Practical
Numerical Methods
for Chemical Engineers

Using Excel with VBA
Fourth Edition

Richard A. Davis

Department of Chemical Engineering
University of Minnesota Duluth

Duluth, Santa Barbara, Provo, Paris, Cupertino, Palo Alto

Practical Numerical Methods for Chemical Engineers
Using Excel® with VBA, 4th Edition
By Richard A. Davis

Copyright © 2018. Richard A. Davis. All rights reserved.

No part of this book or accompanying *Excel* workbooks and VBA code may be reproduced, stored in a retrieval system or transmitted in any form or by any means – electronic, mechanical, photocopying, recording, scanning, or otherwise – without prior written permission of the Publisher, except as permitted under Section 107 or 108 of the 1976 United States Copyright Act. Requests to the Publisher for permission should be sent to the address below

> Dr. Richard A. Davis, Publisher
> Department of Chemical Engineering
> University of Minnesota Duluth
> Duluth, Minnesota 55812
>
> Email: rdavis@d.umn.edu

Excel® is a registered trademark of Microsoft Corp. (www.microsoft.com). Other company and product names mentioned herein are the trademarks of their respective owners.

Disclaimer: Use the examples, programs, or algorithms in this book and associated web site at your own risk! No warranties are expressed or implied, that the examples, programs, or algorithms in this book and associated web site are free of error, or will meet your requirements for any particular application. The publisher disclaims all liability for direct or consequential damages resulting from any use of the examples, programs or algorithms in this book or associated web site, including, but not limited to, *Excel* worksheets and *VBA* programs.

9 8 7 6 5 4 3 2 1

ISBN-13: 978-1790143092

ISBN-10: 1790143098

Practical Numerical Methods, 4th Edition is available at: https://sites.google.com/d.umn.edu/pnm4

Contents

PROLOGUE IX

CHAPTER 1 INTRODUCTION TO NUMERICAL METHODS AND MODELING 1
- 1.1 NUMERICAL METHODS 3
  - 1.1.1 Linear versus Nonlinear Equations 6
  - 1.1.2 Common Features of Numerical Methods 6
  - 1.1.3 Significant Figures and Computer Round-off Error 7
- 1.2 MATHEMATICAL MODEL BUILDING 9
  - 1.2.1 Laws of Conservation 11
  - 1.2.2 Steady-state versus Equilibrium 12
  - 1.2.3 Generation 14
  - 1.2.4 Lumped versus Distributed Systems 15
  - 1.2.5 Degrees of Freedom 21
  - 1.2.6 What Do We Mean by "Order?" 22
  - 1.2.7 Dimensionless Equations* 22
  - 1.2.8 Model Verification and Validation* 23
  - 1.2.9 Stochastic versus Deterministic Modeling* 24
- 1.3 EXPERT PROBLEM SOLVERS 25
- 1.4 APPENDIX: PRIMER ON REACTION ENGINEERING 31

CHAPTER 2 ENGINEERS USE EXCEL 35
- 2.1 GETTING STARTED WITH EXCEL 36
  - 2.1.1 Backup Your Work 36
  - 2.1.2 Calculators versus Excel Spreadsheets 37
  - 2.1.3 Shortcuts for Managing an Excel Worksheet 38
  - 2.1.4 Relative vs Absolute Cell Referencing and Named Cells 40
  - 2.1.5 Auto Fill a Range of Cells 41
  - 2.1.6 Array Operations with CNTL SHIFT ENTER 42
  - 2.1.7 Worksheet Cell and Formula Errors 43
  - 2.1.8 Avoid Division by Zero Errors 43
  - 2.1.9 Troubleshoot a Worksheet with Formula Auditing Tools* 44
  - 2.1.10 Circular Reference Errors 45
- 2.2 FUNCTION LIBRARY 45
  - 2.2.1 Functions for Rounding and Significant Figures 48
  - 2.2.2 Unit Conversions 50
  - 2.2.3 Logical, Lookup, Match and Index Reference Functions 51
  - 2.2.4 Indirect Function 52
- 2.3 DOCUMENT YOUR DATA AND CALCULATIONS 53
  - 2.3.1 Presenting Tabulated Results 53
  - 2.3.2 Never Type Variable or Data Values Directly Into a Formula and Other Recommendations for Formatting Worksheets 54
  - 2.3.3 Drawing Tools 55
  - 2.3.4 Charts and Graphs 55
  - 2.3.5 Pivot Tables and Pivot Charts 63
  - 2.3.6 Text Wrapping and Breaking Text* 64
  - 2.3.7 Use Conventional Engineering Symbols and Equations* 65
  - 2.3.8 Comments, Data Validation and Conditional Formatting* 65

2.4	"What-if" Analysis Using Goal Seek and Data Table	67
2.5	Excel Add-Ins: Analysis Tools and Solver	70
2.6	Epilogue on Excel	74

CHAPTER 3 VISUAL BASIC FOR APPLICATIONS 75

3.1	"Macro Enabled" Workbooks	76
3.2	Visual Basic Editor (VBE) Projects and Modules in the PNM Suite	76
3.2.1	Objects and the Properties Window	78
3.2.2	Code and Immediate Windows	79
3.3	User-defined Functions, Sub Procedures, and Macros	79
3.3.1	Keywords, Text, Breaks, and Multiple Statements per Line	82
3.3.2	Descriptive Comments in Code	82
3.4	Variable Data Types and Constants	82
3.4.1	Implicit Variable Declaration and Default Variable Type	84
3.5	Structured Program Flow Control	85
3.5.1	Loops and GoTo	85
3.5.2	If-Then-Else Logic	88
3.5.3	Case Structure*	90
3.5.4	GoSub...Return*	96
3.6	VBA Math and Worksheet Functions	97
3.6.1	Speed Up VBA Computations	99
3.6.2	Pseudo Random Number Generator	101
3.6.3	Evaluate Worksheet Cell Formulas in VBA*	104
3.7	Arrays	106
3.7.1	Range Variables and Worksheet References	107
3.7.2	Sorting Arrays	109
3.7.3	Worksheet Array Functions and User-defined Functions that Return Arrays	110
3.8	Communicate with the User	111
3.8.1	Macro Controls on the Worksheet	111
3.8.2	MsgBox	116
3.8.3	InputBox	118
3.8.4	User-forms*	119
3.8.5	Charting and Graphing with VBA	124
3.8.6	Read/Write Data To/From Files	131
3.8.7	Status Bar	132
3.8.8	Events to Trigger Macros*	132
3.8.9	Optional Arguments in User-defined Functions*	134
3.8.10	FORMAT Numbers in Output	134
3.8.11	Use VBA to Add Cell Comments and Colors*	134
3.9	Debugging Code and Preventing Errors	135
3.9.1	Immediate, Locals, and Watch Windows	138
3.9.2	Debug Menu	138
3.10	PNM Suite of Macros and User-defined functions	139
3.11	Excel Add-In Macros*	141
3.12	Epilogue on VBA	142

CHAPTER 4 LINEAR ALGEBRAIC SYSTEMS 143

4.1	Linear Algebraic Functions	143
4.2	Linear Algebraic Systems	144
4.3	Gaussian Elimination	145
4.3.1	Matrix Solution Improvement by Maximum Column Pivoting and Error Estimation	146
4.3.2	Crout Reduction Method of LU Decomposition*	148
4.3.3	Thomas Algorithm for Tridiagonal Matrix Inversion	151
4.3.4	Cholesky Decomposition of a Symmetric Matrix*	151

4.4	Excel Worksheet Array Functions	152
4.5	Eigenvalues and Eigenvectors*	154
4.6	Successive Substitution: Jacobi and Gauss-Seidel Methods	157
4.6.1	Iteration on Circular References in Excel	162
4.6.2	Relaxation*	163
4.7	Epilogue on Linear Equations	165

CHAPTER 5 TAYLOR SERIES ANALYSIS AND DERIVATIVE APPROXIMATION ... 167

5.1	Taylor Series	167
5.2	Accelerate Series Convergence*	172
5.3	Derivative Approximations	178
5.3.1	Finite-difference Derivative Approximations	179
5.3.2	Higher-order Derivative Approximations*	182
5.3.3	Derivatives from Unequally-Spaced Data*	183
5.3.4	Richardson's Extrapolation and Ridders' Algorithm	188
5.4	Sensitivity Analysis and Maximum Uncertainty*	194
5.5	Epilogue on Taylor Series and Derivatives	198

CHAPTER 6 ROOT-FINDING METHODS ... 199

6.1	Graphical Solution	200
6.2	Numerical Roots of a Single Equation	203
6.2.1	Fixed-point Ordinary Iteration	204
6.2.2	Wegstein's Acceleration of Fixed-point Iteration*	204
6.2.3	Bisection	205
6.2.4	Regula Falsi Linear Interpolation*	208
6.2.5	Excel's Goal Seek	211
6.2.6	Newton's Method	212
6.2.7	Secant and Muller's Methods	219
6.2.8	Padé Interpolation (or Extrapolation)*	223
6.3	Roots of Simultaneous Equations	225
6.3.1	Ordinary Iteration (Successive Substitution)	225
6.3.2	Aitken's Δ^2 and Steffensen's Methods of Accelerating Series Convergence*	225
6.3.3	Equation Sequence and Arrangement	226
6.3.4	Quasi-Newton's Method for Simultaneous Equations	231
6.3.5	Excel's Solver for Simultaneous Equations	235
6.4	Bairstow's Method of Polynomial Factorization*	240
6.5	Scaling and Transformation*	245
6.6	Initializing Iterative Methods and Equation Tearing to Promote Convergence*	247
6.7	Epilogue on Nonlinear Equations	249

CHAPTER 7 OPTIMIZATION ... 251

7.1	Objective Function	252
7.2	Define the Optimization Problem	253
7.3	Calculus Maximum versus Minimum	254
7.4	Indirect Search and Newton's Method	256
7.5	Direct Search Methods	258
7.5.1	Graphical Direct Search for One or Two Dimensions	258
7.5.2	Bracketing with Quadratic Interpolation	260
7.5.3	Response Surfaces*	262
7.5.4	Golden Sections	263
7.5.5	Powell's Method for Multiple Dimensions	266
7.5.6	Downhill Simplex Method of Nelder and Mead*	268
7.6	Stochastic Global Optimization*	270
7.6.1	Simulated Annealing*	270

- 7.6.2 *Luus-Jaakola Global Optimization**......275
- 7.6.3 *Particle Swarm Optimization (PSO)*......280
- 7.6.4 *Firefly Algorithm for Multimodal Optimization**......283
- 7.6.5 *Evolutionary Genetic Algorithm**......286
- 7.7 EXCEL'S SOLVER TOOLKIT FOR OPTIMIZATION......290
- 7.8 CONSTRAINTS ON THE SEARCH REGION......294
  - 7.8.1 *Lagrange Multipliers*......294
  - 7.8.2 *Slack Variables for Inequality Constraints*......298
  - 7.8.3 *Variable Transformation for Removing Constraints*......298
  - 7.8.4 *Penalty Functions for Inequality Constraints**......301
  - 7.8.5 *Redundancy and Separation**......302
- 7.9 MULTI-OBJECTIVE DESIRABILITY FUNCTIONS*......304
- 7.10 VARIABLE SCALING*......305
- 7.11 SENSITIVITY ANALYSIS (REVISITED)*......306
- 7.12 EPILOGUE ON OPTIMIZATION......312

CHAPTER 8 UNCERTAINTY ANALYSIS......314

- 8.1 MODELS OF MEASUREMENT UNCERTAINTY......316
  - 8.1.1 *Precision versus Accuracy*......317
  - 8.1.2 *Type A Standard Uncertainty from Random Errors*......319
  - 8.1.3 *Confidence Intervals, Outliers, and the Bootstrap Method*......319
  - 8.1.4 *Type B Fixed Uncertainty from Systematic Errors*......324
  - 8.1.5 *Uncertainty in Quoted Values and Single or Small Sets of Measurements*......326
  - 8.1.6 *How (Not) to Report Uncertainty*......328
- 8.2 UNCERTAINTY PROPAGATION......329
  - 8.2.1 *The Law of Propagation of Uncertainty*......330
  - 8.2.2 *How (Not) to Handle Unit Conversions*......334
  - 8.2.3 *Correlation**......336
  - 8.2.4 *Expanded Uncertainty and Degrees of Freedom Using the Welch-Satterthwaite Formula*......338
  - 8.2.5 *Automate Propagation of Uncertainty in Excel with Jitters*......340
  - 8.2.6 *Uncertainty in Replicated Experiments**......349
- 8.3 MONTE CARLO SIMULATION OF UNCERTAINTY*......350
  - 8.3.1 *Latin Hypercube Sampling**......354
  - 8.3.2 *Correlation in Monte Carlo Uncertainty Analysis**......356
- 8.4 TRIANGULAR AND LOG NORMAL DISTRIBUTIONS*......358
- 8.5 EPILOGUE ON UNCERTAINTY ANALYSIS......360

CHAPTER 9 REGRESSION......361

- 9.1 LEAST-SQUARES REGRESSION......363
  - 9.1.1 *Ordinary Least-squares Regression*......366
  - 9.1.2 *Multivariable Linear Least-squares Regression*......367
  - 9.1.3 *Nonlinear Least-squares Regression*......372
- 9.2 SELECT GOOD INITIAL GUESSES FOR PARAMETERS......373
- 9.3 MODEL FIDELITY: VERIFICATION, VALIDATION, ASSESSMENT, AND UNCERTAINTY ANALYSIS......379
  - 9.3.1 *Linearization for Model Validation*......380
  - 9.3.2 *Coefficient of Determination: R^2 and Adjusted R^2*......387
  - 9.3.3 *Regression Uncertainty with Parameter Correlation**......388
  - 9.3.4 *F-test, P-Value and t-Stat for Hypothesis Testing**......389
  - 9.3.5 *Residual Plots for Model Verification*......390
  - 9.3.6 *Residual Outliers**......394
- 9.4 WEIGHTED LEAST-SQUARES REGRESSION*......402
- 9.5 JACKKNIFE METHOD OF UNCERTAINTY ANALYSIS*......404
- 9.6 GAUSS-NEWTON LEAST-SQUARES REGRESSION WITH UNCERTAINTY ANALYSIS......405
- 9.7 LEVENBERG-MARQUARDT REGRESSION......418

9.8	MONTE CARLO ANALYSIS OF MODEL UNCERTAINTY*	421
9.9	GRAPH CONFIDENCE INTERVALS AND ERROR BARS	426
9.10	CHOOSING AN EMPIRICAL MODEL	429
9.10.1	*Rational Least-squares*	*429*
9.10.2	*Recommendations for Effective Modeling*	*436*
9.11	LEAST ABSOLUTE DEVIATION REGRESSION*	437
9.12	EPILOGUE ON MODELING AND REGRESSION	439

CHAPTER 10   INTERPOLATION **441**

10.1	POLYNOMIAL INTERPOLATION	443
10.1.1	*Linear Interpolation*	*443*
10.1.2	*Newton's Method of Divided Difference*	*445*
10.1.3	*Lagrange Interpolating Polynomials*	*447*
10.1.4	*Inverse Polynomial Interpolation*	*449*
10.2	CUBIC SPLINE POLYNOMIAL INTERPOLATION	450
10.3	HERMITE POLYNOMIALS*	463
10.3.1	*Akima's Hermite Cubic Spline Polynomial Interpolation**	*464*
10.3.2	*Kruger's Constrained Hermite Cubic Spline Interpolation**	*466*
10.4	RATIONAL INTERPOLATION*	470
10.4.1	*Bulirsch-Stoer Algorithm for Rational Interpolation**	*473*
10.4.2	*Stineman Piecewise Rational Interpolation**	*474*
10.5	BIVARIATE (TWO-DIMENSIONAL) GRID INTERPOLATION*	476
10.6	DATA SMOOTHING AND B-SPLINES*	479
10.7	EPILOGUE ON INTERPOLATION	482

CHAPTER 11   INTEGRATION **485**

11.1	GRAPHICAL INTEGRATION (RECTANGLE RULE)	486
11.2	TRAPEZOIDAL AND MIDPOINT RULES	488
11.2.1	*Improper Integrals and the Midpoint Rule*	*494*
11.2.2	*Integration Accuracy and Precision*	*496*
11.2.3	*Data Integration with Uncertainty Propagation*	*497*
11.2.4	*Romberg Integration with Error Control*	*499*
11.3	SIMPSON'S RULES WITH ERROR CONTROL	503
11.4	CUBIC SPLINE POLYNOMIAL INTEGRATION	512
11.5	GAUSS QUADRATURE WITH ERROR CONTROL	513
11.6	MULTIPLE INTEGRATION	519
11.6.1	*Multiple Integration by Trapezoidal and Simpson's Rules, and by Cubic Spline Interpolating Polynomials*	*520*
11.6.2	*Multiple Integration by Gauss-Kronrod Quadrature**	*525*
11.7	MONTE CARLO INTEGRATION*	527
11.8	CONTINUOUS LEAST-SQUARES APPROXIMATION*	530
11.9	EPILOGUE ON INTEGRATION	532

CHAPTER 12   INITIAL-VALUE PROBLEMS **534**

12.1	TAYLOR SERIES METHOD	536
12.2	EULER METHOD	537
12.2.1	*Stability and Stiffness*	*540*
12.2.2	*Simultaneous First-order Differential Equations*	*542*
12.3	TRAPEZOIDAL RULES FOR IVPS	543
12.4	IMPROVED EULER'S METHOD WITH EXTRAPOLATION*	549
12.5	RUNGE-KUTTA HIGHER ORDER METHODS	553
12.5.1	*Integration Accuracy versus Precision*	*557*
12.5.2	*Variable Step Runge-Kutta Methods*	*560*
12.6	ADAMS-BASHFORTH-MOULTON MULTI-STEP METHOD*	567

12.7	Differential-Algebraic Systems*	570
12.8	Newton's Method with Continuation for Root-finding (Homotopy)*	573
12.9	Variable Step Quadrature*	577
12.10	Nonlinear Parameter Estimation in IVPs*	578
12.11	Epilogue on Initial-value Problems	582

CHAPTER 13 BOUNDARY-VALUE PROBLEMS 584

13.1	Shooting Method	585
13.2	Finite-difference Method	593
13.3	Orthogonal Collocation on Finite Elements*	598
13.4	Nonlinear Parameter Estimation in BVPs*	606
13.5	Epilogue on Boundary-value Problems	607

CHAPTER 14 PARTIAL DIFFERENTIAL EQUATIONS 608

14.1	Finite Difference Solutions of Elliptic Equations	609
14.2	Parabolic Equations	612
14.2.1	Method of Lines	612
14.2.2	Crank-Nicolson and Dufort-Frankel Implicit Methods	615
14.2.3	Orthogonal Collocation for Parabolic PDEs*	618
14.3	Epilogue on Partial Differential Equations	624

CHAPTER 15 REVIEW 625

15.1	Appendix A: Excel Worksheet Functions	631
15.2	Appendix B: VBA Functions and Keywords	633
15.3	Appendix C: VBA User-defined Functions	636
15.4	Appendix D: VBA Sub Procedures (Macros)	640
15.5	Appendix E: VBA User-forms	647

REFERENCES 648

INDEX 652

Advanced Topics

Prologue

Learn numerical methods and develop technical problem-solving skills using the spreadsheet software *Excel* with the built-in object-oriented programming language *VBA*. Scientists and engineers from all disciplines will find the coverage of computational methods applicable to a variety of problems not specific to chemical engineering. Note the top three emerging engineering work activities: design, **computer applications**, and management (Burton 1998). According to a *CACHE* survey of chemical engineering graduates (Edgar 2003):

- **100%** *use a computer every day.*
- **98%** **use Excel** *(88% for data analysis; 47% for numerical analysis, 25% for material and energy balances, 24% for financial accounting).*
- **83%** said **computing enhanced their problem solving skills** and wanted more exposure to computational methods.
- **78%** indicated that a **programming language should be required** at the undergraduate level (the specific language is less important than just knowing how to code).
- **75%** characterized their work as "technical" with their highest priorities being research and development, plant/process support, and **process design and analysis**.
- **73%** responded that their employers expected them to be competent in a **programming language** (38% were required to write a computer program.)

Excel has become the *de facto* standard computational tool for engineering and scientific calculations across many fields. Our colleagues use *Excel* for a variety of engineering, analysis, and management tasks. The object-oriented programming language *Visual Basic for Applications* (*VBA*, the *Basic* programming language in *MS Office* applications) further enhances *Excel*'s capabilities. We recommend the combination of *Excel* with *VBA* as tools to begin developing the computing skills valued by practicing chemical engineers. Consequently, this book has four primary goals to help students:

1. Learn practical numerical methods for solving engineering problems,
2. Become proficient using *Excel* for engineering problem solving,
3. Gain skills for enhancing the capabilities of *Excel* with *VBA* user-defined functions and macros, and
4. Develop good habits of documenting solutions for archiving and collaboration.

"For the things we have to learn before we can do them, we learn by doing them." – Aristotle

Learning numerical methods for engineering is a bit like the paradox of which came first, the chicken or the egg. A strong background in engineering science and design is important to apply many numerical techniques; yet many basic problems in modern engineering courses require the use of numerical methods to arrive at a working solution. This book intends to resolve this issue by introducing essential numerical techniques required in a typical chemical engineering curriculum, accompanied by a variety of examples from the discipline.

"The student needs to develop an understanding, however partial and imperfect, by descriptions rather than definitions, by typical examples rather than grandiose theorems." – Gian-Carlo Rota

Advanced theories of numerical analysis were intentionally sidestepped in favor of concentrating on basic derivations and applications of numerical techniques ... risking the *"howl of the Boeotians."* [1] In the author's experience, students begin using these *practical* computing skills right away in their technical courses and research. Frequently, students returning to campus after completing an engineering internship or a few years after graduation report that their ability to use *Excel* spreadsheets and *VBA* macros for modeling, data collection, and analysis was their most practical skill transferred from college to the workplace.

"Never start teaching or research in a new field of applied mathematics from general concepts, statements, theories, and theorems. Consider some instructive examples and the general theory will come and be cast naturally." – G.I Barenblatt

The coverage of topics in this book assumes readers have a background in topics from calculus, linear algebra, and differential equations. The examples assume some experience with introductory physics and chemistry. Lower division students need not fully understand the science or engineering behind these examples to learn the application of the numerical technique presented. However, they should be able to recognize the different *mathematical formulations* of the problems, *e.g.*, an integral or differential

---

[1] Gauss (1777-1855) in an 1829 letter to a confident, "... for I fear the howl of the Boeotians if I speak my opinion aloud."

equation that requires integration limits, initial conditions, or boundary conditions. We draw a few examples from transport phenomena and engineering unit operations. Some examples involve concepts from chemical reaction engineering, a course typically taken late in the curriculum, with concepts that use simple principles of dynamic material and energy balances. For those who may need it, the appendix to Chapter 1 has a primer on basic reaction engineering principles used throughout the book. For more examples and experience with the numerical methods in *Excel*, the companion website has practice problems at all levels for novice to advanced users.

Although several "canned" software applications are commercially available for solving problems numerically[2], we must avoid treating them like a "black box". The combination of *Excel* and *VBA* forms a powerful platform for learning and implementing numerical methods (Coronell and Hariri 2008). The associated software help files are indispensable. *Excel* and *VBA* help files include examples of the syntax and some information about the mathematics and statistics. As previously noted, *Excel* comes equipped with *VBA*, a versatile, object-oriented programming language used to implement the algorithms presented throughout the text and enhances *Excel* as an engineering tool. We may modify the macros and example files for our particular needs. Take care to use appropriate *VBA* programming syntax when modifying these programs for personal use. The appendices to Chapter 15 contain tables of the macro titles that accompany the textbook.

> *"The more I study, the more insatiable do I feel my genius for it to be."* – Ada Lovelace (first computer programmer[3])

We use commercial numerical software applications as computational tools, not necessarily teaching tools. They rarely illustrate the numerical algorithms through logically programmed steps required by this teaching text (Kuku and Karamani 2011). We encourage students to learn to use the various versions of commercial computational software once they have mastered the underlying fundamentals, and have a feel for when different methods work or fail. Otherwise, we risk overreliance on the software tools and miss an opportunity to develop our critical thinking skills. Without an understanding of the underlying numerical methods, we may become frustrated when the software does not work in the way we anticipate.

> *"If the only tool you have is a hammer, every problem begins to look like a nail."* — Abraham Maslow

Furthermore, process simulators, such as *Aspen Plus*, incorporate several of the basic numerical methods presented in this book, as well as advanced techniques not covered here. Often, the user is required to supply initial conditions or guesses to "jumpstart" the solution. Users need to be familiar with the different solution methods to make the proper specifications and coax a solution from the software. Perhaps even more important is having the background to find alternative routes to a solution when a canned program fails.

With our modern access to vast amounts of information on the internet, encyclopedic coverage of numerical analysis is not necessary in a textbook such as this. Rather, we introduce practical tools and methods to develop the basic concepts in numerical techniques (such as requirements for initial guesses to start iterative solutions), Taylor series analysis, derivative approximations, Monte Carlo techniques, etc. We also include useful tools for students and practicing engineers for use immediately in class or on the job. The book includes several practical, yet powerful *VBA* macros and functions and provides a foundation of essential numerical concepts that are common to advanced numerical methods.

By design, the arrangement of topics follows an intuitive structure that builds upon itself using a spiral-learning model by which the learner revisits a topic at increasing levels of breadth and depth. This book has five principal parts:

I. Introduction to numerical methods and tools for problem solving
   a. Mathematical modeling
   b. *Excel* for Engineers
   c. *VBA* for enhancing *Excel* with custom programming
II. Numerical methods for solving equations
   a. Roots to simultaneous linear algebraic equations
   b. Taylor series function approximations to transition from linear to nonlinear problems
   c. Roots to single or simultaneous nonlinear equations
   d. Optimization
III. Numerical methods for working with data
   a. Uncertainty analysis
   b. Regression and data modeling for interpretation and functionalizing noisy experimental results
   c. Interpolation of smooth data for functionalizing experimental results or for replacing complicated equations
IV. Numerical methods for engineering mathematics: Calculus and Differential Equations
   a. Integral equations or numerical quadrature
   b. Initial-value problems involving first order differential equations.
   c. Initial-Boundary-value problems involving ordinary and partial differential equations
V. Summary review of numerical methods and software tools

---

[2] e.g., *Mathcad, Mathematica, Maple, Matlab, Polymath*, etc.
[3] https://www.famousscientists.org/ada-lovelace/

PROLOGUE

We first introduce modeling concepts to reinforce the need for numerical methods, as well as establish the pattern for problem solving used for many of the example problems. Introductory chapters present practical features of *Excel* and *VBA* for engineering problem solving. We then review methods of solving systems of linear equations followed by a transition chapter on the Taylor series, which serves as the foundation for numerical methods presented in the remainder of the book. Derivative approximations derived from Taylor series function approximations are useful for accelerating the search for roots to nonlinear functions. The same procedures carry over into optimization and uncertainty analysis. We positioned uncertainty analysis as the centerpiece of the text to stress the importance of thinking critically about the reliability of our numerical results.

> *"It is absolutely necessary, for progress in science, to have uncertainty as a fundamental part of our inner nature."* – Richard Feynman

Least-squares regression naturally follows optimization as a special application. We use regression techniques to model, smooth, and interpolate data. A separate chapter covers interpolation of smooth data. The last few chapters of the book present numerical solutions to integrals (which are based on interpolating functions) and differential equations that require methods previously introduced for solving systems of equations, derivative approximations, and solutions to nonlinear functions. The final chapter includes appendices with summary tables of useful *Excel* and *VBA* functions for implementing numerical methods.

To maximize the return on your investment, we recommend the **SQ3R** study method (Harrington and Zakrajsek 2017). Before removing the cap from your yellow highlighter, work your way through the following steps for each chapter of the book:

- **S** — **Survey** the contents of a chapter first to get a feel for what types of problems and numerical methods are introduced.
- **Q** — From the initial survey of the contents, develop a few **Questions** about the topics and look for answers in the chapter. To help you, each chapter starts with a set of *SQ3R Focused Reading Questions*. Add a few of your own questions you have after surveying the chapter.
- **R^1** — **Read** a few manageable sections in the chapter with your questions in mind to focus your study of the material as you search for answers.
- **R^2** — Close the book and **Recite** a few notes (in writing) from your recollection of the contents of the sections. Keep the book closed for this step - don't peek!
- **R^3** — **Reread** the section and add any missing key point to your notes. It is OK to use your yellow highlighter now that you have a better feel for the important concepts in a section. Try to limit your highlights to no more than two key points in a paragraph. This helps focus your thinking and learning.

> *"Let no one be deluded that knowledge of the path can substitute for putting one foot in front of the other."* - M.C. Richards

We strongly encourage working through the examples to get the most out of the book. The examples use the computer spreadsheet application Microsoft Windows *Excel*® version 2007 or later and Visual Basic for Applications (*VBA*). The *Excel* example files and *VBA* function and macro files are available from the author's web site:

https://sites.google.com/d.umn.edu/pnm4.

Lewis Carroll (1939) gave the following helpful advice to students of mathematics (the same applies to engineers):

> The learner, who wishes to try the question fairly, whether this little book does, or does not, supply the materials for a most interesting mental recreation, is earnestly advised to adopt the following rules:
>
> 1. Begin at the beginning, and do not allow yourself to gratify a mere idle curiosity by dipping into the book, here and there. This would very likely lead to your throwing it aside, with the remark 'This is much too hard for me!', and thus losing the chance of adding a very large item to your stock of mental delights.
> 2. Don't begin any fresh Chapter, or Section, until you are certain that you thoroughly understand the whole book up to that point, and that you have worked, correctly, most, if not all of the examples which have been set. Otherwise, you will find your state of puzzlement get worse and worse as you proceed, till you give up the whole thing in utter disgust.
> 3. When you come to a passage you don't understand, read it again; if you still don't understand it, read it again: if you fail, even after three readings, very likely your brain is getting a little tired. In that case, put the book away, and take to other occupations, and the next day, when you come to it fresh, you will very likely find that it is quite easy.
> 4. If possible, find some genial friend, who will read the book along with you, and will talk over the difficulties with you. Talking is a wonderful smoother-over of difficulties. When I come upon anything … that entirely puzzles me, I find it a capital plan to talk it over, aloud, even when I am all alone. One can explain things so clearly to one's self! And then, you know, one is so patient with one' self: one never gets irritated at one's own stupidity!

This book's contribution extends the trail blazed by several excellent engineers and scientists who have recognized early on the power of *Excel* and *VBA* for learning and applying numerical methods. We hope this book will inspire you to push ahead to find new ways to use these powerful tools to tackle your own engineering problems.

> *"If I have seen further it is only by standing on the shoulders of Giants."* – Isaac Newton

Suggestions for Instructors

Practical Numerical Methods may be used as the primary textbook for a course in numerical methods for chemical engineers, as a supplementary reference for courses in computational methods, or as an independent study guide for learning scientific computing with *Excel* and *VBA*. The content ranges from introductory material for a first course on numerical methods to advanced topics for senior and first-year graduate students or researchers. The advanced topics are indicated with an asterisk * after the section heading and are included in the textbook for readers interested in using *Excel* with *VBA* for solving more challenging problems and analysis.

To help chemical engineering students see the relevance of *Excel* as a problem-solving tool, several of the examples are drawn from chemical engineering principles – however, students in their first or second year of an undergraduate program may not have sufficient training to understand the engineering concepts in the examples. Students in the early stages of their academic careers should focus on the mathematics and corresponding solution methods, and then refer back to the examples as they progress in their academic and professional careers to review and apply the numerical methods as needed. Instructors may need to provide additional introduction of the chemical engineering principles in the applied examples to help students appreciate the application of *Excel* with *VBA* for solving real chemical engineering problems.

At the author's home institution, the University of Minnesota Duluth, we cover the sequence of topics in the following table for primarily 2nd and 3rd year students in a 15-week course that meets in a computer lab for three hours per week. An *Excel* workbook design project is assigned to integrate many of the topics and show the power of *Excel* as an engineering tool. We also intersperse process simulation exercises throughout the course using a commercial software application. We recommend reserving the starred (*) sections noted in the **Table of Contents** for advanced-undergraduate or graduate level courses in computational methods where we introduce additional programming tools for scientific computing in addition to *VBA* and *Excel*.

Week	Topics	Chapters
1	Introduction to Numerical Methods and Mathematical Modeling (All sections) Introduction to Excel (All sections)	1 - 2
2	Introduction to *VBA* (All sections) Simultaneous Linear Equations • Gauss Elimination with Pivoting and Matrix Operations in Excel • Successive Substitution (ordinary iteration) Taylor Series and Derivative Approximations • 1st and 2nd Order Finite Difference Derivative Approximations • Richardson' Extrapolation	3 - 4
3	Roots of single and simultaneous Nonlinear Equations • Graphical Solution and Fixed-point Iteration • Bisection and Goal Seek • Newton's Method and Solver (single and simultaneous roots) • Bairstow's method for roots of polynomials	5 6
4	Introduction to Process Simulation 1	Instructor Notes
5	Review and Exam 1 (Comprehensive to date)	1 - 6
6	Optimization • Objective Functions and the Calculus Optimum • Simplex (Linear and Nonlinear Programming Methods) • Solver options with constraints	7
7	Uncertainty Analysis • Models of Measurement Uncertainty • Law of Propagation of Uncertainty • Expanded Uncertainty and the JITTER Macro	8
8	Process Simulation 2 (Physical Properties, Thermodynamic Models and Flash calculations	Instructor Notes

9	Least-squares Regression
	• Single and Multivariable Linear Regression
	• Nonlinear Regression
	    ○ Verification, Validation and Assessment with Linearization and Residual Plots
	    ○ Levenberg-Marquardt Method of least-squares minimization
		9

10	Review and Exam 2 (Comprehensive to date)	1 - 9
11	Interpolation • Linear Interpolation • Lagrange Polynomials • Cubic Splines and Constrained Splines	10

Week	Topics	Chapters
12	Integration or Quadrature • Graphical and Trapezoidal Rule • Simpson's rules • Romberg Method with Richardson's Extrapolation Initial-value Problems • Euler's and Implicit Trapezoidal Methods • Fixed and Variable Step Runge-Kutta Methods	11 - 12
13	Boundary-value Problems and Partial Differential Equations • Shooting Method • Finite Difference (one and two dimensions) • Method of lines	13-14
14	Process Simulation 3 (Reactors and Process Systems Integration)	Instructor Notes
15	Review and Exam 3 (Comprehensive)	1 - 14

Each new edition continues in the same tradition of the first, using expert problem solving strategies to highlight practical numerical methods employed by students and practicing engineers. A few highlights include:

What's New in the 4th Edition?

Much of the text was revised for clarity and conciseness. Several additional numerical methods and associated macros with examples were added to further develop the versatility of the *PNMSuite* workbook and expand the coverage of *Excel* with *VBA* as a tool for solving scientific and engineering problems. These include:

- Each chapter starts with a list of reading questions to focus the student's study of the material.
- Now more than 200 examples and illustrations of *Excel* and *VBA* for numerical problem solving throughout the book.
- Hundreds of practice problems available from the book's website with new problems added annually.
- *MS Office* 2016 updates useful for implementing numerical methods.
- Recommendations for presenting tabulated data in reports.
- Several updates to current macros to minimize user error and improve computational efficiency.
- More than 180 *VBA* user-defined functions, sub procedures, macros, and forms.
- New section in Chapter 2 introducing **Pivot Tables** and **Charts**.
- Expanded coverage of *Excel* and *VBA* debugging tools.
- Roulette wheel algorithm and user-defined function for selecting from a list proportional to probability of selection.
- User-defined function for improved calculations of the standard deviation of a range of data.
- Sub procedure for swapping the contents of two equal sized ranges on a worksheet.
- Simple user-defined function for calculating the convergence criterion in terms of relative residuals.
- Introduction to using *VBA* for creating charts in *Excel*.
- Updated QYXPLOT macro to select more appropriate axis limits.
- Macro for generating right-triangle ternary phase diagrams from equilibrium data.
- Macros for Padé and Shanks transformations for accelerating series convergence.*
- Tearing method and macro for solving nonlinear equations.
- Example of Linear Programming using *Excel's* **Solver** option.
- Simulated Annealing Metropolis algorithm for global optimization of multimodal objective functions.
- Introduction to *Excel's* **Solver** Evolutionary Algorithms.
- Particle swarm optimization (PSO) with a *VBA* sub procedures for implementation in *Excel*.*
- *VBA* implementation of the Genetic algorithm for evolutionary minimization with constraints.
- Macros for generating spider and tornado sensitivity plots of an optimized objective function versus percent change in the variables.
- Sorted *z*-score plotting macro for validation of least squares models using Filliben's criteria.*
- A macro for calculating and graphing "studentized" regression residuals to detect possible outliers in the data.
- Sub procedure for double integration by Gauss-Kronrod with error control.
- Improved Euler's method for initial-value problems with global extrapolation.*
- Nonlinear parameter estimation involving models using numerical solutions to differential equations (IVP and BVP).*
- Improved coverage of orthogonal collocation for boundary-value problems.*
- Separate chapters for numerical solution methods for BVPs and PDEs.

Topics Added in the 3rd Edition

Builds on the content of the Second edition with the following additions:
- Color formatting of cells used by some macros to indicate unconverged solutions
- Introduction to the *Microsoft Equation* object for placing images of equations on the worksheet
- Included a table of formula errors and fixit tips
- Added descriptions to user-defined functions and sub procedures for selection from menus
- Simple bubble sorting macros available in the *PNM3Suite*
- Macros to create y vs x scatter plots of data following the recommended graphing guidelines
- User-defined functions that evaluate worksheet formulas now accept named cells in formulas or variables
- Example of *VBA* driven animation of function plots
- Pivoting and scaling for Gauss Elimination and Crout's method of LU decomposition to ensure non-zero terms on the diagonal of a coefficient matrix for linear equations
- User-form for selecting either Crout reduction of Gaussian elimination to solve a linear system of equations
- Macros for calculating eigenvalues and eigenvectors of a real, square matrix for linear homogeneous systems by the power, Jacobi, and interpolation methods
- User-defined array function for solving a linear system of equations arranged in matrix format, which also checks for consistent arrays and the determinant using *Excel* worksheet functions
- Macro for solving implicit equations by fixed-point or Gauss-Seidel iteration with relaxation
- Accurate approximations of higher order derivatives with Richardson's extrapolation using Ridders' algorithm
- Bisection macro for root finding in a single equation
- User-defined function for finding the root of a formula on an *Excel* worksheet by the regula falsi, secant, Muller, and quasi-Newton methods
- Imbedded a reduced step loop in the algorithm for Newton's method
- User-defined function for finding a root to a nonlinear equation by rational Padé interpolation
- User-defined array function **POLYROOTS** for finding real and imaginary roots of a polynomial by Bairstow's method
- Nelder and Mead downhill simplex method for nonlinear programming to minimize a multivariable objective function
- Additional options for Gauss-Newton, Levenberg-Marquardt, and Powell's methods on the user-form **ROOTS_UsrFrm** for finding roots to systems of nonlinear algebraic equations
- Downhill Simplex method of multivariable minimization, *a.k.a.* "The Blob"
- Firefly Algorithm and macros for multivariable, multimodal optimization
- Triangular distributions for random and fixed uncertainty
- Macros add comments to output cells to define terms
- Macro for bootstrap resampling methods with replacement for calculating confidence intervals for statistics of a data set
- Macro to implement nonlinear rational least-squares regression
- Derivation of continuous least-squares approximation of functions
- User-defined function for Lagrange polynomial interpolation
- User-defined function for B-spline interpolation with fixed ends.
- User-defined function with the Bulirsch-Stoer algorithm for rational interpolation
- Hermite interpolating polynomials using information about the first derivative at the data points, including Akima, Bica, and constrained cubic spline interpolation methods
- Inverse polynomial interpolation
- User-defined functions for two-dimensional (bivariate) interpolation by linear, Stineman, and cubic spline.
- User-defined functions for integrating data sets by the composite trapezoidal rule and Simpson's methods with Richardson's extrapolation where possible.
- Simple macros for demonstrating Gauss-Legendre and Gauss-Kronrod 10 and 15 point single integrals.
- Adaptive quadrature with Simpson's 1/3 rule employing Richardson's error estimates for interval size control.
- Introduction to quasi-linearization for solving finite-difference equations in two-point boundary-value problems
- Optional arguments for the limits of integration in the user-defined array function for orthogonal collocation.
- Application of the method of orthogonal collocation to parabolic partial differential equations.
- User-defined function for expanding the capability of unit conversions in an *Excel* worksheet.
- User-defined function that returns a value for the ideal gas constant in various unit systems.

Topics Added in the 2nd Edition

Builds on the content of the First edition with the following additions:
- User-defined functions for number rounding to the correct number of significant figures.
- A macro for generating an *yx* scatter plot of an *Excel* worksheet formula.
- A new collection of sub procedures for matrix inversion, multiplication, and transposing for use in macros operating on arrays that exceed Excel's limits for similar worksheet functions.
- User-defined functions for finding a root of a *formula* in an Excel worksheet cell.
- An option in the Levenberg-Marquardt macro for nonlinear regression to supply the analytical values of the Jacobian.
- A macro for finding good staring parameters for rational least squares curve fitting.
- Newton's method of divided difference for local polynomial interpolation and extrapolation.
- Stineman's method with piece-wise rational interpolation of smooth data that eliminates poles.
- An introduction to data smoothing methods.
- An option for clamped end conditions with cubic splines.
- User-defined functions for differentiating and integrating *formulas* in Excel worksheets.
- Spline method for numerical solution of differential equations.
- Variable change in Romberg method using trapezoidal rule for improper integrals.
- Multiple integration by Simpson, Gauss-Kronrod, and Monte Carlo methods of quadrature.
- Taylor series method for initial-value problems with Padé approximation for series acceleration.
- Macros for solving systems of stiff ordinary differential equations by implicit single step methods.
- Table of common unit conversion factors and ideal gas constants on the last page for convenience.

We expended much energy to correct errors in this text. Please email corrections, comments, and inquiries to rdavis@d.umn.edu.

Acknowledgements

I am indebted to my long-time friends, colleagues, and mentors: Professors Owen Hanna and Orville Sandall at UCSB for their creative genius, my UMD colleague Professor Keith Lodge for his helpful reviews, program testing, and suggestions, and Ronald Visness with the Minnesota Department of Natural Resources, who inspired me to take my first steps on this project. Each has provided much support, encouragement, enthusiasm, as well as helpful feedback along the way. My students gave several insightful comments and suggestions. Thank you to Professor John Gossage at Lamar University, and Professors James Petersen, Hongfei Lin, and Christian Cuba at Washington State University for their constructive feedback.

    I am most grateful to my eternal companion Elsie for her support of this project. I dedicate this book to my mother the nurse and father the aerospace engineer, who started a typical conversation with the following exchange:

Mom: "It's not *rocket* science!"   *Dad*: "Well, *actually* ..."

Duluth, Minnesota: 2018

Richard A. Davis

Chapter 1   Introduction to Numerical Methods and Modeling

> *"The ability to compute separates the engineer from the technician. An education in engineering mathematics generates an insight into physical things which cannot be attained in any other way and the generation of new qualitative ideas which will work."* – Lawrence Kamm

We begin with a rather candid observation – most of us do not have a future career in fashion modeling.[4] Nevertheless, as engineers and scientists, we may have a future in an alternative modeling career. Mathematical and statistical modeling, simulation, and analysis may not be as glamorous as walking down the fashion runway, but they do provide elegant insights and solutions to important technical problems that benefit society.

The subject of applied mathematics is not just a wardrobe of techniques for solving mathematical problems – it also includes mathematical modeling – using the language of mathematical physics to describe, design, analyze, and optimize engineering systems, processes and operations. Engineers and scientists represent their hypothesis for how a system functions in the form of a mathematical model – then test their model hypothesis through model validation and verification. In this chapter, we introduce working definitions of mathematical models and numerical methods, and offer compelling reasons for mastering these subjects. Additional topics include:

- Linear versus nonlinear models
- Common features of numerical methods
- Computer round-off error
- Laws of conservation
- Steady-state versus equilibrium
- "Order" in mathematical modeling and numerical methods
- Expert problem solving advice

SQ3R Focused Reading Questions

1. What are differences between numerical and analytical solutions?
2. What makes a model rigorous?
3. What are the foundational principles of mathematical models in chemical engineering?
4. What are the different contexts for the term "order"?
5. What are the different meanings of the terms verify and validate?
6. What are good habits of expert problem solvers?
7. How can making a mathematical model dimensionless enhance the analysis?
8. What is a stochastic model?
9. Why do I need to worry about degrees of freedom?
10. What are the differences and similarities of precision, accuracy, and significant figures?

---

[4] There are always exceptions – *Super Model* Cindy Crawford launched a successful modeling career after one term on a chemical engineering scholarship at Northwestern University. Check out the following excerpt from an interview with Crawford on NPR's *Not My Job*:

    NPR: "You were going to be a chemical engineering major."
    CRAWFORD: "Yeah, I know, crazy."
    NPR: "No, it's not crazy. You would've changed the reputation of chemical engineers forever!"
    (http://www.npr.org/2016/10/15/497937670/not-my-job-supermodel-cindy-crawford-gets-quizzed-on-scale-models)

Consider the following responsibilities and desired experience for an actual job description: strong computational experience for optimization, analysis, accuracy, and documentation. This book starts us down that path by providing some practical tools for learning and applying numerical methods for problem solving. However, just as lifelong learning is a lifetime commitment of any competent engineer or scientist, this book is just the beginning – computational tools are continually evolving, becoming increasingly powerful, and the associated skill to apply them is refined over a career of engineering and scientific practice.

Responsibilities:

- Provide process engineering technical assistance to process operations for solving operating problems and optimizing assigned units; develop and design major projects on revamps and new units.
- Review plant operations and laboratory analyses daily for better efficiency and compliance with operating plans.
- Review projects and correspondence for technical accuracy and clarity.
- Provide documentation of engineering and analytical results with recommendations to management.

Required Experience:

- A Bachelor's Degree in chemical engineering or related field.
- Strong computer-simulation experience for simulation of unit operations in order to better optimize their operation.

We may use good mathematical models to guide the design of equipment and processes and optimize expensive bench scale studies. The system of equations that constitute a mathematical model may involve some combination of algebraic, integral, or differential equations derived from the application of principles of mass and energy conservation, thermodynamics, and kinetics.

The modern physicist David Goodstein (1985) paraphrased from Galileo's writings:

> "*Il libro della natura e scritto nella lingua matematica* - The book of nature is written in the language of mathematics."

Goodstein's response to Galileo:

> "*Si, ma deve essere letto nella lingua fisica* – Yes, but it must be read in the *language* of physics."

We use mathematical models to describe, understand, and manipulate our physical world. Nevertheless, our models must make physical sense – natural law, or the laws of physics govern our models. For chemical engineers, the laws of thermodynamics and momentum, mass, and energy conservation are paramount! The "real" world is complex. How we get from "start to finish" is not always obvious. Experiments and process data usually answer the questions, *what*, *where*, and *when*? Mathematical models derived from first principles of physics and chemistry help answer the questions of *why* and *how* process *outputs* result from process *inputs*.

Recall Occam's razor[5]:

> "*Non sunt multiplicanda entia praeter necessitatem* – Things should not be multiplied without good reason."

We should try to eliminate all unnecessary information in our model; the simplest elegant solution is usually the best solution. Einstein qualified Occam's razor:

> "*Make everything as simple as possible, but not simpler.*"

Einstein may be warning modelers not to take Occam's razor to an extreme. We must balance the goal of simplicity with the goal of usefulness. Levenspiel (2002) refers to "$10, $100, and $1000 models" – reminding us that the cost of complex models increases by orders of magnitude, while not *necessarily* providing us with orders of magnitude increases in understanding. However, as computers become faster, and modeling software becomes more sophisticated, we are finding that we can reduce the amount of expensive experiments formerly required, and replace them with relatively inexpensive computer simulations, or *virtual* experiments. Furthermore, as illustrated in Figure 1.1, the cost of computing is decreasing while the cost of experimenting is increasing over time.

---

[5] A razor shaves away unnecessities.

CHAPTER 1: INTRODUCTION

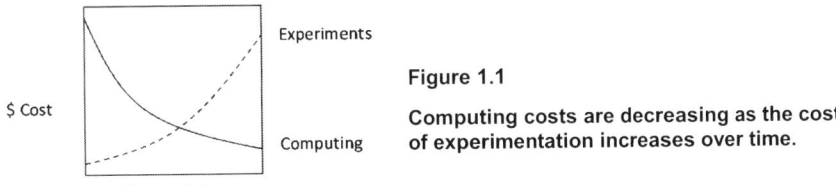

Figure 1.1

Computing costs are decreasing as the cost of experimentation increases over time.

Models range in complexity according to their intended purposes. Simple models may provide enough insight into a problem to facilitate a decision whether to proceed with further development, stop, dig a little deeper, or change engineering directions. Increasing levels of model detail may be required as increased accuracy and precision become necessary for design and decision-making. The cost of developing a mathematical model depends on the degree of uncertainty in the model predictions we are willing to accept. As illustrated in Figure 1.2, generally the higher the uncertainty in the calculated results we are willing to accept, the lower the cost of the model.

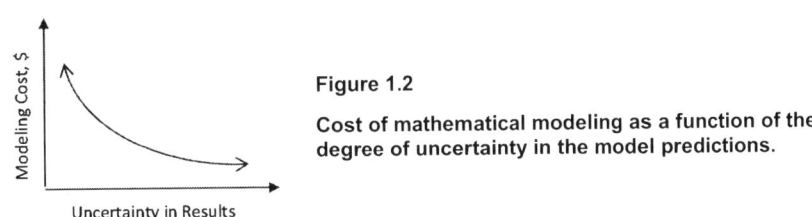

Figure 1.2

Cost of mathematical modeling as a function of the degree of uncertainty in the model predictions.

Part of the increasing cost associated with achieving lower uncertainty in the model predictions comes from the work required to find the solution to the model equations. Preliminary engineering models tend towards the higher end of uncertainty (lower precision) when efforts to lower uncertainty do not justify the exponentially increasing modeling costs. Scientific models tend towards the lower end of uncertainty (higher precision), where the goal is a detailed understanding of a particular natural phenomenon. Engineers compensate for uncertainty with design safety factors. For example, a chemical engineer may design a distillation column by calculating the minimum number of ideal stages, and then apply a heuristic that recommends doubling the minimum stages, then adjusting the reflux to achieve the desired separation (Walas 1990).

1.1 Numerical Methods

"Computers are useless. They can only give you answers." – Pablo Picasso

We define *numerical methods* as computational techniques for approximating solutions to mathematical models that may be inconvenient, difficult, or even impossible, to solve using standard analytical methods. Analytical solutions produce results that have symbolic representations, such as rearranging an equation explicitly for one of the variables. When analytical results are infeasible, we turn to numerical methods to obtain good approximate solutions.[6] To illustrate, consider calculating the molar volume of a gas from models that relate the molar volume to temperature and pressure. Two candidates are the ideal gas law and the Redlich-Kwong equation-of-state for non-ideal behavior. We can rearrange the ideal gas law for the molar volume:

$$V = R_g T / P \qquad (1.1)$$

where R_g is the ideal gas constant, T is temperature, and P is pressure. However, we cannot rearrange the following Redlich-Kwong model *explicitly* for either molar volume *or* temperature:[7]

---

[6] When we say, "approximate," we do not mean low quality. Approximate numerical solutions, though not exact, may have extremely high accuracy and precision to more than 10 significant figures.
[7] Explicit is defined by an expression containing only independent variables.

$$P = \frac{R_g T}{V - b} - \frac{a}{V(V+b)\sqrt{T}} \tag{1.2}$$

The parameters a and b are functions of the critical temperature and pressure of the gas. There is no way to rearrange Equation (1.2) with the molar volume only appearing on one side of the equation. Instead, the molar volume in the Redlich-Kwong equation must be determined from numerical approximation methods for specific cases of T and P.

"It is unworthy of excellent persons to lose hours ... in the labor of calculation." – Gottfried Wilhelm von Leibniz

Many numerical methods require iterative calculations conveniently carried out with the aid of a computer. As computers become faster and software gains in sophistication, numerical methods are fast becoming the norm rather than the exception for solving complex mathematical models that describe engineering processes. Even when a problem has an analytical solution, a numerical solution may be easier to implement, saving us time and effort to work on other problems.[8]

Typical problems encountered in chemical engineering consist of large systems of linear equations – a result from the application of the laws of conservation of momentum, mass and energy, or non-linear equations based on the inherent nonlinear nature of physical and chemical properties, transport phenomena, thermodynamics, and chemical reactions. Some common examples of nonlinear functions in chemical engineering include the temperature dependent Antoine equation for vapor pressure, the modified Arrhenius function for reaction rate constants, and the modified Henri function for enzyme reaction kinetics with substrate inhibition:

- *Antoine*
$$\ln P_v(T) = A - \frac{B}{C + T} \tag{1.3}$$

- *Modified Arrhenius*
$$k(T) = AT^2 \exp\left(-\frac{E_a}{R_g T}\right) \tag{1.4}$$

- *Modified Henri*
$$v(S) = \frac{v_m S}{K_m + aS + bS^2} \tag{1.5}$$

Although linear equations have analytical solutions, it may be tedious to determine the solution of a system larger than three or four equations. Analytical solutions for large systems of linear equations involve extensive algebra and symbolic bookkeeping. For many nonlinear equations, no analytical solution exists. We need methods for finding good approximate solutions to nonlinear functions and large systems of linear equations.

We illustrate the need for numerical methods with a simple model from chemical reaction engineering.[9] The reaction mechanism is elementary, irreversible, and first-order in species A:

$$A \rightarrow B \tag{1.6}$$

Consider the steady-state mole balance for the product species B around a perfectly stirred chemical reactor, illustrated in Figure 1.3, with feed and effluent concentrations of C_{A0}, and unreacted C_A and product C_B, respectively.

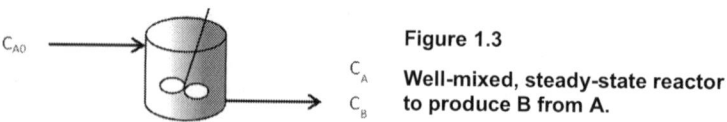

Figure 1.3
Well-mixed, steady-state reactor to produce B from A.

We start with species conservation for mathematically modeling this reactor. With no B in the feed, the concentration of B leaving the reactor equals the product of the residence time, τ, with the rate of generation of B per unit volume in the reactor, r_B:

$$C_B = \tau r_B \tag{1.7}$$

For elementary, first-order reactions, the rate of production of B is a linear function of concentration:

---

[8] Now you can howl, Boeotians!
[9] The appendix to Chapter 1 includes a brief introduction to mathematical models used in chemical reaction engineering.

CHAPTER 1: INTRODUCTION

$$r_B = k(C_{A0} - C_B) \tag{1.8}$$

where $(C_{A0} - C_B)$ is equivalent to the concentration of unreacted A in the reactor in terms of the initial concentration of reactant A, and k is the first order reaction rate constant. We can rearrange Equations (1.7) and (1.8) explicitly for the concentration of product C_B exiting the reactor:

$$C_B = \frac{\tau k C_{A0}}{1 + \tau k} \tag{1.9}$$

However, for higher-order elementary reactions, or non-elementary reactions, we typically describe the rate of generation by a nonlinear function of concentration. For example, a catalyzed reaction may behave according to some form of Langmuir kinetics:

$$r_B = \frac{k(C_{A0} - C_B)}{\left[1 + K_A(C_{A0} - C_B) + \frac{K_B}{C_B}\right]^2} \tag{1.10}$$

where K_A and K_B are reaction equilibrium constants for reversible steps in the reaction mechanism. Substitution of the rate law from Equation (1.10) into the mole balance of Equation (1.7) yields a nonlinear equation implicit[10] in C_B:

$$C_B = \frac{\tau k(C_{A0} - C_B)}{\left[1 + K_A(C_{A0} - C_B) + \frac{K_B}{C_B}\right]^2} \tag{1.11}$$

We cannot rearrange Equation (1.11) explicitly for C_B. Instead, we calculate C_B iteratively, by guessing the solution in a "trial and error" approach[11], until our calculations for C_B converges on the solution.

The previous illustration demonstrates the need for numerical methods to solve nonlinear equations. Other problems encountered in chemical engineering may require solutions to nonlinear integrals,

$$I(x) = \int f(x)\,dx \tag{1.12}$$

or coupled systems of nonlinear, non-homogeneous, "initial-value" first-order differential equations, with f and g representing nonlinear functions of some combination of the independent and dependent variables:

$$\frac{dy}{dx} = f(w, x, y) \quad \text{and} \quad \frac{dw}{dx} = g(w, x, y) \tag{1.13}$$

Our models may involve non-homogeneous, second order "boundary-value" type differential equations:

$$\frac{\partial^2 C}{\partial x^2} + \frac{\partial^2 C}{\partial y^2} + \lambda r(C) = \frac{\rho}{\alpha}\frac{\partial C}{\partial t} \tag{1.14}$$

where C is concentration, $r(C)$ may be a nonlinear rate law, and α, ρ, and λ are model parameters that may be functions of x, y, or t.

The sequence of topics in this book intentionally builds from solutions to equations and on to integration and differential equations. We find that problems involving integrals and differential equations are solved numerically using combinations of simpler techniques presented in the earlier chapters.

---

[10] Implicit is defined by an expression in which the dependent variable and one or more independent variables are not separated on opposite sides of an equation. http://www.merriam-webster.com/dictionary/implicit

[11] Trial and error methods iterate on two steps: (1) Trial = Guess the solution and evaluate the result (2) Error = Use an estimate of the error from the trial to upgrade the guess for a better solution.

1.1.1 Linear versus Nonlinear Equations

We can easily distinguish between linear and nonlinear functions using visual graphical analysis:

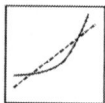

Without a graphical analysis, we recognize nonlinear equations from their terms involving exponents, transcendental functions, or combinations of dependent variables. To illustrate nonlinearity, consider the following, second-order, ordinary differential equation:

$$\frac{d^2y}{dx^2} + \frac{x}{y}\frac{dy}{dx} + x^2y^2 = 0 \qquad (1.15)$$

The first term in Equation (1.15) is linear in the dependent variable y. However, the second and third terms are nonlinear in y because they involve the product of two functions of y. Consequently, Equation (1.15) is a nonlinear ordinary differential equation and may not have an analytical solution.

Of course, we may cite the well-known common exception – the second-order polynomial, or quadratic equation:

$$ax^2 + bx + c = 0 \qquad (1.16)$$

with two roots calculated by the standard quadratic formula:[12]

$$x_1 = \frac{-b - \sqrt{b^2 - 4ac}}{2a} \quad \text{and} \quad x_2 = \frac{-b + \sqrt{b^2 - 4ac}}{2a} \qquad (1.17)$$

Higher order polynomials of order $m > 2$ have m roots.

$$\alpha x^m + \beta x^{m-1} + \cdots ax^2 + bx + c = 0 \qquad (1.18)$$

In principle, an analytical solution for the m roots of a polynomial is possible. In practice, however, analytical solutions of polynomials with m greater than three are not trivial. When analytical solutions are difficult or impossible to achieve, one must resort to numerical approximations.

1.1.2 Common Features of Numerical Methods

Most numerical methods share a few common features. In particular, they:
1. Rely on local linear approximations of nonlinear functions to simplify the calculations,
2. Use error approximation to refine the solution,
3. Employ iteration and successive approximation to upgrade solutions, and are
4. Prone to computational round-off error that may cause instability in the solution method.

Many numerical methods require good linear approximations to the nonlinear equations on a local level. As discussed previously, nonlinear equations can be algebraic, differential, or integral type. For each type of problem, we base solution approximation techniques on local linearization of the nonlinear equations by truncated Taylor series expansions. Chapter 5 introduces the Taylor series for approximating functions with a power series. The novice practitioner of numerical methods will discover that we may solve a wide variety of nonlinear problems with a surprisingly small number of numerical techniques based on the Taylor series.

Linear approximations for nonlinear functions allow for approximate analytical solutions to nonlinear problems. If we can use good linear approximations of nonlinear functions, we then have all of the analytical tools needed for solving linear equations at our disposal. We use estimates of the error in the linear approximation to upgrade the solutions iteratively until the error falls below our required threshold. We consider each event to upgrade the solution

---

[12] See Section 1.1.3 to learn about computing problems with the standard quadratic formula.

CHAPTER 1: INTRODUCTION

as one iteration. On one hand, a crude method of finding an approximate solution is to guess the solution, then plug the guessed values into the function. If our guess does not work, we try a new guess. We should make educated guesses using all of the information available. Using our experience, along with some luck, we may pick a value that is close to the true solution. On the other hand, we design numerical methods to systematically improve the approximation for the solution after each iteration, and remove the guesswork from the process. Once we master these basic, common features of numerical computation, we will find that we can focus on the details of the specific method.

Problems that involve solutions over a range of time or space require a sequence of calculations similar to an iterative solution. We successively approximate the solution at different points in time or space based on the solution at a previous point in time or space. For example, we may approximate the solution throughout the dimensions of space and time by taking discrete steps across the limits of time and space defined by the problem. Whether we are trying to find a good approximate solution at a specific point, or stepping through a function to get a range of solutions, the concepts of iteration and successive approximation apply.

A common source of confusion involves iteration in a computer program where we need to replace the value of a variable with a mathematical operation on that variable. From our mathematical training, we might state that the following equation is only valid when $\phi = 0$:

$$x = x + \phi \tag{1.19}$$

When we state that ϕ must equal zero, we imply that the direction of equality goes both ways (*e.g.*, $x + \phi = x$). However, when setting up a sequence of iterative calculations using a computer, we really mean that the current value of x is *replaced* by $x + \phi$, such that the flow of information is one direction: $x \leftarrow x + \phi$. To illustrate, suppose we need to calculate the sum of a series, $\phi_1, \phi_2 \ldots \phi_n$. We start by initializing $x = 0$ and $i = 0$, then iterate on Equation (1.20) until $i = n$:

$$i \leftarrow i+1 \qquad x_i \leftarrow x_{i-1} + \phi_i \tag{1.20}$$

where the subscript i is an index variable for ϕ and the sum of the values in the series is the final value of x when $i = n$. We make extensive use of this type of operation in numerical methods. To illustrate, if $x_0 = 0$ and $\phi = (1, 2, 3, 4, 5\ldots)$ then $x_1 = 0+1=1$, $x_2 = 1+2=3$, $x_3 = 3+3=6$, $x_4 = 6+4=10$, $x_5 = 10+5=15\ldots$.

1.1.3 Significant Figures and Computer Round-off Error

> *Professor: "Sixty-five million years ago the earth was trod by dinosaurs"*
> *Student: "You mean seventy-three, don't you?"*
> *Professor: "Why do you say that?"*
> *Student: "Because I heard you give this same lecture eight years ago!"* – Levenspiel (2007)

Engineers should report only the significant figures in numerical results. We generally consider only reliable digits in a number as significant. By convention, only the right-most digit in a number should have any significant degree of uncertainty. To illustrate, if we read a value from the Y-axis of the following graph, where the smallest division is 1/10 or 0.1, then a value at $X = 6.4$ may correlate with $Y = 1.46$ (with three significant figures – the last figure "6" is uncertain), certainly not 1.455.

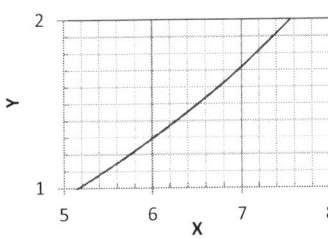

Computers store numbers as single precision with seven significant digits, or double precision, with 15 significant digits. The computer rounds a single-precision number with eight significant digits to a number with seven significant digits.[13] Consider the following simple illustration involving the addition of two numbers in *Excel*'s *VBA* programming environment using single versus double precision values:

- *Single Precision*: $123456789 + 987654321 = 1111111168$
  (round to seven significant figures 1111111000, **incorrect**)
- *Double Precision*: $123456789 + 987654321 = 1111111110$ (**correct**)

The single precision result is accurate to just seven significant figures, whereas the double precision result gives the correct answer. We learn from this example not to trust single precision results involving more than seven significant figures. How far can we go with double precision?

$$123456789123456789 + 987654321987654321 = 1111111111111110000$$

Answer: 15 significant figures! What do we conclude? Use double precision (which is the default in an *Excel* worksheet, but not *VBA*)! *VBA* requires that we specify double precision. However, simply using double precision does not eliminate all round-off errors. Be careful to avoid addition or subtraction of numbers with extreme differences in their orders of magnitude. For example, in double precision, $123456789 \pm 0.00000001 = 123456789$, with no change! As a general practice, we should try to sum small positive numbers, and then sum small magnitude negative numbers before combining them with large magnitude numbers.

Example 1.1   Avoid Round-off Errors in the Quadratic Formula

One of the two standard roots to a quadratic equation, defined by Equations (1.16) and (1.17), potentially loses precision due to round-off error through subtraction. To avoid loss of precision, we keep the root involving addition (x^+), and find the second root (x^-) from the relationships:

$$x^+ \cdot x^- = \frac{c}{a} \quad \text{and} \quad x^- = \frac{c}{a \cdot x^+} \quad (1.21)$$

Thompson's (1987) example of a quadratic root involves a buffer solution of 0.0010 M chloroacetic acid and 0.0100 M sodium chloroacetate. The chemical equilibrium relationship in terms of the hydrogen ion concentration x is

$$1.4 \times 10^{-3} = \frac{x(0.0100 + x)}{0.0010 - x}$$

This reduces to a quadratic function in x:

$$x^2 + 0.0114x - 1.4 \times 10^{-6} = 0$$

The standard roots using 10^{-4} precision are

$$x = \frac{-0.0114 \pm 0.0116}{2}$$

The first root involves a subtraction to give an incorrect answer $x^- = (0.0116 - 0.0114)/2 = 0.0001$. The second root involves an addition (of two terms with the same sign) to give a correct value of $x^+ = (-0.0114 - 0.0116)/2 = -0.0115$. Use x^- in Equation (1.21) to get the other correct root:

$$x^- = \frac{-1.4 \times 10^{-6}}{-0.0115} = 1.22 \times 10^{-4}$$

Press, et al, (1992) recommend the following alternative form of the quadratic rule that eliminates round-off error by avoiding subtraction between two potentially small terms:

$$q = -0.5 \left[b + \text{sign}(b) \sqrt{b^2 - 4ac} \right] \quad (1.22)$$

---

[13] See Section 2.2.1 on significant figures and precision in *Excel* calculations.

Chapter 1: Introduction

where[14] $\quad x_1 = q/a \quad$ and $\quad x_2 = c/q \quad$ (1.23)

and
$$\text{sign}(b) = \begin{cases} -1 & \text{for } b < 0 \\ +1 & \text{for } b \geq 0 \end{cases} \quad (1.24)$$

To illustrate further, compare the calculation in *Excel*, shown in Figure 1.4, with the roots to the following equation:

$$f(x) = 0.321x^2 + 214x + 0.00123 = 0 \quad (1.25)$$

A comparison of the two solution methods for the root in rows five and six reveals the problem with the standard version of the quadratic formula. The bottom two rows show the inaccuracy of the standard solution when we substitute the first root back into Equation (1.25). From this example, we learn to be careful with arithmetic involving the addition or subtraction of an extremely large number and an extremely small number.

	A	B
1	a	0.321
2	b	213
3	c	1.23E-03
4	q =	-212.999998146338000000
5	x_1 standard =	-0.00000577464792771470
6	$x_1 = c/q =$	-0.00000577464793757862
7	x_2 standard =	-663.551396094511000000
8	$x_2 = q/a =$	-663.551396094511000000
9	f(x_1 standard) =	0.00000000000210101511
10	f($x_1 = c/q$) =	0.00000000000000000000

Figure 1.4

Comparison of solutions to the quadratic Equation (1.25) using *Excel* with small parameters *a* and *c*.

□

1.2 Mathematical Model Building

"By a model is meant a mathematical construct which, with the addition of certain verbal interpretations, describes observed phenomena. The justification of such a mathematical construct is solely and precisely that it is expected to work." – John von Neumann

Henri's equation is a classic example of model building, which he discovered for describing the rate of enzyme-catalyzed reactions as a function of the reacting substrate concentration.[15] A typical plot in Figure 1.5 of reaction rate, *r*, versus substrate concentration, [S], reveals the asymptotic nature observed in many reactive systems involving catalysis.

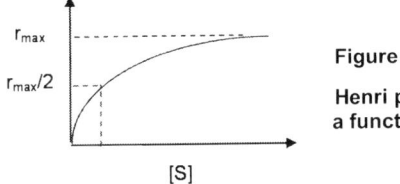

Figure 1.5

Henri plot of enzyme reaction rate, *r*, as a function of substrate concentration *s*.

The curve plotted in Figure 1.5 represents a *graphical* model. We may interpret the behavior of the function directly from the plot. Henri observed that, for small substrate concentrations, the reaction rate is proportional to the substrate concentration, $r \propto [S]$. For large [S], the rate approaches an asymptotic maximum value, $r \to r_{max}$. He proposed the following mathematical equation, or model, to match the observations of the data:

---

[14] *Excel* worksheet formula: sign(b) = if(b<0, -1, 1)
[15] Victor Henri (1872-1940) was a French physical chemist, the 19th or 20th century's prelude to a chemical engineer.

$$r = \frac{a[S]}{b+[S]} \quad (1.26)$$

where a and b are the constants specific to the reaction mechanism. For very small substrate concentration, $[S] \ll b$, the rate expression approaches a linear function in $[S]$ with proportionality constant a/b. In the limit of large $[S] \gg b$, the reaction rate approaches a constant[16], maximum value, $a = r_{max}$:

$$r \cong \begin{cases} a[S]/b & [S] \ll b \\ a & [S] \gg b \end{cases} \quad (1.27)$$

The value of parameter b may also be determined by recognizing that parameter a disappears from Equation (1.26) when $r = a/2$:

$$\frac{a}{2} = \frac{a[S]}{b+[S]} \rightarrow \frac{1}{2} = \frac{[S]}{b+[S]} \rightarrow b = [S] \quad \text{at } r = \frac{a}{2} \quad (1.28)$$

Henri's model successfully fits enzyme reaction data for interpolation, but does not provide any insight into *why* the reaction kinetics behaves this way. Michaelis and Menten (1913) proposed a two-step reaction mechanism to explain how Henri's model works. Their model involves an initial, reversible step that forms an intermediate complex molecule of enzyme-substrate. The enzyme-substrate complex reacts irreversibly to form the reaction product in the second step:

$$S + E \underset{k_{-1}}{\overset{k_1}{\rightleftarrows}} ES \quad (1.29)$$

$$ES \xrightarrow{k_2} E + P \quad (1.30)$$

where E is the free enzyme, ES is the enzyme-substrate complex, and k_1, k_{-1}, and k_2 are the forward and reverse reaction rate constants for the two steps. Assuming that the second step is rate limiting, the reaction rate is proportional to the concentration of the complex $[ES]$:

$$r = k_2[ES] \quad (1.31)$$

while the first step reaches a pseudo-equilibrium state:

$$\frac{[ES]}{[S][E]} = \frac{k_1}{k_{-1}} \quad (1.32)$$

The conservation of enzyme catalyst in the reaction yields the following balance:

$$[E] = [E]_0 - [ES] \quad (1.33)$$

where $[E]_0$ is the initial enzyme concentration in the reactor. Combining Equations (1.31) through (1.33) to eliminate $[E]$ and $[ES]$ gives the well-known Michaelis-Menten expression for the reaction rate:

$$r = \frac{k_2[E]_0[S]}{k_{-1}/k_1 + [S]} = \frac{r_{max}[S]}{K_M + [S]} \quad (1.34)$$

where K_M is the Michaelis equilibrium constant. We should confirm that Equation (1.34) has the same basic functionality of reaction rate with substrate concentration as the Henri Equation (1.26) (i.e. what are a and b?), but also comes with an explanatory mechanism in Equations (1.29) and (1.30). We encounter the alternative Briggs-Haldane mechanism in Section 1.2.3.

Empirical models like Henri's equation do not have a basis in first principles of momentum, energy, or mass conservation. However, empirical models can incorporate mathematical functionality that reproduces the trends in

---

[16] Apply l'Hôpital's rule to Equation (1.26).

CHAPTER 1: INTRODUCTION

the data. Polynomials and rational functions, like Equation(1.26), are examples of convenient models used to interpolate experimental or process data. They are useful for predicting behavior, but may struggle to discern or explain the underlying mechanisms of the system.

In the following sections, we introduce a variety of approaches to mathematical modeling based on underlying assumptions about the system. We start with a review of the inviolable law of conservation. Our models are useless if we cannot make a proper accounting of where all of the mass and energy inputs end up. We follow conservation principles with pairs of opposites (Rasmuson, et al. 2014) with the underlying assumptions in brackets: mechanistic [based on physics] versus empirical, steady [time independent] versus unsteady state, lumped [uniform or well-mixed] versus distributed systems, and stochastic [uncertainty] versus deterministic modeling.

> *"The research rat of the future allows experimentation without manipulation of the real world. This is the cutting edge of modeling technology."* – John Spencer

1.2.1 Laws of Conservation

Mathematical models translate the descriptive word-form of a conservation equation into the symbolic language of mathematics. Although not *absolutely* required, it is *much* more convenient *and* efficient to pose the governing equations in symbolic mathematical form. If you are not convinced, try describing the derivative of the Henri Equation (1.26) for reaction rate with respect to the substrate concentration in a meaningful way, without using any mathematical terms. Ah, you did it, but it wasn't easy!

Development of a mathematical model begins with the laws of conservation for momentum, mass and energy. Start by defining the boundaries on the system control volume. Then, apply rate balances to the system and across the defined boundaries for each of the conservation laws:[17]

$$(\text{Rate of } \xi \text{ in}) - (\text{Rate of } \xi \text{ out}) + (\text{Rate of } \xi \text{ generation}) = (\text{Rate of } \xi \text{ accumulation}) \quad (1.35)$$

where ξ may represent either mass, energy, or momentum. Application of the laws of conservation often yields nonlinear algebraic, integral, or differential equations. Operations on these types of equations are straightforward using standard analytical and numerical techniques. Mathematical models replace each term in Equation (1.35) with its equivalent symbolic mathematical representation of measurable, quantifiable parameters. To illustrate, the product of the volumetric flow rate with density, $Q \times \rho$, may replace the mass (rate in). Check that dimensions of each term in Equation (1.35) are mutually consistent. The dimensions are usually in units of velocity, mass (or moles) or energy per time. Thus, if the (rate *in*) term is described by the flux (rate per unit area), multiply by the normal cross-sectional area to convert the flux into a rate. Table 1.1 contains typical constitutive rates for transport phenomena.

Table 1.1 Constitutive expressions of transport rates. A = area, c_p = specific heat, h = heat transfer coefficient, H = enthalpy, k = mass transfer coefficient, m = mass flow rate, T = temperature, u = velocity, V = volume, x = length dimension, κ = thermal conductivity, ρ = density, μ = dynamic viscosity

	Rate	Diffusive	Advection	Convection
Newton's Law of Viscosity	Momentum Transfer	$\tau = -\mu \dfrac{\partial u}{\partial x}$	$V \rho u$	–
Fourier's Law of Heat Conduction	Heat Transfer	$q = -A\kappa \dfrac{\partial T}{\partial y}$	$q = \dot{m}\left(H^0 + \displaystyle\int_{T^0}^{T} c_p dT\right)$	$q = hA\Delta T$
Fick's Law of Molecular Diffusion	Mass Transfer	$N = -AD \dfrac{\partial C}{\partial y}$	$\dot{m} = A\rho u$	$\dot{m} = kA\Delta C$

---

[17] Stand, face the direction of Philadelphia, place your right hand on a copy of your material and energy balance textbook, raise your calculator in your left hand, and repeat the chemical engineer's motto in Equation (1.35) aloud! When you really need to impress someone, use the alternative form of the motto integrated over time: $\int F_{in} dt - \int F_{out} dt + \int G dt = N_{sys}$.

By definition, a batch reactor has neither inlet nor outlet streams, which leaves only the generation and accumulation terms:

$$0 - 0 + \text{(Product Generation Rate)} = \text{(Product Accumulation Rate)} \tag{1.36}$$

Steady-state systems have no accumulation terms. A batch process at steady state has reached a state of equilibrium where the net rate of generation is zero. For reversible reactions, this leads to one form of the chemical equilibrium constant as a ratio of reaction rate constants. For example, consider the first order reversible reaction $A \rightleftharpoons B$. At steady state for this reaction, Equation (1.36) reduces to the following algebraic expression:

$$-r = k_f C_A - k_r C_B = 0 \tag{1.37}$$

where k_f and k_r represent the forward and reverse reaction rate constants, respectively. Rearrange Equation (1.37) for the reaction equilibrium constant:

$$K = \frac{C_B}{C_A} = \frac{k_f}{k_r} \tag{1.38}$$

Equation (1.32) is another example of an equilibrium constant defined in terms of rate constants.

1.2.2 Steady-state versus Equilibrium

"There exists everywhere a medium in things, determined by equilibrium." – Dmitri Mendeleev

Modeling scenarios often encountered by chemical engineers include some combination of an equilibrium state for temperature, pressure, and composition (phase or chemical), a transient state (changes with time), or a steady state (no net change with time). When a system comes to a state of equilibrium, the values of the dependent variables, represented by ξ, become uniform (homogeneous) throughout the system (or at a boundary), and stop changing with time. In the limit of time going to infinity, the accumulation term in a finite system goes to zero, and the problem reduces to the steady-state case.[18] In the case of a transient problem, we replace the accumulation term in Equation (1.35) by the first-order derivative of ξ with respect to time:

- General form:
$$accumulation = \frac{\partial \xi}{\partial t} \tag{1.39}$$

For mass and energy conservation, use:

- Conservation of moles:
$$accumulation = V \frac{\partial C}{\partial t} \tag{1.40}$$

- Conservation of thermal energy:
$$accumulation = CVc_p \frac{\partial T}{\partial t} \tag{1.41}$$

where V and C are the system volume and concentration, and where c_p and T are the heat capacity and system temperature, respectively. At steady state, the quantity ξ remains constant, and the derivative with respect to time goes to zero. We make distinctions among the conditions of equilibrium, transience, and steady state through a set of illustrations for momentum, mass, and energy conservation.

In the first illustration involving momentum conservation, we fill a rubber balloon from a compressed air tank, as shown in part (a) of Figure 1.6. We measure and control the flow rate of air to the balloon with a rotameter and valve, respectively. During the filling step, the mass of air accumulates in the balloon, the volume of gas expands, and the air pressure inside the balloon increases. These are all dynamic conditions because neither the volume, nor mass of air, nor does air pressure inside the balloon remain constant while filling. When the valve at the tank is closed, as shown in part (b), the air pressure inside the balloon equilibrates with the pressures acting on the balloon based on the elasticity of the rubber and atmospheric pressure acting outside the balloon. We may also consider the equilibrium

---

[18] In chemical engineering practice, we interpret "infinite" time as some *finite* point in time beyond which the relative values of the dependent variable stop changing within the level of precision required for our purposes.

state as a steady-state condition. However, equilibrium is not a requirement for steady-state conditions, as we shall see in part (c). Suddenly, a defect in the balloon causes a small pinhole to form that allows the air to leak slowly out from the balloon. To keep the volume of the balloon constant (steady state), we open the tank valve just enough to fill the balloon at the same rate that air is leaking out. Although the volume remains constant, or at a steady state, the pressure inside the balloon is not in equilibrium with the outside pressure due to the movement of air through the pinhole.

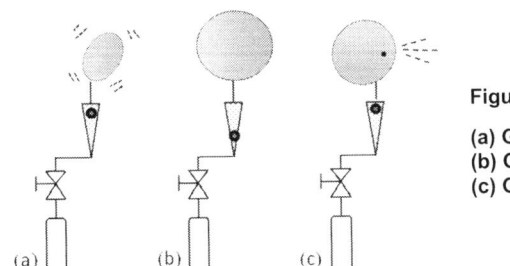

Figure 1.6

(a) Gas on - filling the balloon
(b) Gas off - filled balloon
(c) Gas on - filling rate equals the rate of the steady leak.

In a second illustration involving thermal energy conservation, we analyze heat transfer by conduction through a solid wall with convection to a fluid on the right side. Consider the cross section of a vertical wall that is initially at a uniform temperature, T_i, as shown in Figure 1.7 (a). The left side of the wall is perfectly insulated such that no thermal energy crosses the left boundary, while the right side is exposed to a large reservoir of fluid at a constant temperature, $T_\infty > T_i$. As the wall warms up from contact with the hot fluid, the temperature profile in the wall near the fluid side raises faster (b), although the entire wall eventually warms to the same temperature of the fluid (c). Upon reaching a state of thermal equilibrium, the wall and fluid temperatures are equal.

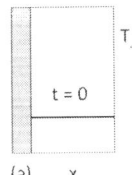 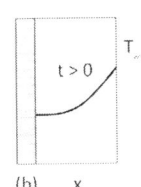 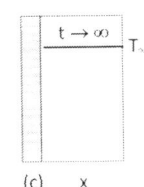

Figure 1.7

(a) Initial, (b) intermediate, and (c) equilibrium temperature profiles in a semi-insulated wall corresponding to (a) time $t = 0$, (b) small time $t > 0$, and (c) large time $t \to \infty$.

We then activate a surface heater between the insulation and the wall at the left side to supply thermal energy at a constant rate, q, to the wall, as illustrated in Figure 1.8 (a). The heater causes the wall temperature to increase even further. The energy from the heating unit initially warms the left side of the wall (b), conducts through the wall to the right side, and then convects away by the fluid until a new steady-state temperature profile develops across the wall. Under this new condition of steady state, the temperature profile in the wall remains unchanging with time, although the wall temperature is not uniform, and exceeds the fluid temperature (c). As shown by this illustration, equilibrium implies steady-state conditions, except the reverse is not true; steady state does not require a temperature equilibrium condition.

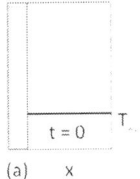 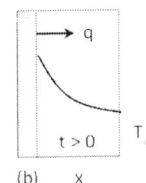 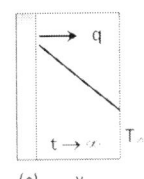

Figure 1.8

Initial, intermediate, and steady state temperature profiles in a heated wall. (a) Initial condition with no heating (b) Heating begins (c) steady-state condition with heating at the left side.

In a third illustration involving mass transfer, a membrane separates two components of a binary gas, such as air, composed predominantly of oxygen and nitrogen. The membrane barrier separates the molecular species by taking advantage of the different rates of diffusion through the barrier. As illustrated in Figure 1.9, a high pressure feed stream passes over one side of the membrane, while the permeating gas collects at the opposite side at low pressure.

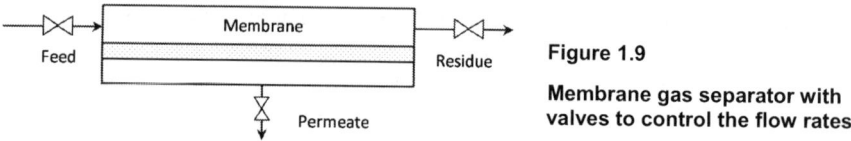

Figure 1.9

Membrane gas separator with valves to control the flow rates.

Initially, the valve controlling the feed stream is closed while the residue and permeate valves are open such that the chambers on either side of the membrane are at ambient pressure. What happens to the pressures and compositions on each side of the membrane if we close the permeate stream valve, and open the feed stream valve? We maintain a high backpressure on the feed side of the membrane at 10 times the ambient pressure by controlling the flow rate with the residue valve. As the high pressure feed gas flows past the membrane surface to the residue stream exit, the molecular species begin to dissolve into the membrane and diffuse over to the permeate side. As shown in the left plot of Figure 1.10, the partial pressure of O_2 in the permeate chamber quickly equilibrates to 21% of the total feed pressure. The slower permeating N_2 gas takes nearly eight times longer to equilibrate to 79% of the feed pressure, shown in the right plot of Figure 1.10. The total pressure in the permeate stream eventually matches the feed pressure.

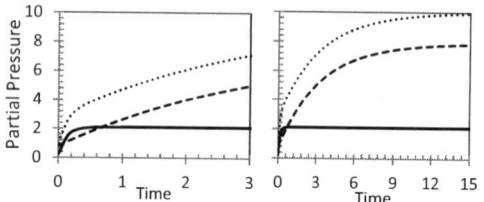

Figure 1.10

Partial pressure of - - N_2 and —O_2 and ... total pressure in the closed permeate chamber over a short and long period.

If we now open the permeate valve, allowing the pressure of the lower chamber to drop back down towards ambient conditions, we see in Figure 1.11 that the permeate stream moves to a new, non-equilibrium, steady-state condition enriched in oxygen at 68% O_2 and 32% N_2.

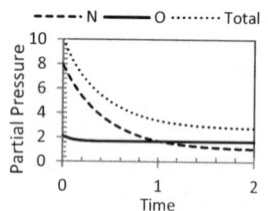

Figure 1.11

Partial pressure of N_2 and O_2 in the open permeate chamber.

An understanding of physical and chemical equilibrium and rate mechanisms is important for chemical engineering model building when defining conditions or relationships at boundaries between adjoining systems.

1.2.3 Generation

The generation term in Equation (1.36) requires a reaction rate law describing the generation of a molecular species. Chemical reactions conserve mass, but not necessarily molecules. The Michaelis-Menten reaction rate law derived in Equation (1.34) is an instance of chemical species generation. Consider an alternative model of enzyme reactions proposed by Briggs and Haldane to develop further the concepts of steady state, equilibrium, accumulation, and generation. Instead of assuming an equilibrium step, as in the case of Michaelis and Menten, Briggs and Haldane proposed conservation equations for each species, including the enzyme-substrate complex:

Rate of Generation = Rate of Accumulation

$$k_1[S]\cdot[E]-k_{-1}[ES]-k_2[ES]=\frac{d[ES]}{dt} \quad \text{and} \quad k_2[ES]=\frac{d[P]}{dt} \quad (1.42)$$

Briggs and Haldane assumed that the conservation of the enzyme-substrate complex concentration quickly reaches a pseudo-steady-state condition, d[ES]/dt ≅ 0. Their model combines Equations (1.33) and (1.42) to eliminate [E] and [ES] from the enzyme-catalyzed reaction rate expression:

$$r = \frac{k_2 [E]_0 [S]}{\frac{k_{-1}+k_2}{k_1} + [S]} \tag{1.43}$$

Equation (1.43) maintains the same functionality as the Henri and Michaelis-Menten equations, but contains an alternative form of the Michaelis constant.

Yet another approach solves the conservation equations without making any assumptions of pseudo-equilibrium or steady state for the reaction steps. This approach does not yield a simple algebraic equation. Instead, we must solve a simultaneous system of differential and algebraic equations:

$$\frac{d[ES]}{dt} = k_1 [S] \cdot [E] - k_{-1}[ES] - k_2[ES] \quad \text{and} \quad \frac{d[P]}{dt} = k_2 [ES] \tag{1.44}$$

$$[E]_0 = [E] + [ES] \quad \text{and} \quad [S]_0 = [S] + [P] \tag{1.45}$$

where the initial concentrations of the enzyme-substrate complex [ES] and product [P] are both assumed zero.

The generation term in a thermal energy conservation equation becomes necessary when accompanied by an endo- or exothermic chemical reaction or electrical heating within the system boundaries. Note that the generation term can be negative or positive. A reaction that consumes the molecular species involves a negative generation term. Similarly, by convention, an exothermic reaction has a negative heat of reaction and positive heat generation. In the case of an endothermic reaction, the overall heat generation term is negative when $r_A < 0$:

- *Chemical Reaction Energy*: $\quad q = V_s r_A \Delta H_{rxn,A}$ (1.46)

where V_s is the system volume, r_A is the reaction rate of species A, and ΔH is the heat of reaction.

Thermal energy generation may involve heat from an electrical resistor:

- *Electrical Energy*: $\quad q = \dfrac{I^2 R}{V}$ (1.47)

where I, R, and V are the electrical current, resistance, and voltage, respectively.

1.2.4 Lumped versus Distributed Systems

> *"May I [suggest] the following modeling strategy: always start by trying the simplest model and then only add complexity to the extent needed. This is the $10 approach."* —Octave Levenspiel *(2002)*

Engineers use mathematical models for design and simulation, as well as understanding and troubleshooting a process. Engineering models typically require simplifying assumptions to overcome complexities in the geometry, flow patterns, or other properties of the system. The assumptions and approximations made in simplifying a model may limit its applicability and introduce error, but may also be necessary for reaching a tractable solution to the mathematical equations. To illustrate, consider the following six examples that demonstrate how simple modeling assumptions may be relaxed to provide additional descriptive capabilities.

1. We begin with the steady state, isothermal operation of a well-mixed reactor, pictured in Figure 1.12. F_{A0} and F_A are the inlet and effluent molar flow rate of species A, respectively, and V is the reactor volume.

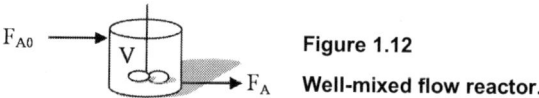

Figure 1.12 Well-mixed flow reactor.

The conservation equation for a molecular species in a steady-state reactor is simply the difference between the molar flow rates in and out of the reactor vessel and generation due to reaction:

$$F_{A0} - F_A + Vr_A = 0 \tag{1.48}$$

where r_A is the reaction rate of formation of A per unit volume of reactor. Define the conversion of the reacting species A in a flow reactor in terms of molar flow rates in and out:

$$X_A = \frac{F_{A0} - F_A}{F_{A0}} \tag{1.49}$$

We solve Equation (1.49) for F_A then substitute into the conservation Equation (1.48) and rearrange for alternative forms of species conservation in terms of conversion of reactant A:

$$\frac{V}{F_{A0}} = \frac{X_A}{-r_A} \quad \text{or} \quad \frac{X_A}{V} = \frac{-r_A}{F_{A0}} \tag{1.50}$$

Numerical methods for finding roots to nonlinear algebraic equations may be required to solve Equation (1.50) for X_A when the reaction rate law for r_A is nonlinear.

The simple design Equation (1.50) for a well-mixed reactor does not reveal anything about the effect of mixing on conversion of a chemical reaction. This is an example of a lumped model that assumes that the composition in the reactor is everywhere the same as the exiting composition. To include mixing affects, we need a model that accounts for spatial or time variations. For instance, we can model a reactor that behaves as if it has two mixing zones, similar to two stirred reactors-in-series, as shown in the flow diagram in Figure 1.13.

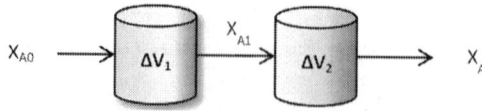

Figure 1.13 Reactor represented by two well-mixed tanks-in-series.

The steady-state material balances for species A around each tank are:

$$\frac{X_{A1}}{\Delta V_1} = \frac{-r_{A1}}{F_{A0}} \quad \text{and} \quad \frac{X_{A2} - X_{A1}}{\Delta V_2} = \frac{-r_{A2}}{F_{A0}} \tag{1.51}$$

where the sum of volumes for the mixing zones equals the total volume of the reactor:

$$V = \Delta V_1 + \Delta V_2 \tag{1.52}$$

Optimization tools may help determine the best split between the two volumes that match model predictions with experimental results for conversion.

Eventually, as the number of tanks-in-series increases to infinity, the volume of each tank becomes infinitely small, and we replace the ratio of conversion to volume with a derivative:

$$\lim_{\Delta V \to 0} \frac{X_{A,V+\Delta V} - X_{A,V}}{\Delta V} = \frac{dX_A}{dV} \tag{1.53}$$

In this case, the model approaches the distributed design equation for a continuous plug flow reactor, pictured in Figure 1.14:

$$\frac{dX_A}{dV} = \frac{-r_A}{F_{A0}} \tag{1.54}$$

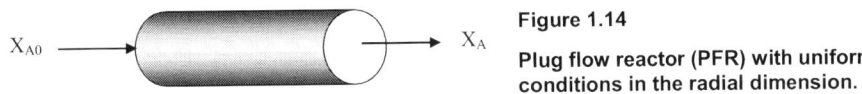

Figure 1.14

Plug flow reactor (PFR) with uniform conditions in the radial dimension.

The solution to Equation (1.54) may require numerical methods designed for approximating solutions to integrals, or for ordinary differential equations when the reaction rate law, r_A is nonlinear such that we cannot separate the variables.

By adding increasing degrees of complexity, we transformed the simple algebraic design equation for a well-mixed reactor into the differential plug-flow-reactor (*PFR*) design equation. The *PFR* equation accounts for changes in the axial direction of flow, but ignores mixing effects in the radial direction (hence, the term, "plug flow"). One method of accounting for deviations from plug flow, without a three-dimensional model, is to introduce a dispersion term into the *PFR* balance. Dispersion accounts for uneven mixing in the radial direction by assuming that the gradient in the concentration along the axis of flow drives the mixture away from a flat concentration profile in the radial direction, in a manner akin to Fick's law. The dispersion flux equation is

$$J_A = -S^2 E \frac{dC_A}{dV} \quad (1.55)$$

where E is the dispersion coefficient, similar to a diffusion coefficient, and S is the cross sectional area of the plug-flow reactor. The molar flow in and out terms now take the form

$$\text{Molar flow rate in or out} = F_{A0} X_A + S J_A \quad (1.56)$$

Substitute the results from Equations (1.55) and (1.56) into the differential *PFR* design Equation (1.54), divide each term by the differential volume, and take the limit as the differential volume goes to zero:

$$\lim_{\Delta V \to 0} \frac{\Delta \left[F_{A0} X_A + C_{A0} S^3 E \left(\frac{dX_A}{dV} \right) \right]}{\Delta V} = F_{A0} \frac{dX_A}{dV} + C_{A0} S^3 E \frac{d^2 X_A}{dV^2} = -r_A \quad (1.57)$$

The second-order differential Equation (1.57) may require appropriate boundary conditions along with a numerical method for approximating the solution to this boundary-value problem.

We observe here that the added levels of rigor transform the original algebraic model into a second-order differential equation, which also adds an order of difficulty to the problem solution. Thus, a simple model of a reactor employing the well-mixed design equation becomes increasingly more rigorous as needed to account for some imperfect mixing effects in a continuous flow reactor. Note, however, that if perfect mixing is a good assumption within our comfort zone of precision, the simple algebraic well-mixed design equation will give the same result as the more complex plug flow or dispersion models, *e.g.*, as the dispersion coefficient, $E \to 0$.

In general, when setting up a mathematical model, make as many reasonable simplifying assumptions as possible. Only add increasing levels of rigor to the model as needed by the requirements of the problem. There is no reason, from a practical engineering standpoint, to create an expensive, yet rigorous, mathematical model to describe a process if the solution to the model equations is intractable *and* a simpler model gives satisfactory results (recall Figure 1.2). We may extract enough information and insight from simple models to meet our needs for preliminary design and economic analysis.

2. The famous Bernoulli equation from fluid mechanics is another classic example of a lumped system that relates the fluid velocity to pressure in a pipe:

$$\frac{\Delta P}{\rho} + g \Delta z + \frac{\Delta u^2}{2} = 0 \quad (1.58)$$

where ΔP is the pressure difference between the fluid inlet and outlet of a system, g is the acceleration due to gravity, Δz is the vertical change in height between the points of fluid inlet and outlet, u is the average fluid velocity, and ρ is the fluid density. When ignoring friction, the only information needed for determining the pipe flow are the conditions at the ends. The model requires neither length nor geometry of the pipe between the inlet and outlet, such as the conduit displayed in Figure 1.15.

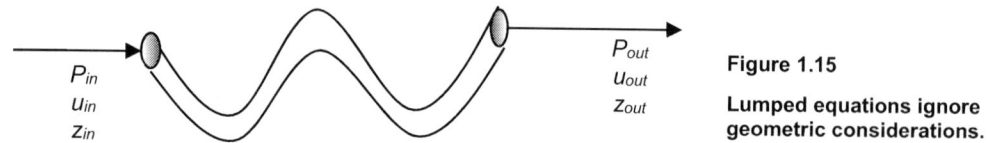

Figure 1.15

Lumped equations ignore geometric considerations.

When we include a term in Equation (1.58) to account for friction, we need to understand the complete geometry to calculate the friction term:

$$\frac{\Delta P}{\rho} + g\Delta z + \frac{\Delta u^2}{2} = F = \frac{2fLu^2}{d} \tag{1.59}$$

where f is the Fanning friction factor, and L and d are the length and diameter of the tube, respectively. The nonlinear Colebrook Equation (6.33) gives the friction factor for flow in rough circular tubes as a function of the Reynolds number, $Re = \rho u D/\mu$, where ρ, u, D, and μ are the fluid density, velocity, pipe diameter, and viscosity, respectively. With this information, we can now calculate the velocity or pressure at any position along the length of the pipe using a numerical root-finding method.

3. Consider the mass and energy balances around a mixing tank shown in Figure 1.16. To simplify our analysis we assume all of the fluid mixing and heat transfer behaviors are lumped together. In this case, we cool a full tank of hot water having volume V and initial temperature T_0 with a stream of cold liquid having a constant volumetric flow rate Q and temperature T_{in}.

Figure 1.16

Mixed tank with volume V and constant inlet and outlet flow rate Q.

We derive an expression for calculating the temperature of the water as a function of time by assuming constant properties for water, a constant feed rate, and well-mixed tank. Our model system is the water in the tank. Because the tank is full, the rate of water exiting the tank must equal the rate entering. There is no mass accumulation or generation in the tank, i.e.:

$$\text{Rate of Mass in} = \text{Rate of Mass out} \tag{1.60}$$

For the case of constant density, the mass of water in the tank remains constant, such that:

$$\rho \frac{dV}{dt} = 0 \tag{1.61}$$

where ρ is the fluid density.

Next, we apply the principle of conservation of energy to the water in the tank. Because the system generates no energy within the water in the tank, the energy balance reduces to

$$(\text{Rate of Energy in}) - (\text{Rate of Energy out}) = (\text{Rate of Energy accumulation}) \tag{1.62}$$

We replace the terms with their mathematical equivalents from thermodynamics:

$$Q\rho c_p (T_{in} - T) = V\rho c_p \frac{dT}{dt} \tag{1.63}$$

where c_p is the average specific heat of the water. We separate variables and integrate Equation (1.63) assuming constant Q and V:

$$\frac{Q}{V}\int_0^t dt = \int_{T_0}^T \frac{dT}{(T_{in} - T)} \tag{1.64}$$

then solve Equation (1.64) for the temperature in the tank at time t:

CHAPTER 1: INTRODUCTION

$$T = T_{in} - (T_{in} - T_0)\exp\left(-\frac{t}{\tau}\right) \tag{1.65}$$

where the residence time is $\tau = Q/V$. This result behaves asymptotically: $T \to T_{in}$ as $t \to \infty$.

4. For our next illustration, consider heat convection from a hot aluminum sphere with diameter, d, that is suddenly plunged into an infinite volume of colder fluid (also known as quenching). Our model system is the solid sphere. If the diameter of the sphere is small, and the thermal conductivity of the material is large (true for aluminum) such that the temperature distribution in the solid is practically uniform, the energy balance gives:

$$0 - (Rate\ of\ Energy\ Out) = (Rate\ of\ Energy\ Accumulation)$$

$$-h(T - T_\infty)\frac{\pi d^2}{4} = \frac{\pi d^3}{6}\rho c_p \frac{dT}{dt} \tag{1.66}$$

with the initial condition $\qquad T = T_0$ at $t = 0$

We assume the temperature in Equation (1.66) is independent of position in the sphere (lumped).

5. On the other hand, it may be necessary to consider variations in the physical space of the system without uniform properties. We treat this type of problem at the local level of the system in terms of the distribution of the properties of viscosity, temperature, or concentration. The modeling approach for a distributed system applies the conservation equations in a shell balance to a local finite volume element. The local in and out terms of a finite volume element are described by constitutive equations, such as Newton's law of viscosity, Fick's law of diffusion, or Fourier's law of heat conduction from Table 1.1. For example, consider the illustration of heat conduction through a wall in Figure 1.8. The left side of the wall is heated. The right side of the wall loses heat by convection to a neighboring fluid. Start with a differential element with thickness Δx at any point in the wall along the x direction, as illustrated in Figure 1.17.

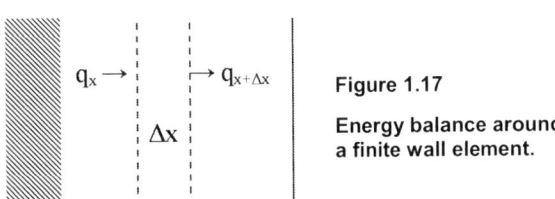

Figure 1.17

Energy balance around a finite wall element.

An energy balance around the differential length Δx gives:

$$(Rate\ of\ Energy\ In)\ -\ (Rate\ of\ Energy\ Out) = (Rate\ of\ Energy\ Accumulation)$$

$$-A\kappa\frac{\partial T}{\partial x}\bigg|_x - A\left(-\kappa\frac{\partial T}{\partial x}\bigg|_{x+\Delta x}\right) = A\Delta x \rho c_p \frac{\partial T}{\partial t} \tag{1.67}$$

where the terms on the left side represent heat conduction in and out by Fourier's law. The term on the right side accounts for accumulation of thermal energy. Divide Equation (1.67) by the differential volume of the wall represented by the product of the cross sectional area and differential thickness, $A\Delta x$, and then take the limit as the differential length goes to zero:

$$\lim_{\Delta x \to 0} \frac{-A\kappa\frac{\partial T}{\partial x}\big|_x - A\left(-\kappa\frac{\partial T}{\partial x}\big|_{x+\Delta x}\right) = A\Delta x \rho c_p \frac{\partial T}{\partial t}}{A\Delta x} \to \frac{\partial}{\partial x}\left(\kappa\frac{\partial T}{\partial x}\right) = \rho c_p \frac{\partial T}{\partial t} \tag{1.68}$$

The result is a second-order partial differential equation subject to the boundary conditions at the left and right sides of the wall where $x = 0$ and $x = L$, respectively.

Apply energy conservation at the walls for the boundary conditions. At the left side, the rate of thermal heating equals the rate of conduction from the surface into the wall. The heat conduction into the right surface of the wall equals the rate of thermal convection to the fluid:

$$\text{Rate of Energy In} = \text{Rate of Energy Out}$$

$$q = -\kappa \frac{\partial T}{\partial x}\bigg|_{x=0} \quad \text{and} \quad -\kappa \frac{\partial T}{\partial x}\bigg|_{x=L} = h(T_L - T_\infty) \tag{1.69}$$

6. In the case of mass transfer, additional constitutive relationships may be necessary for thermodynamic conditions such as vapor-liquid equilibrium at the system boundaries. To illustrate, we derive Fick's *second* law in one dimension from the local species balance around a finite volume element of length Δx, width Δz, and height Δy, as shown in Figure 1.18.

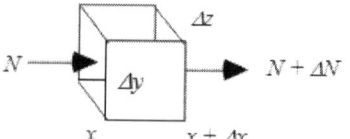

Figure 1.18
Finite volume element in a Cartesian coordinate system with dimensions Δx, Δy, Δz.

N represents the molar flux for the diffusing species. The rate balance for the case in Figure 1.18 accounts for the flux in and out of the element at positions x and $x + \Delta x$. Similar expressions apply to the y and z dimensions:

(Rate In–Out)$_x$ + (Rate In–Out)$_y$ + (Rate In–Out)$_z$ + (Rate In–Out) = (Rate of Accumulation)

$$\Delta y \Delta z (N_x - N_{x+\Delta x}) + \Delta x \Delta z (N_y - N_{y+\Delta y}) + \Delta x \Delta y (N_z - N_{z+\Delta z}) = \Delta x \Delta y \Delta z \frac{dC}{dt} \tag{1.70}$$

where the product of the lengths of two sides $\Delta y \Delta z$ is the area of flux normal to the *x*-direction, etc., and the product of the lengths of all three sides $\Delta x \Delta y \Delta z$ is the finite element volume. The element volume multiplied by the time-derivative of concentration represents the accumulation term. The equation takes the form of a second-order *partial* differential equation (space and time dimensions) in the zero limit of the volume of the finite element $\Delta x \Delta y \Delta z \to 0$:

$$\lim_{\Delta x \to 0} \left\{ \frac{\Delta y \Delta z (N_x - N_{x+\Delta x}) + \Delta x \Delta z (N_y - N_{y+\Delta y}) + \Delta x \Delta y (N_z - N_{z+\Delta z})}{\Delta x \Delta y \Delta z} \right\}$$

$$= -\left(\frac{\partial}{\partial x} N_x + \frac{\partial}{\partial y} N_y + \frac{\partial}{\partial z} N_z \right) = \frac{\partial C}{\partial t} \tag{1.71}$$

If we replace the molar flux terms in Equation (1.71) by Fick's first law from Table 1.1, Equation (1.71) becomes:

$$-\left[\frac{\partial}{\partial x}\left(-D\frac{\partial C}{\partial x}\right) + \frac{\partial}{\partial y}\left(-D\frac{\partial C}{\partial y}\right) + \frac{\partial}{\partial z}\left(-D\frac{\partial C}{\partial z}\right) \right] = \frac{\partial C}{\partial t} \tag{1.72}$$

subject to an initial condition: $\quad C = C_0$ at $t = 0$

and two boundary conditions for each dimension. For example, the *x*-dimension boundary conditions are:

$$\text{Rate In} = \text{Rate Out}$$

$$\beta_2 (C - C_\infty) = \beta_1 \frac{\partial C}{\partial x} \quad \text{at} \quad x = x_0 \quad \text{and} \quad \beta_3 \frac{\partial C}{\partial x} = \beta_4 (C - C_\infty) \quad \text{at} \quad x = x_1 \tag{1.73}$$

where β_1 through β_4 are diffusion and convection parameters. The y and z dimensions require similar boundary conditions. The initial condition is the solution to the equations for the special case at $t = 0$, and is the starting

CHAPTER 1: INTRODUCTION

point for the calculations. Boundary conditions impose the unique, physical nature of the system into the model. We derive boundary conditions from the application of equilibrium thermodynamics or the conservation equations applied to the boundary surfaces[19].

For constant diffusivity, and in the limit $V \rightarrow 0$, Equation (1.72) reduces to

$$D\left(\frac{\partial^2 C}{\partial x^2} + \frac{\partial^2 C}{\partial y^2} + \frac{\partial^2 C}{\partial z^2}\right) = \frac{\partial C}{\partial t} \quad (1.74)$$

To complicate the matter further, reaction rates other than first-order give non-linear rate expressions, which, when substituted for the generation terms, give non-linear differential equations. For example, consider the one-dimensional problem with a nonlinear reaction term for generation:

$$D\frac{\partial^2 C}{\partial x^2} - \frac{kC}{K+C} = \frac{\partial C}{\partial t} \quad (1.75)$$

We can approximate solutions to this type of differential equation by numerical methods introduced in Chapter 13 that are appropriate for parabolic partial differential equations.

1.2.5 Degrees of Freedom

"We cannot choose freedom established on a hierarchy of degrees of freedom, on a caste system of equality like military rank. We must be free not because we claim freedom, but because we practice it." – William Faulkner

A *system* of simultaneous equations consists of:
- *E* number of equations, and
- *V* number of variables

The relationship between the numbers of *E* equations and *V* variables present different issues in finding solutions:

- *E = V*   An independent system of equations has the characteristic that no combination of any subset of the equations may form any single equation in the original system. There exists one unique solution to a set of linear, independent equations. Nonlinear equations, however, may have multiple roots.

- *E ≠ V*   An ill-conditioned system of equations has an unequal number of equations and unknowns

- *E < V*   There are *F = V - E* degrees of freedom in the solution; the values of *F* unknowns must be specified before the system of equations may be solved for a unique solution. An underspecified problem is not necessarily a bad situation because it may provide an opportunity for optimization.

- *E > V*   If there are more equations than unknowns, the system is over-specified, and there is no unique solution to the set of equations. In this case, the set of solutions to the system of equations is infinite. An over-specified system of equations may have a subset of *V* independent equations with a unique solution. We may use uncertainty in the additional equations to find the solution that *mostly* satisfies the system of equations.

Take care to specify properly a large system of equations before attempting a solution. It is not enough that the number of variables match the number of equations. Some sets of linear equations are singular, and do not have a unique solution. Although nonlinear problems often have multiple solutions, we recognize the correct solution by identifying the results that fit within our understanding of the physics of the system. For example, we can eliminate any solution that gives a mole fraction less than zero or greater than one. Recall that, by definition, absolute temperatures and concentrations must be positive.

---

[19] Boundary conditions on models may also involve ethical, legal, environmental, financial, as well as physical requirements, but we will save that for another course.

1.2.6 What Do We Mean by "Order?"

"Ideally a book would have no order to it, and the reader would have to discover his own." – Mark Twain

Before we get too far along, a word of caution about the term, "order" is ... well ... in order. We use the term "order" in a variety of contexts throughout this book. We have already seen several examples using the term "order":

1. *Derivatives and differential equations*: Does it have a first or second order derivative

$$\frac{dy}{dx} \quad \text{or} \quad \frac{d^2y}{dx^2} \quad ...?$$

2. *Reaction order*: Is it first or second order in the reacting species? kC_A or kC_A^2?

3. *Polynomials*: Is it linear (first-order), quadratic (second-order), or cubic (third order) ...?

$$a+bx \quad \text{or} \quad a+bx+cx^2 \quad \text{or} \quad a+bx+cx^2+dx^3 \quad ...?$$

4. *Number order of magnitude*: First order 10 versus second order $10^2 \to O(\Delta x)$ or $O(\Delta x^2)$

Learn to interpret the various meanings of the term "order" from the context of the discussion. To illustrate, we describe the following first-order differential equation as having a second order generation term, C_A^2:

$$\frac{dy}{dx} = kC_A^2$$

We learn about fourth-order accurate methods for solving first-order, ordinary differential equations in Chapter 12. These distinctions become increasingly evident as we develop our numerical skills and start to recognize the meaning of the word "order" from the context of how and where it is used.

1.2.7 Dimensionless Equations*

"The purpose of computing is insight, not numbers." – Richard Hamming

We may recast model equations in dimensionless terms to identify *significant* versus *negligible* terms, and to minimize the effects of round-off error in numerical solutions. Follow Aris' (1993) rules when making equations dimensionless:

1. When the effect of a certain quantity on the model prediction is of interest, the quantity should appear in only one dimensionless parameter, not in a dimensionless variable.
2. Dimensionless dependent variables should range from zero to one.
3. Dimensionless independent variables should range over finite intervals, *e.g.* (0, 1) or (-p, +p).
4. Dimensionless parameters show the relative magnitudes of terms in the model equations. Consider ignoring relatively small terms.

Making the model equations dimensionless usually reduces the number of parameters by at least one. Non-dimensionalization also reveals the minimum number of parameters needed to describe the physical nature of the system. To illustrate, make Equation (1.57) dimensionless following the Aris rules:

$$\frac{dX_A}{d\overline{V}} + \varepsilon \frac{d^2 X_A}{d\overline{V}^2} = \phi \qquad (1.76)$$

where the conversion X_A is already dimensionless, with a range of zero to one. Define the dimensionless independent variable and parameters as follows:

CHAPTER 1: INTRODUCTION

$$\bar{V} = \frac{V}{V_0} \qquad \varepsilon = \frac{C_{A0}A^2E}{F_{A0}V_0} \qquad \phi = \frac{-r_A V_0}{F_{A0}}$$

Note the reduction in the total number of parameters. The three parameter groups F_{A0}, $C_{A0}S^3E$, and $-r_A$ are combined into two parameters ε and ϕ. Aris' fourth rule of dimensionless numbers applies to the parameter ε. For small values of ε, (for cases of large $F_{A0}V_0$ or small $C_{A0}S^3E$) Equation (1.76) reduces to the PFR result in Equation (1.54).

For some problems, there are advantages of scaling the independent variable to range between the integration limits of zero to one:

$$x^* = \frac{x - x_0}{x_n - x_0} \quad \rightarrow \quad x = x_0 + x^*(x_n - x_0) \tag{1.77}$$

where x_0 and x_n are the integration limits, or initial and final values for x. The advantages include:

- Increased computational stability with smaller integration steps
- Improved accuracy with smaller integration steps
- Higher computational efficiency with fewer integration steps
- Ability to integrate in both directions of x (positive from x_0 to x_n or negative from x_n to x_0)

The first-order differential equation becomes dimensionless in the independent variable:

$$\frac{dy}{dx} = \frac{1}{(x_n - x_0)} \frac{dy}{dx^*} = f(x^*, y) \quad \rightarrow \quad \frac{dy}{dx^*} = (x_n - x_0) f(x^*, y) \tag{1.78}$$

Recall that x^* is a function of x according to Equation (1.77). To apply scaling in our numerical methods, we simply multiply each function for the differential equations by the difference in the limits of integration $(x_n - x_0)$ and define x in terms of x^* in the expressions for f.

1.2.8 Model Verification and Validation*

> *"If you put tomfoolery into a computer, nothing comes out but tomfoolery. But this tomfoolery, having passed through a very expensive machine, is somehow ennobled and no one dares criticize it."* – Pierre Gallois

The general usefulness of mathematical models warrants a word of caution. We must not forget that we use models to understand a system, find relationships among operating parameters, process improvement, or design. No amount of effort in mathematical modeling entirely replaces good experimental results. However, as computers and computational tools advance in speed, accuracy, and accessibility, this experimental maxim is shifting. There are already cases in aviation design and chemical property prediction where some computer model results are trusted without experimental validation.[20] Some chemical reactors may involve extremely harsh conditions where probes fail. We use computer simulations as "soft sensors" to infer conditions inside these reactors where experimental measurements are impossible. Despite these advances, the current state-of-the-art of chemical process simulation still requires bench-scale experiments, followed by pilot scale studies before we trust the simulations.

The modeling process is cyclic, as depicted in Figure 1.19. Repeat the following modeling steps until achieving satisfactory results:

1. Define the physical system and identify what you know.
2. Make simplifying assumptions to generate a tractable mathematical model of the system.

---

[20] e.g., the ACS *Journal of Chemical and Engineering Data* accepts article that present the results of computer experiments.

3. Solve the model equations over a range of model parameters to discover relationships.
4. Verify the model analysis and validate the model by comparison to experiments.
5. Refine your understanding of the system and modeling assumptions as necessary.

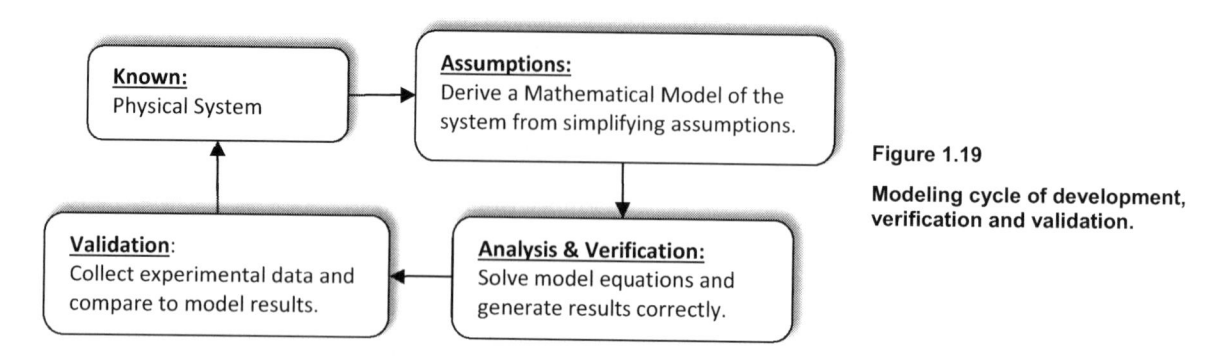

Figure 1.19

Modeling cycle of development, verification and validation.

"The real problem is not whether machines think but whether humans do." – B.F. Skinner

Computers beguile us with their power, graphics, and speed of solution of mathematical models. Be careful not to develop a false sense of security, and make the mistake of placing too much confidence in computer models without careful verification and validation. We should plan to expend considerable effort to verify and validate the correctness of the solution to the model equations (is it true?) and predictive accuracy and usefulness of any mathematical model (does it matter?). Ask the question: "How sensitive are the model results to uncertainties in the variables or parameters?" If the answer is "VERY" then the model may need additional rigor. Simply because it came from a computer is not enough to validate the results. Remember that models are, at the least, only as good as the information supplied to them. Hence the mantra *GIGO*:

"Garbage in = Garbage out!"

The real value of computational tools, such as *Excel*, is the ability to use a validated model to ask a lot of "What if ...?" type questions. We may ask a variety of questions about our process, and use the model repeatedly to get answers quickly and discover insights without disrupting our real process. We get a return on our modeling investment when we use it to explore the process behavior and discover untapped opportunities. If we do not use the model to further our understanding of the parameter space we waste much of our work developing, solving, and validating our model.

1.2.9 Stochastic versus Deterministic Modeling*

"Just because something doesn't do what you planned it to do doesn't mean it's useless." – Thomas Edison

Unlike deterministic models that have an outcome based on a single set of numerical inputs, stochastic models produce a probability distribution of outcomes, like the histograms in Figure 1.20, based on random inputs selected from their own unique probability distributions of uncertainty. Stochastic models predict the expected outcome, and provide confidence intervals of reliability. Smaller confidence intervals indicate better reliability of the probable outcome. Models used in forecasting typically rely on stochastic inputs due to their inherently uncertain nature[21].

---

[21] IBM's Watson is a prime example.

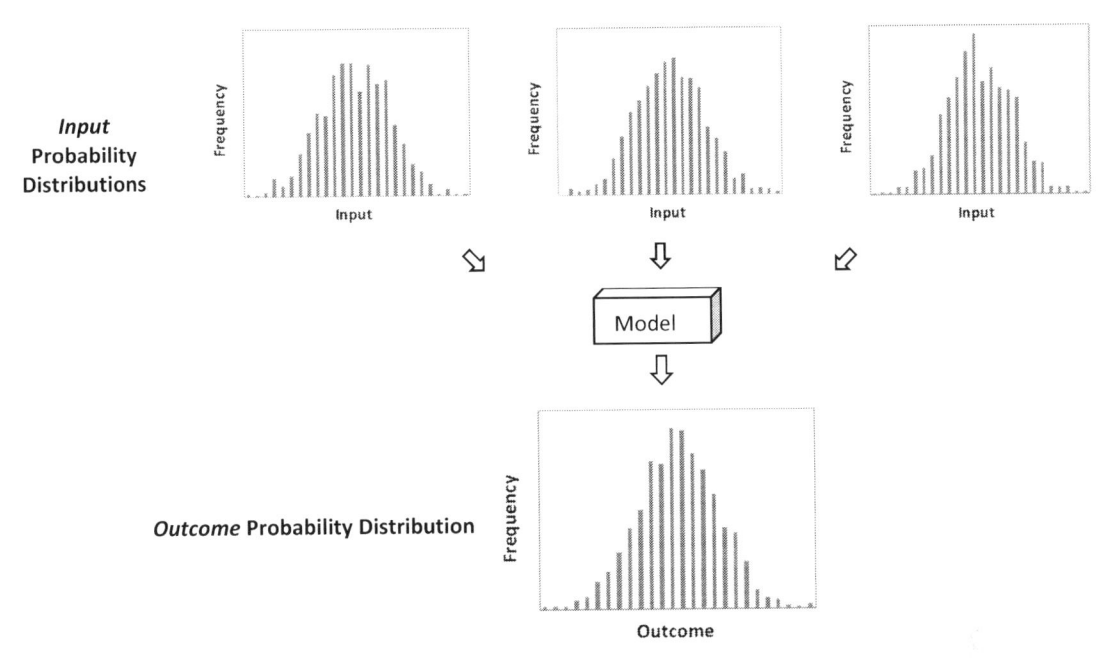

Figure 1.20 Probability distribution of the outcome from a stochastic model with input probability distributions.

Whereas this book deals primarily with numerical methods for solving deterministic type model equations, we also introduce aspects of empirical and stochastic modeling in Chapter 7 on evolutionary optimization, and Chapter 8 and Chapter 9 on uncertainty propagation through data analysis and least-squares regression, respectively. Chapter 11 also presents stochastic Monte Carlo methods for solving difficult integrals.

1.3 Expert Problem Solvers

"You will only be remembered for two things: the problems you solve or the ones you create." - Mike Murdock

Engineers enjoy solving a problem – then moving on to the next one. We must avoid the temptation to skip the critical step in problem solving where we document and reflect on our results. Expert problem solvers have identified several key elements important to the task of creative problem solving that have evolved from Polya's classic text, *How to Solve It* (1985). Our proven problem-solving methodology has four principle components: start with the basic equations, clearly state all assumptions, perform the analysis, and check the results for engineering consistency and common sense. A detailed outline of a formal set of problem solving steps in Table 1.2 incorporates these components.

Adapt these steps to the requirements of your particular problem. For example, an engineer working with toxic solvents may want to add comments or analysis that address safety, environmental impact, waste disposal, and treatment. Beware that simplifying assumptions introduce uncertainty in the results. We present methods for quantifying the reliability of computed results in Chapter 8.

Table 1.2   Sequence of recommended expert problem solving steps. (Compare to Figure 1.19)

Section	Solution Step	Solution Procedure
Known:	Restate the problem in your own words.	• Make sure that you understand the problem. • If you cannot restate the problem in your own words, you likely do not understand it. • What do we know? What do we not know? • Can you provide missing information from estimations or typical conditions? • Pose the problem from a different viewpoint. For example, invert the problem: given A find B becomes given B find A.
Find:	What are you looking for?	• List the desired outcomes of the problem.
Schematic:	Draw and label a simple diagram	• A schematic uses block symbols instead of pictures to represent the elements of a system, including only relevant information. • Label streams, unit operations, conditions, etc., in a simple schematic or block flow diagram (*e.g.*, See Figure 1.3).
Assumptions:	State all assumptions	• "Begin challenging your assumptions. Your assumptions are your windows on the world. Scrub them off every once in a while or the light won't come in." – Alan Alda • Are there any conditions (constraints or boundaries) on the problem? • Make simplifying assumptions necessary to derive a mathematical model of the problem and solve the equations.
Analysis:	Solve the problem	• Plan your solution method. Are there similar problems that have a standard solution method? Can you start from the end and work it backwards? • Break the original problem down into a set of smaller problems. • Collect additional experimental data as needed. • Check your calculations and solution for accuracy. • Verify and validate any modeling results.
Comments:	Critique the solution and make connections to similar problems or techniques	• You may want to ask two questions about the results[22]: What is true? What matters? Then deal with the answers to these questions. • Compare to similar results from your experience or results reported in the literature. • How sensitive is the solution to uncertainty in the inputs? • Do the results violate any basic natural laws, such as conservation of mass and energy? • Are there issues of safety or environment that we should note? • Reflect on what you learned about problem solving to improve your skills.

Example 1.2   Mathematical Model of a ChemE-Car

Chemical engineering students compete in competitions to build a shoebox size car powered by a chemical reaction.[23] Teams of students design a car to travel safely over a range of distances carrying a payload. On the day of the race, officials reveal the travel distance and payload requirements. Student teams adjust the operating parameters for their car to travel the specified distance. The winning car finishes with the smallest difference between the actual travel and target distances. The following example is adapted from the design and analyses of Farhardi, et al (2009) and Lewis, et al (2006).

Known: Early ChemE car designs used pressurized gases generated from liquid phase reactions to power the car. Gases released at high velocity through a nozzle overcome the rolling friction and move the car toward the goal line. Table 1.3 has the parameters needed to model the car's performance. One popular reaction involved the decomposition of hydrogen peroxide to produce oxygen gas:

$$2H_2O_2 \xrightarrow{catalyst} 2H_2O + O_2(g) \tag{1.79}$$

---

[22] Kenneth Solen, Emeritus Professor of Chemical Engineering, Brigham Young University.
[23] http://www.aiche.org/Students/Conferences/chemecar.aspx

CHAPTER 1: INTRODUCTION

Table 1.3 ChemE Car experimental parameters.

Parameter	Symbol	Value	Unit
Ambient pressure	P_{atm}	1.01325×10^5	Pa
Temperature	T	298	K
Nozzle area	A	3×10^{-6}	m^2
Total Volume	V	1×10^{-3}	m^3
Reaction liquid volume	V_{liq}	5×10^{-4}	m^3
Catalyst concentration	C_I-	50	mol/m^3
Total mass	m_{car}	1	kg
Coefficient of friction	μ	0.005	-
Initial reactant concentration	C_{A0}	1000	mol/m^3

Find: (1) Derive a mathematical model of the travel distance from the laws of conservation of mass, energy, and momentum. (2) Use the model to investigate the effect of various parameters on the car's performance.

Schematic: In the following diagram, the reacting liquid partially fills the reaction chamber. Gas escapes through the nozzle to propel the car. The variables that affect the car's movement include the gas escape velocity u_{gas}, car velocity u_{car}, pressure, P, gas volume, V_{gas}, liquid volume, V_{liq}, and temperature, T.

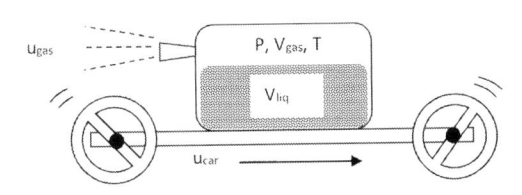

Assumptions: Isothermal, well mixed reactor, negligible air drag force, irreversible reaction, ideal gas, constant temperature and constant liquid volume.

Analysis: Calculate the car velocity from a momentum balance around the car that includes the force generated by the gas escaping through the nozzle and the rolling friction force:

$$\begin{bmatrix} \text{Rate of momentum} \\ \text{accumulation} \end{bmatrix} = \begin{bmatrix} \text{Rate of momentum} \\ \text{in from thrust} \end{bmatrix} - \begin{bmatrix} \text{Rate of momentum} \\ \text{out from friction} \end{bmatrix}$$

$$m_{car} \frac{du_{car}}{dt} = \left(A u_{gas} \rho_{gas} \right) u_{gas} - \mu m_{car} g \quad (1.80)$$

where A = cross sectional area of the nozzle, m^2 u_{car} = car velocity, m/s
g = gravity acceleration = 9.81 m/s^2 u_{gas} = nozzle gas velocity, m/s
m_{car} = total mass of the car and payload, kg ρ_{gas} = density of gas product, kg/m^3
t = rolling time, sec μ = coefficient of rolling friction

The car does not accelerate until the thrust in Equation (1.80) overcomes the static friction. Apply Bernoulli's mechanical energy balance to calculate the instantaneous velocity of the gas through the nozzle (ignoring friction effects due to the relatively short nozzle length):

$$P_a - P_{atm} = \frac{1}{2} \rho_{gas} u_{gas}^2 \quad (1.81)$$

Estimate the density of the gas phase from the ideal gas law:

$$\rho_{gas} = \frac{P_{atm} M_{gas}}{R_g T} \quad (1.82)$$

where M_{gas} = molecular weight of gas product, kg/kmol R_g = ideal gas constant, 8314 m³ Pa/kmol K
P_a = absolute pressure in the reactor, Pa T = temperature, K P_{atm} = atmospheric pressure, Pa

Substitute from Equation (1.82) for the density in Equation (1.81) and solve for the instantaneous gas velocity:

$$u_{gas} = \sqrt{\frac{2R_g T P_g}{M_{gas} P_{atm}}} \tag{1.83}$$

where $P_g = P_a - P_{atm}$ or gauge pressure.[24]

Next, account for the change in pressure with respect to reaction time by differentiating the ideal gas law, assuming negligible temperature change in a dilute reactor:

$$\frac{dP_g}{dt} = \frac{RT}{V_{gas}} \frac{dn_{gas}}{dt} \tag{1.84}$$

where V_{gas} = constant volume of the gas headspace in the reactor, m³.

A mass balance for the gas produced by the reaction accounts for the generation of O_2 and the loss of gas through the nozzle:

$$\frac{dn_{gas}}{dt} = V_{liq} r_{gas} - u_{gas} A \rho_{gas} \tag{1.85}$$

where r_{gas} = rate of gas production, kmol/m³·s and V_{liq} = volume of the aqueous phase, m³.

In the presence of iodide catalyst, the reaction mechanism has two steps that yield a first-order rate law:

$$\begin{aligned} H_2O_2 + I^- &\to IO^- + H_2O \\ H_2O_2 + IO^- &\to I^- + H_2O + O_2(g) \end{aligned} \tag{1.86}$$

where the first step in the catalyzed reaction mechanism limits the rate. A material balance for the reactant around the aqueous phase of the reactor gives the rate of production of gas according to the stoichiometry in Equation (1.79). The mole balance for H_2O_2 (symbolized by A) in the liquid phase includes the pseudo first-order reaction:

[Rate of molar accumulation] = [Rate of molar generation]

$$V_{liq} \frac{dC_A}{dt} = -V_{liq} k C_A \tag{1.87}$$

where V_{liq} is the volume of liquid in the reaction chamber. The rate constant with units of inverse seconds includes the molar concentration of iodide catalyst (Nyasulu and Barlag 2008):

$$k/s = 0.011 C_{I^-} \tag{1.88}$$

Separate variables and integrate Equation (1.87) for a function of the reactant concentration with time:

$$C_A = C_{A0} \exp(-kt) \tag{1.89}$$

For dilute aqueous systems, the combined mass and specific heat is relatively large such that any temperature change due to heat of reaction is negligible.

Combine the results from Equations (1.85) and (1.89) into the pressure Equation (1.84) to yield the following simultaneous first-order differential equations in terms of the liquid volume, V_{liq}, gas volume $V_{gas} = V - V_{liq}$, and gauge pressure, $P_g = P_a - P_{atm}$:

---

[24] For compressible, constant enthalpy, isentropic flow (isentropic exponent $k = 1.4$ for oxygen): $u_{gas} = \sqrt{\frac{2k}{(k-1)}\left(\frac{P_a}{\rho_a}\right)\left[1 - \left(\frac{P_{atm}}{P_a}\right)^{(k-1)/k}\right]}$.

$$\frac{dP_g}{dt} = \frac{V_{liq}RTkC_{A0}\exp(-kt)}{2(V-V_{liq})} - \frac{A}{(V-V_{liq})}\sqrt{\frac{2RTP_g(P_g+P_{atm})}{M_{gas}}} \qquad (1.90)$$

$$\frac{du_{car}}{dt} = \frac{AP_g}{m_{car}} - \mu g \qquad (1.91)$$

subject to the following initial conditions at $t = 0$:

$$P_g = 0 \quad\text{and}\quad u_{car} = 0$$

and constraints:

$$P_g \geq 0 \quad\text{and}\quad u_{car} \geq 0$$

The final travel time t_f is determined at the point when the car stops ($u_{car} = 0$). Integrate the velocity profile to calculate the total travel distance using the results for the car velocity from the solution to the system of Equations (1.90) and (1.91):

$$z_f = \int_0^{t_f} u_{car}\,dt \qquad (1.92)$$

We present numerical methods for solving integrals and first-order differential equations in Chapter 11 and Chapter 12. We treat the coefficient of friction as an adjustable parameter determined by experimentation. Students explore the significance of the other parameters by experimentation with the model.

The students initiate the reaction by adding catalyst to the reaction chamber. As seen in Figure 1.21, the simulated reactor pressure builds up quickly in the first few seconds of reaction, and then gradually decreases as the liquid reactants become depleted and the gases escape the reactor vessel.

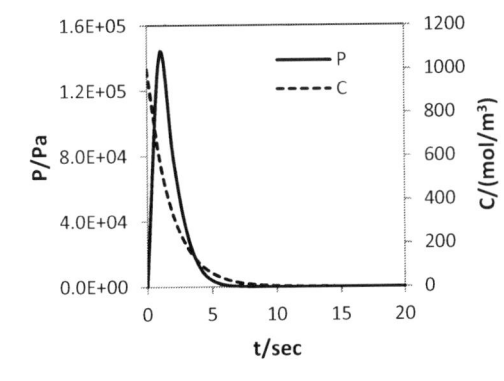

Figure 1.21

ChemE Car reaction chamber gauge pressure and reactant concentration for the conditions in Table 1.3.

The plot of the velocity in Figure 1.22 shows an initial acceleration followed by a constant deceleration as the car coasts to a stop after the reaction is complete, and the pressure drops due to the escaping gas.

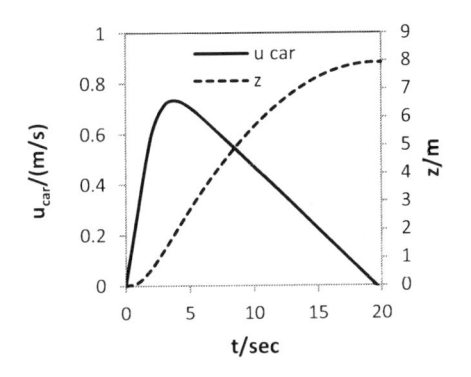

Figure 1.22

ChemE Car velocity and travel distance versus time for the conditions in Table 1.3.

A graph of numerical results in Figure 1.23 has a series of curves for the distance traveled versus friction factor with different payloads. Student teams use pre-race experiments to determine the coefficient of friction from travel distance by reading the value from the graph.

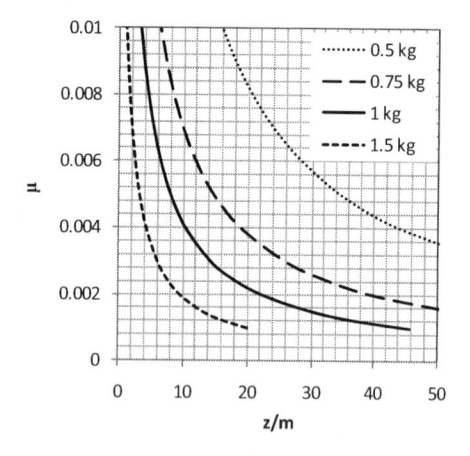

Figure 1.23

Coefficient of friction versus travel distance for the conditions in Table 1.3 and m_{car} = 0.5, 0.75, 1.0, 1.5 kg.

The plot includes gridlines to help read interpolated values from the curves. For instance, when the car has a total mass of 1.0 kg, an experimental travel distance of 10 meters corresponds to a coefficient of friction equal to 0.004. Power law functions provide good models of the calibration data. For example, the following equation fits the calibration data of the friction coefficient versus travel distance for a car with a total mass of one kilogram:

$$\mu = 0.0294 z^{-0.873}$$

Chapter 9 covers methods of least-squares regression for fitting empirical models to data. For the conditions of the race, the only variable under the control of the students is the liquid volume in the reactor.

Comments: The model makes assumptions that require experimental validation. For example, changing temperatures affect reaction rates and pressure in the chamber. In addition, the reactor contents may not be well mixed. We must not ignore safety during the manufacturing and operational stages of the chemical car.

□

> *"The trouble with models is that every one of them has an associated set of assumptions. The most common mistake when dealing with a model is to forget the assumptions."* – Michalewicz and Fogel (2000)

The legacy of the space shuttle program is a sobering reminder of the *power* and *consequences* of both *good* and *bad* problem solving that led to the 1986 Challenger and 2003 Columbia Space Transportation Systems (space shuttle) triumphs and untimely accidents (Gehman 2003). Students occasionally complain about losing points for incorrect calculations in graded assignments and exams, arguing that their method was *mostly* correct. However, history has shown that "there are no small mathematical errors in engineering work." (Kowalczyk 1963). The loss of NASA's Mars Orbiter[25] attributed to overlooked unit conversions and the collapse of the Quebec Cantilevered Bridge[26] because of failure to redesign for added length are additional examples of costly engineering mistakes.

The accuracy of our results and clarity of our presentation are both issues of ethics and professionalism. Lives may depend on the reliability of our results. Goddard (2001) emphasizes that a solution checked by an engineer gains a large measure of credibility and trust; and that we must match the psychological gain in credibility with technical sufficiency. Checking calculations is an important layer of safety protection. We need to realize that verification and

---

[25] http://www.cnn.com/TECH/space/9909/30/mars.metric.02/
[26] https://en.wikipedia.org/wiki/Quebec_Bridge

CHAPTER 1: INTRODUCTION

validation are not trivial parts of the solution process.[27] The problem-solving format recommended here promotes engineering documentation, proper numerical precision, accuracy, sound engineering judgment, and critical evaluation of the results.

> *"Review your work. You will find, if you are honest, that 90% of the trouble is traceable to loafing."* – Ford Frick

Mathematical models allow engineers to explore the parameter space, understand a process, look for relationships, and find opportunities for improvement. Models may come from first principles (laws of conservation) or convenient functions such as polynomials that capture the behavior of the real system. We should start with simple expressions, then add rigor only when necessary. In many cases, a solution to model equations requires numerical approximation techniques. The remaining chapters introduce *Excel* with VBA as a platform for building models and offer a comprehensive tool set for finding solutions to our common modeling problems.

1.4   Appendix: Primer on Reaction Engineering

> *"It is your **reaction** to adversity, **not the adversity itself**, that determines how your life's story will develop."* – Dieter Uchtdorf

Physical chemists commonly study fluid phase reactions in small, well-controlled batches, using well-mixed vessels such as beakers and flasks, or small high-pressure experimental reactors like the one pictured in Figure 1.24.

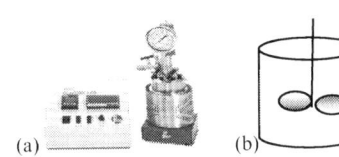

Figure 1.24

(a) Laboratory scale controller and research reactor
(b) Simple schematic of an ideal batch reactor.

We use batch reactors to observe the change in reactant concentration over time. It is tempting to define the reaction rate law, r_A, for species A as the change in concentration of reactant, C_A, with time, t:

$$r_A = -\frac{dC_A}{dt} \tag{1.93}$$

We must be careful to avoid this conceptual trap![28] The expression in Equation (1.93) is just the application of the law of mass conservation in Equation (1.35) to a batch reactor, assuming no influent or effluent, and constant fluid volume:

$$(\text{Rate In} = 0) - (\text{Rate Out} = 0) + \begin{pmatrix} \text{Rate of Generation} \\ Vr_A \end{pmatrix} = \begin{pmatrix} \text{Rate of Accumulation} \\ V\dfrac{dC_A}{dt} \end{pmatrix} \tag{1.94}$$

Analysis of batch reactor data consisting of concentration change with time and temperature reveals the functionality of the reaction rate law, in terms of the reactant concentrations and temperature:

$$r_A = f(C_A, T) \tag{1.95}$$

Elementary reactions follow rate laws involving exponents on concentrations that correspond to the order of reaction, m, typically from the stoichiometry, *e.g.*:

---

[27] Please note that *Verification and Validation* have different meanings in different contexts. For our purposes here, verification and validation are the steps we take in our responses to the questions, "Is it true? Does it matter?" In other words, validation asks, "Are you building the right model?" while verification asks, "Are you modeling it right?" (Boehm 1981)

[28] Personal note from Professor Robert Rinker, University of California Santa Barbara, Department of Chemical Engineering.

$$r_A = kC_A^m \tag{1.96}$$

where k is the rate "constant" with units $s^{-1} \cdot (mol/m^3)^{1-m}$. To illustrate, the rate law for the elementary, irreversible, second-order reaction mechanism $2A \to B$ is:

$$r_A = -kC_A^2 \tag{1.97}$$

Another second-order reaction might involve two different molecular species combining to form a product, $A+B \to C$:

$$r_A = -kC_A C_B \tag{1.98}$$

The rate law in Equation (1.98) is first-order in either species A or B, but second-order overall. For equal starting concentrations of A and B, the reaction behaves as a second order reaction in either A or B. Alternatively, when the initial concentration of component A is large, $C_A \gg C_B$, the reaction rate is pseudo-first-order:

$$r_A = -k' C_B \tag{1.99}$$

where the pseudo-first-order rate constant $k' = kC_A$, assuming a relatively negligible change in C_A.

For small temperature ranges, reaction rate constants typically exhibit an Arrhenius exponential temperature dependence illustrated in Figure 1.25.

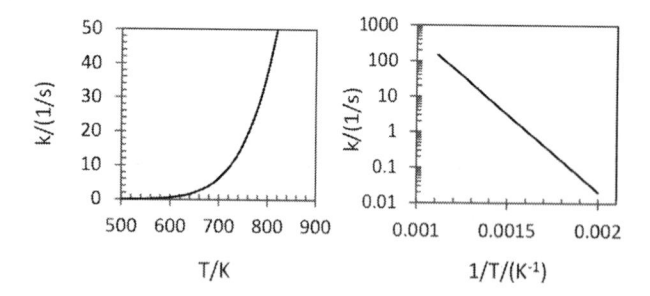

Figure 1.25

Example of Arrhenius temperature dependence of a reaction rate constant.

The Arrhenius function for rate constants involves the exponential of the inverse absolute temperature:

$$k(T) = k_0 \exp(-E_a/R_g T) \tag{1.100}$$

where k_0 is the pre-exponential, or frequency factor with the same units for k, R_g is the ideal-gas constant, and E_a is the activation energy.

When the reaction temperature changes by any appreciable amount, an energy balance is required to include the effects of temperature on the rate constant:

$$\frac{dT}{dt} = \frac{U \cdot a \cdot (T_s - T) + (\Delta H \cdot r)_A}{\rho \cdot c_p} \tag{1.101}$$

where Ua is the product of the overall heat transfer coefficient and specific surface area of heat transfer, T_s is the temperature of the surrounding heating/cooling fluid, $\Delta H \cdot r$ is the product of the heat of reaction and reaction rate, and where ρ and c_p are the density and heat capacity of the reactor contents, respectively. Chapter 12 presents numerical methods for solving systems of first-order, ordinary differential equations like Equations (1.93) and (1.101).

Non-elementary reactions, such as those involving catalysts or complex mechanisms, have rate laws with complex nonlinear functions. For instance, a bacterium, B that grows on substrate S may reproduce according to the Monod rate law:

$$r_B = \frac{\mu_{max} C_S C_B}{K_S + C_S} \tag{1.102}$$

where μ_{max} is the maximum specific growth rate and K_S is the Monod constant, both unique to the type of bacteria and growth substrate. Under conditions of large substrate concentration, $C_S >> K_S$, Equation (1.102) reduces to a simple first-order rate law:

$$r_B = \mu_{max} C_B \tag{1.103}$$

Batch reactor design answers the question of how much reaction time is required to achieve a specified conversion of reactant or product in a given reactor volume. We predict the required holding time from the solution to the differential mass balance in Equation (1.93). With conditions of constant temperature, we may separate variables and integrate for the time:

$$t = \int_{C_A}^{C_{A0}} \frac{dC_A}{r_A} \tag{1.104}$$

where the limits of integration C_{A0} and C_A are the initial and final reactant concentrations, respectively. Chapter 11 presents methods for calculating good approximate solutions to integrals.

Complex rate laws that involve more than one molecular species may require simultaneous solution of coupled differential equations for mass balances around a batch reactor for each species, e.g.:

$$\frac{dC_A}{dt} = -r_A \quad \text{and} \quad \frac{dC_B}{dt} = -r_B \tag{1.105}$$

A continuous flow reactor, operating at steady state as depicted in Figure 1.16, has no net accumulation. The accumulation term in the species balance is set to zero, whereas the influent and effluent rates are nonzero. The result is an algebraic reactor design equation:

$$Q(C_{A0} - C_A) + V r_A = 0 \tag{1.106}$$

where Q is the volumetric flow rate of the feed and exit streams and V is the volume of reacting fluid. We interpret the design for a well-mixed reactor in terms of the residence time, $\tau = V/Q$:

$$\tau = \frac{V}{Q} = \frac{(C_{A0} - C_A)}{-r_A} \tag{1.107}$$

When the rate law involves the concentrations of both reacting species, the model needs two material balance equations around the steady-state reactor:

$$Q(C_{A0} - C_A) - V r_A = 0 \quad \text{and} \quad Q(C_{B0} - C_B) - V r_B = 0 \tag{1.108}$$

The equations are coupled through rate laws that may be functions of the exit concentrations C_A and C_B, such as Equation (1.98). Chapter 6 presents methods for finding roots to nonlinear algebraic equations. An unsteady-state flow reactor involving two molecular species requires a set of mass balances that include the accumulation terms:

$$\frac{dC_A}{dt} = Q(C_{A0} - C_A) - V r_A \quad \text{and} \quad \frac{dC_B}{dt} = Q(C_{B0} - C_B) - V r_B \tag{1.109}$$

We obtain values for parameters in rate laws from experiments. The order of an elementary reaction is determined from the ratio of reaction rates under two different concentration conditions:

$$\frac{r_2}{r_1} = \frac{k}{k} \left(\frac{C_2}{C_1} \right)^m \tag{1.110}$$

With uniform temperature, the reaction rate constant in Equation (1.110) cancels, allowing for the calculation of the reaction order independently from the rate constant:

$$m = \frac{\ln(r_2/r_1)}{\ln(C_2/C_1)} \tag{1.111}$$

Two common methods used to obtain the rate constant include the differential and integral methods. There are two variations of the differential method. In the first variation, the initial rate of reaction in a batch reactor is determined by extrapolating the composition versus time data back to the starting point of reaction at $t = 0$. For example, consider a second-order, elementary reaction:

$$k = -\frac{dC_A/dt\big|_{t=0}}{C_{A0}^2} \qquad (1.112)$$

The alternative version of the differential method uses the derivative over time in a batch reactor:

$$-\frac{dC_A}{dt} = kC_A^2 \qquad (1.113)$$

A plot of $-dC_A/dt$ versus C_{A0}^2 in Figure 1.26 gives a line with slope k.

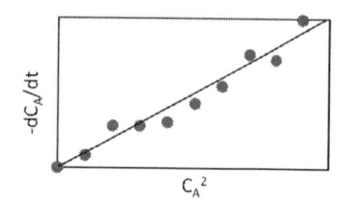

Figure 1.26

Verification of second-order reaction and calculation of the rate constant using the derivative method.

We present methods for approximating derivatives of data in Chapter 5. Chapter 9 presents least squares methods for fitting data with a model function that smooths noisy data.

The integral method for a batch reactor finds the rate constant for an assumed rate law by separating variables and integrating the equation. For the second-order case this gives:

$$-\int_{C_{A0}}^{C_A} \frac{dC_A}{C_A^2} = k\int_0^t dt \quad \text{or} \quad \frac{1}{C_A} - \frac{1}{C_{A0}} = kt \qquad (1.114)$$

If the reaction is truly second-order, a plot of $1/C_A$ versus t, as seen in Figure 1.27, should give a line with slope k and intercept $1/C_{A0}$. Due to random fluctuations in experimental data, a numerical method is needed to find the equation of a line that best fits the transformed data in Equation (1.113) or (1.114).

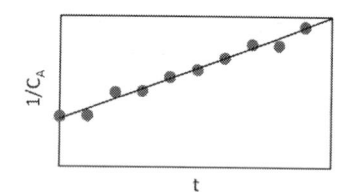

Figure 1.27

Verification of second-order reaction and calculation of the rate constant using the integral method.

Use the method of least squares regression presented in Chapter 9 to model data and obtain kinetic parameters. For more complex rate laws, we may solve the integral on the left hand side of Equation (1.114) using numerical methods presented in Chapter 11. Alternatively, we may solve the differential equations for the batch reactions using methods presented in Chapter 12. Chapter 8 develops methods for quantifying reliability of numerical results from models for reaction kinetics and reactor design.

Chapter 2   Engineers Use Excel

"No crime is so great as daring to excel." — Winston Churchill

Excel has evolved from its original conception as a financial accounting tool into its current version that not only supports business activity, but also has many features of a serious computational tool suitable for solving complex science and engineering problems (Lilley 2010), including:

- Data analysis, and modeling
- Documenting and sharing data and numerical results,
- Built-in math, engineering, and statistical functions,
- Charting and graphing for data visualization,
- Goal seeking for inverse problem solving,
- User-interface for model inputs and results, and
- **Visual Basic for Applications** (*VBA*) programming environment for customization.

SQ3R Focused Reading Questions
1. What is the difference between an *Excel* workbook and a worksheet?
2. What are relative and absolute cell references?
3. How do I rename a cell?
4. What are shortcut methods for moving around a worksheet?
5. What is a circular reference?
6. Where can I find a list of available worksheet functions?
7. What does a good graph look like?
8. What is an array function or array operation?
9. What is a **Data Table** versus a **Pivot Table**?
10. What can I do with the **Solver**?

This chapter does not contain encyclopedic coverage of *Excel*'s capabilities and applications. We merely scratch the surface to expose the power of *Excel* by highlighting features that make it an important engineering tool, as well as give you experience and confidence to use *Excel* to tackle your own problems. We introduce a few additional features of *Excel* needed for implementing various numerical methods throughout the book. All of the example *Excel* workbook files are downloadable from the book's companion website: https://sites.google.com/d.umn.edu/pnm4

"I have been impressed with the urgency of doing. Knowing is not enough; we must apply. Being willing is not enough; we must do." — Leonardo da Vinci

To get the most from this book, we recommend redoing the examples for yourself in *Excel*. Novice *Excel* users should get started by working through Microsoft's on-line tutorials. Click on the help icon located at the top right corner of the worksheet to access the help files that accompany the *Excel* software. Refer to *Excel*'s Help features often to find useful tips and examples. Keep reinventing yourself – moving from competency to expertise. Several web sites post additional *Excel* tips (this same advice applies to numerical methods in general). Many expert *Excel* users are eager to share what they have learned about clever, efficient workarounds for all types of problems. Your expertise will grow as you embrace the wide world of *Excel*.

2.1 Getting Started with Excel

"Many people who excel are self-taught." – Herb Ritts

When we open a new *Excel* workbook, we start with a blank worksheet consisting of a grid of rectangular cells where we add data, text, and formulas. Refer to a cell's location or "address" on an *Excel* worksheet according to column and row intersections, such as **B6**, just like Milton-Bradley's *Battleship* game. Access all of the functionality of *Excel* from the tabs on the ribbon at the top of the worksheet window. As shown in Figure 2.1, the ribbon organizes the capabilities in various groups accessible from tabs. The **Home** tab has groups of the basic operations such as formatting, cutting, pasting, and data manipulation; all features common to most *Windows* and *Mac* programs. Access advanced functionality from the other tabs, including special functions in mathematics, statistics and finance, charting, and data analysis.

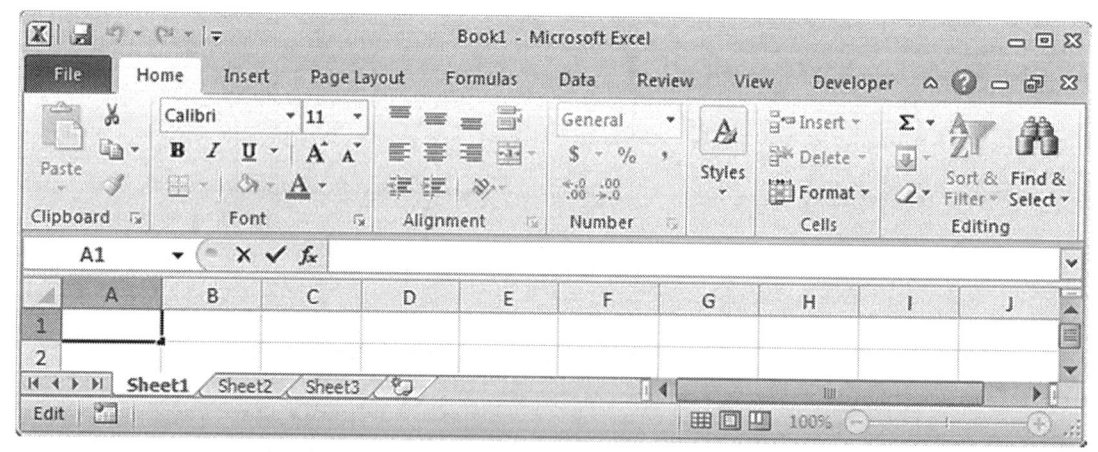

Figure 2.1 *Excel* worksheet showing the Ribbon File tab in the top left corner, tabs for accessing the functionality of *Excel*, and a grid of cells for adding data to the worksheet.

By default, a new *Excel* workbook opens with three blank worksheets, tabbed at the bottom of the window, as seen in Figure 2.1. To add or delete extra *Excel* worksheets, right-click on an *Excel* worksheet tab at the bottom and select **Add** or **Delete**. We can quickly add an *Excel* worksheet by clicking on the **Add Worksheet** button located next to the rightmost worksheet tab. With the left mouse button, double-click on the name of an *Excel* worksheet to select and rename it with a descriptive title to indicate its contents.

2.1.1 Backup Your Work

"As a kid, Terry Bradshaw didn't amaze me. My hero was Steelers backup Terry Hanratty, who nabbed two Super Bowl rings while completing three passes." – Stephen Rodrick

Develop a habit of periodically making backup copies of your digital files. There are few more frustrating experiences with computing than losing a file due to negligence, corruption, a computer "crash", power failure, etc. Create a plan for saving and organizing your workbooks with a logical naming scheme and backup system in case you lose a file. Students like to use small flash drives for convenience in portability. Flash drives have the added advantage that they do not require internet access. The disadvantages of portable flash drives include the possibilities of loss and theft. We also recommend using some version of on-line "cloud" tool for storing files, such as Apple *iCloud*, Google *Drive*, and Microsoft *SkyDrive* (all free). The obvious disadvantage of internet tools is the requirement of internet access. Whatever you use, get into the habit of regularly backing up your important files.

CHAPTER 2: EXCEL

2.1.2 Calculators versus Excel Spreadsheets

Whether you are a student or practicing engineer, you likely have some experience using a hand calculator to solve engineering problems. Most engineering students own powerful graphing calculators that perform some of the numerical methods introduced in this book. So why do we need a spreadsheet program to do our number crunching? The short answer – engineering problems require complex calculations with several intermediate steps and documentation for sharing. One simple analogy depicted in Figure 2.2 for understanding *Excel* compares an *Excel* worksheet to a collection of individual "calculators."

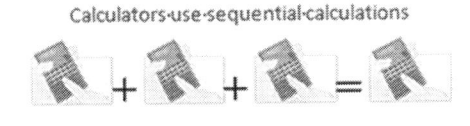

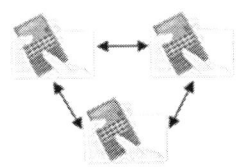

Figure 2.2

Use calculators to perform simple engineering calculations. Spreadsheets act as interactive calculators.

Each cell in the array on a spreadsheet can function independently as a single "calculator." However, unlike individual hand-held calculators, the "calculators," or cells, in an *Excel* worksheet can store information, interact, and share information to conveniently organize and solve complex problems.

A typical scenario with a hand calculator might involve performing a calculation, writing the result by hand on paper, performing a second calculation, then combining the result from the first calculation with the second calculation by re-entering the information from the first step – not terribly inconvenient for simple calculations, although tedious and error prone for complicated calculations involving many steps. This is where spreadsheet software, such as *Excel*, is indispensable for accurate and efficient calculations involving complex problems. We minimize calculation errors by breaking complicated problems into smaller, manageable pieces, and then reassembling the intermediate results for a final answer.

We reference worksheet cells by their locations in lettered columns and numbered rows. In place of buttons on a calculator, *Excel* uses the keyboard and menus for entering information into a cell. Enter formulas for calculations in the formula bar, just above the cells, as shown in Figure 2.3. Once we have our formula set-up in a cell, display the calculated result by pressing the **Enter** key (just like a calculator!), or clicking the check mark ✓, just left of the formula bar. The other two options on the formula tab, represented by the icons ✘ and f_x, function to clear the contents of the cell and open a drop-down list of built-in functions to use in the formula, respectively. Unlike a single calculator, however, we do not have to clear the previous result to do another calculation. We simply start a new calculation in a different cell, keeping the previous result in its original cell. The number of cells in a workbook is *practically* limitless.[29]

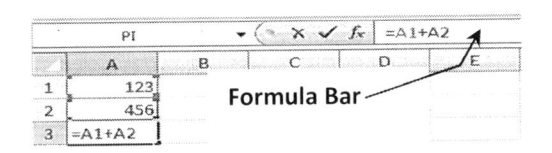

Figure 2.3

Cells in an *Excel* worksheet. Use cell A3 to add the numbers in cells A1 and A3.

By default, *Excel* workbooks automatically recalculate whenever the user makes a change to any cell in a worksheet. Recalculation of large spreadsheets may take a long time. We can turn off automatic calculation in the **File** settings, and manually recalculate the worksheet using the *F9* key.

A number may be stored directly in a cell without further designation or manipulation. Any reference to the cell with the number accesses the value of the number. Initiate a mathematical calculation in a cell with an equal sign. Table 2.1 contains the basic mathematical operations available for building *Excel* formulas.

---

[29] *Excel* 2007 increased the number of rows in a worksheet to over 1 million and the number of columns to 16,384. This makes it much more challenging to find and sink your battle ships!

Table 2.1 Common mathematical operations in *Excel*

Operation	Symbol	Operation	Symbol	Operation	Symbol
Equal	=	Addition	+	Multiplication	*
Exponent	^	Subtraction	-	Division	/

Be careful when entering formulas: *Excel* performs ^, *, and / (power, multiplication, and division) operations before + and – (addition and subtraction) operations. This may be different from how you perform a calculation with your hand calculator. Control the order of mathematical operations using parentheses, ( ). To illustrate, Figure 2.4 shows worksheet formulas and results for the following expression with and without parentheses to control the sequence of calculations. Notice how the negative sign is associated with the number three when squared without parentheses:

$$1 + 2 \times \left(-\frac{3^2}{4}\right)^5 = -114 \qquad (2.1)$$

Figure 2.4 (a) Calculation of Equation (2.1) without parentheses (cell A1) and with parentheses (cell A2) to control the order of calculations. (b) View cell formulas by clicking on Show Formulas in the Formulas Tab/Formula Auditing section (c).

The "calculator" analogy only takes us so far in our understanding of the power of spreadsheets. *Excel* worksheets are in essence multi-dimensional calculators where each cell may use information from other cells (including cells from other worksheets) to construct complex sequences of calculations. The results may be displayed in cell ranges (groups of cells) or graphically. Although it may not seem like it, setting up a worksheet to perform a complex series of calculations is a form of computer programming. Becoming a power user of Excel enhances our coding skills. Several built-in mathematical, statistical, logical functions, and programmed macros extend the capability of *Excel* for performing engineering calculations. The next time your reach for your calculator, you might want to open an *Excel* worksheet instead! Free *MS Excel* apps are now available for use on our smart phones and tablets.

2.1.3 Shortcuts for Managing an Excel Worksheet

Excel has several short cuts for customizing the worksheet display and moving around a worksheet:

a. Select/Activate a cell or range of cell: Use the mouse to click in a new cell to activate it or click, hold, and drag to the final cell to select ranges of cells. A single active cell displays a dark, solid border. An active range of cells is shaded with a solid border.
b. Build formulas in a cell starting with an equal sign, "=". To reference a cell address in a formula, click in the referenced cell, or type the column:row address into the formula. Edit formulas directly in the cell or in the formula bar below the ribbon.
c. Use the tab and arrow keys to move from one active cell to another, row to row or column to column. Press the *SHIFT* key then use the arrow keys to select an active range of cells.
d. To change the active cell to the end of a row or column of contiguous filled cells, double-click on the border facing the direction of movement.
e. To select a block range of cells, click in the cell at the top left corner of the range, hold down the *SHIFT* key, and then click in the cell at the bottom right corner of the range.
f. To select a row or column of contiguous filled cells, press the *SHIFT* key while double-clicking on the cell border (not the corner) facing the fill direction.
g. To select a table of contiguous filled cells, click anywhere in the table and type *CNTL SHIFT ** or *CNTL a*.
h. Move the contents of a cell or block of cells by selecting the cell range and dragging the border to a new location on the worksheet.

CHAPTER 2: EXCEL

i. Press the *CTRL* key while dragging any border of a block of cells to move a *copy* of the selected range and leave the original cells unchanged.
j. To adjust a column width or row height to match the cell contents, double-click on the border of the column or row header. Drag the border of a row or column header to change the dimension.
k. To freeze part of the worksheet, select from the tab menu **View>Freeze Panes** so that table headers do not move while scrolling across or down an *Excel* worksheet. Select the option to **Unfreeze** if needed.
l. To display just the selected block of cells, use the ribbon tab menu option **View>Zoom to Selection** to increase the magnification of the worksheet. Click 100% in the same group to return to regular viewing size.
m. To select multiple non adjacent cells or ranges, select a cell (or range) then hold down the **CTRL** key while selecting additional cells (or ranges). Alternatively, press **SHIFT+F8** before selecting additional cells (or ranges). Press **SHIFT+F8** again to stop selecting additional cells (or ranges).
n. To combine ranges, use a comma between consecutive references to ranges. Leave a space between references to multiple ranges to perform and operation on the intersection of the ranges.

The book's web site has a folder containing all of the example *Excel* worksheets for each chapter. The Chapter 2 *Examples* folder has the worksheet in Figure 2.5 with blocks of cells setup as shown below for practicing the shortcut methods described above.

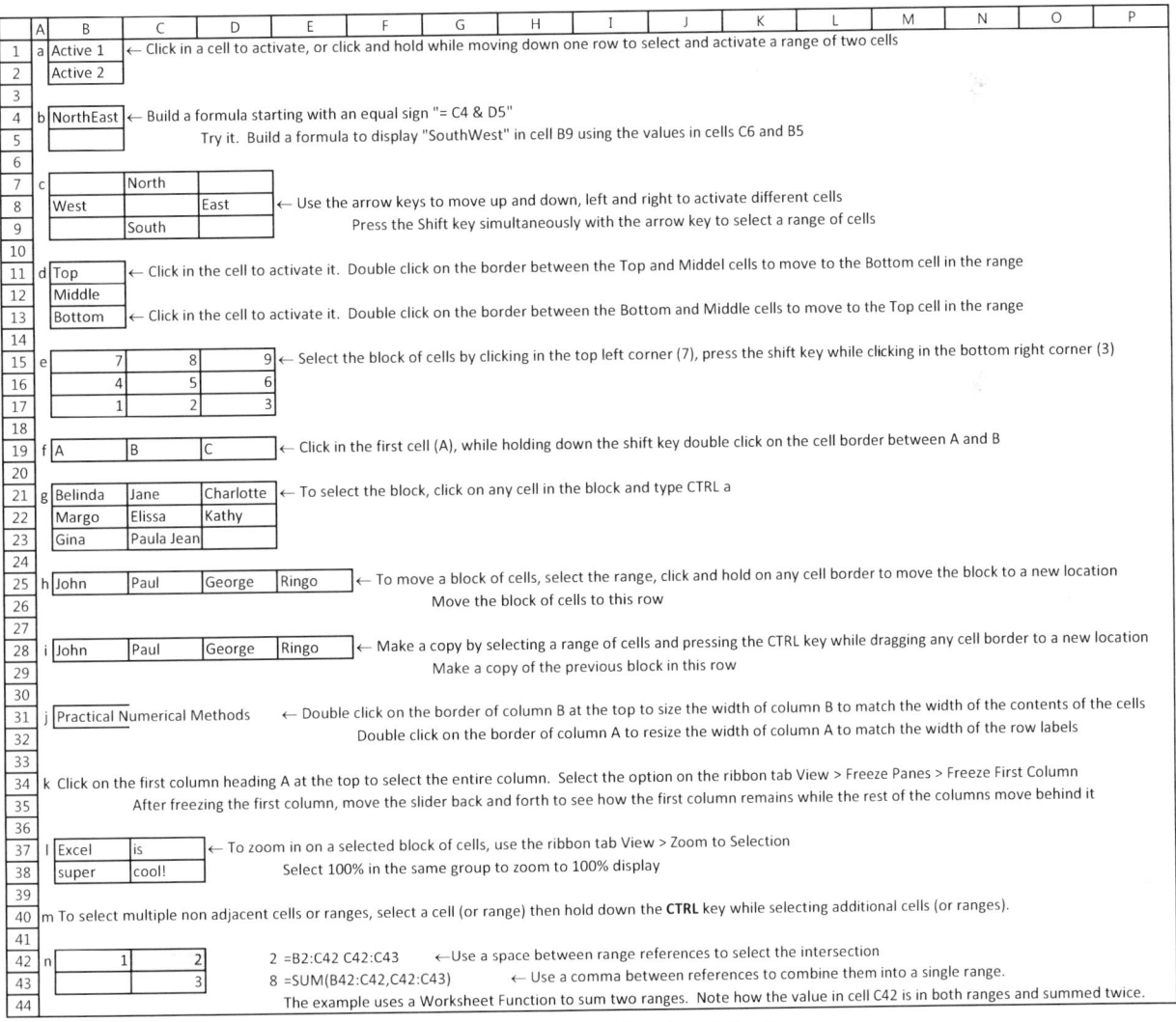

Figure 2.5   Examples of shortcuts for getting around an *Excel* worksheet.

2.1.4 Relative vs Absolute Cell Referencing and Named Cells

We reference a cell by its column and row address, **C#R#**, where **C#** refers to the column header and **R#** refers to the row number. When referencing a cell in a formula we have the choice of using a relative or absolute reference. This becomes useful when copying a formula in one cell to another. If cell references are relative, then the cell addresses in the copied and pasted formula automatically adjusts to the new cell's relative position in the worksheet. Absolute cell referencing uses the dollar sign, **$** in front of either column or row identifier, or both. For example, the reference **$A$1** indicates that both column and row are absolute. **$A1** indicates that the column is absolute and the row is relative to its location on the worksheet. Repeatedly pressing the **F4** function key cycles through combinations of relative and absolute cell references (*e.g.*, click on **A1** to select it, then press the **F4** function key four times to cycle through **$A$1** or **A$1** or **$A1** or back to **A1**).

Occasionally, we may need to make copies of cells without altering relative references. Use the following simple workaround to make a copy of a range of cells that keeps the original relative references in the formulas:

1. Select the range of cells containing the relative references.
2. Open **Find and Replace** by simultaneously typing the keys **CNTL** and **H**. Alternatively, select **Replace** from the **Find & Select** option in the **Editing** group on the ribbon **Home** tab.
3. Use **Find and Replace** to change all = to #. This temporarily inactivates the formulas.
4. Copy and paste the range of cells in a new location on the worksheet.
5. Select the original range of cells, and use **Find** and **Replace** to change all # back to =.

Excel's powerful time saving **Auto Fill** feature, described in Section 2.1.5, conveniently fills the contents of a range of cells with values or formulas based on the address of the starting cell(s). The next illustration compares relative with absolute cell addressing when using **Auto Fill**. Before getting to the illustration, it is often useful to rename a cell when performing engineering calculations. For example, we may need to use the ideal gas constant, temperature, and molar volume to calculate a gas pressure. Rename cells using common engineering terms or symbols that reflect their contents. Use the new names of the cells as absolute references in place of the column-row default name in formulas.

Figure 2.6

Rename cell B1 to Rg, representing the ideal gas constant. Use Name Manager in the Defined Names section on the Formulas tab to change cell names.

We rename cells by typing the new name in the **Name** box, located just left of the formula bar, as shown for **Rg** in Figure 2.6. Type the function key *F3* to get a list the named cells. We may find that we need to change or delete a cell name. Use the **Name Manager** in the **Defined Names** group of the **Formulas** tab to make any cell name changes required. Alternatively, right-click in a cell and select **Name a Range** to change the scope of the cell reference.

Cell naming has some restrictions. We may not rename a cell with a different column-row designation (*e.g.*, we cannot rename cell **A1** as **B2**). In addition, we may not use intrinsic *Excel* function names for cell names (*e.g.*, we cannot use the name **AVERAGE** reserved for the *Excel* function with the same name for calculating the average value of a range of cells). For another example, *Excel* does not permit cell names *C* or *R* (*e.g.*, the ideal gas constant) because upper or lower case **C** and **R** conflicts with *Excel*'s shorthand references to column and row. One way to get around naming conflicts is to incorporate an underscore _ within a cell name, at the beginning, middle, or end, such as **C_**. The *Excel* worksheet in *Example 1.1* used named cells for the constants *a*, *b*, and *c* in a quadratic equation.

A convenient tool for simultaneously naming multiple cells uses the **Create from Selection** command in the **Defined Names** group in the **Formulas** tab, shown in Figure 2.6. Type the names and values into the first three cells of columns **A** and **B**, as shown in Figure 2.7. Select the two-column range of cells containing the names and values. Click **Create from Selection** and check **Create names from values in the Left column** in the dialog box. Cells **B1** and

CHAPTER 2: EXCEL

B2 now have absolute reference names *a* and *b*, respectively. However, *Excel* modified the name for cell **B3** with an underscore **c_**.

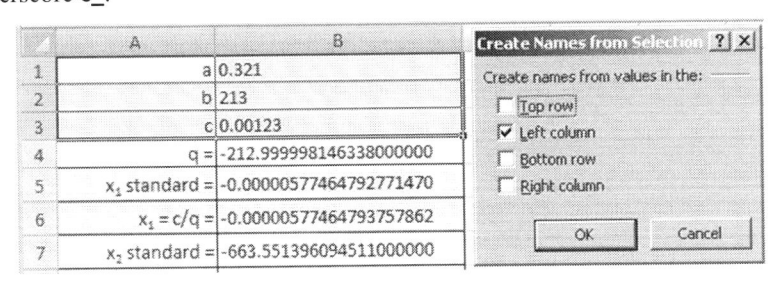

	A	B
1	a	0.321
2	b	213
3	c	0.00123
4	q =	-212.999998146338000000
5	x₁ standard =	-0.00000577464792771470
6	x₁ = c/q =	-0.00000577464793757862
7	x₂ standard =	-663.551396094511000000

Figure 2.7

Create multiple named cells from a selection in an *Excel* worksheet.

The formulas for calculating the roots to the quadratic equation use the new user-defined cell names, as shown in Figure 2.8.

	A	B
1	a	0.321
2	b	213
3	c	0.00123
4	q	=-0.5*(b+signb(b)*SQRT(b^2-4*a*c_))
5	x₁ standard =	=(-b+SQRT(b^2-4*a*c_))/(2*a)
6	x₁ = c/q =	=c_/q
7	x₂ standard	=(-b-SQRT(b^2-4*a*c_))/(2*a)
8	x₂ = q/a =	=q/a
9	f(x₁ standard) =	=a*B5^2+b*B5+c_
10	f(x₁ = c/q)=	=a*B6^2+b*B6+c_

Figure 2.8

Formulas for the quadratic rule use named cells to reference the constants.[30] *Excel* automatically replaces the name c with c_ to avoid conflicts with *Excel*'s shorthand reference for column.

Beware that named cells are equivalent to absolute references. If we use a cell name in place of the relative address in a formula it will *not* automatically change the address relative to the row or column when we use **Auto Fill** to generate a range of cells with the formula along a row or column. Example 2.1 and Example 2.4 demonstrate the difference between relative and absolute cell references, including named cells.

2.1.5 Auto Fill a Range of Cells

Use *Excel*'s **Auto Fill** feature to add a series of data or formulas to a range of cells that follow a pattern using relative cell references. To illustrate, suppose we need a row of cells that contain a series of even integer numbers starting with zero. For a small set, we can type the numbers directly in the cells. For a large set, use the following **Auto Fill** steps.

- Establish the pattern in the first few cells of the row (or column). This illustration requires just the first two integers in the pattern.

	A	B
1	0	2

- Select the cells by clicking on the first cell, holding the left mouse button, and dragging over the last cell. Next, hold-click on the fill handle located at the bottom-right corner of the selected range of cells. Notice that the mouse icon changes from a hollow cross ✛ to a small solid cross ✚ when we select the fill handle.

	A	B
1	0	2

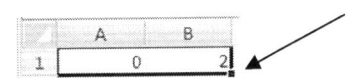

- Drag the fill handle to the right to create a range of cells containing even integer numbers that follows the pattern established in the first two cells.

---

[30] *Excel*'s worksheet **SIGN(x)** function returns 1 for x > 0, 0 for x = 0, and -1 for x < 0. We use a corrected formula in cell B4 **SIGNB(x)** that returns 1 for x = 0.

- *Excel* displays the value of the latest cell in a popup window as we drag through the row. The intersecting row/column icon at the bottom right corner of the fill handle contains a drop-down menu of fill options when right-clicked. **Auto Fill** works in columns and rows.

When using **Auto Fill**, we must be careful not to click away from the fill handle before completing the operation or we will need to restart the process.

Alternatively, we can create a series of values in a row or column using the ribbon tab **Home > Fill** option from the **Editing** group:

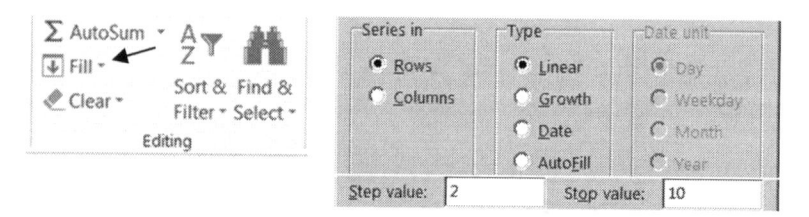

Start by selecting the cell in a worksheet containing the starting value in the series (zero in cell **A1** for this illustration). Then click on the **HOME** tab on the ribbon and select **Series** from the **Fill** option in the Editing group. Select **Series in Rows** or **Columns** to place the series values in a range of cells along a row or down a column. Use the **Type Linear** and set the **Step value** for a uniform series to generate the values on the worksheet that ends with the **Stop value**. The setup in the figure produces the same result created above using a fill handle starting from zero. Using the **Fill** tab option instead of the fill handle is particularly useful when the series involves more than 30 terms, or extends past the bottom or right edge of the visible part of the worksheet.

Auto Fill is an important feature that makes *Excel* a powerful tool for quickly solving complex problems. **Auto Fill** is particularly useful for implementing numerical methods that require iterative calculations that reuse results or formulas from the previous step in the sequence of calculations. *Example 2.1* and *Example 2.4* show how **Auto Fill** works with formulas. We use **Auto Fill** in Chapter 4 and Chapter 6 for finding roots to linear and nonlinear equations, Chapter 9 for least-squares regression, as well as Chapter 11 and Chapter 12 for generating *Excel* solutions to integrals and differential equations that requires several evaluations of the integrand or derivative functions over a range of points between the limits of integration.

2.1.6 Array Operations with CNTL SHIFT ENTER

Array operations perform the same operation on each cell in the selected range. Type the keyboard combination *CTRL SHIFT ENTER* to perform an array operation on a range of cells or a named range. To illustrate, use an array operation to generate a series of results involving the product of the contents of two columns of numbers. Select a range of cells for the results. In this case, select the range of cells in column **C** then type the formula to multiply the range of cells in column **A** with the range in column **B**.

Type the combination *CTRL SHIFT ENTER* simultaneously to generate the results in column **C**, for the product of the elements in each row. Note how *Excel* adds the curly braces to the formula to indicate the result of an array operation. Do not add the braces yourself.

CHAPTER 2: EXCEL

	A	B	C	D
1	1	4	4	
2	2	5	10	
3	3	6	18	

C1 ▾ fx {=A1:A3*B1:B3}

Use the *Esc* key to exit an array formula. To remove an array formula from a range of cells, first select the cells, then delete the formula and type *CTRL SHIFT ENTER*, or use **Clear** from the **Home** tab. Occasionally we become "stuck" in an array operation on an *Excel* worksheet. Type the **Escape** key (esc) to get unstuck, then select the range of cells in the array and modify the worksheet formulas. Once again, type *CTRL SHIFT ENTER* to complete the array operation. *Example 2.4* uses an array formula as an alternative method for calculating gas pressures over a range of temperatures using equations of state.

2.1.7 Worksheet Cell and Formula Errors

Excel returns #_!? error codes in a cell when the cell's formula attempts to perform an illegal operation. Table 2.2 gives definitions of worksheet error codes with suggestions for making corrections.

Table 2.2   Excel worksheet error codes in cells.

# Error Code	Formula Error and Suggested Fix
########	Adjust the width of the column to display a large number. Double click on the right side border of the column heading to match the width of the column to the length of the number.
#DIV/0!	Division by zero. Check for blank cells or that the cell references in denominators do not contain zero. In some cases, it may be appropriate to add a small number to the value in the dominator to avoid division by zero (see Section 2.1.8).
#NAME?	Text in a formula is not recognized. Check for correct names of worksheet formulas, cell addressing (includes both row and column) or a named cell (C and R are not permitted for cell names)
#N/A	Missing required arguments or data in a formula, worksheet or user-defined formula.
#NULL!	May refer to the intersection of two ranges that does not exist.
#NUM!	An incorrect number type. Check the magnitude or sign of the number.
#REF!	Invalid reference in a formula. Check for a valid cell or range address (e.g. $A$3, not AB) or named cell.
#VALUE!	Incorrect type of data for an argument. For example, we cannot mix strings and numbers in an arithmetic operation (e.g. #VALUE! = 2+2+B).

2.1.8 Avoid Division by Zero Errors

Division by zero is a common problem in numerical computing. When doing calculations "by hand", we quickly recognize the problem and avoid the mistake. *Excel* returns the error message **#DIV/0**. A division by zero error halts any further calculations by the application. We may use validation methods introduced in Section 2.3.8 to avoid scenarios that involve division by zero. Alternatively, we may try a couple other computational tricks that sidestep the problem of division by zero:

1. Add a tiny number to the denominator, just large enough to prevent the division-by-zero computation error, but small enough not to affect the required precision in the numerical result. For example, consider the variable x in the denominator:

$$f = \frac{a}{x} \rightarrow f \cong \frac{a}{x+\delta} \quad (2.2)$$

where $0 < \delta \ll 1$. This method carries a slight risk that the values of x and δ in Equation (2.2) are the same, but have opposite signs. We can eliminate this potential issue by multiplying the numerator and denominator by $x \rightarrow ax/(x^2 + \delta)$.

2. Alternatively, invert the equation, if possible, to remove any chance for division by zero, *e.g.*:

$$\text{New } f = \frac{a}{x} \rightarrow \frac{1}{f} = \frac{x}{a} \quad (2.3)$$

3. Another option involves a change of variables. For an example, see Section 11.2.1. The following Chen (1987) approximation to the log mean temperature difference eliminates division by zero when $\Delta T_1 = \Delta T_2$:

$$\Delta T_{lm} = \frac{\Delta T_2 - \Delta T_1}{\ln(\Delta T_2/\Delta T_1)} \cong \left[\Delta T_1 \Delta T_2 \left(\frac{\Delta T_1 + \Delta T_2}{2}\right)\right]^{1/3} \qquad (2.4)$$

We revisit the problem of division by zero when evaluating convergence, iterative calculations, and improper integrals.

2.1.9 Troubleshoot a Worksheet with Formula Auditing Tools*

"A person who never made a mistake never tried anything new."— Albert Einstein

A cell normally displays just the result of a calculation or descriptive text. A formula in a cell remains hidden behind its result. To view the formula, click in the cell and examine the formula in the formula bar or as displayed within the cell on the worksheet. To display a formula in a cell instead of the numerical result, add a space or apostrophe in front of the equal sign, *e.g.*, '=. Beware that a displayed formula becomes inactive. Instead, we recommend copying important formulas and pasting the result after an apostrophe in an adjacent cell to display both formula and result simultaneously. To copy a range of cells, select them, then press the *CTRL* key while dragging the cells to a new location to make a duplicate copy on the worksheet. Then go back and add apostrophes before the equal sign in the formulas of the copies. We may also use the *Excel 2013* (and higher) worksheet function **FORMULATEXT(cell_address)** to display the text of the formula in the range **cell_address**.

Excel has the option of displaying all the formulas in an *Excel* worksheet by selecting **Show Formulas** in the **Formula Auditing** group on the **Formulas** tab, seen in Figure 2.9. Figure 2.4 and Figure 2.8 display worksheets with **Show Formulas** enabled. Unfortunately, we cannot show formulas on part of an *Excel* worksheet – it is all or nothing. There are several examples of worksheets with **Show Formulas** enabled throughout the remainder of the book.

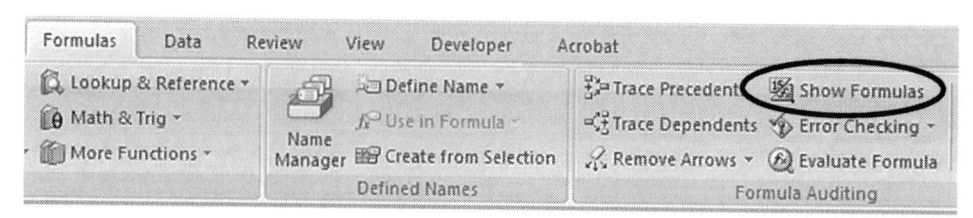

Figure 2.9 View formulas in an *Excel* worksheet by selecting Show Formulas from the Formula Auditing group.

Select from the other options described in Table 2.3 in the **Formula Auditing** group to trace the cells in a formula as precedents, or point to where the worksheet uses the cell as a dependent term. *Excel* draws connecting arrows among the formula cell and other cells used by the formula on the worksheet. Once satisfied with the formula, select **Remove Arrows**. A circular reference error happens when a formula in a cell references itself, as described in Section 2.1.10.

Table 2.3 Formula auditing tools on the Formulas Tab of the ribbon.

Icon	Auditing Tool	Purpose
	Trace Precedents	Draw arrows from the current cell to all cells referenced in the formula.
	Trace Dependents	Draw arrows from the current cell to all cells that reference the current cell.
	Remove Arrows	Clear trace arrows. Options to remove only dependent or precedent arrows.
	Show Formulas	Show all formulas in the cells of the worksheet.
	Error Checking	List cells containing errors. Options to trace errors or locate circular references.
	Evaluate Formula	Evaluate a formula when auto calculation is deactivated.

2.1.10 Circular Reference Errors

Ordinarily, we should not use the address of the current active cell within the formula of the active cell. When a cell formula references its own address, the following circular reference warning pops up:

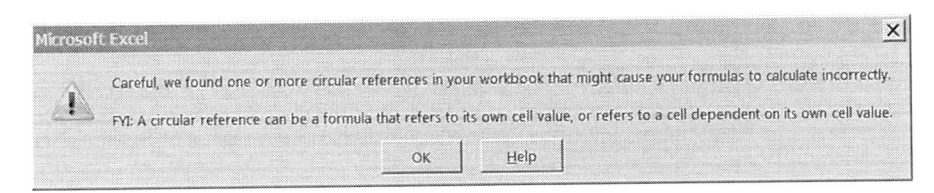

Occasionally, we make intentional circular references. Section 6.2.1 introduces the method of fixed-point iteration for finding roots to systems of equations using circular referencing. To turn on iteration, select from **File > Options > Formulas > Enable iterative calculation**.

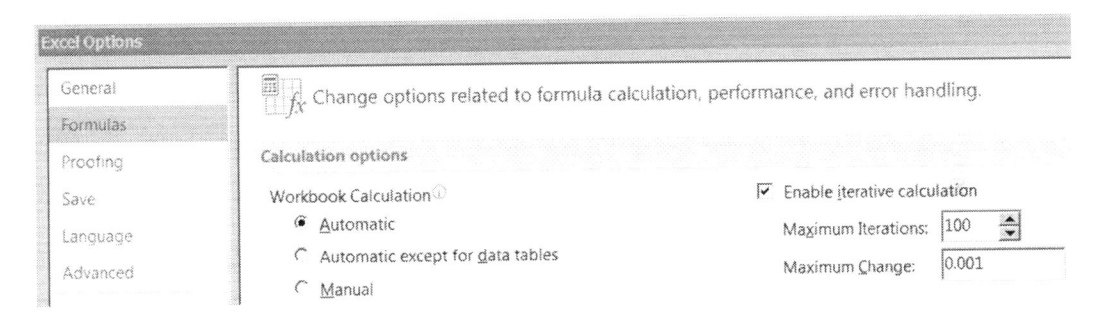

2.2 Function Library

Excel comes with several built-in worksheet functions available from the ribbon **Formulas** tab, shown in Figure 2.10. Click on the icon for a category to view the list of available functions. When we move the mouse pointer over a function name, a popup message box appears with a brief description of the function. We can also find details in *Excel*'s **Help** to become familiar with the different options and use them when needed. For example, *Excel* has financial functions needed to perform an economic analysis.

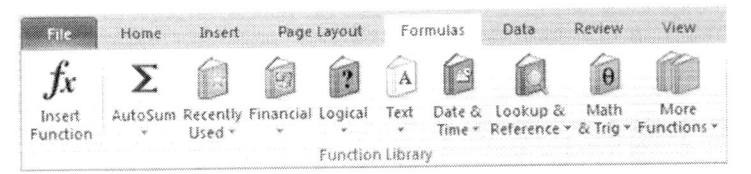

Figure 2.10

The formulas tab has several intrinsic function categories.

SQRT(x) is an example of a frequently used worksheet function for calculating the square root as seen in Equation (2.11) of Example 2.7. We used four unique worksheet functions in Equation (2.7). We can select the function from the list, or type the function name directly in a cell's formula. Functions perform their designated operation on arguments listed between parentheses. For convenience, type the keyboard combination *CNTL SHIFT A* to populate the list of required arguments, and then replace the argument names with your values. Two exceptions are the functions **PI()**, used to return the value for π and **RAND()**, which generates a uniformly distributed random number between zero and one. **PI()** and **RAND()** require the parentheses, but do not use arguments. If we select a function from the list, *Excel* opens a dialog box for the required arguments. Consult the help feature of *Excel* for detailed information on the procedures for implementing other functions. For example, trigonometric functions, such as **COS(r)**, **SIN(r)**, and **TAN(r)** operate on the angle **r** in units of radians. You may use the worksheet functions **RADIANS(d)** to convert **d** from degrees into radians (or the worksheet function **DEGREES(r)** to convert **r** from radians into degrees). The appendix to Chapter 15 lists several important functions from *Excel*'s library used to implement the numerical methods in this book.

Example 2.1 Temperature Profile in a Rectangular Block

Known: The thermal diffusion equation describes the temperature in a homogeneous, two-dimensional rectangular solid (assuming constant properties):

$$\frac{\partial^2 T}{\partial x^2} + \frac{\partial^2 T}{\partial y^2} = 0 \tag{2.5}$$

where T is temperature and x and y are the coordinates. For the special case where the temperatures of three of the four sides are equal (as shown in the schematic), the dimensionless solution to Equation (2.5) for the temperature consists of a Fourier series:

$$\theta = \frac{T - T_1}{T_2 - T_1} = \frac{2}{\pi} \sum_{n=1}^{\infty} C_n \sin\left(\frac{n\pi x}{L}\right) \sinh(n\pi y/L) \tag{2.6}$$

where n is the index for the terms in the summation with

$$C_n = \frac{(-1)^{n+1} + 1}{n \sinh(n\pi W/L)}$$

Schematic: A rectangular solid has width W and length L. The left, right, and bottom surfaces are maintained at temperature T_1; the top surface is maintained at temperature T_2.

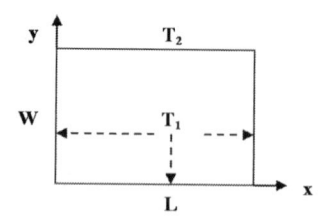

Find: Use built-in worksheet functions accessible from the ribbon **Formulas>Function Library>Math & Trig** to calculate the dimensionless temperature, θ at the center of a rectangle ($x = 0.5L$, $y = 0.5W$) for the following conditions: W/L = 1, 0.5, 0.25.

Analysis: Figure 2.11 displays an *Excel* worksheet showing the result of the Fourier series solution. Equation (2.6) is broken down into smaller parts that are easier to manage. We assemble the parts back together for the desired result. The second row contains headings for the table. Row 3 contains the Fourier series function formulas for the first term in the summation. We used the **Auto Fill** feature to generate the column of values for each summation index n, using the pattern established in rows four and five. We also used **Auto Fill** to create the formulas in the remaining rows. Notice that *Excel* automatically increments the relative cell references. **Auto Fill** treats the named cell WoL (for W over L) in **B1** as an absolute cell reference. After 20 rows, we combined the terms in a product then summed them according to Equation (2.6) using the worksheet function **SUM** to give the dimensionless temperature in cell **F24**. We used **Insert>Symbol** to put the Greek letter theta, θ, in cell **E24**. We can use **Formulas>Formula Auditing>Show Formula** to explore the relationships among cells used in formula calculations, or to display the equations in the cells.

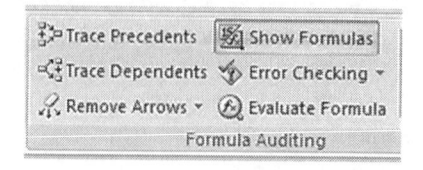

CHAPTER 2: EXCEL

	A	B	C	D	E	F
1	W/L	1				
2	n	n*sinh(n*pi*W/L)	Cn	sin(n*pi*x/L)	sinh(n*pi*y/L)	Product
3	1	1.15E+01	1.73E-01	1.00E+00	2.30E+00	3.99E-01
4	2	5.35E+02	0.00E+00	1.23E-16	1.15E+01	0.00E+00
5	3	1.86E+04	1.08E-04	-1.00E+00	5.57E+01	-5.99E-03
6	4	5.74E+05	0.00E+00	-2.45E-16	2.68E+02	0.00E+00
7	5	1.66E+07	1.21E-07	1.00E+00	1.29E+03	1.55E-04
8	6	4.61E+08	0.00E+00	3.68E-16	6.20E+03	0.00E+00
9	7	1.24E+10	1.61E-10	-1.00E+00	2.98E+04	-4.79E-06
10	8	3.29E+11	0.00E+00	-4.90E-16	1.43E+05	0.00E+00
11	9	8.56E+12	2.34E-13	1.00E+00	6.90E+05	1.61E-07
12	10	2.20E+14	0.00E+00	6.13E-16	3.32E+06	0.00E+00
11	9	8.56E+12	2.34E-13	1.00E+00	6.90E+05	1.61E-07
12	10	2.20E+14	0.00E+00	6.13E-16	3.32E+06	0.00E+00
13	11	5.60E+15	3.57E-16	-1.00E+00	1.60E+07	-5.70E-09
14	12	1.41E+17	0.00E+00	-7.35E-16	7.68E+07	0.00E+00
15	13	3.55E+18	5.64E-19	1.00E+00	3.69E+08	2.08E-10
16	14	8.84E+19	0.00E+00	8.58E-16	1.78E+09	0.00E+00
17	15	2.19E+21	9.13E-22	-1.00E+00	8.55E+09	-7.80E-12
18	16	5.41E+22	0.00E+00	-9.80E-16	4.11E+10	0.00E+00
19	17	1.33E+24	1.50E-24	1.00E+00	1.98E+11	2.97E-13
20	18	3.26E+25	0.00E+00	1.10E-15	9.51E+11	0.00E+00
21	19	7.96E+26	2.51E-27	-1.00E+00	4.58E+12	-1.15E-14
22	20	1.94E+28	0.00E+00	-1.23E-15	2.20E+13	0.00E+00
23					Sum =	0.393
24					θ =	0.25

Figure 2.11

Excel **worksheet for calculating terms in Fourier series.**

Comments: The results for the dimensionless temperature reveal several important features of the series solution. First, the terms in the summation quickly become negligible for large values of n. When $W/L = 0.5$, only 20 terms are needed to get a precise result. Extra significant terms are required as $W/L \rightarrow 0$. As we squash the rectangle, the dimensionless temperature at the midpoint approaches the average temperature of the top and bottom sides. The temperatures of the vertical sides play a diminishing role due to the longer distance from the midpoint relative to the upper and lower surfaces. As W/L decreases, the problem changes from two dimensions to one dimension. This example also emphasizes the advantage of breaking problems into smaller, more manageable parts to minimize potential for formula errors and troubleshoot mistakes.

☐

"If your experiment needs statistics, you ought to have done a better experiment." – Ernest Rutherford

Experimental data are subject to some natural variability that must be quantified using statistical methods. *Excel* has extensive functionality for statistical analysis. Access *Excel*'s statistical functions from the ribbon **Formulas>More Functions** tab shown in Figure 2.12. We may type the syntax for the function directly in an *Excel* worksheet cell, or insert the active function into an *Excel* worksheet cell from the list. As previously noted, inserting a function brings up a dialog box for the required arguments.

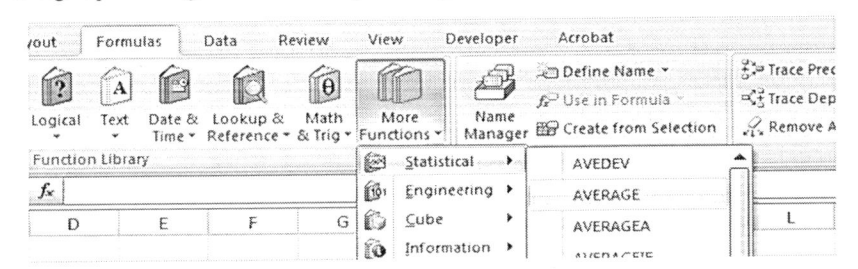

Figure 2.12

Excel's **statistical functions. Move the mouse pointer over a function name for a brief description of the function.**

We use a few of *Excel*'s descriptive statistical functions in Chapter 8 to analyze uncertainty in experimental data. However, because the primary focus of this book is numerical methods, we recommend consulting a reference on statistics that uses *Excel*'s capabilities for a thorough coverage of statistical analysis (Donnelly 2004) (Kirkup 2002).

2.2.1 Functions for Rounding and Significant Figures

Calculators and spreadsheet software often report numbers with higher precision than warranted by the precision of the input values. Be careful to adjust the display of significant figures, or digits in a calculated result to match the precision justified by the analysis.[31] We determine the number of significant figures from the location of the decimal and nonzero numbers. For example, 10 and 100 each have just one significant figure. By contrast, 10.00 and 100.0 each have four significant figures. We report the number 12.34 as accurate to the second decimal place, or ± 0.005. However, the number 123.4 is only accurate to the first decimal place, or ± 0.05, i.e., 12.34 + 123.4 = 135.7± 0.05. Scientific notation is useful for displaying the correct number of significant figures, e.g., 1.3×10^1 and 1.3×10^2 each have two significant figures. *Excel* and VBA use E notation for numbers in scientific notation, e.g., 1.3E1 or 1.3E2 are equivalent to 13 and 130 (or 1.3×10^1 and 1.3×10^2), respectively.

There are exceptions to the rule of significance. Exact numbers do not influence the number of significant figures in a result. Some examples of essentially exact numbers include the ideal gas constant, π, or other engineering constants known with a high degree of precision. The number of members in a small population, such as the size of an engineering team, or the number of unit operations in a process, is an exact number with infinite precision.

Consider the calculation of the average pipe length in a pipe network. The pipes have the following lengths: 12.3 m, 101 m, 62 m, 1.45 m, and 6.7 m. The number of pipes is not approximately 5 ± 1, nor rounded to 5.0 ± 0.1, but exactly 5. For example, we would not round up to 10 pipes, and certainly not down to zero pipes! Here are some basic rules to follow when carrying significant figures through a sequence of calculation steps.

- *Addition/Subtraction*: Report the result with a precision matching the higher order of significant decimals between the two numbers. The average pipe length is

$$\frac{12.3 + 101 + 62 + 1.45 + 6.7}{5} m = 37 m \quad (Not\ 36.7\ m\ or\ 40\ m)$$

  In this example, the least significant digits in the numbers 101 and 62 control the precision, which are precise to the first order.

- *Multiplication/Division*: Report the final value with a precision that matches the least number of significant figures of the multipliers or divisors. Report the average pipe length with two significant figures because the number of five pipes is an exact number. The number of 1.45 m pipes combined to equal the length of the 101 m pipe is:

$$\frac{101\,m}{1.45\,m} = 69.7 \quad (Not\ 69.66\ or\ 70)\ \text{pipes}$$

  However, the number of 62 m pipes that equals the length of one 101 m pipe is

$$\frac{101\,m}{62\,m} = 1.6 \quad (Not\ 1.63)\ \text{pipes}$$

  This result is correct only in terms of the rules for determining significant figures. However, if we used 1+6/10 shorter pipes, the total length would be short by about 1.8 m. In some cases, it is appropriate to round up instead of rounding off. Therefore, to be safe, replace a 101 m pipe with 1.7 lengths of 62 m pipe. Always use good engineering judgment when displaying significant figures in a result.

- Do not round off intermediate numbers during a sequence of calculations – report the final calculated result rounded to the correct number of significant figures determined from the input numbers.

Formatting a number in a cell controls the display of significant figures. *Excel* offers a choice between scientific notation and decimals displayed. When the column width is too small for the number, *Excel* displays **##**. Double-click the column heading divider to size the column automatically to fit the cell contents. To control the display of decimal places for a single cell, or range of cells, select the appropriate format from the ribbon **Home tab>Number** group.

---

[31] Some engineers do not trust any number with more than two significant digits.

CHAPTER 2: EXCEL

Figure 2.13 displays the same nine-digit number in four different cells using four different formatting options. Cell **A2** contains the number displayed with the default format in the worksheet. Alternatively, use scientific notation to change the number of significant figures displayed in the values with higher orders of magnitude. To input a number with scientific notation, add the letter E±## after the significant digits to indicate the order of magnitude as ##. For example:

$$1234.5 = 1.2345E3 \quad \text{or} \quad 0.0123 = 1.23E\text{-}2$$

Access scientific number formatting by clicking the right mouse button in the cell and selecting **Format Cells**. The same formatting option is accessible from the **Numbers** group on the **Home** tab.

Excel also has the worksheet function **ROUND(x, n)** for rounding **x** to the desired integer number of decimals, **n**, depending on the value of the first truncated term, *e.g.*, **2 = ROUND(1.51, 0)** and **1 = ROUND(1.49, 0)**. Worksheet functions **ROUNDUP** and **ROUNDDOWN** control the direction of rounding. Use a negative integer value for **n** to round a number left of the decimal place, e.g., **20 = ROUND(15,-1)**. To illustrate, use the worksheet function **ROUND** in cell **A3** to control the number of decimal places. In this case, only two decimal places are displayed using the syntax **1234.57 = ROUND(1234.56789, 2)**. We may accomplish the same functionality as **ROUND** in cell **A4** by clicking on the **Decrease Decimal** button in the **Number** section of the **Home** tab. The worksheet function **MROUND(x, m)** rounds the value of **x** to the nearest multiple **m**, such that **1230 = MROUND(1234.56, 10)**.

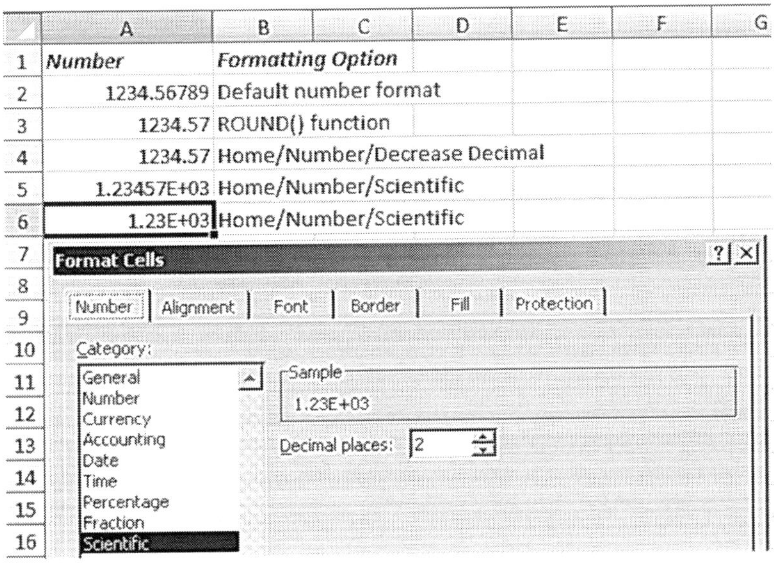

Figure 2.13
Excel options for formatting numbers in cells.

Unfortunately, there are no built-in worksheet functions for rounding numbers to the desired number of *significant figures*. For this, we use a combination of embedded worksheet functions in the following formula:

$$= \text{ROUND}(x, SF - 1 - \text{INT}(\text{LOG10}(\text{ABS}(x)))) \tag{2.7}$$

where **x** is the number to round, and **SF** is the integer value of the number of significant digits desired in the result. The worksheet functions **LOG10** returns the logarithm of **x** in base 10 for the order of magnitude of the value of **x**. The worksheet function **ABS** returns the absolute value of **x** to avoid errors when evaluating logarithms of negative numbers. Use the worksheet function **INT** to get the integer portion of the value of the logarithm. To illustrate, apply the method to the following cases using two significant figures:

120 = ROUND(123.45, 2-1-INT(LOG10(123.45)))   0.0012 = ROUND(0.0012345, 2-1-INT(LOG10(0.0012345)))

We use this simple algorithm for rounding to significant figures for uncertainty analysis in Chapter 8 and Chapter 9.

2.2.2 Unit Conversions

CONVERT is a particularly useful worksheet function for engineers that need to convert the units of a number from one system of measurement to another. It may be necessary to activate this function as an *Excel* add-in to make it available in the workbook. The function syntax is

CONVERT (number, from_unit, to_unit)

where **number** is the value to convert, **from_unit** is the original unit(s) for **number**, and **to_unit** is the unit(s) for the converted result. **CONVERT** accepts strings of text (in quotation marks) for **from_unit** and **to_unit**. Table 2.4 lists some examples of units important for chemical engineers. Search the *Excel* help file for a complete list of unit conversions. Note that unit names and prefixes are case-sensitive. For incorrect input data types, **CONVERT** returns the error message #VALUE!. If the unit is not supported, or the units are for different types of quantities (*e.g.*, kg versus atm), **CONVERT** returns the #N/A error message.

Table 2.4 Some unit conversions available with *Excel*'s CONVERT function.

Measurement	from_unit or to_unit	Measurement	from_unit or to_unit	Measurement	from_unit or to_unit
Gram	"g"	Minute	"min"	BTU	"BTU"
Pound mass	"lbm"	Second	"sec"	Horsepower	"HP" or "h"
Meter	"m"	Pascal	"Pa" or "p"	Watt	"W" or "w"
Mile	"mi"	Atmosphere	"atm" or "at"	Degrees C	"C" or "cel"
Inch	"in"	mm of Hg	"mmHg"	Degrees F	"F" or "fah"
Foot	"ft"	Newton	"N"	Kelvin	"K" or "kel"
Year	"yr"	Pound force	"lbf"	Gallon	"gal"
Day	"day"	Joule	"J"	Liter	"l" or "lt"
Hour	"hr"	Calorie	"cal"		

To illustrate, the following conversion from 300 K to degrees Fahrenheit in a cell gives 80.33 °F:

=CONVERT(300,"K","F")

We may pre-pend the abbreviated unit prefixes in Table 2.5 to any metric **From_unit** or **To_unit**. For example, convert 1.200 inches to 30.48 mm in an *Excel* worksheet cell:

30.48=CONVERT(1.2, "in", "mm")

Table 2.5 Unit prefixes to append to from_unit or to_unit.

tera	1E+12	"T"	deci	1E-01	"d"	milli	1E-03	"m"
giga	1E+09	"G"	centi	1E-02	"c"	micro	1E-06	"u"
mega	1E+06	"M"	kilo	1E+03	"k"	nano	1E-09	"n"

To convert the units for a quantity that has combinations of units, use **CONVERT** on the value of the quantity for the first unit in the numerator, then multiply or divide the result with **CONVERT** on a value of one for the remaining units. To illustrate, convert volumes from 1.20 m^3 to 42.4 ft^3 and velocity from 3.40 m/s to 7.61 mile/hr using the following formulas:

42.4=CONVERT(1.2,"m","ft")*CONVERT(1,"m","ft") ^ 2   7.61=CONVERT(3.4,"m","mi")/CONVERT(1,"sec","hr")

Unfortunately, the units available for the **CONVERT** worksheet function do not include many of the units required to quantify engineering calculations. We solve this problem using *VBA* to create user-defined functions to extend the conversion factors to those commonly encountered in engineering. The *VBA* user-defined functions **UNITS** and **RGAS** are described in the table of conversion factors at the back of this book.

2.2.3 Logical, Lookup, Match and Index Reference Functions

Engineers often work with large data sets. *Excel* has a variety of logical, lookup, and reference functions that conveniently compare and access information from ordered tables. The most useful are **IF**, **IFS**, **VLOOKUP** (or **HLOOKUP**), **MATCH**, and **INDEX**. The next example uses *Excel* functions to compare and retrieve a value from a table. We use several of these functions in Chapter 10 for data interpolation. Section 10.1.1 has an example using **MATCH** and **INDEX**. Consult the on-line *Excel* **Help** for more details and examples about working with logical, lookup and reference functions.

Example 2.2   Selection of a Pipe Size for a Specified Flow Rate

Known/Find: Use **IF** and **VLOOKUP** worksheet functions to match a pipe size to a volumetric flow rate. The data listed in the *Excel* worksheet of Figure 2.14 give the nominal sizes for various ranges of flow rates. For example, nominal three-inch pipe has a flow capacity ranging from 10.45 to 23 gallons per minute.

Analysis: Set up an *Excel* worksheet to find the pipe size from the table for a specified flow rate. Use an **IF** function to compare pipe sizes. The **IF** function has three arguments separated by commas:

= IF(logical_test, value_if_true, value_if_false)

The first argument is a logical test. When the result of the logical test is true, the **IF** function returns the result of the second argument, otherwise it returns the result of the third argument. In this example, we nested the **IF** function in cell **E4** eight times to cover the whole data set of pipe sizes. Cell **E3** contains the value for the flow rate.

E4 = IF(E3<A3, B2, IF(E3<A4, B3, IF(E3<A5, B4, IF(E3<A6, B5, IF(E3<A7, B6, IF(E3<A8, B7, IF(E3<A9, B8, IF(E3<A10, B9, IF(E3<A11, B10, B11)))))))))

MS Office 2016 introduced a new worksheet function **IFS** that eliminates the need for nested **IF** functions. The arguments consists of a series of paired arguments separated by commas. Each pair requires a logical expression, followed by the action expected if the logic is **TRUE**:

E4 = IFS(E3<A3, B2, E3<A4, B3, E3<A5, B4, E3<A6, B5, E3<A7, B6, E3<A8, B7, E3<A9, B8, E3<A10, B9, E3<A11, B10, E3>=A11, B11)

Alternatively, use the simpler **VLOOKUP** function (vertical lookup) to give the same result:

= VLOOKUP(lookup_value, table_array, col_index_num)

VLOOKUP also requires three arguments, separated by commas.
- The first argument is the lookup value, the next argument is the multi-column range of table values, and the third, optional argument is an integer (1 or higher) identifying the column that contains the matched value returned by the lookup function.
- The values in the first column of the lookup table must be in ascending order.
- The lookup value is matched with the entries in the first column of the lookup table. Note that we create the range of the lookup table by typing the top-left cell reference and bottom-right cell references, separated by a colon. It is often faster to enter the range references by placing the curser inside the parentheses of the function and selecting the range from the worksheet using the mouse.

In this example, the function matches the flow rate to the next lower value in the left column, and then returns the corresponding value for the pipe schedule from the second column of the lookup table, corresponding to the nominal sizes in Column **B**.

E4 = VLOOKUP(E3, A2:B6, 2)

	A	B	C	D	E
1	Capacity/gpm	Nominal Size/in			
2	0	1			
3	2.69	2		Flow Rate/gpm =	20
4	10.45	3		Nominal Size/in =	3
5	23	4			
6	39.6	5			

Figure 2.14
VLOOKUP worksheet function for extracting information from an *Excel* table.

Comments: Although this example is trivial due to the small size of the database, these reference functions play an important role in managing larger data sets. The function **HLOOKUP** performs a similar function for accessing data arranged horizontally in rows.

□

Be careful when comparing values using **IF** statements. For example, we should avoid comparing exact numbers. *Excel* stores numbers to 15 digits, and rounding errors may cause a logical statement to return a **FALSE** result when lower precision values should return a **TRUE** result. A preferred method uses less than (<) or greater than (>) instead of equals (=) to avoid this problem. When we need exact comparisons, we can control the rounding using the built-in **INT, ROUND, ROUNDUP,** or **ROUNDDOWN** functions that allow us to specify the decimal place for rounding. For example, consider what appears to be the same calculation in an *Excel* worksheet that return false and true results, respectively:

FALSE = 0.29*100 − INT(0.29*100)   TRUE = ROUND(0.29*100, 0) − ROUND(0.29*100, 0)

2.2.4 Indirect Function

The **INDIRECT(ref)** worksheet function returns the cell or range reference identified by the argument text string **ref**. We use **INDIRECT** in formulas or lookup type functions where we want to change the reference to a cell or range without creating a different formula. We illustrate the use of **INDIRECT** in Example 2.3.

Example 2.3   Diffusivity of Gases in Water

Known/Find: Create an *Excel* worksheet to calculate the diffusivity of gas species in water using the following Wilke-Chang equation:

$$D = \frac{5.0645 \times 10^{-7} T}{\mu v^{0.6}} \qquad (2.8)$$

where D is the diffusivity in cm²/s, T is the temperature in degrees K, μ is the dynamic viscosity of water in cP units calculated from the following correlation:

$$\mu = 1.792 \exp\left[-1.94 - 4.8\left(\frac{273.16}{T}\right) + 6.74\left(\frac{273.16}{T}\right)^2\right] \qquad (2.9)$$

and v is the molar volume of the gas species in cm³/mol tabulated in the Excel worksheet for several gas species.

Analysis: Rename the cells in column **H** according to the labels in column **G** (note the underscore is needed in the new names to distinguish them from the column/row cell addresses). Apply **Data Validation** from the tab **Data>Data Tools group** to cell **B4** to allow the user to select the gas species from the list in column **G**. Provide the temperature in cell **B5**. Use the **INDIRECT** function in cell **B8** to get the molar volume for the selected gas species:

B8=INDIRECT(B4)

Calculate the viscosity according to Equation (2.9) in cell **B9** and the diffusivity in cell **B10** according to the Wilke-Change Equation (2.8):

B9=1.792*EXP(-1.94-4.8*(273.16/B5) +6.74*(273.16/B5) ^2)   B10=0.00000050645*B5/ (B9*B8^0.6)

	A	B	C	D	E	F	G	H
1	Diffusivity of Gas in Water by Wilke-Change Equation							
2							Gas	v/(cm^3/mol)
3	User Specified Values						Air	29.9
4	Gas	CO2_					O2_	25.6
5	T/K	298					N2_	31.2
6							CL2_	48.4
7	Calculated Values						CO2_	34
8	v/(cm^3/mol)	34					H2S	32.9
9	μ/cP	0.9109					NH3_	25.8
10	D/(cm^2/s)	2.00E-05					SO2_	44.8

Comments: The result for the diffusivity of CO_2 in water agrees with experimental results. The **INDIRECT** worksheet function works like the combination of worksheet functions **MATCH** and **INDEX** to extract the molar volume of the gas species from the tabulated values in column **H**.

□

2.3 Document Your Data and Calculations

"An idea NOT coupled with action will never get any bigger than the brain cell it occupied." – Arnold H. Glasow

Engineers generally do not work in isolation, but in teams. A team might consist of a manager, engineers, technicians, operators, accountants, marketers, and scientists. Engineers need to document their calculations to share their work with teams of collaborators, archive their work for future reference, or provide information to external groups or agencies. For example, we might work with vendors, consultants, or government agencies that regulate our process. We may find ourselves in a meeting where we need to back up our recommendation with persuasive data and sound engineering analysis.

Computer spreadsheets are ubiquitous throughout universities, industry, business, and government – they provide a common platform for collecting, archiving, modeling, analyzing, and sharing information. Since its introduction, *Excel* has quickly risen to the top and held its position as the *de facto* standard spreadsheet application. When documenting calculations, consider the following features of *Excel* for displaying information:

- Tabulates results organized with row and column headings.
- Uses conventional names, symbols, and units.
- Displays numbers with the appropriate level of significant figures.
- Uses graphs to show trends and relationships in data.
- Includes an editor to add equation objects to the worksheet.

We may add text as well as numerical information to cells on an *Excel* worksheet to provide descriptive information about our calculations. As we type text into various cells, *Excel* uses **Auto Complete** to finish spelling text it recognizes from previous entries on the worksheet. This is often more problematic than useful. To turn off **Auto Complete**, uncheck this option from the ribbon tab **File >Options >Advanced >Enable Auto Complete**.

2.3.1 Presenting Tabulated Results

Excel worksheets are inherently tabular. We must take appropriate steps to arrange our results for improved readability for future reference and collaboration with users of our workbooks. As a general principle: keep tables and figures simple and uncluttered. The following recommendations for preparing tables were culled from author instructions in research journals:

- Arrange results in columns to help the user identify trends in the data.
- Use row and column headings to identify the data/variable type – include units for dimensional quantities.

- Order the data according to magnitude. For multiple columns, order the data by magnitude of the corresponding dependent variable. Use multiple tables if necessary to eliminate complexity.
- Do not use lines to separate columns. Use lines to separate rows only when each row contains a unique type of variable. Excel uses faded lines to box a cell. Add darker lines between rows of the table wherever needed.
- Round off numbers to their significant figures. Where feasible, limit the number of decimal places to two, at most. Use scientific notation if necessary to indicate the magnitude of small and large numbers.
- Line up numbers in a column at their decimal place.
- Try to keep the width of each row uniform by using row headings with a single line. If you need to use abbreviations, spell them out in a footnote to the table (this does not apply to units or chemical formulas).
- Tables should have a concise heading placed above the first row. Use distinct number in the title when the document contains multiple tables. Summarize the content of the table in the heading.
- Tables and figures should contain enough information in the headings to 'stand-alone' without the need to refer to the text of the document for understanding.
- Provide references for any cited data.
- Use lines, bold text, spacing, highlights, and/or shading where appropriate to help the reader interpret the contents of the table.

Use the preceding recommendations to compare and contrast the reader's experience with the following two tables that present the same data of binary diffusion coefficients of several gas species in air, at the given temperature, and one atmosphere pressure. How does Table 2.6 improve the reader's access to the information?

Gas Diffusivities

Gas	T (K)	$D \times 10^5$ (m^2/s)
Acetone	273	1.1
Benzene	298	.88
CO2	298	1.6
H2	298	4.1
H2O	298	2.6
Naphthalene	300	0.62
NH3	298	2.8
O2	298	2.1

Table 2.6 Binary diffusion coefficients in air at one atmosphere. (Incropera and DeWitt 2002)

Species	T/K	D/(m^2/s)
Naphthalene	300	6.2×10^{-6}
Benzene	298	8.8×10^{-6}
Acetone	273	1.1×10^{-5}
CO$_2$	298	1.6×10^{-5}
O$_2$	298	2.1×10^{-5}
H$_2$O	298	2.6×10^{-5}
NH3	298	2.8×10^{-5}
H$_2$	298	4.1×10^{-5}

2.3.2 Never Type Variable or Data Values Directly Into a Formula and Other Recommendations for Formatting Worksheets

Koomey (2008) makes several more recommendations for documenting calculations in a spreadsheet that make them more accessible to collaborators and archives of work for future reference:

- Never type data values directly into a formula. Enter numerical data into cells on the worksheet, and then reference the data cells in the formulas. Following this practice keeps all of the data exposed to the user to help with readability, trouble-shooting, and further worksheet development. If we need to change the data, we do not have to modify formulas. We follow this advice in the worksheet set up for *Example 2.4* and throughout the book.
- Designate different sections of the worksheet for data, constants, and results. An organized worksheet helps us quickly catch up with what we were doing with the spreadsheet.
- Define all units for dimensional quantities and be careful about converting between unit systems. Place the units in adjoining cells or in cell comments.
- Include references to sources of data, formulas, or methods (put references in cells, text boxes, or **Comments**).
- Provide dates for creation of the sections of the spreadsheet to allow users to reconstruct the sequence of analysis if necessary.

CHAPTER 2: EXCEL

2.3.3 *Drawing Tools*

MS Office applications, including *Excel*, have drawing tools for creating simple illustrations. Click on the **Insert>Shapes** option in the **Illustrations** group, shown in Figure 2.17, to expose menus for adding text boxes, lines, blocks, and other drawing features to charts or worksheets. We created all of the schematics in the example problems for this book using the drawing tools displayed in Figure 2.15. The first icon on the **Basic Shapes** section allows us to put a text box on the worksheet. Use text boxes to type notes, or add explanatory information to clarify the contents of an *Excel* worksheet. To assign macros and add text to shapes, right click on the shape and select the desired property from the drop-down menu.

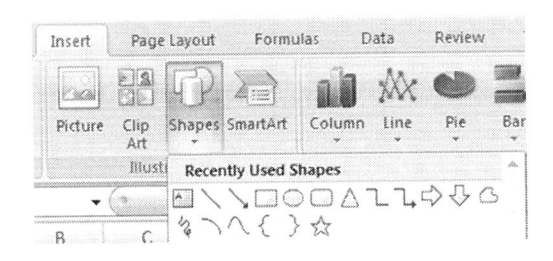

Figure 2.15

Drawing shapes and text box for annotating charts or creating simple schematics in worksheets.

2.3.4 *Charts and Graphs*

"Of all of our inventions for mass communication, pictures still speak the most universally understood language." – Walt Disney

Charts and graphs are important tools for conveying information in compact, concise formats. Engineers must learn to create proper graphs with maximum utility that are clear and readable. Graphs must be true to the data and avoid any potential for hiding or distorting relationships among variables. *Excel* offers a variety of graph types and styles for engineering and business needs. David Letterman[32] would be proud of the pie chart in Figure 2.16; however, engineers rarely use such displays because they do not demonstrate relationships among variables.

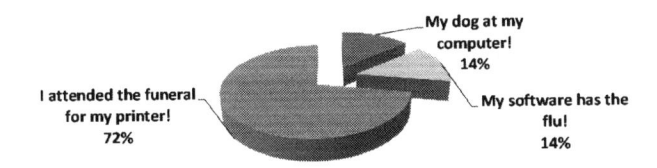

Figure 2.16

Pie chart created in *Excel* representing the distribution of student excuses for missing assignments.

www.cis.gsu.edu/~dstraub/Courses/Grandma.htm

Tufte (1983) recommends preparing scientific or engineering graphical displays that follow the experimental principles of making "controlled comparisons" for the following purposes:

- Documenting the characteristics of data
- Quantitatively demonstrating cause and effect
- Evaluating alternative explanations

"Data graphics should draw the viewer's attention to the sense and substance of the data, not to something else. The data graphical form should present the quantitative contents. Occasionally artfulness of design makes a graphic worthy of the Museum of Modern Art, but essentially ... graphics are instruments to help people reason about quantitative information" - Tufte

Consider the following two images of the same scenario in terms of Tufte's recommendations for preparing visual displays. How does the left image violate Tufte's principles to misrepresent the facts?

---

[32] https://www.youtube.com/watch?v=P5PSTldcWnw

Unfortunately, *Excel* has several wizards to make it easy for us to create a *bad* graph with one click. We then must put considerable effort into modifying the cartoonish result to display a proper graph (Su 2008).

The most common chart for representing engineering information is a scatter plot (or *y* versus *x* plot). Access the **Chart** features from the **Insert** tab, as shown in Figure 2.17.

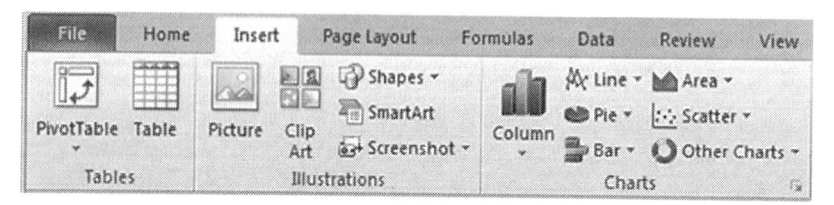

Figure 2.17 Examples of chart types available in *Excel*.

Refer to the example worksheet with data and graph in Figure 2.18 displaying vapor-liquid equilibrium data for a water-acetic acid system at one atm pressure. The plot shows the vapor (*y*) and liquid (*x*) mole fractions of water and corresponding temperatures.

	A	B	C	D	E	F
1	Experimental Data			Ideal System		
2	x	y		x	y	T/K
3	0.02	0.035		0	0.000	391.1
4	0.04	0.069		0.1	0.169	388.5
5	0.06	0.103		0.2	0.312	386.1
6	0.08	0.135		0.3	0.437	384.0
7	0.1	0.165		0.4	0.546	382.1
8	0.2	0.303		0.5	0.643	380.3
9	0.3	0.425		0.6	0.729	378.7
10	0.4	0.531		0.7	0.807	377.2
11	0.5	0.627		0.8	0.877	375.8
12	0.6	0.715		0.9	0.941	374.4
13	0.7	0.796		1	1.000	373.2
14	0.8	0.865				
15	0.9	0.929				
16	0.94	0.957				
17	0.98	0.985				

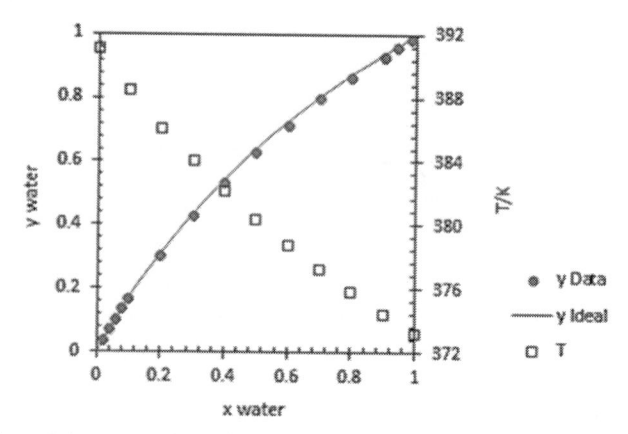

Figure 2.18 Sample graph of VLE data for Water-Acetic Acid system plotted from data in an *Excel* worksheet (Cornell and Montonna 1933).

To create the graph, first select the experimental data for *x* and *y* in the range **A3:B17** on the worksheet. Then select the ribbon tab **Insert>Scatter>Scatter** with only markers. *Excel* creates an *yx* scatter plot, similar to Figure 2.19:

CHAPTER 2: EXCEL

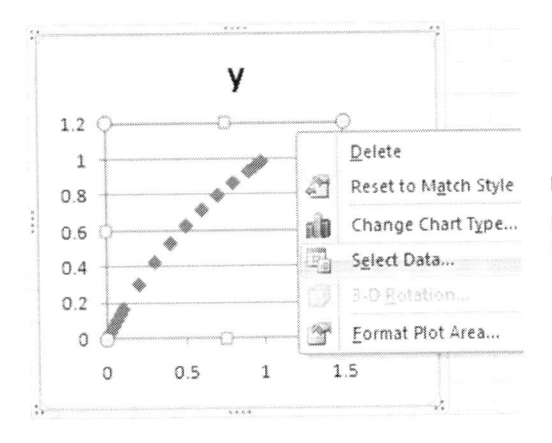

Figure 2.19

Right-click on a chart and Select Data for editing the series.

Click on the graph to activate it. Right-click and chose **Select Data** from the popup menu. A **Select Data Source** dialog box opens as seen in Figure 2.20 with options for editing and adding data to the plot.

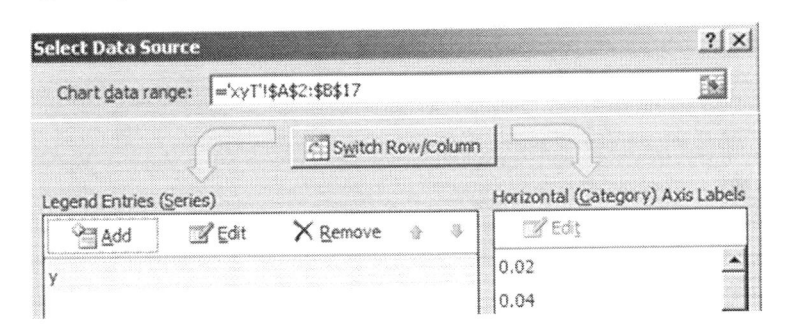

Figure 2.20

Select Data Source dialog box for adding or removing data from a chart.

Click **Add** to create a new data series for the ideal VLE system y versus x data. Select the range **E3:E13** for the **Series X values** and **F3:F13** for the **Series Y values**. Add another data series for the T versus x data. The T data distorts the plot because the temperature values are outside of the range on the y-data. We need to show the temperature data on a secondary y-axis. To add a secondary axis, right-click on the T data in the plot to activate it on the worksheet. Select **Format Data Series** from the popup menu, shown in Figure 2.21. Then click **Secondary Axis** under **Series** options.

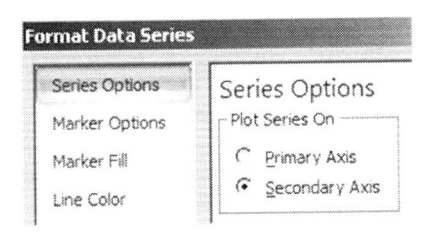

Figure 2.21

Add a secondary axis to a plot.

The remaining formatting options allow us to modify the size and colors of lines and markers. Experiment with the options to find a combination of formatting that produces a meaningful graph according to Tufte's principles. We can also access formatting options from the **Chart Tools** tab that appears on the ribbon when the graph is active on the worksheet. Consider the following guidelines when creating a proper y versus x graph in *Excel*:[33]

- **Axis Titles**: All graphs should have axis titles, with appropriate units for dimensional quantities. To add axis titles, click on the graph and select from the ribbon tab **Chart Tools>Layout>Axis Titles**. In the **Labels** group, modify the **Primary Vertical** and **Primary Horizontal Axis Titles**. With a secondary axis, select the

---

[33] Professor Keith Lodge, University of Minnesota Duluth, private communication. These are recommended graphing guidelines – be careful to follow the conventions established by your individual instructors, company, or publisher.

Secondary Axis Title. Starting with *Excel 2013*, we can access formatting options from the plus ✚ option that appears next to the active chart.

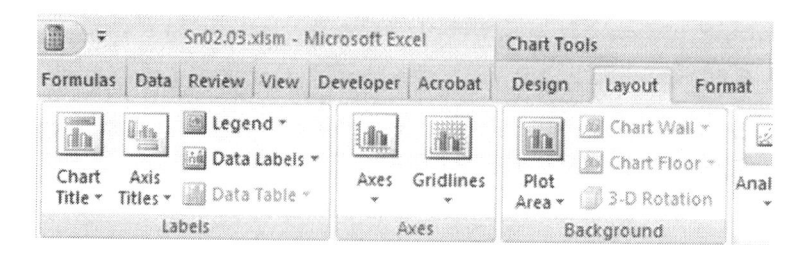

- **Axis Scales**: Adjust the scale of the abscissa (horizontal *x*-axis) to match the range of the independent data. To adjust the scale, select the axis, then right click and select **Format Axis** from the popup menu. Change the minimum and maximum values to fit the range of data. Adjust the scale of the ordinate (vertical *y*-axis) to convey the real story. As seen in the following plots of the *same* data, where the *y*-scale is too large, the data may appear flat, and the graph may not reveal any potential correlation between the dependent and independent data. If the scale is too narrow, the graph may not display all of the data, or mislead the viewer into seeing a strong correlation between the *y* vs *x* data, where there may be none (Vesiland 1999).

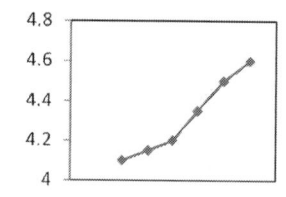

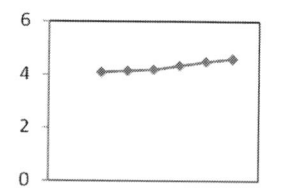

The same issue applies to the abscissa. In the following plots of the same data, the shifted limits on the scale of the plot at the right tend to conceal the left-most data. The addition of a line adds even more visual emphasis, or weight, to the two data points at the far right, further concealing the relatively significant influence of the data at the left.

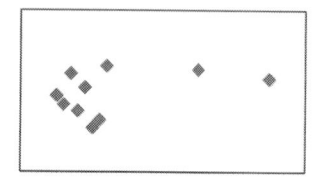

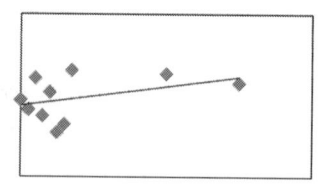

- **Log Scales**: Use log scales on the axis wherever the range of values extends beyond two orders of magnitude. For example, consider the plots of a reaction rate constant (k) with inverse temperature ($1/T$) in Figure 2.22. The first plot hides the temperature effects at low temperature (high $1/T$). The second plot uses a log scale on the ordinate axis to reveal the temperature dependence at all temperatures. To format an axis, right-click on the axis and select the desired formatting options, such as log scale. See Section 9.3.1 for additional recommendations for reading log plots.

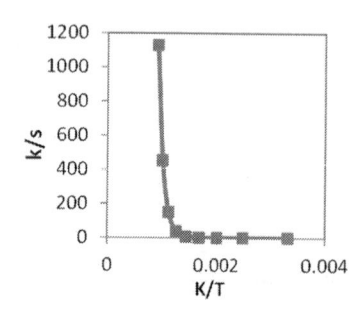

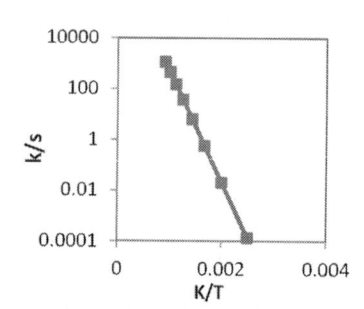

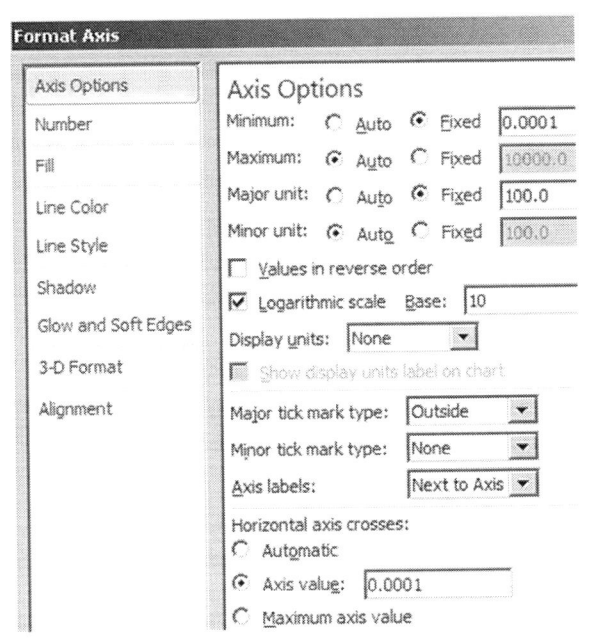

Figure 2.22
Axis formatting options in *Excel*.

- **Secondary Y Axis**: Plot any data series with different dimensions or ranges of values on a secondary *y*-axis; note the second *y*-axis in Figure 2.18 with its temperature scale on the right. Avoid too much clutter in a single graph. Instead, use a series of smaller single-series graphs for each unique data set. *Excel* has a feature called **Sparkline** that permits small line and bar graphs of a single series within cells on an *Excel* worksheet. To create **Sparklines**, select from the ribbon tab **Insert>Line** in the **Sparkline** group. A dialog box asks for the data to plot and cell location for the **Sparkline**. When we click in the cell with the graph, a new **Sparkline Tools Design** tab appears with options for formatting the line.

- **Tic Marks**: Always use appropriate tick marks on the axis to help the viewer interpolate the data in the graph. Major tick marks point outside the plot area, whereas minor tick marks should point inside the plot area (This may be a matter of personal opinion). Use minor tick marks particularly where the viewer needs to read values from the graph.

- **Grid Lines**: We usually omit grid lines unless we need to read values from our plots. Log plots are an important exception where grid lines may be important for signaling the presence of a log scale on an axis. Without proper care gridlines may be confused with function or data curves, or create an optical illusion that distorts the appearance of relationships in the data represented by markers or plotted curves. We must eliminate extraneous lines or shapes on graphs that cause the viewer to see misleading trends or unintentional patterns in data. In the following graphs, the left error bar *appears* larger than the right one, the top line *appears* longer than the bottom one, the left inner circle *appears* larger than the right outer circle, and the points of the angles facing inward appear closer than the points of the angles facing outward. See Figure 9.18 for another example (Vesiland 1999). In actuality, the identified elements have equal dimensions or positions.

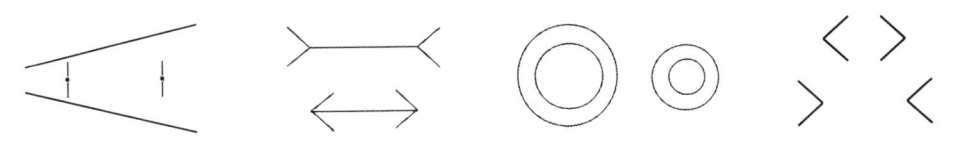

Remove the default shading and gridlines from the graph background. Delete the default horizontal grid lines by selecting them and pressing the delete key. Right-click on the graph to change the border of the plot. Right-click in the space surrounding the plot to change the border of the object. To help the user read values from a plot, first try using minor tic marks on the axis, then add both horizontal and vertical gridlines only if necessary, using gridlines with faint colors that don't obscure the rest of the plot.

- **Markers**: Use markers to represent experimental (measured) data. Markers show the presence or lack of any "noise" or variance in the data. It is acceptable to use lines without markers for points generated from functions or equations. However, we normally do not use markers alone for plots of equations or functions. As scientists or engineers, we are trained to see markers as experimental data points, and lines as smooth, continuous functions that interpolate experimental data or models.

- **Curves and Lines**: Draw smooth curves *through* the data points to help the viewer recognize real trends in the correlation between the function and independent variable; note that the y-curve in Figure 2.18 does not *connect* experimental data points. Do not use the default setting whereby markers with a line joining them are used – the result is a discontinuous function (our measurements usually represent a sampling from a continuous function). Chapter 9 presents methods for fitting data with functions by the method of least-squares regression.

- **Color**: When using color for markers and lines, be sure also to include other distinguishing characteristics in case of black and white printing or color-blind viewers. For instance, use distinguishable markers (square versus circle) and lines (solid versus dashed). It is better to remove color from yx graphs entirely.

- **Significant Figures**: Use an appropriate number of significant figures on axis labels (*e.g.*, 385 instead of 385.000, or 3.85E5 instead of 385000). Right-click on the axis labels to change the formatting of the numbers, modify the scale, *etc*.

- **Legend**: Only include a legend in a graph with more than one data series. Position the legend where it does not obstruct the objects in the graph. Click and drag the legend to reposition.

- **Titles**: Delete the default graph title placed above the plot. By convention, scientists and engineers place descriptive captions below or next to a graph in a report or manuscript (tables use titles above.). Refer to the technical literature for examples of figure captions.

- **Size and Properties**: On occasion when you have several graphs in the same document, it may be appropriate to set the size of the graphs for uniformity in the height and width. Property options allow you to fix the size of the graph so that it does not scale with the sizes of cells or changing numbers of rows or columns.

Fortunately, the latest versions of *Excel* allows us to create custom graph templates according to these guidelines. Once you have a graph in an *Excel* worksheet with the basic formatting features described above, right-click on the graph object and select **Save as Template**. To reuse the template, select it from the ribbon tab **Input>Charts>Templates** folder.

A *VBA* macro named **GRAPHYXGUIDE** (pronounced, "Graphics guide") is also available in this book's companion *Excel* workbook *PNMSuite* that formats ranges of data for plotting according to these graphing guidelines. See Section 3.10 for more information about accessing and using the macro. The macro requires the data arranged in contiguous columns in an *Excel* worksheet.

- The cells in the first row contain text for the labels of each data series in contiguous columns.
- The first column contains the range of data for the independent variable (x).
- The remaining columns contain one, or more, ranges of independent data (y).
- Format the graph generated by the macro **GRAPHYXGUIDE** as necessary. For instance:
    - Use different line styles or dash types with multiple data series on the same chart,
    - Remove lines from data series,
    - Remove markers from function series (series created from equations),
    - Add a secondary axis for values with different units or different orders of magnitude, or
    - Alter the axis scales to match the data ranges, or use log scales, etc.

The next example illustrates most of the *Excel* features discussed to this point in the chapter.

CHAPTER 2: EXCEL

Example 2.4 Ideal versus Non-ideal Gas Pressure

Known/Find: Compare the ideal gas law with the Redlich-Kwong (*RK*) equation of state for calculating the pressure of ammonia, *P*, at a specific molar volume, *V*, over a range of temperatures, *T*.

- Ideal Gas:
$$P = \frac{R_g T}{V} \qquad (2.10)$$

- Redlich-Kwong:
$$P = \frac{R_g T}{V-b} - \frac{a}{V(V+b)\sqrt{T}} \qquad (2.11)$$

where $a = 0.42747\left(\dfrac{R_g^2 T_c^{5/2}}{P_c}\right)$ and $b = 0.08664\left(\dfrac{R_g T_c}{P_c}\right)$

Properties of Ammonia	Value
Ideal gas constant/(L·atm/mol·K)	R_g = 0.08206
Molar volume/(L/mol)	V = 0.5
Critical Temperature/K	T_c = 405.5
Critical Pressure/atm	P_c = 111.3

Analysis: Use the following steps to set up the comparison between the two equations of state in *Excel*.

1. Provide a description for each parameter and cell name in columns **A** and **B**, respectively. Rename the cell addresses in column **C** with the names in column **B**. Enter the values for the constants (Rg, V, Tc, Pc, a, b) in column **C**. We can enter names in the **Name Box**, located just left of the **Formula Bar**.

	A	B	C	D	E	F	G
1	Parameter	Name	Value	Units		RK Constants	
2	Ideal Gas Constant	Rg	0.08206	L atm/ mol K		a	85.63487
3	Molar Volume	V	0.5	L/mol		b	0.025903
4	Critical T	Tc	405.5	K			
5	Critical P	Pc	111.3	atm			

Alternatively, name the range cells in column **C** using the **Name Manager**. Select the range of cells that includes the names and values, then chose from **Create Names from Selection** on the **Formulas** tab and click **OK**.

	A	B	C
1	Parameter	Name	Value
2	Ideal Gas Constant	Rg	0.08206
3	Molar Volume	V	0.5
4	Critical T	Tc	405.5
5	Critical P	Pc	111.3

Create Names from Selection
Create names from values in the:
- ☐ Top row
- ☑ Left column
- ☐ Bottom row
- ☐ Right column

Check the names by clicking in a named cell. The new name appears in the **Name Box**. We can also edit names in the **Name Manager**, if needed.

2. Calculate the constants *a* and *b* for the Redlich-Kwong equation of state in cells **G2** and **G3**. Use parentheses to control the order of operations in the calculations. Reference the other parameters by the new cell names. Create names *a* and *b* for cells **G2** and **G3**, respectively.

	F	G
1	RK Constants	
2	a	=0.42747*((Rg^2)*(Tc^2.5)/Pc)
3	b	=0.08664*(Rg*Tc/Pc)

3. Set up a column of temperatures in 50 K increments from 200 to 1000 K. Type the first two temperatures 200 and 250 in consecutive cells in column **A**. Select both temperature cells, then use the fill handle to create the temperature range automatically by dragging down the column until reaching the value 1000 in the last cell of the range. Alternatively, use the **Fill** option to create the series of temperatures in column **A**.

4. Calculate the pressure at each temperature using the ideal gas and Redlich-Kwong equations of state in columns **B** and **C**, respectively. At this point, we have not named any cells for T. Be sure to reference the temperature in the formula using a relative reference to the corresponding cell in the row for temperature from column **A**. Use parentheses to control the sequence of operations in the mathematical formula.

	A	B	C
7	T/K	IG P/atm	RK/atm
8	200	=Rg*T/V	=Rg*A8/(V-b)-a/(V*(V+b)*SQRT(A8))

5. Use the fill handle to put the formula in a column of cells corresponding to the column of temperatures. We can also double-click on the fill handle to achieve the same result obtained by manually dragging the fill handle. *Excel* creates formulas for the pressures corresponding to each temperature. Note that the named cells for the parameters are absolute references, whereas the relative temperature reference automatically increments in each formula for the temperature in the corresponding row.

B8 f_x =Rg*A8/V

	A	B	C	D	E	F	G
1	Parameter	Name	Value	Units		RK Constants	
2	Ideal Gas Constant	Rg	0.08206	L atm/ mol K		a	85.63487
3	Molar Volume	V	0.5	L/mol		b	0.025903
4	Critical T	Tc	405.5	K			
5	Critical P	Pc	111.3	atm			
6							
7	T/K		IG P/atm	RK/atm			
8		200	32.824	11.58916			
9		250					

6. Select the range of cells for ideal and non-ideal pressure. From the **HOME** tab, change the number format to **Scientific Notation** with three significant figures (2 decimal places).

7. Plot the results in a *y* versus *x* **Scatter** plot to compare the results. First, select the contiguous columns containing the temperatures and pressures. Then select the scatter chart on the **INSERT** tab. Click on the plot to reveal the tabs with the **Chart Tools**. Format the plot according to the guidelines in this section using the options listed on the **Layout** and **Format** tabs.

	A	B	C	D	E	F	G	H
1	Parameter	Name	Value	Units		RK Constants		
2	Ideal Gas Constant	Rg	0.08206	L atm/ mol K		a	85.6349	
3	Molar Volume	V	0.5	L/mol		b	0.0259	
4	Critical T	Tc	405.5	K				
5	Critical P	Pc	111.3	atm				
6								
7	T/K		IG P/atm	RK/atm				
8		200	3.28E+01	1.16E+01				
9		250	4.10E+01	2.27E+01				
10		300	4.92E+01	3.31E+01				
11		350	5.74E+01	4.32E+01				
12		400	6.56E+01	5.30E+01				
13		450	7.39E+01	6.25E+01				
14		500	8.21E+01	7.20E+01				

Comments: The results of the comparison indicate that ammonia gas approaches ideal behavior at high temperatures and pressures.

CHAPTER 2: EXCEL

An alternative method for creating the columns of gas pressures uses an array operation on the formula. Name the range of temperature data **A8:A24** by selecting the range of cells and typing **T** in the name box, then *Enter*. Select the entire range **B8:B24** for the results for pressure, and type the formula for the ideal gas pressure, followed by the keyboard combination *CTRL SHIFT ENTER* to put the pressure values corresponding to each temperature in column **B**. Repeat this step for the Redlich-Kwong results in column **C**.

7	T/K	IG P/atm	RK/atm
8		200	=Rg*T/V	1.16E+01
9		250		
10		300		
11		350		

□

2.3.5 Pivot Tables and Pivot Charts

When working with large amounts of categorical data tabulated in an *Excel* worksheet, consider using *Excel's* **Pivot Table** and **Pivot Chart** tools to summarize quickly your data for analysis and presentation. To use *Pivot* operations, the data must be arranged in contiguous columns with column labels in the row at the top of the data. Be careful not to mix the *types* of data in a column (e.g., strings versus numbers) – the default **Pivot Table** calculates sums of numerical data, or counts the number of entries of non-numerical data. Once created, **Pivot Tables** are easily customized for your particular analytical, exploration, or presentation purposes. We demonstrate *Pivot* tools in the next example.

Example 2.5   Pivot Table and Pivot Chart Summary of Building Utility Data

Known: The energy data tabulated in the *Excel* worksheet in Figure 2.23 shows the type of utility requirements for three different batch-processes located in separate buildings in a chemical plant.

	A	B	C	D
1	Bldg	Utility	MWh/ton	MWh/month
2	C	Cooling	0.23	0
3	A	Steam	0.96	759.8
4	B	Steam	0.73	773.5
5	A	Cooling	0.1	9
6	C	Steam	0.6	706.3
7	B	Electricity	0.23	137
8	B	Cooling	0.16	7.3
9	A	Electricity	0.28	130
10	C	Electricity	0.41	277.5

Figure 2.23

Energy requirements for batch processing facilities in a chemical plant (Bieler, Fisher and Hungerbuhler 2003).

Find: Create a **Pivot Table** to summarize the energy requirements for each building and type of utility.
Analysis: Click on any cell in the contiguous range of tabulated data on the *Excel* worksheet. Click on the **Insert** tab of the ribbon and select **Pivot Table**, from the **Tables** group:

FILE   HOME   INSERT

PivotTable   Recommended   Table
             PivotTables
                Tables

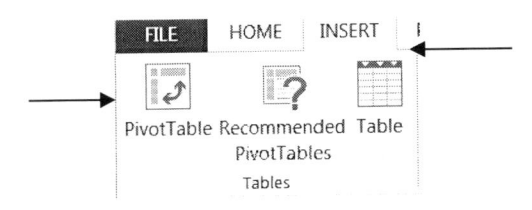

From the dialog, select the data labels for Utility, MWh/t, and MWh/month to generate the **Pivot Table** in Figure 2.24.

Row Labels	Sum of MWh/t	Sum of MWh/mo
Cooling	0.49	16.3
Electricity	0.92	544.5
Steam	2.29	2239.6
Grand Total	3.7	2800.4

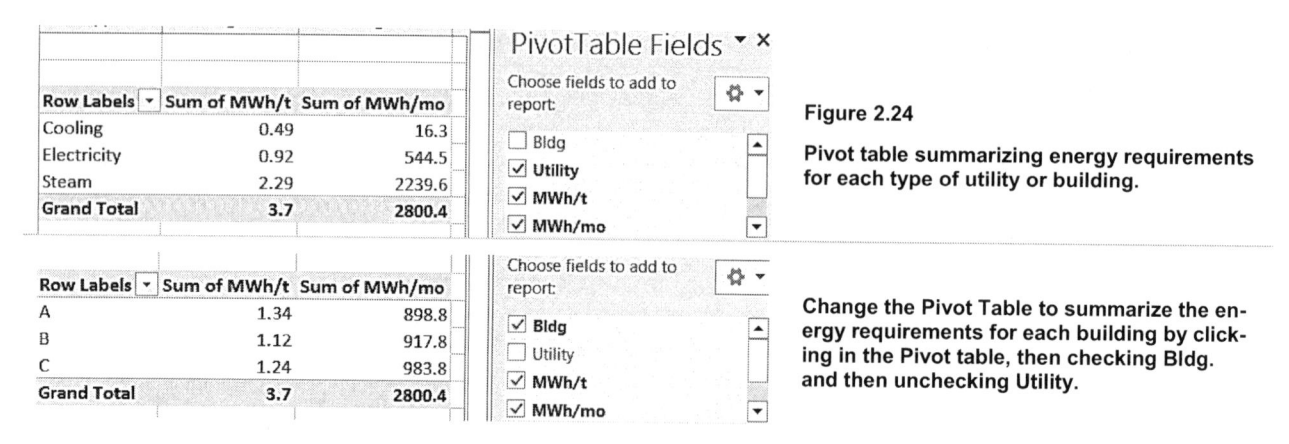

Figure 2.24

Pivot table summarizing energy requirements for each type of utility or building.

Row Labels	Sum of MWh/t	Sum of MWh/mo
A	1.34	898.8
B	1.12	917.8
C	1.24	983.8
Grand Total	3.7	2800.4

Change the Pivot Table to summarize the energy requirements for each building by clicking in the Pivot table, then checking Bldg. and then unchecking Utility.

To change the summary value of the field from the default **SUM** to a different calculation, click on the down arrow next to a field listed under Σ **VALUES**, and select **Value Field Settings** from the popup window. Choose the type of calculation from the list of options. To illustrate, we may decide to calculate the Average MWh/ton for each type of utility:

Row Labels	Average of MWh/t
Cooling	0.163333333
Electricity	0.306666667
Steam	0.763333333
Grand Total	0.411111111

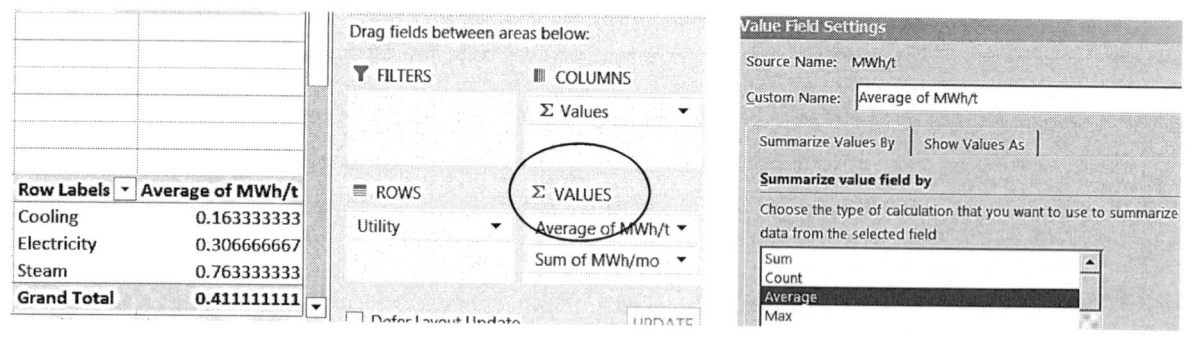

To create a **Pivot Chart**, click in the **Pivot Table** and select the **Analyze** tab that appears on the ribbon. Then select **Pivot Chart** and customize as needed in a fashion similar to **Pivot Tables**. In this illustration, a column chart displays the sum of MWh/t for each building:

Row Labels	Sum of MWh/t
A	1.34
B	1.12
C	1.24
Grand Total	3.7

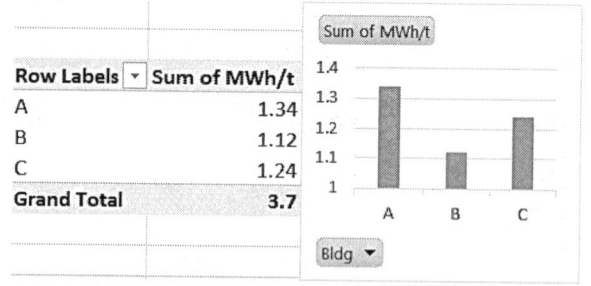

Comments: We used a relatively small data table in this example for illustrating how to work with **Pivot Tables**. The utility of **Pivot Tables** and **Charts** becomes much more apparent when working with larger data sets.

2.3.6 Text Wrapping and Breaking Text*

When the contents of a cell exceeds the height of the row, automatically adjust the row size. Use **Text Wrapping** in the **Alignment** group on the **Home** tab to expand the row height to show all of the contents of the cell, as shown in Figure 2.25. Unfortunately, *Excel* does not always split the words at the best location. Use the keyboard combination **Alt-Enter** to break the text at the desired location and position the remaining text in the next line within the same cell.

CHAPTER 2: EXCEL

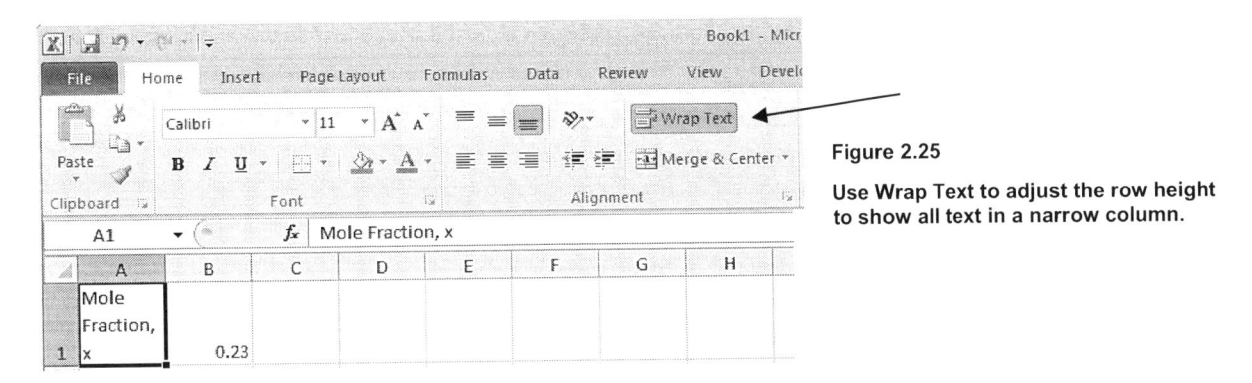

Figure 2.25
Use Wrap Text to adjust the row height to show all text in a narrow column.

2.3.7 Use Conventional Engineering Symbols and Equations*

Engineering math uses Greek symbols for a variety of physical and chemical constants. Improve the clarity of worksheet documents by inserting symbols into cells used for headings or labels. Find the desired symbol from the **Text** group on the ribbon tab **Insert>Symbol**:

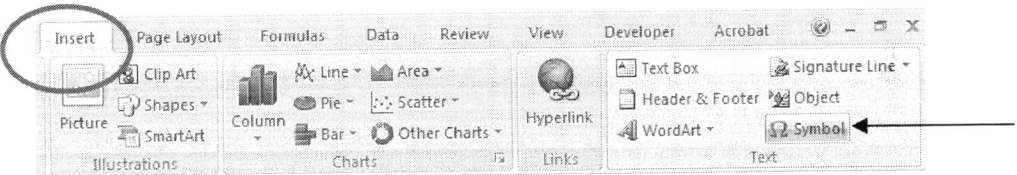

As engineers and scientists, we communicate in the language of mathematics. Starting with *Excel* 2010, we have access to a convenient object for creating images of equations in selected areas on the worksheet. To access the object named **Microsoft Equation**, select from ribbon tab **Insert > Object**. Scroll to find **Microsoft Equation** then click **OK**. *Excel 2013* added **Equation** to the tab **Insert>Symbol** group to expose a new pair of tabs to access the equation editor. Build up an equation, such as the Hagen-Poiseuille law shown here, by selecting the options for fractions, Greek symbols, and exponents from the various groups in the **Equation** toolbar:

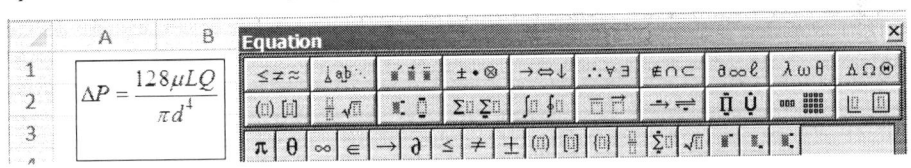

$$\Delta P = \frac{128 \mu L Q}{\pi d^4}$$

2.3.8 Comments, Data Validation and Conditional Formatting*

"You validate people's lives by your attention." - Unknown

We may consider adding a descriptive comment tied to a cell that pops up whenever the user moves the mouse over that cell. To add a comment, right-click on the cell and select **Insert Comment**. Add descriptive text in the box. To make changes to a comment, right-click in a cell or range of cells, and select **Edit** or **Delete Comment**. A cell with a comment has a small red triangle in the upper right corner.

Excel has tools to help us minimize or eliminate numerical errors by controlling the types of data entry in an *Excel* worksheet. Click in a single cell or select a range of cells then chose from the ribbon tab **Data>Data Validation** to open a dialog box. *Data validation* controls the scope and range of data entry to a cell. For example, if a cell contains a value for mole fraction, select **Decimal** then enter the upper and lower limits for mole fraction of zero and one, respectively (see Figure 2.26). We may also add input and error messages on additional tabs in the dialog box to display pop-up boxes that inform and guide the user to make correct input decisions.

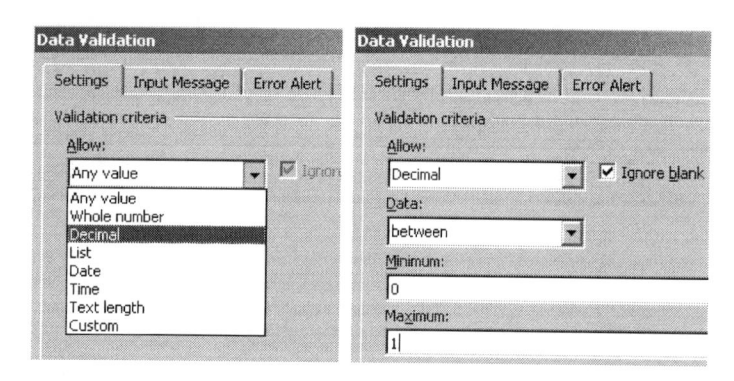

Figure 2.26

Data validation options. Decimal data allows for minimum and maximum allowed entry to a cell.

We may use conditional formatting accessed from the tab **Home>Styles** group to change the appearance of a cell to accentuate the cell contents:

With conditional formatting, we may highlight cells, add data bars, or color ranges depending on the conditions we impose. To add more than one condition, simply apply multiple conditions, one at a time from the **Conditional Formatting** drop-down menu. To remove conditional formatting, go to **Manage Rules** in the drop-down menu. We use conditional formatting in Example 14.1 to indicate the relative magnitude of temperature with a color scale.

Example 2.6 Table of Ideal Gas Constants

Known/Find: Create an *Excel* worksheet for obtaining the ideal gas constant for various combinations of units.

Analysis: For this example, we use a default value for the gas constant of R_g = 8314 m³ Pa/kmol·K. The value of the gas constant for different unit systems requires unit conversions:

$$R_g = \left(8314 \frac{m^3 \cdot Pa}{kmol \cdot K}\right)\left(\frac{f_V \cdot f_P}{f_N \cdot f_T}\right) \quad (2.12)$$

where *f* is the conversion factor to change from the default unit of volume (*V*), pressure (*P*), moles (*N*), or temperature (*T*) to the new units for that quantity. Employ the worksheet reference functions **INDEX(array, row_num, optional col_num)** with imbedded **MATCH(lookup_value, lookup_array, optional type)** to pull conversion factors from a table. Use validation to limit the selection of units in the worksheet.

1. Set up an *Excel* worksheet with units and conversion factors for each dimensional quantity in the gas constant:

	A	B	C	D	E	F	G	H	I	J	K
1	Parameter	Units		Conversions							
2	Volume	L		1000	=INDEX(B10:B13,MATCH(B2,A10:A13,0))						
3	Pressure	psia		0.000145038							
4	Mole	lbmol		2.20462							
5	Temperature	R		1.8							
6	R_gas =	3.039E+02									
9	V Unit	Convert Factor		P Unit	Convert Factor		N Unit	Convert Factor		T Unit	Convert Factor
10	m3	1		Pa	1		kmol	1.0000E+00		K	1
11	L	1000		atm	9.8692E-06		gmol	1.0000E+03		R	1.8
12	cm3	1.000E+06		psia	1.4504E-04		lbmol	2.2046E+00			
13	ft3	3.531E+01		mmmHg	7.5006E-03						

CHAPTER 2: EXCEL

2. To validate the choice of units in the range **B2:B5**, activate each cell in turn and choose from the tab **Data>Data Validation**, then select **List** from the validation criteria shown in Figure 2.26. In the **Settings/Source** box, type or select the range address on the worksheet containing the optional units. For example, the range **$A$10:$A$13** contains the available units for volume. This creates a drop down list to select the value for cell **B2**. Use similar drop down lists for the units of the remaining quantities in the range **B3:B5**.

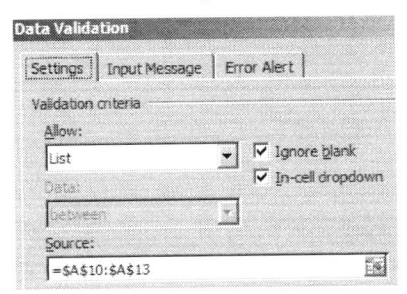

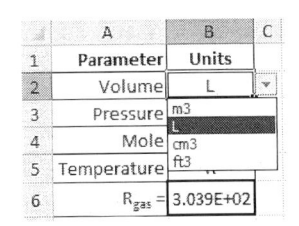

3. In the range **D2:D5** use **INDEX** with imbedded **MATCH** worksheet functions to locate the correct conversion factors from the tables according to the units selected in the range **B2:B5**:

    D2=INDEX(B10:B13, MATCH(B2, A10:A13, 0))   D4=INDEX(H10:H12, MATCH(B4, G10:G12, 0))
    D3=INDEX(E10:E13, MATCH(B3, D10:D13, 0))   D5=INDEX(K10:K11, MATCH(B5, J10:J11, 0))

    The third (optional) argument in **MATCH** determines if the match is (1) the largest value less than or equal to the lookup value with the data in ascending order, (0) equal to the look up value, or (-1) the smallest value greater than or equal to the lookup value with the data in descending order.

4. Calculate the gas constant in cell **B6** according to Equation (2.12):

    B6=8314*D2*D3/(D4*D5)

5. Click in any of the cells for units in column **B** and select from the drop-down list to recalculate the gas constant in the new units.

Comments: To clean up the appearance of the worksheet, hide the rows and columns on the worksheet that contain the data used for the unit list, conversion factors, and reference functions. To hide and unhide rows and columns, select the desired rows (or columns), then select from the tab **Home/Format>Hide & Unhide**.

□

2.4 "What-if" Analysis Using Goal Seek and Data Table

"What if the hokey-pokey isn't what it's all about?" – Bumper Sticker

Imagine a scenario where we know the value of a function, but not the value of the input variable needed to get the known functional result. We may try several ways of finding good input values, including graphing the function and simple "trial and error" where we try a few values to see what happens. **Goal Seek** uses a numerical algorithm built into *Excel* for adjusting values in calculations to meet specified criteria. **Goal Seek** is available from the **Data** tab, in the group **Data Tools>What-if Analysis**. We illustrate how **Goal Seek** works with an example. Be aware that **Goal Seek** stops its search for a solution when the "**Set cell**" value stops changing within a specified tolerance. The default tolerance is 10^{-3}. If the change in the magnitude of the function is smaller than the default, **Goal Seek** will stop the search prematurely. For example, the viscosities of gases have values of the order 10^{-5} Pa·s. **Goal Seek** will stop after one iteration. Use an appropriate setting to fit the problem. Raise or lower the convergence setting in **File>Options>Formulas >Maximum Change**. Otherwise, try scaling the function value by 10^3 for a value with a magnitude of at least one.

We may explore the effects of parameter changes on functions using the **Data Table** option under **What If Analysis**. **Data Table** is a convenient tool in *Excel* for generating formula values over ranges of input in one or two dimensions. Once we have the data in rows and columns on the worksheet, we may quickly generate line, surface, or contour plots of the data to visualize the effects of changing parameters in the formula. We illustrate how to use **Goal Seek** and **Data Table** in the following example.

Example 2.7   Molar Volume of Ammonia

Known/Find: Refer to the information in *Example 2.4* to find the molar volume of ammonia at a pressure of 10 atm and temperature of 300 K.

Analysis: Set up an *Excel* worksheet as shown in Figure 2.27 with the data and formulas for calculating the pressure from the Redlich-Kwong equation of state in Equation (2.11). The value of cell **B9** is the difference between the pressure in cell **B7** and calculated pressure in cell **B8**.

	A	B	C	D	E	F
1	R_g (L·atm/mol·K)	0.08206				
2	T_c/K	405.5				
3	P_c/atm	111.3				
4	a	85.63487	=0.42747*(Rg^2*Tc^(5/2)/Pc)			
5	b	0.025903	=0.08664*Rg*Tc/Pc			
6	T/K	300				
7	P/atm	10				
8	P/Redlich-Kwong	10.00042	=Rg*T/(V-b) - a/(V*(V+b)*SQRT(T))			
9	ΔP/atm	-0.00042	=B7-B8			
10	V/(L/mol)	2.275195				

Figure 2.27

Molar volume of ammonia from Redlich-Kwong equation of state.

Access **Goal Seek** from the **Data** tab **What-if Analysis** drop-down list, as follows:

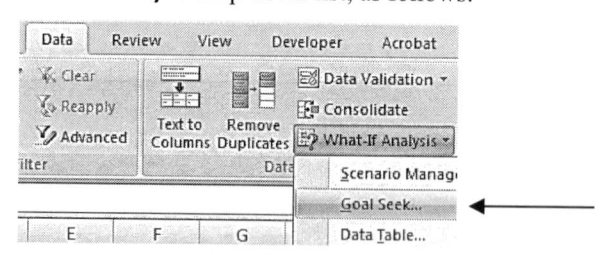

- Select cell **B9** containing the pressure change for the **Set cell**, then specify zero for **To value**.
- Select cell **B10** for the input to the box named **By changing cell**.
- Click **OK** for the **Goal Seek** solution. **Goal Seek** adjusts the volume to achieve a pressure change of approximately zero (i.e., match the target pressure).
- A **Goal Seek Status** message box provides information about the solution compared with the target:

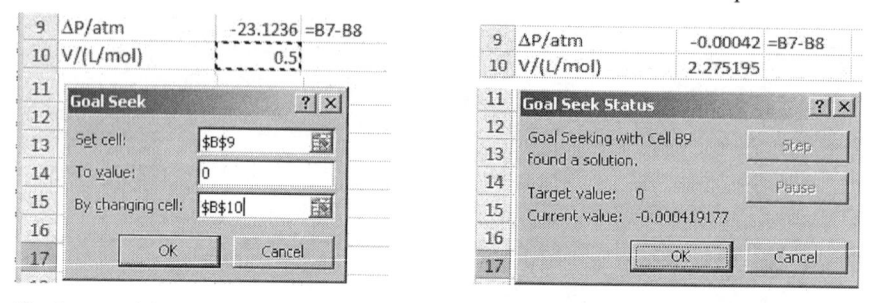

If necessary, specify the precision of **Goal Seek** in **File>Options>Formulas >Maximum Change**, as shown in Figure 2.28. The maximum change represents the change in the function result of **Set cell**. The problem may require scaling to control the precision of the solution.

CHAPTER 2: EXCEL

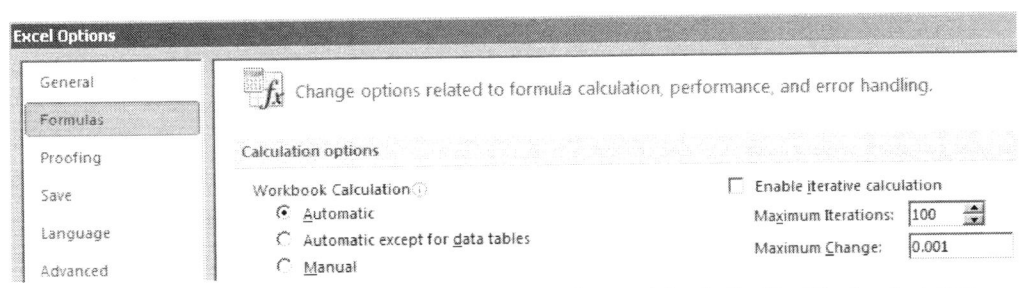

Figure 2.28 Maximum Change for controlling the precision in the Goal Seek calculation.

We use the existing worksheet setup and formulas with **Data Table** to create columns of data for plotting the pressure versus molar volume:

1. Create a column of molar volume values ranging from one to three incrementing by 0.25 in the range **A13:A21**. Set the pattern with 1.00 and 1.25 in cells **A13** and **A14**, select the two cells with the pattern, then drag the fill handle down to generate the column of values. Alternatively, use **AutoFill** to create the volume series in column **A**.
2. Set the value of cell **B12** equal to the result of the formula for pressure in cell **B12=B8**
3. Select the entire range of cells **A12:B21**, including the cell that refers to the pressure calculation.
4. Select from the tab **Data> What if Analysis > Data Table**.
5. Leave the **Row input cell** blank on the **Data Table** input box. The **Column input cell** is **B10** (molar volume used in the formula for pressure in cell **B8**).

6. Click **OK** to generate pressures in column **B** corresponding to each value of molar volume in column **A**. Plot the molar volumes in the range **B13:B21** versus pressures in the range **A13:B21**. We can read the volume form the graph corresponding to 10 atm pressure.

Comments: **Goal Seek** is a robust tool for finding local roots to single variable functions, but is limited to finding a root for a single variable problem. **Data Table** allows us to use our existing worksheet formulas to generate data quickly for analysis by tabulation or plotting.

□

2.5 Excel Add-Ins: Analysis Tools and Solver

> *"42.7 percent of all statistics are made up on the spot."* – Steven Wright

Some advanced features of *Excel* become available only when activated as add-ins. **Analysis Tools** and **Solver** are two popular *Excel* add-ins for engineering problem solving. Activate add-ins on the menus inside the ribbon tab **File>Options>Add-ins>Manage:** *Excel* **Add-ins>Go>Analysis Tool Pak**. These add-in steps are illustrated next.

1. From *Excel* **Options**, select **Add-ins**, then *Excel* **Add-ins** from the **Manage** link to access the add-ins options.

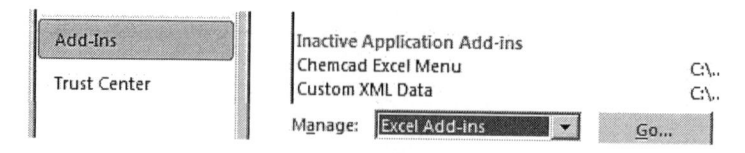

2. Click in the boxes for the desired add-ins. A description of a highlighted **Analysis Tool Pak** add-in appears below the list:

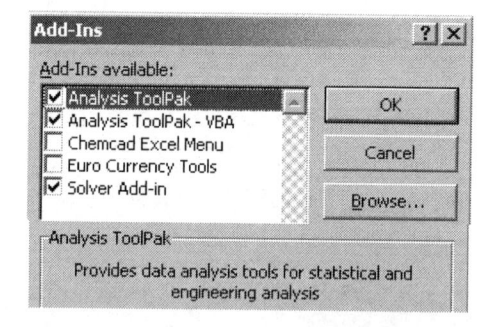

Some Window's software has *Excel* add-ins for interaction between the software and *Excel*. For example, the chemical process simulator *CHEMCAD*®, shown in the available add-ins for this illustration, has features to share data with *Excel*. We normally select an add-in once, and it remains available for future workbooks until deselected.

The data **Analysis Tool Pak** consists of a collection of tools for financial, engineering, and data analysis, including linear least-squares regression for data modeling, Fourier analysis, and statistical functions. Once added, access the **Analysis Tools** or the **Solver** from the **Analysis** group on the **Data** tab, as shown in Figure 2.29.

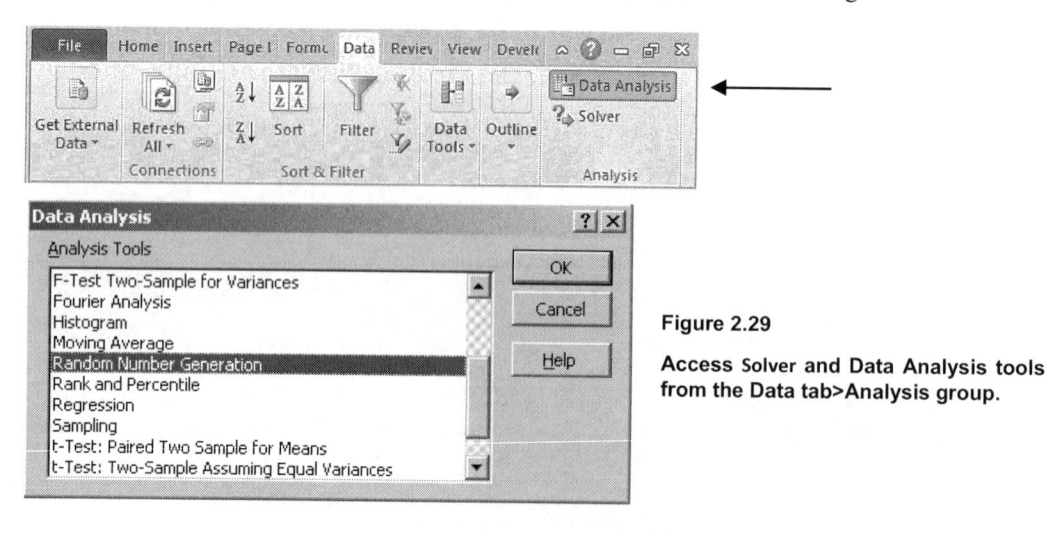

Figure 2.29

Access Solver and Data Analysis tools from the Data tab>Analysis group.

CHAPTER 2: EXCEL

Example 2.8 Descriptive Statistics and Analysis in Excel

Known/Find: Generate a set of 10 normally distributed data points with a mean value of five and a standard deviation of one using the nested worksheet functions **NORMINV(RAND(),5,1)**. Alternatively, we can generate random numbers from the **Data Analysis** add-in. Note that **NORMINV** will give a new normal random number anytime a change is made to the worksheet. Turn off **Automatic Calculation** by changing the setting in **File> Options> Formulas>Calculation Options> Workbook Calculation> Manual**, shown in Figure 2.28. Select **Random Number Generation** from the **Data Analysis>Analysis Tools,** as shown in Figure 2.29. Specify the number of variables, number of random values of the variable, and distribution, as shown in Figure 2.30:

Figure 2.30

Dialog box for Analysis Tool Pak>Analysis Tools> Random Number Generation.

Analysis: Due to the nature of random numbers, if we are trying to duplicate this example we will get different random numbers than those listed in the figures. Use **Data Analysis Tools>Descriptive Statistics** to quickly generate a summary of descriptive statistical information for the data. Specify the range of input values and the location of the output range to contain the results of the descriptive statistics, as seen in Figure 2.31:

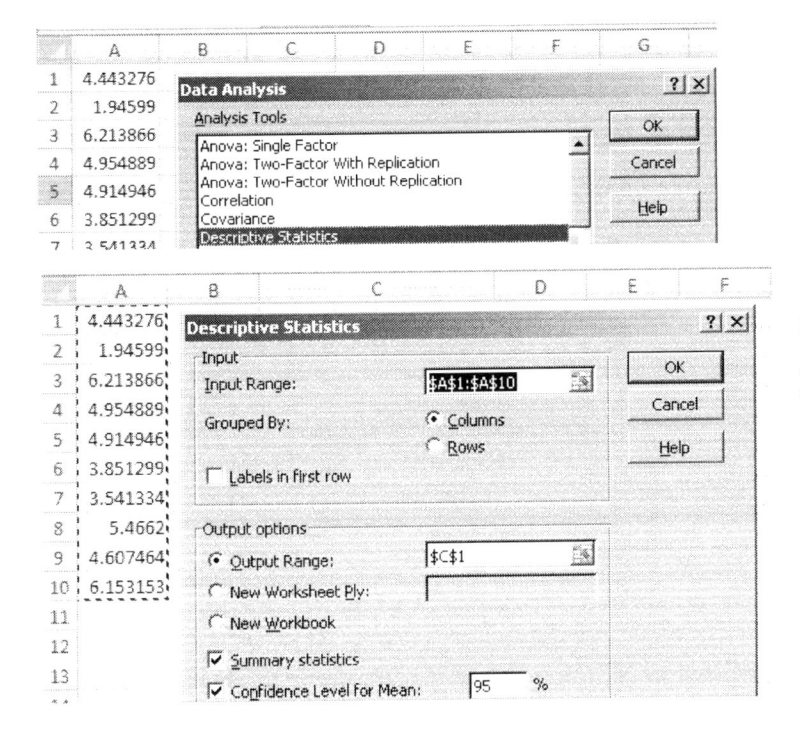

Figure 2.31

Analysis Tool Pak>Descriptive Statistics and output table.

	A	B	C	D
1	4.443276		Column1	
2	1.94599			
3	6.213866		Mean	4.60924174
4	4.954889		Standard Error	0.40471046
5	4.914946		Median	4.761205345
6	3.851299		Mode	#N/A
7	3.541334		Standard Deviation	1.279806846
8	5.4662		Sample Variance	1.637905564
9	4.607464		Kurtosis	0.933709597
10	6.153153		Skewness	-0.79754662
11			Range	4.267876648
12			Minimum	1.945989551
13			Maximum	6.213866199
14			Sum	46.0924174
15			Count	10
16			Confidence Level(95.0%)	0.915518666

Observe how the results shown in Figure 2.31 span 16 rows – so be careful not to overwrite important content on the *Excel* worksheet. We may calculate each of the results from **Descriptive Statistics** using an equivalent worksheet function. For example, calculate the mean, standard deviation, and mode using the worksheet functions **AVERAGE**, **STDEV**, and **MODE**, respectively.

In a second illustration, conduct a two-sample *t*-test. Generate a second set of numbers normally distributed about the same mean, but with different numbers of data points and different standard deviations.

- Use the add-in **Data Analysis>t-Test: Two Sample Assuming Unequal Variances** to compare the means of the two data sets, as seen in Figure 2.32.
- Click in the dialog boxes for the requested values, and then select the corresponding ranges on the worksheet to give the worksheet addresses of the variable ranges.
- Provide the location for displaying the results of the two-sample *t*-test on the worksheet, then click **OK** for calculating the table of results shown in Figure 2.32.

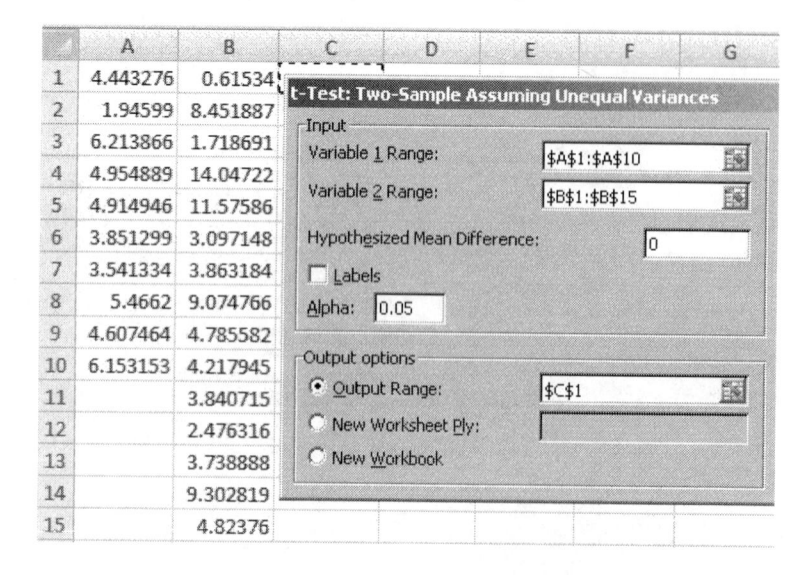

Figure 2.32

Excel add-in Data Analysis>Analysis Tools>*t*-test and table of results.

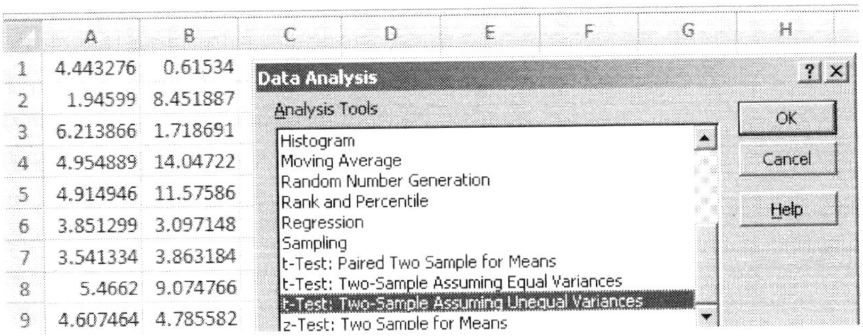

CHAPTER 2: EXCEL

	A	B
1	4.443276	0.61534
2	1.94599	8.451887
3	6.213866	1.718691
4	4.954889	14.04722
5	4.914946	11.57586
6	3.851299	3.097148
7	3.541334	3.863184
8	5.4662	9.074766
9	4.607464	4.785582

Data Analysis

Analysis Tools:
- Histogram
- Moving Average
- Random Number Generation
- Rank and Percentile
- Regression
- Sampling
- t-Test: Paired Two Sample for Means
- t-Test: Two-Sample Assuming Equal Variances
- **t-Test: Two-Sample Assuming Unequal Variances**
- z-Test: Two Sample for Means

	A	B	C	D	E
1	4.443276	0.61534	t-Test: Two-Sample Assuming Unequal Variances		
2	1.94599	8.451887			
3	6.213866	1.718691		Variable 1	Variable 2
4	4.954889	14.04722	Mean	4.609242	5.708674861
5	4.914946	11.57586	Variance	1.637906	14.96369503
6	3.851299	3.097148	Observations	10	15
7	3.541334	3.863184	Hypothesized Mean Di	0	
8	5.4662	9.074766	df	18	
9	4.607464	4.785582	t Stat	-1.020196	
10	6.153153	4.217945	P(T<=t) one-tail	0.160576	
11		3.840715	t Critical one-tail	1.734064	
12		2.476316	P(T<=t) two-tail	0.321153	
13		3.738888	t Critical two-tail	2.100922	

Note that the *magnitude* of the *t*-statistic is smaller than the critical value for two tails (-1.02 < 2.1), which supports the null hypothesis that there is no statistical difference between the two means for 95% confidence ($\alpha = 0.05$). Alternatively, we observe that the *p*-test value is greater than α, again supporting the null hypothesis.

Comments: Using statistical formulas directly in the worksheet instead of generating the results using the **Data Analysis** tools has the advantage that the values for the descriptive statistics and hypothesis tests update automatically on the worksheet for any changes to the data.

☐

Excel's **Solver** is a powerful tool for tackling a variety of engineering problems. The **Solver** add-in does what **Goal Seek** does, and much more (like **Goal Seek** on steroids). For instance, the **Solver** is capable of working with multiple variables. The **Solver** uses numerical methods that iteratively search for the solution to the problem. The numerical methods employed by the **Solver** converge efficiently to a good solution using approximations for the derivatives of the function with respect to the function variables. We reach the solution when the values for the search stop changing between iterations within a specified tolerance.

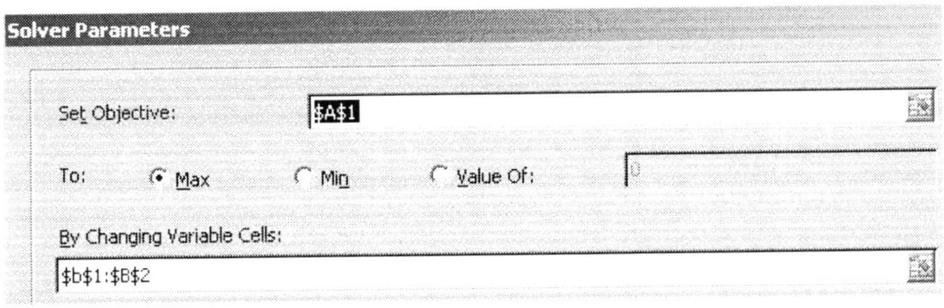

Solver Parameters

Set Objective: $A$1

To: ● Max ○ Min ○ Value Of: 0

By Changing Variable Cells:
b1:$B$2

Figure 2.33

Solver dialog box for setting up the dependent and independent variables.

Table 2.7 lists definitions for the parameters on the *Excel* **Solver** dialog box, shown in Figure 2.33. The first group of parameters including **Set Target Cell**, **Equal To**, and **By Changing Cells** work in the same way as **Goal Seek**. Use the additional options of **Max** and **Min** for optimization problems, described in detail in Chapter 7. As demonstrated in subsequent chapters, we may try using the **Solver** for cracking problems ranging from nonlinear equations to constrained optimization. Refer back to this section of the book for help setting the parameters in the **Solver**, or consult *Excel*'s on-line **HELP**, .

Table 2.7 *Solver* parameters and definitions.[34]

Parameter	Definition
Set Objective	Provide the address of the cell with the objective formula that we want to minimize, maximize, or set to a specified value.
By Changing Variable Cells	Specify the range of cells that contain the problem variables. The adjustable cells must be referenced, either directly or indirectly, by the objective cell formula.
Add, Change, Delete	Set constraints on the values in the **By Changing Cells** references.
Solve	Run the **Solver**.
Close	Close the dialog box and retain any changes made to the **Solver**.
Options	Open the **Solver Options** dialog box to access advanced features and control settings.
Reset All	Reset all parameters to their original values.
Load/Save	Load or save previously defined settings
Make Unconstrained Variables Nonnegative	Causes **Solver** to assume a lower limit of zero for all adjustable cells for which we have not set a lower limit in the **Constraint** box.

2.6   Epilogue on Excel

Excel has risen to the top of the list of standard, collaborative engineering tools for managing data, performing engineering calculations, and collaborating with other engineers, scientists, accountants, *etc*. We turn to *Excel* whenever we need to prepare a graph, crunch some numbers, calculate statistics, or perform "what if" analysis. The Appendix in Section 0 lists several *Excel* worksheet functions for implementing numerical solutions to engineering problems used in the variety of examples throughout this book. Do not stop there, however. The best way to become an expert is to start using *Excel* and find new ways of performing engineering analysis. *Excel* has much more to offer than we can possibly squeeze into a textbook on numerical methods.

---

[34] Click on **Solver Parameters>Help** for more details.

Chapter 3   Visual Basic for Applications

"Most of the fundamental ideas of science are essentially simple, and may, as a rule, be expressed in a language comprehensible to everyone." – Albert Einstein

Computer programs of numerical methods are scripts of mathematical operations performed by the computer. In the very early days of computing, programs consisted of setting switches for ones and zeros to turn transistors on and off – a task reserved for the specialist. Thanks to the brilliance of Grace Hopper – tedious, and error prone, machine-level programming was replaced by high-level programming in familiar human-like languages, such as BASIC[35], that are translated into computer machine language by a compiler or interpreter.[36] *Visual Basic for Applications* (*VBA*) is a dialect of *Microsoft's* object-oriented programming language *Visual Basic*. Microsoft Office applications have *VBA* built in for enhancing their capabilities. *VBA* programs allow us to customize and boost *Excel* beyond its standard worksheet capabilities as an engineering problem solving tool. *VBA* is particularly useful for incorporating advanced numerical methods into *Excel* worksheets. We learned about several useful built-in *Excel* worksheet functions for engineering in Chapter 2. However, we may need a function not available in *Excel*'s library. In this case, we can create custom functions with *VBA* that are accessible just like *Excel* functions in cells on the worksheet.

> **SQ3R Focused Reading Questions**
> 1. What is a macro-enabled workbook?
> 2. What is the difference between a macro, sub procedure, and user-defined function?
> 3. Where does *Excel* store *VBA* code?
> 4. What is the difference between types of variables?
> 5. How does *VBA* display the results of calculations?
> 6. How does *VBA* interact with an *Excel* worksheet?
> 7. Are worksheet functions available for use in *VBA*?
> 8. What tools are available for debugging code?
> 9. How can I control the flow of calculations in a program?
> 10. How can I program to speed up the calculations?

A *VBA* program automates complex calculations. Whereas we can usually perform the same calculations in an *Excel* worksheet, programs simplify the worksheets and are available for reuse, thus reducing our workload and minimizing errors in our formulas. The *VBA* programs developed for the various numerical methods introduced throughout this book provide several examples of *VBA* programming syntax and functionality that we may consult as we learn to use this programming language. The *MS Office* help files are good places to look for functions or code. The internet has many freely accessible web sites devoted to *VBA* programming where we may find tips and examples, as well as macros to enhance our *Excel* worksheet calculations.

This chapter introduces the following features of *VBA* programming that are common to practically all programming languages:
- Editor and syntax for viewing and writing procedures
- User-defined functions and Sub procedures (or macros) that perform a sequence of calculations

---

[35] BASIC (an acronym for Beginner's All-purpose Symbolic Instruction Code) is a family of general-purpose, high-level programming languages whose design philosophy emphasizes ease of use. (https://en.wikipedia.org/wiki/BASIC).

[36] She later said, "No-one thought of that earlier, because they weren't as lazy as I was." In fact, Grace was famous for hard work! (http://www.bbc.com/news/business-38677721)

- Types of variables for storing information in a macro, including arrays
- Logic for program flow control
- Communication between *Excel* and *VBA* using message and input boxes and forms

3.1 "Macro Enabled" Workbooks

The next sentence is important! We must save *Excel* workbooks using the **macro-enabled** option or we will lose our *VBA* code when we close the file! If our workbook contains *VBA* code, *Excel* prompts us with a reminder to specify the file type the first time we save it, as shown in Figure 3.1. Click **No** and save the file again as a macro enabled workbook, otherwise, you will lose all of your *VBA* code. *Excel* appends the file name with the extension *.xlsm*. We only need to save our workbook once as a macro enabled file. We have three options for locations to save our macros:

1. Current workbook
2. New workbook
3. Personal.xlsb Macro Workbook

Whether or not we save the macro in an existing or new workbook, the macro is only available when that workbook is open. However, saving the macro to the **Personal Macro Workbook** makes the macro available any time we use that particular installation of *Excel*. If we use more than one computer, we will need to save our macro to the **Personal Macro Workbook** on each machine. Access the *PERSONAL.XLSB* file in the folder: **C:\Users\ <*username*> \AppData \Roaming \Microsoft ***Excel* **XLSTART**.

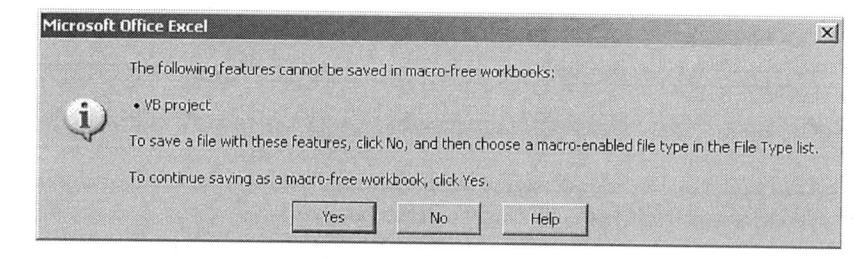

Figure 3.1

Excel prompt to save a workbook with *VBA* as a macro-enabled file.

Unfortunately, nefarious hackers use *VBA* to create computer malware to wreak havoc on our computers. By default, *Excel* disables macros to prevent malicious code from running when we first open a .xlsm file. To activate a *VBA* program in a saved workbook, select **Enable Content** from the **Security Warning** options:

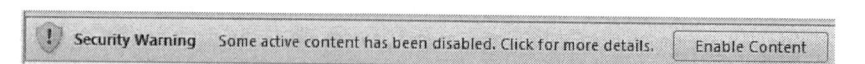

3.2 Visual Basic Editor (VBE) Projects and Modules in the PNM Suite

We write code for *VBA* programs in the **Visual Basic Editor** (*VBE*). We start programming in a few simple steps from an *Excel* worksheet by selecting from the ribbon tab **View>View Macros**. Type a new name for the program and click **Create** to insert a module and add a **Sub Procedure** to the code window of the *VBE*.

To access the *VBE* without creating a program we must first expose the **Developer** tab, as shown in Figure 3.2, by selecting from the ribbon tab **File> Options> Customize Ribbon> Customize the Ribbon> Developer**.

CHAPTER 3: VBA

Figure 3.2

Add the Developer Tab to the Ribbon.

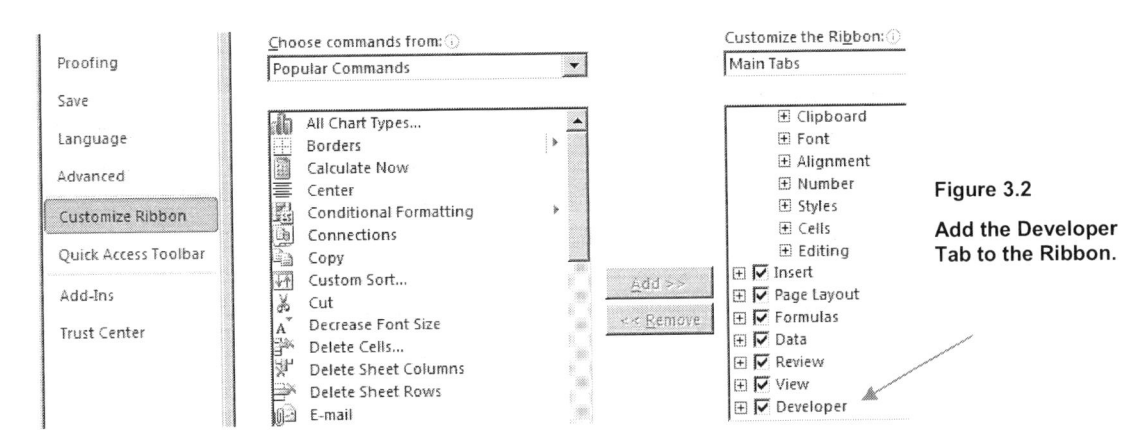

To open the *VBE* from the **Developer** tab, click on the **Visual Basic** icon shown at the left in Figure 3.3.

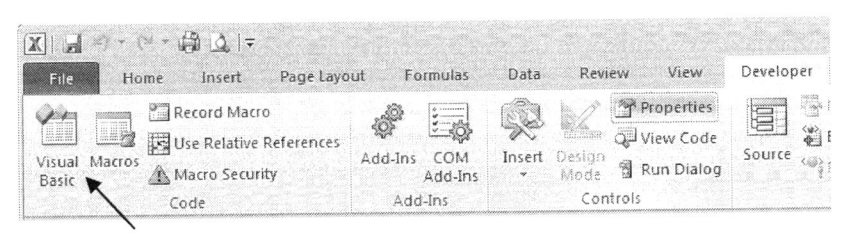

Figure 3.3

Select Visual Basic from the Code group of the Developer tab to access the *VBA* features of *Excel*.

The *VBE* uses a navigation tree (labeled *Project - VBA Project*) to list all *VBA* projects currently open. The tree lists our current project as *VBA Project (Workbook Name)*. Open a project from the **View** menu. The project window contains a navigable tree structure with add-ins, such as the **Solver**; objects, including all worksheets and the workbook; and any modules added by the user to house macros. Worksheet objects may contain sub procedures for event handlers or controls added to an *Excel* worksheet. We use modules to contain any custom macros, sub procedures and user-defined functions developed by the user. The user may change the names of modules in the corresponding **Properties** window, described in Section 3.2.1.

We need a module to contain our code. To add a new module to our project (*i.e.*, workbook), select the menu option **Insert/Module** in the *VBE* as shown in Figure 3.4, or right-click on the bold project name and select the option **Insert/Module**.

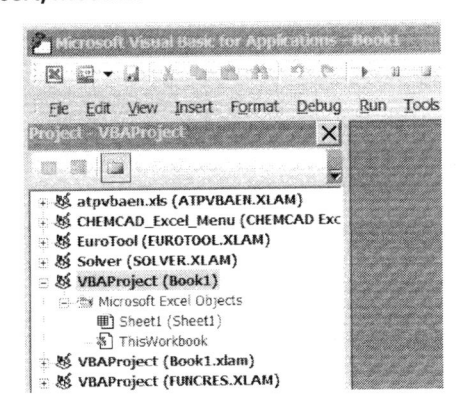

Figure 3.4

Add a module to the project to contain our macro's code. Select the VBA Project for the workbook (name) and select the menu Insert/Module. Be careful not to save your macros in other workbook object modules, such as FUNCRES that deletes custom macros.

To summarize, take the following steps to create a new macro:
1. Open the *VBE* from the Developer tab.
2. Activate the *VBA* Project for the workbook (workbook name in parenthesis).
3. Insert a module from the menu.
4. Insert a **Function** or **Sub** procedure from the **Insert** menu.
5. Add code to the procedure.

As needed, modify the format of the text in the programming window through the **Tools/ Options/ Editor Format** menu, as shown in Figure 3.5.

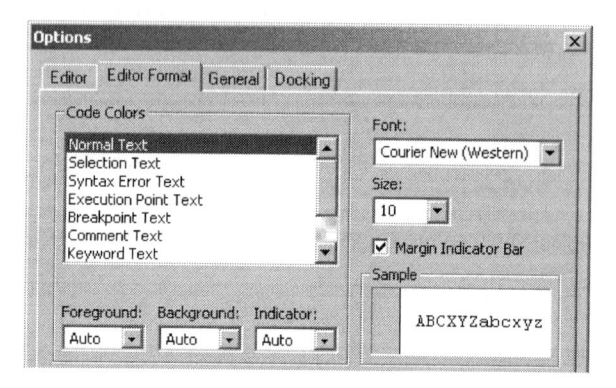

Figure 3.5

VBE formatting options from the Tools/Options menu.

An *Excel* workbook named *PNMSuite* (Practical Numerical Methods Suite) containing a suite of *VBA* macros is available to implement the numerical methods described in this book. All of the macros are sorted into modules by book chapter titles. Download the *PNMSuite* workbook from the companion web site: https://sites.google.com/d.umn.edu/pnm4 To view and edit the codes in the **Visual Basic Editor**:

1. Click on **PNMSuite** in the **Project** window.
2. Provide the password AsteroidB612 in the dialog box, and then open the modules to view the code.

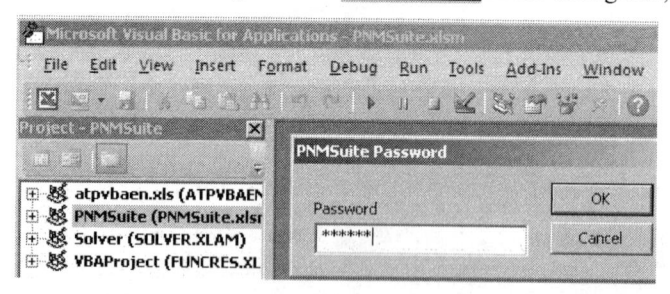

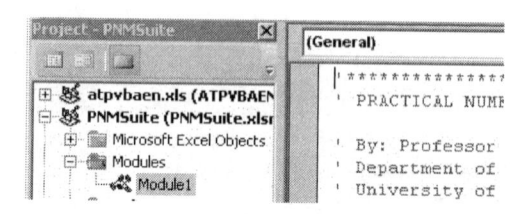

To delete or change the password, in the *VBE*, select the menu **Tools/PNMSuite Properties/Protection**. Unlock, delete, or change the password as needed for your own convenience.

3.2.1 Objects and the Properties Window

VBA is an object-oriented language. Examples of objects include workbooks, worksheets, modules, cell ranges, controls, and charts. We may modify the properties of objects, such as the name, formatting, as well as functionality. To open the **Properties** window, select from the *VBA* menu **View/Properties**, or right click on an object and select **Properties** from the popup window. For example, we may rename a module to reflect its contents better (this is the only property of a module). To rename a module, click on the module name in the **Project** window, open the **Properties** window, and then type the new name. To illustrate, rename **Module1** to **FrictionFactor**:

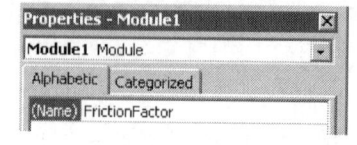

The property options for each object depend on the nature of its functionality. Most default settings are usually fine for our purposes. However, we should explore the options to learn more about different features of objects as we develop our *VBA* skills. We make use of object properties in many of the *VBA* sub procedures created for implementing numerical methods. We identify a modified object with the period syntax between the name of the object and its property, i.e., *Object.Property*. In the following sample code, the object is the active cell on the worksheet. The first property replaces the value of the cell with a string of text "U95% = ±". The second property changes the set of three characters "95%" into a subscript:

```
With ActiveCell
    .Value = "U95% = ±"
    .Characters(start:=2, Length:=3).Font.Subscript = True
End With
```

The active cell on the worksheet now contains the result: $\boxed{U_{95\%} = \pm}$. We do not list all of the properties here. Rather, we scout out the code found in this book, *Excel's* help files, or sample code from the internet for examples of useful properties. To understand the object properties, read the operations in reverse order. To illustrate, we translate the second line after `ActiveCell` above, starting at the right end: "Apply a subscript to the three font characters starting at position two in the string of the active cell."

Select the **Object Browser** from the **View** menu. A new window opens with classes of objects and members of classes. Navigate to a class member and copy the name to the clipboard to paste into your code. For example, if we select the class **WorksheetFunction**, the browser displays all of the members, including the common worksheet functions used for engineering calculations, such as trigonometric functions, etc. We use the **Object Browser** to find the names of other functions available for use in our macros.

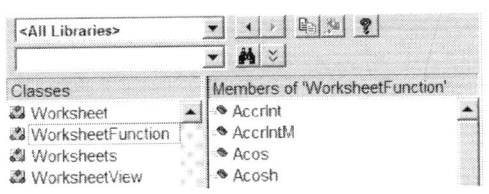

3.2.2 Code and Immediate Windows

The **Code** window displays the text for the *VBA* sub procedures and user-defined functions. To access the **Code** window, select an object, such as a module, or select from the **View** menu. Alternatively, double-click on a module to open up the corresponding code window. Two drop-down menus at the top of the **Code** window contain lists of objects or declarations of sub procedures and user-defined functions. Use the **Declarations** menu to navigate quickly to a macro. The **Immediate** window displays the values of **Debug.Print** variables during the execution of a macro. See Section 3.9 for more on this topic.

3.3 User-defined Functions, Sub Procedures, and Macros

Modules house the code for *VBA* procedures. *VBA* procedures can be user-defined functions that return a single result, array functions that operate on a range of values, or sub procedures that may return multiple results to the calling program. User-defined functions return a single result, such as a number, array, or string when called from an *Excel* worksheet or sub program, and are available for use in an *Excel* worksheet, just like any of the built-in worksheet functions described in Section 2.2. Recalculate a workbook by pressing the function key **F9** after making any changes to a user-defined function. A *VBA* sub procedure may be called from another function or sub procedure, launched using an *Excel* worksheet button, or run from the **View>Macros** menu. However, sub procedures are not accessible directly from cells in worksheets.

"Macro" is another name for a sub procedure, typically generated by the macro recorder. We may generate macros by recording a sequence of steps to set up an *Excel* worksheet. Alternatively, type the desired *VBA* code directly in a module within the *VBE*. To record a macro, select the *Excel* workbook tab option **View>Macro>Record Macro**. A dialog box like that in Figure 3.6 prompts for the macro name, with options to assign a keyboard short cut, store the macro in the **Personal Macro Workbook**, and add a brief description. We should provide a descriptive name for the recorded macro for future reference. Macro names:

- Require a letter at the beginning.
- Are limited to a length of fewer than 80 characters.
- Are limited to letters and numbers and underscore.
- May not contain spaces, special characters, or punctuation

Stop recording by selecting **View>Macro>Stop Recording**. Any actions we take to set up a workbook between pressing the **OK** button and **STOP** recording will be interpreted as *VBA* statements in a new sub procedure placed in a *Module* in our project or personal macro workbook. For example, if we already have a module named *Module1* then *Excel* adds the recorded macro to a new *Module2*. Alternatively, access macro recording controls using the toggle button located at the bottom left of the worksheet.

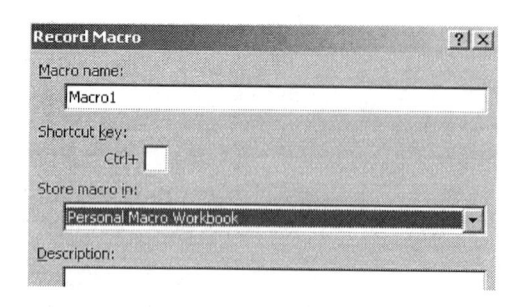

Figure 3.6

Record macro dialog for naming and storing a recorded macro.

To add a function or macro to a module from within the *Visual Basic Editor*:
- Select the menu option **Insert/Procedure** as shown in Figure 3.7.
- Type a name for the procedure
- Select the type from the dialog box.

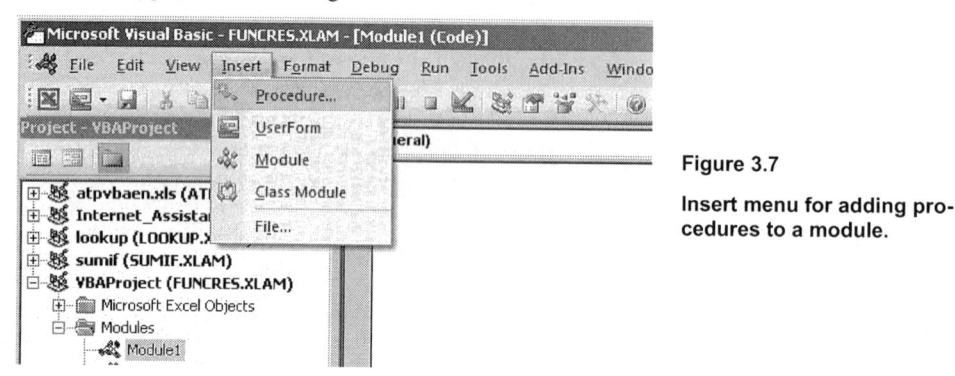

Figure 3.7

Insert menu for adding procedures to a module.

Adding a procedure to a module automatically creates the required opening and closing statements that wrap the rest of the programming statements for the procedure in the **Code** window:

```
Public Function name(arguments)
    :
End Function
```
or
```
Public Sub name(arguments)
    :
End Sub
```

These wrapper statements signal where a unique procedure starts and ends. We may add multiple procedures with unique names in a module. Procedures have the following requirements:
- All of the code for the procedure must be within the two wrapper statements.
- `Public` statements typed at the top of the module, outside of any procedure, are available to all open workbooks. Private statements are only available to procedures in the same module.
- `Public` procedures are accessible to other procedures in the workbook. `Private` procedures may only be accessed by other procedures within the same module.
- Information is passed to a procedure through a list of arguments enclosed in parentheses after the name. Use commas to separate multiple arguments passed to or from a procedure.

CHAPTER 3: VBA

- Be careful to use the arguments in the same order as listed when calling the procedure from an *Excel* worksheet or another function or sub procedure.

Because *VBA* is an interpreted programming language, the sequence of operations executes in the order of left to right, top to bottom. To illustrate, the following code assigns values from the active worksheets cell ranges **A1** and **A2** to two variables a and b, assigns the product of the sum of the variables with the argument *d* to a third variable *c*, then puts the result on the worksheet in cell range **A3**. The value of variable d is returned to the calling procedure.

```
Sub abc(d)
    a = Range("A1").Value: b = Range("A2").Value: c = (a+ b)*d
    Range("A3").Value= c: d = a/b
End Sub
```

By default, the argument (*d*) in the sub program **abc** passes to the macro "by reference," meaning that the value of variable *d* in the program that called the sub program takes on the new value of *d* passed back to the calling program (bi-direction assignment). Alternatively, pass the argument "by value" so that the corresponding variable in the calling program is not affected (one direction assignment). Type **ByVal** before the variable or **By Value** after the variable, or add parentheses around the argument to pass "by value", *e.g.*:

`Sub abc(ByVal d)` **or** `Sub abc(d By Value)` or, the equivalent: `Sub abc((d))`

Note, to use **ByVal** for arrays, they must be **Variant** data types (defined in Section 3.4).

To see the sub procedures in a workbook, use the *Excel* tab option **View>Macro**. This will produce a popup dialog box listing all macros (if any) available to our workbook. We may also access the macros list from the **Developer** tab. Alternatively, from within *VBE* select the menu option **Tools/Macros**. To run a macro, click on the macro name and press the **Run** button. To edit a macro in the *VBE*, select the desired macro, and then click on the **Edit** button. The object of the macro (i.e., workbook, worksheet, form, or module) will be highlighted in the *VBE's* navigation tree window. We can also create a new macro by first naming the macro before selecting the **Edit** button. To delete a macro, select the name in the list box under **View>View Macros** and click the **Delete** button.

User-defined functions are limited to performing calculations and returning a single result, or a range of results using the *CTRL SHIFT ENTER* keyboard combination in the case of an array function. User-defined functions access information from cells on an *Excel* worksheet through the function arguments or from **Cell** or **Range** references, which may include arrays (or ranges as described in Section 3.7.1). The only additional "worksheet" action allowed by a function is **MsgBox** (see Section 3.8.2).

Access the code for a user-defined function from the ribbon tab **View>Macro>View Macros**. Unfortunately, the list only shows macros (sub procedures), not functions. To view a function, type the function name in the **Macro Name** dialog space. Click **Options** to add a limited amount of descriptive information for a procedure, such as definitions of arguments. *Excel* displays the description when the function is selected from the ribbon tab **Formula>Insert Function**. To insert a user-defined function into a cell on an *Excel* worksheet, click on the tab **Formulas>Insert Function**, and then locate the custom function from the **User-Defined** category, as shown in Figure 3.8.

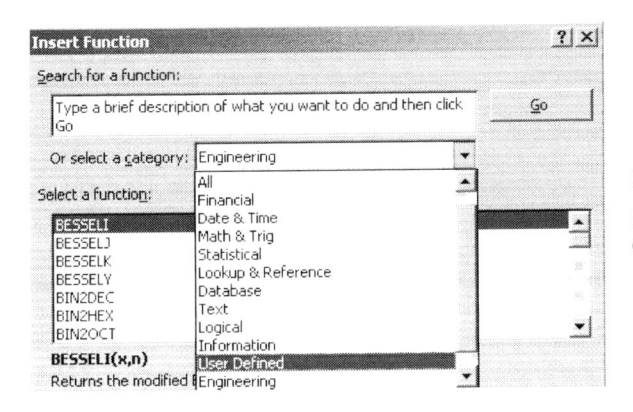

Figure 3.8

Find custom functions in the Insert Function dialog box, User-defined category.

3.3.1 Keywords, Text, Breaks, and Multiple Statements per Line

Keywords are part of the *VBA* language and may not be used for custom names of terms in a statement. The *VBA* editor automatically controls spacing, changes the color of keywords to blue and capitalizes the first letter of the keyword. We may use this to our advantage when creating custom code. *VBA* programmers recommend using lower case letters for all custom code statements to help identify unintentional keywords in a statement. If *VBA* leaves the name untouched, we can usually rest assured that we are not trying to use a reserved *VBA* name.

Use quotes around text strings. For example, when referring to a cell on an *Excel* worksheet, type the worksheet address "A1". Use double quotes inside the string quotes where you want to include quotation marks as part of the text for the string, *e.g.*, " ""Hello, World!"" ", will display, "Hello, World!" with quotation marks.

Lines of code have no length limits; however, we may want to break a single line of code into more than one line in the module for readability. Type a space, followed by an underscore _ at the break point, then move to the next line, and continue adding the rest of the code for that line. To illustrate, the following code puts all of the operations from the previous illustration in a single statement, but uses two lines to display all of the code:

```
Range("A3").Value = (Range("A1").Value _
          + Range("A2").Value)*d
```

A few *VBA* statements do not permit line breaks, such as breaking in the middle of text in quotations. *Excel* will display an error message if we try to add an inappropriate break. Concatenate text and variables with an ampersand. A colon separates distinct statements on the same line.

```
Sub abc(d)
     a = Range("A1").Value: b= Range("A2").Value: c = (a + b)*d
     Range("A3").Value= c: d = a/b: Range("A4") = "Result = " & d
End Sub
```

When the calling program passes $d = 3$ to Sub abc of our illustration, the result on the worksheet is

	A	B	C	D
1	1	2	9	Result = 0.5

3.3.2 Descriptive Comments in Code

Good programmers add descriptive comments to their programs to document the purpose of the code, explain the steps in the calculations, define variables, or include other descriptive information to help the user understand the structure. Comments follow an apostrophe '. For example, add the following comment above the previous *VBA* statements:

```
' Sum two values.  Put the result on the worksheet.
```

The *VBE* automatically changes the color of comments to green. We may place comments in separate lines, or at the end of a line. The program does not interpret the comments as part of the operation. When we put a comment at the start of a line, *VBA* assumes any text after the apostrophe is part of the comment, and not available for interpretation. All of the *VBA* programs in this book make extensive use of comments to describe the purpose of the code.

3.4 Variable Data Types and Constants

VBA uses variables to store alphanumeric or Boolean types of information. Variable names have restrictions similar to naming macros. They must start with a letter and may not contain empty spaces or special characters (with the exception of an underscore). However, a variable name may consist of up to 255 characters. All variables are assigned a type required by various *VBA* functions. Table 3.1 has a list of variable types. The default variable type is VARIANT, which may accept different data types. Overuse of the default VARIANT type slows down the execution of a *VBA* macro because the program requires extra time to determine what type of data is assigned to the VARIANT.

CHAPTER 3: VBA

Table 3.1 Variable data types

Type	Definition	Example
Boolean	Boolean values	F = False
Constant	Constant value for use in a multiple procedures	A = 3.14159326
Currency	Monetary values that range from −922,337,203,685,477.5808 to 922,337,203,685,477.5807	M = 123.45
Date	Dates and times	D=#4/15/14 16:35#
Decimal	Greatest number of significant figures (29), +/−79,228,162,514,264,337,593,543,950,335 (+/−7.9228162514264337593543950335 E+28)	C = 7.93E02
Integer	Integer values ranging from -32,768 to 32,767.	X=2
Long	Integer values ranging from -2,147,483,648 to 2,147,483,647.	Y=1234
Single	Single precision floating-point numbers, ranging in value from -3.402823 E38 to -1.401298 E-45 for negative values and from 1.401298 E-45 to 3.402823E38 for positive values.	Z=12.34
Double	Double precision floating-point numbers ranging in value from -1.79769313486231 E308 to -4.94065645841247 E-324 for negative values and from 4.94065645841247 E-324 to 1.79769313486232 E308 for positive values.	W=123.4
String	Text	T= "Name"
Range	Refers to an address of a cell range on an *Excel* worksheet	R=Range("A1:B2")
Variant	Accepts all data types	All of the above

Declare the variable type in the arguments of a procedure or with a keyword `DIM` statement within a procedure. Each variable in a `Dim` statement must be declared by type, whether or not they are of the same type. The data type for arguments of a procedure are specified in the argument list. For example, the opening line for a user-defined function with four arguments declares the type after each one, separated by commas:

```
Public Function TEST(a() as Double, b as Long, r as Range, t as String) As Double
```

Note that the first argument is an array variable, as indicated by the empty set of parentheses. See Section 3.7 for more details about array variables. This example also declares the data type for the result of the function after the closing parenthesis of the function. The function in this case returns a double precision number.

By default, *VBA* uses `SINGLE` precision for numbers, whereas *Excel* worksheets use double precision numbers. We recommend using the `Dim` keyword to declare numerical data types as `DOUBLE` precision in *VBA* to minimize round-off errors in calculations and speed up transferring values between worksheet cells and *VBA* variables. A `STRING` variable contains text. A `RANGE` variable contains the address to a cell or range of cells in an *Excel* worksheet. Look for examples of variable declarations in each of the macros used to implement the numerical methods in this book. To illustrate, the following dimension statement declares several types of variables, separated by commas in one line.

```
DIM X as Integer, Y as Long, Z as Single, _
    W as Double, D(1 to 5) as Double, T as String, R as Range
```

The double precision variable D is an array with five elements.

Declare variables for local or global use within the module or workbook with **Private** or **Public** statements at the top of the module, with the following conditions:

- Procedures in any module of the *workbook* may access a global variable.
- Constants are different from variables. Once declared, the value of a constant may not be changed within a procedure.
- Variables declared as **Private** are reserved for use in the procedure or module that contains the variable.
- **Public** variables are available to any module within the workbook.
- To make a variable or constant available within a single *procedure* (user-defined function or sub), declare the constant right after the procedure **Sub** or **Function** statement.

- To make a variable or constant available to all procedures in a *single module*, declare the private constant at the top of the module before any procedures.
- To make a variable or constant available to all modules in the *workbook*, use the **Public** keyword and declare the constant in any module before any procedures.

The following example *VBA* statement uses the syntax for **Public** constants:

```
Public Const k as Integer = 10, Rg as Double = 8.314, msg as String = "Converged"
```

where Rg is the gas constant. Any procedure in the workbook may now use Rg without redefinition. Beware that Rg is no longer available as a variable name in the workbook, and may not be assigned a different value in a new declaration statement.

After we assign a type to a variable, we will get an error notice if we try to give it a value with a different data type. Seasoned *VBA* programmers recommend defining all variable types and associated precision using a Dim statement near the beginning of the procedure to help with debugging and minimizing user error. We can enforce variable type declarations with an **Option Explicit** statement at the top of the module. Alternatively, we can set *VBA* to require the program to declare the variable type by checking the box labeled **Require Variable Declaration** under **Tools/Options**, as shown in Figure 3.9.

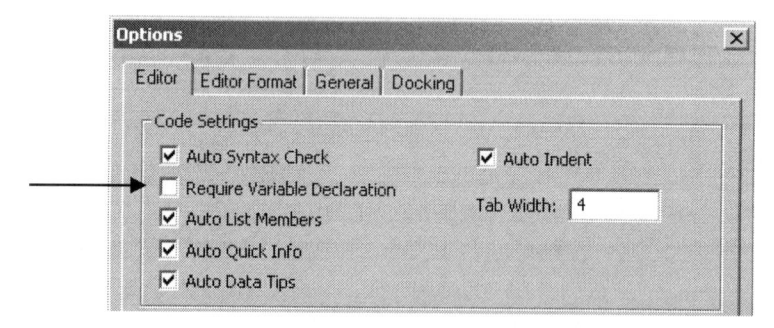

Figure 3.9

VBA programming options. Check the box to Require Variable Declaration.

In addition to the **Dim** keyword, we may also declare variables using the keyword **Static**. Similar to **Public** declarations, a static variable does not change between calls to the procedure, but unlike a public variable, its value is limited to the procedure that declared it.

3.4.1 Implicit Variable Declaration and Default Variable Type

Special characters listed in Table 3.2 allow for implicit declaration of a variable type the first time we use the variable in a *VBA* statement. Note that we only need use implicit declarations once when declaring the variable – thereafter we do not need the special character for the data type.

Table 3.2 Special characters for implicit type declaration.

Special Character	Data Type	Example
%	Integer	n_iter%
&	Long	pop&
@	Currency	payment@
!	Single	t_in!
#	Double	p_out#
$	String	Boxtitle$

By default, *VBA* assigns the **Variant** type to undeclared variables. We may change the default type using the following statement at the module level:

```
DefType Letter1[-Letter2]
```

CHAPTER 3: VBA

where `Type` refers to an abbreviated string representing the data type, `Letter1` is the first letter of the variable name, and the [optional] `Letter2` completes a range of first letters in the variable name. To illustrate, the following statement automatically treats any variable that starts with letters from *I* to *N* as Integer types[37]:

```
DefInt I-N
```

The **DefType** key words are listed in Table 3.3. Be careful to use these statements at the module level. *VBA* does not permit redefinition of data types within a **Sub** or **Function** procedure.

Table 3.3 *VBA* DefType keywords

DefType	Data Type	DefType	Data Type	DefType	Data Type	DefType	Data Type
DefBool	Boolean	DefCur	Currency	DefSng	Single	DefStr	String
DefInt	Integer	DefLng	Long	DefDbl	Double	DefVar	Variant

3.5 Structured Program Flow Control

"I am not a product of my circumstances. I am a product of my decisions." — Stephen Covey

VBA has logical constructs similar to *Excel* worksheet functions **IF** and **Auto Fill**. We learned how to use **IF** functions in *Excel* worksheets to make decisions about operations in a cell formula. *VBA* uses **IF** statements as well as **CASE** statements for making decisions about alternative calculations. In worksheets, think of **Auto Fill** over a range of cells with the same formula as a sequence of calculations. In a similar fashion, *VBA* has the ability to perform a sequence of calculations in a controlled **For Next** loop that increments the value of an expression with each iteration or step. The flow of calculations may require decisions, iterations, or comparisons. To illustrate, Figure 3.10 has a flow chart for a common scenario in numerical techniques where the solution method requires iteration until satisfying specified convergence criteria.[38]

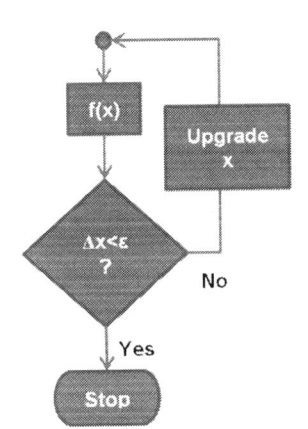

Figure 3.10

Flow chart for program flow control. Loop around the process f(x) until the change in Δx becomes smaller than some convergence tolerance, ε. (flow chart created using MS Office Shapes)

3.5.1 Loops and GoTo

An **If** statement coupled with a **GoTo** statement allows us to direct the flow of the program in many directions. We present examples of **If** statements in Section 3.5.2. Loops are more efficient and compact for repeated calculations or accessing elements of arrays in a series. A **For Next** loop uses an index variable to keep track of the number of iterations. The operations within the loop may use the value of the index to access elements of arrays or in other ways.

---

[37] FORTRAN 77 reserves variable names beginning with letters from I through N for integers.
[38] Program flowcharting (http://en.wikipedia.org/wiki/Flowchart).

Logical statements are often used to branch out of a loop using **Exit For** and **GoTo**. Table 3.4 has descriptions of loop programming statements in *VBA*.

Table 3.4 Loops for controlling iterative calculations in *VBA*. Condition is a logical statement.

Loop	Description
`If condition ... Then GoTo label` `On condition ... GoTo label`	Transfer the program flow to the label statement. Label has a colon, :, at the end of the string name for the label (may be alphanumeric)
`For i = 1 to n: ... : Next i`	Repeat the operations within the loop n specified number of times, similar to **Auto Fill** on an *Excel* worksheet.
`For Each element In group: ... : Next element`	Loop through each *element* in a group, such as an array or range. The variable *element* can be any name.
`Do: ... : Loop Until condition` `Do Until condition: ... : Loop`	Uses logical statements that branch out of the loop when the specified *condition* is **True**. Evaluates the condition at the end or beginning of the loop.

`Do While condition: ... : Loop`
`Do: ... : Loop While condition`
`While condition: ... : Wend`
	Uses logical statements that branch out of the loop when the specified *condition* is **False**. Evaluates the *condition* at the end or beginning of the loop.

Example 3.1 shows how to use a **For...Next** loop in a *VBA* program that calculates the dimensionless temperature according to Equation (2.6). We indented the statements within the loop for convenience in reading the code.

There are examples of looping in many of the *VBA* programs used to implement the numerical methods described in this book. This is not surprising, because looping is one of the strengths of *VBA* programming that extends the intrinsic capabilities of *Excel* as an engineering tool for computation. Study the programs in this book to become familiar with the variety of methods for controlling the flow of a program. Use the keyboard escape key (*Esc*) or the key combination *CTRL BREAK* to terminate a program when it becomes stuck in an "infinite loop".[39] Seasoned programmers prefer to use **For Next** and **For Each** loops that have a *finite* number of iterations to avoid being stuck in a scenario of an "infinite" loop (no end to the number of iterations). As we peruse the macros looking for examples of loops, we find several nested loops (loops within loops). Nested loops become important when working with multidimensional problems or arrays. In the following illustration, notice how the j-loop nests inside the i-loop. The j-loop executes m times for each of the n number of i-loops.

```
For i = 1 to n
    For j = 1 to m
        ⋮
    Next j
Next i
```

Example 3.1   Dimensionless Temperature in a Rectangular Bar

Known/Find: Repeat *Example 2.1* using a *VBA* user-defined function to calculate the dimensionless temperature at the center of a rectangle for various width-to-length ratios.

Analysis: Open an *Excel* worksheet and select from the ribbon tab **View>Macros>View Macros**.

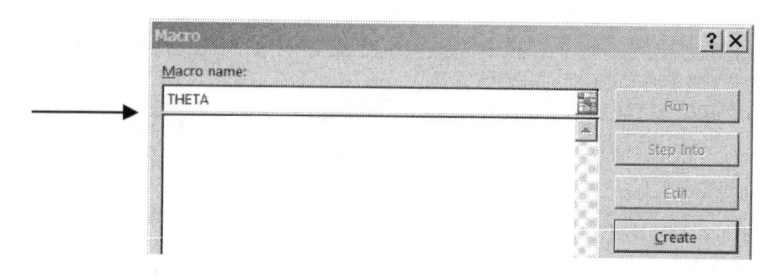

---

[39] For Windows keyboards without a *BREAK* key, press the start button, type *On*, then select *Onscreen Keyboard* from the list. With the onscreen keyboard, click on *CTRL* then *PAUSE* (which is the same as the *BREAK* key).

CHAPTER 3: VBA

1. Type **THETA** in the name box, and then click **Create**. The *VBE* opens with the first and last wrapper statements for a sub procedure named THETA listed in a new module.
2. Replace the keyword **Sub** with **Function** and add the arguments **Infinity as Single, WoL as Double**, as shown in the first line of Figure 3.11. Note how *VBE* automatically changes the **End Sub** to **End Function**. The argument WoL represents the dimensionless ratio of the rectangle width to length. The integer variable Infinity in the **For** ... **Next** loop represents the number of terms in the Fourier summation.
3. Use comments to describe the variables and other operations in the code.
4. Use a loop to calculate the dimensionless temperature by summing the Fourier terms in each iteration. The loop uses a step size of 2 because only the odd terms have non-zero values.

```
Function THETA(Infinity As Integer, WoL As Double) As Double
' Infinity = number of terms in the Fourier series
' THETA = dimensionless temperature at block center
' WoL = ratio of block width to length
' ********************************************************************
' Dimensionless temperature at the center of a rectangular block

' *** Nomenclature ***
Dim c_n As Double ' = coefficient for terms in Fourier series
Dim n As Integer ' = index for terms in the Fourier series
Dim npi As Double ' = n*pi
Dim s As Double '= sum of terms in the Fourier Series
' ********************************************************************
s = 0# ' Initialize the sum of Fourier terms to zero
With WorksheetFunction
For n = 1 To Infinity Step 2 ' Sum the Fourier terms
    npi = n * .Pi() ' Calculate the product of index n and pi
    c_n = 2 / (n * .Sinh(npi * WoL)) ' Calculate coefficient
    s = s + c_n * Sin(npi / 2#) * .Sinh(npi * WoL / 2#) ' Sum of terms in loop
Next n
THETA = 2# * s / .Pi() ' Calculate dimensionless temperature from sum of terms
End With
End Function ' THETA
```

Figure 3.11 *VBA* user-defined function to calculate dimensionless temperature at the center of a rectangle.

5. Try the user-defined function in an *Excel* worksheet for 20 terms in the sum with a width-to-length ratio of WoL = 1. The *VBA* user-defined function makes it easy to compare the solution using different numbers of terms in the series:

	A
1	=theta(20,1)

	A
1	0.25

Comments: The *VBA* user-defined function gives the same result as the worksheet calculations in *Example 2.1*, but is more compact on the worksheet. The *VBA* function code also makes it easier to avoid calculation errors that may be difficult to detect in long worksheet cell formulas.

We can use a logic based **Do While** loop to sum the series until the sum converges to a specified tolerance. Consider the following modification to the code in Figure 3.11 designed to branch out of the loop when the term in the series becomes smaller than a convergence tolerance of 10^{-8}:

```
Function THETA2(WoL As Double) As Double
' THETA2 = dimensionless temperature at block center
' WoL = ratio of block width to length
' ********************************************************************
' Dimensionless temperature at the center of a rectangular block
' *** Nomenclature ***
Dim c_n As Double '= coefficient for terms in Fourier series
Dim n As Integer, = index for terms in the Fourier series
Dim npi As Double '= n*pi
Dim s As Double '= sum of terms in the Fourier Series
Dim t As Double '= *** NEW *** term in series
Const tol As Double = 0.00000001 ' *** NEW *** constant value for tolerance
' ********************************************************************
```

```
s = 0#: n = -1 ' Initialize the sum of Fourier terms to zero and initial index
With WorksheetFunction
Do
    n = n + 2 ' *** NEW statement for the index of the terms
    npi = n * .pi() ' Calculate the product of index n and pi
    c_n = 2 / (n * .Sinh(npi * WoL)) ' Calculate coefficient
    t = c_n * Sin(npi / 2#) * .Sinh(npi * WoL / 2#) ' series term
    s = s + t ' Sum of terms in loop
Loop While .Max(Abs(t), 0) > tol
THETA2 = 2# * s / .pi() ' Calculate dimensionless temperature from sum of terms
End With
End Function ' THETA2
```

The index for the number of terms in the original **FOR NEXT** loop is replaced with a statement inside the loop that increments the value of the variable *n* by a value of two.

□

3.5.2   If-Then-Else Logic

If-Then statements make decisions based on comparisons between values. If the comparison is true, then the next statement is executed, otherwise, the flow of the program skips over the next statement. Comparison operators include combinations of "greater than" >, "less than" <, and "equals" =. For example, the sequence <= means "less than or equal to" and < > means "not equal to." **If Then** logical statements may be nested, contain more than one logical condition, or use a default **Else** option. **If** statements normally require a concluding **End If** statement, unless a single result of a true outcome is in the same line as follows:

```
If ... Then ...
```

or

```
If ... Then
    ...
ElseIf ... Then
    ...
Else
    ...
End If
```

To illustrate, consider the modified version of the *Excel* worksheet **SIGN** function used in Figure 2.8:

```
Public Function SIGNB(b As Double) As Double
' Returns the sign of b: +1 for b>=0, -1 for b<0.
    If b < 0 Then
        SIGNB = -1
    Else
        SIGNB = 1
    End If
End Function
```

Where possible, avoid comparing non-Boolean, non-integer or non-string values as equal =. The precision of the single and double precision numbers may inadvertently cause the comparison to fail. Combinations of comparisons use the terms **AND** for inclusive comparisons, and **OR** for either/or comparisons, *e.g.*:

```
If x < b And x > a Then
    ...
ElseIf y <> c Or y ≥ d then
    ...
End If
```

We cannot put comparisons on both side of a variable. For example, the following expression does not perform what we might think:

- **Incorrect**   `If a < x < b Then ...`

CHAPTER 3: VBA

Example 3.2 Correct Object Names for Worksheets

Known: Names for objects in *VBA* must start with a letter and may contain only alphanumeric characters or underscores. Worksheet names referenced in *VBA* do not permit blank spaces and punctuation marks. The total string length must not exceed 40 characters.

Find: Write and test a *VBA* function that removes any non-alphanumeric characters or spaces from a string.

Analysis: The *VBA* user-defined function in Figure 3.12 takes the following steps to eliminate unwanted characters.
1. Pass the string variable to the function through the argument list.
2. Replace any blank spaces, commas, or ampersands in the string with underscores using the **REPLACE** function.
3. Check the first character in the string for nonnumeric values using **MID** to get the first character, and **LIKE** to compare the first character to the alphabet. If true, keep the first character.
4. Loop through the remaining characters in the string variable using **FOR…NEXT** and compare them to all possible alphanumeric characters or underscore.
5. Create a string variable by adding only the alphanumeric or underscore characters in the order they appear in the original text. Characters are concatenated with an ampersand &.

```
Public Function JUSTABC_123(ByVal s As String) As String
' s = original sheet name, new sheet name after replacement of bad characters
'*************************************************************************
' Replace the value of a string variable with only alphanumeric characters or an
' underscore. First character must be alphabetic or replaced by word "Sheet".
' *** Nomenclature ***
Dim b As String ' = temporary value of character in string
Dim c As String ' = new value for string variable
Dim i As Long ' = number index for characters in string
Dim j As Long ' = index for finding first alphabetic character
Dim k As Long ' = counter for number of characters up to 40
Dim n As Long ' = number of characters in string
'*************************************************************************
10: ' Replace blank, comma, etc. with underscore
s = Replace(Replace(Replace(Replace(s, " ", "_"), ",", "_"), "'", "_"), "&", "_")
n = Len(s): b = VBA.Mid(s, 1, 1) ' get length of "s" and character in first position
If b Like "[a-zA-Z]" Then ' First character must be from alphabet (not numeral)
    c = b ' set the name of new string variable as first character
    j = 2 ' starting point for next character in string
    k = 1 ' initialize counter for number of characters
Else ' Start of new name if first character is not alphabetic
    c = "Sheet": j = 1 ' starting point for next character in string
    k = 5 ' initialize counter for number of characters ("Sheet" has 5 characters)
End If
For i = j To n ' check each character for allowed alphanumeric values
    b = VBA.Mid(s, i, 1) ' Get the ith character in the string
    If b Like "[a-zA-Z_0-9]" Then' Add only alphanumeric/underscore values to new string
        c = c & b: k = k + 1 ' concatenate names & increment counter
        If k >= 40 Then Exit For
    End If
Next i
If s <> c Then ' Warn the user about worksheet name change
    s = Application.InputBox(Type:=2, Default:=c, _
        prompt:="New worksheet name required (use suggestion or type a new name):" & _
        vbNewLine & "(Must start with an alphabetic character (aA-zZ) followed" & _
        vbNewLine & " by alphanumeric characters (aA-zZ, 0-9, or underscore.")
    GoTo 10 ' check the new worksheet name for compatibility
End If
JUSTABC_123 = s ' return the new string value
End Function ' JUSTABC_123
```

Figure 3.12 *VBA* user-defined function JUSTABC_123 to remove all non-alphanumerical characters for a string variable (except underscore). "A B C is easy as one two three …" – Jackson Five

We tested the function in an *Excel* worksheet using the string in cell **A1**. Cell **B1** has the formula with the function and result:

	A	B
1	1,ABC & 123!	=JUSTABC_123(A1)

	A	B
1	1,ABC & 123!	Sheet1_ABC_and_123

Comments: The user-defined function `JUSTABC_123` assumes English language alphanumeric characters for the string. Use the *VBA* help menu to learn more about each *VBA* function used in this code. We use this function in several custom macros that write output to an *Excel* worksheet object.

□

3.5.3 Case Structure*

An alternative to the **If...Then...Else** option uses the **Case** flow-control structure to decide among two or more options. In the following code, if the variable `j` equals one or two then the program only executes the line after case 1 or 2. If the variable `j` does not equal one or two, then neither case is selected. Use **Case Else** for any default options. Otherwise, the program skips over the `Select Case` operations.

```
Select Case j
    Case 1
        ...
    Case 2
        ...
    Case Else ' (optional)
        ...
End Select
```

To assign a case to multiple options in a series, use for example `Case 1 To 3`. To select from a collection of options, separate them with commas, *e.g.*, `Case 7, 9, 11`. We may also use less than, equals, or greater than type arguments, such as `Case Is > 12`, or Boolean, such as `Case True` or `Case False`. Although the logical structure **If Then** executes faster in *VBA*, **Case** control simplifies the structure of a program.

Example 3.3   Darcy Friction Factor in a Circular Duct

Known: We define the Darcy friction factor[40] for laminar flow in a circular duct in terms of the dimensionless Reynolds number:

$$f_D = 64/\text{Re} \tag{3.1}$$

where
$$\text{Re} = \rho u D / \mu$$

and where ρ is the fluid density, u is the average fluid velocity, D is the inside diameter of the duct, and μ is the fluid dynamic viscosity. The explicit Haaland (1983) equation correlates the Darcy friction factor for turbulent flow in terms of Reynolds numbers greater than 2000 and the pipe roughness ratio (ε/D):

$$f_D = \left\{ 0.782 \ln \left[\frac{6.9}{\text{Re}} + \left(\frac{\varepsilon}{3.7D} \right)^{1.11} \right] \right\}^{-2} \tag{3.2}$$

where ε is the roughness factor for the duct.

Find:

1. Create a user-defined function for the Darcy friction factor in terms of the roughness, diameter, and Reynolds number.

---

[40] Darcy's friction factor is four times the Fanning friction factor ($f_D = 4f_F$)

CHAPTER 3: VBA

2. Set up an *Excel* spreadsheet using **Goal Seek** in a *VBA* macro to calculate the flow rate of water in a horizontal pipe, given the length, diameter, roughness, and pressure loss from Bernoulli's equation:

$$\frac{\Delta P}{\rho} = f \frac{L}{D} \frac{V^2}{2} \tag{3.3}$$

where V is the fluid velocity, ΔP is the pressure loss in the pipe, and L and D are the length and inside diameter of the pipe, respectively.

Analysis:

Part 1: Create a User-defined Function

1. Open the *Visual Basic Editor* and add a module. Then add a public function procedure from the **Insert/Procedure** menu (DFF is an acronym for Darcy's Friction Factor):

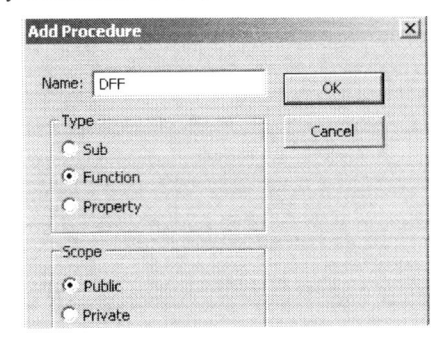

2. Add code between the **Function/End Function** wrapper statements in the module for the Darcy friction factor, as shown in Figure 3.13. Use one parameter for the roughness ratio $RR = \varepsilon/D$. Note the colors used by *VBA* to distinguish among comments (green), reserved words (blue), errors (red), and custom code (black). We get an error message if we try to rename or use a reserved *VBA* word incorrectly. To change the precision from the default single to double precision, use **As Double** after a function name to return a double precision value. The roughness ratio, RR, is listed as an **Optional** Variant type argument because it is only required for turbulent flow. The function checks if the roughness ratio argument is missing when needed for turbulent flow calculations.

```
Public Function DFF(Re As Double, Optional RR as Variant) As Double
' Returns the Darcy friction factor for flow in a circular duct calculated from the
' Haaland equation for turbulent flow. Re = Reynolds number, RR = roughness ratio
    Select Case Re
    Case Is < 2100
        DFF = 64 / Re
    Case Else
        If IsMissing(RR) Then ' Check for roughness ratio in arguments
            MsgBox "Needs roughness ratio.": Exit Function
        End If
        DFF = (0.782 * Log(6.9 / Re + (RR / 3.7) ^ 1.11)) ^ (-2)
    End Select
End Function
```

Figure 3.13 User-defined *VBA* function DFF for the Darcy friction factor using Cases to select from flow conditions.

The user-defined function for the Darcy friction factor in Figure 3.13 uses **Case** program flow control. Notice the use of Case Else as the default case for turbulent flow when **Re** is greater than, or equal to 2100.

It is a good idea to test the user-defined function in an *Excel* worksheet before using it in the analysis. The *VBA* user-defined function in cell **A1**, with formula shown in the formula bar gives the correct result:

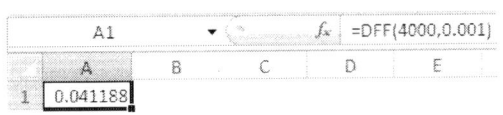

The function is now available for any calculations in the workbook involving pressure drop for fluid flow in a pipe.

Part 2: Use named cells for the parameters.

1. Set up an *Excel* worksheet with the parameters, names, values, and units:

	A	B	C	D
1	Parameter	Name	Value	Units
2	Velocity	v	1	m/s
3	Density	dens	997	kg/m3
4	Viscosity	visc	8.90E-04	N s/m2
5	Length	L	300	m
6	Diameter	D	0.1	m
7	Roughness	eps	2.50E-04	m
8	Pressure Loss	dP	1.00E+04	Pa

2. Name the cells containing the parameter values in column **C** according to the list in column **B**. Select the range of cells containing the names and values, then click on **Create From Selection** in the **Defined Names** group of the **Formulas** tab. Check the box for **Left column** to create names for the cells in column **C** from those in column **B**:

3. Complete the worksheet by calculating the roughness ratio, Reynolds number, and friction loss, using the user-defined function **DFF(Re,eD)** for the Darcy friction factor, developed in part one. Name the additional cells in column **C** using the labels in column **B**. Finally, add the energy balance in cell **C12** according to Bernoulli's Equation (3.3).

	A	B	C	D
1	Parameter	Name	Value	Units
2	Velocity	v	1	m/s
3	Density	dens	997	kg/m3
4	Viscosity	visc	0.00089	N s/m2
5	Length	L	300	m
6	Diameter	D	0.1	m
7	Roughness	eps	0.00025	m
8	Pressure Loss	dP	10000	Pa
9	Roughness Ratio	eD	=eps/D	
10	Reynold's Number	Re	=dens*D*v/visc	
11	Friction Loss	F	=(DFF(Re,eD)*L*v^2)/(2*D)	m^2/s^2
12	Bernoulli's Equation	F-dP/dens	=F-dP/dens	

4. At this stage, we do not have enough experience with *VBA* to program the macro using **Goal Seek**. Use the macro recorder to learn how to call **Goal Seek** from within a macro. There are three ways to start recording a macro. The simplest way is to click on the record button at the bottom of the workbook, . Otherwise, access the recorder from the **View>Macros** tab:

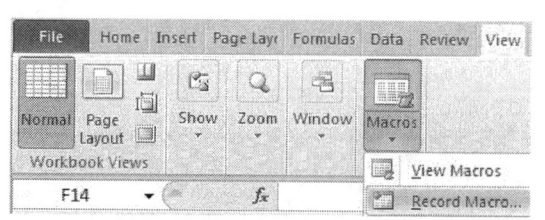

CHAPTER 3: VBA

Alternatively, on the **Developer** tab, click **Record Macro** in the **Code** group to start recording the actions to setup the worksheet. Notice when we click **OK**, the **Code** group replaces **Record Macro** with **Stop Recording**.

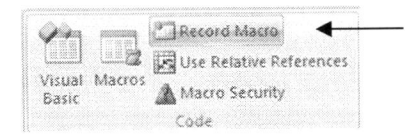

5. The **Record Macro** dialog box prompts us for a name and location to store the macro. We can also add a description and shortcut key if desired. We use the default values for now.

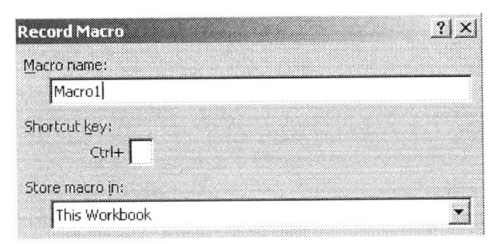

6. While recording the macro, set up **Goal Seek** from the tab **Data>What-if Analysis>Goal Seek** to solve for the velocity on the worksheet:

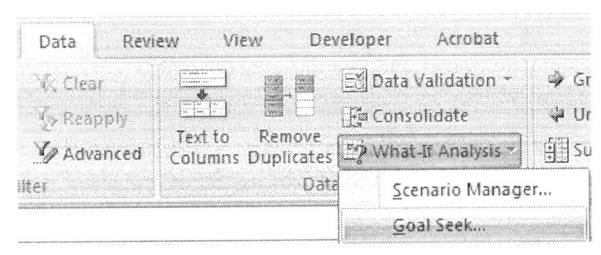

The **Set cell:** corresponds to the Bernoulli energy balance in cell **C11**. This formula returns a value of zero at the correct velocity in the cell address $C$2 provided in the box labeled **By changing cell**.

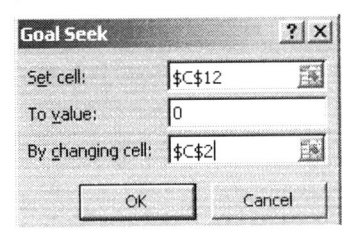

Click **OK** to start **Goal Seek** and find the correct value for velocity.

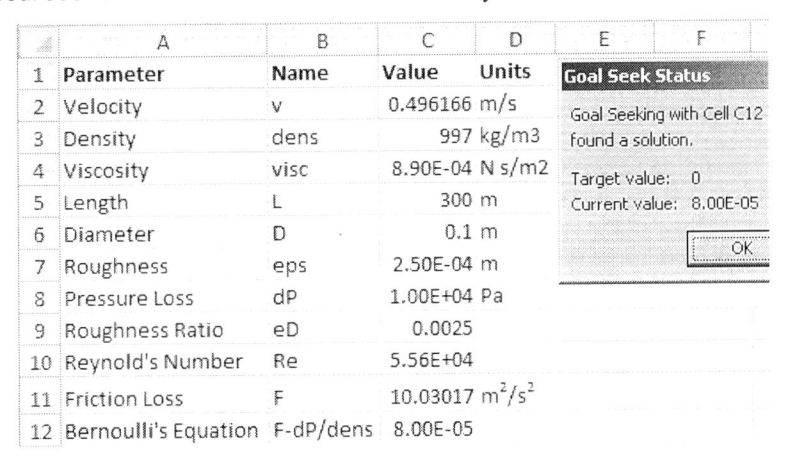

	A	B	C	D
1	Parameter	Name	Value	Units
2	Velocity	v	0.496166	m/s
3	Density	dens	997	kg/m3
4	Viscosity	visc	8.90E-04	N s/m2
5	Length	L	300	m
6	Diameter	D	0.1	m
7	Roughness	eps	2.50E-04	m
8	Pressure Loss	dP	1.00E+04	Pa
9	Roughness Ratio	eD	0.0025	
10	Reynold's Number	Re	5.56E+04	
11	Friction Loss	F	10.03017	m^2/s^2
12	Bernoulli's Equation	F-dP/dens	8.00E-05	

94   PRACTICAL NUMERICAL METHODS

7. End recording by selecting the **View>Macros>Stop Recording** option or from the **Developer** tab on the worksheet. Alternatively, click on the stop recording button at the bottom of the workbook, ■:

8. Examine the recorded *VBA* code by selecting the worksheet menu tab option **View>Macros>View Macros**.
9. Highlight the recorded macro name from the list on the macro dialog box.

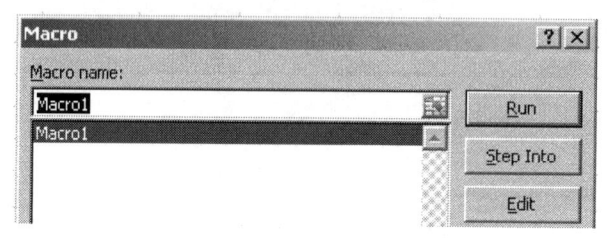

10. Click **Edit** to bring up the *VBE* with the recorded *VBA* code in **Module2**.

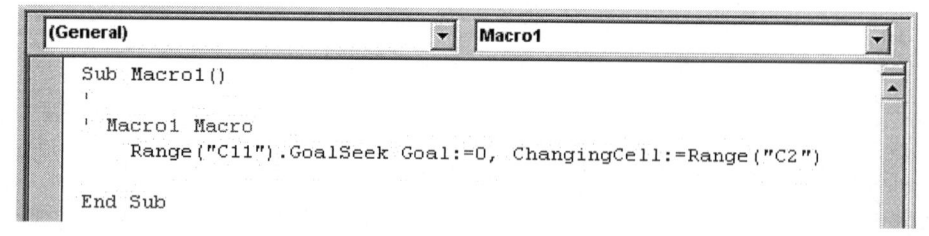

```
Sub Macro1()
'
' Macro1 Macro
    Range("C11").GoalSeek Goal:=0, ChangingCell:=Range("C2")

End Sub
```

Each time we make a change in any of the parameters that define the friction flow we must calculate a new flow velocity. We can now rerun the recorded macro without the need to set up **Goal Seek** by selecting the macro from the **View Macros** list and clicking **Run**.

Comments: By recording the setup, we learned how to access **Goal Seek** from *VBA*. However, reusing the code requires just as many steps as setting up **Goal Seek**.

□

Example 3.4   Macro for Expanded Unit Conversions

Known: The *Excel* worksheet function **CONVERT** has a limited set of unit conversions for pressure and other physical quantities.

Find: Create a *VBA* macro to perform additional unit conversions commonly encountered by chemical engineers

Analysis: To convert from one unit system to another, multiply the value of the physical property by the ratio of the conversion factors for the basis unit system:

$$\genfrac{}{}{0pt}{}{\text{To Property}}{\text{Value}} = \left(\genfrac{}{}{0pt}{}{\text{From}}{\text{Property Value}}\right)\left(\frac{\text{To Units Conversion Factor}}{\text{From Units Conversion Factor}}\right) \qquad (3.4)$$

Create a user-defined function procedure named **UNITS**. We show just the *VBA* code for pressure unit conversion factors in Figure 3.14. To view the other unit conversions, see the code in the *VBE* of the *PNMSuite*. Use the **Select Case** structure to select the unit conversion calculation for pressure. Note how the available pressure units are listed

CHAPTER 3: VBA

as strings after the **Case** statement, separated by commas. A private user-defined function named **PRESSURE** returns the corresponding conversion factors for units selected by the user with units of "atm" for the base system. Also, note how the units named "mmHg" and "torr" or "psi" and "lbf/in2" share the same conversion factors.

Refer to the table at the back of the book for all of the conversion factors available in **UNITS**. The result is rounded to five significant figures using another user-defined function **SIGFIGS**, listed in Figure 3.17.

```
Public Function UNITS(v As Double, from_unit As String, to_unit As String) As Double
' v = value of physical property to convert "from_unit" into "to_unit" value
' from_unit = original units of value
' to_unit = new units of converted value
' UNITS = returns the value converted to the new units
'****************************************************************************
' Add units to the case list below (SI in first line and English in second line).
' Add the case for the conversion value in the corresponding private user-defined
' functions below.
'****************************************************************************
Dim u As Double ' pressure units
Select Case from_unit
Case "atm", "Pa", "bar", "N/m2", "torr", "mmHg", "lbf/in2", "lbf/ft2", "lbm/ft.s2", _
     "inHg", "psi", "inH2O", "ftH2O"
     u = v * PRESSURE(to_unit) / PRESSURE(from_unit) ' pressure units
Case Else
    MsgBox "Unit(s) not available.": u = 0' Error message
End Select
UNITS = SIGFIGS(u, 5) ' Round off to significant figures
End Function ' UNITS

Private Function PRESSURE(unit As String) As Double
' unit = string for pressure units
'****************************************************************************
' Returns conversion factors for pressure units
'****************************************************************************
Select Case unit
Case "atm": PRESSURE = 1#
Case "Pa", "N/m2": PRESSURE = 101325#
Case "bar": PRESSURE = 1.01325
Case "mmHg", "torr": PRESSURE = 760#
Case "psi", "lbf/in2": PRESSURE = 14.696
Case "lbf/ft2": PRESSURE = 2116.2
Case "lbm/ft.s2": PRESSURE = 68087#
Case "inH2O": PRESSURE = 406.8
Case "ftH2O": PRESSURE = 33.9
Case "inHg": PRESSURE = 29.921
Case Else: MsgBox "Unit(s) not available.": PRESSURE = 0
End Select
End Function ' PRESSURE
```

Figure 3.14 User-defined function UNITS for converting physical property values between unit systems.

To apply the user-defined function in an *Excel* worksheet, type the name of the function and provide the strings for the unit systems. For example, to convert from 10^5 Pa to inH$_2$O:

$$401.48 = \text{UNITS}(100000, \text{"Pa"}, \text{"inH2O"})$$

Comments: To customize the **UNITS** user-defined function with additional unit conversion factors, follow the pattern used for pressure units to add new conversion factors for other units. Run the macro named **UNITS_TABLES** from the *PNMSuite* to create lists of units currently available in **UNITS**.

□

3.5.4 GoSub...Return*

The combination of **GoSub** *label*...**Return** statements provide an alternative to creating separate, self-contained **Sub** or **Function** procedures. With **GoSub**, we may create a small subroutine within a procedure that is callable only *within* the calling procedure. To use **GoSub** we first create the sub procedure after the code in the main procedure, starting with a **label:** and ending with a **Return** statement. To call the subroutine, we use the statement **GoSub** followed by the name of the label indicating the beginning of the subroutine. **GoSub** routines should be placed at the end of the procedure code where they do not interfere with the main program. Place an **Exit Sub** or **Exit Function** statement before the label to quit the procedure before reaching the **GoSub** routine.

GoSub routines do not have argument lists inside a set of parenthesis like other **Sub** or **Function** procedures. Instead, the variables in the calling procedure are available to the **GoSub** routine. The code in **GoSub** branches out of the main program to the position of code following the **label** statement, then returns to the procedure line following the **GoSub** statement.

We illustrate the **GoSub** structure in Figure 3.15 with a simple roulette wheel algorithm designed to mimic the random outcome of a roulette wheel for selecting an element from a list proportional to probability. In this illustration, we choose one element out of a list of three elements in column **A** proportional to their probabilities of selection specified as 20%, 50% and 30%, respectively. In other words, we have a 1 in 5 chance of selecting **A2**, a 1 in 2 chance of selecting **A3**, and a 3 in 10 chance of selecting **A4**. For a large number of selections, we expect to select **A2** approximately 20% of the time, and so on.

The roulette wheel algorithm starts by generating a random number from a uniform distribution, $0 < r < 1$, to compare with the cumulative probability distribution of the list of probabilities. An element in the list is selected when the cumulative probability exceeds the value of the random number. The calling sub procedure **SELECTPRO** uses a **For**...**Next** loop to call the **GoSub ROULETTE** subroutine over 1000 iterations and keeps track of the number of times each of the three elements is selected. The selection frequencies are placed in column **B**. We see that the results correspond well with the probability of selection, showing 18% selected as **A1**, 51% as **A2**, and 31% as **A3**. Because these results come from a stochastic selection process, the frequency values will change a little each time we run the macro **SELECTPRO**. A separate user-defined **ROULETTE** function is also available from the *PNMSuite* for proportional probability selections. The required argument for **ROULETTE** is a range of probabilities. The function returns the number of the element from the range. The function has an optional second argument for the index array.

	A	B
1	Probability	Frequency
2	0.2	181
3	0.5	512
4	0.3	307

```
Public Sub SELECTPRO()
' Select elements from an array proportional to their probabilities.
' *** Nomenclature ***
Dim a As Range ' = corresponding probabilities
Dim b As Range ' = range of results for frequency of selection
Dim i As Long, j As Long ' = index for loop in n number of elements
Dim m As Long, n As Long ' = maximum samples and size of list
Dim p As Double ' = cumulative probability
Dim r As Double ' = random number
'*************************************************************************
Set a = Range("A2:A4"): Set b = Range("B2:B4"): m = 1000: n = a.Count
For i = 1 To n: b(i) = 0: Next i ' initialize the range of c to zero
For j = 1 To m
    GoSub ROULETTE ' Select an element from range a
    Select Case i ' add to the count for each element when selected
        Case 1: b(1) = b(1) + 1
        Case 2: b(2) = b(2) + 1
        Case 3: b(3) = b(3) + 1
    End Select
Next j
Exit Sub ' Exit sub before the roulette subroutine
```

CHAPTER 3: VBA

```
'*****************************************************************************
ROULETTE:
p = 0: r = Rnd ' initialize the cumulative probability and random number
For i = 1 To n
    p = p + a(i)
    If p > r Then Exit For
Next i
Return
End Sub
```

Figure 3.15 *VBA* sub procedure SELECTPRO to demonstrate `GoSub Label ... Return` statements for creating subroutines within a procedure. The subroutine implements the roulette wheel algorithm for randomly picking an element from a list proportional to its probability of selection. The frequency results represent 1000 proportional probability selections.

We recommend using **GoSub** routines for calculations used repeatedly within a procedure that do not need to change a variable name. For another example, Figure 7.25 lists the genetic algorithm that employs three **GoSub** subroutines in a sub procedure.

3.6   VBA Math and Worksheet Functions

Several of the mathematical and statistical operations and functions found in a typical scientific calculator or *Excel* worksheet are available in *VBA* programming. The same arithmetic operations used in an *Excel* worksheet apply in *VBA*. For example, use $x \wedge n$ for raising the value of x to some power n. Arithmetic operations in *VBA* use the standard symbols: +, -, *, and / for performing addition, subtraction, multiplication, and division, respectively.[41] *VBA* uses the backslash \ for integer division, rounding the result to the nearest integer. The # sign after a number indicates that the value is not an integer. This is not required for numbers with values in the decimal places. Use **E** notation to represent numbers with powers of 10. For example, use `1.23E4` in place of `1.23*10^4`. The *VBE* changes the appearance of the number to include zeros for the powers of 10, in our example `12300`. Likewise, `1.23E-4` becomes `0.000123`.

As discussed in Section 2.1.2 for *Excel* worksheet formulas, be careful to control the order of operations. For example, the negation operation - is done before a power operations ^ in an *Excel* worksheet. However, it is the other way around in *VBA*. When in doubt, use parentheses to control the sequence of calculations. The *VBA* code for the calculation in Equation (2.1) assigned to variable *a* is

```
a = 1 + 2 * (3 / 4) ^ 5
```

Some useful worksheet functions are noticeably absent from *VBA*, including **Pi()** and **LN()**. *Excel* worksheet functions for log base 10 are **Log(x, 10)** or **Log10(x)**, however *VBA* uses **LOG()** for *natural* logarithms. *VBA* does not have a function for log base 10. Do not despair; we get the logarithm in any base *b* from the quotient $\log_b x = \log_{10} x / \log_{10} b = \ln(x)/\ln b$. Thus, we may evaluate a logarithm in any base with the following code:

```
Log(x)/Log(CInt(base))
```

where base is any positive integer value greater than one. We may also use *Excel* worksheet functions when equivalent functions are missing from *VBA* using one of the following statements preceding the worksheet function name: either **Application.Function_name()** or **WorksheetFunction.Function_name()**. For example, the following statement assigns to the variable b the maximum value of variable a and the constant π using a nested worksheet functions:

```
b = WorksheetFunction.Max(a, WorksheetFunction.Pi())
```

---

[41] The *VBA* editor adds spaces between arithmetic operations as you type. There is a bug in Excel 2010 when typing ^ for exponents; you must type the spaces before and after ^, e.g., x ^ 2, not x^2.

VBA does not permit the use of worksheet functions in place of *VBA* functions that perform the same operation. For instance, a statement `WorksheetFunction.Exp(5)` produces an error because the *VBA* function `EXP(5)` is available to perform the same operation. To see the list of *VBA* Math functions, open a **Module** and select from the menu **View/Object Browser**. Then select the desired category from the *VBA* <All Libraries>. The *VBA* functions are categorized according the themes, such as Financial or Math, as seen in Figure 3.16. Click on the **Help** menu to find definitions and descriptions of the functions. Several of the *VBA* Math functions used throughout this book for implementing numerical methods in *Excel* with *VBA* are described in Appendix 15.2.

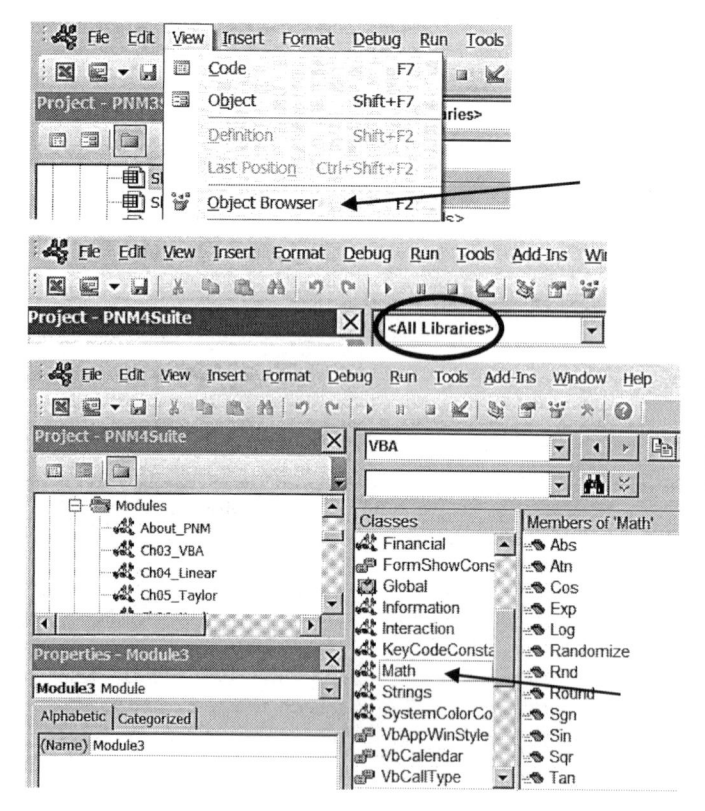

Figure 3.16
VBA library of functions showing members of Math library. To view the VBA functions, open a module and select from the menu View/Object Browser. Click on the drop down arrow next to <All Libraries> and select VBA.

Example 3.5 Round off to Specified Significant Figures

Known/Find: *Excel* does not have an intrinsic worksheet function for rounding a numerical value to a desired number of significant figures. Create a simple user-defined function for this purpose.

Analysis: Use the formula in Equation (2.7). Beware, the *VBA* **Round** function employs a different rounding protocol than the *Excel* worksheet **Round** function. *VBA* rounds numbers that end in five up to even values. *Excel* rounds numbers that end in five up to the next higher value in the decimal place. In addition, the *VBA* **Round** function only rounds values to the right of the decimal place, and does not accept negative values for the optional argument of decimal places. We have also learned that *VBA* does not have a function for log base 10. We may use the worksheet function **Log10** for base 10, or the quotient formula with the *VBA* natural **LOG** function, as seen in Figure 3.17:

```
Public Function SIGFIGS(x As Double, Optional sf As Integer = 2) As Double
' x = number for rounding
' sf = (optional) number of significant figures.  Default is 2
'*****************************************************************************
' Rounds a real number, x, to the selected number of significant figures, sf.
'*****************************************************************************
If x = 0 Then
    SIGFIGS = 0: Exit Function
```

CHAPTER 3: VBA

```
       End If
       SIGFIGS = WorksheetFunction.Round(x, sf - 1 - Int(Log(Abs(x)) / Log(10)))
       End Function ' SIGFIGS
```

Figure 3.17 *VBA* user-defined function SIGFIGS returns a number rounded to a specified number of significant figures.

We tested our *VBA* user-defined function **SIGFIGS** in an *Excel* worksheet using one, two (the default), three, and four significant figures for two numbers: -123.45 and 0.0012345:

	A	B	C	D
1	x/sf	1	2	3
2	-123.45	=sigfigs($A2,B$1)	=sigfigs($A2)	=sigfigs($A2,D$1)
3	0.0012345	=sigfigs($A3,B$1)	=sigfigs($A3)	=sigfigs($A3,D$1)

	B	C	D	E
1	1	2	3	4
2	-100	-120	-123	-123.5
3	0.001	0.0012	0.00123	0.001235

Comments: The user-defined function automates the output of significant figures in *Excel* worksheets. □

3.6.1 Speed Up VBA Computations

As an interpreted programming language, *VBA* executes operations slowly relative to similar expressions in *Excel* worksheets. Using **For Next** or **Do** loops, logical **if...Then** statements, **WorksheetFunction**, and **Range** references between our code and the worksheet, can significantly slow down *VBA* execution. Take steps in long programs to speed up the *VBA* execution where possible. The following suggestions may help speed up macros:

- Avoid looping through arrays by using **Variant** or **Range** variables to replace contents of arrays in one single operation. Be careful not to overuse the **Variant** variable declaration, which slows down program execution because *VBA* must decide on the type for each instance using the variable.

- In default mode, *Excel* recalculates all formulas in an *Excel* worksheet whenever the value of any cell changes on the worksheet, then updates the screen. Suspend and enable visible updates on the worksheet using the following pair of statements:

```
Application.ScreenUpdating = False
       :
Application.ScreenUpdating = True
```

If the user needs to know about the progress of calculations during execution, use the **Status Bar** at the bottom of the worksheet to display information while `ScreenUpdating = False` (see Section 3.8.7). However, overuse of the status bar may also slow down the execution of the program.

- Unlike compiled languages, such as C++, working with double precision increases the speed of execution in *VBA* over single precision calculations. In addition, because the worksheet uses double precision, *Excel* does not require a change in type when accepting double precision values from *VBA* calculations.

- When the worksheet is not involved in the calculations, but merely a repository of information, turn off the automatic calculations so that *Excel* does not recalculate the entire workbook each time a change is made to the contents of a cell or range:

```
' Helps speed up macros
Application.EnableCancelKey = False
Application.Calculation = xlCalculationManual
Application.ScreenUpdating = False
    :
Application.ScreenUpdating = True
Application.Calculation = xlCalculationAutomatic
```

However, most of the *VBA* macros accompanying this book do rely on worksheet calculations. Thus, we do not recommend this particular time saving tip when using any of the macros in the *PNMSuite*.

- Use **True** comparisons before **False**. *VBA* takes longer to test for **False**.

- In *Example 3.5*, we used "With Application ..." to incorporate worksheet functions into the *VBA* code. When accessing several worksheet functions, use the following "**With WorksheetFunction ... End With**" or "**With Application ... End With**" syntax around the functions to speed up execution. The following example nests the worksheet functions **MAX** and **PI** for returning the maximum value from a list of arguments, and π, respectively:

    ```
    With WorksheetFunction: b = .Max(a,.Pi()): End With
    ```

- Unfortunately, repeated reference to ranges or worksheet functions in *VBA* slows down *VBA* execution. Where possible and practical, replace slower worksheet functions with *VBA* code that gives the same result. The ratio of logarithms to change the base is one example from Example 3.5. In another example, we can replace the previous line of *VBA* code with the logical comparison:

    ```
    npi = 4*Atn(1)  ' Calculate the value of pi from arctan trig function
    If a < npi then  ' Use logical if statement to check for the maximum
        b = npi
    Else
        b = a
    End If
    ```

For yet another example, an inverse sine function is not available in *VBA*; nevertheless, we can create a user-defined function for this operation. The sub procedure test in Figure 3.18, compares the execution time for the worksheet function **ASIN()** with a user-defined function for the arcsine in *VBA* by performing one million calculations. The macro writes the output to the **Immediate** window, accessed from the **View** menu in the *VBE*.

```
Public Sub test()
' Test the execution time for an Excel worksheet function
' *** Nomenclature ***
Dim finish as Single  ' = end time
Dim i as Long, n as Long  ' = index for iterations and number of trials
Dim start as Single  ' = start time
Dim totaltime as Single  ' = calculation time
Dim y as Double  ' = temporary variable for result of calculation
'***************************************************************************
n = 1000000  ' set a large number for iterations
starting = Timer  ' Set start time.
With Worksheet  ' Loop to repeat worksheet function for arcsine
    For i = 1 To n : y = .Asin(0#) : Next i
End With
finish = Timer : totaltime = finish - starting  ' Calculate total time. Set end time.
Debug.Print "Worksheet Function Time =", totaltime  ' Print to immediate window
starting = Timer  ' Set start time.
For i = 1 To n : y = Arcsin(0#) : Next I  ' Loop user-defined function for arcsine
finish = Timer: totaltime = finish - starting  ' Set end time * Calculate total time
Debug.Print "User-defined function Time =", totaltime  ' Print to immediate window
End Sub
Public Function Arcsin(x As Double) As Double
' User-defined function for arcsine in terms of the arc tangent, Atn
    Arcsin = Atn(x / Sqr(-(x ^ 2) + 1))
End Function
```

```
Immediate
Worksheet Function Time =    1.4375
User Function Time =         0.234375
```

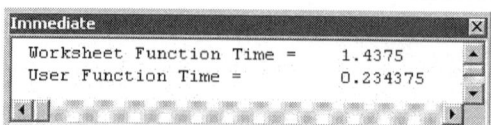

Figure 3.18

VBA code and Immediate Window to test and compare execution speed of an *Excel* worksheet function relative to a user-defined function.

The *VBA* function **TIMER** returns the elapsed time in fractions of seconds after 12:00 AM for the computer system. Two additional *VBA* features **For...Next**, described in the Section 3.5.1, and **Debug.Print**, described in Section 3.9, are used to cycle through the calculations and display the results in the **Immediate** window, respectively. For this test, the user-defined function **ARCSIN** reduced the execution time by a factor of five relative to the time using the worksheet function **ASIN**.

3.6.2 Pseudo Random Number Generator

"The generation of random numbers is too important to be left to chance."— Robert R. Coveyou

We use Monte Carlo techniques throughout this book that require simulations based on large sets of random numbers. Unfortunately, the pseudo random number generators provided with *Excel* (**RAND**) and *VBA* (**RND**) have relatively short cycles of ~ 10^6 values before the random number sequences start to repeat (McCullough 2008).[42] A simple pseudo-random number generator developed by Wichmann and Hill (2005) listed in Figure 3.19, has a cycle of 10^{36}, which should be adequate for any Monte Carlo simulations using *VBA*. The *VBA* user-defined function **RNDWH** employs the *VBA* **Mod** operator and computer **Timer** to seed the generation of pseudo-random numbers.

We show how to use the function **RNDWH** in an *Excel* worksheet to generate 10 random numbers in the range **A1:A10**. We also plot the results to examine the randomness. We find that we get a different set of uniform random numbers each time we recalculate the worksheet.

	A
1	=RND4WH()
2	=RND4WH()
3	=RND4WH()
4	=RND4WH()
5	=RND4WH()
6	=RND4WH()
7	=RND4WH()
8	=RND4WH()
9	=RND4WH()
10	=RND4WH()

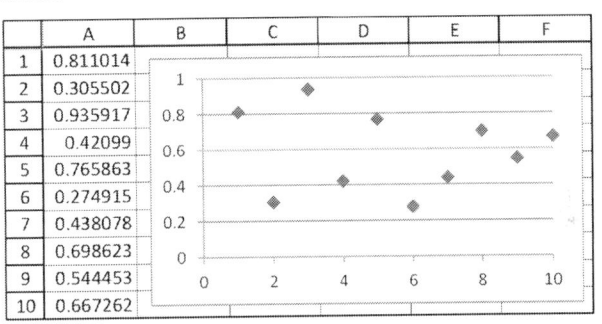

	A
1	0.811014
2	0.305502
3	0.935917
4	0.42099
5	0.765863
6	0.274915
7	0.438078
8	0.698623
9	0.544453
10	0.667262

```
Public Static Function RNDWH() As Double
' Wichmann-Hill Pseudo Random Number Generator (PRNG) with a period of ~ 10^36. The
' PRNG is initialized by the Timer, or internal computer clock. This is an alternative
' for VBA Rnd() function (period is ~10^6). Note: the static function is required to
' preserve the values of the constants between calls to the function.
' *** Nomenclature ***
Dim f As Boolean ' = flag to check for initialization
Dim r As Double ' = random number
Dim s As Double ' = initialization seed for the PRNG
Dim t As Long, x As Long, y As Long, z As Long ' = integer values for PRNG algorithm
'*******************************************************************************
On Error Resume Next ' Skip over random errors
' Note that True conditions run faster in VBA. t, x, y & z must be > 0.
' Mod divides two numbers and returns only the remainder
If f = True Then ' Congruential algorithm
    x = 11600 * (x Mod 185127) - 10379 * (x / 185127)
    y = 47003 * (y Mod 45688) - 10479 * (y / 45688)
    z = 23000 * (z Mod 93368) - 19423 * (z / 93368)
    t = 33000 * (t Mod 65075) - 8123 * (t / 65075)
Else ' Initialize pseudo-random-number-generator (PRNG). Use the computer clock to
    ' generate seed for PRNG. Timer returns number of seconds elapsed since 12 am.
    s = Timer * 60
    x = 11600 * (s Mod 185127) - 10379 * (x / 185127)
    y = 47003 * (s Mod 45688) - 10479 * (y / 45688)
    z = 23000 * (s Mod 93368) - 19423 * (z / 93368)
    t = 33000 * (s Mod 65075) - 8123 * (t / 65075)
    f = True
End If
If x > 0 Then Else x = x + 2147483579 ' Use fast x > 0 Then 'do nothing' Else
If y > 0 Then Else y = y + 2147483543
If z > 0 Then Else z = z + 2147483423
If t > 0 Then Else t = t + 2147483123
' Random number, CDbl returns a double precision value
```

---

[42] Apparently, the Wichmann-Hill algorithm was incorrectly programmed in *Excel* 2007 (and later releases) with a cycle of only 10^6. **RNDWH** uses four linear congruential generators to produce a cycle of 10^{36}.

```
r = CDbl(x) / 2147483579# + CDbl(y) / 2147483543# _
    + CDbl(z) / 2147483423# + CDbl(t) / 2147483123#
RNDWH = r - Int(r)  ' Uniform random deviate (0<r<1)
End Function  ' RNDWH
```

Figure 3.19 *VBA* user-defined function **RNDWH** implementation of the Wichmann-Hill pseudo-random number generator with a cycle of 10^{36}.

The user-defined function **RNDWH** returns a value uniformly distributed between zero and one ($0 < r_u < 1$). Simple scaling formulas for generating uniform random numbers that range from negative one to positive one are:

$$-1 < 2[0.5 - r_u] < 1 \quad \text{or} \quad -1 < (2r_u - 1) < 1 \tag{3.5}$$

We generate normally distributed random numbers from a uniform distribution using the Box-Muller (1958) transformation of a pair of uniformly distributed random numbers:

$$r_{n1} = \cos(2\pi r_{u2})\sqrt{-2\ln(r_{u1})} \quad \text{and} \quad r_{n2} = \sin(2\pi r_{u2})\sqrt{-2\ln(r_{u1})} \tag{3.6}$$

The normal random numbers come from a Gaussian (normal) distribution with mean of zero and standard deviation of one. The Box-Muller method gives a pair of normal random numbers. A user-defined function **NRDWH** in Figure 3.20 returns a normal random deviate. The function **NRDWH** gets the uniform random number required by the Box-Muller transformation from the user-defined function **RNDWH**. Alternatively, the nested *Excel* worksheet functions **NORMSINV(RAND( ))** or **NORMSINV(RNDWH( ))** returns a normally distributed random number with mean zero and standard deviation of one.

```
Public Function NRDWH() As Double
' Calculates a normal random deviate (NRD) with Wichmann-Hill prng. Use the Box-Muller
' transformation. Generates one normal random deviates. Note: VBA Log is natural log.
' *** Required User-defined Functions and Sub Procedures from the PNM Suite ***
' RNDWH = Wichmann-Hill pseudo random number
'*****************************************************************************
NRDWH = Cos(8 * Atn(1) * RNDWH()) * Sqr(-2 * Log(RNDWH()))
End Function  ' NRDWH
```

Figure 3.20 *VBA* user-defined function **NRDWH** that returns normal random deviates using the Box-Muller transformation of uniform deviates.

We tested the pseudo-random number generator *VBA* function **RNDWH** and **NRDWH** with a set of 5000 normal random deviates plotted in a normal probability plot, sorted from low to high, and a histogram using *Excel*'s **Data Analysis Histogram** add-in tool. The **HISTOGRAM** tool has the option of generating a frequency chart. The results plotted in Figure 3.21 shows the telltale bell shaped distribution and sigmoidal shaped cumulative distribution expected for normally distributed values. We used a bin size of 0.5 to collect values in the range -3.75 to 3.75. We put the random numbers in column **A** and the bins in column **B** as seen in Figure 3.22.

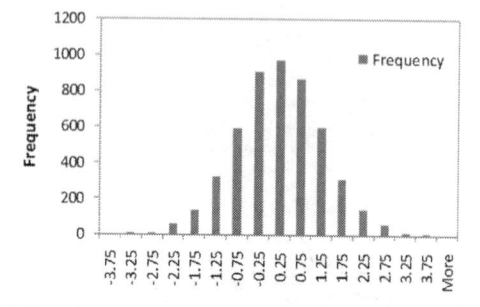

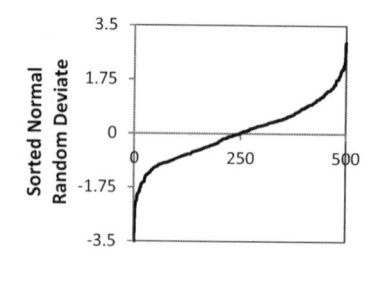

Figure 3.21 Histogram showing the "bell-shaped" curve and probability plot showing the sigmoid shape for normally distributed values from the Box-Muller transformation of uniform random numbers.

CHAPTER 3: VBA

Use the following simple heuristic for selecting the number of bins and bin dimension for *n* data points (Stephen, et al. 2011):

1. Number of bins: **b=ROUNDUP(SQRT(n),0)**

2. Bin size: $$\Delta x = \frac{x_{max} - x_{min}}{b}$$

3. Bin dimension: $$x_{bi} = x_{min} + i\Delta x \quad for \quad i = 0, 1 \ldots b-1, b$$

Adjust the number of bins and the bin dimensions as necessary to create a histogram that best displays the distribution of the data. The PNMSuite has a sub procedure named **BINS** that calculates the number of bins and bin size to generate a series of bin dimensions for use in a frequency chart of histogram.

The alternative *Excel* worksheet *array* function **FREQUENCY** generates the values for a histogram in an *Excel*:

1. Select a range of cells with the same number of cells as bins.
2. Type the worksheet function = **Frequency(A2:A5002, B2:B17)**.
3. Enter the array function in the range of cells by typing the key combination *CTRL SHIFT ENTER*.
4. *Excel* adds the frequency of entries for each bin interval in the corresponding cells on the worksheet.

Beginning with Office 2007, *Excel* added conditional formatting options that fill cells with data bars having length proportional to the relative value of the cell. To create the appearance of a histogram in the range of cells, select the ribbon tab **Home>Conditional Formatting>Data Bars** from the **Styles** group. We adjusted the width of column **C** to show the histogram created by the **Data Bars**. As an additional check on the normality of the distribution of the random deviates, we calculated the average value and standard deviation of the random numbers as seen in Figure 3.22 using the following *Excel* worksheet functions:

           B19 = AVERAGE(A2:A5002)   B20 = STDEV(A2:A5002)

The results agree with a normal distribution having a mean value of zero and standard deviation of one within 0.1% and 0.4% error, respectively. We can reduce the error using additional simulations.

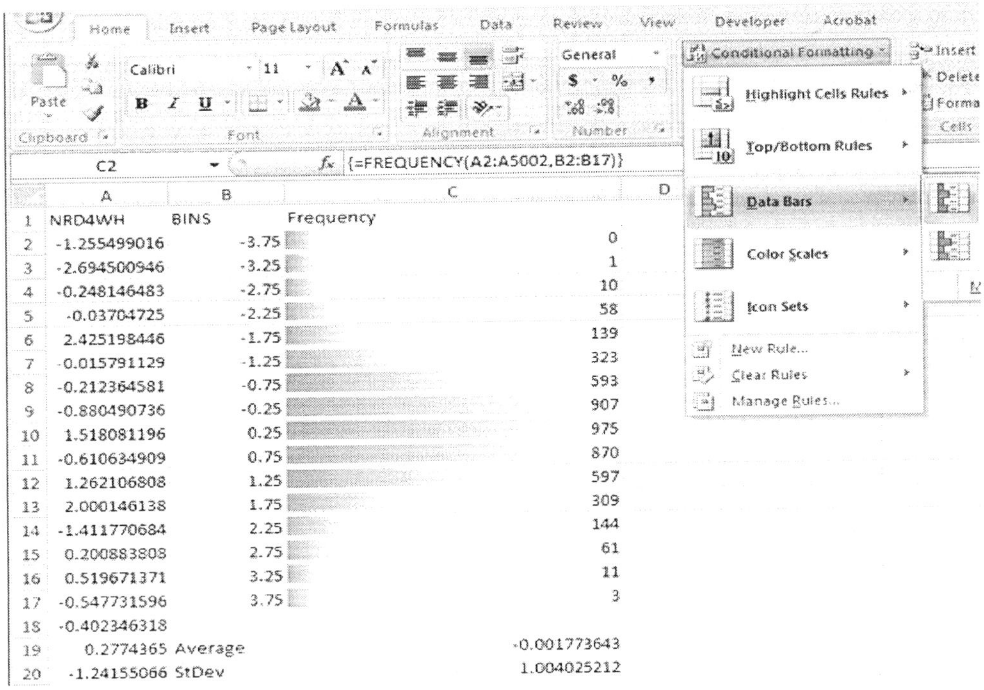

Figure 3.22   Use the array function FREQUENCY(data_array, bins_array) and conditional formatting with Data Bars to generate a histogram in an *Excel* worksheet.

3.6.3 Evaluate Worksheet Cell Formulas in VBA*

On occasion, we may want to use *VBA* to evaluate an *Excel* worksheet formula created in a cell. For example, instead of using **Range** or **Cells** objects to exchange information between our code and the worksheet, which is relatively slow, we can evaluate the formula directly in *VBA*. Capturing formulas from an *Excel* worksheet allows us to create macros that have general utility for the *Excel* user who may not have a background in *VBA* programming. To incorporate worksheet formulas into our numerical methods in *VBA*, we use an improved version of Billo's (2006) clever algorithm programmed in Figure 3.23. The algorithm creates a string variable of the formula with a unique substring for each occurrence of the cell address (or name) for the independent variable. The outline of the steps for the improved algorithm follows:

1. Write a formula in a cell in a worksheet with reference to the cell containing the value of the independent variable. For example, calculate the formula $f(x) = \ln(x)\cos(15x)$ in cell **B2** in terms of the value for x in cell **B1**:

	A	B
1	x	1.5
2	f(x)	=LN(B1)*COS(15*B1)

2. Insert a function or sub procedure in a module in the *VBE*.

3. Assign Variant type names to the formula and independent variable for use in the procedure. For this generic algorithm, use **f** for the formula and **x** for the independent variable.

4. Assign the address for the independent variable and formula to string variables in the procedure, **x_a = x.Address**. We use **x_str = RANGE(x_a).Name.Name** to extract the string for a named cell on the worksheet. Let **x_str = ""**. By design, this produces an error for unnamed cells. The statement **On Error Resume Next** skips to the next statement. For this generic algorithm use **f_str** for the formula string. We plan to use a string "**X_**" as a placeholder for all instances of the values of **x_a** or **x_str** in the formula. Just in case the formula has "**X_**" as any part of the string, we loop using the *VBA* function **InStr** to check and rename the placeholder by adding "**_**" to the end of the string until the **InStr** condition is false. By default, the **.Address** object property returns an absolute reference for x, so we need to convert all references for x in the formula to absolute. For our simple worksheet from step one:

   $$x\_str = \text{``\$B\$1''} \quad \text{and} \quad f\_str = \text{``\$B\$2''}$$

   ```
   On Error Resume Next
   x_a = x.Address: x_str = "":  x_str = Range(x_str).Name.Name
   x_u = "X_": f_str = UCase(f_str.Formula)  ' upper case sensitivity
   Do While InStr(f_str, x_u)  ' check for dummy variable string in formula
       x_u = x_u & "_"  ' add an underscore to end of string
   Loop
   f_str = Application.ConvertFormula(f_str, xlA1, xlA1, xlAbsolute)
   ```

5. Find the number of distinct formula references to **x_a** by dividing the length of the sum of all **x_a** strings in **f_str** by the length of the string in **x_a**:

   ```
   n = (Len(f_str) - Len(Application.Substitute(f_str, x_a,""))) / Len(x_a)
   ```

   A similar calculation applies to **x_str** for named cells.

6. Substitute numerical values for **x_a** into **f_str** for evaluating the formula to check for errors. As Billo (2006) warned, we need to be cautious when locating and substituting values for addresses of the independent variable in the formula string. To illustrate, imagine a formula for a simple function that contains the cell address $B$1 for the independent variable and an address $B$12 for another parameter in the function:

   $$= \$B\$1 + \$B\$12$$

   Suppose the numerical value of the independent variable x is five. If we are not careful, we may mistakenly find and replace the string "$B$1" portion of the address $B$12 that then becomes the incorrect number 52 in our formula:

CHAPTER 3: VBA

$= 5 + 52$ *which is evaluated as 57, instead of the desired formula* → $= 5 + \$B\$12$

To prevent similar mistakes, Billo (2006) devised a simple protocol for avoiding incorrect substitutions by adding a blank space after the substituted value and evaluating the formula to check for errors. If a trailing character is left after the blank space (such as the two in the previous illustration which gives = 5 + 5 2), the evaluation returns an error. Whenever a substitution returns an error, the formula string remains unchanged.

We have improved Billo's code in two ways. First, we enhance computational efficiency by creating a new formula string **f_x** with a substring "**X_**" in place of **x_a**. In this way, we only need to replace the **x** addresses once, and then use the new formula string in successive combinations of value substitution and evaluation. The other improvement shown in step four allows for named cells in the worksheet formula by replacing a string for a named cell with a row-column cell address for the variable in the formula. We use the temporary string variable **t** to evaluate the formula to check for errors. To accomplish this, we use two loops, one for "**A1**" type addresses, and another for named cells:

```
f_x = f_str
For j = nx To 1 Step -1 ' Loop to check for cell address in formula
    t = Application.Substitute(f_str, x_a, x.Value & " ", j)
    If Not IsError(Evaluate(t)) Then
        f_string = t: f_x = Application.Substitute(f_x, x_a, x_u, j)
    End If
Next j
ns = Len(x_str)
if ns > 0 then ' Check for division by zero
n = (Len(f_str) - Len(Application.Substitute(f_str, x_str,""))) / ns
For j = n To 1 Step -1 ' Loop to check for cell name in formula
    t = Application.Substitute(f_str, x_str, x.Value & " ", j)
    If Not IsError(Evaluate(t)) Then
        f_string = t: f_x = Application.Substitute(f_x, x_str, x_u, j)
    End If
Next j
End If
```

The **Substitute** operation replaces only the **j** index occurrence of **x_a** or **x_str** in **f_str**.

7. Whenever we need to evaluate the formula in our procedure, we simply substitute a numerical value for each **x_u** in the formula string and evaluate the result using imbedded operations:

```
= Evaluate(Application.Substitute(f_x, x_u, x_new))
```

If we leave out the last (optional) argument in **Substitute**, the operation replaces all instances of **x_u** in the string **f_x** with **x_new**.

We put the steps together in a template listed in Figure 3.23 of a user-defined function for evaluating an *Excel* worksheet formula. The template replaces the cell address for **x** in the formula with the value of **x_new**. The user-defined function evaluates the result and returns the value to the worksheet. The template looks for "**x1**" addresses first, then checks if the variable is in a named cell.

```
Public Function UDF(f As Variant, x As Variant, x_new As Double) As Double
' f = cell with formula in terms of cell with x
' x = cell with variable in formula f
' x_new = cell with value of x for evaluating the formula
'*****************************************************************************
' Template for evaluating Excel worksheet formulas in VBA.
' *** Nomenclature ***
Dim j As Integer ' = loop index variable
Dim n As Integer ' = number of occurrences of x variable in the formula string
Dim ns As Integer ' = length of string for named cell with independent variable x
Dim f_str As String ' = formula string for evaluation in VBA
Dim f_x As String ' = formula string with string for independent variable
Dim t As String ' = temporary formula string with substitution for x for evaluation
Dim x_a As sting ' = address of the independent variable
Dim x_str As String ' = named cell on a worksheet
```

```
Dim x_u As String ' = independent variable string with underscore
'************************************************************************
On Error Resume Next
x_a = x.Address: x_str = "":  x_str = Range(x_a).Name.Name
x_u = "X_": f_str = UCase(f.Formula) ' upper case sensitivity
Do While InStr(f_str, x_u) ' check for dummy variable string in formula
    x_u = x_u & "_" ' add an underscore to end of string to create a unique string
Loop
With Application
    f_str = .ConvertFormula(f.Formula, xlA1, xlA1, xlAbsolute): f_x = f_str
    n = (Len(f_str) - Len(.Substitute(f_str, x_a, ""))) / Len(x_a)
    For j = n To 1 Step -1 ' check for variables in the formula string
        t = .Substitute(f_str, x_a, x.Value & " ", j)
        If Not IsError(Evaluate(t)) Then
            f_str = t: f_x = .Substitute(f_x, x_a, x_u, j)
        End If
    Next j
    ns = Len(x_str) ' repeat for x-variable in named cells
    If ns > 0 Then ' replace named variables in the formula string
        n = (Len(f_str) - Len(.Substitute(f_str, x_str, ""))) / ns
        For j = n To 1 Step -1 ' check for variables in the formula string
            t = .Substitute(f_str, x_str, x.Value & " ", j)
            If Not IsError(Evaluate(t)) Then
                f_str = t: f_x = .Substitute(f_x, x_str, x_u, j)
            End If
        Next j
    End If
    UDF = Evaluate(.Substitute(f_x, x_u, x_new))
End With
End Function
```

Figure 3.23 Template of a user-defined function for evaluating an *Excel* worksheet formula within *VBA*.

Test the template function listed in Figure 3.23. For $x = 1.5$ in cell **B1**, the user-defined function returns the correct answer of -0.35409 in cell **D1**:

	A	B	C	D
1	x	1.5		=UDF(B2,B1,5)
2	f(x)	=LN(B1)*COS(15*B1)		

Example 3.11 uses the template in Figure 3.23 to create a macro for plotting formulas in a *yx*-scatter plot. This book has several examples of worksheet formula evaluations in user-defined functions for numerical approximation of roots of equations, derivatives, and integrals of worksheet formulas. The only restriction is that the formula must be contained in a single cell on the worksheet; it may not consist of a combination of formulas from multiple cells in the workbook.

3.7   Arrays

An array in *VBA* is analogous to a range of cells arranged in rows or columns on an *Excel* worksheet. An array is a variable with a collection of values of the same data type. The *VBA* syntax for an array is the name followed by the number of elements in parentheses, *e.g.*, A(0 to 5). Access the elements of an array using an index number for the row and column of the array, similar to cell addressing on an *Excel* worksheet. Arrays may have a single dimension, such as `B(3)`, like an *Excel* worksheet row, or two dimensions, such as `D(0 to 2, 0 to 3)`, like a matrix of rows and columns in an *Excel* worksheet. By default, *VBA* arrays start with an index value of zero. We may change the default starting-index to 1 with the statement `Option Base 1` at the top of the module, or through a dimension statement (although we most often begin with a zero or one, we may start the indexing at any positive integer value):

```
DIM a(1 to 5) as Double
```

A redimension statement may change the size of an array after initializing the array:

```
REDIM a(1 to 10) as Double: REDIM PRESERVE a(1 to 5) as Double
```

We must be careful when redimensioning arrays; we will lose any information previously stored in the array. Use the keyword **PRESERVE** to save the array contents before redimensioning.

When passing arrays between a *VBA* macro and an *Excel* worksheet, use the **VARIANT** or **RANGE** data types. We must use the `Variant` type for user-defined functions that operate on arrays. We take advantage of the convenience of arrays extensively for a variety of numerical methods involving solutions to systems of algebraic or differential equations.

To illustrate, the following simple user-defined function returns the value of a specified arithmetic operation on two numeric values in an array. The arguments for user-defined function **ARITHMETIC** consist of a string in quotes "add" or "multiply" and an array of two numerical values. The **ARITHMETIC** function combines the argument string with "2NUMBERS" to create a new string for the name of one of two user-defined functions **ADD2NUMBERS** or **MULTIPLY2NUMBERS**. The *VBA* application function **RUN** uses the string to call the function by name followed by the values for the string-named function's arguments, separated by commas. Finally, the user function returns the value of the corresponding arithmetic operation to the worksheet:

```
Public Function ARITHMETIC(m As String, a As Variant) as Double
    ARITHMETIC = Application.Run(m & "2NUMBERS", a(1), a(2))
End Function

Public Function ADD2NUMBERS(a as Double, b as Double) as Double
    ADD2NUMBERS = a + b
End Function

Public Function MULTIPLY2NUMBERS(a as Double, b as Double) as Double
    MULTIPLY2NUMBERS = a * b
End Function
```

We demonstrate the functions in an *Excel* worksheet with arrays of values (2, 4) in the cell ranges **A1:B1** and **A2:B2**:

	A	B	C
1	2	4	=arithmetic("add",A1:B1)
2	2	4	=arithmetic("multiply",A2:B2)

	A	B	C
1	2	4	6
2	2	4	8

3.7.1 Range Variables and Worksheet References

Excel workbooks may have multiple worksheets. Worksheets have default names **Sheet1**, **Sheet2**, etc. Double-click on the name in the worksheet tab to edit the worksheet name. To activate an *Excel* worksheet with *VBA*, use the following statement, with the name of the worksheet in quotes. Worksheet names referenced in *VBA* may not start with a number nor have spaces or special characters:

```
Worksheets("Sheet1").Activate
```

If we do not activate a specific worksheet in *VBA*, or include the worksheet object name, *VBA* references ranges of cells in the current active worksheet.

Range type variables pass information between cells on an *Excel* worksheet and variables in a sub procedure. To assign a range address to a range variable use syntax that follows the pattern: `Set A = Range("A1")`. The `Range` object accesses the contents or address of a range of cells on an *Excel* worksheet using strings for column-row addresses. For example, the following line assigns the values in the worksheet range **A1:B2** to a two-by-two array:

```
Dim a(2, 2) as Double: a = Range("A1:B2").Value
```

This assignment of a value to a variable passes in one direction from the worksheet to the sub procedure, such that any subsequent changes made to the values in array **a** by the sub procedure are not reflected in the cells on the worksheet. The **.Value** object property is optional.

The next line uses the statement **"Set"** to assign the worksheet range address to a range variable **r**:

```
Dim r as Range: Set r = Range("A1:B2")
```

Now we can use the range variable to access the contents of the worksheet range **A1:B2**, as well as program the sub procedure to modify the contents of the range of cells at that location on the worksheet.

Concatenate parts of a cell address in the **Range** object. For example, we may want to fill the contents of a range of cells with integer values 1 to 10 in column **A** using a **For Next** loop:

```
For i = 1 to 10: Range("A" & i) = i: Next i
```

The statement **Cells(row, column)** provides an alternative *VBA* statement for passing information to and from an *Excel* worksheet. The **Cells** object uses integer values for the row and column indices. We use the **Cells** object with row/column number addressing, such as **Cells(1, 1)** for the address **A1**. The following line assigns the value of the worksheet cell located in row 1 column 2 to the variable **a**, then replaces the value of the worksheet cell with twice the value of the variable **a**:

```
a = Cells(1, 2): b = 2: Cells(1, 2) = a*b
```

We can also use a **Cells** object inside of **Range** objects when we need to use integer row and column addressing rather than string "column: row" addressing. The following fills an array with the values in the range **A1:B2**:

```
a = Range(Cells(1, 1), Cells(2, 2)).Value
```

Use quotations in *VBA* to place a string of text on the worksheet, for example:

```
Range("A2") = "x = "
```
or
```
Cells(2,1) = "x = "
```

To work with named cells, use the option on the worksheet tab **Formulas>Define Name**. Type a name for the cell and that name will serve as the cell identifier. These identifiers are only available for operations within *Excel* (e.g., formulas defined in cells). The identifier's names are not passed to the *VBA* code as variables. However, we can refer to the cell or range name in quotations in a **Range** object. For example, we may access a cell renamed *x* using **Range**:

```
a = Range("x")
```

The macros throughout the book provide a variety of examples for **Range** and **Cells** objects. Consult the *VBA Help* for additional information and examples.

By default, one-dimensional *VBA* arrays contain data in a row. The **Transpose** object property changes a row array into a column array. The *VBA* macro in Figure 3.24 checks for rows versus columns and transposes the array if necessary.

```
Public Sub ROWCOL(n, y, yr, yrows)
' n = number of elements in the array
' y = array of values to place on an Excel worksheet
' yr = range of cells on an Excel worksheet
' yrows = number of rows in the range yr
'*********************************************************************
' Checks row versus column for dependent variable in worksheet
'*********************************************************************
If yrows = n Then
    yr = Application.Transpose(y)
Else
    yr = y
End If
End Sub ' ROWCOL
```

Figure 3.24 *VBA* sub procedure ROWCOL checks one-dimensional arrays for rows or columns.

We use sub **ROWCOL** extensively in the *PNMSuite* of macros that accompany this book to allow users the flexibility of setting up their inputs and outputs from or to worksheet ranges in either rows or columns of contiguous cells.

3.7.2 Sorting Arrays

If we need to sort data in an array, we have two options. We may write the contents of an array onto an *Excel* worksheet, use the worksheet sort operation, and then read the sorted results back into our array. Alternatively, we may write a simple macro for sorting arrays. The macro **HEAPSORT** listed in Figure 3.25 uses the comparison-based heap sort method to replace the original contents of an array with the same values in ascending order (Floyd 1964). The *PNMSuite* has a similar macro **HEAPSORTXY** for sorting two arrays according to the contents of the first array. To sort the elements in a column of an array, pass the array to the sub procedure **HEAPSORT**, and include the corresponding column number of the array for sorting. The *PNMSuite* also includes macros using the bubble sort algorithm.

We illustrate how to use **HEAPSORT** with a simple example involving a small series of random numbers listed in column **A** of an *Excel* worksheet. We write a simple macro that gets the range of values from the worksheet and assigns them to an array variable x. Our macro then calls **HEAPSORT**, which returns the values in the array rearranged in ascending order. Our little macro then puts the results back onto the worksheet in column **B**.

```
Public Sub SORT()
    x = Range("A1:A5")
    Call HEAPSORT(x, 1)
    Range("B1:B5") = x
End Sub
```

	A	B
1	2	1
2	4	2
3	3	3
4	5	4
5	1	5

```
Public Sub HEAPSORT(x, k)
' x() = array for sorting     k = column of sort array x
'*****************************************************************
' Sort column array x in ascending order using the method heap sort.
' *** Required User-defined Functions and Sub Procedures from the PNM Suite ***
' HEAPIFY = regional sort with array element exchange
' *** Nomenclature ***
Dim i As Long ' = index of array x
Dim i_min As Long, i_max As Long ' = first and last index number of array x
Dim temp As Double ' = temporary save value for exchange
'*****************************************************************
i_min = LBound(x, 1): i_max = UBound(x, 1) ' index for lower/upper array values
' Note back slash \ performs division and rounds the result to nearest integer
For i = (i_max + i_min) \ 2 To i_min Step -1: Call HEAPIFY(x, i, i_min, i_max, k): Next i
For i = i_max To i_min + 1 Step -1
    temp = x(i, k): x(i, k) = x(i_min, k): x(i_min, k) = temp
    Call HEAPIFY(x, i_min, i_min, i - 1, k)
Next i
End Sub ' HEAPSORT

Public Sub HEAPIFY(x, ByVal i, i_min, i_max, k)
'' i = index variable              i_min, i_max = minimum and maximum elements of x
' k = column of sorting array      x() = array for sorting
'*****************************************************************
' Regional array sort with element exchange
' *** Nomenclature ***
Dim j As Long   ' = index variable
Dim temp As Double ' = temporary value for exchanging array elements
'*****************************************************************
Do ' Loop through array elements and replace element j with i
    j = i + i - (i_min - 1)
    Select Case j
        Case Is > i_max: Exit Do
        Case Is < i_max
            If x(j + 1, k) > x(j, k) Then j = j + 1
    End Select
    If x(i, k) > x(j, k) Then Exit Do
    temp = x(i, k): x(i, k) = x(j, k): x(j, k) = temp: i = j
Loop
End Sub ' HEAPIFY
```

Figure 3.25 *VBA* sub procedures **HEAPSORT** and **HEAPIFY** used for sorting an array in ascending order.

3.7.3 Worksheet Array Functions and User-defined Functions that Return Arrays

We may use worksheet array functions in *VBA*, such as **MINVERSE** that inverts a square matrix. Array functions require the `Variant` data type. To use an *Excel* worksheet array function, modify the `Application` or `Worksheet` object with the worksheet array function. For example:

```
B = WorksheetFunction.MINVERSE(A)
```

where the array variable B that accepts the result is a `Variant` data type.

A user-defined array function may return an array of values similar to worksheet array functions. Select a range of cells with the same dimensions as the output array, type " =NAME()", where NAME is the name of the user-defined array function, then type the keyboard combination *CTRL SHIFT ENTER* to place the array of results in the selected range of cells on the worksheet. To illustrate, we created a *VBA* user-defined array function to generate a random subset of size k from a larger set of size n. Engineers use simple random samples taken from larger populations for quality control or experimental trials. The user-defined array function **SAMPLE** in Figure 3.26 accepts a range of values from the worksheet in the argument, and returns a smaller random subset to the selected cells in the output range. The size of the output range must be smaller than the size of the population.

The user-defined function uses the object property `Application.Caller.Rows.Count` to determine the size of the range of cells selected for output on the worksheet. The function checks for the output as a row or column range, and then uses the object `Application.Transpose` to convert the default row array into a column array for output to the worksheet, if necessary. Note that we also include parentheses ( ) after the name of the sample array **s** to signal that the result is an array.

```
Public Function SAMPLE(p As Variant) As Variant
' p = population range selected from the worksheet.
'****************************************************************************
' Returns array of simple random samples from population. Array-user-defined function
' requires CTRL SHIFT ENTER. Select range of cells smaller/equal to the population.
' *** Nomenclature ***
Dim i As Long, j As Long ' = index variables for samples and population
Dim k As Long, kcol As Long ' = number of samples, and number of samples in columns
Dim n As Long ' = number of elements in the population array
Dim s() As Variant ' = sample array
'****************************************************************************
With Application ' Get size of the selected range of sample cells.
    k = .Caller.Count: kcol = .Caller.Columns.Count
End With
ReDim s(1 To k) As Variant: n = p.Count: i = 0
For j = 1 To n
    If Rnd() < (k - i) / (n - j + 1) Then
        i = i + 1: s(i) = p(j)
    End If
Next j
If kcol = 1 Then
    SAMPLE = Application.Transpose(s())
Else
    SAMPLE = s()
End If
End Function ' SAMPLE
```

Figure 3.26   User-defined array function SAMPLE for generating a random sample from a larger population.

We tested the user-defined function **SAMPLE** on a range of five values in column **A** to generate a random subset of three values in column **B** on an *Excel* worksheet. To generate the simple random sample, select three empty cells on the worksheet, type "=SAMPLE(A1:A5)", then type the keyboard combination *CTRL SHIFT ENTER* to obtain a result similar to the one below. Repeat these steps to generate a different random sample.

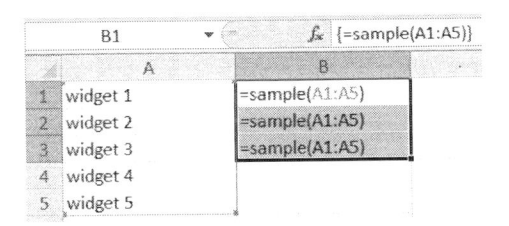

3.8 Communicate with the User

VBA in *Excel* for *Windows* has keywords for reading aloud strings of text. For example, we may call the whimsical user-defined function in Figure 3.27 to alert a stoked surfer of an *Excel* worksheet error:

```
Public Function KOWABUNGA(Optional s As Variant)
' s = (optional) Variant for text string
'*****************************************************************
With Application.Speech ' Read the text string argument (MS Windows only)
    If IsMissing(s) Then
        .Speak "Yo, dude. Gnarly, wipeout. Try again, brah."
    Else
        .Speak s
    End If
End With
End Function ' KOWABUNGA
```

Figure 3.27 VBA user-defined function KOWABUNGA for reading a string aloud (Windows only)

Engineers work in teams. As team members, we frequently create *Excel* workbooks to collect or disseminate information. We stress the importance of clear organization and documentation throughout this book. In this section, we survey a variety of techniques for making our workbooks user-friendly with functionality that provides additional access to our calculations and results. Careful use of these tools not only provides the user with access, but also validates user input to reduce errors. As we have seen, perhaps the simplest method of communication collects and reports information on an *Excel* worksheet using `Range` and `Cells` objects in procedures, as described in Section 3.7.1. We have other options, including:

- Controls
- Message and Input dialog boxes
- User-forms for organizing input
- Reading and writing data to or from files
- **Status Bar** for displaying progress

3.8.1 Macro Controls on the Worksheet

> *"If it keeps up, man will atrophy all his limbs but the push-button finger."* – Frank Lloyd Wright

We can add controls to an *Excel* worksheet for manipulating the contents of cells or macros.

- Macro controls, such as virtual buttons, are accessed from the ribbon tab **Developer>Insert>Form controls** or **ActiveX Controls**.
- The **Controls** group contains a **Design Mode** toggle and **Insert** buttons.
- The **ActiveX Controls** include check, toggle, spin, and command buttons, list and combo boxes, a scroll bar, label, and an image placeholder (similar **Form Controls** are legacy from earlier versions of *Excel*).

Use **Command** buttons to launch a program directly from an *Excel* worksheet. Place a command button on an *Excel* worksheet by clicking on the command button icon in **ActiveX Controls**. Click on the worksheet to add the button, then dimension it by dragging the corner to the desired size.[43] To modify the properties of a control, select **Design Mode,** then right-click on the control and select the **Properties** option. A window opens where we modify the control's properties such as the name, caption, font, etc. We can also access the properties by activating the control while in **Design Mode,** then selecting the **Properties** option from the **Developer** tab.

Double-click on the control to assign *VBA* code. This sends us to the code window for the worksheet in the *VBE*. Type the code using *VBA* programming statements. For example, we may want to add a button to an *Excel* worksheet that runs a macro when clicked. Add the line `Call Macro_name()` to the button code (the syntax `Call` is optional), where `Macro_name` can be the name of any macro or sub procedure in the workbook.

Example 3.6   Add a Button to an Excel Worksheet to Run a Macro

Known/Find: We take another short cut in our pipe friction flow problem by adding a "button" to the worksheet that launches the macro in *Example 3.4*.

Analysis: An outline of the steps follows.

1. Add an **ActiveX Control** button to the worksheet from the ribbon tab **Developer>Insert>ActiveX Controls.**

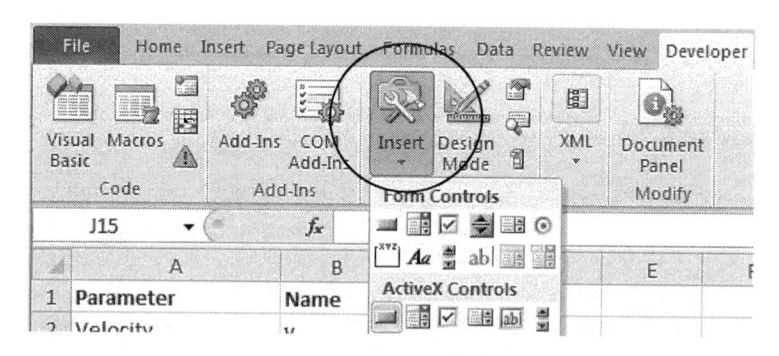

2. To add the button, select it from the **ActiveX Controls**, move the mouse pointer to the desired location on the worksheet, click and hold the left mouse button, and drag to create the dimension of the button:

3. Adding a button activates the **Design Mode,** which deactivates the worksheet while formatting the button. Double-click on the button to open the *VBE*. Then add the code for running **Goal Seek** (or *Right-click* on the button and select **View Code**):

---

[43] A bug in *Excel 2007* and higher may cause the text in an **ActiveX Control Button** to shrink with repeated clicks. If this happens, click **Design Mode** and move the control on the worksheet. Otherwise, consider using a **Form Control Button** instead.

CHAPTER 3: VBA

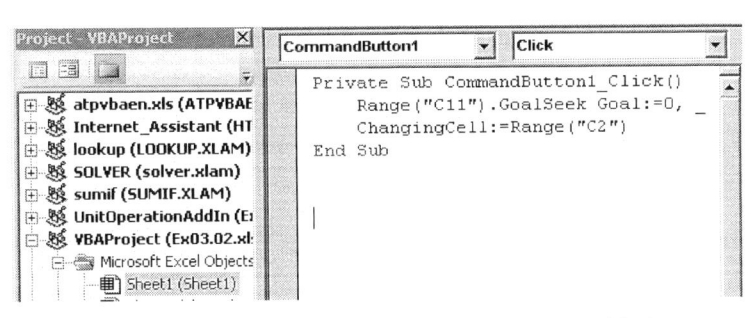

```
Private Sub CommandButton1_Click()
    Range("C11").GoalSeek Goal:=0, _
    ChangingCell:=Range("C2")
End Sub
```

4. Be sure to toggle out of **Design Mode** before using the worksheet with the macros. Try the worksheet macro with different pipe lengths. Put the worksheet in **Design Mode** again to change the button properties. To bring up the properties window, right-click on the button and select **Properties**.

5. Change button properties by clicking in the property box in the second column and edit as desired. Make the following changes: **BackColor** = Green, **Caption** = Run, **Font** = 18 pt, Bold, **Shadow** = True.

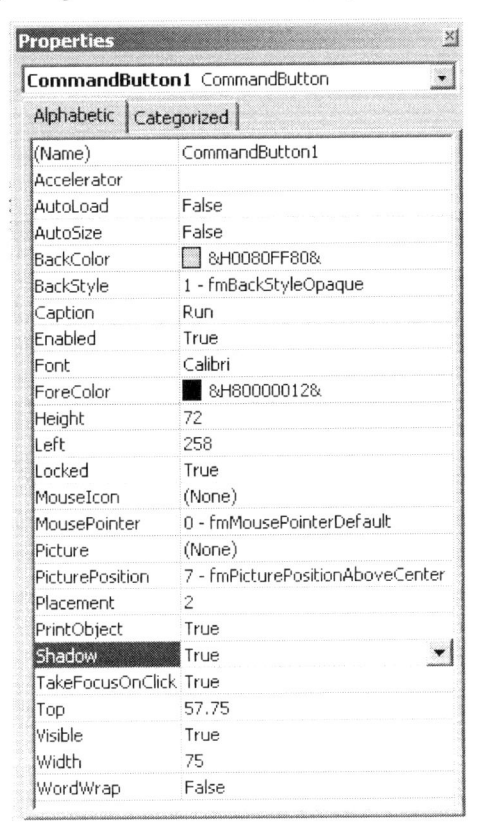

Comments: Consistent button properties help the user identify buttons on our worksheets and become familiar with our style of documentation. A button replaces all of the steps for finding, selecting and running the macro from the **View Macros** list, with one click!

Example 3.7 Animation of Breakthrough from an Adsorption Column

Known: We may use Klinkenberg's (1948) model to simulate the mass transfer zone of an adsorbate solute through a packed bed adsorber in terms of operating time ($1 < t < 150$ min) and axial position ($0 < x < 100$ cm):

$$c = \frac{1}{2}\left[1+\text{erf}\left(\sqrt{t}-\sqrt{x}+\frac{1}{8\sqrt{t}}+\frac{1}{8\sqrt{x}}\right)\right] \quad (3.7)$$

where c is the dimensionless concentration of the solute in the column at position x and time t.

Find: Use an *Excel* plot with an **Active X** scroll-bar control on the worksheet to animate the mass transfer zone along the axial length of the column and breakthrough curve at the exit of the packed bed.

Schematic: A packed bed used for separating a solute from the bulk fluid adsorption has different feed and outlet concentrations of solute. A moving mass transfer zone tracks in the x length dimension. C and C_{in} are the local and feed solute concentrations.

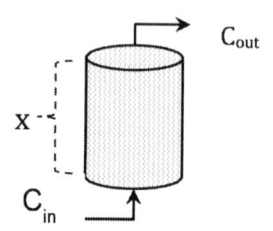

Assume: uniform solute concentration in the radial direction of the packed column.

Analysis: Create two y vs x plots: one for the dimensionless concentration wave front along the length of the column, and a second for the dimensionless concentration at the exit point of the column, as seen in Figure 3.28. The concentration formula references cell **B1**, renamed **t** for time. Use **AutoFill** to add values for x from 1 to 100 in the range **A4:A103**. Rename the ranges **A4:A103** for **x** and **C4:C103** for the breakthrough time **tb**, respectively. With named ranges, use array formulas to calculate the dimensionless concentration along the length of the column and column exit point. To perform an array operation, select the range of cells for the result, then type the formula followed by typing the keyboard combination *CTRL SHIFT ENTER* (note that *Excel* adds the curly brackets):

B4:B103= {0.5*(1+ERF(SQRT(t)-SQRT(x) + (1/SQRT(t) +1/SQRT(x))/8))}

D4:D103= {0.5*(1+ERF(SQRT(tb)-10+ (1/SQRT(tb) +1/10)/8))}

Calculate the breakthrough time range from the value for time in cell **B1** to plot the exit concentration with 100 points over 150 minutes:

C5 = C4+t/100 ...

Add an **Active X** scroll bar control to the worksheet and link it to cell **B1**. Get the scroll bar control from the **Controls** group on the **Developer** tab: **Developer>Insert>ActiveX Controls>Scroll Bar**. Place the control on the worksheet, then click and drag the top right corner of the control to size as shown below the graphs in Figure 3.28. The next step requires linking the control to the worksheet. Toggle on **Design Mode** in the **Developer** tab. Then right-click on the scroll bar to select **Properties** from the popup window. Change the values for **LinkedCell** to **B1**, and set **Max** to 150 and **Min** to 1 (Warning: Klinkenberg's formula is indeterminate at $t = 0$ or $x = 0$).

To use the scroll bar, toggle off **Design Mode** on the **Developer** tab. Click on the slider bar and drag left or right to watch the plot of the mass transfer zone move along the length of the column, and the break through curve develop with time.

CHAPTER 3: VBA

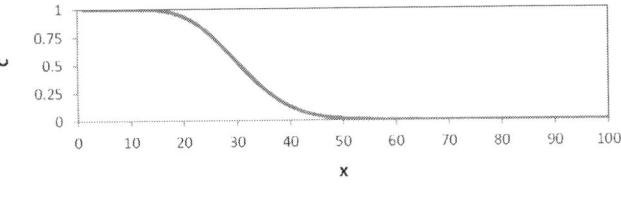

	A	B	C	D
1	t scroll bar	30		
2				
3	x	c(x)	t_b	c_b
4	1	1	1.3	0
5	2	1	1.6	0
6	3	1	1.9	0
7	4	1	2.2	0
8	5	0.999999	2.5	0

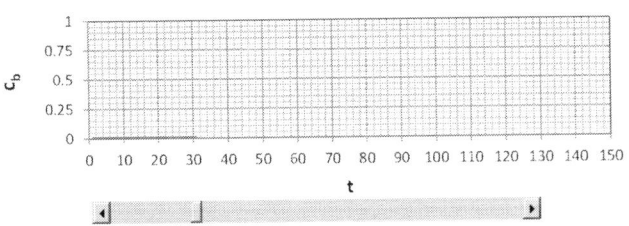

Figure 3.28

Mass transfer zone and dimensionless breakthrough concentrations in an adsorption column after 30 minutes of operation.

The plots in Figure 3.29 show the results of Klinkenberg's model for 75, 100, and 120 minutes. By convention, the breakthrough time occurs when the exit concentration reaches 5% of the feed value. In this example, the breakthrough time is 75 minutes. The mass transfer zone is defined as the region of the column where the concentration drops from the inlet value to zero. When comparing the shape of the mass transfer zones at small times in Figure 3.28 with the mass transfer zone at larger times in Figure 3.29, we see how the zone width spreads due to dispersion as the wave front moves through the column.

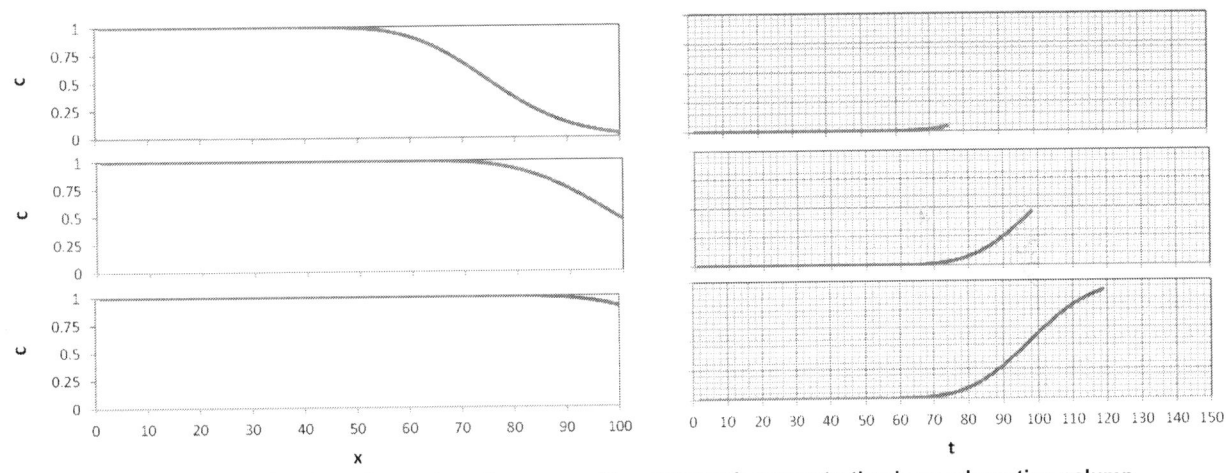

Figure 3.29 Animation of mass transfer zone and breakthrough concentration in an adsorption column.

We may alternatively use a control button with a simple *VBA* macro to automate the simulation. The macro loops through time values, assigns them to the **t** cell on the worksheet, then recalculates the worksheet for each loop to replot the model in the graphs at each point in time.

```
Public Sub simulate()
For i = 1 To 150
    Range("t") = i: ActiveSheet.Calculate
Next i
End Sub
```

Simulate

Comments: Engineers use the complete breakthrough curve plotted after 150 minutes to design an adsorption column and account for the length of unused bed. Animated worksheet calculations promote better understanding of functional relationships and enhance presentations when sharing results. In this example, the model shows how the mass transfer zone begins to spread as the concentration wave front moves through the column.

□

3.8.2 MsgBox

A message box displays information generated by a *VBA* function or sub procedure, whereas an input box prompts the user for information required by the macro. The macros in this book make extensive use of **MsgBox** and **InputBox** for interacting with the user to set up the numerical methods and provide information about the progress of calculations. There are two ways to use a message box. The first simply displays information, *e.g.* :

```
MsgBox "Iteration = " & i
```

The ampersand, &, concatenates combinations of strings and numbers. In this example, the message box displays the word "Iteration =" followed by the numerical value of the variable *i*.

A **MsgBox** keyword followed by arguments listed in parentheses returns an integer value depending on the user's response to the dialog box. The following syntax assigns the value to a variable v:

```
[v =] MsgBox(Prompt[, Buttons][, Title])
```

where [ ] indicates an optional argument. The **Prompt** and **Title** arguments are text statements that prompt the user for a response and provide a title to the message box, respectively. The **Buttons** argument defines the possible user responses to the message box. For example, the following statement prompts the user to continue calculating, or not.

```
v = MsgBox("Continue?", vbYesNo, "Calculate")
```

The **Buttons** argument has a few options listed in Table 3.5. Add buttons to the default options using an addition sign +. For example, **vbOkCancel + vbDefaultButton2** shows the **OK** and **Cancel** buttons with the cancel highlighted as the default action if the user presses the *Enter* key instead of clicking on a button.

Table 3.5   MsgBox button arguments.

Buttons	Description	Default	Description
vbOKOnly	OK button only (Default)	vbDefaultButton1	First button is default (default)
vbOKCancel	OK and Cancel buttons	vbDefaultButton2	Second button is default
vbYesNoCancel	Yes, No, and Cancel buttons	vbDefaultButton3	Third button is default
vbYesNo	Yes and No buttons	vbDefaultButton4	Fourth button is default

For this illustration, the integer return value assigned to variable **v** is 6 for **Yes**, or 7 for **No**. We may then use a conditional *VBA* statement described in Sections 3.5.2 and 3.5.3, such as **If a = 6 then** ... , to control the next step in the program depending on the user's response to the message box. Table 3.6 lists other button return values.

Table 3.6   MsgBox return-values for button selection.

Return Value	Description	Return Value	Description
1	OK button pressed	6	Yes button pressed
2	Cancel button pressed	7	No button pressed

We can change the order of parameters for the **MsgBox** using named parameters. Rewrite the previous example as:

```
i = 3: MsgBox Prompt:= "Iteration = " & i
a = MsgBox(Buttons:=vbYesNo, Title:= "Calculate", Prompt:="Continue?")
```

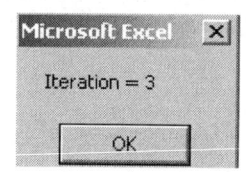

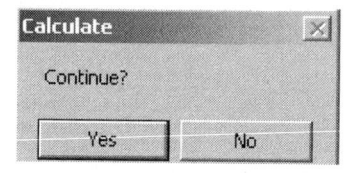

CHAPTER 3: VBA

Example 3.8 Message Boxes in a Sub Procedure for Swapping Cell Contents

Known: *Excel* users frequently have the need to swap the contents of two cell ranges. The manual way requires moving, or cutting and pasting the contents of the first cell range to another location on the worksheet then moving, or cutting and pasting the contents of the second cell range into the location of the first cell range. Finally, move, or cut and paste the first cell range into the location of the second cell range. Is there a better, more efficient way?

Find: Create a simple *VBA* sub procedure for swapping the contents of two cell ranges.

Analysis: We solved this problem by programming a sub procedure, listed in Figure 3.30 that operates on a pair of active cell ranges on the active worksheet. To select a pair of cell ranges for swapping, first select a cell range, then, while holding down the **CTRL** key, select the second cell range.

The sub procedure first checks that only two different cell ranges are active (selected) on the worksheet and displays a message box if more or less than two cell ranges are selected. The **SELECTION** object now has two **Areas** for the two cell ranges that must have the same number of cells for swapping – if not, the code displays an error message box, and then exits the sub procedure. A **For...Next** loop iterates through the two areas of the cell ranges to swap the cells in order (left to right and top to bottom) by using a temporary variable to hold the contents from the first cell range while it is replaced by the contents of the second range. Finally, the temporary contents from the first cell range is placed into the cells of the second cell range.

```
Public Sub SWAP()
' Swap the contents of two Cell Ranges of the same length.  The two Cell Ranges may
' be non-adjacent.  To activate two cell ranges: Select the first Cell Range, then
' hold the CTRL key while selecting the second Cell Range. Run the macro SWAP.
' *** Nomenclature ***
Dim i As Long ' = index variable for elements of selected Cell Ranges
Dim n As Long ' = length of a Cell Range
Dim t As Variant ' = temporary content for first swapped Cell Range selected
'*************************************************************************
With SELECTION
    If .Areas.Count <> 2 Then
        MsgBox "Only the first two selected cell ranges swap.": Exit Sub
    End If
    n = .Areas(1).Count ' get the size of the Cell Ranges for swapping
    If .Areas(2).Count <> n Then ' check for equal sized swapping Cell Ranges Areas
        MsgBox "The swapped Cell Ranges must have the same number of cells.": Exit Sub
    End If
    For i = 1 To n ' swap the first two selected Cell Ranges
        t = .Areas(1)(i): .Areas(1)(i) = .Areas(2)(i): .Areas(2)(i) = t
    Next i
End With ' SELECTION
End Sub ' SWAP
```

Figure 3.30   VBA sub procedure SWAP for exchanging the contents of two selected cell ranges.

To illustrate the performance of the sub procedure **SWAP**, create two different cell ranges on an Excel worksheet, as seen in the following left side image. Select the cells containing the numbers 1 to 4, hold down the **CTRL** key and select the cells containing the letters A to D. With the two cell ranges selected (highlighted) as seen in the left side, run the sub procedure **SWAP** from the ribbon **VIEW>Macros>View Macros**, as seen in the popup form. The result of the swap is shown on the right side with the letters replaced by the numbers, and vice versa.

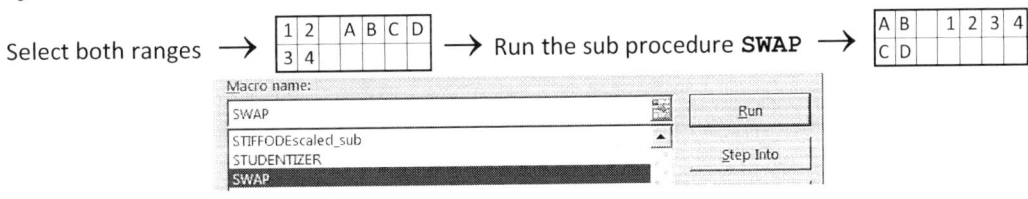

Select both ranges → [1 2 / 3 4] [A B C D] → Run the sub procedure **SWAP** → [A B / C D] [1 2 3 4]

Macro name: SWAP
STIFFODEscaled_sub
STUDENTIZER
SWAP

Run
Step Into

Comments: The macro only requires two cell ranges of the same length. However, the two cell ranges may have different orientations and be non-adjacent on the worksheet, or different worksheets.

3.8.3 InputBox

A sub procedure may collect information *from* the user with an **InputBox**. An input box has an appearance similar to a message box, but includes a text box for entering the requested information. The syntax is:

```
[Set] v = Application.InputBox(Prompt[, title][, default][, type])
```

where the square brackets [ ] indicates an optional argument. Instead of returning an integer value like MsgBox, an input box assigns the user input to the variable v. To set up an input box, we must provide the arguments in the order listed and put quotations around text strings. We may change the order of the arguments by defining an argument type with a colon-equal sign, *e.g.* :=. Table 3.7 is a summary of the options for variable **Type**. Use the optional keyword **Set** before the variable assignment with **Type:=8** to input a range variable. Otherwise, omit **Set** and use the statement **Type:= 1, 2, or 4** for numbers, strings, or Boolean data types (**True** or **False**), respectively. The statement **Type:= 64** checks for an array of values.

Table 3.7 Useful InputBox variable type declarations.

Type:=	Definition	Type:=	Definition
1	Numbers	8	Range (Requires Set)
2	String	64	Array of values
4	Logical (True or False)		

Input boxes provide a simple way to validate input data. By defining the type of input variable, *VBA* checks for proper input and gives an error message when the user provides the wrong type. With an input error, *Excel* redisplays the input box for the user to try again.

Use the optional argument **Default:=** to display a default value in the input box. To illustrate, the following statement prompts the user for a tolerance, with a default value of 0.00001. The user can click **OK** to accept the default value displayed, or replace the default value with a different number.

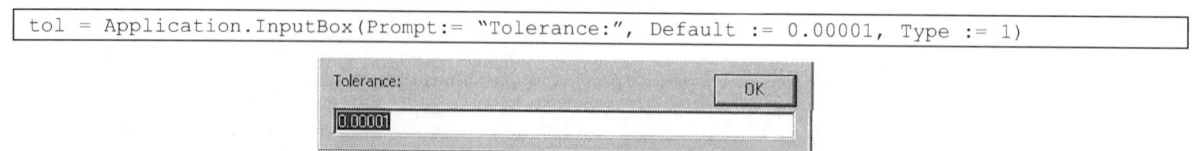

```
tol = Application.InputBox(Prompt:= "Tolerance:", Default := 0.00001, Type := 1)
```

Example 3.9 Use Input Boxes to Set Up a Friction Calculation

Known/Find: Use input boxes to set up **Goal Seek** in Example 3.4 and Example 3.6.

Analysis: To illustrate how input boxes work, we incorporate them in the pipe flow "button" macro for selecting the locations of the guess for the velocity and the result of Bernoulli's energy balance. Replace the ranges in the code for the **Goal Seek** line with the range variable names set by the input boxes:

```
Private Sub CommandButton1_Click()
    Dim set_cell As Range, changing_cell As Range
    Set changing_cell = Application.InputBox("Click on cell for velocity", Type:=8)
    Set set_cell = Application.InputBox("Click on cell for Bernoulli equation", Type:=8)
    set_cell.GoalSeek Goal:=0, ChangingCell:=changing_cell
End Sub
```

Click the **Run** button on the worksheet to launch the macro and display the input boxes. Click on the cell for velocity, and then click **OK**, as seen in Figure 3.31. In the next input box, provide the cell reference for Bernoulli's equation in **C11** to complete the input and calculate the velocity that satisfies the mechanical energy balance in **C11**.

CHAPTER 3: VBA

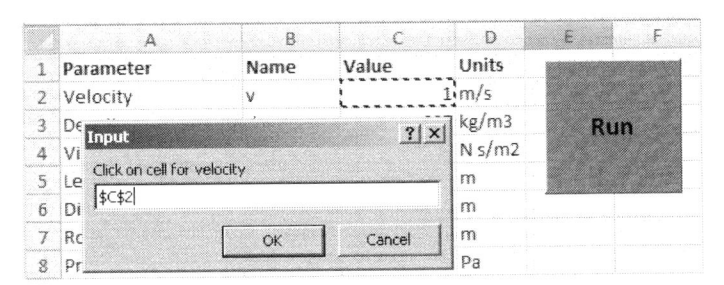

Figure 3.31

Input box prompt for the location of the cell containing the fluid velocity.

Comments: Using input boxes to interact with the macro may feel like a step backward because we must select cell ranges and click three times instead of just one click, as required by the **Goal Seek** dialog box. However, the cells for velocity and Bernoulli's equation are no longer hard coded into the *VBA* macro. We now have descriptive titles for the inputs in place of the generic titles in **Goal Seek**. The input boxes give us the freedom to move the cells around on the worksheet without the need to go back into the code and reassign cells addresses required by **Goal Seek**.

□

3.8.4 User-forms*

We experienced *Excel* forms when we provided input to the dialog boxes for **Goal Seek** and the *Excel* add-ins **Solver** and **Descriptive Statistics**. User-forms permit multiple inputs in a single dialog box and allow for additional controls beyond the limited capabilities of input boxes. *Excel VBA* has features for creating custom user-forms to provide information and instructions for our own *VBA* macros.

To create a **User-form**, open the *VBE* and select from the menu **Insert/UserForm**:

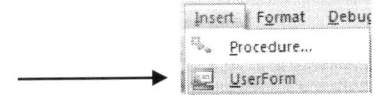

This opens a blank user-form with a grid for laying out the controls on the form with format properties and controls available from the **Toolbox**:

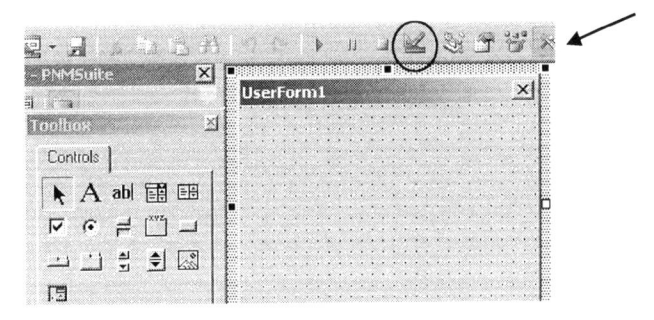

When we click on a form, it becomes active and the **Properties** window appears in the *VBE* for specifying the **Name**, **Caption**, **Font**, and other formatting options. If the properties window does not come to the front, open it from the **View** menu or right-click on the **User-form** and select **Properties**. Use the **Select Object** control to resize the **User-form** by clicking and dragging on the handles at the sides and corners of the frame. Usually, the **Toolbox** appears when we click in the **User-form**. Otherwise, we can open it from the **View** menu. Click on the toolbox icon (circled in the row of icons in the previous image) located below the menus for accessing controls:

For our purposes, the most important tools are **Label**, **Textbox**, **RefEdit**, **Toggle**, and **CommandButton**. Add controls to the **User-form** to collect input and control the execution of any affiliated *VBA* code. Table 3.8 summarizes the form control options.

Table 3.8 User-form Toolbox control options.

Toolbox Control	Name	Description
	SelectObject	Move or resize tools on the user-form with select object.
	Label and Image	Use labels to place descriptive comments about controls on the user-form. Change the label directly in the control on the form, or Caption in the Properties window. Use Image to put a picture on the form.
	TextBox and RefEdit	Input or output box for strings or numerical values. Use the Name in the Properties window to access the value in the text box. Use RefEdit when the input is a range on an *Excel* worksheet (Very important - do not use RefEdit inside another control, such as a Frame or Multipage). Set default Values in the properties.
	Combo or List Boxes	Combo boxes show the first element of a list for selection in a drop-down box. A list box shows all elements in a list (no drop-down box). The RowSource in the Properties window requires an *Excel* worksheet range for the elements of the list.
	CheckBox, Option, or Toggle Button	When checked, the value of the Boolean Name becomes True; otherwise, the value is False. The user may set the default value in the Properties for the control.
	Frame	Graphic for grouping controls on the user-form, such as mutually exclusive **CheckBox** and **Option** buttons (*VBA* takes care of the exclusivity of the checkboxes or buttons).
	CommandButton	When clicked, runs the assigned procedure.
	TabStrip and MultiPage	Tab strips and multi-page controls organize user input options when we need to organize the controls.
	ScrollBar and SpinButton	Cycle through selections in a list on a scroll bar or spin button.

In general, take the following steps to create a user-form:
1. Open the Visual Basic Editor (*VBE*).
2. Create a **User-form** from the *VBE* **Insert** menu.
3. Lay out the controls on the form as needed. To expedite form creation, use **Copy** and **Paste** from the **Edit** menu to duplicate controls with similar formats.
4. Provide descriptive names and captions in each control's **Properties** window. Right-click on the control to select **Properties** or **View Code** from the popup menu. Specify formatting in the properties window as desired to make the form readable and understandable to users. The value of a control is available for use by sub procedures through the name of the control.
5. Write *VBA* code in `Private` sub procedures to show and unload the user-form and manipulate the inputs selected by the user. Double-click on controls to access the programming environment of the user-form.

We list here the most common coding requirements for manipulating user-forms with *VBA* sub programs, using the generic name **UserForm1** in the examples. We certainly have the freedom to rename our user-forms with more descriptive titles that convey their intended use.

- Create a simple macro in a module to open a **UserForm** (the name and location of the sub procedure is not specific in this operation):

```
Sub OpenForm1()
    UserForm1.Show
End Sub
```

We can now call this macro from the tab **View>View Macros**, just like our other macros.

- Use the following *VBA* code template to open and display a **UserForm** automatically when opening the host workbook:

```
Private Sub Workbook_Open()
    UserForm1.Show
End Sub
```

The specific sub procedure **WorkBook_Open** is required for this operation. Place this sub procedure in the object named **ThisWorkbook** accessed from the **Project** window.

As a matter of convenience in this text, we refer to the user forms by the name of the calling sub procedure (with `_UsrFrm` appended to the end of the name).

- Close a **UserForm** with either of the following codes in a sub procedure, such as a button on a form or at the end of the code called by the form:

```
Unload UserForm1
```
or
```
UserForm1.Hide
```

The **Unload** statement closes the form and clears it from memory in the computer. The **Hide** statement closes the form, but leaves its settings in memory – the form retains the last settings and contents of controls as long as the workbook remains open.

- Initialize values of controls in a **UserForm** with the **UserForm_Initialize** sub procedure associated with the form's module. Right-click on the user-form and select **View Code** from the popup menu. Type the name of the procedure, or create the procedure by selecting the names from the drop down lists above the module:

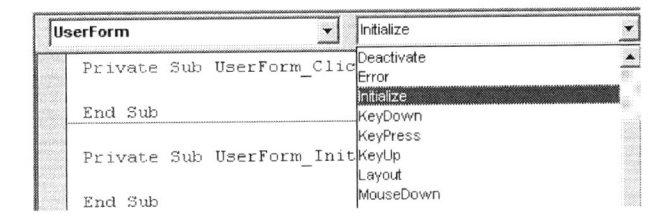

In the following illustration, we initialize the choices in `ComboBox1` with the four musicians in the "best band ever,"[44] and assign the current date to `TextBox1`.

```
Private Sub UserForm_Initialize()
    With UserForm1.ComboBox1
        .AddItem "Jimmy Page" : .AddItem "Robert Plant"
        .AddItem "John Paul Jones" : .AddItem "John Bonham"
    End With
    UserForm1.TextBox1 = Date
End Sub
```

Several of the macros in the *PNMSuite* workbook have corresponding versions with user-forms for input in place of input boxes. The names of the user-forms in the project end with `_frm` whereas the names of the show user-

---

[44] Jack Black's induction speach of *Led Zeppelin* at the 2012 Kennedy Center Honors Ceremony. (http://www.youtube.com/watch?v=Ta0gDfGb9u0) ← Be sure to watch the last tribute cover by *Heart*! It will blow your mind, and your head, and your brain, too!

form macros end with `_UsrFrm`. We do not make a distinction when referring to user-forms to implement the numerical methods. Unfortunately, the resolution of the user's computer screen may cutoff text in user forms created with higher resolution. For this reason, we recommend generous use of the "real estate" on the user form and controls to create user forms that show all of the text for screens with different resolutions. The next example builds a simple user-form for the pipe flow examples.

Example 3.10 User-form for Inputs to Bernoulli's Equation

Known/Find: Create a user-form to replace the input boxes in Example 3.9.

Analysis: Open the workbook created for Example 3.9, and access the *VBE*. This workbook contains *VBA* code for a user-defined function to calculate the Darcy friction factor in **Module1** as well as input boxes and a control button to launch **Goal Seek** in the code for **Sheet1** of the **Project** window. Modify these features of the workbook for the user-form in the following steps:

1. Create a blank user-form from the **Insert** menu in the *VBE*.

2. Add two **RefEdit** controls for the ranges of input required by **Goal Seek**. Add labels and two **CommandButton** controls from the **Toolbox**.

3. In their respective **Properties** windows, rename the user-form to **FlowVelocity**, then rename the **RefEdit** controls to **rV** and **rB**, respectively. Add captions in the properties for the user-form, labels, and button controls to create a user-form shown in Figure 3.32.

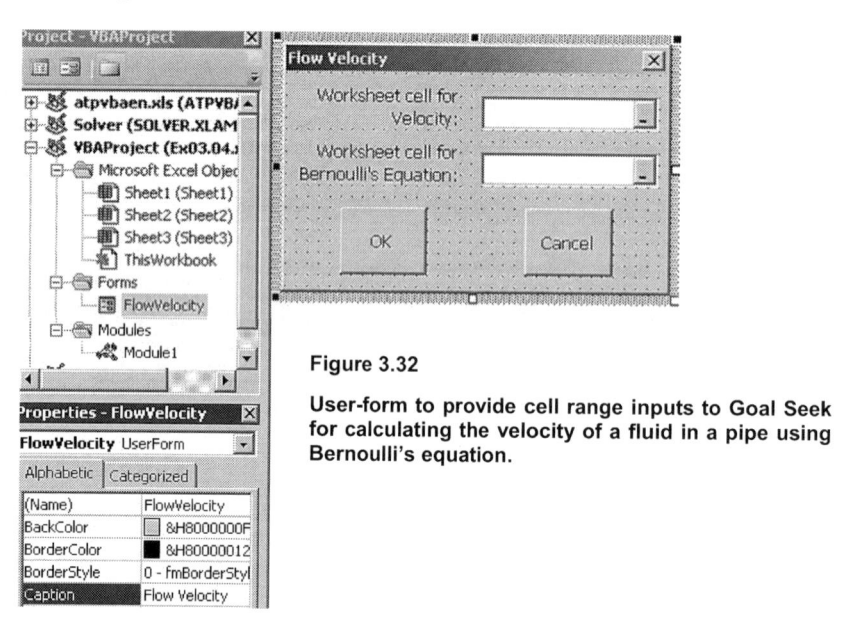

Figure 3.32
User-form to provide cell range inputs to Goal Seek for calculating the velocity of a fluid in a pipe using Bernoulli's equation.

4. Set up the **Tab** key to move among the control objects on the live form. Select the menu **View/Tab Order** or right-click on the form and select **Tab Order**. Arrange the sequence of active controls with the **Move Up** and **Move Down** buttons. When the user-form opens, the input for velocity becomes active on the form. The **Tab** key cycles through the list of controls in the tab order.

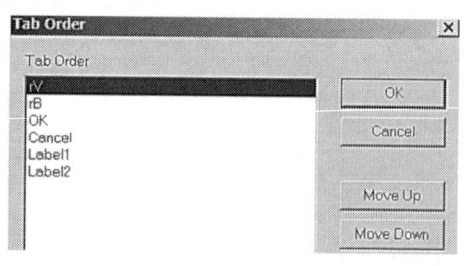

CHAPTER 3: VBA

5. Double-click on the **Cancel** button on the form and add code to unload the user-form. Double-click on the **OK** button on the form and add code for **Goal Seek**, followed by a statement to unload the user-form. Modify the line for accessing **Goal Seek** from Example 3.9 to use the ranges from the user-form **RefEdit** controls. Notice that the range variables use the `Set` statement in their assignments.

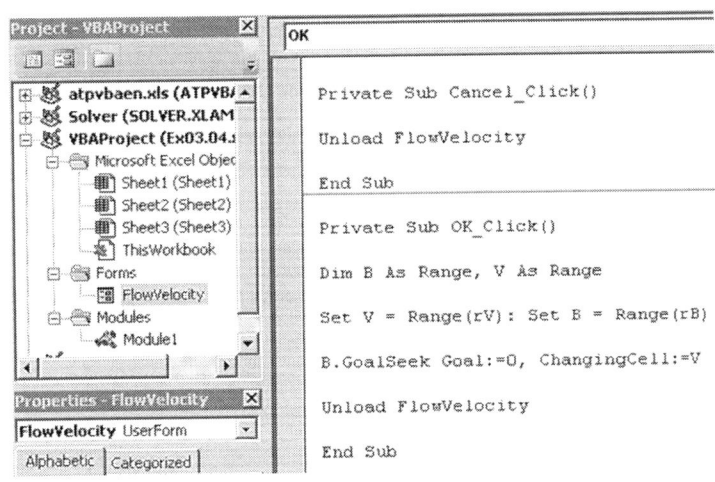

```
Private Sub Cancel_Click()

Unload FlowVelocity

End Sub

Private Sub OK_Click()

Dim B As Range, V As Range

Set V = Range(rV) : Set B = Range(rB)

B.GoalSeek Goal:=0, ChangingCell:=V

Unload FlowVelocity

End Sub
```

6. Finally, open the code for **Sheet1** and replace the lines in the sub procedure for the **Run** button to make the user-form visible:

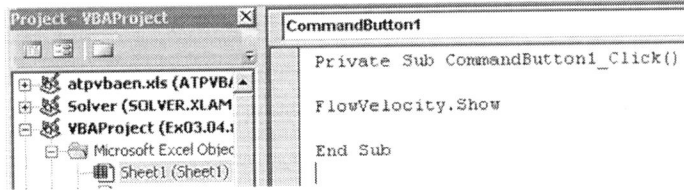

```
Private Sub CommandButton1_Click()

FlowVelocity.Show

End Sub
```

Test the user-form. The velocity may be already determined from the previous example. Change the initial value in the cell for velocity to a different value, such as one, and then click the **Run** button on the worksheet. In the user-form that pops up on the screen, select the input boxes and click on the corresponding cells for the velocity and Bernoulli's equation. Then click **OK** to calculate the velocity of 0.496 m/s in the worksheet using **Goal Seek**.

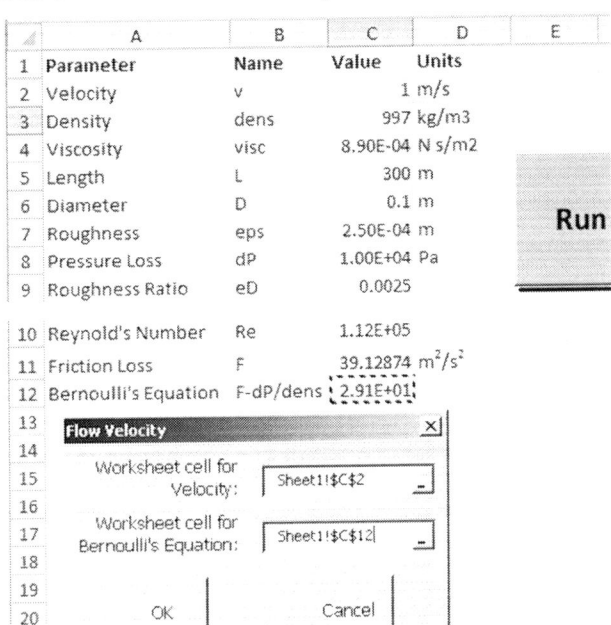

	A	B	C	D
1	Parameter	Name	Value	Units
2	Velocity	v	1	m/s
3	Density	dens	997	kg/m3
4	Viscosity	visc	8.90E-04	N s/m2
5	Length	L	300	m
6	Diameter	D	0.1	m
7	Roughness	eps	2.50E-04	m
8	Pressure Loss	dP	1.00E+04	Pa
9	Roughness Ratio	eD	0.0025	
10	Reynold's Number	Re	1.12E+05	
11	Friction Loss	F	39.12874	m^2/s^2
12	Bernoulli's Equation	F-dP/dens	2.91E+01	

Flow Velocity

Worksheet cell for Velocity: Sheet1!$C$2

Worksheet cell for Bernoulli's Equation: Sheet1!$C$12

OK   Cancel

Comments: User-forms are tools for organizing required input to an *Excel* worksheet or macro and provide a powerful way to interact with the user. In this case, we have come full circle by essentially duplicating the **Goal Seek** dialog box with descriptive language specific to the problem setup on our worksheet. However, user-forms require considerably more effort than input and message boxes, which are convenient when we do not need this level of sophistication.

□

3.8.5 Charting and Graphing with VBA

Several numerical methods presented in this book generate series of numerical results for graphical display. As engineers and scientists, we also encounter tabulated experimental data that require graphical interpretation. For convenience, we use *VBA* to organize numerical and experimental data into charts and graphs in *Excel* workbooks. We focus on y versus x scatter plots in most of our examples as the more common chart type, with an understanding that *VBA* has statements for generating any of the charts available in *Excel*. The conventional practice for graphing uses the symbols y and x to represent the ordinate (vertical) and abscissa (horizontal) axes, respectively.

The easiest and fastest way to generate *VBA* code for creating an *Excel* chart uses the *VBA* macro recorder to translate the steps taken to create a chart in an *Excel* worksheet into *VBA* code. The recorded macro is stored in a new module where we can find the *VBA* code statements used to create and format a new chart object. If we are satisfied with the macro, we can give it a memorable name and reuse it as is when needed. Otherwise, we can modify the macro to create general-purpose procedures for plotting numerical results. The following steps outline typical charting steps for scatter plots in *VBA*:

1. Add a chart object to the active worksheet.
2. Modify the chart object for the desired type. For example, the following statement modifies a chart object as a scatter plot for displaying a data series as a smooth curve without markers: `.ChartType = xlXYScatterSmoothNoMarkers`.
3. Format the display of the chart as desired, including the title, border, axis labels, tic marks, and grid lines.
4. Create x and y series for the data displayed in the chart. The data may come from *VBA* calculations or from data ranges on the worksheet. Note that *VBA* refers to the y-axis as `xlvalue` and the x-axis as `xlcategory`.

The next two examples show how to use *VBA* to create scatter plots for a single series or multiple y-series. In the first example, we create a general-purpose procedure for plotting numerical values generated from a worksheet formula. The second example uses input boxes to select ranges of experimental data tabulated in a worksheet used to create a phase diagram for a ternary system of solvents used for extraction. The object property statements used to format the charts are relatively self-explanatory in terms used by *Excel* to create and format charts. A macro **GRAPHYXGUIDE** (pronounced "Graphics Guide") is available from the *PNMSuite* for creating y versus x scatter plots from ranges of experimental data tabulated in an *Excel* worksheet. **GRAPHYXGUIDE** formats the chart according to the guidelines in Section 2.3.4

Example 3.11 Generate Quick Y versus X Scatter Plots in Excel Worksheets

Known/Find: We occasionally need to generate a simple scatter plot of a function to observe its behavior with respect to the independent variable. Create a *VBA* macro in a user-form to generate a y vs x scatter plot of a cell formula in an *Excel* worksheet. Test the macro by graphing the following damped function in an *Excel* worksheet.

$$y = \ln(x)\cos(15x)$$

Analysis: Our show user-form macro **QYXPLOT_UsrFrm** (pronounced "Quick Plot") uses the formula evaluation method from the template of Figure 3.23 to create arrays of points for a scatter plot of an *Excel* worksheet formula. The user-form prompts for the cell addresses on the worksheet containing the formula and independent variable, as well as the minimum and maximum scale of the x-axis. The formula is evaluated 1000 times in a **For** loop to generate arrays for the x and y values in the plot. The macro adds a chart on the worksheet using conventional formatting for

CHAPTER 3: VBA

a *y* vs *x* scatter plot with smooth lines. The user may modify the format or delete a **QYXPLOT** chart just like any other *Excel* chart on the worksheet.

To create a user-form named **QYXPLOT_frm**, follow the steps outlined in Section 3.8.4. Open the *Visual Basic Editor*, then:

1. From the Insert menu, select **UserForm**.
2. In the **Properties** window for the user-form, change the name to **QYXPLOT_frm**.
3. Arrange the form with **Labels**, **RefEdit** for the formula and variable, and **Textbox** for the axis scales, as shown in Figure 3.33 (Note: never user **RefEdit** control inside another control, such as a **Frame** or **MultiPage**).

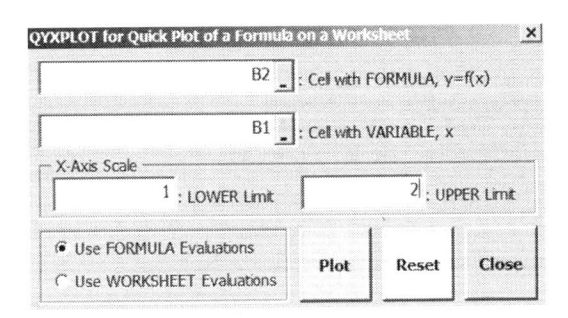

Figure 3.33
User-form for QYXPLOT_frm.

4. In each control **Properties** window, change the font to **Tahoma** with size 9, and name the controls as follows:

Control	Input Label	Properties Name
RefEdit	Cell with formula for dependent variable	f_rfed
RefEdit	Cell with independent variable	x_rfed
TextBox	x-axis lower scale limit	xlow_txtbx
TextBox	x-axis upper scale limit	xup_txtbx
OptionButton	Use FORMULA Evaluations	f_optbtn
OptionButton	User WORKSHEET Evaluations	ws_optbtn
CommandButton	Plot	RUN_cmdbtn
CommandButton	Reset	RESET_cmdbtn
CommandButton	Close	CANCEL_cmdbtn

5. Add a frame to contain two option buttons with mutual exclusivity. Use the option buttons to select among formula evaluations or worksheet evaluations of the function.
6. Add three **CommandButtons**. Provide names in the properties windows: RESET_QYXPLOT, RUN_QYXPLOT and CANCEL_QYXPLOT. Add **Captions** "Reset", "Plot" and "Close", then change the font size to nine in the corresponding properties windows.
7. Right-click on the Plot command button and select **View Code** to open the code window. Add the *VBA* code in Figure 3.34 to the **Sub RUN_cmdbtn_Click** for the Plot button. The private sub **QYXPLOT** uses the control **RefEdit** values for the formula and variable to generate a series of points in an array for plotting. The chart displays the text string of the formula in the title above the graph. The user-form is unloaded when errors occur. The graph's axis scale parameters are determined from the lower and upper limits of the data, using the recommendations of Peltier[45] in the sub **AXISCALE**, available in the *PNMSuite*.

```
Private Sub RUN_cmdbtn_Click()   ' call the macro to create the plot
    Call QYXPLT
End Sub

Private Sub QYXPLT()
' Create a quick yx smooth line plot of a formula in a worksheet cell.  The macro uses
' the cell locations from the user form of the y=f(x) function and independent
' variable x.  The user specifies the minimum and maximum values on the x-axis scale.
```

---

[45] https://peltiertech.com/calculate-nice-axis-scales-in-excel-vba/

```vb
' The macro generates a generic plot with gridlines. To change the appearance of the
' plot, change the settings in the VBA code.  The default number of points = 1000.
' *** Controls ***
' f_optbtn        (OptionButton) = use formula evaluations in VBA
' f_rfed          (RefEdit) = cell with formula for plotting
' ws_optbtn       (OptionButton) = evaluate the formula in the worksheet
' xlow_txtbx      (TextBox) = lower limit on x scale
' x_rfed          (RefEdit) = cell with independent variable in formula
' xup_txtbx       (TextBox) = upper limit on x scale
' *** Required sub procedure ***
' AXISCALE = get axis scaling parameters for the graph
' *** Nomenclature ***
Dim bxmax As Double, bxmin As Double ' = boundaries on x-axis scale
Dim bymax As Double, bymin As Double ' = boundaries on y-axis scale
Dim dx As Double ' = x scale division
Dim i As Long, j As Long ' = index counters
Dim n As Long ' = number of function evaluations (data pairs) for generating the plot
Dim noc As Long ' = number of occurrences of xr in yr (default = 1000)
Dim ns As Long ' = number of occurrences of x cell name in yr formula
Dim s As Series ' = existing series in chart for removal
Const small As Double = 0.000000000001 ' = small number avoids division by zero
Dim t As String ' = temporary string for formula with substitutions for x address
Dim uxmaj As Double, uxmin As Double ' = major and minor units on x-axis scale
Dim uymaj As Double, uymin As Double ' = major and minor units on y-axis scale
Dim x As Double ' = independent variable
Dim xa As String ' = string for address of x variables on worksheet
Dim xdata() As Double, ydata() As Double ' = arrays of x and y data
Dim xmax As Double, xmin As Double ' = scale limits for independent variable x on plot
Dim xr As Range ' = range variable for independent variable
Dim xs As String ' string for x address
Dim xu As String ' = string placeholder for x substitution/evaluation
Dim xychart As Chart ' = chart object for chart (plot)
Dim xyseries As Series ' = series object for x y data
Dim ya As String ' = string for address of y = f(x) formula on worksheet
Dim yf As String ' = string for y function
Dim ymax As Double, ymin As Double ' = scale limits for dependent variable y axis
Dim yr As Range ' = range variable for dependent variable function of x
'************************************************************************
' Check for complete form
If f_rfed = vbNullString Or x_rfed = vbNullString Or xlow_txtbx = vbNullString _
   Or xup_txtbx = vbNullString Then
    MsgBox "Incomplete setup. Check form for missing information.":Exit Sub
End If
n = 1000 ' Number of divisions on x axis
Set yr = Range(f_rfed) ' Worksheet Cell with FUNCTION formula, y = f(x)
Set xr = Range(x_rfed) ' Worksheet Cell with INDEPENDENT variable, x
xmin = xlow_txtbx: xmax = xup_txtbx ' LOWER UPPER x-axis scale
If Not xmax > xmin Then
    MsgBox "Incorrect X-axis limits. Upper limit must be greater than lower limit."
    Exit Sub
End If
ReDim xdata(1 To n + 1) As Double, ydata(1 To n + 1) As Double
On Error Resume Next ' Check for errors in cell address naming
xa = xr.Address: xs = "": xs = Range(xa).Name.Name ' Get strings for x address/named
xu = "X_": ya = VBA.UCase(yr.Formula) ' make upper case to eliminate case sensitivity
Do While InStr(ya, xu) ' check for dummy variable string in formula
    xu = xu & "_" ' add an underscore to end of string
Loop
With Application ' length of characters in formula
ya = .ConvertFormula(yr.Formula, xlA1, xlA1, xlAbsolute): yf = ya
noc = (Len(ya) - Len(.Substitute(ya, xa, ""))) / Len(xa)
For j = noc To 1 Step -1 ' Create formula string for substitution for x
    t = .Substitute(ya, xa, xmin & " ", j)
    If Not IsError(Evaluate(t)) Then
        ya = t: yf = .Substitute(yf, xa, xu, j)
    End If
Next j
ns = Len(xs) ' repeat for x variable in a named cell
If ns > 0 Then
```

```
        noc = (Len(ya) - Len(.Substitute(ya, xs, ""))) / ns
        For j = noc To 1 Step -1 ' Create formula string for substitution for x
            t = .Substitute(ya, xs, xmin & " ", j)
            If Not IsError(Evaluate(t)) Then
                ya = t: yf = .Substitute(yf, xs, xu, j)
            End If
        Next j
    End If
    dx = (xmax - xmin) / n: x = xmin: xdata(1) = x
    If f_optbtn = True Then ' use formula
        ydata(1) = Evaluate(.Substitute(yf, xu, x))
        For i = 2 To n + 1 ' Evaluate data
            x = x + dx: xdata(i) = x: ydata(i) = Evaluate(.Substitute(yf, xu, x))
        Next i
    Else ' evaluate the formula in the worksheet
        xr = x: ydata(1) = yr
        For i = 2 To n + 1 ' Evaluate data
            x = x + dx: xr = x: xdata(i) = x: ydata(i) = yr
        Next i
    End If
    Call AXISCALE(bxmax, bxmin, xmax, xmin, uxmaj, uxmin) ' get x-axis scale parameters
    ymin = .Min(ydata): ymax = .Max(ydata) ' Get the y axis scale
    Call AXISCALE(bymax, bymin, ymax, ymin, uymaj, uymin) ' get x-axis scale parameters
End With ' Application
' Create a smooth line xy plot on the active worksheet using the data arrays
Set xychart = ActiveSheet.Shapes.AddChart.Chart
With xychart
    For Each s In .SeriesCollection: s.Delete: Next s ' clear any series in chart
    .ChartType = xlXYScatterSmoothNoMarkers: .Legend.Delete
    Set xyseries = .SeriesCollection.NewSeries
    xyseries.Values = ydata: xyseries.XValues = xdata
    xyseries.Format.Line.Weight = 0.5
    With .Axes(xlCategory) ' rescale the x axis & add gridlines (optional)
        .MinimumScale = bxmin: .MaximumScale = bxmax
        .MajorUnit = uxmaj: .MinorUnit = uxmin
        .HasMajorGridlines = False: .HasMinorGridlines = False
        .MinorTickMark = xlInside ' put minor tic marks on inside
    End With
    With .Axes(xlValue) ' rescale y axis and add grid lines (optional)
        .MinimumScale = bymin: .MaximumScale = bymax
        .MajorUnit = uymaj: .MinorUnit = uymin
        .HasMajorGridlines = False: .HasMinorGridlines = False
        .MinorTickMark = xlInside ' put minor tic marks on inside
    End With
    .PlotArea.Select ' add box around plot area
    With SELECTION.Format.Line
        .Visible = msoTrue: .ForeColor.ObjectThemeColor = msoThemeColorText1
    End With
    With .Axes(xlCategory)
        .HasTitle = True: .AxisTitle.Caption = "x"
    End With
    With .Axes(xlValue)
        .HasTitle = True: .AxisTitle.Caption = "y"
    End With
    With .ChartArea ' remove box from chart area and resize
        .Format.Line.Visible = msoFalse: .Width = 225: .Height = 200
    End With
    .SetElement (msoElementChartTitleAboveChart) ' Use formula for chart title
    With .ChartTitle
        .Text = "y" & Application.Substitute(yf, xu, "x")
        .Format.TextFrame2.TextRange.Font.Size = 10: .Bold = msoFalse
    End With
End With
bummer: xr = xmin: Application.ScreenUpdating = True: beep
MsgBox prompt:=Err.Description & ": Check setup.": Unload QYXPLOT_frm
End Sub ' QYXPLT
```

Figure 3.34 VBA sub procedures QYXPLOT_frm and QYXPLT for generating a scatter plot of a formula.

8. Right-click on the Cancel command button, select **View Code** to open the code window, and fill in the line to close the user-form. Repeat for the Reset command button:

```
Private Sub CANCEL_cmdbtn_Click()  ' unload the user-form for QYXPLOT
    Unload QYXPLOT_frm
End Sub
Private Sub RESET_cmdbtn_Click()  ' reset the form
    Unload QYXPLOT_frm: QYXPLOT_frm.Show
End Sub
```

9. Create a sub procedure in a module that launches the user-form from the **View Macros** ribbon:

```
Public Sub QYXPLOT_UsrFrm()  ' Launch the user-form QYXPLOT_FRM
    QYXPLOT_frm.Show
End Sub
```

Set up an *Excel* worksheet with the formula for the damped function in terms of the independent variable in Cell **B2**, starting with a value of one for *x* in Cell **B1**. Run the macro **QYXPLOT_UsrFrm** from the ribbon **View>View Macros>QXYPLOT_UsrFrm>Run** and provide the cell references for *y* and *x* as requested. Change the scale for *x* to range from 1 to 2. Observe from the plot that the trial function oscillates about the abscissa with increasing amplitude.

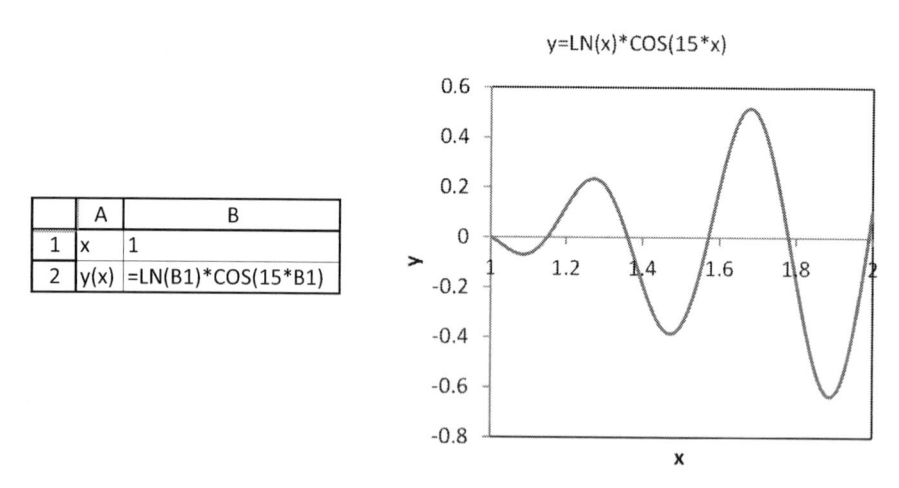

	A	B
1	x	1
2	y(x)	=LN(B1)*COS(15*B1)

Comments: A macro **QYXPLOT** with input boxes is also available in the *PNMSuite* for creating quick plots of worksheet formulas. The *VBA* user-form **QYXPLOT** is convenient for quickly generating a plot of an *Excel* worksheet formula without creating columns of data for graphing in an *Excel* worksheet.

□

Example 3.12   Create a Right-triangle Plot of Liquid-liquid Extraction Data

Known: A 50 kg feed mixture of a 35 wt% chloroform (C) and 65 wt% acetic acid (A) is extracted with 25 kg of water solvent (*W*). The following table contains the weight fractions for the equilibrium tie lines of the ternary system on the phase miscibility curve (Wankat 2017).

Find: Create a right-triangle phase diagram from the data to determine the compositions and weights of raffinate and extract phases in a single stage extraction.

CHAPTER 3: VBA

Raffinate Phase Weight Fractions			Extract Phase Weight Fractions		
C	W	A	C	W	A
0.9901	0.0099	0	0.0084	0.9916	0
0.9185	0.0138	0.0677	0.0121	0.7369	0.251
0.8	0.0228	0.1772	0.073	0.4858	0.4412
0.7013	0.0412	0.2575	0.1511	0.3471	0.5018
0.6715	0.052	0.2765	0.1833	0.3111	0.5056
0.5999	0.0793	0.3208	0.252	0.2539	0.4941
0.5581	0.0958	0.3461	0.2885	0.2328	0.4787

Analysis: A right-triangle graph used for analysis of liquid-liquid extraction consists of a collection of tie lines plotted between phase miscibility curves. The tie lines connect the points in the extract and raffinate in each row in the data table. We can display the equilibrium data for only two of the components in a right triangle. We get the weight fraction of the third component by difference. We use *VBA* to create a simple sub procedure to automate the process of plotting the miscibility curves for the extract and raffinate phases and the phase equilibrium tie lines. The *VBA* sub procedure **LLETRIANGLE** listed in Figure 3.35 prompts the user for the axis labels and the ranges of data for plotting. The *VBA* code creates a chart object in the worksheet then modifies the chart to a scatter plot with a series of curves for the tie lines.

```
Public Sub LLETRIANGLE()
' Creates a right-triangle phase diagram for ternary data systems in liquid-liquid
' extraction.  The macro prompts for the axis labels and the ranges of equilibrium
' data for two of the three components in the extract and raffinate phases.
' *** Nomenclature ***
Dim i As Long, ii As Long ' = index for component
Dim n As Long ' = number of LLE data pairs
Dim p As Double ' percent multiplier (1 or 100)
Dim s As Double ' size of graph
Dim tx(1 To 2) As Double, ty(1 To 2) As Double ' tie-line points
Dim xlabel As String, ylabel As String ' axis labels
Dim x1r As Range, x2r As Range ' raffinate data ranges
Dim y1r As Range, y2r As Range ' extract data ranges
' ***********************************************************************************
With Application
    xlabel = .InputBox(prompt:="Label Component on x-Axis:", Default:="Comp. A/wt.%")
    ylabel = .InputBox(prompt:="Label Component on y-Axis:", Default:="Comp. B/wt.%")
    Set x1r = .InputBox(prompt:="Data Range of " & xlabel & " wt. frac. in the " & _
                        "Raffinate:", Type:=8)
    Set y1r = .InputBox(prompt:="Data Range of " & xlabel & " wt. frac. in the " & _
                        "Extract:", Type:=8)
    Set x2r = .InputBox(prompt:="Data Range of " & ylabel & " wt. frac. in the " & _
                        "Raffinate:", Type:=8)
    Set y2r = .InputBox(prompt:="Data Range of " & ylabel & " wt. frac. in the " & _
                        "Extract:", Type:=8)
End With
n = x1r.Count ' number of data pairs
With WorksheetFunction
    If .Max(x1r) > 1 Or .Max(x2r) > 1 Or .Max(y1r) > 1 Or .Max(y2r) > 1 Then
        p = 100 ' check for mass fraction or mass%
        ' xlabel = xlabel & " wt%": ylabel = ylabel & " wt%"
    Else
        p = 1
        ' xlabel = xlabel & " wt frac.": ylabel = ylabel & " wt frac."
    End If
    s = WorksheetFunction.Min(500, .Max(200, n * 50)) ' scale the size of the plot
End With ' WorksheetFunction
ActiveSheet.Shapes.AddChart2(, xlXYScatterSmoothNoMarkers).Select ' add a scatter plot
With ActiveChart
    .ChartTitle.Delete: .Parent.Height = s: .Parent.Width = s ' resize
    .Line.Visible = msoFalse ' remove border
```

```
        With .Axes(xlCategory) ' format the x-axis
            .MajorTickMark = xlOutside: .MinorTickMark = xlInside
            .MinimumScale = 0: .MaximumScale = 1 * p
            .MajorUnit = 0.1 * p: .MinorUnit = 0.02 * p
            .HasTitle = True: .AxisTitle.Characters.Text = xlabel
        End With ' Axes
        With .Axes(xlValue) ' format the y-axis
            .MajorTickMark = xlOutside: .MinorTickMark = xlInside
            .MinimumScale = 0: .MaximumScale = 1 * p
            .MajorUnit = 0.1 * p: .MinorUnit = 0.02 * p
            .HasTitle = True: .AxisTitle.Characters.Text = ylabel
        End With ' Axes
        .SeriesCollection.NewSeries ' plot the raffinate immiscibility curve
        With .FullSeriesCollection(1)
            .Name = "Raffinate": .XValues = x1r: .Values = x2r: .Format.Line.Weight = 0.5
            .Format.Line.ForeColor.ObjectThemeColor = msoThemeColorText1
        End With ' FullSeriesCollection
        .SeriesCollection.NewSeries ' plot the extract immiscibility curve
        With .FullSeriesCollection(2)
            .Name = "Extract": .XValues = y1r: .Values = y2r: .Format.Line.Weight = 0.5
            .Format.Line.ForeColor.ObjectThemeColor = msoThemeColorText1
        End With 'FullSeriesCollection
        For i = 1 To n ' plot the tie lines
            ii = i + 2
            tx(1) = x1r(i): ty(1) = x2r(i): tx(2) = y1r(i): ty(2) = y2r(i)
            .SeriesCollection.NewSeries
            With .FullSeriesCollection(ii)
                .Name = "Tie Line " & i: .XValues = tx: .Values = ty
                With .Format.Line
                    .Weight = 0.5: .ForeColor.ObjectThemeColor = msoThemeColorText1
                    .DashStyle = msoLineSysDash
                End With ' Format.Line
            End With ' FullSeriesCollection
        Next i
        tx(1) = p: ty(1) = 0 ' plot the diagonal line
        tx(2) = 0: ty(2) = p
        .SeriesCollection.NewSeries
        With .FullSeriesCollection(n + 3)
            .Name = "Diagonal": .XValues = tx: .Values = ty
            With .Format.Line
                .Weight = 0.1
                .ForeColor.ObjectThemeColor = msoThemeColorBackground1
                .ForeColor.Brightness = -0.2
            End With ' Format.Line
        End With ' FullSeriesCollection
End With ' ActiveChart
End Sub ' LLETRIANGLE
```

Figure 3.35 *VBA* sub procedure **LLETRIANGLE** for generating a right-triangle scatter plot of liquid-liquid extraction data with tie lines for conducting graphical mass balance analysis.

Create an *Excel* worksheet with the tabulated data in columns. Run the macro **LLETRIANGLE**. To generate the right-triangle ternary phase diagram in Figure 3.36, use the symbols **W** for the *x*-axis label and **A** for the *y*-axis label. Select the ranges of data for the raffinate and extract **W** and **A** in the input boxes. In the first step of the analysis, the mixing point is located from the overall composition of the ternary system. The overall weight fractions of acetic acid (**A**) and water (**W**) are calculated from the feed and solvent flow rates: Total = 50 + 25 = 75 kg. **W** = 25/75 = 33%. **A** = (0.65 × 50)/75 = 43%. The weight fractions give the coordinates of the mixing point on the triangle plot. The equilibrium compositions of the extract and raffinate phases are read from the ends of the tie line passing through the mixing point: Extract (water layer) = 40% **W**, 48% **A**, 12% **C**. Raffinate (chloroform layer) = 2% **W**, 22% **A**, 76% **C**.

CHAPTER 3: VBA

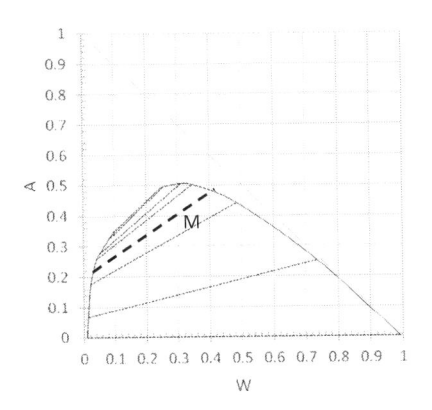

Figure 3.36

Right-triangle plot of acetone-water in a ternary system with chloroform generated by the *VBA* macro LLETRIANGLE.

The mixing point (M) is located at the coordinates of the overall composition. The compositions of the extract and raffinate phases are read from the ends of the interpolated tie line - - - passing through the mixing point.

We use the inverse lever rule to get the amounts of extract and raffinate from the relative position of the mixing point on the tie line: Raffinate ~ 20 kg, Extract ~ 80 kg.

Comments: The *VBA* macro **LLETRIANGLE** is a convenient tool for creating right-triangle phase diagrams useful for analysis of liquid-liquid extraction in ternary systems. Interpolation methods in Chapter 10 are recommended for calculating the equilibrium compositions instead of reading from a graph.

□

3.8.6 Read/Write Data To/From Files

Engineers use *VBA* programs with *Excel* to collect data from sensors, access database files, such as the industrial plant historian, as well as format and generate summary reports of the contents of an *Excel* workbook for archival records or distribution.

Before reading from, or writing to a file, we must first create or access the file using an **Open** statement with the following syntax and parameter definition:[46]

```
Open pathname For mode As #
```

pathname	String expression that specifies the file name, and may include the directory path to the location of the file on a drive. Without the path, the file shows up in the same directory as the *Excel* workbook.
mode	**Input, Output,** or **Append** (used to add output to an existing file)
#	File number, requires an integer in the range 1 to 511

Use the following *VBA* statements to write and read sequential data in files.

`Print #,`	Write values to file **#** without commas, separated by tabs.
`Line Input #,`	Access data from file **#** created by a **Print** statement, separated by tabs.
`Write #,`	Write data in quotations in file **#**, separated by commas.
`Input #,`	Access data separated by commas from file **#**, such as data created with **Write**.
`Close #,`	Close data file **#**.

The simple macro in Figure 3.37 first opens a file named **T_Data** in the folder containing the current workbook, creates a heading, and writes the values from the worksheet range **A2:A6** to the file, preceded by a label string. If the file does not yet exist, the Open statement first creates the file in the folder that contains the *Excel* workbook. Figure 3.37 also shows the *Excel* worksheet and data file results using the Windows system application *Notepad*. Use

---

[46] *VBA* **Help** gives the following warning: do not use the Open statement "to open an application's own file types. For example, don't use Open to open ... a Microsoft *Excel* spreadsheet ... Doing so will cause loss of file integrity and file corruption."

of the back-slash \ in the path, quotation marks " ", ampersand, &, and **Tab** in Figure 3.37, to control the format of the output, and distinguish between text strings and numerical values. Use a semi-colon to continue with the next statement on the same output line, or use a comma to separate values on the same line. Inputting data from a file follows the same general syntax. We may import and export a range of data using a range variable or worksheet address.

```
Public Sub Temp()
' Print a temperature report
    Dim i As Integer
    Open ThisWorkbook.Path & "\T_Data" _
        For Output As #1
    Print #1, "Thermocouple"; Tab; " T/°C"
    Print #1, "------------"; Tab; " ----"
    For i = 2 To 6
        Print #1, i - 1; Tab; Cells(i, 1)
    Next i
    Close #1
End Sub
```

Figure 3.37 Simple *VBA* macro to write data from an *Excel* worksheet to a data file named T_Data.

3.8.7 Status Bar

Excel uses the **Status Bar** located at the bottom of a workbook to display information about the current state of calculations in the active worksheet. We use the **Status Bar** extensively to display information about the progress of our *VBA* macros. For example, we use the **Status Bar** to show the current state of convergence for iterative calculations or percent completion of a sequence of operations.

```
Application.StatusBar = "Residual = " & res
```

For $res = 10^{-5}$, the previous line shows the value in the **Status Bar** below the worksheet sheet tab, as follows:

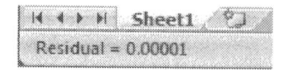

3.8.8 Events to Trigger Macros*

> *Winners make a habit of manufacturing their own positive expectations in advance of the event."* – Brian Tracy

User-defined functions behave just like other worksheet functions – they recalculate whenever a change occurs on the workbook (this requires **Workbook Calculation** set to **Automatic**). On the other hand, sub procedures only execute when called. However, we can place code into workbook objects or worksheet objects in the *VBE* to run sub procedures automatically when triggered by specific events, such as a change to the value of a cell. For example, we may add the following code in the object `ThisWorkbook` to run a macro each time the workbook opens:

```
Private Sub Workbook_Open()
    Call name() ' name is generic for any sub procedure specified by the user
End Sub
```

Customize the following simple macro as needed. Place the result in the code window of an *Excel* worksheet *object* to launch the sub procedure `name1()` or `name2()` whenever the value in the specified cell $A$1 or the range $A$2:$B$3 changes on the worksheet (*Note*: the cell addresses in the code must be absolute and capitalized, *i.e.* "$A$1", not "a1"):

```
Private Sub Worksheet_Change(ByVal Target as Range)
    Select Case Target.Address
        Case "$A$1"
```

CHAPTER 3: VBA

```
                Call name1()
        Case "$A$2", "$A$3", "$B$2", "$B$3", Range("na_me").Address
            Call name2()
    End Select
End Sub
```

To use named ranges after Case, replace the range address "$A$1" with Range("na_me").Address, where the string na_me can be any range name. Other Case selection criteria are allowed, including Is >, Is <, as well as If Then logic. To add an event trigger, open the object from the project window in *VBE* and select the object type and property from the drop-down lists at the top of the code window, as seen in Figure 3.38 of the following example.

Example 3.13 User-form for Inputs to Bernoulli's Equation

Known/Find: Add an event trigger in the worksheet object from Example 3.4 to run the **Goal Seek** macro to recalculate automatically the fluid velocity whenever the value for the pipe length or pipe diameter changes in the worksheet.

Analysis: Open the *Excel* workbook created for Example 3.4, and access the *VBE*. This workbook contains *VBA* code for a user-defined function to calculate the Darcy friction factor in **Module1** as well as code to run **Goal Seek** in **Sheet1** of the **Project** window. Open the *VBE* and take the following steps shown in Figure 3.38 to add an event trigger:

1. Double click on the Microsoft *Excel* Object **Sheet1** (highlighted in the project window).

2. Select **Worksheet** from the **(General)** dropdown list above the code window. Delete the new private **Sub** procedure named Worksheet_SelectionChange that was automatically added in the code window.

3. From the **(Declarations)** drop down list to the right, select **Change**. Another **Sub** procedure is added to the code window named Worksheet_Change.

4. Modify the new **Sub** procedure to select cases for the cell addresses with the pipe dimensions of length and diameter, as shown in Figure 3.38.

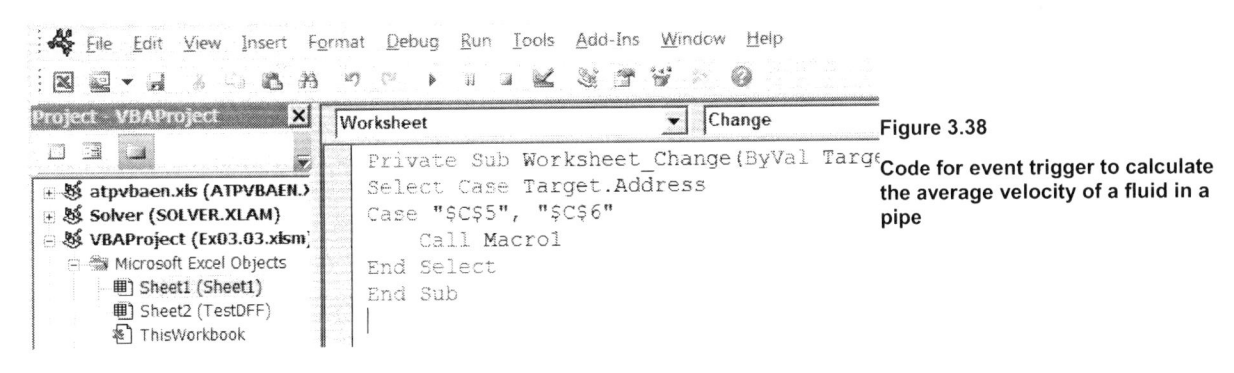

Figure 3.38
Code for event trigger to calculate the average velocity of a fluid in a pipe

Test the procedure by changing the values for **L** and **D** on the worksheet. Changing either of these values triggers the launch of Macro1, which employs **Goal Seek** to recalculate the corresponding fluid velocity. For example, using the original parameters, when we change the pipe length to 30 m, the worksheet automatically recalculates the velocity to 1.6 m/s.

Comments: Event triggers should be well-documented and advertised when sharing workbooks with colleagues to avoid any confusion when changing values in spreadsheet simulations.

□

3.8.9 Optional Arguments in User-defined Functions*

We can use optional arguments in our user-defined functions to control their functionality. We demonstrate the syntax in the following trivial example of a user-defined function that returns the surface area or volume of a sphere using the radius:

```
Public Function BALL(r As Double, Optional v as Variant)
' Returns a result related to the radius of a circle or sphere
p = Application.Pi() ' get the value for the constant pi
If IsMissing(v) Then ' return the surface area of a sphere
    BALL = 4 * p * r ^ 2
Else ' return the volume of a sphere
    BALL = (4 / 3) * p * r ^ 3
End If
End Function
```

Note that the optional argument requires the default **Variant** type for the statement **IsMissing**. Alternatively, we may assign a default value to the optional argument, such as the value 2 for optional b in the following code:

```
Public Function BALL2(r As Double, Optional b As Integer = 2)
' Returns a result related to the radius of a circle or sphere
p = Application.Pi() ' get the value for the constant pi
Select Case b
Case 2 ' Area
    BALL2 = (4 * p * r ^ 2)
Case 3 ' Volume
    BALL2 = (4 * p * r ^ 3) / 3
Case Else
    MsgBox "Optional Argument must be 2 or 3"
End Select
End Function
```

Test the functions in an *Excel* worksheet to calculate the surface area and volume of a sphere with radius one:

	A	B
1	=BALL(1)	=ball2(1)
2	=BALL(1,2)	=ball2(1,3)

	A	B
1	12.56637	12.56637
2	4.18879	4.18879

3.8.10 FORMAT Numbers in Output

Use the keywords **VBA.FORMAT(number,"#.##")** or **VBA.FORMAT(number,"0.0E+00")** as needed to control the display of a numerical value. **Format** uses a combination of symbols for numerical display. For example, in a message box or input box, format the number assigned to the variable p:

```
Prompt:= "% = " & Format(p,"0.#%")
```

Use # in place of 0 to display only non-zero digits in the least decimal place, E+ for scientific notation, or % for percentage. Add or remove extra "#" symbols or "0" to change the number of digits displayed, or "E" and "%" as needed to format the number with scientific notation or as a percent. We may need to replace **FORMAT** with **VBA.FORMAT** or with the worksheet formula **TEXT** in Macintosh versions of *Excel VBA*:

```
Prompt:= "% = " & WorksheetFormula.Text(p,"#.#E+##")
```

Use a backslash "\" before E or % to display the text characters E or % without the corresponding formatting action.

3.8.11 Use VBA to Add Cell Comments and Colors*

We recommend cell comments (described in Section 2.3.8) and cell colors to document the worksheet setup. The *VBA* code in Figure 3.39 provides the syntax for removing and adding a cell comment and sizing the frame. The code first clears any existing comment from the active cell before adding a new comment. The next line sets the visibility to false so that the comment only becomes visible when the mouse curser passes over the cell. The rest of the code adds the desired text for the comment and sizes the frame to fit the text. To move the text entry to the next line, use **vbNewLine** for a "carriage return". Use the **Format** function, if needed, to set the number of figures in the display of a number. Several of the macros in this book add comments to cells containing the output from numerical results to document the name of the macro and ranges on the worksheet for the macro inputs.

```
With ActiveCell
    .ClearComments: .AddComment
    With .Comment
        .Visible = False
        .Text Text:="Comment text" & vbNewLine & "more text" _
                & VBA.Format(number, "0.#E+#0%")
        .Shape.TextFrame.AutoSize = True
    End With
End With
```

Figure 3.39

VBA code for adding a comment to the active cell on a worksheet with formatting.

We may also color a cell to communicate to the user the relative meaning of the cell contents. For example, we may use red to indicate if the content of a cell represents an intermediate result. To color the interior of a cell or its contents, use the following self-explanatory VBA statements:

```
ActiveCell.Interior.Colorindex = 3: ActiveCell.Font.Colorindex = 1
```

The integer index values in the previous line are 1 = **Black** and 3 = **Red**. To view the colors for the rest of the 56 index values, run the simple macro **CLRNDX** from the *PNMSuite* to generate a range of shaded cells and corresponding index values in an *Excel* worksheet named **Color_Index**.

3.9 Debugging Code and Preventing Errors

"It is only when they go wrong that machines remind you how powerful they are." – Clive James

The value of modeling and numerical solutions strongly depends on what we put into them. As we sit in front of the computer monitor – frustrated with unsatisfactory results, we should humbly admit that solving problems using computational tools is a *PICNIC*:

"Problem in the Chair – Not In the Computer"

By design, when given the same set of instructions and conditions, computers are supposed to give precisely the same results; otherwise, computers would be useless for engineering calculations. If there is a problem with the programming, we can bet we have a syntax or calculation error somewhere.

We need to begin developing our debugging skills if we hope to survive serious *VBA* programming. A programming "bug" is an error in our code that causes the code to malfunction or return incorrect results.[47] Debugging code refers to the process of hunting down, trapping, and exterminating errors. We begin by classifying errors into three general types, and then provide some possible fixes in Table 3.9:

- *Syntax Errors* – The *Excel VBE* colors syntax errors red to indicate incomplete statements, misspelled keywords, missing parenthesis, etc.
- *Run-time Errors* – Excel/VBE uses message boxes to indicate errors that occur while our code is running. Table 3.9 lists some common run-time errors and suggestions to remedy the problem
- *Logical Errors* – Our code may produce an error due to our own implementation of incorrect logic in the programming steps. Some examples of bad logic include cases that permit division by zero or "infinite" loops that prevent the program from stopping.

---

[47] While Grace Hopper was working on a Mark II Computer at Harvard University in the 1940s, the technicians found a moth stuck in a relay that impeded operation. She noted that they were literally "debugging" the system and pasted the offending bug in her lab notebook (https://en.wikipedia.org/wiki/Debugging).

Table 3.9 Common mistakes, *VBA* error messages, and potential bug fixes.

Error Message	Potential Solution Fixes
Variable not defined	With **Option Explicit**, use a **Dim** statement to declare a variable type. Remember that each variable must be declared individually, even if all of the variables in a **Dim** statement have the same type.
Compile error: Variable not found	To use a worksheet function in *VBA*, prepend **WorksheetFunction.** or **Application.**, including the periods, to the worksheet function. Be careful to include ( ) and any arguments as required.
#Name!	Missing procedure. The procedure does not exist. Check the spelling of the name. Do not use a name that is a reserved keyword or a range or cell address. The procedure is not in a module (check if you misplaced your code in a worksheet object of the workbook object. If you have multiple workbooks open, be sure that your code is in a module of the workbook calling the procedure, or prepend the path to the workbook with the procedure.
Ambiguous name detected: *name*	We cannot have two macros with the same name in open workbooks. Check the list of macros in the workbook and rename a macro that has a duplicate name. Remember, macros in a workbook may be located within worksheet objects or in different modules.
Type mismatch	Declared variable type does not match the variable value
Reference Error	Avoid using names reserved for cells, such as row/column names.
#Value!	Missing an argument in an *Excel* worksheet or user-defined function, or an incomplete reference (missing row or column, or improperly named cell)
Can't find function	User-defined functions for use in an *Excel* worksheet must be located in modules, not worksheet or project objects. Sub procedures cannot be used in an *Excel* worksheet cell.
Function/Sub Error	Do not open the same workbook more than once. Avoid opening workbooks that contain procedures with the same name.
VBA code missing	Be sure to macro enable your workbook. Put custom *VBA* functions or macros in a module in the project object with the name of the workbook. If you add a module to another object, such as FUNCRES, it is removed when the workbook is closed.

Chemical engineering professors Kamaruddin and Hamid at Universiti Teknologi Malaysia give the following tips[48] to minimize programming bugs in your code:

1. Use **Option Explicit** at the top of each module to require declaration of variable types in your code. Declaring variables catches spelling errors in variable names and enforces compatibility with types of values assigned to variables.
2. Declare the variable type assigned to values for **InputBox** to validate user-supplied values.
3. Use structured programming logic, such as **If...Then** and **MsgBox** to check for erroneous input or calculated results and provide the user with information to correct errors.
4. Trap errors with the keyword **On Error** followed by an appropriate keyword statement for handling the error, such as **Resume Next**, **Go To**, **Exit**, *etc.*
5. Use indentation to format code segments for readability. By aligning code within **If...Then** or **Loop** structures we can better follow our programming logic.
6. Add comments liberally throughout the code that describes each segment of code to help us remember why we included the code and what are the intended purposes of each statement.
7. Organize the code into **Sub** and **Function** procedures that have specific purposes. We can reduce errors when we are able to reuse well-crafted and thoroughly debugged procedures.
8. Record macros of *Excel* worksheet activity to see how to use *VBA* to code steps that we have correctly implemented in an *Excel* worksheet.
9. Take advantage of *Excel's* worksheet validation tools to prevent improper user data on the worksheet that gets used by the *VBA* programs.
10. Finally, learn how to use *Excel's* debugging tools for tracking errors in your code.

We employ several of the previous tips in the *VBA* user-defined function listed in Figure 3.40 for calculating the volume of a continuously stirred tank reactor. The first line invokes the explicit option to require variable type declaration. Several comments define the variables and intended actions in the code segments. The code uses logical

---

[48] DKK 3352 Computer Aided Chemical Engineering (Excel/VBA)

CHAPTER 3: VBA

statements to check that all parameters in the CSTR design equation are positive. In the case of conversion, by definition it must be greater than zero and less than one. A message box reports the input errors to the user, and then steps over the volume calculations to return an error message to the cell with the user-defined function.

```
Option Explicit
Public Function CSTRVolume(C As Double, F As Double, k As Double, X As Double)
' Calculate the volume of a CSTR with a first order reaction
' C = initial concentration, kmol/m3
' F = molar flow feed rate, kmol/s
' k = reaction rate constant, 1/s
' r = reaction rate, kmol/m3 s
' X = reactant conversion
Dim r As Double
On Error GoTo 10 ' catch errors in the input or calculations
Debug.Print "C = ", C, "F = ", F: Debug.Print "k = ", k, "X = ", X
If C <= 0 Or F <= 0 Or k <= 0 Then ' validate C, F, k
    MsgBox "Concentration, molar flow rate, and rate constant must be > 0."
    GoTo 10
End If
If X < 0 Or X >= 1 Then ' validate the conversion
    MsgBox "Conversion must be 0 < X <= 1.": GoTo 10
End If
r = k * C * (1 - X) ' reaction rate
CSTRVolume = F * X / r ' reactor volume
Exit Function ' return to the worksheet if no errors
' Return a string to report errors in the calculations
10: CSTRVolume = "Invalid CSTRVolume Arguments": End Function
```

Figure 3.40 A simple *VBA* user-defined function with error handling features to validate inputs and check for errors before calculating the volume of a CSTR.

The *VBA* expression **On Error Resume Next** is also useful for trapping errors to prevent the program from a disastrous conclusion. For example, we modified the code for the flow velocity in *Example 3.9* as shown in Figure 3.41 to avoid displaying an error message, such as Figure 3.42 when we click the *Cancel* button associated with an **InputBox**. We added an **Exit Sub** statement to step over any following code and stop the program when an error occurs.

```
Option Explicit
Private Sub CommandButton1_Click()
    Dim set_cell As Range, changing_cell As Range
    On Error Resume Next
    Set changing_cell = Application.InputBox("Click on cell for velocity", Type:=8)
    If Err.Number = 424 Then Exit Sub
    Set set_cell = Application.InputBox("Click on cell for Bernoulli equation", Type:=8)
    If Err.Number = 424 Then Exit Sub
    set_cell.GoalSeek Goal:=0, ChangingCell:=changing_cell
    Exit Sub
End Sub
```

Figure 3.41 Error handling in *VBA* using On Error GoTo statements. Error code 424 indicates that the Cancel button was selected, and then exits the sub procedure.

Skip over an error using **On Error Resume Next** if potential errors do not affect the results of the program.

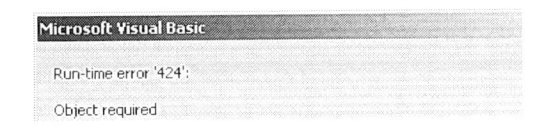

Figure 3.42

Error message when selecting Cancel for InputBox.
Click Debug to go to the line where the error originated.

Test the program. Click **Cancel** with and without the error trapping code to observe the differences between the results.

Good programmers take steps to prevent errors from happening due to faulty user input. Some of the programs in this book include lines of code to check the validity and consistency of inputs. For example, when selecting ranges of *y* vs *x* data, a programmer may want to ensure that the number of *x*'s matches the number of *y*'s. Use flow control and message boxes to alert users to any potential problems with their inputs and reroute the flow of the program execution to allow the user to make changes to fix problems. Look for examples of this practice in the macros provided throughout the remaining chapters. For instance, in a few macros, the name of the worksheet causes problems when it includes spaces between characters. We created the sub procedure **JUSTABC_123** in Figure 3.12 to mitigate this problem.

3.9.1 Immediate, Locals, and Watch Windows

When we want to know the value of a variable without using a `MsgBox`, or without assigning the values into a range on an *Excel* worksheet, select the menu option **View/Immediate Window** then use `Debug.Print` in the code to display what follows the `Print` statement to the **Immediate** window. In the immediate window, we can check for current values of variables by typing a question mark (?) followed by the variable name. To check for intermediate values of variables, insert a break point at a position in the code where you want the program to pause during execution. To insert a break point in the code, click in the vertical bar at the left side of the code to insert a red, circular marker and highlight the line with red shading. Click on the break point marker again to remove the break point. The user-defined function in Figure 3.40 uses `Debug.Print` to display the values of the variables in the arguments to the **Immediate Window**. We also added a break point in the next figure to check the value of the reaction rate by typing `? r` directly in the **Immediate Window**.

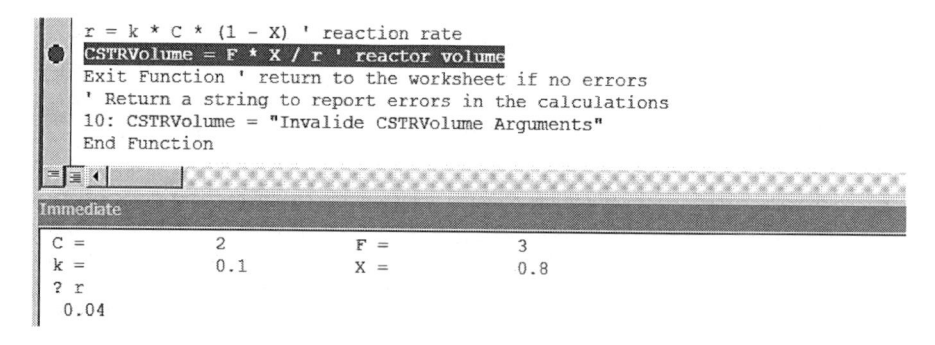

```
r = k * C * (1 - X)    ' reaction rate
CSTRVolume = F * X / r ' reactor volume
Exit Function          ' return to the worksheet if no errors
' Return a string to report errors in the calculations
10: CSTRVolume = "Invalide CSTRVolume Arguments"
End Function
```

```
Immediate
C =         2           F =         3
k =         0.1         X =         0.8
? r
 0.04
```

While debugging our code, we can execute *VBA* statements in the **Immediate Window**. For example, we can change the value of a variable while in break mode.

The **Locals Window** displays the current values of the variables in the procedure at module levels. However, if we need to change a value of a variable, we must execute the statement in the **Immediate window**. The **Watches Window** allows us to monitor the values of specific variables or expression that we select. Open the **Watch Window** from the **View** menu. Add watch variables from the **Debug** menu in the *VBE*. Right click on the **Watches Window** to edit, add, or delete variables or expressions for watching during program execution. To help debug the program, we can edit the watches to pause when a watch variable is true or changes value.

3.9.2 Debug Menu

We may access the *VBE* debugging tools form the **Debug** menu or the **Debug** toolbar. To display the **Debug** toolbar, select from the menu **View/Toolbars/Debug**. Table 3.10 describes each of the debugging tools available from the menu or toolbar.

Table 3.10 VBA debugging options from the menu or Debugging Toolbar.

Debug Tool Bar	Icon	Shortcut Keys	Description
Design Mode			Toggle on or off design mode while creating forms or inserting controls.
Run Sub/User Form		F5	Starts the macro from the insertion point (location of the cursor)
Break		CNTL Break	Pauses the macro and switches to break mode
Reset			Clears all variables and breakpoints and resets the macro
Toggle Breakpoint		F9	Adds or removes a breakpoint that stops program execution
Step Into		F8	Steps in to a macro one line at a time
Step Over		SHIFT F8	Execute the lines of code one line or procedure at a time.
Step Out		CNTL SHIFT F8	Executes the remaining lines of code in a procedure following the current execution point.
Locals Window			Open the Locals Window. Displays all of the values of all variables in a running macro.
Immediate Window		CNTL G	Open the Immediate Window. Use ? to display values of variables. Execute statements or change values of variables.
Watch Window			Open the Watch Window. Edit the watches to monitor the values of specific variables or expressions during program execution.
Quick Watch		Shift F9	Display the dialog box with the current value of the selected variable or expression.
Call Stack			Display the list of calls to procedures that have not completed execution.

Place the cursor in the code and press the shortcut key **F8** to step through the code, line by line, to view the action and result of each statement in the **Immediate** or **Watch Windows**. Type **Shift F8** to step over a line and **CTRL Shift F8** to step out of the code. Click in the text of a macro to place the cursor at a location after the point of the suspected error. Then select the menu option **Debug/Run to Cursor**. We might find an improperly assigned variable. Move the mouse over variables in the module to display the current value of the variable in a small popup box.

3.10 PNM Suite of Macros and User-defined functions

"The idea is to have a toolbox so that you can deal with whatever job you come across." - Stephen King

An *Excel* workbook containing a suite of *VBA* macros is available to implement the numerical methods described in this book. All of the macros are sorted into modules by book chapter titles within the companion *Excel* workbook named *PNMSuite* (Practical Numerical Methods Suite). Download the *PNMSuite* workbook from the companion web site: https://sites.google.com/d.umn.edu/pnm4. To view and edit the codes in the **Visual Basic Editor**:

1. Click on *PNMSuite* in the **Project** window, as shown in Section 3.2.
2. Provide the password AsteroidB612 in the dialog box, and then open the modules in the *PNMSuite* project to view the code.

There are several options for accessing and using the book's macros:
- First, the simplest method: Set up the worksheet calculations directly in the workbook *PNMSuite*. Access the list of available macros from the ribbon **View>Macros** tab menu. To use a macro, click on the macro name in the list, and then select **Run**. Notice the short description of the selected macro at the bottom of the macro window.

- To access user-defined functions, open the *PNMSuite* workbook, type the name of the function with the required arguments directly in an *Excel* worksheet cell, or insert the user-defined function from the **Formulas** tab. Select from the list in the category **User-defined**, as shown in Figure 3.43.
- To see the required arguments in the formula, type the name of the function and the opening parenthesis in an *Excel* worksheet cell after an equal sign, *e.g.*, "**= Function_Name(...**", then type the keyboard combination **CTRL SHIFT A** to fill in the list of required arguments for the function. Replace the symbol for each argument in the list with the required argument value (as either a cell reference or direct input).
- We may also access the macros from other workbooks. To use any of the functions and macros that accompany this book, open the workbook *PNMSuite* along with a new macro-enabled workbook. Be sure to enable the workbook *PNMSuite* when opening the file. While the workbook containing the macros is open, the public macros are available to any other macro-enabled workbook. To use the functions from the suite in an *Excel* worksheet, prepend the path to the workbook *PNMSuite* to the function *name*, *e.g.*:

    C:\'*PNMSuite*.xlsm'!*name*( )

- Alternatively, copy the code for the functions or sub procedures from *PNMSuite* into a module in the new macro-enabled workbook.
- We also have the option to save the workbook *PNMSuite* as an *Excel* add-in.

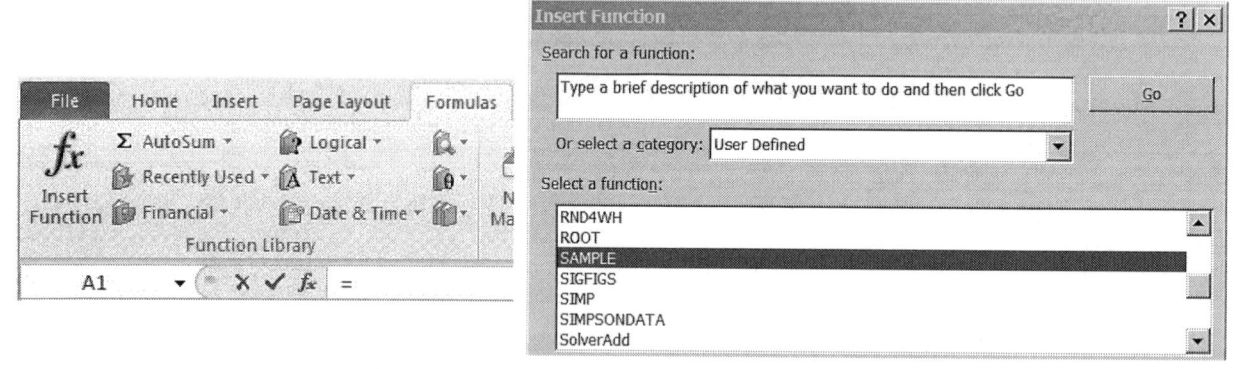

Figure 3.43 Insert a user-defined function into a cell by selecting the function from the ribbon Formulas>Insert Function>User Defined category. In this example, the functions are in the workbook *PNMSuite*.

The macros in the *PNMSuite* were designed from the philosophy of *Excel* worksheet *enhancement*. The application of the numerical methods uses the *Excel* worksheet computational engine as much as possible to define and calculate algebraic equations, integral functions, and derivative functions. Using *Excel* to perform function calculations eliminates *VBA* programming required by the user to make the routines available to *Excel* users who may not have the inclination to write code.

In general, the macros depend on an *Excel* worksheet for storing information and performing much of the calculations. *Excel* worksheets serve as calculators as well as interfaces to the *VBA* macros. The macros do what macros do best by taking over the implementation of repetitive calculations. We use input boxes to select the regions of the worksheet that contain the variables, parameters, and functions required by the numerical method, as well as the location for any output back to the worksheet. This approach allows the user to implement the numerical techniques on a wide variety of problems, all while avoiding extra *VBA* programming.

We may modify the macros as needed for our own purposes. For instance, we may decide to replace the input boxes with hard-coded references to cells on the worksheet to eliminate the input steps when using the macro repeatedly for the same problem. We may consider adding a control button to launch the macro directly from our worksheet.

The macros follow the general program flow structure displayed in Figure 3.44:

1. Input data to set up the problem.
2. Process the data using the numerical method(s) required to obtain the desired result.
3. Output the results to the worksheet, including a comment in an output cell with the name of the macro, setup ranges, date, and time.

Most of the *PNMSuite* workbook macros use input boxes to assign worksheet cells to program variables, which allow the macros to interact with the active worksheet. Input boxes also allow for input validation by requiring a specific type of input. Each of the macros has some error control statements to help the user make proper input decisions and avoid mistakes that disrupt execution. Some macros have **User-form** versions. The codes listed in the *PNMSuite* validate the location selected for the output to avoid overwriting cells on the worksheet. The macros assume that formulas in the worksheet automatically recalculate for any changes to dependent or precedent cells referenced by the formula. Be sure to select **Automatic Calculation** in the menu **File>***Excel* **Options>Formulas>Calculation Options>Automatic**, as shown in Figure 2.28. Alternatively, add a line of *VBA* code `Application.Calculation = xlCalculationAutomatic` (or `xlCalculationManual`) near the top of the code in the procedure.

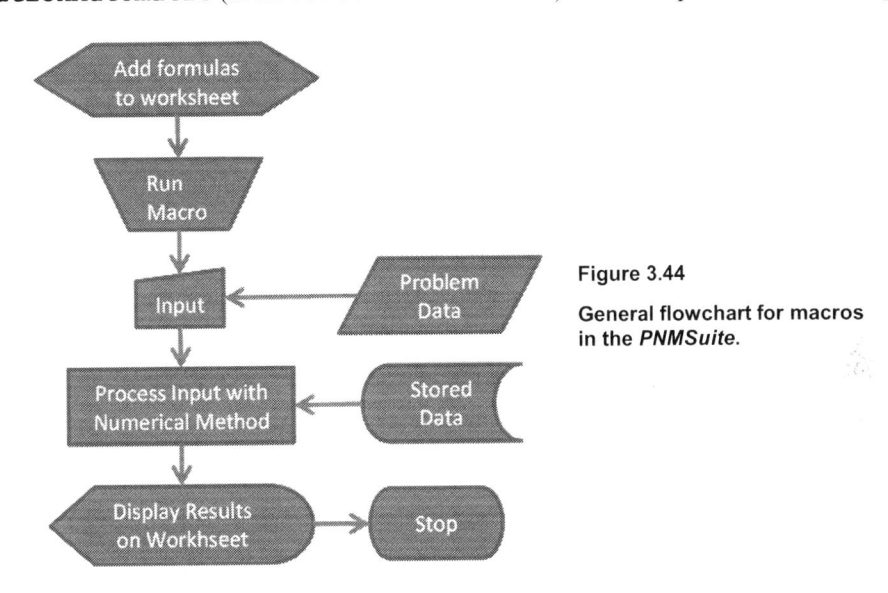

Figure 3.44

General flowchart for macros in the *PNMSuite*.

A few of the macros in the *PNMSuite* for *MS Windows* use the **Solver**. Before we can access the **Solver** from within a *VBA* macro, we need to check that **VBE** references the **Solver**. First activate the **Solver** add-in from the **File>***Excel* **Options> Add Ins**. In the **VBE** window, select the menu item **Tools/References**, then check **Solver**. Chapter 7 and Chapter 9 have examples of *VBA* **Solver** use for optimization and regression uncertainty analysis.

3.11 Excel Add-In Macros*

To make the *VBA* user-defined functions and sub procedures available to any *Excel* workbook without opening the *PNMSuite* workbook, save the workbook as an *Excel* **Add-in**. For a workbook with *VBA* macros, select from the ribbon tab **File>Save As**. Select **Save as type:** *Excel* **Add-In** from the drop down box:

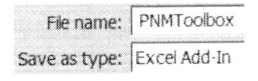

Follow these steps to manage and install add-ins:
1. Click **File>Options>Add-Ins**.
2. Select an add-in type then click **Go**.
3. Select add-ins to add, remove, load, or upload. Alternatively, browse to locate add-ins to install.

When we open a workbook saved as an *.xla* file, all of its macros and user-defined functions are available in any other open workbook.

3.12 Epilogue on VBA

Excel with *VBA* is a powerful and versatile tool for performing engineering computations. The appendix in Section 15.2 lists several *VBA* functions and objects for implementing numerical solutions to engineering problems. The other appendices in Chapter 15 summarize the user-defined functions, sub procedures, and user-forms developed for this book. As we work through the rest of the book, we encounter several user-defined functions and sub procedures programmed to implement advanced numerical methods in *Excel* worksheets. Take time to study the structures of the programs and develop your own *VBA* programming skills and style. Table 3.11 contains a summary of general-purpose user-defined functions developed for this chapter.

Table 3.11   Summary of *VBA* user-defined functions (UDF) and macros, and their methods of input.

Macro/Function	Inputs	Operation
BINS	Sub procedure with input boxes	Calculates a series of bin dimensions for a histogram chart.
CLRNDX	Macro (no inputs)	Create color index in an *Excel* worksheet named Color_Index
GRAPHYXGUIDE	Sub procedure with input boxes	Plot data in a scatter plot following the guidelines in Section 2.3.4
HEAPSORT HEAPSORTXY	Sub arguments vector of numerical values	Sort columns in array or vectors in ascending order.
JUSTABC_123	UDF arguments	Removes non alphanumeric characters from a string
KOWABUNGA	UDF argument	*VBA* speech for speaking string from text (MS Windows only)
LLETRIANGLE	Sub procedure with input boxes	Generates a right-triangle for ternary liquid-liquid extraction analysis
NRDWH	UDF with no arguments	Normal random deviate with mean zero and standard deviation of one.
QYXPLOT QYXPLOT_UsrFrm	Sub procedure with input boxes Show user-form macro	Quick y vs x graph of an *Excel* worksheet formula.
RGAS	UDF arguments	Returns values for the ideal gas constant for different unit systems.
RNDWH	UDF arguments	Pseudo-random number generator with a cycle of 10^{36}
ROULETTE	UDF argument	Selects an element proportional to its probability of selection.
ROWCOL	UDF arguments	Checks for vector in row or column and transposes to column
SAMPLE	UDF arguments	Simple random sample from a population.
SELECTPRO	Sub procedure with input boxes	Select elements from an array according to their probability
SIGFIGS	UDF arguments	Round to the number of significant figures.
SWAP	Sub procedure (no inputs)	Swaps the contents of two active cell ranges on a worksheet
UDF	UDF arguments	Template to evaluate a worksheet function in *VBA*
UNITS	UDF arguments	Convert the value to a new unit system
UNITS_TABLES	Sub procedure (no inputs)	Create tables of unit strings available to the UDF UNITS.

Chapter 4 Linear Algebraic Systems

"Linear thinking typifies a highly developed industry. It starts to get these patterns built into it somehow. I'm not sure how that happens, but certainly you take a look at dinosaurs." – Michael Nesmith

Methods for solving coupled systems of linear equations are important for implementing several numerical methods. Systems of linear equations arise from material and energy balances around chemical processes involving multiple branching and connecting streams, as well as linear least-squares regression, and solution approximations for boundary value problems. We introduce the following methods for solving linear systems with some examples drawn from the practice of chemical engineering:

- Gauss elimination and Crout reduction with maximum column pivoting and error reduction
- Thomas, and Cholesky methods for sparse or symmetric matrices
- Matrix algebra using *Excel* worksheet functions
- Interpolation, power, and Jacobi methods for linear Eigen systems
- Jacobi and Gauss-Seidel iteration with relaxation for accelerating convergence

We begin with a review of the concept of linearity and solution methods for small systems before introducing numerical methods for efficiently finding solutions to larger systems of linear equations.

> **SQ3R Focused Reading Questions**
> 1. What makes a function linear?
> 2. What is a system of simultaneous equations?
> 3. What *Excel* worksheet functions are available for solving a linear system?
> 4. When do I need the key combination CONTROL SHIFT ENTER?
> 5. Why do I need *VBA* macros for solving linear equations?
> 6. What are the advantages and disadvantages of iterative methods for solving linear equations?
> 7. How do I fix an error in a worksheet array function?
> 8. What is an eigenvalue problem?
> 9. What does the determinant of a matrix tell me?
> 10. What is a sparse matrix?

4.1 Linear Algebraic Functions

"We may always depend on it that algebra, which cannot be translated into good English and sound common sense, is bad algebra." – William Kingdon Clifford

A linear system of equations has a unique solution when the number of variables and equations match (zero degrees of freedom[49]), as discussed in Section 1.2.5. We define a function, $f(x)$, as a mathematical relationship between independent variables x and dependent variable y:

$$y = f(x) \tag{4.1}$$

---

[49] See concepts of *existence* and *uniqueness* in more advanced treatments.

Linear functions involve first-order terms in all variables. Consider the function:

$$f(x) = ax + \frac{b}{x+c} + d\ln(x) + e \tag{4.2}$$

Equation (4.2) is linear (or first-order) in variable x in the first term, as well as the parameters a, b, d, and e in all of the terms, but nonlinear in parameter c and variable x in the second and third terms. Some simple transformations yield a multivariable linear function in terms of the transformed terms x_0, x_1, and x_2:

$$f(x) = ax_0 + bx_1 + dx_2 + e \tag{4.3}$$

where $\quad x_0 = x \quad\quad x_1 = \dfrac{1}{x+c} \quad\quad x_2 = \ln(x)$

Observe, however, that we cannot linearize Equation (4.2) in terms of x without involving parameter c.

4.2 Linear Algebraic Systems

We easily solve small systems of linear equations by "hand calculations" involving algebraic rearrangement and substitution. To illustrate, consider the simple system of two linear equations in two unknowns:

$$5x_1 + 6x_2 = 7 \tag{4.4}$$

$$2x_1 - 3x_2 = 4 \tag{4.5}$$

Divide each term in Equations (4.4) and (4.5) by the coefficients of the respective first terms:

$$x_1 + \frac{6}{5}x_2 = \frac{7}{5} \tag{4.6}$$

$$x_1 - \frac{3}{2}x_2 = \frac{4}{2} \tag{4.7}$$

Subtract Equation (4.6) from Equation (4.7) to eliminate x_1:

$$(1-1)x_1 - \left(\frac{3}{2} + \frac{6}{5}\right)x_2 = \frac{4}{2} - \frac{7}{5} \tag{4.8}$$

or

$$0 - \frac{27}{10}x_2 = \frac{3}{5}$$

Then, solve Equation (4.8) for x_2 by dividing the right-hand-side by the remaining coefficient of x_2:

$$x_2 = -\frac{30}{135} = -0.2222 \tag{4.9}$$

Finally, substitute the result for x_2 into Equation (4.6) and solve for x_1:

$$x_1 = \frac{7}{5} - \frac{6}{5}x_2 = \frac{7}{5} - \frac{6}{5}\left(-\frac{30}{135}\right) = 1.6667 \tag{4.10}$$

For larger systems of equations, repeat this basic approach with each variable until a single linear equation remains with a single variable. Find the values of the remaining unknowns in like manner. This is the basis of the method of Gaussian elimination. A large system of n linear equations, such as Equation (4.11), becomes difficult to solve by "hand calculation" due to the extensive bookkeeping involved. In the following system of linear equations, the first subscript references the equation whereas the second subscript references the variable:

$$a_{1,1}x_1 + a_{1,2}x_2 + \ldots a_{1,n}x_n = b_1$$

CHAPTER 4: LINEAR EQUATIONS

$$a_{2,1}x_1 + a_{2,2}x_2 + \ldots a_{2,n}x_n = b_2$$
$$\vdots$$
$$a_{i,1}x_1 + a_{i,2}x_2 + \ldots a_{i,n}x_n = b_i \quad (4.11)$$
$$\vdots$$
$$a_{n,1}x_1 + a_{n,2}x_2 + \ldots a_{n,n}x_n = b_n$$

where $a_{i,j}$, b_i, and x_j are the coefficients, constants, and variables, respectively. In compact matrix notation, Equation (4.11) becomes:

$$\begin{bmatrix} a_{11} & a_{12} & \ldots & a_{1n} \\ a_{21} & a_{22} & \ldots & a_{2n} \\ \vdots & \vdots & \ddots & \vdots \\ a_{n1} & a_{n2} & \ldots & a_{nn} \end{bmatrix} \cdot \begin{bmatrix} x_1 \\ x_2 \\ \vdots \\ x_n \end{bmatrix} = \begin{bmatrix} b_1 \\ b_2 \\ \vdots \\ b_n \end{bmatrix} \quad (4.12)$$

Alternatively, use shorthand notation for the matrices and vectors:

$$A \cdot x = b \quad \rightarrow \quad x = A^{-1} \cdot b \quad (4.13)$$

where the matrix A contains the coefficients of the variables and the vector b contains only constants. We present a few methods of solution for solving intermediate to large systems of equations. Most engineering students take separate courses in linear algebra, or learn these methods in a course on differential equations. Thus, much of this chapter may serve as a review. However, some useful examples using *Excel* and *VBA* further develop our skills with these computing tools and set the stage for numerical methods for nonlinear problems.

4.3   Gaussian Elimination

The Gaussian elimination method with backward substitution solves a properly specified system of linear equations based on the standard technique demonstrated for the small system in Section 4.2. The following steps outline the method of Gaussian elimination applied to a system of linear equations arranged like Equation (4.11):[50]

1. Divide each of the coefficients in the first equation by the coefficient of the first variable:

$$x_1 + \frac{a_{1,2}}{a_{1,1}}x_2 + \ldots \frac{a_{1,n}}{a_{1,1}}x_n = \frac{b_1}{a_{1,1}} \quad (4.14)$$

2. Multiply the first equation by the coefficient of the first term in the second equation. Then subtract the first equation from the second equation to eliminate the first term in the second equation:

$$0 + \left(a_{2,2} - \frac{a_{2,1}a_{1,2}}{a_{1,1}}\right)x_2 + \ldots \left(a_{2,n} - \frac{a_{2,1}a_{1,n}}{a_{1,1}}\right)x_n = b_2 - \frac{a_{2,1}b_1}{a_{1,1}} \quad (4.15)$$

3. Repeat step 2 for each subsequent equation to eliminate the first term from the remaining equations $i = 2, 3 \ldots n$.

$$0 + \left(a_{i,2} - \frac{a_{i,1}a_{1,2}}{a_{1,1}}\right)x_2 + \ldots \left(a_{i,n} - \frac{a_{i,1}a_{1,n}}{a_{1,1}}\right)x_n = b_i - \frac{a_{i,1}b_1}{a_{1,1}} \quad (4.16)$$

---

[50] Gauss and Newton seem to "compete" for prominence in the early field of numerical methods (~250 references to Newton and ~200 references to Gauss). Gauss elimination is a case of Gauss getting a bit more credit for *Newton's* ground work due to some historical confusion related to the advent of computing in the 1950's. Many other eminent mathematicians played prominent roles in the advancement of the elimination method as well. However, in the ancient record, it appears Chinese mathematicians beat them both by a millennium (Grcar 2011).

4. Repeat the procedure in the first three steps, beginning with the first non-zero term in the second equation, and continuing until the set of equations take the form of an upper triangle:

$$x_1 + a'_{1,2}x_2 + \ldots a'_{1,n}x_n = b'_1$$

$$0 + x_2 + a'_{2,3}x_3 \ldots a'_{2,n}x_n = b'_2$$

$$\vdots$$

$$0 + 0 + \ldots x_i + a'_{i,i+1}x_{i+1} \ldots a'_{i,n}x_n = b'_i \qquad (4.17)$$

$$\vdots$$

$$0 + 0 + \ldots x_n = b'_n$$

where a' and b' are the groups of terms from combining equations. Alternatively, in compact matrix notation, Equation (4.17) becomes:

$$\begin{bmatrix} 1 & a'_{12} & a'_{13} & \ldots & a'_{1n} \\ 0 & 1 & a'_{23} & \ldots & a'_{2n} \\ 0 & 0 & 1 & \ldots & a'_{3n} \\ \vdots & \vdots & \vdots & \ddots & \vdots \\ 0 & 0 & 0 & \ldots & 1 \end{bmatrix} \cdot \begin{bmatrix} x_1 \\ x_2 \\ x_3 \\ \vdots \\ x_n \end{bmatrix} = \begin{bmatrix} b'_1 \\ b'_2 \\ b'_3 \\ \vdots \\ b'_n \end{bmatrix} \qquad (4.18)$$

5. The last row in Equation (4.18) gives the solution for $x_n = b'_n$. Substitute this result back into the next-to-last equation and solve for $x_{n-1} = b'_{n-1} - a'_{n-1,n}b'_n$. Solve for the remaining unknown terms in like manner by back substitution with the known values in reverse order of the equations.

4.3.1 Matrix Solution Improvement by Maximum Column Pivoting and Error Estimation

For successful implementation of the method of Gauss elimination, the relative positions of the equations in the matrix of Equation (4.12) may need rearrangement, or ranking, to ensure that the ith term of the ith equation on the diagonal a_{ii} is non-zero. In other words, we may need to interchange rows to avoid division by any zero diagonal elements. Additionally, we scale the terms in the coefficient matrix to reduce the effects of round-off errors during computation. We simultaneously minimize computational round-off errors if we interchange the rows to position the elements from each column with the largest magnitude along the diagonal of the coefficient array.

We define the error in the approximate solution in terms of the residual vector r:

$$r = b - Ax_c \qquad (4.19)$$

where x_c is the vector of calculated roots to the system of linear equations. We then multiply each side of Equation (4.19) by A^{-1} such that the error vector becomes the difference between the calculated and true solution (recall $A^{-1}A = I$, the identity matrix):

$$E = x_c - x \cong A^{-1}r \qquad (4.20)$$

An improved solution to the linear system of equations becomes:

$$x = x_c - E \qquad (4.21)$$

If necessary, we may repeat the previous steps for additional solution improvement; however, one error correction generally suffices (Hanna and Sandall 1995).

A macro **GAUSSELIM**, with input boxes listed in Figure 4.1, is available in the *PNMSuite*. This macro requires the square coefficient matrix and constant vector in a range on an *Excel* worksheet. The *PNMSuite* also has a user-form named **LINSYS** with the same worksheet setup and inputs with an option for Gaussian elimination that uses maximum column pivoting and one iteration of solution improvement.

```
Public Sub GAUSSELIM()
' Solution to a linear system of equations by Gauss elimination with maximum column
' pivoting with one iteration of error correction. The equations must be defined by
' |A||x|=|b|. Define the constant coefficient matrix and constant vector in ranges on
' an Excel worksheet. The constant coefficient matrix must be in contiguous cells.
' Each row represents an equation. The vector must be in contiguous cells in a column.
' The user selects a range of cells for the solution vector.
' *** Required User-defined Functions and Sub Procedures from the PNM Suite ***
' GAUSSPIVOT = Gaussian elimination with maximum column pivoting.
' ROWCOL = transpose row to column vector
' *** Nomenclature ***
Dim a() As Double   ' = square coefficient array
Dim ar As Range     ' = coefficient matrix range
Dim b() As Double   ' = constant vector
Dim boxtitle As String  ' = string for macro name
Dim br As Range     ' = constant vector range
Dim i As Long, j As Long  ' = index for row and column in coefficient array
Dim n As Long, nv As Long ' = number of equations and variables
Dim nvrow As Long   ' = number of rows
Dim x() As Double   ' = solution array
Dim xr As Range     ' = solution vector range
'*********************************************************************************
boxtitle = "GAUSSELIM"
With Application ' Identify the solution method
a_input: ' Label to get the coefficient matrix for linear system
Set ar = .InputBox(prompt:="Range of Square COEFFICIENT matrix, Ax=b:" & vbNewLine _
        & "(must be in contiguous cells)", Title:=boxtitle, Type:=8)
n = ar.Rows.Count ' Get number of variables (rows should match columns)
If n <> ar.Columns.Count Then ' Check for a square coefficient matrix
    MsgBox Title:=boxtitle, _
    prompt:="Range of Coefficient matrix must be square (rows = columns."
    GoTo a_input
End If
b_input: ' Label to get the constant vector for linear equations
Set br = .InputBox(Title:=boxtitle, prompt:="Range for CONSTANT vector, b:", Type:=8)
If br.Count <> n Then ' Check consistency between sizes of vector & matrix
    MsgBox Title:=boxtitle, prompt:="Constant vector must match coefficient matrix."
    GoTo b_input
End If
x_input: ' Get the location for the solution vector
Set xr = .InputBox(Title:=boxtitle, prompt:="RANGE for SOLUTION vector, x:", Type:=8)
nv = xr.Count: nvrow = xr.Rows.Count ' Get number of variables and rows
End With
If nv <> n Then ' Check for consistency between solution vector size & size of system
    MsgBox Title:=boxtitle, prompt:="Solution vector must match constant vector."
    GoTo x_input
End If
ReDim a(1 To n, 1 To n) As Double, b(1 To n) As Double, x(1 To n) As Double
For i = 1 To n ' Get values of ranges to transfer to solution sub procedures
    For j = 1 To n: a(i, j) = ar(i, j): Next j
    b(i) = br(i)
Next i
Call GAUSSPIVOT(a, b, n, x) ' use Gauss elimination with maximum column pivoting
Call ROWCOL(nv, x, xr, nvrow) ' Check row/column solution vector. rows by default.
End Sub ' GAUSSELIM
```

Figure 4.1 *VBA* macro GAUSSELIM that calls sub procedure GAUSSPIVOT for solving a linear system by Gauss elimination with maximum column pivoting and one error correction.

To use **GAUSSELIM**, set up the matrix of coefficients and vector of constants in ranges of cells on an *Excel* worksheet. The macro **GAUSSELIM**:

- Requires a range of ordered cells for the square coefficient matrix, *A*.
- Requires a range of cells ordered in a column for the vector of constants, *b*.

- Checks for consistency in the input values for the sizes of arrays.
- Calls a sub procedure **GAUSSPIVOT** to solve the system of equations by Gaussian elimination with maximum column pivoting and one iteration of error improvement.
- Reports the results to either a row or column vector of contiguous cells selected by the user.

Illustration: Use the macro **GAUSSELIM** to solve the simple linear system in Equations (4.4) and (4.5).
1. Put the square coefficient matrix in the range **A1:B2** and the constant vector in the column range **C1:C2**.
2. Open the workbook *PNMSuite* and run the macro **GAUSSELIM**.
3. Select the ranges of cells for the coefficient matrix and constant vector as prompted by the input boxes.
4. Select the column range **D1:D2** for the solution vector.
5. Run the macro **GAUSSELIM** again with the same inputs, but this time select the range **E1:F1** for the solution vector in a row:

	A	B	C	D	E	F
1	5	6	7	1.666667	1.666667	-0.22222
2	2	-3	4	-0.22222		

4.3.2 Crout Reduction Method of LU Decomposition*

Crout reduction is an efficient method of LU decomposition for systematically solving simultaneous linear equations without the necessity of ranking *equations* (Perry 2003). Crout's method replaces the coefficient matrix A with a decomposed matrix consisting of lower (L) and upper (U) triangular matrices. Crout's method has the advantages of fewer computational steps compared with Gauss elimination, and a reduced matrix A independent from the constant vector b. For problems that maintain the coefficient matrix but change the constants, we can recycle the transformed matrix. The following steps outline Crout's method:

1. Arrange the set of linear equations in the ordered format of Equation (4.11) or (4.12).
2. Reduce the coefficient array according to Equation (4.22) or (4.23). Initially, set the elements of $c_{i,j}$ to zero. Then replace the elements $c_{i,j}$ with reduced terms:

$$c_{i,j} = a_{i,j} - \sum_{k=1}^{j-1} c_{i,k} c_{k,j} \quad \text{for } i \geq j \tag{4.22}$$

$$c_{i,j} = \frac{1}{c_{i,i}} \left[a_{i,j} - \sum_{k=1}^{i-1} c_{i,k} c_{k,j} \right] \quad \text{for } i < j \tag{4.23}$$

Notice that the coefficient matrix A is transformed independent of the constant vector, b.
3. For $i = 1, 2 \ldots n$, calculate the reduced constant array:

$$d_i = \frac{1}{c_{i,i}} \left[b_i - \sum_{k=1}^{i-1} c_{i,k} d_k \right] \quad i < j \tag{4.24}$$

The reduced system of linear equations becomes $c \cdot x = d$.
4. Calculate the solution for x by back substitution into the reduced equations as $i = n, n-1 \ldots 1$:

$$x_i = d_i - \sum_{k=i+1}^{n} c_{i,k} x_k \tag{4.25}$$

The macro **CROUT** is available in the *PNMSuite* for implementing the method of Crout reduction with scaled row pivoting via the sub procedures **CROUTLU** that replaces the coefficient matrix A with a lower/upper triangle matrix LU, and **CROUTLUX** that finds the solution vector x. Use the same *Excel* worksheet setup required by **GAUSSELIM**. Crout's method is an option in user-form named **LINSYS**.

CHAPTER 4: LINEAR EQUATIONS

Example 4.1   Material Balances Around a Two-stage Flash Separation

Known: Two equilibrium flash drums in series separate a mixture of hexane, octane, and decane as shown in the schematic. The stream compositions are tabulated below in terms of mole fraction:

Compound	Stream Mole Fraction			
Streams	1	2	5	6
C_6H_{14}	0.3300	0.4716	0.1584	0.0310
C_8H_{18}	0.3400	0.3549	0.3832	0.2211
$C_{10}H_{22}$	0.3300	0.1735	0.4584	0.7479

Schematic: A two-stage flash separation employs an inter-stage-heater to create two-phase flow into the second stage. The diagram shows the solution for the stream molar flow rates as percentages of F_1.

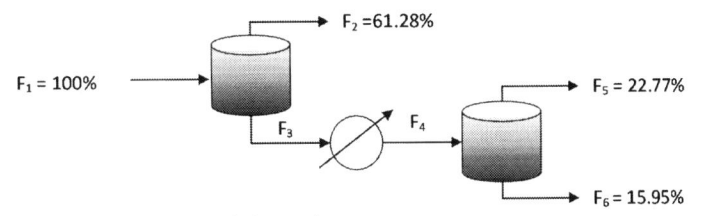

$F_1 = 100\%$, $F_2 = 61.28\%$, $F_5 = 22.77\%$, $F_6 = 15.95\%$

Find: Calculate the total relative flow rates of the outlet streams.

Assumptions: Steady state, ideal equilibrium stages.

Analysis: Set up steady-state material balances (*rate out = rate in*) in the worksheet for each component. Three material balances (one for each component, or two component balances and an overall balance in this case) give the following system of independent linear equations:

$$0.4716F_2 + 0.1584F_5 + 0.0310F_6 = 0.3300 \times 100 \qquad (4.26)$$

$$0.3549F_2 + 0.3832F_5 + 0.2211F_6 = 0.3400 \times 100 \qquad (4.27)$$

$$F_2 + F_5 + F_6 = 100.0 \qquad (4.28)$$

For convenience in an *Excel* worksheet, use matrix notation for the system of equations:

$$\begin{vmatrix} 0.4716 & 0.1584 & 0.0310 \\ 0.3549 & 0.3832 & 0.2211 \\ 1.000 & 1.000 & 1.000 \end{vmatrix} \begin{vmatrix} F_2 \\ F_5 \\ F_6 \end{vmatrix} = \begin{vmatrix} 33.00 \\ 34.00 \\ 100.0 \end{vmatrix} \qquad (4.29)$$

Put the arrays in ranges of cells on an *Excel* worksheet:

	A	B	C	D
1	0.4716	0.1584	0.031	33
2	0.3549	0.3832	0.2211	34
3	1	1	1	100

Divide the first row by its first term, then use the reduced form of the first row to eliminate the first terms in each of the remaining two rows. Using array operations, select a range of cells **A5:D5** for the results of the row reduction. Type "=" then select the range **A1:D1**. Then type **/A1** in the formula bar, followed by the keyboard combination *CNTL SHIFT ENTER* to simultaneously perform the array operation of division on each cell in the range.

	A	B	C	D
1	0.4716	0.1584	0.031	33
2	0.3549	0.3832	0.2211	34
3	1	1	1	100
5	=A1:D1/A1			

Perform similar operations to eliminate the coefficients for the second and third equations:

	A	B	C	D
5	=A1:D1/A1	=A1:D1/A1	=A1:D1/A1	=A1:D1/A1
6	=A2:D2-A2*A5:D5	=A2:D2-A2*A5:D5	=A2:D2-A2*A5:D5	=A2:D2-A2*A5:D5
7	=A3:D3-A5:D5	=A3:D3-A5:D5	=A3:D3-A5:D5	=A3:D3-A5:D5

	A	B	C	D
5	1	0.3358779	0.0657337	69.974555
6	0	0.2639969	0.1977711	9.1660305
7	0	0.6641221	0.9342663	30.025445

Repeat the process beginning with the second term in the second row to eliminate the second term from each of the first and third rows.

	A	B	C	D
9	=A5:D5-B5*A10:D10	=A5:D5-B5*A10:D10	=A5:D5-B5*A10:D10	=A5:D5-B5*A10:D10
10	=A6:D6/B6	=A6:D6/B6	=A6:D6/B6	=A6:D6/B6
11	=A7:D7-B7*A10:D10	=A7:D7-B7*A10:D10	=A7:D7-B7*A10:D10	=A7:D7-B7*A10:D10

	A	B	C	D
9	1	0	-0.1858864	58.312803
10	0	1	0.7491417	34.720214
11	0	0	0.4367447	6.9669824

Finally, reduce the third row and eliminate the third terms from the first and second rows:

	A	B	C	D
13	=A9:D9-C9*A15:D15	=A9:D9-C9*A15:D15	=A9:D9-C9*A15:D15	=A9:D9-C9*A15:D15
14	=A10:D10-C10*A15:D15	=A10:D10-C10*A15:D15	=A10:D10-C10*A15:D15	=A10:D10-C10*A15:D15
15	=A11:D11/C11	=A11:D11/C11	=A11:D11/C11	=A11:D11/C11

	A	B	C	D
13	1	0	0	61.278077
14	0	1	0	22.769854
15	0	0	1	15.952069

The result is the solution to the system of linear equations arranged in row-reduced echelon format in the range **D13:D15**. Alternatively, solve the system of linear equations with the macro **CROUT**. Open the macro-enabled workbook *PNMSuite*. Select **Enable Macros**, and set up the coefficient matrix and constant vector in an *Excel* worksheet in the *PNMSuite* workbook. Launch the macro **CROUT** from the tab option **View>View macros**. Click **OK** to bring up three input boxes for selecting the range of the coefficient matrix, constant vector, and solution vector. Select the ranges of cells on the worksheet by clicking in the top left cell and holding down the left mouse button while dragging to the bottom right cell. Note that the range of cells for the solution vector must contain the same number of cells as the number of unknowns, arranged in a row or column. In this example, we selected the range **G2:G4** for the results, as shown in Figure 4.2.

	A	B	C	D	E	F	G
1	A				b		x
2	0.4716	0.1584	0.031		33		61.27807671
3	0.3549	0.3832	0.2211		34		22.76985402
4	1	1	1		100		15.95206927

Figure 4.2
Solution to the system of equations represented by the matrix form using Crout's method.

Comments: The macro **CROUT** gives the same result as Gaussian elimination, but requires less effort on the part of the user. Neither Gaussian elimination nor Crout reduction efficiently uses computer storage for large systems of equations where round-off errors may creep into the solution.

4.3.3 Thomas Algorithm for Tridiagonal Matrix Inversion

Matrix inversion and multiplication are inefficient methods for solving sparse systems of equations. The Thomas algorithm efficiently inverts tridiagonal matrices like Equation (4.30) that we find in material and energy balances in staged separations processes, cubic spline interpolation, and finite-difference solutions to boundary value problems:

$$\begin{bmatrix} a_{11} & a_{12} & 0 & 0 & 0 & 0 & 0 \\ a_{21} & a_{22} & a_{23} & 0 & 0 & 0 & 0 \\ 0 & a_{32} & a_{33} & a_{34} & 0 & 0 & 0 \\ 0 & 0 & \ddots & \ddots & \ddots & 0 & 0 \\ 0 & 0 & 0 & a_{n-2,n-3} & a_{n-2,n-2} & a_{n-2,n-1} & 0 \\ 0 & 0 & 0 & 0 & a_{n-1,n-2} & a_{n-1,n-1} & a_{n-1,n} \\ 0 & 0 & 0 & 0 & 0 & a_{n,n-1} & a_{n,n} \end{bmatrix} \cdot \begin{bmatrix} x_1 \\ x_2 \\ \vdots \\ x_n \end{bmatrix} = \begin{bmatrix} b_1 \\ b_2 \\ \vdots \\ b_n \end{bmatrix} \qquad (4.30)$$

The coefficient matrix consists of three diagonal bands centered on the main diagonal.

The Thomas algorithm follows the pattern of Gaussian elimination by first transforming the coefficients in the upper diagonal and constant vector:

$$c_1' = \frac{a_{12}}{a_{11}} \quad \text{and} \quad c_i' = \frac{a_{i,i+1}}{a_{i,i} - c_{i-1}'a_{i,i-1}} \qquad \text{for } i = 2 \text{ to } n-1 \qquad (4.31)$$

$$b_1' = \frac{b_1}{a_{11}} \quad \text{and} \quad b_i' = \frac{b_i - b_{i-1}'a_{i,i-1}}{a_{i,i} - c_{i-1}'a_{i,i-1}} \qquad \text{for } i = 2 \text{ to } n \qquad (4.32)$$

The solution then comes from back substitution, row by row:

$$x_n = b_n' \quad \text{and} \quad x_i = b_i' - c_i' x_{i+1} \qquad \text{for } i = n-1 \text{ to } 1 \qquad (4.33)$$

The *PNMSuite* has a VBA implementation of the Thomas algorithm in a sub procedure named **THOMAS**. The sub procedure requires one-dimensional arrays for the upper, lower and middle diagonal terms in the coefficient matrix, as well as an array of the constant vector. Section 10.2 introduces the method of cubic spline interpolation that employs the Thomas algorithm for efficiently solving the banded linear system of equations. The finite-difference method presented in Section 13.2 generates sparse banded matrices for solving linear boundary-value type second-order differential equations.

4.3.4 Cholesky Decomposition of a Symmetric Matrix*

Cholesky's method decomposes a symmetric positive-definite square matrix into a lower triangular matrix L:

$$A = LL^T \qquad (4.34)$$

Solve a system of linear equations involving a symmetric coefficient matrix by first solving for y then x:

$$Ly = b \quad \rightarrow \quad L^T x = y \qquad (4.35)$$

The algorithm for decomposing a symmetric matrix stems from Gaussian elimination. We use Cholesky's method for correlated variables in Monte Carlo simulations for uncertainty analysis in Section 8.3.2. The *PNMSuite* has a macro named **CHOLESKY** for implementing Cholesky decomposition. The macro first checks for positive-definiteness then returns the lower triangular matrix L.

4.4 Excel Worksheet Array Functions

In Equation (4.13), A is the $n \times n$ square coefficient matrix and b is the $n \times 1$ constant vector. The $n \times n$ square identity matrix I is defined such that

$$AI = IA = A \qquad (4.36)$$

The *inverse* matrix, A^{-1} is defined such that

$$AA^{-1} = A^{-1}A = I = \begin{bmatrix} 1 & 0 & \cdots & 0 \\ 0 & 1 & 0 & \vdots \\ \vdots & 0 & \ddots & 0 \\ 0 & \cdots & 0 & 1 \end{bmatrix} \qquad (4.37)$$

In a manner analogous to the algebraic solution of a single linear equation (although computationally intensive), transform Equation (4.13) to give the solution vector, x, through the following sequence of operations on A and b:

$$A^{-1}Ax = A^{-1}b \quad \rightarrow \quad Ix = A^{-1}b \quad \rightarrow \quad x = A^{-1}b$$

To find the inverse of a square $n \times n$ matrix A, solve n systems of n linear equations for n number of z vectors with the constant b vector replaced by the corresponding column from the identity matrix:

$$A \cdot \begin{bmatrix} z_1 \\ z_2 \\ \vdots \\ z_n \end{bmatrix}_1 = \begin{bmatrix} 1 \\ 0 \\ \vdots \\ 0 \end{bmatrix} \qquad A \cdot \begin{bmatrix} z_1 \\ z_2 \\ \vdots \\ z_n \end{bmatrix}_2 = \begin{bmatrix} 0 \\ 1 \\ \vdots \\ 0 \end{bmatrix} \qquad \cdots \qquad A \cdot \begin{bmatrix} z_1 \\ z_2 \\ \vdots \\ z_n \end{bmatrix}_n = \begin{bmatrix} 0 \\ 0 \\ \vdots \\ 1 \end{bmatrix}$$

The columns of the inverted matrix consist of the solution vectors z as follows:

$$A^{-1} = \begin{bmatrix} \begin{bmatrix} z_1 \\ z_2 \\ \vdots \\ z_n \end{bmatrix}_1 & \begin{bmatrix} z_1 \\ z_2 \\ \vdots \\ z_n \end{bmatrix}_2 & \cdots & \begin{bmatrix} z_1 \\ z_2 \\ \vdots \\ z_n \end{bmatrix}_n \end{bmatrix} \qquad (4.38)$$

Excel has capabilities for performing several operations of matrix algebra including matrix multiplication and inversion, both convenient for finding the solution to a system of linear equations. The *Excel* worksheet array functions for determinant, inverse, and matrix multiplication functions are **MDETERM(M_1:N_n)**, **MINVERSE(M_1: N_n)** and **MMULT(M_1: N_n, R_1:S_n)**, respectively. For linear systems:

- The range **M_1: N_n** must be a square matrix of cells (equal number of rows and columns).
- The range **R_1: S_n** must have the same number of rows as columns in the range **M_1:N_n**.
- To perform the worksheet array function, first select a range of cells to contain the result of the function, then type an equal sign followed by the worksheet matrix function. Note that the formula appears in the first cell of the selected range only.
- Type the simultaneous keyboard combination *CTRL SHIFT ENTER*. Excel adds curly braces { } around the formula in each cell of the solution range. Do not type the braces.

Illustration: Solve the following linear system of equations for x_1, x_2, and x_3 using matrix functions in Excel.

$$-2x_1 + 3x_2 + x_3 = 9$$
$$3x_1 + 4x_2 - 5x_3 = 0$$
$$x_1 - 2x_2 + x_3 = -4$$

CHAPTER 4: LINEAR EQUATIONS

Put the coefficients in the range A1:C3 and constant vector in the range D1:D3. Select the range E1:E3 for the solution, type the imbedded worksheet function formula as shown, and then type *CNTL SHIFT ENTER* for the answer shown in the range E1:E3.

	A	B	C	D	E
1	-2	3	1	9	=MMULT(MINVERSE(A1:C3),D1:D3)
2	3	4	-5	0	
3	1	-2	1	-4	

	A	B	C	D	E
1	-2	3	1	9	-1
2	3	4	-5	0	2
3	1	-2	1	-4	1

To correct formula errors in array operations on the worksheet, first select the cells containing the array operation and then select from the **Editing** group on the **Home** tab **Home>Clear>Clear All**. For large systems (>50 equations), round-off errors start to creep into solution methods based on Gauss elimination without error correction. For this reason, *Excel* limits the input size for worksheet array functions to 53 rows by 53 columns. Use one of the *VBA* macros **GAUSSELIM** or **CROUTRED** for larger systems of linear equations.

A singular matrix has a zero determinant (or close to zero). Before attempting to find the roots to a linear system, we may want to first calculate the determinant of the coefficient matrix in *Excel* using the worksheet array function **MDETERM(A)**, where **A** represents a square range of cells. The *PNMSuite* also has simple *VBA* sub procedures for inverting a square matrix (**MATINV**), matrix multiplication (**MATMLT**), and transposing a matrix (**MATTRN**).

The user-defined *array* function **LSOLVE(A, b)** is available in the *PNMSuite* for performing matrix inversion and multiplication to solve a linear system of equations. The arguments of **LSOLVE** are the coefficient matrix A in a square range of cells and constant vector b in a column range on the worksheet. The function also calculates the determinant. To use the function, select a range of cells on the worksheet for the solution, then type **=LSOLVE(A, B)** followed by the combined keystrokes *CTRL SHIFT ENTER*.

Example 4.2   *Material Balances Around a Two-stage Flash Separation*

Known/Find/Assumptions: Refer to *Example 4.1*.

Analysis: First, enter the coefficients of the A and b matrices into ranges on the worksheet. The worksheet is set-up with borders and titles to show where the calculations occur. Calculate the inverse of the A matrix by first selecting the range of cells that will contain the result for the inverse matrix, A^{-1}, entering the matrix inverse function in the formula bar, **MINVERSE(B2:D4)**, then typing the combined keystrokes: *CTRL SHIFT ENTER*.

	A	B	C	D	E	F	G	H
1	A							
2		0.4716	0.1584	0.031				
3		0.3549	0.3832	0.2211				
4		1	1	1				
6	A^{-1}					b		x
7	=minverse(B2:D4)					33		
8					x	34	=	
9						100		

Notice that the formula is now inside a pair of braces, {}, indicating that this group of cells is now considered an array in *Excel*. *Excel* inserts the braces automatically – do not attempt to type braces into the formula. Calculate the determinant of the square matrix to check for a value of zero using the array function **MDETERM**:

$$0.05437514 = \text{MDETERM(B2:D4)}$$

To multiply the constant-vector b by the inverse matrix A^{-1}, select the cells that will contain the solution matrix and enter the **MMULT** function, as shown in the worksheet.

	A	B	C	D	E	F	G	H	I
6	A^{-1}					b		x	
7		2.981142	-2.34298	0.425618		33		=mmult(B7:D9,F7:F9)	
8		-2.46068	8.102968	-1.71528	x	34	=		
9		-0.52046	-5.75999	2.289667		100			

Finally, get the solution matrix in the selected range with the simultaneous keystrokes, *CTRL SHIFT ENTER*.

	B	C	D	E	F	G	H
6	A^{-1}				b		x
7	2.981142	-2.34298	0.425618		33		61.27808
8	-2.46068	8.102968	-1.71528	x	34	=	22.76985
9	-0.52046	-5.75999	2.289667		100		15.95207

Comments: The results agree with the solution obtained by Gaussian elimination. Note that we may find the solution in a single step by nesting the matrix functions or with the user-defined array function (in either case, we must simultaneously strike the keys *CTRL SHIFT ENTER*):

H7:H9 = MMult(MInverse(B2:D4), F7:F9)   H7:H9 = LSOLVE(B2:D4, F7:F9)

□

4.5 Eigenvalues and Eigenvectors*

We encounter Eigen problems in linear algebra with application to the solution of homogeneous differential equations, among other applications. In linear Eigen problems, we search for the eigenvalues, λ, and corresponding nontrivial eigenvectors, $x \neq 0$, for linear systems with the following form:[51]

$$Ax = \lambda x \qquad \text{or} \qquad (A - \lambda I)x = 0 \qquad (4.39)$$

where A represents the real, square matrix of linear coefficients and I is the identity matrix. The nontrivial solution requires a singular coefficient matrix $(A - \lambda I)$, or the vanishing determinant:

$$|A - \lambda I| = \begin{vmatrix} a_{11} - \lambda & a_{12} & \cdots & a_{1n} \\ a_{21} & a_{22} - \lambda & \cdots & a_{2n} \\ \vdots & \vdots & \ddots & \vdots \\ a_{n1} & a_{n2} & \cdots & a_{nn} - \lambda \end{vmatrix} = 0 \qquad (4.40)$$

The expanded determinant produces the characteristic equation for matrix A, which is an nth degree polynomial in λ:

$$\alpha_0 + \alpha_1 \lambda + \alpha_2 \lambda^2 + \cdots + \alpha_{n-1} \lambda^{n-1} + \alpha_n \lambda^n = 0 \qquad (4.41)$$

The n roots of the characteristic equation are the $i = 1 \ldots n$ eigenvalues (or characteristic values), λ_i. With each eigenvalue, we may solve for their corresponding eigenvectors by arbitrarily reducing the Eigen problem in Equation (4.39) by one row, setting the removed value to one, and solving for the vector x of missing terms in the eigenvectors. Note that the eigenvectors found in this manner represent scaled vectors by any constant (positive or negative).

Several numerical methods are available that range in complexity, ease-of-implementation, and robustness for solving Eigen problems. For its convenience of implementation in *VBA*, we recommend a modified version of the method of interpolation (Fadeev and Fadeeva 1963) for real square matrices programmed in the macro **EIGENVI** in the *PNMSuite*. The macro uses the method of Gaussian elimination with pivoting from Section 4.3.1 to solve the linear systems, and Bairstow's method, described in Section 6.4, for finding the roots to the characteristic Equation

---

[51] "Eigen" is old German for peculiar, special, or unique.

Chapter 4: Linear Equations

(4.41). The macro randomly selects the row for reduction in the solution for the eigenvectors. If the selected term from the eigenvector has a relative value of zero, then a different term is selected at random and the calculation is repeated. Due to the random selection of rows in the calculation of the eigenvectors, we must remember to normalize the values for comparison between repeated application of the macro **EIGENVI**.

To use **EIGENVI**, provide the coefficients for the matrix of the eigenproblem in a square range of cells on an *Excel* worksheet. The macro prompts for the address of the range of cells for the matrix and the cell address for the top, left corner of the results output to the worksheet. The results are arranged on the worksheet with the eigenvalues in a row above the columns of the eigenvectors.

Example 4.3   Eigen Problem Test Cases

Known/Find: Find the eigenvalues and eigenvectors for the following two real, square matrices (Carnahan, Luther and Wilkes 1969):

$$\begin{vmatrix} 11 & 6 & -2 \\ -2 & 18 & 1 \\ -12 & 24 & 13 \end{vmatrix} \qquad \begin{vmatrix} 10 & 9 & 7 & 5 \\ 9 & 10 & 8 & 6 \\ 7 & 8 & 10 & 7 \\ 5 & 6 & 7 & 5 \end{vmatrix}$$

Assumptions: Non-repeating, real eigenvalues

Analysis: Set up the matrices in an *Excel* worksheet and run the macro **EIGENVI**. Select the worksheet ranges for the matrices in the input box and provide the location for the results on the worksheet.

11	6	-2	EIGENVALUES:		
-2	18	1	21	7	14
-12	24	13	EIGENVECTORS in corresponding columns:		
			-0	0.5	1
			1	-0	0.5
			3	1	-0

10	9	7	5	EIGENVALUES:			
9	10	8	6	3.858	30.289	0.010	0.843
7	8	10	7	EIGENVECTORS in corresponding columns:			
5	6	7	5	-1.578	0.986	1.000	-0.747
				-0.685	1.044	-1.686	1.000
				1.551	1.000	-4.055	-0.397
				1.000	0.719	6.714	0.123

Carnahan, Luther, and Wilkes (1969) also included a third sparse test matrix. The macro **EIGENVI** successfully solved the sparse matrix set up in the following worksheet for all eight eigenvalues and eigenvectors:

2	-1	0	0	0	0	0	0	EIGENVALUES:							
-1	2	-1	0	0	0	0	0	2.347	3.000	1.653	3.532	1.000	3.879	0.121	0.468
0	-1	2	-1	0	0	0	0	EIGENVECTORS in corresponding columns:							
0	0	-1	2	-1	0	0	0	1.532	1.000	-1.532	-0.653	1.000	-0.395	1.000	0.742
0	0	0	-1	2	-1	0	0	-0.532	-1.000	-0.532	1.000	1.000	0.742	1.879	1.137
0	0	0	0	-1	2	-1	0	-1.347	0.000	1.347	-0.879	0.000	-1.000	2.532	1.000
0	0	0	0	0	-1	2	-1	1.000	1.000	1.000	0.347	-1.000	1.137	2.879	0.395
0	0	0	0	0	0	-1	2	1.000	-1.000	-1.000	0.347	-1.000	-1.137	2.879	-0.395
								-1.347	0.000	-1.347	-0.879	0.000	1.000	2.532	-1.000
								-0.532	1.000	0.532	1.000	1.000	-0.742	1.879	-1.137
								1.532	-1.000	1.532	-0.653	1.000	0.395	1.000	-0.742

Comments: The results agree with the values reported in the reference. Due to the random selection of rows in the calculation of the eigenvectors, we must remember to scale the values for comparison between repeated application of the macro **EIGENVI**.

□

Example 4.4 System of Homogeneous First-order Differential Equations

Known/Find: Find the nontrivial solution to the following system of homogenous linear first-order differential equations (Zill 1982):

$$\frac{dx}{dt} = x + 2y + z$$
$$\frac{dy}{dt} = 6x - y$$
$$\frac{dz}{dt} = -x - 2y - z$$

Assumptions: Non-repeating, real eigenvalues

Analysis: The particular solution takes the form:

$$x = Ke^{\lambda t}$$

where K is the eigenvector and λ is the corresponding eigenvalue. Set up the matrix for the right-hand-side in the range A1:C3 of an *Excel* worksheet and run the macro **EIGENVI** to calculate the eigenvalues and eigenvectors of the characteristic equation. Provide the information requested by the input boxes with the square matrix in the range **A1:C3**, output to cell **E1**, and the default relative convergence tolerance. The solution for the eigenvalues is 0, 3, and -4. By convention, normalize the values in the eigenvectors with the first row set to one.

	A	B	C	D	E	F	G	H	I	J	K	L
1	1	2	1		EIGENVALUES:					Normalize the eigenvectors		
2	6	-1	0		-1.7E-10	3	-4			using the values in the first row		
3	-1	-2	-1		EIGENVECTORS in corresponding columns:							
4					-0.07692	-1	-1			1	1	1
5					-0.46154	-1.5	2			6	1.5	-2
6					1	1	1			-13	-1	-1

The general solution to the system of first-order differential equations is (note that $e^{0 \cdot t} = 1$ for the first eigenvalue):

$$x = c_1 + c_2 e^{3t} + c_3 e^{-4t}$$
$$y = 6c_1 + 1.5c_2 e^{3t} - 2c_3 e^{-4t}$$
$$z = -13c_1 - c_2 e^{3t} - c_3 e^{-4t}$$

Determine the values for the integration constants c_1, c_2, and c_3 from the solution to the linear system of equations defined at the initial conditions for x, y, and z at $t = 0$:

$$x_0 = c_1 + c_2 + c_3$$
$$y_0 = 6c_1 + 1.5c_2 - 2c_3$$
$$z_0 = -13c_1 - c_2 - c_3$$

Use any of the methods, such as Gaussian elimination derived earlier in the chapter for solving the linear system.

Comments: The macros **EIGENVI** and **GAUSSELIM** are convenient for solving homogeneous linear systems of first-order differential equations, such as those derived from mass balances involving first-order reactions.

□

4.6 Successive Substitution: Jacobi and Gauss-Seidel Methods

"The definition of insanity is doing the same thing over and over and expecting different results." – Benjamin Franklin

There are alternatives to the Gaussian elimination class of methods for finding solutions to large, or near singular, systems of equations. In this section, we introduce an iterative method that employs successive substitution beginning with an initial guess for the roots followed by recursive substitution of the upgraded results until the solution converges. Iterative methods are easy to implement in *Excel* and efficiently use computing resources. However, iterative methods require good initial guesses for the solution, and still may diverge away from the root.

Consider linear algebraic equations arranged in the following form:

$$f(x) = 0 \tag{4.42}$$

Rearrange the linear equations implicitly to isolate the variable from the remaining part of the function (that may include the variable). Use an upper case F to indicate an implicit equation:

$$f(x) = x - F(x) = 0 \tag{4.43}$$

Solve Equation (4.43) in terms of x for the fixed-point form:

$$x = F(x) \tag{4.44}$$

Initialize an iterative solution with a guess near the root, $g \approx x$. Find an upgraded solution by substituting the initial guess into the right side of Equation (4.44) and solving for an upgraded approximation for x:

$$x \cong F(g) \tag{4.45}$$

Replace the guessed value with the upgraded value for x recursively, $g \leftarrow x$:

$$x_i = F(x_{i-1}) \qquad i = 1, 2 \ldots \tag{4.46}$$

where x_{i-1} and x_i are the approximations to the solution at successive iterations $i-1$ and i. Use the subscript i to keep track of the approximations for the solution from previous trials, or iterations. Increase the counter for index i by one for each iteration.

Jacobi's method iterates on Equation (4.46) by successively replacing the guess for x in F with the most recent upgraded value from the previous iteration until the solution converges according to the following criterion:

$$\left| \frac{x_i - x_{i-1}}{x_i + \delta} \right| < tol \tag{4.47}$$

The term δ in the denominator of Equation (4.47) is a small number (e.g., 10^{-8}) used to avoid division by zero in the event that the solution for x is actually zero. The convergence tolerance, *tol*, should be smaller than the floating-point precision of the computer (W. Press, S. Teukolsky and S. Vetterling, et al. 1992). To avoid computer round-off issues, use a tolerance no smaller than the square root of the machine precision. The precision of *Excel* calculations is 10^{-16}, which leads to a tolerance no smaller than 10^{-8}.

Jacobi's method is simple to apply. However, convergence to the solution is often slow, requiring several iterations. If the iterative value for the solution diverges, or becomes unstable (oscillating between increasingly more positive and negative values), try repeating the calculations with a different initial guess for the solution, or an alternative arrangement of the implicit equations in terms of x.

We recommend different adaptations of successive substitution for solving large *systems* of linear equations, as well as nonlinear equations. Consider a system of *n* equations in *n* unknowns:[52]

$$x_1 = F_1(x_1, x_2, x_3, \ldots x_n) \tag{4.48}$$

$$x_2 = F_2(x_1, x_2, x_3, \ldots x_n) \tag{4.49}$$

$$x_3 = F_3(x_1, x_2, x_3, \ldots x_n) \tag{4.50}$$

$$\vdots$$

$$x_n = F_n(x_1, x_2, x_3, \ldots x_n) \tag{4.51}$$

Get an approximate solution by making *n* initial guesses, one for each unknown value, and iterating on these equations.

In Gauss-Seidel iteration, the rate of convergence to the true solution is *usually* increased (requiring fewer iterations) by using the updated solutions in the remaining equations as soon as they become available:

$$x_{1,i} = F_1(x_{1,i-1}, x_{2,i-1}, x_{3,i-1}, \ldots x_{n,i-1}) \tag{4.52}$$

$$x_{2,i} = F_2(x_{1,i}, x_{2,i-1}, x_{3,i-1}, \ldots x_{n,i-1}) \tag{4.53}$$

$$x_{3,i} = F_3(x_{1,i}, x_{2,i}, x_{3,i-1}, \ldots x_{n,i-1}) \tag{4.54}$$

$$\vdots$$

$$x_{n,i} = F_n(x_{1,i}, x_{2,i}, x_{3,i}, \ldots, x_{n-1,i}, x_{n,i-1}) \tag{4.55}$$

Note how Equation (4.53) uses $x_{1,i}$ instead of $x_{1,i-1}$, Equation (4.54) uses $x_{1,i}$ and $x_{2,i}$ instead of $x_{1,i-1}$ and $x_{2,i-1}$, and so on. A convenient criterion for checking convergence uses the maximum *relative* difference between iterations of all the variables:

$$\max_j \left| \frac{x_{j,i} - x_{j,i-1}}{x_{j,i} + \delta} \right| < tol \tag{4.56}$$

where *j* is the variable index, *i* is the iteration number, and δ is a small number (typically 10^{-8}) used to prevent possible division by zero in the event that $x_{j,i} \to 0$. Alternatively, use the residual vector norm to check for convergence:

$$\sqrt{\sum_{j=1}^n (x_{j,i} - x_{j,i-1})^2} < tol \tag{4.57}$$

In *Excel*, we calculate the residual vector norm from the imbedded worksheet functions, **SQRT(SUMXMY2(x_i, x_{i-1}))**. The *PNMSuite* has a user-defined function **VNORMRR(x_{i-1}, x_i)** that returns vector norm of the *relative* residuals for the ranges of values for *x* from two successive iterations:

$$\sqrt{\sum_{j=1}^n \left(\frac{x_{j,i} - x_{j,i-1}}{x_{j,i} + \delta} \right)^2} < tol \tag{4.58}$$

We summarize the Gauss-Seidel algorithm with the following steps applied to a simple system of two simultaneous equations:

1. Arrange the system of equations to give diagonally dominant terms, such that the magnitude of the diagonal coefficient is generally larger than the other terms.

---

[52] An explicit form does not involve the iterating variable in the function: $x_1 = F_1(x_2, x_3 \ldots x_n)$; the implicit form may include the iterating variable in the function: $x_1 = F_1(x_1, x_2, x_3 \ldots x_n)$.

CHAPTER 4: LINEAR EQUATIONS

$$\begin{bmatrix} a_{11} & a_{12} \\ a_{21} & a_{22} \end{bmatrix} \cdot \begin{bmatrix} x_1 \\ x_2 \end{bmatrix} = \begin{bmatrix} b_1 \\ b_2 \end{bmatrix} \quad (4.59)$$

$$a_{11} > a_{12} \quad \text{and} \quad a_{22} > a_{21} \quad (4.60)$$

2. Write the system of equations in explicit form, one equation for each variable, with the row in Equation (4.59) corresponding to the variable:

$$x_1 = F_1(x_2) = \frac{b_1 - a_{12}x_2}{a_{11}} \quad \text{and} \quad x_2 = F_2(x_1) = \frac{b_2 - a_{21}x_1}{a_{22}} \quad (4.61)$$

3. Use good engineering judgment to make the best initial guess for the solution:

$$g_1 \rightarrow x_1 \quad \text{and} \quad g_2 \rightarrow x_2 \quad (4.62)$$

4. Input the guesses into the right side of the explicit equations to calculate upgraded values for the solution:

$$x_1 = F_1(g_2) \quad \text{and} \quad x_2 = F_2(g_1) \quad (4.63)$$

5. Successively substitute the upgraded solutions for the guessed values until the solution converges according to the criteria in Equation (4.56) or (4.57):

$$x_1 \leftarrow F_1(x_2) \quad \text{and} \quad x_2 \leftarrow F_2(x_1) \quad (4.64)$$

Gauss-Seidel iteration also works for systems of non-linear equations. In either case, the sensitivity of the solution to the set of initial guesses increases with n, the number of variables. *Excel* is capable of implementing Gauss-Seidel iteration for the solution of a system of equations directly in an *Excel* worksheet using iteration on circular references, as described in Section 4.6.1.

Example 4.5   Temperature Profile in a Conducting Solid

Known: Consider heat conduction in a two-dimensional plane, one unit length square, as seen in the schematic. The north and west sides are maintained at a constant temperature of 100 and 75°C, respectively. The east and south side temperatures are 25 and 0°C, respectively.

Find: Calculate the temperature at four equally-spaced nodes inside the plane.

Assumptions: steady-state heat transfer, uniform and constant thermal conductivity

Schematic: Interior temperature nodes are located at the grid intersections. The temperature at a node is the average temperature of the grid space surrounding the node, depicted in the dashed square around node one shown in the right image.

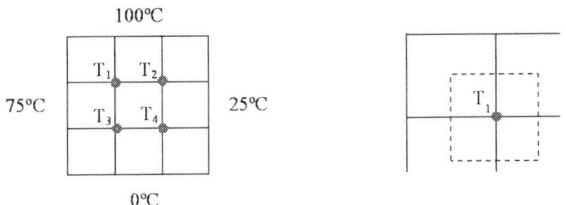

Analysis: This gives a system of four explicit linear equations in four unknowns: T_1, T_2, T_3, and T_4:

$$T_1 = \frac{T_3 + 75 + 100 + T_2}{4} \qquad T_2 = \frac{T_4 + 75 + T_1 + 0}{4} \qquad T_3 = \frac{25 + T_1 + 100 + T_4}{4} \qquad T_4 = \frac{25 + T_2 + T_3 + 0}{4}$$

Use an *Excel* worksheet to solve these equations by Gauss-Seidel iteration. Setup explicit functions for Gauss-Seidel iterative solution at the temperature nodes: (a) T_1 in terms of initial conditions, (b) T_2 uses T_1 from the current iteration and T_4 from initial conditions, (c) T_3 uses T_1 from the current iteration and T_4 from the initial condition, and (d) T_4 uses T_2 and T_3 from the current iteration.

(a) Formula bar: `=(C2+D2+75+100)/4`

	A	B	C	D	E
1	Iteration	T_1	T_2	T_3	T_4
2	0	50	50	50	50
3	1	=(C2+D2+75+100)/4			

(b) Formula bar: `=(B3+E2+75)/4`

	A	B	C	D	E
1	Iteration	T_1	T_2	T_3	T_4
2	0	50	50	50	50
3	1	68.75	=(B3+E2+75)/4		

(c) Formula bar: `=(B3+E2+25+100)/4`

	A	B	C	D	E
1	Iteration	T_1	T_2	T_3	T_4
2	0	50	50	50	50
3	1	68.75	48.4375	=(B3+E2+25+100)/4	

(d) Formula bar: `=(C3+D3+25)/4`

	A	B	C	D	E
1	Iteration	T_1	T_2	T_3	T_4
2	0	50	50	50	50
3	1	68.75	48.4375	60.9375	=(C3+D3+25)/4

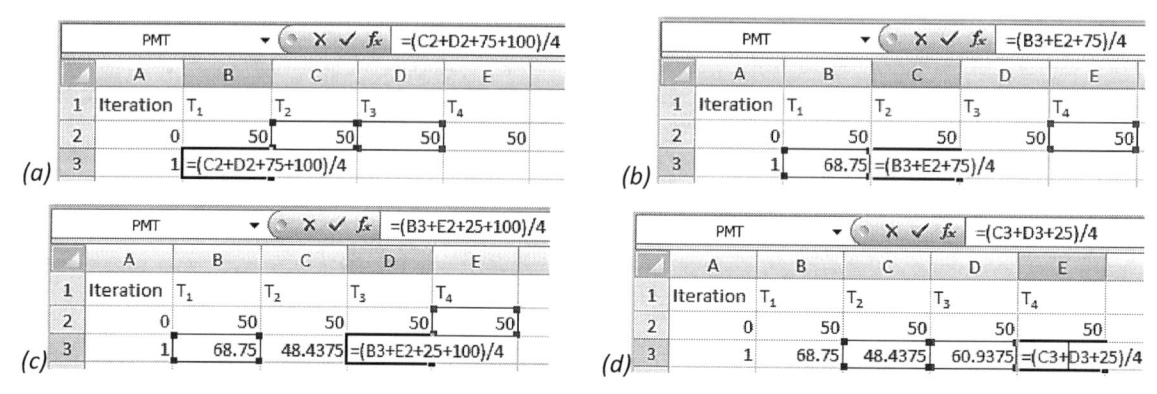

Start with an initial guess of 50°C for each temperature. Column **E** calculates the maximum relative residual of the variables between iterations with a *VBA* user-defined function to check for convergence according to Equation (4.57). After adding the formulas in row **3** using relative cell references, select the cells in the range **B3:F3** and drag the fill handle down several rows to fill out the worksheet with the formulas. After 12 iterations, the solution converges within a relative tolerance of 1×10^{-7}.

	A	B	C	D	E	F
1	Iteration	T_1	T_2	T_3	T_4	Max Res
2	0	50	50	50	50	
3	1	68.75	48.4375	60.9375	33.59375	0.488372
4	2	71.09375	44.92188	57.42188	31.83594	0.078261
5	3	69.33594	44.04297	56.54297	31.39648	0.025352
6	4	68.89648	43.82324	56.32324	31.28662	0.006378
7	5	68.78662	43.76831	56.26831	31.25916	0.001597
8	6	68.75916	43.75458	56.25458	31.25229	0.000399
9	7	68.75229	43.75114	56.25114	31.25057	9.99E-05
10	8	68.75057	43.75029	56.25029	31.25014	2.5E-05
11	9	68.75014	43.75007	56.25007	31.25004	6.24E-06
12	10	68.75004	43.75002	56.25002	31.25001	1.56E-06
13	11	68.75001	43.75	56.25	31.25	3.9E-07
14	12	68.75	43.75	56.25	31.25	9.75E-08
15	13	68.75	43.75	56.25	31.25	2.44E-08

Comments: The solution converges to two decimal places after seven iterations. The solution converges to the precise solution after just 12 iterations.

□

Successful convergence of iterative methods depends on the choice of initial guesses for the solution, or the selection of equations arranged explicitly for the unknown terms. We promote convergence by judiciously choosing equations for explicit rearrangement based on the dominant terms, as illustrated in the next example.

Example 4.6 Analysis of Gas Absorption in a Packed Column

Known/Schematic: The gas and liquid streams in an absorption column have flow rates and compositions as shown in the schematic. The variables y and x are the vapor and liquid mole fractions, and V and L are the vapor and liquid molar flow rates, respectively.

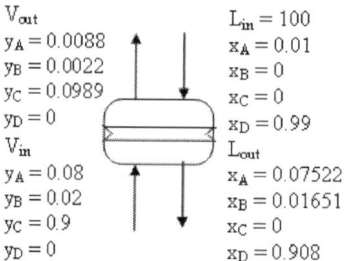

V_{out}
$y_A = 0.0088$
$y_B = 0.0022$
$y_C = 0.0989$
$y_D = 0$

V_{in}
$y_A = 0.08$
$y_B = 0.02$
$y_C = 0.9$
$y_D = 0$

$L_{in} = 100$
$x_A = 0.01$
$x_B = 0$
$x_C = 0$
$x_D = 0.99$

L_{out}
$x_A = 0.07522$
$x_B = 0.01651$
$x_C = 0$
$x_D = 0.908$

Find: Calculate the unknown flow rates, V_{out}, V_{in}, and L_{out}.

CHAPTER 4: LINEAR EQUATIONS

Assumptions: steady-state operation.

Analysis: Simultaneously solve the species mole balances for solutes A, B, and C for the unknown molar flow rates V_{in}, V_{out} and L_{out}.

- Species A $\qquad (0.01)(100) + 0.08V_{in} = 0.07522L_{out} + 0.0088V_{out}$ \hfill (4.65)

- Species B $\qquad 0.02V_{in} = 0.01651L_{out} + 0.0022V_{out}$ \hfill (4.66)

- Species C $\qquad 0.9V_{in} = 0.989V_{out}$ \hfill (4.67)

Divide each equation by its dominant term:

$$\frac{(0.01)(100)}{0.08} + V_{in} = \frac{0.07522}{0.08}L_{out} + \frac{0.0088}{0.08}V_{out} \tag{4.68}$$

$$V_{in} = \frac{0.01651}{0.02}L_{out} + \frac{0.0022}{0.02}V_{out} \tag{4.69}$$

$$\frac{0.9}{0.989}V_{in} = V_{out} \tag{4.70}$$

This gives a simultaneous system of three equations with three unknowns:

$$12.5 + V_{in} = 0.94025L_{out} + 0.11V_{out} \tag{4.71}$$

$$V_{in} = 0.8255L_{out} + 0.11V_{out} \tag{4.72}$$

$$0.91001V_{in} = V_{out} \tag{4.73}$$

Equation (4.73) is the clear candidate for the solution to variable V_{out}. Equations (4.71) and (4.72) both appear to be candidates for the variable V_{in}. There is no clear candidate for L_{out}. Upon further inspection, the coefficient for L_{out} in Equation (4.71) is slightly larger than that in Equation (4.72). Therefore, select Equation (4.71) for L_{out} and Equation (4.72) for V_{in}:

$$L_{out} = \frac{12.5 + V_{in} - 0.11V_{out}}{0.94025} \tag{4.74}$$

$$V_{in} = 0.8255L_{out} + 0.11V_{out} \tag{4.75}$$

$$V_{out} = 0.91001V_{in} \tag{4.76}$$

Solve the system of linear equations by the method of Gauss-Seidel iteration. The left side plot in Figure 4.3 shows the converged results. We found that the solution diverges for other arrangements. For example, consider the following alternative set of divergent explicit equations using Equation (4.72) for L_{out}.

$$L_{out} = \frac{V_{in} - 0.11V_{out}}{0.8255} \tag{4.77}$$

$$V_{in} = 0.94025L_{out} + 0.11V_{out} - 12.5 \tag{4.78}$$

$$V_{out} = 0.91001V_{in} \tag{4.79}$$

The rearranged equations (4.77) to (4.79) quickly diverge away from the solution. By plotting the iterative results for the convergent and divergent forms of explicit equations in Figure 4.3, we observe that the divergent form of the equations has problems even when starting from good initial guesses for the roots.

162   PRACTICAL NUMERICAL METHODS

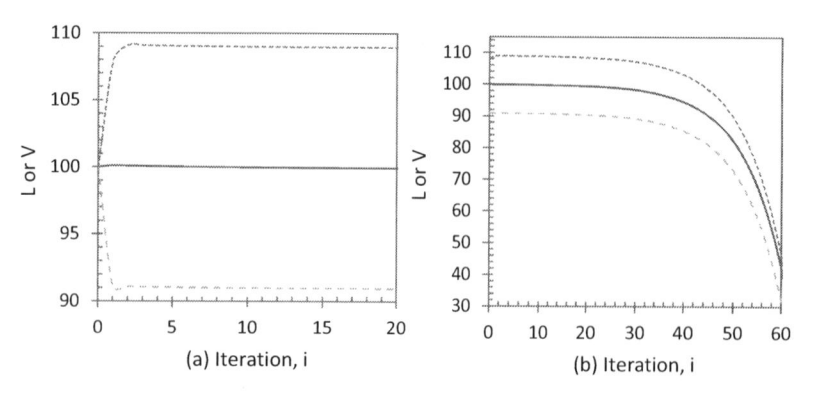

Figure 4.3

(a) Diagonally dominant converging versus (b) diverging solutions to a system of simultaneous equations by iteration.

The selection using diagonally dominant terms converged after four iterations, for L_{out} = 108.93, V_{in} = 99.93, and V_{out} = 90.93. The alternative arrangement appears to head initially towards convergence, and then rapidly diverges after 25 iterations.

Comments: The selection of equations for explicit functions and initial guesses for the solution may be critical for obtaining a converged solution by ordinary iteration.

□

4.6.1 Iteration on Circular References in Excel

"There's no point in stepping up to the ... platform if you're going to repeat yourself." – Robert Plant

A circular reference warning means that *Excel* has detected that a cell containing a formula refers to itself, either directly or indirectly, through the formula of another cell. However, we may want to iterate on the cell reference to solve an implicit function by successive substitution. We can use circular referencing to implement Jacobi iteration directly in an *Excel* worksheet by enabling iteration from the ribbon tab **File>*Excel* Options> Formulas**. We illustrate the method of circular references by example.

Example 4.7   Temperature Profile in a Conducting Solid

Known/Find: Demonstrate iteration on circular references in *Excel* for the problem in *Example 4.5*.

Analysis:

1. Open a new *Excel* worksheet. Add variable names for each of the four temperature nodes in column **A**.

2. Name the adjacent cells in column **B** using the names in column **A**. First, select the range **A1:B4** on the worksheet. Then use the tab **Formulas>Create from Selection>Create names from values in the Left column**.

3. Enter the explicit cell formulas for the temperatures into column **B** on the worksheet using the named cells for each temperature variable. The initial guesses for the solution consist of the original values of zero in each cell. While entering the cell formulas, a warning pops up indicating the presence of a circular reference.

CHAPTER 4: LINEAR EQUATIONS

4. Click **OK** and follow the instructions. To turn on automatic iterations in *Excel*, select **File>*Excel* Options>Formulas>Enable iterative calculations**. We may need to adjust the values set for **Maximum Iterations** and **Maximum Change** to control the precision of the converged result.

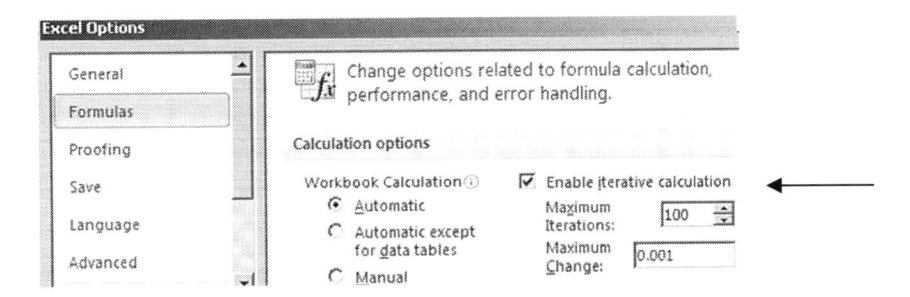

[Excel Options dialog showing Formulas tab with "Enable iterative calculation" checked, Maximum Iterations: 100, Maximum Change: 0.001]

The default **Maximum Change** convergence criterion is not a relative change, but refers to changes smaller than the third decimal place. Lower the value of **Maximum Change** for higher precision. If our result is a number smaller than the maximum change, the iterations will stop after the first iteration. For example, a value of 0.0003 is already lower than the maximum change. Be careful to adjust **Maximum change** to fit the desired precision of the magnitude of the smallest variable in the problem.

5. Complete entering the explicit Gauss-Seidel iteration formulas for the remaining temperatures. *Excel* iterates on the values in the range **B1:B4** until they converge within the maximum change.

	A	B
1	T_1	=(T_3+75+100+T_2)/4
2	T_2	=(T_4+75+T_1)/4
3	T_3	=(25+T_1+100+T_4)/4
4	T_4	=(25+T_2+T_3)/4

	A	B
1	T_1	68.75
2	T_2	43.75
3	T_3	56.25
4	T_4	31.25

Comments: Notice that the precision of the results satisfies the default convergence criteria set by **Maximum Change**. Setting up iteration on circular references makes the calculation "live" in the sense that any changes to the problem parameter automatically restarts the iterative calculations for a new solution. □

4.6.2 Relaxation*

> *"Man is so made that he can only find relaxation from one kind of labor by taking up another."* — Anatole France

At the ith iteration in the Gauss-Seidel method, we can approximate errors in the solution to a system of $j = 1...n$ equations from the residual differences between the current iteration values and explicit functions as follows:

$$r_1 = F_1\left(x_{1,i-1},...x_{j,i-1},...x_{n,i-1}\right) - x_{1,i-1}$$
$$\vdots$$
$$r_j = F_j\left(x_{1,i},...x_{j-1,i},x_{j,i-1},...x_{n,i-1}\right) - x_{j,i-1} \tag{4.80}$$
$$\vdots$$
$$r_n = F_j\left(x_{1,i},x_{2,i},...x_{n-1,i},x_{n,i-1}\right) - x_{n,i-1}$$

As the iterations converge on the true solution, the residual goes to zero: $r \to 0$. To upgrade the solution after each iteration, sum the previous result with some fraction of the error estimate:

$$x_{j,i} = x_{j,i-1} + \omega_j r_j \qquad \text{or} \qquad x_{j,i} = \left(1 - \omega_j\right)x_{j,i-1} + \omega_j F_j \tag{4.81}$$

where ω_j is a relaxation factor for equation (or variable) j. A good value for ω stabilizes the iterative calculations and promotes convergence.

Equation (4.81) represents a weighted average of the long-term averages of previous iterations and the upgraded solution from the current iteration. When the relaxation factor is set to one, ($\omega = 1$), the method reduces to Gauss-Seidel iteration. For $\omega < 1$ Equation (4.81) is under-relaxed, with generally slower convergence, but higher stability for poor initial guesses for the solution. If $\omega > 1$, the method is considered over-relaxed with the potential for faster convergence, but also the strong potential for instability. Although there are no foolproof rules for choosing the best relaxation factor, we recommend a value limited to the range $0 < \omega < 2$.

The concept of relaxation becomes most useful for large systems of equations, particularly when the convergence stability shows sensitivity to the quality of the initial guesses for the solution. In some problems, not all of the iterative equations require relaxation for convergence. Consider selectively applying various levels of relaxation to individual equations as needed to achieve a converged solution. A macro **RELAXIT** is available in the *PNMSuite* that applies relaxation to the Gauss-Seidel method. To use the macro **RELAXIT**, set up the implicit equations in contiguous cells on an *Excel* worksheet in terms of a different range of cells containing the variables. A separate range contains the relaxation factors for each equation.

Example 4.8   Temperature Profile in a 2-D Bar

Known/Find: Rework Example 4.5 using over-relaxation to achieve faster convergence. We employ a uniform relaxation factor between variables and equations.

Analysis: Applying Equation (4.81) to the system of equations in *Example 4.5* gives

$$T_{1,i} = T_{1,i-1} + \frac{\omega}{4}\left(175 + T_{3,i-1} + T_{2,i-1} - 4T_{1,i-1}\right)$$

$$T_{2,i} = T_{2,i-1} + \frac{\omega}{4}\left(75 + T_{4,i-1} + T_{1,i} - 4T_{2,i-1}\right)$$

$$T_{3,i} = T_{3,i-1} + \frac{\omega}{4}\left(125 + T_{1,i} + T_{4,i-1} - 4T_{3,i-1}\right)$$

$$T_{4,i} = T_{4,i-1} + \frac{\omega}{4}\left(25 + T_{2,i} + T_{3,i} - 4T_{4,i-1}\right)$$

Set up an *Excel* worksheet with formulas for the iterative equations using relaxation:

1. Provide the relaxation factor in a named cell **G1**. The next row displays the results from each successive-iteration. Increment the iteration number in column **A**.

2. Provide the initial guesses of 50 for each temperature in row **2**.

3. Place the explicit formulas for the temperature equations in columns **B:E** beginning in row 3.

	A	B	C
1	Iteration	T_1	T_2
2	0	50	50
3	=A2+1	=B2+w*(175+D2+C2-4*B2)/4	=C2+w*(75+E2+B2-4*C2)/4
4	=A3+1	=B3+w*(175+D3+C3-4*B3)/4	=C3+w*(75+E3+B3-4*C3)/4

	D	E	F	G
1	T_3	T_4		w = 1.1
2	50	50	Max Resid	
3	=D2+w*(125+B3+E2-4*D2)/4	=E2+w*(25+C3+D3-4*E2)/4	=maxres(B2:E2,B3:E3)	
4	=D3+w*(125+B4+E3-4*D3)/4	=E3+w*(25+C4+D4-4*E3)/4	=maxres(B3:E3,B4:E4)	

4. Calculate the maximum residual in column **F** using the user-defined function **MAXRES**.

CHAPTER 4: LINEAR EQUATIONS

5. Fill the formulas in subsequent rows using relative referencing down the page until reaching convergence.

6. After some experimentation with values for the relaxation factor, we find that for ω = 1.1, the solution converged after eight iterations (compared with 13 for ω = 1) within a tolerance of 10^{-5}.

	A	B	C	D	E	F	G
1	Iteration	T_1	T_2	T_3	T_4	w =	1.1
2	0	50	50	50	50	Max Resid	
3	1	70.62500	43.12500	62.54688	30.93477	6.16E-01	
4	2	70.12227	44.24144	55.91100	31.32344	1.19E-01	
5	3	68.65469	44.09843	56.27789	31.34614	2.14E-02	
6	4	68.86302	43.71539	56.30473	31.24592	8.76E-03	
7	5	68.74423	43.78342	56.24182	31.25735	1.73E-03	
8	6	68.75752	43.74709	56.25491	31.24981	8.30E-04	
9	7	68.74980	43.75231	56.24940	31.25049	1.19E-04	
10	8	68.75049	43.74985	56.25033	31.25000	5.62E-05	
11	9	68.75000	43.75015	56.24997	31.25003	7.14E-06	
12	10	68.75003	43.74999	56.25002	31.25000	3.57E-06	
13	11	68.75000	43.75001	56.25000	31.25000	4.56E-07	
14	12	68.75000	43.75000	56.25000	31.25000	2.25E-07	
15	13	68.75000	43.75000	56.25000	31.25000	2.94E-08	

Figure 4.4 *Excel* worksheet for implementing iteration in successive rows with relaxation.

We may also use the macro **RELAXIT** using the same set up with the variables in the range **B2:E2** and implicit equations in the range **B3:E3**. We add relaxation factors of 0.5 in the range **H1:K1**. The macro iterated on the solution in the range **B2:E2** until reaching a converged solution in the same range within a tolerance of 10^{-8} after 35 iterations.

Comments: Unfortunately, we do not know the best value for the relaxation factor a priori – it must be determined from experiments with the solution. We should try some degree of under relaxation whenever Gauss-Seidel iterative solutions diverge. Wegstein's method introduced in Section 6.2.2 and Steffensen's method in Section 6.3.2 provide estimates for good relaxation factors to promote convergence.

□

The Gauss-Seidel iterative solution method has several features in common with many of the numerical methods used in the remaining chapters. In particular, concepts of successive approximation and substitution, convergence, equation rearrangement, and requirements for guesses to initialize the solution method are important for many other numerical techniques. Make sure that you understand these concepts before moving on to the rest of the book.

4.7 Epilogue on Linear Equations

Excel with *VBA* has convenient tools for performing linear algebraic operations. Solution methods for linear systems of equations fall into two categories:

1. Matrix methods
   - Inversion and multiplication using *Excel* worksheet functions.
   - Gaussian Elimination and Crout's reduction for medium sized systems of equations
   - Maximum column pivoting and error improvement to reduce computational round-off errors
   - Thomas' algorithm for sparse banded matrices
   - Cholesky's method for symmetric matrices
   - Modified interpolation method for Eigen problems (eigenvalues and eigenvectors)

2. Iterative methods
   - Gauss-Seidel successive substitution (ordinary iteration)
   - Relaxation to control convergence

Methods from either category require effort setting up the equations for solution. Matrix inversion and multiplication has the advantage of producing a result without the need for initial guesses for the roots. Crout's method is more efficient for larger systems than those allowed by *Excel*'s worksheet functions, whereas Thomas' algorithm applies to banded systems. For much larger systems (>100), we resort to iteration on implicit functions, with relaxation to control convergence. Wegstein's method and Steffensen's modification of Aitken's Δ^2 algorithms described in Chapter 6, accelerate convergence by a form of relaxation. The modified interpolation method for Eigen problems tends to work well for small systems involving square matrices. Table 4.1 summarizes the macros introduced in this chapter.

Table 4.1 Summary of VBA macros, user-forms, and user-defined functions (UDF) and methods of input for linear equations.

Macro/Function	Inputs	Operation
CHOLESKY	Sub arguments	Inverts a symmetric square positive-definite matrix.
CROUT	Macro input boxes	Solves a linear system by Crout reduction.
EIGENVI	Macro input boxes	Method of interpolation for finding eigenvalues and vectors for a symmetric, real matrix.
GAUSSELIM	Macro input boxes	Solves a linear system by Gauss elimination with maximum column pivoting.
LINSYS_UsrFrm	Show user-form	Solution to linear systems in matrix format by Gauss elimination or Crout reduction.
LSOLVE	UDF arguments	Solution to a linear system by matrix inversion and multiplication.
MATINV	Sub arguments	Invert a matrix.
MATMLT	Sub arguments	Product of matrices.
MATTRN	Sub arguments	Transpose a matrix.
RELAXIT	Macro input boxes	Solve a system of algebraic equations arranged in implicit form by iteration with relaxation.
THOMAS	Sub arguments	Inverts a banded tridiagonal square matrix.
VNORMRR	UDF arguments	Returns the vector norm of the relative residuals between two ranges of equal length.

Chapter 5   Taylor Series Analysis and Derivative Approximation

"If everything that exists has a place, place too will have a place, and so on ad infinitum." – Aristotle's Paradox

Despite the extensive training engineers receive in mathematics, only a few nonlinear problems in chemical engineering have analytical or closed-form solutions. We rely on numerical methods for obtaining good approximate results to problems that do not have exact answers. The Taylor series expansion is the starting point for finding good linear approximate solutions to a variety of nonlinear problems, including algebraic equations, derivatives, differential equations, and integrals. For example, process control engineers rely on truncated Taylor series expansions to linearize nonlinear differential equations for developing control strategies around process set points. We also need some idea about the quality of our numerical approximations. A Taylor series is useful for estimating errors in numerical results. This chapter transition from solving linear problems to nonlinear problems by introducing five important concepts:

1. Derivation of the Taylor series
2. Derivative approximations derived from Taylor series
3. Richardson's extrapolation to improve the accuracy of derivative approximations
4. Analysis of sensitivity to variables
5. Uncertainty analysis

> **SQ3R Focused Reading Questions**
> 1. What is a truncated Taylor series?
> 2. How does a finite difference approximate a derivative of a function?
> 3. When can I use finite differences to approximate derivatives of data?
> 4. What is the "order" of an approximation?
> 5. What is the meaning of "order" of a derivative?
> 6. What is the "order" of error in a derivative approximation?
> 7. How does extrapolation improve the accuracy of a derivative approximation?
> 8. What is a sensitivity coefficient?
> 9. Can you derive the formula for a two-point central difference derivative approximation?
> 10. What changes in finite difference approximations when data are not equally spaced?

5.1   Taylor Series

The Taylor series approximation of a function is the keystone of numerical methods and deserves our brief consideration before we develop specific numerical methods. Students new to this subject have limited experience with Taylor series from a calculus course. The following derivation is adapted from the fun little book by Lyon (1998).

Consider the arbitrary function plotted as the solid line in Figure 5.1, along with a linear Taylor series approximation for the function, plotted as the heavy dashed line.

Figure 5.1

First-order function approximation near x = g.

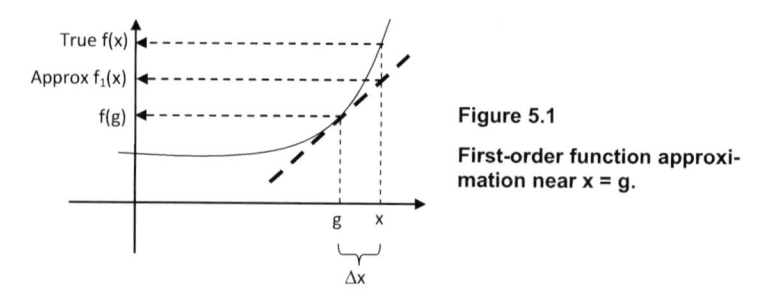

A *linear* approximation for the function of x near the point or expansion $x \cong g$ is a tangent line to the function at g with slope $\partial f / \partial x$ and intercept $f(g)$:

$$f_1(x) \cong f(g) + \Delta x \left. \frac{df}{dx} \right|_g \tag{5.1}$$

where $\Delta x = x - g$ and the derivative (or slope) and function are evaluated at the expansion point, $x = g$.[53] The approximation for the function improves as the region near g shrinks, or $\Delta x \to 0$. We may improve the approximation by adding higher order terms in Δx to the linear approximation in Equation (5.1) to form a polynomial equation:

$$f_\infty(x) = f(g) + \Delta x \left. \frac{df}{dx} \right|_x + c_2 \Delta x^2 + c_3 \Delta x^3 + \ldots \tag{5.2}$$

For example, we see how a second-order (quadratic) approximation shown in Figure 5.2 provides a better approximation to the function over a larger range of Δx.

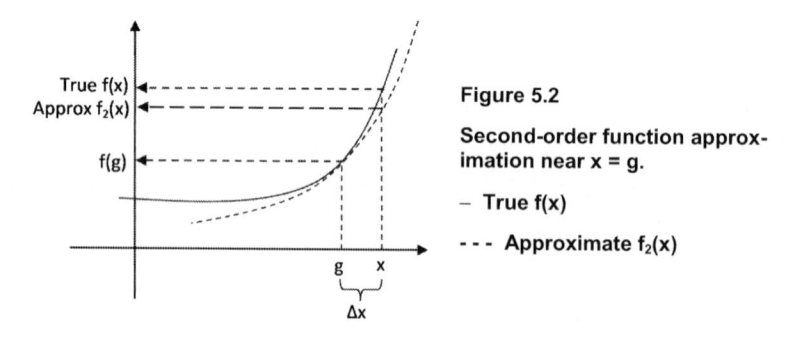

Figure 5.2

Second-order function approximation near x = g.

— True f(x)

- - - Approximate f₂(x)

We find the series of coefficients c by differentiating Equation (5.2) m times with respect to x, then setting $\Delta x = 0$ as follows. Consider the first derivative:

$$\frac{df(x)}{dx} \cong \Delta x \left. \frac{d^2 f}{dx^2} \right|_g + 2c_2 \Delta x + 3c_3 \Delta x^2 + \ldots = \Delta x \left. \frac{d^2 f}{dx^2} \right|_g + \sum_{n=2}^{\infty} n c_n \Delta x^{n-1} \tag{5.3}$$

A second derivative gives:

$$\frac{d^2 f(x)}{dx^2} \cong \Delta x \left. \frac{d^3 f}{dx^3} \right|_g + 2c_2 + 2 \cdot 3 c_3 \Delta x + \ldots = \Delta x \left. \frac{d^3 f}{dx^3} \right|_g + 2c_2 + \sum_{n=3}^{\infty} n(n-1) c_n \Delta x^{n-2} \tag{5.4}$$

After m derivatives, we get:

---

[53] The symbol g is arbitrary. In the next chapter on solutions to nonlinear equations, g stands for the *guess* for the root to a nonlinear function.

$$\frac{d^m f(x)}{dx^m} = \Delta x \frac{d^{m+1} f}{dx^{m+1}}\bigg|_g + m! c_m + \sum_{n=m+1}^{\infty} \left[c_n \Delta x^{n-m} \prod_{\ell=1}^{m}(n-\ell+1) \right] \qquad (5.5)$$

When we set Δx to zero, we eliminate the last $\sum$ term in Equation (5.5):

$$c_m = \frac{\left(d^m f / dx^m \big|_g\right)}{m!} \qquad (5.6)$$

Replacing the coefficients in Equation (5.2) with the result from Equation (5.6) for $m = 2, 3 \ldots$ gives the Taylor series:

$$f(x) = f(g) + \Delta x \frac{df}{dx}\bigg|_g + \frac{\Delta x^2}{2}\frac{d^2 f}{dx^2}\bigg|_g + \frac{\Delta x^3}{3!}\frac{d^3 f}{dx^3}\bigg|_g \cdots \sum_{i=4}^{\infty} \frac{\Delta x^i}{[i]!}\frac{d^i f(x)}{dx^i}\bigg|_{x=g} \qquad (5.7)$$

where the functions and derivatives are evaluated at the expansion point $x = g$. When $g = x$, $\Delta x = 0$, and Equation (5.7) becomes exact.

The Maclaurin series is a special case of the Taylor series expansion about $g = 0$. This special case appears often when equations are made dimensionless and the dependent variable has a limit at $x = 0$. In some cases, we center variables about the mean value by subtracting the mean value, to give a transformed variable:

$$x^* = x - \bar{x} \qquad (5.8)$$

We obtain a linear approximation to the function like Equation (5.1) by truncating the second-order and higher-order terms, leaving a linear equation in x:

$$f(x) \cong f(g) + [x-g]\frac{df(x)}{dx}\bigg|_{x=g} \pm O\left([x-g]^2\right) \qquad (5.9)$$

The remainder has an order of magnitude proportional to the first truncated term $(x-g)^2 d^2f/dx^2$. The result is a local linear approximation to the non-linear function near $x = g$. The error in this linear approximation is the sum of the truncated terms, or the first truncated term evaluated at some value of $\tilde{x}$ where $g < \tilde{x} < x$ such that it forms an exact remainder:

$$e(x) = \sum_{i=2}^{\infty} \frac{[x-g]^i}{[i]!}\frac{d^i f(x)}{dx^i}\bigg|_{x=g} = \frac{[\tilde{x}-g]^2}{2}\frac{d^2 f(x)}{dx^2}\bigg|_{x=g} \qquad (5.10)$$

As the value for x approaches the expansion point, $x \to g$, the error approximated by the remainder goes to zero.

The plot shown in Figure 5.3 illustrates the exponential behavior of the truncation error. Notice that higher order powers give smaller values for $\pm \Delta x < 1$, and then increase dramatically for $\pm \Delta x > 1$.

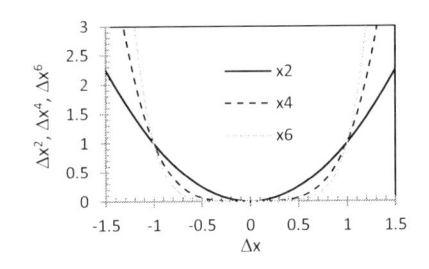

Figure 5.3

Comparison of orders of magnitude $(\Delta x)^n$ for Δx. $\Delta x^n \ll 1$ for $n \to \infty$.

To illustrate, we use a second-order Taylor series expanded around zero to approximate the fourth-order polynomial in Equation (5.11):

$$f(x) = 1 + 0.25x + 0.5x^2 + 0.15x^3 + 0.1x^4 \qquad (5.11)$$

$$f_{Taylor}(x) = f(0) + xf_0' + 0.5x^2 f_0'' \tag{5.12}$$

where $f(0) = 1$ $f_0' = 0.25 + x + 0.45x^2 + 0.4x^3$ $f_0'' = 1 + 0.9x + 1.2x^2$

The plot of the function f versus the Taylor series approximation f_{Taylor} in Figure 5.4 shows two important characteristics of Taylor series approximations of functions:

1. As we get closer to the expansion point, $x \to g$, Taylor series approximations of functions improve. In this case, a visual inspection indicates that the Taylor series performs well for $x < 0.3$.
2. The first truncated term provides an estimate of the error in the truncated Taylor series. In this case, the error is the vertical distance between the function and the Taylor approximation. The error increases nonlinearly with x, or proportional to $(x - g)^3$.

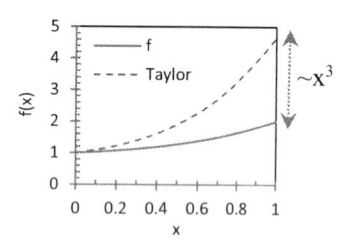

Figure 5.4

Comparison of the fourth-order polynomial in Equation (5.11) with a second-order Taylor series approximation in Equation (5.12).

Example 5.1 Linearize the Vapor-Liquid Equilibrium Curve

Known: The molecular species in a two-component, or binary, fluid mixture are separable using a flash vessel. A flash vessel operates under specific conditions of temperature and pressure that maintain a vapor phase in equilibrium with a liquid phase. The two phases separate easily with the liquid draining from the bottom of the vessel and vapor floating out the top.

Schematic: A single stage flash vessel has labeled streams including the molar feed (F), effluent vapor (V) and liquid (L) flow rates, and corresponding mole fractions of the light component, z, y, and x.

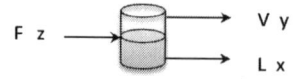

Find: For an equimolar feed ($z = 0.5$) with $F = 100$ kmol/s, $V = 50$ kmol/s, and $L = 50$ kmol/s, calculate the mole fractions in the vapor and liquid streams for a relative volatility, $\alpha = 2.5$.

Assumptions: Streams exit the flash vessel at equilibrium according to Raoult's law, steady-state operation

Analysis: A component material balance for the more volatile species around the flash vessel gives

$$zF = xL + yV \tag{5.13}$$

Solve (5.13) for the vapor mole fraction, y:

$$y = \frac{zF - xL}{V} \tag{5.14}$$

Use Raoult's law to model vapor-liquid equilibrium for the binary system:

- Species 1 $\qquad yP = xP_1^v \tag{5.15}$

- Species 2 $\qquad (1-y)P = (1-x)P_2^v \tag{5.16}$

where P and P^v are the total and pure component vapor pressures, respectively. We define the relative volatility for conditions obeying Raoult's law as the ratio of the vapor pressures in terms of the mole fractions in each phase:

CHAPTER 5: TAYLOR SERIES & DERIVATIVES

$$\alpha = \frac{P_1^v}{P_2^v} = \frac{\frac{y_1 P}{x_1}}{\frac{(1-y_1)P}{(1-x_1)}} = \frac{y_1(1-x_1)}{x_1(1-y_1)} \qquad (5.17)$$

By convention, number the components as $P_1^v > P_2^v$, such that $\alpha > 1$. Rearrange Equation (5.17) for y in terms of x. The vapor-liquid equilibrium curve for an ideal binary system that obeys Raoult's law has the general form:

$$y = \frac{\alpha x}{1+x(\alpha-1)} \qquad (5.18)$$

Find a locally linear approximation to this curve by expanding this function in a truncated first-order Taylor series about an expansion point g:

$$y = \frac{\alpha g}{1+g(\alpha-1)} + (x-g)\frac{d}{dx}\left[\frac{\alpha x}{1+x(\alpha-1)}\right]_{x=g} \qquad (5.19)$$

Equation (5.19) reduces to a linear equation in y and x:

$$y = \frac{\alpha x}{\left[1+g(\alpha-1)\right]^2} + \frac{\alpha g^2(\alpha-1)}{\left[1+g(\alpha-1)\right]^2} \qquad (5.20)$$

where the slope of the linearized expression is the coefficient of x in the first term and the intercept of the dependent variable is the second term:

$$Slope = \frac{\alpha}{\left[1+g(\alpha-1)\right]^2} \qquad Intercept = \frac{\alpha g^2(\alpha-1)}{\left[1+g(\alpha-1)\right]^2} \qquad (5.21)$$

Set $g = 0$, such that Equation (5.20) simplifies to the following single term (a good approximation for dilute systems where Henry's law applies):

$$y = \alpha x \qquad (5.22)$$

At the other extreme for a concentrated system, the expansion point is $g = 1$ and Equation (5.20) reduces to the following linear expression:

$$y = \left(\frac{1}{\alpha}\right)x + \left(\frac{\alpha-1}{\alpha}\right) \qquad (5.23)$$

Combine Equations (5.14) and (5.20) to eliminate y:

$$\frac{zF-xL}{V} = \frac{\alpha x}{\left[1+g(\alpha-1)\right]^2} + \frac{\alpha g^2(\alpha-1)}{\left[1+g(\alpha-1)\right]^2} \qquad (5.24)$$

Solve for x:

$$x = \frac{zF\left[1+g(\alpha-1)\right]^2 - Vg^2\alpha(\alpha-1)}{L\left[1+g(\alpha-1)\right]^2 + V\alpha} \qquad (5.25)$$

Equation (5.25) is exact when $g = x$ at the flash point. Determine the value of g and x by an iterative "trial and error" solution method demonstrated next.

To simplify the calculations, program simple *VBA* user-defined functions for the vapor mole fraction in the vapor stream, at equilibrium, and the Taylor series approximation, respectively:

```
Public Function y(x As Double) As Double
    F = 100: V = 50: L = 50: Z = 0.5: y = (Z * F - x * L) / V
End Function

Public Function ye(x As Double) As Double
```

```
        alpha = 2.5: ye = alpha * x / (1 + x * (alpha - 1))
End Function

Public Function yt(g As Double, x As Double) As Double
    alpha = 2.5
    yt = (alpha * (x + (alpha - 1) * g ^ 2)) / (1 + g * (alpha - 1)) ^ 2
End Function
```

Start by setting the expansion point equal to the feed composition: $g = z = 0.5$. This is a reasonable starting point midway in the allowable range for x: $0 < x < 1$. Plot the y vs x component mass balance in Equation (5.14), vapor-liquid equilibrium relationship in Equation (5.18), and linearized vapor-liquid approximation in Equation (5.20), as seen in Figure 5.5.

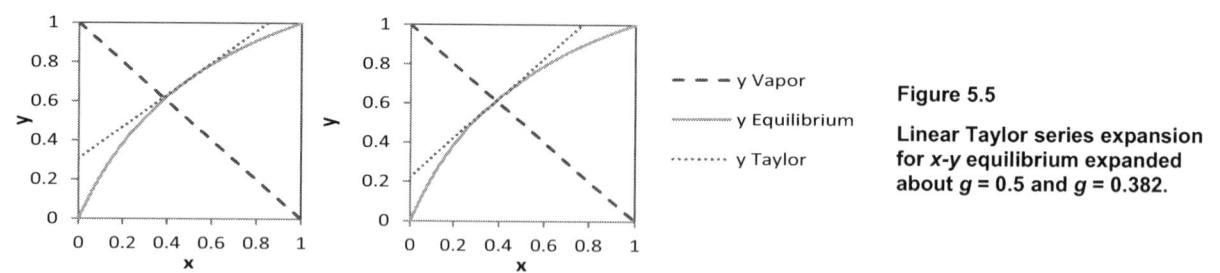

Figure 5.5

Linear Taylor series expansion for x-y equilibrium expanded about $g = 0.5$ and $g = 0.382$.

The first-order Taylor series expansion for the vapor-liquid equilibrium function provides a good approximation near the expansion point $g = 0.5$, but becomes a poor approximation at the extremes of $x \to 0$ and $x \to 1$. Calculate the liquid phase mole fraction then vapor phase mole fraction from the equilibrium and mole balance equations (5.25) and (5.14), respectively, for $x = 0.382$ and $y = 0.618$.

The result for x is close to the initial expansion point, as required for a good Taylor series approximation. Improve the solution by choosing an expansion point closer to the true solution. In the second graph of Figure 5.5 we plot the volatile species mass balance with the vapor-liquid function and the linear approximation for the vapor-liquid function expanded about the previous result setting $g = 0.382$.

Reset the expansion point to the value of $x = 0.382$ and recalculate the mole fractions from Equations (5.25) and (5.14) for $x = 0.387$ and $y = 0.613$. Additional iterations do not provide any practical improvement to the solution.

Comments: We found a good solution after just two iterations on the first-order Taylor series approximation. Try different starting values for the expansion point. For instance, what happens when we start with $g = 0.001$?

□

5.2 Accelerate Series Convergence*

We may try to accelerate convergence of a Taylor series using Shanks transformation or Padé rational function approximation (Bender and Orszag 1999). Richardson's extrapolation presented in Section 5.3.4 is another important (if not the most important) method of sequence transformation and acceleration. The method of Shanks transformation converts a Taylor series into a new series of partial sums as follows:

$$f_n = \sum_{i=0}^{n} a_i \qquad S(f_n) = f_{n+1} - \frac{(f_{n+1} - f_n)^2}{f_{n+1} - 2f_f + f_{n-1}} \qquad (5.26)$$

$$f_0 = a_0 \qquad f_1 = a_0 + a_1 \qquad f_2 = a_0 + a_1 + a_2 \qquad \ldots$$

where a_i represents the terms with exponent i in the case of a Taylor series approximation of Equation (5.7)

$$a_0 = f(g) \qquad a_1 = \Delta x \left.\frac{df}{dx}\right|_g \qquad a_2 = \frac{\Delta x^2}{2} \left.\frac{d^2 f}{dx^2}\right|_g$$

Repeated application of Shanks transformations on the new series may provide further acceleration of series convergence:

$$S^2(f_n) = S(S(f_n)) \qquad S^3(f_n) = S(S(S(f_n))) \tag{5.27}$$

The *PNMSuite* has a *VBA* user-defined function named **SHANKS**, listed in Figure 5.6, for accelerating a truncated Taylor series. Shanks transformation is a special case of Aitkin's method introduced in Section 6.3.2.

```
Public Function SHANKS(b As Variant) As Double
' b = variant array of series for calculating the new series of partial sums
'****************************************************************************
' Shanks transformation accelerates convergence of a series of partial sums.
' *** Nomenclature ***
Dim a() As Double, a1 As Double  ' = vector of partial sums
Dim i As Integer, j As Integer, j1 As Integer  ' = index variable
Dim m As Integer, n As Integer   ' = size of b array
Dim s() As Double, s1 As Double  ' = Shanks transformed partial sums
Const small As Double = 0.000000000001  ' = small number to avoid division by zero
n = b.Count ' size of b array
If n < 3 Then
    MsgBox "Requires a minimum of 3 terms in the series!": Exit Function
End If
ReDim a(1 To n) As Double, s(1 To n, 2 To n - 1) As Double
a(1) = b(1) ' create a new series of partial sums from the original series
For i = 2 To n: a(i) = a(i - 1) + b(i): Next i
m = n - 1
For i = 2 To m ' first Shanks transformation
    a1 = a(i + 1)
    s(1, i) = a1 - ((a1 - a(i)) ^ 2) / (a1 - 2 * a(i) + a(i - 1) + small)
Next i
For j = 1 To n ' if possible, perform additional Shanks transformations on s(j)
    If m - j < 3 Then
        SHANKS = s(j, m): Exit Function
    End If
    m = m - 1: j1 = j + 1
    For i = j + 2 To m
        s1 = s(j, i + 1)
        s(j1, i) = s1 - ((s1 - s(j, i)) ^ 2) / (s1 - 2 * s(j, i) + s(j, i - 1) + small)
    Next i
Next j
SHANKS = s(j1, m)
End Function ' SHANKS
```

Figure 5.6 VBA user-defined function SHANKS for series convergence acceleration.

We may also try accelerating convergence of a Taylor series with a Padé rational function approximation (Baker and Graves-Morris 1996):

$$y(x) \cong \frac{p_0 + p_1 x + p_2 x^2 + \ldots}{1 + q_1 x + q_2 x^2 + \ldots} \tag{5.28}$$

In a Padé approximation, the sum of the orders of polynomials in the numerator and denominator match the order of the truncated Taylor series expansion polynomial (Press, Teukolsky and Vetterling 2007). Experience tells us that the Padé rational function often works best when the orders of the polynomials in the numerator and denominator are equal, or differ by no more than one. To illustrate the method, we find a Padé approximation for a second-order Taylor series, using prime notation for the derivative ($f' = df/dx$). The sum of the orders of the polynomials in the numerator and denominator of the Padé approximation (1+1=2) matches the order of the Taylor expansion (2):

$$y = y_0 + \Delta x f_0 + \frac{\Delta x^2}{2!} f_0' \cong \frac{p_0 + p_1 \Delta x}{1 + q_1 \Delta x} \qquad (5.29)$$

where $\Delta x = x - x_0$. We then calculate the Padé parameters by cross-multiplication with the denominator and equate terms in powers of x on each side of the equation (Hanna and Sandall 1995):

$$\left(y_0 + \Delta x f_0 + \frac{\Delta x^2}{2!} f_0' \right)(1 + q_1 \Delta x) = p_0 + p_1 \Delta x$$

$$y_0 + (y_0 q_1 + f_0) \Delta x + \left(q_1 f_0 + \frac{f_0'}{2!} \right) \Delta x^2 + \frac{q_1 f_0'}{2!} \Delta x^3 = p_0 + p_1 \Delta x \qquad (5.30)$$

$$\Delta x^0 : y_0 = p_0$$

$$\Delta x^1 : y_0 q_1 + f_0 = p_1$$

$$\Delta x^2 : q_1 f_0 + \frac{f_0'}{2!} = 0$$

Since we only have three Padé parameters, we ignore terms involving Δx^3, or higher. We solve for the p's and q's starting from the last equation for terms in powers Δx^2:

$$p_0 = y_0 \qquad p_1 = \frac{y_0 f_0'}{2 f_0} + f_0 \qquad q_1 = \frac{f_0'}{2 f_0} \qquad (5.31)$$

After substitution into Equation (5.29) and rearrangement, our Padé rational approximation takes the following form:

$$y = \frac{2 y_0 f_0 + \left(y_0 f_0' + 2 f_0^2 \right) \Delta x}{2 f_0 + f_0' \Delta x} \qquad (5.32)$$

To improve the accuracy, we may extend this method to higher order Taylor series and Padé approximations as far as we can differentiate the function of the Taylor series. A *VBA* macro **PADE** listed in Figure 5.7 is available in the *PNMSuite* to calculate the Padé coefficients p's and q's in Equation (5.28).

```
Public Sub PADE()
' Accelerate a power series with a Pade rational function.
'
'                                  p0 + p1*x + p2*x^2 ...
' y = a0 + a1*x + a2*x^2 + a3*x^3 ... = --------------------
'                                  1  + q1*x + q2*x^2 ...
' *** Required User-defined Functions and Sub Procedures from the PNM Suite ***
' GAUSSPIVOT = solve a system of linear equations by Gauss elimination with pivoting
' JUSTABC_123 = convert worksheet name for compatibility with VBA
'
' *** Nomenclature ***
Dim a() As Double  ' = vector of polynomial coefficients, starting at lowest order term
Dim ar As Range    ' = range of polynomial coefficients, starting at lowest order term
Dim bxtl As String ' = macro name
Dim c() As Double  ' = matrix of coefficients in a*q product array
Dim i As Long, j As Long
Dim k As Long, m As Long  ' = orders of polynomials in numerator and denominator
Dim n As Long      ' = order of polynomial
Dim outrange As String ' = range for output
Dim ow As Long     ' = check for overwrite
Dim pq() As Double ' = vector of Pade coefficients
Dim pr As Range    ' = location of Pade coefficient results on worksheet
Dim s As Long      ' = counter for rows
```

CHAPTER 5: TAYLOR SERIES & DERIVATIVES

```
Dim ws As String   ' = worksheet name
'************************************************************************
bxtl = "PADE"
With Application
Set ar = .InputBox(Prompt:="Select range of polynomial COEFFICENTS:" & vbNewLine & _
                "y = a0 + a1*x + a2*x^2 ...", Type:=8)
Set pr = .InputBox(Prompt:="Select cell for Pade P & Q RESULTS:", Type:=8)
n = ar.Count
ws = JUSTABC_123(pr.Worksheet.Name)   ' Check worksheet name
pr.Worksheet.Name = ws   ' Check output cells to avoid overwrite
ReDim a(1 To n) As Double, c(1 To n, 1 To n) As Double, pq(1 To n) As Double
' Check order of polynomial for even or odd number of terms and set orders of Pade
If n Mod 2 = 0 Then   ' even
    k = n / 2: m = n / 2
Else   ' odd, set order of denominator one higher than numerator
    m = (n - 1) / 2: k = n - m
End If
For i = 1 To n   ' Set the values of diagonal of product array
    a(i) = ar(i): c(i, i) = 1
Next i
For j = k + 1 To n   ' Set values of product array for q columns
    s = 0
    For i = j - k + 1 To n
        s = s + 1: c(i, j) = -a(s)
    Next i
Next j
Call GAUSSPIVOT(c, a, n, pq)   ' solve the linear system for Pade coefficients
With pr   '
    For i = 1 To k   ' P numerator results on the worksheet
        .Offset(i - 1, 0) = "P(" & i - 1 & ") =": .Offset(i - 1, 1) = pq(i)
    Next i
    .Offset(k, 0) = "Q(0) =": .Offset(k, 1) = 1   ' Q denominator results on worksheet
    For i = 1 To m
        .Offset(k + i, 0) = "Q(" & i & ") =": .Offset(k + i, 1) = pq(i + k)
    Next i
End With   ' pr
End Sub   ' PADE
```

Figure 5.7 VBA sub procedure PADE for calculating the p's and q's of the numerator and denominator of a Padé rational approximation of a polynomial power series.

Example 5.2 Accelerate Convergence of a Taylor Series Numerical Solution of an Ordinary Differential Equation

Known: Consider the following linear first order, ordinary differential equation with initial condition:

$$\frac{dy}{dt} = y + t \qquad y(t=0) = 1 \tag{5.33}$$

Find: (a) Approximate the solution with a Taylor series expanded about the initial condition for $0 \leq t \leq 1$. (b) Accelerate the series solution using Shanks and Padé transformations. (c) Compare the numerical solutions with the following analytical solution (Gerald and Wheatley 2004):

$$y = 2e^t - t - 1 \tag{5.34}$$

Analysis: A fifth-order Taylor series expansion for the solution about $t = 0$ has the following form of a Maclaurin series:

$$y = y(0) + t\frac{dy}{dt}\bigg|_{t=0} + \frac{t^2}{2}\frac{d^2y}{dt^2}\bigg|_{t=0} + \frac{t^3}{6}\frac{d^3y}{dt^3}\bigg|_{t=0} + \frac{t^4}{24}\frac{d^4y}{dt^4}\bigg|_{t=0} + \frac{t^5}{120}\frac{d^5y}{dt^5}\bigg|_{t=0} \tag{5.35}$$

Evaluate the derivatives in the expansion by repeated differentiation of the first derivative in Equation (5.33):

$$y(0) = 1$$
$$\left.\frac{dy}{dt}\right|_{t=0} = y + t = 1 + 0 = 1$$
$$\left.\frac{d^2y}{dt^2}\right|_{t=0} = \left.\frac{dy}{dt}\right|_{t=0} + 1 = 1 + 1 = 2 \qquad (5.36)$$
$$\left.\frac{d^5y}{dt^5}\right|_{t=0} = \left.\frac{d^4y}{dt^4}\right|_{t=0} = \left.\frac{d^3y}{dt^3}\right|_{t=0} = \left.\frac{d^2y}{dt^2}\right|_{t=0} = 2$$

Substitute the derivatives from Equation (5.36) into the Maclaurin series of Equation (5.35) and simplify for the following first six terms in a power series:

$$y = 1 + t + t^2 + \frac{t^3}{3} + \frac{t^4}{12} + \frac{t^5}{60} \qquad (5.37)$$

Higher order terms become increasingly smaller as $2/n!$, which converges slowly. We used the *VBA* macro **PADE** from the *PNMSuite* to generate the coefficients for the following Padé transformation of Equation (5.37):

$$y \cong \frac{1 + (5/8)t + (213/320)t^2}{1 - (3/8)t + (13/320)t^2 + (1/960)t^3} \qquad (5.38)$$

The results of the numerical results of the McLauren series, Padé, and Shanks transformations are compared with the analytical solution displaying four decimals in Table 5.1. Shanks transformation extends the agreement with the analytical solution over the full range of *t*. Padé's transformation gives third decimal agreement over the full range. Padé transformation works best when with alternating signs in the terms of the series.

Table 5.1 Comparison of the numerical Taylor series, Padé and Shanks transformations with the analytical solution for the differential Equation (5.33).

t	Maclaurin	Padé	Shanks	Analytical
0	1.0000	1.0000	1.0000	1.0000
0.2	1.2428	1.2428	1.2428	1.2428
0.4	1.5836	1.5836	1.5836	1.5836
0.5	1.7974	1.7974	1.7974	1.7974
0.6	2.0441	2.0442	2.0442	2.0442
0.7	2.3271	2.3274	2.3275	2.3275
0.8	2.6503	2.6509	2.6511	2.6511
0.9	3.0175	3.0189	3.0192	3.0192
1	3.4333	3.4359	3.4365	3.4366

The results for Shanks transformation were generated using a **Data Table** in an *Excel* worksheet with the user-defined function **SHANKS** from the *PNMSuite*. Figure 5.8 compares the calculation results for each of the numerical methods.

CHAPTER 5: TAYLOR SERIES & DERIVATIVES

	A	B	C	D	E	F	G	H
1	t	0						
2								
3	Taylor			Pade			Shanks	
4	1	1		P(0) =	1		y	=shanks(B4:B9)
5	1	=B1		P(1) =	0.625		0	=TABLE(,B1)
6	1	=B1^2		P(2) =	0.665625		0.1	=TABLE(,B1)
7	=1/3	=(B1^3)/3		Q(0) =	1		0.2	=TABLE(,B1)
8	=1/12	=(B1^4)/12		Q(1) =	-0.375		0.3	=TABLE(,B1)
9	=1/60	=(B1^5)/60		Q(2) =	0.040625		0.4	=TABLE(,B1)
10	y =	=SUM(B4:B9)		Q(3) =	0.00104166666666667		0.5	=TABLE(,B1)
11				y =	=(E4+E5*B1+E6*B1^2)/(E7+E8*B1+E9*B1^2+E10*B1^3)		0.6	=TABLE(,B1)
12							0.7	=TABLE(,B1)
13							0.8	=TABLE(,B1)
14							0.9	=TABLE(,B1)
15							1	=TABLE(,B1)

	A	B	C	D	E	F	G	H
1	t		0					
2								
3	Taylor (Maclaurin) Series			Pade			Shanks	
4		1	1	P(0) =	1		y	1
5		1	0	P(1) =	5/8		0	1
6		1	0	P(2) =	213/320		0.1	1.1103
7	0.333333333		0	Q(0) =	1		0.2	1.2428
8	0.083333333		0	Q(1) =	- 3/8		0.3	1.3997
9	0.016666667		0	Q(2) =	13/320		0.4	1.5836
10		y =	1	Q(3) =	1/960		0.5	1.7974
11				y =	1		0.6	2.0442
12							0.7	2.3275
13							0.8	2.6511
14							0.9	3.0192
15							1	3.4365

Figure 5.8

Excel worksheet setup for the numerical solution of the differential equation by Taylor series with Padé and Shanks transformations.

We used custom number formatting to display the Padé coefficients as fractions. To apply **Fraction**, select the range of cells and select **Fraction** from the ribbon **Home>Number group** drop down menu, as follows:

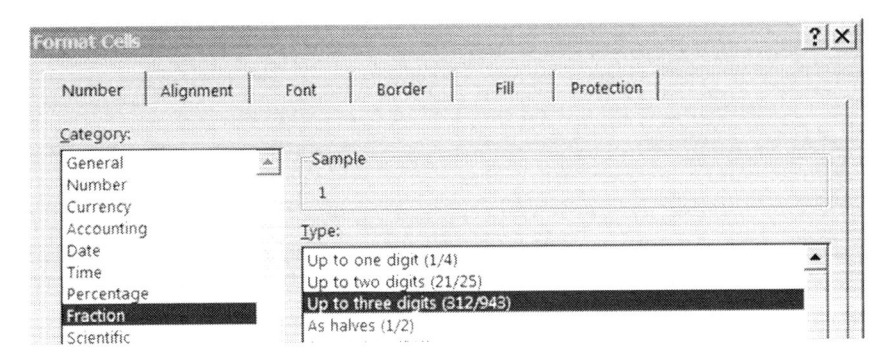

Comments: Shanks transformation proved to be a superior method of series acceleration for this example. We can add higher order terms to the McLauren series to increase the precision of the numerical results.

☐

5.3 Derivative Approximations

The Taylor series is a useful tool for generating approximations of derivatives from tabulated data, or for complicated functions that are difficult or tedious to differentiate analytically. For example, suppose we need to determine the rate of a chemical reaction. We propose to accomplish this by monitoring the concentration of the reactant at fixed increments of time in a well-mixed, constant volume batch reactor, depicted in Figure 5.9.

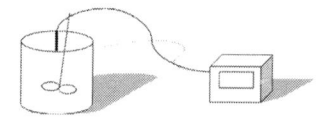

Figure 5.9

Schematic of a well-mixed batch reactor with concentration probe and data recorder.

Table 5.2 lists the transient concentration data obtained from our experiment.

Table 5.2 Batch reactor data with initial concentration of one mol/L.

i	t/min	C/(mol/L)	i	t/min	C/(mol/L)	i	t/min	C/(mol/L)	i	t/min	C/(mol/L)
1	0.5	0.86	6	3	0.41	11	5.5	0.19	16	8	0.09
2	1	0.74	7	3.5	0.35	12	6	0.17	17	8.5	0.08
3	1.5	0.64	8	4	0.30	13	6.5	0.14	18	9	0.07
4	2	0.55	9	4.5	0.26	14	7	0.12	19	9.5	0.06
5	2.5	0.47	10	5	0.22	15	7.5	0.11	20	10	0.05

A plot of the reactor data in Figure 5.10 shows the change in concentration C at each time step of $\Delta t = 0.5$ minutes.

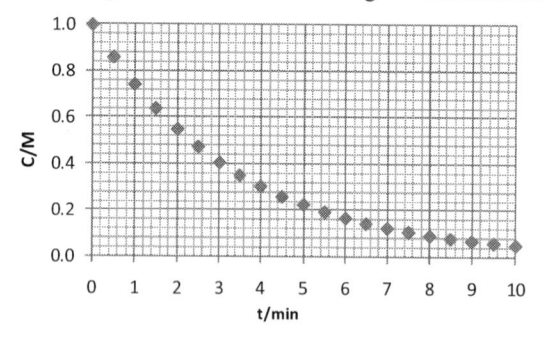

Figure 5.10

Plot of batch reactor concentration data.

The material balance for this batch reaction process is:

(Rate In = 0) − (Rate Out = 0) + Rate of Generation = Rate of Accumulation

For a constant volume system, the material balance reduces to a first order differential equation:

$$-r = \frac{dC}{dt} \qquad (5.39)$$

where r is the rate of reaction (generation term per unit volume), C is the concentration of the reacting molecular species, and t is time. At any point in time, the reaction rate is proportional to the slope of a smooth function of the concentration that passes through the data (The rate law function of concentration is unknown).

From calculus, we define a derivative as the limit of the ratio of changes in the dependent to independent variable as the denominator goes to zero:

$$\frac{dy}{dx} = \lim_{\Delta x \to 0} \frac{y(x+\Delta x) - y(x)}{\Delta x} \qquad (5.40)$$

To illustrate, substitute the function $y = x^2$ into Equation (5.40):

$$\frac{dy}{dx} = \lim_{\Delta x \to 0} \frac{(x+\Delta x)^2 - x^2}{\Delta x} = \frac{x^2 + 2x\Delta x - x^2}{\Delta x} = 2x \qquad (5.41)$$

In the next section, we show how to use the discrete concentration versus time data in Table 5.2 to approximate the slope, or derivative, required in Equation (5.39) for the reaction rate, r.

5.3.1 Finite-difference Derivative Approximations

For the data in Table 5.2, employ a first-order Taylor series expansion of a function for concentration, C, about time, t, to obtain an approximate concentration value at an *earlier* time step $t - \Delta t$, for $C(t - \Delta t)$:

$$C(t - \Delta t) \cong C(t) - \Delta t \left. \frac{dC}{dt} \right|_t \mp O(\Delta t^2) \tag{5.42}$$

where the order of magnitude in the error in the truncated expansion is proportional to Δt^2. Upon rearrangement for the derivative, obtain a first-order *backward* difference approximation of the first derivative using two adjacent data pairs:

$$\left. \frac{dC}{dt} \right|_{t-} \cong \frac{C(t) - C(t - \Delta t)}{\Delta t} \pm O(\Delta t) \tag{5.43}$$

Another first order Taylor series expansion about t for the concentration at a *later* time step for $C(t + \Delta t)$ gives

$$C(t + \Delta t) \cong C(t) + \Delta t \left. \frac{dC}{dt} \right|_t \pm O(\Delta t^2) \tag{5.44}$$

Upon rearrangement, obtain a first-order *forward* difference approximation of the first derivative of concentration with respect to time in terms of two adjacent data pairs, which looks like the derivative definition in Equation (5.40) *without* the limit $t \to 0$, hence the name *finite* difference derivative approximation:

$$\left. \frac{dC}{dt} \right|_{t+} \cong \frac{C(t + \Delta t) - C(t)}{\Delta t} \pm O(\Delta t) \tag{5.45}$$

Following a similar procedure, obtain a linear approximation for the second derivative of concentration with respect to time. Add the second-order terms to each of the Taylor expansions in Equations (5.42) and (5.44):

$$C(t - \Delta t) \cong C(t) - \Delta t \left. \frac{dC}{dt} \right|_t + \frac{\Delta t^2}{2} \left. \frac{d^2 C}{dt^2} \right|_t \mp O(\Delta t^3) \tag{5.46}$$

$$C(t + \Delta t) \cong C(t) + \Delta t \left. \frac{dC}{dt} \right|_t + \frac{\Delta t^2}{2} \left. \frac{d^2 C}{dt^2} \right|_t \pm O(\Delta t^3) \tag{5.47}$$

Sum Equations (5.46) and (5.47) then solve for the second-order, *central* difference, second derivative approximation in terms of three data pairs near t:

$$\left. \frac{d^2 C}{dt^2} \right|_t \cong \frac{C(t + \Delta t) - 2C(t) + C(t - \Delta t)}{\Delta t^2} \pm O(\Delta t^2) \tag{5.48}$$

Obtain a *central* difference, second-order accurate approximation of the *first* derivative by subtracting Equation (5.46) from (5.47) to eliminate the second derivative terms, then solving for the *first* derivative:

$$\left. \frac{dC}{dt} \right|_{t_i} = \frac{C(t + \Delta t) - C(t - \Delta t)}{2\Delta t} \pm O(\Delta t^2) \tag{5.49}$$

Derive second-order accurate *forward* and *backward* difference derivative approximations in a similar fashion. For a forward approximation, expand truncated Taylor series about t for C at $t + \Delta t$, as in Equation (5.47), and $t + 2\Delta t$:

$$C(t + 2\Delta t) \cong C(t) + 2\Delta t \left. \frac{dC}{dt} \right|_t + 2\Delta t^2 \left. \frac{d^2 C}{dt^2} \right|_t \pm O(\Delta t^3) \tag{5.50}$$

Solve Equations (5.47) and (5.50) simultaneously for the first and second derivatives at t in terms of three data pairs:

$$\left.\frac{dC}{dt}\right|_{t+} = \frac{-3C(t) + 4C(t+\Delta t) - C(t+2\Delta t)}{2\Delta t} \pm O(\Delta t^2) \quad (5.51)$$

Follow the same approach to derive a three-point backward derivative approximation from Equation (5.46) and an expansion for C at $t - 2\Delta t$:

$$\left.\frac{dC}{dt}\right|_{t-} = \frac{3C(t) - 4C(t-\Delta t) + C(t-2\Delta t)}{2\Delta t} \pm O(\Delta t^2) \quad (5.52)$$

Following conventional notation, use a subscript index to reference the data points in the finite-difference approximations. Using an index i for the first point, rewrite the forward difference approximation for the first derivative in Equation (5.51):

$$\left.\frac{dC}{dt}\right|_i = \frac{-3C_i + 4C_{i+1} - C_{i+2}}{2\Delta t} \quad (5.53)$$

Use Equation (5.53) to calculate the reaction rate at time zero from the data in Figure 5.10:

$$\left.\frac{dC}{dt}\right|_i = \frac{-3C_0 + 4C_1 - C_2}{2\Delta t} \quad (5.54)$$

or

$$r = \left.\frac{dC}{dt}\right|_{t=0\,\text{min}} \cong \frac{-3\cdot(1.00\text{ M}) + 4\cdot(0.86\text{ M}) - 1\cdot(0.74\text{ M})}{0.5\,\text{min}} = -0.6\frac{\text{M}}{\text{min}}$$

Use various combinations of Taylor series expansions to get other forms of forward, backward, and central differences, as well as higher order derivative approximations.

In summary, we may obtain finite-difference approximations for derivatives of any order by solving linear systems of equations derived from Taylor series expansions for the derivative terms. For example, second order derivatives require at least two equations consisting of second order Taylor series expansions for a coupled system of two equations in two unknowns. The following equations use shorthand symbols f' and f'' for the first and second-order derivatives:

$$f_{i+1} = f_i + \Delta x f_i' + \frac{\Delta x^2}{2} f_i'' \qquad f_{i-1} = f_i - \Delta x f_i' + \frac{\Delta x^2}{2} f_i'' \quad (5.55)$$

where f_i is the function evaluated at x_i and f_{i+1} is the function evaluated at x_{i+1}, and so forth. The variable for position x_{i+1} is a distance Δx from x_i, or $x_{i+1} = x_i + \Delta x$. The variable f_{i-1} is the function evaluated at x_{i-1} where $x_{i-1} = x_i - \Delta x$. We evaluate all of the derivatives at x_i. Note also that in the expansions in the negative direction from x_i all odd terms are negative because the $(x - x_i)$ terms are replaced by $(x_{i-1} - x_i)$, etc. where $x_{i-1} < x_i$ such that $(x_{i-1} - x_i)$ raised to an odd power is negative or positive when raised to an even power. We usually use even-order expansions, which require an odd number of uniformly spaced points, such as those shown in Figure 5.11.

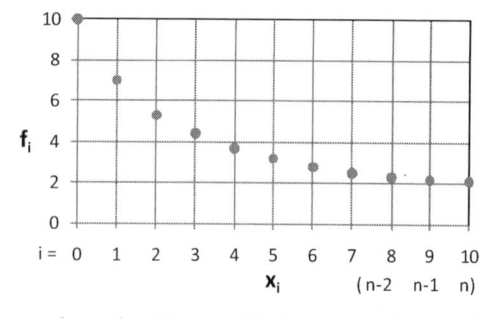

Figure 5.11

Example of nodes for finite-difference derivative approximations.
- Node 0 requires a forward difference approximation.
- Node n requires a backward difference approximation.
- Nodes 1 through n-1 use central difference approximations.

Once again, solve the two Taylor expansions in Equations (5.55) for the first and second derivatives to obtain the three-point central-difference derivative approximations in Table 5.3. We obtain similar results for three-point finite-difference derivative approximations at the beginning and end points ($i = 0$ or $i = n$) from Taylor expansions in each direction:

Chapter 5: Taylor Series & Derivatives

$$f_{i+1} = f_i + \Delta x f_i' + \frac{\Delta x^2}{2!} f_i'' \qquad\qquad f_{i+2} = f_i + 2\Delta x f_i' + \frac{(2\Delta x)^2}{2!} f_i'' \qquad (5.56)$$

$$f_{n-2} = f_n - 2\Delta x f_n' + \frac{(-2\Delta x)^2}{2!} f_n'' \qquad\qquad f_{n-1} = f_n - \Delta x f_n' + \frac{(-\Delta x)^2}{2!} f_n'' \qquad (5.57)$$

Table 5.3 Three point finite-difference first and second derivative approximations for i=0...n points in ascending order: $x_0, x_1, x_2 \ldots x_{i-1}, x_i, x_{i+1} \ldots x_{n-2}, x_{n-1}, x_n$.

Forward Difference ($i = 0$)	$\left.\dfrac{df}{dx}\right\|_0 = \dfrac{-3f_0 + 4f_1 - f_2}{2\Delta x}$	$\left.\dfrac{d^2f}{dx^2}\right\|_0 = \dfrac{f_2 - 2f_1 + f_0}{\Delta x^2}$
Central Difference ($i = 2 \ldots n-1$)	$\left.\dfrac{df}{dx}\right\|_i = \dfrac{f_{i+1} - f_{i-1}}{2\Delta x}$	$\left.\dfrac{d^2f}{dx^2}\right\|_i = \dfrac{f_{i+1} - 2f_i + f_{i-1}}{\Delta x^2}$
Backward Difference ($i = n$)	$\left.\dfrac{df}{dx}\right\|_n = \dfrac{3f_n - 4f_{n-1} + f_{n-2}}{2\Delta x}$	$\left.\dfrac{d^2f}{dx^2}\right\|_n = \dfrac{f_n - 2f_{n-1} + f_{n-2}}{\Delta x^2}$

Example 5.3 Rate of Change in Concentration in a Batch Reactor

Known/Find/Analysis: Approximate the first derivatives of the batch reactor data in Table 5.2 by second-order accurate finite-difference methods in an *Excel* worksheet. The time increment in the data is 0.5 minutes. Use forward and backward derivative approximations at the ends, and central difference approximations for the interior points.

	A	B	C	D
1	i	t/min	C/M	dC/dt, Δt = 0.5
2	0	0	1	0.300
3	1	0.5	0.86	0.260
4	2	1	0.74	0.220
5	3	1.5	0.64	0.190

	A	B	C	D
1	i	t/min	C/M	dC/dt, Δt = 0.5
2	0	0	1	=-(-3*C2+4*C3-C4)/(2*0.5)
3	1	0.5	0.86	=-(C4-C2)/(B4-B2)
4	2	1	0.74	=-(C5-C3)/(B5-B3)
5	3	1.5	0.64	=-(C6-C4)/(B6-B4)

Plot the results for the first derivative over the range $0 < t < 10$ min in Figure 5.12 to show that the reaction is fastest (steeper slope) at the beginning when the concentration of reactant is higher.

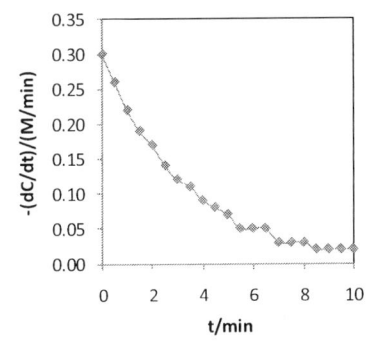

Figure 5.12
Time profile for the derivative of concentration in a batch reactor.

Comments: As seen in Figure 5.12, the derivative calculations for the batch reactor reveal a lack of smoothness in the data at larger reaction times. Derivative approximations tend to magnify any noise inherent in the data. However, we expect the rate of reaction to change smoothly over the course of the batch reactor operation. Chapter 9 introduces data smoothing techniques using regression methods to interpolate the data with model equations.

□

5.3.2 Higher-order Derivative Approximations*

We derive finite-difference approximations for fourth-order accurate derivatives using *five points* by solving the following system of linear equations for the derivatives:

$$f_{i+2} = f_i + (2\Delta x)d_1 + \frac{(2\Delta x)^2}{2!}d_2 + \frac{(2\Delta x)^3}{3!}d_3 + \frac{(2\Delta x)^4}{4!}d_4 \qquad (5.58)$$

$$f_{i+1} = f_i + \Delta x d_1 + \frac{\Delta x^2}{2!}d_2 + \frac{\Delta x^3}{3!}d_3 + \frac{\Delta x^4}{4!}d_4 \qquad (5.59)$$

$$f_{i-1} = f_i - \Delta x d_1 + \frac{\Delta x^2}{2}d_2 - \frac{\Delta x^3}{3!}d_3 + \frac{\Delta x^4}{4!}d_4 \qquad (5.60)$$

$$f_{i-2} = f_i - (2\Delta x)d_1 + \frac{(2\Delta x)^2}{2}d_2 - \frac{(2\Delta x)^3}{3!}d_3 + \frac{(2\Delta x)^4}{4!}d_4 \qquad (5.61)$$

Note that the number of equations required for the derivation must equal the order of the derivative in the last term, and that we must expand the Taylor series to an order that matches the order of the derivative approximation. The error in the derivative approximations matches the order of the Taylor series expansions. In this example, we expand four Taylor series to terms with fourth-order derivatives, which yields central-difference derivative approximations with errors of $O(\Delta x^4)$:

$$\left.\frac{df}{dx}\right|_x \cong \frac{f(x-2\Delta x) - 8f(x-\Delta x) + 8f(x+\Delta x) - f(x+2\Delta x)}{12\Delta x} \pm O(\Delta x^4) \qquad (5.62)$$

$$\left.\frac{d^2 f}{dx^2}\right|_x \cong \frac{-f(x-2\Delta x) + 16f(x-\Delta x) - 30f(x) + 16f(x+\Delta x) - f(x+2\Delta x)}{12\Delta x^2} \qquad (5.63)$$

We need forward and backward finite difference approximations for the first two and last two points. Using the subscript-indexing notation, we rewrite Equations (5.62) and (5.63) along with the forward and backward formulas for the first and second derivatives in Table 5.4.

Table 5.4 Fourth-order accurate, five-point finite-difference first and second derivative approximations for i=0...n points in ascending order: $x_0, x_1, x_2 \ldots x_{i-1}, x_i, x_{i+1} \ldots x_{n-2}, x_{n-1}, x_n$.

Forward ($i = 0$)	$\left.\dfrac{df}{dx}\right\|_0 \cong \dfrac{25 f_0 - 48 f_1 + 36 f_2 - 16 f_3 + 3 f_4}{12\Delta x}$	$\left.\dfrac{d^2 f}{dx^2}\right\|_0 \cong \dfrac{35 f_0 - 104 f_1 + 114 f_2 - 56 f_3 + 11 f_4}{12\Delta x^2}$	
Forward ($i = 1$)	$\left.\dfrac{df}{dx}\right\|_1 \cong \dfrac{3 f_0 + 10 f_1 - 18 f_2 + 6 f_3 - f_4}{12\Delta x}$	$\left.\dfrac{d^2 f}{dx^2}\right\|_1 \cong \dfrac{11 f_0 - 20 f_1 + 6 f_2 + 4 f_3 - f_4}{12\Delta x^2}$	
Central ($2 \le I \le n-2$)	$\left.\dfrac{df}{dx}\right\|_i \cong \dfrac{f_{i-2} - 8 f_{i-1} + 8 f_{i+1} - f_{i+2}}{12\Delta x}$	$\left.\dfrac{d^2 f}{dx^2}\right\|_i \cong \dfrac{-f_{i-2} + 16 f_{i-1} - 30 f_i + 16 f_{i+1} - f_{i+2}}{12\Delta x^2}$	
Backward ($i = n-1$)	$\left.\dfrac{df}{dx}\right\|_{n-1} \cong \dfrac{3 f_n + 10 f_{n-1} - 18 f_{n-2} + 6 f_{n-3} - f_{n-4}}{12\Delta x}$	$\left.\dfrac{d^2 f}{dx^2}\right\|_{n-1} \cong \dfrac{11 f_n - 20 f_{n-1} + 6 f_{n-2} + 4 f_{n-3} - f_{n-4}}{12\Delta x^2}$	
Backward ($i = n$)	$\left.\dfrac{df}{dx}\right\|_n \cong \dfrac{25 f_n - 48 f_{n-1} + 36 f_{n-2} - 16 f_{n-3} + 3 f_{n-4}}{12\Delta x}$	$\left.\dfrac{d^2 f}{dx^2}\right\|_n \cong \dfrac{35 f_n - 104 f_{n-1} + 114 f_{n-2} - 56 f_{n-3} + 11 f_{n-4}}{12\Delta x^2}$	

CHAPTER 5: TAYLOR SERIES & DERIVATIVES

In *summary*, the steps for deriving finite-difference derivative approximations of any order include:
1. Expand the Taylor series about the point of interest for approximating derivatives.
2. Use the Taylor series expansion to find an approximate solution to the function at several uniformly spaced points near the expansion point.
3. The number of points required is one more than the highest order of derivative approximations (*e.g.*, the fourth derivative in Equation (5.58) requires five points).

Following the same procedure for deriving lower order finite difference derivative approximations, we use higher order Taylor series expansions to derive formulas for the higher order finite differences with accuracy of $\pm O(\Delta x^2)$. We summarize the results in compact notation using binomial coefficients:

$$\Delta x^m \frac{d^m f}{dx^m}\bigg|_i \cong \sum_{k=0}^{m}(-1)^k \binom{m}{k} f_k \qquad (5.64)$$

where m is the order of the derivative, and

Forward   **Central (1 ≤ i ≤ n - 1)**   **Backward**

$$f_k = f\left[x_0 + (m-k)\Delta x\right] \qquad f_k = f\left[x_i + \left(\frac{m}{2} - k\right)\Delta x\right] \qquad f_k = f\left[x_n - k\Delta x\right]$$

We get the binomial coefficients from the formula:

$$\binom{m}{k} = \frac{m!}{k!(m-k)!} \qquad (5.65)$$

For example, we use Equation (5.64) dividing by Δx^3 for the following formula of a third-order derivative:

$$\frac{d^3 f}{dx^3}\bigg|_i = \frac{f(x_i + 1.5\Delta x) - 3f(x_i + 0.5\Delta x) + 3f(x_i - 0.5\Delta x) - f(x_i - 1.5\Delta x)}{\Delta x^3} \pm O(\Delta x^2) \qquad (5.66)$$

Note how a third order derivative requires four function approximations, and so on. The *Excel* work sheet function **COMBIN(m, k)** returns the numerical value for the binomial coefficient, given the integer values for m and k, subject to the condition $m \geq k$. Thus far, these finite-difference formulas only work for uniformly spaced values of x. With uniform spacing, we may use Richardson's extrapolation to obtain higher accuracy, as described in Section 5.3.4.

5.3.3 Derivatives from Unequally-Spaced Data*

The previous derivations of the finite-difference derivative approximations assume a *uniform* spacing in the independent variable. Occasionally, approximating derivatives from data involves unequally-spaced data points. Let $\Delta x_{i-1} = x_i - x_{i-1}$, and $\Delta x_i = x_{i+1} - x_i$, where $\Delta x_{i-1} \neq \Delta x_i$. In this case, we must keep track of the spacing among all points. Second-order Taylor series expansions for the function $f(x_{i-1})$ and $f(x_{i+1})$ about x_i give:

$$f(x_{i-1}) \cong f(x_i) - \Delta x_{i-1} \frac{df}{dx}\bigg|_{x_i} + \frac{\Delta x_{i-1}^2}{2} \frac{d^2 f}{dx^2}\bigg|_{x_i} \quad \text{and} \quad f(x_{i+1}) \cong f(x_i) + \Delta x_i \frac{df}{dx}\bigg|_{x_i} + \frac{\Delta x_i^2}{2} \frac{d^2 f}{dx^2}\bigg|_{x_i} \qquad (5.67)$$

The solution to this linear system of equations for df/dx and d^2f/dx^2 yields the following central-difference derivative approximations for unequally spaced nodes:

$$\frac{df}{dx}\bigg|_{x_i} = -\frac{\Delta x_i f(x_{i-1})}{\left[\Delta x_{i-1}^2 + \Delta x_{i-1}\Delta x_i\right]} + \frac{\left[\Delta x_i - \Delta x_{i-1}\right] f(x_i)}{\left[\Delta x_{i-1}\Delta x_i\right]} + \frac{\Delta x_{i-1} f(x_{i+1})}{\left[\Delta x_{i-1}\Delta x_i + \Delta x_i^2\right]} \qquad (5.68)$$

$$\left.\frac{d^2 f}{dx^2}\right|_{x_i} = 2\left\{\frac{f(x_{i-1})}{\left[\Delta x_{i-1}^2 + \Delta x_{i-1}\Delta x_i\right]} - \frac{f(x_i)}{\left[\Delta x_{i-1}\Delta x_i\right]} + \frac{f(x_{i+1})}{\left[\Delta x_{i-1}\Delta x_i + \Delta x_i^2\right]}\right\} \qquad (5.69)$$

Note that Equations (5.68) and (5.69) reduce to the central-difference approximation of Equation (5.48) with equally-spaced points (Δx = constant). Lagrange interpolating polynomials introduced in Section 10.1.3 are also useful for deriving derivative approximations of unequally-spaced points, particularly for forward and backward derivative approximations. We also recommend models from least-squares regression in Chapter 9 for noisy data, or cubic splines presented in Section 10.2 for smooth data, to approximate derivatives involving unequally-spaced points.

Example 5.4   Titration End-point from Unequally-Spaced Data

Known: A 20 ml sample of phosphoric acid was titrated with sodium hydroxide (NaOH). The pH of the solution was recorded and plotted over a range of titration volumes in Figure 5.13.

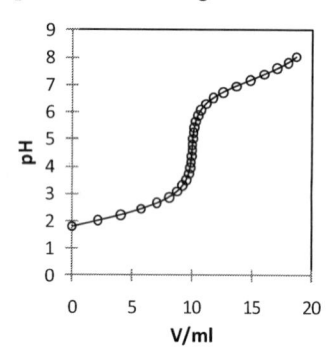

Figure 5.13

Titration pH versus volume. The inflection point in the graph indicates the titration endpoint.

Find: Locate the titration end-point from the derivatives of the data.

Assumptions: smooth data

Analysis: By visual inspection of the curve in Figure 5.13, we observe the titration endpoint occurred after the addition of ~10 ml. To confirm the endpoint numerically, we approximate the first and second derivatives using second-order finite-difference formulas for unequally-spaced data in an *Excel* worksheet, as shown in Figure 5.14.

	A	B	C	D	E
1	V/ml	pH	ΔV/ml	dpH/dV	d²pH/dV²
2	0.000	1.806	2.157		
3	2.157	2.020	1.939	0.1053	0.0055
4	4.096	2.235	1.658	0.1207	0.0104
5	5.754	2.449	1.323	0.1476	0.0220
			:		
12	9.817	3.951	0.076	2.4118	10.3719
13	9.894	4.165	0.051	3.6629	22.3829
14	9.944	4.380	0.037	5.1222	35.1783
15	9.982	4.594	0.033	6.2034	23.0417
16	10.014	4.809	0.036	6.2817	-18.2384
17	10.050	5.023	0.048	5.3108	-35.6536
18	10.098	5.238	0.072	3.8645	-24.3769

	C
1	ΔV/ml
2	=A3-A2
3	=A4-A3
4	=A5-A4
5	=A6-A5
12	=A13-A12
13	=A14-A13
14	=A15-A14
15	=A16-A15
16	=A17-A16
17	=A18-A17
18	=A19-A18

Figure 5.14

Titration data showing the end-point in rows 15 and 16.

Formulas use uneven data spacing for finite-difference calculations in rows **14** to **17** of columns **D** and **E** for the first and second derivatives according to Equations (5.68) and (5.69):

	D
14	=-C14*B13/(C13^2+C13*C14)+(C14-C13)*B14/(C13*C14)+C13*B15/(C13*C14+C14^2)
15	=-C15*B14/(C14^2+C14*C15)+(C15-C14)*B15/(C14*C15)+C14*B16/(C14*C15+C15^2)
16	=-C16*B15/(C15^2+C15*C16)+(C16-C15)*B16/(C15*C16)+C15*B17/(C15*C16+C16^2)
17	=-C17*B16/(C16^2+C16*C17)+(C17-C16)*B17/(C16*C17)+C16*B18/(C16*C17+C17^2)

CHAPTER 5: TAYLOR SERIES & DERIVATIVES

	E
14	=2*(B13/(C13^2+C13*C14)-B14/(C13*C14)+B15/(C13*C14+C14^2))
15	=2*(B14/(C14^2+C14*C15)-B15/(C14*C15)+B16/(C14*C15+C15^2))
16	=2*(B15/(C15^2+C15*C16)-B16/(C15*C16)+B17/(C15*C16+C16^2))
17	=2*(B16/(C16^2+C16*C17)-B17/(C16*C17)+B18/(C16*C17+C17^2))

Plot the first and second derivative results, as seen in Figure 5.15 with pH on the secondary axis, to locate the titration endpoint by inspection.

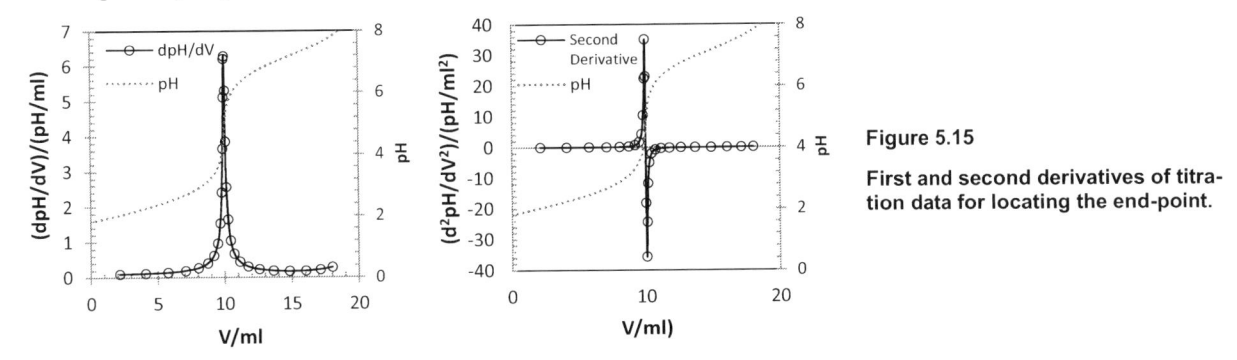

Figure 5.15

First and second derivatives of titration data for locating the end-point.

Rows 15 and 16 in Figure 5.14 indicate where the second derivative crosses the horizontal axis, changing from positive to negative values. The endpoint is between 9.982 and 10.014 ml of NaOH titrated. A linear interpolation gives the endpoint at (9.982 + 10.014)/2 = 9.998 ~ 10 ml. Visit Chapter 10 for more advanced methods of interpolation.

Comments: A comparison of the plots of the first and second derivatives shows the advantage of the second derivative for determining the titration end-point. The second-derivative plot crosses the V axis at the endpoint because the first derivative changes from a positive to a negative value.

□

Finite-difference formulas for unequally-spaced data are complicated and prone to errors when entering in a spreadsheet. We use *VBA* to create functions of derivatives that are less prone to error. The *VBA* user-defined function **FDERIV(f, x, d)**, with arguments defined in Figure 5.16, is available in the *PNMSuite* for approximating first, and second-order derivatives using three consecutive points. The user-defined function allows for uneven spacing between points of the independent variable. To use **FDERIV**, set up an *Excel* worksheet with three-point ranges for x and $f(x)$ in the arguments. The optional last argument determines the finite difference approximation for a first derivative as forward (1), central (0), backward (–1), or second derivative (2). The default is zero for central derivative approximations.

```
Public Function FDERIV(f, x, Optional d As Integer = 0) As Double
' f = range of 3 cells containing three dependent variables
' x = range of 3 cells containing three independent variables
' d = (Optional) -1 = backward first-order, 0 = central default first-order,
'      1 = forward first-order, 2 = second-order derivative
'*********************************************************************
' First and second derivative from 2nd order finite difference approximation using
' three equally-or unequally-spaced values. x and f must be in ascending order according
' to vector x.
' *** Nomenclature ***
Dim dxlow As Double, dxlow2 As Double ' = lower interval size in xd, dxlow^2
Dim dxup As Double, dxup2 As Double   ' = upper interval size in xd, dxup^2
Dim dxlowdxup As Double ' = dxlow*dxup
'*********************************************************************
If Not x.Count = 3 And Not f.Count = 3 Then
    MsgBox "x and f ranges require three elements.": Exit Function
End If
dxlow = x(2) - x(1): dxup = x(3) - x(2): dxlowdxup = dxlow * dxup
dxlow2 = dxlow ^ 2: dxup2 = dxup ^ 2
Select Case d ' Select Forward (1) , Central (0), Backward (-1)
```

```
Case 2 ' Return the 3-point second derivative approximation
    FDERIV = 2 * (f(1) * dxup - f(2) * (dxlow + dxup) + f(3) * dxlow)
Case 1 ' Forward difference approximation of first derivative
    FDERIV = (-f(1) * (dxup2 + 2 * dxlowdxup) _
        + f(2) * (dxlow2 + dxup2 + 2 * dxlowdxup) - f(3) * dxlow2)
Case -1 ' Backward difference approximation of first derivative
    FDERIV = (f(1) * dxup2 - f(2) * (dxlow2 + dxup2 + 2 * dxlowdxup) _
        + f(3) * (dxlow2 + 2 * dxlowdxup))
Case Else ' Default central difference approximation of first derivative
    FDERIV = (-f(1) * dxup2 + f(2) * (dxup2 - dxlow2) + f(3) * dxlow2)
End Select
FDERIV = FDERIV / (dxlowdxup * (dxlow + dxup) + 0.0000000001)
Function ' FDERIV
```

Figure 5.16 *VBA* user-defined function **FDERIV** for calculating first and second derivatives from three data pairs by the finite difference method.

Illustration: Use **FDERIV** in an *Excel* worksheet using the first three and last three points in Table 5.2:

1. The derivative at the first point in cell **C2** requires a forward difference approximation with the optional third argument set to 1.
2. The derivatives in cells **C3** and **F3** use central difference approximations, with the optional third argument set to 0, or omitted.
3. Cell **F4** uses a backward difference approximation for the last point, with the third argument set to −1:

	A	B	C	D	E	F
1	x	f	df/dx	x	f	df/dx
2	0.5	0.86	=FDERIV(B2:B4,A2:A4,1)	9	0.07	
3	1	0.74	=FDERIV(B2:B4,A2:A4,0)	9.5	0.06	=FDERIV(E2:E4,D2:D4,0)
4	1.5	0.64		10	0.05	=FDERIV(E2:E4,D2:D4,-1)

	A	B	C	D	E	F
1	x	f	df/dx	x	f	df/dx
2	0.5	0.86	-0.26	9	0.07	
3	1	0.74	-0.22	9.5	0.06	-0.02
4	1.5	0.64		10	0.05	-0.02

An alternative *VBA* user-defined function **FINDIF(f, x, xi, d)** is available in the *PNMSuite* that performs three-point finite difference derivative approximations with uniform or non-uniform spacing in the data. The function **FINDIF** automatically selects forward and backward difference formulas as required by the first and last point in the data, respectively. It uses central difference approximations for the interior points. The arguments in **FINDIF** are similar to **FDERIV**, except the **x** and **f** refer to the ranges for the complete set of data. **FINDIF** includes an additional argument **xi** to indicate the location of the derivative approximation in the **x** data. The value of **xi** must match a value in the range for **x**. **FINDIF** uses the worksheet function **MATCH** to locate the points needed for the derivative approximation. The optional argument **d** (=1 or 2) is used to indicate first or second derivative approximations. If *d* is missing, the function returns a first derivative. If necessary, the procedure sorts the data in ascending order according to the values of the *x* data using the macro **HEAPSORTXY**. We illustrate how to use the function **FINDIF** using the first five points from Table 5.2 in an *Excel* worksheet.

1. The result in cell **C2** uses the forward difference derivative approximation whereas the result in cell **C6** uses the backward difference approximation.
2. Use absolute references to the ranges of *y* and *x* data for **AutoFill**.
3. The rest of the results in column **C** use the central difference approximation:

	A	B	C
1	x	f	df/dx
2	0.5	0.86	=FINDIF($B$2:$B$6,$A$2:$A$6,A2)
3	1	0.74	=FINDIF($B$2:$B$6,$A$2:$A$6,A3)
4	1.5	0.64	=FINDIF($B$2:$B$6,$A$2:$A$6,A4)
5	2	0.55	=FINDIF($B$2:$B$6,$A$2:$A$6,A5)
6	2.5	0.47	=FINDIF($B$2:$B$6,$A$2:$A$6,A6)

	A	B	C
1	x	f	df/dx
2	0.5	0.86	-0.26
3	1	0.74	-0.22
4	1.5	0.64	-0.19
5	2	0.55	-0.17
6	2.5	0.47	-0.15

Chapter 5: Taylor Series & Derivatives

Example 5.5 Specific Growth Rate

Known: An *Excel* worksheet in Figure 5.17 contains data for the change in biomass concentration with time in a batch fermentation reactor.

Find: Calculate the specific growth rate for biomass.

Analysis: Get the specific growth rate of biomass during the exponential growth phase from the mass balance in the fermentation reactor (rate of generation = rate of accumulation):

$$\mu X = \frac{dX}{dt} \tag{5.70}$$

where μ is the specific growth rate and X is the biomass concentration. Use the *VBA* user-defined functions **FDERIV** and **FINDIF** to calculate the first derivatives of biomass concentration with time. The arguments for the *VBA* function **FDERIV** include a three-point range of contiguous cells on the worksheet for the independent variable (t) and a three-point range of contiguous cells containing values for the dependent variable (X). The third argument indicates whether we are using a forward (1), central (0), or backward (-1) finite-difference formula. The arguments for the function **FINDIF** include the complete range of t and X data and the point of interpolation. The following table compares the arguments for the user-defined functions:

FDERIV	FINDIF
C2 = FDERIV(B2:B4,A2:A4,1)	C2 = FINDIF($B$2:$B$15,$A$2:$A$15,A2)
C3 = FDERIV(B2:B4:A2:A4)	C3 = FINDIF(($B$2:$B$15,$A$2:$A$15,A3)
:	:
C15 = FDERIV(B13:B15,A13:A15,-1)	C15 = FINDIF(($B$2:$B$15,$A$2:$A$15,A15)

A plot of $(dX/dt)/X$ from column **D** on a secondary axis, versus time from column **A** in Figure 5.17, reveals the range of data that corresponds to the exponential growth phase.

	A	B	C	D
1	t/hr	X/(kg/m³)	dX/dt	(dX/dt)/X
2	0	0.2	0.0115	0.05740
3	0.33	0.21	0.0491	0.23394
4	0.5	0.22	0.1969	0.89509
5	0.75	0.32	0.5000	1.56250
6	1	0.47	0.7533	1.60284
7	1.5	1	1.6300	1.63000
8	2	2.1	3.4200	1.62857
9	2.5	4.42	6.9067	1.56259
10	2.8	6.9	10.8067	1.56618
11	3	9.4	14.1667	1.50709
12	3.1	10.9	11.0000	1.00917
13	3.2	11.6	5.3333	0.45977
14	3.5	11.7	-0.1667	-0.01425
15	3.7	11.6	-0.8333	-0.07184

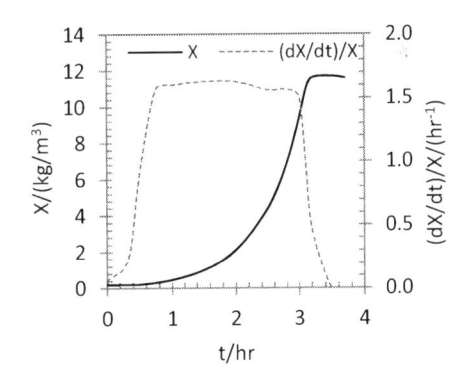

Figure 5.17
Biomass concentration and derivative showing the exponential growth period.

It appears from the relatively flat part of the curve for the specific derivative that the exponential growth phase covers the data in the range of 0.75 hours to 2.8 hours. We calculate the specific growth rate from the average values of $(dX/dt)/X$ in the exponential growth range of data using the *Excel* worksheet function:

$$1.59 \text{ hr}^{-1} = \text{AVERAGE(D5:D10)}$$

Comments: Perhaps a better value for the specific growth rate comes from the slope of the line formed by plotting dX/dt versus X. Right-click on the data in a chart and select **Add Trend line** from the popup menu, as seen in Figure 5.18. Select **Linear** under **Trend line Options** and **Display Equation** on the chart. The slope gives $\mu = 1.57$ hr^{-1}, which agrees with the averaged result in the first decimal place.

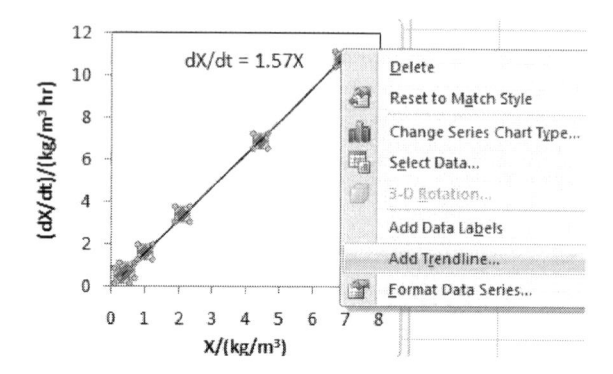

Figure 5.18

Plot of Equation (5.70) to obtain the specific growth rate from the slope.

Consult Chapter 9 for details of the method of least squares regression used to generate a trend line and fit linear equations to data.

□

5.3.4 Richardson's Extrapolation and Ridders' Algorithm

"Expecting something for nothing is the most popular form of hope." – Arnold H. Glasow

To improve the accuracy of finite-difference derivative approximations, we may consider using smaller spacing between the points. This may not be possible for experimentally derived data. We also need to be careful to control computer round-off errors that creep into the results when the spacing is too small. Richardson (1910) developed an extrapolation method that provides significant improvements in accuracy and an estimate of the numerical error starting from relatively large interval spacing (Hanna and Sandall 1995).

From Taylor series analysis, we find that errors in derivative approximations are proportional to the spacing between points for the independent variable, Δx. Consider two derivative approximations using different spacing in the independent variable (using prime notation for the derivatives):

$$f' = f'(\Delta x_1) + E_1 = f'(\Delta x_2) + E_2 \tag{5.71}$$

where E_1 and E_2 represent exact errors in each approximation for $\Delta x_2 < \Delta x_1$. Approximate the error functions from the first truncated term in a Taylor series for f':

$$E_i \cong \frac{\Delta x_i^{n-1}}{n!} \left.\frac{d^n f}{dx^n}\right|_i \tag{5.72}$$

Determine the order of magnitude of the truncated term from the order of the finite-difference derivative approximation. For example, the order of the first derivative approximation in Equation (5.62) is four. Evaluate the ratio of error approximations for two different levels of node spacing to eliminate the higher derivatives:

$$\frac{E_1}{E_2} = \left(\frac{\Delta x_1}{\Delta x_2}\right)^{n-1} \tag{5.73}$$

Substitute this expression into Equation (5.71) for E_1 (for convenience, we drop the Δx functionality in the f' derivative notation):

$$f'_1 + E_2 \left(\frac{\Delta x_1}{\Delta x_2}\right)^{n-1} = f'_2 + E_2 \tag{5.74}$$

Solve Equation (5.74) for an estimate of the error using Δx_2:

CHAPTER 5: TAYLOR SERIES & DERIVATIVES

$$E_2 = \frac{f_2' - f_1'}{\left(\frac{\Delta x_1}{\Delta x_2}\right)^{n-1} - 1} \tag{5.75}$$

Now, Equation (5.71) becomes:

$$f' = f_2' + \frac{f_2' - f_1'}{\left(\frac{\Delta x_1}{\Delta x_2}\right)^{n-1} - 1} = \frac{f_2'\left(\frac{\Delta x_1}{\Delta x_2}\right)^{n-1} - f_1'}{\left(\frac{\Delta x_1}{\Delta x_2}\right)^{n-1} - 1} \tag{5.76}$$

For example, using two first-order forward derivative approximations, where $n = 1$ with $\Delta x_1 = 2\Delta x_2$:

$$f_1' = \frac{f(x + \Delta x_1) - f(x)}{\Delta x_1} \quad \text{and} \quad f_2' = \frac{f(x + \Delta x_2) - f(x)}{\Delta x_2} \tag{5.77}$$

such that

$$f' = \frac{2f_2' - f_1'}{2 - 1} = 2f_2' - f_1' \pm O(\Delta x^2) \tag{5.78}$$

The error estimate for Equation (5.78) according to Equation (5.75) is $E \cong \left| f_2' - f_1' \right|$.

Now consider the *central* difference approximation for a first-order derivative, using two different derivative approximations with $\Delta x_1 = 2\Delta x_2$.

$$f_1' = \frac{f(x + \Delta x_1) - f(x - \Delta x_1)}{2\Delta x_1} \tag{5.79}$$

and

$$f_2' = \frac{f(x + \Delta x_2) - f(x - \Delta x_2)}{2\Delta x_2} = \frac{f(x + 0.5\Delta x_1) - f(x - 0.5\Delta x_1)}{\Delta x_1} \tag{5.80}$$

Recall the derivation of the central difference derivative approximation in Equation (5.49) from second-order Taylor series expansions, truncating the *third*-order terms. The error of the derivative approximation is $O(\Delta x^2)$. Therefore, $n = 3$ for Richardson's extrapolation and the formula in Equation (5.76) becomes:

$$f' = \frac{4f_2' - f_1'}{3} \pm O(\Delta x^4) \tag{5.81}$$

Richardson's extrapolation is a weighted average of the derivative approximations, with more weight applied to the more accurate term that employs smaller spacing in x. Weighted averaging allows some terms to influence the result more than others do. Richardson's extrapolation improves the approximation by two orders of magnitude with relatively little effort. In Equation (5.81), the extrapolated derivative approximation becomes accurate to $O(\Delta x_2^4)$.

We gain two orders of magnitude in accuracy each time we perform Richardson's extrapolation.

Repeated application of Richardson's extrapolation to extrapolated results by successively reducing the interval size Δx by half gives the following recursive formula using second-order accurate finite difference derivative approximations:

- *Richardson's Extrapolation*
$$f_{j,k}' \cong \frac{4^{k-1} f_{j+1,k-1}' - f_{j,k-1}'}{4^{k-1} - 1} \tag{5.82}$$

where the index $j \geq 1$ indicates the relative size of the finite-difference interval, $\Delta x_j = \Delta x_0 / (2j)$, and the index $k = 2, 3 \ldots$ signifies the level of the Richardson's extrapolation, beginning with derivative approximations of order $O(\Delta x^2)$. Richardson's extrapolation has the advantage of obtaining approximations with higher order accuracy from relatively larger spacing in the independent variable. We may apply Richardson's extrapolation to any method derived from Taylor series expansions for reducing error, including higher order derivatives, integrals, and differential equations.

Example 5.6 Derivative Approximations for Reaction Rate

Known: Table 5.2 contains equally-spaced data for concentration with time in a batch reactor.

Find: Calculate the central-difference derivative approximations using the data spaced $\Delta t = 1$ and $\Delta t = 0.5$ minutes apart. Use Richardson's extrapolation to improve the derivatives.

Analysis: Use an *Excel* worksheet to calculate the results using fourth-order approximations without extrapolation from Equation (5.62) for comparison. Columns **D** and **E** contain the second-order accurate finite difference formulas.

	A	B	C	D	E
1	i	t/min	C/M	dC/dt, Δt = 1	dC/dt, Δt = 0.5
2	0	0	1	=-(-3*C2+4*C4-C6)/(2*1)	=-(-3*C2+4*C3-C4)/(2*0.5)
3	1	0.5	0.86	=-(-3*C3+4*C5-C7)/(2*1)	=-(C4-C2)/(B4-B2)
4	2	1	0.74	=-(C6-C2)/(B6-B2)	=-(C5-C3)/(B5-B3)
5	3	1.5	0.64	=-(C7-C3)/(B7-B3)	=-(C6-C4)/(B6-B4)

⋮

	A	B	C	D	E
19	17	8.5	0.08	=-(C21-C17)/(B21-B17)	=-(C20-C18)/(B20-B18)
20	18	9	0.07	=-(C22-C18)/(B22-B18)	=-(C21-C19)/(B21-B19)
21	19	9.5	0.06	=-(3*C21-4*C19+C17)/(2*1)	=-(C22-C20)/(B22-B20)
22	20	10	0.05	=-(3*C22-4*C20+C18)/(2*1)	=-(3*C22-4*C21+C20)/(2*0.5)

We put the Richardson extrapolated results and a fourth-order accurate finite difference derivative approximation in columns **F** and **G**, for comparison.

	F	G
1	Richardson's Extrapolation	4th Order Finite Difference
2	=(4*E2-D2)/3	
3	=(4*E3-D3)/3	
4	=(4*E4-D4)/3	=-(C2-8*C3+8*C5-C6)/(12*0.5)
5	=(4*E5-D5)/3	=-(C3-8*C4+8*C6-C7)/(12*0.5)
6	=(4*E6-D6)/3	=-(C4-8*C5+8*C7-C8)/(12*0.5)
7	=(4*E7-D7)/3	=-(C5-8*C6+8*C8-C9)/(12*0.5)
8	=(4*E8-D8)/3	=-(C6-8*C7+8*C9-C10)/(12*0.5)
9	=(4*E9-D9)/3	=-(C7-8*C8+8*C10-C11)/(12*0.5)
10	=(4*E10-D10)/3	=-(C8-8*C9+8*C11-C12)/(12*0.5)
11	=(4*E11-D11)/3	=-(C9-8*C10+8*C12-C13)/(12*0.5)
12	=(4*E12-D12)/3	=-(C10-8*C11+8*C13-C14)/(12*0.5)
13	=(4*E13-D13)/3	=-(C11-8*C12+8*C14-C15)/(12*0.5)
14	=(4*E14-D14)/3	=-(C12-8*C13+8*C15-C16)/(12*0.5)
15	=(4*E15-D15)/3	=-(C13-8*C14+8*C16-C17)/(12*0.5)
16	=(4*E16-D16)/3	=-(C14-8*C15+8*C17-C18)/(12*0.5)
17	=(4*E17-D17)/3	=-(C15-8*C16+8*C18-C19)/(12*0.5)
18	=(4*E18-D18)/3	=-(C16-8*C17+8*C19-C20)/(12*0.5)
19	=(4*E19-D19)/3	=-(C17-8*C18+8*C20-C21)/(12*0.5)
20	=(4*E20-D20)/3	=-(C18-8*C19+8*C21-C22)/(12*0.5)
21	=(4*E21-D21)/3	
22	=(4*E22-D22)/3	

The results of the calculations show dramatic improvement in precision from Richardson's extrapolation that matches the 4th order accuracy of the five-point finite-difference derivative approximation displayed in column **G**:

	A	B	C	D	E	F	G
1	i	t/min	C/M	dC/dt, Δt = 1	dC/dt, Δt = 0.5	Richardson's	4th Order Finite
2	0	0	1	0.295	0.300	0.30167	
3	1	0.5	0.86	0.245	0.260	0.26500	
4	2	1	0.74	0.225	0.220	0.21833	0.21833
5	3	1.5	0.64	0.195	0.190	0.18833	0.18833
6	4	2	0.55	0.165	0.170	0.17167	0.17167

⋮

	A	B	C	D	E	F	G
18	16	8	0.09	0.025	0.030	0.03167	0.03167
19	17	8.5	0.08	0.025	0.020	0.01833	0.01833
20	18	9	0.07	0.02	0.020	0.02000	0.02000
21	19	9.5	0.06	0.015	0.020	0.02167	
22	20	10	0.05	0.02	0.020	0.02000	

Comments: The improved derivatives from Richardson's extrapolation agree with the fourth-order finite-difference approximations, despite using low-order approximations.

□

CHAPTER 5: TAYLOR SERIES & DERIVATIVES

Unfortunately, computer round-off error in the finite difference derivative approximations may cause Richardson's extrapolation to misbehave when applied to finite difference approximations. Repeated application of Richardson's extrapolations starting from a poor value of Δx in the finite difference calculation tends to diverge away from an accurate result. The effects of computer round-off error are pronounced for third order, and higher, derivative approximations. Ridders (1982) developed a practical algorithm that takes the guesswork out of selecting a good value for Δx. Ridders' method terminates successive extrapolations when the error estimates from Richardson's extrapolation start to increase. The *PNMSuite* implementation of the algorithm takes the following steps:

- Use the central-difference derivative approximations in Equation (5.64) for n^{th}-order derivatives.
- Estimate the initial value of the interval size from the order of magnitude of the value for the independent variable:

$$\Delta x = 0.2 \, Integer \left[Log_{10} \left(|x| \right) \right] \qquad (5.83)$$

- Apply repeated application of Richardson's extrapolation.
- Terminate when relative error estimates according to Equation (5.75) *increase* between successive extrapolations.
- Following the recommendation of Press, et al. (2007), use $1.4^{2(k-1)}$ in place of 4^{k-1} for the central difference formula of Equation (5.64).

The *VBA* macros **PDERIV** and **RIDDERS**, available in the *PNMSuite*, use Ridders (1982) algorithm to approximate the derivative of a function. Both macros prompt the user for the cells containing the formula and independent variable. **RIDDERS** allows the user to calculate derivatives up to order six with respect to several variables by cycling back through the input boxes. Both macros display the estimated error of the approximation in the **Status Bar** located at the bottom of the worksheet. Numerical integration by Romberg's method, presented in Section 11.2.4, also uses Richardson's extrapolation in a dramatic way to improve integral solutions and control the error in the result.

The *VBA* user-defined function **DREX(f, x, dv)**, with arguments defined in Figure 5.19, returns an approximation of a derivative of a *formula* in an *Excel* worksheet using finite differencing with Richardson's extrapolation (**DREX** does not apply to a range data points). The function uses the string-to-*VBA* conversion algorithm from Section 3.6.3 based on the template in Figure 3.23 for evaluating the formula from the worksheet. To apply the user-defined function, provide cell references for the formula and independent variable in the argument list, separated by commas. For second derivatives (or higher order), include the optional integer value of two (2), or higher, at the end of the list of arguments. Do not use **DREX** if components of the formula involve calculations in other cells on the worksheet – instead, use either of the macros **RIDDERS** or **PDERIV**.

We use a simple quadratic function to illustrate how to set up an *Excel* worksheet for calculating the value of a derivative function at $x = 1$ with the user-defined function **DREX**:

$$f(x) = 3x^2 - 2x + 1 \quad \text{with} \quad \left. \frac{df(x)}{dx} \right|_{x=1} = 6x - 2 = 6 \cdot 1 - 2 = 4 \quad \text{and} \quad \left. \frac{d^2 f(x)}{dx^2} \right|_{x=1} = 6$$

1. Fill cell **A1** with the formula of the function in terms of the value for x in cell **A2**.
2. Use **DREX** in cell **A3** and **A4** to calculate the values of the first and second derivatives (accurate to 10^{-8}):

	A	B	C	D
1	=3*B1^2-2*B1+1	1	=DREX(A1,B1)	=DREX(A1,B1,2)

	A	B	C	D
1	2	1	4.000000007	6.000000005

```
Public Function DREX(f, x, Optional m As Integer = 1)
' f = cell containing the formula for the dependent variable (function)
' x = cell containing the independent variable
' m = (optional) integer value for derivative type: 1 = first, 2 = second
'****************************************************************
' Use Ridders' method of approximating derivatives of worksheet formulas by finite
' difference with Richardson's extrapolation.  Stopping criteria adapted from Press,
' et. al. _Numerical Recipes in Fortran_, Cambridge University Press, New York,
' 2nd Ed., 1992, pp 182-3.
' *** Nomenclature ***
Dim a As Long ' = check for calculating another derivative
```

```vb
Dim big As Double ' = big number
Dim c As Double, c2 As Double ' = step size reduction factor
Dim cfd() As Double ' = coefficients in finite difference derivative formulas
Dim d(1 To 98, 1 To 98) As Double ' = array for extrapolated derivatives
Dim dx As Double ' = derivative step size
Dim er As Double, ert As Double ' = error approximation and temporary error
Dim fa As String ' = string of formula with absolute address and dummy variable xu
Dim fi As String ' = string of formula with value for x inserted
Dim fj As Double ' = value of function at j position in finite difference formula
Dim j As Long, j1 As Long, k As Long, k1 As Long ' = variables of index, j1, k1 = j, k-1
Dim m2 As Double ' = order of derivative, m2 = m/2
Dim noc As Long ' number of occurrences of variable string in formula
Dim ns As Long ' = number of occurrences of x cell name in f formula
Dim sf As Double ' = safety factor
Dim small As Double ' = small number
Dim t As String ' = string of formula used to find occurrences of x address in f
Dim x0 As Double ' = initial value for variable x at location of derivative
Dim xj As Double ' = x value at j position in finite difference formula
Dim xa As String ' = address on worksheet for x
Dim xi As String ' = value of independent variable inserted into string of formula
Dim xs As String ' = string address for independent variable x
Dim xu As String ' = string placeholder for x substitution/evaluation
Dim w As Double ' = weight for derivative estimates
'****************************************************************************
big = 1E+30: small = 0.000000000001
sf = 2 ' safety factor for error control
c = 1.4: c2 = c * c: m2 = m / 2 ' step size reduction factor and m/2 for fin. dif.
If Abs(x) < small Then
    dx = 0.002
Else
    dx = 0.002 * 10 ^ Int(small + Log(Abs(x)) / Log(10)) ' initial step size
End If
If dx = 0 Then dx = 0.002 ' check for zero
x0 = x.Value ' save initial values variable at x value of derivative approximation
With Application
ReDim cfd(0 To m) As Double ' Get the coefficients in the finite difference formulas
For j = 0 To m: cfd(j) = .Combin(m, j) * (-1) ^ j: Next j
xu = "X_": fa = VBA.UCase(f.Formula) ' make upper case to eliminate case sensitivity
Do While InStr(fa, xu) ' check for dummy variable string in formula
    xu = xu & "_" ' add an underscore to end of string
Loop
' Get strings for formula evaluations at inserted values of the independent variable
On Error Resume Next ' skip xa setting if Name.Name returns an error due to no name
xa = x.Address: xs = "": xs = Range(xa).Name.Name
fa = .ConvertFormula(f.Formula, xlA1, xlA1, xlAbsolute): fi = fa
noc = (Len(fa) - Len(.Substitute(fa, xa, ""))) / Len(xa)
For j = noc To 1 Step -1
    t = .Substitute(fi, xa, x0 & " ", j) ' add space to value to evaluate
    If Not IsError(Evaluate(t)) Then
        fa = .Substitute(fa, xa, xu, j): fi = t
    End If
Next j
ns = Len(xs)
If ns > 0 Then
    noc = (Len(fa) - Len(.Substitute(fa, xs, ""))) / ns
    For j = noc To 1 Step -1
        t = .Substitute(fi, xs, x0 & " ", j) ' add space to value to evaluate
        If Not IsError(Evaluate(t)) Then
            fa = .Substitute(fa, xs, xu, j): fi = t
        End If
    Next j
End If
d(1, 1) = 0 ' evaluate the finite difference derivative order m with initial interval
For j = 0 To m
    xj = x0 + (m2 - j) * dx: fj = Evaluate(.Substitute(fa, xu, xj))
    d(1, 1) = d(1, 1) + cfd(j) * fj
Next j
d(1, 1) = d(1, 1) / dx ^ m
er = big ' initialize error to arbitrary large value
```

```
For k = 2 To 98
    dx = dx / c   ' adjust interval size for next extrapolation
    d(1, k) = 0   ' evaluate finite difference derivative order m with new interval dx
    For j = 0 To m
        xj = x0 + (m2 - j) * dx: fj = Evaluate(.Substitute(fa, xu, xj))
        d(1, k) = d(1, k) + cfd(j) * fj
    Next j
    d(1, k) = d(1, k) / dx ^ m
    w = c2: k1 = k - 1  ' reset the factor for the next set of extrapolations
    For j = 2 To k  ' Richardson's extrapolations
        j1 = j - 1
        d(j, k) = (d(j1, k) * w - d(j1, k1)) / (w - 1)
        w = w * c2
        ' estimate errors in derivative approximation
        ert = (.Max(Abs(d(j, k) - d(j1, k)), Abs(d(j, k) - d(j1, k1)))) _
            / (Abs(d(j, k)) + small)
        If ert < er Then  ' save the result with the smallest error
            er = ert: RDREX = d(j, k)  ' set the best values for error and the derivative
        End If
    Next j
    ' exit if the errors increase with higher order extrapolations
    If Abs(d(k, k) - d(k1, k1)) / (Abs(d(k, k)) + small) > sf * er Then Exit For
Next k
End With  ' Application
End Function  ' DREX
```

Figure 5.19 *VBA* user-defined function DREX for approximating derivative approximations of worksheet formulas with Richardson's extrapolation.

Example 5.7 Partial Derivative of Pressure with Respect to Temperature

Known/Find: Calculate the partial derivative of pressure with respect to temperature, $\partial P/\partial T$, at constant volume for CO_2 at 350 K using the Soave-Redlich-Kwong (*SRK*) equation of state:

$$P = \frac{R_g T}{V - b} - \frac{a\left[1 + m\left(1 - \sqrt{T_r}\right)\right]^2}{V(V+b)}$$

where $\quad T_r = T/T_c \qquad m = 0.48508 + 1.5517w - 0.1561w^2$

$$a = 0.42747\frac{(RT_c)^2}{P_c} \qquad \text{and} \qquad b = 0.08664\frac{RT_c}{P_c}$$

Analysis: Set up an *Excel* worksheet with the *SRK* parameters for CO_2:

	A	B	C	D
1	T_c/K	304.2	T/K	350
2	P_c/atm	72.9	V/(L/mol)	2.6
3	R/(L atm/mol K)	0.08206	w	0.225
4	a/(atm L^2)	=0.42747*(B3*B1)^2/B2	T_r	=D1/B1
5	b/(L/mol)	=0.08664*B3*B1/B2	m	=0.48508+1.5517*D3-0.1561*D3^2
6	P/atm	=B3*D1/(D2-B5)	($\partial P/\partial T$)/(atm/K)	0.031925772352983

Use the *VBA* macro **PDERIV** to calculate the partial derivative in cell **D6**.

PDERIV

Click on cell of INDEPENDENT variable (x):
$D$1

Click on cell of DEPENDENT variable (y):
$B$6

	A	B	C	D
1	T_c/K	304.2	T/K	350
2	P_c/atm	72.9	V/(L/mol)	2.6
3	R/(L atm/mol K)	0.08206	w	0.225
4	a/(atm L²)	3.65392	T_r	1.15056
5	b/(L/mol)	0.02967	m	0.82631
6	P/atm	11.17404	(∂P/∂T)/(atm/K)	0.0319

Use the default value in the macro for the relative error tolerance. We may also try the user-defined function **DREX** directly in the worksheet:

$$0.0319 = \text{DREX}(B6, D1)$$

Comments: The result for $\partial P/\partial T$ in cell **D6**, or from the function **DREX**, is now available for use in other calculations, such as predicting the heat capacity for the gas.

☐

5.4 Sensitivity Analysis and Maximum Uncertainty*

"My biggest weakness is my sensitivity. I am too sensitive a person." – Mike Tyson

A complete engineering analysis of functional relationships among variables usually includes the response (sensitivity) to disturbances (perturbations) in the input variables and model parameters. We may also need to consider the sensitivity of our model to *process* parameters.[54] In the case of independent variables, we can analyze the sensitivity of the response to perturbations by generating a response surface, or contour plot.

Alternatively, we may analyze the model's sensitivity to perturbations in the independent variables from the "rate of change" in the function as the independent variables deviate from a base value. Partial derivatives of the function with respect to an independent variable *near* the solution describe the rate of change. The first-order finite-difference approximation to the derivative (or slope) at some location x provides a linear approximation of the sensitivity:

$$c = \left.\frac{\partial f}{\partial x}\right|_x \cong \frac{\Delta f}{\Delta x} = \frac{f(x+\Delta x) - f(x)}{\Delta x} \tag{5.84}$$

The partial derivatives, c, also go by the names sensitivity coefficients or process gains. We may describe the *relative* change in f with x in terms of logarithms:

$$\frac{x}{f}\frac{\partial f}{\partial x} = \frac{\partial \ln f}{\partial \ln x} \tag{5.85}$$

We can also conduct a sensitivity analysis of the problem parameters by evaluating the partial derivatives of the model function with respect to the model or process parameters. We illustrate the procedure using a simple function with two parameters, a and b:

$$y = a(x-b)^2 \tag{5.86}$$

Find the sensitivity of the value for the function to changes in the parameters a and b by differentiating the function with respect to each parameter:

$$\frac{\partial y}{\partial a} = (x-b)^2 \quad \text{and} \quad \frac{\partial y}{\partial b} = -2a(x-b) \tag{5.87}$$

In this simple illustration, the sensitivity of the function to parameter a depends only on parameter b and the value for the independent variable. However, the sensitivity to parameter a is quadratic in parameter b, whereas the sensitivity

---

[54] Notice the subtle distinction between response (independent variables) and sensitivity (parameters).

CHAPTER 5: TAYLOR SERIES & DERIVATIVES

to parameter b is linear in both parameters a and b. Thus, small changes in b can potentially have a greater effect on the sensitivity to a, than b, itself.

In terms of relative sensitivity, use the natural logarithm to transform Equation (5.86) by evaluating the natural log of both sides:

$$\ln y = \ln a + 2 \ln(x - b) \tag{5.88}$$

The evaluation of the partial derivative of $\ln(y)$ with respect to $\ln(a)$ is straightforward:

$$\frac{\partial \ln y}{\partial \ln a} = 1 \tag{5.89}$$

A 10% change in the parameter a gives a 10% change in y. Obtaining the partial derivative with respect to $\ln(b)$ requires extra effort. Use the log form in Equation (5.85) to explore the relative sensitivity:

$$\frac{\partial \ln y}{\partial \ln b} = \frac{\partial\left[2\ln(x-b)\right]}{\partial \ln b} = \frac{2}{(x-b)} \frac{\partial(x-b)}{\partial \ln b} = \frac{-2}{(x-b)} \frac{\partial b}{\partial \ln b} = \frac{-2b}{(x-b)} \frac{\partial b}{b \partial \ln b} = \frac{-2b}{(x-b)} \frac{\partial \ln b}{\partial \ln b} = \frac{-2b}{(x-b)} \tag{5.90}$$

The magnitude of parameter b relative to x has asymptotic limits in Equation (5.90):

$$\frac{\partial \ln y}{\partial \ln b} = -\frac{2b}{(x-b)} = \begin{cases} 2 & b \gg x \\ -\dfrac{2b}{x} & b \ll x \end{cases} \tag{5.91}$$

Thus, for large b, a 10% change in b yields a 20% change in y. For small b, the sensitivity is a function of parameter b as well as the dependent variable x.

A *VBA* macro **SENSITIVITY** in the *PNMSuite* employs Richardson's extrapolation of finite-difference derivative approximations to calculate the sensitivity coefficients and relative sensitivity for a function of several variables or parameters. The macro prompts the user for the cell containing the function, and the range of cells containing the variables and parameters used by the function. The code listed in the *PNMSuite* validates the location selected for the output to avoid overwriting cells on the worksheet. The output is a table of sensitivity and relative sensitivity coefficients. We illustrate how to set up an *Excel* worksheet to evaluate the sensitivity coefficients for Equation (5.86) at $x = 0$ with $a = 5$ and $b = 1$:

1. Add a formula in cell **B1** for the function in terms of the variable x, and the parameters in the range **B2:B4**.

    B1=B3*(B2 - B4) ^ 2

2. Run the macro **SENSITIVITY** and select cell **B1** for the function input box and the range **B2:B4** for the variables input box.
3. Select cell **C1** for the location of the results.
4. According to the relative sensitivity results in column **F**, the function is most sensitive to the third "variable" corresponding to parameter b:

	A	B	C	D	E	F
1	f		5	$c_1 =$	-10	rel $c_1 =$	0.00E+00
2	x		0	$c_2 =$	1	rel $c_2 =$	1.00E+00
3	a		5	$c_3 =$	10	rel $c_3 =$	2.00E+00
4	b		1			

Example 5.8 *Flow through a Packed Bed*

Known: Levenspiel (1984) describes a simple experiment to study the effects of friction in fluid flow with a superficial velocity of 0.1 m/s through a one meter packed bed of 0.01 m diameter spheres.

Schematic: The apparatus consists of a tube in the shape of a U with one longer branch. The packed bed is located between points 2 and 3.

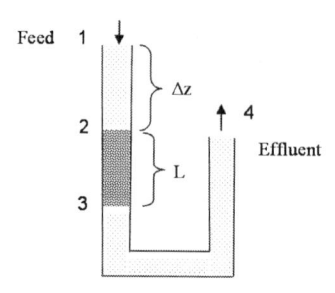

The left branch of the tube is longer than the right branch to create a liquid head of pressure above the packed bed. The liquid stream enters the apparatus at point 1. The height of the liquid above the bed between points one and two controls the liquid flow rate through the tube. The liquid exits the tube at point 4, located at the same vertical position as the top of the packed bed. Each end of the tube is open to the atmosphere.

Find: Investigate the sensitivity of the friction to porosity. Explore the effect of variable bed porosity on Δz, the height of the liquid column for water.

Assumptions: Under steady-state conditions, the change in pressure between points 1 and 4 is negligible.

Analysis: The fluid does no shaft work so that the mechanical energy balance in Equation (1.58) reduces to terms involving only the static head and friction, F:

$$F - g\Delta z = 0 \tag{5.92}$$

where g and Δz are the gravity acceleration and fluid height above the bed, respectively. We ignore the effects of wall friction due to the relatively large surface area of the particles in the packed bed.

The Ergun equation correlates friction with several bed characteristics, including: particle diameter (d), fluid properties of density (ρ) and viscosity (μ), bed length (L), and bed porosity (ε):

$$F = au^2 + bu \tag{5.93}$$

where
$$a = \frac{1.75(1-\varepsilon)L}{\varepsilon^3 d} \quad \text{and} \quad b = \frac{150(1-\varepsilon)^2 \mu L}{\rho \varepsilon^3 d^2}$$

The superficial velocity, u, corresponds to flow through an empty tube.

Combining Equations (5.92) and (5.93) gives the height of the liquid column above the bed for a fixed velocity:

$$\Delta z = \frac{au^2 + bu}{g} \tag{5.94}$$

We use the partial derivative of bed height with respect to bed porosity to get the *relative* or *percent* change:

$$c_{relative} = \frac{\varepsilon}{\Delta z} \frac{\partial \Delta z}{\partial \varepsilon} = \frac{\partial \ln(\Delta z)}{\partial \ln(\varepsilon)} \tag{5.95}$$

A log-log plot accomplishes the same thing. Figure 5.20 has a log-log plot of the liquid height versus bed porosity for a dramatic picture of the sensitivity of bed height to porosity. An order of magnitude change in ε_b yields three to five orders of magnitude change in Δz.

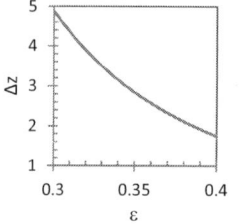

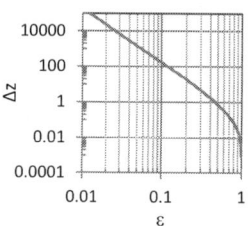

Figure 5.20

Regular and log-log plots showing the sensitivity of the liquid length to porosity.

We can also calculate the sensitivity coefficients for this problem using the *VBA* macro **SENSITIVITY**. The function is located in the cell **B10**, according to Equation (5.93). The variables and parameters are located in the range **B2:B7**. The results at $\varepsilon = 0.4$ for relative sensitivity coefficients in column **F** indicate that the head is most sensitive to the porosity ("variable" 2), followed by the velocity ("variable" 1). The head in this case is relatively insensitive to the properties of density ("variable" 5) and viscosity ("variable" 6).

	A	B	C	D	E	F
1	Parameter	Value	c			
2	u	0.1	$c_1 =$	34.21566	rel $c_1 =$	1.96E+00
3	ε	0.4	$c_2 =$	-16.162	rel $c_2 =$	-3.70E+00
4	d	0.01	$c_3 =$	-182.594	rel $c_3 =$	-1.04E+00
5	L	1	$c_4 =$	1.749168	rel $c_4 =$	1.00E+00
6	ρ	997.1	$c_5 =$	-7.7E-05	rel $c_5 =$	-4.39E-02
7	μ	0.00089	$c_6 =$	86.25917	rel $c_6 =$	4.39E-02
8	a	1640.625				
9	b	7.531216				
10	Δz	1.749171				

Comments: Small uncertainties in the void fraction may lead to large uncertainties in the effects of friction on flow through a packed bed. In this example, a 5% change in the void fraction has the effect of 20% change in the column height of liquid required to maintain the same liquid velocity.

□

"Uncertainty is the refuge of hope." – Henri Frederic Amiel

We use Taylor series approximations to provide estimates of errors (uncertainties) in functions due to uncertainties in the independent variables. To illustrate, linearize a multivariable function with a first-order Taylor series expansion about g, which represents our best estimates for the values of x:

$$f(x_1, x_2, \ldots) = f(g_1, g_2, \ldots) + \Delta x_1 \left.\frac{\partial f}{\partial x}\right|_{g_1} + \Delta x_2 \left.\frac{\partial f}{\partial x}\right|_{g_2} + \ldots \qquad (5.96)$$

where Δx is the uncertainty in g. The error in the first term of the approximation is proportional to the sum of the terms involving the first-derivatives, or sensitivity coefficients. Each term involving a derivative may have a positive or negative sign, with potential for error cancelation in the sum of error terms. Thus, we find an upper limit or maximum value of the uncertainty from the inequality:

$$\left|\sum_{i=1}^{n} \xi_i\right| \leq |\xi_1| + |\xi_2| + \ldots |\xi_n| \qquad (5.97)$$

Application of the inequality in Equation (5.97) to Equation (5.96) gives the following cumulative maximum uncertainty in the approximation for the function f:

$$u_f = |f(x) - f(g)| = |\Delta f(x)| \leq u_{f\max} = \sum_{i=1}^{n} \left|\Delta x_i \left.\frac{\partial f}{\partial x_i}\right|_g \right| \qquad (5.98)$$

where u_f is the approximation for uncertainty in f, or an estimate of the region containing the difference between the true value of $f(x)$ and the approximation $f(g)$. Chapter 8 presents details of experimental uncertainty analysis that we must not ignore when presenting experimental results, including methods for estimating uncertainties in measurements (Hall, Kirwin and Updike 1975).

Example 5.9 Maximum Uncertainty in the Ideal Gas Law

Known/Find: Consider the maximum uncertainty in the calculation of the mass of a volume of air at a specific pressure and temperature from the ideal gas law. The values of the parameters and uncertainties in the variables are listed in the *Excel* worksheet. Note that the gas constant and molecular weight also have uncertainty.

Analysis: First, create a *VBA* user-defined function for mass in terms of pressure, temperature, molecular weight, and volume according to the ideal gas law:

```
Public Function masst(P As Double, V As Double, M As Double, R As Double, T As Double) As Double
    masst = P * V * M / (R * T) 'Ideal gas law. P[=]amt, V[=]L, M[=]g/mol, R[=]L atm/mol K, T[=]K
End Function
```

Use the *VBA* sub procedure **PDERIV** to evaluate the partial derivatives required by Equation (5.98). The independent variables are in the range **B2:B6**. The dependent variable is in cell **B7**. Cycle through the calculation of the derivatives for the three variables and place the results in column **C**. Finally, calculate the absolute value of the product of the derivative and uncertainty for each variable and sum the results as shown in column **E**. The worksheet shows the cell formulas and the results of the calculations. The ideal gas law in cell **B7** and **E7** gives 2.36 ± 0.23 grams or more conservatively 2.4 ± 0.3 (±13%) grams of gas.

	A	B	C	D	E
1		x	d(mass)/dx	Δx	$|(dmass/dx) \times \Delta x|$
2	P/atm	0.99	2.38667	0.01	0.02387
3	V/L	2	1.18140	0.1	0.11814
4	T/K	296	-0.00798	1	0.00798
5	R/(L atm/mol K)	0.0821	-28.77963	0.0001	0.00288
6	M/(g/mol)	29	0.0814761	1	0.08148
7	mass/g	2.3628		Δmass	0.23

Comments: We can reduce the uncertainty using higher precision measurements for temperature, volume, and pressure. We observe from column **E** that the primary contribution to the uncertainty comes from the uncertainties in volume and molecular weight. This analysis does not address the uncertainty in the ideal gas *model*.

□

5.5 Epilogue on Taylor Series and Derivatives

Taylor series is the foundation for developing numerical methods and error analysis. A linear truncated Taylor series expanded near a point of solution provides a good approximation of a nonlinear function and is much easier to work with. We use the Taylor series approximation of functions as the foundation for many of the numerical methods developed in the following chapters. Of most importance are the approximations for derivatives using function values at discrete intervals in the range of the independent variable. Derivative approximations from finite-difference methods require smooth data. We should model noisy data by an appropriate differentiable function, as presented in Chapter 9 on data modeling. Table 5.5 summarizes the methods of derivative approximation developed in this chapter.

Table 5.5 Summary of *VBA* macros and user-defined functions for derivative approximations.

Macro/Function	Inputs	Operation
DREX	Function arguments	Derivative approximation of an *Excel* worksheet *formula* using Ridder's algorithm.
FDERIV	Function arguments	Three-point finite difference derivative approximations of first and second-order derivatives.
FINDIF		FINDIF determines forward, central, and backward formulas from the data.
PADE	Macro input boxes	Calculates the Padé rational function coefficients from a range of polynomial coefficients.
PDERIV	Macro input boxes	Finite difference derivatives of an *Excel* worksheet function with Richardson's extrapolation
RIDDERS	Macro input boxes	Ridders' algorithm for finite difference derivatives order 1 to 6.
SENSITIVITY	Macro input boxes	Partial derivatives and relative sensitivity of a function with respect to variables/parameters
SHANKS	Function argument	Performs repeated Shanks transformations to accelerate convergence of a series.

Chapter 6   Root-finding Methods

> *"If you hit a tuning fork twice as hard it will ring twice as loud but still at the same frequency. If you hit a person twice as hard they're unlikely just to shout twice as loud. That property lets you learn more about the person than the tuning fork."* -Neil Gershenfeld

The need for solving nonlinear algebraic and transcendental equations arises frequently in chemical engineering due to the nonlinearity of models for thermodynamic phase equilibrium, reaction kinetics, and transport phenomena. In this chapter, we introduce several methods for finding roots to nonlinear equations, including:

- Successive substitution (ordinary iteration) employing Wegstein and Steffensen's variation of Aitkin's relaxation by method
- Bisection and Regula Falsi bracketing methods for finding a root to a continuous function
- Newton and Secant root-finding methods using derivative approximations
- Padé interpolation or extrapolation to a root
- Bairstow's method for roots of polynomials
- *Excel*'s **Goal Seek** and **Solver**
- Method of equation tearing

> **SQ3R Focused Reading Questions**
> 1. What makes a function nonlinear?
> 2. What is a system of simultaneous nonlinear equations?
> 3. What *Excel* add-ins are available for solving single and simultaneous equations?
> 4. Why do I need *VBA* macros for solving nonlinear problems?
> 5. What methods are available to guarantee finding a root?
> 6. How can I find all of the roots to a polynomial?
> 7. Why do I need initial guesses for the root?
> 8. How does the arrangement of an equation affect the solution for the root?
> 9. What is needed when the roots have different orders of magnitude?
> 10. How can I use a circular reference in *Excel* to find a root to an equation?

A function manifests *linearity* when its output is directly proportional to its input. A *nonlinear* function is proportional to another function of the input variable. Nonlinear equations have at least one term involving some combination of the variables with an exponent other than one. Consider the following equation with four terms:

$$f(x) = a + \frac{b}{x} + \frac{cx}{1+d} + ex^{1/2} \qquad (6.1)$$

The exponents (orders) of each x-term in the equation are 0, -1, 1, and ½. The second and fourth terms are nonlinear in x as proportional to $1/x$ and $x^{1/2}$, respectively. Polynomials, on the other hand, are linear in the coefficients of each term, even when they are nonlinear in the independent variable. Consider a third-order polynomial:

$$f(x) = a + bx + cx^2 + dx^3 \qquad (6.2)$$

Set $f(x) = 0$ and solve Equation (6.2) for a, b, c, d, and x:

$$a = -\left(bx + cx^2 + dx^3\right) \qquad b = \frac{-\left(a + cx^2 + dx^3\right)}{x} = \frac{-a}{x} - \left(cx + dx^2\right)$$

$$c = \frac{-\left(a + bx + dx^3\right)}{x^2} = \frac{-a}{x^2} - \frac{b}{x} - dx \qquad d = \frac{-\left(a + bx + cx^2\right)}{x^3} = \frac{-a}{x^3} - \frac{b}{x^2} - \frac{c}{x} \qquad (6.3)$$

$$x = \frac{-\left(a + cx^2 + dx^3\right)}{b} \qquad (6.4)$$

The four solutions in Equation (6.3) are explicit in the polynomial coefficients on the left-hand-side. In other words, the explicit parameter only appears on one side of the equation. However, Equation (6.4) in this example is implicit in x, meaning the variable x appears on both sides of the equation. It is challenging to arrange the implicit function in Equation (6.4) to solve for x in terms of only a, b, c, or d.

Most nonlinear functions require numerical methods to find approximate roots. A good approximate solution has a value close enough to the true solution that is adequate for engineering design or analysis. Unlike linear equations that have one root, nonlinear equations may have multiple roots. An nth-order polynomial may have up to n roots. Often, only one root has any real value of practical significance for the scientist or engineer.

In general, we find the roots to nonlinear equations by systematically guessing the value for the root and calculating an approximate error in the guess. We use the error estimate to upgrade our guess for the root closer to the true solution, and repeat the cycle by "trial and error" until we reduce the approximate error below a tolerable limit. We have already seen an example of a nonlinear problem in *Example 5.1* where we found the root iteratively by solving a linear system of equations that included a linear approximation to the vapor-liquid equilibrium equation. We upgraded the guess for the root using the result from the previous iteration. All of the solution algorithms presented in this chapter require iteration starting from an initial guess for the root, then working towards improved solutions after each iteration.[55] Iterative methods converge on a root when the results between iterations stop changing within a defined level of precision. We specified the criteria for convergence in Equation (4.57). The iterative process terminates once the tolerance for convergence is satisfied.

As we learned in Example 4.6 for linear problems, iterative methods may diverge away from a root starting from poor initial guesses for the solution, or for poorly arranged problems. Diverging solutions tend to oscillate between increasingly positive and negative values, or "explode" to an extreme value on either side of the root. If allowed to iterate indefinitely, a diverging solution may cause the algorithm to "crash" when the magnitude of a number becomes too large for the computer to handle. We present methods for controlling and checking on the level of convergence along with each algorithm.

Excel's **Goal Seek** and **Solver** are easy-to-use, general purpose tools for root finding that incorporate many of the features and techniques illustrated in this chapter. It is important to learn the basic methods and requirements for solving nonlinear equations to help set up and trouble-shoot solutions using general-purpose computational tools like **Goal Seek** and **Solver**.

6.1   Graphical Solution

We may find the roots of a one-dimensional nonlinear function by inspecting the graph of the function for the location where the curve intersects the horizontal axis for the independent variable. The plot of a single nonlinear equation arranged in the following form changes sign as it crosses the horizontal axis at the root as illustrated in Figure 6.1:

$$f(x) = 0 \qquad (6.5)$$

---

[55] Now you see why we chose to use the letter g (guess) for our expansion point in Chapter 5.

CHAPTER 6: ROOT-FINDING METHODS

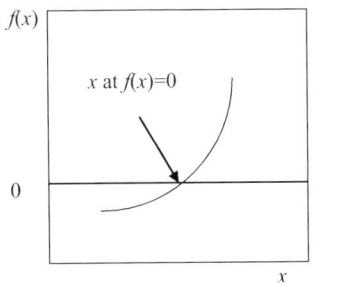

Figure 6.1 Graphical solution of a non-linear equation.

We may also solve for the roots of a pair of coupled nonlinear equations in two unknowns by plotting the two solution variables against each other:

$$x = F(x, y) \quad \text{or} \quad y = G(x, y) \tag{6.6}$$

where F and G are implicit non-linear functions of y or x. The solutions are the values of x and y at the intersection of the curves, as illustrated in Figure 6.2. This graphical procedure becomes difficult to implement for three equations, and impossible for four equations or more. The accuracy of a graphical solution also depends on the precision of the readability of the plot.

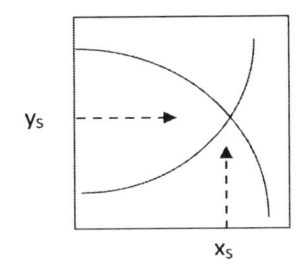

Figure 6.2 Graphical solution of two, coupled non-linear equations.

Example 6.1   Economic Rate of Return

Known: We expect a yield of $350 from three monthly investment payments of $100.

Find: Use *Excel* to determine the rate of return (*ROR*) on our investment.

Analysis: Calculate the interest factors for the present value (*PV*) in terms of the monthly payments ($A = -\$100$) and the future worth after three months ($F = \$350$):

$$PV = 0 = \frac{F}{(1+i)^n} + A\left[\frac{(1+i)^n - 1}{i(1+i)^n}\right] \tag{6.7}$$

Determine the rate of return, which is the root to *Excel*'s financial worksheet function **PV** for Equation (6.7):

$$PV(i, n, A, F)$$

where
- i = interest rate
- n = number of periods
- A = payment
- F = future value (optional)

Note that **F** is an optional argument. If we omit **F**, then **PV** calculates the present value for **n** payments with an interest rate **i**. In this problem, we include **F** to find the unknown interest rate that satisfies Equation (6.7). In other words, at what point do we break even? *Excel* uses the conventional format with a $ sign for currency by adding parentheses and red colored fonts to the content of cells containing negative currency values, as shown in Figure 6.3.

	A	B
1	i	PV
2	0	$50.00
3	0.02	$41.42
4	0.04	$33.64
5	0.06	$26.57
6	0.08	$20.13
7	0.1	$14.27
8	0.12	$8.94
9	0.14	$4.08
10	0.16	($0.36)
11	0.18	($4.41)
12	0.2	($8.10)

	B
1	PV
2	=PV(A2,3,100,-350)
3	=PV(A3,3,100,-350)
4	=PV(A4,3,100,-350)
5	=PV(A5,3,100,-350)
6	=PV(A6,3,100,-350)
7	=PV(A7,3,100,-350)
8	=PV(A8,3,100,-350)
9	=PV(A9,3,100,-350)
10	=PV(A10,3,100,-350)
11	=PV(A11,3,100,-350)
12	=PV(A12,3,100,-350)

Figure 6.3

Present value versus interest rate. Use auto-fill to generate the formulas in column B.

A plot of **PV** versus the interest rate in Figure 6.4 shows a rate of return for this problem of 0.155 or 15.5%.

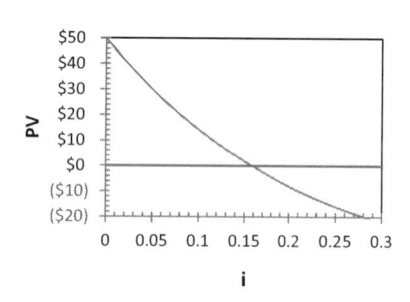

Figure 6.4

Present value versus interest rate. Locate the root where the curve crosses the abscissa.

Comments: The worksheet function **RATE** performs the same calculation. The result is not a bad *ROR*. When can I expect a dividend check?

The next example illustrates the graphical solution of a pair of simultaneous equations.

Example 6.2 Ignition Temperature

Known: A fuel mixture in a reactor has the following properties shown in an *Excel* worksheet:

	A	B	C
1	Heat of Reaction/(J/mol)	dH	500
2	Pre-exponential Factor/(1/s)	k	3.23E+03
3	Activation Energy/(J/mol)	Ea	4.57E+04
4	Gas Constant/(J/mol K)	Rg	8.314
5	Concentration/(mol/m^3)	C	0.3
6	Volume/m^3	V	1
7	Heat Transfer Area/m^2	A	0.5
8	Surrounding Temperature/K	Ts	300
9	Heat Transfer Coefficient/(W/m^2 K)	h	15

Find: Use *Excel* to determine the ignition temperature of the fluid mixture.

Assumptions: Steady state, Arrhenius rate constant, first-order rate law.

Analysis: The auto ignition temperature occurs when the rate of heat generation due to the exothermic reaction exceeds the rate of heat removed by convection in the reactor.

$$q_{conv.} = hA(T - T_s) \quad \text{and} \quad q_{rxn} = \Delta HVCk \exp\left(-\frac{E_a}{R_g T}\right)$$

Generate ranges of heat rates for graphing over a 1500 K temperature range to locate the intersection between the q-curves corresponding to the ignition temperature. To simplify the calculations, use named cells for the values in column **C** according to the labels in column **B** (note that **C** changes to **C_** because **C** is reserved). Name the range **A12:A25** as T for array operations on the following formulas in columns **B** and **C**, using CNTL SHIFT ENTER:

B12:B25 = h*A*(T-Ts)   C12:C25 = dH*V*C_*k*EXP(-Ea/(Rg*T))

	A	B	C	D	E
11	T/K	Q Convection	Q Reaction		
12	200	-750	5.688E-07		
13	300	0	0.0053921		
14	400	750	0.5249811		
15	500	1500	8.1879213		
			⋮		
20	1000	5250	1991.7449		
21	1100	6000	3282.0733		
22	1200	6750	4976.344		
23	1300	7500	7077.2244		
24	1400	8250	9571.173		
25	1500	9000	12433.319		
26	1326	7697.866957	7697.867	-4.9E-07	<-- Goal Seek

If we click on the plot to make it active, and then move the mouse pointer along the curve, a small popup box displays the local values of the points. In this example, the temperature points are spaced at 100 degree increments, so we locate the intersection for the ignition temperature between $1300 < T < 1400$ K. We can refine the value using smaller increments in T.

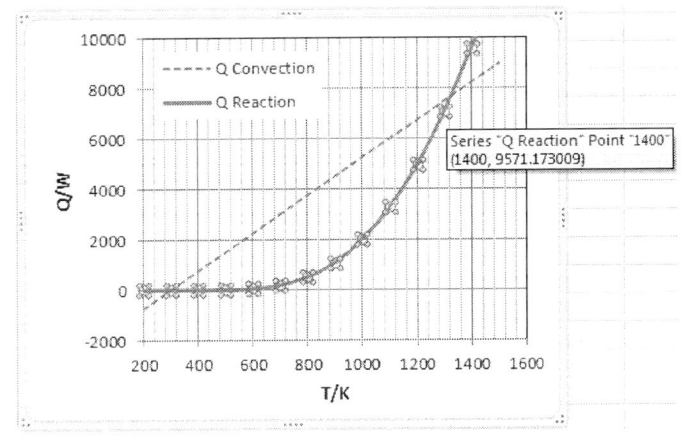

Comments: When the temperature exceeds the auto ignition temperature, the reaction rate "runs away" and we lose temperature control on the reactor. Compare the trace approximation with the precise solution using *Excel*'s **Goal Seek** for $T = 1326$ K and $Q = 7698$ W. Graphical methods have limited usefulness. Problems involving more than two unknowns are difficult to visualize in two-dimensional graphs. □

6.2 Numerical Roots of a Single Equation

Numerical methods for solving single nonlinear equations include bisection and regula falsi that guarantee convergence, as well as the simple method of successive substitution, which we have already seen as an alternative solution method for solving linear systems. We also cover indirect methods that accelerate convergence using derivatives (or derivative approximations), such as Wegstein's method, Newton's method, and the secant method.

6.2.1 Fixed-point Ordinary Iteration

Rearrange a nonlinear function from the form of Equation (6.5) to an implicit form in terms of the dependent variable, similar to the Jacobi or Gauss-Seidel methods for linear functions in Section 4.4:

$$x = F(x) \tag{6.8}$$

We re-introduce the symbolic convention where we use lower case f for functions set to zero and upper case, or capitalized F for implicit functions set to x, like Equation (6.8). The last expression in Equation (6.3) is an example of a candidate for successive substitution, also known as ordinary iteration.

Start with a guess for x, and then evaluate the implicit expression F for x according to Equation (6.8). Substitute the result for x back into the implicit expression F iteratively until achieving convergence in x:

$$x_i = F(x_{i-1}) \quad \text{for} \quad i = 0, 1, 2\ldots \tag{6.9}$$

The subscript index i keeps track of the iteration number.

Though simple to implement, the method of ordinary iteration suffers from the same potential problems encountered with Gauss-Seidel iteration. Successful convergence may be sensitive to the initial guess for the root as well as the arrangement of the implicit expression. Try variable transformation or equation rearrangement to avoid computational problems such as logarithmic and exponential terms, division by zero or calculating the square root of a negative number while iterating on the implicit function.

6.2.2 Wegstein's Acceleration of Fixed-point Iteration*

"There is nothing wrong with change if it is in the right direction." – Winston Churchill

Ordinary iteration may converge slowly depending on the form of the implicit expression, and the selection of the initial guess for the root. Wegstein's (1958) method attempts to accelerate convergence by linearly extrapolating the results from two ordinary iterations to estimate an optimal relaxation factor (see Section 4.6.2). Consider two successive ordinary iterations, starting with the initial guess x_0, leading to the converged root x:

$$x_{i-1} \cong F(x_{i-2}) \to x_i \cong F(x_{i-1}) \ldots \to x = F(x) \tag{6.10}$$

A linear extrapolation to the root of the function by the point-slope form of a line gives:

$$F(x) - F(x_i) = \left[\frac{F(x_i) - F(x_{i-1})}{x_i - x_{i-1}} \right] (x - x_i) \tag{6.11}$$

Define the Wegstein acceleration parameter in terms of the previous two iterations for x and $F(x)$:

$$w = \frac{F(x_i) - F(x_{i-1})}{x_i - x_{i-1}} \tag{6.12}$$

Substitute from Equation (6.10) for the $F(x)$ terms into Equation (6.11) and rearrange for the following compact, weighted form of Wegstein's recursive formulas for x:

- Wegstein
$$x_{i+1} = \frac{F(x_i) - wx_i}{1-w} = \left(\frac{1}{1-w}\right) F(x_i) - \left(\frac{w}{1-w}\right) x_i \tag{6.13}$$

Steffensen's application of the Aitken Δ^2 method described in Section 6.3.2 extends the Wegstein approach to coupled systems of equations. Wegstein's method is popular for use in process simulators to solve large systems of equations derived from material and energy balances around unit operations and flow networks.

Example 6.3   Accelerated Series Convergence with Wegstein's Method

Known/Find: Find a root to the following equation by ordinary iteration and Wegstein's method.

$$4\cos(x) = \exp(x) \tag{6.14}$$

Analysis: Use *Excel* to perform the iterations for each method. Rearrange Equation (6.14) for fixed-point iteration:

$$x = \ln\left[4\cos(x)\right] \tag{6.15}$$

Start with an initial guess of $x_0 = 1$, which is close to the root. The formulas used by ordinary iteration in columns **B** and **C** of the worksheet in Figure 6.5 are

                 B4=C3   C4=LN(4*COS(B4))

After two ordinary iterations, implement Wegstein's method in columns **F**, **G**, and **H** beginning in row **5** with the following worksheet formulas:

       F5=G4/(1-H4)-H4*F4/(1-H4)   G5=LN(4*COS(F5))   H5=(G5-G4)/(F5-F4)

Calculate the relative residuals in columns **D** and **I** according to Equation (4.56). The method of ordinary iteration produces results that oscillate about the root, as seen in Figure 6.5. Wegstein's method converges after five iterations within an error of $\sim 10^{-4}$ %. The relaxation factor in cell **H13** after 10 iterations involves division by zero when the iterative results converge exactly to $x_9 = x_{10}$ within the precision of the computer.

	A	B	C	D	E	F	G	H	I
1	Ordinary Iteration					Wegstein			
2	i	x	F(x)	Res%		x	F(x)	w	Res%
3	0	1	0.770668			1	0.770668		
4	1	0.770668	1.054236	2.98E+01		0.770668	1.054236	-1.2365	2.98E+01
5	2	1.054236	0.680856	2.69E+01		0.897459	0.914045	-1.10568	1.41E+01
6	3	0.680856	1.134023	5.48E+01		0.905336	0.904091	-1.2638	8.70E-01
7	4	1.134023	0.525953	4.00E+01		0.904786	0.904791	-1.27334	6.08E-02
8	5	0.525953	1.241091	1.16E+02		0.904788	0.904788	-1.27262	2.55E-04
9	6	1.241091	0.258556	5.76E+01		0.904788	0.904788	-1.27263	8.06E-08
10	7	0.258556	1.35249	3.80E+02		0.904788	0.904788	-1.2	1.23E-13
11	8	1.35249	-0.14352	8.09E+01		0.904788	0.904788	-2	2.45E-14
12	9	-0.14352	1.37596	1.04E+03		0.904788	0.904788	-2	2.45E-14
13	10	1.37596	-0.25564	1.10E+02		0.904788	0.904788	#DIV/0!	0.00E+00

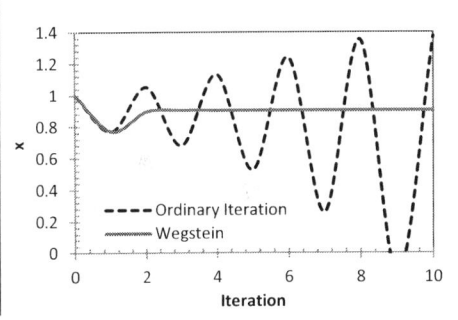

Figure 6.5   Comparison of ordinary iteration with Wegstein's method.

Comments: For this problem, Wegstein's method accelerates convergence where ordinary iteration diverges due to a poor implicit arrangement of the nonlinear equation.

□

6.2.3   Bisection

As with fixed-point iteration, the bisection method is relatively unsophisticated in its approach to finding a root to a nonlinear equation. However, unlike ordinary iteration, bisection has the following requirements that guarantee convergence to a root:

- Arrange the nonlinear function in the standard form of Equation (6.5), $f(x) = 0$.
- Start with *two* initial guesses that bracket the root.
- Apply to smooth and continuous functions within the solution bracket.

     A root is considered bracketed when the functions evaluated at the bracket points have opposite signs. A bracketed root guarantees that the function crosses the horizontal axis somewhere between the brackets at the point where $f(x) = 0$. As illustrated in Figure 6.4, the interest rates $i = 0.1$ and $i = 0.2$ give PV(i=0.1) = 14.3 and PV(i=0.2) = -8.1 that bracket the root at $i = 0.155$ where PV(i=0.155) = 0. To implement the method of bisection, cycle through the following three steps:

1. Bracket the nonlinear function $f(x) = 0$ in the interval

$$x_{low} < x < x_{up} \tag{6.16}$$

where

$$f(x_{low}) \cdot f(x_{up}) < 0$$

2. Upgrade the search for the root by bisecting the x-interval:

$$x_{new} = \frac{x_{low} + x_{up}}{2} \tag{6.17}$$

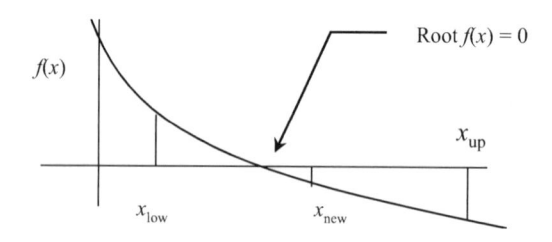

3. If the product of the functions evaluated at the lower bracket and midpoint is negative, $f(x_{low}) \cdot f(x_{new}) < 0$, the root lies in the lower half of the initial interval. Replace the *upper* bracket point with the midpoint from Equation (6.17):

$$x_{up} = x_{new} \tag{6.18}$$

Otherwise, the root is located in the upper half of the interval, $x_{new} < x < x_{up}$. Replace the *lower* bracket point with the midpoint from Equation (6.17):

$$x_{low} = x_{new} \tag{6.19}$$

After each iteration, the bracket size decreases by a factor of two according to Equation (6.17). We determine the number of iterations required from the convergence tolerance by rearranging $\Delta x_0 = tol \cdot 2^{n+1}$ for n:

$$n = \frac{\ln\left[\Delta x_o / (2 \cdot tol)\right]}{\ln 2} \tag{6.20}$$

where *tol* is the desired precision of the root and $\Delta x_o = |x_1 - x_0|$ is the absolute value of the original bracket width for the independent variable.

The bisection method converges linearly because the solution procedure upgrades the guess for the unknown solution by halving the bracketing interval between the current upper and lower bounds on the solution. We implement the bisection method in *Excel* using the worksheet function **IF** introduced in Section 2.2.3. A *VBA* macro **BISECTION** is available in the *PNMSuite*. To use **BISECTION**, put an initial guess for the root into a worksheet cell. In another cell, add a formula for the function $f(x)$ that references the cell with the independent variable x. The macro prompts the user for the cells that contain the variable and function, as well as a tolerance for convergence. **BISECTION** employs an accelerated search for the other bracket on the root before implementing the bisection algorithm.

Example 6.4 Liquid Level in a Spherical Tank

Known/Find: A 1.2 m diameter spherical tank stores liquid hydrogen. Determine the height of liquid in the tank for a total liquid volume of 0.45 m^3.

Schematic: A spherical tank has a diameter D (radius $r = D/2$) and liquid level h.

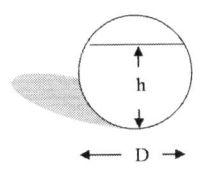

Analysis: Calculate the volume of a partially filled spherical tank in terms of the liquid height and radius:

$$V = \frac{\pi}{3}\left[4r^3 - (2r-h)^2 (r+h) \right] \tag{6.21}$$

Rearrange Equation (6.21) in the form of Equation (6.5), set to zero at the root:

$$f(h) = 0 = 3V - \pi\left[4r^3 - (2r-h)^2 (r+h) \right] \tag{6.22}$$

A simple *VBA* user-defined function for Equation (6.22) simplifies the implementation of the bisection method.

```
Public Function fsphere(h As Double) As Double ' Spherical tank function of liquid height
Dim V As Double, D As Double, r As Double
V = 0.45: D = 1.2: r = D / 2# ' Set the values for tank volume, diameter, and radius
fsphere = 3# * V - WorksheetFunction.pi() * (4 * r ^ 3 - ((2 * r - h) ^ 2#) * (r + h))
End Function
```

We use the *VBA* macro **QYXPLOT** from the *PNMSuite* to graph the function for locating brackets on the root by inspection [where x = h, y = f(h)]:

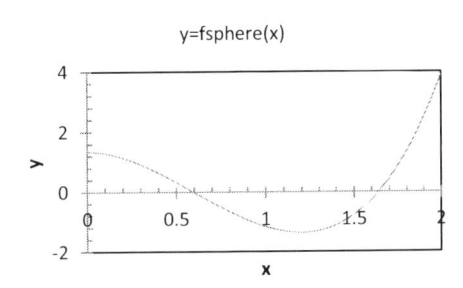

The maximum possible height corresponds to the diameter of the tank. The graph shows two roots, with only one feasible root below x = 1. Thus, after inspecting the graph, we use $h = 0.01$ m and $h = 0.99D = 1.188$ m to bracket the root. The solution converges within a precision of 10^{-3} after 10 iterations as predicted by Equation (6.20):

$$n = \frac{\ln\left[(1.188 - 0.01)/(2 \cdot 10^{-3})\right]}{\ln 2} = 10 \tag{6.23}$$

Use the logical worksheet function **IF()** to control the bisection method in an *Excel* worksheet. The formulas for the cells in row four use relative referencing:

B4=IF(E3*G3<0, B3, D3)	C4=IF(F3*G3<0, C3, D3)	D4= (B4+C4)/2	
E4=f(B4)	F4=f(C4)	G4=f(D4)	H4=ABS(D4-D3)

Select the cells containing the formulas in row four then use the fill handle to drag down 10 rows to auto fill the bisection formulas for each successive iteration that eventually converges on the root.

	A	B	C	D	E	F	G	H
1	Diameter	1.20		Iterations	9.2021238			
2	Iteration	h_{low}	h_{up}	h_{bisect}	f_{low}	f_{up}	f_{bisect}	$|\Delta h|$
3	0	0.010000	1.188000	0.599000	1.349438	-1.363527	-0.003775	
4	1	0.010000	0.599000	0.304500	1.349438	-0.003775	0.914377	0.294500
5	2	0.304500	0.599000	0.451750	0.914377	-0.003775	0.485596	0.147250
6	3	0.451750	0.599000	0.525375	0.485596	-0.003775	0.244723	0.073625
7	4	0.525375	0.599000	0.562188	0.244723	-0.003775	0.120957	0.036813
8	5	0.562188	0.599000	0.580594	0.120957	-0.003775	0.058653	0.018406
9	6	0.580594	0.599000	0.589797	0.058653	-0.003775	0.027447	0.009203
10	7	0.589797	0.599000	0.594398	0.027447	-0.003775	0.011837	0.004602
11	8	0.594398	0.599000	0.596699	0.011837	-0.003775	0.004031	0.002301
12	9	0.596699	0.599000	0.597850	0.004031	-0.003775	0.000128	0.001150
13	10	0.597850	0.599000	0.598425	0.000128	-0.003775	-0.001824	0.000575

The bisection method finds the liquid height of 0.598 m in the spherical tank. Column **H** compares the solution at the midpoint between successive iterations (or rows) to check for convergence on the root. We may use the variable and function in row 3 of the worksheet for the **BISECTION** macro to produce the same root.

Comments: The bisection solution converged on the root without difficulty. The bisection method is preferred when other methods fail to locate the root. Note that Equation (6.22) is a cubic polynomial in h. We may also use the VBA macro **BAIRSTOW** presented in Section 6.4 to calculate the height of liquid.

□

6.2.4 Regula Falsi Linear Interpolation*

The regula falsi method (or method of false position), like bisection, requires an initial bracket of the root to begin the search process. Sometimes referred to as the method of linear interpolation, the regula falsi method upgrades the guesses for the root from the intersection of the line formed between the points for the brackets with the horizontal axis, as illustrated in Figure 6.6. However, the regula falsi method generally converges faster than bisection when starting from a *good* selection of the initial bracket around the root.

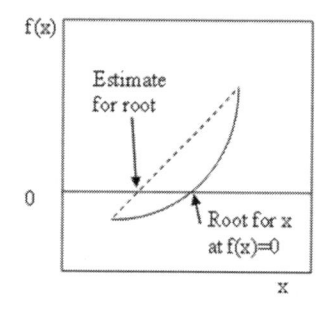

Figure 6.6

Regula falsi estimate for the bracketed root of a non-linear equation.

Start with a point-slope form of a line in terms of two points that bracket the solution:

$$0 - f(x_{up}) = \left[\frac{f(x_{up}) - f(x_{low})}{x_{up} - x_{low}}\right](x_{new} - x_{up}) \tag{6.24}$$

Rearrange Equation (6.24) for the upgraded approximation of x_{new} for the root to give the following recursion formula:

- *Regula Falsi*

$$x_{new} = \frac{f(x_{up})[x_{low} - x_{up}]}{f(x_{up}) - f(x_{low})} \tag{6.25}$$

After each iteration, upgrade the bracketing interval in the same fashion as the bisection method in Section 6.2.3.

Although the regula falsi method has guaranteed convergence for smooth, continuous functions, it may converge slower than the bisection method if the root is close to either end of the initial bracket. A simple modification to the regula falsi method promotes faster convergence by monitoring the values of the bracket after each iteration. When either bracket point remains unchanged after two iterations, decrease the value of the function at that point by a factor of two (Chapra and Canale 2002):

$$x_{new} = x_{up} + \frac{f(x_{up})(x_{low} - x_{up})}{f(x_{up}) - \frac{f(x_{low})}{2}} \tag{6.26}$$

Otherwise, double the function value if the upper bracket value remains unchanged after two iterations:

$$x_{new} = x_{up} + \frac{f(x_{up})(x_{low} - x_{up})}{f(x_{up}) - 2f(x_{low})} \tag{6.27}$$

A *VBA* user-defined function **REGFAL(f, x, xup, tol)** in Figure 6.7 is available in the *PNMSuite* for solving a single nonlinear function in the form $f(x) = 0$ with a bracketed root. The user-defined function **REGFAL** follows the template of Section 3.6.3 for evaluating an *Excel* worksheet formula in *VBA*. The *PNMSuite* also has a macro **REGULAFALSI** with input boxes for the locations of the function and brackets on an *Excel* worksheet. The macro reports the number of iterations to the **Status Bar**. To use either **REGFAL** or **REGULAFALSI** in an *Excel* worksheet, add the value of the lower bracket to a cell in an *Excel* worksheet. Define the nonlinear function in terms of the cell with the lower bracket for the variable. Place the value for the upper bracket in another cell on the worksheet. Both user-defined function **REGFAL** and macro **REGULAFALSI** employ a default value for the relative convergence tolerance of $|(x_{up} - x_{low})/x_{new}| < 10^{-8}$. We illustrate how to set up an *Excel* worksheet for the macro **REGULAFALSI** to find the root to a simple nonlinear function:

$$f(x) = 5x^2 - e^x \tag{6.28}$$

1. By inspection, locate points that bracket the root at $f(x=0) = -1$ and $f(x=1) = 2.28$.
2. Set up an *Excel* worksheet with the lower bracket in cell **A1** and the upper bracket in cell **B1**.
3. Add the formula for the function of Equation (6.28) in cell **A3** with reference to cell **A1** for the variable x.
4. Run the macro **REGULAFALSI** and select the cells for the brackets and functions as requested by the input boxes. When converged, cell **A1** contains the solution for the root.

	A	B	C
1	0	1	=5*A1^2-EXP(A1)

	A	B	C
1	0.605267	1	8.34E-13

5. Alternatively, set up the user-defined function **REGFAL** in a cell and select the cells for the brackets and functions for the arguments:

0.605267 = REGFAL(C1, A1, B1)

```
Public Function REGFAL(f, x, xup As Double, Optional tol As Double = 0.0000001) As Double
' f = worksheet address of function formula in terms of address for x
' x = variable worksheet address and initial lower bracket value
' xup = initial upper bracket value
' tol = (optional) relative convergence tolerance, default = 1E-8
'***************************************************************************
' Solve a single nonlinear equation f(x) = 0 by the modified regula falsi method.  The
' user must put the value of the initial lower bracket into a cell on the worksheet
' and provide a value for the upper bracket, and then formulate the nonlinear function
' in a cell in terms of the cell containing the lower bracket.  If a bracket does not
' change after two iterations, the function value at that bracket is halved.
' *** Nomenclature ***
Dim dx As Double    ' = relative convergence criteria
Dim fa As String, fx As String  ' = formula string
Dim ffup As Double  ' = product of f and fup
Dim fr As Double    ' = regula falsi value of function
```

```vba
    Dim flow As Double, fup As Double ' = lower and upper bracket function values
    Dim j As Long ' = for-loop index
    Dim n As Long, ns As Long ' = occurrences of x in f, number of regfal iterations
    Dim nlow As Long, nup As Long ' = iterations w/o change to lower/upper limit
    Dim small As Double ' = small number to avoid division by zero
    Dim t As String ' = temporary string for function formula
    Dim xa As String ' = variable address
    Dim xlow As Double ' = lower bracket point
    Dim xs As String, xu As String ' string for variable name in formula
    '*****************************************************************************
    On Error Resume Next: small = 0.000000000001 ' check for input errors
    xlow = x.Value ' LOWER bracket value and variable address string
    xa = x.Address: xs = "": xs = Range(xa).Name.Name
    xu = "X_": fa = VBA.UCase(f.Formula) ' make upper case to eliminate case sensitivity
    Do While InStr(fa, xu) ' check for dummy variable string in formula
        xu = xu & "_" ' add an underscore to end of string
    Loop
    With Application ' Get formula string for substitution and evaluation
        fa = .ConvertFormula(f.Formula, xlA1, xlA1, xlAbsolute): fx = fa
        n = (Len(fa) - Len(.Substitute(fa, xa, ""))) / Len(xa)
        For j = n To 1 Step -1
            t = .Substitute(fa, xa, xlow & " ", j)
            If Not IsError(Evaluate(t)) Then
                fa = t: fx = .Substitute(fx, xa, xu, j)
            End If
        Next j
        ns = Len(xs)
        If ns > 0 Then
            n = (Len(fa) - Len(.Substitute(fa, xs, ""))) / ns
            For j = n To 1 Step -1
                t = .Substitute(fa, xs, xlow & " ", j)
                If Not IsError(Evaluate(t)) Then
                    fa = t: fx = .Substitute(fx, xs, xu, j)
                End If
            Next j
        End If
        flow = Evaluate(.Substitute(fx, xu, xlow)) ' Evaluate function at initial brackets
        fup = Evaluate(.Substitute(fx, xu, xup))
        If flow * fup > 0 Then ' Validate the initial brackets around the root
            MsgBox "REGFAL: Root is NOT bracketed!": Exit Function
        End If
        nlow = 0: nup = 0 ' Initialize counters for modified regula falsi
        Do ' regula falsi loop
            ' Get upgraded x value and evaluate the function
            REGFAL = (xlow * fup - xup * flow) / (fup - flow)
            fr = Evaluate(.Substitute(fx, xu, REGFAL)): ffup = fr * fup
            ' Check for new reduced bracket location
            If ffup > 0 Then ' replace upper bracket
                xup = REGFAL: fup = fr: nup = 0: nlow = nlow + 1
                If nlow > 1 Then ' modify lower function value
                    flow = 0.5 * flow: nlow = 0
                End If
            ElseIf ffup < 0 Then ' replace lower bracket
                xlow = REGFAL: flow = fr: nlow = 0: nup = nup + 1
                If nup > 1 Then ' modify upper function value
                    fup = 0.5 * fup: nup = 0
                End If
            ElseIf ffup = 0 Then
                xlow = xup
            End If
            dx = Abs(xup - xlow) / (Abs(REGFAL) + small)
        Loop While dx > tol
    End With ' Application
End Function ' REGFAL
```

Figure 6.7   VBA user-defined function REGFAL for finding a root to a single equation by regula falsi.

6.2.5 Excel's Goal Seek

As described in Section 2.4, **Goal Seek** uses a robust, linear search algorithm, similar to modified regula falsi, which begins with a guess for the root in the **By Changing Cell**. *Excel* uses a small change in the initial guess to recalculate the formula in the **Set cell**. Whichever value for the guess brings the formula result closer to the targeted result (**To Value**) defines the direction (positive or negative) in which **Goal Seek** heads to reach the target value. Recall that, unlike the **Solver**, **Goal Seek** only searches for the root of a single variable. Pay attention to the magnitude of the convergence tolerance in **File>Options>Formulas >Maximum Change**. The maximum change is not relative, but absolute. **Goal Seek** may stop searching for the root prematurely if the magnitude of |**By Changing**| < **Maximum Change**.

Example 6.5   Temperature and Viscosity using Goal Seek

Known/Find: Use **Goal Seek** with the following equation to find the temperature in degrees K when the viscosity of CO_2 is 2×10^{-5} Pa·s.

$$\mu = \frac{2.148 \times 10^{-6} T^{1.46}}{T + 290} \tag{6.29}$$

Analysis: Put an initial guess of 1000 for the temperature in cell **B1** and calculate the viscosity in cell **B2**.

B2=(0.000002148*B1^1.46)/(B1+290)

	A	B
1	T/K	1000
2	μ/Pa·s	3.99434E-05

Run **Goal Seek** from the ribbon tab **Data>What-if Analysis** to change the temperature in cell **B1** to set the viscosity in cell **B2** to a value of 2E-5.

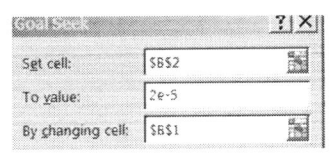

The **Goal Seek Status** box indicates a found solution. However, there is no change in the value for temperature and the current value does not match the target value. The problem is that the value of viscosity has a magnitude of 10^{-5}, which is smaller than the default maximum change value in the *Excel* Options:

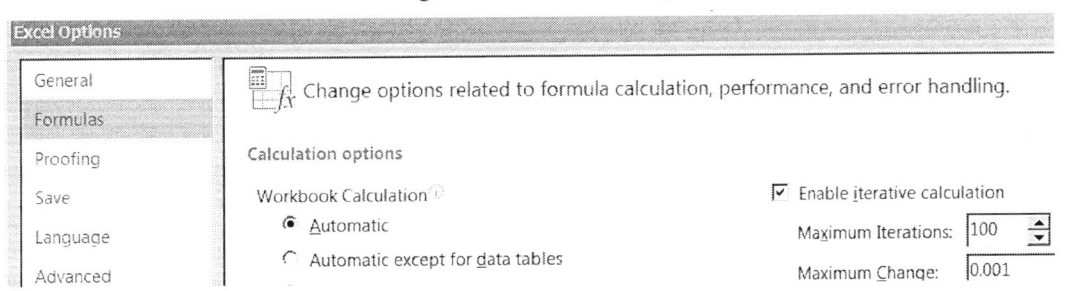

To find the solution, set the value for **Maximum Change** to 0.00000001 and rerun **Goal Seek** to calculate the temperature of 409.4 K. Note that **Goal Seek** now reaches the target value within the maximum change:

	A	B
1	T/K	409.3835169
2	μ/Pa·s	2E-05

Comments: Alternatively, we can scale the viscosity values by a factor of 10^3 to use the default maximum change. □

6.2.6 Newton's Method

"No great discovery was ever made without a bold guess" – Isaac Newton

Newton's recursive method of root finding:
- Requires smooth, continuous functions
- Requires just one guess for the root to initialize the method
- Uses information from the derivative of the function to point towards the location of the root
- Converges quickly to the root from good initial guesses

We derive Newton's method by expanding the function in Equation (6.5) in a linear Taylor series about an initial guess for the solution, truncating second order, and higher terms.

$$f(x) = 0 \cong f(g) + [x-g]\left.\frac{df}{dx}\right|_g \tag{6.30}$$

where we evaluate the derivative at the initial guess g for the root x. Solve Equation (6.30) for x:

$$x \cong g - \frac{f(g)}{df/dx\big|_g} \tag{6.31}$$

The second term on the right side corrects the initial guess to give an upgraded approximation for the root.

Figure 6.8 illustrates the first two steps in Newton's method for one iteration based on the initial guess g. The upgraded solution for x is the intersection of the line tangent to the function with the abscissa.

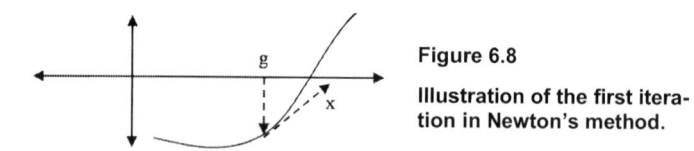

Figure 6.8 Illustration of the first iteration in Newton's method.

Repeat the process by replacing g_0 with the upgraded approximation for the solution x from Equation (6.31) for the recursive formula of Newton's method:

- *Newton's Method*

$$x_i = x_{i-1} - \frac{f(x_{i-1})}{df/dx\big|_{x_{i-1}}} \tag{6.32}$$

where the subscript i is the iteration counter. Iterate on Equation (6.32) until achieving satisfactory convergence on the root. Table 6.1 is a simple example of Newton's method applied to the function $f(x) = x^2 - 4$ with derivative $f'(x) = 2x$ starting with an initial guess for $x = 0.5$. The first iteration gives

$$x_1 = 4.25 = 0.5 - \frac{(-3.75)}{1}$$

Table 6.1 Newton's method steps for $f(x) = x^2 - 4$, $f'(x) = 2x$.

Iteration	x_i	$f(x_i)$	$(df/dx)_i$	Abs[$(x_i - x_{i-1})/x_i$]
0	0.5	-3.75	1	-
1	4.25	14.0625	8.5	88.235%
2	2.595588	2.737078	5.191176	63.739%
3	2.068332	0.277999	4.136665	25.492%
4	2.001129	0.004516	4.002258	3.358%
5	2	1.27×10^{-6}	4.000001	0.056%

Figure 6.9 is a plot of the iterative steps in Newton's method from Table 6.1. The solution converges quickly to the root after just three iterations, and finds the true root $x = 2$ within 0.1% relative error after five iterations.

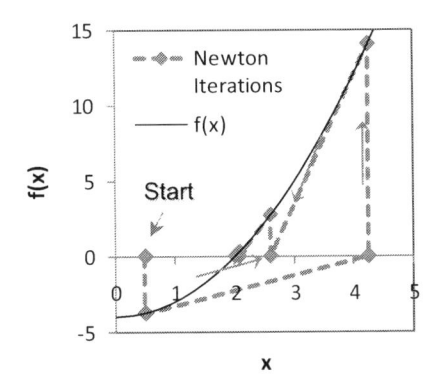

Figure 6.9

Newton's method for finding the root to $f(x) = x^2 - 4$. Iterations start at $x = 0.5$.

A user-defined function **NEWTON(f, x, tol)** is available in the *PNMSuite* for finding the root of a single equation defined by a formula in an *Excel* worksheet. To use **NEWTON**, define the formula of the function in a cell in terms of an initial guess in another cell. The arguments consist of the worksheet addresses for the function, **f**, and initial guess for the root **x**. The user may change from the default relative convergence tolerance by including the optional third argument, **tol**.

Example 6.6 *Friction factor from the Colebrook Equation*

Known: We use the implicit Colebrook equation to calculate a friction factor for turbulent pipe flow: [56]

$$0 = \frac{1}{\sqrt{f}} + 1.737 \ln\left(\frac{\varepsilon}{3.7D} + \frac{1.255}{\text{Re}\sqrt{f}}\right) \qquad (6.33)$$

where f is the friction factor, ε/D is the ratio of surface roughness to inside pipe diameter, and Re is the dimensionless Reynolds number.

Find: Calculate the friction factor when $Re = 10^5$ and $\varepsilon/D = 0.0002$.

Analysis: Solve the nonlinear Colebrook equation by trial and error techniques such as Newton's method. To avoid the possibility of calculating the square root of a negative number, transform the iteration variable: $\phi = 1/\sqrt{f}$. The transformed Colebrook equation becomes

$$\xi(\phi) = \phi + 1.737 \ln\left(\frac{\varepsilon}{3.7D} + \frac{1.255}{\text{Re}}\phi\right) \qquad (6.34)$$

Calculate the derivative of the Colebrook equation (6.33) with respect to ϕ:

$$\frac{d\xi}{d\phi} = 1 + \frac{2.180}{\left(\frac{\varepsilon \text{Re}}{3.7D} + 1.255\phi\right)} \qquad (6.35)$$

Substitute these results into Equation (6.32) for Newton's recursive formula in ϕ:

---

[56] Note that the Haaland Equation (3.2) is explicit in the friction factor, and agrees with the Colebrook equation within 1%:

$$\phi_i = \phi_{i-1} - \frac{\left[\phi_{i-1} + 1.737 \ln\left(\frac{\varepsilon}{3.7D} + \frac{1.255}{Re}\phi_{i-1}\right)\right]}{\left[1 + \frac{2.180}{\left(\frac{\varepsilon Re}{3.7D} + 1.255\phi_{i-1}\right)}\right]} \quad (6.36)$$

Implement Equation (6.36) in an *Excel* worksheet, with the results from each iteration placed in successive rows:

	A	B	C	D	E
1	Iteration	φ	ξ	ξ'	|(φ_i - φ_{i-1})/φ_i|
2	0	10	=B2+1.737*LN(0.0002/3.7+1.255*B2/100000)	=1+2.18/(0.0002*100000/3.7+1.255*B2)	
3	=A2+1	=B2-C2/D2	=B3+1.737*LN(0.0002/3.7+1.255*B3/100000)	=1+2.18/(0.0002*100000/3.7+1.255*B3)	=ABS((B3-B2)/B3)
4	=A3+1	=B3-C3/D3	=B4+1.737*LN(0.0002/3.7+1.255*B4/100000)	=1+2.18/(0.0002*100000/3.7+1.255*B4)	=ABS((B4-B3)/B4)
5	=A4+1	=B4-C4/D4	=B5+1.737*LN(0.0002/3.7+1.255*B5/100000)	=1+2.18/(0.0002*100000/3.7+1.255*B5)	=ABS((B5-B4)/B5)
6	=A5+1	=B5-C5/D5	=B6+1.737*LN(0.0002/3.7+1.255*B6/100000)	=1+2.18/(0.0002*100000/3.7+1.255*B6)	=ABS((B6-B5)/B6)
7	=A6+1	=B6-C6/D6	=B7+1.737*LN(0.0002/3.7+1.255*B7/100000)	=1+2.18/(0.0002*100000/3.7+1.255*B7)	=ABS((B7-B6)/B7)

For an initial guess of 10, the solution converges to $\phi = 14.50609$ after four iterations within a convergence tolerance of 10^{-8}. This gives a friction factor of $f = (1/\phi)^2 = 0.004752$.

	A	B	C	D	E
1	Iteration	φ	ξ	ξ'	|(φ_i - φ_{i-1})/φ_i|
2	0	10	-4.981684507	1.121411906	
3	1	14.44233246	-0.069653994	1.092645587	0.307591067
4	2	14.50608048	-1.01932E-05	1.092331659	0.004394572
5	3	14.50608981	-2.59064E-11	1.092331613	6.43291E-07
6	4	14.50608981	0	1.092331613	1.63491E-12
7	5	14.50608981	0	1.092331613	0

Comments: Newton's method converges quickly to the root, particularly for nearly linear functions. We may use our current worksheet setup with the user-defined function **NEWTON** in any available cell on the worksheet:

14.50609 = NEWTON($C$2, $B$2)

□

We may be tempted to check for convergence using the value of the function shrinking towards zero after each iteration:

$$|f(x)| < \varepsilon \quad (6.37)$$

However, in some cases, the function may *approach* zero without crossing the abscissa in a region away from the root, as illustrated in Figure 6.10. In this case, a convergence criterion based on the magnitude of the function may prematurely stop the search far from the actual solution where the value of the function is relatively small, though not a true root. Under these circumstances, the derivative of the function in the denominator of Equation (6.32) may go to zero causing a singularity in the iterative solution. Avoid singularities by adding an arbitrarily small number to the denominator of the correction term in the following alternative form of Newton's method:

$$x_i = x_{i-1} - \frac{f_{i-1} f'_{i-1}}{\left(f'_{i-1}\right)^2 + \delta} \quad (6.38)$$

The squared term in the denominator of equation (6.38) also prevents the unusual case of division by zero when a negative derivative f' cancels the small value δ. For these reasons, we recommend the convergence criterion defined by Equation (4.57) that compares the successive estimates for the root.

CHAPTER 6: ROOT-FINDING METHODS

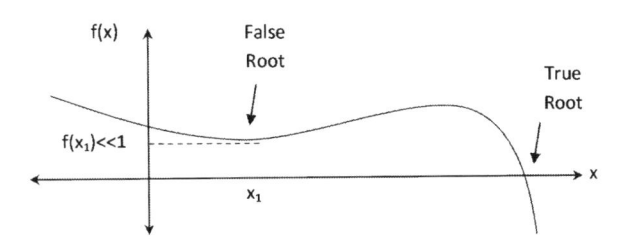

Figure 6.10

Function with a small value of its derivative near the root.

An analysis of the error in Newton's method reveals the potential cause for diverging solutions. Expand the function in a second order Taylor series:

$$0 \cong f(x_{i-1}) + \left[x_i^* - x_{i-1}\right] \frac{df}{dx}\bigg|_{x_{i-1}} + \frac{\left[x_i^* - x_{i-1}\right]^2}{2} \frac{d^2 f}{dx^2}\bigg|_{x_{i-1}} \tag{6.39}$$

where x^* is the approximation for the root in Equation (6.39). Rearrange the linear expansion from Newton's method in Equation (6.32):

$$0 = f(x_{i-1}) + \left[x_i - x_{i-1}\right] \frac{df}{dx}\bigg|_{x_{i-1}} \tag{6.40}$$

Subtract the corresponding terms in Equation (6.40) from those in Equation (6.39):

$$0 \cong \left[x_i^* - x_i\right] \frac{df}{dx}\bigg|_{x_{i-1}} + \frac{\left[x_i^* - x_{i-1}\right]^2}{2} \frac{d^2 f}{dx^2}\bigg|_{x_{i-1}} \tag{6.41}$$

Let $E_i = x^* - x_i$ and $E_{i-1} = x^* - x_{i-1}$ represent estimates for the errors in the two approximate solutions after iteration i and $i - 1$, respectively, then rearrange Equation (6.41) for the corresponding error term, E_i:

$$E_i \cong \frac{E_{i-1}^2}{2} \left(\frac{d^2 f / dx^2\big|_{x_{i-1}}}{df / dx\big|_{x_{i-1}}} \right) \tag{6.42}$$

According to Equation (6.42), a relatively large error in the previous iteration may generate even larger errors in the current iteration. A flat function, where $df/dx \to 0$, produces an infinite error!

The problem of a "flat" function near the root leads us to another problem encountered by Newton's method. A poor initial guess too far from the root may cause Newton's method to point away from the root, or oscillate between two points without converging, as shown in Figure 6.11. A particularly poor guess where the slope of the function is relatively flat sends the search off to extreme values of x between iterations.

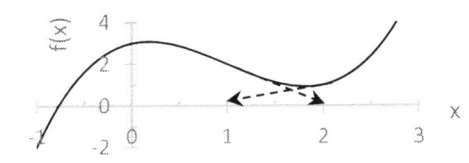

Figure 6.11

Function $f = 3 + x - 3x^2 + x^3$ with a "flat" derivative away from the root near $x = 1$ causing Newton's iterations to oscillate without convergence (True root at $x = -0.77$). Try Newton's method to find the root of the function starting at $x = -11$ (Fausett 2002).

A simple way to handle a divergent solution uses a reduced Newton "correction" step. If the magnitude of the *function* at the upgraded value for the root *increases* after a single Newton step, iteratively reduce the correction term by a factor of two until the *function* at x_i decreases:

$$x_i = x_{i-1} - \left(\frac{f_{i-1}}{2^m}\right) \frac{f'_{i-1}}{\left[\left(f'_{i-1}\right)^2 + \delta\right]} \qquad m = 1, 2 \ldots 6 \text{ or until } |f_i| < |f_{i-1}| \tag{6.43}$$

Example 6.7 Rachford-Rice Multicomponent Flash Calculation

Known: Flash a mixture of seven hydrocarbons in a single equilibrium stage. The *Excel* worksheet shown in Figure 6.13 lists the feed mole fractions, z, and component K-values.

Find: Calculate the vapor-to-feed ratio, Ψ, using Newton's method on the Rachford Rice flash function, and an alternative form of the flash function

Rachford-Rice $$f(\Psi) = 0 = \sum \frac{z(1-K)}{1+\Psi(K-1)} \tag{6.44}$$

Alternative $$f(\Psi) = 0 = 1 - \sum \frac{zK}{1+\Psi(K-1)} \tag{6.45}$$

Assumptions: Steady-state, isothermal

Analysis: Employ *VBA* user-defined functions for the *Rachford-Rice* and *Alternative* Equations, (6.44) and (6.45), respectively, to simplify the calculations in an *Excel* worksheet:

```
Public Function RR(psi As Double, z, K) As Double  ' Rachford-Rice flash function
    Dim i As Integer, n As Integer
    ' Get the size of the data vectors & initialize RR sum
    n = z.Count: RR = 0
    For i = 1 To n: RR = RR + z(i) * (1 - K(i)) / (1 + psi * (K(i) - 1)): Next i
End Function

Public Function AF(psi As Double, z, K) As Double  ' Alternative flash function
    Dim i As Integer, n As Integer
    ' Get the size of the data vector & initialize the AF sum
    n = z.Count: AF = 1
    For i = 1 To n: AF = AF - z(i) * K(i) / (1 + psi * (K(i) - 1)): Next i
End Function
```

Plot the Rachford Rice and alternate functions to locate the region that contains the root, as shown in Figure 6.12.

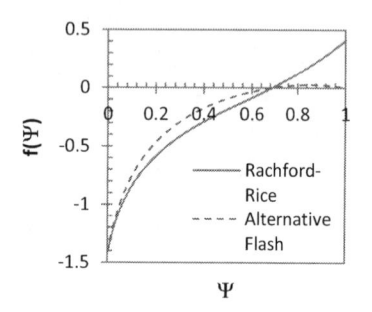

Figure 6.12

Rachford Rice function compared with the alternative form using either just $\Sigma x\text{-}1$ or $\Sigma y\text{-}1$.

The plot reveals a potential problem with the alternative flash function that levels off near the root such that the derivative in Equation (6.31) is small, causing Newton's method to diverge from the root. A poor initial guess of $\Psi > 0.8$ has the potential to converge on the wrong root, or cause divergent iterations.

Newton's method requires the first derivative of the flash functions:

Rachford-Rice $$\frac{df}{d\Psi} = \sum \frac{z(1-K)^2}{[1+\Psi(K-1)]^2} \tag{6.46}$$

Alternative $$\frac{df}{d\Psi} = \sum \frac{zK(K-1)}{[1+\Psi(K-1)]^2} \tag{6.47}$$

Program *VBA* user-defined functions of the derivatives of the flash functions:

```
Public Function dRR(psi As Double, z, K) As Double
' First derivative of the Rachford-Rice flash function
    Dim i As Integer, n As Integer
    n = z.Count: dRR = 0 ' Get the size of the vector & initialize the sum
    For i = 1 To n
        dRR = dRR + (z(i) * (K(i) - 1) ^ 2) / (psi * (K(i) - 1) + 1) ^ 2
    Next
End Function

Public Function dAF(psi As Double, z, K) As Double
' First-derivative of the alternative function
    Dim i As Integer, n As Integer
    n = z.Count: dAF = 0 ' Get the size of the vector & initialize the sum
    For i = 1 To n
        dAF = dAF + K(i) * z(i) * (K(i) - 1) / (1 + psi * (K(i) - 1)) ^ 2
    Next i
End Function
```

Implement Newton's method in an *Excel* worksheet, as seen in Figure 6.13. Put the results from each Newton iteration in succeeding rows.

	A	B	C
1	Component	z	K
2	1	0.01	16.2
3	2	0.13	5.2
4	3	0.09	2.6
5	4	0.27	2.1
6	5	0.06	9.9
7	6	0.13	0.7
8	7	0.31	0.3

	A	B	C	D	E	F
10	Newton Iteration	Rachford-Rice Ψ		Alternative Ψ		
11	1	0.5		0.5	0.85	0.9
12	2	0.681368561		0.633091377	-2.36565	1.134736
13	3	0.686485995		0.680150231	-6.87915	1.05362
14	4	0.686477155		0.686367343	-65.6725	1.010667
15	5	0.686477155		0.686477121	-8472.07	1.000509
16				0.686477155	-1.5E+08	1.000001

	A	B	C	D
10	Newton Iteration	Rachford-Rice Ψ		Alternative Ψ
11	1	0.5		0.5
12	2	=B11-RR(B11,$B$2:$B$8,$C$2:$C$8)/dRR(B11,$B$2:$B$8,$C$2:$C$8)		=D11-AF(D11,$B$2:$B$8,$C$2:$C$8)/DAF(D11,$B$2:$B$8,$C$2:$C$8)
13	3	=B12-RR(B12,$B$2:$B$8,$C$2:$C$8)/dRR(B12,$B$2:$B$8,$C$2:$C$8)		=D12-AF(D12,$B$2:$B$8,$C$2:$C$8)/DAF(D12,$B$2:$B$8,$C$2:$C$8)
14	4	=B13-RR(B13,$B$2:$B$8,$C$2:$C$8)/dRR(B13,$B$2:$B$8,$C$2:$C$8)		=D13-AF(D13,$B$2:$B$8,$C$2:$C$8)/DAF(D13,$B$2:$B$8,$C$2:$C$8)
15	5	=B14-RR(B14,$B$2:$B$8,$C$2:$C$8)/dRR(B14,$B$2:$B$8,$C$2:$C$8)		=D14-AF(D14,$B$2:$B$8,$C$2:$C$8)/DAF(D14,$B$2:$B$8,$C$2:$C$8)

Figure 6.13 *Excel* worksheet implementation of Newton's method for solving the Rachford-Rice equation.

Beginning with an initial guess for the root of 0.5, the Newton's iterations for the Rachford-Rice function converge quickly to the correct root of 0.6865. The alternative method diverges quickly away from the root starting from and initial guess of $\Psi = 0.85$. Starting from $\Psi = 0.9$, the alternate equation converges to an *incorrect* root.

Comments: The Rachford-Rice flash function is preferred to the alternative flash function for Newton's method because it is less sensitive to the starting point.

□

Newton's method has the disadvantage of requiring the analytical evaluation of the derivative, which may be difficult to obtain depending on the form of the nonlinear equation. We circumvent this problem with finite-difference approximations of the derivative. Substitute a simple two-point forward-difference approximation like Equation (5.45) for the derivative in Equation (6.32) for a quasi-Newton's recursive formula:

$$x_i = x_{i-1} - \frac{f(x_{i-1})\Delta x}{\left[f(x_{i-1}+\Delta x)-f(x_{i-1})\right]} \tag{6.48}$$

where Δx is a small perturbation in x. In the quasi-Newton method, we simply replaced a derivative function evaluation in Newton's method with a second function evaluation, thus maintaining a similar computational requirement. Why don't we use a more accurate three-point finite difference derivative approximation? Because we can control the spacing between our two points, we can select a sufficiently small value for Δx to control the magnitude of the error in the derivative approximation. However, a central difference approximation like Equation (5.49) does not need three points, so we achieve higher accuracy in the derivative with one additional function evaluation:

$$x_i = x_{i-1} - \frac{2\Delta x f(x_{i-1})}{\left[f(x_{i-1}+\Delta x) - f(x_{i-1}-\Delta x)\right]} \qquad (6.49)$$

The magnitude of the perturbation value should be small relative to the absolute value of the root, or $\Delta x \cong |x_i| \times 10^{-3}$. It may seem counterintuitive, but we do not need higher accuracy in the derivative approximation in Newton's method because the value of the function in the numerator goes to zero as the iterations converge on the root.

Example 6.8 Newton's Method Using Circular References

Known: We can use *Excel*'s capability for automating circular iteration to incorporate Newton's method within a single cell. This is useful for creating an *Excel* worksheet that automatically finds the new root of a nonlinear function whenever the function parameters are changed.

Find: Use the quasi-Newton method to solve the modified form of the Colebrook equation for friction factors in Equation (6.34) of *Example 6.6*.

Analysis: Enable circular iteration by selecting from the ribbon tab **File>*Excel* Options>Formulas>Enable Iterative Calculations**. Adjust the **Maximum iterations** and **Maximum Change** under *Excel* options to control convergence:

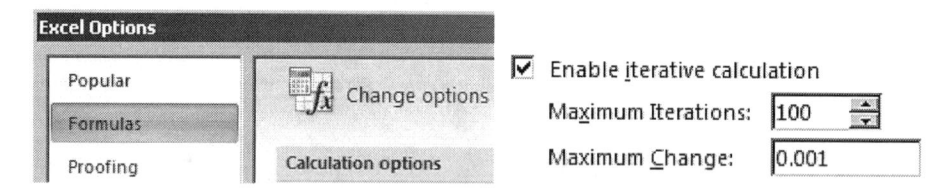

Set up an *Excel* worksheet for the quasi-Newton solution. Define a small perturbation value in cell **B5** for approximating the derivative of the function by a first-order finite-difference approximation in cell **B10**.

	A	B
1	Parameters	Values
2	Re =	1.00E+05
3	ε =	0.0002
4	D =	1
5	δ =	1.00E-05
6	ϕ =	1.45E+01
7	$\phi+\delta$ =	1.45E+01
8	$\xi(\phi)$ =	0.00E+00
9	$\xi(\phi+\delta)$ =	1.09E-05
10	dξ/dϕ=	1.09E+00
11	f =	4.75E-03

	B
1	Values
2	100000
3	0.0002
4	1
5	0.00001
6	=B6-B8/B10
7	=B6+B5
8	=B6+1.737*LN(B3/(3.7*B4)+1.255*B6/B2)
9	=B7+1.737*LN(B3/(3.7*B4)+1.255*B7/B2)
10	=(B9-B8)/B5
11	=1/B6^2

Apply Newton's recursive formula to converge on the root in cell **B6**. Cell **B11** contains the final solution for the friction factor.

Comments: With any changes to the parameters in the range **B2:B4** on the worksheet, the circular calculation automatically iterates to converge on the new solution for the friction factor. This approach has the advantage that we can now perform "what-if" analysis to explore the effects of changing parameters without manually restarting the search for the solution.

□

CHAPTER 6: ROOT-FINDING METHODS

A serious problem with automatic iteration in *Excel* arises when convergence is sensitive to the initial guess for the solution. Circular iteration in a cell uses the cell's default value of zero to initiate the iterative calculations. An initial guess other than zero is possible using a simple modification to the implementation of Newton's method:

1. Guess the root x_0 in a cell.
2. Set the value of the first iteration $x_i = 0$.
3. Calculate the difference between x after an iteration:
$$\Delta x_i = x_i - x_0 \tag{6.50}$$
4. Apply Newton's recursive formula:
$$\Delta x_{i+1} = \Delta x_i - \frac{f(x_i)}{f'(x_i)+\delta} \quad \rightarrow \quad x_{i+1} = \Delta x_{i+1} + x_0 \tag{6.51}$$
5. Iterate through Equations (6.50) and (6.51) until converging on the root.

The following simple example illustrates how to control the selection of the initial guess when using circular cell referencing in an *Excel* worksheet.

Example 6.9   Initial Guesses Using Circular References in Excel Worksheets

Known/Find: We use a simple quadratic function to illustrate a method for specifying an initial guess for Newton's method using circular referencing in an *Excel* worksheet:

$$f(x) = x^2 - b^2 \tag{6.52}$$

Analysis: For this simple equation the roots are $x = \pm b$. Newton's method requires the first derivative of the function in Equation (6.52):

$$f'(x) = 2x \tag{6.53}$$

Apply the algorithm in Equations (6.51) to find a root to Equation (6.52) using an *Excel* worksheet with iteration enabled, b set to four, and an initial guess of three for the root. Change the initial guess to 7 and the worksheet automatically recalculates for a new, but also correct, root.

	A	B
1	b	4
2	x_0	3
3	x_i	=B6+B2
4	$f(x_i)$	=B3^2-B1
5	$f'(x_i)$	=2*B3
6	Δx_i	=B3-B2-B4/(B5+0.00000001)

	A	B
1	b	4.0
2	x_0	3.0
3	x_i	-2.0
4	$f(x_i)$	0.0
5	$f'(x_i)$	-4.0
6	Δx_i	-5.0

	A	B
1	b	4.0
2	x_0	7.0
3	x_i	2.0
4	$f(x_i)$	0.0
5	$f'(x_i)$	4.0
6	Δx_i	-5.0

Comments: Selecting initial guesses with circular references in *Excel* becomes important when the root is far from zero. This becomes particularly important in problems with multiple roots where we need to find different roots by starting Newton's iteration from different initial conditions.

□

6.2.7   Secant and Muller's Methods

The secant and Muller root-finding methods do not require functional derivative evaluations, but possess similar convergence speed as the quasi-Newton method. The secant and Muller methods find the zero of a function by linear and quadratic function approximation using two and three points for each iteration, respectively.

The secant method has the following characteristics:

- *Like* regula falsi, the secant method requires two points to start the search and has a similar recursive formula, but uses an algorithm that eliminates the need for brackets around the root.
- *Unlike* regula falsi, the secant method does not guarantee convergence to a root of a continuous function.
- The secant method requires just one function evaluation after each iteration, which reduces the computational burden when compared with Newton's method.

This derivation of the secant method is an extension of the quasi-Newton method that approximates the derivative from two points. Expand a Taylor series approximation in the linear terms for the function of x_{i-2} about x_{i-1}.

$$f(x_{i-2}) \cong f(x_{i-1}) + [x_{i-2} - x_{i-1}] \left.\frac{df}{dx}\right|_{x_{i-1}} \tag{6.54}$$

Expand a second linear Taylor series approximation of the function at x_i about x_{i-1}.

$$f(x_i) \cong f(x_{i-1}) + [x_i - x_{i-1}] \left.\frac{df}{dx}\right|_{x_{i-1}} \tag{6.55}$$

Solve Equations (6.54) and (6.55) each for the first derivative and set them equal:

$$\frac{f(x_i) - f(x_{i-1})}{[x_i - x_{i-1}]} = \frac{f(x_{i-2}) - f(x_{i-1})}{[x_{i-2} - x_{i-1}]} \tag{6.56}$$

Finally, solve Equation (6.56) for x_i by setting $f(x_i) = 0$ for an approximation of the root to give the recursive formula for the secant method (we use a small number in the denominator to avoid division by zero):

- *Secant Method*

$$\begin{aligned} x_i &= x_{i-1} - \frac{f(x_{i-1})[x_{i-2} - x_{i-1}]}{f(x_{i-2}) - f(x_{i-1})} \\ &\quad x_{i-1} - \frac{f(x_{i-1})[x_{i-2} - x_{i-1}][f(x_{i-2}) - f(x_{i-1})]}{[f(x_{i-2}) - f(x_{i-1})]^2 + \delta} \end{aligned} \tag{6.57}$$

Note that Equation (6.57) is in form similar to Equation (6.25) of the regula falsi method. Implementation of the secant method requires two initial guesses for the solution, x_{i-2} and x_{i-1}. After each iteration, upgrade the guesses and functional evaluations in Equation (6.57) to the most recent values for the guessed solution.

$$x_{i-2} \leftarrow x_{i-1} \quad \text{and} \quad x_{i-1} \leftarrow x_i \tag{6.58}$$

$$f(x_{i-2}) \leftarrow f(x_{i-1}) \quad \text{and} \quad f(x_{i-1}) \leftarrow f(x_i) \tag{6.59}$$

Figure 6.14 illustrates graphically how the secant method progressively converges on the root. The same convergence criterion for Newton's method applies to the secant method. The choice and sequence of initial guesses required to initialize the secant method can affect the speed of convergence, or the stability of the iterative solution. Poor initial guesses may lead to divergence. The *VBA* sub procedure **SECANT** in the *PNMSuite* employs the secant method and requires the function in a worksheet cell formulated in terms of another cell with an initial guess for the root.

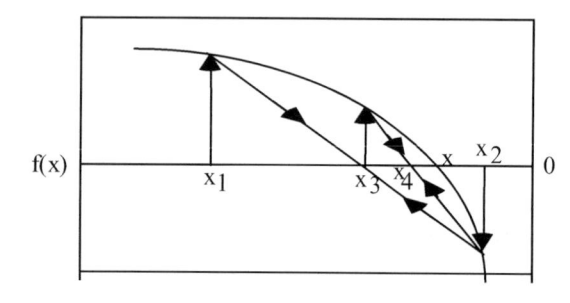

Figure 6.14
Graphical illustration of the steps towards convergence on the root using the secant method.

CHAPTER 6: ROOT-FINDING METHODS

Muller's method interpolates the function near the root with a quadratic equation, which requires three initial guesses for the root to start the search. We calculate the three coefficients in the quadratic equation from the simultaneous solution of three linear equations in terms of the three starting values for the independent variable and function evaluated at each guess. A convenient form of the quadratic solution gives the following recursive formulas for $i = 4\ldots$ with the same convergence criteria as Newton's method:

$$p = \frac{f_{i-2} - f_{i-3}}{x_{i-2} - x_{i-3}} \qquad q = \frac{f_{i-1} - f_{i-2}}{x_{i-1} - x_{i-2}} \qquad r = \frac{q - p}{x_{i-1} - x_{i-3}} \qquad s = q + r(x_{i-1} - x_{i-2}) \qquad (6.60)$$

- Muller's Method
$$x_i = x_{i-1} - \frac{2 f_{i-1}}{s + sign(s)\sqrt{s^2 - 4 f_{i-1} r}} \qquad (6.61)$$

The advantage of reduced sensitivity of Muller's method to the starting values (Fausett 2002) offsets the disadvantage of the requirement for three initial guesses when compared with other root-finding methods.

The *VBA* user-defined function **ROOT(f, x1, [x2, x3, tol])** in the *PNMSuite* employs the secant or Muller's methods to find the root of a *formula* in an *Excel* worksheet. **ROOT** follows the method of Section 3.6.3 for evaluating an *Excel* worksheet formula. The arguments in the user-defined function consist of the cell references for the formula, **f**, and independent variable in the formula, **x1**. The starting value in the cell for the independent variable also serves as the first initial guess for the root. The user may provide a optional third and fourth arguments **x2** and **x3** for the second and third guesses of the root, and optional fifth argument **tol** for the relative convergence tolerance (default is 10^{-8}). When the optional third argument is missing, the program uses the quasi-Newton method. Otherwise, **ROOT** employs the secant or Muller methods when provided additional guesses for the root in the third and fourth arguments, respectively. For an initial guess of $\phi = 10$ in the worksheet in *Example 6.6*, the user-defined function **ROOT** finds the solution after five iterations, **14.50609 = ROOT(C2, B2)**, which is comparable to the performance of Newton's method.

Example 6.10   Dew Point Temperature of a Binary Mixture

Known: Phase equilibrium between a condensing vapor and a liquid is first established at the dew point temperature. We define the relationship between the compositions in the gas and liquid phases for each species, in terms of a partition coefficient, or *K*-value:

$$y_j = K_j x_j \qquad (6.62)$$

where y_j, x_j, and K_j are the gas-liquid mole fractions and partition coefficient, respectively, for species *j*.

Find: Calculate the dew-point temperature for an equimolar mixture of propane and butane at one atmosphere.

Assumptions: Raoult's law for ideal vapor-liquid equilibrium

Analysis: By definition, the sum of the mole fractions in the liquid phase is one, $\Sigma x = 1$. Substitute for *x* in terms of *y* from Equation (6.62):

$$\frac{y_B}{K_B} + \frac{y_P}{K_P} = 1 \qquad (6.63)$$

Rearrange Raoult's law for the *K*-values in terms of the ratio of the vapor pressure (P_{vj}) of pure species *j* to the total pressure (P_T):

$$K_j = \frac{P_{v,j}}{P_T} \qquad (6.64)$$

Use Antoine's equation to model the temperature function for vapor pressure of each component:

$$P_{v,j} = \exp\left(A_j - \frac{B_j}{C_j + T}\right) \qquad (6.65)$$

where A_j, B_j, and C_j are the Antoine constants unique to species j. Substitute Equations (6.64) and (6.65) into the dew-point Equation (6.63) for a single nonlinear equation in T.

$$f(T) = 1 - \frac{y_B P_T}{\exp\left(A_B - \dfrac{B_B}{C_B + T}\right)} - \frac{y_P P_T}{\exp\left(A_P - \dfrac{B_P}{C_P + T}\right)} = 0 \tag{6.66}$$

For convenience, we create two *VBA* user-defined functions for the vapor pressures of propane and butane:

```
Public Function PvC3H8(T As Double) As Double
' Returns the vapor pressure of Propane, mmHg, T is degrees K
    Dim A As Double, B As Double, C As Double
    A = 15.726: B = 1872.46: C = -25.16 : PvC3H8 = Exp(A - B / (C + T))
End Function

Public Function PvC4H10(T As Double) As Double
' Returns the vapor pressure of Butane, mmHg, T is degrees K
    Dim A As Double, B As Double, C As Double
    A = 15.6782: B = 2154.9: C = -34.42 : PvC4H10 = Exp(A - B / (C + T))
End Function
```

We may consider Newton's method for this problem; however, we prefer the secant method to avoid evaluating the derivative. The recursive formula of the secant method for this example becomes

$$T_i = T_{i-1} - \frac{f(T_{i-1})[T_{i-1} - T_{i-2}]}{f(T_{i-1}) - f(T_{i-2})} \tag{6.67}$$

We need two initial guesses for the dew point temperature of the mixture to initiate the secant method. We know that the dew point of the binary system lies between the normal boiling points of pure propane and pure butane of 231 K and 273 K, respectively.

To set up the secant method in an *Excel* worksheet, place the formulas for the successive iterations in column **C** and **D**. Calculate the local convergence criteria in column **E** according to Equation (4.47).

	A	B	C	D	E
1	P_T/mmHg	760	T/K	f(T)	ε
2	y_{C3H8}	0.5	231	2.90249529	
3			273	-0.398853618	0.15385
4			267.925755	-0.273408256	0.01894
5			256.866436	0.132348723	0.04305
6			260.473735	-0.02498084	0.01385
7			259.900967	-0.001870562	0.0022
8			259.854606	2.9109E-05	0.00018
9			259.855317	-3.3284E-08	2.7E-06
10			259.855316	-5.89639E-13	3.1E-09

The formulas for the secant method use absolute references for the parameters. Use auto-fill to generate the formulas for each iteration in successive rows.

	A	B	C	D	E
1	P_T/mmHg	760	T/K	f(T)	ε
2	y_{C3H8}	0.5	231	=$B$1*($B$2/PvC3H8(C2)+(1-$B$2)/PvC4H10(C2))-1	
3			273	=$B$1*($B$2/PvC3H8(C3)+(1-$B$2)/PvC4H10(C3))-1	=ABS((C3-C2)/C3)
4			=C3-D3*(C3-C2)/(D3-D2)	=$B$1*($B$2/PvC3H8(C4)+(1-$B$2)/PvC4H10(C4))-1	=ABS((C4-C3)/C4)
5			=C4-D4*(C4-C3)/(D4-D3)	=$B$1*($B$2/PvC3H8(C5)+(1-$B$2)/PvC4H10(C5))-1	=ABS((C5-C4)/C5)

For initial guesses of 231 K and 273 K, the secant method finds the dew point at 259.86 K after eight iterations with a relative error smaller than 10^{-6}%. We may also try the user-defined function **ROOT** for this setup in cell **F1**:

F1 = ROOT(D2, C2, C3)

	F
1	259.8553

CHAPTER 6: ROOT-FINDING METHODS

Comments: The secant method converges quickly and does not require derivative evaluations. The secant method is preferred over the quasi-Newton method when derivatives are computationally expensive to calculate. Muller's method requires three initial guesses. For this example, we recommend the average of the pure-component boiling points for the fourth argument required by Muller's method:

$$259.86 = \text{ROOT}(D2, C2, C3, (C2 + C3)/2)$$

□

6.2.8 Padé Interpolation (or Extrapolation)*

Rational functions, described in Sections 9.10.1 and 10.4, have relatively good properties for interpolation or extrapolation to the root of a nonlinear equation. We use the following simplest Padé acceleration of Muller's quadratic interpolating function to interpolate (or extrapolate) our nonlinear function to a root:

$$f(x) \cong \frac{p_0 + p_1 x}{1 + q_1 x} \tag{6.68}$$

Estimate the root by setting Equation (6.68) to zero and solving for x:

$$0 \cong \frac{p_0 + p_1 x}{1 + q_1 x} \rightarrow x \cong -\frac{p_0}{p_1} \tag{6.69}$$

To implement the rational Padé method of root-finding:

1. Provide two initial guesses for the root, x_1 and $x_2 = x_1 + \delta$, where δ is a small perturbation.
2. Use the two initial guesses in the secant method of Equation (6.57) to find a third point. Evaluate the nonlinear function at each point to obtain three pairs (x_1, f_1), (x_2, f_2), and (x_3, f_3). Iterate on the following steps until converging on a root to the nonlinear function.
3. Use the three initial guesses for the root and corresponding function values to calculate the parameters for the polynomials in the numerator and denominator of the rational function.

$$f(x_1) = \frac{p_0 + p_1 x_1}{1 + q_1 x_1} \qquad f(x_2) = \frac{p_0 + p_1 x_2}{1 + q_1 x_2} \qquad f(x_1) = \frac{p_0 + p_1 x_3}{1 + q_1 x_3} \tag{6.70}$$

Solve the system of simultaneous equations for p_0, p_1, and q_1 – we only need the p's in Equation (6.69):

$$p_0 = x_3(x_1 - x_2) f(x_1) f(x_2) + x_2(x_3 - x_1) f(x_1) f(x_3) + x_1(x_2 - x_3) f(x_2) f(x_3)$$
$$p_1 = (x_1 - x_2) f(x_1) f(x_2) + (x_3 - x_1) f(x_1) f(x_3) + (x_2 - x_3) f(x_2) f(x_3) \tag{6.71}$$

4. Replace the initial guesses and function values with the upgraded values for x and f:

$$\begin{array}{ccc} x_1 \leftarrow x_2 & x_2 \leftarrow x_3 & x_3 \leftarrow x \\ f_1 \leftarrow f_2 & f_2 \leftarrow f_3 & f_3 \leftarrow f(x) \end{array} \tag{6.72}$$

5. Check for convergence in the root after each iteration i according to Equation (4.57):

A *VBA* user-defined function named **PADEROOT** is available in the *PNMSuite* for finding a root to a nonlinear equation by the Padé interpolation method in an *Excel* worksheet.

The coefficient in the denominator of Equation (6.68) plays a significant role in the Padé approximation when x is far from the root. As the successive values for x approach the root, they come close together, and the linear numerator becomes the dominant term in the interpolation or extrapolation to the root. The Padé method has convergence properties similar to Newton's method without derivative function evaluations, but at the inconvenience of requiring three initial guesses for the root (compared with one guess for Newton's method). However, for most problems, successful convergence of the Padé method is less sensitive to initial guesses for the root compared with Newton's method.

We may try accelerating convergence even further by incorporating each additional function evaluation into the rational Padé interpolation function:

$$f(x) \cong \frac{p_0 + p_1 x + p_2 x^2 + \ldots}{1 + q_1 x + q_2 x^2 + \ldots} \quad (6.73)$$

After each iteration, we may add the next higher order term to the polynomial in the numerator and denominator and solve for the new set of p and q parameters using the methods described in Section 10.4. However, the accelerated convergence to the root usually does not compensate for the additional computational effort to find the parameters p and q in Equation (6.73).

Example 6.11 Find the Root by Padé Interpolation

Known/Find: Use the method of Padé interpolation to find the root of the following nonlinear function:

$$f(x) \cong x^3 - x - 1 \quad (6.74)$$

Analysis: Use the *VBA* user-form **QYXPLOT** to graph the function for inspection:

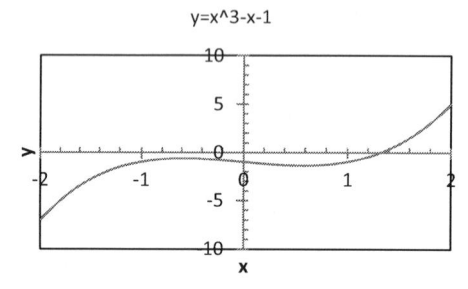

Set up an *Excel* worksheet in the *PNMSuite*, with the formula for the function in cell **A2** in terms of the variable in cell **A1**. Start with an initial guess of $x = 0.5$ near the root in cell **A1**. Put the *VBA* user-defined function **PADEROOT** in cell **A3** to find the root $x = 1.325$:

	A	B
1	x	0.5
2	f(x)	=B1^3-B1-1
3	PADE	=paderoot(B2,B1)

	A	B
1	x	0.5
2	f(x)	-1.375
3	PADE	1.324718

Comments: A comparison in Table 6.2 among quasi-Newton, secant, and Padé methods shows how the secant and Padé methods have similar convergence properties, but more efficient computational requirements in terms of fewer function evaluations per iteration. Note how the secant and Padé methods usually recover faster from poor initial guesses smaller than one.

Table 6.2 Comparison of convergence on a root for Equation (6.74) using Newton, Secant, and Padé methods of root finding with a convergence tolerance of 10^{-8}.

Initial Guess for the root	Function Evaluations/ Quasi-Newton Iterations	Function Evaluations/ Secant Iterations	Function Evaluations/ Padé Iterations
5	16/8 = 2	12/11 = 1.09	10/8 = 1.25
3	14/7 = 2	10/9 = 1.11	9/7 = 1.29
1	12/6 = 2	8/7 = 1.14	7/5 = 1.40
0.5	42/21 = 2	23/22 = 1.05	12/10 = 1.20
0	44/22 = 2	5/4 = 1.25	28/26 = 1.08
-1	40/20 = 2	20/19 = 1.05	24/22 = 1.09
-2	96/48 = 2	27/26 = 1.04	11/9 = 1.08
Average	38/19 = 2	15/14 = 1.07	14/12 = 1.17

□

6.3 Roots of Simultaneous Equations

Simultaneous equations are coupled through their common variables:

$$f_1(x_1, x_2, \ldots x_n) = 0$$
$$\vdots$$
$$f_n(x_1, x_2, \ldots x_n) = 0 \qquad (6.75)$$

We must search simultaneously for the roots to the system of equations. As in the case for a single nonlinear equation, multiple sets of roots may exist for a system of simultaneous nonlinear equations. In this section, we extend the root-finding methods for single equations to coupled systems.

6.3.1 Ordinary Iteration (Successive Substitution)

We may use ordinary iteration from Section 6.2.1 (also referred to as successive substitution, or Picard's method) for systems of non-linear equations. This method is practically identical to Gauss-Seidel iteration, applied in Chapter 4 to systems of linear equations. We solve nonlinear equations *implicitly* (as opposed to explicitly for linear functions), one equation for each unknown variable. Recall that we use capitalized F for a fixed-point implicit style function, e.g., $f_1 = x_1 - F_1$, etc.:

$$x_1 = F_1(x_1, \ldots x_n)$$
$$\vdots$$
$$x_n = F_n(x_1, \ldots x_n) \qquad (6.76)$$

The procedure requires initial guesses for the solution, and successive substitution of the solution calculated from the left-hand side of Equation (6.76) back into the implicit functions on the right-hand side to (hopefully) converge to a set of roots when $|x - F| < tol$. For difficult convergence, consider relaxation as described in Section 4.6.2. Aitken's method introduced in the next section may promote and accelerate convergence by finding better relaxation parameters for troublesome problems.

6.3.2 Aitken's Δ^2 and Steffensen's Methods of Accelerating Series Convergence*

Aitken's (1926) method of accelerating series convergence, like Wegstein's method, assumes that the values from successive iterations converge linearly. We describe linear convergence in terms of relatively constant ratios of errors between iteration values:

$$\frac{x - x_{i+2}}{x - x_{i+1}} \approx \frac{x - x_{i+1}}{x - x_i} \qquad (6.77)$$

where x represents the root. Solve Equation (6.77) for an improved value from the previous three values in the form of Aitken's Δ^2 correction to the first of two previous iterations:

- *Aitkin's Method*
$$x \approx x_i - \frac{(x_{i+1} - x_i)^2}{x_{i+2} - 2x_{i+1} + x_i} \qquad (6.78)$$

Steffensen's application of Aitken's Δ^2 method accelerates two consecutive ordinary iterations with one Δ^2 acceleration step. As with all iterative methods, successful convergence requires good initial guesses and proper arrangements of the implicit functions. A *VBA* macro **STEFFENSEN** implements Aitken's method for single or simultaneous equations in an *Excel* worksheet. To use the macro, supply initial guesses in a range of cells for the variables on an *Excel* worksheet. Add formulas in another range for the implicit functions in terms of values in the range of cells for the variables. **STEFFENSEN** displays the maximum residual in the **Status Bar** and uses a default maximum relative convergence tolerance of 10^{-8}. We illustrate how to set up an *Excel* worksheet for the **STEFFENSEN** macro in *Example 6.12*.

Example 6.12   Divergent Set of Explicit Equations

Known/Find: The solution to the explicit Equations (4.77) to (4.79) using ordinary iteration in *Example 4.6* diverges. Calculate the solution to the system of fixed-point equations by Steffensen's method.

Analysis: Use the *VBA* macro **STEFFENSEN**, to solve the system of explicit equations.

1. Add names for the variables in column **A** of an *Excel* worksheet.
2. Provide initial guesses for the solution in the range **B2:B4**. Name the cells in the range **B2:B4** according to the values in the range **A2:A4**.
3. Define the explicit formulas in column **C** in terms of the values for each variable in column **B**.

	A	B	C
1	Variables	Named Cells	Explicit Functions
2	Lout	100	=(Vin-0.1*Vout)/0.8255
3	Vin	100	=0.94025*Lout+0.1*Vout-12.5
4	Vout	100	=0.91*Vin

4. Launch the *VBA* macro **STEFFENSEN**, which prompts the user for the cells containing the variables and implicit equations:

	A	B	C
1	Variables	Named Cells	Explicit Functions
2	Lout	100	109.0248334
3	Vin	100	91.525
4	Vout	100	91

STEFFENSEN
Select range of INITIAL guesses for roots, x=F(x): (must be in contiguous cells)
$B$2:$B$4

	A	B	C
1	Variables	Named Cells	Explicit Functions
2	Lout	100	109.0248334
3	Vin	100	91.525
4	Vout	100	91

STEFFENSEN
Select range of fixed-point implicit FUNCTIONS, F(x): (must be in contiguous cells)
$C$2:$C$4

5. The solution converged after 14 iterations within a relative convergence criterion of 10^{-8}, compared with 40 iterations needed for the alternative arrangement of the explicit equations.

	A	B	C
1	Variables	Named Cells	Explicit Functions
2	Lout	108.9324639	108.9324641
3	Vin	98.92601663	98.92601669
4	Vout	90.02267539	90.02267513

Comments: Aitken's method finds local relaxation factors for each variable that promote convergence without needing to rearrange the implicit equations. Although Aitken's method improved the rate of convergence for this example, this is not a guaranteed outcome for all problems.

□

6.3.3   Equation Sequence and Arrangement

As we experienced previously in Example 4.6 for linear systems, the proper selection of equation sequence and arrangement may also promote convergence in root finding methods applied to systems of nonlinear equations. This is particularly true of ordinary iteration where we must solve one equation implicitly for each unknown. The Jacobian matrix for a system of nonlinear equations provides some useful information in this regard. The Jacobian matrix

CHAPTER 6: ROOT-FINDING METHODS

consists of partial derivatives of each function in the form $f(x) = 0$ with respect to each variable. For ordinary iteration, we choose the equations for the implicit variables from the corresponding dominant terms in the Jacobian.

To illustrate, consider the following simple 2×2 system of nonlinear equations:

$$f_1 = (x_1 + x_2 - a) - (x_1^2 + x_2^2) = 0 \qquad (6.79)$$

$$f_2 = (x_1 - b) - (cx_1 x_2^2) = 0 \qquad (6.80)$$

The Jacobian matrix for this system is:

$$\begin{bmatrix} \dfrac{\partial f_1}{\partial x_1} & \dfrac{\partial f_1}{\partial x_2} \\ \dfrac{\partial f_2}{\partial x_1} & \dfrac{\partial f_2}{\partial x_2} \end{bmatrix} = \begin{bmatrix} (1-2x_1) & (1-2x_2) \\ (1-cx_2^2) & (-2cx_1 x_2) \end{bmatrix} \qquad (6.81)$$

Find the dominant terms by comparing the relative magnitudes of the terms in each row:

- In the first row, as $x_1 \to \infty$, $\qquad \left|\dfrac{1-2x_1}{1-2x_2}\right| \gg 1 \quad or \quad \left|\dfrac{1-2x_2}{1-2x_1}\right| \ll 1 \qquad (6.82)$

- In the second row, as $x_2 \to \infty$, $\qquad \left|\dfrac{-2cx_1 x_2}{1-cx_2^2}\right| \ll 1 \quad or \quad \left|\dfrac{1-cx_2^2}{-2cx_1 x_2}\right| \gg 1 \qquad (6.83)$

The results in Equations (6.82) and (6.83) indicate that we should consider using Equation (6.80) to form an implicit equation for x_1, leaving Equation (6.79) to form an implicit equation for x_2.

Figure 6.15 contains the code for the *VBA* macro **JACOBIAN** for calculating the matrix of partial derivatives in an *Excel* worksheet using central-difference approximations from Table 5.3. Because we evaluate the derivatives at specific values for the variables, the initial guesses for the roots influence the value of the Jacobian derivatives. To use the calling procedure **JACOBIAN** for a system of n equations in n variables:

1. Put values for x in a range of contiguous cells on an *Excel* worksheet.
2. Add formulas for f in terms of the values for x from step 1.
3. Decide on the location for the square $n \times n$ Jacobian matrix on the worksheet.

```
Public Sub JACOBIAN()
' Calculate partial derivatives of functions required for the Jacobian matrix.
' *** Required User-defined Functions and Sub Procedures from the PNM Suite ***
' JACOBI = calculates the Jacobian square matrix
' *** Nomenclature ***
Dim boxtitle As String ' = string with title for message and input boxes
Dim df() As Double ' = partial derivative of f(i) with respect to x(j)
Dim dx As Double ' = small change in x for finite difference derivative approximation
Dim f() As Double ' = non-linear function in x
Dim fm1 As Double, fp1 As Double ' = function at x-dx and x+dx for fin dif derivatives
Dim fr As Range ' = range of functions on worksheet (contiguous cells)
Dim i As Long, n As Long ' = index for variables, number of unknowns
Dim pcol As Long, prow As Long ' = column and row of print out to worksheet
Dim small As Double ' = small number to avoid division by zero (module level)
Dim tol As Double ' = convergence tolerance (set by user)
Dim x() As Double ' = solution to nonlinear equation
Dim xr As Range ' = range of independent variables (contiguous cells)
'*****************************************************************************
small = 0.000000000001: boxtitle = "JACOBIAN"
With Application
x_input: 'Get initial guesses for variables. Input initial parameters
Set fr = .InputBox(prompt:="Range of FUNCTIONS, f(x):" & vbNewLine & _
            "(contiguous cells)", Title:=boxtitle, Type:=8)
Set xr = .InputBox(prompt:="Range of INDEPENDENT variables, x:" & vbNewLine & _
            "(contiguous cells):", Title:=boxtitle, Type:=8)
n = fr.Count ' Get number of variables and functions
```

```
If n <> xr.Count Then ' Check for consistency in number of variables & equations
    MsgBox Title:=boxtitle, prompt:="Match variables & equations": GoTo x_input
End If
prow = output.row: pcol = output.Column
tol = 0.00000001 ' Relative Convergence TOLERANCE
ReDim df(1 To n, 1 To n) As Double, f(1 To n, 1 To 1) As Double, x(1 To n) As Double
For i = 1 To n: x(i) = xr(i): f(i, 1) = fr(i): Next i ' guesses & function values
Select Case n
Case 1 'Find partial derivative of a single nonlinear equation. Perturbation in x
    dx = .Max(0.0001 * xr, small): xr = x(1) + dx: fp1 = fr: xr = x(1) - dx: fm1 = fr
    df(1, 1) = (fp1 - fm1) / (2 * dx) ' 1st-order finite difference derivatives
Case Else ' Get Jacobian matrix
    Call JACOBI(df, fr, xr, n, x)
End Select
Range(outrange) = df ' write Jacobian to worksheet
End With ' Application
End Sub ' JACOBIAN

Public Sub JACOBI(df() As Double, fr As Range, xr As Range, n As Long, x() As Double)
' df() = partial derivative of f(i) with x(j), forward difference
' fr = range of functions       xr = range of variables on worksheet
' n = number of unknowns        x() = values of variables
'*************************************************************************
' Approximates the partial derivatives in the square Jacobian by 2nd-order finite
' difference derivative approximation.
' *** Nomenclature ***
Dim dx As Double ' = difference in x for finite difference derivative calc.
Dim fp1 As Double ' = function values at x + dx for finite difference derivatives
Dim i As Long, j As Long ' = indexes for number of variables and equations
'*************************************************************************
For i = 1 To n ' Loop through rows
    For j = 1 To n ' Loop through columns. Initial perturbation in x
        dx = 0.00001 * (Abs(x(j)) + 0.00001)
        xr(j) = x(j) + dx: fp1 = fr(i): xr(j) = x(j) - dx
        ' 2nd-order finite difference estimates of partial derivatives
        df(i, j) = (fp1 - fr(i)) / (2 * dx): xr(j) = x(j) ' reset x
    Next j
Next i
End Sub ' JACOBI
```

Figure 6.15 *VBA* macros JACOBIAN and JACOBI for approximating the Jacobian matrix of a system of nonlinear functions by finite-difference partial derivative approximations.

When rearranging equations in preparation for a particular method of solution, there are a few techniques worth trying:
1. Identify *groups* of variables that are common to more than one equation. Solve for the unknowns out of the common groups when formulating the implicit forms of the equations.
2. Rearrange the equations to minimize or eliminate the mathematical operation of division by the variables.
3. Rearrange equations such that each term has approximately the same order of magnitude.
4. Combine terms with the same order of magnitude, where possible.

Example 6.13 Steady-state Reactor Concentrations (J. Riggs 1994)

Known: The following complex reaction mechanism occurs in a 100 L, well-mixed, continuous flow reactor.

$$A \xrightarrow{1} 2B \tag{6.84}$$

$$A \underset{3}{\overset{2}{\rightleftarrows}} C \tag{6.85}$$

$$B \xrightarrow{4} D + C \tag{6.86}$$

The rate laws for each species in terms of rate constants and concentrations are:

CHAPTER 6: ROOT-FINDING METHODS

$$r_A = -k_1 C_A - k_2 C_A^{3/2} + k_3 C_C^2 \tag{6.87}$$

$$r_B = 2k_1 C_A - k_4 C_B^2 \tag{6.88}$$

$$r_C = k_2 C_A^{3/2} - k_3 C_C^2 + k_4 C_B^2 \tag{6.89}$$

$$r_D = k_4 C_B^2 \tag{6.90}$$

where the reaction rate constants have values: $k_1 = 1 \text{ s}^{-1}$, $k_2 = 0.2 \text{ L}^{1/2} \cdot \text{mol}^{-1/2} \cdot \text{s}^{-1}$, $k_3 = 0.05 \text{ L} \cdot \text{mol}^{-1} \cdot \text{s}^{-1}$, $k_4 = 0.4 \text{ L} \cdot \text{mol}^{-1} \cdot \text{s}^{-1}$.

Schematic: A well-mixed continuous flow reactor has a volume V and volumetric flow rates, Q. C_i is the molar concentration of component i:

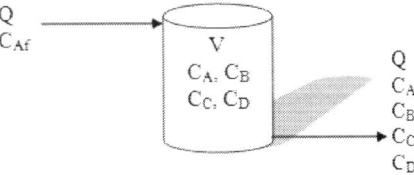

Find: Calculate the steady-state exit concentrations for a volumetric feed rate of 50 L·s^{-1} with a feed concentration of 1.0 mol·L^{-1} of A.

Assumptions: steady-state operation, homogeneous mixtures, constant temperature.

Analysis: At steady state, there is no accumulation of the reaction species in the rector such that the accumulation term is zero. Material balances for each species in the continuous reactor give:

$$\text{Rate In} - \text{Rate Out} + \text{Rate of Generation} = 0$$

Species A: $$Q(C_{A0} - C_A) + V(-k_1 C_A - k_2 C_A^{3/2} + k_3 C_C^2) = 0 \tag{6.91}$$

Species B: $$Q(0 - C_B) + V(2k_1 C_A - k_4 C_B^2) = 0 \tag{6.92}$$

Species C: $$Q(0 - C_C) + V(k_2 C_A^{3/2} - k_3 C_C^2 + k_4 C_B^2) = 0 \tag{6.93}$$

Species D: $$Q(0 - C_D) + V k_4 C_B^2 = 0 \tag{6.94}$$

Calculate the Jacobian in an *Excel* worksheet using initial guess of 0.25 M for each variable. Use named cells for the reaction parameters. First, put the initial guesses into cells on the worksheet in column **C**. Then add the formulas for the mole balances from Equations (6.91) to (6.94) in column **D**:

D2 =Q*(CAf-C2) +V*(-k_1*C2-k_2*C2^ (3/2) +k_3*C4^2)   D3 =-Q*C3+V*(2*k_1*C2-k_4*C3^2)

D4 =-Q*C4+V*(k_2*C2^ (3/2)-k_3*C4^2+k_4*C3^2)   D5 =-Q*C5+V*k_4*C3^2

Use the *VBA* macro **JACOBIAN** to calculate the matrix of partial derivatives in the range **E2:H5**.

	A	B	C	D	E	F	G	H
1	k_1		1	Variable	f(x)=0	Jacobian		
2	k_2	0.2	0.25	10.3125	-165	0	3	0
3	k_3	0.05	0.25	35	200	-70	0	0
4	k_4	0.4	0.25	-7.8125	15	20	-53	0
5	V		100	0.25	-10	0	20	0	-50
6	Q		50					
7	CAf		1					

With the exception of the second equation, the dominate terms line up with the equations. However, the dominant terms in Equation (6.92) differ by just one order of magnitude, whereas the dominant terms in Equation (6.91) differ by two orders of magnitude. Therefore, we use Equation (6.91) for C_A and Equation (6.92) for C_B.

Solve the system of equations for the exit concentrations by ordinary iteration using implicit forms:

$$C_A = C_{A0} + \frac{V}{Q}\left(-k_1 C_A - k_2 C_A^{3/2} + k_3 C_C^2\right) \tag{6.95}$$

$$C_B = \frac{V}{Q}\left(2k_1 C_A - k_4 C_B^2\right) \tag{6.96}$$

$$C_C = \frac{V}{Q}\left(k_2 C_A^{3/2} - k_3 C_C^2 + k_4 C_B^2\right) \tag{6.97}$$

$$C_D = \frac{V}{Q} k_4 C_B^2 \tag{6.98}$$

Note that the mole balance for C_D is the only explicit equation. The other expressions do not include C_D. It is not necessary to solve for C_D simultaneously with the other unknown concentrations because we can calculate C_D from the results for C_B directly. Solve the system of equations by ordinary iteration in an *Excel* worksheet using the following *Excel* formulas with relative references for concentration. Start with initial guesses of 0.25 for each variable. Use subsequent rows for new iterations.

CA =CAf + (V/Q)*(-k_1*B10-k_2*B10^ (3/2) +k_3*D10^2)   CB = (V/Q)*(2*k_1*B10-k_4*C10^2)

CC = (V/Q)*(k_2*B10^ (3/2)-k_3*D10^2+k_4*C10^2)   CD = (V/Q)*(k_4*C10^2)

	A	B	C	D	E
9	Iteration	C_A	C_B	C_C	C_D
10	0	0.25	0.25	0.25	0.25
11	1	0.45625	0.95	0.09375	0.05
12	2	-0.03489	1.103	0.844393	0.722
13	3	#NUM!	-1.11286	#NUM!	0.973
14	4	#NUM!	#NUM!	#NUM!	0.991

Despite our rearrangement of the equations for the dominant terms, the solution diverges quickly after just two iterations. Follow the suggestion of grouping terms with the same order of magnitude for each variable to promote convergence. For example, combine all first order terms for C_A in Equation (6.91):

Species A:
$$C_A\left(Q + Vk_1 + k_2 C_A^{1/2}\right) = QC_{A0} + Vk_3 C_C^2 \tag{6.99}$$

Create similar groups of terms in the remaining equations and solve them for each variable:

$$C_A = \frac{QC_{A0} + Vk_3 C_C^2}{Q + V\left(k_1 + k_2 C_A^{1/2}\right)} \qquad C_B = \frac{2Vk_1 C_A}{Q + Vk_4 C_B} \qquad C_C = \frac{V\left(k_2 C_A^{3/2} + k_4 C_B^2\right)}{Q + Vk_3 C_C}$$

For convenience in the worksheet, use *VBA* user-defined function **VNORMRR** from the *PNMSuite* to calculate the residual vector norm in Equation (4.58) for checking relative convergence. This new equation arrangement reaches a converged solution after 25 iterations when starting from the original guesses.

CHAPTER 6: ROOT-FINDING METHODS

	G	H	I	J	K	L
9	Iteration	C_A	C_B	C_C	C_D	Convergence
10	0	0.25	0.25	0.25	0.25	
11	1	0.314453	0.833333	0.097561	0.05	0.711481553
12	2	0.31044	0.754688	0.62004	0.555556	0.627294855
13	3	0.322211	0.774284	0.494188	0.455643	0.151013801
14	4	0.317448	0.795865	0.526742	0.479613	0.041347095
15	5	0.318644	0.775828	0.549329	0.506721	0.036136157
			⋮			
35	25	0.318866	0.783884	0.534982	0.491579	1.6915E-07

Comments: Although the Jacobian result helped us identify the equations for iteration on each variable, we still needed to rearrange the equations for convergence by ordinary iteration. Aitken's method may accelerate convergence in ordinary iteration for a system of coupled non-linear equations. However, the method also requires good initial guesses for the solution and proper equation sequence and arrangement. Aitken's method does not guarantee convergence, although with some luck, it takes some of the guesswork out of finding a good relaxation factor. When we use the *VBA* macro **STEFFENSEN** to solve the original implicit set in Equations (6.95) to (6.98), the solution converges in 20 iterations within a relative tolerance of 10^{-8}.

□

6.3.4 Quasi-Newton's Method for Simultaneous Equations

We may apply Newton's method to finding roots to simultaneous equations. Consider a pair of arbitrary non-linear functions in two variables, represented by $f_1(x_1, x_2) = 0$ and $f_2(x_1, x_2) = 0$. Expand each nonlinear function in a first-order Taylor series about initial guesses for the roots:

$$0 \cong f_1(g_1, g_2) + \Delta x_1 \left.\frac{\partial f_1}{\partial x_1}\right|_{g_1, g_2} + \Delta x_2 \left.\frac{\partial f_1}{\partial x_2}\right|_{g_1, g_2} \tag{6.100}$$

$$0 \cong f_2(g_1, g_2) + \Delta x_1 \left.\frac{\partial f_2}{\partial x_1}\right|_{g_1, g_2} + \Delta x_2 \left.\frac{\partial f_2}{\partial x_2}\right|_{g_1, g_2} \tag{6.101}$$

where $\Delta x_1 = (x_1 - g_1)$ and $\Delta x_2 = (x_2 - g_2)$. Equations (6.100) and (6.101) are linear in the unknown variables Δx_1 and Δx_2. Find the approximate solution for x_1 and x_2 by first solving for the residuals Δx_1 and Δx_2 using Gaussian elimination from Chapter 4. Get the upgraded values for the roots:

$$x_1 = g_1 + \Delta x_1 \qquad x_2 = g_2 + \Delta x_2 \tag{6.102}$$

As with all "trial and error" type methods for nonlinear equations, one iteration does not necessarily give the true solution, but a better approximation to the true solution if we start from good initial guesses. Iterate on the equations, replacing g_1 and g_2 with the upgraded approximations, x_1 and x_2, until reaching satisfactory convergence on the roots.

The partial derivatives in Equations (6.100) and (6.101) form a 2 × 2 Jacobian matrix. We may extend Newton's method to larger systems of *n* equations in *n* unknowns by calculating the *n* × *n* Jacobian and solving the linear system of equations for the vector Δx_i, $i = 1$ to *n*:

$$\begin{bmatrix} \dfrac{\partial f_1}{\partial x_1}\bigg|_g & \dfrac{\partial f_1}{\partial x_2}\bigg|_g & \cdots & \dfrac{\partial f_1}{\partial x_n}\bigg|_g \\ \dfrac{\partial f_2}{\partial x_1}\bigg|_g & \dfrac{\partial f_2}{\partial x_2}\bigg|_g & \cdots & \dfrac{\partial f_2}{\partial x_n}\bigg|_g \\ \vdots & \vdots & \ddots & \vdots \\ \dfrac{\partial f_n}{\partial x_1}\bigg|_g & \dfrac{\partial f_n}{\partial x_2}\bigg|_g & \cdots & \dfrac{\partial f_n}{\partial x_n}\bigg|_g \end{bmatrix} \cdot \begin{bmatrix} \Delta x_1 \\ \Delta x_2 \\ \vdots \\ \Delta x_n \end{bmatrix} = \begin{bmatrix} -f_1(g) \\ -f_2(g) \\ \vdots \\ -f_n(g) \end{bmatrix} \qquad (6.103)$$

where the Jacobian matrix of partial derivatives and vector of negative functions are evaluated at the current guesses for the solution, $g = (g_1, g_2 \ldots g_n)$.

The number of derivative evaluations in Newton's method increases exponentially as we add equations for additional variables, making it computationally inefficient for large systems of equations. In the quasi-Newton's method, we use finite-difference approximations from Section 5.2 for the partial derivatives.

Newton's method converges when the maximum, relative difference in the upgraded values stop changing within a specified convergence criterion, such as Equation (4.58). As in the case of single equations, Newton's method applied to systems of equations is sensitive to initial guesses for the solution. A reduced step modification to Newton's algorithm may promote convergence in some cases where the elements of the step change Δx are too large and do not produce a new vector for x closer to the root (Beers 2007). The reduced-step method imbeds a loop inside the Newton iterations that cycles through a systematic reduction in Δx until the norm of the function vector starts to decrease:

$$x = g + \alpha_n \Delta x \qquad (6.104)$$

where the reduction factor is $\alpha_n = 1/2^n$, for the nth iteration of the step reduction cycle. In Section 9.7 we introduce the Levenberg-Marquardt method to apply relaxation to the quasi-Newton method.

Macros **QUASINEWTON**, with input boxes, and user-form **ROOTS_frm** are available in the *PNMSuite*. **ROOTS_frm** calls the *VBA* macro **RUN_QUASINEWTON**, listed in Figure 6.16, to solve a system of equations for a set of roots using the quasi-Newton method. The macro uses the method of Gaussian elimination with maximum column pivoting to solve the system of linear expansions for Δx. **RUN_QUASINEWTON** calls the sub procedure **JACOBI** to calculate the Jacobian matrix for each iteration. The **Status Bar** shows the current convergence level and number of Newton iterations. To implement Newton's method in an *Excel* worksheet, provide initial guesses for each variable in a range of cells, and then add formulas in a separate range of cells for the nonlinear functions of the form in Equation (6.75) using the cells with the initial guesses for the variables.

We use a simple system of nonlinear equations to illustrate how to set up an *Excel* worksheet for finding roots with the macros:

$$f_1(x_1, x_2) = 5x_2^2 - x_1^2 - 25 \qquad \text{and} \qquad f_2(x_1, x_2) = x_1 x_2^2 - 5 \qquad (6.105)$$

1. Provide a vector of initial guesses for the roots in the range **B1:B2** on the worksheet. The corresponding labels are in column **A**.
2. Add formulas for the nonlinear equations in the range **E1:E2**. The formulas reference the cells in the range **B1:B2** for the variables. The corresponding labels are in column **D**.
3. Run the user-form macro **ROOTS_UsrFrm** and select the option for the Quasi-Newton solution method.
4. Select the ranges for the variables and functions in the input boxes.
5. The range **B1:B2** contains the converged result for the roots:

	A	B	C	D	E
1	x1 =	1		f1 =	=5*B2^2-B1^2-25
2	x2 =	1		f2 =	=B1*B2^2-5

	A	B	C	D	E
1	x1 =	0.96415		f1 =	0
2	x2 =	2.277261		f2 =	0

```
Private Sub RUN_QUASINEWTON()
' Reduced step quasi Newton's method for solving coupled nonlinear equations
' using finite difference derivative approximations. A linearized system of equations
```

CHAPTER 6: ROOT-FINDING METHODS

```vba
' is solved by Gauss elimination with maximum column pivoting & one error correction.
' *** Required User-defined Functions and Sub Procedures from the PNM Suite ***
' GAUSSPIVOT = solution to linear system of equations
' JACOBI    = Jacobian matrix of partial derivatives
' *** Nomenclature ***
Dim boxtitle As String ' = string with title for message and input boxes
Dim df() As Double ' = partial derivative of f(i) with respect to x(j)
Dim dx As Double ' = small change in x for finite difference derivative approximation
Dim f() As Double ' = non-linear function in x
Dim f1 As Double ' = function evaluated at small perturbation in x
Dim fnorm As Double, fnorm1 As Double ' = vector norms of function range
Dim fr As Range ' = range of functions on worksheet (contiguous cells)
Dim xr As Range ' = range of solution vector guesses(contiguous cells)
Dim i As Long, j As Long, k As Long ' = indexes for variables at Newton iterations
Dim it As Long ' = Newton iteration
Dim m As Long ' index for reduced step iterations
Dim maxit As Long '= maximum number of iterations
Dim n As Long ' = number of unknowns
Dim nit As Long ' = maximum number of Newton iterations
Dim q As Double ' = quotient for step size in Newton's formula
Dim res As Double ' = sum of residuals
Dim sc As Integer ' = show calculations or continue calculations check variable
Dim small As Double ' = small number to avoid division by zero (module level)
Dim tol As Double ' = convergence tolerance (set by user)
Dim x1() As Double ' = xr(j) - x(j)
Dim x() As Double ' = solution to nonlinear equation
'**********************************************************************************
' Set counter for of Newton iterations, small number, title for input boxes
nit = 999: small = 0.000000000001: boxtitle = "QUASINEWTON"
With Application
.Calculation = xlCalculationAutomatic ' turn on worksheet automatic calculation
x_input: 'Get initial guesses for variables
Set fr = .InputBox(prompt:="Range of FUNCTIONS, f(x)=0:" & vbNewLine & _
                "(must be in contiguous cells)", Title:=boxtitle, Type:=8)
Set xr = .InputBox(prompt:="Range of INITIAL guesses, x:" & vbNewLine & _
                "(must be in contiguous cells)", Title:=boxtitle, Type:=8)
n = fr.Count ' Get number of variables and functions
If n <> xr.Count Then ' Check for consistency between number of variables & equations
    MsgBox Title:=boxtitle, prompt:="Numbers of variables & equations must match."
    Exit Sub
End If
tol = 0.00000001 ' relative convergence tolerance
ReDim df(1 To n, 1 To n) As Double, f(1 To n) As Double, x(1 To n) As Double, _
      x1(1 To n) As Double
Select Case n
Case 1 'Find root of a single nonlinear equation. Initial perturbation
fnorm1 = Abs(fr)
For k = 1 To nit ' Newton iterations. Count iterations & reset sum of residuals
    For i = 1 To n: x(i) = xr(i): f(i) = fr(i): Next i ' x & function values for subs
    dx = .Max(Abs(0.0001 * x(1)), small): xr = x(1) + dx: f1 = fr
    df(1, 1) = (f1 - f(1)) / dx ' Finite difference derivatives
    q = f(1) * df(1, 1) / (df(1, 1) ^ 2 + small) ' Newton step size for x
    xr = x(1) - q: res = Abs(q) / (Abs(x(1)) + small) ' Upgrade & x residual
    If res < tol Then Exit For ' Check for solution convergence
    For m = 1 To 6 ' reduced step iterations
        fnorm = Abs(fr)
        If fnorm < fnorm1 Then
            fnorm1 = fnorm: Exit For
        End If
        xr = x(1) - q / (2 ^ m)
    Next m
    .StatusBar = boxtitle & " iterations = " & k _
            & "; Relative Residual = " & VBA.Format(res, "0.#E+#0%")
Next k
Case Else ' Get Jacobian matrix & solve for the roots with Gauss elimination
fnorm1 = Sqr(.SumSq(fr)) ' calculate vector norm of function values
For k = 1 To nit ' Newton iterations. Count iterations & reset sum of residuals
    For i = 1 To n: x(i) = xr(i): f(i) = fr(i): Next i ' x & function values for subs
    Call JACOBI(df, fr, xr, n, x): Call GAUSSPIVOT(df, f, n, x1): res = 0
```

```
            For j = 1 To n  ' calculate the maximum residual & upgraded x
                res = .Max(Abs(x1(j)) / (Abs(x(j)) + small), res)
            Next j
            For j = 1 To n: xr(j) = x(j) - x1(j): Next j  ' update solution
            If res < tol Then Exit For  ' Check for solution convergence
            For m = 1 To 6  ' reduced step iterations
                fnorm = Sqr(.SumSq(fr))  ' vector norm of functions
                If fnorm < fnorm1 Then
                    fnorm1 = fnorm: Exit For
                End If
                For j = 1 To n: xr(j) = x(j) - x1(j) / (2 ^ m): Next j
            Next m
            .StatusBar = boxtitle & " iterations = " & k & "; Relative Residual = " _
                & VBA.Format(res, "0.#E+#0%")
    Next k
    End Select  ' Display progress at bottom of worksheet
    End With  ' Application
End Sub  ' RUN_QUASINEWTON
```

Figure 6.16   *VBA* macro RUN_QUASINEWTON used by the user-form ROOTS for solving a system of nonlinear equations by the quasi-Newton method with finite-difference derivative approximations. See Figure 6.15 for sub procedure JACOBI and the *PNMSuite* for sub procedure GAUSSPIVOT.

Example 6.14   Solve the Mass Balances in a Steady-state Reactor using Newton's Method

Known/Find/Analysis: Find the steady-state concentrations in the well-mixed reactor described in *Example 6.13* employing the *VBA* user-form **ROOTS**. Formulate the system of nonlinear functions according to Equations (6.91) through (6.94) in the range **C9:C12** on an *Excel* worksheet using named cells for the reaction parameters in the range **B1:B7** and variables in the range **B9:B12**:

	A	B	C
1	k_1	1	
2	k_2	0.2	
3	k_3	0.05	
4	k_4	0.4	
5	V	100	
6	Q	50	
7	CA_f	1	
8		Concentrations	Mole Balances
9	A	0.318865812256048	=Q*(CAf-B9)+V*(-k_1*B9-k_2*B9^1.5+k_3*B11^2)
10	B	0.783883977224619	=-Q*B10+V*(2*k_1*B9-k_4*B10^2)
11	C	0.534981835031428	=-Q*B11+V*(k_2*B9^1.5-k_3*B11^2+k_4*B10^2)
12	D	0.491579271799572	=-Q*B12+V*k_4*B10^2

To compare Newton's method with other successive approximation methods in Example 6.13, begin with the same initial guesses for the solution and use a relative convergence tolerance of 10^{-8}. The *VBA* user-form **ROOTS** prompts for the ranges of variables and functions:

CHAPTER 6: ROOT-FINDING METHODS

Newton's method converged to the solution within the relative convergence tolerance after just five iterations (Steffensen's method required 20 iterations).

	A	B	C
8		Concentrations	Mole Balances
9	A	0.318865812	-8.52651E-14
10	B	0.783883977	-9.166E-13
11	C	0.534981835	9.98313E-13
12	D	0.491579272	9.20153E-13

Comments: We retained a large number of decimal places to show the precision of the solution. Normally we round off to fewer significant figures (two in this example). Although the quasi-Newton's method required just five iterations, this corresponds to 16 function evaluations for each iteration to calculate the finite-difference approximations, reducing the computational efficiency of the method to a level comparable to other iteration methods, such as Aitken's Δ^2 method.

□

Newton's method generally converges in fewer iterations than other methods of successive approximation, but may quickly diverge away from the solution without good initial guesses. Section 12.8 introduces the robust Method of Continuation for root finding that combines Newton's method with methods for solving differential equations from Chapter 12.

6.3.5 Excel's Solver for Simultaneous Equations

Excel has two intrinsic features for solving nonlinear problems: **Goal Seek** and **Solver**. We have some experience with **Goal Seek** for finding a root of a single nonlinear function from Section 6.2.5. Section 2.5 introduces the add-in **Solver**. *Excel*'s **Solver** has additional capabilities beyond **Goal Seek** that make it the first method of choice for solving single or multivariable systems of equations in *Excel*. For instance, the **Solver** has the ability to constrain the search region to avoid searching for values where the function is indeterminate. See Section 7.8 for a discussion of numerical methods for handling constraints.

We can control advanced features of the **Solver** solution process. Each option has default settings, shown in Figure 6.17, that are appropriate for most problems. The default solution method is *General Reduced Gradient* (*GRG*), which combines the constraints with the objective function for solution by an extension of the quasi-Newton's method. Table 6.3 lists definitions for a few of the **Solver Options**.

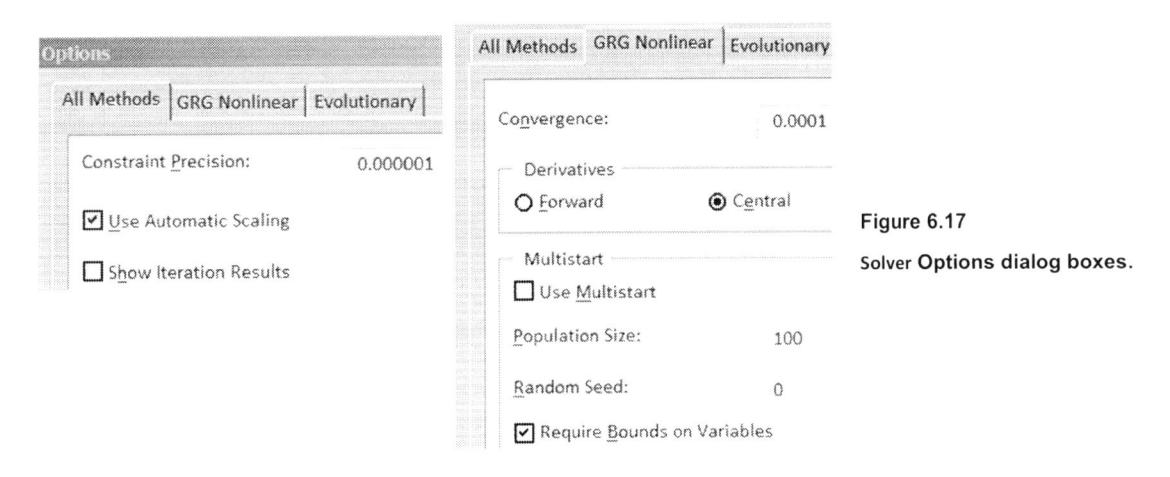

Figure 6.17

Solver **Options** dialog boxes.

Table 6.3   Solver parameters and definitions.[57]

Parameter	Definition
Constraint Precision	Control the precision of solution to determine whether the value of a constraint cell meets a target or satisfies a lower or upper bound. Precision must be a fractional number between 0 (zero) and 1.
Convergence	**Solver** stops when the relative change in the variables is smaller than the number in the **Convergence** box.
Use Automatic Scaling	Use automatic scaling when the variables have large differences in magnitude. Automatic scaling has no affect when the values are smaller than 10^{-3}. Refer to methods of manual scaling described in Section 7.9.
Show Iteration Results	Select this option to pause the solution after each iteration in the search. The **Status Bar** displays the current values of the **Target** cell and iteration number at the bottom of the worksheet.
Derivatives	Approximates the slope of the function in the **Set Target Cell** using finite-difference formulas at the current values in the **By Changing Cells** references. Chapter 5 has the derivations of the finite-difference approximation methods. We recommend **Central** derivatives for all applications of the **Solver**.
Use Multi Start	Look for global solutions by randomly starting the search around the initial values in the **Changing Cells**. The **Population Size** limits the number of searches, and the optional **Seed** initializes the pseudo random number generator. Use bounds (constraints) to set the size of the random search region.

Like **Goal Seek**, the **Solver** only allows for a single **Target Cell**. To implement the **Solver** for a coupled, multivariable system of equations, combine the functions into a single, multidimensional equation as a sum of squares in the form of Equations (6.75):

$$f_s = \sum_i (f_i)^2 = 0 \qquad (6.106)$$

Equation (6.106) works on the following principle: if $f_1(x_1, x_2) = 0$ and $f_2(x_1, x_2) = 0$ at the roots, then the sum $f_1 + f_2 = 0$. However, we square the functions to make each term in the summation positive, which avoids any potential for inadvertently canceling negative with positive function values, and returning a false root before converging. *Excel*'s **Solver** searches for the values of the independent variables that satisfy Equation (6.106). We also exploit the **Solver** in Chapter 7 for optimization and Chapter 8 for nonlinear least-squares regression of data. Use the following steps to set up the **Solver** for simultaneous nonlinear equations:

- Set the **Objective** cell as the sum of squares of the functions according to Equation (6.106).
- Because we recognize that the zero of the sum of squares is also the minimum, check **Equal to: Min** instead of **Value** of 0 because the **Solver** objective function in Equation (6.106) does not cross the horizontal axis, but remains positive for all *x*, as illustrated in Figure 6.18 for a two variable problem.

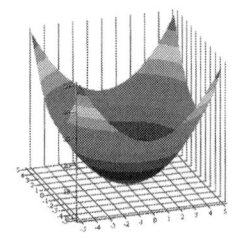

Figure 6.18

Sum-squared function for finding the roots of a system of two nonlinear equations.

- Fill in the input box for **By Changing Cells** with the range of cells containing the variables.
- We may need to uncheck the box labeled **Make Unconstrained Variables Non-Negative**.
- For the solution method, select from the quasi-Newton or conjugate gradient methods (*GRG* Nonlinear, or Generalized Reduced Gradient).[58] In either case, the **Solver** approximates the derivatives using finite-differences. The more accurate centered finite-difference approximations are preferred to forward-difference derivative approximations.
- Control the precision of the solution by adjusting the value of **Convergence** in **Solver Options**. Unlike **Goal Seek**, the **Solver** uses relative values of the variables to check for convergence.

---

[57] Click on **Solver Options>Help**, as shown in Figure 6.17, for more details.
[58] The conjugate gradient method uses derivatives to pick the search direction, similar to Powell's method.

CHAPTER 6: ROOT-FINDING METHODS

We may use any of the optimization methods presented in Chapter 7 to find the roots of a system of equations with an objective function defined according to Equation (6.106). The macro **ROOTS** includes Powell's method from Section 7.5.5 as an option, but calculates the sum of squared functions internally so that the worksheet setup is the same as required by either quasi-Newton or the method of continuation. The user-form **ROOTS** also includes the methods Gauss-Newton and Levenberg-Marquardt from Sections 9.6 and 9.7, which are designed to minimize the sum of squared residuals according to Equation (6.106). The *PNMSuite* also has macros **ROOTSGN** and **ROOTSLM** for finding roots to equations by the Gauss-Newton and Levenberg-Marquardt methods using input boxes. In all cases, the worksheet setup is the same as **QUASINEWTON**.

Example 6.15 Use the Solver to Find the Steady-state Reactor Concentrations

Known/Find: Using *Excel*'s **Solver**, find the roots to the mole balances for the reactor in Example 6.14.

Analysis: Start with the same *Excel* worksheet used for Newton's method shown in Example 6.14 and add one line for the sum of the squares of the functions:

$$C14 = SUMSQ(C9:C12)$$

Open the **Solver** and specify the objective and changing cells as shown in Figure 6.19. Select **To: Min**. This has the same effect as setting **Equal To: Value of 0** in the sum of squares. Modify the convergence control parameters under **Options** if necessary. Click **Solve** to locate the values for the roots that match those in Example 6.14.

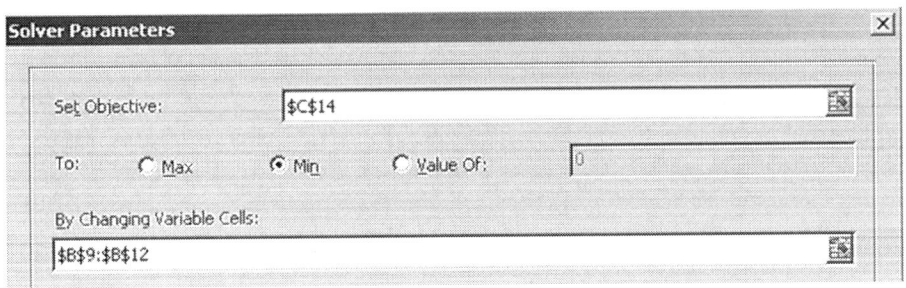

Figure 6.19

Solver options for finding the roots to a system of nonlinear equations in Example 6.14.

Comments: The choice of initial guesses for reactant concentrations in this problem is critical. If we initially specify zero concentrations for B, C, and D the **Solver** will only search on the concentration of A. To illustrate the importance of summing the squares of the functions, replace the formula for the objective from the sum of squared functions to the sum of the functions:

$$C15 = Sum(C9:C12)$$

Starting with initial guesses for the steady-state concentrations of one in the range **B9:B12**, set up the **Solver Parameters** to **Set Target Cell: $C$15 Equal to: Value of: 0** and **Solve** to obtain the results in Figure 6.20. Notice that while the sum of the functions in cell **C15** is approximately zero (as specified for the objective) the values of the functions in the range **C9:C12** are not zero, giving false roots in the range **B9:B12**. Using the *sum of the squares* of the functions in cell **C14** solves the problem.

	A	B	C
8		Concentrations	Mole Balances
9	A	1.059223155	-126.2609138
10	B	1.035533988	117.174706
11	C	0.940776845	13.23182395
12	D	0.940776845	-4.145616656
13			
14		Sum of Squares	29863.99739
15		Sum	-5.56455E-07

Figure 6.20

Incorrect Solver solution to nonlinear reactor mole balances using the sum of functions in place of the sum of squared functions.

Example 6.16 Batch Distillation of a Binary System

Known: A single-stage batch still operating at atmospheric pressure is charged with a binary mixture of 15 mol% butane and 85 mol% pentane. The distillation stops when 90 mol% of the butane is removed from the pot.

Find: How much pentane ends up in the distillate?

Assumptions: Raoult's law

Schematic: A batch still contains a liquid mixture with W total moles and mole fraction x of butane in a mixture of pentane and a distillate D with mole fraction y:

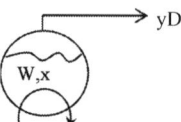

Analysis: Calculate the butane in the distillate from a mass balance:

$$0.9 x_0 W_0 = x_0 W_0 - xW \tag{6.107}$$

where x_0 and W_0 are the initial mole fraction of the butane and total molar charge to the still pot, respectively. The Rayleigh equation using an average relative volatility, α, relates the total moles in the still to the composition of the light component (butane):

$$\ln\left(\frac{W_0}{W}\right) = \frac{1}{(\alpha-1)}\left[\ln\left(\frac{x_0}{x}\right) + \alpha \ln\left(\frac{1-x}{1-x_0}\right)\right] \tag{6.108}$$

Get the relative volatility at the initial boiling point temperature of the mixture. Find the bubble-point temperature from the sum of the product of K-values and liquid mole fractions for the two components, where B = butane, P = pentane:

$$K_B x + K_P (1-x) = 1 \tag{6.109}$$

and where the K-value is the ratio of the pure component saturation vapor pressure to total pressure:

$$K_i = P_i^s / P_T \tag{6.110}$$

The relative volatility is defined as the ratio of vapor pressures, or ratio of K-values. We typically use the Antoine equation to correlate the saturation vapor pressure with temperature (Reklaitis 1983):

$$\ln\left(P_B^s / kPa\right) = 13.9836 - \frac{2292.44}{(T/K) - 27.8623} \quad \text{and} \quad \ln\left(P_P^s / kPa\right) = 13.9778 - \frac{2554.6}{(T/K) - 36.2529}$$

Set up the problem in an *Excel* worksheet, as shown in Figure 6.21.

	A	B
1	W_0	100
2	W	45.9492531053598
3	P_T/kPa	101.325
4	T/K	300.058960142641
5		x_0
6	Butane	0.15
7	Pentane	=1-B6
8	Bubble Pt	=E6+E7-1
9	α	=KB/KP
10	Recovery	=0.9*x_0*W_0-x_0*W_0+x*W
11	Rayleigh	=LN(W_0/W)-(1/(alpha-1))*(LN(x_0/x)+alpha*LN((1-x)/(1-x_0)))
12	SUMSQ	=SUMSQ(B11:B11)
13		
14	Pentane in Distillate	=((1-x_0)*W_0-(1-x)*W)/((1-x_0)*W_0)

CHAPTER 6: ROOT-FINDING METHODS

	C	D	E	F
5	P^s/kPa	K-value	x$_0$*K	x
6	=EXP(13.9836-2292.44/(T-27.8623))	=C6/PT	=x_0*KB	0.0326446950625057
7	=EXP(13.9778-2554.6/(T-36.2529))	=C7/PT	=(1-x_0)*KP	=1-F6

	A	B	C	D	E	F
1	W$_0$	100				
2	W	45.94925311				
3	P$_T$/kPa	1.01E+02				
4	T/K	300.0589601				
5		x$_0$	P^s/kPa	K-value	x$_0$*K	x
6	Butane	0.15	260.2391454	2.568360675	0.385254101	0.032644695
7	Pentane	0.85	73.27145122	0.723133	0.61466305	0.967355305
8	Bubble Pt	-8.28488E-05				
9	α	3.551712721				
10	Recovery	-6.44026E-07				
11	Rayleigh	3.07786E-07				
12	SUMSQ	9.47319E-14				
13						
14	Pentane in Distillate	48%				

Figure 6.21 *Excel* worksheet results and formulas to solve for the bubble point temperature, batch distillation charge, and composition.

Use **Goal Seek** in an *Excel* worksheet to find the initial bubble point temperature:

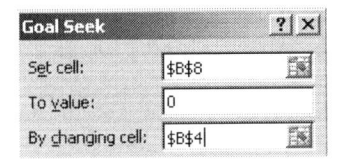

Goal Seek
Set cell: $B$8
To value: 0
By changing cell: $B$4

Then calculate the relative volatility from the ratio of vapor pressures or *K*-values:

$$\alpha = K_B/K_P = P_B^v/P_P^v \tag{6.111}$$

Now solve Equations (6.107) and (6.108) simultaneously for *x* and *W*. Use *Excel*'s **Solver** to find the values of *x* and *W* that minimize the sum of squared residuals according to Equation (6.106). For convenience, use the worksheet formula **SUMSQ** in cell **B12**. Note that *x* and *W* appear in the denominators of the natural logarithms, which may pose a problem for the **Solver** when searching for the roots. To avoid division by zero or negative numbers, impose constraints on $x > 0$ and $W > 0$. However, the default precision of the constraints is 10^{-5}. Therefore, pick a lower constraint of 10^{-3} that still prevents the values of *x* and *W* from going below 10^{-5}.

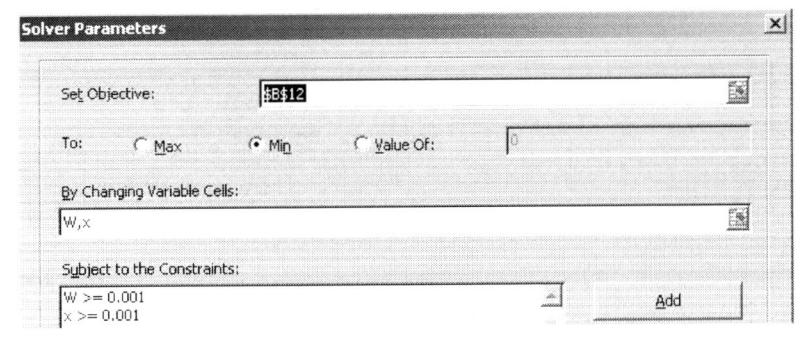

Solver Parameters
Set Objective: $B$12
To: Min
By Changing Variable Cells: W,x
Subject to the Constraints:
W >= 0.001
x >= 0.001

The results for *x* and *W* are 0.0326 and 45.9 in Cells **F6** and **B2**, respectively. These values are larger than the lower limits imposed by the constraints. Finally, calculate the amount of pentane in the distillate according to Equation (6.112) in cell **B14**:

$$yD = \frac{(1-x_0)W_0 - (1-x)W}{(1-x_0)W_0} \times 100 \qquad (6.112)$$

Comments: The relative volatility changes slightly with temperature. We calculate an improved solution by finding the relative volatility at the final composition of the still using the geometric average of the initial and final values in the Rayleigh Equation (6.108):

$$\alpha = \sqrt{\alpha_{initil} \times \alpha_{final}} \qquad (6.113)$$

We must iteratively upgrade the relative volatility to improve the result. Alternatively, we may include Equation (6.113) and the bubble-point temperature calculation in our system of nonlinear equations for solution by the **Solver**.

□

6.4 Bairstow's Method of Polynomial Factorization*

We encounter polynomial functions as models of physicochemical properties, in designed experiments, and transfer functions in process control. We may find a single root to a polynomial using any of the methods in this chapter. However, our choice of guesses for initializing the iterative methods strongly influences where the solution converges. To locate all of the roots using the methods presented to this point requires repeated searches starting from different initial guesses.

There are several alternative methods for efficiently finding all roots to polynomials of the form:

$$f_n(x) = a_0 + a_1 x + a_2 x^2 + \ldots a_N x^N \qquad (6.114)$$

where N is the order of the polynomial. Most polynomial root-finding methods are based on the idea of deflation, which iteratively reduces the order of the polynomial by performing synthetic division to factor the function into the product of first-order (linear) terms:

$$f_n(x) = (x - r_1)(x - r_2)\ldots(x - r_N) \qquad (6.115)$$

where r_i represents the vector of N roots ($i=1, 2\ldots N$). To factor a polynomial in *Excel*, we chose Bairstow's method with the following qualities:

- Relatively simple algorithm for implementation in VBA,
- Ability to produce real and complex roots by deflation with a quadratic divisor ($x^2 - bx - c$), and
- Employment of the now familiar Newton's method for solving simultaneous equations.

We derive Bairstow's method for polynomial deflation following the general outline of Chapra and Canale (2002):

1. Guess values for b and c in the following quadratic divisor (*e.g.*, we may start with $b = c = 0$):

$$x^2 - bx - c \qquad (6.116)$$

2. Divide the original N^{th} order polynomial in Equation (6.114) by the quadratic factor in Equation (6.116) to give a new polynomial with a reduced order $N-2$ and remainder R:

$$f_{n-2}(x) = p_2 + p_3 x + \ldots p_{n-2} x^{N-2} \qquad (6.117)$$

$$R = p_0 + p_1(x - b) \qquad (6.118)$$

Reset $N \leftarrow N - 2$, then recursively calculate the coefficients of the new lower order polynomials:

$$p_N = a_N \qquad (6.119)$$

$$p_{N-1} = a_{N-1} + b p_N \qquad (6.120)$$

$$\text{for } j = N-2, N-1 \ldots 0: \qquad p_j = a_j + bp_{j+1} + cp_{j+2} \tag{6.121}$$

3. The remainder in Equation (6.118) is zero when p_0 and p_1 are each zero. Apply Newton's method for simultaneous equations to upgrade the values for b and c in the quadratic factor to obtain a zero remainder.

$$b_{new} \leftarrow b + \Delta b \qquad c_{new} \leftarrow c + \Delta c \tag{6.122}$$

To set up Newton's method, expand the new polynomial coefficients p_0 and p_1 in truncated Taylor series like Equations (6.100) and (6.101), and set them to zero for a pair of linear equations in Δb and Δc:

$$0 \cong p_0 + \Delta b \left.\frac{\partial p_0}{\partial b}\right|_{b,c} + \Delta c \left.\frac{\partial p_0}{\partial c}\right|_{b,c} \tag{6.123}$$

$$0 \cong p_1 + \Delta b \left.\frac{\partial p_1}{\partial b}\right|_{b,c} + \Delta c \left.\frac{\partial p_1}{\partial c}\right|_{b,c} \tag{6.124}$$

4. The challenge is finding the partial derivatives in Equations (6.123) and (6.124). Bairstow[59] found the partial derivatives from synthetic division of the new polynomial by the quadratic factor in Equation (6.116):

$$q_N = p_N \tag{6.125}$$

$$q_{N-1} = p_{N-1} + bq_N \tag{6.126}$$

$$\text{for } j = N-2, N-1 \ldots 1: \qquad q_j = p_j + bq_{j+1} + cq_{j+2} \tag{6.127}$$

The partial derivatives become:

$$\frac{\partial p_0}{\partial b} = q_1 \qquad \frac{\partial p_0}{\partial c} = q_2 \qquad \frac{\partial p_1}{\partial b} = q_2 \qquad \frac{\partial p_1}{\partial c} = q_3 \tag{6.128}$$

5. Replace the guesses for b and c in the quadratic factor according to Equations (6.122) and iterate on steps three and four until the values for b and c converge within a specified relative tolerance:

$$\left|\frac{\Delta b}{b}\right| < tol \qquad \text{and} \qquad \left|\frac{\Delta c}{c}\right| < tol \tag{6.129}$$

6. Solve the quadratic factor for its two roots by the alternative form of the quadratic formula presented in Section 1.1.3:

$$q = 0.5\left[b + \text{sign}(b)\sqrt{b^2 + 4c}\right] \qquad x_1 = q \qquad x_2 = -c/q \tag{6.130}$$

7. If $N - 2 \geq 3$ for the new polynomial in Equation (6.117), repeat Bairstow's method using the previous results for b and c as initial guesses for Newton's method until arriving at a first or second order quotient. For a first-order quotient, the remaining root is simply $x = -c/b$, using the latest values for b and c found for the last quadratic divisor. In the case of a second order quotient, apply the quadratic formula in Equation (6.130) for the remaining two roots.

A *VBA* macro named **BAIRSTOW**, listed in Figure 6.22, uses input boxes to select the range of polynomial coefficients. Alternatively, try the user-defined array function **POLYROOTS**. The worksheet setup requires the polynomial coefficients in a range of contiguous cells, arranged by increasing order of terms in the polynomial (starting with the constant whether or not it is zero). The arrangement of the rows of output corresponds to the order of the polynomial. The maximum number of Newton iterations and relative convergence tolerance are set to 10^3 and 10^{-8}, respectively. We may change the number of iterations and tolerance in the code if necessary. The code listed in the *PNMSuite* validates the location selected for the output to avoid overwriting cells on the worksheet.

---

[59] British aviation researcher and mathematician.

```vb
Public Sub BAIRSTOW()
' Finds the real and imaginary roots to a polynomial by Bairstow's method given the
' coefficients ordered: f(x) = a0+a1*x+a2*x^2 + ... an*x^n.  The polynomial is
' deflated into the product of factors f(x) = (x-r1)(x-r2)...(x-rn).
' *** Required User-defined Functions and Sub Procedures from the PNM Suite ***
' QUADRADF = solves for the roots of the quadratic Bairstow factor
' *** Nomenclature ***
Dim a() As Double    ' = vector of coefficients (from low to high order terms)
Dim ar As Range ' = polynomial coefficients on worksheet, from low to high order terms
Dim b As Double ' = first-order term in quadratic factor
Dim boxtitle As String ' = string of name of macro used in dialog boxes
Dim c As Double    ' = zero-order term in quadratic factor
Dim db As Double ' = correction factor for b
Dim dc As Double ' = correction factor for c
Dim den As Double ' = denominator
Dim eb As Double, ec As Double ' = relative errors in factor coefficients b and c
Dim im() As Double ' = vector of imaginary parts to roots
Dim i1 As Double, i2 As Double ' = imaginary roots
Dim j As Long, j_2 As Long ' = index for ordered terms in polynomial, j_2 = j-2
Dim nc As Long ' = number of coefficients in polynomial (order - 1)
Dim nc_1 As Long ' = nc-1
Dim nci As Long ' = initial value of nc
Dim p() As Double ' = new deflated polynomial coefficients
Dim pcol As Long, prow As Long ' = print row and column
Dim q() As Double ' = Bairstow polynomial coefficients used for partial derivatives
Dim re() As Double ' = vector of real roots
Dim r1 As Double, r2 As Double ' = real roots
Dim small As Double ' = small number to avoid division by zero
Dim tol As Double ' = convergence tolerance for quadratic divisor
'**************************************************************************
small = 0.0000000001: boxtitle = "BAIRSTOW Polynomial Factorization"
tol = 0.00000001 ' Convergence tolerance for divisor factors
Set ar = Application.InputBox(Title:=boxtitle, Type:=8, _
    Prompt:="Polynomial COEFFICIENTS (n <= 8): f(x) = a0 + a1*x + x2*x^2 ... xn*x^n)")
prow = output.row: pcol = output.Column ' Get row & column for output
ReDim a(1 To nc) As Double, p(1 To nc) As Double, q(1 To nc) As Double, _
    im(1 To nc) As Double, re(1 To nc) As Double
' Initial guesses and errors for divisor factor coefficients b and c
b = 0: c = 0: eb = 1: ec = 1 '<--- adjust b and c if necessary
For j = 1 To nc: a(j) = ar(j): Next j ' create vector of coefficients
Do Until nc < 4 ' Loop through deflation steps, skip if polynomial is order 2 or fewer
    Do ' Newton iterations for b and c
        nc_1 = nc - 1 ' Get new deflated polynomial coefficients
        p(nc) = a(nc): p(nc_1) = a(nc_1) + b * p(nc)
        q(nc) = p(nc): q(nc_1) = p(nc_1) + b * q(nc)
        For j = nc - 2 To 1 Step -1
            p(j) = a(j) + b * p(j + 1) + c * p(j + 2)
            q(j) = p(j) + b * q(j + 1) + c * q(j + 2)
        Next j
        den = q(3) ^ 2 - q(4) * q(2) + small ' Newton upgrade step
        db = (-p(2) * q(3) + p(1) * q(4)) / den: b = b + db
        dc = (-p(1) * q(3) + p(2) * q(2)) / den: c = c + dc
        ' b and c relative errors
        eb = Abs(db / (b + small)): ec = Abs(dc / (c + small))
        Application.StatusBar = boxtitle & ": Newton Error = " _
            & VBA.Format((eb + ec) / 2, "#E+##")
    Loop Until eb <= tol And ec <= tol
    ' Get roots (real and imaginary) to quadratic divisor factor
    Call QUADRADF(b, c, r1, i1, r2, i2)
    re(nc) = r1: im(nc) = i1: re(nc - 1) = r2: im(nc - 1) = i2
    nc = nc - 2 ' Get reduced coefficients for next Bairstow application
    For j = 1 To nc: a(j) = p(j + 2): Next j
Loop
' Solve a quadratic equation for roots to quadratic divisor factor
If nc = 3 Then ' Second order polynomial
    b = -a(2) / a(3): c = -a(1) / a(3): Call QUADRADF(b, c, r1, i1, r2, i2)
    re(nc) = r1: im(nc) = i1: re(nc - 1) = r2: im(nc - 1) = i2
```

CHAPTER 6: ROOT-FINDING METHODS

```
Else ' First order polynomial
    re(nc) = -a(1) / a(2): im(nc) = 0
End If
Worksheets(ws).Activate ' Output real & imaginary roots to worksheet output cells
Cells(prow, pcol).Activate ' Print labels & uncertainty in f
With ActiveCell
    For j = 2 To nci
        j_2 = j - 2
        .Offset(j_2, 0).Value = "r" & j - 1 & " ="
            .Offset(j_2, 0).Characters(Start:=2, Length:=1).Font.Subscript = True
        .Offset(j_2, 1).Value = re(j)
        If im(j) < 0 Then
            .Offset(j_2, 2).Value = VBA.Format(im(j), "0.0###E+##") & "  i"
        ElseIf im(j) > 0 Then
            .Offset(j_2, 2).Value = "+" & VBA.Format(im(j), "0.0###E+##") & "  i"
        End If
        .Offset(j_2, 2).HorizontalAlignment = xlLeft
    Next j
End With
End Sub ' BAIRSTOW

Public Sub QUADRADF(b, c, r1, i1, r2, i2)
'***************************************************************************
' b = first order coefficient
' c = zero order coefficient (constant)
' r1, r2 = real parts for roots 1 and 2
' i1, i2 = imaginary parts for roots 1 and 2
'***************************************************************************
' Calculate the real and imaginary roots of a quadratic function: x^2 - bx - c = 0.
' *** Nomenclature ***
Dim d As Double, s As Double ' = term under the radical and sign of b
'***************************************************************************
If b = 0 Then ' get sign of coefficient b
    s = 1
Else
    s = Sgn(b)
End If
d = b ^ 2 + 4 * c
If d > 0 Then ' real roots
    r1 = 0.5 * (b + s * Sqr(d)): r2 = -c / r1: i1 = 0: i2 = 0
Else ' real and imaginary root parts
    r1 = 0.5 * b: r2 = r1: i1 = 0.5 * Sqr(Abs(d)): i2 = -i1
End If
End Sub ' QUADRADF
```

Figure 6.22 VBA macro BAIRSTOW for factoring a polynomial. The macro returns the real and imaginary roots. The sub procedure QUADRADF solves for the roots of the quadratic factor.

We illustrate how to use the macro **BAIRSTOW** with a simple example in an *Excel* worksheet using the following fifth-order polynomial:

$$f(x) = x^5 - 19x^4 + 119x^3 - 281x^2 + 180x$$

1. Note that the constant a_0 for this polynomial is not missing – it's just zero – and must be included.
2. Fill contiguous cells in a row (or column) on an *Excel* worksheet with the coefficients, ordered left to right (or top to bottom), starting with the lowest order terms. In this illustration the coefficients are in a row on the worksheet in the range **A1:F1**:

	A	B	C	D	E	F
1	0	180	-281	119	-19	1

3. Run the macro **BAIRSTOW**, or apply the user-defined array function **POLYROOTS**. Select the range of coefficients **A1:F1**, and cell **A2** for the location of the output on the worksheet, as requested by the input boxes. For the user-defined function, select a range of cells for the results in **D2:D6**, type the function name with the range of coefficient: **= POLYROOTS(A1:F1)**, then type the simultaneous key combination *CNTL SHIFT ENTER*. Recall that the number of roots matches the order of the polynomial.

4. The results for the five roots for this fifth-order polynomial are $x = 0, 1, 4, 5, 9$:

	A	B	C
2	$r_1 =$	9	
3	$r_2 =$	4	
4	$r_3 =$	5	
5	$r_4 =$	-1.2E-36	
6	$r_5 =$	1	

Try the Bairstow method to find the roots of the volume function in terms of liquid height in a spherical tank, described in Example 6.4. The expanded form of Equation (6.22) is a third-order polynomial in h:

$$f(h) = 0 = 3V - 3\pi r h^2 + \pi h^3 \tag{6.131}$$

a0	1.35	$r_1 =$	1.640285
a1	0	$r_2 =$	-0.43817
a2	-5.65487	$r_3 =$	0.597887
a3	3.141593		

Root 3 gives the feasible answer that lies within the diameter of the tank. Roots 1 and 2 are unrealistic results for the height of liquid in a tank with a diameter of 1.2.

Example 6.17 Molar Volume from the Beattie-Bridgeman Equation

Known/Find: Calculate the molar volume (L/mol) of N_2O at *STP* (273.15 K and 1 atm) using the Beattie-Bridgeman equation of state. The Beattie-Bridgeman constants for N_2O are (Van Wylen and Sonntag 1965):

$A_0 = 5.0065$ $a = 0.07132$ $B_0 = 0.10476$ $b = 0.07235$
$C_0 = 66 \times 10^4$ $R_g = 0.08206$ atm L/mol K

Analysis: The Virial form of the Beattie-Bridgeman equation is a fourth-order polynomial in terms of the inverse molar volume $1/V$:

$$0 = P - \frac{R_g T}{V} - \frac{\beta}{V^2} - \frac{\gamma}{V^3} - \frac{\delta}{V^4} \tag{6.132}$$

where $\beta = B_0 R_g T - A_0 - \dfrac{R_g C_0}{T^2}$ $\gamma = A_0 a - R_g T B_0 b - \dfrac{B_0 R_g C_0}{T^2}$ $\delta = \dfrac{R_g B_0 b C_0}{T^2}$

Use the macro **QYXPLOT** to graph the virial equation to identify the regions near the roots.

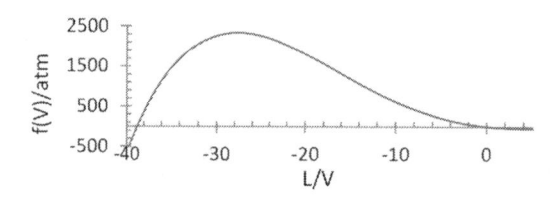

The plot reveals two real roots near $1/V = -40$ and 0 L^{-1}. We may change the scales on the axes to refine our estimate of the roots, or use Bairstow's method to calculate the roots of the polynomial.

Let $x = 1/V$, then compare Equation (6.132) to Equation (6.114) for $a_0 = P$, $a_1 = -R_g T$, $a_2 = -\beta$, $a_3 = -\gamma$, $a_4 = -\delta$. Calculate the coefficients for the fourth-order polynomial in an *Excel* worksheet and use the *VBA* macro **BAIRSTOW** to factor the polynomial for the roots.

CHAPTER 6: ROOT-FINDING METHODS

	A	B	C	D	E	F	G	H	I
1	T	273.15	K		Polynomial				
2	P	1	atm		$a_0 =$	1			
3	Rg	0.08206	L atm/mol K		$a_1 =$	-22.4147			
4	A_0	5.0065			$a_2 =$	3.384231			
5	a	0.07132			$a_3 =$	-0.11113			
6	B_0	0.10476			$a_4 =$	-0.0055			
7	b	0.07235			Bairstow Results				
8	C_0	6.60E+05			$r_1 =$	9.265737	+	-4.30156	i
9	β	-3.38423			$r_2 =$	9.265737	+	4.301558	i
10	γ	0.111129			$r_3 =$	0.044918	+	0	i
11	δ	0.005502			$r_4 =$	-38.775	+	0	i

BAIRSTOW Polynomial Factorization   [?][X]
Select range of polynomial coefficients (n <= 8).
Ordered: f(x) = a0 + a1*x + x2*x^2 ... xn*x^n)
$F$2:$F$6

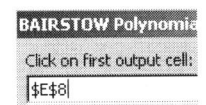

BAIRSTOW Polynomia
Click on first output cell:
$E$8

The results from Bairstow's method indicate just one significant root (the other three roots are negative or imaginary, which makes no physical sense). Calculate the molar volume from the inverse of the root, $V = 1/(0.044918 \text{ L}^{-1})$, for $V = 22.263$ L.

Comments: The molar volume is slightly smaller than the value of 22.4 L/mol for an ideal gas by 0.62%, indicating that N_2O is nearly ideal at *STP*.

□

6.5   Scaling and Transformation*

Systems of equations with roots that differ in value by several orders of magnitude may pose problems for the **Solver**. We can scale relatively large or small variables to have similar orders of magnitude, as demonstrated in *Example 6.18*. To scale a variable in an *Excel* worksheet, use two additional cells: one for the scaling factor, and the other for the product of the changing cell and the scale factor. Use the product result in the formula for the objective function. The *PNMSuite* macro **SCALIT** described in 7.10 automates the scaling method. To illustrate, we use a scale factor in cell **B2** for the calculation of the diffusion coefficient for CO_2, which has a value of 2.5×10^{-9} m²/s in cell **C2**. Cell **A2** contains the value manipulated by the **Solver**:

	A	B	C
1	Changing Cell	Scale Factor	Diffusivity
2	2.5	0.000000001	=A2*B2

	A	B	C
1	Changing Cell	Scale Factor	Diffusivity
2	2.5	1.00E-09	2.50E-09

Example 6.18   Single Stage Multicomponent Flash Separation

Known: A feed stream to a single stage isothermal flash unit consists of six components with the following molar flow rates and *K*-values:

	Component	Feed Rate/(lbmol/hr)	K-value
1	H_2	1900	29.18
2	CH_4	215	11.71
3	Ethane	17	5.406
4	Benzene	577	0.4546
5	Toluene	1349	0.2682
6	p-Xylene	1349	0.1623

246   PRACTICAL NUMERICAL METHODS

Schematic: The feed is separated into vapor and liquid with mole fractions y and x, respectively.

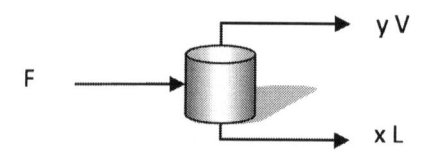

Find: Calculate the equilibrium mole fractions of the vapor and liquid effluent streams.

Analysis: Set up the mole balances and equilibrium relationships for each component:

$$F - xL - yV = 0 \qquad \text{and} \qquad y - Kx = 0 \qquad (6.133)$$

By definition, the mole fractions of the effluent streams sum to one:

$$1 - \sum x = 0 \qquad \text{and} \qquad 1 - \sum y = 0 \qquad (6.134)$$

Define the parameters and equations in an *Excel* worksheet using named ranges for F, K, x, y, L and V, as shown in Figure 6.23. Use initial guesses for x, y, V, and L in column **D**. Use array operations to create the system of equations in column **E**. Select the range **E2:E7** and type the formula for the mole balances followed by the keyboard combination *CTRL SHIFT ENTER*. Select the range **E8:E13** and type the formula for the vapor-liquid equilibrium relationship followed by *CTRL SHIFT ENTER*. Finally, add the formulas for the sums of the mole fractions in cells **E14** and **E15**.

	A	B	C	D	E
1	Feed Flow			Variables	Functions
2	F1	0.351396	x1	0.023287	=(F-x*L-y*V)
3	F2	0.039763	x2	0.006257	=(F-x*L-y*V)
4	F3	0.003144	x3	0.000981	=(F-x*L-y*V)
5	F4	0.106713	x4	0.146724	=(F-x*L-y*V)
6	F5	0.249491	x5	0.393451	=(F-x*L-y*V)
7	F6	0.249491	x6	0.429298	=(F-x*L-y*V)
8	Total F	=SUM(F)	y1	0.679522	=y-K*x
9	K-values		y2	0.073271	=y-K*x
10	K1	29.18	y3	0.005306	=y-K*x
11	K2	11.71	y4	0.066700	=y-K*x
12	K3	5.406	y5	0.105523	=y-K*x
13	K4	0.4546	y6	0.069675	=y-K*x
14	K5	0.2682	V	0.499986	=1-SUM(x)
15	K6	0.1623	L	0.500013	=1-SUM(y)
16			SSR		=SUMSQ(E2:E15)

Figure 6.23
Initial guesses and functions for isothermal flash calculations.

Add a formula in cell **E16** on the worksheet for the sum of the squares of the values in column **E** and try the **Solver** to find the values of the variables in column **D** to minimize the objective function **SUMSQ(E2:E15)** in cell **E16**. The **Solver** is unable to locate the roots from the starting points.

Employ the quasi-Newton option in the *VBA* user-form **ROOTS** to get the solution after six iterations. Observe that Newton's method appears to diverge during the initial iterations, and then manages to converge to the roots in this case. Alternatively, with some effort, we can use the **Solver** to find a solution by scaling the feed rates. Divide each feed rate term by the total feed rate for scaled values with order of magnitude one. Using scaled values for F and minimizing the sum of the squares of the functions, the **Solver** arrives at the correct results for x and y in Figure 6.24.

	A	B	C	D	E
1	Feed Flow			Variables	Functions
2	F1	0.351396338	x1	0.023287	1.32E-13
3	F2	0.03976327	x2	0.006257	-3.89E-13
4	F3	0.003144072	x3	0.000982	-5.39E-13
5	F4	0.10671352	x4	0.146724	-6.31E-13
6	F5	0.2494914	x5	0.393452	1.37E-13
7	F6	0.2494914	x6	0.429298	3.38E-14
8	Total F	1	y1	0.679523	-6.71E-13
9	K-values		y2	0.073271	-1.39E-12
10	K1	29.18	y3	0.005307	1.37E-12
11	K2	11.71	y4	0.066701	-4.01E-13
12	K3	5.406	y5	0.105524	2.92E-13
13	K4	0.4546	y6	0.069675	7.2E-14
14	K5	0.2682	V	0.499987	-4.04E-14
15	K6	0.1623	L	0.500013	1.33E-13
16			SSR		5.4E-24

Figure 6.24

Scaled values for *F*. The roots in column D were found by minimizing the sum of the squares of the functions.

Multiply the dimensionless flow rates in **D14:D15** by the total feed rate for $V = 2703$ lbmol/hr and $L = 2704$ lbmol/hr.

Comments: The quasi-Newton method in the macro **ROOTS** finds the result, whereas the **Solver** needs help with manual scaling of the variables to find the solution.

□

Occasionally, nonlinear functions need rearrangement, or variable transformation for the root finding methods to work. For example, a function that has a gentle slope near the root may cause Newton's method to diverge far from the root. Try logarithmic transformations, $\ln(y) = \ln[f(x)]$, inversion, $1/y = 1/f(x)$, or scaling, $z = c \cdot x$ to promote convergence, if necessary. We may need to remove discontinuities near the root. As an example, the terms $\ln(x)$ and $1/x$ are undefined as $x \to 0$. If possible, isolate the term with the discontinuity and transform the equation using $\exp[\ln(x)]$ or inversion. We may need to define a new variable, such as $z = \sqrt{x}$. Alternatively, we can add constraints to the solution method using methods from Section 7.8 (Shacham and Brauner 2002).

6.6 Initializing Iterative Methods and Equation Tearing to Promote Convergence*

Iterative, "trial and error," root-finding methods require initial guesses for the roots to begin the recursive calculations. Convergence by some methods may be more or less sensitive to the choice of initial guess and sequence and arrangement of equations. When confronted with difficult problems, take steps to alleviate some frustration by preparing the problem for the method of solution. Try the following techniques to narrow the region of good initial starting points:

1. Use initial guesses that make physical sense for the model. For example, if the variables represent mole fractions, start with initial guesses between zero and one, *e.g.*, 0.5.

2. As illustrated at the beginning of the chapter, try graphing the function then scanning the result to locate where it crosses the abscissa (for one or two-dimensional problems).

3. In case of nonlinear, higher-order polynomial functions, apply the method of truncation to solve for an approximate root (Constantinides and Mostoufi 1999). To illustrate, consider a second-order polynomial $0 = a + bx + cx^2$. We may truncate the constant where $|a| \ll |bx + cx^2|$, then rearrange for an approximation to the root: $x \cong -b/c$. In a similar manner, if $|a + bx| \gg |cx^2|$, then $x \cong -a/b$. Following this pattern, we

may truncate terms before and after adjacent ordered terms to solve for a root approximation: $x \cong -a_n/a_{n+1}$ where n indicates the order or exponent of the independent variable associated with the coefficient, a.

4. If the root may be close to zero, expand the nonlinear function in a McLauren series, then perform one Newton step with the root initialized at zero, or at some reasonable value from the first suggestion:

$$g_{init} = x = -\frac{f(0)}{df/dx|_{x=0}} \qquad (6.135)$$

5. Use variable transformation and equation changes to eliminate possible discontinuities, e.g. replace logarithms with exponential operations or invert equations to avoid division by zero.

6. Where possible, reduce the number of unknowns and equations by solving for any explicit variables followed by substitution of the explicit functions for these variables into the remaining equations.

7. Consider the relative magnitudes of each term in the function. Preserve the terms that dominate the function, while temporarily ignoring the less important terms. To illustrate, consider the following simple, non-linear function in x (which is just a different form of a quadratic equation):

$$f(x) = ax + \frac{b}{x} + c = 0 \qquad (6.136)$$

For large x, we may neglect the second term. At the opposite extreme, for small x, the first term in Equation (6.136) becomes negligible. Approximate solutions obtained under limiting conditions give reasonable choices for initializing an iterative solution method.

Limit	Equation (6.136) Approximate f(x)	Approximate x
Large x>>1	$f(x) \cong ax + c = 0$	$x \cong -c/a$
Small x<<1	$f(x) \cong \frac{b}{x} + c = 0$	$x \cong -b/c$

Tearing Method: To promote convergence, we may consider applying the tearing method that isolates a problematic nonlinear equation that tends to prevent Newton's method from converging on the roots of a simultaneous system of equations. The tearing method iterates on the following steps until all roots converge:

1. Identify the problematic nonlinear equation in the system and tear it away from the remaining equations.

2. Identify the torn *variable* in the most nonlinear term(s) of the torn equation, such as exponential, power, or logarithmic functions.

3. Solve for the single root of the torn equation by a robust single root-finding method, such as regula falsi, while holding the values of the non-torn variables constant.

4. Between successive regula-falsi iterations for the solution of the torn equation, solve the remaining system of $n-1$ equations for the non-torn roots by the faster quasi-Newton's method.

The *VBA* sub procedure **TEARING**, available in the *PNMSuite*, employs the quasi-Newton method to solve the non-torn equations within an iteration of the method of regula falsi for the torn equation. To use **TEARING**, setup a worksheet for solving a system of equations by Newton's method, with the equations and variables in two different ranges of contiguous cells, respectively. The torn equation and corresponding variable should be first or last in the range. The initial guess for the root of the torn variable serves as the value of the lower bracket in the regula falsi method used to find the root of the torn equation. The macro prompts for a value of the upper bracket on the torn variable.

Example 6.19 Solution to Nonlinear Equations by Tearing Method

Known/Find: Solve the following system of equations by the tearing method.

$$x_1^2 + 10x_2^2 - x_3 - 30 = 0$$
$$\frac{5}{x_1} + x_2^3 + x_3^{1.2} - 17 = 0 \qquad (6.137)$$
$$\frac{x_1 x_2}{3} + \frac{e^{-x_3^2}}{1000} - 9 = 0$$

Analysis: Choose x_3 and the third equation for the torn variable and equation, respectively. Setup a worksheet with initial guesses for the variables in column **B** and equations in column **C**. Row **4** contains the torn variable and equation.

	A	B	C
1	Variable	x	f(x)
2	1	1	=B2^2+10*B3^2-B4-30
3	2	1	=5/B2+B3^3+B4^1.2-17
4	3	1	=0.001*EXP(-B4^2)+B3*B2/3-9

Run the sub procedure **TEARING** with the following inputs:

    Torn Function = C4   n-1 Equations = C2:C3   Tolerance = 1E-8

    Torn Variable = B4   Upper Bracket = 5   n-1 Variables = B2:B3

The solution converges on a set of roots after 50 regula falsi iterations:

	A	B	C
1	Variable	x	f(x)
2	1	0.677909016	0
3	2	1.804167233	0
4	3	3.00975468	6.52101E-12

Comments: The tearing method converges slowly. Resort to equation tearing when other root finding methods fail.

□

6.7 Epilogue on Nonlinear Equations

> *"People assume that time is a strict progression of cause to effect. Actually, from a non-linear, non-subjective viewpoint, its more like a big ball of wibbly wobbly timey wimey...stuff."* — David Tennant (Dr. Who)

We have a smorgasbord of methods for finding roots to nonlinear equations. As we consider our options, we should note that each of these methods for finding roots to nonlinear equations share some common features:

- Methods use iteration on linear approximations for the function.
- Nonlinear problems may have multiple roots.
- Convergence may require good initial guesses for the root(s). We rely on our experience and engineering judgment to identify good starting points.
- Different equation sequences and arrangements may promote convergence.
- Convergence criteria are based on the change in variables between iterations (not function values).
- It may become necessary to scale variables to achieve convergence.
- Other than Bairstow's method, the methods presented here do not locate imaginary roots.

When problems arise while searching for the roots to a system of equations, try different solution methods. Table 6.4 compares the various methods according to their respective strengths and weaknesses. The secant or Padé extrapolation methods have advantages of computational efficiency when compared with ordinary iteration and the quasi-Newton method. We recommend the quasi-Newton method for small systems and Steffensen's method for large systems. The relatively robust methods of optimization in Chapter 7 may be used to find roots to nonlinear problems. In Chapter 12 we present a powerful version of Newton's method with continuation that transforms the system of equations into a system of differential equations for solution by numerical integration methods. Try Newton's method with continuation when other methods fail. Table 6.5 summarizes the user-defined functions and sub procedures or macros developed for solving nonlinear equations in the *PNMSuite* workbook.

Table 6.4   Comparison of methods for root finding.

Method	Strengths	Weaknesses
Bisection	• Guaranteed solution for continuously smooth functions	• Slow convergence • Requires brackets around the root.
Regula falsi (modified)	• Guaranteed solution for continuously smooth functions with a bracketed root • Faster convergence than Bisection	• May converge slowly for poor selection of a starting bracket around the root.
Secant	• Faster convergence • No derivatives required	• Requires two initial guesses • May diverge from poor initial guesses
Newton	• Requires just one initial guess • Fastest convergence • Solves systems of equations	• Requires good initial guesses • Requires derivative evaluations
Padé interpolation	• Fast convergence • No derivatives required	• Requires three starting points
Bairstow	• Finds real and imaginary roots for polynomials • Relatively fast convergence using a Newton style recursive method of factorization	• Not guaranteed to converge on every root of the polynomial.
Ordinary iteration with Relaxation	• Only requires one initial guess. • Easy to set up • Solves systems of equations • May promote convergence	• Slow convergence with small relaxation factors. • May diverge from poor starting points or improper equation arrangement.
Wegstein	• Usually controls divergence • May Accelerate convergence	• No guarantee of convergence • Primarily limited to single equations
Aitken's Δ^2 (Steffensen)	• Automates relaxation • Controls convergence	• Works best for linear or nearly linear problems.
Method of Continuation (See Section 12.8)	• Robust convergence for poor equation arrangement and initial guesses	• Requires derivative approximations • Computationally expensive
Method of Tearing	• Focus the solution on a single difficult nonlinear equation in a system of equations.	• Slows down the rate of convergence on the roots.

Table 6.5   Summary of VBA sub procedure macros, user-forms, and user-defined functions (UDF) for solving nonlinear equations, including the input methods for initial guesses and functions.

Macro/Function	Inputs	Operation
BAIRSTOW POLYROOTS	Macro input boxes UDF array argument	Bairstow's method of polynomial factorization for finding n real and imaginary roots of an nth order polynomial
BISECTION	Macro input boxes	Bisection method of root finding between brackets
JACOBIAN	Macro input boxes	Calculates the Jacobian using finite difference derivative approximation
NEWTCON	Macro input boxes	Newton's method of continuation for nonlinear systems (See Section 12.8)
NEWTON	UDF arguments	Find a root of a formula by the quasi-Newton method.
PADEROOT	UDF arguments	Find a root of a function by Padé rational interpolation.
REGFAL REGULAFALSI	UDF arguments Macro input boxes	Find a root of a formula by the method of regula falsi.
ROOT	UDF arguments	Newton, Secant, or Muller's methods for finding the root of an *Excel* worksheet *formula*
ROOTS_UsrFrm	Show user-form macro	Options for quasi-Newton, Gauss-Newton, Levenberg-Marquardt, method of continuation, or Powell with either Quadratic of Golden Section search for simultaneous equations
ROOTSGN ROOTSLM	Macro input boxes	Gauss-Newton and Levenberg-Marquardt methods of finding roots to single or systems of nonlinear equations.
SECANT	Macro input boxes	Find a root to a function by the secant method.
STEFFENSEN	Macro input boxes	Steffensen's ordinary iteration method accelerated by Aitkin's delta square method.
TEARING	Macro input boxes	Method of equation tearing that finds the root of the torn nonlinear equation by regula falsi and the roots of the remaining equations by Newton's method after each iteration of regula falsi.

Chapter 7   Optimization

"... alas! alas! life is full of disappointments; as one reaches one ridge there is always another and a higher one beyond which blocks the view."
— Fridtjof Nansen (in "The First Crossing of Greenland")

While ascending a mountain, a climber may choose to reach the top of any of several peaks in a range. However, only one peak may claim the distinction of the pinnacle, with the maximum height in the range. This chapter introduces optimization methods for locating maxima or minima, i.e., the *best*, or *optimum*, solution to underspecified model equations that may have an infinite number of *viable* solutions. We refer to mathematical optimization models as objective functions that may consist of a single equation or a system of underspecified equations.[60] Methods of optimization go by other names, including linear and nonlinear programming.[61] We typically need to optimize a process to maximize profit, minimize cost, or maximize production, within limitations that constrain the feasibility of the process. Optimization methods presented in this chapter fall into two principle categories:

1. Direct search methods for single and multivariable objective functions, including:
   - Graphical search for one or two dimensional problems
   - Powell's and the downhill Simplex method for multidimensional nonlinear optimization
   - Simulated Annealing, pattern search, Luus-Jaakola, and the bio-inspired Particle Swarm and Firefly stochastic search algorithms for the global optimum of nonsmooth or multimodal objective functions
   - Genetic evolutionary algorithm in *Excel*'s **Solver**
2. Indirect search methods that rely on derivative approximations in *Excel*'s **Solver**:
   - Quasi Newton's method
   - Conjugate gradient or generalized reduced gradient methods

> **SQ3R Focused Reading Questions**
> 1. What is the calculus minimum or maximum?
> 2. What is the difference between direct and indirect search?
> 3. What is the difference between a local and global optimum?
> 4. When should I use a stochastic search method?
> 5. Is there any guarantee for finding the optimum?
> 6. What optimization tools are available in *Excel*?
> 7. How does a graph of the objective function help with optimization?
> 8. Why do I need to restart the search for the optimum after finding a solution?
> 9. When do I need to constrain the search for the optimum?
> 10. Why is sensitivity analysis important for optimization?

   A simple illustration compares a direct to an indirect approach to finding the optimum of an objective function. When an engineer moves to a new city, an objective function defines the viable travel routes between her home and

---

[60] See Degrees of Freedom in Section 1.2.5.
[61] *Programming* in this sense has no direct connection to computer coding, though we use computation to implement our solution strategy. Rather, *programming* refers to the selection of steps to reach an optimum solution. For example, consider television *programming* where producers match advertising with a show's content to maximize exposure to target audiences.

work site. She may look on a map for the shortest distance. On the other hand, she may not be concerned with the distance to work, but instead want to find the route that gets her home from work in the shortest amount of time. After a few days of trying her initial route, she notices potential short cuts not identified on the map. She tries a few short cuts and discovers a path that shortens the distance or time of travel. Perhaps her travel route has constraints. There may be a tollbooth, school zones, road construction, or potholes in a road that she wants to avoid, thus minimizing time and expense associated with getting stuck in slower traffic or damage to her vehicle. After a few weeks of direct search by trial and error, she determines the optimum route that satisfies her travel requirements. Alternatively, the engineer can collect information about the various routes and indirectly determine the optimum route without the need of actually traveling any alternative routes. The indirect approach may be more efficient depending on the effort required to collect and use information about the alternative routes. We use these two basic approaches to develop computational techniques for optimizing mathematical functions within constraints.

It may not be possible to quantify an optimization problem in terms of numbers and equations. In the introductory example of finding the best route to and from work, it is difficult to quantify the value of pleasant scenery, a favorite restaurant, or light traffic on a particular route. Fortunately, most problems of interest to chemical engineers *are* quantifiable. We present methodologies for optimizing objectives described in terms of a mathematical function of adjustable independent variables.

Some examples of optimization related to chemical engineering include finding the economic pipe diameter for fluid transport and distribution, reactor design conditions to minimize unwanted side reactions, chemical equilibrium, or optimizing energy integration in a chemical process. In Chapter 9, we present the method of least-squares regression for optimizing the fit of a model to a data set. Often, the objective is to minimize cost or maximize profit through savings in time, capital expenditures, or energy.

7.1 Objective Function

Methods of optimization apply equally whether or not the objective is to find the maximum or minimum. Objective functions are "reversible" from a minimum to maximum by simply changing the sign on the objective function:

$$f(x)_{min} \rightarrow -f(x)_{max} \tag{7.1}$$

In the case of optimization in several dimensions, combine a system of objective functions into a single function by summing their squares:

$$f_T(x_1, x_2, x_3, \ldots) = \sum_i f(x_1, x_2, x_3, \ldots)_i^2 \tag{7.2}$$

We used the sum of squares in Section 6.3.5 with the **Solver** feature of *Excel* to find the roots of a system of nonlinear equations. We use the sum of squares in Chapter 9 to find model parameters that minimize the sum of squared residuals between experimental and model values.

Example 7.1   Optimum Team Size [Adapted from Goldberg (2006)]

Known: A project manager forms a team of n engineers to implement a complex project.

Find: What team size minimizes the time needed to complete the project?

Analysis: The manager invites the engineers one at a time to propose solutions – this is the creative step. The engineers and manager then vote for the best idea – this is the critical thinking step. Each engineer has the same time to propose a solution for a total proposal time of $n \cdot t_1$. The time needed to implement the best solution is t_2. If the project manager divides the proposal and project work equally among the n team members, the total time needed to complete the project becomes:

$$t = nt_1 + \frac{t_2}{n} \tag{7.3}$$

Figure 7.1 shows a plot of the model function for $t_1 = 5$ and $t_2 = 50$ time units.

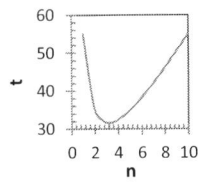

Figure 7.1

Plot of time, *t*, versus team size, *n*.

By inspection, the plot of the model indicates an optimal team size of three engineers.

Comments: The total time is slightly more sensitive to the team size below three members, as seen from a comparison of the slope of the curve on either side of the optimum.

7.2   Define the Optimization Problem

> *"To define it rudely but not inaptly, engineering is the art of doing that well with one dollar which any bungler can do with two."* — Arthur M. Wellington

The mathematical objective function for optimization may consist of a single equation, or a coupled system of equations that require simultaneous solution. Indirect search methods require continuous objective functions, whereas some direct search methods work for discrete functions, as well. Optimization is the process of finding the combination of values for the independent variables that minimizes or maximizes the objective. Objective functions may have local optima, and a global optimum value. Unimodal objective functions have only one optimum value within the constraints, such as the model plotted in Figure 7.1. Bimodal, or multimodal objective functions have local optimum solutions, and a global (best) optimum. Figure 7.2 is an example of a bimodal objective function for maximization.

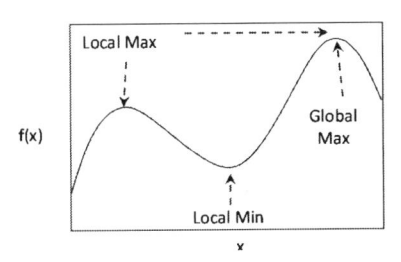

Figure 7.2

Example of bimodal maxima in an objective function.

The development of numerical methods guaranteed to find the true overall global optimum is an unmet challenge. For complex problems, we are never entirely sure that the true global optimum has been located without searching through every possible combination of variables – this can take a long time. Despite this challenge, we present several optimization methods that find a local optimum. We attempt to locate the global optimum by searching over the region of feasibility for the problem and comparing the solutions for local optima.

Before we can solve an optimization problem, we need to define the problem by identifying the objective function and variables. The following steps simplify the process:

1. Define the objective function by answering the question, "What are we trying to minimize or maximize?" For example, minimize costs, maximize profits, minimize time, or minimize waste and pollution.

2. Determine the optimization variables available in the problem, *e.g.*, space or time dimensions, mass rates or concentration, pressure, and temperature. We typically do not use physical properties, such as density, viscosity, or heat capacity as optimization variables because they are not controlled. However, we may use properties when selecting different materials. It is easier to identify the variables of optimization once we have the objective function. For example, suppose we are interested in maximizing profit. The objective function includes the selling price of our product and the cost of manufacturing:

$$\text{Profit} = (\$\,\text{Product}) \cdot (\text{Product}) - (\$\,\text{Materials}) \cdot (\text{Materials}) - (\$\,\text{Utilities}) \cdot (\text{Utilities})$$

In this example, the variables of optimization may be the amount of product and materials.

3. Impose any constraints on the problem parameters, such as those required to satisfy laws of conservation, chemical process safety, or limitations on resources.
4. Find the operating conditions of rates of production and time of production form the variables of optimization, e.g., Product = Rate × Time.

7.3 Calculus Maximum versus Minimum

Recall from calculus that we can locate an optimum value at the point where the derivative of the function is zero:

$$\left.\frac{df(x)}{dx}\right|_{extremum} = 0 \tag{7.4}$$

We identify the optimum value as a minimum, maximum or saddle point by evaluating the second derivative, as shown in Table 7.1.

Table 7.1 First and second derivative optimization information. *f* = objective function, *x* = variable.

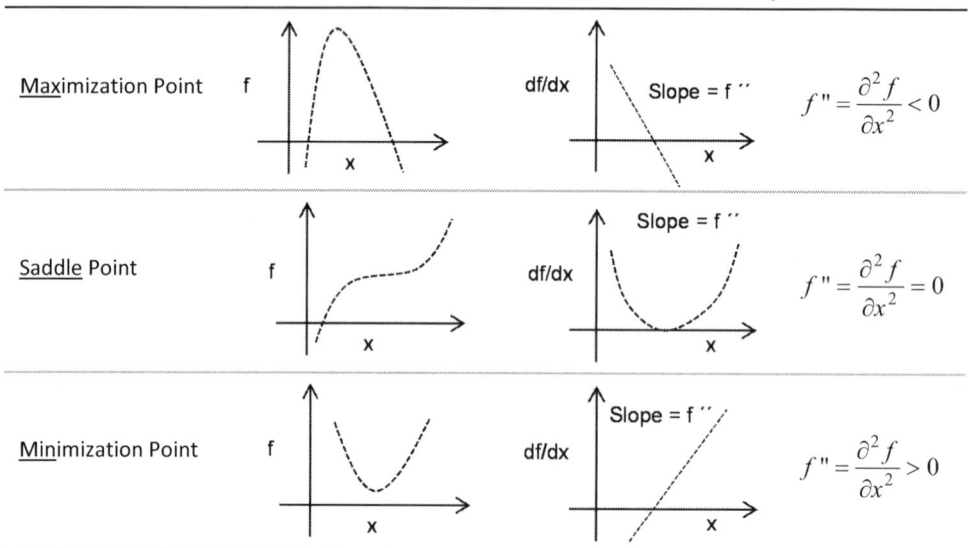

Maximization Point	*f* vs *x*	*df/dx* vs *x*, Slope = f″	$f'' = \dfrac{\partial^2 f}{\partial x^2} < 0$
Saddle Point	*f* vs *x*	*df/dx* vs *x*, Slope = f″	$f'' = \dfrac{\partial^2 f}{\partial x^2} = 0$
Minimization Point	*f* vs *x*	*df/dx* vs *x*, Slope = f″	$f'' = \dfrac{\partial^2 f}{\partial x^2} > 0$

Using derivatives for indirect search provides an alternative definition of optimization as the process of finding the values for *x* where the first derivative of the objective function in Equation (7.4) goes to zero, which satisfies the desired second derivative from Table 7.1.

The derivative of Equation (7.3) gives an analytical solution for the team size that minimizes the solution time:

$$\frac{\partial t}{\partial n} = t_1 - \frac{t_2}{n^2} = 0 \tag{7.5}$$

Rearrange Equation (7.5) for the optimum team size *n*:

$$n_{min} = \sqrt{t_2/t_1} \tag{7.6}$$

Equation (7.6) provides additional information about the sensitivity of the team size to the ratio of solution and implementation times, as shown in the plots in Figure 7.3. The log-log plot at the right shows how the minimum team size increases by one order of magnitude with two order-of-magnitude changes in the solution time ratio.

CHAPTER 7: OPTIMIZATION

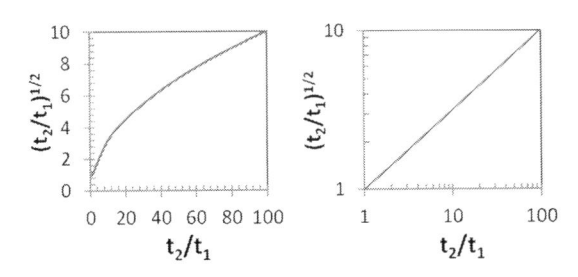

Figure 7.3

YX and log-log plots showing the sensitivity of team size to variation in the time ratio in standard and log plots.

The second derivative confirms that the result is a minimum for all *implementation* times and team sizes:

$$\frac{\partial^2 t}{\partial n^2} = \frac{2t_2}{n^3} > 0 \tag{7.7}$$

Example 7.2 *Multimodal Optimization by Graphical Analysis*

Known: Deb and Saha (2010) used the following multimodal optimization function to test various search methods of optimization.

$$f(x) = 1 - \exp(-x^2)\sin^2(2\pi x) \tag{7.8}$$

Find: Locate any local and global maxima and minima in the range $0 \leq x \leq 1$.

Analysis: Calculate the derivatives as follows:

$$\frac{df(x)}{dx} = 2\exp(-x^2)\left[x\sin^2(2\pi x) - 2\pi\cos(2\pi x)\sin(2\pi x)\right] \tag{7.9}$$

Generate points for the graphs using **Data> What-if Analysis> Data Table** in *Excel*. Graph the function in Equation (7.8) and its derivatives in Equation (7.9) as seen in Figure 7.4

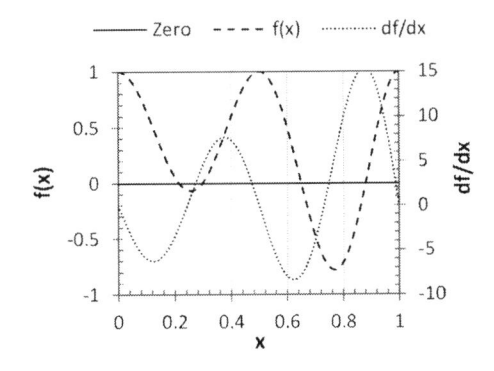

Figure 7.4

Function and derivative showing the locations of the global and local maxima and minima for Equation (7.8).

By inspection of the *f* plot, we identify three maxima and two minima in $0 \leq x \leq 1$. The *df/dx* plot also indicates the location of optima where the curve crosses the zero axis. The second derivative, or slope of the *df/dx* curve is negative at $x = 0, 0.5$, and 1, indicating three maxima with equivalent values of $f(x) = 1$. The slope of the *df/dx* curve is positive for local and global minima at $x = 0.245$ and 0.735, respectively.

Comments: We recommend starting with the graphical search method for one-dimensional problems.

Example 7.3   Optimal Cyclone Diameter from the Calculus Minimum

Known: The annual operating cost of a multiple cyclone system for particulate emission control is a function of the cyclone body diameter, D:

$$AOC = \frac{K_1}{D} + K_2 D^2 + \frac{K_3}{D^3} \quad (7.10)$$

Find: Determine the optimal diameter for a system to remove 10 μm particles with 95% efficiency where $K_1 = \$1000$ m/yr, $K_2 = \$1 \times 10^6$ /yr m², and $K_3 = \$100$ m³/yr.

Analysis: Set the derivative of Equation (7.10) to zero and multiply each term in Equation (7.11) by $-D^4$ to give a 5th order polynomial:

$$\frac{dAOC}{dD} = 0 = -\frac{K_1}{D^2} + 2K_2 D - \frac{3K_3}{D^4} \quad \rightarrow \quad 3K_3 + K_1 D^2 - 2K_2 D^5 = 0 \quad (7.11)$$

Use the VBA macro **BAIRSTOW** described in Section 6.4 to find the diameter from the roots: $D = 0.175$ m.

	A	B	C	D	E	F	G	H	I
1	K_1	1000		Polynomial Coefficients			$r_1 =$	0.175260944	
2	K_2	1.00E+06		D_0	-300		$r_2 =$	-0.138005963	-1.0425E-1 i
3	K_3	100		D_1	0		$r_3 =$	-0.138005963	+1.0425E-1 i
4				D_2	-1000		$r_4 =$	0.050375491	-1.6147E-1 i
5				D_3	0		$r_5 =$	0.050375491	+1.6147E-1 i
6				D_4	0				
7				D_5	2.00E+06				

Comments: The polynomial in Equation (7.11) has just one real, practical root at $D = 0.175$.

□

7.4   Indirect Search and Newton's Method

> *"Engineering problems are under-defined, there are many solutions, good, bad and indifferent. The art is to arrive at a good solution. This is a creative activity, involving imagination, intuition and deliberate choice."* – Ove Arup

In Section 7.3 we gave the calculus definition for optimization as the method of finding the values for x that satisfies Equation (7.4) and the criteria in Table 7.1. We refer to optimization methods that involve derivative evaluations as indirect searches, as opposed to direct search methods that compare values for the objective function. A function is considered smooth when it is twice differentiable at all points in the domain of optimization. Indirect search methods make extensive use of derivatives to locate the optimum point – as such, they do not work for nonsmooth functions.

Example 7.4   Indirect Optimization of Rosenbrock's in Two-dimensions

Known/Find: Use first derivative evaluations to find the values of x_1 and x_2 that minimize the following function:

$$f(x_1, x_2) = 10\left(x_2 - x_1^2\right)^2 + (1 - x_1)^2 \quad (7.12)$$

Analysis: Solve the following partial derivative equations simultaneously for $x_1 = 1$ and $x_2 = 1$.

$$\frac{\partial f}{\partial x_1} = x_1\left[2 - 40\left(x_2 - x_1^2\right)\right] - 2 = 0 \quad \text{and} \quad \frac{\partial f}{\partial x_2} = 20\left(x_2 - x_1^2\right) = 0 \quad (7.13)$$

Comments: Complicated engineering problems may require finite-difference derivative approximations.

□

CHAPTER 7: OPTIMIZATION

We may use any of the nonlinear equation solving techniques presented in Chapter 6 to find solutions to a one dimensional optimization problem. We recognize that the first derivative of an objective function equals zero at a local optimum point, as defined in Equation (7.4). An optimum of an objective function is just the root of the equations for the first derivative of the function. For example, the recursive form of Newton's method applied to the optimization problem $df(x)/dx = 0$ starts with a linear Taylor series expansion about a guess for the location of the optimum:

$$0 = \left.\frac{df}{dx}\right|_g + (x-g)\left.\frac{d^2f}{dx^2}\right|_g \tag{7.14}$$

where g is an approximation, or "guess," for the optimum solution. Solve Equation (7.14) for x, and then use the new value for x as the guess for the next iteration of the following recursive formula:

$$x_i = x_{i-1} - \frac{\left.df/dx\right|_{x_{i-1}}}{\left.d^2f/dx^2\right|_{x_{i-1}}} \tag{7.15}$$

We may replace the derivatives with finite-difference approximations, derived in Chapter 5. Use the sign of the second derivative to check for minimum versus maximum results. For example, the sign of the second derivative is positive at a location for the minimum. The location of the initial guess may be important whether searching for the minimum or maximum. Consider the function plotted in Figure 7.5 where the plot of the first derivative shows two roots; the lower value corresponds to the local minimum and the upper value corresponds to the local maximum. Depending on the location of the initial guess for the optimum, Newton's method may converge to one optimum (*min*) or the other (*max*).

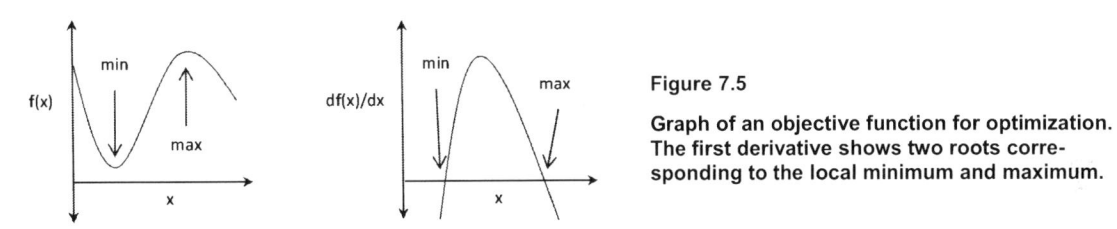

Figure 7.5

Graph of an objective function for optimization. The first derivative shows two roots corresponding to the local minimum and maximum.

The extension of Newton's method to multidimensional optimization problems requires the iterative solution to a system of Newton equations:

$$x_i = x_{i-1} - \left(\left.\frac{\partial^2 f}{\partial x^2}\right|_{x_{i-1}}\right)^{-1}\left(\left.\frac{\partial f}{\partial x}\right|_{x_{i-1}}\right) \tag{7.16}$$

where x represents a vector of variables, and where the matrices of derivatives become:

$$\frac{\partial f}{\partial x} = \begin{vmatrix} \partial f/\partial x_1 \\ \partial f/\partial x_2 \\ \vdots \\ \partial f/\partial x_n \end{vmatrix} \quad \text{and} \quad \frac{\partial^2 f}{\partial x^2} = \begin{vmatrix} \frac{\partial^2 f}{\partial x_1^2} & \frac{\partial^2 f}{\partial x_1 \partial x_2} & \cdots & \frac{\partial^2 f}{\partial x_1 \partial x_n} \\ \frac{\partial^2 f}{\partial x_1 \partial x_2} & \frac{\partial^2 f}{\partial x_2^2} & \cdots & \cdots \\ \vdots & \vdots & \ddots & \vdots \\ \frac{\partial^2 f}{\partial x_n \partial x_1} & \frac{\partial^2 f}{\partial x_n \partial x_2} & \cdots & \frac{\partial^2 f}{\partial x_n^2} \end{vmatrix} \tag{7.17}$$

The **Solver** has an option for a variant of Newton's method called *Generalized Reduced Gradient* (GRG), described in Section 7.7.

Example 7.5 Newton's Method of Optimization

Known: The objective function for this example is a simple quadratic equation, $f = x^2 - x$, with first and second derivatives.

Find: Calculate the minimum using Newton's method.

Analysis: We employ Newton's method of one-dimensional optimization to arrive at the solution for the minimum after one iteration:

$$f = x(x-1) \qquad \frac{df}{dx} = 2x - 1 \qquad \frac{d^2 f}{dx^2} = 2 \qquad x - \frac{df/dx}{d^2 f/dx^2} = \frac{1}{2} \qquad (7.18)$$

Comments: Because the original objective function is quadratic, Newton's method gives the exact result in a single step. Typical nonlinear objective functions approach quadratic behavior near the optimum solution.

□

7.5 Direct Search Methods

> *"If each of your time steps is one week long, you are not modeling the stock price terribly well over a one-week time period, because you are saying that there are only two possible outcomes."* – John Hull

The simplest direct search method discretizes the dimensions of the objective function into a large grid (with many nodes at intersections), as shown in Figure 7.6 for a two-dimensional problem, and then compares the solution of the objective function at each node to determine the vicinity of the optimum. Each node consists of a different combination of values for the independent variables within any constraints. This type of direct search approach may miss the optimum if the node spacing is too coarse. Therefore, the node spacing determines the precision of the result. When selecting the node spacing, we seek an acceptable balance between precision of solution and speed of computation. Typically, the node spacing is set to half the desired solution precision to avoid missing the optimum point. We systematically compare the results for the objective function at each node. We retain the node that generates the local optimum value of the objective function for future comparisons as the search continues, until we have compared all node results. Depending on the problem, the grid search for the optimum may require an unreasonably large set of nodes and corresponding objective function evaluations.

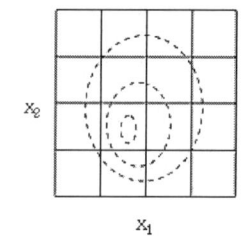

Figure 7.6

Discretization of the variable space for objective function evaluation. The dashed lines are contours of function values.

Techniques employed to refine the solution include the use of successively smaller node spacing and quadratic interpolation of the function values near a suspected optimum. A uniform grid may bias the search. To eliminate any unintended bias, select random points within the constraints on the objective function.

7.5.1 Graphical Direct Search for One or Two Dimensions

We may find a quick route to the general vicinity of the optimum through graphical analysis in one or two-dimensional grids. This technique works best for one-dimensional objective functions, and does not apply when the dimension of the problem exceeds two independent variables. We used graphical analysis to find the optima in Example 7.1 and Example 7.2. Where possible, use the approximate graphical solution as the starting point for more sophisticated numerical search methods. In the case of two-dimensional functions, use surface and contour plots in concert to locate the optimum. A surface plot distinguishes between global and local optima, but may be difficult to read for identifying the coordinates of the optimum. A contour plot shows the existence as well as location of the optimum, but may not so easily differentiate between minimum and maximum points.

A well-constructed contour plot gives the same information as a three dimensional plot. A typical example is a topographical map that shows elevation, steepness of the peaks, *etc*. Contour lines close together indicate steeper terrain. Figure 7.7 shows the topography of Chester Bowl between the author's home and the campus of the University of Minnesota Duluth.[62] *Excel* adds a color scale to the contours to indicate the relative magnitudes of each contour.

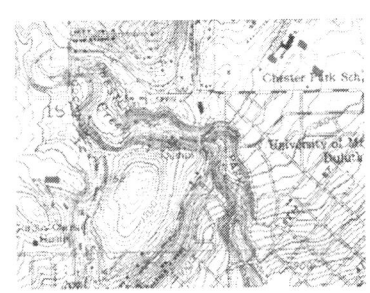

Figure 7.7

USGS topographical map of Chester Creek Park near the campus of the University of Minnesota Duluth.

Example 7.6 Graphical Search for the Global Minimum of a Test Function

Known: Michalewicz (2000) proposed a two-dimensional function to test methods of global optimization:

$$f(x_1, x_2) = -\sin(x_1)\sin^{20}\left(\frac{x_1^2}{\pi}\right) - \sin(x_2)\sin^{20}\left(\frac{2x_2^2}{\pi}\right) \tag{7.19}$$

Find: Use surface and contour plots to find the location of the global minimum.

Analysis: We generate points for plotting by creating a **Data Table** from Equation (7.19). The values for x_1 and x_2 range from one to three. To get a sense of the relative local minima and maxima, we used **Home>Styles>Conditional Formatting>Color Scales** on the cells in the table. We find the approximate location of the global minimum at $x_1 = 2.2$, $x_2 = 1.55$.

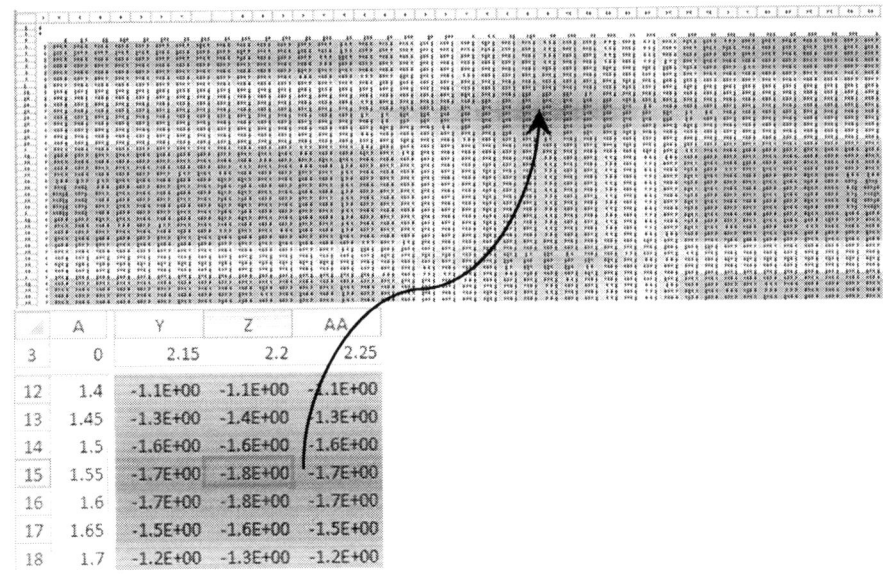

---

[62] It is an uphill climb to work in the mornings followed by a relaxing downhill stroll through the park in the evenings (except in the middle of an infamous northern Minnesota winter!).

A surface plot in Figure 7.8 created in an *Excel* worksheet reveals the presence of several local minima, and the global minimum. To change the fine scale size for either surface or contour plots, adjust the major scale of the vertical axis.

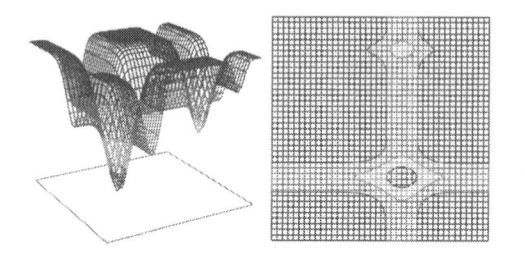

Figure 7.8

Surface and contour plots of Michalewicz's multimodal objective function showing local and global minima. See the example file on the book's website for a color view.

We locate the minimum in the contour plot by inspection at $x_1 \sim 2.2$, $x_2 \sim 1.6$. *Excel* uses colors to show the relative values of the contours, as seen in the electronic file for this example (available from the book's companion web site). The contour plot shows the location of the local optima.

Comments: Use surface and contour plots in concert to identify minima versus maxima and to locate the global optimum. The surface plot shows the relative sensitivity of the function to changes in the variables. Narrow color bands indicate a steeper gradient in the function. Wider bands farther apart indicate a region of relatively constant value of the function. The contour plot becomes a useful tool in sensitivity analysis as discussed in Section 7.11. Regions with steeper gradients, as indicated by tightly spaced contour lines, are more sensitive to changes in the variables than regions with small gradients.

□

7.5.2 Bracketing with Quadratic Interpolation

We learned in Chapter 5 that a Taylor series does a good job of approximating a function for values of the independent variable close to the expansion point. Let us use a simple truncated second-order Taylor series to approximate our objective function:

$$f(x) \cong f(g) + \frac{(x-g)}{1!}\frac{df}{dx}\bigg|_g + \frac{(x-g)^2}{2!}\frac{df}{dx}\bigg|_g \qquad (7.20)$$

where g is our initial guess for the value of x that optimizes $f(x)$. We recognize that Equation (7.20) as a quadratic equation in x, which we simplify to the form:

$$f \cong a + bx + cx^2 \qquad (7.21)$$

We may use a quadratic function to approximate any continuous and twice differentiable objective function near the optimum. We take advantage of this feature by cycling through two stages:

1. Perform a systematic search to bracket the optimum values of x and $f(x)$.
2. Use a quadratic function to predict the location for the optimum.

Bracketing the optimum has the advantage that it limits the search region for quadratic interpolation and guarantees finding an optimum point for smooth, continuous functions. A smaller search region ultimately requires fewer objective function evaluations. To launch a search for the bracket, begin at appropriate starting points in the independent variable, then evaluate the objective function at increasing values for the independent variable until finding values that bracket the optimum. For example, beginning with two points, we may double the step size in the search after each step, evaluating the objective function at each point as we go:

$$x_i = x_{i-1} + 2(x_{i-1} - x_{i-2}) = 3x_{i-1} - 2x_{i-2} \qquad (7.22)$$

where the subscript i is the index of the search step. It may become necessary to take smaller steps or switch search direction if a step moves away from the location of the optimum, i.e., $f(x_i) > f(x_{i-1})$ for minimization. At each

CHAPTER 7: OPTIMIZATION

step, compare the value of the objective function with the result from the previous step. We recognize a bracket around the optimum when the sign of the slope of the objective-function changes between x brackets:

$$[f(x_i) - f(x_{i-1})][f(x_{i-1}) - f(x_{i-2})] < 0 \qquad (7.23)$$

When three consecutive points satisfy Equation (7.23), the bracketing search procedure stops and the last two points determine the bracket. The starting point for the search or length of search steps may need adjustment to find the solution bracket. If necessary, repeat the search within the new bracket, starting with a smaller step size to reduce the bracket size further.

To speed up the search, we predict the location of the optimum value within the bracket by fitting the last three search points to a quadratic interpolating polynomial like Equation (7.21). Solve the following three quadratic equations evaluated at each point simultaneously for the coefficients a, b, and c:

$$f_i = a + bx_i + cx_i^2 \qquad f_{i-1} = a + bx_{i-1} + cx_{i-1}^2 \qquad f_{i-2} = a + bx_{i-2} + cx_{i-2}^2 \qquad (7.24)$$

Parabolic functions are only twice differentiable, and conveniently have just one maximum or minimum point. We find an improved prediction for the optimum location from the calculus definition of a minimum by first evaluating the derivative of the quadratic function, then solving for x to give an interpolated value for the location of the optimum point:

$$\frac{df}{dx} = 0 \cong b + 2cx \quad \rightarrow \quad x \cong \frac{-b}{2c} \qquad (7.25)$$

We do not need the second derivative because the bracket search identified the point as a minimum or maximum. The result from the three points in Equations (7.24) gives the following estimate for the location of the optimum value for the objective function in the next iteration:

$$x = \frac{(f_{i-2}) \cdot (x_i^2 - x_{i-1}^2) + (f_{i-1}) \cdot (x_{i-2}^2 - x_i^2) + (f_i) \cdot (x_{i-1}^2 - x_{i-2}^2)}{2[(f_{i-2}) \cdot (x_i - x_{i-1}) + (f_{i-1}) \cdot (x_{i-2} - x_i) + (f_i) \cdot (x_{i-1} - x_{i-2})]} \qquad (7.26)$$

After each iteration, update the values for the next iteration: $x_{i-2} = x_{i-1}$, $x_{i-1} = x_i$, and $x_i = x$ until converging on the optimum. We may use the interpolated result as the starting point for further refinement of the location for the optimum by iterating between the bracketing search and the quadratic method. We employ the accelerated quadratic interpolation method in the VBA macro **POWELL** described in Section 7.5.5.

Example 7.7 Maximum Specific Growth Rate of Yeast

Known: The specific growth rate of yeast is a function of the substrate concentration, S:

$$\mu = \frac{3S}{5 + 0.9S + 2S^2 + 0.3S^3} \qquad (7.27)$$

Find: Use accelerated bracket search with quadratic interpolation to calculate the substrate level that yields the maximum specific growth rate, μ_{max}.

Analysis: Start with the accelerated search method of Equation (7.22) to bracket the location of the maximum. The *Excel* worksheet in Figure 7.9 shows three trajectories of the bracketing method with quadratic acceleration for locating the optimum. The first search begins with $S = 0.01$ and an initial step size of 0.005 in rows three and four. The *Excel* worksheet formulas for columns **A** and **B** use relative referencing:

A5 =3*A4-2*A3   B4 =3*A5/(5+0.9*A5+2*A5^2+0.3*A5^3)

Columns **A** and **B** contain the results for the substrate concentration and specific growth rate. The shaded cells contain the bracket for the optimum according to Equation (7.23), $0.645 < S < 2.565$. Approximate the location and value of the maximum specific growth rate at $S = 1.6$ using a quadratic interpolating function:

B16 = 0.5*((A12^2-A11^2)*B10+ (A10^2-A12^2)*B11+ (A11^2-A10^2)*B12)/ ((A12-A11)*B10+ (A10-A12)*B11+ (A11-A10)*B12)

$$B17 = 3*B16/(5+0.9*B16+2*B16^2+0.3*B16^3)$$

Column **C** uses Equation (7.23) to locate where the sign of the slope changes to indicate the location of a pair of bracket points around the maximum. In the first trajectory, the sign changes in row **11**:

$$C11 = \text{SIGN}((B11-B10)*(B12-B11))$$

In the second trajectory, we refine the search by restarting at the lower end of the bracket from the first trajectory. The new, narrower bracket becomes $0.96 < S < 1.92$. The quadratic interpolation gives $S_{max} = 1.41$. The third trajectory starting at $S = 0.96$ predicts the maximum at $S_{max} = 1.35$. We repeat the process until the solution converges within a specified tolerance. After several trajectories, we arrive at the maximum $S_{max} = 1.34$.

	A	B	C	D	E	F	G	H	I
1	Iteration 1			Iteration 2			Iteration 3		
2	S	µ	Sign	S	µ	Sign	S	µ	Sign
3	0.01	0.005989		0.645	0.298011		0.96	0.361236	
4	0.015	0.008975	1	0.65	0.299429	1	0.965	0.361851	1
5	0.025	0.014929	1	0.66	0.302223	1	0.975	0.363047	1
6	0.045	0.026761	1	0.68	0.30764	1	0.995	0.365313	1
7	0.085	0.050087	1	0.72	0.317798	1	1.035	0.369354	1
8	0.165	0.095114	1	0.8	0.335495	1	1.115	0.375598	1
9	0.325	0.176821	1	0.96	0.361236	1	1.275	0.381716	-1
10	0.645	0.298011	1	1.28	0.381788	-1	1.595	0.375563	1
11	1.285	0.381853	-1	1.92	0.355026	1	2.235	0.329464	1
12	2.565	0.301414	1	3.2	0.251372	1	3.515	0.229726	1
13	5.125	0.14996	1	5.76	0.12908	1	6.075	0.120266	1
14	10.245	0.056216	1	10.88	0.051167	1	11.195	0.048912	1
15									
16	S_{max}	1.613778		S_{max}	1.410717		S_{max}	1.354701	
17	$µ_{max}$	0.374665		$µ_{max}$	0.381545		$µ_{max}$	0.382126	

Figure 7.9

Excel **worksheet implementation of accelerated search with quadratic interpolation for the maximum of Equation (7.27).**

The first plot in Figure 7.10 uses the points of the first accelerated search and the quadratic fit of the last three points bracketing the maximum for three iterations. The second graph has plots of the three trajectories, showing how the method converges on the optimum location and value.

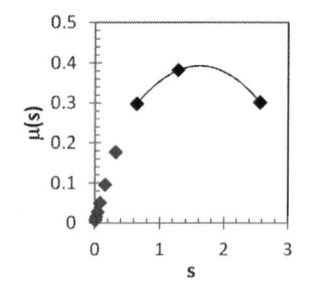

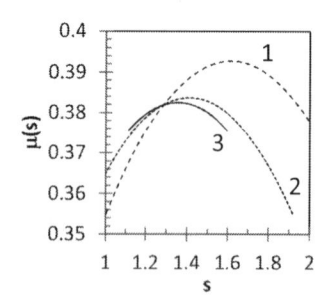

Figure 7.10

First iteration steps and comparison of three trajectories of the quadratic direct search with accelerated search for the bracket.

Comments: Quadratic interpolation quickly narrows the region containing the optimum for objective functions that behave quadratically near the optimal point.

□

7.5.3 Response Surfaces*

Response surfaces have application in statistical design of experiments and sensitivity analysis. Figure 6.18 is an example of a two dimensional response surface. We generate a multidimensional response surface by fitting the function to a second-order multivariable Taylor expansion. For example, over a small range of x, we can approximate a two variable function with a second order Taylor expansion about a point near the optimum:

CHAPTER 7: OPTIMIZATION

$$f(x_1, x_2) \cong a + b_1 x_1 + c_1 x_1^2 + b_2 x_2 + c_2 x_2^2 + d x_1 x_2 \tag{7.28}$$

This is essentially the same approach used for quadratic interpolation in Section 7.5.2. We get the values of the parameters a, b_1, b_2, c_1, c_2, and d from six unique sets of data points f, x_1, x_2. Equation (7.28) includes nonlinear and confounding effects of the two variables in the second order terms. The calculus optimum gives two equations in two unknowns, x_1 and x_2:

$$\frac{\partial f(x_1, x_2)}{\partial x_1} = 0 \cong b_1 + 2c_1 x_1 + d x_2 \quad \text{and} \quad \frac{\partial f(x_1, x_2)}{\partial x_2} = 0 \cong b_2 + 2c_2 x_2 + d x_1 \tag{7.29}$$

Solve Equations (7.29) simultaneously for the roots to obtain an approximate location for the optimum:

$$x_1 = \frac{2b_1 c_2 - b_2 d}{d^2 - 4c_1 c_2} \quad \text{and} \quad x_2 = \frac{2b_2 c_1 - b_1 d}{d^2 - 4c_1 c_2} \tag{7.30}$$

We improve our search for the location of the optimum iteratively using the latest values as our expansion points in our Taylor series until the solution converges, similar to the method of quadratic interpolation. We may extend this approach to multidimensional problems with more than two variables. Chapter 9 introduces least-squares regression methods for fitting data to multivariable linear problems, such as Equation (7.28). Statistical analysis of the variance and uncertainty in the coefficients indicate relative significance of effects and interaction among the variables in designed experiments.

We use second derivatives to confirm that the location of the optimum corresponds to a minimum or maximum. For example, a one-dimensional problem compares the sign of the coefficient of the x^2 term to the second derivatives in Table 7.1:

$$\frac{\partial f^2(x_1, x_2)}{\partial x_1^2} \cong 2c_1 \quad \text{and} \quad \frac{\partial f^2(x_1, x_2)}{\partial x_2^2} \cong 2c_2 \tag{7.31}$$

7.5.4 Golden Sections

The Golden Section search method is to optimization what the bisection and regula falsi methods are to root finding. Golden Section:

- Applies to one-dimensional, unimodal objective functions
- Accelerates the process of locating an optimum
- Guarantees an optimum solution for smooth continuous functions
- Systematically narrows the region of search for the optimum in successive iterations by identifying the section of the search region that contains the optimum

Figure 7.11 contains a diagram of the sequence of Golden Section search steps. In the first step, we identify the initial bracketed region for the independent variable Δx_i that contains the optimum solution. We recommend the accelerated search procedure described in Section 7.5.2 for finding the starting points that bracket the optimum. For bookkeeping purposes at iteration i, the lower bound is labeled x_{Li}, and the upper bound is labeled x_{Ui}, where $\Delta x_i = x_{Ui} - x_{Li}$. Cycle through the following three steps of the Golden Section method until the solution converges on the optimum value:

1. Use two points inside the bounds located at a distance $\phi \Delta x_i$ from the upper or lower bounds:

$$x_1 = x_L + \phi \Delta x \quad \text{and} \quad x_2 = x_U - \phi \Delta x \tag{7.32}$$

2. Compare the objective functions at these points with the criteria for the optimum. Delete the section of length $\phi \Delta x_i$ at one end that does not contain the optimum. This leaves a new search region for the next iteration related to the previous search region:

$$\Delta x_{i+1} = (1 - \phi) \Delta x_i \tag{7.33}$$

3. In subsequent iterations, only one objective function evaluation is required at a distance $\phi\Delta x_{i+1}$ from the remaining boundary of the previous iteration, x_{Li}, or x_{Ui}.

$$\phi\Delta x_{i+1} = (1-2\phi)\Delta x_i \qquad (7.34)$$

Obtain a quadratic equation in ϕ from the combination of Equations (7.33) and (7.34) to eliminate Δx:

$$1-3\phi+\phi^2 = 0 \qquad (7.35)$$

The two roots of the quadratic Equation (7.35) are $\phi = 0.38197$ and $\phi = 2.61803$. Only the root with a value smaller than one has any significance in this application. Note that, according to Equation (7.33), after n iterations, the length of the nth interval is

$$\Delta x_n = \Delta x_1 (1-\phi)^n = \Delta x_1 (0.61803)^n \qquad (7.36)$$

We may use Equation (7.36) to determine the maximum number of iterations required to achieve a prescribed tolerance in Δx:

$$n = \frac{\ln\left(\dfrac{\Delta x_1}{\Delta x_n}\right)}{0.48122} \qquad (7.37)$$

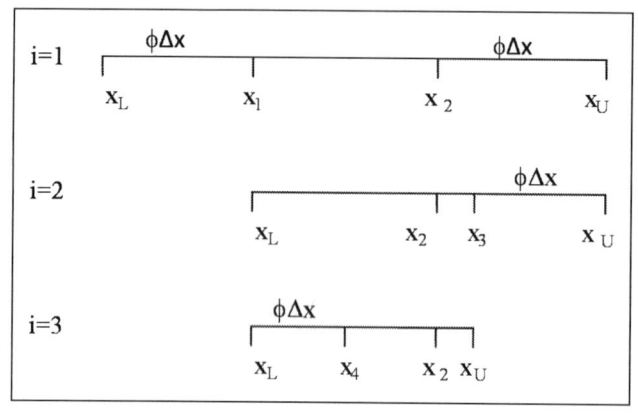

Figure 7.11 Three Golden-Section-Search iterations.

In Figure 7.11, the objective function evaluated at x_2 in the first iteration is closer to the optimum. Remove the section next to the lower bound for a new, smaller search interval for the next iteration. In the second iteration, the objective function evaluated at x_2 is closer to the optimum. Remove the upper section next to the upper bound for a new smaller search interval for the third iteration. Continue with this procedure until the length of the search region becomes smaller than a specified convergence tolerance. We incorporate the Golden Section search method into Powell's method in Section 7.5.5. Unfortunately, the Golden Section method has a tendency to converge slowly near the optimum.

Example 7.8   Golden Section Search for the Maximum Specific Growth Rate

Known/Find: Use Golden Sections to find the maximum for the model of the specific growth rate in Example 7.7.

Analysis: Perform the calculations in an *Excel* worksheet. Start with $S = 0$ in cell **B2** and a step size of 0.1 in cell **B3**. Search for a set of substrate concentrations that bracket the optimum according to Equation (7.22). Apply the Golden Section method to sequentially reduce the bracket size and converge on the local optimum solution.

The formulas in Figure 7.12 show how to use logical **IF** worksheet functions to select the upper and lower brackets for the first three iterations. For convenience, we show the successive iterations of Golden Section search in groups in the *Excel* worksheet.

CHAPTER 7: OPTIMIZATION

	A	B	C
1	Iteration	S	μ
2		0	=3*B2/(5+0.9*B2+2*B2^2+0.3*B2^3)
3		0.1	=3*B3/(5+0.9*B3+2*B3^2+0.3*B3^3)
4		=3*B3-2*B2	=3*B4/(5+0.9*B4+2*B4^2+0.3*B4^3)
5	Bracket	=3*B4-2*B3	=3*B5/(5+0.9*B5+2*B5^2+0.3*B5^3)
6		=3*B5-2*B4	=3*B6/(5+0.9*B6+2*B6^2+0.3*B6^3)
7		=3*B6-2*B5	=3*B7/(5+0.9*B7+2*B7^2+0.3*B7^3)
8			
9	1	=IF(C5>C6,B4,B5)	=3*B9/(5+0.9*B9+2*B9^2+0.3*B9^3)
10		=(B12-B9)*0.38197+B9	=3*B10/(5+0.9*B10+2*B10^2+0.3*B10^3)
11		=B12-(B12-B9)*0.38197	=3*B11/(5+0.9*B11+2*B11^2+0.3*B11^3)
12		=IF(C5>C6,B6,B7)	=3*B12/(5+0.9*B12+2*B12^2+0.3*B12^3)
13			
14	2	=IF(C10>C11,B9,B10)	=3*B14/(5+0.9*B14+2*B14^2+0.3*B14^3)
15		=(B17-B14)*0.38197+B14	=3*B15/(5+0.9*B15+2*B15^2+0.3*B15^3)
16		=B17-(B17-B14)*0.38197	=3*B16/(5+0.9*B16+2*B16^2+0.3*B16^3)
17		=IF(C10>C11,B11,B12)	=3*B17/(5+0.9*B17+2*B17^2+0.3*B17^3)
18			
19	3	=IF(C15>C16,B14,B15)	=3*B19/(5+0.9*B19+2*B19^2+0.3*B19^3)
20		=(B22-B19)*0.38197+B19	=3*B20/(5+0.9*B20+2*B20^2+0.3*B20^3)
21		=B22-(B22-B19)*0.38197	=3*B21/(5+0.9*B21+2*B21^2+0.3*B21^3)
22		=IF(C15>C16,B16,B17)	=3*B22/(5+0.9*B22+2*B22^2+0.3*B22^3)

Figure 7.12

Bracketing with Golden Section optimization.

	A	B	C
1	Iteration	S	μ
2		0.000000000000	0.000000000000
3		0.100000000000	0.058704968397
4		0.300000000000	0.164892545025
5	Bracket	0.700000000000	0.312830520341
6		1.500000000000	0.379346680717
7		3.100000000000	0.258712059042
8			
9	1	0.700000000000	0.312830520341
10		1.616728000000	0.374519444600
11		2.183272000000	0.333827113434
12		3.100000000000	0.258712059042

	A	B	C
14	2	0.700000000000	0.312830520341
15		1.266565405840	0.381578838936
16		1.616706594160	0.374520502568
17		2.183272000000	0.333827113434
18			
19	3	0.700000000000	0.312830520341
20		1.050154417771	0.370720005863
21		1.266552176389	0.381578609753
22		1.616706594160	0.374520502568

A plot of the bracket points in Figure 7.13 shows convergence on $S = 1.336$ for a maximum specific growth rate of $\mu = 0.3822$ after 10 iterations.

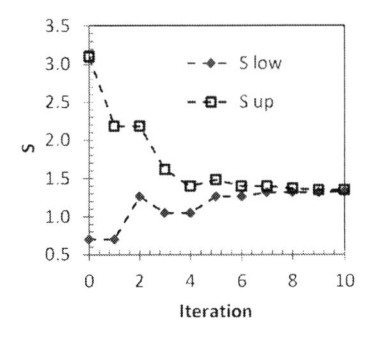

Figure 7.13

Golden Section iterations converging on the optimum location.

Comments: The Golden Section set-up in an *Excel* worksheet is tedious, and we do not recommend this manual approach for general use. Instead, use the *VBA* macro **POWELL** that automates the Golden Section search in an *Excel* worksheet for single or multivariable objective functions.

7.5.5 Powell's Method for Multiple Dimensions

Powell's multidimensional optimization method selects the search directions moving along a vector toward the optimum using a line search procedure such as quadratic interpolation or Golden Section presented in Sections 7.5.2 and 7.5.4. Initially, Powell's method cycles through each dimension, finding the local optimum in each dimension. These single-dimension optimum points form a vector pointing toward the multidimensional optimum, as illustrated in Figure 7.14. Powell's method searches for the optimum along the conjugate directions to a local minimum.[63]

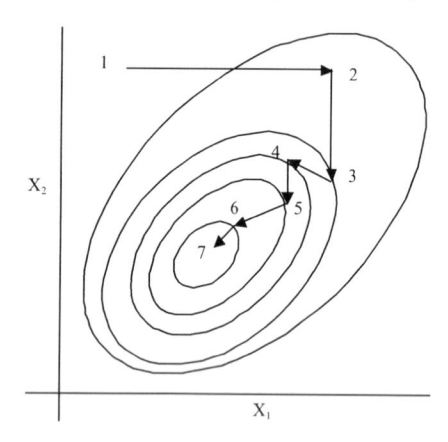

Figure 7.14

Optimum search steps by Powell's method graphed on a contour plot of the objective function.

We find the conjugate directions as follows (Press, Teukolsky and Vetterling 2007):
1. Given an n-dimensional objective function, set the starting point of n variables with an initial set of n search directions.
2. In the first Powell iteration, find a new local optimum by searching in each direction sequentially.
3. Replace the initial optimum point with the new optimum along each conjugate direction.
4. The vector connecting the previous optimum point to the new optimum point defines the new line search direction, as illustrated in Figure 7.14. Identify the search direction that produced the largest step change in the objective function for replacement by the new search direction. We base this replacement choice on the observation that the old direction that produced the largest step change has the most influence on the new direction.
5. Add the new average direction to the set of directions only if we can show that it improves the search for the optimum.

Although this approach is simple, and works well for most problems, there is at least one class of problems where this direct search approach becomes slow and cumbersome. The search for the optimum may require an inordinate amount of small steps if the optimum exists in a long, extended trough or peak that is not parallel to either axis of the independent variables.

The plot in Figure 7.14 demonstrates Powell's method for a two-dimensional objective function represented by the contour plot. The first two search directions are parallel to the two axes for X_1 and X_2, respectively. Note that the terminal points are the local optima in these search directions. The third search direction follows the vector path between the starting point 1 and the terminal point 3, until the local optimum terminal point 4 is found in that direction. The initial search direction parallel to the X_1 axis produced the greatest change from the starting point, which is consequently abandoned. We retain the second and third search directions to give points 5 and 6 in sequence. The next search direction follows the vector from point 4 to point 6 until the terminal point 7 is located. Further refinements are then determined based on the desired precision in the result. Although there are only seven steps in the hypothetical search in Figure 7.14, each Powell step involves several objective function evaluations to find the optimum along the search direction employing either quadratic or Golden section line-search methods.

---

[63] Excel's **Solver** has the optional method of conjugate gradients using a similar approach with derivative approximations to determine the search direction.

CHAPTER 7: OPTIMIZATION

The Chapter 7 Examples folder on the book's companion web site contains an *Excel* workbook with the macro **POWELL_ANIMATION** that uses random starting points in a search for the minimum of the following objective function in the region $-2 < x_1 < 2$ and $-2 < x_2 < 2$:

$$f(x_1, x_2) = x_1 e^{-(x_1^2 + x_2^2)} + 0.05(x_1^2 + x_2^2) \tag{7.38}$$

An example animation in Figure 7.15 shows 10 Powell-steps on the contour plot starting near $x_1 = 1.4$ and $x_2 = -1$ that move along gradients down and around a local maximum into the global minimum near $x_1 = -0.6$ and $x_2 = 0$. The surface plot gives a three-dimensional perspective of the objective function in Equation (7.38).

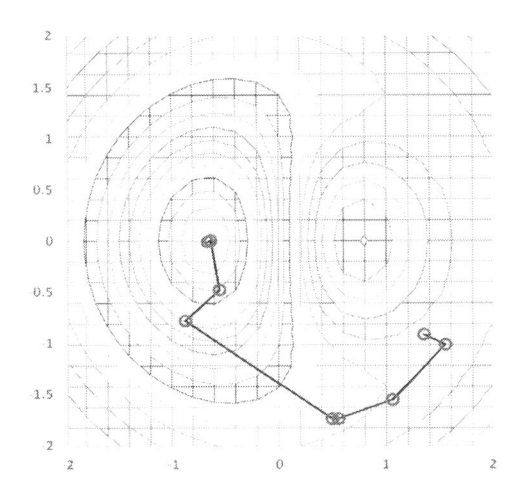

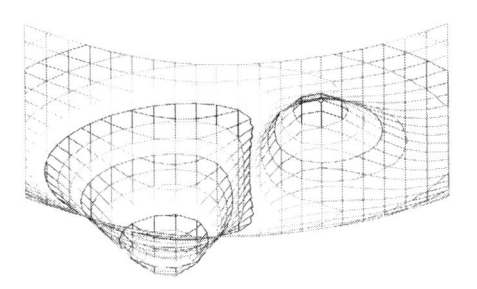

Figure 7.15

Minimization search steps in Equation (7.38) of ANIMA-TION_POWELL with a random starting point at (1.36, -0.91) and global minimum at (-0.61, 0). The surface plot shows the location of the global minimum.

The *PNMSuite* has a *VBA* macro **POWELL**, adapted from the algorithm of Press, et al. (1992). The user-form **OPTIMIN** has an option for Powell's method of minimization with the same worksheet setup requirements for **POWELL**. The macro **POWELL** incorporates a quadratic interpolation method with accelerated bracketing from Section 7.5.2 for the line searches to find the location for the optimum quickly. **POWELL** has the option of polishing the solution for higher precision, if necessary, with the robust Golden Section method from Section 7.5.4 that has a guaranteed local optimum when starting inside a bracket around the location of the optimum. **POWELL** requires a contiguous range of cells for the optimization variables and a cell for the objective function with a formula referencing the cells with the optimization variables. **POWELL** displays the total number of objective function evaluations in the *Status Bar*. The user-form **ROOTS** also includes an option for Powell's method to find the roots to a system of simultaneous equations by minimizing the sum of squares according to Equation (6.106).

Example 7.9   Minimize the Butterfly Function

Known: The following two-dimensional Rosenbrock (1960) function is often used as a test case for optimization strategies:[64]

$$f(x_1, x_2) = 100(x_2 - x_1^2)^2 + (1 - x_1)^2 \tag{7.39}$$

Find: Solve for the minimum using Powell's method.

Schematic: Examine the test function in a surface plot. A **Data Table** generated the data for the plot:

---

[64] Due to its shape, Rosenbrock's function is referred to as the "butterfly" or "banana" function.

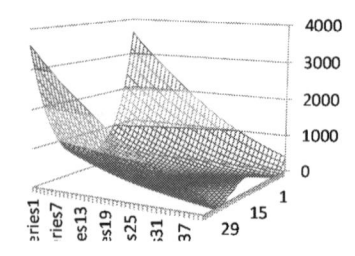

Analysis: By inspection, the minimum occurs at $x_1 = 1$ and $x_2 = 1$. To illustrate Powell's method, enter the formula for the objective function in terms of the variables in cell **C2** of an *Excel* worksheet. Start with initial guesses for the location of the optimum at $x_1 = 2$ and $x_2 = 3$:

	A	B	C
1	x₁	x₂	f
2	2	3	=100 * (B2 -A2 ^ 2) ^ 2 + (1 - A2) ^ 2

Run the *VBA* macro **POWELL** and provide the required input:

Select cell for MINIMIZATION OBJECTIVE function: $C$2

Select range of VARIABLES: (must be in contiguous cells) $A$2:$B$2

Convergence TOLERANCE: 0.00001

Starting from the same initial guesses for comparison, the quadratic bracketing line search method requires 207 function evaluations:

	A	B	C
1	x₁	x₂	f
2	1.00000	0.999999	8.95E-12

Comments: We recommend quadratic interpolation with Powell's method unless the function does not behave like a parabola near the optimum point.

☐

7.5.6 Downhill Simplex Method of Nelder and Mead*

"Beware of the blob, it creeps and leaps, and glides and slides across the floor, Right through the door, And all around the wall, a splotch, a blotch Be careful of the blob!" – Lyrics by Burt Bacharach[65]

The Nelder and Mead (1965) multivariable minimization method uses a geometric simplex with linear polygonal surfaces to model the objective function (not to be confused with the linear programming simplex option in the **Solver**). The downhill simplex method searches for the minimum of the objective function by "morphing" the shape of the simplex according to a simple set of rules that move the vertices of the simplex "downhill" from positions of higher to lower values of the objective function. The robust simplex behaves like a mobile, gelatinous "blob" that squishes through narrow valleys, expands into flatter regions, and contracts itself around the location for the minimum. We illustrate the rules for downhill simplex minimization using the simplest multivariable example of a two-dimensional objective function:

1. Start with an initial guess for the coordinates of the location for the minimum in *n*-dimensions (independent variables). Select *n* additional random guesses for the location of the minimum to create a two-dimensional simplex with *n*+1 vertices. The macro **SIMPLEXNLP**, listed in the *PNMSuite*, selects the random points from a uniform distribution based on one tenth of the order of magnitude of each of the coordinates.

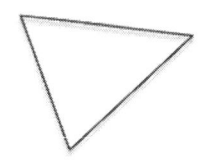

---

[65] Theme song for a B-movie, starring the teenage Steve McQueen, who discovers a meteor with an alien amoeba possessing the ability to change its shape in order to optimize the consumption of the residents of a small rural community (yuck!).

2. Determine the vertices on the current simplex corresponding to the highest and second highest values for the objective function. In the simplex shown at right, the high point is the top left vertex. Reflect the simplex about the opposite face to create a new vertex (hopefully with a lower value of the objective function).

3. If the value of the objective function at the new reflected vertex from step 2 is lower than the previous best value, extrapolate the reflected vertex further to explore if any additional improvement is made towards the minimum. If the second extrapolation does not produce an improvement, keep the vertex from the first reflection and start again at rule 2.

4. If the first reflection from rule 2 does not improve the result, perform a shorter reflection across the opposite face and compare with the current best value to see if the reflected vertex gives and improvement in terms of a lower value of the objective function.

5. If the shorter reflection also gives no improvement, contract the simplex from the high vertex towards the opposite face and compare with the current best value.

6. If the contraction from the high vertex fails to produce an improvement, contract the highest and second highest vertices towards the low vertex, and then start at rule 2. Cycle through these six rules to converge on the minimum.

The simple figure at right illustrates how a two-dimensional simplex may transform in four steps from a higher value of the objective function, located at the top left vertex, towards a lower region of the objective function at the bottom right. The steps follow rules 2 (two steps of a single reflection), 5 (contraction from the high vertex), and 6 (contraction of a face towards the low vertex).

Higher order minimization problems require three or higher dimensional simplexes. The three-dimensional tetrahedron at the right shows one reflection across the two-dimensional face (shaded) located opposite the high vertex. We leave it to the reader to imagine a three-(or higher) dimensional simplex for each of the six rules of the downhill simplex method.

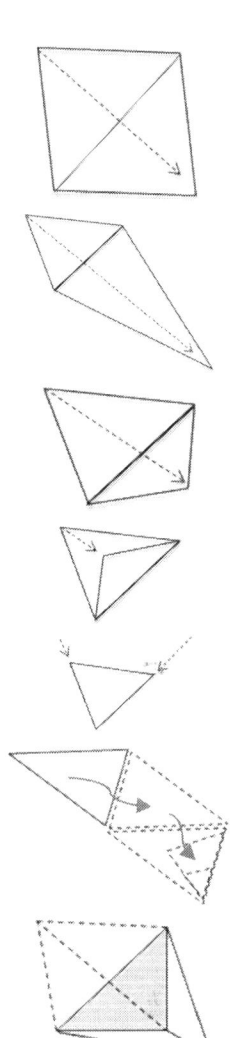

The rules of the downhill simplex method are coded in the *PNMSuite* as a *VBA* macro named **SIMPLEXNLP** (NLP for nonlinear programming), adapted from Press, et al. (1992). To use the macro (*a.k.a.*, "BLOB"), set up an *Excel* worksheet with the formula of the objective function in one cell in terms of the independent variables in another range of cells, just like the setup for **POWELL**. Edit the *VBA* code to change the convergence tolerance, or other search parameters for reflection, extrapolation, or contraction, as needed to arrive at the minimum. The macro displays the progress in the status bar. For comparison, **SIMPLEXNLP** required 188 function evaluations to minimize the butterfly function in Example 7.9. Simplex works about as well as Powell's method because the simplex method does not require bracketing as required by the line-search routines of Golden Section or quadratic programming. The computational cost is a large number of function evaluations to move the simplex around the problem space.

An Excel workbook named **SIMPLEX_ANIMATION** is available in the examples folder for Chapter 7 from the book's web site. To use the animation, open the workbook, enable the macros, and then click on the **Run** button. The animation typically runs for approximately 30 seconds. The animation graphs the simplex for the minimization problem in Example 7.9. The animated search starts from a random coordinate in the ranges $0 < x_1 < 2$ and $0 < x_2 < 2$ and plots the two-dimensional, triangular simplex as it moves downhill towards the minimum located at (1, 1). The animation also tracks the coordinates of the lowest vertex in the simplex after each reflection. Figure 7.16 shows the trajectory of a dozen steps of the simplex movement down two isoclines of the objective function before it eventually encompasses the minimum, marked with cross hairs at the center. Try different random starting points by clicking the **Run** button again after each animation to get a feel for how the simplex morphs along towards the location for the minimum, according the rules of expansion and contraction.

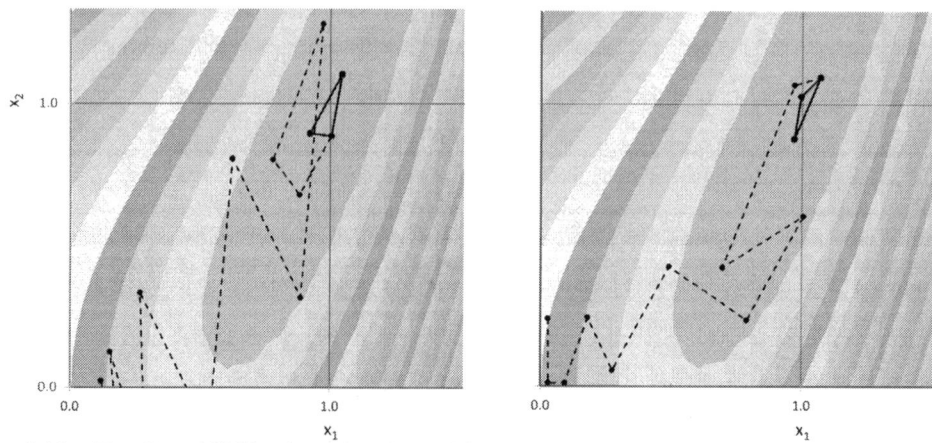

Figure 7.16 Two Downhill Simplex animations of the search for the minimum in *Example 7.9* starting from random points. In each case, the simplex eventually converges and encloses the coordinates for the minimum (located at the cross hairs) after ~15 iterations

7.6   Stochastic Global Optimization*

"I think you can say a lot of evil behavior by companies is short-term optimization." – Sam Altman

The optimization methods presented to this point were designed to find a local optimum near the starting point of the search. How do we know if a better optimum does not exist somewhere within the constraints on the search domain? Although it may not always be possible to locate the true global optimum of a complicated objective function, there are some simple techniques for getting close. First, we notice from our previous experience with optimization that where we end up in our search for a local optimum strongly depends on where we start looking. The key to global optimization is to compare solutions located from several different starting points. We explore four global search methods implemented in `VBA`:

1. Simulated Annealing imitates a metallurgical process of softening metals.
2. Luus-Jaakola uses cycles of random numbers in a shrinking search region.
3. Particle Swarm Optimization (PSO) models the behavior of a social population.
4. Firefly algorithms model the behavior of fireflies attracting mates.
5. Genetic algorithms emulate evolution of genes through crossover and mutation within chromosomes.

We start with one of the earliest successful methods of global optimization named simulated annealing and finish this section with an implementation of the genetic algorithm. The stochastic direct search method of Luus and Jaakola (1973) in Section 7.6.2 and the Firefly algorithm, introduced in Section 7.6.3, iteratively select random sets of values for the optimization variables within a specified search region to find the proximity of a local optimum. Random selection eliminates any unintended procedural bias in the direct search and does not require a continuous function.

7.6.1   Simulated Annealing*

The method of simulated annealing derives its unusual name from the metallurgical annealing process of rapid initial heating followed by a gradual reduction in temperature to release mechanical stresses. Annealed metals are generally softer and more malleable for shaping. In direct-search optimization, we turn to the numerical method of simulated annealing to locate the global minimum of a multimodal objective function where the line-search methods from the Section 7.5 become stuck in a local minimum. Simulated annealing also has application to non-smooth, discrete functions.

Simulated annealing cycles through two basic operations:
1. Perturb the values of the optimization variables in the neighborhood of the current best solution.
2. Evaluate the new location for improvement in the optimization.

There are many different implementations of simulated annealing. We present a commonly used version based on the Metropolis algorithm that employs a Boltzmann probability function for accepting new coordinates of the minimization variables. In the first operation, we calculate an error term as the difference between the current and new value of the objective function:

$$\Delta E = f_{perturbed} - f_{current} \quad (7.40)$$

In the second operation, we accept new perturbed coordinates according to the Boltzmann probability:

$$P = \begin{cases} 1 & \Delta E \leq 0 \\ e^{-\frac{\Delta E}{\sigma_k T}} & \Delta E > 0 \end{cases} \quad (7.41)$$

where ΔE is the error term from Equation (7.40), T is the annealing "temperature", and σ_k is the standard deviation of the errors from the perturbations that normalizes the error term in the numerator. We always accept the new result when an improvement in the minimization is achieved. Otherwise, we accept perturbed values with random probability. The probability test allows the search method to escape from local minima.

The initial and final annealing "temperatures" are calculated from the following initial and final probabilities of accepting a worse value of the objective function:

$$T_0 = -\frac{1}{\ln(P_0)} \qquad T_f = -\frac{1}{\ln(P_f)} \quad (7.42)$$

We recommend an initial probability $P_0 = 0.8$ and final probability $P_f = 10^{-8}$. The simulated annealing algorithm iteratively reduces the temperature according to the following schedule:

$$C = \left(\frac{T_f}{T_0}\right)^{1/(m-1)} \qquad T_k = C \cdot T_{k-1} \quad (7.43)$$

where m is the maximum number of cooling steps and C is the temperature contraction factor. As we cycle through the operations at the current temperature, we calculate the standard deviation of the error in Equation (7.41) incrementally using Welford's (1962) formula:

$$\mu_1 = \Delta E_1 : v_1 = 0$$
$$For\ k = 2 \ldots q$$
$$\mu_k = \mu_{k-1} + (\Delta E_k - \mu_{k-1})/k \quad (7.44)$$
$$v_k = v_{k-1} + (\Delta E_k - \mu_{k-1})(\Delta E_k - \mu_k)$$
$$\sigma_k = \sqrt{\frac{v_k}{k-1}}$$

where q is the parameter for the maximum number of equilibrium cycles at the current temperature, and k in the index variable of the current cycle. A user-defined function **STDEW** for calculating the standard deviation of a range of values by Welford's formula is also available in the *PNMSuite*. To implement the method of simulated annealing for minimizing an objective function:

1. Select initial values for the variables of optimization, x, and calculate the value of the objective function.
2. Specify initial values of the standard deviations used in the next step for perturbing the optimization variables. When selecting values of s for each variable, recall that 99.7% of values from a normal distribution lie within $\pm 3s$. The size of $\pm 3s$ is similar to the contracting search region in the Luus-Jaakola method presented in Section 7.6.2. This becomes important for unconstrained searches where randomly selected variables may not be appropriate for the objective function.

3. Randomly select new values for each variable in the neighborhood of their current values using normal random deviates:

$$x_j = x_{j-1} + s_j \cdot NRD \tag{7.45}$$

where x_j is the value of variable j, s_j is the standard deviation of variable j, and *NRD* is a normal random deviate defined in Section 3.6.2. Evaluate the objective function, f_j, at the new coordinates.

4. Calculate the error $\Delta E = f_j - f$ where f is the currently accepted value of the objective function.
5. Compare the error in Equation (7.41) to determine the current "best" solution.
6. Cycle through the previous steps until the temperature equilibrates.
7. Reduce the temperature according to the schedule in Equation (7.43) and start again at step 2 from the overall best result.

The macro **SIMANN** listed in Figure 7.17 is available in the *PNMSuite* for global minimization by the method of simulated annealing in an *Excel* worksheet. The macro uses input boxes to specify the cell with the objective function and ranges of variables and corresponding standard deviations. The macro provides default parameter values, but these may require experimentation to determine better combinations of annealing parameters.

```
Sub SIMANN()
' Global minimization by the method of simulated annealing. The user specifies initial
' and final probability of accepting a poor result, the number of temperature cooling
' steps, the number of random perturbation cycles at each temperature, and the initial
' standard deviation for each variable.
' *** Required User-defined Functions and Sub Procedures from PNM Suite ***
' RNDWH = pseudo-random number from a uniform distribution
' *** Nomenclature ***
Dim a1 As Double, a2 As Double ' = average errors for standard deviation calculations
Dim bxtl As String ' = title of input and message boxes
Dim co As Long ' = counter for averaging the energy difference
Dim cf As Double ' = contraction factor for temperature
Dim de As Double, dea As Double ' = activation energy for f, average energy
Dim f As Double, fbest ' = save value of the objective function and best value
Dim flag As Integer ' = check for improved solution
Dim fr As Range ' = range for minimization function
Dim i As Long, j As Long, k As Long ' = index variable
Dim m As Long ' = maximum number of temperature steps
Dim n As Long, nr As Long ' = number of optimization variables & rows in the range
Dim p As Double ' = Boltzmann probability
Dim pi As Double, pf As Double ' = initial and final probability
Dim q As Long ' = number of equilibrium cycles
Dim sx() As Double ' = vector of perturbation standard deviations for variables x
Dim sxr As Range ' = range of perturbation in x
Dim t As Double, ti As Double, tf As Double ' = annealing temperature, initial & final
Dim v1 As Double, v2 As Double ' = incremented error variances
Dim x() As Double, xbest() As Double ' = current and best values for variables x
Dim xr As Range ' = range of optimization variables
' *************************************************************************
bxtl = "Simulated Annealing"
With Application
Set fr = .InputBox(prompt:="Cell with Objective FUNCTION, f:", Type:=8)
Set xr = .InputBox(prompt:="Range of VARIABLES, x:", Type:=8)
n = xr.Count: nr = xr.Rows.Count ' get the number of variables and rows in xr
Set sxr = .InputBox(prompt:="Range of Variable STANDARD DEVIATIONS, s:", Type:=8)
m = .InputBox(prompt:="Number of Temperature STEPS:", Default:=100, Type:=1)
q = .InputBox(prompt:="Number of Equilibrium Cycles:", Default:=m, Type:=1)
pi = .InputBox(prompt:="INITIAL Probability:", Default:=0.8, Type:=1)
pf = .InputBox(prompt:="FINAL Probability:", Default:=0.00000001, Type:=1)
ReDim sx(1 To n) As Double, x(1 To n) As Double, xbest(1 To n) As Double
flag = 0 ' initialize the improvement flag
ti = -1 / Log(pi): tf = -1 / Log(pf) ' initial & final "temperatures"
cf = (tf / ti) ^ (1 / (m - 1)) ' temperature contraction factor
tf = tf * 10 ' final t for Boltzmann probability
10: ' label if no improvement is found, start cycles again
f = fr: t = ti ' save the value of the objective function and temperature
```

```
For j = 1 To n
    sx(j) = sxr(j): x(j) = xr(j) ' initialize the st dev and variables
    ' xr(j) = x(j) + sx(j) * (2 * RNDWH() - 1)
    xr(j) = x(j) + sx(j) * NRDWH() ' perturb from a normal distribution
Next j
If f < fr Then ' save the current best result
    a1 = f - fr: fbest = f: xbest = x
Else
    a1 = fr - f: f = fr: fbest = f
    For j = 1 To n: x(j) = xr(j): Next j
    xbest = x
End If
v1 = 0 ' initialize error average for incremental st dev calculations
For i = 1 To m ' loop of temperature cooling steps
    co = 2 ' initialize the count of error terms at a temperature
    For k = 1 To q ' temperature equilibrium cycles
        For j = 1 To n: xr(j) = x(j) + sx(j) * NRDWH(): Next j
        de = fr - f
        a2 = a1 + (de - a1) / co ' increment the average
        v2 = v1 + (de - a1) * (de - a2) ' increment the variance
        If de < 0 Then ' improvement in minimization
            a1 = a2: v1 = v2 ' update for next cycle
            f = f + de: co = co + 1 ' set new f and increment the counter
            For j = 1 To n: x(j) = xr(j): Next j
            If f < fbest Then
                xbest = x: fbest = f
            End If
            flag = 1 ' signal improved solution located
        ElseIf t > tf Then ' stop Boltzmann probability near end of temperature steps
            p = Exp(-de / (Sqr(v2 / co) * t)) ' probability of acceptance
            If RNDWH() < p Then
                a1 = a2: v1 = v2: co = co + 1 ' update next cycle & increment counter
                ' dea = Abs(dea + de) / co ' average de
                For j = 1 To n: x(j) = xr(j): Next j
            End If
        End If
    Next k
    x = xbest ' reset x to best values for current minimization
    t = cf * t ' contract the temperature
    For j = 1 To n: sx(j) = cf * sx(j): Next j ' contraction of standard deviations
    If i Mod 10 = 0 Then .StatusBar = bxtl & " Progress = " & VBA.Format(i / m, "0%")
Next i
If Not flag = 1 Then GoTo 10 ' check for improvement in the steps during t steps
If n = nr Then ' put the overall best result on the worksheet
    xr = .Transpose(xbest)
Else
    xr = xbest
End If
End With ' Application
End Sub ' SIMANN
```

Figure 7.17 *VBA* sub procedure SIMANN for multimodal global minimization of an objective function in an *Excel* worksheet by the Boltzmann method of SIMulated ANNealing.

Application of the method of simulated annealing requires some experimentation with parameters for the number of temperature cycles, annealing steps, and standard deviation of perturbation. As with all numerical algorithms, Simulated Annealing has several advantages and disadvantages:

Advantages
- Simple programming
- Guaranteed to find an optimal solution
- Does not require derivatives of the objective function
- Applies to discontinuous objective functions

Disadvantages

- Need to experiment with the parameters in the algorithm (standard deviation, number of temperature cycles, equilibrium steps, *etc.*)
- Slow convergence
- No simple criteria for stopping without proceeding through all of the steps and cycles
- Not guaranteed to locate the global optimum

Example 7.10 Reflux Ratio to Minimize the Purchase Cost of a Distillation Column

Known: The height and diameter of a continuous distillation column are functions of the external reflux ratio. The conditions of the column are tabulated below:

Distillation Rate, D = 492   Minimum Number of Trays, N_{min} = 10   Minimum Reflux Ratio, R_{min} = 1.33

The purchase cost is proportional to a function of the vapor rate in the rectifying section V and the number of trays N according to the following relationship

$$PC \propto \left(N\sqrt{V}\right)^{0.85}$$

Find: Determine the combination of dimensions that minimize the column purchase cost as a function of the external reflux ratio.

Assume: Steady-state operation, negligible operating costs for this example.

Analysis: We use the Davis correlation of Gilliland's data to calculate the number of trays in the column from the minimum number of trays:

$$X = \frac{R - R_{min}}{R + 1} \quad \rightarrow \quad Y = \frac{\left(1 + 40.6 X^{0.71}\right)\left(1 - X^{0.71}\right)}{1 + 65.4 X^{0.71}} \quad \rightarrow \quad N = \frac{N_{min} + Y}{1 - Y}$$

We set up the calculations for the purchase cost as a function of the reflux ratio in an *Excel* spreadsheet and use the **QYXPLOT** user form to plot 1000 values of the purchase cost versus reflux ratio in the range: $2 < R < 5$.

	A	B
1	D	492
2	N_{min}	10
3	R_{min}	1.33
4	R	2
5	S_R	0.1
6	X	0.2233
7	Y	0.4293
8	N	19
9	L	984
10	V	1476
11	PC	271.52

	A	B
1	D	492
2	N_{min}	10
3	R_{min}	1.33
4	R	2
5	S_R	0.1
6	X	=(B4-B3)/(B4+1)
7	Y	=0.75*(1-B6^0.5668)
8	N	=ROUNDUP((B2+B7)/(1-B7),0)
9	L	=B1*B4
10	V	=B1+B9
11	PC	=(B8*SQRT(B10))^0.85

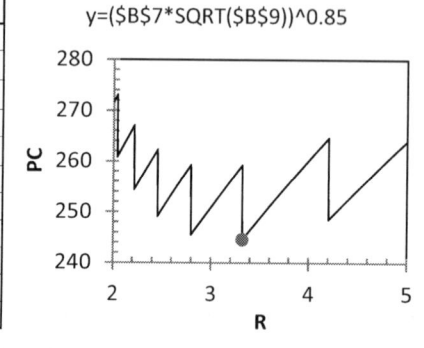

The plot of the purchase cost reveals a discontinuous function that passes through a global minimum. The direct search optimization methods require derivative evaluations, which are undefined at the sharp points on the plot. An attempt at locating the global minimum with the **Solver** produced a run-time error message. We turn to simulated annealing using the macro **SIMANN** starting the search at $R = 2$ with a standard deviation of ±0.1, to find the value of the reflux ratio that minimizes the purchase cost at $R = 3.32$, which agrees with the graphical result. The following cell ranges were used for the macro inputs.

   B11 = Cell with objective function

   B4 = Range of optimization variables

   B5 = Range of variable standard deviations

We used the default values for the rest of the simulated annealing parameters.

CHAPTER 7: OPTIMIZATION

Comments: For simple, one-dimensional problems, the graphical solution works well for locating a minimum. The method of simulated annealing available in the *PNMSuite* does not require derivative approximations, as such, it works well for non-smooth, or discontinuous objective functions like the one in this example. Be aware that stochastic algorithms require a large number of iterations and slowly converge on the global optimum.

□

7.6.2 Luus-Jaakola Global Optimization*

"There has to be a global mission of human progress" – A.P.J. Abdul Kalam

Luus and Jaakola's stochastic pattern search algorithm gradually reduces the size of the search domain around the global optimum. By starting with a relatively wide berth, the algorithm avoids becoming stuck in a region occupied by a local optimum. The Luus-Jaakola method systematically contracts the region of search over several iterations until settling on the global optimum.

Luus-Jaakola optimization uses a remarkably simple algorithm for achieving results for complex problems. This technique has found application in a variety of important problems, including parameter estimation and process control. Table 7.2 lists the definitions of the parameters required by the Luus-Jaakola method.

Table 7.2 Luus-Jaakola parameter definitions.

f	= Objective function	p	= Pass index
i	= Iteration index	r_j	= Uniform random number for variable j
j	= Variable index	s_j	= Search region for variable j
n	= Number of variables	x_j	= Variable j
n_I	= Number of iterations per pass	x^*	= Random initialization values for the set of variables
n_P	= Number of passes	γ	= Contraction factor
n_R	= Number of random variable sets		

Luus-Jaakola uses nested cycles that refine the region of search after each pass, as shown in Figure 7.18. The user specifies n_R random sets of variables for each iteration, and n_I iterations for each of the n_p passes.

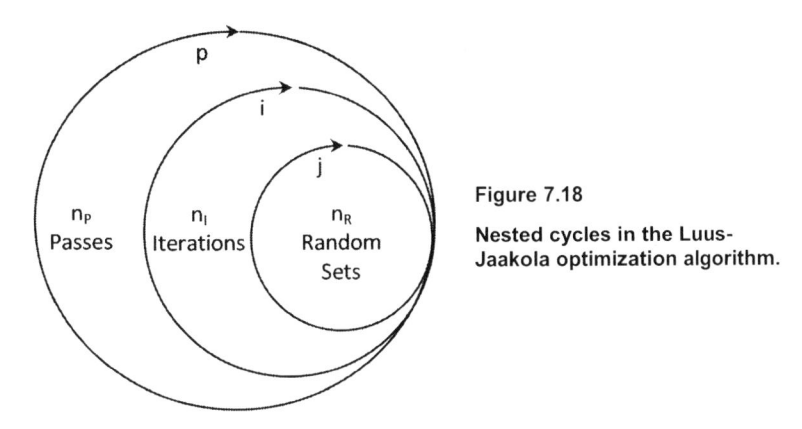

Figure 7.18

Nested cycles in the Luus-Jaakola optimization algorithm.

To implement the Luus-Jaakola method:
1. Specify the initial size of the search regions for each variable:

$$x_j^* - s_j \leq x_j \leq x_j^* + s_j \qquad for \quad j = 1, 2 \ldots n \qquad (7.46)$$

2. Randomly initialize values for the n variables of optimization, x^*, within the defined search space and calculate the objective function at this starting point, $f(x^*)$.

3. Specify the contraction factor. A typical value is $\gamma = 0.95$. Following the alternative suggestion of Englezos and Kalogerakim (2001), try a contraction factor that gives 10% reduction of the current n-variable space:

$$\gamma = (0.9)^{1/n} \tag{7.47}$$

Repeat steps four to nine for n_P passes.

4. Specify a maximum number of passes, for example, $n_P = 100$. Specify the number of iterations per pass in terms of the number of variables, for example, $n_I = 50 \times n$. This approach uses more iterations for larger values of the contraction factor. Likewise, specify the number of random variable sets per iteration, e.g., $n_R = 50 \times n$.

Cycle through the next steps five to seven for n_I iterations, while retaining the set of variables with the smallest objective function value after each iteration.

5. Calculate n_R random sets of values for n optimization variables that are uniformly distributed within their respective search regions:

$$x_j = x_j^* + (2r_j - 1)s_j \quad \text{for} \quad j = 1, 2 \ldots n \tag{7.48}$$

where r_j is a uniformly distributed random number in the range $0 \le r \le 1$.

6. Calculate the objective function for each set of n variables. Retain the variable set that minimizes the objective function after each iteration.

$$x^* \leftarrow x^{\min} \tag{7.49}$$

Alternatively, modify the original Luus-Jaakola optimization method by updating the current best variable set within an iteration whenever finding an improved set.

7. Reduce the size of the search region in the current pass for the next iteration according to the contraction factor from Equation (7.47):

$$s_{i+1}^p = \gamma s_i^p \tag{7.50}$$

Prepare for the next pass and check for a conclusion according to the following steps:

8. Calculate new initial search regions for the next pass for each variable from the change in x between the first and last iteration:

$$s_j^{p+1} = \left| x_j^{\min} - x_j^* \right| \quad \text{for all } j = 1 \ldots n \tag{7.51}$$

9. Stop when the sizes of the relative search regions for all variables converge to a level below the specified tolerance:

$$\frac{s_j^p}{\left| x_j \right| + \delta} < tol \quad \text{for all } j = 1 \ldots n \tag{7.52}$$

where δ represents a small number added to avoid division by zero.

The primary advantages of Luus-Jaakola optimization include:
- The simplicity of the algorithm,
- Higher potential for global optimization, and
- Application to discontinuous functions.
- Stopping criteria when the solution converges on the optimum.

The requirement for selecting appropriate search parameters presents a disadvantage of the Luus-Jaakola optimization method. We recommend the values in the algorithm from experience. The VBA macro **LUUSJAAKOLA**, listed in

CHAPTER 7: OPTIMIZATION

Figure 7.19 employs the Luus-Jaakola algorithm for global optimization. **LUUSJAAKOLA** uses the user-defined function **RNDWH** from Section 3.6.2 to generate the pseudo random numbers. To set up an *Excel* worksheet, add the objective function in one cell, formulated in terms of the optimization variables in a separate range of contiguous cells. Also, include a range of contiguous cells that contain the initial ±size of the search region around the starting values for the optimization variables. The user-form **OPTIMIN** has an option for the method of Luus-Jaakola.

The examples folder on the book's website contains an *Excel* workbook named **LJ_ANIMATION** that animates the Luus-Jaakola method for optimizing Michalewicz's test function from *Example 7.6*. The random points in the first pass are located on the local minima located at the top and center of the isometric plot. As the search region contracts, the random points land around the global minimum in the center of the graph.

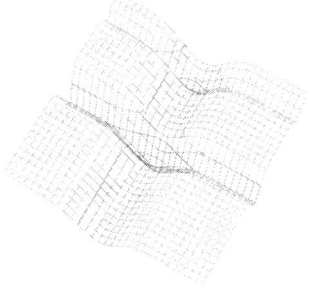

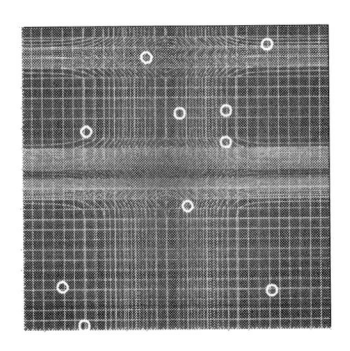

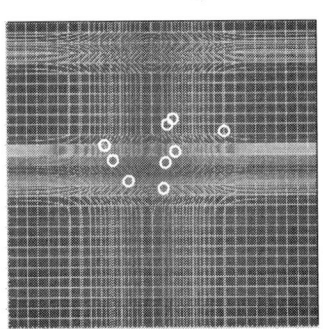

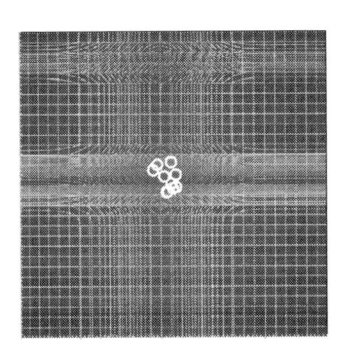

```
Public Sub LUUSJAAKOLA()
' Modified Luus-Jaakola Optimization method.  Macro finds the global minimum within a
' search region for each variable. Search regions are systematically contracted until
' convergence in the variables is reached. User specifies initial values for variables
' and search regions numbers of passes, iterations, random variable sets, and relative
' tolerance for convergence.
' *** Required User-defined Functions and Sub Procedures from the PNM Suite ***
' RNDWH = uniform random number generator
' ROWCOL = change from row to column vector
' *** Nomenclature ***
Dim boxtitle As String ' = string with title for message and input boxes
Dim dx As Double ' = small change in x
Dim fr As Range ' = range of objective function
Dim fs As Double ' = saved objective function value
Dim g As Double ' = contraction factor
Dim i As Long, j As Long, k As Long ' = index for iterations and variables
Dim n As Long ' = number of variables
Dim ni As Long ' = number of iterations
Dim np As Long ' = number of passes
Dim nr As Long ' = number of random variable sets
Dim p As Long ' = index for passes
Dim r As Long ' = index for random samples
Dim s As Double ' = maximum relative variable search region
Dim sc As Integer ' = show calculations
Dim small As Double ' = small number
Dim sr As Range ' = range of variable search regions
Dim srows As Long ' = number of rows in array for search region
Dim ss() As Double ' = saved search regions
Dim tol As Double ' = relative convergence tolerance
Dim ub As Integer ' = use best variable set within an iteration
```

```vba
Dim x() As Double    ' = local best variable set
Dim xr As Range      ' = range of optimization variables
Dim xrows As Long    ' = number of rows in xr
Dim x_save() As Double ' = saved variable values
'************************************************************************
small = 0.000000000001: boxtitle = "LUUS JAAKOLA MINIMIZATION"
With Application
Set fr = .InputBox(Title:=boxtitle, Type:=8, _
                   Prompt:="Cell with OBJECTIVE FUNCTION f(x):")
Set xr = .InputBox(prompt:="Range of VARIABLES x:" & vbNewLine & _
                   "(contiguous cells)", Title:=boxtitle, Type:=8)
xr.Interior.COLORINDEX = 3
' Get default values for numbers of passes, iterations, sets.
tol = 0.00001: n = xr.Count: nr = 50 + 10 * n: g = 0.9 ^ (1 / n)
ni = CInt(Log(0.1) / Log(g)): np = 100 ' user specified number of passes
srinput: ' Label for search region input
Set sr = .InputBox(prompt:="Range of Variable ±SEARCH REGIONS:" & vbNewLine & _
                   "(contiguous cells)", Title:=boxtitle, Type:=8)
' Check for consistency in the size of the search region and
If sr.Count <> n Then ' number of variables
    MsgBox prompt:="Match search regions to variables", Title:=boxtitle: GoTo srinput
End If
xrows = xr.Rows.Count: srows = sr.Rows.Count
ReDim ss(1 To n) As Double, x(1 To n) As Double, x_save(1 To n) As Double
fs = fr: s = 0 ' Save current x's & objective function
For j = 1 To n
    ss(j) = sr(j): x(j) = xr(j): x_save(j) = x(j)
    s = .Max(s, Abs(sr(j)) / (Abs(x(j)) + small))
Next j
.ScreenUpdating = False
For k = 1 To np ' Loop through passes
    np = np + 1
    For i = 1 To ni ' Loop through iterations
        For r = 1 To nr ' Loop through random variable sets
            For j = 1 To n ' Calculate a set of random variables
                xr(j) = x(j) + (2 * RNDWH - 1) * sr(j)
            Next j
            If Not IsError(fr) Then
                If fr < fs Then ' Refine local best location
                    .ScreenUpdating = True
                    fs = fr ' save local best variable set
                    For j = 1 To n: x(j) = xr(j): Next j
                    .ScreenUpdating = False
                End If
            End If
            .StatusBar = boxtitle & ": Pass = " & VBA.Format(k, "#") & _
                        ": Iteration = " & VBA.Format(i / ni, "#%") & _
                        ": Relative Error = " & VBA.Format(s, "0.0E+#0")
        Next r
        ' Contract the search regions for the next iteration & display progress
        For j = 1 To n: sr(j) = sr(j) * g: Next j
    Next i
    s = 0
    For j = 1 To n ' Set new search space for next pass, save best x
        dx = Abs(x(j) - x_save(j)): x_save(j) = x(j)
        If dx > sr(j) Then sr(j) = dx
        s = .Max(s, sr(j) / (g * Abs(x(j)) + small))
    Next j
    If s < tol Then GoTo addcomment ' Check for convergence
Next k
MsgBox "Not converged after maximum Luus-Jaakola passes = " & k & vbNewLine & _
       "Try different guesses, increase convergence tolerance, " & vbNewLine & _
       "or increase number of passes."
End With
End Sub ' LUUSJAAKOLA
```

Figure 7.19 *VBA* sub procedure LUUSJAAKOLA for global minimization in an *Excel* worksheet by the Luus-Jaakola stochastic direct search method.

Example 7.11 Multimodal Objective Test Function for Luus-Jaakola

Known: Chong and Zak (1996) proposed the following multimodal objective test function:

$$f(x_1, x_2) \cong 3(1-x_1)^2 e^{\left[-x_1^2-(1+x_2)^2\right]} - 10\left(\frac{x_1}{5} - x_1^3 - x_2^5\right) e^{\left(-x_1^2 - x_2^2\right)} - \frac{e^{\left[-(1-x_1)^2 - x_2^2\right]}}{3} \tag{7.53}$$

Find: Use Luus and Jaakola's method to locate the global minimum in the range $-3 < x_1 < 3$, $-3 < x_2 < 3$

Analysis: To simplify the analysis, create a VBA user-defined function for Equation (7.53):

```
Public Function f(x1 As Double, x2 As Double) As Double
    f = 3 * ((1 - x1) ^ 2) * Exp(-x1 ^ 2 - (1 + x2) ^ 2) _
      - 10 * (x1 / 5 - x1 ^ 3 - x2 ^ 5) * Exp(-x1 ^ 2 - x2 ^ 2) _
      - Exp(-(1 - x1) ^ 2 - x2 ^ 2) / 3
End Function
```

Figure 7.20 shows a surface plot that reveals several local minima and maxima within the specified ranges of constraints for x_1 and x_2. The contour plot gives a better idea of the locations of the local optima. Match the contour plot with the surface plot to discern the differences among minima and maxima.

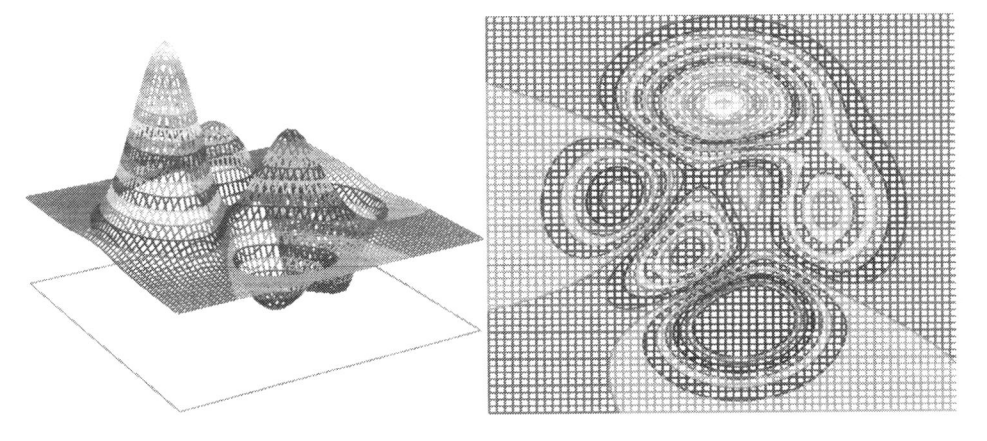

Figure 7.20

A Surface plot (left) indicates whether the optimum is a maximum or minimum value. The contour plot (right) shows the locations of the local optimum values at the center of the contours.

Setup an *Excel* worksheet with the objective function in cell **B3=f(B1, B2)** in terms of the variables in the range **B1:B2**. Initialize the values for x_1 and x_2 to zero in the range **B1:B2**. Specify the ± search region for x_1 and x_2 in the range **D1:D2** according to Equation (7.46). The **Status Bar** displays the current values of the Luus-Jaakola parameters from Table 7.2. The Luus-Jaakola direct search method finds the value of the global minimum $f = -6.551$ located at $x_1 = 0.228$ and $x_2 = -1.626$.

	A	B	C	D
1	x₁	0.228279	s₁	3
2	x₂	-1.625535	s₂	3
3	OF	-6.558966		

Comments: The Luus-Jaakola method of global optimization is simple to program, but may be less efficient than indirect search methods. The trick is picking the best values for the Luus-Jaakola parameters in Table 7.2. □

7.6.3 Particle Swarm Optimization (PSO)

"The fiercest serpent may be overcome by a swarm of ants." – Isoroku Yamamoto

Particle swarm optimization (PSO) methods are designed to imitate the cognitive and social behavior of animals, such as a flock of birds, that moves towards the global optimum position in sync according each member's historically best position and the paramount position of the swarm.

A particle has a unique set of values for each variable of optimization in an objective function. The current values are the coordinates of the particle in the domain of the search. A population of distinct particles makes up the swarm. The coordinates of each particle are chosen randomly to initiate the search and sample the search space without bias. The PSO algorithm moves each particle with a velocity and direction according to information about its best position from previous moves and the overall best particle's position. The following PSO algorithm tracks the best position of each particle and the overall best particle in terms of their contributions towards minimizing the objective function.

1. Define the objective function in terms of the *n* variables of optimization, x_j.
2. Provide lower and upper search domain boundary limits on the values of each variable *j*: $x_j^L < x_j < x_j^U$. Use information about the problem to select the tightest limits on the variables, as possible, to improve the computational efficiency of the search.
3. Specify a maximum number of iterations and size of the population of particles, *m*. A population size of 50, or fewer, is recommended (Marini and Walczak 2015).
4. Initialize the coordinates of each particle using Latin hypercube sampling randomly (see Section 8.3.1).

The PSO algorithm cycles through the next steps to converge on the global optimum:

5. The particle's velocity components for the next iteration are calculated from the current velocity according to the particle's best condition from all of the previous iterations and the global best of the population of particles:

$$v_{i+1,j} = w \cdot v_{i,j} + r_1 c_1 \left(p_j - x_j\right) + r_2 c_2 \left(g_j - x_j\right) \qquad i = 0,1,2\ldots \qquad (7.54)$$

where Δt is the "time" step (equal to one iteration), *w* is an adaptive inertia weighting factor with particle success feedback (Nickabadi, Ebadzadeh and Safabakhsh 2011), c_1 and c_2 are the relative particle and global acceleration constants with values typically in the range of 1.5 to 2, and r_1 and r_2 are random numbers drawn from a uniform distribution (0 < *r* < 1). We use values of *w* = 0.5 and $c_1 = c_2 = 1.5$ in our implementation of PSO per the recommendation of Hassan, et al. (2005).

6. The trajectory of a particle in the next iteration is calculated from the previous position and velocity:

$$x_{i+1,j} = x_{i,j} + \Delta t v_{i+1,j} \qquad (7.55)$$

To prevent variables from exceeding their boundary limits, the coordinates calculated from Equation (7.55) are compared with the boundary limits and adjusted, if necessary, using the harmonic mean distances from the particle to the boundary at $x_{i+1,j}$ and $x_{i,j}$, as follows:

$$x_{i+1,j} = \begin{cases} x_j^L + \left(1 - \dfrac{x_{i,j} - x_j^L}{|v_{i+1,j}|}\right)\left(x_{i,j} - x_j^L\right) & \text{if } x_{i+1,j} < x_j^L \\[2ex] x_j^U - \left(1 - \dfrac{x_j^U - x_{i,j}}{|v_{i+1,j}|}\right)\left(x_j^U - x_{i,j}\right) & \text{if } x_{i+1,j} > x_j^U \end{cases} \qquad (7.56)$$

Figure 7.21 contains a graphical illustration of a two-dimensional particle's trajectory after one iteration from position X_i to X_{i+1} shows the influence of the particle's inertia, memory, and the swarm on its new velocity and position. The dashed lines show the vectors pointing toward the current best particle, P_i, and global best coordinates P_g. The dot-dashed lines show the initial and final velocity vectors determined by Equation (7.54). The solid lines represent the results of the parallel components of the particle movement in Equation (7.55), calculated from the weighted inertia, memory, and swarm influence:

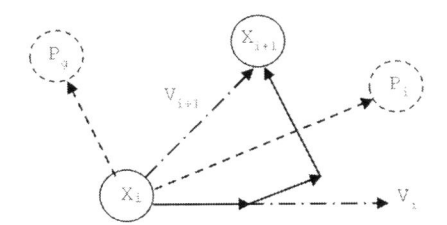

Figure 7.21

A particle trajectory influenced by its own interia and the coordintates of the particle's best position and the best postion of the swarm.

7. Compare the value of the objective functions to determine the local and global best particles. Retain the historically best position of each particle and the overall best position.
8. Terminate the cycle when the value of the global optimum stops changing after a specified number of iterations. The number of termination cycles varies by problem. We recommend $50n$ termination cycles.

A *VBA* sub procedure named **PSO**, listed in Figure 7.22, is available in the *PNMSuite* for implementing PSO in an *Excel* worksheet. To use of the **PSO** macro, formulate the objective function in a cell on the worksheet in terms of the variables of optimization in a separate range of cell. Provide corresponding ranges for the lower and upper limits on the variables. The user must also specify the maximum number of cycles, the number of cycles for convergence, and the values of the velocity parameters. Track the progress in the status bar at the bottom of the worksheet.

```
Public Sub PSO()
' Particle Swarm Optimization method for minimization of a multimodal objective
' function with Latin Hypercube Initialization to explore the domain of the variables.
' An adaptive inertia factor uses particle success feedback.
' *** Required VBA procedures from PNMSuite ***
' RNDWH = function for random number generation
' *** Nomenclature ***
Dim bxtl As String ' = input box title
Dim c As Long ' = convergence counter index to check for change between cycles
Dim cmax As Long ' = maximum cycles without change for convergence
Dim c1 As Double, c2 As Double ' = acceleration constants (typically 1.5)
Dim dj() As Double ' = dx for Latin hypercube sampling
Dim dx() As Double ' = size of domain of variable between lower and upper limits
Dim dxl As Double, dxu As Double ' = lower and upper distance to boundaries
Dim f() As Double, fg As Double ' = particle function values and best function value
Dim fr As Range ' = range address of the objective function
Dim g As Long, j As Long, k As Long ' = index for fittest, variables (j) and particles (k)
Dim i As Long, imax As Long ' = cycles and maximum cycles for PSO
Dim m As Long, n As Long ' = number of particles (each with dimension n) number of variables
Dim p() As Double ' = individual particle best position
Dim s As Double, ps As Double ' = success sum and particle success factor
Dim small As Double ' = small number to avoid division by zero
Dim t As Double ' temporary value for shuffling initial coordinates
Dim v() As Double ' = particle "velocity" vector
Dim w As Double ' = inertia weight factor (typically 0.5)
Dim x() As Double ' = array of particle coordinates (optimization variables)
Dim xg() As Double ' = global best particle coordinates
Dim xr As Range ' = range address of variables on the worksheet
Dim xlr As Range, xur As Range ' = range of lower and upper bounds on variables
Dim xl() As Double, xu() As Double ' = lower and upper limits on variables
'*****************************************************************************
small = 0.00000001: bxtl = "PSOL: Particle Swarm Optimization (Minimize)" ' boxtitle
With Application
Set fr = .InputBox(prompt:="Cell of Objective FUNCTION, f(x):", Title:=bxtl, Type:=8)
Set xr = .InputBox(prompt:="Range of VARIABLE(S), x:", Title:=bxtl, Type:=8)
n = xr.Count ' number of variables of optimization
Set xlr = .InputBox(prompt:="Range of Variable LOWER Bounds:", Title:=bxtl, Type:=8)
Set xur = .InputBox(prompt:="Range of Variable UPPER Bounds:", Title:=bxtl, Type:=8)
imax = .InputBox(prompt: "Maximum CYCLES:", Title:=bxtl, Default:=1000, Type:=1)
cmax = .InputBox(prompt:="Cycles with NO CHANGE:", Title:=bxtl, Default:=20 * n, Type:=1)
m = .InputBox(prompt:="Particle POPULATION Size:", Title:=bxtl, _
            Default:=.Min(15 * n, 50), Type:=1)
w = .InputBox(prompt:="Inertia WEIGHT:", Title:=bxtl, Default:=0.5, Type:=1)
c1 = .InputBox(prompt:="Particle ACCELERATION Factor:", Title:=bxtl, Default:=1.5, Type:=1)
```

```
        c2 = .InputBox(prompt:="Global Particle ACCELERATION Factor:", Title:=bxtl, _
                Default:=1.5, Type:=1)
        ReDim dj(1 To n) As Double, f(1 To m) As Double, p(1 To m, 1 To n) As Double, _
            v(1 To m, 1 To n) As Double, x(1 To m, 1 To n) As Double, dx(1 To n) As Double, _
            xg(1 To n) As Double, xl(1 To n) As Double, xu(1 To n) As Double
        For j = 1 To n ' check for relative lower and upper limits on variables
            xu(j) = .Max(xur(j), xlr(j)): xl(j) = .Min(xur(j), xlr(j))
            dx(j) = xu(j) - xl(j): dj(j) = dx(j) / m
        Next j
        For j = 1 To n ' Check for initial variable consistency with limits and first particle
            If xr(j) < xl(j) Or xr(j) > xu(j) Then xr(j) = xl(j) + RNDWH() * dx(j)
        Next j
        For k = 1 To m ' Use Latin Hypercube Sampling to initialize particles
            For j = 1 To n: x(k, j) = xl(j) + (k - 1 + RNDWH()) * dj(j): Next j
        Next k
        For j = 1 To n ' Random shuffle of variables for initial particles
            For k = m To 1 Step -1
                i = Int(k * RNDWH + 1): t = x(i, j): x(i, j) = x(k, j): x(k, j) = t
            Next k
        Next j
        For k = 1 To m ' Initialize the population of particles and velocities
            For j = 1 To n ' variable initialization loop
                p(k, j) = x(k, j): xr(j) = p(k, j): v(k, j) = 0 ' initial particle & velocity
            Next j
            f(k) = fr
            If Not k = 1 Then
                If f(k) < fg Then ' record the global best particle function and index
                    fg = f(k): g = k: For j = 1 To n: xg(j) = x(g, j): Next j
                End If
            Else
                fg = f(k): g = 1 ' use the first particle as the initial global best
            End If
        Next k
        c = 0: ps = 1 ' initialize counter for checking convergence & particle success factor
        For i = 1 To imax ' overall search PSO loop
            c = c + 1:s = 0  ' search convergence counter, reset the success factor
            For k = 1 To m ' particle loop
                For j = 1 To n ' variable loop
                    v(k, j) = ps * w * v(k, j) + RNDWH() * c1 * (p(k, j) - x(k, j)) _
                        + RNDWH() * c2 * (p(g, j) - x(k, j)) ' velocity
                    dxl = Abs(x(k, j) - xl(j)): dxu = Abs(x(k, j) - xu(j)) ' distance to bound
                    x(k, j) = x(k, j) + v(k, j) ' new x
                    If x(k, j) < xl(j) Then ' prevent x from exceeding limits
                        x(k, j) = xl(j) + dxl * Abs((x(k, j) - xl(j)) / v(k, j))
                    ElseIf x(k, j) > xu(j) Then
                        x(k, j) = xu(j) - dxu * Abs((x(k, j) - xu(j)) / v(k, j))
                    End If
                    xr(j) = x(k, j) ' set the variables on the worksheet to calculate f
                Next j
                If fr < f(k) Then ' update particle best coordinates and fitness value
                    f(k) = fr: s = s + 1 ' upgrade the best particle function value
                    For j = 1 To n: p(k, j) = x(k, j): Next j ' save local best particles
                    If f(k) < fg Then ' new global best particle
                        c = 0: fg = f(k): g = k ' reset counter & save global best part. index
                    End If
                End If
            Next k
            If Not c = cmax Then ' check for reaching convergence cycle limit
                ps = s / m ' normalize particle success factor
            Else
                Exit For
            End If
        Next i
        For j = 1 To n: xr(j) = p(g, j): Next j ' report the best particle coordinates
    End With ' Application
End Sub ' PSO
```

Figure 7.22 *VBA* macro PSO for particle swarm optimization of a minimization problem with lower and upper search limits on the variables of optimization.

Example 7.12 Multimodal Objective Test Function for PSO

Known: Test the PSO method with Rastigrin's function:

$$f(x,y) \cong 20 + x^2 + y^2 - 10\left[\cos(2\pi x) + \cos(2\pi y)\right] - 80 \tag{7.57}$$

Figure 7.20 shows a surface plot that reveals several local minima and maxima within the specified ranges of constraints for x and y with the global minimum located at the origin.

Figure 7.23

Surface plot of Equation (7.57)

Find: Use PSO to locate the global minimum of Equation (7.57) in the domain: $-5 < x < 5$ and $-5 < y < 5$.

Analysis: *Excel's* **Solver** with the *GRG* option fails to locate the global minimum after starting the search from a random location in the domain of x and y. To apply the **PSO** macro, provide ranges for the variables and boundary limits in the ranges **B2:B3**, **C2:C3**, and **D2:D3**, as seen in the worksheet. Define the objective function in cell **B4**:

B4=20+B2^2+B3^2-10*(COS(2*PI()*B2)+COS(2*PI()*B3))-80

	A	B	C	D
1	Variables		Lower Limits	Upper Limits
2	x	-1.29337E-09	-5	5
3	y	-5.33785E-09	-5	5
4	f	-80		

Using a population of 40 particles with the default settings for the search parmaters, the sub procedure **PSO** found the global optimum $f = -80$ at $x = 0$ and $y = 0$ after 137 cycles using Latin hypercube initialization and an adaptive inertia factor.

Comments: PSO seems to be computationally expensive. However, the PSO method was able to locate the global minimum of an objective function that challenges other methods. To reduce the computational burden, try reducing the size of the population and search domain. For example, **PSO** found the solution after ~70 cycles with a population of 25 particles.

☐

7.6.4 Firefly Algorithm for Multimodal Optimization*

> Book: "She don't look like much."
> Kaylee: "Oh, she'll fool ya. You ever sailed in a Firefly?" – Firefly Series

The Firefly algorithm, based on the natural attraction among fireflies, incorporates movement of candidate solutions towards the optimum according to the intensity of a fly's position determined by the local and global optima. The natural behaviors of fireflies that attract their mates with bioluminescent flashing inspired the Firefly Algorithm of Yang (2009) for multivariable, multimodal numerical optimization of continuous functions. The Firefly Algorithm is simple to implement and outperforms many competing methods based on evolutionary approaches to optimization. Yang proposed the following simple rules to mimic fireflies applied to numerical optimization:

1. A unique firefly is represented by a set of values for the variables of optimization. The relative magnitude of the objective function evaluated at the values determines the firefly's "brightness".

2. All fireflies are attracted to each other at a level proportional to their "brightness", which is a function of the Cartesian distance between the fireflies:

$$r_{ij}^2 = \sum_{k=1}^{d}\left(x_{i,k} - x_{j,k}\right)^2 \qquad (7.58)$$

where d is the number of dimensions in the objective function, and i, j, and k are the indexes for the two comparison flies and variable coordinates, respectively.

3. For minimization, the less bright fireflies move towards brighter fireflies when $f(x_j) < f(x_i)$. The new position of the firefly is determined from three parts:

$$x_i = x_i + \beta e^{-\gamma r_{ij}^2}\left(x_j - x_i\right) + \alpha_i\left(2RAND - 1\right) \qquad (7.59)$$

or

$$x_i = x_i + \beta e^{-\gamma r_{ij}^2}\left(x_j - x_i\right) + \alpha_i NRD \qquad (7.60)$$

where β is the attractiveness at $r = 0$, γ is an absorption coefficient, and α_i is a randomization parameter. Typically, the absorption factor is defined from a characteristic length scale for the problem as $\gamma = 1/L$, where L may be the order of magnitude of the variable. *RAND* is a random number *uniformly* distributed between zero and one. *NRD* is a normal random deviate defined in Section 3.6.2. Equation (7.60) uses an alternative normal random deviate drawn from a *normal* distribution centered at zero with a standard deviation of one.

A *VBA* macro named **FIREFLY** is available in the *PNMSuite* for the original firefly algorithm. The Firefly algorithm is also an option in the user-form **OPTIMIN**. To use the macro:

1. Specify the initial brightness level, absorption coefficient, and randomization parameter. Typical values or ranges of parameters are $\beta_0 = 1$ and $\alpha_i = 0.01 L_j$. We may use *VBA* statements similar to Figure 3.17 to evaluate $\gamma = 1/L$ from the order of magnitude of the variable of optimization, $L = O(x)$:

```
Gamma(k) = 10 ^ (-Int(Log(Abs(xbest(k))) / Log(10)))
```

2. Specify the number of fireflies and the initial search region. The macro randomly initializes the coordinates for the independent variables of optimization within the specified search region.

The macro automates the remaining steps:

3. Evaluate the objective function at the current coordinates of each firefly.
4. Compare the brightness levels between each pair of fireflies and move the less bright fireflies towards the more bright fireflies according to Equation (7.59) or (7.60).
5. Contract the random factor α after each generation by a factor of 0.97 (or similar value specified by the user).
6. Iterate on the previous two steps for the maximum number of generations, saving the current best fly coordinates.

We added a contraction factor to the randomization parameter to promote faster convergence near the optimum location in a manner similar to Equation (7.47) for Luus and Jaakola's method. McCaffrey[66] noted that the original Firefly algorithm tends to organize smaller swarms of "fireflies" around local minima simultaneous with convergence on the global minimum.

Fister, et al. (2013) reviewed firefly algorithms and discussed how researchers continue to adapt and modify the algorithm to solve a variety of optimization problems. For specific instances of the parameters β and γ the Firefly algorithm is equivalent to three competing methods: Particle Swarm Optimization, Simulated Annealing, and Differential Evolution. As Fister observed, by exploiting in part some advantage from each of these methods, "it is no surprise that the firefly algorithm can perform very efficiently." Research into other randomization schemes has produced improvements over the Gauss distribution. However, we use uniform and normal randomization for convenience in programming with the *PNMSuite*. We demonstrate the method with an example.

---

[66] https://msdn.microsoft.com/en-us/magazine/mt147244.aspx

Example 7.13 Firefly Algorithm Applied to the Michalewicz's Test Function

Known/Find: Use the **FIREFLY** macro in the *PNMSuite* to locate the global minimum of Michalewic's test function for multimodal minimization in Example 7.6.

Analysis: Setup an *Excel* worksheet with initial values for the variables in the range **B2:B3** and search region in the range **C2:C3**. Define the formula from Equation (7.19) in cell **B4** in terms of the named cells for the variables.

	A	B	C
1	Variables		Search Regions
2	x_1	2	1
3	x_2	2	1
4	f	=-SIN(x_1)*SIN((x_1^2)/PI())^20-SIN(x_2)*SIN(2*(x_2^2)/PI())^20	

Run the macro **FIREFLY** from the *PNMSuite* workbook. As prompted by the input boxes, select the function in cell **B4**. Select the variable and search ranges as defined above. Use the default firefly algorithm parameters suggested by the macro. **FIREFLY** locates the global minimum of $f = -1.80$ at $x_1 = 2.20$ and $x_2 = 1.57$.

	A	B	C
1	Variables		Search Regions
2	x_1	2.202894687	1
3	x_2	1.570785796	1
4	f	-1.801303404	

Comments: The solution agrees with the results from inspecting the graphs of the test function in Figure 7.8. We may elect to polish the solution for higher precision using the **Solver**, or one of the macros **SIMNPLEXNLP** and **POWELL**, starting from the best **FIREFLY** result.

□

One drawback of the original firefly algorithm is the lack of convergence criteria. To promote convergence on the global optimum, we modified the algorithm with an additional random term to move weaker fireflies in the direction of the current brightest firefly:

$$x_i = x_i + \beta e^{-r_{ij}^2}\left(x_j - x_i\right) + (1-c)\alpha_i NRD + c\delta RAND\left(x_{best} - x_i\right) \quad (7.61)$$

where the two randomized terms are weighted by a contraction factor calculated from a logistic equation in terms of the current generation g and maximum number of generations g_{max}:

$$c = \left\{1 + \exp\left[-10\left(\frac{2g}{g_{max}} - 1\right)\right]\right\}^{-1} \quad (7.62)$$

The contraction factor starts near zero and increases to one through a sigmoidal curve as the number of generations approaches a maximum set by the user. Thus, in the early stages of the search, the normal random term dominates the search for the global optimum. When the number of search generations passes the halfway point, the influence of the normal random term declines and the swarm of fireflies starts to move in the direction of the current best solution with increasing rapidity, ultimately converging on the global optimum. Our modified firefly algorithm replaces the absorption coefficient with a scaled Cartesian distance between fireflies to remove the effects of different orders of magnitude in the variables of optimization:

$$r_{ij}^2 = \sum_{k=1}^{d}\left[\gamma_k\left(x_{i,k} - x_{j,k}\right)\right]^2 \quad (7.63)$$

Unlike the original firefly algorithm, Equation (7.63) is dimensionally consistent. Two macros using the Davis modifications are available in the *PNMSuite*. The sub procedure **FIREFLYDM** requires the same setup as the original **FIREFLY**, but also requires an additional randomization factor for the best term. The macro uses input boxes with recommended values for the brightness and randomization factors. A **FIREFLY_ANIMATION** macro is available in the Chapter 7 examples folder on the companion web site using the problem from *Example 7.11*. The macro **FIREFLY_ANIMATION** animates the modified search method by updating a plot of the random search coordinates of the swarm of fireflies after each generation, as shown in the sequence of contour plots in Figure 7.24.

Generations: 0   15   30   45   60

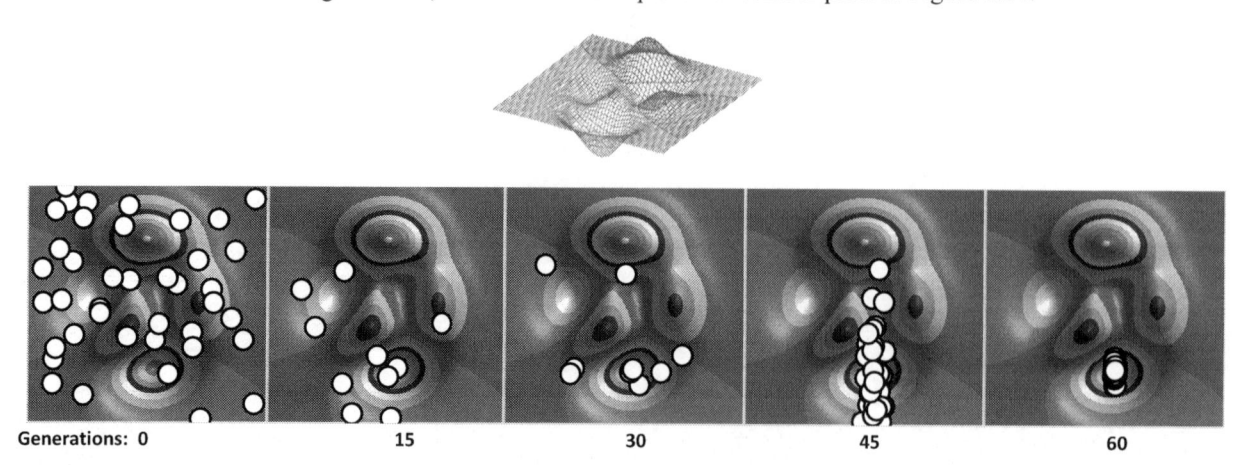

Figure 7.24   Firefly animation of random swarm coordinates in Example 7.11 at increasing generations.

The surface plot shows the peaks and valleys of the landscape. Early generations show the swarm of fireflies "buzzing" around two different local minima near the left and bottom regions of the contour plots. In later generations, the fireflies congregate along a trough between the global minimum and maximum, finally settling on the global minimum in the lower section of the plot. Explore the parameter space with the animation macro to acquire a feel for how the various parameters influence global optimization by the Davis modified Firefly Algorithm.

7.6.5 *Evolutionary Genetic Algorithm**

"Patience is a virtue." – Prudentius

Evolutionary algorithms use randomly generated populations of candidate solutions to search for the location of the global optimum of nonsmooth or multimodal objective functions, as the one graphed in Figure 7.8. The evolutionary algorithm described here was designed to mimic biological processes of natural selection for iteratively upgrading the population of candidates to converge on the overall best candidate. Unlike direct search algorithms that require derivatives, the evolutionary algorithm presented here works with nonsmooth objective functions that may not be twice differentiable everywhere. We develop a *VBA* macro for an evolutionary method and follow up in Section 7.7 on the **Solver's** capabilities for optimization that also includes an evolutionary genetic search option.

The genetic algorithm mimics the process of natural selection by iterating through several generations involving gene-crossover between parent chromosomes and mutation of genes within populations of chromosomes. Evolutionary genetic algorithms are also useful for finding the global optimum in multimodal problems, such as Figure 7.24, or nonsmooth functions.[67] We outline the following steps for a typical implementation of the genetic algorithm:

1. Formulate the objective function, $f(x)$ in a cell on a worksheet in terms of a separate range of cells containing the variables of optimization, x. Each unique variable in the objective function is a *gene*, and a distinct set of optimization variables makes up a *chromosome*. The two-dimensional Michaelewicz problem in *Example 7.13* has two genes (x_1 and x_2) per chromosome.

2. Specify the number of chromosomes, n_p, in the population of the candidate pool. A larger population improves our chances of locating the optimum, but increases the computational burden on the search process, which may superfluously lengthen the time to reach a solution. We may need to experiment with the size of the population to find the globally best solution.

---

[67] Pattern search methods such as Luus-Jaakola and Firefly tend to outperform genetic algorithms for nonsmooth functions.

3. Initialize an array containing the population of chromosomes by randomly selecting values for each variable, or gene, within constraints on the search domain for each variable. For a variable constrained by $x_{min} \leq x \leq x_{max}$, the values for the genes of each chromosome are selected to fall within the domain of the constraints:

$$x = x_{min} + r \cdot (x_{max} - x_{min}) \quad (7.64)$$

where r is a uniformly distributed random number, $0 \leq r \leq 1$.

4. Evaluate the objective function for each chromosome to determine the initial population fitness in terms of the relative values for the objective function. In the case of minimization, a chromosome that produces a smaller value of the objective function has a higher level of fitness relative to chromosomes that produce larger values of the objective function. The fitness score for each chromosome is normalized to values that range from zero to one in terms of the objective function in the population ($0 \leq Z \leq 1$):

$$\text{Fitness Score, } z_j = \frac{f_{max} - f}{f_{max} - f_{min}} \quad \rightarrow \quad \text{Normalized Fitness Score, } Z_j = \frac{z_j}{\sum_{i=1}^{n_p} z_i} \quad (7.65)$$

The normalized fitness score Z_j represents the probability of selecting chromosome j for gene crossover. In this way, genes from the "more fit" chromosomes tend to endure in subsequent generations, whereas genes from "less fit" chromosomes become extinct from the gene pool.

5. We use a simple roulette wheel algorithm[68] for selecting parent chromosomes proportional to their fitness score:

Generate a uniform random number, r, from a pseudo-random number generator, such as RNDWH
Initialize the cumulative probability, p = 0
For i = 1 to np (size of the chromosome population)
    p = p + Fitness Score i
    If p > r then select chromosome i and exit the For loop

The roulette wheel algorithm works on the principle that larger fitness scores cover a larger span of the cumulative probability distribution, thus endowing them with a higher likelihood of selection. A pair of parent chromosomes is randomly selected from the current population proportional to their fitness to generate the offspring chromosomes for the next generation. There is a variety of crossover methods. In the simplest form, we simply swap half of the genes in a pair of parent chromosomes. Alternatively, we may replace the genes in a chromosome with random weighted-average values from two parent chromosomes:

$$x = r \cdot x_{Mother} + (1-r) \cdot x_{Father} \quad (7.66)$$

6. Specify the mutation rate, r_m. The mutation rate is simply the probability of mutation in a chromosome. To implement random mutation in a chromosome, we first generate a random number for comparison with the mutation rate. If the mutation rate exceeds the random number, then we randomly mutate a single gene in the chromosome by random selection with equal probability:

Generate a uniform random number, r, from a pseudo-random number generator, such as RNDWH
If $r_m > r$ then
    Generate a new random number, r, from a pseudo-random number generator to select the gene for mutation
    Gene index j = ROUNDUP(r·number of genes, 0)
    $x_j = x_{j\,min} + r \cdot (x_{j\,max} - x_{j\,min})$

As with the size of the population, we may need to experiment with the level of the mutation rate to improve the solution. A smaller mutation rate slows down the rate of convergence on the optimum. However, a larger mutation rate may *prevent* the algorithm from converging on the global optimum by frequently replacing genes having superior fitness with inferior genes (Bourg 2006).

---

[68] https://en.wikipedia.org/wiki/Fitness_proportionate_selection

After specifying the population size and mutation rate, we iterate through successive generations of mutation and crossover as the population of chromosomes converges on the location for the global optimum. One of the disadvantages of the genetic algorithm is finding an appropriate stopping criterion. Unlike Luus-Jaakola and the firefly algorithms, we cannot use the relative values of the chromosome population to check for convergence attributed to random mutations. Instead, we allow the algorithm to cycle through a specified number of generations and save the best chromosome as the optimal result.

To illustrate, the *PNMSuite* has a *VBA* sub procedure named **GENETIC**, listed in Figure 7.25, for demonstrating a simple version of the evolutionary genetic algorithm to minimize an objective function with minimum and maximum constraints on each of the variables. Note that the number of variables must be two, or more to allow for crossover between genes. For single variable optimization of non-smooth objective functions with discontinuities, try the method of Luus and Jaakola or the firefly algorithm. The code uses **GoSub**... **Return** statements for subroutines within the procedure to handle repeated calls for fitness normalization, probability proportionate fitness selection by the roulette wheel algorithm for crossover, and random gene mutation within the chromosome. We replace the previously most-fit chromosome with new offspring only when it improves the fitness in order to retain the global best result in each subsequent generation. After the last generation, the program applies one application of the **Solver's GRG Nonlinear** method, described further in Section 7.7, to sharpen up the precision of the best solution at the global optimum.

```
Public Sub GENETIC()
' Evolutionary genetic algorithm for minimizing a multimodal objective function. The
' user formulates the objective function in the worksheet in terms of the variables of
' optimization located in a contiguous range of cells on the worksheet.  GENETIC
' requires ranges of minimum and maximum values that constrain the search regions for
' the variables.  The result from the genetic algorithm is refined for the local
' minimum with the Solver GRG Nonlinear method.
' *** Nomenclature ***
Dim bxttl As String ' = string for title in input and message boxes
Dim c() As Double, c1() As Double, cmin() As Double ' = chromosomes for two generations
Dim dx() As Double ' = span of constraints on x
Dim fmin As Double ' = minimum value of the objective function
Dim fr As Range ' = range of objective function
Dim ft As Double ' total fitness
Dim i As Integer ' = index for chromosomes in a population
Dim j As Integer, jj As Integer, k As Long ' = index for genes and generations
Dim m As Double ' = mutation rate (probability)
Dim nmax As Integer ' = maximum number of generations
Dim ng As Integer, np As Integer ' = number of genes & size of chromosome population
Dim p As Double ' = cumulative probability
Dim p1() As Double, p2() As Double ' parent chromosomes 1 and 2
Dim r As Double ' = pseudo-random number drawn from a uniform 0 to 1 distribution
Dim s As Integer, s1 As Integer, s2 As Integer ' = selected chromosome for crossover
Const small As Double = 0.000000000001 ' = small number avoids division by zero
Dim xr As Range ' = range of variables of optimization and min/max constraints
Dim xminr As Range, xmaxr As Range ' = range of variables min/max constraints
Dim tol As Double ' = relative convergence tolerance
Dim z() As Double, zmin As Double, zmax As Double ' = normalized fitness scores
'******************************************************************************
bxttl = "GENETIC Algorithm for Minimization": tol = 0.00000001
With Application
Set fr = .InputBox(prompt:="OBJECTIVE Function, f(x):", Title:=bxttl, Type:=8)
Set xr = .InputBox(prompt:="Range of Optimization VARIABLES, x:", Title:=bxttl, Type:=8)
Set xminr = .InputBox(prompt:="Range of MINIMUM Constraints, x MIN: ",Title:=bxttl, Type:=8)
Set xmaxr = .InputBox(prompt:="Range of MAX Constraints, x MAX:", Title:=bxttl, Type:=8)
np = .InputBox(prompt:="POPULATION Size:", Title:=bxttl, Default:=100, Type:=1)
m = .InputBox(prompt:="MUTATION Rate:", Title:=bxttl, Default:=0.1, Type:=1)
nmax = .InputBox(prompt:="Maximum GENERATIONS:", Title:=bxttl, Default:=100, Type:=1)
ng = xr.Count ' number of genes in a chromosome
If ng <> xminr.Count Or ng <> xmaxr.Count Then
    MsgBox "Length of ranges of x, xmin, or xmax don't match": Exit Sub
End If
ReDim c(1 To np, 1 To ng) As Double, c1(1 To np, 1 To ng) As Double, _
      cmin(1 To ng) As Double, dx(1 To ng) As Double, _
      p1(1 To ng) As Double, p2(1 To ng) As Double, z(1 To np) As Double
For j = 1 To ng ' get the span for the constraints and save the initial values for xr
```

```
            dx(j) = xmaxr(j) - xminr(j): cmin(j) = xr(j)
    Next j
    fmin = fr ' initial value of the objective function
    For i = 1 To np ' Initialize the chromosomes with random values within the constraints
        For j = 1 To ng
            c(i, j) = xminr(j) + RNDWH() * dx(j)
            xr(j) = c(i, j) ' put the chromosome on the worksheet to calculate f
        Next j
        z(i) = fr ' save the function value for each chromosome
        If z(i) < fmin Then ' save the current best result for minimization
            fmin = z(i)
            For j = 1 To ng: cmin(j) = c(i, j): Next j
        End If
    Next i
    For k = 1 To nmax ' Evolutionary loop for natural selection
        GoSub FITNESS ' normalize the fitness scores
        For i = 1 To np ' offspring and mutation loop
            GoSub ROULETTE: s1 = s: GoSub ROULETTE: s2 = s ' select parent chromosome
            For j = 1 To ng ' offspring from cross-over with averaged values
                r = RNDWH: c1(i, j) = r * c(s1, j) + (1 - r) * c(s2, j) ' weighted average
            Next j
        Next i
        For i = 1 To np
            GoSub MUTATE ' randomly mutate a gene in a chromosome
            For j = 1 To ng
                xr(j) = c1(i, j): c(i, j) = c1(i, j)
            Next j
            z(i) = fr
            If z(i) < fmin Then ' save the current best result
                fmin = z(i)
                For j = 1 To ng: cmin(j) = c(i, j): Next j
            End If
        Next i
    Next k
    For j = 1 To ng: xr(j) = cmin(j): Next j ' put global best chromosome on worksheet
    ' Use the Solver's GRG Nonlinear method to improve the final solution
    solverok SetCell:=fr.Address, MaxMinVal:=2, ByChange:=xr.Address, _
        Engine:=1, EngineDesc:="GRG Nonlinear"
        SolverOptions Convergence:=tol, Scaling:=True, AssumeNonNeg:=False, Derivatives:=2
        SolverSolve UserFinish:=True
    beep: Exit Sub
    FITNESS: ' subroutine for normalizing fitness scores to range from 0 to 1
        zmax = .Max(z): zmin = .Min(z) ' get the range of fitness values
        For i = 1 To np: z(i) = (zmax - z(i)) / (zmax - zmin + small): Next i
        ft = .sum(z)
        For i = 1 To np: z(i) = z(i) / ft: Next i
    Return
    ROULETTE: ' subroutine selects parent chromosome for crossover by roulette algorithm
        p = 0: r = RNDWH
        For s = 1 To np ' cumulative probability for fitness scores
            p = p + z(s): If p > r Then Exit For
        Next s
    Return
    MUTATE: ' subroutine for random gene mutation in a chromosome
        If RNDWH < m Then ' criteria to mutate a gene
            jj = .RoundUp(ng * RNDWH, 0) ' randomly select a gene in a chromosome
            c1(i, jj) = xminr(jj) + RNDWH() * dx(jj) ' mutate the gene
        End If
    Return
    End With ' Application
End Sub
```

Figure 7.25 *VBA* sub procedure GENETIC for locating the global minimum of a multimodal objective function in an *Excel* worksheet using the genetic algorithm.

7.7 Excel's Solver Toolkit for Optimization

Perhaps without recognizing it, we were optimizing when we used *Excel*'s **Solver** to find the roots to nonlinear equations in Chapter 6. Section 2.5 shows how to access the **Solver**. The **Solver** offers three options for optimization in terms of either a maximum, set value, or minimum of an objective function in an *Excel* worksheet:

1. Direct search *Simplex* linear programming (Bunday 1984) for finding solutions to linear objective functions with linear constraints (not to be confused with Nelder and Meade's nonlinear downhill simplex described in Section 7.5.6). We illustrate linear programming with an example.
2. *Generalized Reduced Gradient* (*GRG*) indirect search for continuous, smooth, nonlinear problems. *GRG* combines constraints with the objective function using Lagrange multipliers described in Section 7.8.1 for solution by a version of the quasi-Newton method of unconstrained problems.
3. *Evolutionary* solver using a genetic algorithm for optimizing multimodal and nonsmooth objective functions, or functions with discontinuities that are not amenable to indirect methods that require derivative approximations

Example 7.14 Solver Optimization Using Simplex Linear Programming

Known: A chemical company produces two different products, *W* and *Z*, in three process lines. The profits per pound of product *W* and *Z* are $5 and $3 respectively. The available weekly production times of three process lines are 40 hours, 36 hours, and 36 hours, respectively.

Find: Given the matrix of processing times in the *Excel* worksheet, calculate the weekly production rates of products *W* and *Z* to maximize the profit.

Analysis: Derive the profit-maximization objective function in terms of the weekly amounts of products:

$$\text{Total Profit}/\$ = P_W \cdot \sum_1^3 R_W + P_Z \cdot \sum_1^3 R_Z \tag{7.67}$$

where P_W and P_Z are product profits, and R_W and R_Z are the weekly production rates of each product, respectively. Note that the objective function in Equation (7.67) is a linear function of the variables of optimization. The linear programming option does not require iteration and finds the solution exactly.

Set up the worksheet to calculate the process times for each line subject to the weekly processing time constraints. Provide initial guesses of 50 for the weekly production rates in column **B**:

	A	B	C	D	E	F	G	H	I	J	K
1	Product	Rate	Time	Time	Process	Time	Constraint	Product	Unit Profit	Profit	
2	Line	lb/wk	hr/lb	hr/wk	Line	hr/wk	hr/wk		$/lb	$/wk	
3	W1	50	0.5	=B3*C3	1	=D3+D6	40	W	5	=I3*SUM(B3:B5)	
4	W2	50	0.4	=B4*C4	2	=D4+D7	36	Z	3	=I4*SUM(B6:B8)	
5	W3	50	0.2	=B5*C5	3	=D5+D8	36		Total Profit	=SUM(J3:J4)	$/wk
6	Z1	50	0.25	=B6*C6							
7	Z2	50	0.3	=B7*C7							
8	Z3	50	0.4	=B8*C8							

Launch the add-in **Solver** from the **Data** tab on the *Excel* ribbon, select the *Simplex LP* solving method, and setup the input boxes for the objective function, variables, and constraints:

CHAPTER 7: OPTIMIZATION

[Solver dialog box screenshot showing Set Objective: $J$5, Max selected, By Changing Variable Cells: $B$3:$B$8, Subject to the Constraints: $F$3 <= $G$3, $F$4 <= $G$4, $F$5 <= $G$5, Make Unconstrained Variables Non-Negative checked, Select a Solving Method: Simplex LP (circled).]

The *Simplex LP* method finds the distribution of production rates among the process lines exáctly within the constraints. The solution for the production rates in column **B** indicate that process line 1 produces all of product Z and lines 2 and 3 share the production of product W.

	A	B	C	D	E	F	G	H	I	J	K
1	Product	Rate	Time	Time	Process	Time	Constraint	Product	Unit Profit	Profit	
2	Line	lb/wk	hr/lb	hr/wk	Line	hr/wk	hr/wk		$/lb	$/wk	
3	W1	0	0.5	0	1	40	40	W	5	1350	
4	W2	90	0.4	36	2	36	36	Z	3	480	
5	W3	180	0.2	36	3	36	36		Total Profit	1830	$/wk
6	Z1	160	0.25	40							
7	Z2	0	0.3	0							
8	Z3	0	0.4	0							

Comments: The objective function guides the solution setup by determining the variables of optimization. It is critical to understand the difference between the variables of optimization and operation for this example. The set points for the operating conditions that we can control are determined from the values of optimal production rates.

□

The default setup for the **Solver** finds a *local* value of the optimum by *GRG*. Try searching for the optimum from different starting points to find the global optimum. **Solver** has a multi-start option in *GRG* for initiating the search from random points, similar to the method of Luus and Jaakola, but without shrinking the search domain. By default, the random search begins at the starting points ±1. To reset the random search region, set constraints on the variables. For example, if we constrain $-10 \leq x \leq 10$, then the **Solver** restarts from random points between -10 and 10.

Starting with *Excel* 2010, the **Solver** added a genetic evolutionary algorithm[69] to expand its capability for handling discontinuous, discrete, or multimodal objective functions. In theory, the **Solver's** implementation of the evolutionary genetic algorithm works on both constrained and unconstrained problems. In our experience, the **Solver** implementation of the evolutionary genetic algorithm requires constraints. The **Solver** has a default population size of 100 and mutation rate of 0.075. We illustrate potential problems and workarounds using the *GRG Nonlinear* and *Evolutionary* **Solver** methods with three cases of the following dual mode objective function in one variable:

$$f(x) = 1 - \sin(x+1)\sin\left[\frac{2}{\pi}(x+1)^2\right]^{20} \qquad 0 \leq x \leq 2 \qquad (7.68)$$

To begin, we set up the problem in an *Excel* worksheet as seen in Figure 7.26 and plotted in Figure 7.27. For all cases, we used the **Solver** with lower and upper limit constraints on *x*.

---

[69] http://www.solver.com/genetic-evolutionary-introduction

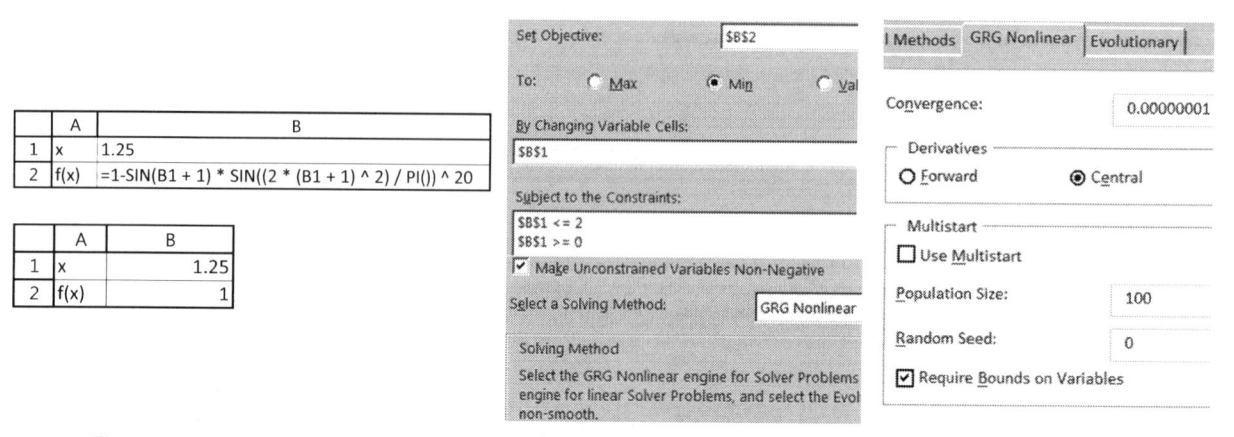

Figure 7.26 Implementation of Equation (7.68) in an Excel worksheet for minimization using the Solver with constraints on *x*. The *GRG* Nonlinear options show the convergence tolerance and central derivative approximation used by Newton's indirect search method.

Case 1: We start from an initial value for $x = 1.25$ midway between the two minima shown in Figure 7.27, and select the faster *GRG* Nonlinear solving method. Unfortunately, the **Solver** is unable to find an improvement to the initial value due to the starting location in a flat region where the derivative of the function in Newton's method is ~ zero, which fools the algorithm that it has found the calculus minimum where $df/dx = 0$.

Case 2: We initialize the value for $x = 1.5$ and try *GRG* Nonlinear again. The **Solver** finds quickly the *local* minimum nearby at $x = 1.71$. However, from the plots in Figure 7.27, we see the true global minimum at $x = 0.57$. We recommend selecting the multi-start option for *GRG* Nonlinear to locate the global minimum within the constraints in this case.

Case 3: We start from the **Solver** solution for Case 2 at the local minimum $x = 1.71$, and select the *Evolutionary* solving method with the default options. After a relatively large number of trials when compared with *GRG*, the **Solver** jumps out of the local minimum and locates the true global minimum at $x = 0.57$!

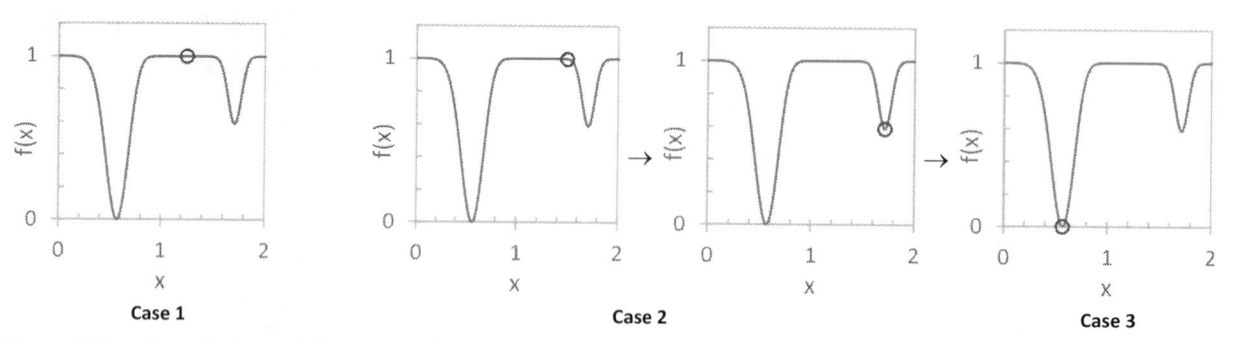

Figure 7.27 Three Solver solution methods for the minimum of the objective function in Equation (7.68): (1) *GRG*, Initial $x = 1.25$, fails to locate a minimum at the final $x = 1.25$ (2) *GRG*, Initial $x = 1.5$, finds a local minimum at the final $x = 1.71$ (3) *Evolutionary*, Initial $x = 1.71$, finds the global minimum at the final $x = 0.57$.

Also by default, the **Solver** assumes non-negative variables of optimization. If the solution allows for negative variables, be sure to uncheck the box: **Make Unconstrained Variables Nonnegative**. Despite the **Solver**'s sophistication and variety of tools for handling linear, nonlinear, and discontinuous functions, we may find that some problems need alternative optimization methods, such as those presented earlier in this chapter. We recommend trying the **Solver** first to take advantage of its speed, then turning to the *VBA* macros if needed.

CHAPTER 7: OPTIMIZATION

Example 7.15 Solver Optimization using Multi-Start and Evolutionary methods

Known: Consider the following test function for multimodal minimization (Bourg 2006):

$$f(x,y) = 1 - \frac{\cos(x^2+y^2)}{x^2+y^2+0.5} \tag{7.69}$$

Find: Calculate the global minimum using the multi-start feature of the **Solver**. (a) Start from $x = 5$ and $y = 5$ without bounds. (b) Start again after setting constraints $-5 \le x \le 5$ and $-5 \le y \le 5$ and checking **Required Bounds**.

Analysis: Notice that the objective function is symmetric in variables x and y. Use the macro **QYXPLOT** to plot the objective function in variable x at $y = 0$. By inspection, we observe that the global minimum is located at the origin.

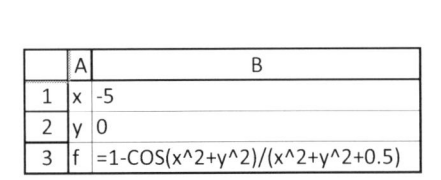

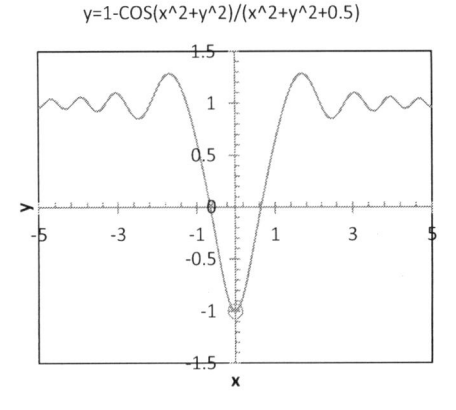

	A	B
1	x	-5
2	y	0
3	f	=1-COS(x^2+y^2)/(x^2+y^2+0.5)

If we use the **Solver** with multi-start, but without setting the constraints on x and y, it finds a local minimum near (5.012, 5.012) using random starting points in the range $4 \le x \le 6$ and $4 \le y \le 6$. With constraints on the cells for $-5 \le x \le 5$ and $-5 \le y \le 5$, the multi-start option finds the true global minimum at (0, 0).

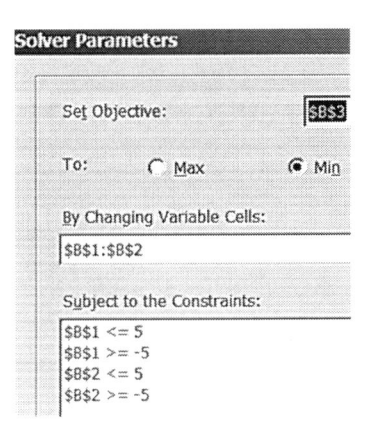

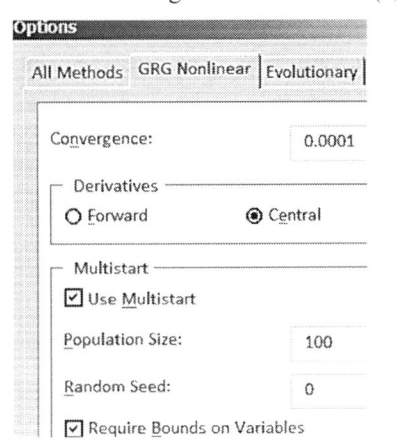

When we selected the *Evolutionary* solving method *without* constraints, the **Solver** appeared to make no change from the initial condition. After applying the constraints, the *Evolutionary* genetic algorithm using the default options locates the true global optimum. We recommend one additional application of the *GRG* Nonlinear solving method staring from the *Evolutionary* solution to refine the result for the location of the optimum. The sub procedure **GENETIC** arrives at the same conclusion using the search region defined by the constraints.

Comments: When having trouble finding the optimum, try options for central derivatives or increasing the population size for multi-start. For comparison, the macro **LUUSJAAKOLA** finds the global minimum with the default settings when starting from (5, 5) with search regions of 5 for each variable. However, the **Solver** is much faster.

□

7.8 Constraints on the Search Region

Engineers tackle a variety of problems with specific conditions that constrain the variables for feasible solutions. Examples of working constraints include safety, ethical, legal, environmental, economic, and physical limitations. However, mathematical models of these kinds of problems need *numerical* constraints on the variables of optimization. Occasionally, combinations of inequalities constrain the solution to the objective function. For example, consider an *n*-dimensional objective function, *f*, with constraints, *c*:

$$f_1(x_1, x_2, \ldots, x_n) = 0 \tag{7.70}$$

$$c_1(x_1, x_2, \ldots, x_n) = 0 \tag{7.71}$$

$$c_2(x_1, x_2, \ldots, x_n) \le 0 \tag{7.72}$$

We may reverse the order of any inequality simply by multiplying by negative one. To illustrate, consider the second constraint in Equation (7.72):

$$-c_2(x_1, x_2, \ldots, x_n) \ge 0 \tag{7.73}$$

Try removing at least one constraint by rearrangement and substitution. For example, consider the following objective function and constraint:

$$f(x) = Ax_1^2 + Bx_2^2 \tag{7.74}$$

$$c(x) = 0 = Cx_1^2 + Dx_2^2 - E \tag{7.75}$$

Solve Equation (7.75) for x_2 and substitute into Equation (7.74) to eliminate both the constraint and variable x_2:

$$x_2 = \sqrt{(E - Cx_1^2)/D} \tag{7.76}$$

$$f(x) = Ax_1^2 + \frac{B}{D}(E - Cx_1^2) = \left(A - \frac{BC}{D}\right)x_1^2 + \frac{BE}{D} \tag{7.77}$$

The unimodal optimization strategies presented in this book require *unconstrained* objective functions. Several techniques are available for eliminating constraints from the optimization problem by transforming the objective function, constraints, or variables.

7.8.1 Lagrange Multipliers

> *"When we ask for advice, we are usually looking for an accomplice."* – Count Joseph-Louis Lagrange

Apply the method of Lagrange multipliers to problems with equality constraints. *Excel*'s **Solver** uses Lagrange multipliers to enforce constraints. To illustrate the technique, consider an objective function *f* in two independent variables *x* and *y*, subject to an equality constraint function *c*:

$$f(x, y) \qquad \text{and} \qquad c(x, y) = 0 \tag{7.78}$$

A Taylor series expansion of the objective function in Equation (7.78) yields:

$$f(x + \Delta x, y + \Delta y) = f(x, y) + \Delta x \frac{\partial f}{\partial x} + \Delta y \frac{\partial f}{\partial y} + \ldots \tag{7.79}$$

The first-order terms in Equation (7.79) must sum to zero at the optimum point for *x* and *y* according to Equation (7.78):

CHAPTER 7: OPTIMIZATION

$$\Delta x \frac{\partial f}{\partial x} + \Delta y \frac{\partial f}{\partial y} = 0 \qquad (7.80)$$

Similarly, for the constraint function:

$$\Delta x \frac{\partial c}{\partial x} + \Delta y \frac{\partial c}{\partial y} = 0 \qquad (7.81)$$

Use an arbitrary multiplier λ to permit the summation of Equations (7.80) and (7.81) while combining terms in Δx and Δy:

$$\Delta x \left(\frac{\partial f}{\partial x} + \lambda \frac{\partial c}{\partial x} \right) + \Delta y \left(\frac{\partial f}{\partial y} + \lambda \frac{\partial c}{\partial y} \right) = 0 \qquad (7.82)$$

An undetermined value of λ allows us to separate the terms in Equation (7.82) as follows:

$$\Delta x \left(\frac{\partial f}{\partial x} + \lambda \frac{\partial c}{\partial x} \right) = 0 \qquad \text{and} \qquad \Delta y \left(\frac{\partial f}{\partial y} + \lambda \frac{\partial c}{\partial y} \right) = 0 \qquad (7.83)$$

Locate the optimum point from the simultaneous solution of Equations (7.83) with the equality constraint in Equation (7.78) for x, y, and λ. The multiplier λ has no particular significance other than to reflect the constraint on the optimum.
    Functions in more than two independent variables, or multiple equality constraints, follow this same pattern of solution. For example, an objective function with three independent variables, $f(x, y, z)$, and two equality constraints, gives a system of equations in five unknowns that includes two Lagrange multipliers, λ_1 and λ_2:

$$\frac{\partial f}{\partial x} + \lambda_1 \frac{\partial c_1}{\partial x} + \lambda_2 \frac{\partial c_2}{\partial x} = 0 \qquad \frac{\partial f}{\partial y} + \lambda_1 \frac{\partial c_1}{\partial y} + \lambda_2 \frac{\partial c_2}{\partial y} = 0 \qquad \frac{\partial f}{\partial z} + \lambda_1 \frac{\partial c_1}{\partial z} + \lambda_2 \frac{\partial c_2}{\partial z} = 0 \qquad (7.84)$$

$$c_1(x, y, z) = 0 \qquad \text{and} \qquad c_2(x, y, z) = 0 \qquad (7.85)$$

Use the methods of solving simultaneous equations from Chapter 4 and Chapter 6 to find the optimum values for x, y, and z, as well as the Lagrange multipliers, λ_1 and λ_2. For convenience, approximate the partial derivatives in the expansions using the finite-difference forms derived in Chapter 5.

Example 7.16   Gibbs Reaction Free Energy Minimization

Known: Production of synthesis gas from char involves the water-gas shift reaction (Lwin 2000):

$$CO + H_2O \rightleftharpoons CO_2 + H_2 \qquad (7.86)$$

Schematic: The reactor feed consists of water and carbon monoxide. The exit stream consists of equilibrium concentrations of CO, H_2O, CO_2, and H_2.

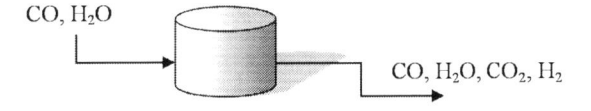

CO, H$_2$O → [reactor] →   CO, H$_2$O, CO$_2$, H$_2$

Find: Given the ratio of water to carbon monoxide in the feed, determine the exit-stream mole fractions at 1000 K and one atm.

Assumptions: chemical equilibrium with excess char.

Analysis: Use *Excel*'s **Solver** to minimize the Gibbs free energy with respect to exit composition:

$$G = \sum_i (n_i \Delta G_i) + RT \sum_i \left[n_i \ln(y_i) \right] \qquad (7.87)$$

where G = total Free Energy, J/mol y_i = mole fraction of species i

R = ideal gas constant, 8.314 J/(mol·K) T = temperature, K

The standard Gibbs free energies of formation at 1 atm and 1000 K for each species are:

$\Delta G_{H2O} = -1.9242 \times 10^5$ J/mol $\Delta G_{CO} = -2.0024 \times 10^5$ J/mol

$\Delta G_{H2} = 0$ J/mol $\Delta G_{CO2} = -3.9579 \times 10^5$ J/mol

The reaction stoichiometry constrains the mole balances. For a given feed ratio r of H_2O to CO, and a basis of one mole CO, elemental species balances give:

- Carbon $1 = n_{CO} + n_{CO2}$ (7.88)

- Hydrogen $r = n_{H2O} + n_{H2}$ (7.89)

- Oxygen $r + 1 = n_{CO} + 2n_{CO2} + n_{H2O}$ (7.90)

where n_i is the moles of component i. We use transformed variables to prevent the search method from selecting a negative term in the logarithms:

$$n_{CO} = z_{CO}^2 \quad n_{H2O} = z_{H2O}^2 \quad n_{CO2} = z_{CO2}^2 \quad n_{H2} = z_{H2}^2 \quad (7.91)$$

Alternatively, we can check the **Solver** option box **Make Unconstrained Variables Nonnegative**. Calculate the mole fraction for each species from the total moles exiting the reactor:

$$y_i = \frac{n_i}{n_{CO} + n_{H2O} + n_{CO2} + n_{H2}} \quad (7.92)$$

Figure 7.28 shows the **Solver** optimized solution for a feed ratio of two moles of water per mole of CO.

	A	B	C	D	E
1	R_g	8.314472	J/(mol K)	Ratio mols H₂O/mols CO	2
2	T	1000	K		
3	Species	ΔG/(J/mol)	z	n/mol	
4	H₂O	-192420	1.13064635431532	=C4^2	
5	CO₂	-395790	0.849493387230104	=C5^2	
6	H₂	0	0.849493387176119	=C6^2	
7	CO	-200240	-0.52759972230682	=C7^2	
8	Total	=A15+Rg*T*D15		=SUM(D4:D7)	
9					
10	n*ΔG	y	ln(y)	n*ln(y)	
11	=B4*D4	=D4/$D$8	=LN(B11)	=D4*C11	
12	=B5*D5	=D5/$D$8	=LN(B12)	=D5*C12	
13	=B6*D6	=D6/$D$8	=LN(B13)	=D6*C13	
14	=B7*D7	=D7/$D$8	=LN(B14)	=D7*C14	
15	=SUM(A11:A14)	=SUM(B11:B14)	=SUM(C11:C14)	=SUM(D11:D14)	
16					
17	Constraints				
18	Carbon	=D5+D7-1			
19	Hydrogen	=D4+D6-ratio			
20	Oxygen	=D4+2*D5+D7-ratio-1			

CHAPTER 7: OPTIMIZATION

	A	B	C	D	E
1	R_g	8.314472	J/(mol K)	Ratio mols H_2O/mols CO	2
2	T	1000	K		
3	Species	ΔG/(J/mol)	z	n/mol	
4	H_2O	-192420	1.1	1.2784	
5	CO_2	-395790	0.8	0.7216	
6	H_2	0	0.8	0.7216	
7	CO	-200240	-0.5	0.2784	
8	Total	-619006		3.0000	
9					

10	n*ΔG	y	ln(y)	n*ln(y)
11	-245982	0.426	-0.8530	-1.0905
12	-285618	0.241	-1.4248	-1.0282
13	0	0.241	-1.4248	-1.0282
14	-55739	0.093	-2.3774	-0.6618
15	-587339	1.000	-6.0802	-3.8087
16				
17	Constraints			
18	Carbon	4.82E-07		
19	Hydrogen	1.93E-07		
20	Oxygen	6.75E-07		

Figure 7.28 *Excel* **worksheet for Gibbs free energy minimization (formulas and results).**

The **Solver** parameters shown in Figure 7.29 include the conservation of mass constraints.

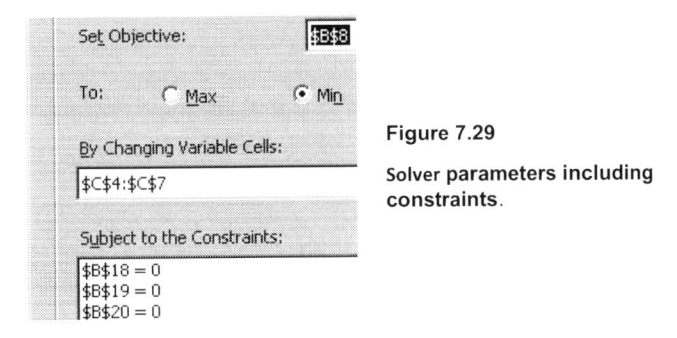

Figure 7.29

Solver parameters including constraints.

Set Objective: $B$8
To: Max ● Min
By Changing Variable Cells:
$C$4:$C$7
Subject to the Constraints:
$B$18 = 0
$B$19 = 0
$B$20 = 0

Select **Sensitivity** in the **Solver Results** message box and click **OK**. The **Solver** creates a new worksheet that contains the selected reports. For this problem, the **Solver** generates a report that gives the Lagrange multipliers, as shown in Figure 7.30.

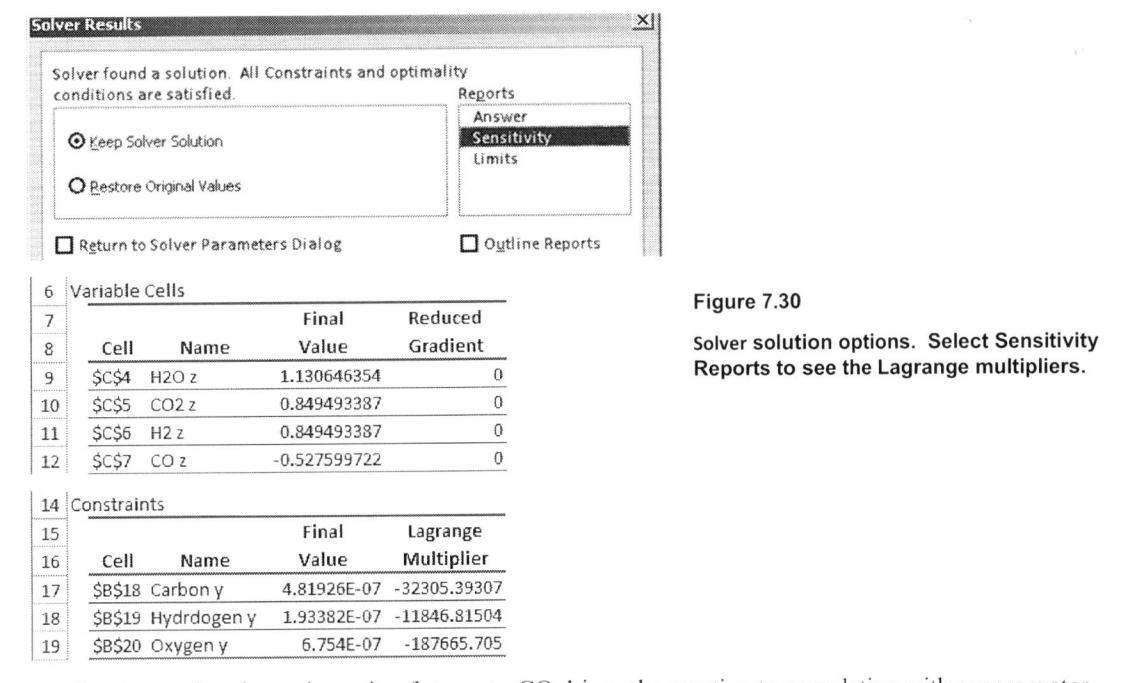

Figure 7.30

Solver solution options. Select Sensitivity Reports to see the Lagrange multipliers.

Variable Cells

Cell	Name	Final Value	Reduced Gradient
$C$4	H2O z	1.130646354	0
$C$5	CO2 z	0.849493387	0
$C$6	H2 z	0.849493387	0
$C$7	CO z	-0.527599722	0

Constraints

Cell	Name	Final Value	Lagrange Multiplier
$B$18	Carbon y	4.81926E-07	-32305.39307
$B$19	Hydrdogen y	1.93382E-07	-11846.81504
$B$20	Oxygen y	6.754E-07	-187665.705

Comments: Increasing the molar ratio of steam to CO drives the reaction to completion with excess water.

7.8.2 Slack Variables for Inequality Constraints

Transform an *inequality* into an *equality* by inserting a new "slack" variable. In our illustration, the constraint in Equation (7.72) becomes:

$$c_3(x_1, x_2, \ldots, x_n, x_{n+1}) = x_{n+1}^2 + c_2(x_1, x_2, \ldots, x_n) = 0 \tag{7.93}$$

where x_{n+1} is the new slack variable. Modifications to the objective function reflect new additional variable in the manner of Equation (7.2). For our illustration:

$$f_2(x_1, \ldots x_{n+1}) = f_1(x_1, \ldots x_n) \pm \mu \left[c_1^2(x_1, \ldots x_n) + c_3^2(x_1, \ldots x_n, x_{n+1}) \right] \tag{7.94}$$

where μ is some arbitrarily large number (*e.g.*, 10^5) such that the magnitude of the $\mu[\ldots]$ term is large relative to the original objective function, f_1:

$$|\mu(\ldots)| > |f_1| \tag{7.95}$$

Subtract the $\mu[\ldots]$ term for *maximization*. Add the $\mu[\ldots]$ term for *minimization*. This technique has the disadvantage of adding new independent variables to the search, which increases the computational requirements for finding the optimum.

7.8.3 Variable Transformation for Removing Constraints

Consider the general form of a two-sided constraint on variable *i*, where the upper and lower bounds are constants:

$$a_i \leq x_i \leq b_i \tag{7.96}$$

Remove a two-sided constraint by introducing a transformation variable, z_i, in one of the following forms involving the trigonometric sin, hyperbolic tangent, or its equivalent sigmoid functions (Urbaniec 1986):

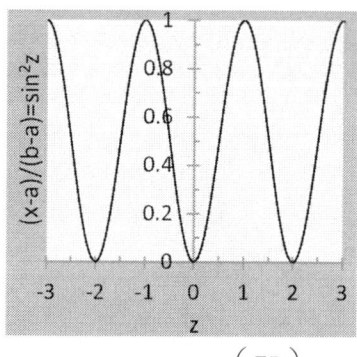

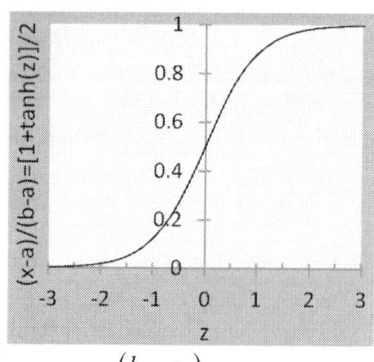

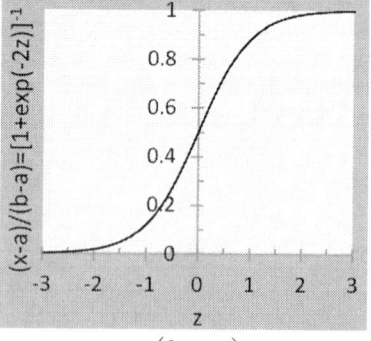

$$x_i = a_i + (b_i - a_i)\sin^2\left(\frac{\pi z_i}{2}\right) \qquad x_i = a_i + \frac{(b_i - a_i)}{2}\left[1 + \tanh(z_i)\right] \qquad x_i = a_i + \frac{(b_i - a_i)}{1 + e^{-2z_i}} \tag{7.97}$$

To transform a one-sided constraint of the following form:

$$a_i \leq x_i \qquad \text{or} \qquad x_i \leq a_i \tag{7.98}$$

select from the following transformations:

CHAPTER 7: OPTIMIZATION

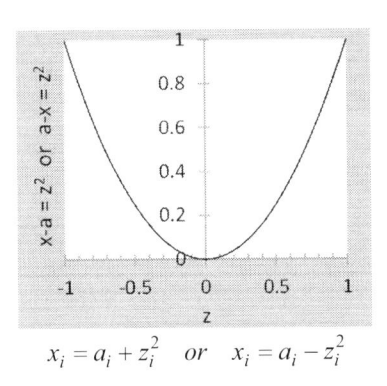

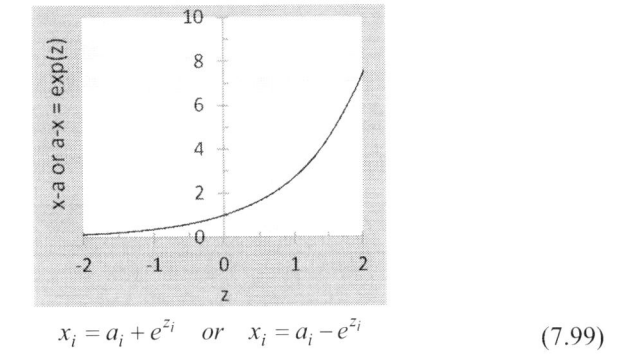

$$x_i = a_i + z_i^2 \quad \text{or} \quad x_i = a_i - z_i^2 \qquad\qquad x_i = a_i + e^{z_i} \quad \text{or} \quad x_i = a_i - e^{z_i} \tag{7.99}$$

Note that a one-sided constraint employing an exponential transformation cannot return the value of the constraining parameter a_i, because $\exp(z) > 0$ for all z. However, optimization search schemes will stop when the difference falls below the convergence tolerance: $\exp(z) < tol$.

We now use our algorithms of optimization to search for values of the *transformed* variables z that optimize the objective function:

1. Specify z using any search method (direct or indirect)
2. Calculate x in terms of z
3. Evaluate $f(x)$ according to the search criteria

Variable transformation does not increase the total number of variables. However, variable transformation has the *disadvantage* of increasing the nonlinearity of the objective function, which may significantly add to the number of iterations needed to locate the optimum.

- For two-sided transformations like those in Equation (7.97), the search regions for the *sin* transformed variables should not exceed $0 < z < 1$. Values for z outside of this region may put the *sin* search process into a non-converging loop.
- In the case of a *sigmoid* transformation, limit the search region: $-3 < z < 3$.
- For indirect search methods using the *exp* or *tanh* transformations, the derivative of the objective function with respect to z goes to zero at positive or negative extreme values for z, which may cause an indirect search procedure using derivatives to mistake the extreme location for an optimal point.

Example 7.17 Optimizing Conversion in Reactors in Series

Known: The cost of operating two, well-stirred, reactors in series involves the reaction conversion in each reactor as follows (Chapra and Canale 2002):

$$OF = C\left[\left(\frac{X_1}{X_2(1-X_1)^2}\right)^{0.7} + \left(\frac{X_2 - X_1}{X_2(1-X_2)^2}\right)^{0.7} + 25\left(\frac{1}{X_2}\right)^{0.7}\right] \tag{7.100}$$

where C is a constant. By definition, conversion of a reactant must fall between zero and one:

$$0 \leq X_1 \leq 1 \qquad \text{and} \qquad 0 \leq X_2 \leq 1 \tag{7.101}$$

However, we also note that for irreversible reactions, the overall extent of conversion in the second reactor is constrained to be larger than the conversion in the first reactor:

$$X_2 \geq X_1 \tag{7.102}$$

Find: Calculate the optimal conversions to minimize operating costs.

Schematic: Two well-mixed reactors are connected in series. The conversion from the first reactor represents the feed conditions to the second reactor.

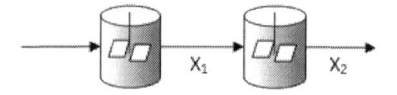

Analysis: Although conversions may vary from zero to one, we observe that the objective function becomes indeterminate at $X_1 = 0$ or 1 and $X_2 = 0$ or 1. To avoid division by zero, limit the range for either conversion between 0.1 and 0.9. Transform the variables to eliminate the constraints in the objective function according to Equation (7.97):

$$X_1 = 0.1 + (X_2 - 0.1)\sin^2 Z_1 \qquad X_2 = X_1 + (0.9 - X_1)\sin^2 Z_2 \qquad (7.103)$$

Equations (7.103) represent a system of linear equations in X_1 and X_2. We ignore the constant in Equation (7.100) in our analysis because it does not affect the location of the optimum. We now solve Equations (7.103) for X_1 and X_2:

$$X_1 = 0.1 \frac{\sin^2 Z_1 \left[9\sin^2 Z_2 - 1\right] + 1}{\sin^2 Z_1 \left[\sin^2 Z_2 - 1\right] + 1} \qquad X_2 = 0.1 \frac{\sin^2 Z_1 \left[\sin^2 Z_2 - 1\right] + 8\sin^2 Z_2 + 1}{\sin^2 Z_1 \left[\sin^2 Z_2 - 1\right] + 1} \qquad (7.104)$$

Create *VBA* user-defined functions for X_1 and X_2, and the objective function for minimizing cost in terms of Z_1 and Z_2.

```
Public Function X_1(Z1 As Double, Z2 As Double) As Double
    X_1 = 0.1 * (Sin(Z1) ^ 2 * (9 * Sin(Z2) ^ 2 - 1) + 1) / _
          (Sin(Z1) ^ 2 * (Sin(Z2) ^ 2 - 1) + 1)
End Function

Public Function X_2(Z1 As Double, Z2 As Double) As Double
    X_2 = 0.1 * (Sin(Z1) ^ 2 * (Sin(Z2) ^ 2 - 1) + 8 * Sin(Z2) ^ 2 + 1) / _
          (Sin(Z1) ^ 2 * (Sin(Z2) ^ 2 - 1) + 1)
End Function

Public Function OF(X1 As Double, X2 As Double) As Double
    OF = (X1 / (X2 * (1 - X1) ^ 2)) ^ 0.7 + ((X2 - X1) / _
         (X2 * (1 - X2) ^ 2)) ^ 0.7 + 25 * (1 / X2) ^ 0.7
End Function
```

Surface and contour plots help us locate the vicinity of the optimum. We use **Data Table** from the ribbon tab **Data>What If Analysis** to generate the data for the plots in an *Excel* worksheet. The values for Z_1, Z_2, X_1, X_2, and the formula for the objective function are in cells **B1**, **B2**, **E1=X_1(B1, B2)**, **E2=X_2(B1, B2)**, and **A3=OF(E1, E2)**, respectively. The data table ranges for Z_1 and Z_2 are in the row **B4:Q4** and column **A4:A19**, respectively. We can use **Conditional Formatting>Color Scales** to shade the cells in the table with colors to indicate their relative magnitudes.

	A	B	C	D	E	F	G	H	I	J	K	L	M	N	O	P	Q
1	Z_1	0.96		X_1	0.55												
2	Z_2	0.93		X_2	0.78												
3	35.68445602	0	0.1	0.2	0.3	0.4	0.5	0.6	0.7	0.8	0.9	1	1.1	1.2	1.3	1.4	1.5
4		0	126	126	126	126	126	126	126	126	126	126	126	126	126	126	126
5		0.1	120	120	120	119	119	118	117	116	114	111	108	102	93	79	59	39
6		0.2	105	105	104	103	102	100	98	95	91	86	79	71	61	50	40	39
7		0.3	88	88	87	86	84	82	79	76	71	66	60	54	47	41	37	43
8		0.4	73	73	73	71	70	68	65	62	58	54	49	45	41	39	38	46
9		0.5	62	62	61	60	59	57	55	52	49	46	43	40	38	37	39	48
10		0.6	54	53	53	52	51	49	48	46	44	41	39	38	37	37	41	49
11		0.7	47	47	47	46	45	44	43	41	40	38	37	36	36	38	42	50
12		0.8	43	43	42	42	41	40	39	38	38	37	36	36	37	39	44	50
13		0.9	40	39	39	39	39	38	37	37	36	36	36	36	37	40	45	50
14		1	38	38	38	37	37	37	36	36	36	36	36	37	38	41	45	51
15		1.1	37	37	37	37	37	37	36	36	36	36	37	37	39	42	46	51
16		1.2	38	38	38	38	38	38	38	37	37	37	38	38	40	42	47	51
17		1.3	41	41	41	40	40	40	40	39	39	39	39	39	40	43	47	51
18		1.4	45	45	45	44	44	43	43	42	41	40	40	40	41	43	47	51
19		1.5	49	49	49	48	47	46	45	44	42	41	40	40	41	43	47	51

We first generate a surface plot and adjust the scale of the vertical axis. We then create a duplicate of the surface plot (copy and paste) to create the contour plot. Right-click on the surface plot and select change plot type, then select contour and pick a format.

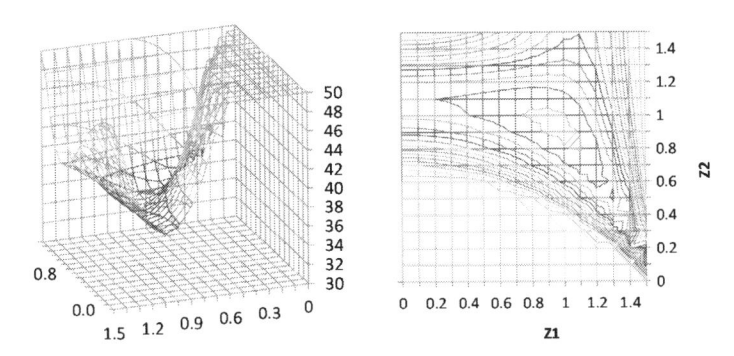

From the data table and plots, we observe that the minimum is located near $Z_1 \approx 1$ and $Z_2 \approx 1$. We use the **Solver** to minimize the unconstrained objective function and calculate the conversions from the result for the transformed variables. Cells **B1** and **B2** hold the **Solver** results for Z_1 and Z_2. The corresponding results for X_1, X_2, and the minimum are shown in cells **E1**, **E2**, and **A3**, respectively.

Comments: The objective function appears to be relatively sensitive to conversion in terms of the transformed Z variables. We confirm this observation by plotting the objective function versus conversions. We observe from the steep gradient in the X_2 direction (contour lines are close together) that the cost function is most sensitive to high conversion in the second reactor.

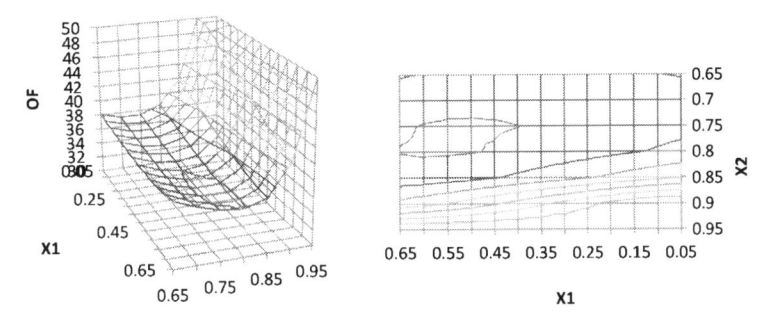

☐

7.8.4 Penalty Functions for Inequality Constraints*

In its simplest form, a constrained minimization function consists of the original objective function plus the addition of penalty functions that significantly increase the magnitude of the objective function when the search violates any constraints on the variables:

$$f_p = f + \sum \delta \mu \tag{7.105}$$

where μ is a large positive constant relative to the magnitude of the minimum value of the objective function, $C \gg |f_{min}|$, and where

$$\delta = \begin{cases} 1, \text{if the constraint is violated} \\ 0, \text{if the constraint is satisfied} \end{cases} \tag{7.106}$$

Alternatively, *interior* or *exterior* penalty functions scale the penalty according to the distance from a constraint on either side of the boundary. An interior function increases the penalty on the objective function as the optimization variable approaches the boundary. An exterior function penalizes the objective when the variable crosses the boundary condition. Three examples of penalty functions include the interior logarithm, inverse barrier, and the quadratic exterior functions for *inequality* constraints like Equation (7.72):

- *Logarithm* $\qquad f_p = f - \mu \sum \ln(-c) \tag{7.107}$

- *Inverse*
$$f_p = f - \mu \sum \frac{1}{c} \qquad (7.108)$$

where μ represents a large penalty number relative to f, and c represents the constraint in Equation (7.72).

A popular *exterior* penalty function uses a quadratic expression:

- *Quadratic*
$$f_p = f + \sum \mu \cdot \max\left[0, c\right]^2 \qquad (7.109)$$

We illustrate the quadratic penalty function approach with the inequality constraint, $a \leq x \leq b$:

$$f_p = f + \mu_a \cdot \max\left(0, a-x\right)^2 + \mu_b \cdot \max\left(0, x-b\right)^2 \qquad (7.110)$$

Select the constants μ_a and μ_b such that $\mu_a|a-x|^2 \gg |f|$ and $\mu_b|x-b|^2 \gg |f|$ for values of x outside of the constraining limits a and b.

7.8.5 Redundancy and Separation*

"With half the race gone, there is half the race still to go." – Murray Walker

In some cases, it helps to decompose the optimization problem into discrete parts, if possible, by first identifying redundant constraints, then looking for separability. Redundancy in constraints on the objective function occurs when any combinations of the other constraints already satisfy at least one of the original constraints. Separability refers to the ability to represent the objective function as a summation of independent one-dimensional objective functions:

$$f(x_1, x_2, \ldots x_n) = \sum_{i=1}^{n} g_i(x_i) \qquad (7.111)$$

When separable, the optimum of the multivariable objective function, f, is the combined optima of the one-dimensional functions, g. The search for the optimum reduces to a series of one-dimensional searches. One-dimensional searches are less computationally intensive than higher dimensional searches.

Example 7.18 Redundancy and Separation in Constrained Optimization

Known: Consider the following set of constraints on a two-dimensional Rosenbrock objective function:

$$y = 10\left(x_2 - x_1^2\right)^2 + \left(1 - x_1\right)^2 \qquad (7.112)$$

$$-5 \leq x_1 \leq 10 \qquad -5 \leq x_2 \leq 10 \qquad x_1 + x_2 \leq 25$$

Find: An equivalent *unconstrained* objective function.

Schematic: Plot the first two constraints as a box and a line for the upper limit.

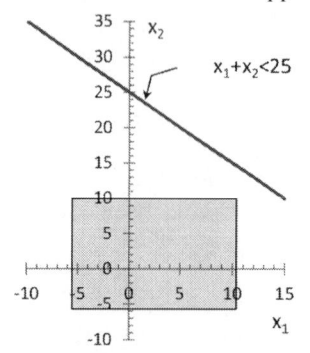

CHAPTER 7: OPTIMIZATION

Analysis: As shown graphically in the schematic, the third constraint represented by the line is redundant given that the range of possible optimum points is tightly contained within the first two constraints (illustrated by the shaded area).

For one or two-dimensional problems, a graphical analysis of the constraints is the simplest approach to identify redundancy. For higher dimensional objective functions, try decomposing the problem into subsets of the original problem. Consider the separability of the objective function in Equation (7.112). From inspection, we find the second term is one-dimensional in x_1. The first term is a function of each dimension, x_1 and x_2.

We employ the methods of transformation and insertion to transform the original objective function into a separable function. Start with an expansion of the first term of the objective function:

$$\left(x_2 - x_1^2\right)^2 = x_2^2 - 2x_1^2 x_2 + x_1^4 \tag{7.113}$$

Introduce new transformation variables:

$$z_1 = x_1^2 + bx_2 \quad \text{and} \quad z_2 = x_2 \tag{7.114}$$

where b is an arbitrary constant. Invert the relationships in Equations (7.114):

$$x_1^2 = z_1 - bx_2 \quad \text{and} \quad x_2 = z_2 \tag{7.115}$$

Substitute the results from Equations (7.115) into the expanded form of the two-dimensional term of the objective function, as found in Equation (7.113):

$$x_2^2 - 2x_1^2 x_2 + x_1^4 = z_2^2 - 2(z_1 - bz_2)z_2 + (z_1 - bz_2)^2 \tag{7.116}$$

Expand the products as follows:

$$x_2^2 - 2x_1^2 x_2 + x_1^4 = z_1^2 + z_2^2\left(1 + 2b + b^2\right) - 2z_1 z_2 (1 + b) \tag{7.117}$$

Eliminate the two-dimensional term in Equation (7.117) by setting the arbitrary constant b to -1. Coincidentally, we also eliminate the other term containing the constant. The result of this transformation and insertion is a separable objective function in three one-dimensional terms:

$$y = 10z_1 + (1 - x_1)^2 = 10\left(x_1^2 - x_2\right) + (1 - x_1)^2 = 10x_1^2 - 10x_2 + (1 - x_1)^2 \tag{7.118}$$

At first, apparently the minimums of the first two terms should be zero. However, $x_1 = 1$ minimizes the third term in Equation (7.118). Substituting this result into the other terms gives $x_2 = 1$ at the minimum value of zero for the objective function. This example also illustrates how to find the optimum points sequentially in a separable function.

Finally, change variables to eliminate the constraints according to the transforms defined by Equation (7.97):

$$x_1 = -5 + 15\sin^2 z_1 \quad \text{and} \quad x_2 = -5 + 15\sin^2 z_2 \tag{7.119}$$

The transformed, unconstrained, two dimensional objective function becomes:

$$y = 10\left(15 - \sin^2 z_1\right)^2 + 50 - 150\sin^2 z_2 + \left(6 - 15\sin^2 z_1\right)^2 \tag{7.120}$$

Optimize Equation (7.120) to find the location of the minimum at z_1 and z_2, and then calculate the solution for the optimal values of x_1 and x_2 from Equations (7.119).

Comments: A little effort to reduce redundant constraints at the beginning of a problem may prevent unnecessary computational effort by reducing sensitivity to initial guesses, and speedup the search for the optimum.

□

7.9 Multi-objective Desirability Functions*

Multiple objectives arise when performing statistically designed experiments to determine any effects of factors on process optimization. We use combinations of desirability functions when dealing with multiple objectives and constraints (Derringer and Suich 1980). Individual objective functions and their corresponding constraints are described in terms of desirability functions, d, with the property: $0 \leq d \leq 1$ (with $d = 1$ as the most desirable state). Given the lower and upper limits (L or U) or target values (T), we may define the desirability functions to hit a target value (set point), or locate a minimum or maximum:

Target:
$$d = \begin{cases} 0, f < L \\ \left|\dfrac{f-L}{t-L}\right|^{\alpha_L}, L \leq f \leq t \\ \left|\dfrac{f-U}{t-U}\right|^{\alpha_U}, t \leq f \leq U \\ 0, f > U \end{cases} \tag{7.121}$$

Minimum:
$$d = \left|\dfrac{f-U}{U-L}\right|^{\alpha}, L \leq f \leq U \tag{7.122}$$

Maximum:
$$d = \left|\dfrac{f-L}{U-L}\right|^{\alpha}, L \leq f \leq U \tag{7.123}$$

where the exponents, α_L and α_U, control the spread, or shape, of the desirability function. For example, we use a small value of α for the case of large uncertainty in the target value, T, or cases where it is not important to meet the target value precisely. Conversely, we use a large value of α in the case of small uncertainty in the target value, or where we need to find results as close to T as possible. In the case of minimum or maximum, the lower and upper limits become values below which, or above which, the values of the objective functions are respectively considered unacceptable.

Exponential functions are useful for handling one-sided constraints when the lower and upper limits on the function are not well defined:

Target:
$$d = \begin{cases} \exp\left(-c_L \left|f-t\right|^{\alpha_L}\right), -\infty \leq f \leq T \\ \exp\left(-c_U \left|f-t\right|^{\alpha_U}\right), t \leq f \leq \infty \end{cases} \tag{7.124}$$

Minimum:
$$d = \exp\left(-c_U \left|f-L\right|^{\alpha_U}\right), L \leq f \leq \infty \tag{7.125}$$

Maximum:
$$d = \dfrac{1-\exp\left(-cf^{\alpha}\right)}{\exp\left(-cL^{\alpha}\right)}, L \leq f \leq \infty \tag{7.126}$$

Use the constants c and α to control the shape of the desirability function.

We combine desirability functions for simultaneous objective functions in a geometric mean (as opposed to an arithmetic mean). Exponents may weigh the importance of each desirability function where $0 < w_i < 1$ and the weighting exponents sum to one:

$$d = \left(\prod_{i=1}^{n} d_i^{w_i}\right)^{1/n} \quad \text{where} \quad \sum_{i=1}^{n} w_i = 1 \tag{7.127}$$

With the geometric mean, when any one desirability function is in the undesirable low position, or zero, the combined functions is also small, or zero. When we combine objective functions this way, we may use any of our techniques for multivariable optimization presented in this chapter.

Multi-objective Optimization (MOO) refers to search methods designed to locate the Pareto frontier, which consists of a set of conditions that each satisfy competing objective functions without one dominating the other. Evolutionary optimization methods, such as genetic algorithms, are becoming important for finding the Pareto front (Deb 2001). Ultimately, engineers must establish criteria, such as the balance of time, resource utilization, and costs, for determining which conditions along the front are most desirable given the overarching goals and constraints of the optimization analysis.

7.10 Variable Scaling*

It may become necessary to scale variables to minimize computer round-off errors when any two or more variables differ by at least one order of magnitude. Without scaling, optimizing techniques tend towards minimizing objective functions with respect to the effects of large variables, while ignoring relatively small ones.[70]

The following techniques scale a variable to range from zero to one in magnitude:

$$x^* = \frac{x - x_{min}}{x_{max} - x_{min}} \qquad \begin{cases} x_{min} \leq x \leq x_{max} \\ 0 \leq x^* \leq 1 \end{cases} \qquad (7.128)$$

$$x^* = \frac{x}{x + x_r} \qquad \begin{cases} 0 \leq x < \infty \\ 0 \leq x^* \leq 1 \end{cases} \qquad (7.129)$$

$$x^* = 1 - \exp\left(-\frac{x}{x_r}\right) \qquad \begin{cases} 0 \leq x < \infty \\ 0 \leq x^* \leq 1 \end{cases} \qquad (7.130)$$

where x_{min} and x_{max} are the lower and upper limits on the range of variable x, and x^* and x_r are the scaled and reference variable values, respectively. We are free to select any value for x_r, however, we recommend using a scaling value that is characteristic of the magnitude in the range of values of the variable.

The **Solver** has the option of automatic scaling of variables with values greater than 10^{-3}. We can apply manual scaling in *Excel* by redefining the search variables using the methods in Equations (7.128) to (7.130), or in terms of the products of a range of scale factors and a range of adjustable parameters used as **Changing Cells** in the **Solver**. For example, for a variable with a value $x = 2.3 \times 10^{-5}$, we may use the scale factor $x = (x^*) \times 10^{-5}$ and direct the **Solver** to adjust the value of x^* in **Changing Cells** to optimize the problem.

To use scaling factors, cycle through the following steps until the **Solver** returns values with an order of magnitude of 1.0 in each of the **Changing Cells**.

1. Fill the range for the **Changing Cells** with the value 1.0 on the worksheet.
2. Specify scale factors for each variable in another range of cells on the worksheet. Start with good guesses for the variable values.
3. Fill a new range of cells with the product of the values for the **Changing Cells** and the scale factors. This range of products becomes the variables used by the objective function.
4. Calculate the objective function in a target cell on the worksheet in terms of the cells containing the values of the variables from step 3.
5. Run the **Solver** to optimize the value of the objective function in the target cell.
6. Replace the values for the scale factors with the values in the range of products of **Changing Cells** with the scale factors, and reset the **Changing Cells** values to 1.0.

A *VBA* macro **SCALIT**, listed in Figure 7.31 automates steps five and six of the cycle for variable scaling in an *Excel* worksheet for minimization using the **Solver**. This macro has application in the solution to systems of equations

---

[70] Review Section 1.2.7 for scaling recommendations.

and nonlinear least-squares regression where we normalize independent or dependent variables using relative scaling (Civan 2011). Example 9.6 uses the macro **SCALIT** to obtain the best model parameters in least-squares regression.

```
Public Sub SCALIT()
' Iterative minimization using the Solver with parameter scaling. The user creates a
' range of CHANGING CELLS starting with values of ~1. Create a range of scale factors
' (best guesses for the model parameters, a range of model parameters defined as the
' product of the CHANGING CELLS and SCALE FACTORS, and a cell containing the objective
' *** Nomenclature ***
Dim ar As Range  ' = range of model parameters
Dim boxtitle As String  ' = string for title in input and message boxes
Dim cr As Range  ' = range of changing cells for Solver
Dim da As Double  ' = difference in parameters
Dim j As Long  ' = index for variables
Dim k As Long  ' = number of parameters
Dim of As Range  ' = range of objective function
Dim small As Double  ' = small number to avoid division by zero
Dim sr As Range  ' = range of scaling factors
Dim tol As Double  ' = relative convergence tolerance
'******************************************************************************
boxtitle = "SCALIT (Minimization with Iterative Scaling)"
With Application
Set of = .InputBox(Type:=8, Title:=boxtitle, prompt:="Cell with OBJECTIVE FUNCTION:")
Set cr = .InputBox(Type:=8, Title:=boxtitle, prompt:="Range of CHANGING values (~1):")
Set sr = .InputBox(Type:=8, Title:=boxtitle, prompt:="Range of SCALE factors:" & _
                    vbNewLine & "(initial guesses for parameters)")
Set ar = .InputBox(Type:=8, Title:=boxtitle, prompt:="Range of PARAMETERS:" & _
                    vbNewLine & " = (CHANGING cell)*(SCALE factor)")
tol = .InputBox(prompt:="Relative Convergence TOLERANCE:", Type:=1, _
                    Title:=boxtitle, Default:=0.000000000001)
End With
k = ar.Count: small = 0.000000000001 ' Get the number of parameters
' Set Solver options
SolverOptions MaxTime:=1000, Iterations:=1000, _
    Estimates:=2, derivatives:=2, Scaling:=False, SearchOption:=2, Convergence:=tol
SolverOk SetCell:=Range(of.Address()), MaxMinVal:=2, ByChange:=Range(cr.Address())
With WorksheetFunction
Do ' Adjust scale factors to same order of magnitude. Minimize SSE using Solver
    SolverSolve UserFinish:=True
    da = 0 ' initial parameter difference
    For j = 1 To k ' Calculate convergence and reset the scale factors
        da = .Max(Abs(cr(j) - 1), da): sr(j) = ar(j): cr(j) = 1
    Next j
Loop While da > tol ' Check for convergence
End With
End Sub ' SCALIT
```

Figure 7.31 *VBA* macro **SCALIT** for scaling variables in optimization using the Solver in an *Excel* worksheet.

7.11 Sensitivity Analysis (Revisited)*

"The sensitivity of men to small matters, and their indifference to great ones, indicates a strange inversion." – Blaise Pascal

A complete optimization analysis considers the sensitivity of the result to perturbations in the variables and parameters (This is a good time to review Section 5.4). In Example 7.2 we may want to explore the sensitivity of the optimum concentration to the prices of each chemical. We should also consider the solution sensitivity to parameters associated with any constraints on the objective function. A response surface and contour plot show the behavior of the objective function relative to perturbations in the independent variables. We have previously seen examples of sensitivity analysis in Sections 7.2 and 7.3. At the location of the optimum, the partial derivatives of the objective function with respect to the variables should be zero. However, sensitivity to variables, parameters, or constraints may shift the location of the optimum in practice.

CHAPTER 7: OPTIMIZATION

Example 7.19 Optimum Reactor Temperature [adapted from Riggs (2001)]

Known: A series of elementary, irreversible reaction steps involves first order rate laws for reacting species A and B:

$$A \xrightarrow{k_A} B \xrightarrow{k_B} C \quad (7.131)$$

$$r_A = k_A C_A \quad \text{and} \quad r_B = k_B C_B \quad (7.132)$$

where C_i is the concentration of species i. The reaction rate constants have Arrhenius temperature dependence:

$$k_A/s^{-1} = 4 \times 10^6 \exp\left[-5{,}000/(T/K)\right] \quad \text{and} \quad k_B/s^{-1} = 2 \times 10^{13} \exp\left[-10{,}000/(T/K)\right] \quad (7.133)$$

where T is the reaction temperature. Use the following economic parameters, operating conditions, and reaction parameters for a continuous (steady state), well-stirred reactor:

Volumetric feed flow rate: $Q/(L/s) = 100$	Price (value) of A in feed:	P_{A0} = \$0.15/kmol
Working reactor volume: $V/L = 1000$	Price (value) of A in effluent:	P_A = \$0.10/kmol
Feed concentration of A: $C_{A0}/(mol/L) = 1$	Price (value) of B in effluent:	P_B = \$0.50/kmol
	Price (value) of C in effluent:	P_C = \$0.20/kmol

Find: Calculate the operating temperature of the reactor to maximize profit.

Assumptions: steady-state, irreversible reaction kinetics, and perfect mixing

Schematic: The diagram uses symbols for the reactor feed and effluent streams volumetric flow rates and concentrations:

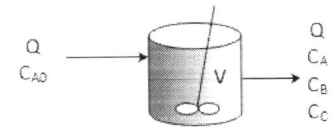

Analysis: Define the objective function in terms of the production and consumption rates, and price-values of chemicals:

$$\frac{\text{Profit}}{(\$/day)} = Q(C_A P_A + C_B P_B + C_C P_C - C_{A0} P_{A0})\left(\frac{\$}{\sec}\right)\left(\frac{3600 \sec}{h}\right)\left(\frac{24h}{day}\right) \quad (7.134)$$

The mole balances around the reactor constitute a system of three linear constraints in terms of the concentrations C_A, C_B, and C_C.

Rate in − Rate Out + Rate of Generation = 0

Species A $\quad Q(C_{A0} - C_A) - k_A C_A V = 0 \quad (7.135)$

Species B $\quad -Q C_B + V(k_A C_A - k_B C_B) = 0 \quad (7.136)$

Species C $\quad -Q C_C + V k_B C_B = 0 \quad (7.137)$

Let $\tau = V/Q$, (residence time), then solve the mole balances for the effluent concentrations:

$$C_A = \frac{C_{A0}}{1 + \tau k_A} \qquad C_B = \frac{\tau k_A C_A}{1 + \tau k_B} \qquad C_C = \tau k_B C_B \quad (7.138)$$

Set up an *Excel* worksheet with the problem parameters in named cells and calculations for the effluent concentrations and profit function (worksheets show formulas and results).

	A	B	C	D	E	F
1	T	294.701260196053	K	tau	=V/Q	s
2	Q	100	L/s	CA	=CA_0/(1+tau*kA)	M
3	V	1000	L	CB	=tau*kA*CA/(1+tau*kB)	M
4	CA_0	1	M	CC	=tau*kB*CB	M
5	kA	=4000000*EXP(-5000/T)	1/s	Profit	=Q*(CA*PA+CB*PB+CC*PC-CA_0*PA_0)*3600*24	$/day
6	kB	=20000000000000*EXP(-10000/T)	1/s			
7	PA_0	=0.15/1000	$/mol			
8	PA	=0.1/1000	$/mol			
9	PB	=0.5/1000	$/mol			
10	PC	=0.2/1000	$/mol			

	A	B	C	D	E	F
1	T	294.7012602	K	tau	10	s
2	Q	100	L/s	CA	0.368639911	M
3	V	1000	L	CB	0.461974185	M
4	CA_0	1	M	CC	0.169385904	M
5	kA	0.171267427	1/s	Profit	1310.932205	$/day
6	kB	0.036665664	1/s			
7	PA_0	$ 0.00	$/mol			
8	PA	$ 0.00	$/mol			
9	PB	$ 0.00	$/mol			
10	PC	$ 0.00	$/mol			

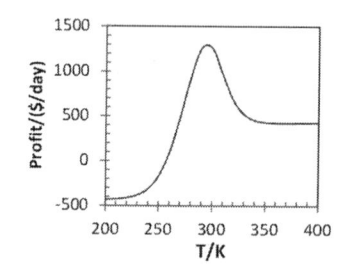

Normally we round off the solution to one or two significant figures. We show more decimal places here to demonstrate the high degree of precision in the numerical solution. Plot the objective function to locate the region of the maximum. Use a **Data Table** to generate the data for the plot in a column, or use the macro **QYXPLOT**. The maximum profit occurs near $T = 300$ K, which we use as an initial guess. To pinpoint the optimum temperature, use the **Solver** to maximize profit at $T = 295$ K for a profit of $1311/day.

Comments: We observe that the objective function flattens off at low and high temperatures where the profit becomes insensitive to changes in temperature. The profit function may go negative if the temperature drops below 260K. At large T, the profit reaches a constant when the second reaction producing C consumes all of B. The indirect search method relies on the derivative, or slope of the objective function to control the search direction. A poor initial guess for the optimum temperature in the flat regions causes the **Solver** to fail to locate the optimum. The optimization problem in this example becomes a good candidate for Powell's method with Golden Section search. Note that the macro **POWELL** is set up for minimization. To use this macro, multiply the objective function in cell **E5** by -1 to convert the maximization problem into a minimization problem.

Next, we explore the sensitivity of the optimum temperature and profit to perturbations in the feed rate. Use a **Data Table** to generate data and plot the objective function in surface and contour plots with variable T and Q to observe that the optimum temperature is relatively insensitive to the feed rate, although the profit does increase with increasing feed rate, as expected.

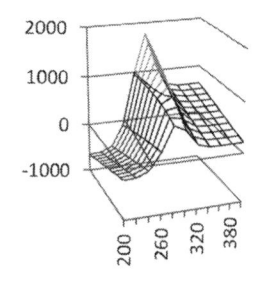

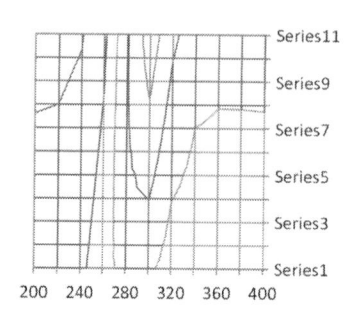

The contour plot shows horizontal isoclines close together, indicating steeper gradients, or higher sensitivity to temperature. Isoclines in the vertical direction are spaced farther apart, showing lower sensitivity to the flow rate.

YX, surface, and contour plots are limited to one or two-dimensional problems. To analyze sensitivity graphically for multi-parameter problems, calculate the objective function at plus or minus a percentage or fraction of the variable or parameter's average value (*e.g.*, a ±10% change in *x* is $x \pm 0.1x$).

$$OF\left[x(1-\Delta)\right] \qquad OF\left[x\right] \qquad OF\left[x(1+\Delta)\right] \qquad (7.139)$$

Then plot the three points for the objective function (-%, 0%, +%) in a line plot similar to Figure 7.32 to inspect the effect of changing the value of the variable or parameter. Do this for each variable or parameter in the objective function and plot the results in the same line chart. The slopes indicate the relative sensitivity of the objective function (*OF*) to changes in parameters or variables. We may want to keep the horizontal grid lines in the sensitivity spider plot because we are interested in how the objective function deviates from the horizontal due to sensitivities in the parameters. In Figure 7.32, the objective function is most sensitive to variable y.

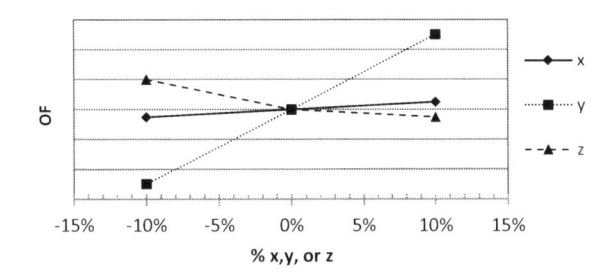

Figure 7.32

Sensitivity spider plot using percent changes in variables or parameters (x, y, z) of an objective function (OF).

The *PNMSuite* includes sub procedures **SPIDER** and **TORNADO** for generating sensitivity charts in an *Excel* worksheet. **SPIDER** and **TORNADO** require the ranges for the cell with the optimized objective function in terms of the variables of optimization located in contiguous cells on the worksheet. The macros prompt for the ranges of cells containing headings or labels for the function and variables as well as the percent change in the variables. The *VBA* code cycles through the variables, perturbing the values of each variable according the specified *Δ* change in Equation (7.139) and records the corresponding change in the objective function. The percent change in the function is tabulated in the worksheet for generating a line chart of sensitivity curves. The macro limits the number of variables to five, or less. The following example uses the macro to generate a sensitivity spider plot.

Spider plots appear cluttered with sensitivity curves for more than five variables. A tornado chart is recommended for a graphical display of sensitivity to a large number of variables. A tornado chart creates a horizontal bar chart with colors and lengths showing the relative sensitivities of the function to the variables, with the most sensitive variable at the top, descending down to the least, which gives the appearance of a funnel, or tornado.

Example 7.20   Economic Pipe Diameter and Sensitivity to Interest Rate

Known: Construct a steel pipeline to transport oil 500 miles with a service life of 20 years and interest rate of 10%, with the parameters specified in the *Excel* worksheet. The pipe wall thickness is a function of the diameter:

$$w = 0.015D \qquad (7.140)$$

Schematic: A pipe has length *L*, inside diameter, *D*, and wall thickness, *W*.

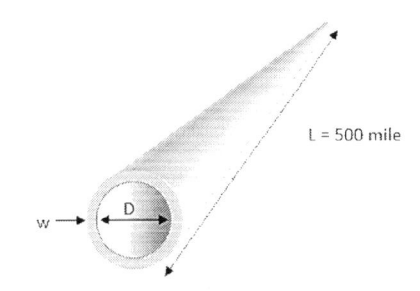

Find: Calculate the economic pipe diameter and its sensitivity to the interest rate.

Assumptions: Constant properties, steady state, laminar, viscous, incompressible flow, no salvage value at the end of the life of the pipe system. Ignore elevation changes in potential energy and annual operating costs.

Analysis: Calculate the power requirements assuming Hagen-Poiseuille flow, and a mass of steel pipe from the volume and density of steel in terms of the pipe diameter:

$$Power = \frac{128 Q^2 \mu L}{\pi D^4} \quad (7.141)$$

$$mass = \frac{\pi}{4} \rho L \left[(D + 2w)^2 - D^2 \right] \quad (7.142)$$

Use *Excel*'s finance function **PMT(i, n, PV)** to calculate the annual operating cost of the pipe and pump. The objective function for annual profit is the net sum of annual revenues and payments:

$$Profit = R_{Oil} + \sum AOC \quad (7.143)$$

where $R_{Oil} = Q \cdot P_{Oil}$ and $AOC_{Power} = -Po \cdot P_{Power}$

$AOC_{Pipe} = PMT\left[0.1, 20, (m \cdot P)_{pipe}\right]$ and $AOC_{Pump} = PMT\left[0.1, 20, (Po \cdot P)_{Pump}\right]$

	A	B	C	D	E	F	G
1	**Pipe Parameters**				**Economic Parameters**		
2	Diameter/m	D	0.7290		Interest Rate	i	0.1
3	Density of Steel/(kg/m³)	ρ	8010		Power/($/J)	P_{power}	1.11E-08
4	Wall thikness/m	w	0.0109		Oil/($/m³)	P_{oil}	31.7
5	Length/km	L	805		Steel Pipe/($/kg)	P_{Pipe}	10.36
6					Pump/($/W)	P_{Pump}	0.42
7	**Oil Parameters**						
8	Flow Rate/(m³/min)	Q	47		**Annual Costs**		
9	Oil Viscosity/(kg/m s)	μ	3.5		Operating Cost/($/yr)	AOC	-2.20E+06
10					Pipe/($/yr)	AOC_{Pipe}	($199,478,700.68)
11	**Calculated Parameters**				Pump/($/yr)	AOC_{Pump}	($12,300,703.35)
12	Power/W	Po	2.49E+08		Power/($/yr)	AOC_{Power}	-8.74E+07
13	mass/kg	m	1.64E+08		Revenue/($/yr)	R_{oil}	7.83E+08
14					Profit/($/yr)		4.82E+08

A plot of the profit function in Figure 7.33 shows the location of the maximum and sensitivity of the profit to variations in pipe diameter and interest rate.

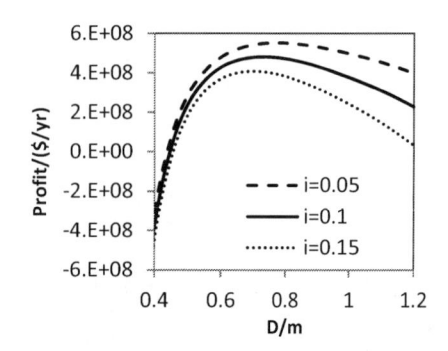

Figure 7.33

Sensitivity of the profit function to changes in pipe diameter and interest rate.

The profit is highly sensitive to changes in the diameter below the economic diameter, however relatively insensitive to changes in diameter above the optimum value. We decide to select the pipe size with a nominal inside diameter just larger than the economic pipe diameter to minimize the sensitivity to upsets in the operating parameters.

Comments: Due to a large profit margin, the project appears to be economically viable. We should explore the effects of fluctuations in price on the profit. We get the relative sensitivity coefficients using the macro **SENSITIVITY**. The function is the profit in cell **G14** and the variables are the prices in the range **G3:G6**.

	I	J	K	L	M	N	O	P	Q
3	c_1 =	-7863173380	rel c_1 =	-0.2156827413	-0.1	=M3*$L$3	=M3*$L$4	=M3*$L$5	=M3*$L$6
4	c_2 =	24703154.47	rel c_2 =	1.93163218087	0.1	=M4*$L$3	=M4*$L$4	=M4*$L$5	=M4*$L$6
5	c_3 =	-26189030.10	rel c_3 =	-0.6692554646					
6	c_4 =	-39834859.88	rel c_4 =	-0.0412691325					

We use the relative sensitivities for slopes in a plot in Figure 7.34 to reveal a relatively insensitive variation of profit to ±10% changes in price of power, pipe, and pump. The profit is most sensitive to fluctuations in the selling price of oil. We used the formulas in columns **N** to **Q** to generate the data for the plot. The independent variable for fractional change in price is located in column **M**.

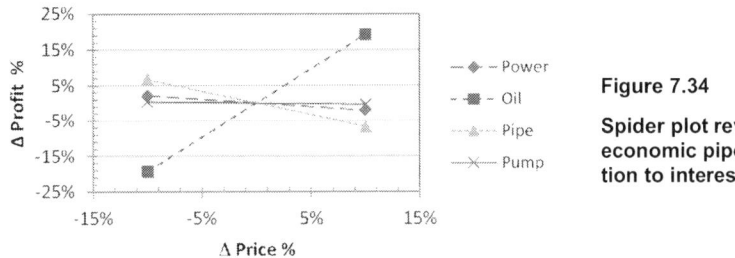

Figure 7.34

Spider plot revealing the sensitivity of the economic pipe diameter and profit function to interest rate.

Alternatively, use the sub procedures **SPIDER** and **TORNADO** from the *PNMSuite* to generate a similar spider sensitivity plot and a tornado sensitivity chart using the following inputs from the worksheet:

Function Heading = **E14**

Function Cell = **G14**

Variable Headings = **F3:F6**

Variable Range = **G3:G6**

Default Percent Change = 10

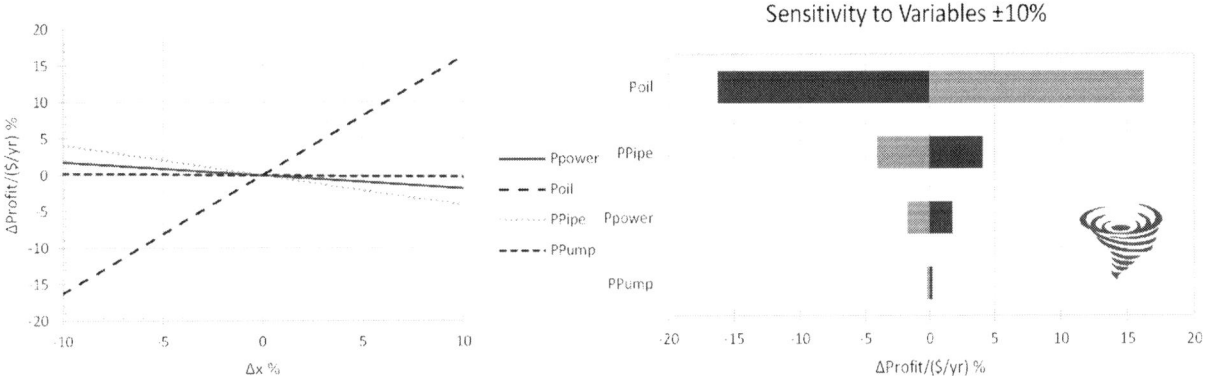

Figure 7.35 Spider and tornado sensitivity plots reveal that profit is most sensitive to the price of oil. The price of the pump has the least influence on profit.

7.12 Epilogue on Optimization

Table 7.3 has a summary of the variety of optimization tools at our disposal. Table 7.4 summarizes the macros for optimization developed in this chapter. These tools help us solve many practical problems of interest to engineers. In general, we find that indirect search methods, such as those employed by *Excel*'s **Solver**, find the optimum in fewer steps because they use additional information about the problem. However:

- Indirect methods require derivative evaluations, or approximations for derivatives, which may not be possible for all problems, particularly objective functions with discontinuities or discrete values.
- Some problems may require direct methods. For instance, some direct methods, such as Simulated Annealing, Luus-Jaakola and the Firefly algorithm, work with discontinuous, discrete functions.
- Global optimization by the random multi-start and *Evolutionary* genetic algorithms in the **Solver**, or the VBA macros for Simulated Annealing, Luus-Jaakola, PSO, Firefly, or the Genetic algorithm use stochastic search patterns in an effort to locate the true optimum.
- The solution for the location of a local optimum depends on the starting point of the search. Use multiple starts from different points to identify a global optimum.
- Constraints can save computational effort by limiting the search to regions of feasibility.
- Some problems require variable scaling to deal with convergence and round-off errors.

Table 7.3 Comparison of methods of optimization.

Method	Strengths	Weaknesses
Graphical	• Visual identification of local and global optima for one or two dimensions	• Low precision • Does not work well for more than two dimensions
Quadratic Interpolation	• Faster convergence • Simple algorithm • Applies to most objective functions encountered by engineers	• Poor approximation when the objective function does not behave quadratically • Requires bracketing the optimum point • Requires continuous, smooth objective functions
Golden Section	• Guaranteed convergence for bracketed, smooth, and continuous objective functions	• Requires bracketing the optimum point
Powell	• Applies to multidimensional problems • Does not require derivative evaluations	• Slow convergence • Requires continuous functions
Simulated Annealing, Luus-Jaakola, PSO, and Firefly Algorithms	• Applies to multidimensional problems • Simple algorithm • Application to discontinuous, discrete objective functions • Does not require derivative evaluations	• Requires experimentation to determine appropriate number of cycles and stochastic function evaluations. • Slow convergence relative to the **Solver**.
Newton (Solver)	• Faster convergence	• Requires derivative evaluations • Sensitive to starting points
Downhill Simplex	• No derivative evaluations • No bracketing required • Lower sensitivity to starting points	• Slower convergence • Large number of function evaluations required • Requires continuous objective functions
Evolutionary Genetic Algorithm	• Applies to multidimensional problems. • Applies to multimodal problems • Works with discontinuous, discrete objective functions	• May not find the global optimum. • Requires experimentation with the parameters for mutation rate, number of generations, and size of the population.

Table 7.4 Summary of VBA macros and user-forms for optimization, including the methods of input.

Macro	Inputs	Operation
FIREFLY	Macro input boxes	Firefly algorithm for multivariable, global minimization
FIREFLYDM	Macro input boxes	Davis modified firefly algorithm for global minimization
GENETIC	Macro input boxes	Genetic algorithm for global optimization with constraints
LUUSJAAKOLA	Macro input boxes	Luus-Jaakola method of multivariable, multimodal minimization
OPTIMIN_UsrFrm	Show user-form macro	Powell, SIMPLEX, Firefly, or Luus-Jaakola minimization methods
POWELL	Macro input boxes	Powell's method with options for quadratic or Golden Section line search
PSO	Macro input boxes	Particle swarm optimization with Latin hypercube initialization and adaptive inertia
SCALIT	Macro input boxes	Scaled variables/parameters for solution using *Excel*'s Solver.
SIMANN	Macro input boxes	Simulated annealing for global minimization
SIMPLEXNLP	Macro input boxes	Downhill simplex for multivariable minimization
SPIDER	Macro input boxes	Generates a sensitivity spider plot of an optimized objective function of variables
STDEW	User-defined function arguments	Standard deviation of the input range by Welford's formula
TORNADO	Macro input boxes	Generate a tornado sensitivity bar chart of an optimized function of variables

Chapter 8   Uncertainty Analysis

"What is not surrounded by uncertainty cannot be the truth." – Richard Feynman

In this chapter, we consider numbers with a little fuzz on them. A "fuzzy" number lacks a high degree of precision, as characterized by a probability distribution about the expected value. For example, we may refer to a results as having an unknown true solution that is within a range of "plus or minus" some percentage of our best estimate. We use uncertainty analysis to assess risk in decision-making. As an example, Branan (2012) lists typical ranges of tolerable uncertainty for cost estimates in the various stages of process design: *Conceptual ±35%* → *Preliminary ±30%* → *Detailed ±20%* → *Definitive ±10%.* In this chapter, we present methods for the analysis of uncertainty in experimental results and answer the question about how experimental uncertainty propagates through engineering calculations. In this regard, we attempt to answer two related questions:
1. How reliable are our experimental measurements?
2. How does the uncertainty in our response to the first question affect the reliability of what we do with these measurements?

Bell (1999) provides simple definitions to distinguish between measurement error and uncertainty:
- The uncertainty of a measurement tells us something about its quality.
- Error is the difference between the measured value and the 'true value' of the thing being measured.
- Whenever possible we try to correct for any known errors: for example, by applying corrections from calibration certificates. However, any error whose value we do not know is a source of uncertainty.
- Uncertainty then is a quantification of the doubt about the measurement result.

The topics presented in this chapter cover a few simple, yet powerful, tools developed to promote uncertainty analysis by engineers and scientists. The primary topics include:
- Models of measurement uncertainty
- Law of propagation of measurement uncertainty through calculations of design or analysis
- Monte Carlo simulations of uncertainty propagation

With few exceptions[71], all experimental measurements and calculated results are subject to some degree of uncertainty. Engineers can choose to ignore uncertainty – commit an act of "disestimation"[72] and hope that it does not influence their analysis – or they can incorporate the effects of uncertainty into their analysis to inform their decisions.

Chemical engineers may find it interesting that the U.S. Atomic Energy Commission in the decade of the 1970's conducted the earliest large-scale application of uncertainty analysis in their reactor safety study (Morgan and Henrion 1990). We must consider uncertainty when evaluating risk for chemical process safety analysis (CCPS 2009). We discussed the ethical responsibility engineers have to report the quality of data as part of the expert problem-solving format in the introductory Section 1.3, and derived an expression for maximum uncertainty propagation in Section 5.4. We introduce a formal process for quantifying uncertainty in experiments, and show how to determine the effects of uncertainty in calculations involving uncertain information.

Uncertainty analysis is a conventional method of estimating, quantifying, and reporting the reliability of a numerical value. Reported values for measurements and calculated results are only complete when accompanied by a

---

[71] One exception is an exact number, such as the count of a small population, e.g., number of students in a classroom.
[72] "Disestimation is the act of taking a number too literally, underestimating or ignoring the uncertainties that surround it. [It] imbues a number with more precision than it deserves, dressing a measurement up as absolute fact instead of presenting it as the error-prone estimate that it really is." (Seife 2010)

CHAPTER 8: UNCERTAINTY

statement of reliability. Nevertheless, busy engineers tend to avoid uncertainty analysis, dreading the tedious calculations they experienced in their chemistry and physics laboratory courses. A shortcut method used by many practicing engineers incorporates safety factors for managing uncertainty in conceptual design calculations (Walas 1990). For example, one heuristic for distillation column design recommends doubling the minimum number of ideal stages. Several design heuristics, or "rules of thumb," have evolved from years of practical experience with processes and unit operations. Yet, even a trusted heuristic may fall short if we are careless about uncertainty in experimental results and fail to account for the propagation of uncertainty through our calculations.

The reported precision and accuracy of a measurement are often just as important as the value of the measurement itself. To illustrate, consider a chemical engineer faced with the problem of designing the volume of a continuously stirred reactor from a steady-state mole balance involving a first-order reaction:

$$V = \frac{QX}{k(1-X)} \quad (8.1)$$

where V is the volume of fluid, k is the first order rate constant, and X and Q represent the reactant conversion and volumetric flow rate, respectively.

The scale-up for the reactor volume in Equation (8.1) needs accurate information about the reaction rate constant. If the chemist reports a rate constant accurate to $k \pm 50\%$, the engineer may design the reactor with a safety factor that doubles the volume to account for the possibility that the reaction may run 50% slower than the reported average value. Otherwise, the engineer risks under sizing the reactor when operating at a lower conversion than required by the process.

When we report the uncertainty in a measurement, engineers and scientists using our results can incorporate the uncertainty into their own analysis and design calculations to inform their decision making process. For the problem of reactor scale up in Equation (8.1) we need reliable kinetic and transport data for our analysis. Reactor design usually requires careful economic analysis before making a decision to go ahead with the implementation. In addition to financial consequences, there may also be ethical consequences of not properly analyzing and reporting uncertainty. If we undersize our reactor, the cooling system may not have the capacity to control runaway exothermic reactions, which may result in catastrophic failure of the vessel, or worse.

It may seem reasonable to conduct an uncertainty analysis only after completing all of the experiments and data collection. A better approach uses existing process data, and information about precisions of measurements to plan the experiment *before* running costly trials. There is no point in conducting a series of expensive experiments if we expect that large uncertainties in the measurements will lend no confidence in the results, which in turn cannot be trusted for use in design or decision-making. Thus, uncertainty analysis in experimental design is an iterative process. We refine the experimental procedure repeatedly until we are comfortable with the precision and expected accuracy of the potential results. Only then, should we perform the experiment, analyze the data, and report the results with uncertainty estimates.

SQ3R Focused Reading Questions

1. What is the difference between error and uncertainty?
2. What is the difference between accuracy and precision?
3. What is the difference between Type A and Type B uncertainties?
4. Why do I need to pay attention to significant figures in the results of uncertainty analysis?
5. How do I report uncertainty in a result?
6. What are different sources of uncertainty?
7. How do I estimate the uncertainty of a digital instrument?
8. What is the coverage factor for expanded uncertainty?
9. What are the degrees of freedom in a fixed uncertainty?
10. What are the advantages and disadvantages of Monte Carlo analysis of uncertainty?

8.1 Models of Measurement Uncertainty

"It can only be attributable to human error." – Hal 9000, "200: A Space Odyssey"

Our objective is to estimate the uncertainty in the dependent variable y from the effects of uncertainty in the independent variables x. Given the measurand output y in terms of the function of the independent variables or input measurements x, we represent the uncertainty in y in terms of a range of reliability:

$$[f(x) - u_y] \leq y \leq [f(x) + u_y] \quad \text{or} \quad y = f(x) \pm u_y \quad (8.2)$$

where $f(x)$ is the functional relationship between y and x, and u_y is the uncertainty in the function result for y. We interpret Equation (8.2) with some degree of confidence to state that the true value of y lies between the lower and upper limit of uncertainty.

The conventional formula for combining the uncertainties of multiple inputs into the analysis of uncertainty propagation has the following form (Kirkup and Frenkel 2006):

$$u_y = \sqrt{\sum (cu)^2} \quad (8.3)$$

with the terms summarized in Table 8.1. In a fashion, Table 8.1 provides a road map for our introduction to uncertainty analysis by identifying the important milestones along the way. The rest of this chapter provides details about each term in Table 8.1 and explores methods for estimating uncertainties and the consequences of uncertainty in engineering calculations. Numerous examples are included to demonstrate each step of the process.

Table 8.1 Definitions of terms in uncertainty analysis and Excel worksheet formulas with functions.

Symbol	Definition	Formula	Excel Worksheet Formulas
n	Number of measurements	-	COUNT(range of x)
m	Number of variable parameters, $m = 1$ for average, $m = 2$ for standard deviation	-	1 or 2
$\bar{x}_i$	Average (mean) value of variable i	$(\sum x)/n$	AVERAGE(range of x)
s_i	Sample standard deviation in x_i	$\sqrt{\sum (x - \bar{x}_i)^2 / (n-1)}$	STDEV.S(range of x)
u_{xi}	Standard error or random uncertainty in x_i	$s_i / \sqrt{n}$	CONFIDENCE(range of x)
u_{zi}	Systematic (fixed) uncertainty	Readability$/\sqrt{3}$	δ / SQRT(3)
u_i	Combined uncertainty in variable/parameter i	$\sqrt{u_{xi}^2 + u_{zi}^2}$	SUMSQ(ux, uz)
c	Sensitivity coefficient	$\partial y / \partial \bar{x}_i$	(y(x + δ) − y(x)) / δ
u_y	Standard uncertainty in y	$\sqrt{\sum_{i}^{n} (c_i u_i)^2}$	SQRT(SUMSQ(range of c, range of u)) CNTL SHIFT ENTER
v_i	Variable degrees of freedom ($m = 1$ for average, $m = 2$ or higher when the average model has two constants).	$v_i \cong (n-m)\left(\dfrac{u_i}{u_{\bar{x}i}}\right)^4$	(n - m) * (u / ux) ^ 4
v_y	Overall degrees of freedom (DOF) in the model with k input variables determined by the Welch-Satterthwaite formula	$v_y = \dfrac{u_y^4}{\sum_{i=1}^{k} \dfrac{(c_i u_i)^4}{v_i}}$	(uy ^ 4)/SUMSQ(((range of c) * (range of u)) ^ 2 / (range of v)) CNTL SHIFT ENTER
t	Coverage factor based on the combined degrees of freedom v at a significance level α	Student-t-statistic	TINV(α, v)
U_y	Expanded uncertainty in y for a specified confidence interval	$t \cdot u_y$	t * uy

CHAPTER 8: UNCERTAINTY

"Statistician—someone who insists on being certain about uncertainty."

There are established guidelines for estimating uncertainty (Kirkup and Frenkel 2006). In this chapter, we adhere primarily to *GUM*, or the *Guide to the Expression of Uncertainty in Measurements* (1997). Conventions for estimating and reporting uncertainty are evolving as the international scientific and engineering communities converge on accepted practices. Nonetheless, engineers have a responsibility for reporting a reasonable approximation of the reliability of any measurements and calculations based on uncertain data.

We begin by showing how to quantify random (Type A) and one form of systematic (Type B) uncertainties in measurements, then follow up with two methods for estimating the effects of variable or parameter uncertainty in our models or design calculations:[73]

- Law of propagation of uncertainty
- Monte Carlo simulations

Uncertainties in measurements come from a variety of sources that include the measured object, the instrumentation, the physical environment, or the person making and recording the measurement. We estimate uncertainties using statistical analysis (Type A), or from other kinds of information about the measurement (Type B). There are generally four types of experimental errors (Holman 2001):

1. Mistakes made by the person performing the experiment (human error).
2. Procedural errors from the method of measuring and data collection.
   - Parallax errors from reading analog instruments from different angles of viewing.
   - Reaction time errors when recording from a timepiece, such as a stop watch.
3. Random errors because of our inability to measure exactly and precisely. Sources of random error include (Kirkup 2002):
   - Random molecular motion on surfaces.
   - Johnson noise, or random voltages, that are the result of thermal effects on materials.
   - Radioactive decay.
4. Systematic errors from the imprecision of the measuring device or lack of calibration.

Mistakes are the results of human error, *e.g.*, recording the wrong number in a lab notebook or inadvertently switching labels on measurement probes. We have no practical way of incorporating the effects of unreported mistakes into uncertainty analysis. However, we should make every effort to eliminate or minimize mistakes by careful experimental planning and implementation. For example, a good experimenter prepares a procedure that includes tables for data collection or other mechanisms for recording data before starting the experiment. When and where possible, we should randomize the sequence of experiments to reduce procedural bias errors. Once we have eliminated, or at least minimized mistakes and procedural errors, we must account for random and systematic errors.

8.1.1 Precision versus Accuracy

"Then there was the man who drowned crossing a stream with an average depth of six inches." – W. I. E. Gates

Levenspiel (2007) recognized one of the earliest examples of error analysis in Galileo's dialog on physics:

"Aristotle claims that 'an iron ball of 100 pounds falling from a height of 100 cubits reaches the ground before a one-pound ball has fallen a single cubit,' I say that they arrive at the same time. You find, on making the experiment, that the larger outstrips the smaller by two fingerbreadths. Now you would not hide behind these two finger-breadths the ninety-nine cubits of Aristotle, nor would you mention my small error and at the same time pass over in silence his very large one?" – *Galilei* (1914)

---

[73] We may interchange the terms: variables and parameters, or the expressions: functions and models, because the methods for quantifying uncertainty in each case are similar.

Experimental measurements have two characteristics of reliability that define the quality of the numerical results: accuracy and precision. In Galileo's experiment, the accuracy comes from observing that the different masses repeatedly fall the same distance in the same approximate period. The differences in the large falling distance among repeated experimental measurements lie within the range of precision of the relatively small breadth of two fingers.

We define the characteristics of quality of data in terms of accuracy versus precision. Figure 8.1 is the classic illustration of the relationship between measurement precision and accuracy in target practice:

Accuracy = how close a measurement is to the target's center, or true value.

Precision = how closely measurements are clustered together around their mean (average) value.

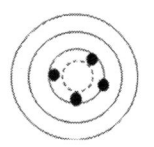

Higher Accuracy / Lower Precision
Lower accuracy / Higher Precision
High Accuracy / High Precision

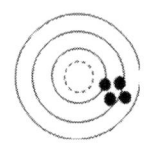

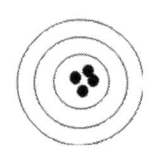

Figure 8.1

Accuracy versus precision in replicated shots on target, or measurements.

From a statistical perspective illustrated in Figure 8.2, accuracy describes the fixed or bias error represented by the distance z between the mean value of a series of measurements and their true value. Precision describes the distribution (standard deviation) of measurement results about the measurement mean.

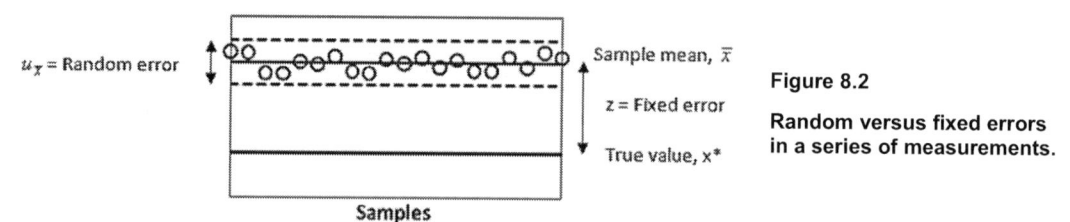

$u_{\bar{x}}$ = Random error
Sample mean, $\bar{x}$
z = Fixed error
True value, x^*
Samples

Figure 8.2

Random versus fixed errors in a series of measurements.

Our model of a *measurement* includes the expected value ($\bar{x}$) combined with a fixed error (z):

$$x = \bar{x} + z \qquad (8.4)$$

where x represents the *best* value. The expected value for the variable is the mean, or average, value of n replicated measurements:

- *Average*
$$\bar{x} = \frac{1}{n}\sum_{j=1}^{n} x_j \qquad (8.5)$$

Use *Excel*'s worksheet function **AVERAGE** to calculate the mean value of the numerical contents of a range of cells. We reduce the fixed error by careful calibration of our instruments to drive $z \to 0$.

Our model of *uncertainty* in x must also include the random uncertainty in the sample mean, $u_{\bar{x}}$, and any uncertainty in the elimination of fixed error, u_z. (Note that our model of the measurement value is not the same as our model of the uncertainty in the value). To obtain an estimate of error in x, we apply Equation (8.3) to the measurement model in Equation (8.4). Because there is no correlation between the sample mean and fixed error, the partial derivatives for $\bar{x}$ and z in Equation (8.3) are simply one, which gives the combined random and fixed uncertainties for a new model of our measurement:

$$x = \bar{x} \pm u \qquad (8.6)$$

where the combined uncertainty in the measurement x includes random and fixed measurement uncertainties:

- *Combined Uncertainty*
$$u = \sqrt{u_{\bar{x}}^2 + u_z^2} \qquad (8.7)$$

The next few sections cover conventions for calculating random and fixed uncertainties in measurements.

8.1.2 Type A Standard Uncertainty from Random Errors

"Measure what is measureable, and make measurable what is not." – Galileo Galilei

In this section, we show how to calculate the standard (random) uncertainty, u_x, required by Equation (8.7). We rely on descriptive statistics to quantify random uncertainty by assuming that random errors are normally (Gaussian) distributed about the expected mean value in Equation (8.5).[74] We define the standard uncertainty in terms of the standard deviation in the measurements. For example, Figure 8.3 has a plot of the normal probability density function for a variable with a mean value of zero and standard deviation of one. The curve shows the distribution of measurements for x concentrated about the average value of zero in this case. At $\pm 2\sigma$, the area under the curve represents ~95.5% probability, the conventional level of confidence reported in uncertainty analysis[75]. This is Type A uncertainty.

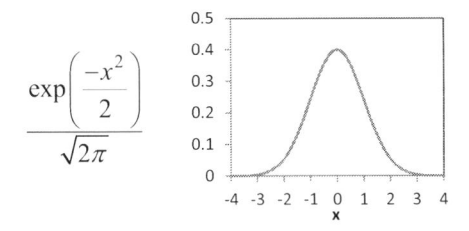

$$\frac{\exp\left(\frac{-x^2}{2}\right)}{\sqrt{2\pi}}$$

Standard Deviations
$\pm 1\sigma = 68.2\%$
$\pm 2\sigma = 95.4\%$
$\pm 3\sigma = 99.6\%$
$\pm 4\sigma = 99.8\%$

Figure 8.3

Plot of a normal distribution with mean 0 and standard deviation 1.

Excel's worksheet function **STDEV.S** calculates the sample standard deviation of n replicate measurements:

- Standard Deviation
$$s = \sqrt{\frac{\sum_{j=1}^{n}(x_j - \bar{x})^2}{n-1}} \quad (8.8)$$

where j is the index of n replicated measurements for x and $n-1$ is the degrees of freedom in the calculation. The standard uncertainty in the random values for x_i is just the standard deviation of the sample mean:

- Standard Uncertainty
$$u_{\bar{x}} = \frac{s}{\sqrt{n}} \quad (8.9)$$

Note how the standard uncertainty goes to zero as the number of replicate measurements increases: $n \to \infty$. One obvious way to reduce random uncertainty is to make more measurements, where practical.

Excel includes a convenient add-in for calculating the standard deviation, mean, degrees of freedom, and standard uncertainty for a data set. Use *Excel*'s **Descriptive Statistics** feature available in the add-in **Data Analysis**, as described in Example 2.8.

8.1.3 Confidence Intervals, Outliers, and the Bootstrap Method

"It is only prudent never to place complete confidence in that by which we have even once been deceived." – Rene Descartes

A small set of observations taken from a larger population represents just one realization of a variable drawn from a "statistical universe of data sets." (W. Press, S. Teukolsky and S. Vetterling, et al. 1992) Different subsets of observations from the same statistical universe may lead to different estimates of "true" mean values of model variables. Confidence intervals from that "statistical universe" provide additional information about the degree of reliability in the reported mean value to account for variability among subsets of observations.

---

[74] Substitute other distributions as warranted.
[75] $1\sigma \sim 68.3\%$ and $3\sigma \sim 99.7\%$ probability, respectively

A confidence interval for an expected value has bounds of plus or minus the product of the standard uncertainty and two-tailed Student's t-statistic. In this book, we use the conventional 95% confidence interval, but we may select from any level of confidence:

- Confidence Interval $\quad \left(\bar{x} - t_{95\%,v}\dfrac{s}{\sqrt{n}}\right) \leq x \leq \left(\bar{x} + t_{95\%,v}\dfrac{s}{\sqrt{n}}\right)$ (8.10)

where v is the degrees of freedom. Get the t-statistic (or coverage factor) from Table 8.2 for $(1 - \alpha) \times 100\%$ or 95% probability that x falls within the confidence interval (a significance level of $\alpha = 0.05$) for different degrees of freedom. Calculate the degrees of freedom from the difference between the size of the data set, n, and the number of parameters, e.g., $v = n - 1$ for the calculation of the mean and $v = n - 2$ for the standard deviation.

Table 8.2 Two-tailed t-statistic for 95% coverage.

v	t₉₅%	v	t₉₅%	v	t₉₅%	v	t₉₅%	v	t₉₅%	v	t₉₅%	v	t₉₅%	v	t₉₅%
1	12.71	5	2.57	9	2.26	13	2.16	17	2.11	22	2.07	33	2.03	69	1.99
2	4.3	6	2.45	10	2.23	14	2.14	18	2.1	24	2.06	38	2.02	96	1.98
3	3.18	7	2.36	11	2.2	15	2.13	19	2.09	27	2.05	45	2.01	159	1.97
4	2.78	8	2.31	12	2.18	16	2.12	21	2.08	30	2.04	54	2	473	1.96

We used the *Excel* worksheet function **TINV(α, v)** to generate the t-table. The curve in Figure 8.4 represents the coverage factors listed in Table 8.2 showing how the t-statistic rapidly approaches the asymptotic limit of 1.96 for large degrees of freedom. Many practitioners of uncertainty analysis conservatively round up to $t_{95\%} = 2$ for data sets larger than 50. We use the *Excel* worksheet function **CONFIDENCE.T(α, s, n)** for calculating the confidence interval $\pm t s/\sqrt{n}$ from a t-distribution only when $v = n - 1$, for $n > 50$. When $v \neq n - 1$ and $n \leq 50$, calculate the confidence interval according to Equation (8.10) using the worksheet function **TINV** for the coverage factor, t.

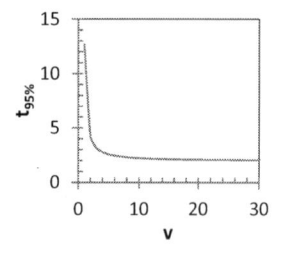

Figure 8.4

Coverage factors t for 95% confidence (α=0.05) with v degrees of freedom.

How should we interpret confidence intervals? Suppose we repeated the set of experiments to calculate a confidence interval 100 times. We then have 100 confidence intervals at a given significance level. For the significance level $\alpha = 0.05$, or 95% confidence, we say with confidence that 95 out of the 100 intervals probably contain the true population mean. Note that this is not the same as a 95% probability that the true population mean falls within each confidence interval (Donnelly 2004). We use the Welch-Satterthwaite formula from Section 8.2.4 to calculate the degrees-of-freedom needed to get the coverage factor for the combined uncertainties in Equation (8.7).

Points that fall far outside the confidence interval suggest outliers in the data. A macro named **OUTLIER** and two user-defined functions **CHAUVENET(x)** and **GRUBBS(x)** are available in the *PNMSuite* to test for a single outlier in a range of data **x** with a normal distribution (Taylor 1982) (NIST 2012). The **OUTLIER** macro colors the outlier cells red in the range of data according to Tukey's (1977) method. The functions return the value of the greatest outlier, or "**No Outlier**" if no data point fails the statistical test. We urge caution when eliminating outliers that may be indications of other problems with the experiment or data. Always repeat the statistical analysis on the reduced data set after removing a single outlier.

For data sets with unknown statistics for its probability distribution, the *PNMSuite* includes a simple macro listed, in Figure 8.5, to apply the **BOOTSTRAP** resampling method (Efron 1979) to estimate the confidence interval for a statistic, such as the mean or median of a data set in an *Excel* worksheet[76]. Bootstrapped Monte Carlo resampling methods simulate the uncertainty using a large number of random subgroups of the data with replacements drawn from the original data to make up a subgroup with the same size as the original data set. On its surface, bootstrapping seems like a "free lunch." However, the bootstrap assumes that a sample of data points drawn from a larger population approximates the distribution of the population. The bootstrap fails when the sample does not approximate the population because it is too small, or was not selected randomly, independently, or without bias (Good and Hardin 2006).

The macro **BOOTSTRAP** selects a large number of random subgroups from a numbered data set using an inner loop that generates random integers for the numbered data. A sorted array saves the values of the descriptive statistical function evaluated for each subgroup. To use the macro, put the formula for a statistic in a cell on a worksheet in terms of the range of data. The macro calculates a smoothing parameter from the standard error of the data set, or the user may specify another value. **BOOTSTRAP** reports the standard deviation, and the margins at a specified significance level, and then plots the distribution of the simulations in a histogram on a new worksheet for analysis. The expanded uncertainty is determined from the range of function values between the lower $\alpha/2$ and upper $(1-\alpha/2)$ quantiles of the sorted array. A **BOOTSTRAP** user-form is also available for calculating the confidence interval for the average value of a data set. See (Hesterberg 2014) for more details about bootstrapping and permutation tests.

```
Public Sub BOOTSTRAP()
' Bootstrap resampling method with substitution and optional smoothing for calculating
' the expanded uncertainty of a data set.  Use a worksheet function to calculate the
' statistic for a range of cells containing the data set.
' *** Required User Functions ***
' HEAPSORT = sort the vector of simulated statistical functions
' NRDWH = normal random deviate
' RNDWH = uniform pseudo random number
' *** Nomenclature ***
Const bxttl = "BOOTSTRAP Expanded Uncertainty" As String
Dim b As Long, db As Double ' = number of bins and bin size
Dim fb() As Double ' array of bin sizes for simulated values of f
Dim f_low As Double, f_up As Double ' = low and upper values of confidence interval
Dim fr As Range, fi As Double ' = range of statistical function and initial value
Dim freq() As Long ' vector of frequencies
Dim fs() As Double ' = array of simulated values for a statistic
Dim i As Long, j As Long, il As Long ' = loop index
Dim m As Long ' = number of resamples
Dim n As Long, nr As Long ' = size of data set & number of rows in vector
Dim output As String, outputr As Range ' = location of output on the worksheet
Dim outrange As String, ow As Integer ' = address of output & overwrite check
Dim p As Double, pi As Integer ' = fraction and percent confidence level
Dim r() As Long ' = random integers from data size
Dim s As Double ' = standard deviation of the mean
Dim u As Double ' = uncertainty
Dim ws As String ' = name of worksheet for output
Dim x() As Variant, xr As Range, xs() As Double ' = range of data set & saved data set
'***************************************************************************
With Application ' use worksheet functions
Set fr = .InputBox(Prompt:="Cell with FUNCTION, f(x):", Type:=8, Title:=bxttl)
Set xr = .InputBox(Prompt:="Range of DATA set, x:", Type:=8, Title:=bxttl)
n = xr.Count: nr = xr.Rows.Count
s = USIGFIGS(.StDev(xr) / Sqr(n), 2)
s = .InputBox(Prompt:="SMOOTHING Parameter:", Default:=s, Type:=1, Title:=bxttl)
p = .InputBox(Prompt:="% CONFIDENCE Level:", Default:=95, Type:=1, Title:=bxttl)
m = .InputBox(Prompt:="Number of SIMULATIONS:", Default:=10000, Title:=bxttl, Type:=1)
pi = .Min(99, .Max(CInt(p), 50)): p = pi / 100 ' must be between 50% & 100%
m = .Max(1000, CLng(m)) ' number of simulations, minimum = 1000
fi = fr.Value ' save initial value of the statistic
ReDim fs(1 To m, 1 To 1) As Double, r(1 To n) As Long, _
      x(1 To n) As Variant, xs(1 To n) As Double
For i = 1 To n: xs(i) = xr(i): Next i ' save original data set
For j = 1 To m ' bootstrap simulation loop
    For i = 1 To n ' random data point selection loop with smoothing
```

---

[76] The name *Bootstrap* comes from phrase "Pull yourself up by your bootstraps."

```
            r(i) = CLng(.RoundUp(RNDWH() * n, 0)): x(i) = xs(r(i)) + NRDWH() * s
    Next i
    If n = nr Then ' replace the data with the resampled vector
        xr = .Transpose(x) ' put in the resample in a column
    Else
        xr = x ' put in the resample in a row otherwise
    End If
    fs(j, 1) = fr ' save local simulation
Next j
Call HEAPSORT(fs, 1) ' sort the array of simulated statistical function evaluations
f_low = fs(CLng((1 - p) * m / 2), 1): f_up = fs(CLng(0.5 * (1 + p) * m), 1)
outputr.Value = "-U" & CStr(pi) & "% = "
outputr.Offset(0, 1).Value = USIGFIGS(Abs(fi - f_low), 2)
outputr.Offset(1, 0).Value = "+U" & CStr(pi) & "% = "
outputr.Offset(1, 1).Value = USIGFIGS(Abs(f_up - fi), 2)
outputr.Offset(2, 0).Value = "s = " ' standard deviation
outputr.Offset(2, 1).Value = SIGFIGS(Application.StDev(fs),3)
If n = nr Then ' replace the data with the original save vector
    xr = .Transpose(xs) ' put in the original column
Else
    xr = xs ' put in the original row
End If
Worksheets.Add ' add a worksheet for a histogram of the bootstrap results
b = .RoundUp(Sqr(m), 0): db = (fs(m, 1) - fs(1, 1)) / b ' number of bins & size
ReDim fb(1 To b) As Double, freq(1 To b) As Long
For i = 1 To b ' get bins for histogram
    fb(i) = fs(1, 1) + i * db: Cells(i, 1) = fb(i)
Next i
i1 = 1 ' fill bins with frequency
For j = 1 To b
    freq(j) = 0
    For i = i1 To m
        If fs(i, 1) <= fb(j) Then
            freq(j) = freq(j) + 1
        Else
            Exit For
        End If
    Next i
    i1 = i + 1: Cells(j, 2) = freq(j)
Next j
With Range(Cells(1, 2), Cells(b, 2)) ' create histogram with conditional formatting
    .FormatConditions.AddDatabar
    With .FormatConditions(.FormatConditions.Count)
        .ShowValue = True: .SetFirstPriority
    End With
    With .FormatConditions(1)
        .MinPoint.Modify newtype:=xlConditionValueAutomaticMin
        .MaxPoint.Modify newtype:=xlConditionValueAutomaticMax
        .BarColor.Color = 13012579: .BarFillType = xlDataBarFillSolid
    End With
End With
Columns("B:B").ColumnWidth = b ' adjust the width of the histogram to size of data
With Cells(b + 1, 2).SparklineGroups
    .Add Type:=xlSparkColumn, SourceData:="B1:B" & CStr(b)
    .Item(1).SeriesColor.Color = 9592887
End With
Rows(b + 1).RowHeight = 2 * b: End With ' Application
End Sub ' BOOTSTRAP
```

Figure 8.5 *VBA* macro BOOTSTRAP for calculating the confidence interval of a statistic for a data set.

"A statistician is a person whose lifetime ambition is to be wrong 5% of the time."

CHAPTER 8: UNCERTAINTY

Example 8.1 Confidence Interval for Temperature in Ideal Gas Concentration

Known: Consider the calculation of concentration, C, for an ideal gas in terms of the pressure and temperature:

$$C = \frac{P}{R_g T} \quad (8.11)$$

where P, T, and R_g are the pressure, temperature, and gas constant, respectively. The temperature was measured with a thermocouple using a digital display and tabulated in column **A** of an *Excel* worksheet.

Find: Calculate the average value for the temperature, standard uncertainty, and 95% confidence interval.

Assumptions: Ideal gas.

Analysis: Use *Excel* to tabulate the temperatures. Use the **Data Analysis** add-in **Descriptive Statistics** to perform the calculations. Run the macro **BOOTSTRAP** for an alternative simulation of the uncertainty. Based on these results, we report the temperature of the gas as $436.6 \pm 0.1°C$ (with 95% confidence).

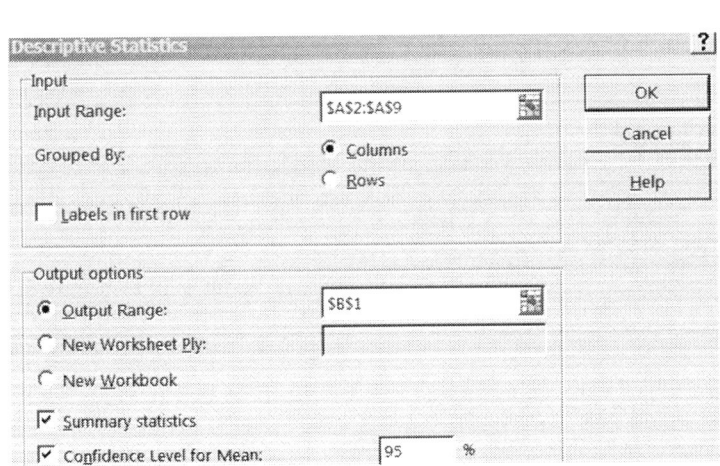

	A	B	C
1	T/°C	Column1	
2	436.7		
3	436.5	Mean	436.55
4	436.6	Standard Error	0.0423
5	436.4	Median	436.55
6	436.5	Mode	436.7
7	436.6	Standard Deviation	0.1195
8	436.4	Sample Variance	0.0143
9	436.7	Kurtosis	-1.456
10		Skewness	-9E-13
11		Range	0.3
12		Minimum	436.4
13		Maximum	436.7
14		Sum	3492.4
15		Count	8
16		Confidence Level(95.0%)	0.0999

Comments: We arrive at the same set of results if we use the individual worksheet functions for the average (mean), standard deviation of a sample, square root, count, and confidence level with student's t-distribution:

Average T/°C	436.6 = AVERAGE(A2:A9)
Standard Error/°C	0.0423 = STDEV.S(A2:A9)/SQRT(COUNT(A2:A9))
$t_{95\%}$	2.364 = TINV(0.05, 7)
Expanded Uncertainty U/°C	0.1 = 2.364 × 0.0423 = CONFIDENCE.T(0.05, STDEV.S(A2:A9), COUNT(A2:A9))

The following bootstrapped result agrees with the expanded uncertainty in the confidence interval. The frequency plot shows a normal distribution for the uncertainty:

Bootstrap Expanded U/°C   0.084 to 0.086 using 1000 simulations of the AVERAGE(A2:A9) with α = 0.05. When rounded to one significant figure, these results agree with the confidence interval of expanded uncertainty.

$-U_{95\%}$ =	8.4E-2
$+U_{95\%}$ =	8.6E-2
s =	4.30E-2

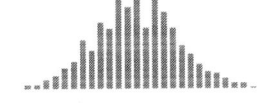

□

8.1.4 Type B Fixed Uncertainty from Systematic Errors

"Not everything that can be counted counts, and not everything that counts can be counted." – Albert Einstein

In this section, we learn how to estimate the systematic (fixed) uncertainty, u_z in Equation (8.7). Effects of systematic errors are practically impossible to eliminate; they are artifacts of our general inability to measure exactly. Manufacturers of measuring devices may provide guidelines for systematic errors in measurements, or we may know these from experience with the measurement technique. Without such information, we base the uncertainty on the precision of the instrument. As illustrated in Figure 8.6, fixed and bias errors may arise from a zero-offset, slope deviation (sensitivity), hysteresis, or nonlinearity in a response. A unique problem with reading the needle position on a dial involves a phenomenon called parallax where the angle of viewing the gauge affects the perception of the position of the needle. For example, if we view the gauge from the right, we perceive the reading to be slightly to the left of the hash mark. Likewise, viewing the gauge from the left side gives the perception that the needle lines up on right side of the hash mark. When we suspect such bias errors, we should take steps to quantify them and include them in our uncertainty analysis.

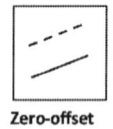

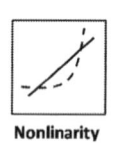

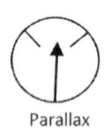

Figure 8.6

Graphical depictions of sources of fixed errors.

Before the digital age, we became accustomed to estimating fixed uncertainty whenever we read the time on the face of an analog clock. We might report the time on the dial in Figure 8.7 as "about ten 'till noon", when we really mean that the clock displays a time between 11:45 and 11:55. To illustrate the concept of fixed uncertainty further, consider the rulers in Figure 8.7. We must interpolate between graduation marks on a ruler when measuring length. Of course, we can spend more money to get a ruler with finer spacing between graduation marks. However, we cannot create a ruler with an infinitesimal spacing. As the marks get closer together, we end up with a ruler consisting of one long solid mark, which is the same as no marks at all.

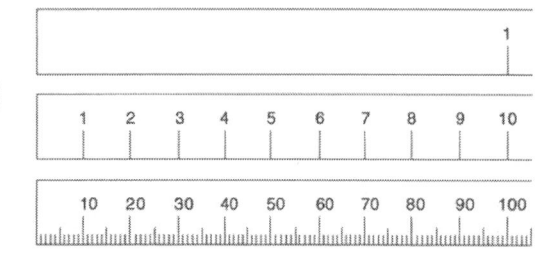

Figure 8.7

Analog clock face, digital readout, and three rulers comparing different levels of readability for 1 dm, 10 cm, and 100 mm.

Referring to a measurement for $\bar{x}$ in Figure 8.8, an engineer records a value, "$\bar{x}$ plus or minus δ", where the range $\pm\delta$ is the readability of the instrument of measurement. For larger spacing between marks, readability is determined from our ability to interpolate between marks. For instance, we may be able to interpolate the top ruler in Figure 8.7 to within one cm and the middle ruler to within 0.2 cm. However, the bottom ruler has readability no better than ± 0.5 mm (at most half the distance between graduation marks on the scale). Readabilities of analog devices may vary from user to user. One person may be more or less comfortable with a higher precision of readability than another person reading from the same analog scale. The specification of readability becomes less important when the random uncertainty is relatively large. However, if the major source of uncertainty comes from the lack of precision in the analog scale, we may need to negotiate the value for the readability among the members of our engineering team.

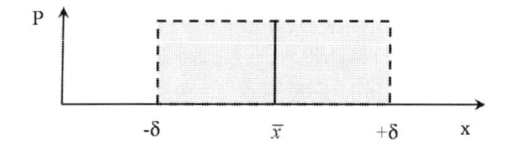

Figure 8.8

Uniform probability distribution about the mean: $2 \cdot \delta \cdot P = 1$. Any value between the lower and upper limits of readability has equal probability.

CHAPTER 8: UNCERTAINTY

Unlike analog scales, there is no ambiguity in the readability of a digital meter where the readability is fixed at 50% of the smallest decimal place in the readout (*e.g.*, the digital readout of **2.46** in Figure 8.7 has a resolution of 0.01 for a readability of $\delta = \pm(0.01)/2 = \pm 0.005$).

Although proper calibration may force the bias error toward zero, $z \to 0$, we can never eliminate the uncertainty in z because we cannot make measurements with infinite precision. This is an example of a Type B uncertainty. Without additional information, we conservatively assume a uniform probability density function for the uncertainty in the systematic error, as illustrated in Figure 8.8.

We calculate the variance of the uniform distribution by integrating the square of the difference with the mean value multiplied by the probability density function, $p = 1/(2 \cdot \delta)$:

$$u_z^2 = \int_{-\infty}^{\infty} (x - \bar{x})^2 \, p \, dx = \int_{\bar{x}-\delta}^{\bar{x}+\delta} (x - \bar{x})^2 \frac{1}{2\delta} dx = \frac{\delta^2}{3} \tag{8.12}$$

We must use our best engineering judgment when assuming a distribution. In some cases where we believe that the extreme uncertainties of a rectangular distribution are unlikely, such as measuring devices manufactured to controlled specifications, we may assume a less conservative triangular distribution (Ellison and Williams 2012) where $u_z^2 = \delta^2/6$. A volumetric flask is one example where a triangular distribution may be appropriate. We then estimate the fixed, systematic uncertainty in the measurement from the standard deviation in terms of the resolution (or readability) of the instrument, $\pm \delta$:

$$\text{Rectangular Distribution} \quad u_z = \sqrt{\frac{\delta^2}{3}} = \frac{\delta}{\sqrt{3}} \qquad \text{Triangular Distribution} \quad u_z = \sqrt{\frac{\delta^2}{6}} = \frac{\delta}{\sqrt{6}} \tag{8.13}$$

We use infinite degrees of freedom for Type B fixed uncertainty because we assume that the standard deviation is zero (Kirkup and Frenkel 2006). Consequently, systematic fixed uncertainty has a coverage factor of 1.96.

Whereas increasing the number of replicate measurements has the effect of reducing the standard uncertainty in Equation (8.9), the fixed uncertainty in Equation (8.13) is independent of the number of replicates such that the combined uncertainty in Equation (8.7) never becomes smaller than the systematic or fixed uncertainty:

$$\lim_{n_i \to \infty} u \to u_z \tag{8.14}$$

Thus, when our estimates for uncertainty in a measurement prove bad, this is most likely *not* due to statistical error, but instead may be the result poor estimates for systematic errors, or ignoring fixed type errors altogether.

Example 8.2   Readability of Analog versus Digital Pressure Gauges

Known/Schematic: The pressure of the ideal gas from *Example 8.1* was measured using analog and digital devices. Compare fixed uncertainties for pressure readings between (a) analog and (b) digital meters. The readability (resolution) of the analog meter is subject to the spacing between graduation marks on the scale, and the user's comfort level with interpolating between the marks. The readability (resolution) of the digital meter is the precision of the smallest decimal place in the display.

(a) [analog pressure gauge showing PSI scale 0–100, needle near 35]   (b) [digital display showing 34.9]

Find: systematic fixed uncertainty for the analog and digital gauges.

Assumptions: Uniform uncertainty distribution in fixed errors.

Analysis: The analog pressure gauge (a) reads approximately 35 psi. The precision of the analog graduation marks is 10 psi. However, resolution of the analog gauge is approximately 2 psi, (for a readability of $\delta = \pm 1$ psi), yielding the following fixed uncertainty:[77]

(a) *Analog*:
$$u_z = \frac{2\,psi}{\sqrt{12}} = \frac{1\,psi}{\sqrt{3}} = 0.6\,psi$$

The digital gauge (*b*) reads exactly 34.9 psi, a value rounded to the first decimal place that can be any value in the range: $34.85 < P < 34.95$ psi. For example, the pressure reading may represent the average of a series of fluctuating pressures, such as:

$$P = \frac{34.91 + 34.86 + 34.88 + 34.93 + 34.94}{5} = 34.9\,psi$$

Thus, the resolution of the digital meter is $2\delta = |34.95 - 34.85| = 0.1$ psi, or a readability of $\delta = 0.05$ psi. The resolution *and* precision of the digital gauge have the same value of 0.1 psi, for a fixed uncertainty:

(b) *Digital*:
$$u_z = \frac{0.1\,psi}{\sqrt{12}} = \frac{0.05\,psi}{\sqrt{3}} = 0.03\,psi$$

Comments: Eliminate systematic errors using carefully calibrated instruments and minimize systematic uncertainty using a measuring device with finer precision for higher resolution in the measurements.

□

8.1.5 Uncertainty in Quoted Values and Single or Small Sets of Measurements

"First get your facts; then you can distort them at your leisure." – Mark Twain

If we only have time or resources for a single measurement, we use the readability of the measurement tool as our source of uncertainty. For a few measurements (< 10) that prevent a proper statistical analysis, we should consider using the more conservative maximum absolute deviation (*MAD*) or slightly less conservative average absolute deviation (*AAD*) from the mean of as estimates of uncertainty:

$$u_{MAD} = Max|x - \bar{x}| \qquad u_{AAD} = \frac{\sum |x - \bar{x}|}{n} \qquad (8.15)$$

A quoted experimental value may come from a variety of sources, such as a table of physical properties in the scientific or engineering literature, equipment vendors, handbooks, etc. If available, use the uncertainty reported in the source. Unfortunately, quoted values often come without uncertainty information. If the manufacturer provides the control tolerance for the value, we can assume a normal distribution. For $\pm 2\sigma$ (standard deviations from the mean with 95% of the distribution) we approximate the standard uncertainty as $\pm tol/2$. The 95% expanded uncertainty with a coverage factor of ~2 becomes $U_{x,95\%} \cong \pm 2 \times tol/2 = \pm tol$. Otherwise, assume a zero standard deviation, which implies infinite degrees of freedom in Equation (8.8).

For single measurements, we may assume that the precision in the number reflects the readability in the measurement and compute the fixed uncertainty from half the possible range of the value that gives back the original value, when rounded to the least significant figure. To illustrate, the value of the variable $x_i = 1.23$ has a readability range of $(0.01)/2$ or 0.005, such that any number in the range $1.225 < \bar{x} < 1.235$ rounds to 1.23. Without additional information, we assume infinite degrees of freedom yielding a coverage factor of $t = 1.96$ that gives a 95% confidence interval. For uniform (rectangular) or triangular distributions, the fixed uncertainty for this example becomes:

---

[77] You may read the analog scale differently than the author, which is OK.

CHAPTER 8: UNCERTAINTY

- *Rectangular Distribution* $\quad 1.23 - 1.96\left(\dfrac{0.005}{\sqrt{3}}\right) \leq \bar{x} \leq 1.23 + 1.96\left(\dfrac{0.005}{\sqrt{3}}\right)$

- *Triangular Distribution* $\quad 1.23 - 1.96\left(\dfrac{0.005}{\sqrt{6}}\right) \leq \bar{x} \leq 1.23 + 1.96\left(\dfrac{0.005}{\sqrt{6}}\right)$

We may further illustrate with the value of the ideal gas constant required by *Example 8.1*, $R_g = 0.08206$ L·atm/mol·K, which has a readability range of $\pm(0.00001)/2$ or ± 0.000005, and a fixed uncertainty $u = 0.000005/\sqrt{3} = 2.9 \times 10^{-6}$ L·atm/mol·K. With "infinite" degrees of freedom, the expanded uncertainty becomes $1.96 \cdot 2.9 \times 10^{-6}$ L·atm/mol·K = 5.7×10^{-6} L·atm/mol·K.

Example 8.3 Uncertainty of Concentration in a Flow Reactor

Known: A transient mole balance for catalytic hydrogenation of olefins in a flow reactor gives

Rate of Accumulation = Rate In − Rate Out + Rate of Generation):

$$V\frac{dC_A}{dt} = Q(C_{A0} - C_A) + Vr_A \tag{8.16}$$

where V and Q are the reactor volume and flow rate, respectively, and C is the concentration of olefin. The reaction rate law is (Parulekar 2006):

$$-r_A/(mol/L \cdot s) = \frac{C_A}{(1+C_A)^2} \tag{8.17}$$

Find: Calculate the steady-state reactant concentration from an initial concentration $C_{A0} = 13$ mol/L, reactor volume $V = 10$ L, and volumetric flow rate $Q = 0.2$ L/s.

Assumptions: constant density, isothermal, well-mixed reactor

Analysis: At steady state, the accumulation term (dC_A/dt) in Equation (8.16) is zero. Substitute for the reaction rate from Equation (8.17) into Equation (8.16):

$$0 = Q(C_{A0} - C_A) - V\frac{C_A}{(1+C_A)^2} \tag{8.18}$$

Rearrange for a cubic polynomial:

$$0 = C_{A0} + \left(2C_{A0} - \frac{V}{Q} - 1\right)C_A + (C_{A0} - 2)C_A^2 - C_A^3 \tag{8.19}$$

Set up an *Excel* worksheet with the parameters and combination of terms of the polynomial coefficients. Use the macro **QYXPLOT** to graph the function and locate the vicinity of the roots. Observe that this reactor has multiple steady states where $f = 0$. The formulas for the results in column **E** correspond to the polynomial coefficients in Equation (8.19). Use the *VBA* macro **BAIRSTOW** from the *PNMSuite* to find three real roots in column **H**. Assume the uncertainty in the volumetric flow rate is uniform at $u = \pm 0.05$, based on the precision of the quoted value.

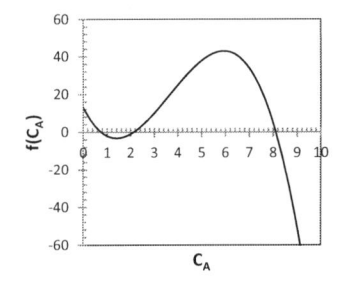

	A	B	C	D	E	F	G	H
1	Parameters			Polynomial Coefficients			Roots for CA	
2	CA_0	13		C_{A0}	13		$r_1 =$	8.117532
3	V	10		$2C_{A0}$-V/Q-1	-25		$r_2 =$	0.751536
4	Q	0.2		C_{A0}-2	11		$r_3 =$	2.130933
5					-1		-1	

Calculate the fixed uncertainty in the volumetric flow rate: $u = 0.1/\sqrt{12} = 0.05/\sqrt{3} = 0.029$. How does the systematic fixed uncertainty in the reactor parameters affect the solution? Plot the steady-state function of Equation (8.18) for the two cases $Q \pm u$ and report the values for the real roots, $C_{A+} = 9.11$, $C_{A-} = 0.45$.

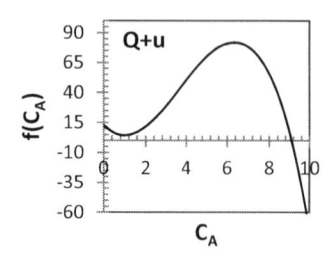

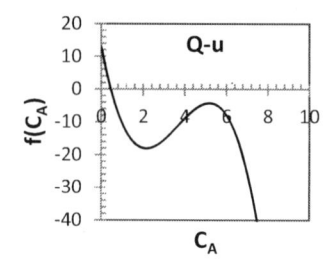

Comments: The steady-state operation of the reactor is sensitive to the volumetric flow rate within the uncertainty in the quoted value. At each of the limits of uncertainty, the reactor has only one real steady-state solution that differs significantly from the base case.

□

8.1.6 How (Not) to Report Uncertainty

> *"There are known knowns. These are things we know that we know. There are known unknowns. That is to say, there are things that we know we don't know. But there are also unknown unknowns. There are things we don't know we don't know."* – Donald Rumsfeld

We must be careful to define uncertainties in our reporting practices to give the users of our results a complete set of information. Generally, we report an expected value plus-or-minus an estimate of the error. The expected value usually refers to the average value of replicated measurements in Equation (8.5). Without additional qualifying information, a reported uncertainty may refer to the standard deviation, standard uncertainty, combined uncertainty, or an expanded uncertainty. Whatever the definition, a simple reporting practice includes parenthetical information with the numerical values. For example,

"$T = 7.8 \pm 0.9°C$ (mean ± standard uncertainty)."

"$C = 1.23 \pm 0.04$ M (mean ± 95% confidence interval)."

As seen above, the levels of precision in the expected value and its uncertainty must agree. The following examples show incorrect alignment of least significant digits between the value and its uncertainty:

Incorrect "$T = 7.83 \pm 0.9°C$" **Incorrect** "$C = 1.2 \pm 0.04$ M."

The first case has too much precision in T. The second example has too little precision in C.

One way to ensure alignment of the uncertainty with the result uses percent uncertainty. We normally retain one or two significant figures in the uncertainty or percent uncertainty:

"$T = 7.8°C \pm 12\%$ (mean ± standard uncertainty)."

"$C = 1.23$ M $\pm 3\%$ (mean ± 95% confidence interval)."

We always round *up* uncertainty. Rounding up the magnitude of uncertainty gives a more conservative answer (meaning we err on the high side of uncertainty). We should never round down the value of uncertainty because this has the effect of increasing the level of precision in the result beyond the specified level of confidence. For example, if we determine that the 95% confidence interval is greater than, or equal to ±11.4% but round the value down to 11%, we are in effect lowering our confidence level below 95%. On the other hand, by rounding the value up to 12%, we are still at 95% confidence, or higher.

Excel has worksheet functions for rounding numbers up or down (**ROUNDUP, ROUNDDOWN**). The first argument in the worksheet function is rounded according to the decimal place specified by second argument:

$$0.04 = \text{ROUNDUP}(0.0321, 2)$$

We use a negative value for the second argument if we are rounding to a position left of the decimal place. Instead of specifying the location of the least significant figure, we recommend the method of imbedded worksheet functions introduced in Section 2.2.1 to round automatically a value *up,* with the specified number of significant figures:

$$=\text{ROUNDUP}(x, \textit{SigFigs} - 1 - \text{INT}(\text{LOG10}(\text{ABS}(x))))$$

To illustrate, the result for 0.0321 with one significant figure gives:

$$0.04 = \text{ROUNDUP}(0.0321, - \text{INT}(\text{LOG10}(\text{ABS}(0.0321))))$$

For uncertainty results, we modified the user-defined function **SIGFIGS** using the **ROUNDUP** worksheet function in user-defined function **USIGFIGS**. The user-defined function also has an optional argument for the number of significant figures set to a default value of one. The function restricts the number of significant figures to at least one or two at most:

$$0.04 = \text{USIGFIGS}(0.0321,1) \qquad 0.033 = \text{USIGFIGS}(0.0321,2)$$

We should include a brief discussion of uncertainty in our reports, including a description of any suspected bias in our results.[78] For example, we may have a bias error that is one-sided, such that we do not get absurd negative results when calculating extreme values in confidence intervals. For instance, our uncertainty analysis may yield an impractical design volume of 1.2 ± 3 m^3 (or a volume in the range: $-1.8 < V < 3.2$ m^3). Clearly, we cannot have a negative volume.

8.2 Uncertainty Propagation

"For my part I know nothing with any certainty, but the sight of the stars makes me dream." – Vincent Van Gogh

Data reduction is the process of extracting order and meaning from experimental results. The propagation of the uncertainty in the inputs through complex data-reduction calculations compounds the effects of uncertainty on the final calculated result. Figure 8.9 illustrates a travel analogy of the potential dangers associated with ignoring uncertainty. A small deviation from the correct direction at the start grows to increasingly large deviations from the desired trajectory of travel.

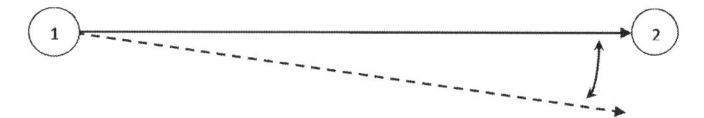

Figure 8.9

A small deviation in the direction at the starting point 1 may have the result of a large deviation from the desired endpoint 2.

In analogous fashion, the more complex the combination of calculations, the more the initial input error can influence the final uncertainty in the calculated result. Engineering design equations typically involve several variables. The uncertainty in each of the variables in the design equation propagates through the calculations to influence the uncertainty in the final design result.

Reconsider the *CSTR* example in Equation (8.1). The design for the volume must include the uncertainties in the expected values of each of the design parameters:

$$\bar{V} = \frac{\left(\bar{Q} \pm u_Q\right)\left(\bar{X} \pm u_X\right)}{\left(\bar{k} \pm u_k\right)\left(1 - \bar{X} \pm u_X\right)} \qquad (8.20)$$

One approach combines the effects of uncertainties in each variable in a manner that gives the extreme limits for the design result. We gave an example of this in Equation (5.98). In the reactor volume equation, we adjust the variables

---

[78] The 1897 Indiana State House considered biased legislation to round up the value of pi to 3.2 for *convenience* when calculating the circumference of a circle (Seife 2010). Wait…what?!

according to the uncertainties to give the greatest volume. Replace Equation (8.20) with the conservative, or "worst case," design scenario:

$$\overline{V} = \frac{(\overline{Q}+u_Q)(\overline{X}+u_X)}{(\overline{k}-u_k)\left[1-(\overline{X}+u_X)\right]} \tag{8.21}$$

In general, this approach tends to overestimate the combined effects of uncertainty, and does not account for the fact that most uncertainties are normally distributed about their expected values. We should also note that the worst-case approach does not allow for the possibility that some combined uncertainties may have a cancelling effect on the overall uncertainty in the calculated result.

8.2.1 The Law of Propagation of Uncertainty

"In this world, nothing is certain except death and taxes." – Benjamin Franklin

The great mathematician Carl Freidrich Gauss[79] proposed a first-order measure of "uncertainty importance" as the product of the sensitivity of the function with the uncertainty in the variable (Morgan and Henrion 1990):

$$u_{Gi} = c_i \cdot u_i \tag{8.22}$$

where i represents the variable index for multivariable problems, c_i is the sensitivity coefficient, and u_i is the combined uncertainty from Equation (8.7) for standard and systematic error estimates. As defined in Section 5.4, we may estimate the sensitivity coefficient c_i from a first-order finite difference model using the partial derivative of the function with respect to the variable x_i, evaluated at the expected (average) values $\overline{x}_i$:

$$c_i = \left.\frac{\partial f}{\partial x_i}\right|_{\overline{x}_i} \cong \frac{f(\overline{x}_i + \Delta x) - f(\overline{x}_i)}{\Delta x} \tag{8.23}$$

What role does the sensitivity coefficient play in uncertainty analysis relative to the uncertainty in x? The product of the sensitivity coefficient with uncertainty for a variable, u_x, gives a relative measure of its contribution to the uncertainty in the function:

$$u_f = \left|\frac{\partial f}{\partial x} u_x\right| \cong \left|\frac{\Delta f}{\Delta x} u_x\right| \tag{8.24}$$

Figure 8.10 compares four cases of the sensitivity of a function to the uncertainty in its variables. We recognize a high level of sensitivity by a relatively steep slope in the plot of the model function. A function with a gradual slope has lower sensitivity to the variable. Thus, as illustrated in the shaded graphs in Figure 8.10, a variable with low sensitivity and large uncertainty may have the same importance in the overall uncertainty of the function, u_f, as a variable with high sensitivity and small uncertainty. For a linear Taylor series expansion about x, we assume that the sensitivity is proportional to the slope of the tangent line to the function at x. The change in the function value on the vertical axis gives a relative measure of the effect of the product of sensitivity with uncertainty. We need low uncertainty to reduce the effects of high sensitivity. Conversely, we may get away with larger uncertainty in a variable when the model has a low sensitivity to perturbations in that variable.

---

[79] Gauss published only his fully developed theories. His personal motto was *pauca sed matura* ("few, but ripe").

CHAPTER 8: UNCERTAINTY

Figure 8.10

Comparison of the effects of sensitivity and uncertainty of a parameter on the propagation of uncertainty in a function. Note how the sensitivity coefficient in the graph above may change from low to high within the region of uncertainty in a variable for larger u_x.

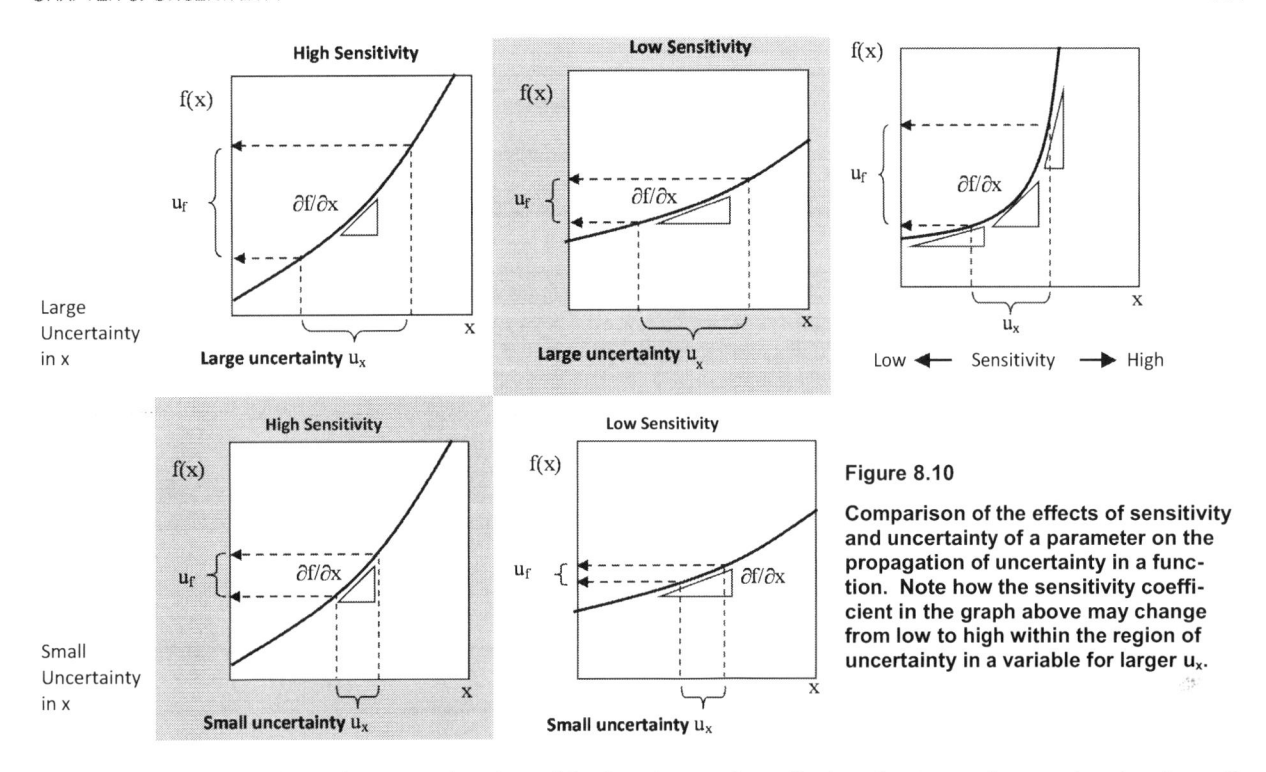

A first-order Taylor series approximation of the function works well when the range of uncertainty in x is small. For large variable uncertainty, the local linear assumption may break down for the sensitivity coefficient in Equation (8.24) where the model function is highly nonlinear. The third graph at the right side of the top row in Figure 8.10 illustrates how the slope of the tangent line along the curve changes over a larger range of uncertainty in x. Section 8.3 presents an alternative method using Monte Carlo simulations to handle uncertainty calculations when a truncated linear Taylor series does not well represent the sensitivity coefficients.

Equation (5.98) estimates the maximum propagated error assuming that the effects of error from each variable are cumulative and concurrent. In practice, the likelihood of all errors occurring at the same time (and in the same direction) is small. We expect that some of the effects of the errors will partially cancel. The variance in the function evaluated for N data sets is

$$\sigma_f^2 = \lim_{N \to \infty} \frac{1}{N} \sum_{i=1}^{N} \left[f(x_1, x_2, \ldots) - f(g_1, g_2, \ldots) \right]_i^2 \tag{8.25}$$

Substitute Equation (5.96) into Equation (8.25) for the special case of two input variables:

$$\sigma_f^2 = \lim_{N \to \infty} \frac{1}{N} \sum_{i=1}^{N} \left[\left(\Delta x_1 \left.\frac{\partial f}{\partial x}\right|_{g_1} \right)^2 + \left(\Delta x_2 \left.\frac{\partial f}{\partial x}\right|_{g_2} \right)^2 + 2 \Delta x_1 \Delta x_2 \left.\frac{\partial f}{\partial x}\right|_{g_1} \left.\frac{\partial f}{\partial x}\right|_{g_2} \right] \tag{8.26}$$

The variances of the two variables in Equation (8.26) are:

$$\sigma_{x_1}^2 = \lim_{N \to \infty} \frac{1}{N} \sum_{i=1}^{N} \left[(\Delta x_1)_i \right]^2 \quad \text{and} \quad \sigma_{x_2}^2 = \lim_{N \to \infty} \frac{1}{N} \sum_{i=1}^{N} \left[(\Delta x_2)_i \right]^2 \tag{8.27}$$

The last term in Equation (8.26) involves the covariance that accounts for any correlation between the two variables:

$$\sigma_{12}^2 = \lim_{N \to \infty} \frac{1}{N} \sum_{i=1}^{N} \left[\Delta x_1 \Delta x_2 \right]_i \tag{8.28}$$

We employ the squared uncertainties as estimates for the variances (Dunn 2005) in Equation (8.26):

$$u_f^2 = \left(u_1 \left.\frac{\partial f}{\partial x_1}\right|_{g_1,g_2}\right)^2 + \left(u_2 \left.\frac{\partial f}{\partial x_2}\right|_{g_1,g_2}\right)^2 + 2u_{12} \left.\frac{\partial f}{\partial x_1}\right|_{g_1,g_2} \left.\frac{\partial f}{\partial x_2}\right|_{g_1,g_2} \tag{8.29}$$

where
$$u_{12} = \frac{1}{n}\sum_{i=1}^{n}(x_{1i}-u_1)(x_{2i}-u_2)$$

Note that a negative value for u_{12} reduces the model uncertainty u_f. We may neglect covariance when the data show no correlation, or have a uniform, or normal, distribution about their mean values. Parameters obtained from least-squares regression typically *do* exhibit correlation. Consult Section 8.2.3 for specific calculations of correlation and Section 9.3 for information on correlation in uncertainty of regression. Equation (8.29) is formally the *Law of Propagation of Uncertainty* applied to a two-variable problem. In the case of multivariable uncertainty analysis, the *Law* becomes:

- Law of Propagation
$$u_f^2 = \sum_i \left(u_i \left.\frac{\partial f}{\partial x_i}\right|_g\right)^2 + 2\sum_i \sum_{j \neq i} u_{ij} \left.\frac{\partial f}{\partial x_i}\right|_g \left.\frac{\partial f}{\partial x_j}\right|_g \tag{8.30}$$

For functions involving *simple* addition, subtraction, multiplication, or division, it is relatively easy to calculate derivatives for the sensitivity coefficients. For these simple cases, the *Law of Propagation of Uncertainty* reduces to the following straightforward mathematical expressions:

- Addition/Subtraction
$$y(x) = a \cdot x_1 \pm b \cdot x_2 \pm c \cdot x_3 \pm \ldots$$
$$u_y^2 = a^2 \cdot u_{x_1}^2 + b^2 \cdot u_{x_2}^2 + c^2 \cdot u_{x_3}^2 + \ldots \tag{8.31}$$

- Multiplication/Division
$$y(x) = k \cdot x_1^a \cdot x_2^b \cdot x_3^c \ldots$$
$$\left(\frac{u_y}{y}\right)^2 = \left(\frac{au_{x_1}}{x_1}\right)^2 + \left(\frac{bu_{x_2}}{x_2}\right)^2 + \left(\frac{cu_{x_3}}{x_3}\right)^2 \ldots \tag{8.32}$$

- Logarithmic Differentiation
$$y(x) = \log x \quad \rightarrow \quad dy = \frac{dx}{x}$$
$$u_y = u_{\log x} = \frac{u_x}{x} \tag{8.33}$$

A simple illustration shows how to the handle uncertainty using logarithms with the law of propagation of uncertainty:

$$y(x) = \frac{k \cdot x_1 \cdot x_2^2}{\sqrt{x_3}}$$

$$\ln y = \ln\left[\frac{k \cdot x_1 \cdot x_2^b}{\sqrt{x_3}}\right] = \ln k + \ln x_1 + 2\ln x_2 - \frac{1}{2}\ln x_3 \quad \left(\frac{u_y}{y}\right)^2 = a^2\left(\frac{u_{x_1}}{x_1}\right)^2 + b^2\left(\frac{u_{x_2}}{x_2}\right)^2 + c^2\left(\frac{u_{x_3}}{x_3}\right)^2 + \ldots$$

$$d\ln y = \frac{dy}{y} = \frac{dx_1}{x_1} + 2\frac{dx_2}{x_2} - \frac{1}{2}\frac{dx_3}{x_3}$$

$$\left(\frac{u_y}{y}\right)^2 = \left(\frac{u_{x_1}}{x_1}\right)^2 + \left(2\frac{u_{x_2}}{x_2}\right)^2 + \left(\frac{1}{2}\frac{u_{x_3}}{x_3}\right)^2 \tag{8.34}$$

To illustrate further, apply the law of propagation in Equation (8.32) to our reactor design Equation (8.1):

$$\left(\frac{u_V}{V}\right)^2 = \left(-\frac{u_Q}{Q}\right)^2 + \left(-\frac{u_k}{k}\right)^2 + \left[\frac{u_X}{\overline{X}(1-\overline{X})}\right]^2 \tag{8.35}$$

The relative uncertainty in the reactor volume increases for relatively large flow rates and rate constants, and for the extremes of low and high conversion, $\overline{X} \to 0$ or $\overline{X} \to 1$. To reduce the contributions from these parameters to the design uncertainty, we need small measurement uncertainties in Q, k, and X.

The law of uncertainty propagation has some important advantages, as well as a few disadvantages to consider when comparing other methods of uncertainty analysis:

Advantages
- Simple implementation using finite difference derivative approximation for the sensitivity coefficients
- Quick solutions
- Applies to linear models and works well for near linear models
- Computes the relative magnitudes for sources of uncertainty to identify which variables contribute more or less to the propagated uncertainty

Disadvantages
- Limited to Gaussian distributions of uncertainty
- Requires derivative calculations
- Does not apply to variables with large random Type A uncertainty in highly nonlinear models (see Monte Carlo uncertainty analysis in Section 8.3 for an alternative method)

Example 8.4   Heat Transfer Coefficients from Correlations

Known: A heuristic for uncertainty in heat transfer coefficients derived from dimensionless correlations claims they are reliable within ±20% of their values. Engineers use the Dittus-Boelter equation to calculate the dimensionless Nusselt number (Nu) in terms of the dimensionless Reynolds (Re) and Prandtl (Pr) numbers for heating turbulent flow in smooth tubes (Incropera and DeWitt 2002):

$$Nu = 0.023 \, \mathrm{Re}^m \, \mathrm{Pr}^n \tag{8.36}$$

Find: Implement the law of propagation of uncertainty to calculate the uncertainty in the prediction for the dimensionless heat transfer coefficient.

Assumptions: Reynolds number of 10^5 for turbulent flow.

Analysis: An *Excel* worksheet contains the remaining parameters and corresponding uncertainties. Calculate the Nusselt number and products of the sensitivity coefficients with uncertainties. We use the macro **SENSITIVITY** from *PNMSuite* to calculate the sensitivity coefficients shown in column **E** in the *Excel* worksheet.

	A	B	C	D	E	F	G	H
1		Variables	u	c				(c*u)^2
2	Re	100000	10000	c_1 =	0.0015953507494268	rel c_1 =	0.799998418891522	=(C2*E2)^2
3	Pr	0.7	0.01	c_2 =	113.953640154725	rel c_2 =	0.399999262785487	=(C3*E3)^2
4	C	0.023	0.001	c_3 =	8670.38566420063	rel c_3 =	0.99999815699303	=(C4*E4)^2
5	m	0.8	0.01	c_4 =	2295.8945838314	rel c_4 =	9.21032337329781	=(C5*E5)^2
6	n	0.4	0.1	c_5 =	-71.1277143761875	rel c_5 =	-0.142669714633631	=(C6*E6)^2
7	Nu	=B4*(B2^B5)*(B3^B6)	=SQRT(H7)					=SUM(H2:H6)
8	u/Nu	=C7/B7						

	A	B	C	D	E	F	G	H
1		Variables	u	c				(c*u)^2
2	Re	1.00E+05	1.00E+04	$c_1 =$	0.001595	rel $c_1 =$	8.00E-01	254.5144
3	Pr	0.7	0.01	$c_2 =$	113.9536	rel $c_2 =$	4.00E-01	1.298543
4	C	0.023	0.001	$c_3 =$	8670.386	rel $c_3 =$	1.00E+00	75.17559
5	m	0.8	0.01	$c_4 =$	2295.895	rel $c_4 =$	9.21E+00	527.1132
6	n	0.4	0.1	$c_5 =$	-71.1277	rel $c_5 =$	-1.43E-01	50.59152
7	Nu	199.4192	30.14454					908.6932
8	u/Nu		15%					

Get the combined uncertainty and percent uncertainty in cells **C7** and **B8**, respectively from the *Law of Propagation of Uncertainty* in Equation (8.29), ignoring correlation.

Comments: The result shows a relative uncertainty of 15%, which supports the general heuristic that heat transfer coefficients are accurate within ±20%. An inspection of each term in column **H** in the propagation equation indicates that the terms involving the Reynolds number, *Re*, and its exponent, *m*, contribute most to uncertainty in the Nusselt number.

□

8.2.2 How (Not) to Handle Unit Conversions

"Being a statistician means never having to say you're certain." – Unknown

When we convert between unit systems, the perceived level of precision in our measurand may change if we are not careful. For example, a 1.2 lb mass is equivalent to 0.54 kg if we maintain the same number of two significant figures[80]. However, we lose information about uncertainty if we try to determine the fixed uncertainty from the magnitude of the least significant figure. Compare our two masses in different unit systems having the same number of significant figures:

Measurand ± Fixed Uncertainty	Range	% Uncertainty
1.2 ± 0.05 lb$_m$	1.195 < 1.2 < 1.25 kg	(0.05/1.2) × 100 = 4.2%
0.54 ± 0.005 kg	0.535 < 0.54 < 0.545	(0.005/0.54) × 100 = 0.93%

Unfortunately, we cannot justify a 78% *decrease* in uncertainty by simply converting units from pounds mass to kilograms. To fix this discrepancy, we use percent uncertainty from the start to determine the level of precision in our converted values. In our example of mass units, start with 1.2 pound mass ±4.2%. Convert the mass to units of kilograms carrying extra figures in the conversion factor:

$$1.2 \text{ lb mass} \times 0.453593 \text{ kg/lb mass} = 0.544312 \text{ kg}$$

Then, calculate the uncertainty in units of kilograms from the percent uncertainty in the original mass in pounds:

$$\% \text{ uncertainty} = \pm 0.544312 \times 4.2\% \text{ kg} = \pm 0.022268 \text{ kg}$$

Finally, round the mass in units of kg according to the level of precision in the converted results retaining two significant figures in the uncertainty (recall that we round *off* the value of the mass, but must round *up* the value for the uncertainty):

$$0.544 \pm 0.023 \ (\pm 4.2\%) \text{ kg}$$

We may lose a small amount of information in our uncertainty estimates if we do not carry more significant figures during the conversion process.

---

[80] Review Section 1.1.3 for a discussion of significant figures and round-off error.

CHAPTER 8: UNCERTAINTY

Alternatively, apply the law of propagation of uncertainty in Equation (8.29) to the expression for unit conversion. Let x and y represent the value of the measurand in the original and new unit systems, respectively, and m is the conversion factor: $y = mx$. If we assume that the uncertainty in the conversion factor is negligible, then the uncertainty in the new unit system is just the product of the unit conversion factor with the original uncertainty:

$$u_y = \pm \sqrt{\left(\frac{\partial y}{\partial x} \cdot u_x\right)^2} = \pm |m \cdot u_x|$$

For our example of mass unit conversion, the uncertainty in units of kilograms is the same as the result on a percent basis:

$$u_y = (0.454 \text{ kg/lb mass}) \times (\pm 0.05 \text{ lb mass}) = \pm 0.023 \text{ kg} (\pm 4.2\%)$$

The uncertainty in a temperature measurement does not change when we convert from °C to K, because $\Delta°C = \Delta K$, and likewise for °F versus R. However, $\Delta°C \neq \Delta°F$. Therefore, be careful when using temperature measurements in your calculations for uncertainty. For example, use an absolute temperature with percent uncertainty.

The best approach to handling uncertainty propagation with unit conversions leaves our measurements and associated uncertainty estimates in their original values and units. The unit conversion step is just part of the overall sequence of calculation of the function using the input variables in their original units. This approach forms the basis of the *Jitter* method in Section 8.2.5.

Example 8.5   Standard Uncertainty in Ideal Gas Concentration

Known/Find: Calculate the concentration and associated standard uncertainty for an ideal gas in terms of the pressures and temperature from *Example 8.1* and *Example 8.2*. This example illustrates how to handle uncertainty in single values and constants.

Analysis: Use the fixed uncertainty for R_g from Section 8.1.4 to calculate the standard uncertainty in C using *Excel* worksheet functions. Equation (8.11) for concentration is a candidate for the simple formula for uncertainty propagation in Equation (8.32):

$$u_C = C \sqrt{\left(\frac{u_T}{T}\right)^2 + \left(\frac{u_P}{P}\right)^2 + \left(\frac{u_{R_g}}{R_g}\right)^2} \qquad (8.37)$$

Be careful to use temperature on an absolute scale in Equation (8.37).

Set up an *Excel* worksheet with the calculations for uncertainty propagation. Assume zero standard uncertainties for the gas constant and pressure, which implies infinite degrees of freedom for these variables. Specify the readabilities from the precision of the values for P, T, and R_g to calculate their fixed uncertainties.

	A	B	C	D
1	Gas Concentration, C/M	=CONVERT(C3,"psi","atm")/(D3*B3)		
2	Variable	T/K	P/psi	R_g/(L atm/mol K)
3	Average Value	=436.55+273.15	34.9	0.08206
4	Standard Uncertainty, u_x	0.0423	0	0
5	Fixed Uncertainty, u_z	=0.05/SQRT(3)	=0.05/SQRT(3)	=0.000005/SQRT(3)
6	Combined Uncertainty, u_i	=SQRT(B4^2+B5^2)	=SQRT(C4^2+C5^2)	=SQRT(D4^2+D5^2)
7	(u/x)^2	=(B6/B3)^2	=(C6/C3)^2	=(D6/D3)^2
8	Standard Uncertainty in C	=B1*SQRT(SUM(B7:D7))		

The results for the propagation of uncertainty include the values for the product of the sensitivity coefficients and combined uncertainty for each variable.

	A	B	C	D
1	Gas Concentration, C/M	4.078E-02		
2	Variable	T/K	P/psi	R$_g$/(L atm/mol K)
3	Average Value	709.7	34.9	0.08206
4	Standard Uncertainty, u$_x$	0.0423	0	0
5	Fixed Uncertainty, u$_z$	0.028867513	0.028867513	2.88675E-06
6	Combined Uncertainty, u$_i$	0.051211555	0.028867513	2.88675E-06
7	(u/x)^2	5.20698E-09	6.84176E-07	1.23753E-09
8	Standard Uncertainty in C	3.39E-05		

The result for concentration must match the precision of the standard uncertainty. For the case where we retain one significant figure in the uncertainty we get $C = 0.04078 \pm 0.00004$ M, which represents $\pm 0.1\%$ uncertainty. The result for $(u/x)^2$ in cell **C7** reveals that the largest source of uncertainty comes from the error in the pressure measurement (where x symbolizes a variable).

Comments: If we use the higher precision pressure reading from the digital meter, we may now use a readability for pressure of 0.05 psi, and calculate $C = 0.04089 \pm 0.00004$ M (or $\pm 0.09\%$) for a slight improvement.

□

8.2.3 Correlation*

"The pure and simple truth is rarely pure and never simple." − Oscar Wilde

Covariance is a statistical measure of how two variables change relative to each other, defined here as the product of the correlation coefficient and standard uncertainties:

$$COV_{ij} = R_{ij} u_i u_j \qquad (8.38)$$

where $(-1 < R_{ij} < 1)$ represents the correlation coefficient of the model input variable i with respect to the model input variable j:

$$R_{ij} = \frac{\sum (x_i - \bar{x}_i)(x_j - \bar{x}_j)}{\sqrt{\sum (x_i - \bar{x}_i)^2 \sum (x_j - \bar{x}_j)^2}} \qquad (8.39)$$

and where the summations are over all values of the paired inputs x_i and x_j. The bar above the variable indicates the average value. The correlation coefficient is negligible when the data for the variables are normally distributed about their mean values. Some engineers and scientists rely too much on the correlation coefficient to determine goodness of fit between data and a model, as discussed in Chapter 9 on regression.

The complete law of uncertainty propagation must account for any correlation between variables:

$$u_y^2 = \sum_{i=1}^{k} (c_i u_i)^2 + 2 \sum_{i=1}^{k-1} \sum_{j=i+1}^{k} COV_{ij} c_i c_j \qquad (8.40)$$

where k is the number of variables in the function for y. The correlation between two input variables may cause an increase or decrease in the result from uncertainty propagation, depending on the $\pm$ signs of the correlation and sensitivity coefficients. *Excel* has two methods for calculating the correlation coefficient for two sets of data. Use the spreadsheet function **CORREL(Data$_i$, Data$_j$)** or the **Data Analysis** add-in tool. **Data$_i$** and **Data$_j$** represent ranges of data that have a suspected correlation. *Excel*'s **CORREL** function requires the same number of points in each data set.

Our usual approach is to randomize the order of experiments to reduce any procedural bias and eliminate time dependent correlation between variables. We may use *Excel* to randomize the number of the experiments. However, the covariance between two variables may be ignored when no evidence of correlation exists, or when the input variables were measured independently, not simultaneously. In Example 8.4, the Reynolds number potentially correlates with the Prandtl number because each parameter is a function of the fluid viscosity, which is also a function of the fluid temperature. However, the Prandtl number for air is relatively constant over a wide range of fluid temperatures, so any significant correlation of Re with Pr is unlikely.

We may handle issues of correlated variables by finding any functional relationships among them. To illustrate, consider a measurand with two correlated input variables, $y = f(x_1, x_2)$. For a simple function that correlates the input variables, such as:

$$x_2 = Ax_1, \qquad (8.41)$$

the law of propagation of uncertainty becomes

$$u_y^2 = (c_1 u_1)^2 + (c_A u_A)^2 \qquad (8.42)$$

where u_A is the standard uncertainty in the linear function parameter, A. Instead of calculating the covariance, we now require the uncertainty in parameter A. The sub procedure **JITTER**, developed in Section 8.2.5, includes the option for providing correlation coefficients between input variables.

Example 8.6   Cross Sectional Area of a Circular Pipe

Known: Consider the uncertainty in the calculation of the cross sectional area of a pipe.

Schematic: A cross-section of a circular pipe has an inside diameter d, outside diameter $D = 6.00 \pm 0.01$ inches, and wall thickness $t = 0.50 \pm 0.01$ inches.

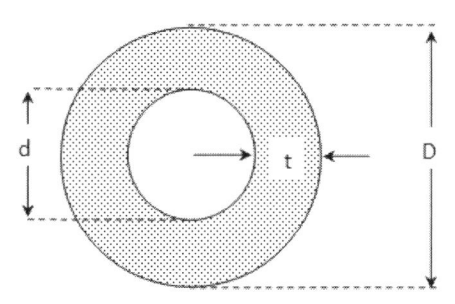

Find: Explore any correlation between geometric parameters.

Analysis: The geometry of the cylinder involves three possible measurements: the inside diameter, d, the wall thickness, t, and the outside diameter, D. Calculate the cross sectional area of the pipe from the differences in the areas of circles with diameters D and d:

$$A(D,d) = \frac{\pi}{4}(D^2 - d^2) \qquad (8.43)$$

However, if we limit our measurements to the outside diameter and wall thickness, the diameters are not independent, but related through the wall thickness:

$$d = D - 2t = 0.127 \text{ in} \qquad (8.44)$$

Eliminate the dependency by substituting for d in Equation (8.43) in terms of Equation (8.44):

$$A(D,t) = \frac{\pi}{4}\left[D^2 - (D-2t)^2\right] \qquad (8.45)$$

Approximate the uncertainty in the area from the uncertainties in the two uncorrelated measurements for D and t. Perform a simple numerical comparison of uncertainties using Equations (8.43) and (8.45). Calculate the sensitivity coefficients using the VBA macro **SENSITIVITY**. First, calculate the uncertainty in the inner diameter by propagating the uncertainty in D and t through Equation (8.44) for the result in cell **C4**. With this result, we get the uncertainty in A calculated from the *Law of Uncertainty Propagation* for Equation (8.43) without and with correlation in the range **C8:C9**. Finally, calculate the uncertainty in A from Equation (8.45) in cell **C13**.

	A	B	C	D	E	F	G	H
1		Variable	u		c			(c*u)^2
2	D	6	0.01	c₁ =	1	rel c₁ =	1.20E+00	1.00E-04
3	t	0.5	0.01	c₂ =	-2	rel c₂ =	-2.00E-01	4.00E-04
4	d(D,t)	5	0.022361					
5								
6	D	6	0.01	c₁ =	9.4248	rel c₁ =	6.55E+00	8.88E-03
7	d	5	0.022361	c₂ =	-7.854	rel c₂ =	-4.55E+00	3.08E-02
8	A(D,d)	8.63938	0.199311	<-- Without correlation				
9			0.081372	<-- With correlation				
10								
11	D	6	0.01	c₁ =	1.5708	rel c₁ =	1.09E+00	2.47E-04
12	t	0.5	0.01	c₂ =	15.708	rel c₂ =	9.09E-01	2.47E-02
13	A(D,t)	8.63938	0.157863					

The calculation for the area in terms of diameters over-estimates the uncertainty by not accounting for the partial cancellation of errors due to correlation between the inside and outside diameters.

Comments: Correlation among the input variables does not always produce larger overall uncertainty. The effect of correlation on the overall uncertainty is a function of the sign and magnitude of the sensitivity and correlation coefficients, and may reduce the magnitude of uncertainty for some problems.

□

When we suspect correlation between any two variables, but lack information required to calculate the correlation coefficient, use Equation (5.98) for the correlated terms to give a conservative, estimate of uncertainty. For example, the maximum contribution of two correlated variables j and k to the overall standard uncertainty involves the maximum uncertainty for just the correlated terms:

$$u_y^2 = \sum_{\substack{i \neq j \\ i \neq k}} (c_i u_i)^2 + (|c_j u_j| + |c_k u_k|)^2 \quad (8.46)$$

However, by including the maximum effect we risk over-estimating the uncertainty if in fact the covariance actually reduces the overall uncertainty. Just as ignoring uncertainty is risky, overestimation of uncertainty may lead to poor design choices when faced with economic consequences of unreliable results.

8.2.4 Expanded Uncertainty and Degrees of Freedom Using the Welch-Satterthwaite Formula

"There's no sense in being precise when you don't even know what you're talking about." – John von Neumann

The expanded uncertainty for 95% confidence is the product of the coverage factor, or Student t-statistic, and the standard propagated uncertainty from Equation (8.40):

$$U_{y,95\%,v} = t_{95\%,v} u_y \quad (8.47)$$

where $t_{95\%}$ is the two tailed coverage factor for v_i degrees of freedom obtained from Table 8.2, or *Excel* function **TINV(α,v)** with $\alpha = 0.05$. Use the following Welch-Satterthwaite formula for approximating the degrees of freedom of the pooled variances in u_y required for the t-statistic (Kirkup and Frenkel 2006):

CHAPTER 8: UNCERTAINTY

$$v_y = \frac{u_y^4}{\sum_{i=1}^{k} \frac{(c_i u_i)^4}{v_i}} \tag{8.48}$$

where k is the number of variables.

Define the degrees of freedom for variable i from the application of the Welch-Satterthwaite formula to the combined random and systematic uncertainties formula in Equation (8.7):

$$v_i \cong \frac{u_i^4}{\frac{u_{\bar{x}i}^4}{n-k} + \frac{u_{zi}^4}{\infty}} = (n-m)\left(\frac{u_i}{u_{\bar{x}i}}\right)^4 \tag{8.49}$$

where we assume infinite degrees of freedom for the systematic (fixed) uncertainty, and m is the number of parameters in the model for the variable x. For example, when calculating the degrees of freedom for the standard deviation from the mean, we set $m = 1$.

The Welch-Satterthwaite formula in Equation (8.48) does not apply to uncertainty problems with correlation in the independent variables, nor does the use of combined random and systematic uncertainties fit into the calculation of covariance (Hall and Willink 2001). Castrup (2010) developed a general form of the Welch-Satterthwaite formula with a correction term in the denominator for correlated variables:

$$v_y = \frac{u_y^4}{\sum_{i=1}^{k} \frac{(c_i u_i)^4}{v_i} + \sum_{i=1}^{k-1}\sum_{j>i}^{k} r_{ij}^2 \frac{(c_i u_i)^2 (c_j u_j)^2}{v_i v_j}\left(v_i + v_j + \frac{1}{2}\right) + 2\sum_{i=1}^{k-1}\sum_{j>i}^{k} r_{ij} c_i c_j u_i u_j \left[\frac{(c_i u_i)^2}{v_i} + \frac{(c_j u_j)^2}{v_j}\right]} \tag{8.50}$$

Note how Equation (8.50) reduces to the original Welch-Satterthwaite formula in Equation (8.48) without covariance. We must round the degrees of freedom down to the next lowest integer value to maintain the specified level of coverage, e.g. 95%. Use the worksheet function **ROUNDDOWN(v)** in *Excel*.

When our only source of error comes from Type B uncertainties, such as fixed uncertainty from quoted values, we skip past Equations (8.48), (8.49), or (8.50) and assume infinite degrees of freedom for the coverage factor. In this case $t_{95\%,\infty} = 1.96$ and we approximate the expanded combined uncertainty from the readability of the measurement, δ defined in Section 8.1.4:

$$u_i = 1.96 u_{zi} = 1.96 \frac{\delta_i}{\sqrt{3}} \cong \delta_i \tag{8.51}$$

The expanded propagation of fixed uncertainty becomes simply:

Special Case of Fixed Uncertainty Only $$U_y \cong \sqrt{\sum_i (c_i \delta_i)^2} \tag{8.52}$$

Example 8.7 Expanded Uncertainty in Ideal Gas Concentration

Known/Find: Calculate the expanded uncertainty for an ideal gas in terms of the pressures and temperature from *Example 8.5* using the digital pressure reading.

Analysis: We need the degrees of freedom for the combined uncertainties for each variable: T, P, and R_g. Use the Welch-Satterthwaite formula in Equation (8.49) for each variable in row **9**. In the case where the standard (random) uncertainty is zero, we use a small number in the denominator to avoid division by zero. Calculate the squares of the products of the sensitivity coefficients and combined uncertainties of variables i in row **10**:

$$\left(\frac{\partial C}{\partial x_i} u_i\right)^2 = \left(C \frac{u_i}{x_i}\right)^2 \tag{8.53}$$

Use the Welch-Satterthwaite formula from Equation (8.48) to get the degrees of freedom in cell **B11** for the expanded uncertainty in C. Finally, get the coverage factor for a 95% confidence level in cell **B12** and combine with the standard uncertainty for the expanded uncertainty in cell **B13**.

	A	B	C	D
9	DOF for combined u	=7*(B6/(B4+0.00000001))^4	=999*(C6/(C4+0.00000001))^4	=999*(D6/(D4+0.00000001))^4
10	(c*u)^2	=$B$1^2*B7	=$B$1^2*C7	=$B$1^2*D7
11	DOF for Expanded U	=(B8^4)/(SUM(B10:D10^2/B9:D9))		
12	t95%	=TINV(0.05,B11)		
13	U95%	=B12*B8		

The results of the calculations in the worksheet give $C = 0.04078 \pm 0.00007$ M (95% confidence) when we retain just one significant figure in the expanded uncertainty.

	A	B	C	D
9	DOF for combined u	15.03862055	6.9375E+28	6.9375E+12
10	(c*u)^2	8.658E-12	1.138E-09	2.058E-12
11	DOF for Expanded U	264554.0766		
12	t95%	1.959972952		
13	U95%	6.64E-05		

Comments: For such a large number of degrees of freedom, some practitioners use a conservative value of $t_{95\%} = 2$ for the coverage factor.

□

8.2.5 Automate Propagation of Uncertainty in Excel with Jitters

> *"As long as there's continued jitters in Nasdaq, we'll have to wait to see what happens."*-Kathy Matsui

We may experience a sense of trepidation when starting down the path of uncertainty analysis involving the law of propagation and Welch-Satterthwaite Equations (8.23) or (8.50). We may recall our anxieties about error analysis as a tedious and mysterious statistical exercise in our chemistry and physics laboratory courses. Before throwing up our hands in surrender, remember that, by its true nature, uncertainty analysis is not an exact science. We have some wiggle room. For instance, we only need good approximations for the derivatives in Equation (8.40). For complicated functions, we may approximate the sensitivity coefficients with a simple, first-order, finite-difference approximations derived in Equation (5.45):

$$c = \left.\frac{\partial y}{\partial x}\right|_{\bar{x}} \cong \frac{y(\bar{x} + \Delta x) - y(\bar{x})}{\Delta x} \tag{8.54}$$

where y may represent the end-result from a complicated sequence of calculations starting from the input variables, x. We do not usually require derivative approximations with higher order precision, such as those presented in Section 5.2, because the very nature of uncertainty does not command higher precision (de Levie 2000). The beauty of Equation (8.54) for uncertainty analysis is that we eliminate the need for calculations of uncertainty in any of the intermediate steps in our calculation of the uncertainty in y from the uncertainty in x.

Our goal is to estimate the expanded uncertainty in the output function (measurand) attributed to the propagation of the standard uncertainties in the input variables, or measurements, through a sequence of complex calculations. We get there following a 10-step recipe:

1. Calculate the average, or mean values of the input measurements from replicated experiments performed in random order using Equation (8.5).
2. Get the random and fixed uncertainties from the standard deviation and readabilities for all input variables using Equations (8.9) and (8.13). We assume that random errors in replicated measurements are *normally* distributed (however, we may use any appropriate distribution for our specific problem). We assume *uniformly* distributed *fixed* uncertainties.
3. Combine the random and systematic uncertainties according to Equation (8.7).

4. Calculate the degrees of freedom for all variables using the special case of the Welch-Satterthwaite formula in Equation (8.49).
5. Calculate the overall sensitivity coefficients at the mean values of the inputs from finite-difference derivative approximations, like Equation (8.54).
6. Calculate the propagation of uncertainty in the output to give the standard uncertainty in the measurand using Equation (8.40).
7. Calculate the total degrees of freedom from the Welch-Satterthwaite formula in Equation (8.50). Round the result for the degrees of freedom down to the next lowest integer value.
8. Look up the *t*-statistic from Table 8.2. Alternatively, use *Excel*'s worksheet function **TINV(α, ν)** to get the (1 – α)×100% or 95% coverage factor, *t*, for a significance level α = 0.05.
9. Calculate the expanded uncertainty, or 95% confidence interval using Equation (8.47).
10. Round up the expanded uncertainty and report the result with parenthetical information defining the confidence interval.

Algorithms for uncertainty analysis that use finite-difference derivative approximations are called jitter methods, analogous to observing the function's behavior when the variables are slightly perturbed, or "have the jitters" (Coleman and Steele 1989). The *PNMSuite* macro **JITTER** automates steps 4 to 10 from our recipe directly in an *Excel* worksheet. The macro listed in Figure 8.11 employs input boxes. The user-form **UNCERTAINTY** has an option for the jitter method. The user must provide the following information in ranges of cells on an *Excel* worksheet:

- Average values for the variables in a contiguous range of cells on the worksheet
- Function formula evaluated in terms of the average values for the variables (directly or indirectly linked to the range of average values on the worksheet). This is critically important – without a direct or indirect link on the worksheet between the dependent function and independent variables, the **JITTER** macro is unable to calculate the derivatives using finite difference approximations.
- Standard uncertainties in the average values from Equation (8.9) in a range of contiguous cells.
- Optional correlation coefficients.
- Degrees of freedom for each variable in a range of contiguous cells. Use a large number, such as 1000 degrees of freedom when the variable uncertainty is determined only from the fixed uncertainty.
- One of the following two options (in a range of contiguous cells):
  1. Readabilities for the precision of the measured variables, or
  2. Systematic (fixed) uncertainties for the measurements of each variable (Equation (8.13).

The **JITTER** macro allows for the inclusion of covariance (de Levie 2004). Use the default values of zero for no correlation between any pairs of parameters.

```
Public Sub JITTER()
' Propagates the uncertainty in the variables to the uncertainty in the function.
' Calculates the sensitivity coefficients by a second-order finite difference first
' derivative approximation. The result is the confidence interval based on expanded
' uncertainty.  The user may set the level of confidence. Covariance is also an option
' for including correlation among variables. The output includes the expanded and
' standard uncertainty, DOF, t, sensitivity coefficients, and percent contributions of
' uncertainty from each variable. The Function (f) equation is defined in terms of x
' cells. Uncertainties in x (u) are the calculated from the combined standard error
' and the systematic errors in x (assuming uniform error distribution). The Welch-
' Satterthwaite formula is used to get the overall degrees of freedom from the
' variable degrees of freedom to calculate the t-statistic, or coverage factor.
' *** Nomenclature
Dim alpha As Double ' = 1-confidence level
Dim ar As Range ' = range of x variables for Richardson's extrapolation
Dim boxtitle As String ' = string for title in message and input boxes
Dim c() As Double ' = sensitivity coefficients
Dim co As Double ' = combined effect of correlation
Dim co_j As Double ' = covariance of parameter j
Dim co_max As Double ' = maximum sum of covariance terms
Dim co_v As Double ' = sum of covariance terms in Welch-Satterthwaite formula
```

```vb
Dim co_2 As Double ' = sum of squares of covariance terms in Welch-Satterthwaite
Dim cu() As Double ' = product of sensitivity coefficient and u
Dim cu_2() As Double ' = square of cu
Dim dx As Double ' = small change in x for derivative approximation
Dim dy As Double ' = vector of derivatives dy/dx
Dim f As Double ' = function
Dim f_ave As Double ' = function evaluated at average x
Dim fr As Range ' = range for function
Dim i As Long, j As Long ' = parameter/variable index
Dim k As Long ' = number of variables
Dim pci As Double ' ' = percent confidence interval (e.g., 95%)
Dim pcol As Long, prow As Long, prowi As Long, _
    prow3 As Long ' = location of print output column and row
Dim pu_f As Double ' = percent standard uncertainty
Dim pu_ft As Double ' = relative expanded uncertainty
Dim pu_max As Double ' = percent maximum uncertainty
Dim r() As Double ' = correlation coefficients for data i and j
Dim rdblt As Integer ' = use readabilities for fixed uncertainties
Dim re As Integer ' = Richardson's derivative extrapolation
Dim s As Double ' = sum denominator of Welch-Satterthwaite formula
Dim sc As Integer ' = correlation coefficient calculation
Dim small As Double ' = perturbation of variable in derivative approximation
Dim t_95 As Double ' = 95% probability coverage factor, t-statistic
Dim tol As Double ' = derivative convergence tolerance
Dim uc() As Double ' = vector of combined uncertainties in variables
Dim u_f As Double ' = uncertainty in f
Dim u_ft As Double, u_f2 As Double ' = expanded uncertainty u_f & t_95, ,u_f^2
Dim u_fco2 As Double ' = square standard uncertainty with correlation
Dim u_max As Double ' = maximum combined uncertainty
Dim ur As Range ' = range of random uncertainties
Dim vc() As Double ' = DOF of combined random and systematic uncertainties
Dim vf As Long ' = Welch-Satterthwaite degrees of freedom
Dim vr As Range ' = range of degrees of freedom
Dim x_ave() As Double ' = average values for variables
Dim x_formula() As String ' = string array of formulas for variables
Dim xr As Range ' = range of variables
Dim xrows As Long ' = number of rows in x vector
Dim zr As Range ' = range of systematic (fixed) uncertainties in variable x
'**********************************************************************
small = 0.000000000001: boxtitle = "JITTER: Propagation of Uncertainty"
With Application
frin: Set fr = .InputBox(Title:=boxtitle, Type:=8, prompt:="FUNCTION cell, f(x):")
If fr.Count <> 1 Then
    MsgBox "Function must be in a single cell.": GoTo frin
End If
GoTo xrskip
xrin: MsgBox "Match ranges for x, u, v, and d.", , boxtitle
xrskip:
Set xr = .InputBox(Title:=boxtitle, Type:=8, prompt:="Range of AVERAGE VARIABLES, x:")
k = xr.Count: xrows = xr.Rows.Count ' Get the number of variables and rows
Set ur = .InputBox(Title:=boxtitle, Type:=8, _
                prompt:="Range of STANDARD (RANDOM) UNCERTAINTIES, ±u_r:")
If Not ur.Count = k Then GoTo xrin
Set vr = .InputBox(prompt:="Range of DEGREES of FREEDOM for x, v:", _
                Title:=boxtitle, Type:=8)
If Not vr.Count = k Then GoTo xrin
rdblt = MsgBox(prompt:="Use READABILITY, ±d ?" & vbNewLine & vbNewLine & _
    "[u_z = ±d/SQR(3)]", Buttons:=vbYesNo, Title:="Systematic (FIXED) Uncertainty")
Select Case rdblt
Case 6 ' Yes
    Set zr = .InputBox(Title:=boxtitle, Type:=8, prompt:="Range of READABILITIES, ±d:")
Case Else ' No
    Set zr = .InputBox(Title:=boxtitle, Type:=8, _
                    prompt:="Range of SYSTEMATIC (FIXED) UNCERTAINTIES, ±u_z:")
End Select
If Not zr.Count = k Then GoTo xrin
ReDim c(1 To k) As Double, cu(1 To k) As Double, cu_2(1 To k) As Double, _
    vc(1 To k) As Double, r(1 To k, 1 To k) As Double, x_ave(1 To k) As Double, _
    x_formula(1 To k) As String, uc(1 To k) As Double
```

```
' Save variables and initial function value for derivative approximations
For i = 1 To k: x_ave(i) = xr(i): x_formula(i) = xr(i).Formula: Next i ' Save variable
' Range("B8").GoalSeek Goal:=0, ChangingCell:=fr ' Example 9.6
f_ave = fr
sc = MsgBox(prompt:="Specify CORRELATION Coefficients?", _
    Buttons:=vbYesNo + vbQuestion + vbDefaultButton2, Title:=boxtitle)
If sc = 6 Then
    For i = 1 To k - 1
        For j = i + 1 To k
            r(i, j) = .InputBox(Title:=boxtitle, Type:=1, _
                            prompt:="R(" & i & "," & j & ")", Default:=0)
        Next j
    Next i
End If
tol = 0.00000001: re = 6
pci = .InputBox(Title:=boxtitle, Type:=1, _
            prompt:="Set CONFIDENCE Interval (50-99) %:", Default:=95)
pci = .Max(.Min(CInt(pci), 99), 50): alpha = 1 - pci / 100 ' Convert to integer value
prow = output.row: pcol = output.Column ' Get row & column for output
For i = 1 To k ' Calculate combined uncertainties & DOF for x
    Select Case rdblt
    Case 6 ' Yes
        uc(i) = Sqr(ur(i) ^ 2 + (zr(i) ^ 2) / 3) ' zr = readability
    Case Else
        uc(i) = Sqr(ur(i) ^ 2 + zr(i) ^ 2) ' zr = fixed uncertainty
    End Select
    If ur(i) > small Then ' Welch-Satterthwaite formula
        vc(i) = .Max(1, vr(i) * (uc(i) / ur(i)) ^ 4)
    Else
        vc(i) = 1000 ' approximates infinite DOF
    End If
Next i
u_f2 = 0: u_max = 0: s = 0 ' Calculate partial derivatives & sum uncertainty terms
For i = 1 To k
    dx = 0.00001 * (Abs(x_ave(i)) + 0.00001) ' perturbation value
    ' sensitivity coefficients from 1st-order FD approximation
    ' Range("B8").GoalSeek Goal:=0, ChangingCell:=fr ' Example
    xr(i) = x_ave(i) + dx: c(i) = fr: xr(i) = x_ave(i) - dx
    c(i) = (c(i) - fr) / (2 * dx): cu(i) = c(i) * uc(i): cu_2(i) = cu(i) ^ 2
    xr(i) = x_formula(i) ' Reset variable for next iteration
    If c(i) = 0 Then ' if derivative is zero, the function of x not set up correctly
      MsgBox "df/dx = 0! Formula for f(x) must reference cells for x." & vbNewLine & _
                "Check your worksheet set up.", , boxtitle
        Exit Sub
    End If
    ' Sum uncertainty terms and DOF
    u_f2 = u_f2 + cu_2(i): s = s + (cu_2(i) ^ 2) / vc(i): u_max = u_max + Abs(cu(i))
Next i
' Range("B8").GoalSeek Goal:=0, ChangingCell:=fr ' Example
co = 0: co_max = 0: co_2 = 0: co_v = 0 ' Covariance for propagation and DOF
If sc = 6 Then
    For i = 1 To k - 1
        For j = i + 1 To k
            co_j = r(i, j) * cu(i) * cu(j)
            co = co + co_j: co_max = co_max + Abs(co_j)
            co_2 = co_2 + (co_j ^ 2) * (vc(i) + vc(j) + 0.5) / (vc(i) * vc(j))
            co_v = co_v + co_j * (cu_2(i) / vc(i) + cu_2(j) / vc(j))
        Next j
    Next i
End If
' Welch-Satterthwaite degrees of freedom,95% coverage factor & Uncertainty propagation
vf = CLng((.RoundDown((u_f2 ^ 2) / (s + co_2 + 2 * co_v), 0)))
vf = .Min(.Max(vf, 1), 1000): t_95 = .TInv(alpha, vf) ' alpha = 0.05 = 95% confidence
' Uncertainties (standard includes covariance, maximum, and expanded)
u_fco2 = u_f2 + 2 * co: u_f = Sqr(u_fco2): u_max = u_max + co_max: u_ft = u_f * t_95
pu_f = Abs(u_f) / (Abs(fr) + small) ' Percent uncertainties
pu_max = Abs(u_max) / (Abs(fr) + small): pu_ft = Abs(u_ft) / (Abs(fr) + small)
' Round up uncertainties for 1 or 2 significant figures
u_f = USIGFIGS(u_f, 2): u_max = USIGFIGS(u_max, 2): u_ft = USIGFIGS(u_ft, 2)
```

```
pu_f = USIGFIGS(pu_f, 2): pu_max = USIGFIGS(pu_max, 2): pu_ft = USIGFIGS(pu_ft, 2)
Worksheets(ws).Activate
Cells(prow, pcol).Activate ' Print labels & uncertainty in f
With ActiveCell
    .Offset(0, 0).Value = "U" & CStr(pci) & "% = ±"
        .Offset(0, 0).Characters(Start:=2, Length:=3).Font.Subscript = True
    .Offset(0, 1).Value = u_ft: .Offset(0, 1).NumberFormat = "0.0E+#0"
    .Offset(0, 2).Value = " = ±": .Offset(0, 3).Value = pu_ft
        .Offset(0, 3).NumberFormat = "###0.0###%"
    .Offset(1, 0).Value = "umax = ±"
        .Offset(1, 0).Characters(Start:=2, Length:=3).Font.Subscript = True
    .Offset(1, 1).Value = u_max: .Offset(1, 1).NumberFormat = "0.0E+#0"
    .Offset(1, 2).Value = " = ±": .Offset(1, 3).Value = pu_max
        .Offset(1, 3).NumberFormat = "###0.0###%"
    .Offset(2, 0).Value = "u = ±"
    .Offset(2, 1).Value = u_f: .Offset(2, 1).NumberFormat = "0.0E+#0"
    .Offset(2, 2).Value = " = ±"
    .Offset(2, 3).Value = pu_f: .Offset(2, 3).NumberFormat = "###0.0###%"
    .Offset(3, 0).Value = "t" & CStr(pci) & "% = "
        .Offset(3, 0).Characters(Start:=2, Length:=3).Font.Subscript = True
    .Offset(3, 1).Value = t_95: .Offset(3, 1).NumberFormat = "#0.0#"
    .Offset(3, 2).Value = "DoF = " : .Offset(3, 3).Value = vf
End With
prow3 = prow + 3
For i = 1 To k
    prowi = prow3 + i: Cells(prowi, pcol).Activate
    With ActiveCell
        .Value = "c" & i & " =": .Characters(Start:=2, Length:=Len(CStr(i))) _
        .Font.Subscript = True
        .Offset(0, 1).Value = c(i): .Offset(0, 1).NumberFormat = "0.0#E+##"
        .Offset(0, 2).Value = "(c*u" & i & ")2 ="
        .Offset(0, 2).Characters(Start:=5, Length:=Len(CStr(i))).Font.Subscript = True
            .Offset(0, 2).Characters(Start:=6 + Len(CStr(i)), _
                                    Length:=1).Font.Superscript = True
        .Offset(0, 3).Value = cu_2(i) / u_f2 : .Offset(0, 3).NumberFormat = "##0.0%"
    End With
Next i
' Reset average values of variables. Note, sub ROWCOL does not work for formulas
For i = 1 To k: xr(i) = x_formula(i): Next i
.Calculate ' Recalculate the workbook after redoing formulas
End With
End Sub ' JITTER
```

Figure 8.11 *VBA* **JITTER macro for applying the law of uncertainty propagation in** *Excel*.

Table 8.3 contains definitions for the output from **JITTER** for a two-variable problem. For purposes of comparison, the **JITTER** macro reports the coefficients for sensitivity analysis defined by Equation (5.85), the contributions from each parameter to the total error, $(c \cdot u)^2$, and the maximum combined uncertainty according to Equation (5.98).

Table 8.3   Map of JITTER macro output for a two-variable problem.

$U_{95\%}=\pm$	Expanded uncertainty for the 95% confidence interval	=±	Percent expanded uncertainty	DOF=	Overall degrees of freedom from the Welch-Satterthwaite equation
$u_{max}=\pm$	Maximum standard uncertainty	=±	Percent maximum uncertainty	$t_{95\%}=$	Coverage factor from the Student t-distribution for 95% confidence
u=±	Standard uncertainty	=±	Percent standard uncertainty		
$u_1=$	Combined uncertainty for variable 1	$c_1=$	Sensitivity coefficient for variable 1	$(c \cdot u_1)^2=$	Percent uncertainty contribution from variable 1
$u_2=$	Combined uncertainty for variable 2	$c_2=$	Sensitivity coefficient for variable 2	$(c \cdot u_2)^2=$	Percent uncertainty contribution from variable 2

CHAPTER 8: UNCERTAINTY	345

Example 8.8   Expanded Uncertainty in an Ideal Gas Concentration Using Jitter

Known/Find: Calculate the expanded uncertainty for an ideal gas in terms of the pressures and temperature from *Example 8.7* using the **JITTER** method in the **UNCERTAINTY** user form.

Analysis: The worksheet is already setup for **JITTER**. We only need to add a row for the degrees of freedom of each of the average values of the variables using 7 for T, and 1000 for the single, quoted values of P and R_g. Access **JITTER** through the **UNCERTAINTY** user form, as follows.

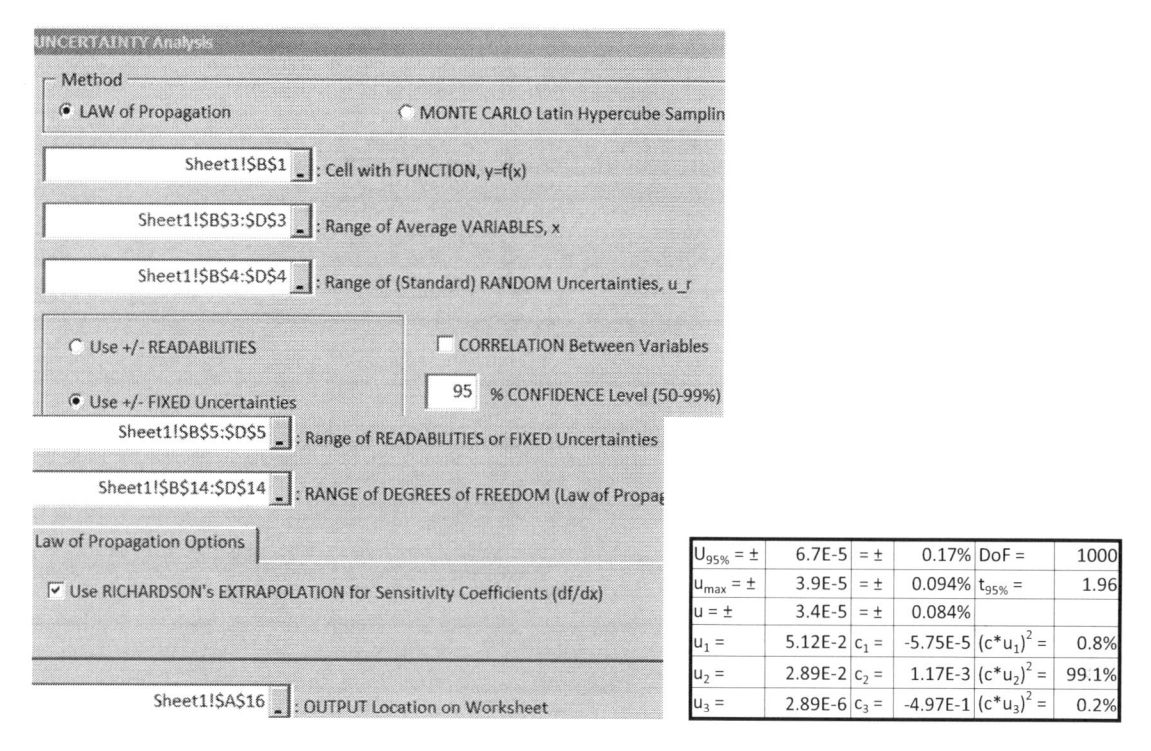

Comments: The **JITTER** method returns the same results calculated "by hand." The results also indicate that the pressure uncertainty continues to contribute the bulk of the uncertainty in the analysis.

□

Example 8.9   Uncertainty in Chemical Equilibrium Experiments

Known: We conducted an experiment to determine the chemical equilibrium constant for the esterification of ethanol by ascetic acid at room temperature:

$$C_2H_5OH + CH_3COOH \rightleftharpoons CH_3COOC_2H_5 + H_2O \tag{8.55}$$

We added measured amounts of acid and alcohol that equilibrated in a batch reactor. After a sufficient period, three 5 mL samples from the reactor were titrated with sodium hydroxide to determine the final acid concentration, and by mole balances, the final concentrations of all reactants and products. We standardized the concentration of the titrant previously with replicated experiments. Table 8.4 lists all other random and fixed systematic uncertainties.

346 PRACTICAL NUMERICAL METHODS

Table 8.4 Standard and fixed uncertainties for a batch reaction.

Variable	Value	Standard Random Uncertainty	Degrees of Freedom (DOF)	Readability ±δ
Initial mass of CH_3COOH	23.37 g	—	∞	0.005 g
Initial mass of C_2H_5OH	11.72 g	—	∞	0.005 g
Reactor volume	37.4 mL	—	∞	0.05 mL
Titration sample volume	5.00 mL	—	∞	0.005 mL
NaOH Concentration	0.948	0.003	3	0.0005
Replicated Titrant Volumes	24.8 mL 24.9 mL 24.6 mL	—	2	0.05 mL

Find: Use the batch reaction data to calculate the equilibrium constant:

$$K = \frac{[CH_3COOC_2H_5][H_2O]}{[C_2H_5OH][CH_3COOH]} \tag{8.56}$$

Assumptions: reaction equilibrium, uncorrelated variables

Analysis: The *Excel* worksheet contains the data and numerical solution. Set up the *Excel* worksheet with the uncertainty information for use by the **JITTER** macro. The sample weights and volumes were only measured once, and are without random uncertainty. For these inputs, we set the degrees of freedom to a relatively large value of 1000 to simulate infinity. Calculate the average value of the titrant volume and its standard uncertainty in cells **B11** and **C11**, respectively. The average moles of NaOH titrated gives the moles of unconverted acid by difference. Determine the equilibrium concentrations from mole balances and the reaction stoichiometry.

	A	B	C	D	E
1	Experimental Data	Titration/ml			
2		24.8			
3		24.9			
4		24.6			
5			Standard		Systematic
6	Variables	Mean	U	DOF	U
7	mass Acid, gm	23.37	0	1000	=0.005/SQRT(3)
8	mass Ethanol, gm	11.72	0	1000	=0.005/SQRT(3)
9	Volume, ml	37.4	0	1000	=0.05/SQRT(3)
10	Sample V, ml	5	0	1000	=0.005/SQRT(3)
11	Titration, ml	=AVERAGE(B2:B4)	=STDEV(B2:B4)/SQRT(3)	2	=0.05/SQRT(3)
12	NaOH Conc., M	0.948	0.003	3	=0.0005/SQRT(3)
13					
14	Calculations		Initial	Initial	Final
15	Reactant	MW	Moles	C/M	C/M
16	Acid	60.052	=B7/B16	=C16*1000/$B$9	=B11*B12/B10
17	Ethanol	46.06844	=B8/B17	=C17*1000/$B$9	=D17-(D16-E16)
18	Acetate		0	=C18*1000/$B$9	=D16-E16
19	Water		0	=C19*1000/$B$9	=D16-E16
20	K =	=E18*E19/(E16*E17)			

We show the values provided to the **JITTER** input boxes and the results in Figure 8.12:

JITTER: Propagation of Uncertainty

Select range of AVERAGE (MEAN) VARIABLES, x:
$B$7:$B$12

Select range of STANDARD (RANDOM) UNCERTAINTIES, ±u_r:
$C$7:$C$12

Select range of mean DEGREES of FREEDOM:
$D$7:$D$12

Select range of SYSTEMATIC (FIXED) UNCERTAINTIES, ±u_z:
$E$7:$E$12

Click on the first OUTPUT cell:
$A$21

CHAPTER 8: UNCERTAINTY

	A	B	C	D	E	F
1	**Experimental Data**	Titration/ml				
2		24.8				
3		24.9				
4		24.6				
5			Standard		Systematic	
6	Variables	Mean	U	DOF	U	
7	mass Acid, gm	23.37	0	1000	0.002886751	
8	mass Ethanol, gm	11.72	0	1000	0.002886751	
9	Volume, ml	37.4	0	1000	0.028867513	
10	Sample V, ml	5	0	1000	0.002886751	
11	Titration, ml	24.8	0.088	2	0.028867513	
12	NaOH Conc., M	0.948	0.003	3	0.000288675	
13						

	A	B	C	D	E	F
14	**Calculations**		Initial	Initial	Final	
15	Reactant	MW	Moles	C/M	C/M	
16	Acid	60.052	0.3892	10.405	4.696	
17	Ethanol	46.06844	0.2544	6.802	1.093	
18	Acetate		0	0.000	5.710	
19	Water		0	0.000	5.710	
20	K =	6.35E+00				
21	$U_{95\%} = \pm$	5.7E-1	= ±	9.0%	DoF =	5
22	$u_{max} = \pm$	3.9E-1	= ±	6.1%	$t_{95\%} =$	2.57
23	$u = \pm$	2.3E-1	= ±	3.5%		
24	$u_1 = \pm$	2.89E-3	$c_1 =$	3.58E+0	$(c*u_1)^2 =$	0.2%
25	$u_2 = \pm$	2.89E-3	$c_2 =$	-3.38E+0	$(c*u_2)^2 =$	0.2%
26	$u_3 = \pm$	2.89E-2	$c_3 =$	-1.18E+0	$(c*u_3)^2 =$	2.4%
27	$u_4 = \pm$	2.89E-3	$c_4 =$	8.82E+0	$(c*u_4)^2 =$	1.3%
28	$u_5 = \pm$	9.28E-2	$c_5 =$	-1.78E+0	$(c*u_5)^2 =$	55.8%
29	$u_6 = \pm$	3.01E-3	$c_6 =$	-4.65E+1	$(c*u_6)^2 =$	40.1%

Figure 8.12

JITTER macro setup and output on an *Excel* worksheet for calculating chemical equilibrium.

The expanded uncertainty in the equilibrium constant represents ±9% (with 95% confidence). The input variables numbered one to six correspond to the variables in the range **B7:B12**. Titration uncertainties make the largest contributions to the combined uncertainty. Additional titrations and higher precision on the titrant concentration will decrease the overall uncertainty in the equilibrium constant.

Comments: The law of propagation of uncertainty uses information about relative contributions to the overall uncertainty. The batch reaction experiment was replicated five times giving the results in column **A** of Figure 8.13. The weighted average values for the reaction equilibrium constant and uncertainty were calculated from the expanded uncertainties according to Equation (8.59) and (8.60) for replicated experiments. The uncertainty results are rounded up with one significant figure:

	A	B	C	D	E	F	G
1	K	U	U^2	$1/U^2$	K/U^2	Weighted Ave K =	6.37
2	6.35	0.57	0.3249	3.0779	19.5445	Weighted U =	0.3
3	6.36	0.6	0.36	2.7778	17.6667	%U =	0.05
4	6.38	0.49	0.2401	4.1649	26.5723		
5	6.34	0.62	0.3844	2.6015	16.4932		
6	6.39	0.59	0.3481	2.8727	18.3568		
7			Sum	15.4948	98.6334		

Figure 8.13

Weighted expanded uncertainty in equilibrium constants (95% confidence).

As expected, the expanded uncertainty in the weighted average of the replicated experiments decreased to ±0.3 due to higher degrees of freedom.

□

Example 8.10 Solvent Flux across an Ultrafiltration Membrane

Known: A model for solvent flux by ultrafiltration has an implicit term to account for concentration polarization:

$$N = P\left[\Delta p - \Delta \pi \exp\left(\frac{N}{kC}\right)\right] \tag{8.57}$$

where N, C, and P, are the solvent flux, concentration, and permeance, respectively, k is the solute mass transfer coefficient, and Δp and $\Delta \pi$ are the transmembrane and osmotic pressure differences, respectively. The *Excel* worksheet in Figure 8.14 shows the values of the parameters in column **B** and fixed uncertainties in column **C** calculated from the precision of the quoted values divided by $\sqrt{12}$. No information is provided for the random uncertainties, which are consequently ignored in the analysis by setting them to zero.

	A	B	C	D	E
1	Parameter	Value	Fixed U	Random U	DOF
2	P/(mol/m^2·s·Pa)	3.2E-08	2.9E-10	0	1000
3	Δp/Pa	7.00E+06	2.9E+03	0	1000
4	$\Delta \pi$/Pa	4.1E+05	2.9E+03	0	1000
5	k/(m/s)	3.1E-06	2.9E-08	0	1000
6	C/(mol/m^3)	5.56E-02	2.9E-05	0	1000

Figure 8.14

Ultrafiltration membrane parameters and associated fixed uncertainties from the precision of the parameter.

Find: Calculate the expanded uncertainty in the solvent flux.

Assumptions: Neglect random uncertainties and assume infinite degrees of freedom.

Analysis: The *Excel* worksheet in Figure 8.14 contains values of zero for the random uncertainties, and 1000 for the degrees of freedom needed for the inputs required by the *VBA* **JITTER** macro.

At first glance, we notice that Equation (8.57) is already arranged to solve for the solvent flux by the method of ordinary iteration. Unfortunately, this problem is highly nonlinear, and ordinary iteration fails to converge to a solution, even when starting from a good initial guess. Next, we try *Excel*'s **Goal Seek** to find the solution by placing a guess for the flux in cell **B7** and the difference between the left and right sides of Equation (8.57) in cell **B8**. **Goal Seek** finds the solution for the solvent flux as a small value with magnitude of order 10^{-7}. The small result causes numerical round-off problems for the *VBA* macro **JITTER**, which approximates the derivatives for the sensitivity coefficients by the method of finite-differences. To circumvent the round-off issue, we scale the variable in the function by 10^{-7} in cell **B8**. Launch **Goal Seek** from the tab **Data>What-If Analysis** menu and set cell **B8** to a value of zero by changing cell **B7**:

	A	B
7	N/(mol/m^2·s)	4.89072548806656
8	f(N)	=B7*0.0000001-B2*(B3-B4*EXP(B7*0.0000001/B5/B6))

The solution for flux is the result in cell **B7** multiplied by the scale factor: $N = 4.89 \times 10^{-7}$.

We need to modify the *VBA* macro **JITTER** in Figure 8.11 to solve the propagation equation for the solvent flux each time a variable is perturbed for the finite-difference derivative approximations. We use the following *VBA* statement just before the calculation of the sensitivity coefficient for variable i:

```
Range("B8").GoalSeek Goal:=0, ChangingCell:=fr
```

where the range variable **fr** in **JITTER** is assigned to the cell **B7**.

Click on propagation FUNCTION cell, f(x):
$B$7

Use the **Goal Seek** *VBA* statement once more just below the **FOR NEXT** loop to restore the solution for the solvent flux to the original value. The location in the worksheet for the implicit function is hard coded to cell **B8** for this example. Be careful to change the setting for **Maximum Change** under **File>Options>Formulas** to a number smaller than 10^{-8} in order to capture with **Goal Seek** the impact of perturbations in the permeance P on the flux.

Finally, run the modified *VBA* macro **JITTER** to calculate the expanded uncertainty in the solvent flux calculation. Provide the additional required user input information as follows:

CHAPTER 8: UNCERTAINTY

Skip the correlation coefficients and assign the output in cell **C7** on the *Excel* worksheet. Figure 8.15 has the results for uncertainty.

	A	B	C	D	E	F	G	H
7	N/(mol/m²·s)	4.89E+00	$U_{95\%} = \pm$	9.3E-2	= ±	1.9%	DoF =	1000
8	f(N)	-2.1E-16	$u_{max} = \pm$	6.1E-2	= ±	1.3%	$t_{95\%} =$	1.96
9			$u = \pm$	4.8E-2	= ±	0.97%		
10			$u_1 = \pm$	2.89E-10	$c_1 =$	1.18E+2	$(c^*u_1)^2 =$	0.0%
11			$u_2 = \pm$	2.89E+3	$c_2 =$	2.46E-7	$(c^*u_2)^2 =$	0.0%
12			$u_3 = \pm$	2.89E+3	$c_3 =$	-4.2E-6	$(c^*u_3)^2 =$	6.6%
13			$u_4 = \pm$	2.89E-8	$c_4 =$	1.58E+6	$(c^*u_4)^2 =$	93.1%
14			$u_5 = \pm$	2.89E-5	$c_5 =$	8.8E+1	$(c^*u_5)^2 =$	0.3%

Figure 8.15

Jitter output for uncertainty in solvent flux through a membrane.

Report the result for solvent flux with its expanded 95% confidence interval:

$$N/(mol/m^2 s) = (4.9 \pm 0.1) \times 10^{-7} \qquad (\text{mean} \pm 95\% \text{ confidence}) \qquad (8.58)$$

Comments: The primary contributor to the propagated uncertainty in the solvent flux is the uncertainty in the solute mass fraction (parameter 4). The jitter method, coupled with a **Goal Seek** solution to the implicit propagation function is a powerful tool for uncertainty propagation. Numerical derivative approximations eliminate the tedious and error prone evaluations of sensitivity coefficients. We may use our user-defined function **ROOT** to calculate the molar flux in the worksheet, instead of **Goal Seek** in the code.

□

8.2.6 Uncertainty in Replicated Experiments*

"The demand for certainty is one which is natural to man, but is nevertheless an intellectual vice." – Bertrand Russell

In cases where we need to combine the results from different experiments with unequal uncertainties, use the uncertainties in the replicates to weight the average value of the replicated experiments:

$$\bar{y} = \left(\sum_{j=1} y_j U_{y_j}^{-2}\right) \bigg/ \left(\sum_{j=1} U_{y_j}^{-2}\right) \qquad (8.59)$$

where *j* is the index for distinct sets of experiments. The uncertainty in the weighted average of the replicated sets of experiments becomes:

$$\bar{U}_{\bar{y}} = \left[\sum_j U_{y_j}^{-2}\right]^{-1/2} \qquad (8.60)$$

We may substitute other estimates of uncertainty, such as standard deviation or standard uncertainty of the mean for U_y in the weighted formulas. Be sure to provide a definition of the uncertainty in parenthetical information for end users, as described in Section 8.1.6.

8.3 Monte Carlo Simulation of Uncertainty*

> *"Anyone who attempts to generate random numbers by deterministic means is, of course, living in a state of sin."* – John von Neumann

As illustrated in Figure 8.10, the *Law of Propagation of Uncertainty* uses a linear Taylor series approximation of a function expanded about the mean values of the variables. Although the combination of bias and random errors is accepted in practice, there is no strict scientific basis for combining them in the sum of squared errors. This approach may provide appropriate estimates of uncertainty propagation for small uncertainties in the model parameters, or for smooth, nearly linear functions. However, it can lead to poor estimates for large uncertainties in the variables of highly nonlinear or discontinuous functions (Morgan and Henrion 1990). We recommend the method of Monte Carlo uncertainty propagation for highly nonlinear functions when the linear approximation of uncertainty breaks down.[81]

The Monte Carlo method applied to uncertainty propagation simulates a large number of function evaluations using random deviates for each variable. We generate the random deviates from probability density functions for the random uncertainty in variable i:

$$x_i = \bar{x}_i + u_i r_N \tag{8.61}$$

where r_N is a random number taken from an appropriate distribution of uncertainty in the measurement. When we estimate the uncertainty from descriptive statistics for a set of replicated measurements, we assume a normal distribution about zero for the random deviates with a standard deviation of one.[82] We assume a uniform distribution to generate the random deviates for fixed uncertainty. The mean value for r_N in Equation (8.61) is approximately zero by design such that the average value of the m number of Monte Carlo simulated results returns the expected value of the original variable:

$$\bar{x}_i = \frac{\sum (\bar{x}_i + u_i r_N)}{m} \tag{8.62}$$

Random deviates for multivariable functions should also include the property of any correlation among variables, usually involving asymmetric probability density functions for the distribution of the random deviates about the mean. As discussed in Section 8.2.3, we can take steps to reduce or eliminate correlation among measured variables by careful experimental planning and implementation. We then apply descriptive statistics to the simulated results to calculate the expanded uncertainty.

Excel has several useful worksheet and *VBA* functions for implementing Monte Carlo simulation of uncertainty propagation. Figure 8.16 shows a histogram of normal random deviates plotted using the *Excel* **Data Analysis** add-in tool from the ribbon tab **Data>Analysis/ Data Analysis> Histogram**. The *Excel* functions for generating 2000 random normal deviates in column **A** use the nested worksheet functions **NORMSINV(RAND())**. Bins represent ranges of values for sorting the data according to frequency of occurrence. We specified the bins with a spacing of 0.5 in column **B**. By visual inspection of the histogram, we see that the mean value for the normal random deviates is approximately zero. After generating a large number of simulations, we calculate the 95% confidence interval by sorting the simulation results from low to high values, and eliminating 2.5% of the values from the bottom and 2.5% of the values from the top of the sorted list. The remaining range of simulated values comprise the 95% confidence interval.

---

[81] The Monte Carlo simulation method gets its name from the famed Monte Carlo casino's games of random chance.
[82] Any appropriate distribution may be used in place of the conventional normal distribution.

CHAPTER 8: UNCERTAINTY

Figure 8.16

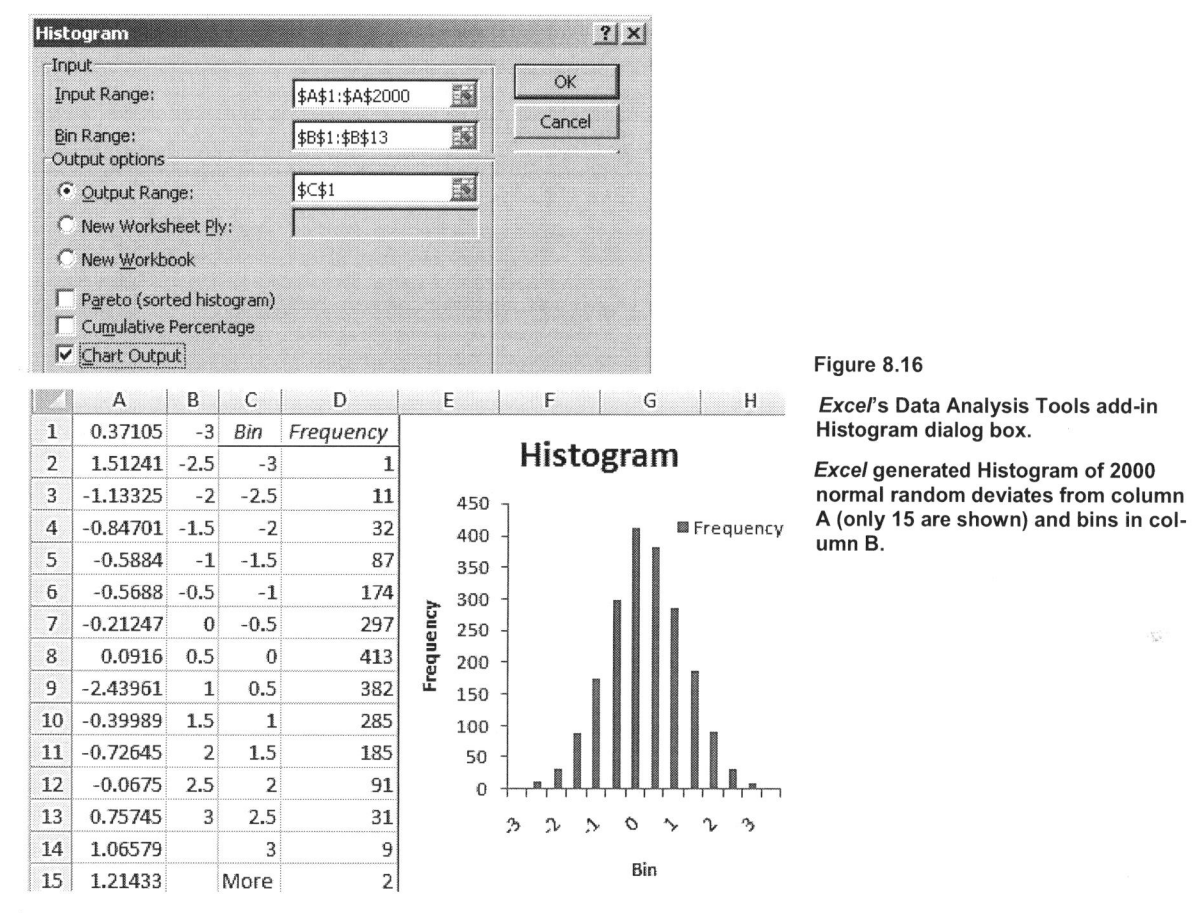

Excel's Data Analysis Tools add-in Histogram dialog box.

Excel generated Histogram of 2000 normal random deviates from column A (only 15 are shown) and bins in column B.

	A	B	C	D
1	0.37105	-3	Bin	Frequency
2	1.51241	-2.5	-3	1
3	-1.13325	-2	-2.5	11
4	-0.84701	-1.5	-2	32
5	-0.5884	-1	-1.5	87
6	-0.5688	-0.5	-1	174
7	-0.21247	0	-0.5	297
8	0.0916	0.5	0	413
9	-2.43961	1	0.5	382
10	-0.39989	1.5	1	285
11	-0.72645	2	1.5	185
12	-0.0675	2.5	2	91
13	0.75745	3	2.5	31
14	1.06579		3	9
15	1.21433		More	2

The Monte Carlo method of uncertainty propagation has many advantages over the law of propagation, but is not without its disadvantages:

Advantages

- Not limited to Gaussian distributions
- May use any combination of distributions for different sources of uncertainty
- Accounts for correlation among variables
- Applies to complex models
- Does not require derivative calculations
- Appropriate for highly nonlinear models
- Handles discontinuous functions

Disadvantages

- Good Monte Carlo analysis requires a large number of simulations.
- It can take a long time to arrive at the results depending on the complexity of the calculations and speed of the computer.
- It may be difficult to identify the main contributors to the overall uncertainty.

How many Monte Carlo simulations does good uncertainty analysis require? A practical answer is "just enough to get a reasonable estimate of uncertainty." A brute, though practical approach (because the computer is doing most of the work) repeats the analysis with 20% additional simulations. As we add more simulations, we accept the results for uncertainty when they stop changing within two significant figures.

Example 8.11 Monte Carlo Uncertainty Propagation in CSTR Design

Known: A first-order reaction takes place in a continuous stirred tank reactor:

$$-r_A = kC_A \tag{8.63}$$

Schematic: A well-mixed reactor for a first-order reaction has a fluid volume V, volumetric feed rate Q, and feed and effluent concentration, C_{A0} and C_A, respectively:

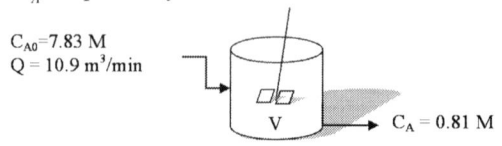

$C_{A0} = 7.83$ M
$Q = 10.9$ m^3/min
$C_A = 0.81$ M

Find: Use Monte Carlo uncertainty analysis to calculate the reactor volume and corresponding 95% confidence interval for 75% conversion according to the following parameters and uncertainties.

Parameter	Mean Value	Standard Uncertainty	Degrees of Freedom	Fixed Uncertainty
Volumetric flow rate, Q/(m³/min)	10.9	0.2	27	0.1
Feed concentration, C$_{A0}$/M	7.83	0.07	3	0.01
Effluent concentration, C$_A$/M	0.81	0.06	3	0.01
Rate constant, k/(1/min)	12.5	0.2	-	0

Assumptions: steady-state operation, well-mixed reactor

Analysis: Calculate the volume of the reactor from a mass balance:

$$V = \frac{Q(C_{A0} - C_A)}{kC_A} \tag{8.64}$$

Implement the Monte Carlo method with a run of 1000 simulations using random deviates:

$$V \pm u_V = \frac{(Q + u_Q r_Q)\left[(C_{A0} + u_{C_{A0}} r_{C_{A0}}) - (C_A + u_{C_A} r_{C_A})\right]}{(k + u_k r_k)(C_A + u_{C_A} r_{C_A})} \tag{8.65}$$

Generate random deviates from a combination of two terms; one involving a standard normally distributed uncertainty contribution and a second, uniformly distributed fixed uncertainty contribution (Vasquez and Whiting 2006):

$$u_i = u_{i,\text{standard}} r_{\text{normal}} + u_{i,\text{fixed}} r_{\text{uniform}} \tag{8.66}$$

Get a uniform random deviate from a pseudo-random number generator that returns values in the range $0 < r < 1$:

$$u_{\text{fixed}} r_{\text{uniform}} = u_{\text{fixed}}(2r - 1) \tag{8.67}$$

for

$$-u_{\text{fixed}} < (u_{\text{fixed}} r_{\text{uniform}}) < +u_{\text{fixed}}$$

Figure 8.17 shows some of the Monte Carlo simulation results. The simulated reactor parameters were calculated in each row according to the following *Excel* worksheet formulas using absolute cell references for the mean values and uncertainties:

Simulated Q	B9=$B$2+NORMSINV(RAND())*$C$2+(2*RAND()-1)*$D$2
Simulated C$_0$	C9=$B$3+NORMSINV(RAND())*$C$3+(2*RAND()-1)*$D$3
Simulated C	D9=$B$4+NORMSINV(RAND())*$C$4+(2*RAND()-1)*$D$4
Simulated k	E9=$B$5+NORMSINV(RAND())*$C$5+(2*RAND()-1)*$D$5
Simulated V	F9=B9*(C9-D9)/(E9*D9)
Average V	F109 = AVERAGE(F9:F108)

CHAPTER 8: UNCERTAINTY

	A	B	C	D	E
1	Variable	Mean	Random Uncertainty	Fixed Uncertainty	DOF
2	$Q/(m^3/min)$	10.9	0.2	0.1	27
3	$C_o/(kmol/m^3)$	7.83	0.07	0.01	3
4	$C/(kmol/m^3)$	0.81	0.06	0.01	3
5	$k/(1/min)$	12.5	0.2	0	1000
6	V/m^3	7.56	0.70		

	A	B	C	D	E	F
8	MC Run	Q	C_o	C	k	V
9	1	10.630	7.941	0.706	12.755	8.535
10	2	10.930	7.758	0.804	12.868	7.341
11	3	11.031	7.897	0.699	12.433	9.131
12	4	10.871	7.861	0.756	12.466	8.194
13	5	10.933	7.929	0.778	12.720	7.898
1003	995	11.052	7.911	0.799	12.488	7.874
1004	996	11.051	7.852	0.896	12.467	6.884
1005	997	10.858	7.947	0.740	12.482	8.467
1006	998	10.646	7.729	0.844	12.601	6.893
1007	999	10.822	7.842	0.726	12.537	8.464
1008	1000	10.695	7.865	0.825	12.526	7.288
1009	Average	10.903	7.830	0.809	12.503	7.623

Figure 8.17

Excel **worksheet displaying ten of 1000 Monte Carlo simulations of reactor volume with random and fixed uncertainty in the design variables.**

Calculate the 95% confidence interval, for *V* by sorting the Monte Carlo simulation results in column **F** in ascending order and eliminating 2.5% of the results from the top and 2.5% from bottom, respectively, as shown in Figure 8.18:

$$[6.4 < (V = 7.6) < 9.2] m^3 \qquad \text{(mean in 95\% confidence interval)} \tag{8.68}$$

For a conservative estimate, we round down the lower limit and round up the upper limits in the confidence interval, while retaining two significant figures. Notice that the confidence interval in this example is asymmetric: 7.6 – 6.4 = 1.2 m^3, 9.2 – 7.6 = 1.6 m^3. The expanded uncertainty falls within the range:

$$[1.2 < U_{V, 95\%} < 1.6] m^3 \tag{8.69}$$

The Monte Carlo results indicate a greater uncertainty on the high side of the reactor volume.

	A
1	5.965609
2	6.021519
3	6.032746
4	6.092178
5	6.132323
22	6.404743
23	6.410402
24	6.4174
25	6.426508
26	6.426936

	A
975	9.100943
976	9.104978
977	9.158304
978	9.174939
979	9.183323
996	9.60561
997	9.683965
998	9.688748
999	9.881273
1000	9.918998

Figure 8.18

First 2.5% and last 2.5% of sorted simulation results for the confidence interval.

Comments: The mean values for each of the deviation variables and simulated volumes in row 1009 of Figure 8.17 show that the Monte Carlo simulations return values distributed around the average values for the variables. The histogram in Figure 8.19 shows the distribution of 1000 Monte Carlo simulated reactor volumes to demonstrate the method in an *Excel* worksheet. However, 10^6 or more simulations are typically required to get a reasonable result that is independent from the number of simulations. We use *VBA* programming to generate a larger number of simulations.

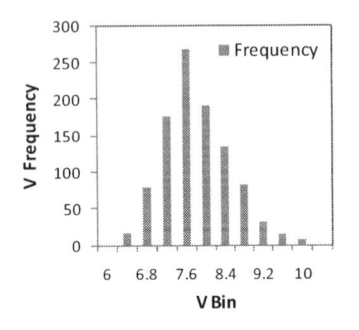

Figure 8.19

Histogram of 100 Monte Carlo simulations of reactor volume calculations with parameter uncertainty.

8.3.1 Latin Hypercube Sampling*

"Heisenberg slept somewhere near here!" – Sign at a historic inn

The Latin hypercube sampling method dramatically reduces the required number of Monte Carlo simulations by at least one order of magnitude. Latin hypercube sampling gets its name from the Latin square, an array of symbols with no two alike. A hypercube is a multidimensional surface. Like the Latin square, Latin hypercube sampling divides the normal distribution for random uncertainty and the uniform distribution for fixed systematic uncertainty into discrete blocks of equal probability. The method draws a unique random sample from each block for use in simulations. Thus, Latin Hypercube ensures sampling of the entire distributions for each variable used in Monte Carlo simulations, and eliminates any potential for clustering of random numbers. Table 8.5 lists the typical number of simulations required for coverage of uncertainty using Latin hypercube sampling.

Table 8.5 Number of simulations required for the method of Monte Carlo uncertainty analysis with Latin hypercube sampling.

Number of Monte Carlo Simulations	10^3	10^4	10^5	10^6
Latin Hypercube Sampling Region	$\pm 3.1 \times U_i$	$\pm 3.7 \times U_i$	$\pm 4.3 \times U_i$	$\pm 4.8 \times U_i$
Percent coverage	99.6%	99.8%	99.9%	99.99%

Figure 8.20 plots blocks of volumes with equal probability drawn from a two-dimensional probability distribution for a variable with normal random and uniform fixed uncertainty. For normally distributed errors, the Latin hypercube sampling method concentrates most of the sampling around the ridge of the normal distribution, which represents the mean value (zero in this case).

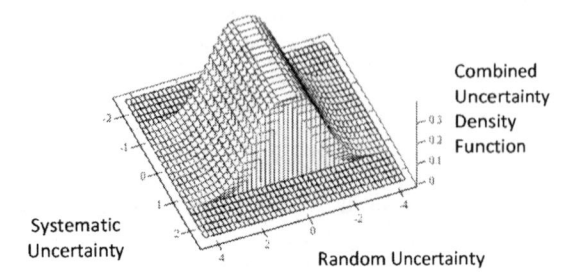

Figure 8.20

Example of a two-dimensional probability density function for the combined random (normal) and systematic (uniform) uncertainty distributions used for Monte Carlo simulations with Latin Hypercube sampling.

Excel and *VBA* have the functionality needed to implement Latin Hypercube sampling in an *Excel* worksheet. We outline the algorithm for Monte Carlo uncertainty analysis with Latin Hypercube sampling in the following steps:

1. Specify the number of intervals, n, to calculate the cumulative probability from Blom's (1958) formula:

$$p_i = \frac{8i+5}{8n+10} \qquad for \qquad i = 0, 1 \ldots n \qquad (8.70)$$

CHAPTER 8: UNCERTAINTY

2. Calculate limits for the normal distribution interval ranges using the worksheet function with zero mean and a standard deviation of one:

$$x = \text{NORMSINV}(p) \tag{8.71}$$

where p is the cumulative probability ranging from zero to one. To illustrate, for $n = 1000$ intervals, start with $p = 0.001$ calculated from Equation (8.70). Figure 8.21 displays the first five results. We remove the cumulative probability of zero, which returns an error message. Column **A** has calculations for the cumulative probability, confirming that each interval has equal probability. Column **B** gives the lower value of the range in the variables for each interval. Column **C** contains a randomly selected value from the interval for use by the Monte Carlo simulations. Column **D** gives the probability of x.

	A	B	C	D	E
1	i	Cumulative p	x	Random x	p
2	0	=(8*A2+5)/(8*$A$1001+10)	=NORMSINV(B2)		=NORMDIST(C2,0,1,FALSE)
3	1	=(8*A3+5)/(8*$A$1001+10)	=NORMSINV(B3)	=C2+RAND()*(C3-C2)	=NORMDIST(C3,0,1,FALSE)
4	2	=(8*A4+5)/(8*$A$1001+10)	=NORMSINV(B4)	=C3+RAND()*(C4-C3)	=NORMDIST(C4,0,1,FALSE)
5	3	=(8*A5+5)/(8*$A$1001+10)	=NORMSINV(B5)	=C4+RAND()*(C5-C4)	=NORMDIST(C5,0,1,FALSE)
6	4	=(8*A6+5)/(8*$A$1001+10)	=NORMSINV(B6)	=C5+RAND()*(C6-C5)	=NORMDIST(C6,0,1,FALSE)

	A	B	C	D	E
1	i	Cumulative p	x	Random x	p
2	0	0.000624844	-3.227289946		0.002183909
3	1	0.001624594	-2.943123501	-3.0162642	0.005247904
4	2	0.002624344	-2.79136111	-2.92619448	0.008108949
5	3	0.003624094	-2.685220917	-2.76194329	0.010843991
6	4	0.004623844	-2.602759201	-2.66959698	0.013485823

Figure 8.21

Calculation of interval limits and probabilities for a normal distribution.

The probability plots in Figure 8.22 illustrate how Latin Hypercube sampling from intervals with uniform probability concentrates the samples about the mean value of zero in a normal distribution. Intervals near the mean value have a narrow base and higher density, giving the same probability as intervals far from the mean that have a wider base with low density. The uniform distribution for systematic fixed-uncertainty uses intervals with equal spacing and density (not shown).

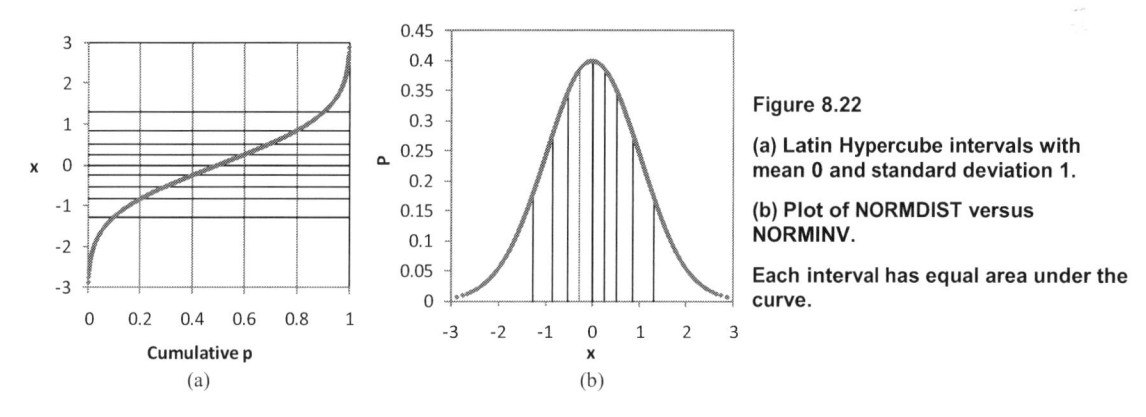

Figure 8.22

(a) Latin Hypercube intervals with mean 0 and standard deviation 1.

(b) Plot of NORMDIST versus NORMINV.

Each interval has equal area under the curve.

3. Sample each interval in the normal distribution using a random value selected between the lower and upper limit of the interval, as shown in column **C** of Figure 8.21. Figure 8.22 graphically illustrates the sampling intervals in the probability density function with **NORMDIST(x, 0, 1, FALSE)** plotted verses $x =$ **NORMSINV(p)**:

$$P(x) = \frac{1}{\sigma\sqrt{2\pi}} \exp\left[-\frac{(x-\mu)^2}{2\sigma^2}\right] \tag{8.72}$$

where $\sigma = 1$ is the standard deviation about the mean, $\mu = 0$. The intervals for our problems use the mean values of the variables and expanded uncertainty in place of the standard deviation.

4. Sample once from each of the n intervals for random uncertainty and again from each of the n intervals of fixed uncertainty.
5. Combine the normal and uniform values from step four for each variable:

$$x_i = \bar{x}_i + r_i u_i + r_{iz} u_{zi} \tag{8.73}$$

6. Repeat steps one to four for each of the m variables in the simulation. The number of intervals for combined uncertainty with equal probability requires $n^2 \times m$ simulations.
7. Randomly shuffle the results from the (n^2 simulations) × (m variables) random samples to remove any experimental procedural bias.
8. Simulate the model using the shuffled sets of variables.
9. Sort the $n^2 \times m$ simulated results in ascending order.
10. Determine the 95% confidence interval about the mean value for variable x by removing the lower 2.5% and upper 2.5% of the sorted results.

8.3.2 Correlation in Monte Carlo Uncertainty Analysis*

"A little nonsense now and then is cherished by the wisest men." — Roald Dahl, *Charlie and the Great Glass Elevator*

To account for correlation among random variables, we employ the symmetric, positive-definite correlation coefficient matrix R_{ij}, in the following steps:

1. Calculate the square root of the correlation coefficient matrix using Cholesky decomposition to yield the lower triangular matrix L, which satisfies the following criterion:

$$|R| = LL^T \tag{8.74}$$

2. Obtain a vector of correlated unit normal random numbers for the variables from the product of the Cholesky decomposition matrix L and a vector of unit normal random numbers (mean 0, standard deviation 1):

$$r_c = L \cdot r \tag{8.75}$$

3. Calculate the correlated normal random deviates for each simulation from the new unit normal random numbers and the standard random uncertainties.
4. Get the fixed systematic uncertainties using uniform random numbers ($-1 < r_{zi} < 1$) for each variable i.
5. Generate correlated variables for use in Monte Carlo simulations with Latin Hypercube sampling:

$$x_i = \bar{x}_i + r_{ci} u_i + r_{zi} u_{zi} \tag{8.76}$$

A *PNMSuite* VBA macro **UNMCLHS** calculates the propagation of uncertainty by the Monte Carlo method using Latin hypercube sampling, while allowing for correlation among the variables. The code listed in the *PNMSuite* validates the location selected for the output to avoid overwriting cells on the worksheet. A user-form **UNCERTAINTY** is also available from the *PNMSuite* with options for *Monte Carlo* methods. The macros have the following attributes:

- Calculate the decomposition matrix by Cholesky's method from user provided input for correlation coefficients.
- Employ the same number of simulations in both standard and systematic uncertainties.
- Prompt the user for the cell containing the function in terms of the range of average values for the variables in the worksheet.
- Input the ranges of cells containing the standard and systematic uncertainties.
- Based on the power of Latin hypercube sampling, the macros require a minimum (100 normal) × (100 uniform) intervals for just 10^4 simulations (this is equivalent to $\sim 10^6$ standard Monte Carlo simulations with ordinary random sampling!). The default number of simulations is 10^4.

CHAPTER 8: UNCERTAINTY

- Display the progress of the simulations during program execution in the **Status Bar**.
- Use a default 95% confidence interval. The macro output includes the lower and upper confidence limits with expanded uncertainties and corresponding percent uncertainties.
- Create a histogram in the worksheet cells to show the distribution of simulated results.
- Deploy the user-defined function **RNDWH** for generating random deviates, and uses the sub procedure **HEAPSORT** for sorting the simulated results in ascending order from low to high values.

Example 8.12 Monte Carlo Propagation of Uncertainty in a CSTR

Known/Find: Redo *Example 8.11* using Latin Hypercube sampling.

Analysis: Run the *VBA* user-form **UNCERTAINTY** for Monte Carlo uncertainty analysis with 10^4 simulations. Provide the required input with reference to the *Excel* worksheet in Figure 8.17, and direct the output to cell **C6**. Ignore correlation for this example.

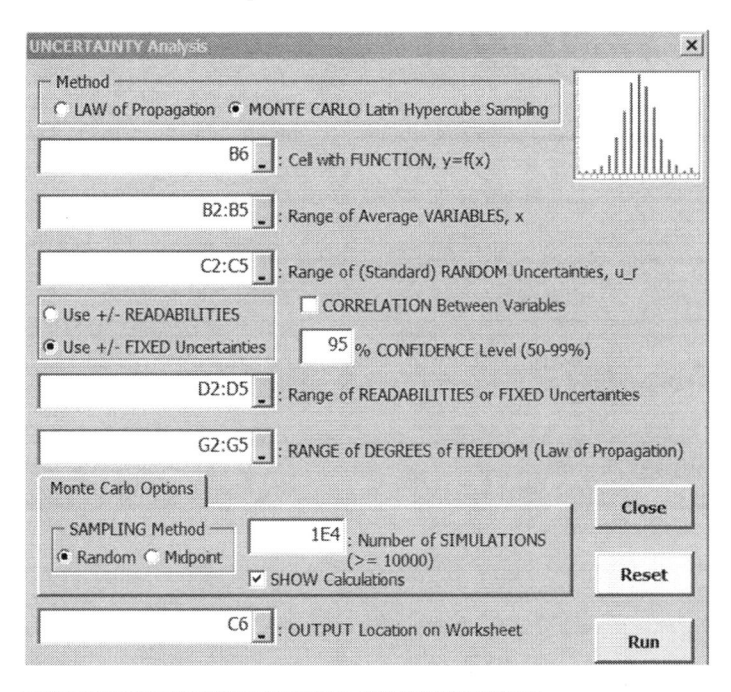

Figure 8.23

Monte Carlo inputs and uncertainty results for the design volume of a CSTR in the user form UNCERTAINTY.

	A	B	C	D
6	V/m³	7.56	$U_{95\%}$ = --	1.2E+00
7			$U_{95\%}$ = +	1.5E+00
8			Bin	Frequency
9			5.385824436	1
10			6.030201821	58
11			6.674579206	652
12			7.318956592	2827
13			7.963333977	3670
14			8.607711363	2010
15			9.252088748	610
16			9.896466133	116
17			10.54084352	40
18			11.1852209	15
19			11.82959829	1

We interpret the results shown in Figure 8.23 as the 95% confidence interval about the mean value:

$$[6.36 < (V = 7.56) < 9.06] m^3 \quad \text{(mean in 95\% confidence interval)} \quad (8.77)$$

or
$$V = 7.56 \pm (1.2 \text{ to } 1.5) m^3 \quad \text{(mean in 95\% confidence interval)}$$

Comments: In this problem, Monte Carlo uncertainty analysis produces an asymmetric confidence interval about the mean value, with a slightly larger uncertainty at the higher end. Monte Carlo methods need large numbers of calculations that require a long time to complete. Fortunately, 10,000 Latin hypercube simulations are adequate for most uncertainty analyses.

Let us compare the Monte Carlo results for uncertainty with uncertainty obtained from the law of propagation of uncertainty using the macro **JITTER**. We do not have information about the degrees of freedom, so we use a large value of 1000. The results for expanded uncertainty in Figure 8.24 fall within the range of uncertainty determined by the Monte Carlo method in Equation (8.77).

	F	G	H	I
6	u = ±	6.7E-01	U$_{95\%}$ = ±	1.3E+00
7	DOF =	1000	t$_{95\%}$ =	1.96
8	c$_1$ =	6.9E-01	(c·u$_1$)^2 =	5.4%
9	c$_2$ =	1.1E+00	(c·u$_2$)^2 =	1.3%
10	c$_3$ =	-1.0E+01	(c·u$_3$)^2 =	90.0%
11	c$_4$ =	-6.0E-01	(c·u$_4$)^2 =	3.3%

Figure 8.24
Uncertainty analysis of a CSTR using the law of propagation implemented by the VBA macro JITTER.

8.4 Triangular and Log Normal Distributions*

"The triangle is a foundation to an offense". – Bill Cartwright

Consider using an asymmetric triangular distribution when the expanded uncertainty exceeds the limitations for allowable physical values of the variables. For example, by definition reaction conversion is limited to the range $0 \leq X \leq 1$, and rate constants or concentrations must be greater than zero (e.g., $k \geq 0$). The probability density function for a triangular distribution has a mode, and lower and upper bounds, as illustrated in Figure 8.25:

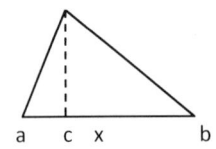

Figure 8.25
Triangular distribution for a<x<b.
The mode is at x = c.

Mean: $\bar{x} = \dfrac{a+b+c}{3} \Rightarrow c = 3\bar{x} - a - b$

Variance: $\sigma^2 = \dfrac{a^2 + b^2 + c^2 - ab - ac - bc}{18}$

We may consider a triangular distribution for a small number of data where we set a and b to the low and high values of the samples and c as the average, mode, or median value of the data. We generate triangular-distributed random deviates in the range $a < x < b$, from uniform random deviates ($0 < p < 1$) as follows:

$$x = \begin{cases} a + \sqrt{p(b-a)(c-a)} & \text{if } 0 < p < (c-a)/(b-a) \\ b - \sqrt{(1-p)(b-a)(b-c)} & \text{otherwise} \end{cases} \quad (8.78)$$

To use a triangular distribution, modify the VBA macro **UNMCLHS** by replacing **WorksheetFunction.NormSInv** with the user-defined function **TRINV** listed in Figure 8.26. Provide the values for a, b, and c as needed for each variable requiring a triangular distribution. The macro then uses the values for the standard uncertainty in x as $\pm u_x(b - c) = \pm u_x(c - a)$.

```
Public Function TRINV(p As Double, a As Double, b As Double, c As Double) As Double
' p = probability, 0 < p < 1
' a, b, c = lower and upper limits and mode of triangular distribution
'*******************************************************************************
' Returns an inverse value (a < x < b) from a triangular distribution.
'*******************************************************************************
If Not p > (c - a) / (b - a) Then ' draw from a triangle distribution
    TRINV = a + Sqr(p * (b - a) * (c - a))
Else
    TRINV = b - Sqr((1 - p) * (b - a) * (b - c))
End If
End Function
```

Figure 8.26 User-defined function TRINV that returns the inverse of a triangular distribution.

We may also consider log normal distributions for one sided density functions. For example, concentration has a lower limit of zero, but not necessarily an upper limit. Variables that represent the product of several normally distributed random deviates typically exhibit log normal distributions. Variables representing physical quantities that must be positive, such as concentration, also have a skewed lognormal distribution for small data sets (Berendsen 2011). Log normal distributions have the distinguishing feature of asymmetry in the tails of the probability density function, exemplified in Figure 8.27:

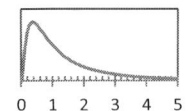

Figure 8.27

Log normal distribution with mean 0 and standard deviation 1.

The lognormal distribution gets its name from the characteristic of the derivative of a log function:

$$\frac{d \ln x}{dt} = \frac{1}{x}\frac{dx}{dt} \tag{8.79}$$

The lognormal probability density function is similar to Equation (8.72) for a normal distribution:

$$P(x) = \frac{1}{x\sigma\sqrt{2\pi}} \exp\left[-\frac{(\ln(x)-\mu)^2}{2\sigma^2}\right] \quad \text{for} \quad x > 0 \tag{8.80}$$

We generate log-normally distributed random numbers from a normal distribution with mean value of zero and standard deviation of one, as follows:

$$r_{LN} = \exp(\mu + \sigma r_N) \tag{8.81}$$

where μ and σ are the normal mean and standard deviations of the natural logs of the data, respectively, and r_N is a normal random deviate. Alternatively, *Excel* has functions for generating lognormal random deviates. If necessary, modify the *VBA* macros **JITTER** and **UNMCLHS** to use log normal distributions in place of normal distributions by replacing **WorksheetFunction.NormSInv** with **WorksheetFunction.LogInv**. Consult Berendsen (2011) and Tyagi and Haan (2001) for descriptions of other probability density functions encountered in science and engineering measurements.

8.5 Epilogue on Uncertainty Analysis

"In theory, there is no difference between theory and practice. But in practice, there is." – Yogi Berra

This chapter introduces uncertainty analysis in engineering experimentation and calculations. Our discussion of uncertainty propagation introduces the basic concepts for most situations encountered in practice. The following steps serve only as a guide for novice uncertainty analysts. We may need to rely on our engineering judgement to tackle problems that do not fit the typical mold.

1. Start by assuming that all input data have some combination of random and fixed uncertainty.
2. Use descriptive statistics, or regression analysis described in the next chapter, to calculate the average value of a set of measurements, the standard uncertainty from the standard deviation, and the degrees of freedom for the average value. If all we have is a single number, we must assume zero standard deviation, which implies infinite degrees of freedom.
3. Use knowledge about the precision of the measurement, from either the reported significant figures or resolution of the instrument, to calculate a fixed uncertainty. Combine the fixed and standard uncertainties for the measurement input.
4. Use the jitter or Monte Carlo methods to propagate the measurement uncertainties through a complex series of calculations.
5. Report the result matched to the significant figures (usually one, but no more than two significant figures) of the expanded uncertainty. Include parenthetical information about the uncertainty type.

Table 8.6 compares the strengths and weaknesses of the methods of uncertainty analysis presented in this chapter. Table 8.7 summarizes the macros and user-defined functions for uncertainty analysis in the *PNMSuite*. Ultimately, we recommend following the convention of uncertainty analysis required by your particular discipline or industry. Consider using both methods of the linear *Law of Propagation* and nonlinear *Monte Carlo* to get the most information about uncertainty.

We caution you not to take uncertainty analysis to unwarranted extremes. Recall that uncertainty analysis is just that, *uncertain*! With too much analysis, we may convince ourselves that we have "real" knowledge of the actual error, when in reality all we have is a set of tools to help us make informed, reasonable decisions based on estimates for reliability of our data and calculations.

Table 8.6 Comparison of methods of uncertainty analysis.

Method	Strengths	Weaknesses
Law of Propagation of Uncertainty	• Widely accepted method of uncertainty analysis • Provides estimates for each variable's contribution to the final uncertainty	• Does not account for asymmetric confidence intervals
Bootstrap	• Applies to models with unknown probability distributions in the data	• Not well known or widely applied.
Monte Carlo	• Asymmetric confidence intervals for nonlinear models • Accounts directly for correlation in the inputs	• Computationally intensive and expensive.

Table 8.7 Summary of VBA macros, user-forms, and user-defined functions (UDF) for uncertainty analysis.

Macro/Function	Inputs	Operation
BOOTSTRAP	Macro input boxes	Bootstrap resampling with replacement for a statistic of a data set
JITTER	Macro input boxes	Jitter method of uncertainty propagation
TRINV	UDF arguments	Inverse value of a triangular distribution
UNMCLHS	Macro input boxes	Monte Carlo method of uncertainty propagation, using Latin hypercube sampling
UNCERTAINTY_UsrFrm	Show user-form macro	Select between the Jitter and Monte Carlo methods
USIGFIGS	UDF arguments	Significant figures in uncertainty

"The government are very fond of statistics. They add them and they multiply them, and they extract the square root and they make very beautiful charts about them but you must always remember, that in the first instance, they are dependent on items put down by the village watchman, who just puts down what he damn pleases." – O. Meredith Wilson

Chapter 9 Regression

"I never went to a modeling school, and I don't suggest to anybody that they go." – Heidi Klum

In Chapter 1, we introduced mechanistic mathematical models of chemical and engineering systems derived from physical principles of momentum, mass, and energy conservation. In this chapter, we introduce methods for *empirical* data modeling as well as techniques for *validating* both mechanistic and empirical models with data. We typically have three goals in modeling data:

1. Discover mechanisms or relationships between dependent variables (model output) and independent variables (model input).
2. Make predictions for future responses to input variables.
3. Determine physicochemical properties by fitting a model of the physical phenomena to experimental data.

Indeed *Excel* excels at data analysis, and has evolved into a popular and convenient software tool for data modeling. Consequently, this chapter highlights *Excel*'s built-in modeling tools as well as a few *VBA* macros designed to enhance *Excel*'s capabilities for data modeling and analysis. Specifically, this chapter includes the following topics:

- Fitting models to data by linear and nonlinear least-squares regression
- Model assessment and validation using residual analysis
- Uncertainty in model predictions and parameters
- Robust regression of least absolute deviation

SQ3R Focused Reading Questions
1. What is the difference between linear and nonlinear regression?
2. What are outliers and how do we deal with them in regression modeling?
3. What *Excel* functions and tools are available for least-squares regression?
4. What model should I use to fit to data?
5. Why do I need good guesses for the model parameters in nonlinear regression?
6. What are the potential hazards with linearizing a nonlinear equation for regression?
7. What are the recommended tools for validating a regression model?
8. How do I estimate the uncertainty in a regression result?
9. What are the advantages of rational functions for modeling data?
10. When should I consider least absolute deviation regression instead of least squares regression?

We begin with a simple illustration to introduce basic concepts of data modeling. Chemists use measurements of concentration changing with time in an isothermal batch reactor to determine reaction kinetics. Consider the experimental batch reactor data for a first-order reaction plotted in Figure 9.1.

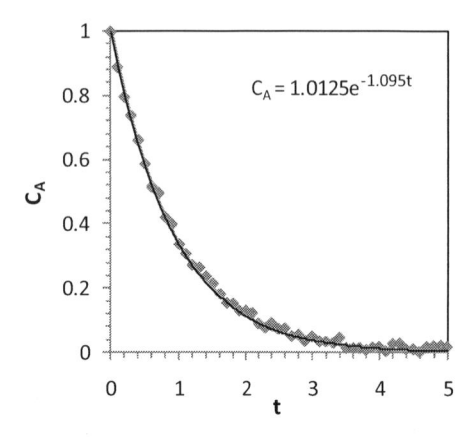

Figure 9.1

Concentration versus time of reactant A in a batch reactor plotted with Equation (9.2).

A simple mole balance begins with the basic expression for the conservation of mass in a batch processes with zero inlet and outlet rates:

$$\text{Rate in} - \text{Rate out} + \text{Rate of generation} = \text{Rate of accumulation}$$

$$0 - 0 + (-kC_A V) = V \frac{dC_A}{dt} \quad (9.1)$$

where V is the fluid volume in the reactor (assumed constant), k is the first order rate constant, C_A is the concentration of reactant A, and t is the batch reaction time. Note that species A is consumed in the reaction, as indicated by the negative rate of generation. Separate variables and integrate Equation (9.1), subject to the initial condition that $C_A = C_{A0}$ at $t = 0$, for the following exponential model of concentration as a function of reaction time:

$$C_A = C_{A0} \exp(-kt) \quad (9.2)$$

We use the experimental data in Figure 9.1 to determine values for the initial concentration C_{A0} and rate constant k in Equation (9.2). Right-click on the data in the plot and select *Excel*'s **Trend line Chart Tool** to produce the exponential fit of the data illustrated by the smooth curve in Figure 9.1 with $C_{A0} = 1.0125$, and $k = 1.095$.

A different approach rearranges the mass balance in Equation (9.1) for the rate constant in terms of the concentration and derivative:[83]

$$k = -\frac{1}{C_{A0}} \frac{dC_A}{dt}\bigg|_{t=0} \quad (9.3)$$

at the initial conditions in the reactor, $t = 0$ and $C_A = C_{A0}$. The slope of the curve in the limit $t \to 0$ corresponds to the derivative of the concentration with respect to time. Use a finite-difference approximation for the derivative from the first two data pairs:

$$\frac{1}{C_{A0}} \frac{dC_A}{dt}\bigg|_{t=0} = \frac{1}{C_{A0}} \left(\frac{C_{A1} - C_{A0}}{t_1 - t_0} \right) = 1.12 \quad (9.4)$$

Alternatively, we linearize Equation (9.2) using logarithms. The rate constant is equivalent to the slope of the line formed from the natural logarithm of the concentration data versus time, plotted in Figure 9.2:

$$-\ln C_A = -\ln C_{A0} + kt \quad (9.5)$$

The intercept corresponds to $\ln[C_{A0}]$. We determine the equation of a line through the logarithmically linearized data using *Excel*'s **Trend line Chart Tool** to calculate $C_{A0} = \exp(-0.0124) = 0.9877$ and $k = 1.0954$.

---

[83] As we learned in Section 5.2, differentiation tends to magnify noise in uncertain data. We should "avoid differentiation when integration will do." (Civan 2011)

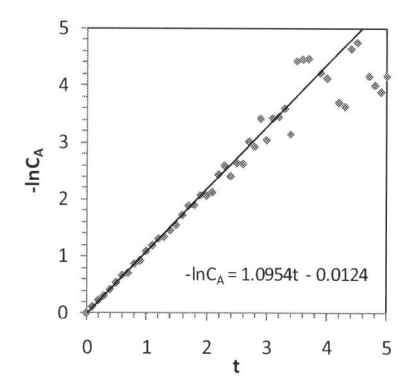

Figure 9.2

Natural log of concentration versus time for reactant A in a batch reactor plotted with Equation (9.5).

$-\ln C_A = 1.0954t - 0.0124$

Our simple example of interpreting batch reactor data distills the essence of data modeling and regression analysis. Each of the three alternative methods of analysis requires finding a best-fit line through the data. We pause here to point out "the elephant in the room." We used three methods and found three slightly different values for the rate constant k, and initial concentration, C_{A0}. Which one is best? Do we have confidence in any of the results? Experimental data can be noisy. The batch reactor data plotted in Figure 9.1 exhibit some random fluctuations (noise) in the data. The finite difference calculation of the slope used just two points of the noisy data. The log plot of Figure 9.2 seems to magnify the fluctuations at large t. This begs the questions: how do we know when we have the best model of the data? Moreover, how reliable is our model? This chapter introduces methods for working with data to accomplish such things as calculating rates from slopes, validating models, and estimating model uncertainties.

9.1   Least-squares Regression

"We dance around the ring and suppose, but the secret sits in the middle and knows" – Robert Frost

Engineers use regression analysis to find functional relationships between independent and dependent data. "Regression" means, "come back to the center." A least-squares regression function passes through a set of data in a way that minimizes the distances between the data points and function. In other words, it passes through the "center" of the data. Notice that the method of least-squares regression also produces models that smooth noisy data for interpolation, as seen in Figure 9.1 and Figure 9.2.

In regression analysis, we first propose a function to model the data, and then perform an optimization calculation for the model parameters (*e.g.*, search for values of k and C_{A0} in Equation (9.2) that minimize the sum of the squared residuals (*SSR*), or differences, between the model and the data pairs):[84]

$$SSR = \sum_{i=1}^{n} \left(y_i - y_{i,\text{model}} \right)^2 \qquad (9.6)$$

where i is the index for a data point, n is the number of y versus x data pairs, and $y_{i,\,model}$ is calculated from the model function at the value of the data for the independent variable x_i:

$$y_{i,\text{model}} = f\left(a_j, x_i\right) \qquad (9.7)$$

The model parameters a_j ($j = 1$ to k) become the variables of optimization.[85] We may use any method of optimization, such as *Excel*'s **Solver**, to find the model parameters, a, that minimize the *SSR* in Equation (9.6).

In general, we have two competing objectives for regression of data (Good and Hardin 2006):

1. Maximize the goodness of fit at the risk of finding the wrong model.

---

[84] Alternatively referred to as Sum Squared Errors (*SSE*).
[85] The great French mathematician and natural philosopher, René Descartes was the first to use the letters x, y, and z for variables and unknowns, and a, b, and c for model parameters and constants.

2. Minimize roughness (residuals) at the risk of over fitting the data.

Depending on the model, we may select from linear or nonlinear least-squares regression methods. A common misconception is that nonlinear functions of the independent variable require nonlinear least-squares solution methods. A second order polynomial, $y = a_0 + a_1 x + a_2 x^2$, is an example of a nonlinear function in x, however linear in the coefficients a_0, a_1, and a_2. For the polynomial, the least-squares problem of determining the coefficients involves the solution to a linear system of equations. Nonlinear regression fits data with a model with nonlinear *parameters*. For example, the rational function $y = (p_0 + p_1 x)/(1 + q_1 x)$ is linear in the parameters p_0 and p_1 in the numerator, however nonlinear in the parameter q_1 in the denominator. Incidentally, this rational function is also nonlinear in the independent variable x in the denominator. We begin with the simpler linear least-squares regression method before introducing the all-purpose method of nonlinear least square regression.

Many software applications, including *Excel*, as well as scientific calculators, have sophisticated linear least-squares regression capabilities for fitting data with common models, such as equations for lines, polynomials, power law, logarithm, and exponential functions. *Excel* has intrinsic plotting capabilities along with the method of linear least-squares regression that allow for a visual confirmation of how well the model fits the data. In *Excel*, use the **Add a Trend line** feature in the **Chart Tools** to fit a canned function, or by right-clicking on the data in a plot and selecting **Trend Line** from the popup menu.

Example 9.1   Reaction Rate Law from Linear Least-Squares Regression

Known: The rate law for an elementary reaction is proportional to the concentration raised to the power of the reaction order:

$$-r = kC^m \tag{9.8}$$

where C is concentration, m is the reaction order, and k is the reaction rate constant. We may transform Equation (9.8) by logarithms to linearize the parameter m in the exponent:

$$\ln(-r) = \ln(k) + m \cdot \ln(C) \tag{9.9}$$

Now, the slope in Equation (9.9) is the reaction order, m, and the intercept is the natural logarithm of the rate constant, $\ln(k)$. Calculate $k = \exp(\text{intercept})$.

Find: Use the linear least squares method applied to the transformed batch reactor data listed in the *Excel* worksheet shown in Figure 9.4 to determine the values of m and k.

Analysis: Plot the data in *Excel*. Right-click on the data markers in the plot and select **Add Trend line** from the popup menu, as shown in Figure 9.3. Select **Linear** and check the box to show the equation on the plot. The linear least squares fit of the logarithmic linearized data gives

$$\ln(-r) = 1.4972 + 1.2273\ln(C) \tag{9.10}$$

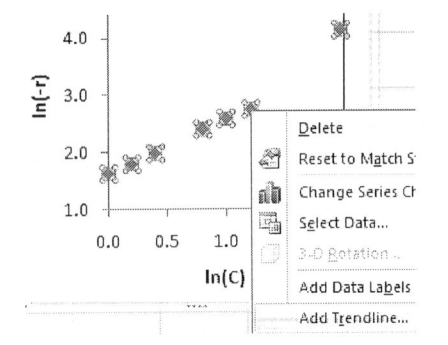

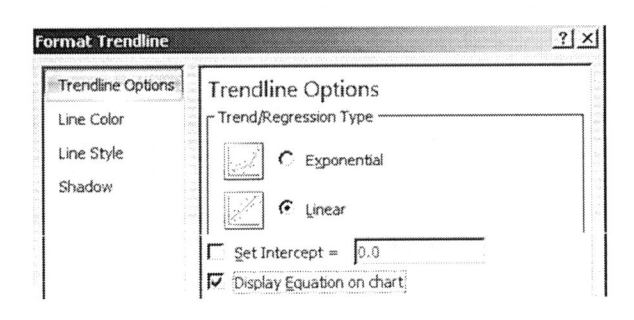

Figure 9.3   Add a trend line from the right-mouse-click menu. Trend line options for showing the regression equation on the chart.

Figure 9.4 is a plot of the least-squares fit along with the logarithmic linearized form of the data. For convenience, we plot $-\ln(-r)$ versus $-\ln C$ to avoid negative values on the axes. This result confirms that the reaction is of order one ($m = 1$) and gives a rate constant $k = \exp(-1.4972) = 0.2238$ s^{-1}.

	A	B	C	D
1	C	-r	-lnC	-ln(-r)
2	1	0.195	0.0000	1.6348
3	0.82	0.165	0.1985	1.8018
4	0.67	0.137	0.4005	1.9878
5	0.45	0.09	0.7985	2.4079
6	0.37	0.075	0.9943	2.5903
7	0.3	0.064	1.2040	2.7489
8	0.14	0.016	1.9661	4.1352
9				
10	Slope	m	1.2273	
11	Intercept	-lnk =	1.4972	

Figure 9.4

Linear trend line fit of logarithmically transformed data.

If we know the reaction order a priori, $m = 1$ in this case, then Equation (9.8) reduces to a linear function of concentration C, with slope k and zero intercept. Now we can perform a linear least squares fit of the data in their original form listed in columns **A** and **B**, shown in Figure 9.5. Select the option **Set Intercept = 0** from the **Trend line** menu to give the rate law:

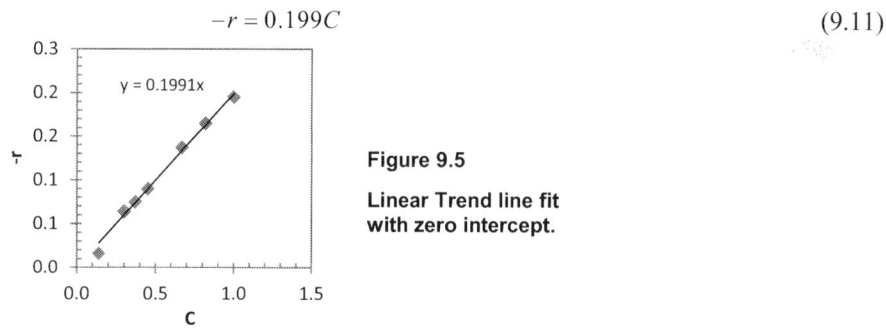

$$-r = 0.199C \tag{9.11}$$

Figure 9.5

Linear Trend line fit with zero intercept.

Comments: Observe that the last point in Figure 9.4 heavily weights the line fit of the linearized data, an indication that transforming the model by linearization may not properly fit the data to the original Equation (9.8).

Transforming the data by logarithmic linearization produces a result with a significant deviation from the best parameter value for the model because of minimizing the errors in the fit associated with the *transformed* data pairs as opposed to fitting the *original* data. To understand how this happens for this example, the linear least-squares minimization function for Equation (9.9) with $m = 1$ becomes:

$$SSR = \sum_{i=1}^{n}\left[\ln(-r_i) - (\ln C_i + \ln k)\right]^2 \tag{9.12}$$

Minimizing Equation (9.12) with respect to $\ln(k)$ gives

$$\frac{\partial(SSR)}{\partial \ln k} = 0 = \sum_{i=1}^{n}\left[\ln(-r_i) - (\ln C_i + \ln k)\right] = \left\{\sum_{i=1}^{n}\left[\ln(-r_i) - \ln C_i\right]\right\} - n\ln k \tag{9.13}$$

The single model parameter is just the exponential of the average value of the sum of the differences in the log-linearized data:

$$\ln k = \frac{1}{n}\sum_{i=1}^{n}\left[\ln(-r_i) - \ln C_i\right] \tag{9.14}$$

This gives $\ln(k) = -1.678$ or $k = 0.187$, which is six percent smaller than the result in Equation (9.11).

□

9.1.1 Ordinary Least-squares Regression

"Eventually you reach the point where you have to draw the line." – Euclid's Last Theorem

A quick, but crude, method for fitting data to a line first plots the data, and then draws a line that appears to give the best fit. If we repeat these two steps, we will get a line with slightly different values for the slope and intercept. The method of single-variable, linear least-squares regression automatically fits n number of y vs x data pairs to a linear equation of the form $y = a_0 + a_1 x$. Linear least-squares regression involves minimizing the least-squares error function, or sum of the squared residuals (SSR) between the data and linear function by substituting an equation of a line for y_{model} in Equation (9.6):

$$SSR = \sum_{i=1}^{n}\left[y_i - (a_0 + a_1 x_i)\right]^2 \tag{9.15}$$

As discussed in Section 7.4 on indirect optimization, we use the calculus definitions to minimize the least-squares error function by solving the partial derivatives of SSR with respect to each parameter:

$$\frac{\partial(SSR)}{\partial a_0} = 0 \quad \text{and} \quad \frac{\partial(SSR)}{\partial a_1} = 0 \tag{9.16}$$

We now have two coupled linear equations in the two unknown fitting parameters, a_0 and a_1. Solve Equations (9.16) simultaneously for the ($k = 2$) model parameters:

$$a_0 = \frac{\left(\sum_{i=1}^{n} x_i^2\right)\left(\sum_{i=1}^{n} y_i\right) - \left(\sum_{i=1}^{n} x_i y_i\right)\left(\sum_{i=1}^{n} x_i\right)}{n\left(\sum_{i=1}^{n} x_i^2\right) - \left(\sum_{i=1}^{n} x_i\right)^2} \tag{9.17}$$

and

$$a_1 = \frac{n\left(\sum_{i=1}^{n} x_i y_i\right) - \left(\sum_{i=1}^{n} x_i\right)\left(\sum_{i=1}^{n} y_i\right)}{n\left(\sum_{i=1}^{n} x_i^2\right) - \left(\sum_{i=1}^{n} x_i\right)^2} \tag{9.18}$$

In cases where the model requires a zero y-intercept, we need only to find a single parameter ($k = 1$):

$$y = ax \tag{9.19}$$

Find the best-fit value of a with zero-intercept by setting to zero the partial derivative of the minimization function with respect to a only:

$$\frac{\partial\left(\sum_{i=1}^{n}[y_i - ax_i]^2\right)}{\partial a} = 0 \tag{9.20}$$

Solve Equation (9.20) explicitly for a:

$$a = \left(\sum_{i=1}^{n} x_i y_i\right)\left(\sum_{i=1}^{n} x_i^2\right)^{-1} \tag{9.21}$$

We may extend this approach to linear regression of models with more than two parameters. For example, a least squares fit of data with a polynomial involves a system of linear equations in the unknown coefficients:

$$SSR = \sum_{i=1}^{n}\left[y_i - (a_0 + a_1 x_i + a_2 x_i^2 + \ldots + a_m x_i^m)\right]^2 \tag{9.22}$$

The solution for the coefficients involves solving the system of ($k = m + 1$) linear equations for the polynomial coefficients using a method from Chapter 4:

$$\frac{\partial (SSR)}{\partial a_j} = 0 \quad j = 0 \ldots (k-1) \tag{9.23}$$

Excel has built-in functions for calculating the slope and intercept of a line by linear least-squares regression. Referring to Figure 9.4 in Example 9.1, use *Excel* worksheet functions **SLOPE(y_data, x_data)** and **INTERCEPT(y_data, x_data)** to get the slope and intercept of the line formed from the log-linearized data:

C10=SLOPE(D2:D8, C2:C8)   C11=INTERCEPT(D2:D8, C2:C8)

Excel's worksheet array function **TREND(y_data, x_data, x_value(s), [intercept])** returns the *y* value from a linear interpolation of a single or multivariable linear regression of **x_data** and **y_data** at the **x_value(s)**. Use **TREND** for a single interpolation at one **x_value**, or the simultaneous key combination *CNTL SHIFT ENTER* to return an array of interpolated results at several **x_values** in a range of cells on the worksheet. Use the optional fourth argument for the intercept in **TREND** to specify a zero *y*-intercept. The default value is **TRUE**. **FALSE** sets the intercept to zero. See Example 10.1 for an illustration of linear interpolation of smooth data using the worksheet function **Trend**.

9.1.2 Multivariable Linear Least-squares Regression

"Gromit was the name of a cat. When I started modeling the cat I just didn't feel it was quite right, so I made it into a dog because he could have a bigger nose and bigger, longer legs." – Nick Park

Multivariable linear regression is an important tool for model building whenever our linear model involves more than one independent variable, such as the following expression:

$$y = a_0 + a_1 x_1 + a_2 x_2 + a_3 x_3 \ldots \tag{9.24}$$

Fitting a polynomial to *y* vs *x* data is one example of multivariable linear regression. Compare the following equation to Equation (9.24):

$$y = a_0 + a_1 x + a_1 x^2 + a_2 x^3 \ldots \tag{9.25}$$

Define each power of *x* as a new "linear" variable:

$$x_1 = x \qquad x_2 = x^2 \qquad x_3 = x^3 \quad \ldots \tag{9.26}$$

We recommend *Excel*'s linear regression tools because they have options to report additional statistical information about the goodness of fit. The *VBA* macro **RATLS** presented in 9.10.1 also performs polynomial least-squares regression and polynomial uncertainty analysis. We demonstrate multiple linear regression in Example 9.2.

Excel comes with an optional add-in for conducting multivariable linear regression that includes information for model assessment and uncertainty analysis. As shown in Section 2.5, to access the **Analysis Tool Pak**, click on the **File** tab and select **Options>Add-Ins>Manage: Excel Add-ins>Go...** Select **Analysis Tool Pak** from the available add-ins and click **OK**. The **Analysis** group on the **Data** tab of the ribbon should now display the **Data Analysis** option. Click on **Data Analysis** and select **Regression** from the list of **Analysis Tools**.

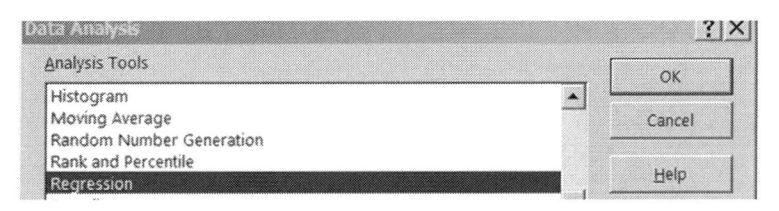

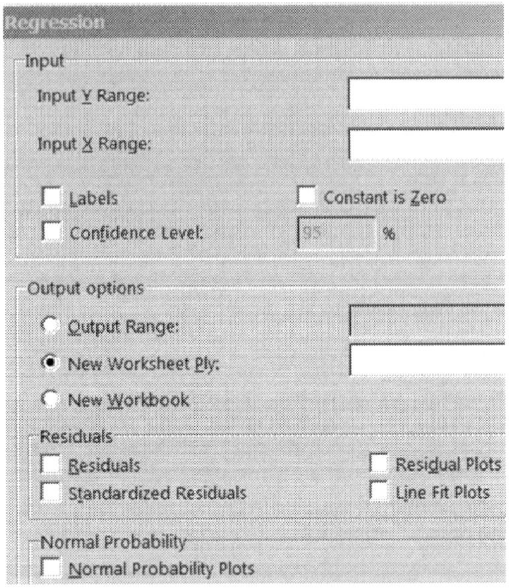

Figure 9.6 Dialogue box for Regression analysis tool with key inputs defined.

Y Range = range of values for the dependent variable

X Range = range of values for the independent variable(s)

Labels = when selected, the first row of data contains labels

Constant is Zero = select for a zero intercept in the model

Confidence Level = specify percent confidence

Output Range = location of top left corner of results on the worksheet

New Worksheet = select to put the results in a new worksheet

New Workbook = select to put the results in a new workbook

Residuals = report the values of the model residuals

Residual Plots = generate residual plots as recommended in 9.3.5.

Standard Least-squares regression makes the following assumptions.

- Error is limited to the dependent variable, y.
- The dependent y-data are normally (Gauss) distributed about their true values

The summary output from linear regression analysis is placed in a worksheet organized into three groups: **Summary Output**, **ANOVA**, and **Residual Output**. We define the more useful output results from **Regression Analysis** in Table 9.1, and indicate where details are covered in the chapter. Example 9.2 performs multivariable linear regression using *Excel's* add-in regression analysis tool.

Table 9.1 Summary output for Regression Analysis Tool

SUMMARY OUTPUT	
R^2	= Coefficient of determination described in Section 9.3.2
Adjusted R^2	= Adjusted coefficient of determination described in Section 9.3.2
Standard Error	= Standard uncertainty of the model prediction for y, defined in Equation (9.68)
Observations	= Number of data pairs used to determine the degrees of freedom
ANOVA	
Significance F	= For 95% confidence, this term should be < 0.05. If > 0.05, check the P-values for variables > 0.05.
Coefficients	= Results for the model parameters from the linear regression
Standard Error	= Standard uncertainty in the coefficients, defined in Equation (9.74)
t Stat	= Model parameter divided by the standard error, defined in Equation (9.78). Significant parameters have t-Stat > 2
P-value	= Probability that the t-Stat is insignificant, described in Section 9.3.4. p-values < 0.05 indicate the model parameter is significant for t-Stat > 2
Lower/Upper %	= % confidence interval for expanded uncertainty in model parameter (coefficient), defined by Equation (9.75)
RESIDUAL OUTPUT	
Residuals	= Differences between data points for the dependent variable and the model defined by Equation (9.79)
Residual Plots	= Plots of the residuals with the data described in Section 9.3.5 (random scatter of the residuals about the zero axis indicate the model captures the data trend)

CHAPTER 9: REGRESSION

Example 9.2 Linearized Reaction Rate Model

Known: Figure 9.7 shows an *Excel* worksheet with the rates of reaction for ethylene (A) polymerization in the presence of a catalyst for a range of temperatures and concentrations.

Find: Use multivariable linear regression to fit a reaction rate law to the data.

Assumptions: Assume a well-mixed reactor and an elementary reaction rate law for the polymerization reaction (*i.e.*, the rate of ethylene disappearance):

$$-r_A = kC_A^m \tag{9.27}$$

The rate "constant" has an Arrhenius function of temperature dependence:

$$k = k_0 \exp\left(-\frac{E_a}{R_g T}\right) \tag{9.28}$$

with pre-exponential factor k_0, activation energy, E_a, and ideal gas constant, R_g. Note T requires units of absolute temperature, such as degrees K.

Analysis: Transform the model by logarithmic linearization to calculate k_0, E/R, and the reaction order, m, using multivariable linear regression on all of the data:

$$\ln(-r_A) = \ln(k_0) - \frac{E_a}{R_g}\frac{1}{T} + m \ln C_A \tag{9.29}$$

The terms $1/T$ and $\ln(C_A)$ are the independent variables, and $\ln(-r_A)$ is the dependent variable. Transform the data in an *Excel* worksheet, as shown in Figure 9.7, for the linearized model in Equation (9.29). *Excel* requires contiguous columns of data for the independent variables.

	A	B	C	D	E	F	G	H
1	T/°C	C_A/(mol/m³)	$-r_A$/(mol/s-g cat)	(1/T)/K⁻¹	lnC_A	ln$(-r_A)$	ln$(-r_A/C_A)$	$-r_A/C_A$
2	22.8	28	63	0.003379	3.332	4.143	0.81093	2.25
3	22.8	142	321	0.003379	4.956	5.771	0.815614	2.2606
4	22.8	256	632	0.003379	5.545	6.449	0.903712	2.4688
5	22.8	415	1025	0.003379	6.028	6.932	0.904169	2.4699
6	41.6	22	121	0.003177	3.091	4.796	1.704748	5.5
7	41.6	58	325	0.003177	4.06	5.784	1.723382	5.6034
8	41.6	92	535	0.003177	4.522	6.282	1.760478	5.8152
9	41.6	149	849	0.003177	5.004	6.744	1.740113	5.698
10	61.1	21	225	0.002992	3.045	5.416	2.371578	10.714
11	61.1	32	358	0.002992	3.466	5.881	2.414797	11.188
12	61.1	46	520	0.002992	3.829	6.254	2.425187	11.304
13	61.1	72	784	0.002992	4.277	6.664	2.387743	10.889
14	91.7	16	520	0.002741	2.773	6.254	3.48124	32.5
15	91.7	22	695	0.002741	3.091	6.544	3.452869	31.591
16	91.7	28	893	0.002741	3.332	6.795	3.462382	31.893
17	91.7	33	1065	0.002741	3.497	6.971	3.474223	32.273

Figure 9.7

Transformed reaction rate data for linear regression (Lynch and Wanke 1991).

Add in the Analysis Tool Pak, then use the ribbon tab **Data>Data Analysis Tool>Regression** to perform a multiple linear regression fit of the transformed data with the linearized model to get the parameters $\ln k_0$, E/R_g, and m.

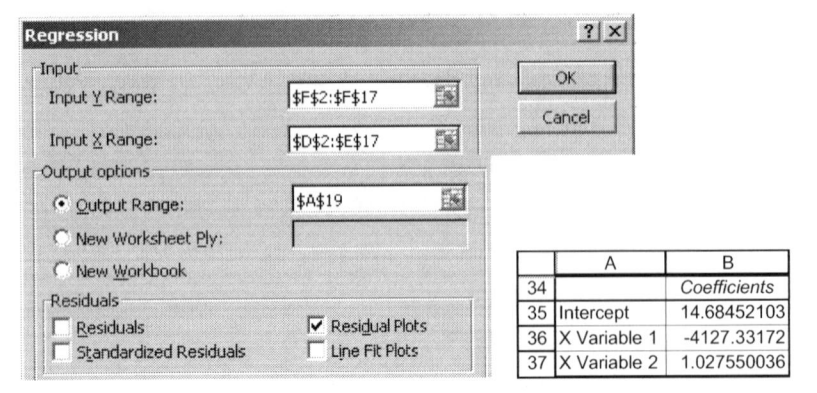

Figure 9.8

Results for the model parameters from multivariable linear regression. The coefficients intercept = $\ln k_0$, x variable 1 = $-E/R$, and x variable 2 = n.

	A	B
34		Coefficients
35	Intercept	14.68452103
36	X Variable 1	-4127.33172
37	X Variable 2	1.027550036

The regression results in Figure 9.8 confirm a first order reaction rate law, $m = 1$. A second linear regression with the order m set to one requires a slight change to the dependent variable by combining the log terms for r_A and C_A:

$$\ln(-r_A) - \ln C_A = \ln\left(\frac{-r_A}{C_A}\right) = \ln(k_0) - \frac{E_a}{R_g}\frac{1}{T} \tag{9.30}$$

Now the y-data become $\ln(-r_A/C_A)$ and the single variable x-data correspond to the inverse temperature $1/T$. Perform a second linear regression directing the output to a new worksheet, as shown in Figure 9.9:

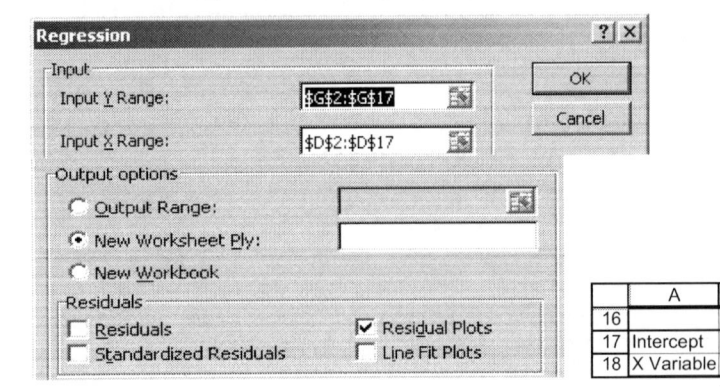

Figure 9.9

Results from linear regression in one variable. The coefficients: Intercept = $\ln k_0$ and x Variable 1 = $-E/R$.

	A	B
16		Coefficients
17	Intercept	14.5590354
18	X Variable	-4050.7014

The value from the intercept for k_0 = EXP(14.5590354) = 2.103×10^6 mol/(s·g cat). The new ratio of the activation energy to the gas constant, E_a/R_g = 4051 K, which gives a better rate law:

$$-r_A = 2.103\times10^6 \exp\left(-\frac{4051}{T}\right)C_A \tag{9.31}$$

Comments: We arrive at the same result in Figure 9.10 using the exponential option of a **Trend line** fit in the y vs x plot of $1/T$ versus $-r_A/C_A$. Excel's **Trend Line** uses variable transformation to linearize non-linear models of logarithmic and exponential functions for regression.

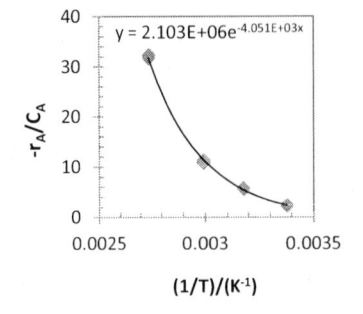

Figure 9.10

Trend line fit of reaction rate to inverse temperature.

□

CHAPTER 9: REGRESSION

The *Excel* worksheet array function **LINEST(y_data, x_data, intercept, statistics)** performs multiple linear regression of the model in Equation (9.24). The arguments of the function have the following definitions:

y_data = range of contiguous cells in a column containing the data for the dependent variable

x_data = range of cells containing the data for the independent variables $x_1, x_2, \ldots$ in contiguous columns ordered left to right.

intercept = *optional* argument. If **TRUE** or omitted, the regression model includes the y intercept, otherwise, **FALSE** sets the intercept to zero.

statistics = *optional* argument. If **TRUE**, **LINEST** reports additional regression statistics. If omitted or **FALSE**, **LINEST** does not report regression statistics to the worksheet.

LINEST is an *Excel* worksheet array function that requires simultaneous *CNTL SHIFT ENTER* key-strokes to return the output to a range of cells. To use **LINEST**, first select a range of cells corresponding to the size of the anticipated results. The dimensions of the **LINEST** output range are based on the number of columns of **x_data** and the value of the intercept argument.

- **LINEST** reports the coefficients and intercept to the first row of the selected range. The number of columns in the output matches the number of parameters in the model.
- When the **statistics** argument is set to **FALSE**, select one row for the output, otherwise, use five rows. The remaining four rows give statistical information about the quality of the fit.
- When the **intercept** argument is set to **FALSE**, the number of columns equals the number of columns in the range of x data columns, otherwise add one column at the right side for the value of the intercept.

The sequence of coefficients is in reverse order of the columns of **x_data**. In reference to Equation (9.24), if the x data are arranged in contiguous columns from left to right: $x_1, x_2, x_3 \ldots$ then the coefficients are ordered right to left in the output range: $\ldots a_3, a_2, a_1, a_0$ (Note that the last term, a_0, is only included when the **intercept** argument is omitted, or set to **TRUE**). Figure 9.11 has a map of the results from the worksheet array function **LINEST**.

a_2	a_1	...	a_0
u_2	u_1	...	u_0
r^2	u_y		
F	DOF		
RSS	SSR		

Figure 9.11

Map of LINEST results. a_i = coefficient of variable x_i. u_i = standard uncertainty in coefficient a_i. r_2 = coefficient of determination (see Section 9.3.2). u_y = standard error in model (see Section 9.3.3). *F* statistic (see Section 9.3.4). *DOF* = degrees of Freedom in *u*. *RSS* = regression sum of squares (see Equation (9.66)). *SSR* = sum of squared residuals (see Equation (9.6)).

Although **LINEST** returns the same results as the **Regression Analysis** tool, it has the significant advantage of automatically updating the results whenever any of the information used by the arguments changes on the worksheet. Automatic updating is convenient for conducting "what-if" analysis of the regression.

Example 9.3   Multiple Linear Regression of Reaction Rate Data

Known/Find: Repeat the problem in *Example 9.2* using **LINEST**.

Analysis: Select three contiguous cells in a row. Type **LINEST** and in the parentheses select the ranges of data for the dependent and independent variables for the worksheet function arguments, separated by commas. Type the key combination *CNTL SHIFT ENTER* to calculate the linear regression coefficients.

19	n	-E/R	ln(k₀)
20	=LINEST(F2:F17,D2:E17)		

19	n	-E/R	ln(k₀)
20	1.02755	-4127.33172	14.684521

Compare the sequence of the **LINEST** results to the coefficients in Equation (9.29), n and $-E/R$, followed by the intercept constant, $\ln(k_0)$.

Comments: *Excel*'s **LINEST** array function gives the same model results as *Example 9.2*.

□

9.1.3 Nonlinear Least-squares Regression

> *"A lot of people ... haven't had very diverse experiences. So they don't have enough dots to connect, and they end up with very linear solutions without a broad perspective on the problem. The broader one's understanding of the human experience, the better design we will have."* – Steve Jobs

Nonlinear least-squares regression requires minimization of the sum of the squared residuals (*SSR*) defined by Equation (9.6) for a function with nonlinear parameters. Methods of nonlinear regression also apply to linear regression. However, due to the iterative methods of solution, nonlinear regression requires initial guesses for the parameter values. We may use any of the multivariable optimization techniques from Chapter 7 to minimize the *SSR* for regression:

- We find *Excel*'s **Solver** particularly convenient and powerful for regression. Unfortunately, the **Solver** does not include uncertainty information about the results of nonlinear regression.
- In Section 9.6, we derive the Gauss-Newton method for implementing linear and nonlinear least-squares regression that includes uncertainty information about the model and parameters.
- In Section 9.7, we extend the Gauss-Newton method into the robust Levenberg-Marquardt method for nonlinear regression problems where convergence is sensitive to the starting point for the model parameters.
- Nonlinear multivariable least-squares regression may require a global search method, such as the **Solver's** multi-start option, Simulated Annealing, Luus-Jaakola, PSO, or Firefly, to find the best-fit parameters that minimize the global *SSR*.

Example 9.4   Nonlinear Reaction Rate Model

Known/Find: Fit the nonlinear reaction rate law to the data in Example 9.2.

Assumptions: Assume a first-order rate law:

$$-r_A = k_0 \exp\left(-\frac{E_a}{R_g T}\right) C_A \tag{9.32}$$

Analysis: Use the results for the parameters k_0 and E_a/R_g from the linearized model as initial guesses for the nonlinear model in the range **F2:F3**. Calculate the model values for the reaction rate in column **D** in terms of the parameters and independent variables, T and C_A. Then, use the worksheet function **SUMXMY2** to calculate the sum of the squared residuals in Cell **F5**:

$$\text{SSR} = \text{SUMXMY2(C2:C17, D2:D17)}$$

Perform the nonlinear least-squares regression with the **Solver** to find the values of the model parameters that minimize the sum of the squared residuals.

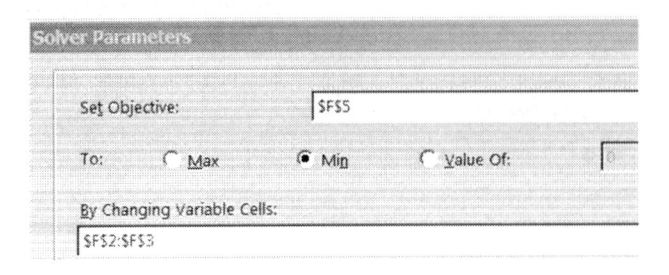

Figure 9.12
Reaction rate data for nonlinear regression of the rate law in Equation (9.32).

CHAPTER 9: REGRESSION   373

	A	B	C	D	E	F
1	T/°C	C_A/(mol/m^3)	-r_A/(mol/s-g cat)	Model	Parameters	
2	22.8	28	63	68.7594	k_0	1.892E+06
3	22.8	142	321	348.7084	E_a/R_g	4012
4	22.8	256	632	628.6574		
5	22.8	415	1025	1019.113	SSR	6189.096
6	41.6	22	121	121.3975		
7	41.6	58	325	320.048		
8	41.6	92	535	507.6624		
9	41.6	149	849	822.1923		
10	61.1	21	225	243.7372		
11	61.1	32	358	371.409		
12	61.1	46	520	533.9005		
13	61.1	72	784	835.6704		
14	91.7	16	520	508.123		
15	91.7	22	695	698.6692		
16	91.7	28	893	889.2153		
17	91.7	33	1065	1048.004		

The nonlinear regression yields a better rate law for the experimental data:

$$-r_A = 1.892 \times 10^6 \exp\left(-\frac{4012}{T}\right) C_A \tag{9.33}$$

Comments: The improved reaction rate model comes from regressing the original data, as opposed to the linear regression of transformed data. We achieve a smaller value for the SSR when we replace the values for E/R and A_0 from the linearized data with the results from nonlinear regression.

□

9.2   Select Good Initial Guesses for Parameters

Iterative methods for performing nonlinear least-squares regression require good initial guesses to achieve convergence in the model parameters at the global minimum *SSR*. There are a few tricks to coax good guesses from the data:

- Judiciously select a subset of data pairs from parts of the complete set that mimic the trends in the data. The size of the subset should equal the number of parameters in the model. Plug each data pair from the subset into the model to give a simultaneous system of equations in the model parameters. The roots become our initial estimates for the model parameters. Use a global minimization method to find the roots.

- As illustrated in Example 9.2 and Example 9.4, try linearizing the model to get good starting values from linear regression of transformed data. Depending on the function, we may use inversion or logarithms as seen in Section 9.3.1, or use first-order Taylor series expansions of the nonlinear part(s) of the model. In cases where we can identify transformed variables that only involve either the independent or dependent variable, we may use linear least-squares regression to get values for the model parameters.

- Use asymptotic analysis of the model to obtain approximations for the parameters from the data at extremes or limiting conditions.

We illustrate these techniques with a simple model consisting of a rational function that is nonlinear in parameter c:

$$y = f(x) = \frac{a + bx}{1 + cx} \tag{9.34}$$

Simultaneous Equations: Apply the model equation to three data pairs (x_1, y_1), (x_2, y_2), and (x_3, y_3):

$$y_1 = \frac{a+bx_1}{1+cx_1} \qquad y_2 = \frac{a+bx_2}{1+cx_2} \qquad y_3 = \frac{a+bx_3}{1+cx_3} \qquad (9.35)$$

Solve the nonlinear system in Equations (9.35) for initial guesses for a, b, and c using any of the nonlinear root-finding methods from Chapter 6, or an optimization method from Chapter 7.

Linearization by Rearrangement: Where possible, rearrange the model by multiplying both sides of Equation (9.34) by the denominator of the right size, then solve for the dependent variable y:

$$y = a + bx - cxy \qquad (9.36)$$

Use good judgement to select three points for x and y. Apply the model Equation (9.36) to the three data pairs (x_1, y_1), (x_2, y_2), and (x_3, y_3) and solve the linear system of equations for initial guesses for the parameters a, b, and c:

$$\begin{vmatrix} 1 & x_1 & x_1 y_1 \\ 1 & x_2 & x_2 y_2 \\ 1 & x_3 & x_3 y_3 \end{vmatrix} \cdot \begin{vmatrix} a \\ b \\ c \end{vmatrix} = \begin{vmatrix} y_1 \\ y_2 \\ y_3 \end{vmatrix} \qquad (9.37)$$

We may also try multivariable linear regression of Equation (9.36) to all of the data.

Asymptotic analysis: Find the limits of the model function at the extreme cases for x:

- For small x $\qquad\qquad\qquad \lim_{x \to 0} f(x) = a = y_0 \qquad (9.38)$

- For large x $\qquad\qquad\qquad \lim_{x \to \infty} f(x) = \frac{b}{c} = y_\infty \qquad (9.39)$

The following examples employ these methods to illustrate how to find good estimates of model parameters to initiate nonlinear least-squares regression.

Example 9.5   Nonlinear Regression of Water Vapor Pressure

Known/Find: Use nonlinear least-squares regression to fit the following Antoine model to the water vapor pressure data listed in the *Excel* worksheet:

$$\ln P_v = A - \frac{B}{T+C} \qquad (9.40)$$

Assumptions: The uncertainty in the data is limited to the vapor pressure.

Analysis: We first attempt to obtain good initial guesses for the model parameters A, B, and C. Inverting the equation does not separate the parameters due to the lack of a common denominator in the two parts on the right-hand-side of Equation (9.40). Instead, we linearize the second term with a truncated Maclaurin series expansion about $C = 0$:

$$\left(\frac{B}{C+T}\right)\bigg|_{c=0} = \frac{B}{T} \quad \text{and} \quad \frac{d}{dC}\left(\frac{B}{C+T}\right)\bigg|_{C=0} = -\frac{B}{T^2} \quad \Rightarrow \quad \frac{B}{C+T} \cong \frac{B}{T} - \frac{BC}{T^2} \qquad (9.41)$$

Substitute the result from Equation (9.41) into Equation (9.40) for the second term to yield the following linearized approximation to the original model:

$$\ln P_v \cong A - \frac{B}{T} + \frac{BC}{T^2} \qquad (9.42)$$

Apply multivariable linear regression to the transformed data to obtain initial guesses for the model parameters. Set up an *Excel* worksheet with transformed data in columns **D**, **E**, and **F**. We converted the temperature into the absolute scale of degree K because we eliminated the sum of terms in the denominators:

CHAPTER 9: REGRESSION

	A	B	C	D	E	F
1	T/°C	P$_v$/mmHg	T/K	1/T	1/T2	ln(P)
2	0	4.579	273.15	0.003661	1.34029E-05	1.521481
3	10	9.209	283.15	0.003532	1.24729E-05	2.220181
4	20	17.54	293.15	0.003411	1.16364E-05	2.864484
5	30	31.82	303.15	0.003299	1.08814E-05	3.460095
6	40	55.33	313.15	0.003193	1.01975E-05	4.013315
7	50	92.51	323.15	0.003095	9.57617E-06	4.527317
8	60	149.4	333.15	0.003002	9.00991E-06	5.006627
9	70	233.7	343.15	0.002914	8.49243E-06	5.454038
10	80	355.1	353.15	0.002832	8.01829E-06	5.872399
11	90	525.8	363.15	0.002754	7.58277E-06	6.264921
12	100	760	373.15	0.00268	7.1818E-06	6.633318

Use the *Excel* worksheet array function **LINEST**, introduced in Section 9.1.2, for multivariable linear regression of Equation (9.42) using **F2:F12** for the **Y Range** and **D2:E12** for the **X Range**. Obtain the approximate values for parameters A, B, and C from the linear regression coefficients according to the formulas in the range **A18:C18**:

	A	B	C
14	BC	-B	A
15	=LINEST(F2:F12,D2:E12)	=LINEST(F2:F12,D2:E12)	=LINEST(F2:F12,D2:E12)
16			
17	C	B	A
18	=A15/B15	=-B15	=C15

	A	B	C
14	BC	-B	A
15	-230716.61	-3748.5325	18.336626
16			
17	C	B	A
18	61.548516	3748.5325	18.336626

Next, use nonlinear least squares regression to refine the solution. Define the *SSR* minimization objective function:

$$SSR = \sum_{i=1}^{n} \left(\ln P_v - A + \frac{B}{T+C} \right)^2 \tag{9.43}$$

Find the model parameters using *Excel*'s **Solver**. To prepare a new worksheet for nonlinear regression, place the initial guesses from the linearized model for the parameters in the range **B1:B3**, as shown in Figure 9.13. Note that we added 273.15 to the estimate for parameter C in order to use temperatures in the Celsius scale. Calculate the model prediction for ln(P) at each data point using the cell references for the initial guesses for the model parameters. Then calculate the square of the differences between the model predictions and transformed data in columns **E** and **F**. Sum the squares of the residuals in cell **F17**.

	A	B	C	D	E	F
1	A	18.3				
2	B	3750				
3	C	335				
4						
5	T/°C	P$_v$/mmHg	ln(P)	A-B/(T+C)	Δln(P)	[Δln(P)]^2
6	0	4.579	=LN(B6)	=$B$1-$B$2/(A6+$B$3)	=C6-D6	=E6^2
7	10	9.209	=LN(B7)	=$B$1-$B$2/(A7+$B$3)	=C7-D7	=E7^2
⋮				⋮		
15	90	525.8	=LN(B15)	=$B$1-$B$2/(A15+$B$3)	=C15-D15	=E15^2
16	100	760	=LN(B16)	=$B$1-$B$2/(A16+$B$3)	=C16-D16	=E16^2
17					Sum Δ^2 -->	=SUM(F6:F16)

Figure 9.13

Excel worksheet showing formulas and initial parameter values for nonlinear regression.

Open the **Solver** dialog box and select the cell with the result for the sum of squared residuals in cell **F17** for **Set Objective** and the range of parameters in **B1:B3** for the **By Changing Variables** cells.

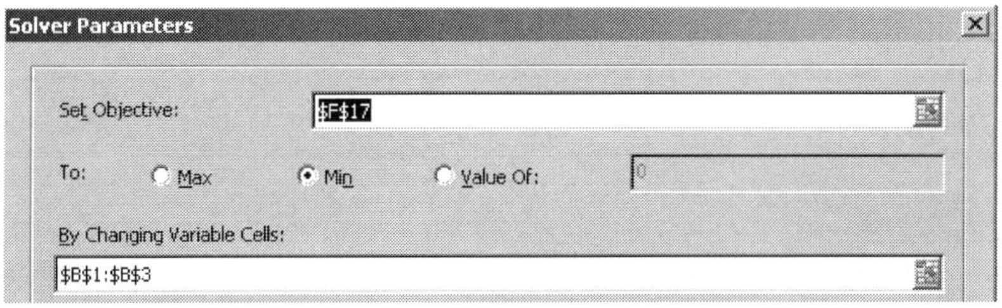

```
Solver Parameters

Set Objective:          $F$17

To:   ○ Max   ● Min   ○ Value Of:  0

By Changing Variable Cells:
$B$1:$B$3
```

If necessary, scale the result for the sum of squared residuals to a value with a magnitude ≈ 1. To scale the sum of squared residuals, simply multiply the value by some number that brings the result up to ≈1. For example, if the sum of squared residuals in Cell **F17** of Figure 9.13 is **1.23E-5**, modify the formula in the cell to **1E6*SUM()**. Just in case we have negative model parameters, uncheck the default option to make unconstrained variables non-negative. The **Solver** finds upgraded parameters for A, B, and C in the range **B1:B3** that minimize the least-squares objective function. We note small changes to the parameters compared to the results from linear regression.

	A	B	C	D	E	F
1	A	18.11059847				
2	B	3728.107842				
3	C	224.5854009				
4						
5	T/°C	P_v/mmHg	ln(P)	A-B/(T+C)	Δln(P)	[Δln(P)]²
6	0	4.579	1.521480634	1.510642165	0.010838469	0.000117472
7	10	9.209	2.220181267	2.218271721	0.001909546	3.64637E-06
8	20	17.54	2.864483987	2.86803768	-0.003553693	1.26287E-05
9	30	31.82	3.460095023	3.46675861	-0.006663587	4.44034E-05
10	40	55.33	4.013315257	4.020222248	-0.006906991	4.77065E-05
11	50	92.51	4.527316747	4.533373209	-0.006056462	3.66807E-05
12	60	149.4	5.006627273	5.010461114	-0.003833841	1.46983E-05
13	70	233.7	5.454038242	5.455158554	-0.001120312	1.2551E-06
14	80	355.1	5.87239944	5.870655813	0.001743627	3.04024E-06
15	90	525.8	6.264920912	6.259737526	0.005183387	2.68675E-05
16	100	760	6.633318433	6.624845165	0.008473268	7.17963E-05
17					Sum Δ² -->	3.80E-04

The nonlinear regression produces the following Antoine function for water vapor pressure:

$$\ln(P_v/mmHg) = 18.111 - \frac{3728.1}{(T/°C) + 224.59} \qquad (9.44)$$

Plot of the model with the data in Figure 9.14 to assess the goodness of the fit.

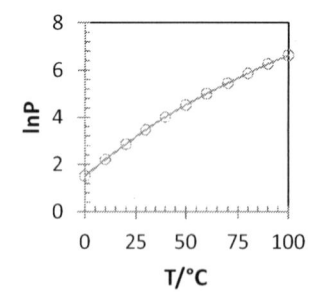

Figure 9.14

Plot comparing the fit of the nonlinear model to vapor pressure data.

Comments: The results for the approximate model parameters from the linear regression agree with the nonlinear results, with the exception of parameter C that was found with a linear expansion about $C = 0$. We may improve the linearized estimate for C by expanding the nonlinear part of Equation (9.40) about a larger value, such as $C = 300$. We can combine the steps in columns **E**, **F** and cell **F17** into one-step using *Excel*'s worksheet function **SUMXMY2(C6:C16, D6:D16)**. We may improve the precision of the nonlinear regression solution by lowering the convergence tolerance in the **Solver Options** for the GRG nonlinear method.

☐

The next example demonstrates alternative methods, including simultaneous equations and linearization by rearrangement, for finding good estimates of model parameters to initialize nonlinear regression.

CHAPTER 9: REGRESSION

Example 9.6 Selecting Initial Guesses for Rational Function Parameters

Known: Figure 9.15 shows an *Excel* worksheet with the molar density of methanol for a range of temperatures in an *Excel* worksheet.

Find: Fit the density ρ versus temperature T data with the following rational function involving four parameters:

$$\rho = \frac{a+bT}{1+cT+dT^2} \tag{9.45}$$

Analysis: Select four points from the data: one at the beginning, two near the middle, and one at the end of the data in T. Use the points in the model Equation (9.45) to generate a system of simultaneous equations in the four parameters, a, b, c, and d:

$$17.5 = \frac{a+b(273)}{1+c(273)+d(273)^2} \qquad 16.3 = \frac{a+b(333)}{1+c(333)+d(333)^2}$$
$$14.3 = \frac{a+b(413)}{1+c(413)+d(413)^2} \qquad 12 = \frac{a+b(473)}{1+c(473)+d(473)^2} \tag{9.46}$$

Use the **Solver** or the macro **NEWTCON** to locate values for a, b, c, and d in the range **B1:B4** that minimize the sum of the squares of the functions in cell **C5**:

	A	B	C
1	a =	22.4412450619747	=17.5-($B$1+$B$2*273)/(1+$B$3*273+$B$4*(273)^2)
2	b =	-0.0395720545808778	=16.3-($B$1+$B$2*333)/(1+$B$3*333+$B$4*(333)^2)
3	c =	-0.000911459562613097	=14.3-($B$1+$B$2*413)/(1+$B$3*413+$B$4*(413)^2)
4	d =	-1.15574183194156E-06	=12-($B$1+$B$2*473)/(1+$B$3*473+$B$4*(473)^2)
5			=SUMSQ(C1:C4)

The result from the simultaneous equations gives initial guesses of $a = 22$, $b = -0.040$, $c = -9.1 \times 10^{-4}$, and $d = -1.2 \times 10^{-6}$.

Equations (9.46) may be recast in linear from for solution by matrix algebra as follows:

$$a+b(273)-17.5\left[c(273)+d(273)^2\right]=17.5 \qquad a+b(333)-16.3\left[c(333)+d(333)^2\right]=16.3$$
$$a+b(413)-14.3\left[c(413)+d(413)^2\right]=14.3 \qquad a+b(473)-12\left[c(473)+d(473)^2\right]=12 \tag{9.47}$$

K1:K4=Mmult(Minverse(E1:H4),I1:I4)   **CTRL SHIFT ENTER**

	E	F	G	H	I	J	K
1	1	273	-4777.5	-1304257.5	17.5	a =	22.44503744
2	1	333	-5427.9	-1807490.7	16.3	b =	-0.039583055
3	1	413	-5905.9	-2439136.7	14.3	c =	-0.000910747
4	1	473	-5676	-2684748	12	d =	-1.15778E-06

Next, try asymptotic values for the initial guesses. For small T, $\rho = a = 17.5$ kmol/m^3. For large T, $b = c = d = 0$. Alternatively, employ linearization of the rational function for linear regression to obtain initial guesses for the parameters. Multiply through Equation (9.45) by the denominator of the right-hand-side:

$$\rho\left(1+cT+dT^2\right)=a+bT \tag{9.48}$$

Rearrange for a linearized form:

$$\rho = a+bT-c\rho T-d\rho T^2 \tag{9.49}$$

Transform the data for multivariable linear regression using the **Regression** tool. The independent variables become T, $-\rho T$, and $-\rho T^2$. Create columns of these combinations in an *Excel* worksheet:

	A	B	C	D
1	ρ/(kmol/m³)	T/K	-ρT	-ρT²
2	17.5	273	-4777.5	-1304257.5
3	17.1	293	-5010.3	-1468017.9
4	16.7	313	-5227.1	-1636082.3
5	16.3	333	-5427.9	-1807490.7
6	15.9	353	-5612.7	-1981283.1
7	15.5	373	-5781.5	-2156499.5
8	15	393	-5895	-2316735
9	14.3	413	-5905.9	-2439136.7
10	13.7	433	-5932.1	-2568599.3
11	12.9	453	-5843.7	-2647196.1
12	12	473	-5676	-2684748

Select **Regression** from the **Data Analysis** tools and specify the ranges of dependent (**Y**) and independent (**X**) variables:

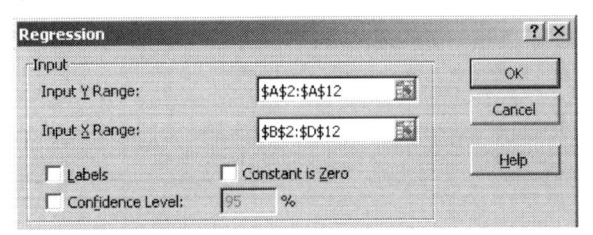

	E	F
16		Coefficients
17	Intercept	22.3455432
18	X Variable 1	-0.0406114
19	X Variable 2	-0.0009545
20	X Variable 3	-1.278E-06

Regression dialog:
- Input Y Range: $A$2:$A$12
- Input X Range: $B$2:$D$12
- Labels, Constant is Zero, Confidence Level: 95%

Multivariable linear regression gives initial guesses for the parameters in the rational function: $a = 22.35$, $b = -0.04061$, $c = 0.0009545$, $d = 1.278 \times 10^{-6}$. Table 9.2 compares the results from these three methods of initializing the nonlinear model parameters. All three methods provide reasonable starting points for implementing nonlinear least-squares regression methods.

Table 9.2 Comparison of the results for model nonlinear parameter initialization using results from three methods.

Initialization Method	a	b	c	d
Simultaneous Equations	22.4	-0.0396	-0.000911	-1.16E-6
Asymptotic Analysis	17.5	0	0	0
Linearization	22.3	-0.0406	-0.000955	-1.28E-6

Figure 9.15 shows an *Excel* worksheet with results from nonlinear regression using the **Solver** with automatic scaling and a convergence setting of 10^{-8}. Columns **A**, **B**, and **C** contain the temperature, density, and model predictions, respectively. The model density calculations in column **C** use Equation (9.45), *e.g.*:

$$C2 = (\$E\$1+\$E\$2*A2)/(1+\$E\$3*A2+\$E\$4*A2^2)$$

The cell ranges **E1:E4** and **E6** contain the initial guesses for the nonlinear model parameters and sum of squared residuals, respectively:

$$E6 = \text{SUMXMY2}(B2:B12, C2:C12)$$

	A	B	C	D	E
1	T/K	ρ/(kmol/m³)	ρ Model	a	22.34554
2	273	17.5	17.4786	b	-0.04061
3	293	17.1	17.1091	c	-0.00095
4	313	16.7	16.7263	d	-1.3E-06
5	333	16.3	16.3254		
6	353	15.9	15.9001	SSR	0.018695
7	373	15.5	15.4419		
8	393	15	14.9389		
9	413	14.3	14.3736		
10	433	13.7	13.7197		
11	453	12.9	12.9351		
12	473	12	11.9482		

Figure 9.15

Excel worksheet for nonlinear least-squares regression of Equation (9.45).

The scaling option in *Excel*'s **Solver** does not work for model parameters smaller than 10^{-3}, such as the values for c and d. We can improve the results for nonlinear regression by manual scaling with the *VBA* macro **SCALIT**, listed in Figure 7.31, for the worksheet and results in Figure 9.16. The macro finds scale factors so that the **Changing** cell values used by the **Solver** have magnitudes of order one.

CHAPTER 9: REGRESSION

	D	E	F	G
1		Model Parameters	Changing Parameters	Scale Factors
2	a	=G2*F2	1	10
3	b	=G3*F3	1	0.01
4	c	=G4*F4	1	0.001
5	d	=G5*F5	1	0.00001
6				
7	SSR	=SUMXMY2(B2:		

	D	E	F	G
1		Model Parameters	Changing Parameters	Scale Factors
2	a	2.16125E+01	1	2.16E+01
3	b	-3.81124E-02	1	-3.81E-02
4	c	-1.08129E-03	1	-1.08E-03
5	d	-8.46659E-07	1	-8.47E-07
6				
7	SSR	0.015518408		

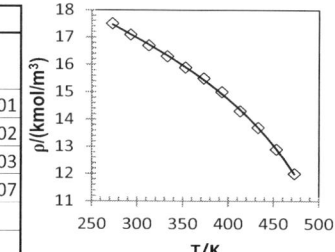

Figure 9.16   Initial and final results for nonlinear regression with a rational model of density using scaled parameters that range over a several orders of magnitude.

The scaled result has a smaller *SSR* with slightly better parameters than those derived without scaling. The results of nonlinear regression give a model density as a function of temperature in degrees K:

$$\frac{\rho}{\left(kmol/m^3\right)} = \frac{21.61 - 0.03811T}{1 - 0.001081T - 8.467 \times 10^{-7} T^2} \tag{9.50}$$

The plot in Figure 9.16 compares the model in Equation (9.50) to the data indicating a good fit over the complete temperature range.

Comments: Linearization with linear least-squares regression provided a good starting point. However, the results for nonlinear regression with manual scaling yield a better fit of the data.

□

9.3   Model Fidelity: Verification, Validation, Assessment, and Uncertainty Analysis

"All theory, dear friend, is gray, but the golden tree of actual life springs ever green." – Goethe

Graphical analysis is the first step in model assessment of fidelity with data. For data with one independent variable, plot the data along with the model for visual inspection. We have two macros, **QYXPLOT** and **GRAPHYXGUIDE** to facilitate graphical analysis. If the model does not appear to match the data, there is no need to go further – we should develop a better model.

To validate a model using a large set of data, randomly divide the data into two subsets of approximately equal size, fit the model to one subset of data, and then compare the model to the complete data set (Good and Hardin 2006). Selecting a random subset of the data is a simple task using *Excel*'s pseudo random number generator:

- Add a column of *random* numbers next to the columns of data. Generate random numbers using worksheet function **RAND(seed)** or **Data Analysis> Random** tool. Use an integer value for the seed to ensure that the values do not change when manipulating the worksheet.

- Sort the two columns according to the values of the random numbers then use the first half of the randomly sorted data in the regression calculations. Alternatively, use the macro **ROWSHUFFLE**, available in the *PNM-Suite* for randomly ordering rows in an *Excel* worksheet.

- Compare the model results to the *complete* set of data.

- Repeat the regression with the complete set of data when satisfied with the model.

We need a sufficiently large data set for this type of validation – a luxury we may not always have. We define *sufficiently large* as large enough for an acceptable level of uncertainty in the regression (see Section 0). We may need to collect more data, or find alternative methods for assessing the fitness of the model.

9.3.1 Linearization for Model Validation

"There's a fine line between fishing and just standing on the shore like an idiot." – Steven Wright

Engineers and scientists prefer to display data in the form of linear relationships. For example, we tend to recognize linear relationships more easily than judging between a quadratic or cubic function. Try the statistician's "trick" of looking along the edge of a sheet of paper at the printed plot, as opposed to viewing directly from above, to check for linearity or curvature in the data, as seen in Figure 9.17. Try this technique with the plot in Figure 9.17 taken from the Gilliland (1940) correlation for the number of distillation trays Y versus reflux ratio X. Looking directly at the data on the chart from a "bird's-eye" view, the correlation appears linear in the region where $X > 0.1$. Rotate the plot, if necessary, to view the data from the bottom right corner to see the full effect of linearity versus nonlinearity in the functional relationship between X and Y.

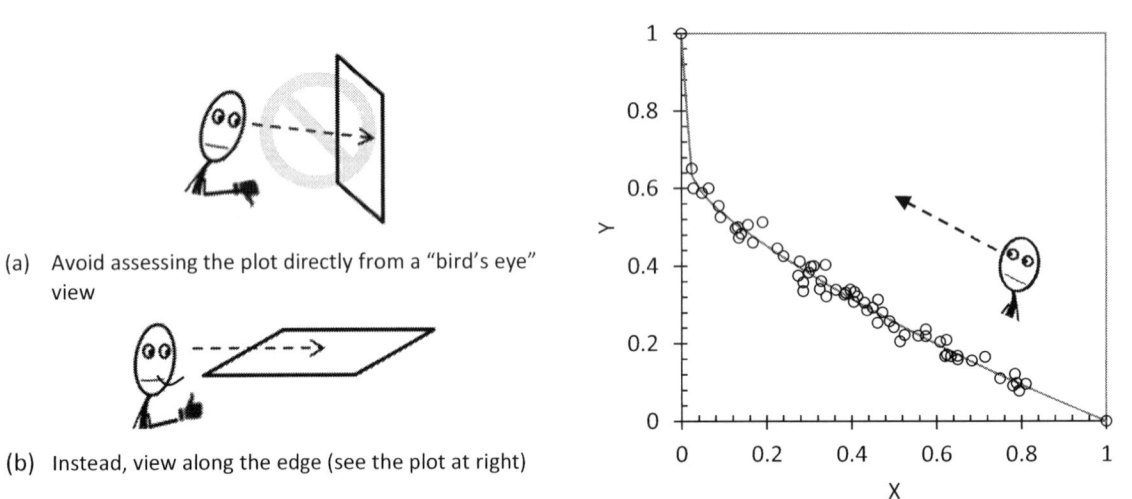

(a) Avoid assessing the plot directly from a "bird's eye" view

(b) Instead, view along the edge (see the plot at right)

Figure 9.17 How to look at a plot to check for linearity. Instead of looking from above (a), hold the plot up to your face at eye level and look along the edge of the paper in the direction of the arrow (b). Try it on the YX Gilliland correlation by holding this page of the book horizontally at eye level to check the curvature.

When preparing graphs, we should look for possibilities of how our plots may be *misinterpreted*. Consider the series of parallel lines in Figure 9.18 with alternating vertical and horizontal short line markers for the data points. The lines in the plot are equally-spaced and parallel. Is that what you see? Try looking at Figure 9.18 again from different edges of the page.

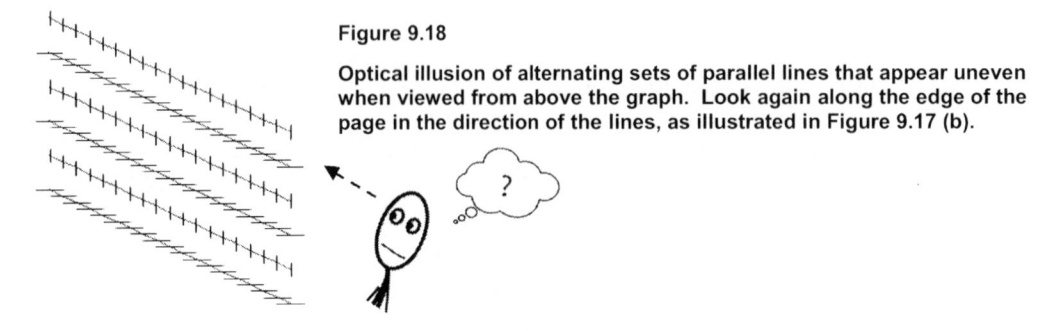

Figure 9.18

Optical illusion of alternating sets of parallel lines that appear uneven when viewed from above the graph. Look again along the edge of the page in the direction of the lines, as illustrated in Figure 9.17 (b).

Use a log plot when the data along either axis ranges over several orders of magnitude. For example, consider the temperature dependence of reaction rate constants. Figure 9.19 is a plot of the rate constant of the dehydrogenation reaction of ethane with oxygen versus inverse temperature (Schmidt 1998). The data have an exponential relationship.

CHAPTER 9: REGRESSION

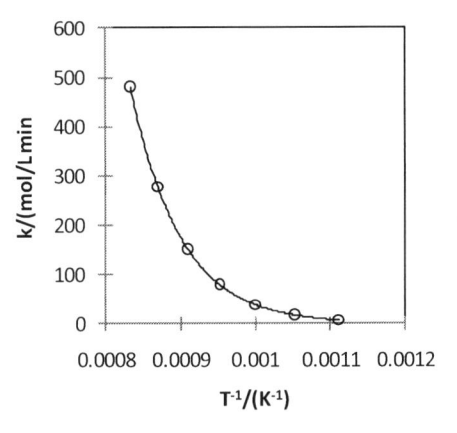

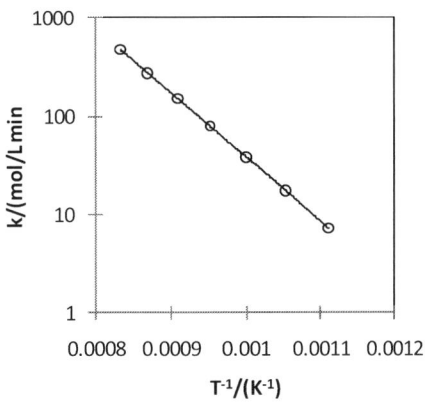

Figure 9.19

Inverse temperature plot of the rate constant for the ethane dehydrogenation reaction.

Semi-log plot of the temperature effect on the rate constant for the dehydrogenation of ethane.

Click on the chart to activate it, and then select **Axes** from the **Layout** tab under **Chart Tools**. Change the axis options for logarithmic scales as shown in Figure 9.20.

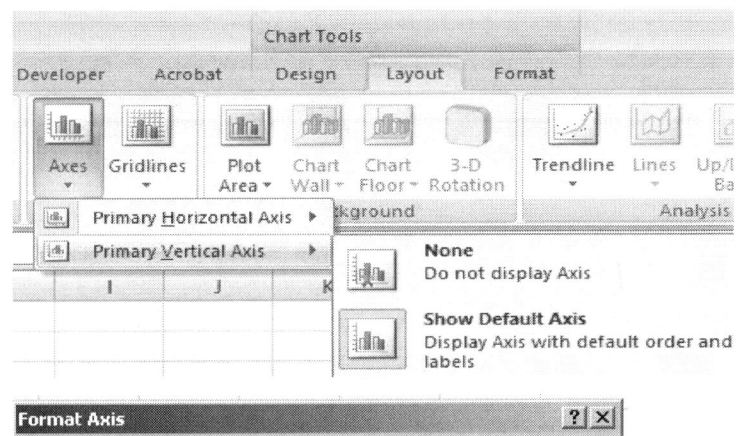

Figure 9.20

Right-click on the axis to modify the format. Check Logarithmic scale for a semi-log plot.

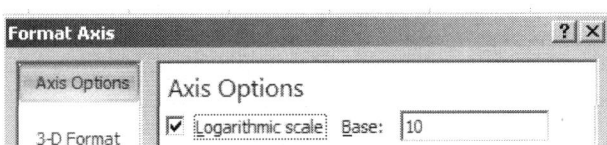

Figure 9.19 also displays the reaction rate constant data in a single cycle of a semi-log graph with inverse temperature on the abscissa and the rate constant on the ordinate. This observation led to the Arrhenius function for the temperature dependence of the reaction rate constant:

$$k = A \exp\left(-\frac{E'}{T}\right) \tag{9.51}$$

where A is the pre-exponential factor, E' is the reduced activation energy, and T is temperature. A straight line on the semi-log plot matches the data for the rate constant.

We recommend log plots when graphing data that ranges over two or more orders of magnitude (Duncan and Reimer 1998). Consider a log-log plot with two-cycles spanning the range 1 to 100 on each axis:

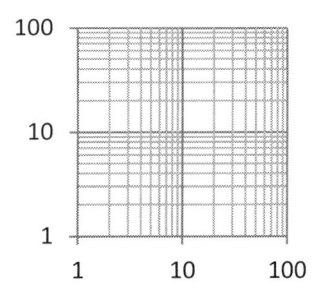

- Log plots have no *y*-intercept in the sense of *y* vs *x* plots because the logarithm of zero is undefined.
- On the *x*-axis, the first gridline to the right of *x* = 1 represents *x* = 2, not 1.1. The grid line just before 10 represents *x* = 9. The grid line just after 10 is not 11 but indicates where *x* = 20. Similarly, the grid line just before 100 indicates where *x* = 90. The same principles apply to the *y*-axis.
- On a log scale, locate the gridline for 10 at the half-way mark between 1 and 100 because the exponent 1 of 10^1 is "half-way" between the exponents of 0 and 2: $10^0 < 10^1 < 10^2$:

$$\log_{10}(10^1) = \frac{\log_{10}(10^0) + \log_{10}(10^2)}{2} = \frac{0+2}{2} = 1$$

- The midpoint between two orders of magnitude is approximately one third of the value of the larger magnitude, or the average of the *log* of each term. For example, the midpoint between 1 and 10 on a log plot is NOT the linear average 11/2 = 5.5, but the midpoint between the logs of 10^0 and 10^1, which is $10^{0.5} \approx 3.2$, in other words, the geometric mean of 1 and 10 defined as $\sqrt{1 \cdot 10}$.

Linear least-squares regression problems consist of equations with only linear terms in the model parameters. We may linearize the parameters in nonlinear functions by transforming the variables and parameters. For example, transform the Arrhenius Equation (9.51) into a linear function of inverse temperature by taking the natural logarithm of each side of the equation:

$$\ln(k) = \ln(A) - \frac{E'}{T} \qquad (9.52)$$

In this form, the transformed dependent variable, ln(*k*), is a linear function of 1/*T*, where the slope of the line is the negative activation energy, *E'*. For the data plotted in Figure 9.21, use **Trend line** to get the slope of -*E'* = 15098 and the intercept of ln(*A*) = 18.757.

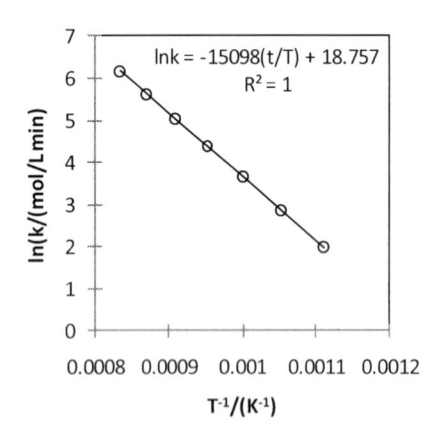

Figure 9.21

Logarithmic linearization of an Arrhenius function.

For another illustration of linearization, consider the Michaelis-Menten rate law for enzyme kinetics:

$$r_P = \frac{kS}{K+S} \qquad (9.53)$$

where *k* and K_M are the rate constant and equilibrium constant, respectively. *S* is the concentration of the reacting substrate. For smooth data, we find the constants from the following observations of the data plotted in Figure 9.22:

$$\lim_{S \to \infty} r_P = k \qquad \text{and} \qquad r_P = k/2 \quad \text{at} \quad S = K \qquad (9.54)$$

CHAPTER 9: REGRESSION

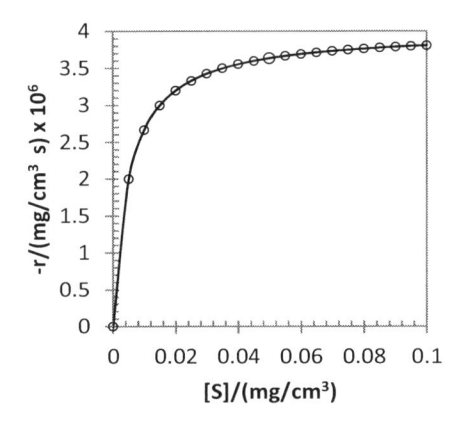

Figure 9.22

Enzyme catalyzed reaction rate data.

Linearize equation (9.53) in terms of $1/S$ by inverting each side of the equality:

$$\frac{1}{r_P} = \frac{K}{k}\frac{1}{S} + \frac{1}{k} \tag{9.55}$$

A Lineweaver-Burk plot of the reciprocal reaction rate versus the reciprocal concentration data in Equation (9.55) has a slope of K/k and an intercept of $1/k$, as shown in Figure 9.23.

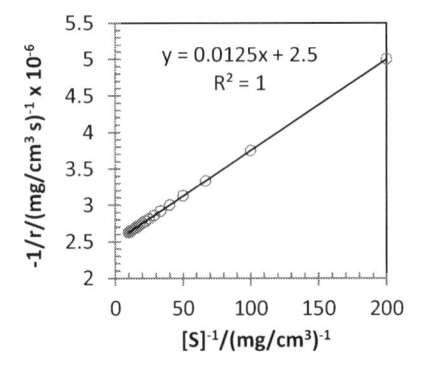

Figure 9.23

Lineweaver-Burk plot of an enzyme catalyzed reaction rate.

The result of the Lineweaver-Burke plot of the data in Figure 9.23 gives $1/k = 250000$ and $K_M/k = 1250$. Note that the data for small S (or large $1/S$) play an overly important role in the linear fit of the data. For this reason, we must follow up with nonlinear regression of the original model and data, using the result from the linearized model as our starting point to initialize the search for the minimum SSR.

The solution of a single linear equation involves straightforward algebraic rearrangement. Table 9.3 has several examples of linear transforms of nonlinear functions. Although these forms help with model validation, we should ultimately avoid data transformation through linearization when obtaining best-fit model parameters from regression. However, we may use linearized regression results for model validation as starting points for a nonlinear regression search for the minimum SSR.

Table 9.3 Examples of common function linearization methods for model validation: y=a₀+a₁x

Nonlinear Function	Linearized Function	Dependent Variable	Independent Variable	a₀	a₁
$y = ax^b$	$\ln(y) = \ln(a) + b\ln(x)$	$\ln(y)$	$\ln(x)$	$\ln(a)$	b
$y = ae^{bx}$	$\ln(y) = \ln(a) + bx$	$\ln(y)$	x	$\ln(a)$	b
$y = a + bx + cx^2$	$\dfrac{y - y_0}{x - x_0} = b + c(x + x_0)$	$\dfrac{y - y_0}{x - x_0}$	x	$b + cx_0$	c
$y = \dfrac{x}{a + bx}$	$\dfrac{1}{y} = \dfrac{a}{x} + b$	$\dfrac{1}{y}$	$\dfrac{1}{x}$	b	a
$y = a + \dfrac{b}{x}$	$y = a + \dfrac{b}{x}$	y	$\dfrac{1}{x}$	a	b
$y = a \cdot b^x$	$\ln(y) = \ln(a) + x\ln(b)$	$\ln(y)$	x	$\ln(a)$	$\ln(b)$
$y = axb^x$	$\ln(y) = \ln(a) + \ln(x) + x\ln(b)$	$\ln(y) - \ln(x)$	x	$\ln(a)$	$\ln(b)$

Example 9.7   Modeling Batch Filtration Data (Shuler and Kargi 2002)

Known: Consider the following tabulated data collected from a study of dead-end filtration of microbe cells with a filter diameter of 20 cm, initial cell concentration of 100 g/L, filtrate viscosity of 1.0 cP, and 1.0 atm pressure drop.

t/min	0	10	20	30	40	50	60	70
V/L	0	2.7	4	5.4	6.8	7.4	7.8	8.3

A flow balance around the volume of filtrate in terms of Darcy's law gives:

$$\frac{dV}{dt} = \frac{A\Delta p}{\mu R} \quad (9.56)$$

where V is the total volume of filtrate, A is the area of the filter, Δp is the pressure drop across the filter, μ is the filtrate viscosity, and R is the combined resistance of the filter and filter cake formed on the filter:

$$R = R_f + R_c \quad (9.57)$$

The cake resistance is a linear function of the volume of filtrate:

$$R_c = \alpha \rho \frac{V}{A} \quad (9.58)$$

where α is the specific resistance of the filter-cake, and ρ is the mass concentration of the solids in the filtered solution.

Find: Determine the values for α and R_f from least-squares regression of the model.

Assumptions: constant properties

Analysis: Separate variables and integrate Equation (9.56) to obtain a quadratic function of t in terms of V:

$$t = \frac{\alpha \rho \mu}{2\Delta p}\left(\frac{V}{A}\right)^2 + \frac{\mu R_f}{\Delta p}\frac{V}{A} \quad (9.59)$$

Linearize Equation (9.59) by dividing each term by V:

$$\frac{t}{V} = \frac{\alpha \rho \mu}{2\Delta p}\frac{V}{A^2} + \frac{\mu R_f}{\Delta p A} \quad (9.60)$$

CHAPTER 9: REGRESSION

A plot of the transformed data in the form of (t/V) versus V should lie about a straight line, as shown in Figure 9.24. Note that we must discard the first point to avoid division by zero. Find the values of the specific cake resistance and the filter resistance using the **Trend Line** least-squares regression on the plot. The results from the linearized form of the equation give:

$$\frac{\alpha\rho\mu}{2\Delta pA^2} = 0.7359 \quad \text{and} \quad \frac{\mu R_f}{\Delta pA} = 1.6892 \quad (9.61)$$

Solve for $R_f = 531$.

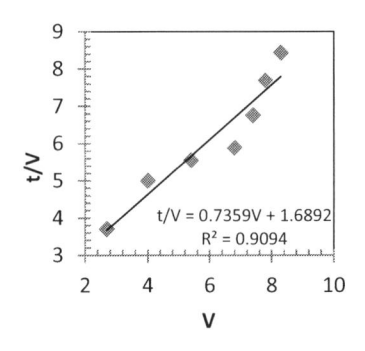

Figure 9.24

Linear least squares fit of transformed filtration data using Trend line.

Excel's charting **Trend Line** feature has the option of fitting the second-order polynomial in Equation (9.59), as shown in Figure 9.25. Force the equation through the origin by selecting zero intercept.

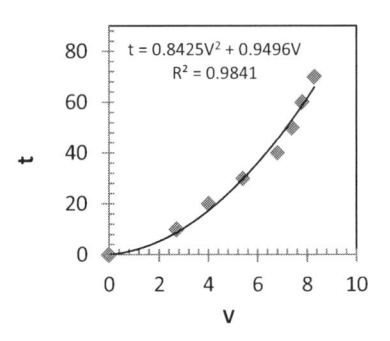

Figure 9.25

Second order polynomial least squares fit of original filtration data using Trend line.

Unfortunately, a **Trend Line** on a chart does not put the coefficients into an *Excel* worksheet for further analysis. We opt for the worksheet function **LINEST** to get the parameter values in a range of cells on the worksheet. Select the range **E2:F2**, type the **LINEST** worksheet function, and then type the keyboard combination *CNTL SHIFT ENTER*, to get the results in Figure 9.26:

E1:F1=LINEST(A2:A9, B2:C9, FALSE)

	A	B	C	D	E	F
1	t	V	V²		0.842532974	0.949631615
2	0	0	0			
3	10	3	7.29		ρ/(g/L) =	100
4	20	4	16		Δp/atm =	1
5	30	5	29.16		A/cm² =	314.1592654
6	40	7	46.24		μ/cP =	1
7	50	7	54.76			
8	60	8	60.84		α =	1663.09343
9	70	8	68.89		R_f =	298.3355705

Figure 9.26

Linear regression of Equation (9.59) using the worksheet function LINEST in the range E1:F1.

The results from the quadratic form of the equation give:

$$\frac{\alpha \rho \mu}{2\Delta p A^2} = 0.8425 \quad \text{and} \quad \frac{\mu R_f}{\Delta p A} = 0.9496 \tag{9.62}$$

Solve for the parameters α and R_f directly in the worksheet:

F8=E1*2*F4*(F5^2)/(F3*F6)   F9=F1*F4*F5/F6

Comments: This example illustrates potential dangers of linearizing functions that involve both dependent and independent data. The linear regression of a second order polynomial (quadratic function) has the advantage of using all of the data points. The loss of the single data point gives dramatically different results for the filtration resistances (compare R_f = 531 to 298)

The next example illustrates how to use linearization in concert with least-squares regression to fit models to appropriate sets of data.

Example 9.8   Growth Phases in Batch Fermentation

Known: Microorganisms growing in a batch fermenter reactor pass through three growth phases: a lag phase where the cells become accustomed to the growth media, followed by an exponential growth phase, then a stationary phase as the growth media becomes depleted. Figure 9.27 shows a plot of batch reactor data for the cell mass density for E-coli. A first-order model for the exponential growth phase accounts for the changes in cell mass X in terms of the specific growth rate, μ:

$$\frac{dX}{dt} = \mu X \tag{9.63}$$

Find: Determine the specific growth rate during the exponential growth phase.

Analysis: Separation of variables followed by integration of Equation (9.63) gives:

$$X = X_0 \exp\left[\mu(t - t_0)\right] \tag{9.64}$$

where X_0 is the cell mass density at time t_0. In the left graph of Figure 9.27 it is difficult to identify where the lag phase ends and exponential growth begins. The location of the stationary phase is obvious. A semi-log plot of the growth data in the graph at the right in Figure 9.27 reveals all three phases.

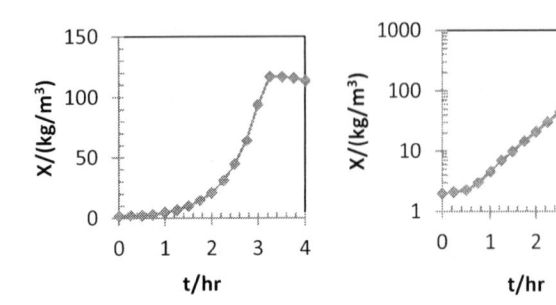

Figure 9.27

Cell mass concentration and semi-log plot showing the period for the exponential growth phase.

Determine the specific growth rate by fitting the data from the exponential growth phase between 1 and 3 hours with Equation (9.64) (try looking at the plot diagonally from the left edge of the paper to identify the time span for the exponential growth phase). The chart **Trend Line** option for an exponential fit is convenient for fitting the subset of data from the exponential growth period. The results shown in Figure 9.28 give a specific growth rate of $\mu \cong 1.5$.

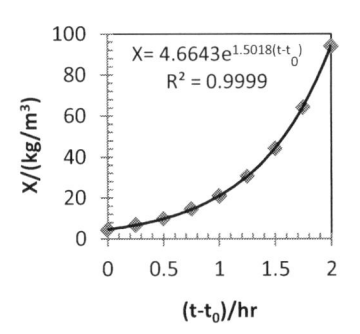

Figure 9.28

Exponential fit of a cell mass time profile during the exponential growth phase in a batch fermenter.

Comments: Although the data in *this* example are smooth, growth data are typically noisy, requiring extra care when performing regression analysis.

□

9.3.2 Coefficient of Determination: R^2 and Adjusted R^2

"The first principle is that you must not fool yourself – and you are the easiest person to fool." – Richard Feynman

The following coefficient of determination is a popular, although limited, measure of the goodness of fit by least-squares regression: [86]

$$R^2 = 1 - \frac{SSR}{RSS} \tag{9.65}$$

where Equation (9.6) defines the *Sum of the Squared Residuals* (*SSR*) used for the least-squares minimization objective function. We define the *Regression Sum of Squares* (*RSS*) in terms of the model predictions for *n* data pairs relative to the average value of the dependent variable:

$$RSS = \sum_{i=1}^{n} \left(\bar{y}_{data} - y_{i,model} \right)^2 \tag{9.66}$$

with

$$\bar{y}_{data} = \frac{1}{n} \sum_{i=1}^{n} y_{i,data}$$

By definition, the coefficient of determination ranges from zero to one, with one being a perfect fit of a model that passes exactly through the data. An R^2 near zero may indicate a poor fit of the data, or no correlation between the dependent and independent variables. R^2 represents the fraction of the variation in the dependent variable explained by the regression curve (Donnelly 2004). In general, linear relationships with $R^2 < 0.9$ are considered a poor fit of the data.

We urge caution when basing decisions about model fidelity on R^2. R^2 may help distinguish between two competing models that use the same number of parameters, but should not discount a model derived from first principles, such as those introduced in Chapter 1. To illustrate, consider a small data set for a linear model shown in Figure 9.29. A higher order polynomial has a fit with $R^2 = 0.9535$ by passing exactly through each endpoint, but does not *interpolate* the data well near the endpoints.

---

[86] The coefficient of determination in Equation (9.65) is the square of the correlation coefficient in Equation (8.39).

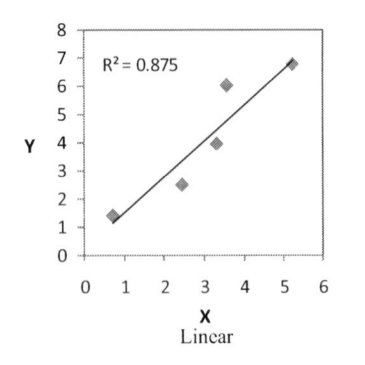

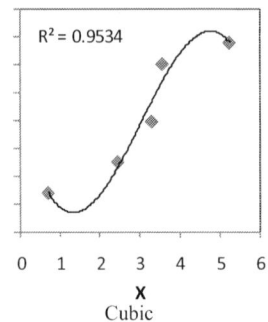

Figure 9.29

Linear and cubic polynomial fits of YX data using Trend line with R^2.

Adding additional parameters to a model increases $R^2 \to 1$, regardless of whether or not the additional terms improve the model predictions. However, we can *adjust* the value of R^2 to account for the effects of additional terms in the regression model (Montgomery 2009):

$$\text{Adjusted } R^2 = 1 - \frac{SSR/(n-k)}{RSS/(n-1)} = 1 - \left(\frac{n-1}{n-k}\right)\left(1 - R^2\right) \tag{9.67}$$

where k is the number of regression parameters and n is the number of data pairs ($k < n$). The adjusted R^2 for the cubic polynomial in Figure 9.29 is 0.818. Note that the adjusted R^2 may actually decrease as the number of regression parameters increases – an indication that the model has unnecessary extra terms that do not enhance its predictive capability. The *adjusted* R^2 increases toward a value of one only when the additional terms justify the improvement in the model fit. See Wisniak and Polishuk (1999) for more examples where R^2 misinforms the validity of the model.

9.3.3 Regression Uncertainty with Parameter Correlation*

Uncertainty analysis in regression refers to the *fit* of the data by the model, not the uncertainty in the data, nor the appropriateness of the model. The following steps for calculating uncertainty in regression models include correlation among model parameters (Billo 2006):

1. Calculate the standard uncertainty in the regression model from the sum of squared residuals (*SSR*) in Equation (9.6):

$$u_r = \sqrt{\frac{SSR}{n-k}} \tag{9.68}$$

where k is the number of regression variables, and $v = n - k$ represents the degrees of freedom.

2. Calculate the systematic (fixed) uncertainties in the regression model. The fixed uncertainty in y comes from the readability of the data for y described in Section 8.1.4:

$$u_z = \pm \frac{2\delta}{\sqrt{12}} = \pm \frac{\delta}{\sqrt{3}} \tag{9.69}$$

3. Incorporate the systematic fixed uncertainty for y into the combined standard uncertainty for y:

$$u_y = \sqrt{u_r^2 + u_z^2} \tag{9.70}$$

4. The expanded uncertainty, or 95% confidence interval, requires the product of the coverage factor, $t_{95\%}$, with the standard uncertainty:

$$U_y = t_{95\%, v_y} u_y \tag{9.71}$$

Get the coverage factor from the worksheet function **TINV(α, ν)**, where α represents the significance level (or 1 - α confidence level) and ν represents the degrees of freedom. Calculate the degrees of freedom from the special case of the Welch-Satterthwaite formula in Equation (8.49) that assumes infinite degrees of freedom for the systematic uncertainty:

$$v \cong (n-k)\left(\frac{u_y}{u_r}\right)^4 \tag{9.72}$$

5. Obtain the covariance matrix from the partial derivatives of the model-predicted values with respect to the model parameters:

$$COV_{j,m} = \sum_{i=1}^{n}\left(\frac{\partial y_{i,\text{model}}}{\partial a_j}\frac{\partial y_{i,\text{model}}}{\partial a_m}\right) \tag{9.73}$$

6. Calculate the standard and expanded uncertainty in the model *parameters* from the diagonal terms of the covariance matrix:

$$u_{aj} = u_r\sqrt{(COV)^{-1}_{jj}} \tag{9.74}$$

The expanded uncertainty in the model *parameters* requires the coverage factor using $v = n - k$ degrees of freedom:

$$U_{a_j} = t_{95\%,v} u_{aj} \tag{9.75}$$

7. Determine R^2 and *adjusted* R^2 according to Equations (9.65) and (9.67). Calculate the linear correlation coefficients among model parameters from the covariance matrix:

$$R_{jm} = \frac{COV_{jm}}{\sqrt{COV_{jj}COV_{mm}}} \tag{9.76}$$

Excel's add-in **Data Analysis/Regression** tool and worksheet function **LINEST** return R^2, *adjusted* R^2, standard uncertainties (errors), and confidence intervals for the model and model parameters derived from linear regression. Unfortunately, **LINEST** does not calculate, nor include correlation coefficients among model parameters. The add-in **REGRESSION** uses a dialog box to prompt the user for the information required by **LINEST**, and formats the output with row and column headings, as well as optional residual plots. *Excel* has no built-in options for nonlinear regression uncertainty analysis. We develop custom *VBA* code for nonlinear regression uncertainty analysis in Sections 0, 9.6, and 0.

We can eliminate correlation among linear model parameters by centering the data on the average value of the independent variable (de Levie 2004). To center the data, change the origin through variable transformation. For example, we can center data for fitting a second-order polynomial:

$$y = a + b(x - \bar{x}) + c(x - \bar{x})^2 \tag{9.77}$$

where $\bar{x} = (\sum x)/n$ is the average value of x such that $\sum(x - \bar{x}) = 0$ in Equation (8.39) for $R = 0$ in Equation (9.76). Example 9.14 illustrates how data centering removes correlation among model parameters.

9.3.4 F-test, P-Value and t-Stat for Hypothesis Testing*

We use the statistical results of multiple linear regression to identify the most significant terms in a model, and conversely, any insignificant terms to consider for elimination from the model.[87] The *Excel* **add-in** tool **REGRESSION**, available from the **Data>Analysis** group, gives *t*-statistics and *P*-values for a hypothesis test of the level of significance of each parameter. Start by checking for the value of *Significance F* < α, where the significance level $\alpha = 0.05$ for

---

[87] This section assumes we have experience with inferential statistics and hypothesis testing. Recall that we may need to add the **Data Analysis Toolpak** to the *Excel* workbook.

95% confidence. If otherwise, look to the *P-values* and eliminate the independent variable with the largest *P*-value > α from the regression model. Rerun the regression and check the *Significance F* again.

Our null hypothesis assumes that an average of values of the dependent variable best represents the data (*i.e.*, the model parameters in Equation (9.24) are $a_0 \neq 0$, $a_1 = 0$, $a_2 = 0$ ...). We reject the null hypothesis for *P*-values below a predetermined level of significance. The alternative to the null hypothesis states that the variables in the model influence the dependent data. We may consider eliminating a term from the linear model that involves a variable with a *P*-value greater than the significance level. It is important to check for only one significant term at a time. We should always redo the regression and validate the new model without the latest rejected term.

We can also use the critical value of the *t*-statistic to test our null hypothesis. The **t-Stat** for model parameter *i* is the ratio of the linear coefficient (model parameter) to its standard error:

$$t\,Stat_i = \frac{a_i}{u_i} \qquad (9.78)$$

Compare the result for **t-Stat** with the critical value calculated from the worksheet function **TINV(0.05, v)** for 95% confidence (or significance level of 0.05), where *v* is the degrees of freedom, or $n - k$ (number of data pairs – number of model parameters). We reject the null hypothesis when |*t-Stat*| > *t-Critical*. A typical value for *t-Critical* is 2 when the degree of freedom exceeds 30. Thus, we *retain* a model parameter when the **t Stat** is larger than 2.

9.3.5 Residual Plots for Model Verification

> *"A theory may be so rich in descriptive possibilities that it can be made to fit any data."* – Phillip Johnson-Laird

Residual plots provide visual confirmation whether-or-not a model captures the true behavior of the phenomena represented by the data, and helps verify the assumption of normally distributed errors. Residuals are the differences between the data and the model predictions in Equation (9.6):

$$R_i = y_i - y_{i,\text{model}} \qquad (9.79)$$

A raw residual plot graphs the residuals on the ordinate axis versus one of the following corresponding values on the abscissa: the numbered position in the data series or the independent variable.

Figure 9.30 shows a raw residual plot for the exponential growth model in Example 9.8. To generate a residual plot in *Excel*, select the range of residuals on the worksheet and insert a scatter plot. Even quicker, use **Sparklines** directly in a cell on an *Excel* worksheet. To create a **Sparkline** plot, select the type from the ribbon tab **Insert>Sparklines** group. Once we have a residual plot, we look for the following features to verify a model:

- Random scatter in the residuals about the zero axis, with approximately 50% of the residuals above, and 50% below the axis.
- A lack of a strong or discernible pattern in the plotted residuals.

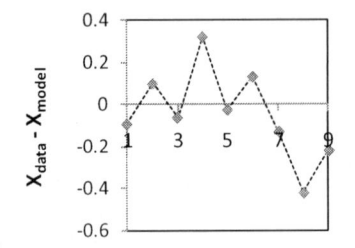

Figure 9.30

Raw residual plot for the exponential growth model in Example 9.8 shows random scatter without a discernable pattern.

Sparkline plot of residuals.

The residuals in Figure 9.30 appear to be randomly scattered about the horizontal axis, with no obvious trend, suggesting that the model correctly captures the behavior of the data and the errors are normally distributed. Contrast Figure 9.30 with the residual plot in Figure 9.31 for the filtration model in Figure 9.7. The residuals from the filter model appear correlated with the independent variable, a warning that the model may not correctly represent the behavior of the filter, or may need additional terms.

CHAPTER 9: REGRESSION

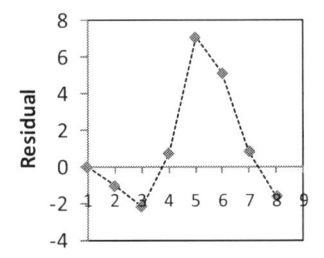

Figure 9.31

Raw residual plot for the filtration model in *Example 9.7* shows a clear pattern, or trend indicating that the model may not properly capture the behavior of the filter.

We also observe the lack of a normal distribution in the random error in the data in Figure 9.31. When the pattern in the residual plot has a clear trend, we might suspect other phenomena unaccounted for in the model, such as a change in the filter resistance over time. Observable patterns in the residual plot may also suggest correlation among different model parameters. Figure 9.32 is the raw residual plot for the reaction kinetic model in Example 9.2. The pattern of residuals has no clear, obvious trend supporting the validity of the kinetic model.

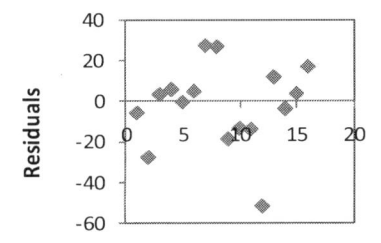

Figure 9.32

Raw residual plot for a kinetic model showing a normal random distribution about the abscissa.

Three secondary forms of the residual plot may also aid in confirming or denying the adequacy of a model (du Plessis 2007). In one form, we generate a parity plot of the observed values for the dependent variable against the predicted values. For a good model, we should see a plot similar to Figure 9.33 with the values randomly scattered about a 45° diagonal line, and with roughly the same number of points above the line as below the line.

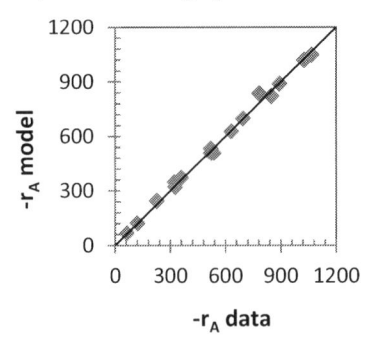

Figure 9.33

Parity plot of experimental versus model results for reaction rate by nonlinear regression.

Another type of residual plot for model assessment uses the residuals sorted from low to high values. A good model of the data has a cumulative normal distribution of residuals. Figure 9.34 shows an example of a cumulative normal distribution of 1000 sorted points with mean value of zero and a standard deviation of one, generated with the nested *Excel* worksheet functions **NORMSINV(RAND())**.

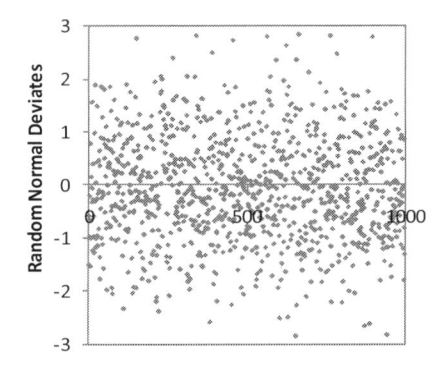

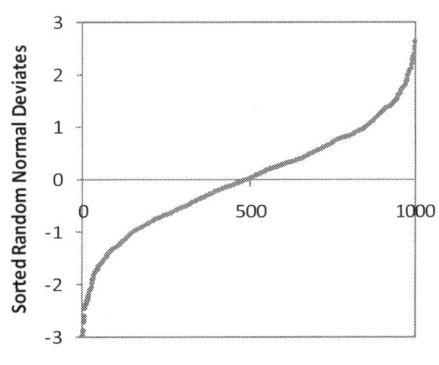

Figure 9.34

Random normal deviates and cumulative normal distribution for 1000 points with zero mean and standard deviation of one.

Raw residual plots show random scatter about the horizontal axis with the residuals concentrated around the zero axis. Sorted residual plots with normally distributed errors show a cumulative normal distribution sigmoid shape. A sorted residual plot provides additional evidence for normally distributed errors about zero, that the model fits the data, and the absence of significant correlation among model parameters. However, cumulative normal distribution plots generally require larger data sets than raw residual plots to judge the goodness of the model fit. Figure 9.35 shows an example of the sigmoid curve in the sorted residual plot for Example 9.2.

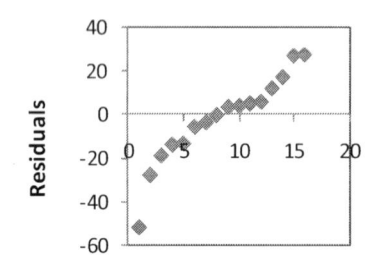

Figure 9.35

Sorted residual plot for nonlinear regression fit of kinetic data showing a sigmoid cumulative normal distribution.

A fourth type of residual plot displays the normalized sorted residuals. To create a normal plot of the sorted residuals, first calculate a cumulative probability for the abscissa from the rank (i) of the sorted residuals using Filliben's (1975) formula:[88]

$$p_i = \begin{cases} 1 - 0.5^{1/n} & i = 1 \\ \dfrac{i - 0.3175}{n + 0.365} & i = 2, 3 \ldots n-1 \\ 0.5^{1/n} & i = n \end{cases} \quad (9.80)$$

Calculate the z-score from the worksheet function z_i =**NORMSINV**(p_i) for the inverse of the normal distribution. The *PNMSuite* macro **ZPLOT**, listed in Figure 9.37, generates a normalized sorted residual plot with a linear trend line and R^2. Normally distributed residuals fall on a line in a plot versus the z-score, as shown in Figure 9.36. **ZPLOT** prompts the user for the range of residuals on the *Excel* worksheet.

Whereas residuals rarely fall precisely on the line, we look for S, U, or C-shaped trends about the line (or rotated forms of these trends) that indicate non-normally distributed residuals. For small data sets ($n < 100$) where residual plots do not provide a complete picture, Filliben proposed a test for residual normality based on the linear correlation coefficient for the sorted residuals with z-scores. Table 9.4 contains a list of critical correlation coefficients recommended by Filliben. According to Filliben's criterion, sorted residuals are normal when $R^2 > R_c^2$. Use Equation (9.81) to interpolate the critical correlation coefficients. The macro **ZPLOT** displays Filliben's critical R^2 in the status bar below the worksheet.

Table 9.4 Critical coefficients of determination for normality test at 95% confidence.

n	10	20	30	40	50	60	70	80	90	100
R_c^2	0.840	0.904	0.929	0.943	0.952	0.958	0.963	0.967	0.971	0.973

$$R_c^2 = \left(\frac{n}{0.7 + 1.036 n^{0.9937}} \right)^2 \quad (9.81)$$

---

[88] Blom's (1958) formula in Equation (8.70) provides an alternative cumulative probability series.

CHAPTER 9: REGRESSION

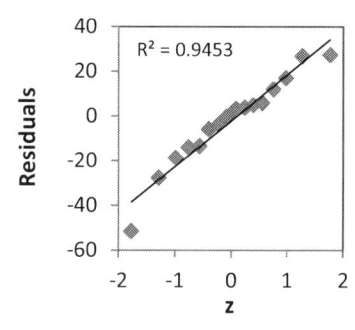

Figure 9.36

Normal sorted residual plot for a nonlinear regression fit of kinetic data showing the linear pattern for a normal distribution. The correlation coefficient R^2 = 0.945 > Critical 0.904 for n = 20.

Be careful when using Filliben's criterion; symmetric normalized residual plots may fool the comparison criterion. Our best tool for model validation besides the plot of the function with the data is a visual inspection of residual plots.

```
Public Sub ZPLOT()
' Create an xy-scatter plot of the sorted residuals versus z-scores created from a
' cumulative probability distribution. The macro uses an input boxe to get the range
' of residuals.  The residuals are sorted using the macro HEAPSORT.  The plot includes
' a linear trendline forced through the origin and the coefficient of determination.
' Filliben's critial R2 is displayed in the statusbar below the worksheet.
' *** Required sub procedure ***
' HEAPSORT = sort the residuals
' *** Nomenclature ***
Dim a As Long ' = new plot
Dim bxtl As String ' = input box title
Dim gx(1 To 2) As Double, gy(1 To 2) As Double ' = cross grid arrays
Dim i As Long ' = index variables
Dim n As Long ' = number of residuals
Dim p As Double ' = uniform cumulative probability
Dim r2 As Double, r2c As Double ' = Filliben's critical coefficient of determination
Dim rr As Range ' = range of residuals
Dim s As Series ' any existing series in plot for removal
Const small As Double = 0.000000000001 ' = small number avoids division by zero
Dim zscore() As Double, rsort() As Double ' = array of x and y data for plot
Dim xychart As Chart ' = chart object for chart (plot)
Dim xyseries As Series ' = series object for xy data
'***************************************************************************
bxtl = "ZPLOT: Sorted Residual vs Z Score"
With Application
Set rr = .InputBox(Title:=bxtl, Type:=8, prompt:="Range of RESIDUALS:")
n = rr.Count ' number of residuals
ReDim res(1 To n, 1 To 1) As Double, zscore(1 To n) As Double, rsort(1 To n) As Double
For i = 1 To n ' calculate the z scores for cumulative probabilities
    Select Case i ' Filliben's cumulative probabilities
    Case 1
        p = 1 - 0.5 ^ (1 / n)
    Case n
        p = 0.5 ^ (1 / n)
    Case Else
        p = (i - 0.3175) / (n + 0.365)
    End Select ' Case
    zscore(i) = .NormSInv(p): res(i, 1) = rr(i) ' create residual array for sorting
Next i
Call HEAPSORT(res) ' sort the residual data
For i = 1 To n: rsort(i) = res(i, 1): Next i
' Create an xy scatter plot on the active worksheet using the sorted residuals and z's
Set xychart = ActiveSheet.Shapes.AddChart.Chart
With xychart
    For Each s In .SeriesCollection: s.Delete: Next s ' clear any series in chart
    .ChartType = xlXYScatter: .Legend.Delete
    Set xyseries = .SeriesCollection.NewSeries
    With xyseries ' fit the data with a linear trendline forced through origin
        .Values = rsort: .XValues = zscore: .Trendlines.Add
        With .Trendlines(1)
            .Intercept = 0: .DisplayRSquared = True
```

```
            End With ' Trendlines(1)
         End With ' xyseries
         With .Axes(xlCategory) ' rescale the x axis & add gridlines (optional)
             .HasMajorGridlines = False: .HasMinorGridlines = False
             .MinorTickMark = xlInside ' put minor tic marks on inside
             .HasTitle = True: .AxisTitle.Caption = "Z-Score"
             .Crosses = xlMinimum ' put the abscissa at the bottom of the plot
         End With ' .Axes
         With .Axes(xlValue) ' rescale y axis and add grid lines (optional)
             .HasMajorGridlines = False: .HasMinorGridlines = False
             .MinorTickMark = xlInside ' put minor tic marks on inside
             .HasTitle = True: .AxisTitle.Caption = "Sorted Residuals"
             .Crosses = xlMinimum ' put the abscissa at the bottom of the plot
         End With ' .Axes
         .PlotArea.Select ' add box around plot area
         With SELECTION.Format.Line
             .Visible = msoTrue: .ForeColor.ObjectThemeColor = msoThemeColorText1
         End With ' Selection.Format.Line
         .ChartArea.Format.Line.Visible = msoFalse ' remove box from chart area and resize
      End With ' xychart
      If n <= 100 Then ' apply Filliben's check for normally distributed residuals
          r2c = (n / (0.7 + 1.036 * (n ^ 0.9937))) ^ 2 ' Filliben's critical R2
          r2 = .RSq(rsort, zscore) ' coefficient of determination
          If r2c - r2 > 0.005 Then ' check for normallcy
              MsgBox "CAUTION: (R2 = " & VBA.Format(r2, "0.00") & ") < (Critical R2 = " _
              & VBA.Format(r2c, "0.00") & ") at 95% Confidence Level. Residuals are NOT normal."
          End If ' Filliben criterion
      End If ' Filliben check
   End With ' Application
End Sub ' ZPLOT
```

Figure 9.37 *VBA sub procedure ZPLOT for generating normalized z-scores plotted versus sorted residuals to check for normally distributed regression residuals.*

9.3.6 Residual Outliers*

An analysis of residuals standardized by the square root of the variance (standard deviation) may suggest potential outliers in the data (Montgomery, Runger and Hubele 1998). A standardized residual divides the raw residual by the standard deviation of the regression residuals:

$$\text{Standardized Residual}_i = \frac{R_i}{s} \qquad (9.82)$$

where s is the standard deviation of residuals. *Excel's* **REGRESSION** analysis tool has the option of listing the standard residuals in the output. For example, according to Figure 8.3, approximately 95% of normally distributed residuals fall within the range of ±2 standard uncertainties calculated using Equation (9.68), or ~99% of normal residuals are within ±3 standard uncertainties:

$$P[-2u_r \leq R_i \leq 2u_r] = 0.95 \qquad (9.83)$$

An analysis of residuals requires sufficient data pairs for statistical significance. However, residuals that fall far outside the interval ±$2u_r$ suggest outliers in the data. A macro named **OUTLIER** and two user-defined functions **CHAUVENET(R)** and **GRUBBS(R)** are available in the *PNMSuite* to test for a single outlier in a range of residuals, **R**, assuming a normal distribution (Taylor 1982) (NIST 2012). The macro **OUTLIER** uses the Tukey (1977) method to color red any cells corresponding to potential outliers. The user-defined functions **CHAUVENET** and **GRUBBS** return the value of the greatest outlier, or "**No Outlier**" if the data fails the statistical test for potential outliers. We urge caution when eliminating outliers that may indicate other problems with the experiment, data, or model. Always repeat the regression analysis after removing an outlier, one at a time.

A plot of standardized residuals like Figure 9.38 may also indicate potential outliers in the data when their corresponding markers fall above or below ±2 (standard deviations).

CHAPTER 9: REGRESSION

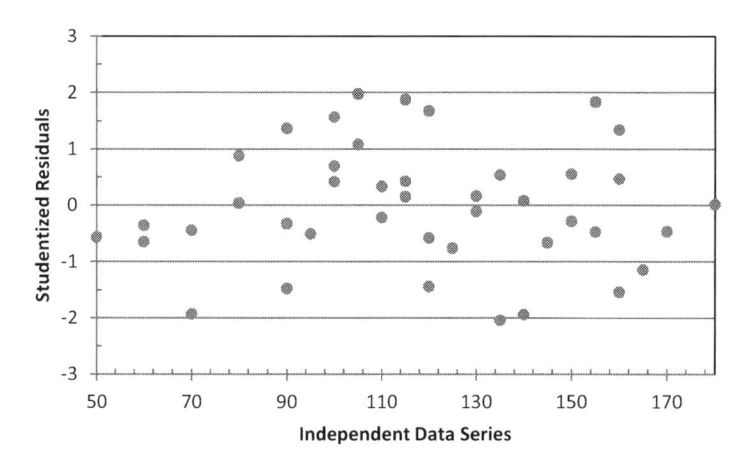

Figure 9.38

"Studentized" residual plot with horizontal gridlines at ±2 and ±3 to indicate potential outliers with 95% or 99% confidence. The data show no strong outliers.

A "studentized" residual does not assume constant variance in the dependent data. A *VBA* sub procedure named **STUDENTIZER**, listed in Figure 9.39, is available in the *PNMSuite* for generating *studentized* residuals according to the following modification to Equation (9.82):

$$t_i^R = \frac{R_i}{s_i \sqrt{1 - \frac{1}{n} - \frac{(x_i - \overline{x})^2}{\sum_j (x_j - \overline{x})^2}}} \quad (9.84)$$

where s_i is the external standard deviation of residuals calculated by removing the potential outlying local residual, as follows:

$$s_i^2 = \frac{\sum_{j \neq i} R_j^2}{n - m - 1} \quad (9.85)$$

and where *n* and *m* are the number of regression residuals and model parameters, respectively. The macro prompts the user for the ranges of the independent variable and regression residuals, the number of regression model parameters, and the location of the output. The macro then plots the studentized residuals in a scatter plot for analysis of model fidelity and any potential outliers in the data relative to the model.

We apply the same analysis used with raw residual plots to studentized residual plots to validate a model, but with the added feature of identifying potential outliers. For a significance level of 5% (95% confidence), one out of 20 standardized/studentized residuals are expected to occur outside of the range of ±1.96. Likewise, we would expect around one out of 100 standardized/studentized residuals should occur outside of the range ±2.58. The macro **STUDENTIZER** is programmed to check for studentized residuals outside of the range ±2 and colors the cells red that contain potential outliers.

```
Public Sub STUDENTIZER()
' Macro to calculate the external "studentized" residuals in a selected column.
' To use STUDENTIZER, create a column of residuals from a regression model.  The macro
' prompts for the ranges of the independent variable, residuals, number of model
' parameters and the location for the column of studentized residuals. The macro plots
' the studentized residuals in a scatter plot.
' *** Required Sub Procedure ***
' AXISCALE = get parameters for axis scales in the graph
' JUSTABC_123 = check name of worksheet for compatability with VBA naming
' *** Nomenclature ***
Dim bxmax As Double, bxmin As Double ' = max and min bounds on x-axis scale in graph
Dim bxtl As String ' = input box titles
Dim i As Long ' = index variable for loop
Dim m As Long ' = number of model parameters
Dim n As Long, n1 As Double, nm1 As Long ' = size of residual vector
Dim o As Range, output As Range, outrange As String ' = output cells and ranges
```

```vba
    Dim ow As Integer ' =  check for overwriting existing worksheet content
    Dim r As Range, r2 As Double ' = range of residuals in a column and sum squared r
    Dim s As Series ' = index for series loop
    Dim uxmaj As Double, uxmin As Double ' = major & minor units for x-axis scale in graph
    Dim v As Double ' = variance of residuals
    Dim x As Range ' = range corresponding independent variables in a column
    Dim xa As Double ' = average value of independent variables
    Dim xma() As Double, xma2 As Double ' = x-xa and sum of squared x-xa
    Dim xmax As Double, xmin As Double ' = max and min values of x-data
    Dim xychart As Chart ' = chart of the residual plot
    Dim xyseries As Series ' = series of residual versus indepndet variable data
    Dim ws As String, ws_temp As String ' = name of the output worksheet
    Dim z() As Double ' = studentized residuals
    '****************************************************************************
    bxtl = "STUDENTIZER: Studentized Residuals": On Error Resume Next ' check for input errors
    With Application
    Set x = .InputBox(prompt:="Range of INDEPENDENT Variables, X:", Title:=bxtl, Type:=8)
        If Err.Number = 424 Then Exit Sub
    Set r = .InputBox(prompt:="Range of Regression RESIDUALS, R:", Title:=bxtl, Type:=8)
        If Err.Number = 424 Then Exit Sub
    m = .InputBox(prompt:="NUMBER of Model Parameters:", Title:=bxtl, Type:=1, Default:=2)
    Set o = .InputBox(prompt:="Output cell:", Title:=bxtl, Type:=8)
        If Err.Number = 424 Then Exit Sub
    n = x.Count ' number of elements of the ranges for r and x
    If n <> r.Count Then ' check for consisent sized for input ranges
        MsgBox "Sizes of residual (r) and independent variable (X) ranges must match": End
    End If
    Set output = Range(o.Address) ' check for overwriting existing content on worksheet
    ws_temp = output.Worksheet.Name: ws = JUSTABC_123(ws_temp): output.Worksheet.Name = ws
    outrange = ws & "!" & output.Address & ":" & ws & "!" & output.Offset(n, 0).Address
    If Not (.CountA(Range(outrange)) = 0) Then
    ow = MsgBox(Title:=bxtl, _
        Buttons:=vbOKCancel + vbExclamation + vbDefaultButton2, _
        prompt:="Output will OVERWRITE " & n + 1 & " rows x 1 column cell range.")
        If ow = 2 Then Exit Sub
    End If
    Range(outrange).Clear ' clear output range contents and formatting
    ReDim xma(1 To n) As Double, z(1 To n) As Double
    xa = .Average(x) ' average x
    For i = 1 To n: xma(i) = (x(i) - xa): Next i ' array of (x - xa)
    r2 = .SumSq(r): xma2 = .SumSq(xma): n1 = 1 - 1 / n: nm1 = n - m - 1
    End With ' Application
    output = "tR" ' label for column of studentized residuals
    output.Characters(Start:=2, LENGTH:=1).Font.Subscript = True
    For i = 1 To n ' studentize the residuals
        v = (r2 - r(i) ^ 2) / nm1 ' external variance of residuals
        z(i) = r(i) / Sqr(v * (n1 - xma(i) ^ 2 / xma2)) ' studentized residuals
        output.Offset(i, 0) = z(i) ' write to worksheet
        If Abs(z(i)) >= 2 Then output.Offset(i, 0).Interior.COLORINDEX = 22 ' Red (3)
    Next i
    xmax = Application.Min(x): xmin = Application.Max(x)
    Call AXISCALE(bxmax, bxmin, xmax, xmin, uxmaj, uxmin) ' get x-axis scale parameters
    ' Create an xy scatter plot on the active worksheet using the studentized residuals
    Set xychart = ActiveSheet.Shapes.AddChart.Chart
    With xychart
        For Each s In .SeriesCollection: s.Delete: Next s ' clear any series in chart
        .ChartType = xlXYScatter: .Legend.Delete
        Set xyseries = .SeriesCollection.NewSeries
        With xyseries ' assign the residiual and independent data to the series
            .Values = z: .XValues = x
        End With ' xyseries
        With .Axes(xlCategory) ' rescale the x axis & add vertical gridlines (optional)
            .HasMajorGridlines = False: .HasMinorGridlines = False
            .MinorTickMark = xlInside ' put minor tic marks on inside
            .MinimumScale = bxmin: .MaximumScale = bxmax
            .MajorUnit = uxmaj: .MinorUnit = uxmin
            .HasTitle = True: .AxisTitle.Caption = "Independent Data Series, X"
        End With ' .Axes
        With .Axes(xlValue) ' rescale y axis and add horizontal grid lines (optional)
```

CHAPTER 9: REGRESSION

```
            .HasMajorGridlines = True: .HasMinorGridlines = False
            .MinorTickMark = xlInside ' put minor tic marks on inside
            .HasTitle = True: .AxisTitle.Caption = "Studentized Residuals, t"
            .MajorUnit = 1: .MinorUnit = 0.5
            .Crosses = xlMinimum ' put the abscissa at the bottom of the plot
        End With ' .Axes
        .PlotArea.Select ' add box around plot area
        With SELECTION.Format.Line
            .Visible = msoTrue: .ForeColor.ObjectThemeColor = msoThemeColorText1
        End With ' Selection.Format.Line
        With .FullSeriesCollection(1)
            .MarkerStyle = 8: .MarkerSize = 5
        End With ' FullSeriesCollection(1)
        .ChartArea.Format.Line.Visible = msoFalse ' remove box from chart area
    End With ' xychart
End SUB ' STUDENTIZER
```

Figure 9.39 *VBA* macro STUDENTIZER for calculating studentized regression residuals and generating a yx scatter plot for identifying potential outliers.

Example 9.9   Assessment of a Model of Heat Capacity Data

Known/Find: Fit the data for heat capacity listed in columns **A** and **D** of the *Excel* worksheet in Figure 9.40 with each of the following multi-parameter models. Assess the models by comparing their R^2, adjusted R^2, residual plots, and check for any potential outliers.

$$c_p = a + bT + cT^2 \tag{9.86}$$

$$c_p = a + bT + cT^2 + dT^3 \tag{9.87}$$

$$c_p = a + bT + \frac{c}{T} + d\ln(T) + eT^2 \tag{9.88}$$

Analysis: Create additional columns of the temperatures raised to powers of two and three.

	A	B	C	D
1	T/K	T^2	T^3	c_p/(kJ/kmol·K)
2	255.37	6.52E+04	1.67E+07	35.06
3	310.93	9.67E+04	3.01E+07	37.675
4	366.48	1.34E+05	4.92E+07	39.995
5	422.04	1.78E+05	7.52E+07	42.057
6	477.59	2.28E+05	1.09E+08	43.898
7	533.15	2.84E+05	1.52E+08	45.537
8	588.71	3.47E+05	2.04E+08	47.01
9	644.26	4.15E+05	2.67E+08	48.336
10	699.82	4.90E+05	3.43E+08	49.533
11	755.37	5.71E+05	4.31E+08	50.638
12	810.93	6.58E+05	5.33E+08	51.614
13	866.48	7.51E+05	6.51E+08	52.516
14	922.04	8.50E+05	7.84E+08	53.326
15	977.59	9.56E+05	9.34E+08	54.063
16	1033.2	1.07E+06	1.10E+09	54.726
17	1088.7	1.19E+06	1.29E+09	55.333

Figure 9.40

Excel worksheet for multivariable linear regression of transformed data using the Data Analysis>Regression tool.

All three equations have linear model parameters, so we use *Excel*'s **Data Analysis>Regression** (add-in tool) to implement multivariable linear regression. Select the option of generating the residuals and a residual plot for assessing each model. Once we have the residuals, we can sort them from low to high values and generate a cumulative normal distribution plot.

Begin with the quadratic model by linear regression of the data tabulated in columns **A** and **B** versus c_p in column **D**, as shown in Figure 9.41.

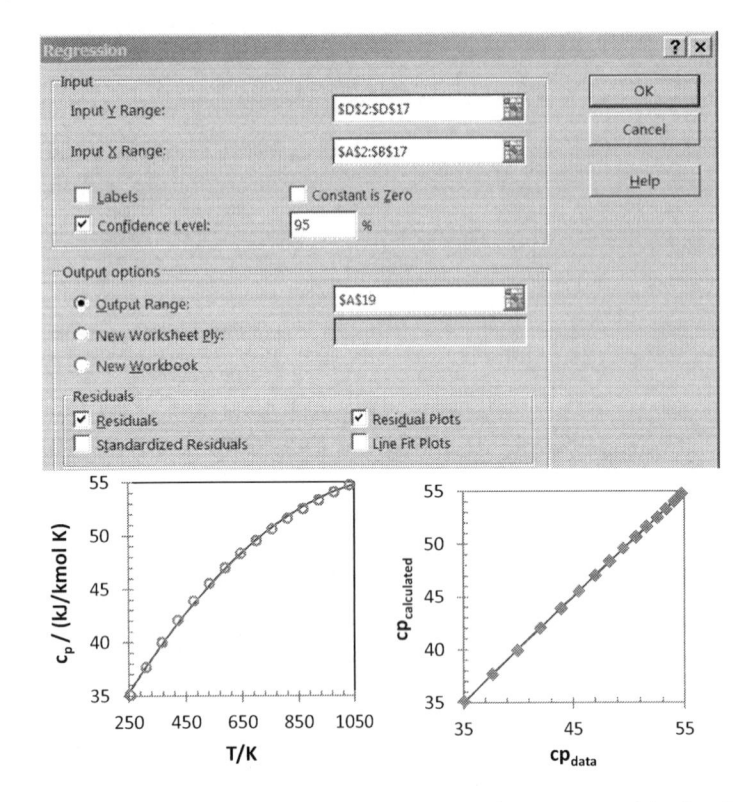

Figure 9.41

Second-order polynomial fit of the c_p data using *Excel*'s regression tool.

Data are plotted with the polynomial for comparison on a y_{data} versus y_{model} parity plot.

The second-order polynomial appears to give a reasonable fit of the data in Figure 9.41. However, raw and sorted residual plots in Figure 9.42 and Figure 9.44 tell a different story. The residuals in Figure 9.42 have a clear discernible pattern that invalidates the model.

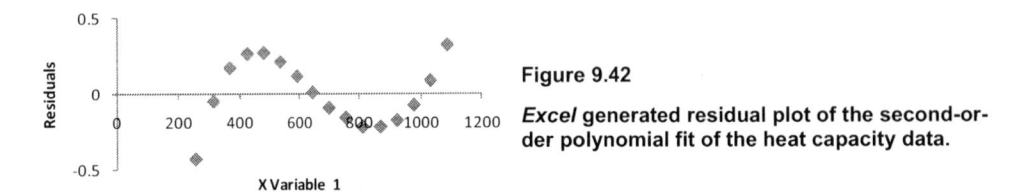

Figure 9.42

Excel generated residual plot of the second-order polynomial fit of the heat capacity data.

To create the sorted residual plot, select the range of residuals in the worksheet, and sort by clicking on the icon for ascending order (A to Z) in the **Sort & Filter** group (on the **Data** tab, as shown in Figure 9.43.).

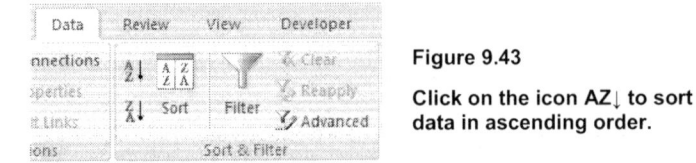

Figure 9.43

Click on the icon AZ↓ to sort data in ascending order.

Figure 9.44 compares the sorted residuals with the shape of a cumulative normal distribution created from sixteen simulated residuals using the following nested *Excel* worksheet functions:

$$=\text{NORMINV}(\text{RAND}(), 1, 0)*u$$

where u is the standard error in the fit generated by the **Regression** tool. The raw residuals in Figure 9.42 show a clear trend, and the sorted residuals in Figure 9.44 do not have the telltale shape of a cumulative normal distribution at the high end of the data when compared with the simulated residuals. The normal plot of residuals reveals the undesirable S-shaped curve about the diagonal line, further denying the validity of the quadratic model of the data.

CHAPTER 9: REGRESSION

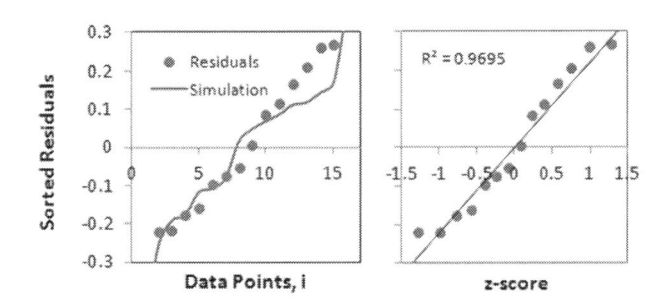

Figure 9.44

Sorted residual plot of the second-order polynomial compared with a simulated cumulative normal distribution indicates a lack of normally distributed residuals.

The normal plot versus the z-score shows an undesirable S-shaped curve with the lower z below the line and upper z above.

Next, fit the data with a third-order polynomial by replacing the **X-range** in the regression tool with the values in the cell range **A2:C17**. The residual and sorted residual plots in Figure 9.45 and Figure 9.46 reveal that the four-parameter, third-order polynomial also does not correctly capture the behavior of the heat capacity data.

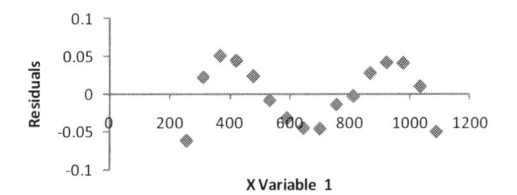

Figure 9.45

Residual plot for the third-order polynomial reveals a pattern that invalidates the model.

Again, we simulated normally distributed residuals from the standard error of the fit. Figure 9.46 has a plot to compare the shapes of the curves formed by a set of simulated residuals with the model residuals. Notice that the simulated residuals in Figure 9.44 and Figure 9.46 are slightly dissimilar based on the different random numbers used for each new simulation. Once again, we do not see the sigmoidal shape of a cumulative normal distribution for the model residuals, although we *do* see the unwanted S-shape in the normalized sorted residual plot. Despite the lack of a sigmoid trend in the sorted residuals, and the presences of an S-shaped trend in the normalized plot about the diagonal line, Filliben's criterion misleads us with $(R^2 = 0.97) > (R_c^2 = 0.88)$ for $n = 14$.

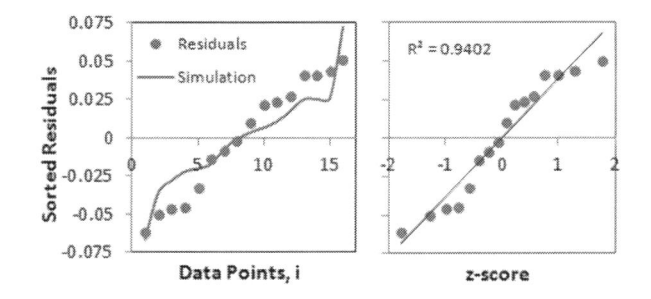

Figure 9.46

Sorted residuals plot (left) for a third-order polynomial compared with a simulated cumulative distribution. The normal plot (right) shows the undesirable S-shaped distribution about the diagonal line, indicating the residuals are not normally distributed.

Finally, fit the last model in Equation (9.88) using the transformed data listed in the *Excel* worksheet in Figure 9.47.

	A	B	C	D	E
1	T/K	1/T	lnT	T²	c_p/(kJ/kmol·K)
2	255.37	3.92E-03	5.54E+00	6.52E+04	35.06
3	310.93	3.22E-03	5.74E+00	9.67E+04	37.675
4	366.48	2.73E-03	5.90E+00	1.34E+05	39.995
5	422.04	2.37E-03	6.05E+00	1.78E+05	42.057
6	477.59	2.09E-03	6.17E+00	2.28E+05	43.898
7	533.15	1.88E-03	6.28E+00	2.84E+05	45.537
8	588.71	1.70E-03	6.38E+00	3.47E+05	47.01
9	644.26	1.55E-03	6.47E+00	4.15E+05	48.336
10	699.82	1.43E-03	6.55E+00	4.90E+05	49.533
11	755.37	1.32E-03	6.63E+00	5.71E+05	50.638
12	810.93	1.23E-03	6.70E+00	6.58E+05	51.614
13	866.48	1.15E-03	6.76E+00	7.51E+05	52.516
14	922.04	1.08E-03	6.83E+00	8.50E+05	53.326
15	977.59	1.02E-03	6.89E+00	9.56E+05	54.063
16	1033.15	9.68E-04	6.94E+00	1.07E+06	54.726
17	1088.71	9.19E-04	6.99E+00	1.19E+06	55.333

Figure 9.47   Transformed data for multivariable linear regression fit of Equation (9.88).

The **X-range** for Equation (9.88) is **A2:D17**. Figure 9.48 shows the results of the multi-linear regression, including a raw residual plot. The *p*-values in Figure 9.48 are all well below the significance level of 0.05, indicating that each variable has some significance in the model (with 95% confidence).

Regression Statistics	
Multiple R	0.99999981
R Square	0.99999962
Adjusted R Square	0.99999949
Standard Error	0.00453491
Observations	16

	Coefficients	Standard Error	t Stat	P-value
Intercept	-90.558451	3.287511952	-27.5462	1.68513E-11
X Variable 1	-0.0031023	0.001027446	-3.019392	0.011668381
X Variable 2	1953.19778	97.03265012	20.12928	4.98732E-10
X Variable 3	21.4517324	0.568896496	37.70762	5.51231E-13
X Variable 4	-2.137E-06	2.87726E-07	-7.425874	1.31643E-05

Lower 95%	Upper 95%
-97.7942156	-83.322686
-0.00536366	-0.0008409
1739.630361	2166.7652
20.19959963	22.703865
-2.7699E-06	-1.503E-06

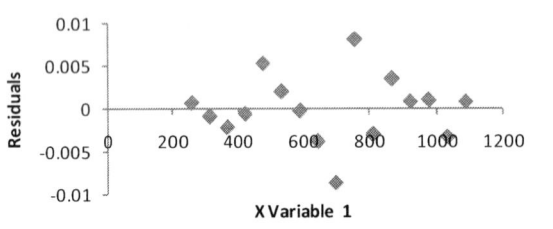

Figure 9.48

Multilinear Regression results and residual plot for Equation (9.88) indicate a good fit of the data.

The raw residual plot shows random distribution about the horizontal axis, with no obvious trend. The sorted and normal residual plots in Figure 9.49 reveal a cumulative normal distribution, similar to the simulated sorted residuals, providing further evidence of a good model. Using Filliben's criterion, $(R = 0.96) > (R_c = 0.88)$ for $n = 14$.

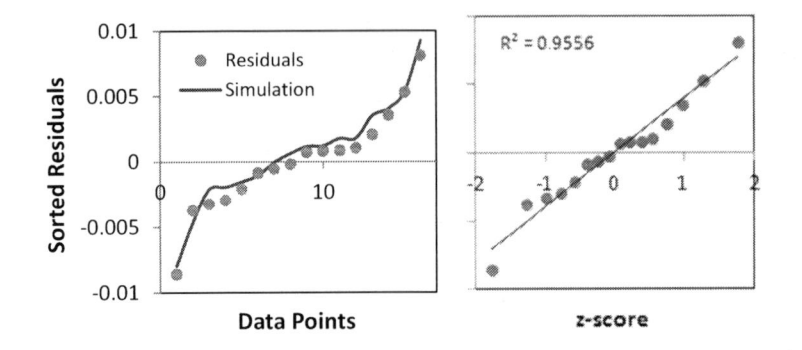

Figure 9.49

Sorted residual plot (left) shows a cumulative normal distribution for the model in Equation (9.88) when compared to simulated normally distributed residuals. The normal plot (right) shows a linear trend confirming normally distributed residuals.

The sub procedure **STUDENTIZER** was used to generate studentized residuals for the plot in Figure 9.50. One of the studentized residuals falls outside of the range of ±2, indicating a possible outlier in the data. The user-defined function **CHAUVENET** also identified the same outlier in the raw residuals. However, the user-defined function **GRUBBS** with a 5% significance level and Tukey methods in the macro **OUTLIER** did not report the outlier. However, we keep the outlier since we can expect 5% of the standardized residuals may exceed ±2. The studentized residuals also show random scatter, indicating a normal distribution.

CHAPTER 9: REGRESSION

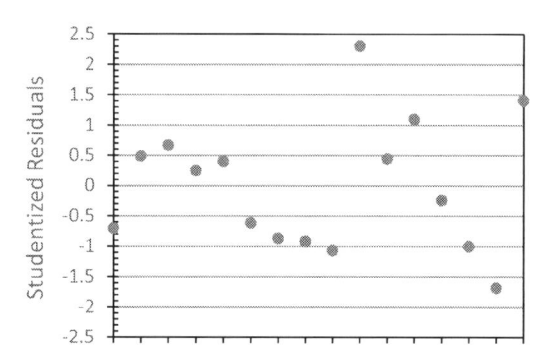

Figure 9.50

Studentized residual plot (left) for the model in Equation (9.88). The plot shows a potential outlier that exceeds 1.96.

In this example, the R^2 and *adjusted* R^2 values increase as we progress from a quadratic ($R^2 = 0.998838$, Adjusted $R^2 = 0.998665$) through a cubic ($R^2 = 0.999964$, Adjusted $R^2 = 0.999955$) to the final equation involving a log term ($R^2 = 0.999999$, Adjusted $R^2 = 0.999999$) indicating that the additional parameters in each successive model are needed to capture the trend in the data.

Comments: For small data sets, the scatter and raw residual plots usually provide a better picture for validating a model than the sorted residual plot. The simulated residuals help, but even small numbers of simulated normal residuals may not give a sigmoid shape. We must be careful when applying our assessment methods. The normalized symmetric S-shaped sorted residual plots fooled Filliben's criterion.

We illustrate the worksheet array function **LINEST** for Equation (9.88) which has five regression parameters. Select a contiguous range of 5 × 5 empty cells on the worksheet in Figure 9.48 for the output. Enter the following worksheet formula including the optional arguments for the intercept constant and statistics:

=LINEST(E2:E17, A2:D17, TRUE, TRUE)

Type the keyboard combination *CNTL SHIFT ENTER* to add the following output to the 5 × 5 range of selected cells:

-2.14E-06	21.45173237	1953.1978	-0.00310226	-90.5584505
2.88E-07	0.568896496	97.03265	0.001027446	3.287511952
1.00E+00	0.004534906	#N/A	#N/A	#N/A
7.31E+06	11	#N/A	#N/A	#N/A
6.02E+02	0.000226219	#N/A	#N/A	#N/A

Compare these results to the results from the **Regression Analysis** tool in Figure 9.48 and Equation (9.88). Note that *Excel* puts **#N/A** in blank cells in the array. The following table maps the results for Equation (9.88):

e	d	c	b	a
Standard Error e	Standard Error d	Standard Error c	Standard Error b	Standard Error a
R^2	SE c_p			
F	DOF			
RSS Eq. (9.66)	SSR Eq. (9.6)			

Unfortunately, **LINEST** does not provide residuals for model assessment.

9.4 Weighted Least-Squares Regression*

Classical least-squares regression assumes homoscedasticity (uniform uncertainty) in the dependent variables of the data. Residual plots that exhibit heteroscedasticity (non-uniform uncertainty) have an hourglass or diamond shape as seen in the residual plot of Figure 9.48, or megaphone shape, as seen in Figure 9.53. A heteroscedastic data set may be a good candidate for some form of weighted least-squares regression (Kirkup 2002). In this case, we fit a model to data by minimizing the weighted sum of squared residuals:

$$WSSR = \sum_{i=1}^{n}(w_i \cdot residual_i)^2 \qquad (9.89)$$

where w_i is a weighting factor for the model residual at point i.

For data displaying heteroscedasticity, try weighting each term in the sum of squared residuals (*SSR*) by the inverse of the uncertainty in each data point, $w_i = 1/u_{yi}$:

$$\text{Uncertainty weighted } SSR = \sum_{i=1}^{n}\left(\frac{y_{i,data} - y_{i,model}}{u_{yi}}\right)^2 \qquad (9.90)$$

With weighted regression, terms with larger uncertainty make a smaller contribution to the *SSR* in Equation (9.68):

$$u_y = \sqrt{\frac{n}{(n-k)}\frac{\sum_{i=1}^{n}\left(\dfrac{y_{i,data} - y_{i,model}}{u_{yi}}\right)^2}{\sum_{i=1}^{n}\dfrac{1}{u_{yi}^2}}} \qquad (9.91)$$

Equation (9.91) reduces to Equation (9.68) for uniform uncertainty in y.

Example 9.10 Weighted Nonlinear Regression to the Logistic Equation[89]

Known/Find: Fit the y vs x data in columns **A** and **B** in the worksheet in Figure 9.51 to the following logistic equation:

$$y = \frac{a}{1+\exp(b+cx)} + d \qquad (9.92)$$

	A	B	C	D	E	F
1	x	y	y logistic	Residuals	a	1
2	-5	9.84	1.982014	-7.85799	b	1
3	-3	10.14	1.880797	-8.2592	c	1
4	-1	9.99	1.5	-8.49	d	1
5	0	9.97	1.268941	-8.70106	SSR =	1046.216
6	1	10.32	1.119203	-9.2008		
7	2	10.56	1.047426	-9.51257		
8	3	10.79	1.017986	-9.77201		
9	4	10.88	1.006693	-9.87331		
10	5	11.03	1.002473	-10.0275		
11	7	10.95	1.000335	-9.94966		
12	9	11.03	1.000045	-10.03		
13	10	11.01	1.000017	-10.01		

Figure 9.51

Excel worksheet set up for nonlinear regression of the logistic function using the Solver in *Excel*.

---

[89] The logistic equation (or leaning "S" curve) is important for modeling natural phenomena such as population growth and business forecasting, including technology adoption (Koomey 2008). The logistic curve models gradual growth initially, followed by exponential growth in the middle, then a tapering off as the upper limit of growth matures.

Analysis: Use *Excel*'s **Solver** to perform nonlinear least-squares regression to determine the parameters in Equation (9.92). Uncheck the box for the option for nonnegative unconstrained variables. Specify initial guesses for the model parameters in column **F,** using named cells according to the labels in column **E**. Column **C** contains the results from the logistic equation using the *x* data in column **A** and initial guesses for the model parameters in the range **F1:F4**:

$$C2 = a/(1 + EXP(b + c\_*A2)) + d$$

Column **D** contains the residual differences between the data and model predictions:

$$D2 = C2 - B2$$

Calculate the sum of squared residuals in cell **F5** using an *Excel* worksheet function:

$$F5 = SUMSQ(D2:D13)$$

The *Excel* worksheet in Figure 9.52 shows the results from the **Solver** for the values of *a, b, c,* and *d* that minimize the sum of the squared residuals.

	A	B	C	D	E	F
1	x	y	y logistic	Residuals	a	1.039294
2	-5	9.84	9.959359	0.119359	b	2.090596
3	-3	10.14	9.962633	-0.17737	c	-1.1895
4	-1	9.99	9.996708	0.006708	d	9.959023
5	0	9.97	10.07336	0.10336	SSR =	0.069122
6	1	10.32	10.2592	-0.0608		
7	2	10.56	10.55309	-0.00691		
8	3	10.79	10.80528	0.015275		
9	4	10.88	10.93084	0.050842		
10	5	11.03	10.97681	-0.05319		
11	7	10.95	10.99629	0.046287		
12	9	11.03	10.99813	-0.03187		
13	10	11.01	10.99826	-0.01174		

Figure 9.52

Results from Solver nonlinear regression of the logistic equation.

Assess the model fit by plotting the results for Equation (9.92) with the data accompanied by a raw residual plot, as shown in Figure 9.53. We see random scatter about the horizontal line in the residual plot, affirming the fitness of the model for the data.

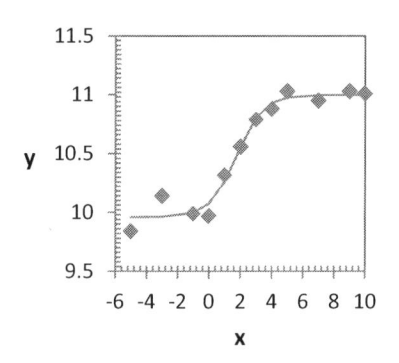

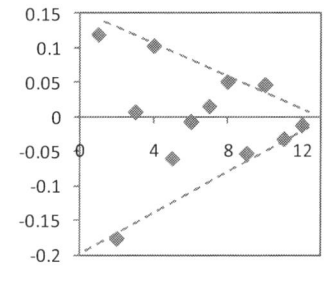

Figure 9.53

Logistic equation fit to YX data and residual plot showing the megaphone shape indicating the need for weighted least-squares regression.

However, the raw residual plot also shows the megaphone shape characteristic of non-uniform uncertainty in the data. An inspection of the residuals in column **D** of the worksheet in Figure 9.52 indicates that the residuals at the lower end are approximately one order of magnitude larger than the residuals at the higher end of *x*. We weight the first four data points by a factor of 1/3 and repeat the regression to obtain the parameter values shown in Figure 9.54.

	A	B	C	D	E	F	G
1	x	y	y logistic	Residuals	Weighted	a	-1.03927
2	-5	9.84	9.95938	0.11938	0.039793	b	-2.09073
3	-3	10.14	9.962653	-0.17735	-0.05912	c	1.189546
4	-1	9.99	9.996722	0.006722	0.002241	d	10.99831
5	0	9.97	10.07337	0.103366	0.034455	SSR =	0.018959

Figure 9.54
Nonlinear regression results with partial weighting of the data.

The weighted residual plot in Figure 9.55 now shows a uniform random distribution about the origin, without the megaphone shape. Only model parameter d changes appreciably after weighting the residuals.

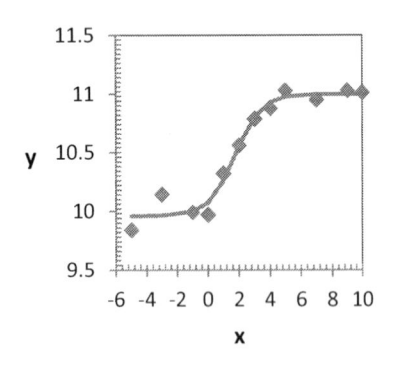

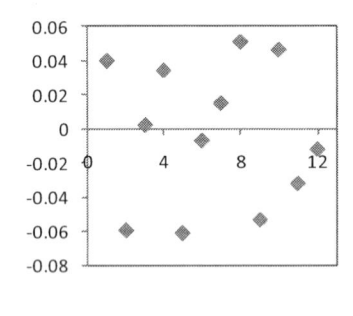

Figure 9.55
Weighted least squares fit of logistic equation to YX data and raw residual plot showing a reasonable fit.

Comments: We may select weights to improve the fit, but should not overdo it at the risk of giving some of the data underserved levels of importance.

9.5 Jackknife Method of Uncertainty Analysis*

"That's not a knife ... this is a knife." – Crocodile Dundee

The jackknife resampling method estimates uncertainty in least-squares regression parameters by repeatedly performing the regression for a set of n data pairs n times, while each time leaving out one subsequent data pair (Bradley and Gong 1983).[90] The cycles of the jackknife method make it a good candidate for implementation in an *Excel* worksheet with a *VBA* macro employing loops. An outline of the procedure follows:

1. Use the **Solver**, or another method such as Levenberg-Marquardt in Section 9.7, to find the model parameters that best fit the data by minimizing the sum of the squared residuals, according to Equation (9.6), or Equation (9.89) for weighted regression. It may be necessary to scale the model parameters depending on their relative magnitudes, as described in Section 6.4.

2. Remove the first data point and perform the least square regression by minimizing the SSR without the corresponding residual at the skipped point:

$$SSR_j = \sum_{i=1}^{j-1}(y_i - y_{i,\text{model}})^2 + \sum_{i=j+1}^{n}(y_i - y_{i,\text{model}})^2 \qquad (9.93)$$

3. Repeat step two n times, once for each data point $j = 2\ldots n$, and record the n sets of model parameters, a_j from each regression.

---

[90] Like a jackknife, we should always keep it handy.

4. Calculate the standard and expanded uncertainty (with $n - 1$ degrees of freedom) for each regressed model parameter generated by the jackknifed data (Harris 1998):

$$u_a = \frac{(n-1)}{\sqrt{n}} s_a \qquad \text{and} \qquad U_a = t_{95\%} u_a \qquad (9.94)$$

where a is the parameter and s_a is the standard deviation of the jack-knifed results for the regression parameter.

A *VBA* macro **JACKKNIFE** with input boxes, and user-form **REGRESSUN** with an option for the jackknife method, are available in the *PNMSuite* for regression uncertainty analysis. The macros use *Excel*'s **Solver** for minimizing the sum of squared residuals. To implement the jackknife method, set up an *Excel* worksheet to perform regression analysis using a range of cells for the model parameters and a cell containing the *SSR*. The macros replace the formula in the worksheet for *SSR* with a different formula that subtracts the residual for the local data point, then cycles (jackknifes) through each data point to obtain the model parameters for calculating standard deviations.

Table 9.5 maps the output from the user-form **REGRESSUN** or macro **JACKKNIFE** for a two-parameter model. *Example 9.11* uses the jackknife method to compare uncertainty in the model fit to the Gauss-Newton method. The jackknife method does *not* provide information about correlation among model parameters, but *does* tend to provide conservative estimates of uncertainty for small data sets.

Table 9.5 Sample map of REGRESSUN or JACKKNIFE macro output using the jackknife method for a two-parameter problem. The macros add comments in the cells with definitions of the terms.

$u_Y = \pm$	Standard uncertainty in model fit	$U_{Y,95\%} = \pm$	Expanded uncertainty in model fit
$u_1 = \pm$	Standard uncertainty in parameter 1	$U_{1,95\%} = \pm$	Expanded uncertainty in parameter 1
$u_2 = \pm$	Standard uncertainty in parameter 2	$U_{2,95\%} = \pm$	Expanded uncertainty in parameter 2
$R^2 =$	Coefficient of determination	Adj $R^2 =$	Adjusted R^2
$SSR =$	Sum of squared residuals	$DOF =$	Degrees of freedom
$AAD =$	Average absolute deviation	$t_{95\%} =$	Coverage factor

9.6 Gauss-Newton Least-squares Regression with Uncertainty Analysis

An alternative technique for least-squares regression combines Gaussian elimination for solving systems of linear equations with Newton's method of solving nonlinear equations. The Gauss-Newton least squares method is similar to Newton's method of optimization, presented in Section 7.4, with the advantage of not requiring second-order, or higher derivatives (Englezos and Kalogeraki 2001). The Gauss-Newton method:

- Uses the same *Excel* worksheet setup for least-squares regression as required for the **Solver**.
- Yields the sensitivity coefficients required for parameter uncertainty analysis.
- Converges quickly from good initial guesses, but can diverge away from the solution for poor initial guesses.

The method of Gauss-Newton linearizes a nonlinear regression function with a first-order Taylor series expansion about the initial guesses for the fitting parameters. This approach transforms the nonlinear minimization problem into a locally linear least-squares regression problem. The solution to the linear regression gives upgraded guesses for the parameters. We repeat the linearization process by successive linear Taylor expansions about of the upgraded model parameters until achieving convergence in the parameter values.

To illustrate the Gauss-Newton method, consider a regression function with two fitting parameters:

$$y = f(x, a_1, a_2) \qquad (9.95)$$

Cycle through the following steps until the solution for the model parameters a_1 and a_2 converges.

1. Linearize the function in a first-order truncated Taylor series expansion about initial guesses for the parameters, $a_1 \approx g_1$ and $a_2 \approx g_2$:

$$f(x, a_1, a_2) \cong f(x, g_1, g_2) + (a_1 - g_1) \frac{\partial f}{\partial a_1}\bigg|_{x,g} + (a_2 - g_2) \frac{\partial f}{\partial a_2}\bigg|_{x,g} \quad (9.96)$$

   Approximate the partial derivatives using the finite-difference methods defined in Section 5.2.

2. Calculate the (optional weighted) sum of squared residuals:

$$SSR = \sum_{i=1}^{n} \left\{ \frac{1}{u_{yi}} \left[y_i - f(x_i, g_1, g_2) - (a_1 - g_1) \frac{\partial f}{\partial a_1}\bigg|_{x_i g} - (a_2 - g_2) \frac{\partial f}{\partial a_2}\bigg|_{x_i g} \right] \right\}^2 \quad (9.97)$$

3. Apply the method of linear regression by setting to zero the partial derivatives of the least-squares SSR error function in Equation (9.97) with respect to each parameter:

$$\frac{\partial(SSR)}{\partial a_1} = 0 = \sum_{i=1}^{n} \left\{ y_i - \left[f(x_i, g_1, g_2) + (a_1 - g_1) \frac{\partial f}{\partial a_1}\bigg|_{x_i g} + (a_2 - g_2) \frac{\partial f}{\partial a_2}\bigg|_{x_i g} \right] \right\} \cdot \left(\frac{1}{u_{yi}} \right)^2 \cdot \frac{\partial f}{\partial a_1}\bigg|_{x_i g} \quad (9.98)$$

$$\frac{\partial(SSR)}{\partial a_2} = 0 = \sum_{i=1}^{n} \left\{ y_i - \left[f(x_i, g_1, g_2) + (a_1 - g_1) \frac{\partial f}{\partial a_1}\bigg|_{x_i g} + (a_2 - g_2) \frac{\partial f}{\partial a_2}\bigg|_{x_i g} \right] \right\} \cdot \left(\frac{1}{u_{yi}} \right)^2 \cdot \frac{\partial f}{\partial a_2}\bigg|_{x_i g} \quad (9.99)$$

4. Rearrange the system of partial differential equations into matrix notation for convenient solution by Gaussian elimination for $(a_1 - g_1)$ and $(a_2 - g_2)$:

$$\begin{bmatrix} \sum \left(\frac{1}{u_{yi}} \cdot \frac{\partial f}{\partial a_1}\bigg|_{x,g} \right)^2 & \sum \left(\frac{1}{u_{yi}} \right)^2 \cdot \left(\frac{\partial f}{\partial a_1} \frac{\partial f}{\partial a_2}\bigg|_{x,g} \right) \\ \sum \left(\frac{1}{u_{yi}} \right)^2 \cdot \left(\frac{\partial f}{\partial a_2} \frac{\partial f}{\partial a_1}\bigg|_{x,g} \right) & \sum \left(\frac{1}{u_{yi}} \cdot \frac{\partial f}{\partial a_2}\bigg|_{x,g} \right)^2 \end{bmatrix} \cdot \begin{bmatrix} a_1 - g_1 \\ a_2 - g_2 \end{bmatrix} = \begin{bmatrix} \sum \left(\frac{1}{u_{yi}} \right)^2 \cdot [y - f(x, g_1, g_2)] \frac{\partial f}{\partial a_1}\bigg|_{x,g} \\ \sum \left(\frac{1}{u_{yi}} \right)^2 \cdot [y - f(x, g_1, g_2)] \frac{\partial f}{\partial a_2}\bigg|_{x,g} \end{bmatrix} \quad (9.100)$$

5. Solve the system of linear equations in (9.100) for the differences $(a_1 - g_1)$ and $(a_2 - g_2)$, then calculate the upgraded values of a_1 and a_2 from the solution for $(a_1 - g_1)$ and $(a_2 - g_2)$ in terms of the values for the initial guesses g_1 and g_2. If necessary, use a relaxation factor to control convergence.

$$a_{m+1} = a_m + w(a - g)_{m+1} \quad (9.101)$$

   where the subscript m indexes the iterations of the Gauss-Newton method.

6. Apply the bisection *rule* to determine the largest relaxation factor in Equation (9.101) that gives a reduced sum of the squared residuals (SSR) as defined by Equation (9.97) (Englezos and Kalogeraki 2001). Begin with $w = 1$, then repeatedly reduce w by a factor of two until the upgraded SSR is smaller than the SSR from the previous Gauss-Newton iteration:

$$\text{Stop when } SSR(a_{m+1}) < SSR(a_m)$$

7. Replace the guesses for the parameters with the upgraded values from step 5: $g_1 \leftarrow a_1$ and $g_2 \leftarrow a_2$.

8. Check for convergence using the vector norm of the relative residuals:

CHAPTER 9: REGRESSION

$$\sqrt{\sum_{j=1}^{k}\left(\frac{a_j - g_j}{a_j + \delta}\right)^2} < tol \qquad (9.102)$$

where k is the number of model parameters in Equation (9.95) and δ is a small number to avoid division by zero in the event that $a_j = 0$.

We may extend the Gauss-Newton method, as outlined above, to any number of parameters in the model. The Gauss-Newton method is programmed in the macro **GAUSSNEWTON** and is an option in the user-form **LSREGRESS**. The *VBA* macro **GAUSSNEWTON**, listed in Figure 9.56, uses finite-difference derivative approximations in step 4. **GAUSSNEWTON** employs the sub procedure **GAUSSPIVOT** for Gaussian elimination with maximum column pivoting and one error correction from Equation (4.21) to solve the linearized system of equations. Table 9.6 maps the output from the macros **GAUSSNEWTON** and **LSREGRESS**. To use the macros:

- Set up an *Excel* worksheet with a range of model parameters, column of *y*-data, and column of the model for *y* evaluated at the corresponding values of the independent data and model parameters.
- The macro has the option for weighted least-squares regression, which requires a column of corresponding weights at each point, per Equation (9.89).
- **LSREGRESS** has an option for adding residual and sorted residual plots to the worksheet.

```
Public Sub GAUSSNEWTON()
' Nonlinear least-squares regression by the Gauss-Newton method. Calculate the
' uncertainties in least-squares regression constants and the sum of the square of the
' residuals between the observed and calculated y values.  Uses Gauss elimination with
' maximum column pivoting and one iteration of error correction to solve the
' linearized regression equations. Derivatives are approximated with 2nd order finite
' difference approximation. A stepping factor is used to control convergence with the
' bisection rule. The macro returns the standard error and expanded uncertainty for
' 95% confidence intervals for each parameter. The macro calculates the standard error
' for y regression, R^2, and adjusted R^2. The user must create ranges of y observed,
' y calculated, and model parameters.
' *** Required User-defined Functions and Sub Procedures from PNM Suite ***
' GAUSSPIVOT = Gaussian elimination solution to linear equations
' JUSTABC_123 = renames worksheet with allowed object naming characters
' MATINV = matrix inversion
' USIGFIGS = significant figures of value
' *** Nomenclature ***
Dim aad As Double ' = average absolute deviation
Dim alpha As Double ' = 1-ci
Dim ar As Range ' = range of least-squares regression constants
Dim ase() As Double ' = standard error for regression constants
Dim b() As Double ' = array of least-squares regression constants
Dim boxtitle As String ' = string title used in input and message boxes
Dim c() As Double ' = array of constants for linear regression
Dim ci As Double ' = confidence interval
Dim da As Double ' = perturbation value for finite difference derivative approximation
Dim df As Long ' = degrees of freedom for regression
Dim df_ws As Long ' = Welch-Satterthwaite degrees of freedom for regression
Dim dy As Double ' = derivative
Dim dyda() As Double ' = partial derivatives of y with respect to ar() at data points
Dim er As Double ' = convergence error
Dim erssr As Double ' = relative error in sum of squared residuals
Dim i As Long ' = data index
Dim j As Long, j1 As Long ' = least-squares regression constant indexes
Dim k As Long ' = number of least-square regression constants
Dim m As Long ' = counter for printout
Dim n As Long ' = number of data pairs
Dim p() As Double ' = partial derivative product matrix
Dim pci As Double ' = percent confidence interval (e.g., 95%)
Dim pcol As Long, prow As Long ' = column and row to start output of results
Dim pinv() As Double ' = inverse of partial derivative matrix
Dim r() As Double ' = array of correlation coefficients
Dim rss As Double ' = regression sum of squares
Dim r2 As Double ' = coefficient of determination
```

```
Dim r2adj As Double   ' = adjusted r^2
Dim s As Double   ' = sum
Dim sc As Integer   ' = show calculation parameter, integer
Dim se As Double   ' = standard error
Dim small As Double   ' = small number for finite difference derivative
Dim ssr As Double, ssri As Double   ' = sum of square residuals, initial ssr
Dim t As Double   ' = t-statistic for 95% coverage
Dim tol As Double   ' = convergence tolerance
Dim tws As Double   ' = 95% coverage for Welch Satterthwaite degrees of freedom
Dim u() As Double   ' = expanded uncertainties in regression constants
Dim uc As Integer   ' = use weights
Dim uexpand As Double   ' = expanded uncertainty
Dim u_rz As Double   ' = combined uncertainty
Dim u_z As Double   ' = fixed uncertainty in y data
Dim v As Double   ' = variance
Dim vu As Double   ' = variance of weightings
Dim w() As Double   ' = vector of weight factors
Dim wprompt As String   ' = string for prompt to use weighted least-squares
Dim wr As Range   ' = range of weight factors
Dim wstep As Double   ' = relaxation factor or stepping parameter
Dim xv() As Double   ' = parameter variables
Dim yave As Double   ' average of y data
Dim ydatar As Range   ' range of y data
Dim ymodel() As Double   ' = array of calculated dependent variable y
Dim yr As Range   ' = range of calculated dependent variable y
'****************************************************************************
small = 0.000000000001: boxtitle = "GAUSS-NEWTON REGRESSION"
With Application
Set ar = .InputBox(Title:=boxtitle, Type:=8, prompt:="MODEL PARAMETERS:" & vbNewLine _
        & "(must be in contiguous cells)")
k = ar.Count: m = 1 ' Get the number of regression constants & output length
For i = 1 To k - 1
    For j = i + 1 To k: m = m + 1: Next j
Next i
m = .Max(m, 5)
y_input: ' Get ranges for Y observed and calculated
Set ydatar = .InputBox(Title:=boxtitle, Type:=8, prompt:="Range of Y DATA:" & _
                    vbNewLine & "(must be in contiguous cells)")
Set yr = .InputBox(Title:=boxtitle, Type:=8, prompt:="Range of Y MODEL:" & _
                    vbNewLine & "(must be in contiguous cells)")
uc = MsgBox(prompt:="Use WEIGHTED Least-Squares?", Title:=boxtitle, _
            Buttons:=vbYesNo + vbQuestion + vbDefaultButton2)
Select Case uc
Case 6
    Set wr = .InputBox(Title:=boxtitle, prompt:="Range of WEIGHTS, w:" & vbNewLine & _
                    "(must be in contiguous cells)", Type:=8)
    wprompt = "*w WEIGHTED "
End Select
n = yr.Count: df = n - k ' Get the number of data pairs & DOF
If ydatar.Count <> n Then ' Check for consistency in number of data pairs
    MsgBox prompt:="Number of y data must match" & vbNewLine & _
                    "number of y model.", Title:=boxtitle: GoTo y_input
End If
' Get fixed uncertainty in y data from readability
u_z = .InputBox(Title:=boxtitle, Type:=1, Default:=0, _
                prompt:="SYSTEMATIC (FIXED) UNCERTAINTY in y:")
tol = 0.000000000001 ' Relative Convergence TOLERANCE
pci = Application.InputBox(Title:=boxtitle, Type:=1, _
        prompt:="Set Confidence Interval (50 to 99)%:", Default:=95)
pci = CInt(pci): pci = .Max(50, .Min(99, pci)): ci = pci / 100: alpha = 1 - ci
If df < 1 Then ' Check for enough data
    MsgBox Title:=boxtitle, prompt:="Too many parameters!": Exit Sub
End If
ReDim ase(1 To k) As Double, b(1 To k) As Double, c(1 To k) As Double, _
    dyda(1 To n, 1 To k) As Double, u(1 To k) As Double, w(1 To n) As Double, _
    ymodel(1 To n) As Double, r(1 To k, 1 To k) As Double, _
    p(1 To k, 1 To k) As Double, pinv(1 To k, 1 To k) As Double, _
    xv(1 To k) As Double
.ScreenUpdating = True ' Show intermediate calculations in the spreadsheet
```

```
If uc = 6 Then ' User provided weight factors, set to 1 if not provided
    For i = 1 To n: w(i) = wr(i): Next i
Else
    For i = 1 To n: w(i) = 1: Next i
End If
ssr = 0
For i = 1 To n: ssr = ssr + ((yr(i) - ydatar(i)) * w(i)) ^ 2: Next i
ssri = ssr ' Save initial sum squared residuals
Do ' Iterate for upgraded parameter solution by Gauss-Newton method
    For j = 1 To k: b(j) = ar(j): Next j ' Save parameters for deriv approximation
    For i = 1 To n: ymodel(i) = yr(i): Next i ' Save y model values
    For j = 1 To k    ' Approximate partial derivatives by 2nd order finite difference
        da = 0.00001 * (Abs(b(j)) + 0.00001): ar(j) = b(j) + da
        For i = 1 To n
            dyda(i, j) = (yr(i) - ymodel(i)) / da
        Next i
        ar(j) = b(j) - da
        For i = 1 To n
            dyda(i, j) = (dyda(i, j) + (ymodel(i) - yr(i)) / da) / 2
        Next i
        ar(j) = b(j)
    Next j
    For j = 1 To k ' Calculate the constant vector for Gauss-method
        c(j) = 0
        For i = 1 To n
            c(j) = c(j) + (ydatar(i) - yr(i)) * dyda(i, j) * w(i) ^ 2
        Next i
    Next j
    For j = 1 To k ' Partial derivative product (covariance) matrix
        For j1 = 1 To k
            p(j, j1) = 0
            For i = 1 To n ' Sum pd evaluated for each data pair
                p(j, j1) = p(j, j1) + dyda(i, j) * dyda(i, j1) * w(i) ^ 2
            Next i
        Next j1
    Next j
    Call GAUSSPIVOT(p, c, k, xv) ' Solve linearized equations
    er = 0: wstep = 1 ' Upgrade fitting parameters and calc. ssr, & convergence
    For j = 1 To k
        ar(j) = wstep * xv(j) + b(j): er = .Max(Abs(xv(j)) / (Abs(ar(j)) + small), er)
    Next j
    ssr = 0 ' Weighted sum of squared residuals
    For i = 1 To n: ssr = ssr + ((yr(i) - ydatar(i)) * w(i)) ^ 2: Next i
    Do While ssr > ssri ' Bisection rule to determine stepping parameter
        wstep = 0.5 * wstep 'upgrade fitting parameters and calculate SSR
        For j = 1 To k: ar(j) = wstep * xv(j) + b(j): Next j
        ssr = 0 ' Weighted sum of squared residuals
        For i = 1 To n: ssr = ssr + ((yr(i) - ydatar(i)) * w(i)) ^ 2: Next i
    Loop
    erssr = Abs(ssr - ssri) / (Abs(ssr) + small): ssri = ssr ' reset SSR
    ' Display progress in status bar at bottom of worksheet
    .StatusBar = "GAUSS-NEWTON residual = " & VBA.Format(er, "0.0E+#0")
Loop While er > tol And erssr > tol ' Check convergence in parameters
If Not k = 1 Then ' Get the inverse of the P array for uncertainty analysis
    Call MATINV(p, pinv)
Else
    pinv(k, k) = 1 / (p(k, k) + small)
End If
yave = .Average(ydatar): rss = 0: aad = 0: vu = 0
For i = 1 To n
    rss = rss + ((ydatar(i) - yave) * w(i)) ^ 2
    aad = aad + Abs(ydatar(i) - yr(i)): vu = vu + w(i) ^ 2
Next i
aad = aad / n: vu = vu / n
For i = 1 To k - 1 ' Calculate the correlation coefficients for fitting parameters
    For j = i + 1 To k
        r(i, j) = pinv(i, j) / Abs(pinv(i, i) * pinv(j, j) + small) ^ 0.5
    Next j
Next i
```

```vba
' Calculate the variance, standard error & (1-alpha)*100%, alpha = 0.05 for 95% conf.
v = ssr / df: se = Sqr(v / vu): t = .TInv(alpha, df)
' Calculate combined uncertainty in model
u_rz = Sqr(se ^ 2 + u_z ^ 2): df_ws = CLng(df * (u_rz / se) ^ 4)
tws = .TInv(0.05, df_ws)
r2 = 1 - ssr / rss: r2adj = 1 - (1 - r2) * (n - 1) / df ' Calculate R^2 & adj R^2
' Get 95% confidence intervals for fitting constants
For j = 1 To k: ase(j) = se * Abs(pinv(j, j)) ^ 0.5: u(j) = ase(j) * t: Next j
prow = output.row: pcol = output.Column ' Get row & column location
u_rz = USIGFIGS(u_rz, 2): uexpand = USIGFIGS(tws * u_rz, 2) ' round up uncertainty
' Print the standard & 95% confidence intervals for parameters
Worksheets(ws).Activate: Cells(prow, pcol).Activate
With ActiveCell ' standard uncertainty in model fit to Y and DOF
    .Value = "uY = ±"
    .Characters(Start:=2, Length:=1).Font.Subscript = True
    .Offset(0, 1).Value = u_rz: .Offset(0, 1).NumberFormat = "0.0E+#0"
    .Offset(0, 2).Value = "UY," & CStr(pci) & "% = "
        .Offset(0, 2).Characters(Start:=2, Length:=5).Font.Subscript = True
    .Offset(0, 3).Value = uexpand: .Offset(0, 3).NumberFormat = "0.0E+#0"
End With
For j = 1 To k ' uncertainties in parameters
    u_rz = USIGFIGS(ase(j), 2): uexpand = USIGFIGS(u(j), 2)
    Cells(prow + j, pcol).Activate
    With ActiveCell
        .Value = "u" & j & " = ±"
        .Characters(Start:=2, Length:=Len(CStr(j))).Font.Subscript = True
        .Offset(0, 1).Value = u_rz: .Offset(0, 1).NumberFormat = "0.0E+#0"
        .Offset(0, 2).Value = "U" & j & "," & CStr(pci) & "% = ±"
            .Offset(0, 2).Characters(Start:=2, _
                Length:=Len(CStr(j)) + 4).Font.Subscript = True
        .Offset(0, 3).Value = uexpand: .Offset(0, 3).NumberFormat = "0.0E+#0"
    End With
Next j
Cells(prow + k, pcol).Activate
With ActiveCell ' Coefficient of determination, r2, adjusted r2, SSR, AAD
    .Offset(1, 0).Value = "R2 ="
        .Offset(1, 0).Characters(Start:=2, Length:=1).Font.Superscript = True
    .Offset(1, 1).Value = r2: .Offset(1, 1).NumberFormat = "0.0000"
    .Offset(1, 2).Value = "Adj R2 ="
        .Offset(1, 2).Characters(Start:=6, Length:=1).Font.Superscript = True
    .Offset(1, 3).Value = r2adj: .Offset(1, 3).NumberFormat = "0.0000"
    .Offset(2, 2).Value = "DoF ="
    .Offset(2, 3).Value = df_ws: .Offset(3, 2).Value = "t" & CStr(pci) & "% ="
        .Offset(3, 2).Characters(Start:=2, Length:=3).Font.Subscript = True
    .Offset(3, 3) = t: .Offset(3, 3).NumberFormat = "0.00"
    .Offset(4, 2).Value = "SSR ="
    .Offset(4, 3).Value = ssr: .Offset(4, 3).NumberFormat = "0.00E+#0"
    .Offset(5, 2).Value = "AAD ="
    .Offset(5, 3).Value = aad: .Offset(5, 3).NumberFormat = "0.00E+#0"
End With
m = 1 ' Print linear correlation coefficients
For i = 1 To k - 1
    For j = i + 1 To k
        m = m + 1
        Cells(prow + k + m, pcol).Activate
        With ActiveCell
            .Value = "R" & i & "," & j & " ="
            .Characters(Start:=2, Length:=Len(CStr(i) & CStr(j)) + 1) _
                .Font.Subscript = True
            .Offset(0, 1).Value = r(i, j): .Offset(0, 1).NumberFormat = "0.0000"
        End With
    Next j
Next i
End Sub ' GAUSSNEWTON
```

Figure 9.56 VBA macro GAUSSNEWTON for least-squares regression with uncertainty analysis.

CHAPTER 9: REGRESSION

Table 9.6 Sample map of GAUSSNEWTON and LSREGRESS output for a two-parameter problem. The macros add comments in the cells with definitions of the terms.

$u_Y = \pm$	Standard uncertainty in model fit	$U_{Y,95\%} = \pm$	Expanded uncertainty in model fit
$u_1 = \pm$	Standard uncertainty in parameter 1	$U_{1,95\%} = \pm$	Expanded uncertainty in parameter 1
$u_2 = \pm$	Standard uncertainty in parameter 2	$U_{2,95\%} = \pm$	Expanded uncertainty in parameter 2
$R^2 =$	Coefficient of determination	Adj $R^2 =$	Adjusted R^2
$R_{1,2} =$	Correlation for parameters 1 and 2	DoF =	Degrees of freedom
		$t_{95\%} =$	Coverage factor
		SSR =	Sum of squared residuals
		AAD =	Average absolute deviation

As with all numerical methods, the Gauss-Newton method for least-squares regression has advantages and disadvantages:

Advantages
- Provides sensitivity coefficients for estimating the uncertainty in the parameters directly from the solution method (de Levie 1999).
- Converges quickly to the solution from good initial guesses.
- Requires just one iteration for linear least-squares regression.
- Useful for finding roots to nonlinear equations (*e.g.*, see macro **ROOTSGN**).
- Apply the Gauss-Newton macro to the converged solution from another optimization method, such as the **Solver**, to obtain regression uncertainty information (Linga, Al-Saifi and Englezos 2006).

Disadvantages:
- Requires large numbers of partial derivative evaluations.
- Convergence can be sensitive to the initial guesses for the parameters.

Example 9.11 Fit Adsorption Isotherm Data by Nonlinear Regression

Known: Figure 9.57 lists isotherm data in columns **A** and **B** of an *Excel* worksheet for adsorption of methylene blue dye on activated carbon at 305 K (Kumar and Sivanesan 2006).

	A	B	C	D
1	C_e	q_e	q model	
2	mg/L	mg/g	mg/g	Residual
3	0	0	0.00E+00	0
4	1.4	43	19.70770127	-23.292299
5	5.4	73	69.49894421	-3.5010558
6	8.5	107.5	101.8404807	-5.6595193
7	11.1	144.5	125.3543536	-19.145646
8	17.2	164	170.0890314	6.0890314
9	24.4	178	208.8640172	30.864017
10	27.9	210.5	223.709745	13.209745
11	34.3	253.5	245.8559154	-7.6440846

12	38.4	258	257.3550039	-0.6449961
13	45.2	274	272.911308	-1.088692
14	51.4	293	284.1316204	-8.8683796
15	59.6	302	295.7494839	-6.2505161
16	68.4	308	305.2043348	-2.7956652
17	76.4	318	311.8119846	-6.1880154
18	85.6	322	317.6762976	-4.3237024
19	95.2	324	322.3128282	-1.6871718
20	105	325	325.8774987	0.8774987
21	115.2	324	328.6430725	4.6430725
22	125	325	330.6075778	5.6075778
23	134.2	329	331.9720472	2.9720472

Figure 9.57 Adsorption data are listed in columns A and B. Results and residuals are in columns C and D.

Find: Determine the isotherm model parameters A, B, and g and their uncertainty for the following Redlich-Peterson empirical model (combines features of Freundlich and Langmuir models):

$$q_e = \frac{AC_e}{1 + BC_e^g} \tag{9.103}$$

Assumptions: equilibrium conditions between the solid and gas phases

Analysis: Get good initial guesses for the parameters from linearized, asymptotic forms of the model. At low concentration, Equation (9.103) reduces to a linear function:

$$\lim_{C_e \to 0} \frac{AC_e}{1+BC_e^g} = AC_e \quad (9.104)$$

Estimate parameter $A \approx 13$ from a linear fit of the concentration data at the low end with the first five data pairs using a linear **Trend line** in *Excel*, and with the intercept set to zero, as seen in Figure 9.58.

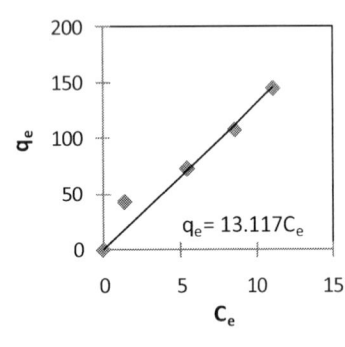

Figure 9.58

Linear least squares fit of a subset of the isotherm data at low loading to find good initial guesses for the parameters, as required for nonlinear least-squares regression.

For large equilibrium concentrations, assuming the exponent parameter $g \approx 1$, Equation (9.103) reduces to a constant:

$$\lim_{C_e \to \infty} \frac{AC_e}{1+BC_e^g} = \frac{A}{B} = 329 \quad (9.105)$$

Calculate $B = A/329 = 13.1/329 \cong 0.04$. Plot the isotherm data with the model using the initial guesses for the parameters to get a general view of the relationship between C_e and q_e as shown in Figure 9.59. The model appears to perform well at small C_e and trends correctly with the data, but under predicts the loading at higher C_e.

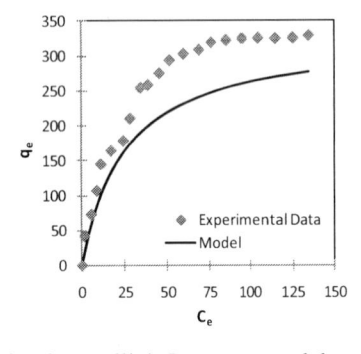

Figure 9.59

Data with adsorption isotherm using initial guesses for the model parameters:

$A \approx 13.1$, $B \approx 0.0398$, $g \approx 1$

Use the estimates for the Redlich-Peterson model parameters for the initial guesses in the user-form **LSREGRESS**:

1. Specify the model parameters in the range of contiguous cells **B25:B27** on the worksheet in Figure 9.60.

2. Calculate model predictions for the equilibrium loading in terms of the experimental equilibrium concentrations and initial guesses for the parameters in the Redlich-Peterson model. For example, insert the formula for the model loading in cell **C3**, then auto fill the column with the formula using relative referencing:

    C3=$B$25*A3/(1+$B$26*A3^$B$27)

3. Calculate the residuals as shown in column **D** of Figure 9.57 and sum of the squares of the residuals shown in Figure 9.60.

    B28=SUMSQ(D3:D23)

4. Employ the *VBA* user-form **LSREGRESS** for nonlinear regression of the Redlich-Peterson model. Supply the data, model, residual ranges, and location for the *SSR* on the worksheet, as required for the input boxes.

CHAPTER 9: REGRESSION

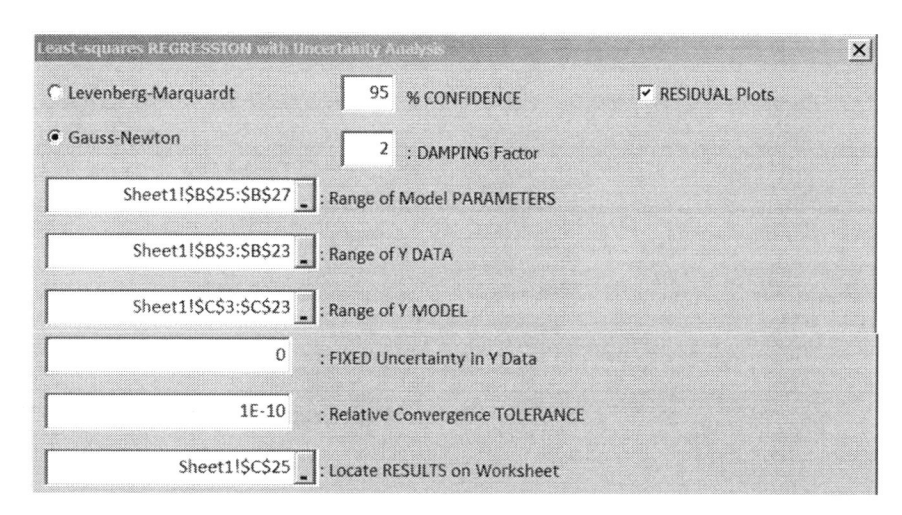

Use the default value of zero for the systematic fixed uncertainty in y, and tolerance of 10^{-10}. Direct the output to cell **C25**. Then click **RUN** for the results displayed in Figure 9.60.

A	14.4439791	$u_y = \pm$	1.2E+1	$U_{y,95\%} =$	2.6E+1
B	1.77E-02	$u_1 = \pm$	1.5E+0	$U_{1,95\%} = \pm$	3.1E+0
g	1.14478703	$u_2 = \pm$	8.2E-3	$U_{2,95\%} = \pm$	1.8E-2
SSR	2.43E+03	$u_3 = \pm$	7.4E-2	$U_{3,95\%} = \pm$	1.6E-1
		$R^2 =$	0.9891	Adj $R^2 =$	0.9879
		$R_{1,2} =$	0.9470	DoF =	18
		$R_{1,3} =$	-0.8929	$t_{95\%} =$	2.10
		$R_{2,3} =$	-0.9655	SSR =	2.43E+3
				AAD =	7.40E+0

Figure 9.60

Model parameters, *SSR*, and uncertainty from *VBA* macro **LSREGRESS** for fitting adsorption data.

5. Figure 9.61 has the updated least-squares model results plotted with the data, which shows a significantly improved fit relative to the initial guesses for the model parameters.

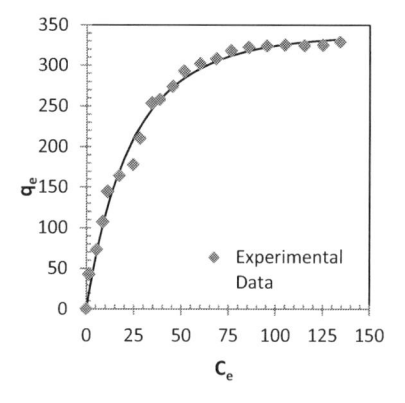

Figure 9.61

Data plotted with the least-squares fit of the adsorption isotherm:

$A = 14.4$, $B = 0.0177$, $g = 1.14$

6. Plot the residuals and cumulative distributions in Figure 9.62 to assess the goodness of fit and appropriateness of the model.

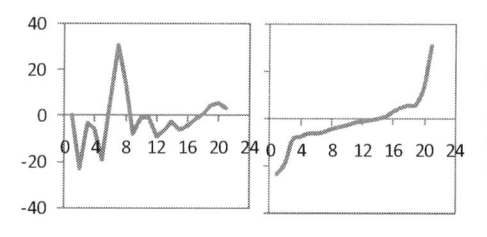

Figure 9.62

Residual and sorted residual plots confirm the model fit with random scatter and a sigmoid shaped cumulative normal distribution, respectively.

It is a simple step to add a residual plot to the *Excel* worksheet from the results for the best fit to the model. The sorted residual plot requires a small amount of extra effort. However, *Excel* has a sort option on the **Data** tab. The residual plot in Figure 9.62 shows random deviations about the abscissa, indicating an appropriate model of the data. Note that two-thirds of the residuals for this small data set are below the abscissa, an indication of potential outliers. The sorted residual plot has a sigmoidal shape indicative of a cumulative normal distribution that confirms the model fit. The results of the nonlinear least squares fit of the data give the following parameters with 95% confidence intervals:

$$A = 14.4 \pm 3.1 \qquad B = 0.018 \pm 0.017 \qquad g = 1.14 \pm 0.15$$

Comments: The large uncertainties in some of the model parameters indicate the need for additional data to produce a higher precision model. Large uncertainties may also identify unnecessary model parameters. Models are less sensitive to parameters with large uncertainty, or may have significant correlation among the parameters, as indicated by the relatively large values for the correlation coefficients R_{12} and R_{13} shown in Figure 9.60. The jackknife resampling method using the macro **REGRESSUN** with similar input requirements to **LSREGRESS** gives conservative estimates of uncertainty, as shown in Figure 9.63. The results from **REGRESSUN** point to the same conclusions.

$u_Y = \pm$	1.2E+01	$U_{Y,95\%} = \pm$	2.4E+01
$u_1 = \pm$	2.1E+00	$U_{1,95\%} = \pm$	4.4E+00
$u_2 = \pm$	9.4E-03	$U_{2,95\%} = \pm$	2.0E-02
$u_3 = \pm$	9.2E-02	$U_{3,95\%} = \pm$	1.9E-01
$R^2 =$	0.9891	Adj $R^2 =$	0.9879
SSR =	7.40E+00	DoF =	18
AAD =		$t_{95\%} =$	2.10

Figure 9.63

Regression parameter uncertainties from the REGRESSUN user-form using the Jackknife method.

1 = A, 2 = B, 3 = g

We employed the user-defined function **DREX** to calculate the partition coefficient of solute loading for small equilibrium concentrations C_e:

$$q = KC_e \tag{9.106}$$

where

$$K = \frac{d}{dC_e}\left(\frac{AC_e}{1+BC_e^g}\right)\bigg|_{C_e \to 0}$$

	A	B
36	C_e	0
37	q	=B25*B36/(1+B26^B27)
38	K	=DREX(B37,B36)

	A	B
36	Ce	0
37	q	0.00E+00
38	K	1.43E+01

For small C_e, a simple Henry's law replaces the adsorption isotherm:

$$q = (14.3 \pm 3.1)C_e \quad (95\% \text{ confidence}) \tag{9.107}$$

□

Large uncertainty in regression parameters may indicate excessive-parameterization of the model, or correlation between data and parameters. In these cases, the values of the regression parameters may become sensitive to the data. Use the adjusted R^2 to help identify unnecessary parameters, or use larger data sets. The *VBA* macro **LSREGRESSS** also computes correlation coefficients between all combinations of pairs of parameters needed for the *Law of Propagation of Uncertainty* in Equation (8.40), as implemented in the **JITTER** macro in Section 8.2.5 or the Monte Carlo macro **UNMCLHS** in Section 8.3.

Example 9.12 Boiling Point Temperature from Vapor Pressure

Known: In this example, we focus on how to use the parameter correlation to improve the result of an uncertainty analysis. An *Excel* worksheet shown in Figure 9.64 lists data for the vapor pressure versus temperature of pure water (Meyer 1997).

Find: Estimate the boiling point temperature of water at 1.00×10^5 Pa from a log linear fit of the data:

$$\ln(P) = \frac{a}{T} + b \tag{9.108}$$

Assumptions: Neglect the effects of linearizing the data on the least-squares regression parameters.

Analysis: Use the *VBA* macro **LSREGRESS** to calculate the uncertainties and correlation coefficient between the model parameters a and b. Referring to Figure 9.64, the *Excel* worksheet calculations in columns **C**, **D**, and **E** use the following formulas:

C2=1/A2   D2=ln(B2)   E2=$H$2*C2+$H$3

	A	B	C	D	E	F
2	T/K	P/Pa	1/T	lnP	lnP model	Residuals
3	353.15	47373	0.0028	10.76581	10.77066387	-0.004856
4	358.15	57815	0.0028	10.965	10.96661537	-0.001612
5	363.15	70117	0.0028	11.15792	11.157171	0.00075
6	368.15	84529	0.0027	11.34485	11.34255059	0.002299
7	373.15	101320	0.0027	11.52604	11.52296221	0.003077
8	378.15	120790	0.0026	11.70181	11.69860294	0.003206
9	383.15	143240	0.0026	11.87228	11.86965954	0.002617
10	393.15	198480	0.0025	12.19844	12.19871997	-0.000276
11	403.15	270020	0.0025	12.50625	12.51145593	-0.005205

	G	H	I	J	K	L
1			Gauss Newton			
2	a	-4956.82	$u_Y = \pm$	3.5E-03	$U_{Y,95\%} =$	8.3E-03
3	b	24.80667	$u_1 = \pm$	1.1E+01	$U_{1,95\%} = \pm$	2.5E+01
4	SSR	8.58E-05	$u_2 = \pm$	2.9E-02	$U_{2,95\%} = \pm$	6.8E-02
5			$R^2 =$	1.0000	Adj $R^2 =$	1.0000
6			$R_{1,2} =$	-0.9992	DoF =	7
7					$t_{95\%} =$	2.36
8					SSR =	8.58E-05
9					AAD =	2.7E-03

Figure 9.64

Water vapor pressure data and linearized model fit using the *VBA* macro LSREGRESS.

Figure 9.65 shows the linearized data plotted as ln(*P*) versus 1/*T*, along with the results from the least-squares regression of the data with Equation (9.108). The coefficient of determination $R^2 = 1$ also supports a good model fit. The residual plot is also included for additional model validation. Despite the appearance of a good fit, the clear trend in the residuals indicates that the model does not capture some nonlinear behavior. Nevertheless, the simple model serves our purpose to illustrate issues with uncertainty and correlation among model parameters.

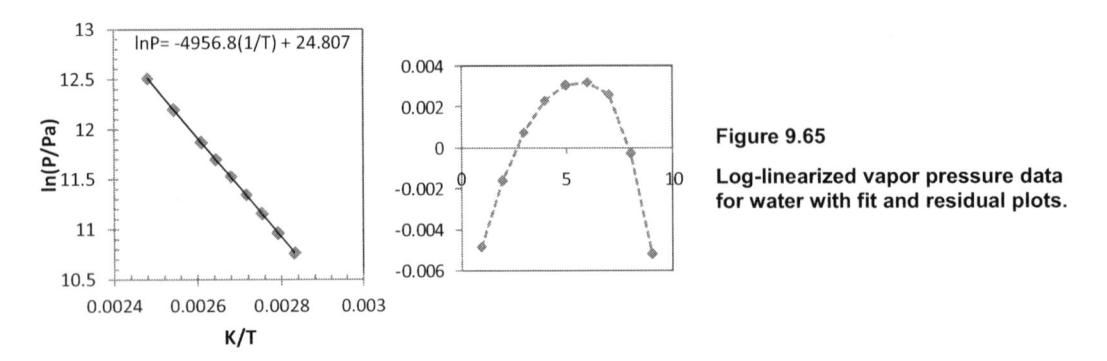

Figure 9.65

Log-linearized vapor pressure data for water with fit and residual plots.

Rearrange Equation (9.108) for T in terms of P:

$$T = \frac{a}{\ln(P) - b} \tag{9.109}$$

Use the `VBA` macro **JITTER** to calculate the uncertainty in the boiling point temperature at 10^5 Pa, ignoring correlation between the model parameters a and b in Equation (9.109). Set up the function, variables, and uncertainty information in an *Excel* worksheet, as shown in Figure 9.66. The temperature calculation uses Equation (9.109) with the mean values shown for P, a, and b from Figure 9.64. The standard uncertainties come from the **LSREGRESS** results in Figure 9.64 for a and b. We assume only systematic uncertainty in the pressure. A linear fit with slope and intercept has degrees of freedom equal to the number of data pairs minus two. We interpret the fixed (systematic) uncertainty in the pressure as a quoted value, set equal to the least significant digit 1000 divided by $\sqrt{12}$.

Variable	Mean	Stand u	DOF	Fixed u		
P/Pa	100000	0	1000	288.6751346		
a	-4957	11	7	0		
b	24.807	0.029	7	0		
T/K	372.8729	$U_{95\%} = \pm$	2.2E+00	$= \pm$	0.60%	
		$u_{max} = \pm$	1.5E+00	$= \pm$	0.41%	
		$u = \pm$	1.0E+00	$= \pm$	0.28%	
		$t_{95\%} =$	2.16	DoF =	13	
		$c_1 =$	2.80E-04	$(c \cdot u_1)^2 =$	0.62%	
		$c_2 =$	-5.68E-02	$(c \cdot u_2)^2 =$	36.91%	
		$c_3 =$	-2.80E+01	$(c \cdot u_3)^2 =$	62.47%	

Figure 9.66

Boiling point temperature calculation and uncertainty information required by the `VBA` macro JITTER ignoring correlation between the parameters a and b.

For comparison, we repeat the calculation of uncertainty propagation including the correlation coefficient of -0.9992 between parameters a and b in **JITTER** to obtain the results shown in Figure 9.67.

$U_{95\%} = \pm$	4.5E-01	$= \pm$	0.12%
$u_{max} = \pm$	2.0E+00	$= \pm$	0.54%
$u = \pm$	2.1E-01	$= \pm$	0.06%
$t_{95\%} =$	2.16	DoF =	13
$c_1 =$	2.80E-04	$(c \cdot u_1)^2 =$	0.62%
$c_2 =$	-5.68E-02	$(c \cdot u_2)^2 =$	36.91%
$c_3 =$	-2.80E+01	$(c \cdot u_3)^2 =$	62.47%

Figure 9.67

`VBA` macro JITTER results from uncertainty propagation in Equation (9.109) including correlation between the parameters a and b.

Cell **J6** in Figure 9.64 contains the correlation coefficient among the regression parameters requested by the **JITTER** macro, as shown in Figure 9.68, where the variables P, a, and b are numbered 1, 2, and 3, respectively. We set all other correlation coefficients to zero.

JITTER:
R(2,3)
=$J$6

Figure 9.68

Input box for the correlation coefficient between parameters a and b. Use the default value 0 for all other values of $R(i, j)$.

CHAPTER 9: REGRESSION

The results for the boiling point temperature and expanded uncertainty with and without correlation give:

- *Ignore Correlation* $\quad\quad T_{bp} = 373 \pm 3 K \quad$ (95% confidence) $\quad\quad$ (9.110)

- *Include Correlation* $\quad\quad T_{bp} = 372.9 \pm 0.5 K \quad$ (95% confidence) $\quad\quad$ (9.111)

Comments: The expanded uncertainties in the boiling temperatures without and with correlation between the fitting parameters *a* and *b* have different results. Ignoring correlation between the regression parameters appears to *overestimate* the uncertainty in the result for the boiling point temperature. The uncertainty value of 0.5 K seems reasonable given that the temperature data have precision to the second decimal place.

We can eliminate correlation by centering the data in Equation (9.112):

$$\ln(P) = \frac{A}{T^*} + B \quad\quad (9.112)$$

where the transformed inverse temperature is:

$$\frac{1}{T^*} = \frac{1}{T} - \frac{1}{n}\sum\left(\frac{1}{T}\right) \quad\quad (9.113)$$

Figure 9.69 has the results from the *VBA* macro **LSREGRESS** using centered data.

$u_Y = \pm$	3.5E-03	$U_{Y,95\%} =$	8.3E-03
$u_1 = \pm$	1.1E+01	$U_{1,95\%} = \pm$	2.5E+01
$u_2 = \pm$	1.2E-03	$U_{2,95\%} = \pm$	2.8E-03
$R^2 =$	1.0000	Adj $R^2 =$	1.0000
$R_{1,2} =$	0.0000	DoF =	7
		$t_{95\%} =$	2.36
		SSR =	8.58E-05
		AAD =	2.7E-03

Figure 9.69

Regression and parameter uncertainty results from the *VBA* macro LSREGRESS using centered data.

Rearrange Equation (9.112) for the temperature:

$$\frac{1}{T} = \frac{\ln(P) - B}{A} + \frac{1}{n}\sum\left(\frac{1}{T}\right) \quad or \quad T = \left[\frac{\ln(P) - B}{A} + \frac{1}{n}\sum\left(\frac{1}{T}\right)\right]^{-1} \quad\quad (9.114)$$

Figure 9.70 contains the results from **JITTER** uncertainty propagation using centered data without correlation.

Variable	Mean	Stand u	DOF	Fixed u	
P/Pa	100000	0	1000	288.6751	
A	-4957	11	7	0	
B	11.560	1.20E-03	7	0	
T/K	372.8689	$U_{95\%} = \pm$	1.7E-01	= ±	0.0463%
		$u_{max} = \pm$	1.2E-01	= ±	0.0315%
		$u = \pm$	8.8E-02	= ±	0.0235%
		$t_{95\%} =$	1.97	DOF =	261
		$c_1 =$	2.80E-04	$(c \cdot u_1)^2 =$	85.1704%
		$c_2 =$	-2.67E-04	$(c \cdot u_2)^2 =$	0.1122%
		$c_3 =$	-2.80E+01	$(c \cdot u_3)^2 =$	14.7174%

Figure 9.70

Uncertainty propagation using the macro JITTER on centered data.

By centering the data to eliminate correlation among model parameters, we get a result for boiling point temperature with the same level of uncertainty that we found when we included correlation in the original model (compare Equation (9.111)):

- *Centered* $\quad\quad T_{bp} = 372.9 \pm 0.2 K \quad$ (95% confidence) $\quad\quad$ (9.115)

□

9.7 Levenberg-Marquardt Regression

Due to its improved convergence stability, the Levenberg-Marquardt method has become a standard method for non-linear least-squares regression in many commercial software applications. The Levenberg-Marquardt method applies a form of relaxation to the Gauss-Newton method, described in Section 9.6, by reducing the sensitivity of convergence to the initial guesses for the model parameters. The method starts from the Gauss-Newton method of finding the model parameters that minimize the sum of squared residuals in Equation (9.6):

$$SSR(a_j) = \sum_{i=1}^{n}\left[y_i - f_i(a_j, x_i)\right]^2 \qquad j = 1\ldots k \qquad (9.116)$$

where a represents the vector of k model parameters, and y_i and f_i are the data and model values for the dependent variable at each value for the independent variable x_i. We expand the model function in a first-order truncated Taylor series about initial guesses for the model parameters to give a linear approximation for the model at each data point:

$$f_i(a + \delta, x_i) \cong f(a) + \sum_{j=1}^{k} \delta_j \frac{\partial f_i}{\partial a_j} \qquad (9.117)$$

where δ is a vector of small perturbations in parameters a. The partial derivatives with respect to each parameter forms a Jacobian matrix of the model function:

$$J_{ij} = \partial f_i / \partial a_j \qquad (9.118)$$

where i and j are the index variables for the data and model parameters, respectively. When we substitute the linear function approximation for f into Equation (9.116) we get:

$$SSR(a + \delta) \cong \sum_{i=1}^{n}\left[y_i - f_i(a, x_i) - J_i \delta\right]^2 \qquad (9.119)$$

where $J\delta$ is the product of the matrices J and δ evaluated at each data point. We now differentiate the SSR in Equation (9.119) with respect to δ and set the result to zero to find the calculus minimum:

$$\frac{\partial SSR(a + \delta)}{\partial \delta_j} = 0 \cong -\sum_{i=1}^{n} 2\left[y_i - f_i(a, x_i) - J_i \delta\right] J_{ij} \qquad (9.120)$$

In vector notation, Equation (9.120) simplifies to

$$(J^T J)\delta = J^T \left[y - f(a, x)\right] \qquad (9.121)$$

where J^T is the transpose of the Jacobian. Equation (9.121) represents a system of $j=1\ldots k$ linear equations in the perturbation parameters δ_j. We iteratively solve Equation (9.121) for the vector of δ values, and then replace the model parameters with $a_j + \delta_j$ until reaching a converged solution.

Levenberg (1944) added a damping factor in Equation (9.121) to control divergence when starting from poor initial guesses for the model parameters:

$$(J^T J + \lambda I)\delta = J^T \left[y - f(a, x)\right] \qquad (9.122)$$

where λ is the damping factor (use a larger value of λ to control convergence; smaller λ to increase rate of convergence by Newton's method) and I is the identity matrix. Unfortunately, for large λ, the method has a slow rate of convergence based on a small gradient of SSR with respect to a. To take larger "steps" towards the minimum along directions of small gradients, and promote faster convergence, Marquardt (1963) proposed scaling the terms and replaced the identity matrix with the diagonal of the gradient:

$$\left[J^T J + \lambda \mathrm{diag}(J^T J)\right]\delta = J^T \left[y - f(a, x)\right] \qquad (9.123)$$

CHAPTER 9: REGRESSION

The *VBA* macro **LEVENBERGM** and its option on the user-form **LSREGRESS** in the *PNMSuite* cycle through the following steps of the Levenberg-Marquardt method until converging on the model parameters that minimize the sum of squared residuals:

1. Guess the values for the model parameters, *a*.
2. Evaluate the model at each data point, *x*, to give the vector $f(a, x)$.
3. Provide analytic expressions for the partial derivatives for the Jacobian matrix, or use second-order finite difference approximations:

$$J_{ij} \cong \frac{f(a_j + \varepsilon, x_i) - f(a_j - \varepsilon, x_i)}{2\varepsilon} \quad (9.124)$$

where ε is a small number relative to *a*.

4. Calculate the coefficient matrix and constant vectors in Equation (9.123):

$$A = \left[J^T J + \lambda \mathrm{diag}(J^T J) \right] \quad \text{and} \quad b = J^T \left[y - f(a, x) \right] \quad (9.125)$$

5. Solve the system of linear equations for the vector of δ values:

$$\delta = A^{-1} b \quad (9.126)$$

6. Replace the values for the model parameters with $a \leftarrow a + \delta$ and calculate the *SSR*.
7. If the *SSR* decreases, reduce the value of the damping factor, λ, by 50%. Otherwise, increase λ by a factor of two, incrementally until SSR decreases.
8. Stop when the *SSR* converges.

The macros require the same *Excel* worksheet set up and inputs needed for least-squares minimization with the **Solver**. We need columns of the data and model predictions using initial guesses for the model parameters. By default, we calculate the Jacobian matrix from finite difference derivative approximations. The user may elect to use analytical values for the Jacobian by changing the *VBA* code in **LEVENBERGM** for the value of the string variable: **jcb = "UD"**. The user-form **LSREGRESS** has the option for the analytical Jacobian formulas on the *Excel* worksheet and an option for generating graphs of the residual and sorted residual plots.

The popularity of the Levenberg-Marquardt method stems from its relative robustness when starting from poor initial guesses for the model parameters. The method of Levenberg-Marquardt usually outperforms the algorithms employed by the **Solver** for nonlinear regression (de Levie 2004). However, we need to be careful when starting from poor guesses when dealing with nonlinear models. We may converge to a local, but not the global minimum. In these cases, we may try an algorithm such as Simulated Annealing, Luus-Jaakola, PSO, or Firefly to locate the global minimum of least-squares regression. We may also use the method of Levenberg-Marquardt to find roots to systems of equations (see the macro **ROOTSLM** or user-form **ROOTS**).

Example 9.13 Levenberg-Marquardt Regression of Antoine's Equation

Known/Find: Repeat Example 9.5 using the Levenberg-Marquardt option in the user-form **LSREGRESS**.

Assume: a first order reaction with $m = 1$.

Analysis: We can use the same worksheet from Example 9.5 with the user-form **LSREGRESS**. We also elect to supply the Jacobian derivatives in the range **G6:I16**, as seen in columns **E**, **F**, and **G** in Figure 9.71.

$$\frac{\partial \ln P_v}{\partial A} = 1 \qquad \frac{\partial \ln P_v}{\partial B} = -\frac{1}{T + C} \qquad \frac{\partial \ln P_v}{\partial C} = \frac{B}{(T + C)^2} \quad (9.127)$$

The worksheet formulas for the columns of partial derivatives are:

E6 = 1   F6 = -1/(A6 + $B$3)   G6 = $B$2/(A6 + $B$3)^2

	E	F	G
5	dlnP/dA	dlnP/dB	dlnP/dC
6	1	-0.00426725	0.072958218
7	1	-0.00409261	0.067108628
8	1	-0.0039317	0.061935346
14	1	-0.00318124	0.040548096
15	1	-0.00308316	0.038086317
16	1	-0.00299094	0.035842109

Figure 9.71

Jacobian matrix of analytical results for the partial derivatives of the regression model with respect to the parameters defined by Equation (9.127).

We start with initial guesses of one for each of the parameters in the range **B1:B3** and run the macro **LSREGRESS** from the *PNMSuite* workbook. The inputs refer to the worksheet in Figure 9.13 and Figure 9.71. The macro replaces the initial guesses in the range **B1:B3** with the results for the model parameters from the regression analysis.

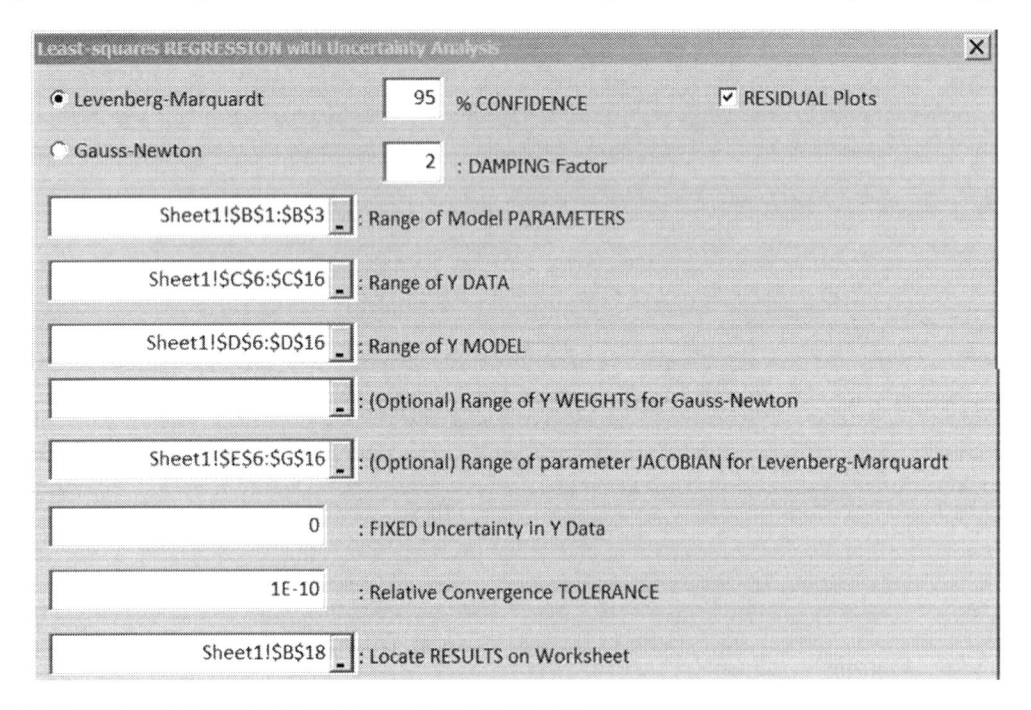

Least-squares REGRESSION with Uncertainty Analysis

- ● Levenberg-Marquardt
- ○ Gauss-Newton

95 % CONFIDENCE ☑ RESIDUAL Plots

2 : DAMPING Factor

Sheet1!$B$1:$B$3 : Range of Model PARAMETERS

Sheet1!$C$6:$C$16 : Range of Y DATA

Sheet1!$D$6:$D$16 : Range of Y MODEL

: (Optional) Range of Y WEIGHTS for Gauss-Newton

Sheet1!$E$6:$G$16 : (Optional) Range of parameter JACOBIAN for Levenberg-Marquardt

0 : FIXED Uncertainty in Y Data

1E-10 : Relative Convergence TOLERANCE

Sheet1!$B$18 : Locate RESULTS on Worksheet

	B	C	D	E
18	$u_Y = \pm$	7.8E-4	$U_{Y95\%} =$	1.8E-3
19	$u_1 = \pm$	2.2E-2	$U_{1,95\%} = \pm$	4.9E-2
20	$u_2 = \pm$	1.2E+1	$U_{2,95\%} = \pm$	2.8E+1
21	$u_3 = \pm$	4.1E-1	$U_{3,95\%} = \pm$	9.4E-1
22	$R^2 =$	1.0000	Adj $R^2 =$	1.0000
23	$R_{1,2} =$	0.9987	DoF =	8
24	$R_{1,3} =$	0.9951	$t_{95\%} =$	2.31
25	$R_{2,3} =$	0.9988	SSR =	4.8E-6
26			AAD =	5.6E-4

	A	B
1	A	18.6179355
2	B	4006.61544
3	C	234.3428581

Comments: The Levenberg-Marquardt result improves the solution obtained with the **Solver** as evidenced by a smaller *SSR* based on the smaller relative convergence tolerance specified for the solution in this example.

☐

9.8 Monte Carlo Analysis of Model Uncertainty*

"Patience is bitter, but its fruit is sweet." – Jean-Jacques Rousseau

We may apply Monte Carlo uncertainty analysis to regression. Before we outline the method, recall that we are calculating *uncertainties* in the model predictions, which do not require a high degree of precision. In Monte Carlo uncertainty analysis, we repeat the least-squares regression many times with simulated random data sets (Delahunty and Mack 1993). We generate the simulated data under the assumption of normal (Gaussian) distributions for the uncertainty in the experimental data for the dependent variable (Feller and Blaich 2001). A large number of Monte Carlo simulations with regressions yield an uncertainty distribution for each model parameter. Monte Carlo analysis of regression uncertainty has the advantage of directly incorporating any effects of correlation between data and fitting parameters (Alper and Gelb 1990). A *VBA* macro **UNSOLVER** and user-form **REGRESSUN** are available in the *PNM-Suite* for implementing Monte Carlo uncertainty analysis of regression employing the following steps:

1. Start from the best-fit parameters determined by nonlinear least-squares regression in an *Excel* worksheet with ranges containing the regression data, calculated (or best fit) data, and sum of the squared residuals.

2. Calculate u_f, the standard uncertainty in the fit according to Equation (9.68):

$$u_f = \sqrt{\frac{SSR}{n-k}} \qquad (9.128)$$

   The Monte Carlo method has the option for including fixed uncertainties in the *y* data.

3. Use the Monte Carlo approach to generate several simulations (≥1000) of "noisy" data, y_{sim}, normally distributed about the best fit, y_{model}. For $i = 1, 2 \ldots n$ data points:[91]

$$y_{i,sim} = y_{i,\text{model}} + r_N u_f + + r_U u_z \qquad (9.129)$$

   where r_N and r_U are random numbers generated from a normal distribution with mean 0 and standard deviation of 1, and a uniform distribution from -1 to 1, respectively. The uncertainties u_f and u_z are the random and fixed uncertainties in *y*.

4. Use the **Solver** to perform a unique least-squares regression of each simulated data set in step 3. The **Solver** runs with auto-scaling. However, we may need to scale the model parameters manually if any parameters have values with magnitude smaller than 10^{-3}. To scale the parameters for the **Solver**, include a multiplier coefficient for each parameter in the model. For example, if a parameter *x* has a value 10^{-3}, replace *x* with the product of a scale factor: $0.001x$. Now the **Solver** searches for a value of *x* with an order of magnitude of one. If necessary, substitute the method of Levenberg-Marquardt for the **Solver**.

5. Find the 95% confidence interval by sorting the results for each regression parameter in ascending order, then eliminating the bottom 2.5% and top 2.5% quantiles of the sorted simulations for each parameter, just like the Bootstrap method introduced in Section 8.1.3. The remaining minimum and maximum values provide the lower and upper limits of the 95% confidence interval for each parameter. We may modify the settings in the macro for any level of confidence.

The *VBA* macro **UNSOLVER** and user-form **REGRESSUN** make extensive use of worksheet functions for statistical analysis. The macros call a simple sub procedure **HEAPSORT** for sorting the simulated results in ascending order. We may elect to use weighted least-squares regression by selecting a range of inverse weight factors corresponding to the dependent data. Table 9.7 maps the output from the *VBA* macro **UNSOLVER** and user-form **REGRESSUN** to the worksheet.

---

[91] We should use an appropriate distribution, as circumstances require, such as lognormal.

Table 9.7 Sample map of user-form REGRESSUN or UNSOLVER macro output using the Monte Carlo method for a two-parameter problem. The macros add comments in the cells with definitions of the terms.

$u_Y = \pm$	Standard uncertainty in the model	$U_{Y,95\%} = \pm$	Expanded uncertainty in the model
$U_{1,95\%} = -$	Lower interval of expanded uncertainty in parameter 1	$U_{1,95\%} = +$	Upper interval of expanded uncertainty in parameter 1
$U_{2,95\%} = -$	Lower interval of expanded uncertainty in parameter 2	$U_{2,95\%} = +$	Upper interval of expanded uncertainty in parameter 2
$R^2 =$	Coefficient of determination	Adj $R^2 =$	Adjusted R^2
$R_{1,2} =$	Correlation coefficient for parameters 1 with 2	DOF =	Degrees of freedom
		$t_{95\%} =$	Coverage factor
		SSR =	Sum of squared residuals
		AAD =	Average absolute deviation

Example 9.14   Monte Carlo Regression Uncertainty in Vapor Pressure

Known/Find: Employ the *VBA* user-form **REGRESSUN** to calculate the uncertainties in the parameters for the vapor pressure model in Example 9.12. Calculate the uncertainty in the boiling point temperature.

Analysis: Add a line to the *VBA* code for the user-form **REGRESSUN** to record the simulated values to the worksheet for the parameters a and b in the original problem and simulated values for A and B in the centered problem.

```
Cells(i+50, j) = a(j)
```

REGRESSUN has the same input requirements as **LSREGRESS**, shown in Example 9.11. Figure 9.72 displays the results from the user-form **REGRESSUN** with *YX* scatter plots for the 1000 simulated parameters for the original equation as well as the modified form for centered data (Ogren, Davis and Guy 2001). We observe that centering the data removes the linear correlation from the regression parameters in the scatter plot of A versus B.

We may use the results from Figure 9.72 in either macro **JITTER** or **UNMCLHS** to calculate the propagation of uncertainty in the boiling point temperature. A better approach calculates the expanded uncertainty directly in the user-form **REGRESSUN**. To do this, we need to calculate the temperature for each simulation of the model parameters and sort the results to eliminate the lower 2.5% and upper 2.5% quantiles. To use the model with centered data for this calculation, we modify the worksheet setup by placing the formula for temperature according to Equation (9.114) in cell **B12** on the worksheet, adding a term to sample the systematic uncertainty in pressure from a uniform distribution:

$$B12 = ((LN(100000 + (2*RAND() - 1)*1000/SQRT(12)) - I3)/I2 + I1) \wedge -1$$

We then add a simple line of *VBA* code in the user-form **REGRESSUN** simulation loop to capture the calculated value for temperature after finding the model parameters for each simulation, and writing the result back to the worksheet in successive rows:

```
Cells(i + 15, 3) = Cells(12, 2)
```

Once we complete the Monte Carlo simulations with the user-form **REGRESSUN**, we sort the results for temperature on the worksheet to get the lower and upper limits on the confidence interval:

$$372.76K \leq \left(T_{bp} = 372.87K\right) \leq 372.98K \; (95\% \; confidence) \tag{9.130}$$

Alternatively, $U_{95\%} = \pm 0.11$ K (which rounds up to ± 0.2K).

CHAPTER 9: REGRESSION

Original Data MC Regression Uncertainty

$u_Y = \pm$	3.5E-03	$U_{Y,95\%} = \pm$	8.3E-03
$U_{1,95\%} = -$	2.1E+01	$U_{1,95\%} = +$	1.8E+01
$U_{2,95\%} = -$	4.9E-02	$U_{2,95\%} = +$	5.5E-02
$R^2 =$	1.0000	Adj $R^2 =$	1.0000
$R_{1,2} =$	-0.9991	DoF =	7
		$t_{95\%} =$	2.36
		SSR =	8.6E-05
		AAD =	2.66E-03

Centered Data MC Regression Uncertainty

$u_Y = \pm$	3.5E-03	$U_{Y,95\%} = \pm$	8.3E-03
$U_{1,95\%} = -$	2.0E+01	$U_{1,95\%} = +$	2.1E+01
$U_{2,95\%} = -$	2.3E-03	$U_{2,95\%} = +$	2.3E-03
$R^2 =$	1.0000	Adj $R^2 =$	1.0000
$R_{1,2} =$	0.0468	DoF =	7
		$t_{95\%} =$	2.36
		SSR =	8.6E-05
		AAD =	2.66E-03

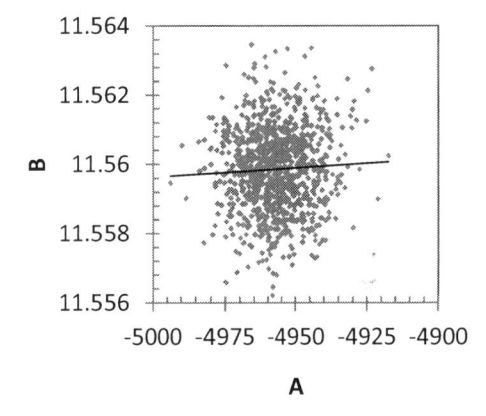

Figure 9.72 Monte Carlo simulations of data for least-squares regression of (*a* vs. *b*) original data showing parameter correlation, and (*A* vs. *B*) centered temperature data with no parameter correlation.

Comments: The Monte Carlo results for regression uncertainty are comparable to the results using the *Law of Propagation of Uncertainty* calculated in *Example 9.12*.

□

Not all least-squares regression analyses allow for centering data. We often regress a model to the original data in order to obtain the values of specific model parameters.

Example 9.15   Monte Carlo Regression Uncertainty in a FOPDT Model

Known: Use a First-Order-Plus-Dead-Time (FOPDT) model for the output plotted in Figure 9.73 using the data shown in the worksheet in Figure 9.74. The data are the result of a unit step change to the input of a process. The *FOPDT* model has two parts (Co 2008):

$$X(t) = \begin{cases} X_0 & t \leq \tau_d \\ X_{ss} + (X_0 - X_{ss})\exp\left(-\frac{t-\tau_d}{\tau}\right) & t > \tau_d \end{cases} \quad (9.131)$$

where *t* is time, X_0 and X_{ss} are the initial and final steady-state values for the output variable, and τ and τ_d are the process time constant and dead time, respectively. The steady-state output relates to the input by the addition of the process gain, *K*:

$$X_{ss} = X_0 + K \quad (9.132)$$

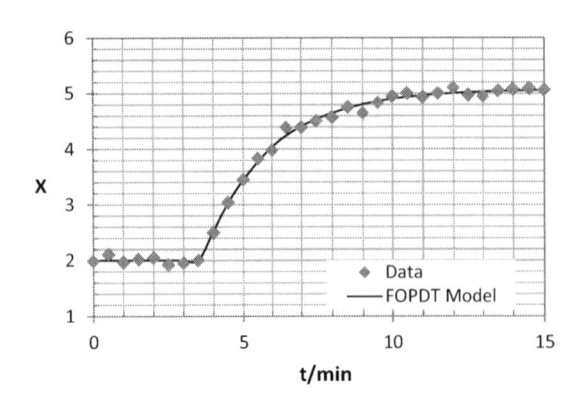

Figure 9.73
Output from a unit step change in the input showing a FOPDT response. The gridlines help estimate the model parameters.

Find: Use nonlinear regression with Monte Carlo uncertainty analysis to find the model parameters τ, τ_d, K and X_0, and their corresponding 95% confidence limits.

Analysis: First, use the **Solver** to perform the nonlinear regression of the model to fit the data. We rely on our engineering judgment to initialize the model parameters in the range **A1:A4**. For example, from the plot of the response in Figure 9.73, we observe the following:

- The initial output value $X_0 \approx 2$
- The steady-state output value approaches $X_{SS} \approx 5$ for large time.
- The dead-time interval from the start ($t = 0$) to where the response begins to increase is $\tau_d \approx 3.5$ min.
- Estimate the process time constant from the time when the response to the step input is approximately two-thirds of the way to the new steady-state value [*i.e.*, $X_{2/3\ SS} = X_0 + (2/3)(X_{SS} - X_0) = 4$], $\tau \approx 5.5 - \tau_d = 2$ min.

Employ the *VBA* user-form **REGRESSUN** to perform 10^4 Monte Carlo simulations of the data with regression. The macro prompts for the model parameters in the range **B1:B4** of Figure 9.74. The data and model predictions for X start in row **16** of columns **B** and **C**. The macro also prompts for a cell to display the progress of the solution, because we are unable to use the **Status Bar** while the **Solver** is running.

Calculate the steady-state output value X_{SS} from the values for the initial output and process gain in cells **B3** and **B4**:

$$B5 = B3 + B4$$

Then, calculate model-predicted values for X using the range **B1:B4** for the parameters in Equation (9.131). The worksheet formula chooses between the two phases of the *FOPDT* model with the *Excel* worksheet **IF** function:

$$C9 = IF(A9 < \$B\$2, \$B\$4, \$B\$5 + (\$B\$4 - \$B\$5) * EXP(-(A9 - \$B\$2)/\$B\$1))$$

	A	B	C	D	E	F
1	τ	2.223	Unsolver			
2	τ_d	3.589	$u_Y = \pm$	6.4E-02	$U_{Y,95\%} = \pm$	1.3E-01
3	K	3.068	$U_{1,95\%} = -$	1.9E-01	$U_{1,95\%} = +$	1.9E-01
4	X_0	2.004	$U_{2,95\%} = -$	1.2E-01	$U_{2,95\%} = +$	1.1E-01
5	X_{SS}	5.072	$U_{3,95\%} = -$	6.3E-02	$U_{3,95\%} = +$	6.7E-02
6	SSR	0.111638	$U_{4,95\%} = -$	4.5E-02	$U_{4,95\%} = +$	4.3E-02
7			$R^2 =$	0.9977	Adj $R^2 =$	0.9975
8			$R_{1,2} =$	-0.7145	DoF =	27
9			$R_{1,3} =$	0.5303	$t_{95\%} =$	2.05
10			$R_{1,4} =$	-0.0432	SSR =	1.12E-01
11			$R_{2,3} =$	-0.4715	AAD =	4.59E-02
12			$R_{2,4} =$	0.3595		
13			$R_{3,4} =$	-0.6934		

CHAPTER 9: REGRESSION

	A	B	C	D
16	t	X	X Model	(X-X Model)
17	0	1.98734	2.00376	0.01642
18	0.5	2.11305	2.00376	-0.10929
19	1	1.97263	2.00376	0.03113
20	1.5	2.02304	2.00376	-0.01927
21	2	2.04900	2.00376	-0.04523
22	2.5	1.92461	2.00376	0.07915
23	3	1.95624	2.00376	0.04753
			⋮	
40	11.5	5.00080	4.98488	-0.01592
41	12	5.09558	5.00248	-0.09310
42	12.5	4.97263	5.01653	0.04390
43	13	4.94518	5.02775	0.08257
44	13.5	5.04905	5.03671	-0.01234
45	14	5.07584	5.04387	-0.03197
46	14.5	5.08867	5.04959	-0.03909
47	15	5.05793	5.05415	-0.00378

Figure 9.74

FOPDT model parameters and 95% confidence limits from *VBA* macro REGESSUN with the Monte Carlo method.

Process output response to a unit step change.

Process data and model predictions in columns **B** and **C**; residuals in column **D**.

Finally, calculate the residuals between the model and data in column **D** and plot as shown in Figure 9.75. We observe no trends in the residual plot and a sigmoid shape in the sorted residual plot indicating a good model of the data.

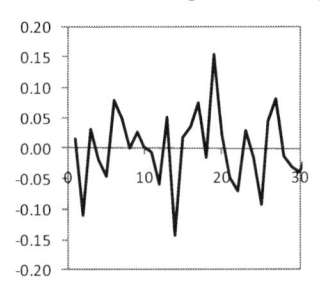

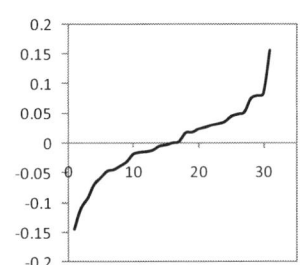

Figure 9.75

FOPDT model raw residual and sorted residual plots show a normal distribution that supports a good model fit of the data.

Comments: The Monte Carlo method of regression uncertainty analysis incorporates correlation among the model parameters, and is thus superior to the uncertainty estimates from Gauss-Newton least-squares regression. However, the method is computationally demanding, and slow using *VBA* with repeated reference to ranges of cells on the worksheet.

Figure 9.76 has the results for the 95% confidence intervals using Gauss-Newton regression and sensitivity coefficients. The Monte Carlo 95% confidence results essentially agree with the Gauss-Newton results for uncertainty propagation, with minor differences due primarily to the effects of correlation among the parameters, and secondarily to any nonlinearity in the model.

$u_Y = \pm$	6.4E-02	$U_{Y,95\%} =$	1.3E-01
$u_1 = \pm$	1.0E-01	$U_{1,95\%} = \pm$	2.1E-01
$u_2 = \pm$	5.9E-02	$U_{2,95\%} = \pm$	1.2E-01
$u_3 = \pm$	3.4E-02	$U_{3,95\%} = \pm$	7.0E-02
$u_4 = \pm$	2.3E-02	$U_{4,95\%} = \pm$	4.7E-02
$R^2 =$	0.9977	Adj $R^2 =$	0.9975
$R_{1,2} =$	-0.7337	DoF =	27
$R_{1,3} =$	0.5559	$t_{95\%} =$	2.05
$R_{1,4} =$	0.0000	SSR =	1.12E-01
$R_{2,3} =$	-0.4727	AAD =	4.59E-02
$R_{2,4} =$	0.2787		
$R_{3,4} =$	-0.6646		

Figure 9.76

Gauss-Newton uncertainty propagation in FOPDT model parameters.

Figure 9.77 shows scatter plots revealing correlation among parameters. We scaled the plotted parameter values to compare the correlations on the same basis. All combinations of parameters show some degree of correlation. However, we find little correlation between the process time constant and the initial output value.

Figure 9.77

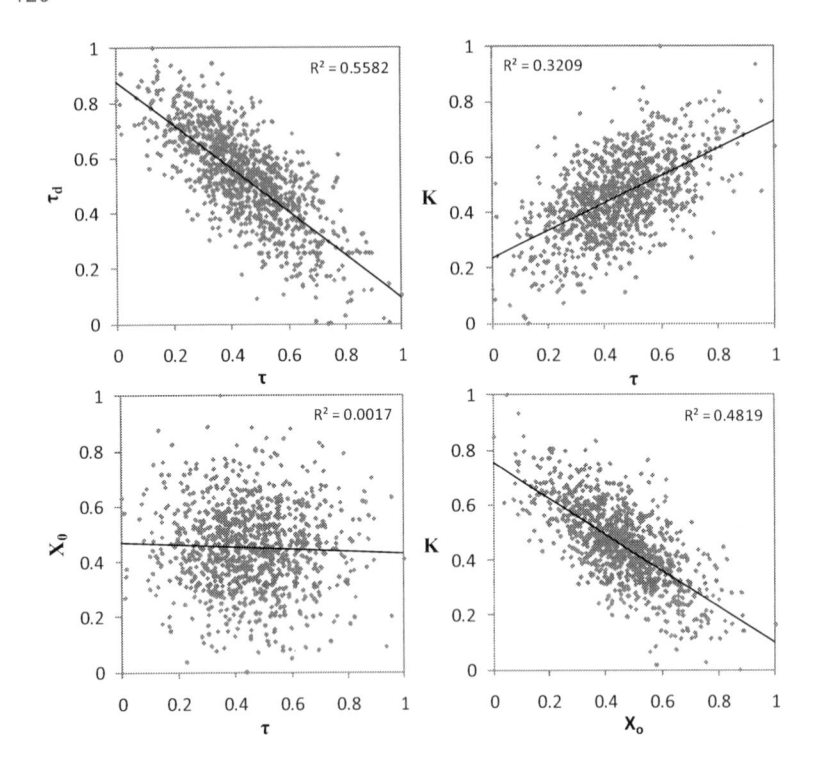

Scaled Monte Carlo simulation data for regression parameters shows some correlation among regression parameters. However, the initial output value is independent of the time constant.

9.9 Graph Confidence Intervals and Error Bars

Confidence intervals for linear calibration curves should account for correlation between the slope and intercept of the best fit to Equation (9.15). The confidence interval is a function of the independent variable, x (Liengme 2009):

$$U_{yx} = \pm U_y \sqrt{\frac{1}{n} + \frac{(x-\bar{x})^2}{SSX}} \qquad (9.133)$$

where U_y is the expanded standard uncertainty (error) in the regression from Equation (9.71), n is the number of data pairs, and SSX is the sum of the squares of the deviations of x from the mean value of the independent variables, $\bar{x}$, defined in Equation (8.5):

$$SSX = \sum_{i=1}^{n}(x_i - \bar{x})^2 \qquad (9.134)$$

Table 9.8 lists the *Excel* worksheet functions for the terms required by Equation (9.133).

Table 9.8 *Excel* worksheet functions for calculating the 95% confidence interval for a line with correlation between the slope and intercept.

Parameter	Excel Worksheet Function	Parameter	Excel Worksheet Function
Coverage factor, $t_{95\%,\nu}$	TINV(0.05,n-2)	SSX	DEVSQ(x_range)
Standard error, $s/\sqrt{n}$	STEYX(y_range, x_range)	n	COUNT(x_range)
Mean, $\bar{x}$	AVERAGE(x_range)		

CHAPTER 9: REGRESSION

Example 9.16 Plot Confidence Intervals for Linear Calibration

Known/Find: Plot the data with the linear fit and confidence intervals for the linearized data in *Example 9.7*.

Analysis: The following steps refer to the *Excel* worksheet and plot in Figure 9.78.

1. Calculate the transformed dependent variable for a linearized filter relationship in column **C** and plot the data using relative cell referencing:

    $$C2 = A2/B2$$

2. Calculate the best fit model values for t/V using the **TREND(y_range, x_range)** worksheet function in column **D**:

    $$D2 = TREND(\$C\$2:\$C\$8, \$B\$2:\$B\$8, B2)$$

3. Calculate the terms in Table 9.8 required for the 95% confidence interval:

    B10 = TINV(0.05, COUNT(B2:B8) - 2)   B11 = STEYX (C2:C8, B2:B8)
    B12 = AVERAGE(B2:B8)   B13 = DEVSQ (B2:B8)   B14 = COUNT(B2:B8)

4. Put the confidence intervals in column **E** according to Equation (9.133):

    $$E2 = \$B\$10*\$B\$11*SQRT(1/\$B\$14 + ((B2 - \$B\$12) \wedge 2)/\$B\$13)$$

5. Add or subtract the confidence interval from the model predictions. Put the results in columns **F** and **G** to plot the curves representing the lower and upper confidence intervals for the best fit, as seen in Figure 9.78:

    F2 =D2-E2   G2 =D2+E2

	A	B	C	D	E	F	G
1	t/min	V/L	(t/V)/(min/L)	(t/V) model	U_V	V-U_V	V+U_V
2	10	2.7	3.703703704	3.67597959	1.0355	2.6405	4.7115
3	20	4	5	4.63258899	0.7552	3.8774	5.3878
4	30	5.4	5.555555556	5.66278372	0.5471	5.1157	6.2099
5	40	6.8	5.882352941	6.69297845	0.5549	6.1381	7.2478
6	50	7.4	6.756756757	7.13449048	0.6302	6.5043	7.7647
7	60	7.8	7.692307692	7.42883183	0.6965	6.7323	8.1253
8	70	8.3	8.43373494	7.79675852	0.792	7.0048	8.5887
10	t_{95}	2.5706					
11	St Error	0.5333					
12	mean time	6.0571					
13	SSX	26.357					
14	n	7					

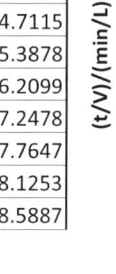

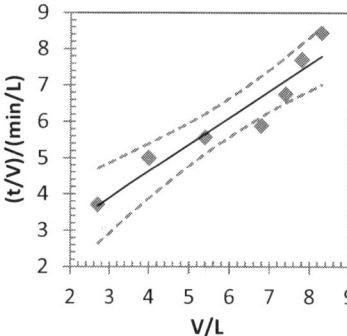

Figure 9.78

Excel worksheet and plot of the confidence interval for the transformed filter data.

Comments: The confidence interval for the linear fit becomes wider at both ends of the data. The confidence interval in this example appears to capture the spread in the data about the fit.

□

We typically use error bars to display the range of uncertainty in the individual data points. In *Excel*, click on the graph, then access error bars from the **Analysis** group of the **Chart Tools**, as shown in Figure 9.79. Use custom error bars to select a range of values on the worksheet that contain the error terms for each point.

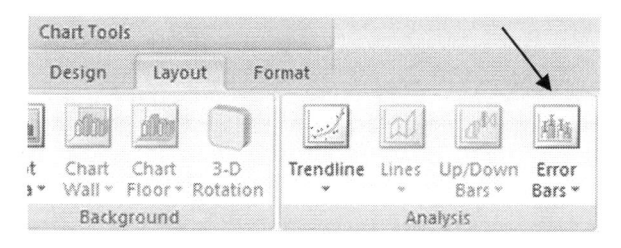

Figure 9.79

Excel worksheet Chart Tools ribbon with option for error bars.

Alternatively, click on the chart and select Error Bars from te Chart Elements (**+**).

Example 9.17 Plot Confidence Intervals and Error Bars for the Logistic Model

Known/Find: Add confidence intervals and error bars to the plot of the data and logistic equation in Example 9.10.

Analysis: Perform an uncertainty analysis of the model regression using the *VBA* macro **LSREGRESS** to obtain the expanded uncertainty in the fit, as shown in the *Excel* worksheet cell **D19** of Figure 9.80.

	A	B	C	D	E	F	G
14	Uncertainty Analysis					Lower CI	Upper CI
15	$u_1 = \pm$	8.1E-02	$U_{1,95\%} = \pm$	1.9E-01		9.74503	10.17373
16	$u_2 = \pm$	6.6E-01	$U_{2,95\%} = \pm$	1.5E+00		9.748303	10.177
17	$u_3 = \pm$	3.2E-01	$U_{3,95\%} = \pm$	7.3E-01		9.782372	10.21107
18	$u_4 = \pm$	4.8E-02	$U_{4,95\%} = \pm$	1.1E-01		9.859016	10.28772
19	$u_Y = \pm$	9.3E-02	$U_{95\%} =$	0.21435		10.04484	10.47354
20	$R^2 =$	0.9709	Adj $R^2 =$	0.9600		10.33874	10.76744
21	$R_{1,2} =$	-0.5789	DOF =	8		10.59093	11.01963
22	$R_{1,3} =$	0.6083	$t_{95\%} =$	2.31		10.71649	11.14519
23	$R_{1,4} =$	-0.6859				10.76246	11.19116
24	$R_{2,3} =$	-0.8824				10.78193	11.21063
25	$R_{2,4} =$	0.1899				10.78378	11.21248
26	$R_{3,4} =$	-0.4272				10.78391	11.21261

Figure 9.80

Uncertainty analysis for logistic equation regression using the *VBA* macro LSREGRESS.

Generate data for the lower and upper confidence intervals of the least squares fit from the model *y*-values in column **C** of Figure 9.54 and the expanded 95% confidence uncertainty in cell **D19**. The calculations of the confidence intervals in columns **F** and **G** of Figure 9.80 use the following formulas:

F15 = C2 - $D$19   G15 = C2 + $D$19

Plot the lower and upper confidence intervals with the data and least squares fit of the model, as shown in Figure 9.81. We can also add error bars to the markers for the data. Figure 9.81 shows vertical error bars representing the standard error in the data. To add error bars to a graph, click on the chart, then select **Chart Tools>Layout>Error Bars** from the **Analysis** group, as seen in Figure 9.79. In Excel 2013, click in the chart and select error bars from the **Chart Elements** (✚). Click on the horizontal error bars and delete them in the plot. Select the desired formatting features

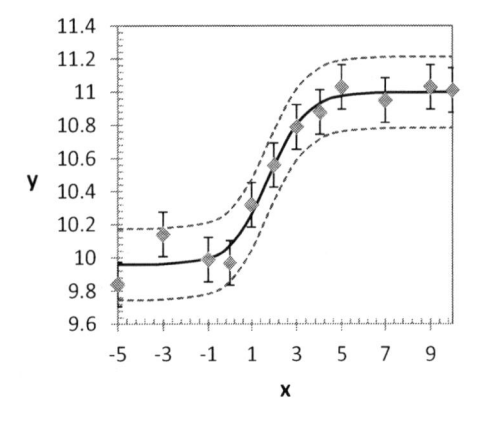

Figure 9.81

Upper and lower confidence intervals (dashed lines) for least-squares regression of the logistic equation.

Vertical error bars representing the standard error (uncertainty) of the y-data.

Comments: Confidence intervals refer to the uncertainty in the model. Generating error bars for the uncertainty in the data requires replication of the experiments and information about the readability of the measuring device. ☐

9.10 Choosing an Empirical Model

> *"Three Rules of Work: Out of clutter find simplicity; from discord find harmony; in the middle of difficulty lies opportunity."* – Albert Einstein

We conclude this chapter with a few guidelines for selecting a model. Least-squares regression is a powerful tool for smoothing and interpolating data, as well as extracting model parameters from experimental results. Where appropriate, polynomial models are easy to obtain, and have the additional benefit of simple differentiation and integration. However, we must be careful when selecting models for interpreting or interpolating data. For example, if our purpose is to extrapolate time dependent data back to time zero, our model must be determinate at the origin. In many problems, we know the value of the prediction is exact at a particular point. For example, when modeling molecular species partitioning between phases, we know that zero concentration in one phase requires zero equilibrium concentration in the adjoining phase. In this case, our least-square regression should force the intercept to pass through the origin. Many physical phenomena exhibit asymptotic behavior where regular polynomials do not perform well. Rational, exponential, or logarithmic functions are just a few examples for consideration when polynomials do not provide satisfactory models of the data:

- *Rational*:[92] $$y = \frac{p_0 + p_1 x + p_2 x^2 + \ldots p_k x^k}{1 + q_1 x + q_2 x^2 + \ldots q_m x^m} \tag{9.135}$$

- *Power*: $$y = a x^b + c \tag{9.136}$$

- *Exponential*: $$y = a e^{bx} + c \tag{9.137}$$

- *Double Exponential*: $$y = a e^{bx} + c e^{dx} \tag{9.138}$$

- *Logistic*: $$y = \frac{a}{1 + b e^{-dx}} + c \tag{9.139}$$

- *Logarithmic*: $$y = a \ln(bx) + c \tag{9.140}$$

- *Combinations*: $$y = a + bx + \frac{c}{x} + d \ln x + g x^2 \ldots \tag{9.141}$$

where y and x are the dependent and independent variables, respectively, with model parameters p, q, a, b, c, d, and g. We may consider using general-purpose interpolation methods, such as cubic spline, presented in Chapter 10, to represent smooth data. Although we found that transforming data for linear regression leads to sub-optimal model parameters, some cases *require* data transformation to eliminate poles, such as infinite derivatives or division by zero.

9.10.1 Rational Least-squares

The rational function in Equation (9.135) is our favorite "generic", utilitarian function for least squares modeling of data, particularly when we lack additional information about the physics of the relationships between dependent and independent variables. Rational functions generally outperform polynomials and are well suited to data that displays asymptotic behavior. In this section, we give some advice for choosing a rational function and derive a simple algorithm for fitting data to a rational function by least squares regression.

When choosing a rational function, consider the following advantages and disadvantages described by the *NIST Engineering Statistics Handbook* (2012):

---

[92] Adjust the orders (degrees) of the polynomials in the numerator (k) or denominator (m) to best fit the data and capture asymptotic behaviors.

Advantages
- Asymptotic properties that accommodate a wider range of shapes than polynomials
- Better interpolation and may be used for extrapolation
- Requires fewer model fitting parameters (coefficients) for complicated shapes

Disadvantages
- It is not clear how to choose the orders of the numerator and denominator polynomials, which may require additional "trial-and-error" effort to find the best form of the rational function.
- Indeterminate poles (vertical asymptotes) form at the roots of the polynomial in the denominator. Poles usually become problematic when they exist within the range of the data for the independent variable (x), or when using the rational function for extrapolation.

Generally, rational functions perform best when the orders of the polynomials in the numerator and denominator are equal, or differ by at most one order (Hanna and Sandall 1995). However, we may judiciously select the form of the rational function by considering the following properties of rational functions (NIST 2012) in reference to the orders of the polynomials in the numerator and denominator of Equation (9.135):

- For $k = m$, a horizontal asymptote for large x is $y = p_k/q_m$.
- For $m > k$, a vertical asymptote is $y = 0$.
- If $m < k$, there are no horizontal asymptotes

Equation (9.135) is nonlinear in the q-parameters and thus requires a nonlinear regression approach for fitting the data with a rational function. Because we need good initial guesses for the p and q parameters, we may first linearize the rational function by multiplying through by the q-polynomial and rearranging into the following linear form in the p's and q's:

$$y = p_0 + xp_1 + x^2 p_2 + \ldots x^k p_k - yxq_1 - yx^2 q_2 \ldots - yx^m q_m \tag{9.142}$$

If we use the x and y data to evaluate the right-hand side of Equation (9.142), we find good starting approximations for the p and q parameters by multivariable linear regression of the linearized data. We then follow up with a nonlinear regression of the data using Equation (9.135) with the **Solver** or one of our macros **GAUSSNEWTON** or **LEVEMBERGM** for nonlinear regression with uncertainty analysis.

Because of the general applicability of rational least squares models, we have included a *VBA* sub procedure **RATLS**, listed in Figure 9.82, that performs nonlinear least-squares regression for the coefficients p and q in Equation (9.135). **RATLS** uses the worksheet array function **LINEST** to perform the multivariable linear regression of the linearized data, with the option for forcing the intercept to zero ($p_0 = 0$). The macro **RATLS** starts from the linearized results from **LINEST** to perform nonlinear least-squares regression of a rational function. For convenience, we have also included a *VBA* user-defined function **RATLSF** for calculating the value of the rational function given x and the values of the polynomial coefficients p and q obtained from regression. The worksheet setup for **RATLS** is similar to the setup required by **GAUSSNEWTON** and **LEVENBERGM**, with two additional inputs and on additional output:

1. The range of values of the independent variable (x).
2. A range of empty cells for the rational function model of the data. **RATLS** populates that range for the rational model with formulas from the user-defined function **RATLSF** to calculate the model at each data point.
3. Creates a formula in a cell on the worksheet to interpolate the data by the rational function **RATLSF**.

Occasionally, a rational least-squares function may form poles at locations where the denominator is zero. We recommend the macro **BAIRSTOW** to find the roots of the polynomial in the denominator and identify any poles where the rational function may become indeterminate due to division by zero. When using rational functions, we recommend starting small and increasing the orders of the polynomials in the numerator or denominator only when necessary to obtain an adequate model. Try changing the orders of the polynomials in the numerator or denominator to eliminate poles. We may use the macro **RATLS** with a zero-order denominator to fit a regular polynomial to our data.

CHAPTER 9: REGRESSION

```vb
Public Sub RATLS()
' Nonlinear regression fit of rational functions by the method of Gauss-Newton:
'             p0 + p1*x + p2*x^2 + ... pk*x^k
'    y(x) =  -------------------------------
'             1  + q1*x + q2*x^2 + ... qm*x^m
' The value of q0 = 1.  Option to require p0 = 0 (y-intercept = 0). The worksheet
' function LINEST is used to solve the linearized system of equations for estimates of
' p's and q's by multivariable linear regression. Use sub procedure RATLSGN to perform
' nonlinear regression for improved p's and q's and uncertainty analysis.  It is
' recommended to use the sub procedure BAIRSTOW to obtain the roots of the polynomial
' in the denominator and check poles in the range of x data.  Set up a worksheet with
' columns of x & y data and a column for y model.  The macro populates the y-model
' range with the formula to calculate the rational function model values at each x.
' *** Required User-defined Functions and Sub Procedures from PNM Suite ***
' JUSTABC_123 = checks for allowed naming of worksheets
' RATLSF = rational function model
' RATLSGN = Gauss Newton least-squares solution for rational function
' *** Nomenclature ***
Dim ar As Range ' = range of least-squares regression parameters
Dim b As Integer ' = prompt to use Bairstow's method to find roots of denominator
Dim boxtitle As String ' = string for name of macro used in titles
Dim c As Integer ' = check for non-zero intercept constant for linear regression
Dim ci As Boolean ' = Boolean for including intercept in linear regression
Dim i As Long, j As Long '  = index for data & polynomial coefficients
Dim k As Long, m As Long ' = order of polynomial in numerator & denominator
Dim n As Long ' = number of data pairs
Dim output As Range, outrange As String ' = range location on worksheet for results
Dim ow As Integer ' = over-write msgbox response
Dim p As String, p0 As String ' = address of range of parameters in numerator
Dim q As String ' = address of range of parameters in denominator
Dim qp As Variant ' = array of results from LINEST
Dim rout As Long ' rows for output
Dim ws As String ' = string for name of worksheet
Dim x As Variant, y As Variant ' = arrays for x and y data
Dim xr As Range, yr As Range, yrtnlr As Range ' = ranges for x and y data and model
' ****************************************************************************
boxtitle = "RATLS"
MsgBox boxtitle & " only works in a PNMSuite worksheet. CANCEL to quit the " & vbCr _
       & " if necessary to setup the worksheet for " & boxtitle & " in PNMSuite."
data_input:
With Application
.Calculation = xlCalculationAutomatic ' turn on worksheet automatic calculation
Set xr = .InputBox(Prompt:="Range of Independent X DATA:", Type:=8, Title:=boxtitle)
Set yr = .InputBox(Prompt:="Range of Dependent Y DATA:", Type:=8, Title:=boxtitle)
Set yrtnlr = .InputBox(Prompt:="Range for Y RATIONAL:", Type:=8, Title:=boxtitle)
n = yr.Count ' get number of data pairs
If Not n = xr.Count Then
    MsgBox "Sizes of X and Y data ranges must match.": Exit Sub
End If
c = MsgBox(Prompt:="Set Y-INTERCEPT = 0?", Buttons:=vbYesNo, Title:=boxtitle)
Select Case c
     Case 6: ci = False
     Case 7: ci = True
End Select
pqset:
k = .Min(2, CInt(n / 2)): m = .Min(k, n - 1 - k) ' recommend order for num/denom
k = .InputBox(Prompt:="ORDER of Numerator:", Type:=1, Title:=boxtitle, Default:=k)
m = .InputBox(Prompt:="ORDER of Denominator:", Type:=1, Title:=boxtitle, Default:=m)
If k + m + 1 > n Then '  validate the orders of polynomials
    MsgBox "The sum of orders of numerator/denominator polynomials" & vbNewLine & _
          "must not exceed the number of data pairs (less one).": Exit Sub
End If
rout = 1 ' calculate the number of rows for the output on the worksheet
For i = 1 To k + m + 1
    For j = i + 1 To k + m + 1: rout = rout + 1: Next j
Next i
rout = .Max(rout, 5)
r_output: Set output = .InputBox(Prompt:="OUTPUT Cell:", Type:=8, Title:=boxtitle)
Set output = Range(output(1, 1).Address) ' first OUTPUT cell
```

```
ws = JUSTABC_123(output.Worksheet.Name)
output.Worksheet.Name = ws   ' Check output cells to avoid overwrite
outrange = ws & "!" & output.Address & ":" _
           & ws & "!" & output.Offset(rout, 5).Address
If Not (.CountA(Range(outrange)) = 0) Then
    ow = MsgBox(Title:=boxtitle, _
    Buttons:=vbOKCancel + vbExclamation + vbDefaultButton2, _
    Prompt:="Output will OVERWRITE " & rout + 1 & " rows x 6 column cell range.")
  If ow = 2 Then Exit Sub
End If
Range(outrange).Clear ' clear output range contents and formatting
If k + m + 1 > n Then ' Check for consistency between number of data pairs and orders
    k = .RoundDown((n - 1) / 2, 0): m = k
End If
ReDim y(1 To n, 1 To 1) As Variant, x(1 To n, 1 To k + m) As Double
For i = 1 To n ' Set the arrays for LINEST
    y(i, 1) = yr(i).Value ' y data array
    For j = 1 To k
        x(i, j) = xr(i) ^ j ' x data array for rational function numerator values
    Next j
    For j = 1 To m
        x(i, j + k) = -yr(i) * xr(i) ^ j ' x data array for ratl function denom values
    Next j
Next i
qp = .LinEst(y, x, ci) ' Linear regression for the rational function parameters
output.Activate ' ORDER OF RESULTS IS q(M), q(M-1), ...q(0), P(K), P(K-1), ... P(0)
With ActiveCell
    .Value = "YRATIONAL =":
    .Characters(Start:=2, LENGTH:=8).Font.Subscript = True
    With .Offset(1, 0) 'Output results to worksheet (parameter and value)
        For j = 1 To k + 1 ' Numerator
            With .Offset(j - 1, 0)
                .Value = "P" & j - 1 & " ="
                .Characters(Start:=2, LENGTH:=Len(CStr(j - 1))).Font.Subscript = True
            End With ' Offset
            .Offset(j - 1, 1) = qp(k + m + 2 - j)
        Next j
        For j = 1 To m ' Denominator
            With .Offset(k + j, 0)
                .Value = "Q" & j & " ="
                .Characters(Start:=2, LENGTH:=Len(CStr(j))).Font.Subscript = True
            End With ' Offset
            .Offset(k + j, 1) = qp(m + 1 - j)
        Next j
        ' Strings of worksheet addresses for parameters in numerator & denominator
        If ci = False Then ' for zero intercept
            p = .Offset(1, 1).Address & ":" & .Offset(k, 1).Address
        Else ' for non-zero intercept
            p = .Offset(0, 1).Address & ":" & .Offset(k, 1).Address
        End If
        q = .Offset(k + 1, 1).Address & ":" & .Offset(k + m, 1).Address
    End With ' Offset
    .Offset(0, 1) = "= RATLSF(" & xr(CInt(n / 2)).Address & ", " & p & "," & q _
                    & "," & ci & ")"
End With ' ActiveCell
For i = 1 To n ' populate the range of y model with function call
    yrtnlr(i) = "= RATLSF(" & xr(i).Address & ", " & p & "," & q & "," & ci & ")"
Next i
Set ar = Range(p, q) ' range of model parameters on the worksheet
Call RATLSGN(ar, output, yr, yrtnlr) ' nonlinear regression by Gauss-Newton method
End With ' Application
End Sub ' RATLS

Public Function RATLSF(x As Double, p, q, optional i as Boolean = True) As Double
' x = independent variable for point of interpolation
' p = range of coefficients of x terms in numerator polynomial: p0+p1*X+p2*x^2 + ...
' q = range of coefficients of x terms in denominator polynomial: 1+q1*x+q2*x^2 + ...
'     (note: only include q1, q2, ...)
' i = Boolean for intercept (False = zero intercept)
```

CHAPTER 9: REGRESSION

```
' *** Nomenclature ***
Dim den As Double, num As Double ' = denominator & numerator polynomials
Dim j As Long, k As Long ' = index of term in polynomial
' ******************************************************************************
num = 0: den = 1 ' get the size of the ranges for q and q
If i = True Then
    For j = 1 To p.Count: num = num + p(j) * x ^ (j - 1): Next j ' sum numerator
Else
    For j = 1 To p.Count: num = num + p(j) * x ^ j: Next j ' sum numerator
End If
For j = 1 To q.Count: den = den + q(j) * x ^ j: Next j ' sum denominator
If den = 0 Then
    MsgBox "Division by zero. Denominator has pole at x.": Exit Function
End If
RATLSF = num / den ' quotient for rational function
End Function ' RATLSF
```

Figure 9.82 VBA macro RATLS for obtaining the coefficients *p* and *q* in a rational function. The function RATLSF evaluates a rational function for values of the independent variable and coefficients *p* and *q*.

Example 9.18   Model Ethanol-water VLE Data using a Modified Rational Function

Known: Figure 9.83 has columns of data for aqueous ethanol vapor and liquid equilibrium (*VLE*) *y* vs *x* mole fractions in an *Excel* worksheet.

Find: Compare the fit of the data using a fourth-order polynomial with zero-intercept to a four-parameter rational function designed to pass through the origin:

$$y = p_1 x + p_2 x^2 + p_3 x^3 + p_4 x^4 \tag{9.143}$$

$$y = \frac{p_1 x + p_2 x^2}{1 + q_1 x + q_2 x^2} \tag{9.144}$$

Analysis: Plot the data in *Excel* and use a chart **Trend line** with a zero-intercept to fit the polynomial in Equation (9.143) to the *VLE* data. We may obtain good initial guesses for the rational function parameters *p* and *q* by transforming Equation (9.144) into a linear form:

$$y\left(1 + q_1 x + q_2 x^2\right) = p_1 x + p_2 x^2 \quad \text{or} \quad y = p_1 x + p_2 x^2 - q_1 yx - q_2 yx^2 \tag{9.145}$$

Figure 9.83 shows the transformed data arranged in contiguous columns **C**, **D**, and **E** for linear regression using *Excel's* **Regression Analysis** tools or the worksheet function **LINEST**.

	A	B	C	D	E
1	y	x	x²	-yx	-yx²
2	0	0	0	0	0
3	0.17	0.019	0.000361	-0.00323	-0.00006137
4	0.3891	0.0721	0.00519841	-0.02805411	-0.002022701
5	0.4375	0.0966	0.00933156	-0.0422625	-0.004082558
6	0.4704	0.1238	0.01532644	-0.05823552	-0.007209557
7	0.5089	0.1661	0.02758921	-0.08452829	-0.014040149
8	0.5445	0.2377	0.05650129	-0.12942765	-0.030764952
9	0.558	0.2608	0.06801664	-0.1455264	-0.037953285

10	0.5826	0.3273	0.10712529	-0.19068498	-0.062411194
11	0.6122	0.3965	0.15721225	-0.2427373	-0.096245339
12	0.6564	0.5079	0.25796241	-0.33338556	-0.169326526
13	0.6599	0.5198	0.27019204	-0.34301602	-0.178299727
14	0.6841	0.5732	0.32855824	-0.39212612	-0.224766692
15	0.7385	0.6763	0.45738169	-0.49944755	-0.337776378
16	0.7815	0.7472	0.55830784	-0.5839368	-0.436317577
17	0.8943	0.8943	0.79977249	-0.79977249	-0.715236538
18	1	1	1	-1	-1

Figure 9.83 Transformed VLE data for multiple linear regression of a linearized rational function.

Use *Excel's* **LINEST** worksheet array function, with *CNTL SHIFT ENTER*, to perform multivariable linear regression of the conditioned data in Figure 9.83. In this problem, the constant is set to zero:

A21:E21 = LINEST(A2:A18, B2:E18, FALSE)

Figure 9.84 has the multi-linear regression results for the coefficients. We do not need to scale the parameters in the regression because all of the parameters have the same order of magnitude.

	A	B	C	D	E
20	q₂	q₁	p₂	p₁	p₀
21	-17.69	27.279	-5.5110891	16.13080553	0

Figure 9.84

Multivariable linear least-squares regression results for a linearized rational function.

The results from the linearized data regression become the initial guesses for the fitting parameters in the non-linear regression. Use the macro **RATLS** to obtain the coefficients directly, using second order polynomials in the numerator and denominator of the rational function, and no *y*-intercept (or constant) in the numerator. The worksheet in Figure 9.85 has the nonlinear regression parameters in the range **F2:F5**, and model and *SSR* formulas in column **C** and cell **F7**, respectively:

C2 = RATLSF(A2, $F$2:$F$3, F4Z:$F$5)   D2 = C2 - B2   F7 = SUMSQ(D2:D18)

	A	B	C	D	E	F
1	x	y	y Rational	Δy	p0	0
2	0	0	0	0	p1	14.03432
3	0.019	0.17	0.185987	-0.01599	p2	-6.18229
4	0.0721	0.3891	0.385682	0.003418	q1	22.49312
5	0.0966	0.4375	0.428878	0.008622	q2	-15.6751
6	0.1238	0.4704	0.463462	0.006938		
7	0.1661	0.5089	0.502025	0.006875	SSR	0.000648
8	0.2377	0.5445	0.54691	-0.00241		
9	0.2608	0.558	0.558557	-0.00056		
10	0.3273	0.5826	0.588249	-0.00565		
11	0.3965	0.6122	0.616118	-0.00392		
12	0.5079	0.6564	0.660237	-0.00384		
13	0.5198	0.6599	0.665114	-0.00521		
14	0.5732	0.6841	0.687786	-0.00369		
15	0.6763	0.7385	0.736927	0.001573		
16	0.7472	0.7815	0.77687	0.00463		
17	0.8943	0.8943	0.886627	0.007673		
18	1	1	1.004347	-0.00435		

	A	B	C	D
20	u_Y = ±	7.1E-3	U_{Y,95%} =	1.6E-2
21	u₁ = ±	6.2E-1	U_{1,95%} = ±	1.4E+0
22	u₂ = ±	5.7E-1	U_{2,95%} = ±	1.3E+0
23	u₃ = ±	1.5E+0	U_{3,95%} = ±	3.1E+0
24	u₄ = ±	7.1E-1	U_{4,95%} = ±	1.6E+0
25	R² =	0.9993	Adj R² =	0.9992
26	R_{1,2} =	0.3007	DoF =	13
27	R_{1,3} =	0.9839	t_{95%} =	2.16
28	R_{1,4} =	-0.8464	SSR =	6.48E-4
29	R_{2,3} =	0.4447	AAD =	5.02E-3
30	R_{2,4} =	0.2130		
31	R_{3,4} =	-0.7736		

Figure 9.85 *Excel* **worksheet set up for nonlinear regression to minimize SSR and LEVENBURGM results.**

Follow up by checking for poles in the rational function by factoring the polynomial in the denominator using the macro **BAIRSTOW**. To set up the worksheet for **BAIRSTOW**, place the number 1 in a cell above the value for q_1. Then run **BAIRSTOW**, selecting the values for $q_0 = 1$, $q_1 = 22.49312$, and $q_2 = -15.6751$ for the following result:

	A	B	C
34	1	r₁ =	-0.04316
35	22.49312	r₂ =	1.47812
36	-15.6751		

The roots to the polynomial in the denominator of the rational function at $x = -0.04316$ and 1.47812 lie outside the range of data for $0 \leq x \leq 1$, thus we can use the rational function for interpolation without concern for poles.

We use plots of the model with the data to compare a fourth-order rational function to a regular fourth-order polynomial (zero intercept) fit of ethanol-water vapor equilibrium data in Figure 9.86. By inspection, we see that the polynomial does a poor job of interpolating the data having several sections that overshoot the data. The rational function looks better. However, scatter and sorted residual cumulative distribution plots for the rational model show a slight pattern and the lack of the desired sigmoid shape at larger *x* indicating that the rational model in Equation (9.144) does not completely capture the behavior of the data.

CHAPTER 9: REGRESSION

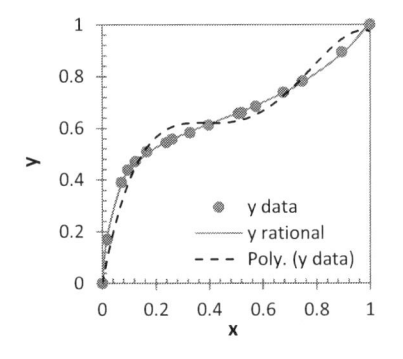

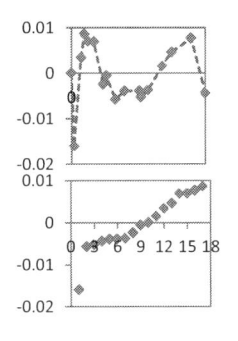

Figure 9.86

Compare polynomial and rational function models of the data. Residual and sorted residual plots indicate a poor fit of the rational function at larger mole fractions.

Comments: The rational function clearly gives a better fit of the data when compared with the fourth-order polynomial. We also note that both models correctly pass through the origin. However, neither of the models pass exactly through $y = 1$ at $x = 1$, as required by the definition of phase equilibrium of pure substances in terms of mole fractions.

Consider a modified arrangement of a rational function that also employs just four fitting constants but passes exactly through endpoints (0, 0) and (1, 1), as required by the definition of phase equilibrium:

$$y = \frac{x\left[1 + p_1(1-x) + p_2(1-x)^2\right]}{1 + x\left[q_1(1-x) + q_2(1-x)^2\right]} \tag{9.146}$$

Use nonlinear regression to obtain the p and q coefficients in the alternative rational function:

$$p_1 = 4.674, \, p_2 = 7.818, \, q_1 = 4.445, \, q_2 = 15.47$$

The plot of the modified rational function with the data in Figure 9.87 reveals a superior fit, particularly at the higher end of the data. Residual and cumulative distribution plots confirm the superiority of the new rational model in Equation (9.146) with slightly more random scatter in the residual plot and the sigmoid shape of the cumulative distribution of the sorted residuals. We suspect Equation (9.146), or similar versions with more terms in the series in the numerator or denominator, may be suitable for general-purpose correlation of y vs x vapor-liquid equilibrium data for many binary systems.

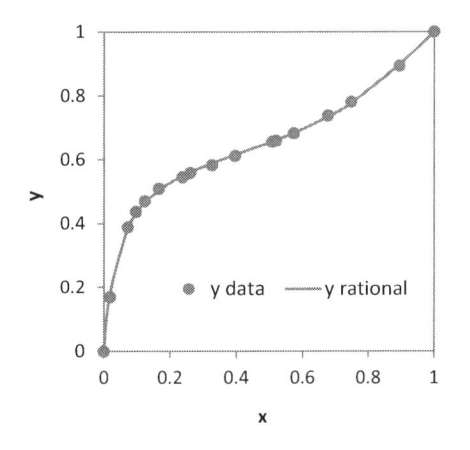

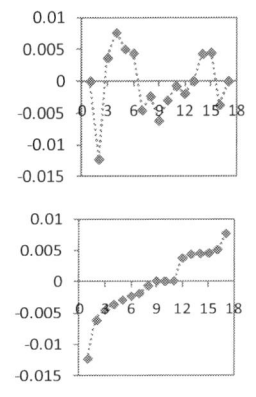

Figure 9.87

Plot of the modified rational function in Equation (9.146) with the data. The residual plots indicate a better model fit to the data compared with a standard rational function.

□

9.10.2 Recommendations for Effective Modeling

Civan (2011) suggests a few simple rules for effective data correlation and analysis. We added a few of our own to the list:

- Measure and correlate data only in terms of the independent variables (*i.e.*, avoid combining independent and dependent data on the same side of an equality, such as Equation (9.60) or (9.145).
- Scale data and model parameters to similar orders of magnitude. For example, when the independent data are large in magnitude, or range over two or more orders of magnitude, we should scale the data for X in the range $[-1,1]$:

$$X = a + bx \quad \text{or} \quad x = \frac{X-a}{b} \quad (9.147)$$

where

$$a = \frac{x_{min} + x_{max}}{x_{min} - x_{max}} \quad \text{and} \quad b = -\frac{2}{x_{min} - x_{max}}$$

To illustrate this point, consider a third-order polynomial with first, second, and third-order monomials plotted in the following graphs. Notice how the unscaled data in the left plot show similar trends, whereas the scaled independent variable in the right plot reveals greater diversity in the behavior of the terms that allow the model to capture the behavior of the data best (Fenton 2010).

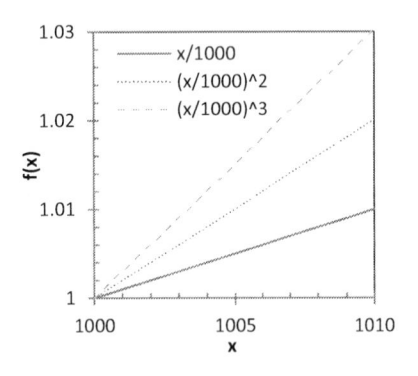

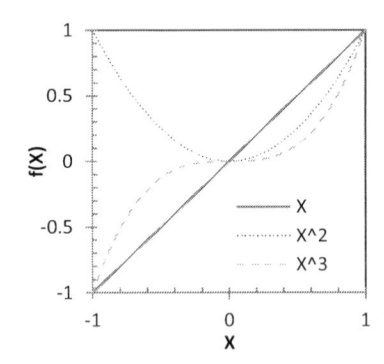

- Select simple models derived from the physics of the problem, when they accurately represent the data.
- Avoid high order polynomials. Instead, use equations that match the nature of the data (*e.g.*, include an exponential, log, or rational terms when the data appear to behave asymptotically). For this very reason, *Excel*'s **Trend Line** function only allows for a polynomial order of six or fewer. To illustrate, compare the linear and fifth order polynomial fits of the data in the following plot. Recall Occam's razor from Chapter 1; without additional information about the nature of the data, we usually choose the simplest model that predicts the trend in the data. Rational function models provide good alternatives to high order polynomials.

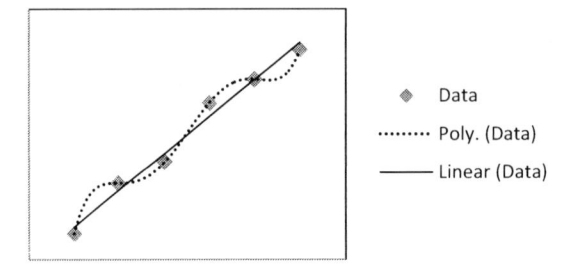

- Evaluate the goodness of fit using a plot of the model with the data, along with residual plots (raw, sorted, and normal) before relying on correlation coefficients or absolute deviations. Use scatter in the data to assess the fit and estimate uncertainty in the model predictions and parameters (see Section 9.3.3).
- Use methods of integration instead of differentiation of data where possible to smooth any noise in the dependent variable.

9.11 Least Absolute Deviation Regression*

The robust method of least absolute deviation[93] is an alternative to least squares regression with the sum of squared residuals in Equation (9.6) replaced with the average of absolute residuals, or deviations:

$$AAD = \frac{1}{n}\sum_{i=1}^{n}|y_i - y_{i,\text{model}}| \qquad (9.148)$$

where n is the number of data pairs in the regression. The *PNMSuite* has a user-defined function **AAD(y, y_model)** for calculating the average absolute deviation for two ranges **y** and **y_model** in an *Excel* worksheet. When comparing competing models of a data set, choose the model with a smaller *AAD*. The popularity of the method of least squares, particularly linear least squares, stems from its compatibility with the calculus minimum. Least squares regression assumes normally distributed errors in the dependent data. The method of least absolute deviation regression has advantages and disadvantages relative to the method of least-squares regression (Thanoon 2015):

Advantages

- Robust methods such as least absolute deviation regression are less sensitive to outliers in the data. By comparison, least squares methods rely on weighting schemes to reduce the sensitivity of the solution to outliers.
- Least absolute deviation regression does not require the assumption of normally distributed errors in the dependent variable.

Disadvantage

- Least absolute deviation has the possibility of more than one solution, particularly for linear models. Thus, least absolute deviation regression is computationally expensive. The search for the model parameters to minimize the *AAD* in Equation (9.148) requires direct stochastic search methods for global minimization, such as Simulated Annealing, Luus-Jaakola, PSO, Firefly, or Genetic algorithms presented in Section 7.6. Use the solution from least-squares regression as the starting point in the global search for the minimum *AAD*.

Example 9.19 Improved Equation for the Gilliland Correlation

Known: A popular shortcut method for multicomponent distillation tower design employs the Gilliland (1940) correlation to estimate the number of equilibrium stages N from the external reflux ratio R in the rectifying section, as shown in Figure 9.88, where

$$Y = \frac{N - N_{\min}}{N+1} \qquad X = \frac{R - R_{\min}}{R+1} \qquad (9.149)$$

where $N_{\min}$ and $R_{\min}$ are the minimum number of stages and reflux ratio, respectively. Several equations are available to represent Gilliland's correlation (Eduljee 1975). However, they remain unsatisfactory because they do not incorporate the theoretical end conditions correctly at the minimum or infinite reflux conditions: $Y=1$ and $dY/dX = -\infty$ at $X = 0$; $Y = 0$ at $X = 1$. The following modified rational equation in terms of a transformed independent X^m variable matches all theoretical end conditions when the exponent $m < 1$.

$$Y = \frac{1 - X^m}{1 - qX^m} \qquad (9.150)$$

Find: Use the method of least absolute deviation regression to determine the three parameters p, q, and m to fit Equation (9.150) to Gilliland's data listed in the *Excel* worksheet.

---

[93] Deviation has alternative names, including residual, error, and value.

Analysis: Robust least absolute deviation regression is recommended in this case where the data does not have a normal error distribution and the model need not be sensitive to outliers. The data are listed in columns **A** and **B** with the model parameters and AAD in column **C**. The transformed X variable, model calculations, residuals and absolute residuals in columns **E** to **H** use the following formulas:

E2 = A2 ^ $D$3   F2 = (1-E2)/(1 + $D$1*E2)   G2 = B2 − F2   H2 = Abs(G2)

Start with the **Solver's** *GRG Nonlinear* option to find approximate values for the parameters p, q, and m to minimize the AAD in cell D6 = Average(H3:H39).

	A	B	C	D	E	F	G	H
1	X	Y	q1	0.91	X^m	Y	Resid	Abs Resid
2	0	1	m	0.041	0.00000	1.00000	0.00000	0.00000
3	0.0244	0.652	SSR	0.020	0.85878	0.64628	0.00572	0.00572
4	0.028	0.6	AAD	0.016	0.86364	0.63693	-0.03693	0.03693
5	0.0476	0.588	MAD	0.040	0.88264	0.59636	-0.00836	0.00836

⋮

	A	B	C	D	E	F	G	H
36	0.75	0.11			0.98827	0.11648	-0.00648	0.00648
37	0.78	0.091			0.98986	0.10215	-0.01115	0.01115
38	0.789	0.098			0.99033	0.09787	0.00013	0.00013
39	0.795	0.078			0.99064	0.09503	-0.01703	0.01703
40	1	0			1.00000	0.00000	0.00000	0.00000

Follow up with the **Solver's** *Evolutionary* genetic algorithm, introduced in Section 7.6.5, to implement the method of least absolute deviation for the global minimum. Constrain the cells for the parameters as shown below in the **Solver** dialogue box.

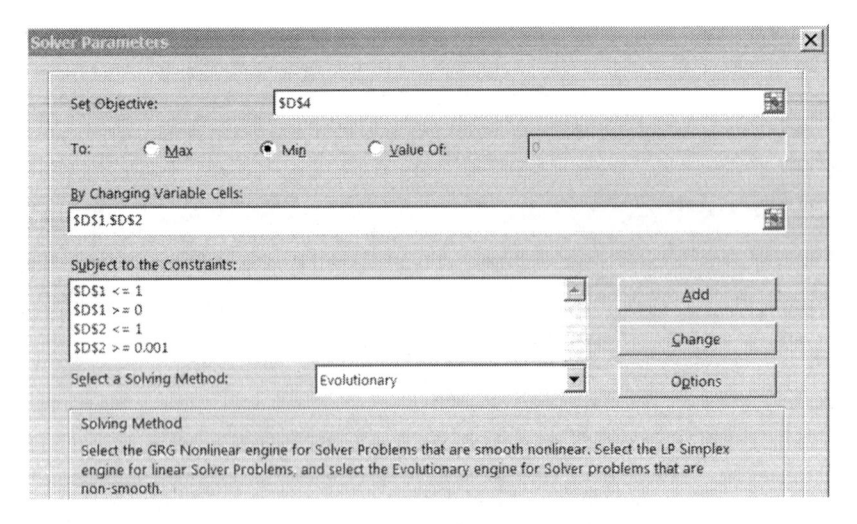

Use the default options to find the global minimum AAD defined in Equation (9.148) and report the maximum absolute deviation (MAD):

$$Y = \frac{1 - X^{0.041}}{1 - 0.91 X^{0.041}}, \qquad \begin{array}{l} AAD = 0.015 \\ MAD = 0.038 \end{array} \qquad (9.151)$$

The MAD is a rough estimate of the uncertainty. The jackknife method, described in Section 0, was used to estimate the 95% confidence interval for Equation (9.151) to be ±0.04.

Figure 9.88 compares Equation (9.151) with Gilliland's original data in a log-log plot showing good correlation. The studentized residual plot, generated with the sub procedure **STUDENTIZER** has a random distribution about the origin, with residuals concentrated inside ± one standard deviation – indications that the model matches the trend in the data, without any outliers.

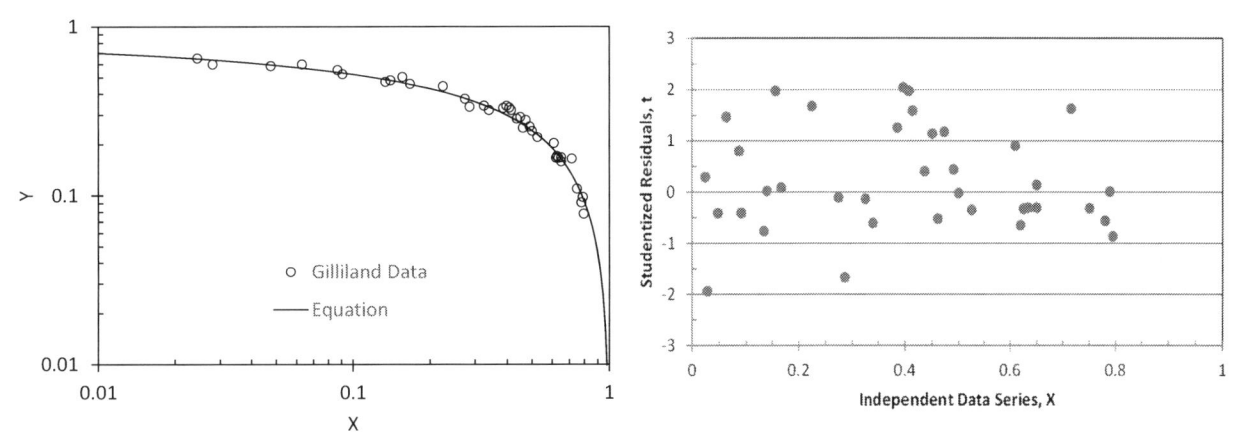

Figure 9.88 Gilliland's correlation data plotted with Equation (9.151) and studentized residual (deviation) plot showing a good fit with normal, random distribution about the origin and no outliers.

Comments: The result of Equation (9.151) is superior to previously published equations for Gilliland's correlation and has the advantage of an explicit inverse function for calculating X in terms of Y:

$$X = \left(\frac{1-Y}{1-0.91Y}\right)^{24} \tag{9.152}$$

Equations (9.151) and (9.152) are recommended for use in computational applications, such as chemical process simulators, for preliminary multicomponent distillation column design and analysis.

□

9.12 Epilogue on Modeling and Regression

"All models are wrong, but some are useful." —George Box

Regression means to come back to the center. At the end of this chapter, we come back around to the important aspects of successful regression. We began the chapter with linear regression, but note that nonlinear least-squares regression methods apply equally to both linear and nonlinear problems. Before computers became commonplace, model linearization was common practice. Today, linearization remains an important tool for validating models by visual inspection and finding good starting points for nonlinear regression, but is no longer the preferred method of finding model parameters for nonlinear regression problems.

Although computing tools such as *Excel* conveniently report the R^2 of a fit, we recommend the more useful and informative residual plots for model assessment, particularly for nonlinear problems. There is no substitute for plotting the model fit with the data to confirm a good fit. We introduced a few practical tools for estimating uncertainty in model fits and model parameters. We must be careful to note that the largest source of uncertainty may come from the bias in the model itself, or, in other words, how well the model represents physical reality.

The emphasis in this chapter is on the use of *Excel* with VBA macros for data modeling, analysis and regression. Table 9.9 compares the strengths and weaknesses of the methods of least squares regression covered in this chapter. Linear regression analysis tools available in *Excel*, coupled with the **Solver,** and macros **GAUSSNEWTON** or **LEVEN-BERGM** or user-forms **LSREGRESS** and **REGRESSUN** (Monte Carlo) for regression uncertainty analysis make *Excel* convenient for manipulating data, plotting results, assessment, and validation.

Table 9.10 summarizes the macros and user-defined functions for regression analysis developed in this chapter. When using the **Solver** for regression minimization, we recommend following up with the macro **LSREGRESS** to evaluate the model uncertainty. **REGRESSUN** includes an option for weighted least-squares regression. Otherwise, our go-to method is Levenberg-Marquardt. Do not forget the macro **SCALIT** from Chapter 7 that uses the **Solver** for minimization with scaled parameters. We recommend rational least-squares functions for general fitting of non-smooth data. Use the regression method of least-absolute deviation when the error is not normally distributed, or you need a robust method to reduce the sensitivity of the regression results to outliers in the data.

Table 9.9   Comparison of least-squares regression methods.

Method	Strengths	Weaknesses
Linear	• No requirement for initial guesses of parameters • Does not over/undershoot the data	• Data may not behave linearly
Gauss-Newton	• Fast computations • Uncertainty analysis of model and parameters for non-linear regression	• Convergence may be sensitive to initial guesses for parameters
Levenberg-Marquardt	• Uncertainty analysis of model and parameters for non-linear regression • Robust convergence	• Slower than Gauss-Newton
Solver	• Available as an *Excel* add-in and fast calculations • Allows for constraints • Random restarts for multimodal optimization	• Sensitive to initial guesses
Jackknife	• Fast computations of uncertainty • Simple implementation	• Not a widely accepted method • May overestimate the uncertainty
Monte Carlo	• Asymmetric confidence intervals for nonlinear models	• Computationally intensive
Rational	• General purpose model that outperforms polynomials • Initial guesses obtained from linearization of the model	• Model may over/undershoot the data trends at poles formed in the denominator
Least average absolute deviation	• Robust to outliers in the data • Does not require normally distributed error in the data • The model passes exactly through a few data points	• May have multiple solutions, which requires minimization by a multi-objective optimization method.

Table 9.10 Summary of *VBA* macros, user-forms, and user-defined functions (UDF) for regression analysis, including the methods of data and model parameter input.

Macro/Function	Inputs	Operation
AAD	UDF arguments	Returns the average absolute deviation from two matched ranges.
CHAUVENET	UDF arguments	Returns the value of an outlier in a data set by Chauvenet's criteria.
GAUSSNEWTON	Macro input boxes	Least-squares regression by the Gauss-Newton method with uncertainty analysis and options for weighted least squares. The worksheet setup is the same as the Solver or LEVENBERGM.
GRUBBS	UDF arguments	Returns the value of an outlier in a data set according to Grubb's criteria.
JACKKNIFE	Macro input boxes	Least-squares regression uncertainty analysis by the Jackknife method. The worksheet setup is the same as the Solver.
LEVENBERGM	Macro input boxes	Nonlinear least-squares regression by the Levenberg-Marquardt method with uncertainty analysis. The worksheet set up is the same as the Solver or GAUSSNEWTON.
LSREGRESS_UsrFrm	Show user-form macro	Nonlinear regression by methods of Gauss-Newton (with option for weighting) or Levenberg-Marquardt (with option for Jacobian). The worksheet setup is the same as the Solver.
OUTLIER	Macro input box	Checks for outliers in a range of data using the Tukey test.
RATLS	Macro input boxes	Rational nonlinear least-squares regression with uncertainty. May be used for polynomials.
RATLSF	UDF arguments	Rational function evaluation given P's and Q's.
REGRESSUN_UsrFrm	Show user-form macro	Regression uncertainty analysis by stochastic Jackknife and Monte Carlo methods.
ROWSHUFFLE	Macro input boxes	Random ordering of the contents of rows in an *Excel* worksheet.
STUDENTIZER	Macro input boxes	Graphs the studentized residuals to check for fit and outliers.
UNSOLVER	Macro input boxes	Regression uncertainty analysis by the Monte Carlo method.
ZPLOT	Macro input box	Normalized sorted residual plot with Filliben's criterion for 95% confidence.

Chapter 10 Interpolation

"Smooth as an android's bottom, eh Data?" – Star Trek's Commander Riker… referring to his clean-shaven face.

Chemical engineers sit in a unique position among the various engineering disciplines, more closely aligned with the natural sciences of chemistry and biology, which rely on physicochemical properties of materials and kinetics. For example, consider physical properties, chemical reaction rate laws, and mass transfer coefficients used to scale up chemical and biological processes. In Chapter 9, we correlated data using mathematical models as doppelgängers of real phenomena for *interpretation* as well as *interpolation*. Mathematical models serve three main purposes, to:

1. Provide physical insights into the nature of a system or process,
2. Smooth noisy data for interpolation, and
3. Functionalize data in the form of compact and portable mathematical expressions that may be more convenient for mathematical manipulation than complex formulas, tabulated data, or graphs.

In this chapter, we extend our modeling options with additional methods for interpolating *smooth* data generated from laboratory experiments, process data, or discrete numerical results. We need good methods of interpolation between data points to get the most information from a limited number of measurements. There is no need for interpolation when we have unlimited time and resources to conduct experiments for every possible configuration of our variables. Fortunately, we can make a few measurement at discrete intervals or obtain numerical results, then interpolate when we need a value in the gaps between our data points. Interpolating smooth data is different from least-squares regression. With regression, we have an over specified problem, and are finding the best parameters in a mathematical model to fit the trend in noisy data. On the other hand, interpolating functions have zero degrees of freedom – they fit the data exactly. As we discovered in Section 9.10, this may not always be the best approach. Nevertheless, we may conveniently rely on methods of interpolation instead of least-squares regression when we lack a good mathematical model that matches the trends in the smooth data.

We introduce the following popular methods of interpolation:

- Newton divided difference and Lagrange polynomials (including linear interpolation)
- Cubic and constrained cubic splines of piecewise polynomials
- Rational functions
- Bivariate interpolation in two dimensions

SQ3R Focused Reading Questions

1. What is required to check the usefulness of an interpolation method or result?
2. What is the difference between divided difference and Lagrange interpolating polynomials?
3. When should I use simplest method of linear interpolation?
4. Why are spline methods of interpolation compared to a Swiss Army knife?
5. What are the differences between regular and constrained spline interpolating methods?
6. How does multidimensional interpolation work?
7. What is the difference between smooth and noisy data?
8. What methods are recommended for interpolating noisy data?
9. What are some advantages of polynomials versus rational functions for interpolation?
10. What are potential complications with rational interpolation?

General-purpose interpolating methods have advantages and disadvantages when compared with least squares models:

Advantages:
- Fit data exactly by passing through all of the points.
- Eliminate the requirement for finding a good model of the data.
- Work just like functions for finding roots, approximating derivatives and integrating data.

Disadvantages:
- Fit data exactly, so are not able to smooth noisy data.
- Give no estimate of uncertainty in the interpolation models.
- In some cases, lack portability or compact formulas (*e.g.*, splines).
- Are unable to validate or correlate data for physicochemical property or parameter estimation.

Consider all of the benefits versus costs of data interpolation when selecting a method of functionalizing and manipulating data. The final section of this chapter presents some methods for smoothing data, primarily used for teasing *qualitative* information from noisy data.

Use extreme caution when applying interpolation methods to extrapolate values beyond the limits of the original data. Novak (2011) makes an analogy of the dangers of extrapolation to the potential hazards of misreading a folded map. We may infer on a folded map that a road traveling in a line from point **A** through points **B** and **C** ending at point **D** will get us from point **B** to **C**. If point **D** lies on the edge of the map, we may be tempted to extrapolate the road to point **E** beyond **D**. However, when we unfold the map to see what lies in between, we discover that our road actually ends at **D**. Although point **E** is in line with the other points on the road, the map reveals that point **E** is located on the other side of a river. Getting to the extrapolated point requires that we follow the road over a bridge crossing the river, and then come back around to point **E**

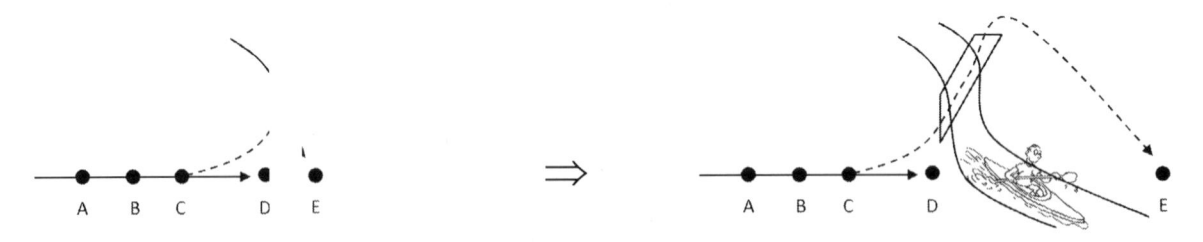

An incomplete data set may not capture the true functionality of its original source. To illustrate, the logistic equation plotted in Figure 9.81 approximates linear behavior in the middle region. Without the rest of the data, we may be tempted to extrapolate far from the true nature of the function at small or large x, as shown in Figure 10.1.

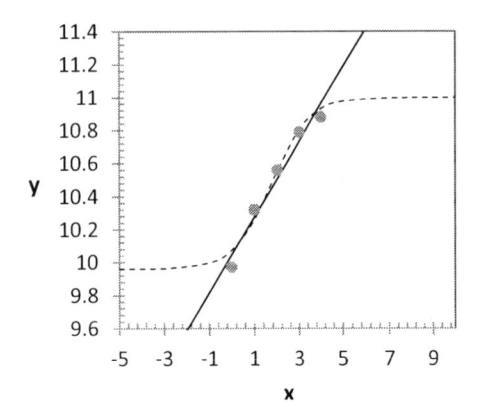

Figure 10.1

The middle subset of data from the logistic equation in Figure 9.81 behaves linearly, as shown by the solid line from linear least-squares regression of the middle five points. Extrapolation of the line outside of the ends of the data takes us far from the true logistic equation indicated by the dashed line.

10.1 Polynomial Interpolation

Polynomials are popular functions for smoothing and interpolating y versus x data based on their ease of use and universal availability in general purpose computational software, including *Excel*:

$$y = a_0 + a_1 x + a_2 x^2 \ldots \quad (10.1)$$

Chapter 9 on least-squares regression presents several tools available in *Excel* for fitting polynomials to data, including charting tools **Trendline**, worksheet array function **LINEST**, and the **Data Analysis** tool **REGRESSION**. Our approach returns once again to *VBA* for programming user-defined functions that accomplish polynomial interpolation of smooth data in in a single command (just like **TREND** for linear interpolation in a single command). Our methods fit the data exactly using polynomials with orders that match the size of the data. For example, two points returns a linear interpolation, three points returns a quadratic interpolation, and four points returns a cubic interpolation, and so on. We rarely have a need for polynomials larger than order five or six. Runge (1901) found that higher order polynomials tend to oscillate around equally spaced points for the independent variable, particularly near the ends. When these polynomials fail to interpolate a complete data set properly, we may piece together lower order polynomials that fit subsets of the data like splines introduced in Section 10.2.

10.1.1 Linear Interpolation

Linear interpolation is a special case of first-order polynomial interpolation that assumes a function behaves as a straight line between two known data points. We may have first encountered formal linear interpolation in our high school chemistry class to get density or solubility values from tables. The concept of readability, introduced in Section 8.1.4, is a matter of performing linear interpolation "mentally". Without realizing it, we subconsciously perform linear interpolation whenever we read from an analog device, such as a map, the hands on a clock, a ruler, or needle gauge. The regula falsi root-finding method from Section 6.2.4 is another example of linear interpolation.

The formula for linear interpolation is simply the point-slope form of an equation of a line between two points (x_i, y_i) and (x_{i+1}, y_{i+1}):

$$y = y_i + \left(\frac{y_{i+1} - y_i}{x_{i+1} - x_i} \right)(x - x_i) \quad (10.2)$$

where $x_i \leq x \leq x_{i+1}$. For convenience, we may use *Excel*'s worksheet function **TREND** for simple linear interpolation in an *Excel* worksheet. A *VBA* user-defined function **LINTERP(ydata, xdata, xnew)**, with arguments defined in Figure 10.2 is available in the *Excel PNMSuite* workbook for linear interpolation between adjacent data pairs. **LINTERP** employs worksheet lookup functions to locate the pair of points out of a larger data set for interpolation. The arguments for **LINTERP** are the ranges of y vs x data and the point of interpolation. The code sorts the data into ascending order according to the independent variable, x. We use the worksheet function **MATCH** to identify the consecutive x points that bracket the interpolant. We may also apply the *VBA* user-defined functions **NEWTONPOLY**, in Figure 10.6, or **LAGRANGE** from the *PNMSuite* for linear interpolation.

```
Public Function LINTERP(ydata, xdata, xnew As Double) As Double
' ydata = range of cells containing dependent variable data
' xdata = range of cells containing independent variable data
' xnew = cell containing the point for interpolation
'**************************************************************************
' Linear interpolation between two data points using the Excel. Data must be in
' ascending order.  Points outside of the data range are extrapolated using the first
' or last two points, accordingly.
' *** Required User-defined Functions and Sub Procedures from the PNM Suite ***
' HEAPSORTXY = sorts pairs of arrays according to the first array values

' *** Nomenclature ***
Dim i As Long, i1 As Long ' = index for first point before xnew, i1 = i+1
Dim n As Long, n1 As Double ' = number of data pairs, n1 = n-1
```

```
Dim x() As Double, y() As Double   ' = vectors of x data and y data
'************************************************************************
n = xdata.Count
If n <> ydata.Count Or n < 2 Then
    MsgBox "x and y data must match, or > 2.": Exit Function
End If
ReDim x(1 To n) As Double, y(1 To n) As Double
For i = 1 To n: x(i) = xdata(i): y(i) = ydata(i): Next i  ' Sort data (just in case)
Call HEAPSORTXY(x(), y())
' Locate points for linear interpolation from data. Check for new points outside range.
If xnew <= x(1) Then ' Use first two points for extrapolation
    LINTERP = (xnew - x(2)) * y(1) / (x(1) - x(2)) _
            + (xnew - x(1)) * y(2) / (x(2) - x(1))
ElseIf xnew >= x(n) Then ' Use last two points for extrapolation
    n1 = n - 1
    LINTERP = (xnew - x(n)) * y(n1) / (x(n1) - x(n)) _
            + (xnew - x(n1)) * y(n) / (x(n) - x(n1))
Else 'Linear interpolation using Lagrange polynomial format for line
    i = WorksheetFunction.Match(xnew, x, 1): i1 = i + 1
    LINTERP = (xnew - x(i1)) * y(i) / (x(i) - x(i1)) _
            + (xnew - x(i)) * y(i1) / (x(i1) - x(i))
End If
End Function ' LINTERP
```

Figure 10.2 *VBA* user-defined function **LINTERP** for linear interpolation.

Example 10.1   Linear Interpolation of Toluene Viscosity

Known: The *Excel* worksheet in Figure 10.3 lists rows of data for the viscosity of liquid toluene over a small temperature range:

	A	B	C	D	E	F	G	H	I	J
1	Data									
2	T/K	295	300	310	320	330	340	350	360	370
3	$\mu \cdot 10^4$/(Pa·s)	5.88	5.54	4.91	4.37	3.93	3.57	3.27	3.01	2.78

Figure 10.3

Viscosity of toluene data in an *Excel* worksheet.

Find: Prepare the *Excel* worksheet for linear interpolation using the worksheet **TREND** function and *VBA* user-defined function **LINTERP**.

Analysis: Plot the viscosity temperature data in Figure 10.4 to check for linearity. The plot suggests that linear interpolation between two data points may yield satisfactory results.

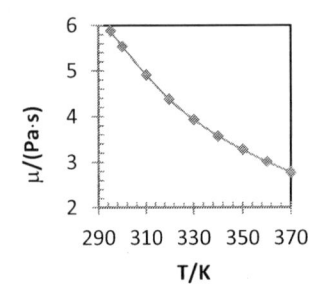

Figure 10.4

Toluene viscosity versus temperature with linear interpolation.

The **TREND** function performs linear interpolation by fitting y vs x data to a line. Follow these steps to set up the worksheet for linear interpolation:

1. Place a value of temperature for interpolation in a cell on the worksheet in Figure 10.5. Use 323 K in cell **B6** for this example.
2. Use *Excel*'s **MATCH(look_up_value, look_up_array)** function to find the number for the column corresponding to the largest temperature smaller than the interpolation value:

$$F6 = MATCH(B6, B2:J2)$$

CHAPTER 10: INTERPOLATION

Note that worksheet function **Match** returns an integer value that corresponds to the index of the column in the range, not the column address in the worksheet. The numbering of columns in the range starts with one. The interpolating points are in columns **E** and **F** (or numbered 4 and 5 in the range).

3. Use *Excel*'s **INDEX(array, row_number, column_number)** function to get temperature and viscosity points on each side of the point of interpolation:

H6 = INDEX(B2:J2, 1, F6)   H7 = INDEX(B2:J2, 1, F6 + 1)   J6 = INDEX(B2:J2, 1, F6)   J7 = INDEX(B2:J2, 1, F6 + 1)

4. Interpolate the viscosity data using *Excel*'s **TREND(known_ys, known_xs, new_x)** function:

B7 = TREND(J6:J7, H6:H7, B6)

	A	B	C	D	E	F	G	H	I	J
5	Interpolation Value				Column		Linear interpolation points			
6	T	323.00			Match	4	T_1 =	320	μ_1 =	4.37
7	$\mu \cdot 10^4$/(Pa·s)	4.24					T_2 =	330	μ_2 =	3.93
8	LINTERP	4.24								

Figure 10.5
Viscosity interpolation using *Excel*'s worksheet function TREND.

The *VBA* user-defined function **LINTERP** accomplishes all of the matching and indexing. Select all of the data, and let the user-defined function **LINTERP** locate the points for interpolation at the value in cell **B6**:

B8 = LINTERP($B$3:$J$3, $B$2:$J$2, B6)

Check the interpolated value with Equation (10.2):

$$\mu = \mu_1 + \left(\frac{\mu_2 - \mu_1}{T_2 - T_1}\right)(x - T_1) = 4.37 + \left(\frac{3.93 - 4.37}{330 - 320}\right)(323 - 320) = 4.24$$

We round the result to two decimal places (or three significant figures) to match the precision of the data.

Comments: *Excel*'s matching functions automate the process of locating the subset of data required for interpolation. Higher order polynomials may provide improved interpolation for nonlinear data. □

10.1.2 Newton's Method of Divided Difference

Taylor series expansions also provide multi-point interpolating formulas near the point of interpolation. For the case of quadratic interpolation, consider three adjacent points (x_{i-1}, y_{i-1}), (x_i, y_i), (x_{i+1}, y_{i+1}), equally-spaced $\Delta x = x_i - x_{i-1} = x_{i+1} - x_i$. A truncated second-order Taylor series for $y(x)$ expanded about the independent variable of the middle point, $x_{i-1} < x_i < x_{i+1}$ gives:

$$y(x) = y_i + (x - x_i)\left.\frac{dy}{dx}\right|_{x_i} + \frac{(x - x_i)^2}{2}\left.\frac{d^2y}{dx^2}\right|_{x_i} \qquad (10.3)$$

Substitute the central difference forms of the finite-difference approximations from Equations 6.24 and 6.27 for the first and second derivatives into Equation (10.3):

$$y(x) = y_i + (x - x_i)\left[\frac{y_{i+1} - y_{i-1}}{2\Delta x}\right] + \frac{(x - x_i)^2}{2}\left[\frac{y_{i+1} - 2y_i + y_{i-1}}{\Delta x^2}\right] \qquad (10.4)$$

Note that the form of Equation (10.4) requires equally-spaced data.

There are other ways to fit three data pairs to the same quadratic function, such as regressing the data by the method of least squares to a second order polynomial: $y = A + Bx + Cx^2$. Perhaps the simplest approach defines the quadratic interpolating function in terms of the first two independent data points:

$$y = a_1 + a_2(x - x_{i-1}) + a_3(x - x_{i-1})(x - x_i) \qquad (10.5)$$

Use the three adjacent data pairs in Equation (10.5) to determine the constant coefficients for the interpolating function in Equation (10.5):

$$a_1 = y_{i-1}$$

$$a_2 = \frac{y_i - y_{i-1}}{x_i - x_{i-1}} = \frac{y_i - a_0}{x_i - x_{i-1}} \tag{10.6}$$

$$a_3 = \frac{\left[\dfrac{y_{i+1} - y_{i-1}}{x_{i+1} - x_{i-1}}\right] - \left[\dfrac{y_i - y_{i-1}}{x_i - x_{i-1}}\right]}{x_{i+1} - x_i} = \frac{\left[\dfrac{y_{i+1} - y_{i-1}}{x_{i+1} - x_{i-1}}\right] - a_1}{x_{i+1} - x_i}$$

We derive higher order interpolating polynomials following the same pattern in Equation (10.6). We call these general forms Newton divided-difference interpolating functions:

$$y = a_1 + a_2(x - x_1) + a_3(x - x_1)(x - x_2) + \ldots a_n(x - x_1)\ldots(x - x_{n-1}) \tag{10.7}$$

The order of the interpolating polynomial matches the size of the data. Newton's divided difference method has several advantages:

- Does not require data in ascending order
- Function passes through all selected data pairs
- Computationally efficient, and
- Works with unequally spaced data.

The primary disadvantage: it does not permit simple differentiation of the interpolating function (unlike the polynomial in Equation (10.1).

A *VBA* user-defined function **NEWTONPOLY(ydata, xdata, xnew)**, with arguments defined in Figure 10.6, is available in the workbook *PNMSuite* for implementing Newton's divided difference method of polynomial interpolation in an *Excel* worksheet. The arguments of the function require ranges of *y* and *x* data, and the point of interpolation.

```
Public Function NEWTONPOLY(ydata, xdata, xnew As Double) As Double
' ydata = range of cells containing dependent variable data
' xdata = range of cells containing independent variable data
' xnew = cell containing the point for interpolation
'***************************************************************************
' Newton divided difference polynomial interpolation.  Data must be in contiguous rows
' or columns.  Uses all data.  Interpolation point should be within the data range.
' *** Nomenclature ***
Dim a() As Double ' = polynomial coefficients
Dim i As Long, j As Long ' = row and column index integer
Dim n As Long ' = number of data pairs
Dim p As Double ' = product of distance between data in interpolation point
'***************************************************************************
On Error Resume Next ' Skip errors due to bad data
n = xdata.Count ' Number of data pairs
If n <> ydata.Count Or n < 2 Then
    MsgBox "x and y data must match, or > 2.": Exit Function
End If
ReDim a(1 To n, 1 To n) As Double
For i = 1 To n: a(i, 1) = ydata(i): Next i ' Set coefficients
For j = 2 To n ' Get coefficients for divided differences
    For i = 1 To n - j + 1
        a(i, j) = (a(i, j - 1) - a(i + 1, j - 1)) / (xdata(i) - xdata(i + j - 1))
    Next i
Next j
p = 1: NEWTONPOLY = a(1, 1) ' Interpolate the polynomial
For j = 2 To n
    p = p * (xnew - xdata(j - 1)): NEWTONPOLY = NEWTONPOLY + a(1, j) * p
Next j
End Function ' NEWTONPOLY
```

Figure 10.6 *VBA* **user-defined function NEWTONPOLY for interpolation by divided difference.**

Example 10.2 Interpolate the Viscosity of Ethylene Glycol with Temperature

Known: An *Excel* worksheet is used to tabulate viscosity data for ethylene glycol with temperature.

Find: Use Newton's divided difference method to interpolate the data and estimate the viscosity at 310 K using the first four points. Repeat using all seven points. Compare the results for viscosity with the experimental value of $\mu = 1.07 \times 10^{-2}$ N·s/m^2.

Analysis: Use an *Excel* worksheet to implement Newton's method of divided difference, as shown in Figure 10.7.

	A	B	C	D	E	F	G	H
1	T/K	µ·10²/(N·s/	a₂	a₃	a₄	a₅	a₆	a₇
2	273	6.51	-0.18296	0.002306	-2E-05	5.2E-07	-6.4E-09	6.11E-11
3	300	1.57	-0.02848	0.0003	-1.1E-05	1.55E-07	-1.7E-09	
4	340	0.431	-0.00655	0.000412	-6.6E-06	6.95E-08		
5	373	0.215	-0.02717	0.000478	-5.9E-06			
6	290	2.47	-0.04773	0.000613				
7	330	0.561	-0.01095					
8	350	0.342						
9								
10	T/K	310						
11	µ (4 pts)	0.816111						
12	µ (7 pts)	1.073277						

Figure 10.7

Newton's divided difference interpolation in an *Excel* worksheet.

Row **2** contains the following formulas for the polynomial coefficients:

C2 = (B2 - B3)/(A2 - A3)   D2 = (C2 - C3)/(A2 - A4)   E2 = (D2 - D3)/(A2 - A5)
F2 = (E2 - E3)/(A2 - A6)   G2 = (F2 - F3)/(A2 - A7)   H2 = (G2 - G3)/(A2 - A8)

Calculate the interpolated values using four and seven points, respectively:

B11 = B2 + C2*(B10 - A2) + D2*(B10 - A2)*(B10 - A3) + E2*(B10 - A2)*(B10 - A3)*(B10 - A4)

B12 = B11 + F2*(B10 - A2)*(B10 - A3)*(B10 - A4)*(B10 - A5) + G2*(B10 - A2)*(B10 - A3)*(B10 - A4)*(B10 - A5)*(B10 - A6) + H2*(B10 - A2)*(B10 - A3)*(B10 - A4)*(B10 - A5)*(B10 - A6)*(B10 - A7)

The *Excel* worksheet has the results of interpolation. The following formulas use the VBA user-defined function **NEWTONPOLY** in the *Excel* worksheet for the same results:

B11 = NEWTONPOLY($B$18:$E$18, $B$17:$E$17, $B$20)   B12 = NEWTONPOLY($B$18:$H$18, $B$17:$H$17, $B$20)

Comments: The seven-point interpolation gives a result that agrees with the experimental value for viscosity. We find the VBA user-defined function **NEWTONPOLY** convenient for polynomial interpolation in an *Excel* worksheet.

□

10.1.3 Lagrange Interpolating Polynomials

Lagrange's[94] method provides an alternative for generating polynomials of smooth data. The Lagrange three-point interpolating formula follows a simple pattern, as shown below for a second-order polynomial:

$$y \cong \frac{(x - x_i)(x - x_{i+1})}{(x_{i-1} - x_i)(x_{i-1} - x_{i+1})} y_{i-1} + \frac{(x - x_{i-1})(x - x_{i+1})}{(x_i - x_{i-1})(x_i - x_{i+1})} y_i + \frac{(x - x_{i-1})(x - x_i)}{(x_{i+1} - x_{i-1})(x_{i+1} - x_i)} y_{i+1} \quad (10.8)$$

---

[94] Chemical engineers have a small connection to the mathematician Lagrange through Lavoisier's law of conservation of mass. Referring to Lavoisier's death by guillotine in the French revolution, Lagrange lamented, "It took them only an instant to cut off that head, and a hundred years may not produce another like it." (https://www.britannica.com/biography/Antoine-Laurent-Lavoisier)

where x_i and y_i are data points and x is the point of interpolation. The result in Equation (10.8) is equivalent to fitting a second-order polynomial to three points by least-squares regression. n^{th}-order Lagrange interpolating polynomials follow the same pattern as the three-point form:

$$y = L_1 y_1 + L_2 y_2 + \ldots + L_n y_n \tag{10.9}$$

where L_i represents the quotient of the products of the differences between the points of the independent variable:

$$L_i = \frac{\prod\limits_{j \neq i}^{n}(x - x_j)}{\prod\limits_{j \neq i}^{n}(x_i - x_j)} \tag{10.10}$$

A user-defined function **LAGRANGE(ydata, xdata, xnew, p)** is available in the *PNMSuite* for polynomial interpolation. The arguments are similar to the user-defined function **NEWTONPOLY**. However, an optional fourth argument **p** is included for **LAGRANGE** to specify the order of the polynomial if we want to conduct a lower order polynomial interpolation using a subset of the x vs y data. If the optional argument is missing, the order of the Lagrange polynomial matches the size of the data set (*e.g.*, 2nd order from 3 data pairs, and so on…). **LAGRANGE** requires data arranged on the worksheet in ascending order according to the independent variable. **LAGRANGE** calls another user-defined function **HEAPSORTXY** that sorts the data array from the arguments into ascending order according to the x data range.

Example 10.3 Composition of Aqueous Methanol by Refractive Index

Known: A common binary distillation involves separating a mixture of methanol and water. The composition is determined by refractive index (*RI*). As shown in Figure 10.8, the refractive index passes through a maximum at 50% by mass, which makes it difficult to use for a control measurement because the same *RI* may correspond to two different compositions. Because distillation involves vapor-liquid equilibrium, we propose the use of the saturated liquid temperature to determine if the mixture is less than or greater than 50% methanol.

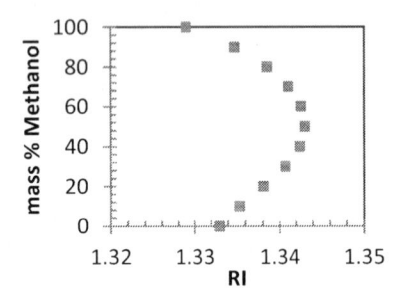

Figure 10.8

Mass fraction of methanol in water versus refractive index (RI).

Find: Organize an *Excel* worksheet to use the Lagrange formula to interpolate data for mass percent versus refractive index of a saturated aqueous mixture of methanol at 1 atm pressure.

Analysis: The boiling points of pure methanol, pure water, and a 50% mixture by mass are 65, 100, and 76°C, respectively. Therefore, if the temperature is greater than 76°C, the mixture is less than 50% methanol, and vice versa. Set up an *Excel* worksheet with the *RI* data and two different Lagrange interpolations for mass percent using data for each case. The formulas for interpolation use the optional argument 2 for second order polynomials:

E8 = LAGRANGE(B7:B12, A7:A12, E2, 2)   E9 = LAGRANGE(B2:B7, A2:A7, E2, 2)

We use an *Excel* **IF** worksheet function to select the correct result based on the temperature:

E3 = IF(E1 < 76, E8, E9)

CHAPTER 10: INTERPOLATION

	A	B	C	D	E
1	RI	Mass %		T/°C	70
2	1.333	0		RI	1.34
3	1.3354	10		mass %	75.1
4	1.3381	20			
5	1.3407	30			
6	1.3425	40			
7	1.3431	50			
8	1.3426	60		T<76°C	75.1
9	1.3411	70		T>76°C	27.0
10	1.3385	80			
11	1.3348	90			
12	1.329	100			

Comments: The data above 50% requires sorting into ascending order in terms of the *RI*. Some experimentation was required to determine that a second order polynomial provided better interpolation in this example.

□

Lagrange's formula is also useful for evaluating derivatives and integrals (see Section 11.3). To illustrate, we show Lagrange 3-point interpolating formulas for approximating derivatives. For any three adjacent points, $x_{i-1} < x_i < x_{i+1}$ (not necessarily equally-spaced) the derivative of the function is found from the Lagrange derivative formula, which is used by the VBA user-defined functions **FDERIV** and **FINDIF** described in Section 5.2:

$$\frac{dy}{dx} = \frac{(2x - x_i - x_{i+1})y_{i-1}}{(x_{i-1} - x_i)(x_{i-1} - x_{i+1})} + \frac{(2x - x_{i-1} - x_{i+1})y_i}{(x_i - x_{i-1})(x_i - x_{i+1})} + \frac{(2x - x_{i-1} - x_i)y_{i+1}}{(x_{i+1} - x_{i-1})(x_{i+1} - x_i)} \quad (10.11)$$

The VBA functions **FDERIV** and **FINDIF** also have options for a second-order derivative approximation:

$$\frac{d^2y}{dx^2} = 2\left\{\frac{y_{i-1}}{(x_{i-1} - x_i)(x_{i-1} - x_{i+1})} + \frac{y_i}{(x_i - x_{i-1})(x_i - x_{i+1})} + \frac{y_{i+1}}{(x_{i+1} - x_{i-1})(x_{i+1} - x_i)}\right\} \quad (10.12)$$

The maximum order of the Lagrange polynomial, or its derivative, matches the size of the data for interpolation. Lagrange's formula is elegant, but computationally less efficient than Newton's method of divided difference for higher order polynomials. We recommend cubic splines presented in Section 10.2 for approximating derivatives involving more than four points from smooth data.

10.1.4 Inverse Polynomial Interpolation*

Consider the polynomial in Equation (10.1) for interpolation of *y* data. If we need to interpolate the *x* data, we may try fitting the data with a similar polynomial for *x* in terms of *y*. However, just because a polynomial performs well for interpolating *y* at *x*, there is no assurance that a polynomial will work just as well for interpolation of *x* at a given value for *y*. In this case, we may need to invert the original polynomial as follows:

$$x = b_1(y - a_0) + b_2(y - a_0)^2 + b_3(y - a_0)^3 \ldots \quad (10.13)$$

Substitute for $y - a_0$ from Equation (10.1) into Equation (10.13):

$$x = b_1\left(a_1x + a_2x^2 + a_3x^3 \ldots\right) + b_2\left(a_1x + a_2x^2 + a_3x^3 \ldots\right)^2 + b_3\left(a_1x + a_2x^2 + a_3x^3 \ldots\right)^3 \ldots \quad (10.14)$$

Equate like terms for each power of *x* on the left and right-hand-side of Equation (10.14). This gives a set of simultaneous equations to solve for the coefficients in the inverted polynomial (Abramowitz and Stegun 1970):

$$b_1 = \frac{1}{a_1} \qquad b_2 = -\frac{b_1 a_2}{a_1^2} \qquad b_3 = \frac{2b_2 a_1 a_2 - b_1 a_3}{a_1^3} \qquad \ldots \quad (10.15)$$

The number of terms is limited such that the order of the inverted polynomial matches the original polynomial. Pay attention to the nature of the polynomial before inversion – there may be more than one value of *x* for a given value of *y*, as we found in Example 10.3:

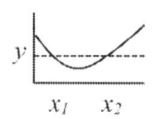

10.2 Cubic Spline Polynomial Interpolation

> Green Hornet: "You know what you are, Kato? You're a human Swiss army knife."
> Kato: "I don't know what that means?"
> Green Hornet: "It's a little thing, and you keep pulling out things, and just when you think there couldn't be any more cool things, a new cool thing comes out, and that's you! You are even dressed like one. You should have a little plus on your chest."

The method of cubic spline interpolation pieces together a series of third-order polynomials (cubic equations) that exactly fit the data points and match the smooth trend of the data. The cubic spline interpolating method takes its unusual name (and associated terminology) from the legacy of the drafting spline tool. Similar to a French curve, drafters used flexible splines to draw smooth curves through data points that did not quite behave as a line or circle. Figure 10.9 shows a picture of a drafting spline in use. The hooked weights hold the flexible edge in place for tracing the curve between specific points or "knots". In mathematical terms, we enforce continuity of interpolating functions between intervals of data pairs to "hold" the cubic spline functions in place. At the endpoints, we may clamp the spline with specific conditions, or let the spline form naturally, free from any weights that constrain the shape.

Figure 10.9
Flexible spline drafting tool for drawing smooth curves through data. The hooked weights hold the drafter's spline in the desired position for tracing the curve.

Each data point is a knot that ties together two cubic polynomial interpolating functions for adjoining intervals. The method of cubic spline is the "Swiss Army Knife" of interpolation.[95] Because the spline is based on a spliced set of third-order polynomials, we can easily perform a variety of operations on the polynomials, including interpolation, differentiation, and integration (see Section 0). In the following applications, the index *k* refers to the coefficients in the polynomials for the interval starting at knot *k*, where $x_k \leq x \leq x_{k+1}$:

- Interpolation $\qquad y_k(x) = A_k + B_k(x - x_k) + C_k(x - x_k)^2 + D_k(x - x_k)^3$ \hfill (10.16)

- First derivative $\qquad y_k'(x) = \dfrac{dy_k}{dx} = B_k + 2C_k(x - x_k) + 3D_k(x - x_k)^2$ \hfill (10.17)

---

[95] Pocket knives with a variety blades and tools for multiple functions used by the Swiss Armed Forces. A registered trademark of the suppliers Wenger and Victorinox.

CHAPTER 10: INTERPOLATION

- Second derivative $$y_k^{\prime\prime}(x) = \frac{d^2 y_k}{dx^2} = 2C_k + 6D_k(x - x_k) \tag{10.18}$$

- Integration $$\int_{x_k}^{x_{k+1}} y\,dx = A_k(x_{k+1} - x_k) + B_k \frac{(x_{k+1}^2 - x_k^2)}{2} + C_k \frac{(x_{k+1}^3 - x_k^3)}{3} + D_k \frac{(x_{k+1}^4 - x_k^4)}{4} \tag{10.19}$$

We use $y_k(x)$ to represent the interpolated value for y at x in the k interval. We use $y_k = y_k(x_k)$ without the parenthesis for the value of the y_k data point at x_k, and so on. We then "tie" cubic spline polynomials together for continuity at common data points, or knots. Cubic spline functions for adjacent data intervals require the following smoothing conditions at common knots:

- Adjoining spline functions must pass exactly through the data:

$$y_{k-1}(x_k) = y_k(x_k) \tag{10.20}$$

- Equal first and second derivatives (continuous) at common points (knots) between polynomials of adjacent intervals:

$$y_{k-1}^{\prime}(x_k) = y_k^{\prime}(x_k) \quad\text{and}\quad y_{k-1}^{\prime\prime}(x_k) = y_k^{\prime\prime}(x_k) \tag{10.21}$$

- Users must specify the method for approximating second derivatives at the first and last points (e.g., $y'' = 0$ for linear or "natural" boundary conditions).

- The independent data must be in ascending order to locate the interval of interpolation: $x_k > x_{k-1}$.

We use the requirements for continuity at the knots to derive the coefficients of the cubic spline polynomials in the following steps:

1. Define the k interval length for consecutive (x, y) data pairs:

$$\Delta x_k = x_{k+1} - x_k \quad\text{and}\quad \Delta y_k = y_{k+1} - y_k \tag{10.22}$$

2. Calculate y and y'' for the interval starting at (x_k, y_k) and ending at (x_{k+1}, y_{k+1}) using Equations (10.16) and (10.18):

$$y_k = A_k \qquad y_{k+1} = A_k + B_k \Delta x_k + C_k \Delta x_k^2 + D_k \Delta x_k^3 \qquad y_k^{\prime\prime} = 2C_k \qquad y_{k+1}^{\prime\prime} = 2C_k + 6D_k \Delta x_k \tag{10.23}$$

3. Solve the four Equations (10.23) for the coefficients A_k, B_k, C_k, and D_k at knot k:

$$A_k = y_k \qquad B_k = \frac{\Delta y_k}{\Delta x_k} - \frac{\Delta x_k}{6}\left(2 y_k^{\prime\prime} + y_{k+1}^{\prime\prime}\right) \qquad C_k = \frac{y_k^{\prime\prime}}{2} \qquad D_k = \frac{y_{k+1}^{\prime\prime} - y_k^{\prime\prime}}{6\Delta x_k} \tag{10.24}$$

4. Polynomials that share a common knot for y and x also require continuous first derivatives across the interval boundary. Apply the continuity requirement at knot k using Equation (10.17) for the derivative with substitutions from Equations (10.24) for the polynomial coefficients:

$$\frac{\Delta y_{k-1}}{\Delta x_{k-1}} - \frac{\Delta x_{k-1}}{6}\left(2 y_{k-1}^{\prime\prime} + y_k^{\prime\prime}\right) + \Delta x_{k-1} y_{k-1}^{\prime\prime} + \frac{\Delta x_{k-1}}{2}\left(y_k^{\prime\prime} - y_{k-1}^{\prime\prime}\right) = \frac{\Delta y_k}{\Delta x_k} - \frac{\Delta x_k}{6}\left(2 y_k^{\prime\prime} + y_{k+1}^{\prime\prime}\right) \tag{10.25}$$

5. Collect the common second derivative terms at each interior knot in Equation (10.25):

$$\Delta x_{k-1} y_{k-1}^{\prime\prime} + 2(\Delta x_{k-1} + \Delta x_k) y_k^{\prime\prime} + \Delta x_k y_{k+1}^{\prime\prime} = 6\left(\frac{\Delta y_k}{\Delta x_k} - \frac{\Delta y_{k-1}}{\Delta x_{k-1}}\right) \tag{10.26}$$

We now have a system of equations like Equation (10.26) for the unknown second derivatives at each knot, one equation for each interval. There are two more unknowns than equations (or two degrees of freedom) for the ends. To solve the equations, we must specify two additional conditions for the interpolating polynomials. Popular cubic spline interpolation methods make specifications about the interpolating functions for the first and last intervals $(x_2 - x_1)$ and $(x_n - x_{n-1})$. We select from the following alternative end-point conditions (and corresponding coefficient matrices) depending on the nature of the data (Rao 2002):

- *Natural end conditions:*[96] For a natural cubic spline we set the second derivatives to zero at the first and last knots:

$$y_1'' = 0 \quad \text{and} \quad y_n'' = 0 \tag{10.27}$$

In matrix notation, this system shown in Equation (10.28) appears tridiagonal. The result is a system of simultaneous equations for the second derivatives at the interior knots ($2 \leq k \leq n - 1$):

$$\begin{bmatrix} 2(\Delta x_1 + \Delta x_2) & \Delta x_2 & 0 & \cdots & & 0 \\ \Delta x_2 & 2(\Delta x_2 + \Delta x_3) & \Delta x_3 & 0 & & \vdots \\ 0 & \ddots & \ddots & \ddots & & 0 \\ \vdots & 0 & \Delta x_{n-3} & 2(\Delta x_{n-3} + \Delta x_{n-2}) & \Delta x_{n-2} \\ 0 & \cdots & 0 & \Delta x_{n-2} & 2(\Delta x_{n-2} + \Delta x_{n-1}) \end{bmatrix} \cdot \begin{bmatrix} y_2'' \\ y_3'' \\ \vdots \\ y_{n-2}'' \\ y_{n-1}'' \end{bmatrix}$$

$$= \begin{bmatrix} 6(\Delta y_2/\Delta x_2 - \Delta y_1/\Delta x_1) \\ 6(\Delta y_3/\Delta x_3 - \Delta y_2/\Delta x_2) \\ \vdots \\ 6(\Delta y_{n-2}/\Delta x_{n-2} - \Delta y_{n-3}/\Delta x_{n-3}) \\ 6(\Delta y_{n-1}/\Delta x_{n-1} - \Delta y_{n-2}/\Delta x_{n-2}) \end{bmatrix} \tag{10.28}$$

The Thomas algorithm, introduced in Section 4.3.3, efficiently solves the tridiagonal system of linear equations for the second derivatives at each knot. Natural (free) boundary conditions are often adequate for most cubic spline data interpolation.

- *Parabolic end conditions*: Set equal the second derivatives at the first and last pairs of knots:

$$y_1'' = y_2'' \quad \text{and} \quad y_n'' = y_{n-1}'' \tag{10.29}$$

Substitute for y_1'' and y_n'' in Equation (10.26) for the $k = 2$ and $k = n - 1$ knots (used for the coefficients and constants in the first and last rows in Equation (10.28)):

$$(3\Delta x_1 + 2\Delta x_2) y_2'' + \Delta x_k y_3'' = 6\left(\frac{\Delta y_2}{\Delta x_2} - \frac{\Delta y_1}{\Delta x_1}\right) \tag{10.30}$$

$$\Delta x_{n-2} y_{n-2}'' + (2\Delta x_{n-2} + 3\Delta x_{n-1}) y_{n-1}'' = 6\left(\frac{\Delta y_{n-1}}{\Delta x_{n-1}} - \frac{\Delta y_{n-2}}{\Delta x_{n-2}}\right) \tag{10.31}$$

- *"Not a knot" end conditions:*[97] Interpolate the first and final pairs of intervals with a single cubic polynomial. We derive the values of the second derivative at the first and last points from linear interpolation. Hence, the second and second-to-last points are not true "knots" in the previous sense. The final two degrees of freedom are satisfied by imposing continuity in the third derivative across the first two and last two intervals:

$$y_1'''(x_2) = y_2'''(x_2) \quad y_{n-1}'''(x_{n-1}) = y_n'''(x_{n-1}) \tag{10.32}$$

---

[96] Referred to as natural or free boundaries because there are no "weights" or imposed conditions securing the curve at the ends.
[97] Used by Mathcad® and Matlab® for their cspline functions.

$$y_2''' = 6D_2 = \frac{y_2'' - y_1''}{6\Delta x_1} = \frac{y_3'' - y_2''}{6\Delta x_2} \qquad \frac{y_{n-1}'' - y_{n-2}''}{6\Delta x_{n-2}} = \frac{y_n'' - y_{n-1}''}{6\Delta x_{n-1}} \qquad (10.33)$$

Then

$$\Delta x_2 \left(y_2'' - y_1'' \right) = \Delta x_1 \left(y_3'' - y_2'' \right) \quad \text{and} \quad \Delta x_{n-1} \left(y_{n-1}'' - y_{n-2}'' \right) = \Delta x_{n-2} \left(y_n'' - y_{n-1}'' \right) \qquad (10.34)$$

The results for the second derivatives at the first and last knots are:

$$y_1'' = \frac{\left[(\Delta x_1 + \Delta x_2) y_2'' - \Delta x_1 y_3'' \right]}{\Delta x_2} \quad \text{and} \quad y_n'' = \frac{\left[(\Delta x_{n-1} + \Delta x_{n-2}) y_{n-1}'' - \Delta x_{n-1} y_{n-2}'' \right]}{\Delta x_{n-2}} \qquad (10.35)$$

After substituting for y_1'' and y_n'' in Equation (10.26) for the $k = 2$ and $k = n - 1$ knots, the corresponding elements of Equation (10.28) become:

$$\frac{(2\Delta x_2 + \Delta x_1)(\Delta x_2 + \Delta x_1)}{\Delta x_2} y_2'' + \frac{(\Delta x_2 + \Delta x_1)(\Delta x_2 - \Delta x_1)}{\Delta x_2} y_3'' = 6\left(\frac{\Delta y_2}{\Delta x_2} - \frac{\Delta y_1}{\Delta x_1} \right) \qquad (10.36)$$

$$\frac{(\Delta x_{n-2} + \Delta x_{n-1})(\Delta x_{n-2} - \Delta x_{n-1})}{\Delta x_{n-2}} y_{n-2}'' + \frac{(2\Delta x_{n-2} + \Delta x_{n-1})(\Delta x_{n-2} + \Delta x_{n-1})}{\Delta x_{n-2}} y_{n-1}'' = 6\left(\frac{\Delta y_{n-1}}{\Delta x_{n-1}} - \frac{\Delta y_{n-2}}{\Delta x_{n-2}} \right) \qquad (10.37)$$

- *Clamped end points*: When we have information about the derivatives (slopes) at the first and last knots, we may specify the values for the derivatives to satisfy the degrees of freedom:

$$y_1' = \alpha \quad \text{and} \quad y_n' = \beta \qquad (10.38)$$

where α and β are specified. The second derivatives at the first and last knots become:

$$y_1'' = \frac{3\left(\frac{\Delta y_1}{\Delta x_1} - \alpha \right)}{\Delta x_1} - \frac{y_2''}{2} \quad \text{and} \quad y_n'' = \frac{3\left(\beta - \frac{\Delta y_{n-1}}{\Delta x_{n-1}} \right)}{\Delta x_{n-1}} - \frac{y_{n-1}''}{2} \qquad (10.39)$$

The elements in Equation (10.28) for knots $k = 2$ and $k = n - 1$ become:

$$\left(\frac{3\Delta x_1 + 4\Delta x_2}{2} \right) y_2'' + \Delta x_2 y_3'' = 6\left(\frac{\Delta y_2}{\Delta x_2} - \frac{\Delta y_1}{\Delta x_1} \right) - 3\left(\frac{\Delta y_1}{\Delta x_1} - \alpha \right) \qquad (10.40)$$

$$\Delta x_{n-2} y_{n-2}'' + \left(\frac{4\Delta x_{n-2} + 3\Delta x_{n-1}}{2} \right) y_{n-1}'' = 6\left(\frac{\Delta y_{n-1}}{\Delta x_{n-1}} - \frac{\Delta y_{n-2}}{\Delta x_{n-2}} \right) - 3\left(\beta - \frac{\Delta y_{n-1}}{\Delta x_{n-1}} \right) \qquad (10.41)$$

Example 13.2 uses clamped end conditions derived from derivatives for interpolating a concentration profile in a catalyst pellet. Natural splines have error $O(\Delta x^2)$ whereas the other options for end conditions produce interpolating polynomials with error $O(\Delta x^4)$.

The *VBA* user-defined function **CSPLINE**, with arguments defined at the top of Figure 10.10, interpolates smooth data with the default method of "not-a-knot" end conditions. The function also has options for returning the first and second derivatives or performing numerical integration. We can also specify parabolic "P" or natural "N" end conditions by including an optional argument in the user-defined function (note: use a comma after each argument, even when an optional argument is omitted, *e.g.*, =CSPLINE(x_data, y_data, x_new, , , "N") for interpolation with natural end conditions). For the case of clamped end conditions, we must also provide values for the optional arguments α or β, each separated by a comma at the end of the argument list. The sub procedure **HEAPSORTXY** sorts the data in ascending order according to the independent variable. The macro uses the worksheet lookup function **MATCH** to locate the interval for interpolation. Thus, the user-defined function does not allow for duplicate values in the *x* dimension. **CSPLINE** may be used to interpolate data located in ranges on a worksheet or data provided in arrays when called from another procedure.

```
Public Function CSPLINE(ydata As Variant, xdata As Variant, xnew As Double, _
                        Optional q As Integer = 0, Optional xup As Double, _
                        Optional cs As String = "K", Optional alpha As Variant, _
                        Optional beta As Variant) As Double
'********************************************************************************
' ydata = range of cells or an array containing dependent variable data (Variant type)
' xdata = range of cells or an array containing independent variable data (Variant type)
' xnew = cell containing point for interpolation. Alternatively used for lower
'        integration limit for quadrature.
' q = (optional argument) -1 = quadrature, 0 = interpolation
'     1 = first derivative, 2 = second derivative)
' xup = (optional argument) upper integration limit
' cs = (optional argument, default is Not-a-knot) for end conditions:
'      "N" = natural, "P" = parabolic, "K" = not-a-knot
' alpha = (optional argument) clamped first derivative (slope) at first data point
' beta = (optional argument) clamped first derivative (slope) at last data point
'********************************************************************************
' Interpolate data by cubic spline with choice of clamped, natural (free), parabolic,
' or Not-a-knot end conditions. Solves for the spline coefficients using the Thomas
' algorithm for a tri-diagonal matrix. The program finds the interpolated value at the
' interpolation point: yi = a + b*(xnew - xk) + c*(xnew - xk)^2 + d*(xnew - xk)^3
' Optional calculation of integral, first or second derivatives. xdata must be in
' ascending order.  Interpolation and derivative approximations are linear outside of
' data range.  Integration must be within the data range.

' *** Required User-defined Functions and Sub Procedures from PNM Suite ***
' HEAPSORTXY = sorts x data in ascending order

' *** Nomenclature ***
Dim dxlo As Double ' = 1st interval size for integration from lower limit to 2nd knot
Dim dxu As Double ' = last interval size for integration from last knot to upper limit
Dim a() As Double, b() As Double, c() As Double, d() As Double ' = spline coefficients
Dim den As Double ' = denominator in Thomas algorithm
Dim dx() As Double ' = interval length for independent data
Dim dxn As Double ' = xnew - xk
Dim dy() As Double ' = change in dependent data across an interval
Dim h() As Double ' = ratio of parameters t_3/t_2 in Thomas algorithm
Dim i As Long ' = loop index
Dim iup As Long ' = match value
Dim k As Long, k1 As Long, kp1 As Long ' = knot for spline interval, k1=k-1, kp1=k+1
Dim n As Long, ny As Long ' number of data pairs in x and y data
Dim n1 As Long, n2 As Long ' = number of data pairs, n1 = n-1, n2 = n-2
Dim p() As Double ' = ratio of parameters t_4/t_2 in Thomas algorithm
Dim s() As Double, st() As Double ' = second derivatives at knots
Dim t_1() As Double, t_2() As Double, _
    t_3() As Double, t_4() As Double ' = coefficients for Thomas algorithm
Dim x() As Double, y() As Double ' = vectors of x-data and y-data
'********************************************************************************
On Error Resume Next ' Skip errors due to bad data
If IsObject(xdata) And TypeOf xdata Is Range Then ' get size of data ranges
    n = xdata.Count
ElseIf IsArray(xdata) Then
    n = UBound(xdata) - LBound(xdata) + 1
Else
    MsgBox "Source of x data must be a range or an array": End
End If
If IsObject(ydata) And TypeOf ydata Is Range Then
    ny = ydata.Count
ElseIf IsArray(ydata) Then
    ny = UBound(ydata) - LBound(ydata) + 1
Else
    MsgBox "Source of y data must be a range object or an array": End
End If
If n <> ny Or n < 2 Then ' check for compatible array size for x and y data
    MsgBox "Size of x and y data ranges or arrays must match, or > 2.": End
End If
n1 = n - 1: n2 = n - 2 ' Number of data pairs for looping and indexing
ReDim a(1 To n) As Double, b(1 To n) As Double, c(1 To n) As Double, _
```

CHAPTER 10: INTERPOLATION

```
            d(1 To n) As Double, dy(1 To n) As Double, dx(1 To n) As Double, _
            s(1 To n), st(1 To n) As Double, t_1(1 To n) As Double, t_2(1 To n) As Double, _
            t_3(1 To n) As Double, t_4(1 To n) As Double, x(1 To n) As Double, _
            y(1 To n) As Double
For i = 1 To n: x(i) = xdata(i): y(i) = ydata(i): Next i ' Sort in ascending order
Call HEAPSORTXY(x(), y())
For k = 1 To n1 ' Set the spline intervals for knots k
    dy(k) = y(k + 1) - y(k): dx(k) = x(k + 1) - x(k)
Next k
' Set constants and coefficients of linear equations for Thomas algorithm
t_1(1) = 0#: t_4(1) = 6 * (dy(2) / dx(2) - dy(1) / dx(1))
t_3(n2) = 0#: t_4(n2) = 6 * (dy(n1) / dx(n1) - dy(n2) / dx(n2))
If Not IsMissing(alpha) Then ' First point clamped
    t_2(1) = 1.5 * dx(1) + 2 * dx(2): t_3(1) = dx(2)
    t_4(1) = t_4(1) - 3 * (dy(1) / dx(1) - alpha)
Else ' First point end condition
    Select Case cs
    Case "N" ' Natural
        t_2(1) = 2 * (dx(1) + dx(2)): t_3(1) = dx(2)
    Case "P" ' Parabolic
        t_2(1) = 3 * dx(1) + 2 * dx(2): t_3(1) = dx(2)
    Case Else ' Not-a-knot
        t_2(1) = (2 * dx(2) + dx(1)) * (dx(1) + dx(2)) / dx(2)
        t_3(1) = (dx(2) + dx(1)) * (dx(2) - dx(1)) / dx(2)
    End Select
End If
If Not IsMissing(beta) Then ' Last point clamped
    t_1(n2) = dx(n2): t_2(n2) = 2 * dx(n2) + 1.5 * dx(n1)
    t_4(n2) = t_4(n2) - 3 * (beta - dy(n1) / dx(n1))
Else ' Last point end condition
    Select Case cs
    Case "N" ' Natural
        t_1(n2) = dx(n2): t_2(n2) = 2 * (dx(n2) + dx(n1))
    Case "P" ' Parabolic
        t_1(n2) = dx(n2): t_2(n2) = 2 * dx(n2) + 3 * dx(n1)
    Case Else ' Not-a-knot
        t_1(n2) = (dx(n2) + dx(n1)) * (dx(n2) - dx(n1)) / dx(n2)
        t_2(n2) = (2 * dx(n2) + dx(n1)) * (dx(n2) + dx(n1)) / dx(n2)
    End Select
End If
For k = 3 To n2 ' Set the tri-diagonal values for interior points
    k1 = k - 1: t_1(k1) = dx(k1): t_2(k1) = 2 * (dx(k1) + dx(k))
    t_3(k1) = dx(k): t_4(k1) = 6 * (dy(k) / dx(k) - dy(k1) / dx(k1))
Next k
' Solve for the second derivatives using the Thomas algorithm for a tri-diagonal matrix
ReDim h(1 To n2) As Double, p(1 To n2) As Double
h(1) = t_3(1) / t_2(1): p(1) = t_4(1) / t_2(1) ' Set ratios
For k = 1 To n2 - 1 ' Forward elimination loop
    kp1 = k + 1: den = t_2(kp1) - t_1(kp1) * h(k)
    h(kp1) = t_3(kp1) / den: p(kp1) = (t_4(kp1) - t_1(kp1) * p(k)) / den
Next k
st(n2) = p(n2) ' Back substitution
For k = (n2 - 1) To 1 Step -1: st(k) = p(k) - h(k) * st(k + 1): Next k
For k = 2 To n1: s(k) = st(k - 1): Next k ' Set values for 2nd derivatives at ends
If Not IsMissing(alpha) Then ' First point clamped
    s(1) = 3 * (dy(1) / dx(1) - alpha) / dx(1) - s(2) / 2
Else ' First point end conditions
    Select Case cs
    Case "N" ' Natural
        s(1) = 0#
    Case "P" ' Parabolic
        s(1) = s(2)
    Case Else ' Not-a-knot
        s(1) = ((dx(1) + dx(2)) * s(2) - dx(1) * s(3)) / dx(2)
    End Select
End If
If Not IsMissing(beta) Then ' Last point clamped
    s(n) = 3 * (beta - dy(n1) / dx(n1)) / dx(n1) - s(n1) / 2
Else ' last point end conditions
```

```
        Select Case cs
        Case "N" ' Natural
            s(n) = 0#
        Case "P" ' Parabolic
            s(n) = s(n1)
        Case Else ' Not-a-knot
            s(n) = ((dx(n1) + dx(n2)) * s(n1) - dx(n1) * s(n2)) / dx(n2)
        End Select
End If
For k = 1 To n1 ' Calculate the cubic spline coefficients for knot k
    a(k) = y(k): b(k) = dy(k) / dx(k) - dx(k) * (2 * s(k) + s(k + 1)) / 6
    c(k) = s(k) / 2: d(k) = (s(k + 1) - s(k)) / (6 * dx(k))
Next k
a(n) = y(n): b(n) = b(n1): c(n) = 0#: d(n) = 0# ' End interval is linear extrapolation
If xnew < x(1) Then ' Find the knot for interval containing the interpolation point
    k = 1
Else
    k = WorksheetFunction.Match(xnew, x, 1)
End If
dxn = xnew - x(k) ' distance of xnew from xk
Select Case q ' Calculate 3rd order spline polynomial
Case 0 ' interpolation
    If xnew >= x(1) Then
        CSPLINE = a(k) + b(k) * dxn + c(k) * dxn ^ 2 + d(k) * dxn ^ 3
    Else
        CSPLINE = y(1) + b(1) * (xnew - x(1)) ' Linear extrapolation
    End If
Case 1 ' Calculate 1st derivative of 3rd order spline polynomial
    If xnew > x(1) Then
        CSPLINE = b(k) + 2 * c(k) * dxn + 3 * d(k) * dxn ^ 2
    Else
        CSPLINE = b(1)
    End If
Case 2 'Calculate 2nd derivative of 3rd order spline polynomial
    If xnew > x(1) Then
        CSPLINE = 2 * c(k) + 6 * d(k) * dxn
    Else
        CSPLINE = 0#
    End If
Case -1 ' Find the intervals containing the integration limits
    iup = WorksheetFunction.Match(xup, x, 1)
    dxu = xup - x(iup)
    dxlo = xnew - x(k) ' Integrate the data between the integration limits
    If Not iup = k Then ' integration limits in different intervals
        CSPLINE = a(k) * (dx(k) - dxlo) + b(k) * (dx(k) ^ 2 - dxlo ^ 2) / 2 + _
            c(k) * (dx(k) ^ 3 - dxlo ^ 3) / 3 + d(k) * (dx(k) ^ 4 - dxlo ^ 4) / 4
        For i = k + 1 To iup - 1
            CSPLINE = CSPLINE + a(i) * dx(i) + b(i) * (dx(i) ^ 2) / 2 _
                + c(i) * (dx(i) ^ 3) / 3 + d(i) * (dx(i) ^ 4) / 4
        Next i
        CSPLINE = CSPLINE + a(iup) * dxu + b(iup) * (dxu ^ 2) / 2 _
            + c(iup) * (dxu ^ 3) / 3 + d(iup) * (dxu ^ 4) / 4
    Else ' integration limits between same knots
        CSPLINE = a(k) * (dxu - dxlo) + b(k) * (dxu ^ 2 - dxlo ^ 2) / 2 + _
            c(k) * (dxu ^ 3 - dxlo ^ 3) / 3 + d(k) * (dxu ^ 4 - dxlo ^ 4) / 4
    End If
Case Else
    MsgBox "Incorrect q argument (must be -1, 0, 1, or 2)!"
End Select
End Function ' CSPLINE
```

Figure 10.10   VBA user-defined function CSPLINE for interpolation of smooth data by cubic splines.

Be careful when using cubic spline functions for *extrapolation*. The default end conditions use "not-a-knot." However, we recommend the natural form for extrapolation to small distances beyond the data, unless additional information suggests one of the other forms. Best practice requires plotting the spline interpolation with several points among the data to check that the interpolation follows the trend in the data. The VBA macro **QYXPLOT** in the *PNM-Suite* is convenient for generating a plot of the spline interpolation at 1000 equally spaced intermediate points. The

CHAPTER 10: INTERPOLATION

"best" spline method of handling the ends for cubic spline interpolation is subjective. As inferred by Shakespeare and Benjamin Franklin, "Beauty is in the eye of the beholder."

Example 10.4 Inverse Interpolation Using the Solver

Known: An experimental adsorption column generated breakthrough data for separating HCFC from air on activated carbon. Chemical engineers use breakthrough data to scale up designs of packed columns. Common scale-up parameters include the breakthrough times when the effluent stream reaches 5, 50, and 95% of the feed concentration (also known as the breakthrough, saturation, and equilibrium times, respectively).

Find: Use the data listed in the *Excel* worksheet to interpolate the breakthrough curve and determine the times for breakthrough, saturation, and equilibrium.

Assumptions: Smooth data

Analysis: Plot the data to observe that the breakthrough curve is symmetrical. Interpolate the data using the *VBA* user-defined function **CSPLINE** available from the *PNMSuite*. Use *Excel*'s **Solver** for inverse interpolation to find simultaneously the corresponding times from the specifications for percentage feed concentration in cells **C20**, **C22** and **C24**. The **Set Cell** for the sum of squared residuals is **D25**.

	A	B	C	D	E	F
1	t/min	%				
2	85	0				
3	90	2.0				
4	94	6.0				
5	98	9.8				
6	101	14.4				
7	107	26.4				
8	111	37.1				
9	113	45.7				
10	119	64.4				
15	141	94.0				
16	145	95.7				
17	149	97.1				
18	155	98.6				
19	160	100.0				
20	t_b	93.003	5	0.0		
21	%	5.0				
22	t_s	114.099	50	0.0		
23	%	50.0				
24	t_e	143.297	95	0.0		
25	%	95.0		1.471E-13		

	A	B	C	D
20	t_b	93.0027508596496	5	=B21-C20
21	%	=cspline(B2:B19,A2:A19,B20,,,"C",0,0)		
22	t_s	114.098531885308	50	=B23-C22
23	%	=cspline(B2:B19,A2:A19,B22,,,"C",0,0)		
24	t_e	143.297439176281	95	=B25-C24
25	%	=cspline(B2:B19,A2:A19,B24,,,"C",0,0)		=SUMSQ(D20,D22,D24)

The breakthrough time for an effluent concentration at 5% of the feed concentration is 93 min. The saturation and equilibrium times for 50% and 95% concentration in the effluent are 114 and 143 min, respectively. We may use alternative end conditions by including the optional argument(s) in the user-defined function. For example, we may want to clamp the first derivatives to zero for the first and last knots where the break through curve has a theoretical slope of zero:

B21=CSPLINE(B3:B18,A3:A18,B20, , ,"C",0,0)   B23=CSPLINE(B3:B18,A3:A18,B22, , ,"C",0,0)

B25=CSPLINE(B3:B18,A3:A18,B24, , ,"C",0,0)

Comments: *Excel* spline interpolations conveniently functionalize the data in ways that are deployable just as regular functions involving algebraic combinations of variables. We must compare the spline interpolation to the data trend by plotting intermediate points between data. The user-form macro **QYXPLOT** is convenient for graphing 1000 points over the range of the data. Add the data to the plot with markers for comparison.

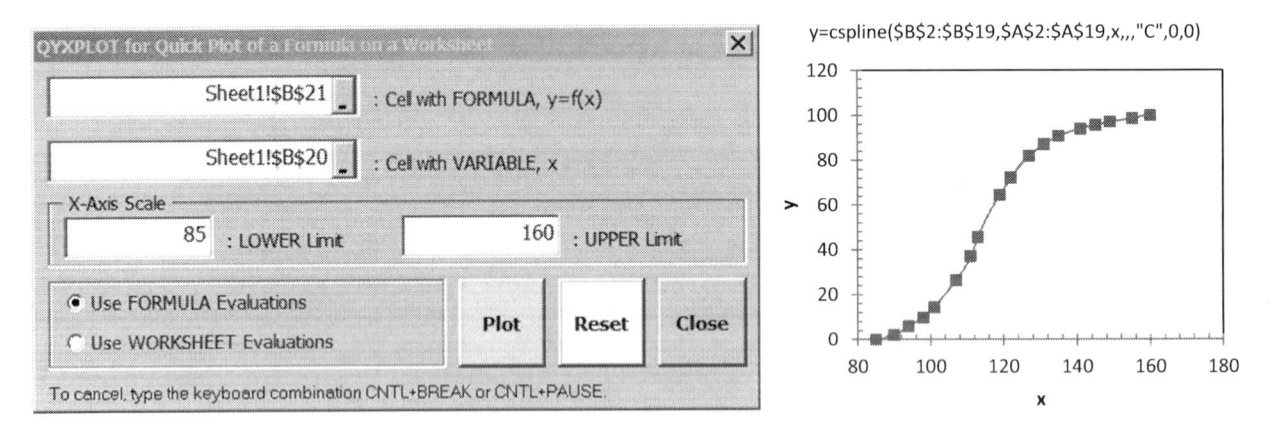

Beware that each form of cubic spline end conditions produces slightly different results, particularly for the breakthrough time. The following table compares the results from the four methods.

End Conditions	t_b/min	t_s/min	t_e/min
Clamped	93.003	114.099	143.297
Natural	93.019	114.098	143.289
Parabolic	92.999	114.099	143.289
Not-a-knot	92.974	114.099	143.288

□

Spline interpolation becomes convenient for "functionalizing" smooth data series when working with equation-based models. In the next example, we use cubic spline interpolation of equilibrium data to perform material balances around a single-stage liquid-liquid extraction process.

Example 10.5 Single-stage Liquid-Liquid-Extraction

Known: Liquid-liquid extraction (*LLE*) of acetic acid from chloroform by water: C = chloroform, W = water, A = acetic acid in a single stage. The *LLE* equilibrium data are listed in the *Excel* worksheet in Figure 10.12.

Find: Compositions and weights of raffinate and extract when 45 kg of a 35 wt% chloroform and 65 wt% acetic acid feed mixture (F) is extracted with 22.75 kg of water (S) in a single stage extraction.

Schematic: A solvent stream (S) extracts a solute (A) from the feed (F). The raffinate stream (R) becomes more pure in solute C. The solvent leaves in the extract stream (E). X_F is the mass fraction of a component in the feed stream; X and Y are the mass fractions of the components in the raffinate and extract phases, respectively.

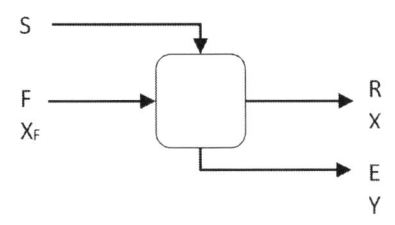

CHAPTER 10: INTERPOLATION   459

Assumptions: Smooth data and an ideal extraction stage with the immiscible exit liquid streams reaching concentration equilibrium according to the data.

Analysis: Calculate the extract and raffinate total mass and mass fractions from the individual species mass balances and equilibrium relationships:

$$X_{FA}F - X_A R - Y_A E = 0 \qquad Y_A - f(X_A) = 0 \qquad (10.42)$$

$$X_{FC}F - X_C R - Y_C E = 0 \qquad Y_C - f(X_C) = 0 \qquad (10.43)$$

$$S - X_W R - Y_W E = 0 \qquad Y_W - f(X_W) = 0 \qquad (10.44)$$

To solve this system of equations, we need mathematical equations to interpolate the equilibrium data. For its ease of use, we first plot the data in *Excel* and use **Trend Line** to fit the data with polynomials. Unfortunately, as seen in Figure 10.11, a single polynomial has trouble fitting the data, whereas the cubic spline performs well.

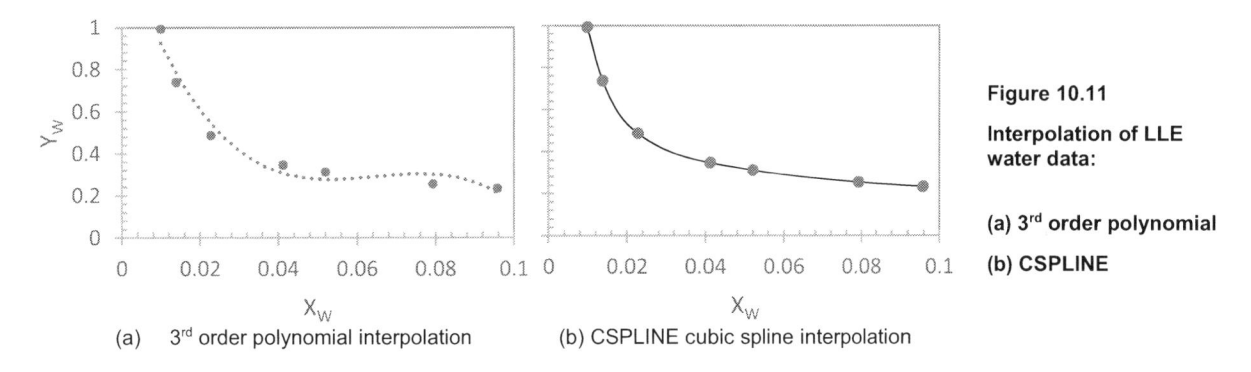

Figure 10.11

Interpolation of LLE water data:

(a) 3rd order polynomial

(b) CSPLINE

(a)   3rd order polynomial interpolation   (b) CSPLINE cubic spline interpolation

We then use the all-purpose user-defined function **CSPLINE** to provide the functional relationship in Equation (10.42) for each of the three components of the extraction. Although the spines are not compact equations, like polynomials, the **CSPLINE** function behaves just like an equation in the *Excel* worksheet. Finally, we must ensure that the mass fractions in each of the exit phases sum to one:

$$1 - \sum Y = 0 \qquad 1 - \sum X = 0 \qquad (10.45)$$

We may solve the system of eight equations for the eight unknowns: X and Y for each component and the masses of E and R. We use *Excel's* **Solver** to search for the values of the exit equilibrium compositions and amounts of E and R that minimize the **SUMSQ** of the eight functions. Initial guesses for the unknowns are specified in the ranges **E12:E14** for X, **F12:F14** for Y, and **H18:H19** for R and E. The mass balances are formulated in the range **G12:G14**, the equilibrium functions are given in the range **H12:H14**, and the sums of mass fractions are in the range **E18:E19**.

To find the solutions using the **Solver**, we must create a cell with the sum of the squares of the eight functions in cell **E21**. The **Solver** is setup to minimize the sum of the function squares by changing the values in the ranges for the variables. The solution is shown in the *Excel* worksheet in Figure 10.12.

	A	B	C	D	E	F	G	H	I
1				X Raffinate LLE Data wt%			Y Extract LLE Data wt%		
2				C	W	A	C	W	A
3				0.9901	0.0099	0	0.0084	0.9916	0
4				0.9185	0.0138	0.0677	0.0121	0.7369	0.251
5				0.8	0.0228	0.1772	0.073	0.4858	0.4412
6				0.7013	0.0412	0.2575	0.1511	0.3471	0.5018
7				0.6715	0.052	0.2765	0.1833	0.3111	0.5056
8				0.5999	0.0793	0.3208	0.252	0.2539	0.4941
9				0.5581	0.0958	0.3461	0.2885	0.2328	0.4787
10									
11				Component	X	Y	Mass Balance	Equilibrium	
12	Feed/kg	45		A	0.221471672248624	0.482104945	=B18-E12*H18-F12*H19	=F12-cspline(I3:I9,F3:F9,E12)	
13	Solvent/kg	22.75		C	0.74880092772988	0.108780377	=B19-E13*H18-F13*H19	=F13-cspline(G3:G9,D3:D9,E13)	
14				W	0.029727400026848	0.409114676	=B20-E14*H18-F14*H19	=F14-cspline(H3:H9,E3:E9,E14)	
15	Feed XA	0.65		SSR			=SUMSQ(G12:G14)	=SUMSQ(H12:H14)	
16	Feed XC	0.35							
17				Mass Fractions			Exit Mass		
18	Feed A/kg	=B15*B12		ΣX-1	=SUM(E12:E14)-1			R/kg	13.0935317676077
19	Feed C/kg	=B16*B12		ΣY-1	=SUM(F12:F14)-1			E/kg	54.6564682324205
20	Feed W/kg	=B13		SSR	=SUMSQ(E18:E19)				
21				Total SSR	=SUM(G15,H15,E20)				

	A	B	C	D	E	F	G	H	I
1				X Raffinate LLE Data wt%			Y Extract LLE Data wt%		
2				C	W	A	C	W	A
3				0.9901	0.0099	0	0.0084	0.9916	0.0000
4				0.9185	0.0138	0.068	0.0121	0.7369	0.2510
5				0.8	0.0228	0.177	0.073	0.4858	0.4412
6				0.7013	0.0412	0.258	0.1511	0.3471	0.5018
7				0.6715	0.052	0.277	0.1833	0.3111	0.5056
8				0.5999	0.0793	0.321	0.252	0.2539	0.4941
9				0.5581	0.0958	0.346	0.2885	0.2328	0.4787
10									
11				Component	X	Y	Mass Balance	Equilibrium	
12	Feed/kg	45		A	0.2215	0.4821	1.78E-13	1.26E-12	
13	Solvent/kg	22.75		C	0.7488	0.1088	-7.43E-13	1.96E-12	
14				W	0.0297	0.4091	5.47E-13	2.40E-13	
15	Feed XA	0.65		SSR			8.82E-25	5.49E-24	
16	Feed XC	0.35							
17				Mass Fractions			Exit Mass		
18	Feed A/kg	29.25		ΣX-1	5.35E-12			R/kg	13.09
19	Feed C/kg	15.75		ΣY-1	-1.80E-12			E/kg	54.66
20	Feed W/kg	22.75		SSR	3.19E-23				
21				Total SSR	3.82E-23				

Figure 10.12

Excel Solver solution for the extract and raffinate mass and component mass fractions.

Comments: *VBA* spline functions present versatile interpolating options for simulating tabulated data as "functions" that may be manipulated like algebraic expressions in numerical routines.

□

In the next example, data passes to the **CSPLINE** function through *VBA* arrays instead of worksheet ranges.

Example 10.6   Interpolation of Liquid Water Density Data

Known: The density of liquid water in the temperature range $0 \leq T \leq 100°C$ passes through local minima and maxima near the freezing point temperature that make it difficult to interpolate with a single equation, such as a polynomial.

Find: Create a *VBA* user-defined function to interpolate water density data as a function of temperature using the method of cubic splines.

CHAPTER 10: INTERPOLATION

Analysis: The problem is solved using two functions available in the *PNMSuite* for spline interpolation and significant figures. Two arrays of density versus temperature data are coded directly in the user-defined function **DENSITY_H2O** in Figure 10.13 The function passes the data arrays through the arguments for the **CSPLINE** user-defined function, which performs the interpolation. The user-defined function **SIGFIGS** rounds the interpolated result to four significant figures to match the precision of the density data. Note that the *VBA* module containing the **DENSITY_H2O** function requires setting `Option Base 1` at the top of the module for compatibility of the arrays with **CSPLINE**.

```
Option Base 1
Public Function DENSITY_H2O(t As Double) As Double
' t = temperature in deg C
' ****************************************************************************
' Returns the density of water at 1 atm pressure in kg/m^3
' ****************************************************************************
' *** Required User-defined Functions from PNM Suite ***
' CSPLINE = interpolation by cubic spline with not-a-knot end conditions
' SIGFIGS = round the density to a specified number of significant figures
' *** Nomenclature ***
Dim d_data() As Variant, t_data() As Variant ' = density vs temperature data arrays
' ****************************************************************************
If t > 100 Or t < 0 Then
    MsgBox "Temperature out of range. Retry with 0 < T < 100 deg C": End
End If
t_data = Array(0.001, 4, 10, 20, 30, 40, 50, 60, 70, 80, 90, 100) ' T data, deg C
d_data = Array(999.8, 1000, 999.7, 999.8, 995.6, 992.2, _
               988, 983.2, 977.8, 971.8, 965.3, 958.4) ' Density data, kg/m3
DENSITY_H2O = CSPLINE(d_data, t_data, t) ' Interpolate the data using cubic spline
DENSITY_H2O = SIGFIGS(DENSITY_H2O, 5) ' return the density with 4 significant figures
End Function ' DENSITY_H2O
```

Figure 10.13 VBA user-defined function for interpolating water density by the method of cubic splines.

The data were plotted as markers in Figure 10.14 with the curve representing 1000 interpolation results over the complete temperature. The second graph zooms in on the region where the data shows local minima and maxima. The spline interpolation performs well in the lower temperature region near the freezing point where the density data reveal local minima and maxima values near the freezing point.

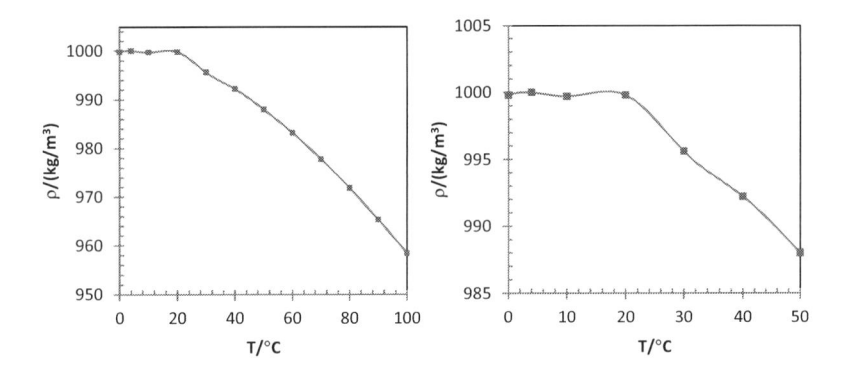

Figure 10.14

Density of liquid water data (markers) plotted with 1000 cubic spline interpolation points (curve) generated from the user-defined function DENSITY_H2O. The second plot zooms in on the data near the freezing point to highlight the unusual trend captured by the spline interpolation.

Comments: The user-defined function **DENSITY_H2O** has the advantage that it does not require worksheet data ranges for interpolating the density versus temperature data.

Approximating derivatives of smooth data is relatively straightforward using cubic spline interpolating functions according to Equations (10.17) and (10.18). Before using this method for approximating derivatives, it is important to recall that splines fit data exactly; they do not smooth noise in experimental data as with least-squares regression for data modeling. The *VBA* user-defined function **CSPLINE** returns a first or second derivative approximation using the optional argument value of 1 or 2 after the argument for the interpolation point. For example,

CSPLINE($C$1:$C$5, $B$1:$B$5, A1, 1) interpolates the data in columns **B** and **C** for the first derivative at the value in cell **A1**. We cannot include the second derivative option in the method of constrained cubic splines developed in Section 10.3 because of the imposed constraints on the second derivative.

Example 10.7 Cubic Spline Derivatives for Titration Endpoint Determination

Known/Find: Repeat Example 5.4 using cubic spline derivative approximations.

Analysis: Figure 10.15 shows the first five points of the titration data and cell formulas containing the cubic spline derivative functions in an *Excel* worksheet.

	A	B	C	D
1	V/ml	pH	dpH/dV	d^2pH/dV^2
2	0.000	1.806	0.096456218	0
3	2.157	2.020	0.103956192	0.005606287
4	4.096	2.235	0.118537729	0.009434391
5	5.754	2.449	0.143244538	0.020370122
6	7.077	2.664	0.186390224	0.044847846

	C	D
1	dpH/dV	d^2pH/dV^2
2	=cspline($B$2:$B$31,$A$2:$A$31,A2,1)	=cspline($B$2:$B$31,$A$2:$A$31,A2,2)
3	=cspline($B$2:$B$31,$A$2:$A$31,A3,1)	=cspline($B$2:$B$31,$A$2:$A$31,A3,2)
4	=cspline($B$2:$B$31,$A$2:$A$31,A4,1)	=cspline($B$2:$B$31,$A$2:$A$31,A4,2)
5	=cspline($B$2:$B$31,$A$2:$A$31,A5,1)	=cspline($B$2:$B$31,$A$2:$A$31,A5,2)
6	=cspline($B$2:$B$31,$A$2:$A$31,A6,1)	=cspline($B$2:$B$31,$A$2:$A$31,A6,2)

Figure 10.15

Cubic-spline interpolated derivative approximations using the *VBA* user-defined function CSPLINE. Formulas use optional fourth arguments 1 or 2 for first or second derivatives, respectively.

The optional fourth argument in the *VBA* user-defined function **CSPLINE** requires an integer value of 1 for returning a first derivative or 2 for returning a second derivative at the point of interpolation, shown in the formulas used in columns **C** and **D**, respectively. Plot the derivative approximations for the complete titration data set to reveal the titration end-point volume at ~10 ml, as shown in Figure 10.16.

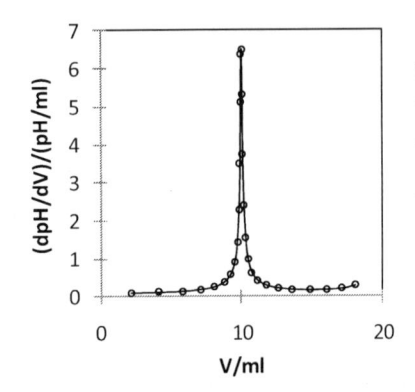

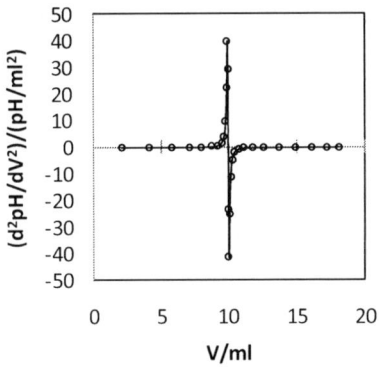

Figure 10.16

Cubic spline interpolation for first and second derivative approximations for locating the endpoint of titration. Compare to Figure 5.15 using finite difference derivative approximations.

Comments: These results agree with the values determined by three-point finite-difference derivative approximations in Example 5.4. The choice of end conditions in this example is probably not important because we are focusing our attention on the results of interpolation at the interior points.

□

CHAPTER 10: INTERPOLATION

10.3   Hermite Polynomials*

Conventional polynomials, such as those generated from Newton divided difference or Lagrange formulas occasionally have problems matching trends in the data. Where possible, Hermite polynomials may improve on regular polynomials by interpolating smooth data with known values of the first derivative, or slope, at each data point. However, Hermite polynomials have the obvious disadvantage of requiring information about the derivatives, so may not have wide application. For example, we may use Hermite interpolation for initial-rate analysis of batch reactor data that consists of the concentration with time and the derivative corresponding to the rate of change of concentration from the reaction rate law. The values of the derivatives at the data points may also come from an approximation, such as finite differences.

As with regular polynomials and some rational functions, cubic spline interpolation may overshoot intermediate points in sparse data, or data with abrupt changes in value or direction. Splines can also have problems with monotonic interpolation of equally spaced data, particularly near the ends. Hermite polynomials use information about the derivative of the function at each point to generate an interpolating polynomial. We exploit Hermite polynomials in piecewise spline interpolation methods created by Akima and Kruger that use local subsets of the data to approximate the derivatives at each knot. Hermite interpolation methods eliminate potential overshooting of the obvious trend in the data. The method of Stineman in Section 10.4.2 employs an approach similar to piecewise rational function interpolation of smooth data.

Example 10.8   Third-order Hermite Polynomial from Two Points

Known: A spherical solid pellet is heated by convection from a hot fluid with a temperature T_f. The average temperature of the pellet is T_a.

Find: Use a second-order Hermite polynomial to estimate the surface temperature from the average temperature.

Schematic: A solid sphere with diameter D, radius r, and surface temperature T_s is immersed in a fluid with heat transfer coefficient h and temperature T_f:

Analysis: Based on symmetry, the change in temperature with radial position across the pellet center is zero:

$$\frac{dT}{dr} = 0 \quad \text{at} \quad r = 0 \tag{10.46}$$

The boundary condition at the pellet surface is determined from a surface energy balance where the rate of heat convection to the surface equals the rate of heat conduction into the sphere:

$$h(T_f - T_s) = \kappa \frac{dT}{dr} \quad \text{at} \quad r = D/2 \tag{10.47}$$

where h and κ are the heat transfer coefficient and pellet thermal conductivity, respectively. A second-order polynomial trial function for the temperature profile in the sphere has three coefficients:

$$T = a_0 + a_1 r + a_2 r^2 \tag{10.48}$$

Application of the boundary conditions to Equation (10.48) gives:

$$\left.\frac{dT}{dr}\right|_{r=0} = a_1 = 0 \tag{10.49}$$

$$\left.\frac{dT}{dr}\right|_{r=D/2} = a_2 D = \frac{h}{\kappa}(T_f - T_s) \tag{10.50}$$

The average temperature is calculated by integrating the polynomial temperature profile over the volume of the pellet:

$$T_a = \frac{\int_0^{D/2} T \cdot 4\pi r^2 dr}{\frac{4\pi}{3}\left(\frac{D}{2}\right)^3} = \frac{24}{D^3}\int_0^{D/2} T \cdot r^2 dr = \frac{24}{D^3}\int_0^{D/2} (a_0 + a_2 r^2) r^2 dr = a_0 + \frac{3D^2}{20}a_2 \tag{10.51}$$

Solve Equations (10.50), (10.49), and (10.51) for the polynomial coefficients:

$$a_0 = \frac{20\kappa T_a + 5DhT_a - 3DhT_f}{20\kappa + 2Dh} \qquad a_1 = 0 \qquad a_2 = \frac{10h(T_f - T_a)}{D(Dh + 10\kappa)} \tag{10.52}$$

Substitute the results for the coefficients from (10.52) into Equation (10.48) at $r = D/2$ and rearrange for T_s:

$$T_s = \frac{10\kappa T_a + DhT_f}{10\kappa + Dh} \tag{10.53}$$

Comments: The result represents a weighted average of the fluid temperature and average pellet temperature. For large κ, $T_S \to T_a$ and for large h, $T_S \to T_f$.

□

10.3.1 Akima's Hermite Cubic Spline Polynomial Interpolation*

Akima (1970) derived a method of cubic spline Hermite polynomial interpolation to prevent overshooting the observed trend in the data where a regular cubic spline method shows undesired behavior. Akima's method uses five points to constrain the derivatives at the knots by a weighted average slope:

$$\bar{m}_k = \frac{|m_{k+1} - m_k| m_{k-1} + |m_{k-1} - m_{k-2}| m_k}{|m_{k+1} - m_k| + |m_{k-1} - m_{k-2}|} \tag{10.54}$$

where the slopes are calculated from first-order forward-difference approximations:

$$m_k = \frac{\Delta y_k}{\Delta x_k} \tag{10.55}$$

The differences in the slopes are indexed according to the lower knot k: $\Delta x_k = x_{k+1} - x_k$ and $\Delta y_k = y_{k+1} - y_k$. We use the average of two adjacent slopes when the denominator of Equation (10.54) is zero for a flat trend in the data:

$$\bar{m}_k = \frac{(m_k + m_{k-1})}{2} \tag{10.56}$$

Because two additional points are required beyond each of the end points, Akima suggested creating image knots before the first data point and after the last data point n by requiring that the first three and the last three points and two additional points at each end are similarly spaced and interpolated by parabolas:

$$m_{-1} = 2m_1 - m_2 \qquad m_{-2} = 2m_{-1} - m_1 \qquad m_{n+1} = 2m_n - m_{n-1} \qquad m_{n+2} = 2m_{n+1} - m_n \tag{10.57}$$

The *VBA* user-defined function **AKIMA** is available in the *PNMSuite* for Akima's method of interpolation. **AKIMA** requires the same arguments as **CPLINE** for interpolation.

CHAPTER 10: INTERPOLATION

Whereas Akima's approach to handling the endpoints works well in some cases, particularly with unequal spacing of the knots, it is subject to oscillations for uniform spacing in x. Bica (2014) derived expressions for the averaged slopes at the ends that reduce the error of interpolation:

$$\bar{m}_1 = \frac{4\left(\Delta x_2^3 + \Delta x_3^3\right)(y_2 - y_1) + 3\Delta x_1\left[3\Delta x_2^3 \bar{m}_3 + \Delta x_1^2(y_2 - y_1) + \Delta x_2^2(y_3 - y_2)\right]}{7\Delta x_1^4 + 16\Delta x_1 \Delta x_2^3} \quad (10.58)$$

$$\bar{m}_2 = \frac{12\Delta x_2^3 \bar{m}_3 + 7\Delta x_1^2(y_2 - y_1) + 4\Delta x_2^2(y_3 - y_2)}{7\Delta x_1^3 + 16\Delta x_2^3} \quad (10.59)$$

$$\bar{m}_{n-1} = \frac{7\Delta x_{n-1}^2(y_n - y_{n-1}) + 12\Delta x_{n-2}^3 \bar{m}_{n-2} + 4\Delta x_{n-2}^2(y_{n-1} - y_{n-2})}{7\Delta x_n^3 + 16\Delta x_{n-1}^3} \quad (10.60)$$

$$\bar{m}_n = \frac{4\left(\Delta x_{n-1}^3 + \Delta x_n^3\right)(y_n - y_{n-1}) + 3\Delta x_{n-1}\left[3\Delta x_{n-2}^3 \bar{m}_{n-2} + \Delta x_{n-2}^2(y_{n-1} - y_{n-2}) + \Delta x_{n-1}^2(y_n - y_{n-1})\right]}{7\Delta x_{n-1}\Delta x_n^3 + 16\Delta x_{n-1}^4} \quad (10.61)$$

Using the rule of continuity for piecewise cubic splines between the knots, we calculate the polynomial coefficients for each k interval required by the interpolation formula in Equation (10.16). We use the optimized slopes from Equations (10.58) to (10.61) for the knots at the ends and Equation (10.54) for the interior knots to calculate the coefficients in the two-point cubic polynomial of Equation (10.16):

$$A_k = y_k \qquad B_k = \bar{m}_k \qquad C_k = \frac{3m_k - 2\bar{m}_k - \bar{m}_{k+1}}{\Delta x_k} \qquad D_k = \frac{\bar{m}_k + \bar{m}_{k+1} - 2m_k}{\Delta x_k^2} \quad (10.62)$$

The *PNMSuite* has a `VBA` user-defined function **ASPLINE** for Akima's method with Bica's end slopes. **ASPLINE** has the same arguments and capabilities as **CSPLINE** for interpolation, first-order derivatives, and integration. However, Akima's method must not be used to extrapolate beyond the ends of the data.

Example 10.9 Comparison of Akima with Cubic Spline Interpolation

Known/Find: Interpolate Bica's (2014) data with Akima's Hermite cubic spline for comparison with regular cubic spline, Kruger's and Stineman's constrained methods, presented in Sections 10.3.2 and 10.4.2.

Analysis: Use the *PNMSuite* macro **QYXPLOT** to generate a plot of the user-defined function **ASPLINE** interpolation. Add the data from the table as markers, and then add the results from other interpolations:

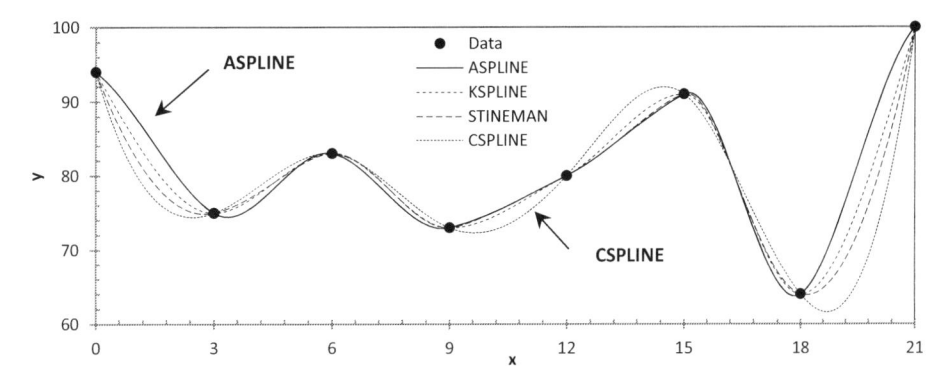

Comments: Akima's Hermite cubic polynomial interpolation method may outperform regular cubic spline by more closely matching the data trends. Akima's method with Bica's end conditions also preserves the pattern in the shapes of the curves of the interpolating polynomials at the ends, where the other methods tend to straighten out.

□

10.3.2 Kruger's Constrained Hermite Cubic Spline Interpolation*

Kruger (2011) developed an alternative Hermite cubic spline interpolation method that constrains the piecewise polynomials to match visual, monotonic trends in smooth data. The method places bounds on the derivative at intermediate points with first-order finite-difference approximations using the three adjacent data pairs for the slope m defined by Equation (10.55). The slope at knot k is constrained to lie between the slopes of the lines formed between successive pairs:

$$Min(m_{k-1}, m_k) \leq y'_k \leq Max(m_{k-1}, m_k) \tag{10.63}$$

We replace the cubic spline requirement for continuity in the second derivative by a function that satisfies the slope constraint in Equation (10.63), at the small expense of smoothness in the interpolating polynomials. The second derivative is now only approximately smooth across a knot. The following simple weighted harmonic mean function for the derivative approximation satisfies the criteria of Equation (10.63):

$$y'_k = \begin{cases} \dfrac{(w_{k-1} + w_k) m_{k-1} m_k}{(w_{k-1} m_k + w_k m_{k-1})} & \text{for } m_{k-1} m_k > 0 \\ 0 & \text{for } m_{k-1} m_k \leq 0 \end{cases} \tag{10.64}$$

with weights: $\quad w_{k-1} = \Delta x_{k-1} + 2\Delta x_k \quad$ and $\quad w_k = 2\Delta x_{k-1} + \Delta x_k$

Equation (10.64) has the property that the derivative at a data point goes to zero as either of the finite-difference approximations go to zero.

The derivation of the constrained cubic spline coefficients employs the following steps outlined for standard cubic spline interpolation with natural boundary conditions:

$$y_k = A_k \tag{10.65}$$

$$y_{k+1} = A_k + B_k \Delta x_k + C_k \Delta x_k^2 + D_k \Delta x_k^3 \tag{10.66}$$

$$y'_k = B_k \tag{10.67}$$

$$y'_{k+1} = B_k + 2C_k \Delta x_k + 3D_k \Delta x_k^2 \tag{10.68}$$

$$y''_k = 2C_k \tag{10.69}$$

$$y''_{k+1} = 2C_k \Delta x_k + 6D_k \Delta x_k \tag{10.70}$$

Solve Equations (10.65) through (10.70) simultaneously for A_k, B_k, C_k, D_k, and the second derivatives, y''_k, y''_{k+1}. The subscripts k and $k+1$ refer to the knots at the beginning and end of an interval, respectively. Calculate the coefficients of Equation (10.66) directly from the derivative approximations at the end of each interval:

$$A_k = y_k \qquad B_k = y'_k \qquad C_k = \frac{-2y'_k + 3m_k - y'_{k+1}}{\Delta x_k} \qquad D_k = \frac{y'_k - 2m_k + y'_{k+1}}{\Delta x_k^2} \tag{10.71}$$

$$y''_k = \frac{4y'_k + 2y'_{k+1} - 6m_k}{\Delta x_k} \qquad y''_{k+1} = \frac{4y'_{k+1} + 2y'_k - 6m_k}{\Delta x_k} \qquad y_{k+1} = y_k + m_k \Delta x_k \tag{10.72}$$

We invoke the natural (free) boundary conditions to get the derivative approximations at the endpoints:

$$y'_1 = \frac{3m_1 - y'_2}{2} \qquad \text{and} \qquad y'_n = \frac{3m_{n-1} - y'_{n-1}}{2} \tag{10.73}$$

Free type boundary conditions usually suffice for constrained splines. However, we may clamp the endpoints similar to Equation (10.38) if we know the values of the derivatives at the knots $k = 1$ and n.

CHAPTER 10: INTERPOLATION

By sacrificing some smoothness in the spline interpolation, we prevent overshooting the data points. Constrained cubic splines have the additional benefit of decoupling the second derivatives of the polynomials between intervals, thus eliminating the requirement for solving the simultaneous system of equations in the standard spline method.

The *VBA* user-defined function **KSPLINE** from the *PNMSuite* workbook implements Kruger's (2011) constrained cubic spline method with linear (natural) end conditions in an *Excel* worksheet. The user-defined function with arguments listed in Figure 10.17 also has the option of returning a first derivative or integral of the interpolating polynomials between integration limits. The sub procedure **HEAPSORTXY** sorts the data in ascending order according to the independent data. **KSPLINE** also does not allow for duplicate values in the *x* dimension.

```
Public Function KSPLINE(ydata, xdata, xnew As Double, _
                   Optional q As Integer = 0, Optional xup As Double, _
                   Optional alpha, Optional beta) As Double
' ydata = range of cells containing dependent variable data
' xdata = range of cells containing independent variable data
' xnew = point for interpolation. Use lower integration limit for quadrature.
' q = (optional argument) -1 = quadrature, 0 = interpolation, 1 = first derivative)
' xup = (optional argument) upper integration limit
' alpha = clamped value of first derivative at first knot
' beta = clamped value of first derivative at last knot
'****************************************************************************
' Interpolate data by constrained linear (natural) cubic spline. The program finds the
' interpolated value at the interpolation point:
'       ynew = a + b*(xnew - xk) + c*(xnew - xk)^2 + d*(xnew - xk)^3.
' Optional calculation of integral or derivative. xdata must be in ascending order.
' *** Required User-defined Functions and Sub Procedures from the PNM Suite ***
' HEAPSORTXY = sorts x data in ascending order
' *** Nomenclature ***
Dim c() As Double, d() As Double ' = cubic spline coefficients
Dim dxlo As Double, dxu As Double ' distance between integration limits and intervals
Dim h() As Double ' = xdata intervals
Dim i As Long ' = loop index
Dim iup As Long ' = index for match
Dim k As Long, k_1 As Long ' = knot for spline interval and k-1
Dim m() As Double ' = finite difference 1st order derivative approximations
Dim mm As Double ' = product of m at a knot
Dim n As Long, n1 As Long ' = number of data pairs and n-1
Dim small As Double ' = small number to avoid division by zero (module level)
Dim wk As Double, wk_1 As Double ' = harmonic mean weights
Dim x() As Double, y() As Double ' = vector of x- and y-data for sorting
Dim yp() As Double ' = constrained 1st derivative approximation at knots
'****************************************************************************
On Error Resume Next ' Skip errors due to bad data
small = 0.000000000001: n = ydata.Count: n1 = n - 1
If xdata.Count <> n Or n < 2 Then
    MsgBox "x and y data must match, or > 2.": Exit Function
End If
ReDim c(1 To n) As Double, d(1 To n) As Double, h(1 To n) As Double, _
      m(1 To n) As Double, x(1 To n) As Double, y(1 To n) As Double, _
      yp(1 To n) As Double
For i = 1 To n: x(i) = xdata(i): y(i) = ydata(i): Next i
Call HEAPSORTXY(x(), y()) ' Sort data in ascending order
For k = 1 To n1 ' Set spline intervals for knots k & derivative approximations
    h(k) = x(k + 1) - x(k): m(k) = (y(k + 1) - y(k)) / (h(k) + small)
Next k
For k = 2 To n - 1 ' Calculate weighted derivative approximations at interior knots
    k_1 = k - 1: mm = m(k_1) * m(k)
    If mm <= 0 Then
        yp(k) = 0
    Else ' Fritsch & Carlson weighted harmonic mean
        wk_1 = h(k - 1) + 2 * h(k): wk = 2 * h(k_1) + h(k)
        yp(k) = (wk_1 + wk) * mm / (wk_1 * m(k) + wk * m(k_1) + small)
    End If
Next k
If IsMissing(alpha) Then ' Get derivative approximations for first & last points
    yp(1) = (3 * m(1) - yp(2)) / 2 ' free
```

```
        Else
            yp(1) = alpha   ' clamped
        End If
        If IsMissing(beta) Then
            yp(n) = (3 * m(n1) - yp(n1)) / 2  ' free
        Else
            yp(n) = beta   ' clamped
        End If
        For k = 1 To n1  ' Calculate C & D cubic spline coefficients for knot k
            c(k) = (-2 * yp(k) + 3 * m(k) - yp(k + 1)) / h(k)
            d(k) = (yp(k) - 2 * m(k) + yp(k + 1)) / (h(k) ^ 2)
        Next k
        c(n) = 0: d(n) = 0  ' Endpoint interval for linear extrapolation beyond last point
        If xnew < x(1) Then  ' Find the knot for interval containing the interpolation point
            k = 1
        Else
            k = WorksheetFunction.Match(xnew, x, 1)
        End If
        Select Case q
        Case 0  ' Interpolation
            If xnew >= x(1) Then  ' Calculate 3rd order constrained spline polynomial
                KSPLINE = y(k) + yp(k) * (xnew - x(k)) _
                    + c(k) * (xnew - x(k)) ^ 2 + d(k) * (xnew - x(k)) ^ 3
            Else
                KSPLINE = y(1) + yp(1) * (xnew - x(1))  ' Linear extrapolation first point
            End If
        Case 1  ' Calculate value of constrained derivative
            If xnew >= x(1) Then
                KSPLINE = yp(k) + 2 * c(k) * (xnew - x(k)) + 3 * d(k) * (xnew - x(k)) ^ 2
            Else
                KSPLINE = yp(1)  ' Linear extrapolation beyond first point
            End If
        Case -1  ' Find the intervals containing the integration limits
            iup = WorksheetFunction.Match(xup, x, 1)
            dxlo = xnew - x(k)  ' Integrate data between integration limits
            dxu = xup - x(iup)
            If Not iup = k Then  ' integration limits in different intervals
                KSPLINE = y(k) * (h(k) - dxlo) + yp(k) * (h(k) ^ 2 - dxlo ^ 2) / 2 _
                    + c(k) * (h(k) ^ 3 - dxlo ^ 3) / 3 + d(k) * (h(k) ^ 4 - dxlo ^ 4) / 4
                For i = k + 1 To iup - 1
                    KSPLINE = KSPLINE + y(i) * h(i) + yp(i) * (h(i) ^ 2) / 2 _
                        + c(i) * (h(i) ^ 3) / 3 + d(i) * (h(i) ^ 4) / 4
                Next i
                KSPLINE = KSPLINE + y(iup) * dxu + yp(iup) * (dxu ^ 2) / 2 _
                    + c(iup) * (dxu ^ 3) / 3 + d(iup) * (dxu ^ 4) / 4
            Else  ' when integration limits are between the same pair of knots
                KSPLINE = y(k) * (dxu - dxlo) + yp(k) * (dxu ^ 2 - dxlo ^ 2) / 2 _
                    + c(k) * (dxu ^ 3 - dxlo ^ 3) / 3 + d(k) * (dxu ^ 4 - dxlo ^ 4) / 4
            End If
        Case Else
            MsgBox "Incorrect q argument for derivative or quadrature!"
        End Select
        End Function  ' KSPLINE
```

Figure 10.17 VBA user-defined function KSPLINE for Hermite constrained cubic spline interpolation of smooth data. See the workbook *PNMSuite* for the Sub HEAPSORTXY.

Example 10.10 Interpolation of Adsorption Isotherm Data

Known: Table 10.1 contains dimensionless adsorption isotherm data for toluene vapor on granulated activated carbon at 25°C and 1 atm pressure. y and x represent the solid and gas phase compositions of the adsorbate toluene.

Table 10.1 Toluene adsorption on carbon isotherm data (Park, Choi and Lee 2007)

x	0.0032	0.0033	0.0158	0.0384	0.0736	0.1139	0.1390	0.1818
y	0.6452	0.6824	0.7147	0.7321	0.7545	0.7819	0.7944	0.8144

Find: Fit the data to two common adsorption isotherm models by nonlinear regression, as described in Section 9.1.3:

- *Freundlich*: $$y = 0.8696 x^{0.04805} \qquad (10.74)$$

- *Langmuir*: $$y = \frac{1323 x}{1 + 1704 x} \qquad (10.75)$$

Then, compare the two models with standard and constrained cubic spline interpolation.

Analysis: A functional fit of the data with a natural cubic spline using the VBA function **CSPLINE** exhibits severe oscillation that overshoots the trends in the data around the bend. The constrained cubic spline interpolation using the VBA function **KSPLINE** provides an interpolation that behaves according to the visible trend in the data, as seen in Figure 10.18.

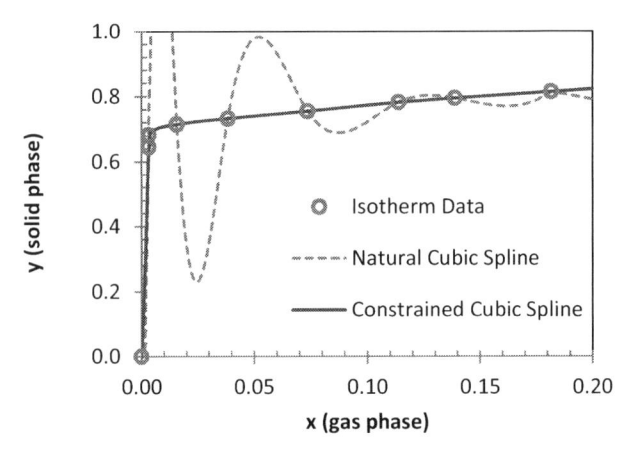

Figure 10.18

Comparison of isotherm data and interpolation results of natural cubic spline and the constrained cubic spline interpolations.

We compare the Langmuir and Freundlich models with constrained cubic spline interpolation for large y in Figure 10.19. Langmuir's model fails to match the trends in the data. Freundlich's model provides a good match, but undershoots the data at the low and high ends of x. The constrained cubic spline appears to match the data well over the entire range for x.

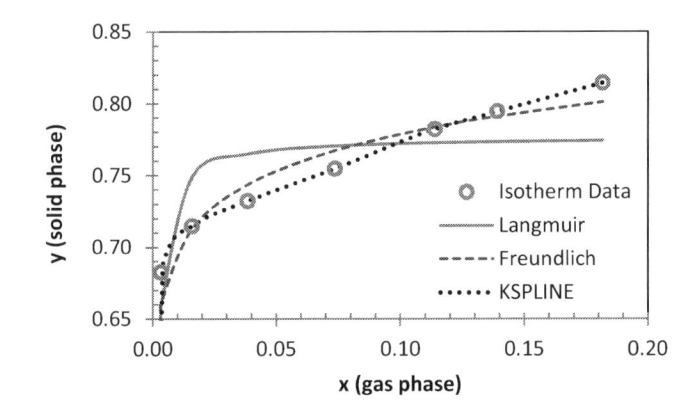

Figure 10.19

Comparison of data with interpolations by Langmuir, Freundlich, and constrained spline models.

Comments: Alternatively, we may use interpolating methods, such as Newton's divided difference interpolating polynomials on local subsets of the data to avoid overshooting.

The constrained Hermite cubic spline interpolation methods avoid oscillations, or data overshooting, that occasionally plague the regular cubic spline piecewise interpolating polynomials for sparse data sets, at the expense of some smoothness in the interpolated values. It is always important to check the fit by plotting interpolated values *between* data points. The *PNMSuite* macro **QYXPLOT** is convenient for generating a graph of an interpolating function with 1000 points over the range of the data. Otherwise, graphing tools such as *Excel* may mask the true behavior of the interpolating function by plotting lines that appear to "connect the dots" of the data points.

For some data, we can avoid constraining the interpolation by inverting the dependent for independent data. Now our regular cubic spline interpolation treats the independent data as dependent, and vice versa, $x \leftrightarrow y$. In this form, we can use our methods of root finding from Chapter 6 to solve for the value of the independent variable given the dependent variable.

10.4 Rational Interpolation*

> *"A man is like a fraction whose numerator is what he is and whose denominator is what he thinks of himself. The larger the denominator, the smaller the fraction."* – Tolstoy

Like rational least squares in Section 9.10.1, a rational Padé interpolating function consists of a ratio of two polynomials with order k in the numerator and order m in the denominator:

$$y_i = \frac{p_0 + p_1 x_i + p_2 x_i^2 + \ldots p_k x_i^k}{1 + q_1 x_i + q_2 x_i^2 + \ldots q_m x_i^m} \tag{10.76}$$

Unlike rational functions for a nonlinear least-squares model, the orders of the polynomials for rational Padé interpolating functions relate directly to n, the number of data pairs for interpolation: $k + m + 1 = n$. Thus, the Padé function represents the "best" rational acceleration of a truncated Taylor series with order $k + m$. See Sections 5.2 and 12.1 for more discussion on Padé series acceleration. The same strengths and weaknesses associated with least squares rational functions also apply to rational Padé interpolation. Experience shows that good rational functions typically have polynomials in the numerator and denominator that differ by no more than one order: $0 \leq |k - m| \leq 1$ (Hanna and Sandall 1995).

Because $n = k + m + 1$, we determine the rational function coefficients exactly by multiplying both sides of Equation (10.76) by the denominator to give a linear expression in the coefficients p and q:

$$p_0 + p_1 x_i + p_2 x_i^2 + \ldots p_k x_i^k - q_1 x_i y_i - q_2 x_i^2 y_i - \ldots q_m x_i^m y_i = y_i \tag{10.77}$$

To find the values for the p's and q's, generate one equation like (10.77) for each data pair to give a linear system of equations in the coefficients. Solve the system using any of the methods presented in Chapter 4. For example, consider the system of equations representing a rational interpolating function for a set of $n = k + m + 1 = 5$ data pairs with $k = 2$ and $m = 2$. Calculate the coefficients using matrix algebra:

$$\begin{bmatrix} p_0 \\ p_1 \\ p_2 \\ q_1 \\ q_2 \end{bmatrix} = \begin{bmatrix} 1 & x_1 & x_1^2 & -x_1 y_1 & -x_1^2 y_1 \\ 1 & x_2 & x_2^2 & -x_2 y_2 & -x_2^2 y_2 \\ 1 & x_3 & x_3^2 & -x_3 y_3 & -x_3^2 y_3 \\ 1 & x_4 & x_4^2 & -x_4 y_4 & -x_4^2 y_4 \\ 1 & x_5 & x_5^2 & -x_5 y_5 & -x_5^2 y_5 \end{bmatrix}^{-1} \cdot \begin{bmatrix} y_1 \\ y_2 \\ y_3 \\ y_4 \\ y_5 \end{bmatrix} \tag{10.78}$$

The rational interpolation method has the advantage that the function passes through every data point. This becomes a disadvantage for noisy data due to large experimental uncertainty. The method of rational interpolation works well for small sets of smooth data, typically fewer than a dozen points. For larger data sets, we typically use a rational interpolation for subsets of data. Otherwise, the rational function may exhibit some unusual behavior, such as poles in the form of vertical asymptotes. A rational function for a few data pairs at either end of the data may be useful for extrapolating better than polynomials beyond the range of data.

A *VBA* user-defined function **RATLIN** is available from the *PNMSuite* for interpolating smooth data in an *Excel* worksheet. **RATLIN** with arguments defined in Figure 10.20 may use any subset of the data specified by the user. The corresponding independent and dependent data must be in two separate ranges on the worksheet. The default orders of the numerator and denominator are set equal for data with an odd number of points. In the case of an even number of points in the data, the order of the denominator is set to one higher than the numerator. Otherwise, the user may specify the order of the numerator to any value between one and one fewer than the number of data pairs.[98]

```
Public Function RATLIN(ydata, xdata, xnew As Double, Optional np) As Double
' ydata = range of cells containing dependent variable data
' xdata = range of cells containing independent variable data
' xnew = cell with point for interpolation
' np = (optional argument) integer for order of polynomial in numerator of rational
' function.  Must be smaller than the sum of orders of numerator and denominator.
'*******************************************************************************
' Rational interpolation.
' ynew = (p0 + p1*x + p2*x^2 ... + pk*x^k)/(1  + q1*x + q2*x^2 ... + pm*x^m)
' User selects smooth data & specifies order of the numerator (optional). If the order
' is not specified, RATLIN uses the same order for numerator and denominator with a
' difference no greater than 1 for the denominator.  The order of numerator must not
' exceed the number of data pairs.  Returns an interpolated value for the dependent
' variable at the specified point of interpolation for the independent variable. The
' linear system of equations is solved by Gaussian elimination.
' *** Nomenclature ***
Dim a() As Double, b() As Double ' = matrix of coefficients, vector of constants
Dim i As Long, j As Long ' = index of data and coefficients
Dim k As Long, m As Long ' = order of numerator/denominator
Dim mi As Long, n As Long ' = counter for denominator terms, number of data pairs
Dim pq() As Double ' = vector of numerator and denominator coefficients
Dim yk As Double, ym As Double ' = numerator/denominator of interpolated value
'*******************************************************************************
n = xdata.Count ' Calculate the orders of the numerator and denominator
If IsMissing(np) Then ' Check np to set numerator order
    k = CInt((n - 1) / 2)
ElseIf CInt(np) < 1 Then
    k = 1
ElseIf CInt(np) > n - 1 Then
    k = n - 1
Else
    k = CInt(np)
End If
m = n - 1 - k ' Set order of denominator
ReDim a(1 To n, 1 To n) As Double, b(1 To n) As Double, pq(1 To n) As Double
For i = 1 To n ' Set up the coefficient matrix and constant vector
    For j = 1 To k + 1: a(i, j) = xdata(i) ^ (j - 1): Next j
    For j = 1 To m: a(i, j + k + 1) = -ydata(i) * xdata(i) ^ j: Next j
    b(i) = ydata(i)
Next i
Call GAUSSPIVOT(a, b, n, pq) ' Solve the linear system of equations for p's and q's
yk = 0 ' Rational interpolation, sum terms in the numerator
For j = 1 To k + 1: yk = yk + pq(j) * xnew ^ (j - 1): Next j
ym = 1: mi = 0 ' Sum the terms in the denominator
For j = k + 2 To m + k + 1
    mi = mi + 1: ym = ym + pq(j) * xnew ^ mi
Next j
RATLIN = yk / ym: Exit Function ' Return the rational interpolated value
End Function ' RATLIN
```

Figure 10.20 *VBA* user-defined function RATLIN for interpolating smooth data with a rational function.

---

[98] Is Equation (10.76) what my grade school teacher meant by, "Mind your p's and q's?"

Occasionally, we need the rational function coefficients for a compact, portable equation representing the data instead of tabulated values for interpolation. The *VBA* macro **RATLINPQ**, available in the *PNMSuite*, requires the same worksheet setup for the user-defined function **RATLIN**, but reports the values of the coefficients in the numerator and denominator of the rational function in Equation (10.76). We may use the user-defined function **RATLSF(x, p, q)** in Figure 9.82 to evaluate the rational function at position x with the results for the coefficients **p** and **q**. The *VBA* macros **RATLIN** and **RATLINPQ** do not require data in ascending order.

The *PNMSuite* also includes a *VBA* user-defined function **STINEMAN**, derived in Section 10.4.2, for interpolating by piecewise rational functions (Stineman 1980). We recommend Stineman's method of interpolation when a rational function overshoots the obvious trend in the data or generates poles.

Example 10.11 Interpolation of Batch Distillation Data

Known: A four-stage batch distillation column separates ethanol from water. The still pot initially consists of 100 moles of a mixture of 20 mol% ethanol in water. The external reflux ratio is fixed at three. Distillation continues until the residue in the pot is three percent ethanol. The Rayleigh equation gives a relationship between the contents and composition of the still:

$$\ln\left(\frac{W_0}{W}\right) = \int_{x_w}^{x_{w0}} \frac{dx_w}{x_D - x_w} \tag{10.79}$$

where W_0 and W are the initial and final moles in the still pot, x_{w0} and x_w are the initial and final mole fractions of ethanol in the still, respectively, and x_D is the ethanol composition in the distillate. A graphical analysis of the integrand in Equation (10.79) produced the data listed columns **A** and **B** in an *Excel* worksheet shown in Figure 10.21.

	A	B	C	D	E
1	x_w	$1/(x_D-x_w)$			
2	0.245	2.02		$P_0 =$	1.038944
3	0.1	1.61		$P_1 =$	-112.049
4	0.04	1.52		$P_2 =$	-176.254
5	0.025	1.6		$Q_1 =$	-87.7177
6	0.02	1.72		$Q_2 =$	36.27943

Figure 10.21
Rational model parameters of batch distillation data calculated from the *VBA* macro RATIONALPQ.

Schematic: A batch still has a liquid charge W with mole fraction x_W and distillate rate D with mole fraction x_D.

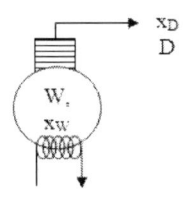

Find: Interpolate the integrand data for several points between 0.2 and 0.03 using a rational function.

Assumptions: Neglect hold up on the trays in the column above the still.

Analysis: Plot the data and try fitting with polynomials using a chart **Trend line**. Figure 10.22 shows the result of a third-order polynomial fit of the data. Although the R^2 value indicates a good fit, clearly the result provides poor interpolation of the data. Though not displayed, a fourth order polynomial does not improve the interpolation. A plot of the rational interpolation of the data created with the macro **QYXPLOT** in Figure 10.22 shows improved interpolation over the entire data range.

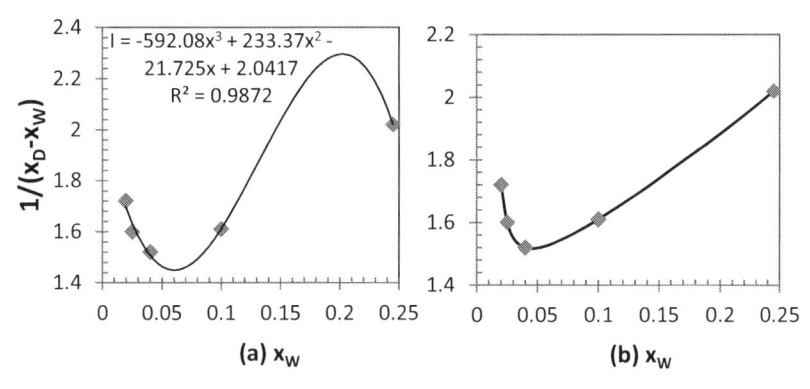

Figure 10.22

(a) Third-order polynomial and (b) k=2, m=2 rational interpolation of batch distillation data from the Rayleigh equation.

Try the *VBA* user-defined function **RATLIN** at $x_W = 0.2$:

$$1.8831 = \text{RATLIN}(\$B\$2:\$B\$6, \$A\$2:\$A\$6, 0.2)$$

The *VBA* macro **RATLINPQ** used the following inputs from the worksheet in Figure 10.21, and the default value for the order of the numerator in the polynomial:

$$\text{y data} = \$B\$2:\$B\$6 \qquad \text{x data} = \$A\$2:\$A\$6 \qquad k = 2$$

The *Excel* worksheet in Figure 10.21 shows the results for the coefficients of the polynomials in the numerator and denominator:

$$\frac{1}{x_D - x_W} = \frac{1.0389 - 112.05 x_W - 176.25 x_W^2}{1 - 87.718 x_W + 36.279 x_W^2} \tag{10.80}$$

Alternatively, use the function **RATLSF(x,p,q)** to evaluate the function in Equation (10.80) at **x** with the ranges for the *p*'s and *q*'s.

$$1.8831 = \text{RATLSF}(0.2, \text{E2:E4}, \text{E5:E6})$$

Comments: Although both the polynomial and rational functions fit the data exactly, the polynomial overshoots the trend in the data at the high end of x_w. In this example, the rational function provides a good fit of the data and appears to match the trend. Chapter 11 describes several methods for numerically evaluating the integral of the function in Equation (10.79).

□

10.4.1 Bulirsch-Stoer Algorithm for Rational Interpolation*

Bulirsch and Stoer (1980) developed an efficient algorithm for rational interpolation, similar to Newton's recursive method of divided difference for polynomials. However, it follows Neville's method that evaluates the interpolant without producing the polynomial coefficients. Their solution is limited to the special case where the order of the numerator and denominator differ by no more than one. For the case where the sum of the orders is odd, the order of the denominator is one greater than the numerator. We illustrate the recursive stages of the Bulirsch-Stoer method for three *y* and *x* data points in Table 10.2 for the general recursive formula for interpolation:

$$y(x) = r_{12\ldots k} = r_{23\ldots k-1} + \frac{r_{23\ldots k} - r_{12\ldots k-1}}{\left(\dfrac{x - x_1}{x - x_k}\right)\left(1 + \dfrac{r_{23\ldots k} - r_{12\ldots k-1}}{r_{23\ldots k} - r_{23\ldots k-1}}\right)} \tag{10.81}$$

A *VBA* user-defined function named **RATLINBS(ydata, xdata, xnew)** is available in the *PNMSuite* workbook for Bulirsch-Stoer rational interpolation of smooth data. The arguments required by **RATLINBS** include the ranges of dependent and independent data and the point of interpolation or extrapolation.

Table 10.2 Bulirsch-Stoer stages of rational interpolation for $y = (p_0 + p_1 x)/(1 + q_1 x)$.

Data	Stage 1	Stage 2	Stage 3
x_1, y_1	$r_1 = y_1$		
x_2, y_2	$r_2 = y_2$	$r_{12} = r_2 + \dfrac{r_2 - r_1}{\left(\dfrac{x - x_1}{x - x_2}\right)\left(1 - \dfrac{r_2 - r_1}{r_2}\right) - 1}$	
x_3, y_3	$r_3 = y_3$	$r_{23} = r_3 + \dfrac{r_3 - r_2}{\left(\dfrac{x - x_2}{x - x_3}\right)\left(1 - \dfrac{r_3 - r_2}{r_3}\right) - 1}$	$r_{123} = r_{23} + \dfrac{r_{23} - r_{12}}{\left(\dfrac{x - x_1}{x - x_3}\right)\left(1 - \dfrac{r_{23} - r_{12}}{r_{23} - r_2}\right) - 1}$

Example 10.12 Rational Interpolation by the method of Bulirsch and Stoer

Known/Find: Compare the results of rational interpolation with the user-defined function **RATLINBS** to the normal probability density function (*PDF*) in Equation (8.72) over the range $-4 \leq x \leq 4$, with a mean of zero and standard deviation of one.

Analysis: First, generate a small set of nine equally spaced points from the density function at $x = -4, -3, -2 \ldots 3, 4$ in an *Excel* worksheet for interpolation. With the user-defined function **RATLINBS**, generate a large number of interpolated values between pairs of PDF function values and compare the results in a graph, as shown in Figure 10.23.

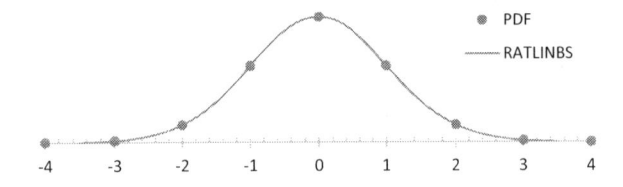

Figure 10.23

Rational interpolation using RATLINBS of points from a normal probability density function (PDF) with standard deviation one and zero mean.

Comments: The interpolated results match the trend in the probability density function well in the range $-3.5 < x < 3.5$. However, the rational interpolation generates small, practically imperceptible oscillations near the ends where the density function flattens out.

□

10.4.2 Stineman Piecewise Rational Interpolation*

Despite the success of rational interpolation in Example 10.11, rational functions are prone to form poles for ill conditioned data, as seen in Figure 10.25. A popular interpolation scheme due to Stineman has properties that prevent poles or overshooting data (Stineman 1980). Stineman's method interpolates smooth data using piecewise low-order rational functions with continuous derivatives at the common points.

Stineman's interpolation method requires data arranged in ascending order of the independent variable. Three consecutive points are indexed with subscripts i, j, and k: $x_i < x_j < x_k$. The formulas for Stineman interpolation between two points at indexes j and k include:

- Slope of a line segment: $$s_j = \frac{y_k - y_j}{x_k - x_j} \tag{10.82}$$

- Line connecting points: $$y_o = y_j + s_j (x - x_j) \tag{10.83}$$

- Vertical distances from (x, y_o) to the line with slope y_j' or y_k':

$$\Delta y_j = y_j + y_j'(x - x_j) - y_o \qquad \Delta y_k = y_j + y_k'(x - x_k) - y_o \tag{10.84}$$

CHAPTER 10: INTERPOLATION

- If $(\Delta y_j)(\Delta y_k) > 0$:

$$y = y_o + \frac{\Delta y_j \Delta y_k}{\Delta y_j + \Delta y_k} \tag{10.85}$$

- If $(\Delta y_j)(\Delta y_k) < 0$:

$$y = y_o + \frac{\Delta y_j \Delta y_k (2x - x_j - x_k)}{(\Delta y_j - \Delta y_k)(x_k - x_j)} \tag{10.86}$$

Stineman's method approximates the first derivatives (slopes) for the interior points from the slope of an arc passing through three adjacent points:

$$y'_j = \frac{(y_j - y_i)\left[(x_k - x_j)^2 + (y_k - y_j)^2\right] + (y_k - y_j)\left[(x_j - x_i)^2 + (y_j - y_i)^2\right]}{(x_j - x_i)\left[(x_k - x_j)^2 + (y_k - y_j)^2\right] + (x_k - x_j)\left[(x_j - x_i)^2 + (y_j - y_i)^2\right]} \tag{10.87}$$

The slopes at the endpoints ($i = 1$ or $k = n$) have special formulas. If $s > 0$ and $s > y_j'$ or $s < 0$ and $s < y_j'$:

$$y' = 2s - y'_j \tag{10.88}$$

Otherwise:

$$y' = s + \frac{|s|(s - y'_j)}{|s| + |s - y'_j|} \tag{10.89}$$

The *VBA* user-defined function **STINEMAN(ydata, xdata, xnew)** interpolates ill-conditioned data in the ranges **ydata** and **xdata** at position **xnew** in an *Excel* worksheet where rational interpolation overshoots the data or forms poles. The function first sorts the data in ascending order as required by the method.

Example 10.13 Stineman interpolation of sieving data in Excel

Known: An *Excel* worksheet shown in Figure 10.24 contains data for the percent of material passing through a sieve.

Find: Compare Stineman with rational interpolation of the complete data set.

Analysis: Recall that interpolating methods pass exactly through the data points. Therefore, it is important to check the interpolation between consecutive data points to verify that the interpolation trends similar to the data. Calculate interpolated sieve values from 2.6% to 100% incremented by 0.2% using both user-defined functions **STINEMAN** and **RATLIN**. The results in columns **E** and **F** shown in Figure 10.24 use the following formulas:

E3 = STINEMAN($B$3:$B$15, $A$3:$A15, D3)   F3 = RATLIN($B$3:$B$15, $A$3:$A15, D3)

	A	B	C	D	E	F
1	Data			Interpolation		
2	%	Sieve		%	Stineman	Rational
3	2.5	0.0026		2.6	0.0034	0.0078
4	2.9	0.0052		2.8	0.0043	0.0017
5	3.3	0.0135		3	0.0072	0.0076
6	4.2	0.0285		3.2	0.0115	0.0117
7	5.8	0.0568		3.4	0.0153	0.0153
8	6	0.063		3.6	0.0186	0.0187
9	8.9	0.09		3.8	0.0218	0.0221
10	23.3	0.125		4	0.0251	0.0253
11	59.2	0.18		4.2	0.0285	0.0285
12	89	0.25		4.4	0.0319	0.0317
13	96.1	0.3		4.6	0.0353	0.0348
14	99.9	0.425		4.8	0.0387	0.0380
15	100	0.5		5	0.0422	0.0412

Figure 10.24

Piecewise rational interpolation of sieve data using the *VBA* user-defined function STINEMAN compared with rational interpolation in an *Excel* worksheet. (Liengme 2009)

As shown in Figure 10.25, both Stineman and rational interpolations performed well above 20% material passing through the sieves. However, poles in the rational interpolation overshoot and undershoot the obvious trend in the data between 5% and 20%, whereas Stineman interpolation appears to perform well over the complete data range.

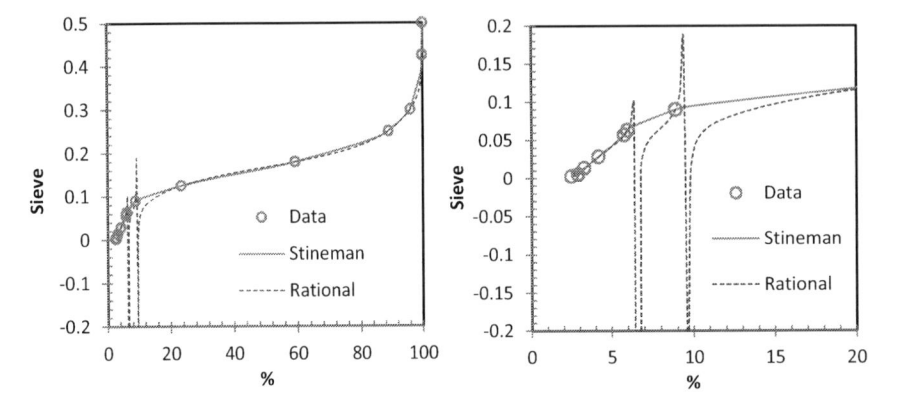

Figure 10.25

Stineman interpolation compared with rational function for sieve data.

Comments: Stineman claims the method never overshoots the trend in the data nor forms poles, at the expense of sacrificing some smoothness in the interpolation.

10.5 Bivariate (Two-Dimensional) Grid Interpolation*

We may apply our methods of single variable interpolation to problems requiring interpolating data on a grid in two or more dimensions. We first consider the simplest case of bilinear interpolation to outline the basic pattern for multivariate interpolation. Consider a two-dimensional rectangular grid of points in Figure 10.26, where x and y are the independent variables arranged in the horizontal and vertical positions on the grid, and where $f(x, y)$ is the value of the dependent variable at x and y.

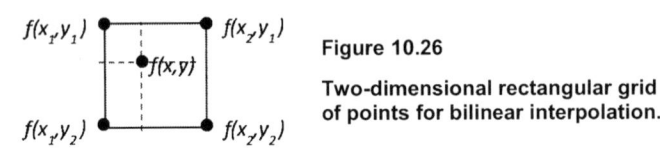

Figure 10.26

Two-dimensional rectangular grid of points for bilinear interpolation.

The general approach starts with linear interpolation in one dimension at the low and high point, followed by linear interpolation of these results in the remaining dimension. For convenience, we employ the Lagrange formula for linear interpolation in the x dimension at positions y_1 and y_2:

$$f_1(x, y_1) = \frac{x - x_2}{x_1 - x_2} f(x_1, y_1) + \frac{x - x_1}{x_2 - x_1} f(x_2, y_1) \tag{10.90}$$

$$f_2(x, y_2) = \frac{x - x_2}{x_1 - x_2} f(x_1, y_2) + \frac{x - x_1}{x_2 - x_1} f(x_2, y_2) \tag{10.91}$$

We finish with the results from Equations (10.90) and (10.91) in Lagrange's linear interpolating formula for the y-dimension:

$$f(x, y) = \frac{y - y_2}{y_1 - y_2} f_1(x, y_1) + \frac{y - y_1}{y_2 - y_1} f_2(x, y_2) \tag{10.92}$$

The extension to three or more dimensions follows the same pattern, but incurs exponential increases in computation.

CHAPTER 10: INTERPOLATION

Interpolating random scattered data that does not fit onto a regular grid involves advanced methods that may be too computationally intensive for *Excel* with *VBA*. We refer you to the reference *Numerical Recipes: The Art of Scientific Computing, 3rd Edition* (Press, Teukolsky and Vetterling 2007) for an introduction to the methods of kriging and radial basis functions for interpolating scattered data in multi-dimensions.

Example 10.14 Henry's Constant for CO$_2$ in Aqueous Diethanolamine

Known: Figure 10.27 contains experimental results of Henry's constant for the solubility of carbon dioxide in aqueous solutions of diethanolamine (DEA) ranging from zero to 30 weight percent over a temperature range of 20 to 40°C (Davis 1992).

Find: Create a *VBA* user-defined function for bilinear interpolation of two-dimensional data. Use the function to estimate Henry's constant at $T = 22°C$ and 5% DEA in water.

Analysis: Comparing the layout of Figure 10.27 with Equation (10.92) and Figure 10.26, x corresponds to weight percent and y corresponds to temperature. We use the worksheet function **TREND** to perform the calculation in an *Excel* worksheet for $H = 3.82$ at 22°C at 5% DEA in water.

	A	B	C	D	E	F	G	H
1		0	10	20	30		T	Henry
2	20	3.27	4.03	5.76	9.56		20	3.65
3	25	3.61	4.52	6.46	10.53		25	4.065
4	30	3.97	5.05	7.21	11.56		22	3.816
5	35	4.35	5.62	8.02	12.65			
6	40	4.75	6.23	8.89	13.8			

	H
1	Henry
2	=TREND(B2:C2,B1:C1,5)
3	=TREND(B3:C3,B1:C1,5)
4	=TREND(H2:H3,G2:G3,22)
5	
6	

Figure 10.27

Worksheet calculations and formulas for bilinear interpolation of Henry's constant at $x = 5$ wt. % DEA in water and $y = 22°C$.

The interpolation formulas in column **H** use the *Excel* worksheet function **TREND** to perform the linear interpolations in Equations (10.90), (10.91), and (10.92).

Our general-purpose *VBA* user-defined function **LINTERP2D**, with arguments defined in Figure 10.28, incorporates worksheet **MATCH** functions to locate the four points for interpolation, followed by Lagrange formulas for x and y-dimension interpolation. Note that *VBA* references elements of arrays by row and column, which feels inverse to x and y positions in $f(x, y)$ shown in Figure 10.26. To use **LINTERP2D**, set up an *Excel* worksheet with the x values arranged along the top of the dependent data, and the y values along the left side. The corresponding values of the dependent variable are located in the rectangular grid, similar to that shown in Figure 10.27. Supply the ranges of values for the arguments. The following formula uses **LINTERP2D** to perform the interpolation in this example, and produces the same result found in the worksheet calculations in cell **H4**:

$$3.82 = \text{LINTERP2D}(\$B\$2:\$E\$6, \$B\$1:\$E\$1, \$A\$2:\$A\$6, 5, 22)$$

```
Public Function LINTERP2D(fdata, xdata, ydata, xnew As Double, ynew As Double) As Double
' fdata = range of x-y grid of dependent values
' xdata = range of x dimension values (in ascending order left to right)
' ydata = range of y dimension values (in ascending order top to bottom)
' xnew = x dimension intermediate value for interpolation
' ynew = y dimension intermediate value for interpolation
'********************************************************************************
' Bilinear interpolation of intermediate value in two-dimensional grid
' (x = horizontal, y = vertical) of f(x,y) values.
'      f(1,1)--f(1,2)      > x-dimension (right direction)
'        |       |         v
'      f(1,2)--f(2,2)      y-dimension (down direction)
' *** Nomenclature ***
Dim f1 As Double, f2 As Double ' = interpolated values in x at y1 and y2
Dim f11 As Double, f12 As Double, f21 As Double, f22 As Double ' = dependent variables
Dim i As Long, j As Long ' = row/column index values in xdata, ydata, fdata arrays
Dim m As Long, n As Long ' = number of terms in xdata and ydata ranges, respectively
Dim p As Long, q As Long ' = offset values in row and column directions
Dim x1 As Double, x2 As Double ' = lower and upper x values for x interpolation
```

```
Dim y1 As Double, y2 As Double ' = lower and upper values for y interpolation
'****************************************************************************
m = xdata.Count: n = ydata.Count ' get length of xdata and ydata ranges
With WorksheetFunction ' locate the 2x2 square grid of interpolants
    xnew = .Max(xnew, xdata(1)): ynew = .Max(ynew, ydata(1))
    j = .Match(xnew, xdata): i = .Match(ynew, ydata) ' column/row index
End With
If i < n Then ' range offsets in x and y dimensions for interior or end values
    p = 1 ' x interpolant is located above last row
Else
    p = -1 ' y interpolant is on last row
End If
If j < m Then
    q = 1 ' y interpolant is located left of last column
Else
    q = -1 ' x interpolant is on last column
End If
' x and y values and corresponding f values at row and column positions
x1 = xdata(j): x2 = xdata(j + q): y1 = ydata(i): y2 = ydata(i + p)
f11 = fdata(i, j): f12 = fdata(i, j + q)
f21 = fdata(i + p, j): f22 = fdata(i + p, j + q)
' Lagrange formulas for linear interpolation in x direction at y1 and y2
f1 = (xnew - x2) * f11 / (x1 - x2) + (xnew - x1) * f12 / (x2 - x1)
f2 = (xnew - x2) * f21 / (x1 - x2) + (xnew - x1) * f22 / (x2 - x1)
' Lagrange formula for linear interpolation in y direction at x
LINTERP2D = (ynew - y2) * f1 / (y1 - y2) + (ynew - y1) * f2 / (y2 - y1)
End Function ' LINTERP2D
```

Figure 10.28 *VBA* user-defined function LINTERP2D for bilinear interpolation of two-dimensional data.

Comments: Linear interpolation is easy to apply, but may not adequately reproduce nonlinear trends in data. □

We follow the pattern of bilinear interpolation to employ other methods of interpolation, such as cubic splines. Stineman's and the constrained Hermite cubic spline methods are particularly useful for general-purpose interpolation of smooth data. These methods eliminate poles, work equally well for both even and unequally spaced data, and are less computationally demanding than regular cubic spline for *VBA* user-defined functions. To illustrate how we may use any of our interpolating routines for two-dimensional interpolation, consider the interpolation in Example 10.14 with the user-defined function **KSPLINE**. The first five formulas interpolate the data in each of the y rows at $x = 5$. The range **A8:A12** contains the results from the x point interpolations. The last formula interpolates the previous five x results at $y = 22$:

        3.59 =KSPLINE(B2:E2,$B$1:$E$1,5)
        4.00 =KSPLINE(B3:E3,$B$1:$E$1,5)
        4.44 =KSPLINE(B4:E4,$B$1:$E$1,,5) → 3.76 =KSPLINE(A8:A12,A2:A6,22)
        4.91 =KSPLINE(B5:E5,$B$1:$E$1,5)
        5.42 =KSPLINE(B6:E6,$B$1:$E$1,5)

For this simple illustration, the size of the data is relatively manageable. For large data sets, we turn to *VBA* to automate the process of two-dimensional nonlinear interpolation. The *VBA* user-defined function **KSLPINE2D(fdata, xdata, ydata, xnew, ynew)** employs a user-defined function **KSPLINEK** in a loop to generate points at x for a final spline interpolation in the y dimension. The *PNMSuite* also has **CSPLINE2D**, **ASPLINE2D**, and **STINEMAN2D** for bivariate interpolation using cubic spline, Akima's, and Stineman's method, respectively. **CSPLINE2D**, **KSLPINE2D**, **ASPLINE2D**, and **STINEMAN2D** return the following results for the interpolation in Example 10.14, which agree with the value from **LINTERP2D** in the first and second decimal places, respectively. Note that the worksheet setup and arguments for the **CSPLINE2D**, **KSLPINE2D**, **ASPLINE2D**, and **STINEMAN2D** user-defined functions are identical to those required by **LINTERP2D**:

        3.76 = CSPLINE2D($B$2:$E$6, $B$1:$E$1, $A$2:$A$6, 5, 22) = KSPLINE2D = ASPLINE2D
        3.82 = STINEMAN2D($B$2:$E$6, $B$1:$E$1, $A$2:$A$6, 5, 22) = LINTERP2D

We created surface plots of the bivariate interpolation of Henry's constant in Example 10.14 from the results of a **Data Table** to check for smoothness of interpolation using **LINTERP2D** and **CSPLINE2D**.

10.6 Data Smoothing and B-Splines*

"The path is smooth that leadeth on to danger." – William Shakespeare

Because of inherent dangers associated with data smoothing, we relegated this topic to the end of the chapter in order to lower its stature as a viable option for interpolation. We refer to random fluctuations in data as "noise." Recall the batch reactor data in Table 5.2. The calculations of the derivatives of the data reveal noise, or a lack of smoothness, in the original data. Instantaneous values may not represent the actual or average state of the process due to large fluctuations in the measurement data. Noisy signals from instrumentation are particularly common in bioprocessing, mineral processing, or other processes involving heterogeneous natural materials. In some cases, we must filter, or smooth, noisy data to obtain better representations of the trends in the data. Beware – smoothing data always involves some loss of information, and may introduce an unintentional bias in the results. It is usually better to smooth the data with a model using least-squares regression or B-splines, if possible. However, data smoothing satisfies aesthetic needs when we just need to *see* the general trend in the data.

Excel provides options for a moving average and exponential smoothing (digital filtering) in the add-in **Data Analysis Tool Pak**. The moving average option uses the worksheet function **AVERAGE**. The method averages n previous data points to forecast the next value of the dependent variable as follows:

$$f_{i,forecast} = \frac{1}{n}\sum_{j=0}^{n-1} f_{i-j} \qquad (10.93)$$

Excel's moving average analysis tool shifts the data ahead by $n-2$ increments. Forecasting is important for filtering noisy data used in process control applications that need the current values and trends in the data to make decisions about controller settings.

Mean smoothing is similar to *Excel*'s moving average:

$$f_{i,mean} = \frac{1}{n}\sum_{j=i-(n-1)/2}^{i+(n-1)/2} f_i \qquad (10.94)$$

Using a moving average centered at the current point eliminates data shifting. The **AVERAGE(x₀, x₁, x₂ ...)** function in *Excel* accomplishes central data smoothing. Average smoothing maintains the integral area under the curve, but tends to shorten the height and broaden the width of peaks in the data. This may influence the numerical calculation of a derivative, but has no effect on numerical integration.

Exponential digital filtering smooths time series data using a weighted average of a set of values that precede the current smoothing point:

$$f_i = wf_{i,s} + (1-w)f_{i-1} \qquad (10.95)$$

where f_s is the current value from the signal, and f_{i-1} is the previous smoothed result. The weight factors are typically in the range: $0.01 \leq w \leq 0.5$, which give averages of the previous 100 to two points, respectively. Digital filtering is particularly important in process control applications where controllers rely on smooth data for adjusting actuators.

Example 10.15 Smoothing Data from Biomass Thermal Decomposition

Known: Thermo-gravimetric analysis (*TGA*) of biomass decomposition yields kinetic information for designing gasification units. The *TGA* procedure measures the change in mass of a sample during a period of constant heating rate. We derive kinetic information from the rate of change in the mass over the course of the experiment. We simulated *TGA* data to illustrate the performance of the smoothing methods. We use a simple first-order model of the reaction in a batch process with a constant heating rate to produce the artificial data:

$$\frac{dm}{dt} = -km \qquad m = 10\,\mathrm{mg}\ at\ t = 0 \qquad T/°C = 10t + 100 \qquad (10.96)$$

where *m* is the mass of the sample of biomass (units are mg), *t* is the reaction time (sec), and *k* is the first-order rate constant (sec^{-1}). The rate constant follows an Arrhenius function for temperature dependence:

$$k/s^{-1} = 4.8 \times 10^5 \exp\left(\frac{-8660}{T+273}\right) \qquad (10.97)$$

The thermal decomposition model was solved in *Excel* using Euler's method from Chapter 12 with time steps of 0.1 seconds. We simulated artificial noisy data using a Monte Carlo approach assuming a normally distributed uncertainty of ±0.1 mg in the model biomass value. The product of the uncertainty and the imbedded *Excel* functions **NORMSINV(RAND())** produced normal random deviates that were added to the model solutions from column **B2** to produce artificial noise in the data for column **B3**:

B3=B2+0.1*NORMSINV(RAND())

Find: Determine the reaction ignition temperature from a plot of the derivative of the mass with respect to time.

Analysis: Figure 10.29 shows the simulated *TGA* mass data plotted with temperature. The plot also includes derivatives of the mass with respect to time. We calculated the derivatives numerically using a three-point finite difference approximation with the *VBA* user-defined function **FDERIV**.

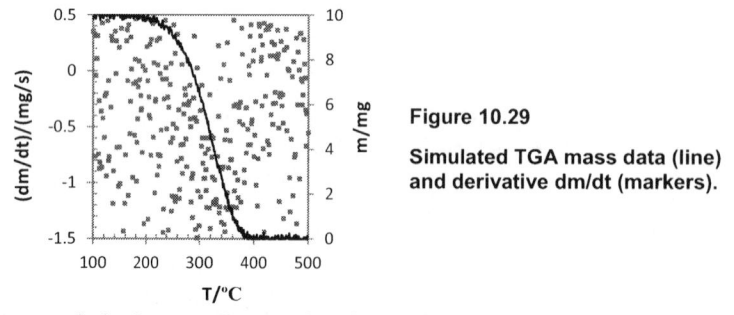

Figure 10.29

Simulated TGA mass data (line) and derivative dm/dt (markers).

The *TGA* mass data appear relatively smooth over the change in temperature. The calculations for the derivatives magnify the noise in the data. Without smoothing, it is difficult to identify the ignition temperature from the plot of the derivatives. We use our various methods to smooth the data for calculating the derivative. Figure 10.30 has two plots for comparing the smoothing results for the derivative *dm/dt* using **Moving Average** and **Exponential** smoothing.

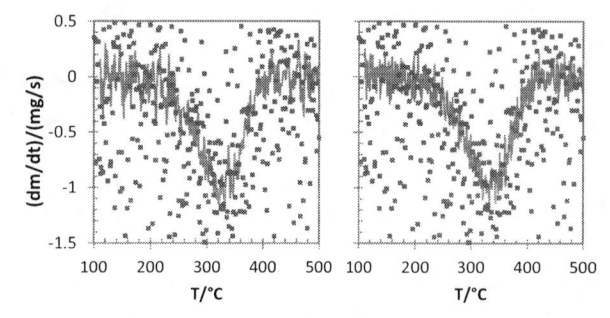

Figure 10.30

Moving average smoothing with 6 points (left) and Exponential smoothing with *w* = 0.9 (right) of simulated TGA derivative data.

Comments: The moving averaging provides enough smoothing to locate the ignition temperature of ~320°C.

B-splines use basis functions that relax the requirement of regular splines to pass exactly through all of the data points. In this sense, a B-spline is not a true interpolation method, but acts to smooth the interpolation of noisy data. B-splines also have application for differentiating and integrating noisy data (Gerald and Wheatley 2004). The cubic B-spline derived here employs weighted values of the independent and dependent variables from a set of four consecutive data points:

$$x = \frac{1}{6}\left[b_{k-1}x_{k-1} + b_k x_k + b_{k+1}x_{k+1} + b_{k+2}x_{k+2}\right] \qquad y = \frac{1}{6}\left[b_{k-1}y_{k-1} + b_k y_k + b_{k+1}y_{k+1} + b_{k+2}y_{k+2}\right] \qquad (10.98)$$

where the b coefficients are basis functions in terms of a weighting factor $0 \le w \le 1$ across a spline interval:

$$b_{k-1} = (1-w)^3 \qquad b_k = 4 - 6w^2 + 3w^3 \qquad b_{k+1} = 1 + 3w + 3w^2 - 3w^3 \qquad b_{k+2} = w^3 \qquad (10.99)$$

Note how Equation (10.98) returns the B-spline knots between intervals at the lower and upper limits for w:

$$x_k^*(w=0) = \frac{x_{k-1} + 4x_k + x_{k+1}}{6} \qquad x_{k+1}^*(w=1) = \frac{x_k + 4x_{k+1} + x_{k+2}}{6} \qquad (10.100)$$

Unlike regular cubic splines, the B-spline knots do not necessarily line up with the data points for uneven spacing of the data. Otherwise, B-splines follow the same rules of continuity at common spline knots employed by regular cubic spline interpolation in Equations (10.20) and (10.21). As with regular splines, B-splines need additional information at the extreme points. Following the derivation of Gerald and Wheatley (2004), we clamp the B-splines to pass exactly through the extreme data points by creating artificial points beyond the first and last points: $x_{-1} = x_0 = x_1$ and $x_{n+2} = x_{n+1} = x_n$. We obtain a cubic polynomial in w by combining like terms for w in Equation (10.98):

$$x = a_0 + a_1 w + a_2 w^2 + a_3 w^3 \qquad (10.101)$$

where

$$a_0 = \frac{x_{k-1} + 4x_k + x_{k+1}}{6} \qquad a_1 = \frac{x_{k+1} - x_{k-1}}{2}$$

$$a_2 = \frac{x_{k-1} - 2x_k + x_{k+1}}{2} \qquad a_3 = \frac{-x_{k-1} + 3(x_k - x_{k+1}) + x_{k+2}}{6} \qquad (10.102)$$

Use the following procedure to interpolate y at a given value for x by the method of cubic B-splines:

1. Create a series of B-spline knots x_k^* by weighting each group of three consecutive data points according to Equation (10.100). Use artificial points to clamp the ends to the data values.
2. Match the interpolation point x with knot k for the smoothed interval $x_k^* \le x \le x_{k+1}^*$.
3. Isolate the values for x and y from the data points required in Equations (10.102) and (10.98).
4. Solve Equation (10.101) at the x point of interpolation for the corresponding weight factor w. For simplicity, we recommend the secant method derived in Section 6.2.7 using initial guesses of $w = 0$ and 1 to initiate the recursive formula in Equation (6.57).
5. Finally, calculate the interpolated value for y in Equation (10.98).

Gossage (2016) recommended B-splines for interpolating vapor-liquid equilibrium data where least squares fitted thermodynamic models struggle to match the trend. Figure 10.31 has a plot of the B-spline interpolation of data for the vapor (y) versus liquid (x) equilibrium mole fractions of ethanol in water. Note how the B-spline smooths the data near x = 0.1 and 0.55, and forces the ends through the exact points y = 0 at x = 0 and y = 1 at x = 1, respectively. To illustrate the smoothing nature of cubic B-splines further, we utilized this method to interpolate Bica's simulated data from *Example 10.9*.

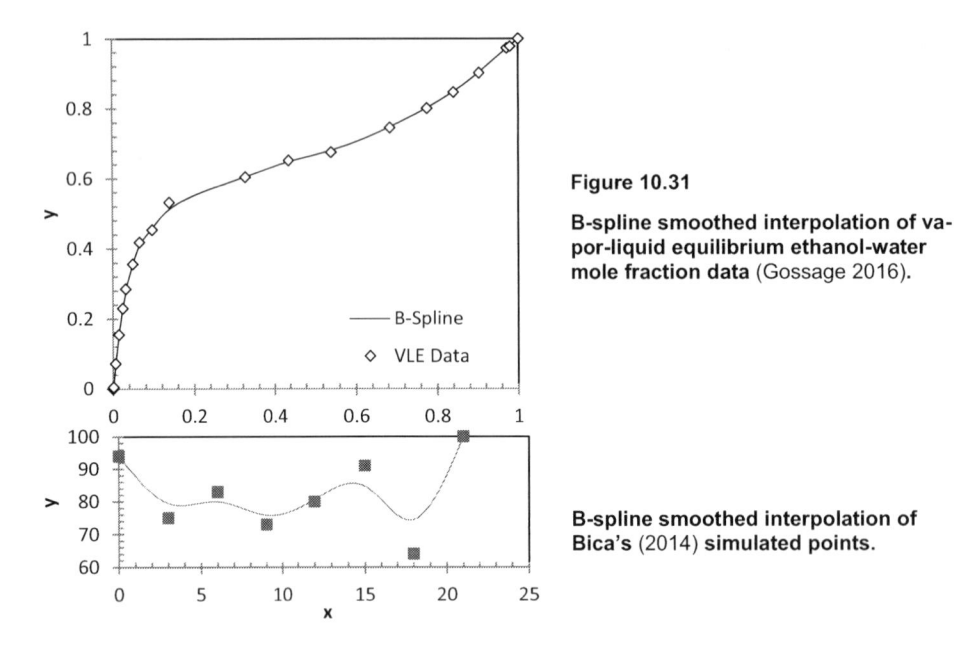

Figure 10.31

B-spline smoothed interpolation of vapor-liquid equilibrium ethanol-water mole fraction data (Gossage 2016).

B-spline smoothed interpolation of Bica's (2014) **simulated points.**

A user-defined function **BSPLINE(y_data, x_data, x_new)** is available in the *PNMSuite* for cubic B-spline interpolation. **BSPLINE** requires the same arguments as **CSPLINE** for interpolation. The user-defined function puts the independent *x* data in ascending order to locate the spline interval of interpolation.

We use data smoothing most often for cosmetic purposes to identify a qualitative view of trends in extremely noisy data. There is no single best *smoothing* technique for all situations. The moving average methods are robust, but may not capture sharp changes in the data and suffer from data shifting. Remember "Good. Better. Best." It is *good* to smooth data to locate a region of a local optimum in the data. It is *better* to collect more data when resources are available. It is *best* to avoid fitting equations to smoothed data, rather use B-splines or the methods of least squares to smooth and fit noisy data with an appropriate model.

10.7 Epilogue on Interpolation

Good interpolation reduces the amount of expensive experiments required to acquire data for analysis. Interpolation methods also allow us to functionalize data without selecting a specific model required by least-squares regression. We have several options summarized in Table 10.3 for interpolating smooth data.

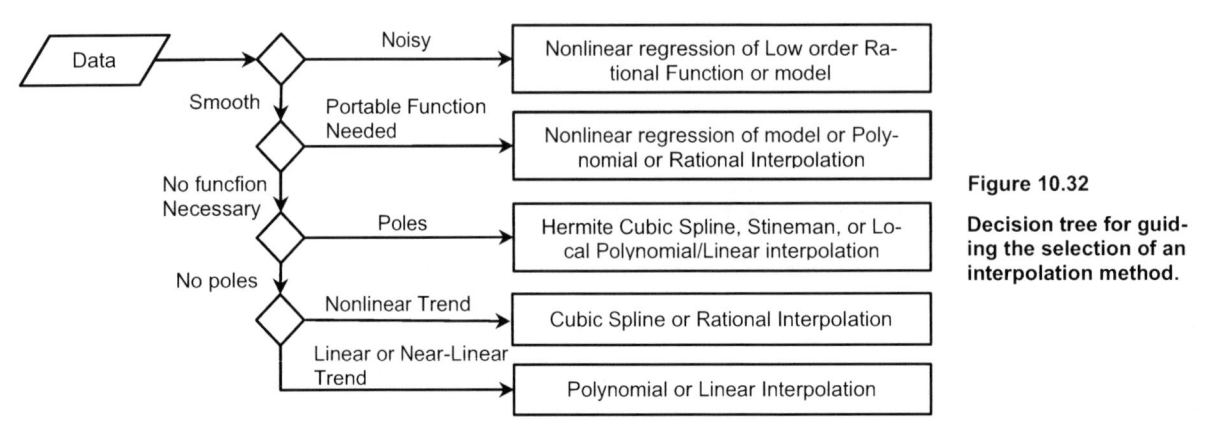

Figure 10.32

Decision tree for guiding the selection of an interpolation method.

Figure 10.32 has a decision tree of heuristics for selecting a method of interpolation. Table 10.4 summarizes the macros and user-defined functions for interpolation developed in this chapter. Generally, we recommend cubic

Chapter 10: Interpolation

spline or constrained spline interpolation for most problems. When portability of the interpolating function is important, rational functions may provide excellent alternatives. For quick results, linear and polynomial interpolations may be generally satisfactory. Use B-splines or regression models for noisy data, or when we need to extrapolate outside the range of data (lower order rational functions make good candidates for noisy data). By design, the interpolation methods presented here fit the data exactly. An important part of assessing interpolation involves plotting interpolation values between data points to check for poles to ensure that the interpolating formula does not overshoot the obvious trend in data. With the possible exception of rational functions, none of these methods is particularly appropriate for extrapolation.

Table 10.3 Comparison of interpolation methods for smooth data.

Method	Strengths	Weaknesses
Linear	• Simplest to implement. • Available in most computational software, including *Excel*. • Guaranteed to follow trends in data. • Will not form poles. • Good results for linear or near linear data. • Easy to differentiate and integrate. • Does not require analysis of fit.	• Poor interpolation for nonlinear data • Requires new linear interpolation formula for each pair of data points. • No second derivatives. • Does not use all data for local interpolation.
Polynomial (Newton Divided Difference or Lagrange)	• Easy to implement. • Most computational software features polynomial regression. • Good results for near linear, quadratic, or cubic trends in data. • Easy to differentiate and integrate. • Works well in many situations. • Portable functions to represent data.	• Requires experimentation to determine the order of polynomial that best fits the data • Frequently over or undershoots the trend in nonlinear data. • May not use all data for local interpolation.
Hermite polynomial	• Uses information about the slope or derivative of the data.	• Derivative information cannot have a lot of noise – must be smooth.
Rational Function	• Good for modeling nonlinear data trends with asymptotic behavior. • Portable function to represent data. • May use all data for sparse data sets. • Best method for extrapolation.	• Requires experimentation to determine the order of numerator/denominator. • May form poles. • Difficult to integrate or differentiate analytically.
Stineman	• Interpolates sparse data that may form poles otherwise.	• No estimate of derivative or integral of data.
Cubic Spline	• Works well for linear and nonlinear data. • No need to experiment with functions. • Easy to differentiate and integrate. • Uses all data.	• May form poles or under/over shoot sparse data. • Does not produce a stand-alone portable function.
Hermite and Constrained Cubic Spline	• Works well for linear and nonlinear data. • No need to experiment with functions. • Easy to differentiate and integrate. • Uses all data. • Guaranteed not to over or undershoot data.	• No second derivatives. • Loses some information in fit compared with natural cubic splines. • Does not produce a single equation for a stand-alone portable function.
Bivariate interpolation	• Linear method is computationally efficient • Reuses our one-dimensional interpolations methods.	• Nonlinear methods are computationally demanding. • Tedious for "hand" calculations.
B-spline	• Smooths noisy data at internal points.	• Passes exactly through the extreme points.

Table 10.4 Summary of VBA macros and user-defined functions (UDF) for interpolation, including the methods of data and parameter input.

Macro/Function	Inputs	Operation
AKIMA ASPLINE ASPLINE2D	UDF arguments	Akima's Hermite cubic spline interpolation method. Options for first derivatives and integration. ASPLINE and ASPLINE2D use Bica's optimized slopes at the ends for interpolation only.
BSPLINE	UDF arguments	Cubic B-spline interpolation of data arranged in ascending order according to the independent variable.
CSPLINE CSPLINE2D	UDF arguments	Cubic spline interpolation with endpoint options: free, clamped, parabolic, not a knot. Options for first or second derivatives and integration. 2D Bivariate cubic spline interpolation in two-dimensions.
KSPLINE KSPLINE2D	UDF arguments	Constrained cubic spline interpolation with free or clamped endpoints. Options for first derivatives and integration. 2D Bivariate constrained spline interpolation in two-dimensions.
LAGRANGE	UDF arguments	Lagrange polynomial interpolation.
LINTERP LINTERP2D	UDF arguments	Linear interpolation in one dimension. 2D Bilinear interpolation in two-dimensions.
NEWTONPOLY	UDF arguments	Newton's method of divided difference polynomial interpolation.
RATLIN RATLINBS	UDF arguments	Rational interpolation. Bulirsch-Stoer method of rational interpolation for Padé functions.
RATLINPQ	Macro input boxes	Rational function parameters.
STINEMAN STINEMAN2D	UDF arguments	Stineman method of rational interpolation eliminates poles. 2D Bivariate Stineman interpolation in a two-dimensional grid.

Chapter 11 Integration

> *"This is a tricky domain because, unlike simple arithmetic, to solve a calculus problem - and in particular to perform integration - you have to be smart about which integration technique should be used ..."* − Marvin Minsky

Many integrals in chemical engineering require numerical approximation because of nonlinear reaction kinetics, thermal and molecular diffusion, and equilibrium thermodynamics. We naturally follow Chapter 10, which covers interpolation, with a presentation in this chapter on numerical approximations of integrals. Our numerical techniques for integration simply rely on analytical integration of interpolating functions. Newton-Cotes methods integrate polynomial interpolations of integrand functions at equally spaced points between the limits of integration. These interpolating integrands range from simple first-order linear functions applied in the trapezoidal rule to higher order polynomials used by Simpson's rules and Gaussian methods of numerical quadrature. We present advanced methods that reduce the computational burden and control the error. Specifically, this chapter covers:

- Graphical methods for quick, but crude, integration
- Trapezoidal rule for evaluating integrals of functions or data (integrating linear interpolation)
- Midpoint rule for improper integrals (integrating rectangles)
- Higher order Simpson's rules for equally-spaced data (integrating polynomial interpolation)
- High precision Romberg integration of functions (applying Richardson's extrapolation to integration of linear interpolation)
- Integration of spline interpolating functions
- Adaptive Gaussian quadrature [99] of functions (integrating Legendre polynomials)
- Multiple integration using Simpsons, Gauss, and Monte Carlo methods

---

SQ3R Focused Reading Questions

1. How can I check the accuracy of a numerical solution for an integral?
2. What is the relationship between numerical methods of interpolation and integration?
3. What is needed to integrate data?
4. When should I use cubic spline for integration?
5. How does extrapolation improve the accuracy of a numerical solution?
6. What is an improper integral?
7. When should I use a midpoint method?
8. How do numerical methods control error?
9. When should I use Monte Carlo methods for approximating the solution of an integral
10. Why are there several different methods for numerical integration?

In graphical terms, the value of an integral is the area between the axis representing the independent variable and the curve of the integral function, as illustrated in Figure 11.1. The value of the integral is the net sum of all areas above the abscissa (or *x*-axis) minus the areas below the abscissa.

---

[99] Quadrature refers to the ancient Greek practice of finding a square with the same area as a geometric construct.

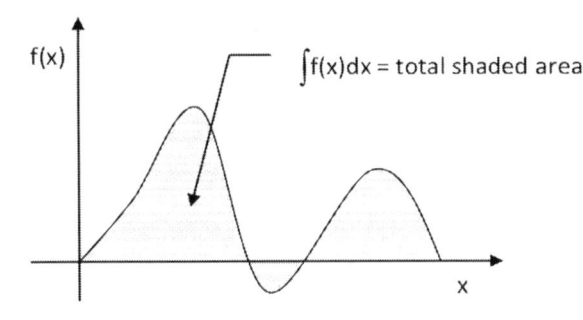

Before digital computing tools became readily available, scientists resorted to clever methods for approximating the solutions to nonlinear integrals. For example, chemists used to plot integrand data on special paper with uniform thickness and density. They carefully cut out the section of paper along the curve of the integrand between the limits of integration. The value of the integral was proportional to the paper's weight. The paper-weighing process was usually repeated several times to find an average result accompanied by statistical analysis.

With the aid of modern computers, powerful numerical methods have been developed for accurately approximating the solution to difficult nonlinear integrals that have no analytical solution. Most numerical methods of integral approximation have a basis in the following expression for the fundamental theorem of calculus by evaluating the integrand function at discrete intervals:

$$\int_a^b f(x)dx = \lim_{n \to \infty} \left\{ \frac{(b-a)}{n} \sum_{i=1}^{n+1} f\left[a + (i-1)(b-a)/n\right] \right\} \tag{11.1}$$

11.1 Graphical Integration (Rectangle Rule)

Graphical integration involves plotting the integral function between the limits of integration on a grid. We approximate the value of the integral by physically measuring the area under the curve of the integral function. A simple technique divides the limits of integration into discrete intervals, as seen in the example in Figure 11.2. We fit rectangles to the curve such that the total area of the rectangles approximately equals the area below the curve.

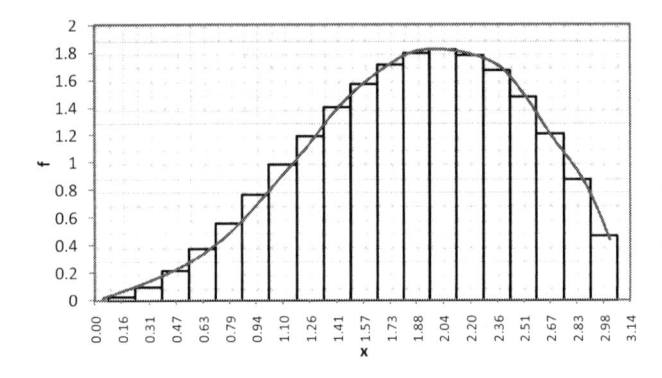

Figure 11.2

Graphical integration of $\int_0^\pi x \sin x \, dx \cong 3.1351$ by approximating the area under the curve with a series of 20 rectangles with width π/20. The true solution is 3.1416.

If the intervals are uniformly spaced, we may approximate the area under the curve as the sum of the areas of the rectangles each having width Δx:

$$\int f(x)dx \cong \Delta x \sum f(x) \tag{11.2}$$

Using more rectangles, each with smaller width, should improve the accuracy of the approximation.

Example 11.1 Graphical Integration of the Packed Column Design Equation

Known: The design equation for determining Z, the height of a packed column for gas absorption involves integrating the number of transfer units:

$$Z = \frac{G}{K_Y a} \int_{Y_i}^{Y_o} \frac{dY}{Y - Y^*} \quad (11.3)$$

where Y = mole ratio of the absorbing species in the gas phase
           Y^* = equilibrium gas phase mole ratio
           G = molar flux of the solute-free gas (10 mole/m² s)
           $K_Y a$ = product of the overall mass transfer coefficient with the specific surface area available for mass transfer (4 mole/m³ s)
           L = liquid feed molar flow rate per cross sectional area of the column (700 mole/m²·s)

Find: Determine the height of packing required for the absorption of solute ClO_2 into pure water at one atm pressure and 25°C to reduce the feed gas stream's mole fraction from 20 to 10% ClO_2.

Assumptions: sparingly soluble gas, steady-state, uniform temperature

Schematic: A packed column for gas absorption conducts counter current molar flows of gas and liquid streams, G and L. The variables y and x are the gas and liquid solute mole fractions, respectively.

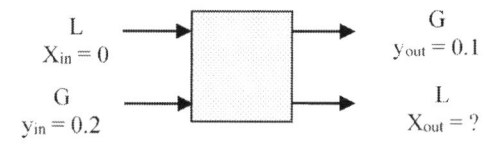

Analysis: Henry's law models the solubility of ClO_2 in water:

$$P = Hx^* \quad (11.4)$$

where P is the partial pressure of ClO_2 in the gas phase, H is the Henry's constant with a value of 5558 kPa at 298 K, and x^* is the equilibrium mole fraction of ClO_2 in the liquid phase.

We require values for the integrand to generate a plot for graphical solution. To begin, choose several values of the gas mole fraction, y between 0.1 and 0.2. Calculate the mole ratio of the solute in the gas phase:

$$Y = \frac{y}{1-y} \quad (11.5)$$

Obtain the corresponding mole ratio and fractions in the liquid phase from the material balance for the dissolving gas species:

$$YG + (XL)_{in} = (YG)_{out} + XL \quad (11.6)$$

or

$$X = \frac{G}{L}(Y - Y_{out}) \quad \text{and} \quad x = \frac{X}{1+X}$$

We get the gas-phase mole fraction at equilibrium from Henry's law, and calculate the corresponding equilibrium mole ratio:

$$y^* = \frac{Hx}{P} \quad \text{and} \quad Y^* = \frac{y^*}{1-y^*} \quad (11.7)$$

Perform each of these steps in an *Excel* worksheet at discrete points in the range $0.1 \leq y \leq 0.2$, and calculate values for the integrand in Equation (11.3) as shown in Figure 11.3. Plot the gas-phase mole ratio versus the integrand function on a grid:

	A	B	C	D	E	F	G
1	G/ mol/m²·s	10		K$_y$a/ (mol/m³·s)	4		
2	L/ mol/m²·s	700		z/m	=(B1/E1)*0.82		
3	y	Y	X	x	y*	Y*	1/(Y-Y*)
4	0.1	=A4/(1-A4)	=($B$1/$B$2)*(B4-$B$4)	=C4/(1+C4)	=CONVERT(5558,"Pa","atm")*D4	=E4/(1-E4)	=1/(B4-F4)
5	0.11	=A5/(1-A5)	=($B$1/$B$2)*(B5-$B$5)	=C5/(1+C5)	=CONVERT(5558,"Pa","atm")*D5	=E5/(1-E5)	=1/(B5-F5)
6	0.12	=A6/(1-A6)	=($B$1/$B$2)*(B6-$B$6)	=C6/(1+C6)	=CONVERT(5558,"Pa","atm")*D6	=E6/(1-E6)	=1/(B6-F6)

	A	B	C	D	E	F	G
1	G/ mol/m²·s	10		K$_y$a/ (mol/m³·s)	4		
2	L/ mol/m²·s	700		z/m		2.05	
3	y	Y	X	x	y*	Y*	1/(Y-Y*)
4	0.1	0.111	0.00E+00	0.00E+00	0.00E+00	0.00E+00	9.00
5	0.11	0.124	1.78E-04	1.78E-04	9.78E-06	9.78E-06	8.09
6	0.12	0.136	3.61E-04	3.61E-04	1.98E-05	1.98E-05	7.33
7	0.13	0.149	5.47E-04	5.47E-04	3.00E-05	3.00E-05	6.69
8	0.14	0.163	7.38E-04	7.38E-04	4.05E-05	4.05E-05	6.14
9	0.15	0.176	9.34E-04	9.33E-04	5.12E-05	5.12E-05	5.67
10	0.16	0.190	1.13E-03	1.13E-03	6.21E-05	6.21E-05	5.25
11	0.17	0.205	1.34E-03	1.34E-03	7.33E-05	7.33E-05	4.88
12	0.18	0.220	1.55E-03	1.55E-03	8.48E-05	8.48E-05	4.56
13	0.19	0.235	1.76E-03	1.76E-03	9.66E-05	9.66E-05	4.26
14	0.2	0.250	1.98E-03	1.98E-03	1.09E-04	1.09E-04	4.00

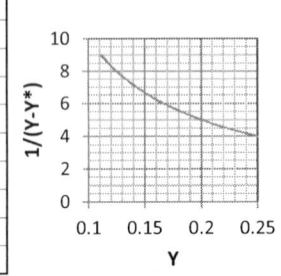

Figure 11.3

Excel **worksheet formulas and results for graphing the integrand of the packed column design equation.**

The area under the curve of the integrand function between the limits of integration corresponds to the value of the integral in Equation (11.3). Estimate the area under the curve by counting the squares and fractions of squares ≈ 165. We get the area of a square from the length and width of a section of the plot:

$$A_{square} = \Delta Y \times \Delta \frac{1}{(Y-Y^*)} = \left(\frac{0.05}{5}\right) \times \left(\frac{2}{4}\right) = 0.005 \qquad (11.8)$$

The value of the integral is then (165) × (0.005) = 0.825. Finally, compute the required height of the packed column from Equation (11.3): Z = (10 mol/m²·s) × (0.825)/(4 mol/m³·s) = 2.1 m.

Comments: Although graphical solutions have low precision, they do provide the foundation for more advanced numerical integration methods.

□

11.2 Trapezoidal and Midpoint Rules

> Andy: "Applicant is attempting to blackmail interviewer, showing low moral character."
> Dwight: "Interviewer is threatening applicant with an arbitrary review process."
> Andy: "Applicant is wasting everyone's time with stupid and inane accusations."
> Dwight: "Interviewer has suspect motives."
> Andy: "Applicant has a head shaped like a trapezoid." – The *Office* TV Sitcom

Every student of calculus learns the trapezoidal rule as a first approximation to the fundamental theorem of calculus in Equation (11.1). A trapezoidal approximation for an integral is simply the integration of linear interpolation between the limits. The trapezoidal rule, illustrated in Figure 11.4, approximates the area under the curve using the graphical method by dividing the region of integration into *n* intervals; the area of each interval is approximately the

CHAPTER 11: INTEGRATION   489

area of the trapezoid formed by a straight interpolating line connecting the integrand functions evaluated at the lower and upper limits of each interval:

- *Trapezoidal Rule* $$\int_{x_i}^{x_{i+1}} f dx \cong (x_{i+1} - x_i)\frac{(f_{i+1} + f_i)}{2} \qquad (11.9)$$

where f_i and f_{i+1} correspond to the integrand evaluated at the integration limits, x_i and x_{i+1}, respectively.

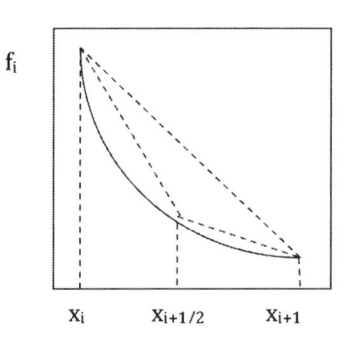

Figure 11.4

Trapezoidal integral approximation.

Figure 11.4 shows the improved area approximation when we divide the interval in half, and apply the trapezoidal rule again to each half:

$$\int_{x_i}^{x_{i+1}} f dx \cong (x_{i+1/2} - x_i)\frac{(f_{i+1/2} + f_i)}{2} + (x_{i+1} - x_{i+1/2})\frac{(f_{i+1} + f_{i+1/2})}{2} \qquad (11.10)$$

Dividing the range of integration into yet smaller increments should produce better approximations of the integral as the shape of the collection of trapezoids approaches the shape of the integrand function. Recall that this is the basic idea with linear Taylor series approximations for curves that give better approximations for a function near the expansion point.

A formal derivation of the trapezoidal rule with error approximation begins with a Taylor series approximation of the integral function. Consider an interval in the independent variable, $x_i < x < x_{i+1}$ between nodes i and $i+1$. Expand a second-order Taylor series about the limits of the interval to get two approximations of the integral function:

$$f \cong f_i + (x - x_i)\frac{df}{dx}\bigg|_{x_i} + \frac{(x - x_i)^2}{2}\frac{d^2 f}{dx^2}\bigg|_{x_i} + O(\Delta x^3) \qquad (11.11)$$

$$f \cong f_{i+1} + (x - x_{i+1})\frac{df}{dx}\bigg|_{x_{i+1}} + \frac{(x - x_{i+1})^2}{2}\frac{d^2 f}{dx^2}\bigg|_{x_{i+1}} + O(\Delta x^3) \qquad (11.12)$$

where $\Delta x = x_{i+1} - x_i$. Sum the two expansions and solve for the integrand function:

$$f = \frac{f_i + f_{i+1}}{2} + \frac{(x - x_i)}{2}\frac{df}{dx}\bigg|_{x_i} + \frac{(x - x_{i+1})}{2}\frac{df}{dx}\bigg|_{x_{i+1}} + \frac{(x - x_i)^2}{4}\frac{d^2 f}{dx^2}\bigg|_{x_i} + \frac{(x - x_{i+1})^2}{4}\frac{d^2 f}{dx^2}\bigg|_{x_{i+1}} + O(\Delta x^3) \quad (11.13)$$

Integrate Equation (11.13) over the limits of the interval and replace the second-order derivatives with simple finite-difference approximations (we use the prime, or f', shorthand notation for the first derivatives to simplify the equations):

$$\int_{x_i}^{x_{i+1}} f dx = \Delta x \frac{(f_i + f_{i+1})}{2} + \frac{3\Delta x^2}{12} f'_i - \frac{3\Delta x^2}{12} f'_{i+1} + \frac{\Delta x^3}{12}\frac{(f'_{i+1} - f'_i)}{\Delta x} + \frac{\Delta x^3}{12}\frac{(f'_{i+1} - f'_i)}{\Delta x} + O(\Delta x^4) \qquad (11.14)$$

Simplify Equation (11.14) by combining like derivative terms to give the trapezoidal rule for a single interval:

$$\int_{x_i}^{x_{i+1}} f dx = \Delta x \frac{(f_i + f_{i+1})}{2} - \frac{\Delta x^2}{12}\left(f'_{i+1} - f'_i\right) + O\left(\Delta x^4\right) \qquad (11.15)$$

Finally, sum over all n intervals to give the composite trapezoidal rule with end-corrections (Moin 2001):

$$\int_{x_1}^{x_{n+1}} f dx = \Delta x \left[\frac{(f_1 + f_{n+1})}{2} + \sum_{i=2}^{n} f_i\right] - \frac{\Delta x^2}{12}\left(f'_{n+1} - f'_1\right) + O\left(\Delta x^4\right) \qquad (11.16)$$

where

$$\Delta x = \frac{x_{n+1} - x_1}{n} \qquad (11.17)$$

Equation (11.16) is fourth-order accurate in Δx and requires integrand function values at equally spaced values for the variable of integration. Note that Equation (11.16) requires the first derivatives of the integrand function at the integration limits. Ignoring the last term involving the first derivatives in Equation (11.16) reduces the expression to a second-order accurate composite trapezoidal rule $\approx O\left(\Delta x^2\right)$. The worksheet in Figure 11.5 solves the test case integral from Figure 11.2 by the composite trapezoidal rule with end corrections using 10 intervals. The numerical result in cell **F3** matches the true solution to the fourth decimal place.

	A	B	C	D	E	F
1		n=	10	Δx=	0.314159	dI/dx(0)=	0
2	x	I		Trapezoid		dI/dx(π)=	-3.14159
3	0	0				Integral=	3.14155
4	0.314159	0.097081	0.015249			
5	0.628319	0.369316	0.073261			
6	0.942478	0.762481	0.177782			
7	1.256637	1.195133	0.307501			
8	1.570796	1.570796	0.434471			
9	1.884956	1.792699	0.528337			
10	2.199115	1.779121	0.56106			
11	2.513274	1.477265	0.511512			
12	2.827433	0.873725	0.369293			
13	3.141593	3.85E-16	0.137244			

Figure 11.5

Numerical integration of $\int_0^{\pi} x \sin x \, dx \cong 3.14155$ by the composite trapezoidal rule using n = 10 intervals with end corrections defined in Equation (11.16). The true solution is 3.14159. The formulas in the worksheet are

A4 = A3+$D$1

B4 = A4*SIN(A4)

C4 =$D$1*(B4+B3)/2

D1 = PI()/10

F3 = SUM(C4:C13)-(D1^2)*(F2-F1)/12

For cases where the interval lengths are not uniform, such as experimental data, apply the trapezoidal rule from Equation (11.15) to each individual interval (without end corrections) and sum all of the interval values for the overall integral approximation. A *VBA* macro **TRAPEZOID** is available in the *PNMSuite* for implementing the composite trapezoidal rule for a function in an *Excel* worksheet, without end corrections. To use the macro **TRAPEZOID**:

- Specify the lower limit of integration in a cell on the worksheet.
- In another cell, formulate the integrand function using references to the cell with the lower limit for the variable of integration.
- Specify the number of uniformly spaced intervals.
- Locate the cell for the result on the worksheet.

Figure 11.6 lists the code for a simple *VBA* user-defined function **TRAP(f, xlow, xup, N)** for integrating an *Excel* worksheet formula by the composite trapezoidal rule in Equation (11.16) with end corrections. **F** is the cell address with the integrand formula in terms of the cell address for **xlow**, and **xup** is the upper limit of integration. **N** is the optional number of intervals where the default is set to 100. The value of **xlow** is set to the lower limit of integration. This procedure uses the template in Figure 3.23 to evaluate the formula for the integrand at each node in the composite trapezoidal rule. To apply the method in *VBA*:

1. Determine the size of each interval according to Equation (11.17) using 100 steps.
2. Convert all relative cell or range addresses in the formula for the integrand to absolute references using the **Application.ConvertFormula** operation.

CHAPTER 11: INTEGRATION

3. Find the number of occurrences of references to the cell containing the independent variable by comparing the length of the formula string to the length of the string of the address for the cell containing the independent variable.
4. Cycle through the formula to replace the address for the independent variable with numerical values corresponding to the nodes between trapezoid intervals.
5. Evaluate the formula for the integrand at each node using the *VBA* function:

   ```
   Evaluate(formula_string)
   ```

6. Use the formula evaluations to estimate the value of the integral according to the composite trapezoidal rule.
7. Use a modified version of the user-defined function **DREX** to calculate the derivatives for the end corrections in Equation (11.16)

We illustrate how to set up an *Excel* worksheet for use with **TRAP** in a simple example of the following integral:

$$\int_0^3 x\sqrt{1+x^2}\,dx = \frac{1}{3}(1+x^2)^{3/2}\Big|_0^3 = 10.2076$$

1. Specify the lower and upper integration limits in cells **B1** and **D1**, respectively.
2. Set the integrand formula in cell **F1** with reference to cell **B1** for the independent variable *x*.
3. Add the user-defined formula **TRAP** in cell **H1** using references to the cells with the integrand formula and limits of integration. The optional argument for the number of integration steps was left to the default =100.
4. The result in **H1** agrees with the true solution to the third decimal place:

	A	B	C	D	E	F	G	H
1	x low	0	x up	3	f(x)	=B1*SQRT(1+B1^2)	Solution	=ftrap(F1,B1,D1)

	A	B	C	D	E	F	G	H
1	x low	0	x up	3	f(x)	0	Solution	10.208

```
Public Function TRAP(f, xlow, xup As Double, Optional n As Long = 100) As Double
' f = address of cell with the formula for the integrand function in terms of x
' xlow = address of cell with the independent variable in f, & lower integration limit.
' xup = value of upper integration limit.
' n = (optional) number of intervals (n+1 nodes), default = 100
'******************************************************************************
' Integration of a worksheet formula by the trapezoidal rule with uniform interval
' size and end corrections evaluated with derivative approximations.
' *** Required User function ***
' TRAPEC = evaluate the derivative of the integrand at the limits of integration
' *** Nomenclature ***
Dim dx As Double ' = interval length
Dim i As Long, j As Long ' = index of node and occurrences of x in f
Dim noc As Long ' = number of occurrences of x in f
Dim ns As Long ' = number of occurrences of cell name in f formula
Dim x As Double ' = variable of integration at node i
Dim fi As String ' = function string at node i
Dim fa As String ' = string of formula with absolute references for x
Dim t As String ' = temp string for function with substituted values for x at node i
Dim xa As String ' = variable address
Dim xs As String ' = integration variable string
Dim xu As String ' = string placeholder for x substitution/evaluation
'******************************************************************************
On Error Resume Next ' formula w/ absolute references & string of variable address
x = xlow.Value: dx = (xup - x) / n ' calculate interval length
xa = xlow.Address: xs = "": xs = Range(xa).Name.Name
xu = "X_": fa = VBA.UCase(f.Formula) ' make upper case to eliminate case sensitivity
Do While InStr(fa, xu) ' check for dummy variable string in formula
    xu = xu & "_" ' add an underscore to end of string
Loop
With Application
    fa = .ConvertFormula(f.Formula, xlA1, xlA1, xlAbsolute): fi = fa
```

```
        noc = (Len(fa) - Len(.Substitute(fa, xa, ""))) / Len(xa)
        For j = noc To 1 Step -1 ' replace occurrences of only xs with new string xu
            t = .Substitute(fi, xa, x & " ", j)
            If Not IsError(Evaluate(t)) Then
                fa = .Substitute(fa, xa, xu, j): fi = t
            End If
        Next j
        ns = Len(xs)
        If ns > 0 Then
            noc = (Len(fa) - Len(.Substitute(fa, xs, ""))) / ns
            For j = noc To 1 Step -1 ' replace occurrences of only xs with new string xu
                t = .Substitute(fi, xs, x & " ", j)
                If Not IsError(Evaluate(t)) Then
                    fa = .Substitute(fa, xs, xu, j): fi = t
                End If
            Next j
        End If
        ' lower integration limit with end correction
        TRAP = Evaluate(fi) / 2# + dx * TRAPEC(f, xlow, x) / 12#
        For i = 2 To n ' interior nodes
            x = x + dx: TRAP = TRAP + Evaluate(.Substitute(fa, xu, x))
        Next i
        fi = fa: x = xup ' upper integration limit with end correction
        TRAP = dx * (TRAP + Evaluate(.Substitute(fa, xu, x)) / 2# _
             - dx * TRAPEC(f, xlow, xup) / 12#)
    End With ' Application
End Function ' TRAP
```

Figure 11.6 *VBA* user-defined function TRAP for evaluating the numerical integral of a worksheet formula for an integrand. See the PNMSuite for the user-defined function TRAPEC that evaluates derivatives at the limits for end-point corrections.

Example 11.2   First-order Batch-reactor Design Equation

Known: The following mole balance in terms of conversion, X, models an isothermal batch reactor with a first-order reaction:

$$\frac{dX}{dt} = kC_0(1-X) \tag{11.18}$$

where k is the reaction rate constant and C_0 is the initial concentration. Separate variables and integrate:

$$kC_0 t = \int_0^{X_f} \frac{dX}{(1-X)} = ln\left(\frac{1}{1-X}\right) \tag{11.19}$$

Find: Numerically approximate the value of the integral for $X_f = 0.9$ or 90% conversion using the trapezoidal rule and compare to the analytical solution in Equation (11.19).

Assume: well-mixed, constant volume reactor

Analysis: Figure 11.7 contains a plot of the function of the integrand between the integration limits. The curve shows a gradual slope at low x values that increases exponentially as x approaches one. We anticipate most of the error in the trapezoidal rule approximation comes from the section where the integrand function transitions to a steep slope. For a conversion of 90%, the exact solution of the integral is 2.3026. We apply the trapezoidal rule in a spreadsheet with nine uniformly spaced intervals, as shown in Figure 11.7, which also displays the *Excel* formulas. Differentiate the integrand for use with end-corrections in the composite trapezoidal rule of Equation (11.16):

$$\frac{d}{dx}\left(\frac{1}{1-X}\right) = \frac{1}{(1-X)^2} \tag{11.20}$$

CHAPTER 11: INTEGRATION

Figure 11.7 shows the *Excel* worksheet formulas for implementing the composite trapezoidal rule with and without end-corrections.

The approximate solution from the trapezoidal rule without endpoint corrections and using nine intervals is 2.3790, with an error of 3.3%. Using nine intervals with endpoint corrections gives an approximate solution for the integral of 2.2965, with an error of 0.27%, which represents a significant improvement. We may reduce the error in the solution by increasing the number of intervals (decreasing the interval size). We solved the integral using the user-defined function **TRAP** with 1000 intervals for the following result with 0.0003% error:

$$2.302591775 = \text{TRAP(B2, A2, A11, 1000)}$$

	A	B	C	D	E	F
1	x	1/(1-x)				
2	0	=1/(1-A2)				
3	0.1	=1/(1-A3)	=(A3-A2)*(B3+B2)/2			
4	0.2	=1/(1-A4)	=(A4-A3)*(B4+B3)/2			
5	0.3	=1/(1-A5)	=(A5-A4)*(B5+B4)/2			
6	0.4	=1/(1-A6)	=(A6-A5)*(B6+B5)/2			
7	0.5	=1/(1-A7)	=(A7-A6)*(B7+B6)/2			
8	0.6	=1/(1-A8)	=(A8-A7)*(B8+B7)/2		End correction	
9	0.7	=1/(1-A9)	=(A9-A8)*(B9+B8)/2		x	1/(x-1)^2
10	0.8	=1/(1-A10)	=(A10-A9)*(B10+B9)/		0	=1/(E10-1)^2
11	0.9	=1/(1-A11)	=(A11-A10)*(B11+B1(		0.9	=1/(E11-1)^2
12		Integral =	=SUM(C3:C11)		Integral =	=C12 - (0.1^2)*(F11-F10)/12

	A	B	C	D	E	F
1	x	1/(1-x)				
2	0	1				
3	0.1	1.111111	0.105556			
4	0.2	1.25	0.118056			
5	0.3	1.428571	0.133929			
6	0.4	1.666667	0.154762			
7	0.5	2	0.183333			
8	0.6	2.5	0.225		End correction	
9	0.7	3.333333	0.291667		x	1/(x-1)^2
10	0.8	5	0.416667		0	1
11	0.9	10	0.75		0.9	100
12		Integral=	2.378968		Integral =	2.296468

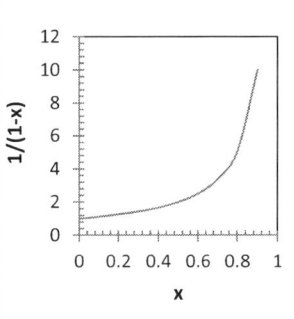

Figure 11.7

Composite trapezoidal rule integration of Equation (11.19). Plot of integrand for trapezoidal rule example.

Comments: Endpoint correction requires accurate evaluation of the derivative of the function at the limits of integration. The precision of the derivative approximations should match the integral result, with an order of magnitude of at least $O(\Delta x^4)$. If analytical derivatives are difficult to obtain, the *VBA* user-defined function **DREX** produces accurate derivative approximations. However, high accuracy derivative approximations are not easy to obtain from experimental data.

To obtain higher accuracy in this example, use more intervals (smaller interval sizes). In Figure 11.8, the formulas for the integrand and integral in column **B** are

$$B3 = 1/(1 - B1) \qquad B5 = \text{TRAP(\$B\$3, \$B\$1, \$B\$2, A5)}$$

The formula in **B5** is filled down the column using absolute references for the integrand function and limits, and relative reference for the number of intervals until $n = 50$ for the plot. At 500 and 1000 intervals, the trapezoidal rule gives 2.302612 and 2.302592, respectively. For $n > 20$, the solution using the trapezoidal rule slowly converges to the true solution of 2.302585 as $n \to \infty$.

	A	B	C	D
1	x low	0		2.55
2	x up	0.9		2.5
3	f(x)	1		2.45
4	n	I Trap	% Error	2.4
5	5	2.524924	0.09656	
6	10	2.365214	0.0272	2.35
7	15	2.331345	0.01249	2.3
8	20	2.318977	0.007119	2.25
9	25	2.313145	0.004586	
10	30	2.309945	0.003196	

Figure 11.8

The trapezoidal rule slowly converges to the true solution as the number of intervals increases.

11.2.1 Improper Integrals and the Midpoint Rule

The trapezoidal rule cannot handle improper integrals with discontinuities at either limit of integration. For an example, consider the solution to Equation (11.19) in the limit $X = 1$. The midpoint rule provides a simple alternative to the trapezoidal rule for improper integrals. To apply the midpoint rule, evaluate the integral function at the midpoint of each interval of integration, as shown in Figure 11.9. This avoids improper integrand evaluations at the limits. Just like the graphical method, we approximate the area under the curve by the area of a rectangle formed from the product of the difference between the limits of integration and the function value at the midpoint:

$$\int_{x_i}^{x_{i+1}} f(x)dx \cong (x_{i+1} - x_i) f(x_{i+1/2}) \tag{11.21}$$

where

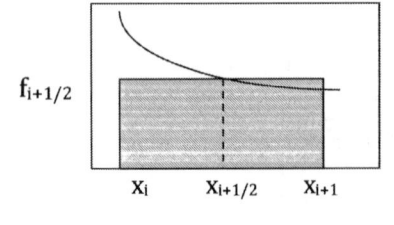

$$x_{i+1/2} = x_i + 0.5(x_i + x_{i+1})$$

Figure 11.9

Midpoint integration approximation.

Although the midpoint rule has similar accuracy to the trapezoidal rule, it has one advantage that is important – it requires only a single function evaluation over each interval of integration. However, this advantage fades when we apply the composite midpoint rule to all n intervals from x_1 to x_{n+1}:

$$\int_{x_0}^{x_n} f(x)dx \cong \Delta x \sum_{i=o}^{n-1} f(x_{i+1/2}) \tag{11.22}$$

Unfortunately, we usually cannot integrate experimental data by the midpoint rule without eliminating ~50% of the data. To improve the accuracy, we usually apply the midpoint rule just to the interval that contains the singular limit of integration. We then use the trapezoidal rule, or higher order integration methods, to integrate the rest of the region (Chapra and Canale 2002). A *VBA* macro **MIDPOINT** is available in the *PNMSuite* for approximating integrals by the composite midpoint rule. The *Excel* worksheet setup is the same as that for the macro **TRAPEZOID**.

CHAPTER 11: INTEGRATION

Example 11.3 Compare the Midpoint Rule with the Trapezoidal Rule

Known: A simple integral function with a singularity at the origin is $1/x$. Integration of this function near the singularity from 0.01 to 1 gives:

$$I = \int_{0.01}^{1} \frac{dx}{x} = \ln x \big|_{0.01}^{1} = 4.605 \tag{11.23}$$

Find: Approximate the solution to the integral using the trapezoidal and midpoint rule for comparison.

Analysis: Use the macro **QYXPLOT** from the *PNMSuite* to graph the function (adjust the axis scales as shown):

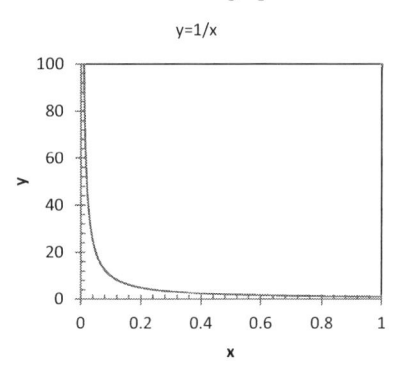

y=1/x

Apply the *trapezoidal rule* with one interval:

$$I = \frac{(1-0.01)}{2}\left[\frac{1}{0.01} + \frac{1}{1}\right] = 49.995 \tag{11.24}$$

Apply the *midpoint rule* with one interval for an improved solution:

$$I = \frac{(1-0.01)}{[0.01 + 0.5(0.01+1)]} = 1.9223 \tag{11.25}$$

Note that the trapezoidal rule overestimates the solution by 986% whereas the midpoint rule underestimates the solution by a significantly smaller amount of 58%.

Comments: Using an interval size of 0.01, the composite trapezoidal rule gives an improved approximation for the integral of 4.682 (absolute error 1.7%), although the composite midpoint rule still gives a more accurate result of 4.569 (absolute error 0.8%).

□

The following clever change of variable proposed by Kahan (1980) eliminates indeterminate limits of integration:

$$\int_{a}^{b} f(x)dx = \frac{3(b-a)}{4} \int_{-1}^{1} (1-\theta^2) f\left[b\left(\frac{3\theta - \theta^3 + 2}{4}\right) - a\left(\frac{3\theta - \theta^3 - 2}{4}\right)\right] d\theta \tag{11.26}$$

At the transformed integration limits of –1 and 1, the integrand becomes $f(a)$ and $f(b)$, respectively, while the term $(1 - \theta^2)$ eliminates the endpoints. The variable change has the added benefit of subduing any resonance stimulated by uniform intervals for periodic integrands, such as trigonometric functions.

We may also transform improper integrals to proper integrals through the following identity:

$$\int_{a}^{b} f(x)dx = \int_{1/b}^{1/a} \frac{f(1/\theta)}{\theta^2} d\theta \tag{11.27}$$

The integral transformation in Equation (11.27) requires a non-zero product of the limits of integration: $a \cdot b \neq 0$. For example, consider the improper integral with the lower limit of $-\infty$. We may divide the integral into two sections where the upper limit of the first section approaches the lower limit:

$$\int_{-\infty}^{b} f(x)dx = \int_{-\infty}^{\alpha} f(x)dx + \int_{\alpha}^{b} f(x)dx \qquad (11.28)$$

We choose the intermediate limit α such that

$$\frac{1}{\alpha^2} \ll 1 \qquad (11.29)$$

We then approximate the integral with the singularity using a combination of the midpoint rule for the interval containing the singularity, and the rest of the intervals by methods such as the trapezoidal rule:

$$\int_{-\infty}^{b} f(x)dx = \int_{-1/\alpha}^{0} \frac{f(1/\theta)}{\theta^2} d\theta + \int_{-\alpha}^{b} f(x)dx \qquad (11.30)$$

11.2.2 Integration Accuracy and Precision

The simplest test for accuracy of numerical integration involves selecting different interval sizes and comparing the precision in the results. The bisection rule is a good place to start. Double the number of intervals to reduce the interval size by half and compare the solution:

$$\left| \frac{I(\Delta x) - I(\Delta x/2)}{I(\Delta x) + \delta} \right| < \varepsilon \qquad (11.31)$$

If the solution does not change within the degree of precision required by the problem using two different interval sizes, we may be inclined to trust the solution. Remember that numerical solutions have no guarantee of arriving at the true solution. We should test our numerical technique on a problem with a known analytical solution to check that it gives the correct result.

At first glance, we may grumble at the extra work required to check for integration accuracy. However, we can use the additional results to squeeze even more accuracy from the solution. When we do not know the order of the accuracy of the solution, we may apply Aitken's Δ^2 method of series convergence acceleration in Equation (6.78) to three approximations for the integral. Alternately, Richardson's extrapolation has the advantage of requiring just two results in the series to improve the solution.

We first encountered Richardson's extrapolation in Chapter 5 as a method of improving the accuracy of finite-difference derivative approximations. The same approach works here to improve the precision of integrals. We find from Taylor series analysis that the truncation error in the derivation of the composite trapezoidal rule without end corrections is proportional to the square of the interval size, Δx^2. The accuracy of the integral approximation increases with diminishing interval size. Consider two applications of the trapezoidal rule with different numbers of intervals, $n_2 > n_1$, or interval sizes $\Delta x_2 < \Delta x_1$. The ratio of the corresponding integration errors E_1 and E_2 is proportional to the ratio of the square of the interval size:

$$\frac{E_1}{E_2} \cong \frac{\Delta x_1^2}{\Delta x_2^2} \qquad (11.32)$$

The integral approximations I_1 and I_2 using the two different interval sizes in the trapezoidal rule are equal when adjusted for error:

$$I \cong I_1 + E_1 = I_2 + E_2 \qquad (11.33)$$

Solve Equations (11.32) and (11.33) simultaneously for the integral I, and errors E_1, and E_2, to obtain an improved integral solution using two different interval sizes:

CHAPTER 11: INTEGRATION

$$I \cong I_2 + \left[\frac{I_2 - I_1}{(\Delta x_1/\Delta x_2)^2 - 1}\right] \quad (11.34)$$

$$E_1 = \frac{4(I_2 - I_1)}{3} \quad \text{and} \quad E_2 = \frac{I_2 - I_1}{3} \quad (11.35)$$

Equation (11.35) gives a conservative estimate of the error of integration. When $n_2 = 2n_1$ or $\Delta x_2 = 0.5\Delta x_1$, Equation (11.34) becomes:

$$I \cong \frac{4I_2 - I_1}{3} \pm O(\Delta x^4) \quad (11.36)$$

We follow the same approach of error estimation to extend Richardson's extrapolation method for higher order approximations from combinations of lower order approximations. The Romberg method of integration presented in Section 11.2.4 automatically finds the interval size required to give an accurate solution using the error estimates in Equation (11.35).

11.2.3 Data Integration with Uncertainty Propagation

We commonly apply the trapezoidal rule to integrate experimental data, such as a time series $f(t)$:

$$I = \int_{t_0}^{t_n} f(t) dt \quad (11.37)$$

If we assume negligible uncertainty in the independent variable for time, the law of propagation of uncertainty gives the uncertainty in the integral, u_I, in terms of the uncertainty in the measurements of the dependent variable, the integration step sizes, Δt, and the number of integration steps, n:

$$u_I = \sqrt{\sum_{i=0}^{n} \left(\frac{\partial}{\partial f} \int_{t_i}^{t_{i+1}} f dt\right)^2 u_f^2} \cong u_f \sqrt{\sum_{i=0}^{n} (\Delta t_i)^2} \quad (11.38)$$

where the time step $\Delta t_i = t_{i+1} - t_i$, and u_f is the uncertainty in the measurement for f. For the case of uniform intervals $\Delta t_i = (t_n - t_0)/n = \Delta t$, common in data collected by computer-based data acquisition systems, we simplify the summation term under the radical in Equation (11.38):

$$u_I = u_f \sqrt{\sum_{i=0}^{n} (\Delta t)^2} = u_f \sqrt{(\Delta t)^2 \sum_{i=0}^{n} 1} = u_f \Delta t \sqrt{n} = u_f \frac{(t_n - t_0)}{\sqrt{n}} \quad (11.39)$$

Recall that the process of integration naturally smooths noisy data, where differentiation tends to magnify the effects of noise in the data. We may approximate the standard uncertainty in the integrand, u_f, from the readability of f or regression analysis methods from Section 9.3. For example, we may fit a subset of the data for the integrand function with a line or polynomial to find the regression uncertainty in f. We then combine the result from Equation (11.39) with the coverage factor to give the expanded uncertainty in the integral, such as $U_I = 1.96 u_I$ for data with 30 or more measurements in the time series.

Doubling the number of intervals reduces the interval size by half, but only reduces the experimental uncertainty by a factor of 0.71 (or 29%):

$$\frac{u_{I \Delta t/2}}{u_{I \Delta t}} = \frac{u_f \frac{\Delta t}{2} \sqrt{2n}}{u_f \Delta t \sqrt{n}} = \sqrt{\frac{1}{2}} \cong 0.71 \quad (11.40)$$

Thus, we see that taking steps to reduce the systematic error, through better instrument calibration or increasing precision, can have a direct effect on reducing the propagation of uncertainty in the integral. However, we can also apply Richardson's extrapolation described in Section 5.3.4 to improve the approximation of the integration of equally spaced data. To use Richardson's formulas, we perform an additional integration using every other data pair, using just half of the data. This gives us two integral approximations in Equation (11.36) with uncertainty estimated from E_2 in Equation (11.35).

The user-defined function **TRAPDATA(ydata, xdata, error)** applies the composite trapezoidal rule to data and attempts an extrapolation to upgrade the result. The arguments **ydata** and **xdata** are the ranges of y (dependent) vs x (independent) data. **Error** is a Boolean type variable set to **True** to display an error estimate, if possible. The user-defined function first checks for uniform spacing in the independent data before conducting an extrapolation.

Example 11.4 TWA of Acetone Concentration in Air

Known: Acetone is a common solvent for cleaning glassware. The Occupational Safety and Health Administration's (*OSHA*) permissible exposure limit (*PEL*) for acetone is 1000 ppm of air (2,400 mg/m^3) as a time-weighted average (*TWA*) over an eight-hour work shift. The concentration of acetone in the air of a research lab was sampled over an eight hour period during which a bottle of acetone was opened several times to access the solvent for cleaning. The sampling data are tabulated and integrated in an *Excel* worksheet.

Find: Calculate the *TWA* and comment on the safety of the lab.

Assumptions: Readability $\delta = \pm 5$ ppm, no error in the time measurements.

Analysis: Plot the data and apply the trapezoidal rule in an *Excel* worksheet. We observe that the *PPM* exceeds the *PEL* for short periods during the eight-hour shift:

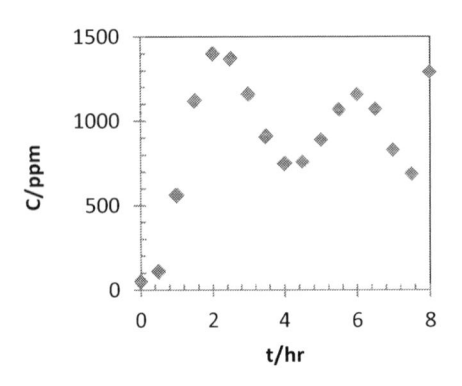

Calculate the time-weighted average (*TWA*) of the toluene concentration using the trapezoidal rule according to the following equation:

$$TWA = \frac{\int_0^{8hr} C \, dt}{\int_0^{8hr} dt} \cong \frac{\Delta t}{8\,hr} \left[\frac{(C_1 + C_{n+1})}{2} + \sum_{i=2}^{n} C_i \right] \quad (11.41)$$

where $\Delta t = 0.5$ hr. Apply the trapezoidal rule to the data in an *Excel* worksheet, and then sum the values in the ranges **C3:C10** and **G3:G10** for the total value of the integral in cell **B13**. We approximate the standard and expanded uncertainty in *TWA* in the range **E13:E14** from the law of uncertainty propagation in Equation (11.39).

E13 = 5*8/SQRT(3*16) E14 = ROUNDUP(TINV(0.05, 1000)*E13, 0)

	A	B	C	D	E	F	G
1	t/hr	C/ppm	Trap		t/hr	C/ppm	Trap
2	0	50			4	750	
3	0.5	110	40		4.5	760	377.5
4	1	560	167.5		5	890	412.5
5	1.5	1120	420		5.5	1070	490
6	2	1400	630.0		6	1160	557.5
7	2.5	1370	692.5		6.5	1070	557.5
8	3	1160	632.5		7	830	475
9	3.5	910	517.5		7.5	690	380
10	4	750	415		8	1290	495
11		Sum	3515			Sum	3745
12							
13	Integral =	7260		u_I =	5.77		
14	TWA =	908		$u_{TWA, 95\%}$ =	12		

We may also try the user-defined function **TRAPDATA** to integrate the concentration data. First, we need to arrange the data in contiguous cells on the worksheet, with the complete time data in column **J** and the complete concentration data in column **K**. Because this represents an odd number of intervals, the user-defined function includes one Richardson's extrapolation:

7206.67 = TRAPDATA(K2:K17, J2:J17)   Relative Error = 0.01 (1%)

Comments: The numerical integration error is negligible in comparison with experimental error. Rounding the result from **TRAPDATA**, $TWA = 901 \pm 12$ ppm (95% confidence) and falls below the PEL of 1000 ppm within the uncertainty of the calculations.

□

11.2.4 Romberg Integration with Error Control

Repeated application of Richardson's extrapolation to quadrature by the trapezoidal rule gives the Romberg's recursive formula:

$$I_{j,k} \cong \frac{4^{k-1} I_{j+2,k-1} - I_{j,k-1}}{4^{k-1} - 1} \tag{11.42}$$

where the index $j = 1, 2, 4, 8\ldots$ indicates the relative number of trapezoidal steps and the index $k = 2, 3, 4\ldots$ signifies the level of the integration error. Each iteration of Romberg extrapolation reduces the error in the integral by two orders of magnitude. Romberg integration approximates the error in the calculation from the difference between successive applications of extrapolation. The *VBA* macro **ROMBERGT**, with required inputs listed in Figure 11.11, solves a single-variable integral by the Romberg integration method with error control using Richardson's extrapolation of the composite trapezoidal rule *without* end correction. The *VBA* macro **ROMBERGI** employs Kahan's change of variable according to Equation (11.26) for improper integrals with indeterminate limit(s).

Since we are doubling the number of intervals, we are able to reuse evaluations of the integrand function from previous iterations to evaluate the integral in the current iteration. For example, if we apply the trapezoidal rule sequentially using two and four intervals, as illustrated in Figure 11.10, we may use the result from two intervals as the starting point for the integral evaluated with four intervals:

- Two Intervals $s_2 = \frac{1}{2}(f_0 + f_4) + f_2$ $\Delta x_2 = \frac{x_4 - x_0}{2}$ $I_2 = \Delta x_2 s_2$ (11.43)

- Four Intervals $s_4 = s_2 + f_1 + f_3$ $\Delta x_4 = \frac{x_4 - x_0}{4}$ $I_4 = \Delta x_4 s_4$ (11.44)

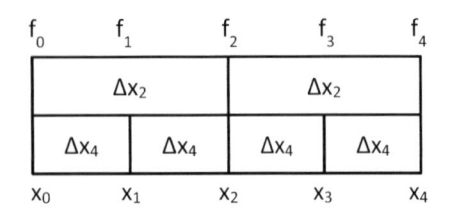

Figure 11.10
Function evaluations at shared nodes for the trapezoidal rule with two and four intervals, respectively.

By reusing the integrand function evaluations from the trapezoidal method at the previous iteration, we cut the number of integrand evaluations in half for a dramatic improvement in computational efficiency. An alternative macro named **ROMBERGM** is available in the *PNMSuite* that uses extrapolation of the composite midpoint rule. However, the midpoint formula does not have the same built-in efficiency of the trapezoidal method for reusing function evaluations. For this reason, we recommend the midpoint version only for improper integrals.

To use the Romberg macros, specify cells for the integration limits, and the integral function evaluated in terms of the value in the cell containing the lower limit of integration. The user-form **INTEGRAL** has options for each of the Romberg methods. We illustrate how to set up an *Excel* worksheet for Romberg integration in *Example 11.5*.

```
Public Sub ROMBERGT()
' Romberg numerical quadrature using Richardson's extrapolation of trapezoidal rule.
' Define the integral function in the worksheet in terms of the cell containing the
' lower limit of integration.  The macro replaces value in the lower integration limit
' cell with values for variable at equally-spaced intervals. The values for integrand
' in the function cell are summed according to the composite trapezoidal rule. The
' number of intervals is doubled and the trapezoidal rule is used to integrate the
' function. Richardson's extrapolation is used to improve the solution iteratively.
' The result is reported to the cell specified by the user.
' *** Nomenclature ***
Dim boxtitle As String ' = string title used in message and input boxes
Dim c As Double ' = extrapolation coefficients
Dim dx As Double ' = width of trapezoid
Dim er As Double ' = relative integration error
Dim fr As Range ' = range variable for integrand function
Dim I(1 To 99, 1 To 99) As Double ' = array of integral solutions
Dim it As Long, it1 As Long ' = iteration counter, it1 = it+1
Dim j As Long ' = counter for managing Richardson's extrapolation
Dim k As Long ' = loop counter for managing Richardson's extrapolation
Dim n As Long ' = number of intervals (requires even integer)
Dim qr As Range ' = range variable for result of quadrature
Dim s As Double ' = sum of integrand function evaluations for trapezoidal rule
Dim sc As Integer ' = show calculations
Dim small As Double ' = small number
Dim tol As Double ' = error control tolerance
Dim x_low As Double, x_up As Double ' = lower/upper integration limit
Dim xr As Range ' = range for integral variable
'****************************************************************************
small = 0.000000000001: boxtitle = "ROMBERG Quadrature (Trapezoidal Rule)"
With Application
Set fr = .InputBox(Title:=boxtitle, Type:=8, _
                Prompt:="INTEGRAND function. f(x):" & vbNewLine & _
                    "(Integrand formula must use cell with x lower limit)")
Set xr = .InputBox(Title:=boxtitle, Prompt:="LOWER integration limit, x:" & _
            vbNewLine & "(Integrand formula must use this cell for x)", Type:=8)
x_up = .InputBox(Prompt:="UPPER integration limit:", Title:=boxtitle, Type:=8)
Set qr = .InputBox(Title:=boxtitle, Type:=8, Prompt:="RESULT cell:")
tol = 0.00000001 ' ERROR Control Tolerance
End With ' Application
x_low = xr ' save value for the lower integration limit
dx = (x_up - x_low) / 2: s = fr / 2 ' size of intervals
xr = x_low + dx: s = s + fr: xr = x_low + 2 * dx: s = s + fr / 2 ' evaluate function
I(1, 1) = s * dx ' trapezoidal rule with two intervals
For it = 2 To 98 ' Loop for Richardson's extrapolation
    it1 = it + 1: n = 2 ^ it: dx = (x_up - x_low) / n
    For k = 1 To n - 1 Step 2: xr = x_low + k * dx: s = s + fr: Next k
    I(it1, 1) = s * dx ' trapezoidal rule integration
```

CHAPTER 11: INTEGRATION

```
        For k = 1 To it ' Richardson's extrapolation
            j = 1 + it - k: c = 4 ^ k: I(j, k + 1) = (c * I(j + 1, k) - I(j, k)) / (c - 1)
        Next k
        er = Abs((I(1, it1) - I(1, it)) / (I(1, it1) + small))' check for convergence
        Application.StatusBar = boxtitle & " % Error = " & VBA.Format(er, "#E+##%")
        If er <= tol Then GoTo 10
    Next it
    MsgBox Prompt:="Maximum Richardson's extrapolations: " & 100 & vbNewLine & _
        ". Adjust error control size or iterations.", Buttons:=0 + 48, Title:=boxtitle
    10: xr = x_low: qr = I(1, it + 1) ' Reset xr and Output solution to worksheet
    End Sub ' ROMBERGT
```

Figure 11.11 *VBA* macro **ROMBERGT** used to solve single integrals using Richardson's extrapolation of the trapezoidal rule. Similar macros **ROMBERGM** and **ROMBERGI** are available in the PNMSuite using the midpoint rule and variable transformation for improper integrals, respectively.

The *PNMSuite* also has two user-defined functions (similar to **TRAP** in Figure 11.6) for integrating a simple worksheet *formula* directly in a cell by the trapezoidal or midpoint methods with Richardson extrapolation.

TREX(f, x_low, x_up, optional tol)   **MPEX(f, x_low, x_up, optional tol)**

where **f** and **x_low** are the cell addresses for the integrand formula and variable of integration, respectively. The value in **x_low** is the lower limit of integration. **x_up** is the upper integration limit and **tol** is an optional argument for the relative error convergence tolerance. The default value for tol = 10^{-8}. The last two arguments may use references to cells containing the values, or specified directly in the argument list of the user-defined functions. The worksheet arrangement for these user-defined functions has the same requirements as the Romberg macros. However, we must not use **TREX** or **MPEX** for integrand functions with calculations involving multiple cells on the worksheet. The integrand formula must be self-contained in one cell.

Example 11.5   Romberg Integration of our Test Function

Known: Test the methods of Romberg integration using an integral with a known analytical solution:

$$\int_0^\pi x\sin(x)\,dx = \pi \qquad (11.45)$$

Find: Solve the integral in Equation (11.45) by the Romberg methods.

Analysis: Plot the integrand function using the macro **QYXPLOT** to observe the behavior of the integral:

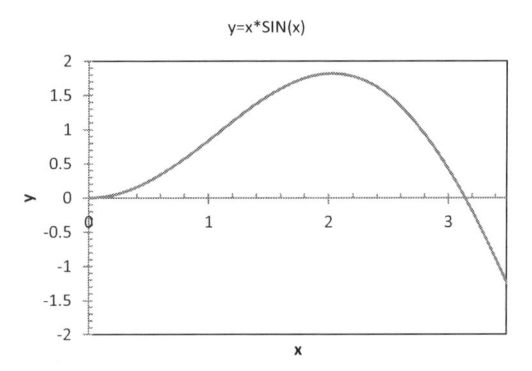

y=x*SIN(x)

Before using the macro **ROMBERGT**, we conducted one Richardson extrapolation in an *Excel* worksheet to show the power of extrapolation. We first solve the integral twice using the composite trapezoidal rule with four and eight intervals, then apply Richardson's extrapolation formula in cell **A13**:

A13 = (4*G13 - C9)/3

	A	B	C	D	E	F	G
1	4 Intervals				8 Intervals		
2	dx	0.785398163			dx	0.39269908	
3	x	I	Trapezoids		x	I	Trapezoids
4	0	0			0	0	
5	0.785398163	0.555360367	0.218089506		0.3926991	0.15027943	0.0295073
6	1.570796327	1.570796327	0.834939781		0.7853982	0.55536037	0.1385521
7	2.35619449	1.666081102	1.271118794		1.1780972	1.08841993	0.3227555
8	3.141592654	3.84892E-16	0.654268519		1.5707963	1.57079633	0.5221359
9		Sum	2.9784166		1.9634954	1.81403322	0.6646097
10					2.3561945	1.6660811	0.6833188
11					2.7488936	1.05195603	0.5336853
12	Richardson's Extrapolation				3.1415927	3.8489E-16	0.2065511
13	3.142015465					Sum	3.1011157

The solutions with four and eight intervals are at best good to one or two significant figures. With one Richardson extrapolation, we improved the result to three significant figures. To see the real power of Richardson's extrapolation, we apply the trapezoidal rule one more time with 16 intervals (not shown) for an approximate solution 3.1349300. We extrapolate the trapezoidal rule solutions for 8 and 16 intervals according to Equation (11.36), then apply Richardson's extrapolation again to the two previously extrapolated results according to Equation (11.42) with $k = 3$:

$$\frac{4 \cdot 3.1314930 - 3.1011157}{3} = 3.1416187 \qquad \frac{16 \cdot 3.1416187 - 3.1420155}{15} = 3.1415923$$

We find the accuracy of the solution slowly improves as we double the number of intervals using the trapezoidal rule (accurate to two significant figures), then accelerates as we apply one Richardson's extrapolation (accurate to four significant figures), but really takes off when we add a second layer of extrapolation (accurate to seven significant figures). Once we have the three trapezoidal rule solutions, the additional effort of extrapolation is minimal for a *substantial* improvement in accuracy:

Intervals	Trapezoidal Rule	1st Extrapolation	2nd Extrapolation
4	2.9784166		
8	3.101115749	3.142015465	
16	3.131492973	3.141618715	3.141592265

To automate Romberg's method, set up a new *Excel* worksheet with formulas for the *Excel* macro **ROMBERGT**. Define the integrand function in a cell in terms of the cell containing the lower limit of integration, as shown in Figure 11.12. Calculate the percent error in cell **B5**. The *VBA* macro **ROMBERGT** uses input boxes to specify the cells containing the integrand function and limits

	A	B
1	Lower Integration Limit x	0
2	Upper Integration Limit	=PI()
3	Integrand function f(x)	=B1*SIN(B1)
4	Romberg Solution	3.14159265803826
5	% Error	=ABS(B4-PI())/PI()

Figure 11.12

Excel worksheet set up for solving an integral using the *VBA* macro ROMBERGT.

Click on cell with LOWER integration limit.
$B$1

UPPER integration limit.
$B$2

Click on cell with INTEGRAND function.
$B$3

Similar input boxes request the cell for the result (**B4**) and integration tolerance (the default tolerance is 10^{-5}). Figure 11.13 contains the results from applying the *VBA* macro **ROMBERGT**.

	A	B
1	Lower Integration Limit x	0
2	Upper Integration Limit	3.141592654
3	Integrand function f(x)	0
4	Romberg Solution	3.141592658
5	% Error	0.00000014%

Figure 11.13

Solution to Equation (11.45) using Romberg's method in an *Excel* worksheet.

CHAPTER 11: INTEGRATION

Alternatively, we may integrate the formula directly in the worksheet using the user-defined function **TREX**:

$$3.141592654 = \text{TREX(B3, B1, B2)}$$

We may also apply Aitken's Δ^2 formula in Equation (6.78) to the three trapezoidal rule results to produce an approximation with accuracy comparable to the first Richardson's extrapolations:

$$I_4 - \frac{(I_4 - I_8)^2}{I_4 - 2I_8 + I_{16}} = 3.14149$$

Comments: We may use smaller tolerances to lower the percent error as desired. Though computationally demanding, Romberg's method takes the guesswork out of the solution method. The computer does the work of integration for us! We may now confidently find accurate solutions to nonlinear integrals. We find similar results from the macros **ROMBERGI**, **ROMBERGM**, and user-defined function **MPEX**.

□

11.3 Simpson's Rules with Error Control

> *"Bart, I want to share three little sentences that will get you through life:*
> *1. Cover for me. 2, Oh, good idea, Boss! 3. It was like that when I got here."* – Homer Simpson

Simpson's rules provide fourth-order accurate approximations for integrals using equally-spaced data by integrating higher order polynomial interpolating functions, such as Lagrange's three-point formula in Equation (10.8). Data may come from experiments or evaluations of the integrand function at discrete points. To derive Simpson's rules for quadrature, expand the integral function in a third-order Taylor series polynomial about the midpoint between the limits of integration.

$$f(x) = f(x_i) + [x - x_i]\frac{df}{dx}\bigg|_{x_i} + \frac{[x - x_i]^2}{2}\frac{d^2f}{dx^2}\bigg|_{x_i} + \frac{[x - x_i]^3}{6}\frac{d^3f}{dx^3}\bigg|_{x_i} \qquad (11.46)$$

Integrate the expansion over two intervals, $(x_{i+1} - x_i) + (x_i - x_{i-1})$, as follows:

$$\int_{x_{i-1}}^{x_{i+1}} f(x)dx = [x_{i+1} - x_{i-1}]f(x_i) + \frac{1}{2}\left\{[x_{i+1} - x_i]^2 - [x_{i-1} - x_i]^2\right\}\frac{df}{dx}\bigg|_{x_i}$$
$$+ \frac{1}{6}\left\{[x_{i+1} - x_i]^3 - [x_{i-1} - x_i]^3\right\}\frac{d^2f}{dx^2}\bigg|_{x_i} + \frac{1}{24}\left\{[x_{i+1} - x_i]^4 - [x_{i-1} - x_i]^4\right\}\frac{d^3f}{dx^3}\bigg|_{x_i} \qquad (11.47)$$

For a uniform interval size $\Delta x = x_{i+1} - x_i$, the terms raised to even powers in Equation (11.46) vanish, leaving the following simple result:

$$\int_{x_{i-1}}^{x_{i+1}} f(x)dx = 2\Delta x f(x_i) + \frac{\Delta x^3}{3}\frac{d^2f}{dx^2}\bigg|_{x_i} \qquad (11.48)$$

Substitute the finite-difference approximation for the second-order derivative from Equation (5.48) into Equation (11.48) to give the following form of Simpson's 1/3 rule:

- *Simpson's 1/3 Rule* $\quad \displaystyle\int_{x_{i-1}}^{x_{i+1}} f(x)dx = 2\Delta x f(x_i) + \frac{\Delta x^3}{3}\left[\frac{f(x_{i-1}) - 2f(x_i) + f(x_{i+1})}{\Delta x^2}\right] \qquad (11.49)$

or
$$\int_{x_{i-1}}^{x_{i+1}} f(x)dx = \frac{\Delta x}{3}\left[f(x_{i-1}) + 4f(x_i) + f(x_{i+1})\right]$$

Simpson's 1/3 rule applies to a region of width $2\Delta x$. Use the following formula for Simpson's 3/8 rule in Equation (11.50) for an odd number of equally spaced intervals applied to a region of width $3\Delta x = (x_3 - x_2) + (x_2 - x_1) + (x_1 - x_0)$:

- Simpson's 3/8 Rule
$$\int_{x_0}^{x_3} f(x)dx \cong \frac{3\Delta x}{8}\left\{f(x_0) + 3\left[f(x_1) + f(x_2)\right] + f(x_3)\right\} \qquad (11.50)$$

We may also derive the Simpson's 1/3 and 3/8 rules by integrating three and four point Lagrange interpolating polynomial formulas for uniformly-spaced points, as shown in Equations (11.51) and (11.52):

$$\int_{x_{i-1}}^{x_{i+1}} \begin{pmatrix} \frac{(x-x_i)(x-x_{i+1})}{(x_{i-1}-x_i)(x_{i-1}-x_{i+1})} f(x_{i-1}) + \\ \frac{(x-x_{i-1})(x-x_{i+1})}{(x_i-x_{i-1})(x_i-x_{i+1})} f(x_i) + \\ \frac{(x-x_{i-1})(x-x_i)}{(x_{i+1}-x_{i-1})(x_{i+1}-x_i)} f(x_{i+1}) \end{pmatrix} dx = \frac{\Delta x}{3}\left[f(x_{i-1}) + 4f(x_i) + f(x_{i+1})\right] \qquad (11.51)$$

and

$$\int_{x_{i-1}}^{x_{i+2}} \begin{pmatrix} \frac{(x-x_i)(x-x_{i+1})(x-x_{i+2})}{(x_{i-1}-x_i)(x_{i-1}-x_{i+1})(x_{i-1}-x_{i+2})} f(x_{i-1}) + \frac{(x-x_{i-1})(x-x_{i+1})(x-x_{i+2})}{(x_i-x_{i-1})(x_i-x_{i+1})(x_i-x_{i+2})} f(x_i) + \\ \frac{(x-x_{i-1})(x-x_i)(x-x_{i+2})}{(x_{i+1}-x_{i-1})(x_{i+1}-x_i)(x_{i+1}-x_{i+2})} f(x_{i+1}) + \frac{(x-x_{i-1})(x-x_i)(x-x_{i+1})}{(x_{i+2}-x_{i-1})(x_{i+2}-x_i)(x_{i+2}-x_{i+1})} f(x_{i+2}) \end{pmatrix} dx$$

$$= \frac{3\Delta x}{8}\left[f(x_{i-1}) + 3f(x_i) + 3f(x_{i+1}) + f(x_{i+2})\right] \qquad (11.52)$$

with interval size $\Delta x = x_{i+1} - x_i$. Composite formulas for Simpson's 3/8 rule require an odd number of intervals in groups of three. To integrate an even number of data points (odd number of intervals), use the 1/3 rule and apply Simpson's 3/8 rule to the last four points (last three intervals).

We may apply Simpson's 1/3 rule to consecutive sets of smaller intervals to increase the accuracy of our solution. For example, if we start with two intervals defined by three equally spaced points, we double the intervals by dividing each of the original intervals in half for five equally spaced points:

$$\int_{x_0}^{x_4} f(x)dx \cong \frac{(x_4 - x_0)}{2}\frac{1}{3}\left[f(x_0) + 4f(x_2) + f(x_4)\right]$$
$$\cong \frac{(x_4 - x_0)}{4}\frac{1}{3}\left\{\left[f(x_0) + 4f(x_1) + f(x_2)\right] + \left[f(x_2) + 4f(x_3) + f(x_4)\right]\right\} \qquad (11.53)$$

Note that each application of Simpson rule uses the common point at x_2, which defines the end of the first set and beginning of the second set, as shown in Figure 11.14. An even number of intervals separates an odd number of points. Each interval of width $\Delta x = (x_n - x_0)/n$ requires $n/2$ applications of Simpson's 1/3 rule, summarized by Simpson's 1/3 composite rule:

$$\int_{x_0}^{x_n} f(x)dx = \frac{\Delta x}{3}\left[f(x_0) + 4f(x_1) + 2f(x_2) + \ldots 4f(x_{n-1}) + f(x_n)\right] \qquad (11.54)$$

CHAPTER 11: INTEGRATION

where $\Delta x = \frac{x_n - x_0}{n}$. In compact form, Simpson's composite 1/3 rule gives:

$$\int_{x_0}^{x_n} f(x)dx = \frac{\Delta x}{3}\left\{f(x_0) + \sum_{i=1}^{n-1}\left[3 + (-1)^{i-1}\right]f(x_i) + f(x_n)\right\} \tag{11.55}$$

Calculate the value of the independent variable between each interval as follows:

$$x_i = x_0 + i\Delta x \tag{11.56}$$

where the index *i* refers to the interval number.

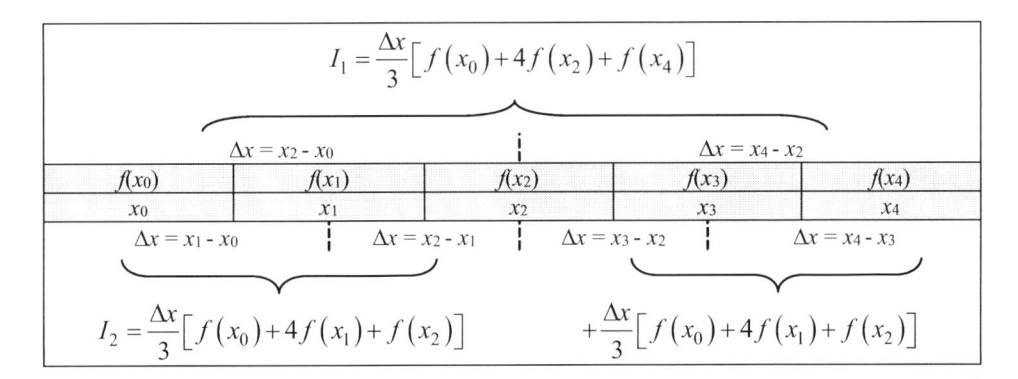

Figure 11.14

Simpson's 1/3 rule applied to sets of three and five points.

The test integral from Figure 11.2 was solved using Simpson's 1/3 rule in an *Excel* worksheet, as seen in Figure 11.15. Note how Simpson's 1/3 rule uses two intervals per application of Equation (11.51) by skipping every other row.

	A	B	C	D	E
1	n= 10			Δx= 0.314159	
2	x	I	Simpson 1/3	Integral=	3.141765
3	0	0			
4	0.314159	0.097081			
5	0.628319	0.369316	0.079339726		
6	0.942478	0.762481			
7	1.256637	1.195133	0.483215848		
8	1.570796	1.570796			
9	1.884956	1.792699	0.970858679		
10	2.199115	1.779121			
11	2.513274	1.477265	1.087666493		
12	2.827433	0.873725			
13	3.141593	3.85E-16	0.520683936		

Figure 11.15

Numerical integration of $\int_0^\pi x\sin x\,dx \cong 3.1418$ by the Simpson's 1/3 composite rule using n = 10 intervals defined in Equation (11.55). The true solution is 3.1416. The formulas in the worksheet follow:

A4 = A3+$D$1

B4 = A4*SIN(A4)

C5=$D$1*(B5+4*B4+B3)/3

D1 = PI()/10

E2 = SUM(C5:C13)

The error in Simpson's composite rule is $O(\Delta x^4)$ and the magnitude of the error in the numerical approximation decreases with diminishing interval size. Both the trapezoidal rule with end-correction and Simpson's composite rule are fourth order accurate. Surprisingly, the trapezoidal rule with end correction is generally more accurate than Simpson's rules, but requires the derivative of the integrand function, which may not always be available when integrating equally spaced data derived from experiments.

Check for accuracy by doubling the number of intervals, repeating the calculation, and comparing the results. Simpson's composite rule approaches the true solution as $n \to \infty$, $\Delta x \to 0$. However, computational rounding error starts to become significant for smaller Δx. Richardson's extrapolation of two Simpson's integrations with $\Delta x_1 = 2\Delta x_2$

increases the precision of our solution by two orders of magnitude. Because Simpson's rule starts with fourth-order accuracy, Richardson's formula starts at the level $k = 3$ in Equation (11.42):[100]

$$I_S \cong \frac{16I_2 - I_1}{15} \pm O\left(\Delta x_1^6\right) \tag{11.57}$$

Lisa: "Dad, just for once don't you want to try something new?"
Homer: "Oh Lisa, trying is just the first step toward failure." — The Simpsons

Example 11.6 Integration of Residence Time Data with Simpson's Rules

Known: Figure 11.16 shows an *Excel* worksheet with data for the distribution of residence time obtained from a tracer experiment in a non-ideal mixed flow reactor.

	A	B			
4	t	E/(1/s)	13	12	0.0253
5	0	0.0000	14	15	0.0164
6	1	0.0030	15	20	0.0079
7	2	0.0220	16	25	0.0038
8	3	0.0540	17	30	0.0018
9	4	0.0630	18	35	0.0009
10	6	0.0608	19	40	0.0004
11	8	0.0454	20	45	0.0002
12	10	0.0339	21	50	0.0001

Figure 11.16
Residence time data for conversion in a batch reactor.

Find: Calculate the conversion of the first-order reaction in the non-ideal reactor with the rate law:

$$-r_A / \left(kmol/m^3 \cdot s\right) = 1.8 C_A \tag{11.58}$$

Assumptions: Isothermal operation, imperfect mixing

Schematic: A mixed-flow reactor has volumetric flow rate Q, volume V, and conversion X_A.

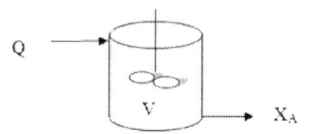

Analysis: Calculate the conversion from the mass balance in terms of the residence time:

$$X_A(\tau) = \frac{k\tau}{k\tau + 1} \tag{11.59}$$

where τ is the ideal residence time calculated from the quotient of the reactor volume (V) to the volumetric flow rate (Q), as follows:

$$\tau_{ideal} = V/Q \tag{11.60}$$

Calculate the average conversion in the non-ideal reactor by integrating the product of the function of conversion with the residence time distribution:

$$\bar{X}_A = \int_0^\infty X_A(\tau) E(\tau) d\tau \tag{11.61}$$

Apply Simpson's rules with extrapolation to obtain the value of the integral in Equation (11.61). Skip the point at $t = 12$ to permit application of Simpson's rules to equally spaced intervals. Use the 1/3 rule for $0 \leq t \leq 4$ and $10 \leq t \leq 50$.

---

[100] This formula is also applicable to upgrade a pair of $O(\Delta x^4)$ trapezoidal rule extrapolations.

CHAPTER 11: INTEGRATION

Use the 3/8 rule for $4 \leq t \leq 10$. Figure 11.17 contains the results on the worksheet and formulas for Simpson's rules in column **E**.

	A	B	C	D	E
1	V/m³	0.96	k/(1/s)	1.8	
2	Q/(m³/s)	0.14	X ideal	0.93	
3	τ/s	6.857	X nonideal	0.603490	
4	t	E/(1/s)	X$_A$	X$_A$·E	Simpson
5	0	0.0000	0	0	0.008310559
6	1	0.0030	0.6429	0.0019	
7	2	0.0220	0.7826	0.0172	0.084928155
8	3	0.0540	0.8438	0.0456	
9	4	0.0630	0.8780	0.0553	0.28629831
10	6	0.0608	0.9153	0.0556	
11	8	0.0454	0.9351	0.0425	
12	10	0.0339	0.9474	0.0321	0.171765698
13	15	0.0164	0.9643	0.0158	
14	20	0.0079	0.9730	0.0077	0.040538874
15	25	0.0038	0.9783	0.0037	
16	30	0.0018	0.9818	0.0018	0.009509239
17	35	0.0009	0.9844	0.0009	
18	40	0.0004	0.9863	0.0004	0.002139443
19	45	0.0002	0.9878	0.0002	
20	50	0.0001	0.9890	0.0001	

	E
4	Simpson
5	=(1/3)*(D5+4*D6+D7)
7	=(1/3)*(D7+4*D8+D9)
9	=2*(3/8)*(D9+3*(D10+D11)+D12)
12	=5*(1/3)*(D12+4*D13+D14)
14	=5*(1/3)*(D14+4*D15+D16)
16	=5*(1/3)*(D16+4*D17+D18)
18	=5*(1/3)*(D18+4*D19+D20)

Figure 11.17

Excel worksheet for integration of residence time data using Simpson's 1/3 rule.

The formula for the value of the integral in cell **D3** uses the worksheet function **SUM**:

D3=SUM(E5:E18)

We apply Richardson's extrapolation to the evaluations for $0 \leq t \leq 4$ and $10 \leq t \leq 50$. First, evaluate the integral for the two ranges using double intervals:

F5 =2*(1/3)*(D5+4*D7+D9) = 0.08279

F12 =10*(1/3)*(D12+4*D14+D16) = 0.21543   F16 =10*(1/3)*(D16+4*D18+D20) = 0.01148

Then apply Richardson's extrapolation and sum the parts to get the integral value:

G20 = 0.60399 = (16*SUM(E5:E7) - F5)/15 + E9 + (16*SUM(E12:E18) - SUM(F12:F16))/15

Comments: The non-ideal conversion is 32% lower than the ideal result. We clearly have an opportunity to improve the mixing, or use a different type of reactor, such as a plug flow reactor to increase the conversion for a similar residence time.

□

The *Excel* workbook *PNMSuite* contains two macros for applying Simpson's rules of integration to equally spaced data or a function evaluated at equally spaced values of the independent variable.
- The sub procedure **SIMPSON** automates Simpson's 1/3 composite rule to integrate functions in an *Excel* worksheet. The macro requires the limits of integration and the formula for the integrand in terms of the variable using the cell for the lower limit of integration.
- The *VBA* user-defined function **SIMPSONDATA(ydata, xdata, error)**, listed in Figure 11.18, uses the 1/3 rule for an even number of intervals (an odd number of data points), where **ydata** is the range of the dependent variable and **xdata** is the range of the dependent variable. For equally spaced data that are factorable by four, **SIMPSONDATA** performs one Richardson's extrapolation according to Equation (11.57) to estimate the error and provide two orders of magnitude increase in accuracy. The optional argument **error** is a Boolean type variable. When the Boolean argument for **error** is set to **True** (the default value is **False**) a message box displays the error estimate, if possible given the number of points in the data.

- For data involving an even number of points (odd number of intervals), the last four points use the 3/8 rule for the integral.
- The macros sort the data in ascending order according the independent variable x, and checks for equally spaced data as required by the method of integration.

```vba
Public Function SIMPSONDATA(ydata, xdata, Optional error As Boolean = False) As Double
' ydata = range of cells containing dependent variable data
' xdata = range of cells containing independent variable data
' error = (optional) Boolean variable to display the error estimate for even intervals
'****************************************************************************
' Solves a single-variable integration of equally-spaced data by Simpson's composite
' rule. Checks for an even or odd number of data pairs and applies 1/3 or 3/8 rule
' accordingly. The data is sorted in ascending order. Assumes that the first and last
' data points are the lower and upper integration limits. For an even number of
' intervals factorable by four, the function performs one Richardson extrapolation to
' improve the accuracy of the integral and estimate the error.
' *** Required User-defined Functions and Sub Procedures from the PNM Suite ***
' HEAPSORTXY = sort data in ascending order by xdata
' *** Nomenclature ***
Dim e As Double ' = estimated error from Richardson's extrapolation
Dim i As Long ' = index for data pair
Dim nx As Long, ny As Long ' = number of x and y intervals
Dim dx As Double, dxi As Double   ' = interval width, interval width for comparison
Dim s As Double, s1 As Double ' = sum for Simpson's composite rules
Const small As Double = 0.000000000001 ' = small number for comparison
Dim x() As Double, y() As Double ' = arrays of x and y data
'****************************************************************************
nx = xdata.Count: ny = ydata.Count ' Get the size of the data set
If ny < 3 Then ' Check size of data
    MsgBox Title:="SIMPSON DATA", prompt:="Range of data must be >= 3": Exit Function
ElseIf nx <> ny Then
    MsgBox Title:="SIMPSON DATA", prompt:="X/Y data ranges must match.": Exit Function
End If
ReDim x(1 To nx) As Double, y(1 To ny) As Double ' Set array size to data set
' Create arrays of data for sorting & integrating. Sort data in ascending order
For i = 1 To nx: x(i) = xdata(i): y(i) = ydata(i): Next i
Call HEAPSORTXY(x(), y()) ' just in case ...
dx = x(2) - x(1) ' Check for equally-spaced data
For i = 3 To nx
    dxi = x(i) - x(i - 1)
    If Abs(dxi - dx) > small Then
        MsgBox "Data must be equally-spaced. Check near data pair: " & i: Exit Function
    End If
    dx = dxi
Next i
s = 0: s1 = 0 ' Simpson's composite rule initial sum total set to zero
Select Case WorksheetFunction.Even(nx) ' round up to nearest even integer
Case nx ' Even points, Use 1/3 rule for all but last three intervals use 3/8 rule
    For i = 1 To nx - 5 Step 2: s = s + y(i) + 4 * y(i + 1) + y(i + 2): Next i
    s = s / 3 ' Use 3/8 rule for last three intervals using last four data pairs
    s = s + 3 * (y(nx - 3) + 3 * (y(nx - 2) + y(nx - 1)) + y(nx)) / 8
Case Else ' Odd points, Use 1/3 rule for even number of intervals
    For i = 1 To nx - 2 Step 2: s = s + y(i) + 4 * y(i + 1) + y(i + 2): Next i
    s = s / 3
    If (nx - 1) Mod 4 = 0 Then ' Repeat with every other data pair.
        For i = 1 To nx - 4 Step 4: s1 = s1 + y(i) + 4 * y(i + 2) + y(i + 4): Next i
        s1 = 2 * s1 / 3: s = (16 * s - s1) / 15 ' Richardson's extrapolation
        e = Abs(s - s1) / Abs(15 * s + small)
        If error = True Then MsgBox "SIMPSONDATA: " & ydata.Address & " ; Error = " _
            & VBA.Format(e, "#.#E+##%")
    End If
End Select
SIMPSONDATA = dx * s ' Return the result to the worksheet
End Function ' SIMPSONDATA
```

Figure 11.18   *VBA* user-defined function SIMPSONDATA for integrating equally-spaced data by Simpson's composite rule.

CHAPTER 11: INTEGRATION

Example 11.7 Number of Transfer Units in a Packed Column

Known/Find: Calculate the number of mass transfer units in a packed column for gas absorption by solving the integral in Equation (11.3):

$$N = \int_{Y_{in}}^{Y_{out}} \frac{dY}{Y - Y^*} \qquad (11.62)$$

where Y is the mole ratio of absorbing solute to solute-free gas and Y^* is the mole ratio that would be in equilibrium with the bulk liquid mole ratio X. Use a liquid-to-gas molar flow ratio of 70, and inlet and outlet solute mole ratios of $Y_{out} = 0.25$ and $Y_{in} = 0.01$, respectively. Henry's constant for this system is 5558 kPa. The column operates at atmospheric pressure.

Analysis: Solve the packed column design integral by Simpson's 1/3 composite rule in an *Excel* worksheet, as shown in Figure 11.19. Specify the fluid flow ratio, Henry's constant, and outlet composition in row one on the worksheet. The values of the integrand must be equally spaced for Simpson's methods. To accomplish this, create several values of the gas-phase mole ratio for the solute Y ranging from the inlet to outlet values in the range **B3:B27**. Then get values of X, x, Y^*, and y^* from the relationships in Equations (11.5) through (11.7):

C4 = (B4 - Y0)/LG   D4 = C4/(1 + C4)   E4 = D4*H/101.325

F4 = E4/(1 - E4)   G4 = 1/(B4 - F4)   H4 = (1/3)*(B4 - B3)*(G3 + 4*G4 + G5)

Select the cells in the range **B4:G4**. Double-click on the fill handle to autofill the formulas down the columns. Select the cells in the range **H3:H4**, then drag down to fill in the formulas for Simpson's 1/3 rule in column **H**. Note how *Excel's* autofill feature uses the established pattern with a blank cell and adds the 1/3 rule formula with relative cell references to every other cell. In this way, the last point in one application of the formula for Simpson's 1/3 rule also serves as the first point in the next application of Simpson's 1/3 rule.

Calculate the value of the integral from the sum of the Simpson's 1/3 rule applied to each group of three consecutive rows:

H28 =SUM(H4:H27)

For comparison, integrate the data using the *VBA* user-defined function **SIMPSONDATA**, which incorporates one Richardson extrapolation for even intervals:

J1 = SIMPSONDATA(G3:G27, B3:B27)

	I	J
1	SIMPSONDATA	11.30423

Error ≈ 0.003%

	A	B	C	D	E	F	G	H
1	L/G =	70	H =	5558	Y₀ =	0.01		Simpson's
2	Intevals	Y	X	x	y*	Y*	1/(Y-Y*)	1/3 Rule
3	0	0.01	0.00000	0.00000	0.00000	0.00000	100.00000	
4	1	0.02	0.00014	0.00014	0.00784	0.00790	82.62505	1.67170
5	2	0.03	0.00029	0.00029	0.01567	0.01592	71.01102	
6	3	0.04	0.00043	0.00043	0.02350	0.02406	62.75318	1.26217
7	4	0.05	0.00057	0.00057	0.03133	0.03234	56.62759	
8	5	0.06	0.00071	0.00071	0.03915	0.04075	51.94647	1.04236
9	6	0.07	0.00086	0.00086	0.04698	0.04929	48.29451	
10	7	0.08	0.00100	0.00100	0.05480	0.05798	45.40700	0.91010
11	8	0.09	0.00114	0.00114	0.06262	0.06680	43.10826	
12	9	0.1	0.00129	0.00128	0.07044	0.07577	41.27816	0.82685
13	10	0.11	0.00143	0.00143	0.07825	0.08490	39.83301	
14	11	0.12	0.00157	0.00157	0.08606	0.09417	38.71406	0.77523

Figure 11.19

Implementation of Simpson's composite rule in an *Excel* worksheet.

15	12	0.13	0.00171	0.00171	0.09388	0.10360	37.88053	
16	13	0.14	0.00186	0.00185	0.10168	0.11319	37.30520	0.74691
17	14	0.15	0.00200	0.00200	0.10949	0.12295	36.97183	
18	15	0.16	0.00214	0.00214	0.11729	0.13288	36.87372	0.73827
19	16	0.17	0.00229	0.00228	0.12510	0.14298	37.01327	
20	17	0.18	0.00243	0.00242	0.13290	0.15326	37.40239	0.74896
21	18	0.19	0.00257	0.00256	0.14069	0.16373	38.06397	
22	19	0.2	0.00271	0.00271	0.14849	0.17438	39.03464	0.78191
23	20	0.21	0.00286	0.00285	0.15628	0.18523	40.36938	
24	21	0.22	0.00300	0.00299	0.16407	0.19627	42.14934	0.84487
25	22	0.23	0.00314	0.00313	0.17186	0.20753	44.49490	
26	23	0.24	0.00329	0.00327	0.17965	0.21899	47.58853	0.95522
27	24	0.25	0.00343	0.00342	0.18743	0.23066	51.71656	
28							Integral	11.30455

Comments: *Excel*'s auto fill is convenient for implementing Simpson's composite rule directly in an *Excel* worksheet. Extrapolation only gives a slight improvement in this case based on the relatively large number of points. □

Adaptive methods adjust the sizes of local integration intervals to control the error and reduce the number of integrand function evaluations for more efficient computation. We use the error estimate from Richardson's extrapolation in Equation (11.57) to adapt the sizes of intervals in Simpson's 1/3 rule. For a local interval $a \leq x \leq b$, we estimate the error from Simpson 1/3 composite rule across two sub intervals:

$$E = \left| \frac{I_2 - I_1}{15} \right| = \frac{1}{15} \left| \int_a^b f(x)dx - \int_a^{(a+b)/2} f(x)dx - \int_{(a+b)/2}^b f(x)dx \right| \tag{11.63}$$

We stop subdividing intervals when the error in Equation (11.63) is smaller than our prescribed tolerance for precision. Otherwise, we continue subdividing intervals with large error until the sum of errors over all pairs of intervals becomes smaller than our tolerance for error. Burden and Faires (1985) developed an algorithm that efficiently keeps track of integrand function evaluations at the ends of intervals for reuse in applications of Simpson's rule to subdivisions. We improved their algorithm in the *VBA* macro **SIMPSONADAPT** with Richardson's extrapolation for sub levels and local relative error. A user-defined function **SIMP** is available in the *PNMSuite* for adaptive integration. The user-form **INTEGRAL** also includes an option for adaptive Simpson's rule. The worksheet setup requirements for the macro **SIMPSONADAPT**, user-defined function **SIMP**, and user-form **INTEGRAL** are the same as those described previously for the integration macro **SIMPSON**:

- Put the lower and upper limits of integration in different cells on the worksheet.
- Add a formula for the integrand function in an *Excel* worksheet cell. The formula references the cell with the lower limit of integration for the integration variable.
- Specify the relative error tolerance if different from the default value of 10^{-8}.

The macro displays the progress in the **Status Bar** then adds a comment to the result cell.

Example 11.8   *PFR Design Equation with an Energy Balance*

Known: The non-isothermal plug-flow reactor (*PFR*) design equation for a first order liquid phase reaction involves the integral of the rate law in terms of reaction conversion:

$$V = \frac{F_0}{k_0 C_0} \int_0^{X_f} \frac{dX}{\exp(-E_a/R_g T)(1-X)} \tag{11.64}$$

CHAPTER 11: INTEGRATION

where F_0 is the molar feed rate, C_0 is the feed concentration, k_0 and E are the reaction rate constant and activation energy, respectively, and X is the conversion. An energy balance provides the relationship between T and X:

$$T = T_0 - X\Delta H_r / c_p \qquad (11.65)$$

where ΔH_r is the heat of reaction and c_p is the heat capacity. Because Equation (11.64) has no analytical solution, it requires a numerical solution.

Find: Use the *VBA* macro **QYXPLOT** from the *PNMSuite* to graph the integrand function for inspection. We find that the function takes a sharp vertical climb after $x = 0.6$:

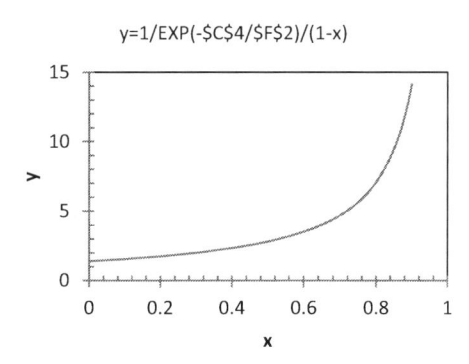

y=1/EXP(-$C$4/$F$2)/(1-x)

Use Simpson's composite rule to evaluate the volume of the *PFR* required for $X_f = 83.4\%$ conversion of A.

Analysis: Add the reactor parameters to an *Excel* worksheet:

	A	B	C	D	E	F
1	Molar Feed Rate	F_0/(mol/s)	100	Heat Capacity	c_p/(J/mol K)	76
2	First-order Rate Constant	k_0/(1/s)	55	Temperature	T/K	298
3	Initial Concentration	C_0/(mol/L)	1	Initial Conversion	X	0
4	Activation Energy	(E/R)/K	103	Final Conversion	X_f	0.834
5	Heat of Reaction	ΔH_r/(kJ/mol)	-12.8		Integrand	1.412890469
6	Initial Temperature	T_0/K	298		Integral Output	2.349479355
7					V/L	4.271780645

Calculate the temperature in cell **F2** and the integrand in cell **F5**, each in terms of the value for conversion in cell **F3**:

F2 = C6 - F3*C5*1000/F1   F5 = 1/EXP(-C4/F2)/(1 - F3)

Use the *VBA* macro **SIMPSON** from the *PNMSuite* to evaluate the integral using 100 intervals. Provide the required information in the input boxes:

Integrand Function = $F$5   Lower Limit = $F$3   Upper Limit = $F$4

Put the result of the integral in cell **E6** and calculate the volume according to Equation (11.64) in cell **E7**.

F7 =$C$1*F6/$C$2/$C$3

Simpson's composite rule approximates the solution of the integral as 2.35. The volume of the *PFR* is then calculated as $V = 4.3$ L.

Comments: If we check the accuracy of the solution by repeating the evaluation with 200 intervals, we find that the solutions with 100 and 200 intervals agree within $10^{-5}\%$. The macro **SIMPSONADAPT** gives a similar result of 2.349479092 for the integral.

□

11.4 Cubic Spline Polynomial Integration

We can use spline-interpolating polynomials to integrate functions of smooth data as described by Equation (10.19). However, like the trapezoidal rule, splines are capable of interpolating unequally spaced data, circumventing limitations imposed by other integration methods, such as Simpson's composite rules presented in Section 11.3. Indeed, splines tend to perform *better* on unequally spaced data. Cubic spline integration, like Simpson's 1/3 rule is based on third-order polynomials, so we expect the accuracy to be comparable to Simpson's 1/3 rule for similar spacing of interpolation points.

Cubic spline integration is not limited to interpolating experimental data. When integrating functions by cubic spline interpolation, we first generate a large number of points from the function, then use cubic spline to integrate the results. To integrate data using our *VBA* user-defined function **CSPLINE**, **KSPLINE**, or **ASPLINE**:

- Set the argument for the interpolation point as the lower limit of integration.
- Use –1 for the first optional argument (intended to indicate an anti-derivative).
- Set the second optional argument as the value of the upper integration limit.
- In the case of **CSPLINE**, include the optional argument to change the interpolation method from "not-a-knot" to natural ("N"), parabolic ("P"), or clamped ("C") end conditions, as desired.
- For clamped conditions, include additional optional arguments for the slopes at the first and last points in either **CSPLINE** or **KSPLINE**.

We may impose integration limits anywhere between the lower and upper extreme points in the data because we are integrating interpolated values. The ability to set integration limits between data points gives the cubic spline an advantage over other methods of integrating experimental data directly, such as the trapezoidal and Simpson's rules in Sections 11.2 and 11.3. Thus, we first identify the upper and lower limits of integration, and their corresponding data intervals. Let the interval that contains the lower integration limit be j and the interval that contains the upper integration limit be k. The solution for the integral over the fraction of the *lower* interval is:

$$I_\ell = \int_{x_\ell}^{x_{j+1}} y(x)\,dx = A_j\left[h_j - (x_\ell - x_j)\right] + \frac{B_j}{2}\left[h_j^2 - (x_\ell - x_j)^2\right] + \frac{C_j}{3}\left[h_j^3 - (x_\ell - x_j)^3\right] + \frac{D_j}{4}\left[h_j^4 - (x_\ell - x_j)^4\right] \quad (11.66)$$

The integral of the cubic spline polynomials for the *interior* intervals, $j < i < k$, is:

$$I_i = \int_{x_{j+1}}^{x_k} y(x)\,dx = \sum_{i=j+1}^{k-1}\left(A_i h_i + \frac{B_i h_i^2}{2} + \frac{C_i h_i^3}{3} + \frac{D_i h_i^4}{4}\right) \quad (11.67)$$

Finally, the integral over the fraction of the *last* interval is:

$$I_u = \int_{x_k}^{x_u} y(x)\,dx = A_k(x_u - x_k) + \frac{B_k}{2}(x_u - x_k)^2 + \frac{C_k}{3}(x_u - x_k)^3 + \frac{D_k}{4}(x_u - x_k)^4 \quad (11.68)$$

The cubic spline approximation for the integral is the sum of the results from Equations (11.66) through (11.68):

$$\int_{x_\ell}^{x_u} y(x)\,dx = I_\ell + I_i + I_u \pm O(\Delta x^4) \quad (11.69)$$

Example 11.9 Conversion with Imperfect Mixing in a Flow Reactor

Known/Find: Repeat Example 11.6 using cubic spline integration.

Analysis: Interpolate the tracer data from Figure 11.16 by the cubic spline method using the *VBA* user-defined function **CSPLINE,** and plot the results as shown in Figure 11.20.

Use the cubic spline integration option in the *VBA* user-defined function **CSPLINE** in cell **D3** of the *Excel* worksheet shown in Figure 11.21. The argument for the interpolation point becomes the lower limit of integration. Set the optional argument to -1 (anti-derivative) for integration, and cell **A21** for the upper limit of integration. This result for the average conversion of 60% in cell **D3** is lower than the ideal conversion of 93% using Equation (11.59) in cell **D2** due to imperfect mixing in the reactor.

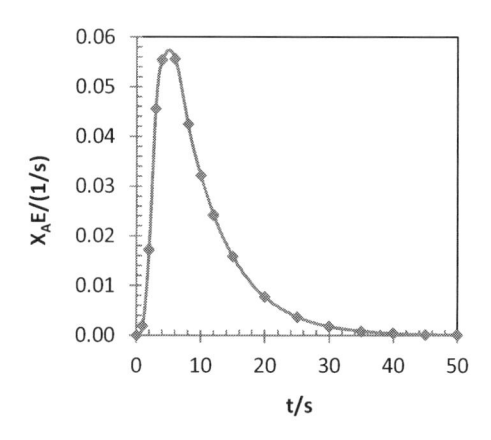

Figure 11.20

Natural cubic spline interpolation of residence time conversion in a flow reactor.

	A	B	C	D
1	V/m³	0.96	k/(1/s)	1.8
2	Q/(m³/s)	0.14	X ideal	=D1*B3/(D1*B3+1)
3	τ/s	=B1/B2	X nonideal	=CSPLINE(D5:D21,A5:A21,A5,-1,A21)
4	t	E/(1/s)	X_A	X_A·E
5	0	0	=$D$1*A5/($D$1*A5+1)	=B5*C5
6	1	0.003	=$D$1*A6/($D$1*A6+1)	=B6*C6
7	2	0.022	=$D$1*A7/($D$1*A7+1)	=B7*C7

	A	B	C	D
1	V/m³	0.96	k/(1/s)	1.8
2	Q/(m³/s)	0.14	X ideal	0.93
3	τ/s	6.857	X nonideal	0.60

Figure 11.21

Excel worksheet setup for solving an integral by cubic spline integration using the *VBA* user-defined function CSPLINE.

Comments: The cubic spline and Simpson results agree. However, the effort required to implement the spline solution is significantly less than the setup requirements for Simpson's rules.

□

11.5 Gauss Quadrature with Error Control

"It is not knowledge, but the act of learning ... which grants the greatest enjoyment." – Carl Friedrich Gauss

Gaussian quadrature approximates an integral as a weighted sum of *n* integrand function values constructed from orthogonal Legendre polynomials that interpolate the integrand function in the normalized domain of the integral. We calculate the weights, *w*, at each of the *n* nodes from the zeroes of orthogonal Legendre polynomials. Gauss quadrature is exact for integrand functions consisting of polynomials with degree $2n - 1$, and works well for most continuous integrand functions. Different integration limits require a change to the normalized integral:

$$I = \int_a^b f(x)\,dx \cong \frac{\Delta x}{2}\sum_{k=1}^n w_k f\left(\bar{x} + \frac{\Delta x}{2}\xi_k\right) \int_{-1}^{1} f(x)\,dx \cong \sum_{i=1}^n w_i f(x_i) \qquad (11.70)$$

where $\bar{x} = \dfrac{a+b}{2}$ and $\Delta x = b - a$

Table 11.1 lists the symmetric nodes, ξ, and corresponding weights, w, for one to five node approximations. Abramowitz and Stegun (1970) have nodes and weights for higher order Gauss-Legendre rules.

Table 11.1 Gauss-Legendre Quadrature abscissa nodes and weights.

n	ξ_k	w_k
1	0	2
2	$-\sqrt{1/3}$, $\sqrt{1/3}$	1, 1
3	$-\sqrt{3/5}$, 0, $\sqrt{3/5}$	5/9, 8/9, 5/9
4	$-\sqrt{\frac{3}{7}+\frac{2}{7}\sqrt{\frac{6}{5}}}$, $-\sqrt{\frac{3}{7}-\frac{2}{7}\sqrt{\frac{6}{5}}}$, $\sqrt{\frac{3}{7}-\frac{2}{7}\sqrt{\frac{6}{5}}}$, $\sqrt{\frac{3}{7}+\frac{2}{7}\sqrt{\frac{6}{5}}}$	$\dfrac{18-\sqrt{30}}{36}$, $\dfrac{18+\sqrt{30}}{36}$, $\dfrac{18+\sqrt{30}}{36}$, $\dfrac{18-\sqrt{30}}{36}$
5	$-\dfrac{1}{3}\sqrt{5+2\sqrt{\dfrac{10}{7}}}$, $-\dfrac{1}{3}\sqrt{5-2\sqrt{\dfrac{10}{7}}}$, 0, $\dfrac{1}{3}\sqrt{5-2\sqrt{\dfrac{10}{7}}}$, $\dfrac{1}{3}\sqrt{5+2\sqrt{\dfrac{10}{7}}}$	$\dfrac{322-13\sqrt{70}}{900}$, $\dfrac{322+13\sqrt{70}}{900}$, $\dfrac{128}{225}$, $\dfrac{322+13\sqrt{70}}{900}$, $\dfrac{322-13\sqrt{70}}{900}$

The midpoint rule, introduced in Section 11.2, is an example of 1-point Gaussian quadrature. Compare the midpoint rule to the following 1-point Gauss-Legendre rule:

$$\int_a^b f(x)\,dx \cong \Delta x f(\bar{x}) \qquad (11.71)$$

Compare the two- and three-point Gauss-Legendre rules to the trapezoidal and Simpson's 1/3 rules (note that Δx for the trapezoidal and Simpsons rules is the spacing between adjacent points, not the distance between the integration limits in the case of Gauss quadrature).

$$\int_a^b f(x)\,dx \cong \frac{\Delta x}{2}\left[f\left(\bar{x} - \frac{\Delta x}{2}\sqrt{\frac{1}{3}}\right) + f\left(\bar{x} + \frac{\Delta x}{2}\sqrt{\frac{1}{3}}\right)\right] \qquad (11.72)$$

$$\int_a^b f(x)\,dx \cong \frac{\Delta x}{18}\left[5f\left(\bar{x} - \frac{\Delta x}{2}\sqrt{\frac{3}{5}}\right) + 8f(\bar{x}) + 5f\left(\bar{x} + \frac{\Delta x}{2}\sqrt{\frac{3}{5}}\right)\right] \qquad (11.73)$$

If we only have resources to collect a few data points, a three or four-point Gauss-Legendre integration of carefully selected values for the variable of integration usually outperforms one application of Simpson's 1/3 or 3/8 rules, respectively. As with Simpson's composite rule, we can divide the range of integration into several sub intervals, apply the 3-point Gauss rules to each sub interval, and then sum the interval results.

The *PNMSuite* has a simple macro **GAUSSLEGENDRE** for a 10-point implementation. The worksheet setup is the same as required for **SIMPSON** and other macros for numerical quadrature. Gauss quadrature has advantages and disadvantages:

Advantages
- Remarkably efficient at finding accurate solutions to most integrals
- Simple algorithm for programming in VBA
- Does not include function evaluations at the limits of integration, so works with improper integrals

CHAPTER 11: INTEGRATION

Disadvantages
- Provides no estimate of error.
- Limited to integrating functions. To integrate data by Gauss quadrature, the data must be collected at appropriate nodes, such as those in Table 11.1. Otherwise, we must first obtain a good interpolating function, such as a least squares fit of data to a rational function or cubic spline.
- Requires smoother integrand functions for higher order Gauss rules (Therefore, most quadrature methods use fewer than 30 nodes. We get around this issue with an adaptive procedure that applies the Gauss quadrature method to several subintervals).

Kronrod (1965) found a set of $n = 15$ Gauss-Legendre nodes for Equation (11.70) that give two approximations to an integral with different order rules, which allows for error approximation. A simple macro **KRONROD** is also available in the *PNMSuite* to demonstrate the method. The macro requires the same *Excel* worksheet setup as **SIMPSON** for single integrals.

Both Kronrod (K) and Gauss (G) integral approximations use Equation (11.70). The absolute value of the difference between the integrals provides an estimate of the integration error:

$$E = \left(200|K - G|\right)^{1.5} \tag{11.74}$$

We adaptively use the error estimate from Kronrod's version of Gauss quadrature to control the error in our numerical approximation for an integral. For large error, apply the Gauss-Kronrod rules to sub intervals of the integration domain. Two sub intervals give the following approximation formula:

$$\int_a^b f(x)dx = \int_a^{\bar{x}} f(x)dx + \int_{\bar{x}}^b f(x)dx \cong \frac{\Delta x}{4} \sum_{k=1}^n w_k \left[f\left(\frac{\bar{x}}{2} + \frac{\Delta x}{4}\xi_k\right) + f\left(\frac{3\bar{x}}{2} + \frac{\Delta x}{4}\xi_k\right) \right] \tag{11.75}$$

The error estimates for all intervals are also summed. We may add subintervals between the integration limits as necessary to control the desired integration error. Adaptive quadrature methods concentrate the subintervals in regions of larger error until the overall error falls below a specified tolerance, as follows:

1. Specify a tolerance for error in the integral approximation.
2. Apply Kronrod's 15-node and Gauss 7-node approximations to the integral and estimate the error with Equation (11.74). Compare the error estimate with the specified tolerance:

$$E < tol? \tag{11.76}$$

3. Stop if the error estimate is smaller than the specified tolerance. Otherwise, divide the range for the limits of integration in half and repeat step two for each half. Compare the combined errors for each half with the specified error tolerance:

$$E_{a\bar{x}} + E_{\bar{x}b} < tol? \tag{11.77}$$

4. Stop if the combined error from step three is smaller than the specified tolerance. Otherwise, subdivide the interval from step three with the largest error estimate and repeat step two for each new half interval. Compare the combined error estimates for all intervals with the specified error tolerance:

$$\sum E_i < tol? \tag{11.78}$$

5. Repeat step 4 by subdividing the interval with the largest error estimate in half until the combined error estimates for all sub intervals is smaller than the specified tolerance.

Figure 11.22 shows the required user input for the *VBA* macro **GAUSSKRONROD** for automating adaptive numerical quadrature in an *Excel* worksheet. The macro uses a default value of 10^{-8} for relative error. The **Status Bar** displays the estimate for the integration error and number of adaptive intervals during execution. We imbed the lower order rule into the higher order rule, and reuse calculations of the integrand function at common nodes to improve computational efficiency. **GAUSSKRONROD** requires the same worksheet setup as the macro **SIMPSON**. To use the macro, set up the integral function in one cell in terms of the lower integration limit defined in another cell. Alternatively, run the user-form **INTEGRAL_UsrFrm** and select the default method Gauss Kronrod. A user-defined function

GKAD(f,xlow,xup) based on the template in Figure 3.23 is available from the *PNMSuite* for integrating an *Excel* worksheet *formula* by the adaptive Gauss-Kronrod method directly in a cell on the worksheet.

```
Public Sub GAUSSKRONROD()
' Solves a single integral Sf(x)dx by adaptive Gauss-Kronrod quadrature with error
' estimate.  Define the integrand f in a cell in terms of a cell for the lower
' integration limit for variable x.  The program adaptively adds intervals for
' additional application of the GK method needed to reduce the total error estimate.
'*** Nomenclature ***
Dim ai(1 To 99) As Double ' = interval x lower limit
Dim b As Double ' = x upper limit
Const boxtitle As String = "GAUSS-KRONROD Quadrature" ' = string for input box title
Dim dxi(1 To 99) As Double ' = interval half distance between x limits
Dim ei(1 To 99) As Double ' = percent integration interval error estimate
Dim f1 As Double ' = integrand function evaluated at lower integration limit
Dim fr As Range ' = range for integrand function
Dim i As Long ' = index for new sub interval
Dim j As Long ' = index for interval with largest error
Dim k As Long ' = index for Kronrod polynomial
Dim qi(1 To 99) As Double ' = interval integral value
Dim qr As Range ' = range for quadrature result
Dim sg As Double ' = sum for integral Gauss 7-point
Dim sk As Double ' = sum for integral in x using Kronrod 15-point
Const small As Double = 0.000000000001 ' = small number
Dim tol As Double ' = error convergence tolerance
Dim total_error As Double ' = total integral error estimate
Dim wg(15) As Double ' = Gauss 7 point weights
Dim wk(15) As Double ' = Gauss-Kronrod 15 point weights
Dim x_ave As Double ' = average x
Dim xl As Double ' = lower limit of integration.
Dim xr As Range ' = range for x variable
Dim z(15) As Double ' = nodes for Kronrod quadrature rules
'************************************************************************************
With Application
Set xr = .InputBox(Title:=boxtitle, Prompt:="LOWER x limit:" & vbNewLine & _
                   "(Integrand formula uses this cell for x)", Type:=8)
b = .InputBox(Title:=boxtitle, Type:=1, Prompt:="UPPER x limit:")
Set fr = .InputBox(Title:=boxtitle, Prompt:="INTEGRAND function f(x):" & vbNewLine & _
         "(Integrand function formula uses" & vbNewLine & "lower x limit)", Type:=8)
xl = xr: f1 = fr: xr = b ' test the relationship between function and variable cells
If f1 = fr Then MsgBox "Warning: Integrand formula may not be function of x cell."
Set qr = .InputBox(Title:=boxtitle, Type:=8, Prompt:="Cell for INTEGRAL result:")
tol = 0.0000000001 ' Relative Convergence TOLERANCE
z(1) = -0.991455371120813: z(15) = 0.991455371120813 ' Set the Kronrod nodes
z(2) = -0.949107912342759: z(14) = 0.949107912342759
z(3) = -0.864864423359769: z(13) = 0.864864423359769
z(4) = -0.741531185599394: z(12) = 0.741531185599394
z(5) = -0.586087235467691: z(11) = 0.586087235467691
z(6) = -0.405845151377397: z(10) = 0.405845151377397
z(7) = -0.207784955007898: z(9) = 0.207784955007898
z(8) = 0#
wk(1) = 0.022935322010529:  wg(1) = 0# ' Set the Kronrod and Gauss weights
wk(2) = 0.063092092629979:  wg(2) = 0.12948496616887
wk(3) = 0.10479001032225:   wg(3) = 0#
wk(4) = 0.140653259715525:  wg(4) = 0.279705391489277
wk(5) = 0.169004726639267:  wg(5) = 0#
wk(6) = 0.190350578064785:  wg(6) = 0.381830050505119
wk(7) = 0.204432940075298:  wg(7) = 0#
wk(8) = 0.209482141084728:  wg(8) = 0.417959183673469
wk(9) = 0.204432940075298:  wg(9) = 0#
wk(10) = 0.190350578064785: wg(10) = 0.381830050505119
wk(11) = 0.169004726639267: wg(11) = 0#
wk(12) = 0.140653259715525: wg(12) = 0.279705391489277
wk(13) = 0.10479001032225:  wg(13) = 0#
wk(14) = 0.063092092629979: wg(14) = 0.12948496616887
wk(15) = 0.022935322010529: wg(15) = 0#
ai(1) = xl: dxi(1) = (b - ai(1)) / 2: x_ave = (ai(1) + b) / 2
sk = 0: sg = 0 ' Initialize sum for x integration
```

CHAPTER 11: INTEGRATION

```
    For k = 1 To 15 ' Loop for x variable summation. Calculate x value for current node
        xr = x_ave + z(k) * dxi(1): sk = sk + fr * wk(k): sg = sg + fr * wg(k)
    Next k
    qi(1) = dxi(1) * sk ' Integral approximation
    ei(1) = (200 * Abs(qi(1) - sg * dxi(1)) ^ 1.5) / Abs(qi(1) + small) ' % error
    For i = 2 To 99 ' Adaptive quadrature loop
        total_error = .sum(ei) * 100 ' Check for total error
        If total_error < tol Then Exit For ' Exit when solution converges within precision
        j = .Match(.Max(ei), ei, 0) ' Identify interval with largest error
        b = ai(j) + dxi(j) ' Gauss-Kronrod quadrature for first half of interval
        dxi(j) = dxi(j) / 2: x_ave = (ai(j) + b) / 2
        sk = 0: sg = 0 ' Initialize sum for x integration
        For k = 1 To 15 ' Loop for x variable summation. Calculate x value for node
            xr = x_ave + z(k) * dxi(j): sk = sk + fr * wk(k): sg = sg + fr * wg(k)
        Next k
        qi(j) = dxi(j) * sk ' Integral approximation
        ei(j) = (200 * Abs(qi(j) - sg * dxi(j)) ^ 1.5) / Abs(qi(j) + small) ' % error
        ai(i) = b ' Gauss-Kronrod quadrature for second half of interval
        b = ai(i) + 2 * dxi(j): x_ave = (ai(i) + b) / 2
        sk = 0: sg = 0 ' Initialize sum for x integration
        For k = 1 To 15 ' Loop for x variable summation. Calculate x value for node
            xr = x_ave + z(k) * dxi(i): sk = sk + fr * wk(k): sg = sg + fr * wg(k)
        Next k
        qi(i) = dxi(i) * sk ' Integral approximation
        ei(i) = (200 * Abs(qi(i) - sg * dxi(i)) ^ 1.5) / Abs(qi(i) + small) ' % error
    Next i
    qr = .sum(qi): xr = ai(1) ' Output integral result & reset lower integration limit
    End With
End Sub ' GAUSSKRONROD
```

Figure 11.22 VBA macro GAUSSKRONROD for solving a single integral using adaptive Gaussian quadrature with Kronrod nodes and weights for error estimation and control.

Illustration: Solve for the final still charge, *W*, in *Example 10.11*. Employ the user-defined functions **RATLINTBS** to interpolate the data in columns **A** and **B** for the integrand function with a rational function like Equation (10.80). Employ the user-defined function **GKAD** to perform the integration of Equation (10.79). The lower limit of integration is x = 0.03 and the upper limit is 0.2. The final still charge calculated in cell **D4** is *W* = 75 moles.

	A	B	C	D
1	x_W	$1/(x_D-x_W)$	x_w interpolant	0.03
2	0.245	2.02	Integrand	=RATLINBS(B2:B6,A2:A6,D1)
3	0.1	1.61	Integral	=gkad(D2,D1,0.2)
4	0.04	1.52	W	=100*EXP(-D3)
5	0.025	1.6		
6	0.02	1.72		

Example 11.10 Probability of Vitamin Deactivation after Heat Sterilization

Known: We need to sterilize a heat labile vitamin. The specific vitamin deactivation rate has an Arrhenius temperature function for the rate constant:

$$k/\text{min}^{-1} = 10^4 \exp\left[-\frac{5000}{T/K}\right] \qquad (11.79)$$

The reactor has an initial temperature of 22°C, a linear heat up rate of r_{HU} = 20°C/min, and a linear cool down rate of r_{CD} = 10°C/min.

Find: Calculate the probability of vitamin activity after heat sterilization by holding the reactor contents for 11 minutes at 122°C.

Assumptions: Well-mixed reactor.

Analysis: Calculate the probability of vitamin activity during the holding time at 122°C:

$$P_{HD} = \exp\left[-k(T_{HD})t_{HD}\right]$$

Next, calculate the heat up and cool down times:

$$t_{HU} = \frac{(122-22)°C}{20°C/\text{min}} = 5\,\text{min} \qquad t_{CD} = \frac{(122-22)°C}{10°C/\text{min}} = 10\,\text{min}$$

Then, calculate the probabilities of vitamin activity during the heat-up and cool-down periods by integrating the specific activity accounting for the changes in temperature:

$$P_{HU} = \exp\left[-\int_0^{t_{HU}} k(T_0 + tr_{HU})\,dt\right] \quad \text{and} \quad P_{CD} = \exp\left[-\int_0^{t_{CD}} k(T_S - tr_{CD})\,dt\right]$$

The overall probability of activity is the product of probabilities for heat-up, holding, and cool-down:

$$P = P_{HU} P_H P_{CD}$$

Set up the problem parameters in columns **A** and **B** of an *Excel* worksheet. The *VBA* macro **GAUSSKRONROD** requires a value in an *Excel* worksheet cell for the lower limit of integration. We use cell **B3** for the lower integration limit for both integrals:

	A	B	C	D	E	F
1	T_o/°C =	22				
2	T_s/°C =	122				
3	t_o/min	0		Inegrand	Integral	P
4	t_{HU}/min =	5		-0.000439	-0.0431	0.9578
5	t_H/min =	11		-0.031966	-0.3516	0.7035
6	t_{CD}/min =	10		-0.031966	-0.0862	0.9174
7					Overall	0.6182

Write a *VBA* user-defined function for the specific vitamin activity in Equation (11.79) as a function of temperature in degrees Celsius:

```
Public Function k(T As Double) As Double ' Specific vitamin activity, min^-1
    k = 10000 * Exp(-5000 / (T + 273.15))
End Function
```

The formulas for the integrand and probabilities in columns **D** and **F** are

D4 =-k(B1+B3*20)   D5 =-k(B2)   D6 =-k(B2-10*B3)
F4 =EXP(E4)   F5 =EXP(E5)   F6 =EXP(E6)

Use the *VBA* macro **GAUSSKRONROD** or access the method from the user-form **INTEGRAL** to solve the integrals in cells **E4** and **E6**. For example, the solution to the integral required for heat up uses the following user input on the **INTEGRAL** user-form:

CHAPTER 11: INTEGRATION

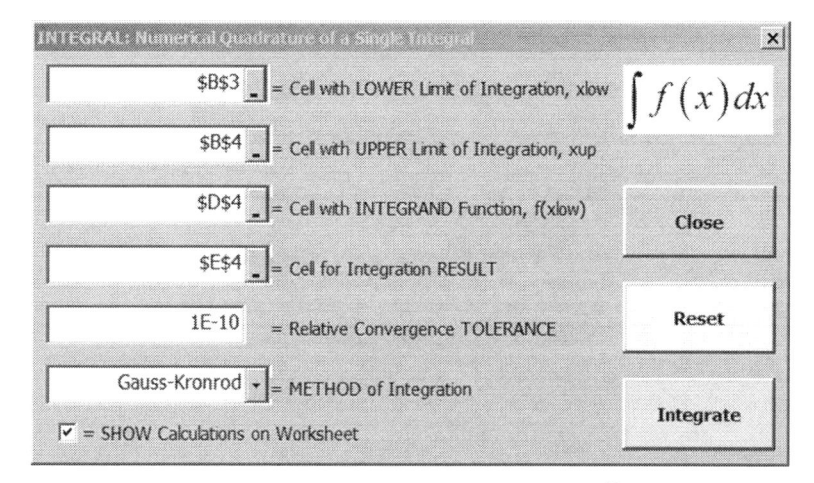

The macro displays the estimate for the integration relative error of $\pm 2 \times 10^{-17}$ in the **Status Bar** at the bottom of the worksheet. Alternatively, evaluate the integrals directly in worksheet cells **E4** and **E6** with the user-defined function
GKAD:

E4 = GKAD(D4, B3, B4)   E6 = GKAD(D6, B5, B6)

In each case, only one Gauss-Kronrod interval was required. The integral for the holding time at a constant sterilization temperature is simply:

E5 =D5*B5

The integral for the cool down period uses the same lower limit of integration, but replaces the upper limit with the cool down time, and the integrand function in cell **C5**. Finally, calculate the overall probability of vitamin activity in cell **F7** as the product of the probabilities for each period of sterilization for approximately 62%:

F7 =F4*F5*F6

Comments: The probability of contamination requires a specific death rate of the microbe as a function of temperature. Any combination of increasing temperature or holding time will reduce the probability of contamination.

□

11.6 Multiple Integration

Graphical analysis of double integrals uses three-dimensional rectangular bars to approximate the volume of the integrand.[101] To illustrate, consider the bar plot of the multivariable function in Equation (7.53). Figure 11.23 compares bar plots for 20 × 20 and 60 × 60 intervals, respectively. The sum of the volumes for the bars gives an approximate solution to the volume integral. Observe how smaller intervals in the two dimensions may give improved accuracy for representing the true volume.

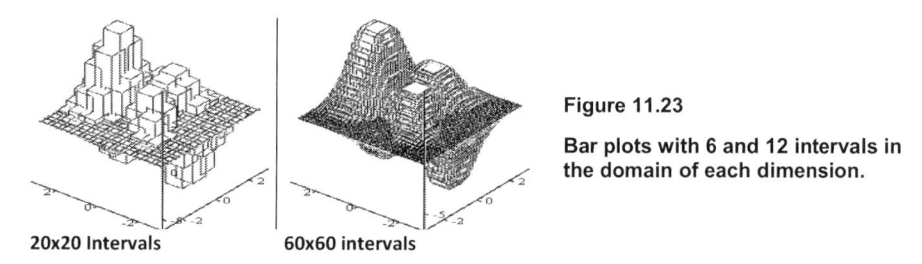

Figure 11.23

Bar plots with 6 and 12 intervals in the domain of each dimension.

20x20 Intervals   60x60 intervals

---

[101] Quadrature refers to the solution of a single integral (area); cubature refers to the solution of a double integral (volume).

To solve multiple integrals numerically, we imbed integration schemes for each dimension into the algorithms of quadrature:

$$I = \int_{y_0}^{y_m} \left(\int_{x_0}^{x_n} f(x,y) dx \right) dy \qquad (11.80)$$

We may approximate solutions to double integrals of the form in Equation (11.80) by nested (imbedded) application of the rules for integration applied to integrals of one dimension. Nesting refers to the solution to Equation (11.80) in the following form:

$$I = \int_{y_0}^{y_m} g(y) dy \qquad (11.81)$$

where

$$g(y) = \int_{x_0(y)}^{x_n(y)} f(x,y) dx$$

For example, we may use the trapezoidal rule to integrate Equation (11.81):

$$I = \Delta y \left\{ \frac{1}{2} [g(y_1) + g(y_{n+1})] + \sum_{i=2}^{n} g(y_i) \right\} \qquad (11.82)$$

where i is the index for the y nodes, and where

$$\Delta y = \frac{y_m - y_0}{m} \qquad \text{and} \qquad y_i = y_{i-1} + \Delta y \qquad (11.83)$$

We imbed the trapezoidal rule for the x-direction within each evaluation of the function $g(y_i)$ in Equation (11.82):

$$g(y_i) = \Delta x_i \left\{ \frac{1}{2} [f(x_1, y_i) + f(x_{n+1}, y_i)] + \sum_{j=2}^{n} f(x_j, y_i) \right\} \qquad (11.84)$$

where j is the index for the x nodes. We calculate the interval size and value of the nodes from the integration limits in the x-direction:

$$\Delta x_i = \frac{x_n(y_i) - x_0(y_i)}{n} \qquad \text{and} \qquad x_j = x_{j-1} + \Delta x_i \qquad (11.85)$$

11.6.1 Multiple Integration by Trapezoidal and Simpson's Rules, and by Cubic Spline Interpolating Polynomials

We first show how to integrate data in two or more dimensions using nested application of the single integral methods and *VBA* macros previously developed in Sections 11.2 to 0. We then automate Simpson's rules for solving a double integral using *VBA*.

Integrating tabulated data by Simpson's rules gives improved precision relative to the trapezoidal method. However, as previously noted, composite Simpson's rules require uniform spacing of the values for the variable of integration. We saw how integration of cubic splines overcomes the uniform-spacing limitation of Simpson's rules to improve the accuracy relative to the trapezoidal method. The idea of imbedded integration applied to the double integral in Equation (11.80) is simple: perform the integration for the x-variable data at each row (or data value) in the y-variable, and then integrate the x integral results in the y-variable. We illustrate the technique for double integration of data by example.

CHAPTER 11: INTEGRATION

Example 11.11 Double Integration of Temperature Data

Known: A thermal imaging camera was used to measure the surface temperatures on a flat, two-dimensional tray dryer used to evaporate water from wet materials at atmospheric pressure.

Find: Calculate the average surface temperature of the drying material.

Schematic: Pictures of the drying surface lying flat on a tray and corresponding thermal image are shown in Figure 11.24. The drying surface is colder than the tray as indicated by the darker tray color on the thermal image.

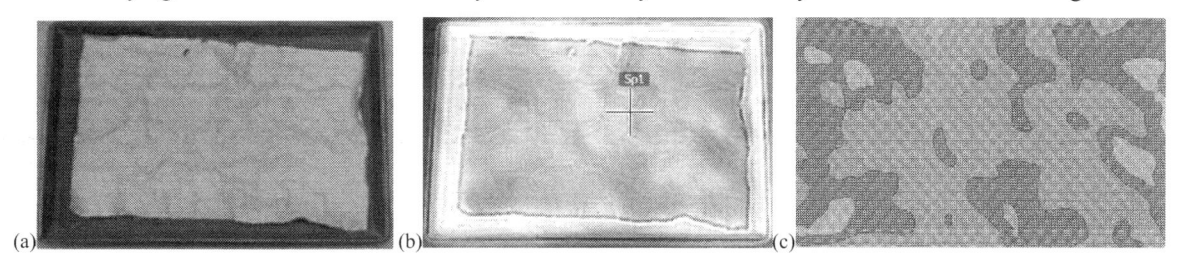

Figure 11.24 (a) Picture of the light colored drying surface on a darker tray and (b) the corresponding thermal image (negative). (c) Contour plot of just the drying surface temperatures.

The drying surface has dimensions 25 cm × 20 cm = 500 cm^2. The colored temperature contour plot shows the temperature isotherms on the drying surface. The temperature scale ranges from 90°C to 98°C. The following surface plot of the two-dimensional temperature profile seen from the long edge reveals the nonlinear nature of the temperature distribution over the drying tray.

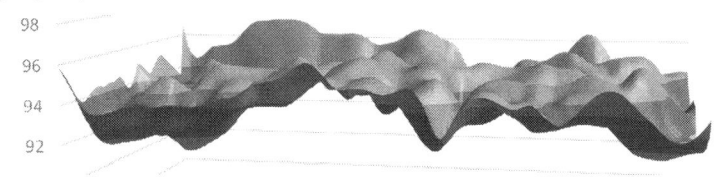

Analysis: We used the user-defined functions **TRAPDATA**, **SIMPSONDATA**, and **CSPLINE** from the *PNMSuite* to perform double integration of the averaging integral for the average temperature as follows:

$$\bar{T} = \frac{1}{A}\int_0^W \int_0^L T(x,y)\,dx\,dy \qquad (11.86)$$

where $A = 500$ cm^2 is the surface area calculated from the product of length and width of the surface.

    x lower limit = 0 x upper limit $L = 25$ cm y lower limit = 0 y upper limit $W = 20$ cm

The surface temperature values are tabulated by x-y coordinate positions in an *Excel* worksheet in a rectangular range of 178 by 104 cells. We added the y-vertical and x-horizontal locations in the left column and top row. The x and y points were calculated by dividing the dimensions by the number of cells in a row for x and a column for y. Due to the large size of the data, only the top left corner of the temperature range is shown in Figure 11.25 with the values of the horizontal x-distance in units of cm in the top row and the values of the vertical y-distance in cm units in the first column.

	A	B	C	D	E	F
1	y/x	0	0.14124	0.28249	0.4237	0.56497
2	0	95.838	95.621	95.382	95.066	94.695
3	0.1942	95.686	95.436	95.153	94.804	94.411
4	0.3883	95.523	95.24	94.935	94.564	94.148
5	0.5825	95.36	95.055	94.728	94.356	93.951

Figure 11.25

Partial view of the *Excel* worksheet containing the temperature data in a xy grid representing the rectangular drying surface.

To integrate the data in the x-dimension, we applied the user-defined functions to the data in each row to create a column of integrals in the y-dimension. The worksheet formulas for the first row are:

```
FX2 = TRAPDATA(B2:FW2, $B$1:$FW$1)
FY2 = SIMPSONDATA(B2:FW2, $B$1:$FW$1)
FZ2 = CSPLINE(B2:FW2, $B$1:$FW$1, 0,-1, 25)
```

Using relative referencing for temperature data in the row and absolute referencing for the x-data in the first row allows us to fill the formulas down a column to give x-integrals at each y position. Now we integrate the new column of x integrals in the y dimension using the user-defined functions:

```
46969.0 = TRAPDATA(FX2:FX105, $A$2:$A$105)
46968.7 = SIMPSONDATA(FY2:FY105, $A$2:$A$105)
46968.7 = CSPLINE(FZ2:FZ105, $A$2:$A$105, 0,-1, 20)
```

All three methods produced similar results for the integral in Equation (11.86) of $46969°C \cdot cm^2$ for an average surface temperature of $93.94°C$ when divided by the surface area of $500 \ cm^2$. Recall that the user-defined functions for the trapezoidal and Simpson's rules use one level of Richardson's extrapolation for improved accuracy.

Comments: The average surface temperature is slightly below the normal boiling point of water due to evaporative cooling.

□

We can automate double integration of a function using *VBA*. Application of Equation (11.55) for Simpson's composite rule to Equation (11.80) gives the following approximation with the x-dimension divided into an even number of n equally spaced intervals:

$$\int_{y_0}^{y_m} \left(\int_{x_0}^{x_n} f(x,y) dx \right) dy \cong \frac{\Delta x}{3} \left\{ \int_{y_0}^{y_m} f(x_0,y) dy + \int_{y_0}^{y_m} \left(\sum_{i=1}^{n-1} \left[3+(-1)^{i-1} \right] f(x_i,y) \right) dy + \int_{y_0}^{y_m} f(x_n,y) dy \right\} \quad (11.87)$$

We then apply Simpson's composite rule to each of the y integrals in Equation (11.87) by dividing the y-dimension into an even number of m equally spaced intervals.

$$\Delta y = \frac{y_m - y_0}{m} \quad (11.88)$$

$$\int_{y_0}^{y_m} f(x_0,y) dy \cong \frac{\Delta y}{3} \left\{ f(x_0,y_0) + \sum_{j=1}^{m-1} \left[3+(-1)^{j-1} \right] f(x_0,y_j) + f(x_0,y_m) \right\} \quad (11.89)$$

$$\int_{y_0}^{y_m} \left(\sum_{i=1}^{n-1} \left[3+(-1)^{i-1} \right] f(x_i,y) \right) dy \cong \frac{\Delta y}{3} \sum_{i=1}^{n-1} \left[3+(-1)^{i-1} \right] \left\{ f(x_i,y_0) + \sum_{j=1}^{m-1} \left[3+(-1)^{j-1} \right] f(x_i,y_j) + f(x_i,y_m) \right\} \quad (11.90)$$

$$\int_{y_0}^{y_m} f(x_n,y) dy \cong \frac{\Delta y}{3} \left\{ f(x_n,y_0) + \sum_{j=1}^{m-1} \left[3+(-1)^{j-1} \right] f(x_n,y_j) + f(x_n,y_m) \right\} \quad (11.91)$$

where
$$y_j = y_0 + j \Delta y$$

The result is four summations. We may extend the same procedure outlined above to triple, or higher dimensional integrals. Figure 11.26 shows the required user input for the *VBA* macro **SIMPSON2D** for solving a double integral. To use the macro:

- Create the double integrand function in a cell on an *Excel* worksheet in terms of the x and y variables in two additional cells.
- Only the x limits of integration may be functions of the cell containing the y variable, not vice versa. For complex geometries, consider using an interpolating scheme, such as cubic spline, for the integration limit functions of $x(y)$ in Equation (11.81).

CHAPTER 11: INTEGRATION

- The macro prompts for the locations of the variable x and y integration limits and the even number of intervals (with the default set to 100) in each of the x and y dimensions.

```
Public Sub SIMPSON2D()
' Numerical solution to a double integral by Simpson's 1/3 composite rule.
' User defines the integrand in a cell in terms of x and y variable cells.
' Only the x-variable limits of integration may be defined in terms of the y-variable.
' *** Nomenclature ***
Dim bxtl As String ' = title used in input boxes
Dim dx As Double, dy As Double ' = width of interval in x and y-dimensions
Dim fr As Range ' = range for integrand function
Dim i As Long, j As Long ' = index for applications of Simpson's rule
Dim nx As Long, ny As Long ' = number of x-dimension and y-dimension intervals
Dim sx As Double ' = x-dimension integral
Dim qr As Range ' = location of output for result to the worksheet
Dim s As Double ' = sum of interval values
Dim xr As Range, yr As Range ' = range for x- and y-variables of integration
Dim xlowr As Range, xupr As Range ' = range for x-dimension lower/upper limits
Dim ylow As Double, yup As Double ' = y-dimension lower/upper limits of integration
'****************************************************************************
bxtl = "SIMPSON2D"
With Application
Set fr = .InputBox(Title:=bxtl, Type:=8, prompt:="INTEGRAND function f(x,y):")
Set xr = .InputBox(Title:=bxtl, Type:=8, prompt:="x-VARIABLE:")
Set xlowr = .InputBox(Title:=bxtl, Type:=8, _
                     prompt:="Cell with LOWER x(y) integration limit:")
Set xupr = .InputBox(Title:=bxtl, Type:=8, _
                     prompt:="Cell with UPPER x(y) integration limit:")
nx = .InputBox(Title:=bxtl, Type:=1, Default:=10, _
     prompt:="Number of x INTERVALS:" & vbNewLine & "(Requires even integer value):")
Set yr = .InputBox(Title:=bxtl, Type:=8, prompt:="y-VARIABLE:")
ylow = .InputBox(Title:=bxtl, Type:=1, prompt:="LOWER y limit of integration:")
yup = .InputBox(Title:=bxtl, Type:=1, prompt:="UPPER y limit of integration:")
ny = .InputBox(Title:=bxtl, Type:=1, Default:=10, _
     prompt:="Number of y INTERVALS:" & vbNewLine & "Requires even integer value):")
Set qr = .InputBox(Title:=bxtl, Type:=8, prompt:="Integral Result OUTPUT cell:")
Set qr = Range(qr(1, 1).Address) ' first OUTPUT cell
nx = .Even(CLng(Abs(.Max(2, nx)))) ' must be at least two intervals and even
ny = .Even(CLng(Abs(.Max(2, ny)))): dy = (yup - ylow) / ny ' integration interval size
yr = ylow
GoSub q_13
s = sx ' Integrate x at ylow ' integrate x variable at ylow
For i = 1 To ny - 1 ' Integrate the y variable at intermediate x
    yr = ylow + i * dy
    GoSub q_13
    s = s + sx * (3 + (-1) ^ (i - 1))
Next i
yr = yup
GoSub q_13
qr = (dy / 3) * (s + sx) ' Integrate y variable at xup
End With ' Application

' ****************************************************************************
q_13: ' ingegrate in the x(y) dimension by Simpson's 1/3 rule
dx = (xupr - xlowr) / nx: xr = xlowr: sx = fr ' Lower integration limit
For j = 1 To nx - 1
    xr = xlowr + j * dx
    sx = sx + fr * (3 + (-1) ^ (j - 1))
Next j
xr = xupr
sx = dx * (sx + fr) / 3 ' Upper integration limit
Return ' q_13
' ****************************************************************************
End Sub ' SIMPSON2D
```

Figure 11.26 *Excel VBA* macro SIMPSON2D for double integration by Simpson's 1/3 composite rule.

Example 11.12 Radiation View Factors

Known/Find: Calculate the view factor for radiation between a small surface, A_1 and a much larger disk of finite area, A_2.

Schematic: A distance L separates two parallel disks. The schematic shows the geometry of a ray of radiation:

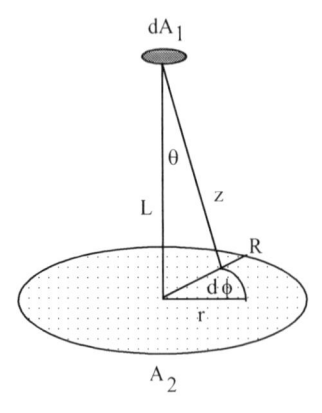

The following double integral defines the view factor in this case:

$$F_{12} = \frac{1}{\pi} \int_0^R \int_0^{2\pi} \left(\frac{r \cos^2 \theta}{z^2} \right) d\phi dr \qquad (11.92)$$

where $\qquad z^2 = L^2 + r^2 \qquad$ and $\qquad \cos \theta = \frac{L}{z}$

The exact solution of this view factor given $L = 1$ and $R = 0.5$ is

$$F_{12} = \frac{R^2}{L^2 + R^2} = 0.2 \qquad (11.93)$$

We get an approximate solution by integrating Equation (11.92) using the *VBA* macro **SIMPSON2D**. Set up an *Excel* worksheet using named cells for the input required by the macro.

	A	B	C	D
1	r	0.5	L	1
2	φ	6.28318530717959	z	=SQRT(L^2+radius^2)
3	f	=radius*(D3/D2)^2	Cos(θ)	=L/D2
4	r_low	0	φ_low	0
5	r_up	0.5	φ_up	=2*PI()
6	I	0.628322201862796	F_12	=B6/PI()

Run the macro **SIMPSON2D** from the *PNMSuite* and provide the information requested in the input boxes:

Function	= $B$3	Output cell	= $B$6
x-variable	= $B$1	y-variable	= $B$2
x-lower integration limit	= $B$4	y-lower integration limit	= $D$4
x-upper integration limit	= $B$5	y-upper integration limit	= $D$5

Cell **B6** has the Simpson's result for 10 r-dimension intervals and 10 ϕ-dimension intervals. The value of the view factor is 0.20000117 in cell **D6** (Note that the extra significant figures were retained for comparison only).

CHAPTER 11: INTEGRATION   525

	A	B	C	D
1	r	0.5	L	1
2	φ	6.283185	z	1.11803399
3	f	0.32	Cos(θ)	0.89442719
4	r_low	0	φ_low	0
5	r_up	0.5	φ_up	6.28318531
6	I	0.628322	F_12	0.20000117

Comments: The results agree with the analytical solution. Though not necessary for this example, more intervals produce higher degrees of precision in the result.

□

11.6.2 Multiple Integration by Gauss-Kronrod Quadrature*

Gauss quadrature has a significant advantage of simplicity when extending the method to multiple integrals. For example, an approximation to a double integral uses nested summations, as follows:

$$\int_c^d \int_{a(y)}^{b(y)} f(x,y) dx dy \cong \frac{\Delta y}{2} \sum_{j=1}^n \left[w_j \frac{\Delta x(y)}{2} \sum_{k=1}^n w_k f(x_k, y_j) \right] \quad (11.94)$$

where
$$\Delta x(y) = b(y) - a(y) \qquad \Delta y = d - c$$
$$x_k = \frac{b(y) + a(y)}{2} + \frac{\Delta x(y)}{2} \xi_k \qquad y_i = \frac{d+c}{2} + \frac{\Delta y}{2} \xi_j$$

We may expand this method to higher order integrals following the pattern of double integration. A *VBA* macro **KRONROD2D** is available for solving double integrals by the method of Kronrod quadrature with error estimates displayed the in the *Status Bar* at the bottom of the worksheet. We simply added a second loop for the *y*-dimension to the **KRONROD** macro. The required user input shown in Figure 11.27 is the same as for the macro **SIMPSON2D** (See Figure 11.22 for the nodes and weights). **KRONROD2D** does not include adaptive interval halving to control the error. However, a macro **GAUSSKRONRON2D** based on the **GAUSSKRONROD** sub procedure described in Section 11.5 is available in the *PNMSuite* for double integration with error control. **GAUSSKRONROND2D** requires the same worksheet setup and inputs as the macro **SIMPSON2D**, with one difference – the input boxes for the numbers of intervals in the two variables of integration are replaced with a single input box for the integration error tolerance. **GAUSS-KRONROD2D** requires the sub procedure **GK_XAD**, which performs adaptive integration with error control in the *x*-dimension, at each point in the *y*-dimension.

```
Public Sub KRONROD2D()
' Solves a double integral Sf(x,y)dxdy by Gauss-Kronrod quadrature. Define the
' integrand in a cell as a function of x and y in separate cells.  Put the limits of
' integration in their own cells.  Only the x variable integration limits may be
' functions of the y variable cell.
'*** Nomenclature ***
Dim ar As Range, br As Range ' = range for x lower limit
Dim boxtitle As String ' = string for input box title
Dim c As Double, d As Double ' = range for y lower and upper limits
Dim dx As Double, dy As Double ' = half distance between x and y limits
Dim er As Double ' = percent integration error estimate
Dim fr As Range ' = range for integrand function
Dim j As Long, k As Long ' = index for y and x summations
Dim qr As Range ' = range for quadrature result
Dim small As Double ' = small number
Dim sx As Double ' = sum for integral in x
Dim sxg As Double ' = sum for integral Gauss 7-point
Dim sy As Double ' = sum for integral in y
```

526   PRACTICAL NUMERICAL METHODS

```
Dim syg As Double ' = sum for integral Gauss 7-point
Dim wg(15) As Double ' = Gauss 7 point weights
Dim wk(15) As Double ' = Kronrod 15 point weights
Dim x_ave As Double, yave As Double ' = average x and y
Dim xr As Range, yr As Range ' = range for x and y variables
Dim z(15) As Double ' = nodes for Kronrod quadrature rules
'*****************************************************************************
small = 0.000000000001: boxtitle = "GAUSS-KRONROD Double Integration"
With Application
Set fr = .InputBox(Title:=boxtitle, Type:=8, Prompt:="INTEGRAND function:")
Set xr = .InputBox(Title:=boxtitle, Type:=8, Prompt:="X VARIABLE:")
Set ar = .InputBox(Title:=boxtitle, Type:=8, Prompt:="LOWER X(Y) integration limit:")
Set br = .InputBox(Title:=boxtitle, Type:=8, Prompt:="UPPER X(Y) integration limit:")
Set yr = .InputBox(Title:=boxtitle, Type:=8, Prompt:="Y VARIABLE:")
c = .InputBox(Title:=boxtitle, Type:=1, Prompt:="LOWER Y integration limit:")
d = .InputBox(Title:=boxtitle, Type:=1, Prompt:="UPPER Y integration limit:")
Set qr = .InputBox(Title:=boxtitle, Type:=8, Prompt:="DOUBLE INTEGRAL result cell:")
End With ' Application
' Half distance between y integration limits & Average y.  Initialize for y integral
dy = (d - c) / 2: yave = (c + d) / 2: sy = 0: syg = 0
For j = 1 To 15 ' Loop for y variable summation
    yr = yave + z(j) * dy: dx = (br - ar) / 2
    sx = 0: sxg = 0: x_ave = (ar + br) / 2 ' Initialize sum for x integration & Average x
    For k = 1 To 15 ' Loop for x variable summation
        xr = x_ave + z(k) * dx: sx = sx + fr * wk(k): sxg = sxg + fr * wg(k) ' x terms
    Next k
    sy = sy + sx * dx * wk(j): syg = syg + sxg * dx * wg(j): Next j ' y terms
' Output double integral result and error estimate
qr = sy * dy: er = (200# * Abs(qr - syg * dy) ^ 1.5) / Abs(qr + small)
Application.StatusBar = boxtitle & " error = " & VBA.Format(er, "#E+##%")
End Sub ' KRONROD2D
```

Figure 11.27 *VBA* sub procedure **KRONROD2D** for Gauss-Kronrod double integration of a worksheet integrand formula. See Figure 11.22 for nodes and weights.

Example 11.13 Double Integration by Gauss-Kronrod Quadrature

Known/Find: Solve the following double integral by the method of Gauss-Kronrod quadrature:

$$I = \int_0^1 \int_0^{x^2} e^{y/x} dy\, dx \qquad (11.95)$$

Analysis: Transform the problem to match the formula in Equation (11.94). Note that the variables *x* and *y* are interchanged when compared to Equation (11.95):

$$I = \int_0^1 \int_0^{y^2} e^{x/y} dx\, dy \qquad (11.96)$$

Set up the integration limits, variables, and integrand function in an *Excel* worksheet:

	A	B	C	D	E	F	G	H
1	x=	0	a =	0		c =	0	f = =EXP(B1/B2)
2	y=	0	b =	=B2^2		d =	1	I =

Notice that the integrand function in cell **H1** is indeterminate at the lower integration limits. However, this does not pose a problem because the Gauss-Kronrod method does not evaluate the integrand function at the integration limits

	A	B	C	D	E	F	G	H
1	x=	0	a =	0		c =	0	f = #DIV/0!
2	y=	0	b =	0		d =	1	I =

Run the *VBA* macro **KRONROD2D** and respond to the input boxes as follows:

CHAPTER 11: INTEGRATION

x-variable	= $B$1	y-variable	= $B$2
x-lower integration limit	= $D$1	y-lower integration limit	= $F$1
x-upper integration limit	= $D$2	y-upper integration limit	= $F$2
Function	= $H$1	Output cell	= $H$2

Specify where **KRONROD2D** writes the solution for the double integral in cell **H2**. The macro displays the estimate of the relative error as 9×10^{-18} in the **Status Bar**.

	A	B	C	D	E	F	G	H
1	x=	0.987237737	a =		0	c =	0	f =	2.695203263
2	y=	0.995727686	b =	0.991473624	d =	1	I =	0.5

Comments: Kronrod's method gives high accuracy for relatively little computational effort, and is easy to modify for higher order multiple integrals. The macro **GAUSSKRONROD2D** with error control returned the same result.

☐

11.7 Monte Carlo Integration*

Monte Carlo methods of analysis employ large sets of random values of the independent variable(s) to observe the corresponding behavior of the dependent variable(s). As discussed in Chapter 8 on uncertainty analysis, a large number of observations is required to sharpen the precision in the Monte Carlo results. The Monte Carlo approach is our option for difficult integrals that are not solvable by the other methods presented in this chapter. We should consider Monte Carlo integration for complicated geometries in multiple dimensions and functions with discontinuities between the limits of integration.

The "dart board" method is a simple example of the application of the Monte Carlo method to integration. With the "dart board" method, we prescribe an area that completely encloses the integrand. We then generate a random set of data pairs for the enclosure space. The fraction of data that fall within the integrand is proportional to the value of the integral. For example, consider a circle with area $A_{circle} = \pi D^2/4$ in a square with area $A_{square} = D^2$, as shown in Figure 11.28. Imagine that darts randomly strike the area of the square, including the area inside the circle. The markers represent the location of each strike. The ratio of the area of the circle to area of the square is equal to the ratio of the number of hits within the circle (n_{in}) to the total number of strikes anywhere within the square enclosure ($n_{in} + n_{out}$):

$$\frac{A_{circle}}{A_{square}} \cong \left(\frac{n_{in}}{n_{in} + n_{out}}\right) \qquad (11.97)$$

The "dart board" method becomes more efficient as the area of the integrand approaches the area of the enclosure, i.e., $A_{circle}/A_{square} \to 1$. The general difficulty here is finding the best enclosure space for the integrand function.

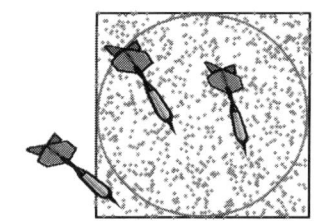

Figure 11.28

"Dart Board" analogy for simple Monte Carlo integration. Each dot represents a random dart, or function evaluation, inside or outside of the integral.

A better approach to Monte Carlo integration transforms the integral from a deterministic problem into a stochastic[102] problem. Consider an arbitrary definite integral with a continuous integrand function: $\int_{x_0}^{x_1} f(x) dx$.

---

[102] *Stochastic*, from the Greek "Στόχος" which means "aim, guess", or characterized by randomness.

A simple probability density function for the integrand in terms of the integration limits is:

$$g(x) = \begin{cases} \dfrac{1}{x_1 - x_0}, & x_0 \leq x \leq x_1 \\ 0, & \text{otherwise} \end{cases} \quad (11.98)$$

where

$$\int_{x_0}^{x_1} g(x)\,dx = 1$$

Integrate the product of the integrand and probability density function to give the average value of the integrand:

$$\bar{f} = \int_{x_0}^{x_1} f(x) g(x)\,dx \cong \frac{1}{n}\sum_{i=1}^{n} f(x_i) \quad x_0 \leq x_i \leq x_1 \quad (11.99)$$

The expected value of the integrand is simply the average function evaluated at a large number of random, uniformly distributed values of the independent variables in the domain of the integration limits, $x_0 \leq x_i \leq x_1$:

$$\bar{I} \cong \int_{x_0}^{x_1} \bar{f}\,dx = (x_1 - x_0)\bar{f} = \frac{(x_1 - x_0)}{n}\sum_{i=1}^{n} f(x_i) \quad (11.100)$$

This approach has the advantage of only evaluating the integrand between the integration limits, and thus arrives at a solution faster than the "dart board" method.

The 95% confidence interval for a Monte Carlo integration result gives an estimate of the expanded uncertainty in the calculation, as follows:

$$U_I = \pm \frac{1.96(x_1 - x_0)}{\sqrt{n}}\sqrt{\frac{1}{n}\left[\sum_{i=1}^{n} f(x_i)^2\right] - \bar{f}^2} \quad (11.101)$$

We may easily extend Monte Carlo integration to multiple integrals where it has a real advantage for complicated integrand functions. The Monte Carlo approach to solving a double integral includes the differences between integration limits in the product with the sum of function evaluations:

$$\int_{y_0}^{y_1}\int_{x_0}^{x_1} f(x,y)\,dxdy \cong \bar{I} = (x_1 - x_0)(y_1 - y_0)\bar{f} \pm U_{\bar{I}} \quad (11.102)$$

where

$$\bar{f} = \frac{1}{n}\sum_{i=1}^{n} f(x_i, y_i) \qquad x_0 \leq x_i \leq x_1 \qquad y_0 \leq y_i \leq y_1$$

and

$$U_I = \pm \frac{1.96(x_1 - x_0)(y_1 - y_0)}{\sqrt{n}}\sqrt{\frac{1}{n}\left[\sum_{i=1}^{n} f(x_i, y_i)^2\right] - \bar{f}^2}$$

We evaluate the function for a large number of random, uniformly distributed values for $x_0 \leq x_i \leq x_1$ and $y_0 \leq y \leq y_1$. In some problems, the integration limits for one variable may be functions of the other variable. In this case, the difference between the integration limits for the second dimension are placed inside the summation, and the Monte Carlo integral becomes

$$\int_{y_0}^{y_1}\int_{x_0(y)}^{x_1(y)} f(x,y)\,dxdy \cong \bar{I} = \frac{(y_1 - y_0)}{n}\sum_{i=1}^{n}\left[(x_1 - x_0)f(x_i, y_i)\right] \pm U_{\bar{I}} \quad (11.103)$$

CHAPTER 11: INTEGRATION

with uncertainty:
$$U_{\bar{I}} = \pm \frac{1.96}{\sqrt{n}} \sqrt{\frac{(y_1 - y_0)^2}{n} \sum_{i=1}^{n} \left[(x_1 - x_0) f(x_i, y_i)\right]^2 - \bar{I}^2}$$

The *VBA* macro **MC2INT**, with required user input listed in Figure 11.29, performs Monte Carlo double integration and allows the *y* integration variable to be a function of the *x* integration variable for the case in Equation (11.103). To use the macro, formulate the integrand function in a cell on the worksheet in terms of all cells containing values for the variables. Define the limits of integration in separate cells. The macro prompts the user for the ranges of inputs. The code listed in the *PNMSuite* validates the location selected for the output to avoid overwriting cells on the worksheet.

```
Public Sub MC2INT()
' Approximate solution to double integral by Monte Carlo method.  This version permits
' the y integration variable to be function of x integration variable. Option to use the
' Wichmann-Hill algorithm for generating pseudo random numbers. Define formula for the
' integrand in terms of cells with the integration limits for each variable, x and y.
'*** NOMENCLATURE ***
Dim boxtitle As String ' = string for title in input boxes
Dim dx As Double, dy As Double ' = range of xr and yr
Dim fe As Double ' = expected function value
Dim fr As Range ' = range for integrand function cell
Dim I As Double ' = real integral value
Dim j As Long ' = integer index for Monte Carlo loop
Dim n As Long ' = integer number of Monte Carlo simulations
Dim pc As Double ' = percent calculated for monitoring progress
Dim sc As Integer ' = variable to check for show calculations
Dim u As Double ' = uncertainty in integral value
Dim xr As Range ' = range for integration variable cell
Dim xlowr As Range, xupr As Range ' = ranges for lower/upper integration x limit cells
Dim yr As Range ' = second integration variable (may be function of xr) cell
Dim ylow As Double, yup As Double ' = lower/upper limit of integration
'*****************************************************************************
boxtitle = "MC2INT Monte Carlo Double Integration"
With Application
n = .InputBox(Title:=boxtitle, Type:=1, _
        Default:=10000, Prompt:="NUMBER of MC simulations:")
n = .Min(n, 10 ^ 35) ' *** for WH PRNG
Set fr = .InputBox(Title:=boxtitle, Type:=8, Prompt:="Integrand FUNCTION, f(x):")
Set xr = .InputBox(Title:=boxtitle, Type:=8, Prompt:="x-variable:")
Set xlowr = .InputBox(Title:=boxtitle, Type:=8, Prompt:="LOWER x(y) limit:")
Set xupr = .InputBox(Title:=boxtitle, Type:=8, Prompt:="UPPER x(y) limit:")
Set yr = .InputBox(Title:=boxtitle, Type:=8, Prompt:="y-variable:")
ylow = .InputBox(Title:=boxtitle, Type:=1, Prompt:="LOWER y-limit:")
yup = .InputBox(Title:=boxtitle, Type:=1, Prompt:="UPPER y-limit:")
' Interval size in xr-dimension, initial Excel PRNG & sum for integral & uncertainty
dy = yup - ylow: I = 0: u = 0: Randomize
For j = 1 To n ' Monte Carlo iterations
    ' Get uniform random values for xr & yr between limits use Wichmann-Hill random
    yr = ylow + RNDWH * dy: dx = xupr - xlowr: xr = xlowr + RNDWH * dx
    ' random expected function value & uncertainty parameter
    fe = dx * fr: u = u + fe ^ 2: I = (I * (j - 1) + fe) / j
    ' Display progress in status bar at bottom of worksheet
    pc = j / n: Application.StatusBar = "MC2INT: " & VBA.Format(pc, "#.#%")
Next j
' Print results for Monte Carlo integral & confidence interval
' (1-alpha)*100%, alpha = 0.05 for 95% confidence
I = dy * I: u = .TInv(0.05, n) * (dy ^ 2 * u / n - I ^ 2) / Sqr(n)
End With ' Application
End Sub ' MC2INT
```

Figure 11.29 *VBA* macro MC2INT for solving double integrals by the Monte Carlo method.

Example 11.14 Double Integration by the Monte Carlo method

Known/Find: Solve the following double integral by the Monte Carlo method using the *VBA* macro **MC2INT** from the *PNMSuite*. Compare the solution using 1000, 10000, and 100000 simulations.

$$I = \int_2^{2.2} \int_{(y)}^{(2y)} \left(y^2 + x^3\right) dx\, dy \qquad (11.104)$$

Analysis: Set up the problem in an *Excel* worksheet.

	A	B	C	D	E	F	G	H
1	y	2		y limits	2	2.2		
2	x	2		x limits	2	4		Simulations
3	f	12		I	16.69136	±	4.1E+00	1000

1. Define the formula for the integrand function in cell **B3** in terms of cells containing the integration variables in the range **B1:B2**:

    B3=B1^2+B2^3

2. Add the integration limits for *x* in cells **E1:F1**.

3. Define the integration limits for *y* in terms of the integration limits for *x*:

    E2=B1   F2=2*B1

4. Run the *VBA* macro **MC2INT** three times, each time changing the default number of Monte Carlo simulations to 1000, 10000, and 100000, respectively. Provide information to the input boxes as requested by clicking on the cells containing the integration variables, integration limits, and integrand function.

5. Direct the output to worksheet cells in the range **E3:E5** for each solution.

	A	B	C	D	E	F	G	H
1	y	2.038188		y limits	2	2.2		
2	x	3.696924		x limits	2.038188	4.076376		Simulations
3	f	54.68097		I	16.69136	±	4.1E+00	1000
4					16.58556	±	1.3E+00	10000
5					16.46385	±	3.9E-01	100000

Comments: The accuracy of the solution as well as computation time increases exponentially with the number of simulations. Compare these results to the true solution accurate to three decimal places of 16.509. Monte Carlo quadrature is simple to implement, but computationally demanding. Monte Carlo should only be used when other methods fail or are inappropriate for a particular integrand function with an unusual geometry. □

11.8 Continuous Least-squares Approximation*

We may use regression to model a complicated equation with a simpler expression. As a matter of convenience in computation, portability, or analysis, we may prefer to work with a simpler model, such as a polynomial or rational function, as an approximation for a much more complicated continuous function. We use the method of continuous least-squares regression to "fit" a simpler model to a complex equation. For continuous least squares, we replace the *summation* of squared residuals over a finite data set in Equation (9.6) with the objective function involving the *integral* of the square of the residual difference between the continuous function $y(x)$ and model equation over the desired limits of application, $a \leq x \leq b$:

CHAPTER 11: INTEGRATION

$$OF = Min\left\{\frac{1}{b-a}\int_a^b \left[y(x) - f_{model}(x)\right]^2 dx\right\} \quad (11.105)$$

We may employ a high-precision adaptive numerical integration method such as Romberg, Simpson, or Gauss-Kronrod to solve the integral in Equation (11.105). We then use an appropriate method of minimization from Chapter 7 to find the model parameters in f_{model} that minimize the objective function of the squared residual integral in Equation (11.105).

For most problems of this nature, we recommend the user-defined function **GKAD** available from the *PNMSuite* in combination with the **Solver** for minimizing Equation (11.105) in an *Excel* worksheet. As a user-defined function, **GKAD** automatically updates the solution for the integral with each change made to the model parameters in the worksheet during minimization. In addition, for the unusual case of an improper integral in Equation (11.105), the Gauss-Kronrod integration method employed by **GKAD** does not evaluate the integrand function at the integration limits. To assess the model fidelity, we evaluate the new model f with the original equation y at a large number of points x between the integration limits a and b and apply the same methods of residual analysis from Section 9.3.5.

Example 11.15 Continuous Least-squares Fit of Molokanov's Equation with a Simple Rational Function

Known: The following Molokanov, et al (1972) equation is popular for estimating the number of stages in Gilliland's correlation for a multicomponent distillation column:

$$Y = 1 - \exp\left[\left(\frac{1+54.4X}{11+117.2X}\right)\left(\frac{X-1}{\sqrt{X}}\right)\right] \quad (11.106)$$

where Y and X give the number of stages and reflux ratio, respectively. The independent variable X ranges from zero to one. Unfortunately, Equation (11.106) is complicated and indeterminate at $X = 0$. Al-ameeri and Said (1985) recommended replacing Molokanov's equation with the following simpler alternative that is continuous over the complete range of X: in terms of a transformed variable, X^m:

$$Y = \frac{1 - X^m}{1 - qX^m} \quad (11.107)$$

Find: Use the method of continuous least-squares to find the parameters m and q in Equation (11.107) to fit the Molokanov Equation (11.106).

Analysis: Substitute for y and y_{model} from Equations (11.106) and (11.107), respectively, into the integral of the objective function in Equation (11.105):

$$OF = Min\int_0^1 \left[Y_{Molokanov}(X) - Y_{Rational}(X)\right]^2 dX \quad (11.108)$$

Integrate the square of the residuals using the adaptive Gauss-Kronrod method in an *Excel* worksheet by the user-defined function **GKAD**. Use the **Solver** to locate values for m and q that minimize the objective function in Equation (11.108). The *Excel* worksheet in Figure 11.30 shows the problem set up and solution by the **Solver** for the following result:

$$Y = \frac{1 - X^{0.11}}{1 - 0.77X^{0.11}} \quad (11.109)$$

	A	B
1	m	0.108609139318945
2	q	0.767515602468436
3	X	0
4	Integrand	=(1-EXP(((1+54.4*B3)/(11+117.2*B3))*((B3-1)/SQRT(B3+0.00000000000001)))-(1-B3^B1)/(1-B2*B3^B1))^2
5	Min	=gkad(B4,B3,1)

	A	B
1	m	0.1085
2	q	0.7678
3	X	0
4	Integrand	0
5	Min	8.93E-05

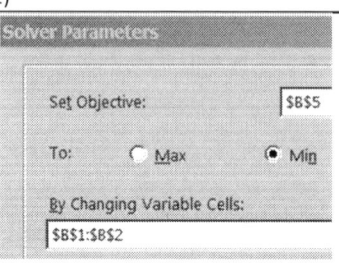

Figure 11.30

Continuous least-squares fit of a rational function to the Molokanov equation using the user-defined function GKAD to integrate the squared residual and the Solver to minimize the objective function.

Comments: The simpler rational equation is continuous over the complete range of *X*, has an explicit analytical inverse for *X* in terms of *Y,* and just two model parameters (compared to four parameters in Equation (11.106).

11.9 Epilogue on Integration

We now have a full menu of integration methods. There is no "one size fits all" best option. In some cases, it is a matter of convenience, and more than one method will do the job. Table 11.2 summarizes the strengths and weaknesses of the variety of methods introduced for numerical quadrature.

Table 11.3 summarizes the macros and user-defined functions for integration, including the methods of input for the integrand function and limits of integration. Consider the following general guidelines when picking an integration method.

The trapezoidal rule is easy to remember and implement on a spreadsheet for a quick solution. However, the trapezoidal rule is low order, and requires small interval sizes between data for good accuracy. In addition to its ease of use, its strength is its ability to handle unequally spaced data.

If we are integrating *equally spaced data*, we should consider Simpson's composite rule, with Richardson's extrapolation if possible. Although not as easy to remember (or derive) as the trapezoidal rule, we get much higher accuracy for little additional computation cost. Simpson's rule is also easy to implement in a spreadsheet. Apply the bisection rule to check the accuracy of a result by doubling the number of intervals and integrate again. When the bisected results stop changing below a prescribed level of precision, we have confidence in our result. When integrating equally spaced data, we may improve the accuracy and estimate the error by integrating every other data point to perform one Richardson's extrapolation of the two results for the integral.

For *unequally spaced data*, our best option is cubic spline integration. We expect the cubic spline integration to have accuracy similar to Simpson's 1/3 rule. We can estimate the accuracy by repeating the spline with a partial set of the data and comparing solutions. Be careful to take the necessary steps to confirm that the cubic splines are properly interpolating the data. Use the constrained cubic spline method when the standard method fails to follow the trend in the data or develops poles between the limits of integration.

Our first choice for integrating continuous integrand *functions* is typically adaptive Gauss-Kronrod quadrature or Simpson rule with error control. However, as good as they are, Legendre polynomials do not necessarily provide the best interpolation for all functions. There are some problems where the adaptive Romberg method may perform better at controlling the integration error. For improper integrals, try the midpoint rule that does not evaluate the integrand function at the limits of integration, or apply variable transformation. Finally, we resort to Monte Carlo integration as a last resort for problems we cannot solve any other way. Fortunately, we have several options at our disposal for tackling a wide spectrum of integral types.

Chapter 11: Integration

Table 11.2 Comparison of methods of quadrature.

Method	Strengths	Weaknesses
Trapezoid	• Easy to remember and implement in an *Excel* worksheet • Works for unequally-spaced data • Best option for integrating noisy data	• Lowest accuracy • Requires accurate derivatives when using endpoint corrections for higher accuracy
Midpoint	• Useful for improper integrals	• Low accuracy • Does not apply easily to data
Romberg	• High accuracy • Estimates error in the integral • Midpoint and transformation options are available for improper integrals	• Does not work with experimental data • Requires a large number of integrand function evaluations
Simpson	• Better accuracy than trapezoidal rule	• Requires equally-spaced values of the integrand (data or function)
Adaptive Simpson	• More efficient than fixed interval methods with an estimate of integration error.	• Does not work with improper integrals without variable transformation
Cubic Spline	• Good for sparse, smooth data sets • Does not require equally-spaced data • Constrained spline method available to prevent over/under shooting or poles.	• No direct estimate of error for solution precision control
Gauss-Legendre Kronrod	• Computationally efficient • Simple extension to multiple integrals • Applies to improper integrals	• Requires functions that are best interpolated by Legendre polynomials.
Adaptive Gauss-Kronrod	• Most efficient method • Adapts to control integration error.	• Requires integrand functions that are approximated by Legendre polynomials
Monte Carlo	• Simple implementation for complex geometries in more than one dimension.	• Lower accuracy with a large computational requirement

Table 11.3 Summary of *VBA* macros, user-forms, and user-defined functions (UDF) for integration, including the methods of input for the integrand function and limits.

Macro/Function	Inputs	Operation

`APLINE`
`CSPLINE`
`KSPLINE`
	UDF arguments	Cubic spline and constrained cubic spline interpolation with endpoint options: free, clamped, parabolic, not-a-knot.

`GAUSSKRONROD` `GAUSSKRONROD2D`	Macro input boxes	Gauss-Kronrod adaptive quadrature of a single integral with error control Guass-Kronrod adaptive quadrature of a double integral with error control.
`GAUSSLEGENDRE`	Macro input boxes	10-point Gauss-Legendre quadrature of a function
`GKAD`	UDF arguments	Adaptive Gauss-Kronrod quadrature applied to an *Excel* worksheet formula
`INTEGRAL_UsrFrm`	User-form	Adaptive Gauss-Kronrod, Adaptive Simpson or Romberg (midpoint or trapezoid, improper trapezoid) integration of a function
`KRONROD` `KRONROD2D`	Macro input boxes	Gauss-Kronrod adaptive method of integrating a function Double integration of a function by the Gauss-Kronrod method
`MPEX`	UDF arguments	Midpoint Romberg integration applied to an *Excel* worksheet formula
`MC2INT`	Macro input boxes	Double integration of a function by the Monte Carlo method
`MIDPOINT`	Macro input boxes	Midpoint rule with uniform spacing of the independent variable

`ROMBERGM`
`ROMBERGT`
`ROMBERGI`
	Macro input boxes	Romberg integration with midpoint (M) and trapezoidal (T) rules. ROMBERGI employs variable transformation for improper integrals.

`SIMP`	UDF arguments	Adaptive Simpson's 1/3 rule with error control.
`SIMPSON` `SIMPSON2D`	Macro input boxes	Integration of a function with Simpson's 1/3 rule Double integration of a formula by Simpson's rules
`SIMPSONADAPT`	Macro input boxes	Adaptive Simpson's 1/3 rule with error control
`SIMPSONDATA`	UDF arguments	Simpson's rules for integrating a set of data (automatically selects 1/3 versus 3/8 rule as necessary). For even data sets with factors of 4, uses one Richardson extrapolation and estimates error.
`TRAP` `TRAPDATA`	UDF arguments	Composite Trapezoidal rule with end-correction applied to an *Excel* worksheet formula Trapezoidal rule applied to data. Performs one Richardson's extrapolation (if possible)
`TRAPEZOID`	Macro input boxes	Composite trapezoidal rule without end correction
`TREX`	UDF arguments	Trapezoidal Romberg integration applied to an *Excel* worksheet formula

Chapter 12 Initial-value Problems

"If everyone is moving forward together, then success takes care of itself." – Henry Ford

There are several dialects of the language of engineering. Technical drawings, for example, convey visual information about a product's specifications, geometry and orientation, or the sequence of unit operations, process instrumentation, and control in process design. *VBA* is another example of an engineering language for logical programming. Calculus is the analysis of rates of change. We use the mathematical language of calculus and differential equations to describe physicochemical and kinetic phenomena, such as thermal or molecular diffusion with chemical reactions.

> **SQ3R Focused Reading Questions**
> 1. Can I derive a simple numerical solution to an ODE from a Taylor series approximation for the function?
> 2. How can I check the accuracy of a numerical solution to an ordinary differential equation?
> 3. How do I solve a simultaneous system of equations numerically?
> 4. What is a stiff first-order differential equation?
> 5. What methods are available for approximating solutions to stiff problems?
> 6. What is the difference between single and multi-step solution methods?
> 7. How do variable step solution methods improve the computational efficiency?
> 8. How can I use numerical integration methods for finding roots to nonlinear equations?
> 9. Can methods for approximating solutions to ordinary differential equations be used to solve integrals?
> 10. Can we use experimental data to find differential model parameters?

We refer to first-order differential equations as initial-value problems (*IVP*). Initial-value problems ensue when we retain the accumulation term in a conservation equation. "Initial" implies some starting point in time or space. In this case, the solution method requires known values for the solution at the starting condition, or at a boundary. We know how to solve linear differential equations using a variety of analytical techniques, such as separation of variables to get the change with time for the concentration of a reactant in a batch reactor of Equation (9.2), plotted in Figure 9.1. However, many nonlinear problems often have no analytical solutions. Suppose that the rate constant in Equation (9.2) is a function of temperature according to the Arrhenius expression of Equation (9.28). Furthermore, suppose that the temperature is a function of the heat of reaction, requiring simultaneous solution with an energy balance:

$$\rho c_p \frac{dT}{dt} = -\Delta H_r kC \tag{12.1}$$

where ΔH_r, ρ, and c_p are the temperature dependent properties of heat of reaction, density, and specific heat of the contents of the reactor, respectively. This problem requires numerical approximation of the solution.

We present numerical methods for solving initial-value problems involving first-order, ordinary differential equations (*ODE*) of the general form:

$$\frac{dy}{dx} = f(x, y) \tag{12.2}$$

where $f(x, y)$ represents an arbitrary function of some combination of the independent and dependent variables. The solution to the differential equation is subject to a known *initial* condition:

$$y = y_0 \quad \text{at} \quad x = x_0 \tag{12.3}$$

CHAPTER 12: INITIAL-VALUE PROBLEMS

Once again, we derive the numerical methods for approximating solutions to initial value problems from truncated Taylor series expansions. We begin with the simple Euler method to lay a common framework for more advanced Runge-Kutta and multistep methods. Specifically, we cover simpler methods for implementation in *Excel* worksheets, as well as advanced methods using the power of *VBA*:

- Euler and trapezoid methods in *Excel* worksheets
- *VBA* macros in *Excel* using higher precision Runge-Kutta methods
- Implicit and multistep methods for mildly stiff problems

Before diving into the numerical methods for obtaining approximate solutions to differential equations, let us review a few characteristics of derivatives to simplify the solution process, where appropriate:

- Apply the chain rule for the derivative of a nested variable or log function:

$$\frac{df(x,y)}{dy} = \frac{\partial f(x,y)}{\partial x}\frac{dx}{dy} \tag{12.4}$$

$$\frac{d\ln[f(x)]}{dx} = \frac{1}{f(x)}\frac{df(x)}{dx} \tag{12.5}$$

- Invert the derivative and function to swap the independent variable for the dependent variable:

$$\frac{dy}{dx} = f(x,y) \quad \rightarrow \quad \frac{dx}{dy} = \frac{1}{f(x,y)} \tag{12.6}$$

- Differentiate an integral to obtain the integrand:

$$y = \int f(x)dx \quad \rightarrow \quad \frac{dy}{dx} = f(x) \tag{12.7}$$

- Convert a higher order differential equation to a system of simultaneous first-order differential equations by introducing new dependent variables:

$$\frac{d^n y}{dx^n} = f \tag{12.8}$$

$$z_1 = \frac{d^{n-1}y}{dx^{n-1}} \quad \rightarrow \quad \frac{dz_1}{dx} = f \tag{12.9}$$

$$z_2 = \frac{d^{n-2}y}{dx^{n-2}} \quad \rightarrow \quad \frac{dz_2}{dx} = z_1 \tag{12.10}$$

$$\vdots$$

$$z_n = \frac{dy}{dx} \quad \rightarrow \quad \frac{dz_n}{dx} = z_{n-1} \tag{12.11}$$

- Solve a differential equation by separation of variables and integration:

$$\int_{y_0}^{y} dy = \int_{x_0}^{x} f(x,y)dx \tag{12.12}$$

Rearrange Equation (12.12) for *y*:

$$y - y_0 = \int_{x_0}^{x} f(x, y^*)dx \tag{12.13}$$

where y^* represents different cases in Table 12.1 for the derivative functionality of *x* and *y*.

Table 12.1 Solution methods for separation of variables.

Case	Solution Method
The function of the derivative does not involve the dependent variable y: $f(x,y) \Rightarrow f(x)$	Separate variables for y and solve the integral by numerical quadrature: $y - y_0 = \int_{x_0}^{x} f(x) dx$
The function of the derivative does not involve the independent variable x: $f(x,y) \Rightarrow f(y)$	Separate variables for x and solve the integral in y by numerical quadrature: $x - x_0 = \int_{y_0}^{y} \frac{dy}{f(y)}$
The derivative function(s) involves both dependent and independent variables, x and y, or just the dependent variables: $f(x,y)$ or $f(x, y_1, y_2, ...)$ or $f(y_1, y_2, ...)$	Solve for y in terms of x using the methods described in this chapter, e.g., Euler's, trapezoidal, Runge-Kutta, Adams-Bashforth-Moulton, etc.

12.1 Taylor Series Method

We may find an approximate solution to an initial value problem represented by Equations (12.2) and (12.3) from a Taylor series expansion about the initial condition:

$$y(x) \cong y_0 + (x - x_0) \frac{dy}{dx}\bigg|_{x_0, y_0} + \frac{(x - x_0)^2}{2!} \frac{d^2 y}{dx^2}\bigg|_{x_0, y_0} + \frac{(x - x_0)^3}{3!} \frac{d^3 y}{dx^3}\bigg|_{x_0, y_0} \ldots \quad (12.14)$$

Substitute $f(x, y)$ for dy/dx from Equation (12.2) into Equation (12.14):

$$y(x) \cong y_0 + (x - x_0) f(x_0, y_0) + \frac{(x - x_0)^2}{2!} \frac{df}{dx}\bigg|_{x_0, y_0} + \frac{(x - x_0)^3}{3!} \frac{d^2 f}{dx^2}\bigg|_{x_0, y_0} \ldots \quad (12.15)$$

The function in Equation (12.2) must be differentiable at each of the higher order terms in Equation (12.15). In general, the function for the first derivative may involve both independent and dependent variables. We must apply the chain rule to evaluate the higher-order derivatives (Hanna and Sandall 1995):

$$\frac{dy}{dx} = f \quad (12.16)$$

$$\frac{d^2 y}{dx^2} = \frac{df}{dx} = \frac{\partial f}{\partial x} \frac{dx}{dx} + \frac{\partial f}{\partial y} \frac{dy}{dx} = \frac{\partial f}{\partial x} + \frac{\partial f}{\partial y} f \quad (12.17)$$

$$\frac{d^3 y}{dx^3} = \frac{d^2 f}{dx^2} = \frac{\partial^2 f}{\partial x^2} + \frac{\partial}{\partial y}\left(\frac{\partial f}{\partial x}\right) f + \left(\frac{\partial f}{\partial y}\right)\left(\frac{\partial f}{\partial x} + \frac{\partial f}{\partial y} f\right) + \left[\frac{\partial}{\partial x}\left(\frac{\partial f}{\partial y}\right) + f \frac{\partial^2 f}{\partial y^2}\right] f \quad (12.18)$$

The expansions of the chain rule for fourth-order and higher derivatives are exponentially more complicated. However, since we evaluate the derivatives at the initial condition, we may use Ridder's method from Section 5.3.4 to obtain numerical values. The accuracy of the approximation in Equation (12.15) is proportional to the order of magnitude of the first truncated term and may require several terms for series convergence.

A good Taylor series approximation for the solution to a differential equation may require more than four terms. We may try accelerating convergence of the Taylor series solution with a Padé rational function approximation (Baker and Graves-Morris 1996) or with Shanks transformations, presented in Section 5.2 and illustrated in Example 5.2.

12.2 Euler Method

> *"The secret of getting ahead is getting started. The secret of getting started is breaking your complex overwhelming tasks into small manageable tasks, and then starting on the first one."* – Mark Twain

As we learned in Chapter 5, a Taylor series approximation of a function improves as the value for x gets closer to the expansion point, x_0, but this limits the range of usefulness of the series solution. To improve the accuracy further, we may try extending this method to higher order Taylor series as far as we can differentiate the function of the differential equation. This approach makes the result specific to each unique problem. For this reason, coupled with the challenge of calculating the complicated derivatives, we rarely use higher-order Taylor methods in practice (Gear 1971). From here, we develop generic solution methods that have broad application not specific to unique problems.

Euler's method is the simplest, two-termed truncated Taylor series for approximating the solution to a first-order, ordinary differential equation. We find an approximate solution to the differential equation for y at a value of x close to the initial condition from a linear, or first-order, truncated Taylor series expansion about the initial condition:

$$y \cong y_0 + (x - x_0) \left.\frac{dy}{dx}\right|_{x_0, y_0} \tag{12.19}$$

Replace the first-order derivative in Equation (12.19) with the function of the initial-value problem on the right side of Equation (12.2), evaluated at the initial condition, as follows:[103]

$$y \cong y_0 + \Delta x_0 f(x_0, y_0) \tag{12.20}$$

where

$$\Delta x_0 = x - x_0$$

We estimate the error in the approximation for y at x from the order of magnitude of the first truncated term in the Taylor series expansion: $\Delta x^2 d^2y/dx^2$, or $O(\Delta x^2)$. For a linear (first-order) approximation, we recommend a sufficiently small $\Delta x_0 < 1$ to reduce the error. We should avoid approximations using $\Delta x_0 > 1$ because the error increases proportional to Δx^2. It may become necessary to scale the variables in the differential equations to ensure a small integration step size.

We approximate the solution to the differential equation (12.2) by successive first-order Taylor series expansions at each integration step between the integration limits $x_0 \leq x \leq x_n$. To illustrate, consider the domain of x divided into n equally spaced intervals:

$$\Delta x = \frac{x_n - x_0}{n} \tag{12.21}$$

We get approximate solutions from linear, first-order Taylor expansions about the value for x and y at the end of each integration step:

$$\begin{aligned}
x_1 &= x_0 + \Delta x & y_1 &\cong y_0 + \Delta x f(x_0, y_0) \\
x_2 &= x_1 + \Delta x & y_2 &\cong y_1 + \Delta x f(x_1, y_1) \\
&\vdots & &\vdots \\
x_n &= x_{n-1} + \Delta x & y_n &\cong y_{n-1} + \Delta x f(x_{n-1}, y_{n-1})
\end{aligned} \tag{12.22}$$

Once we have the approximate solutions at the discrete points ranging from the initial condition $(x_0, y_0), (x_1, y_1)\ldots (x_n, y_n)$, we functionalize the results by interpolation using rational functions, or splines for a continuous representation of the solution, as described in Chapter 10. We recommend the clamped cubic spline that requires values for the derivatives at the starting and endpoints, which we obtain directly from the function of the differential equation evaluated at the lower and upper limits of integration, f_0 and f_n.

---

[103] Note that the derivative function may involve any combination of x and y, e.g. just x or just y, such as $f(x_0)$ or $f(y_0)$ or $f(x_0, y_0)$.

To summarize, Euler's recursive relationship approximates the solution for y between the limits of integration by stepping through the independent variable x, with step sizes of Δx, approximating the value of y at $x + \Delta x$ from the previous solution at x:

- Euler's Method

$$y(x_{i+1}) \cong y(x_i) + \Delta x_i f(x_i, y_i) \quad (12.23)$$

where the index i refers to the discrete positions in x at the local step:

$$x_{i+1} = x_i + \Delta x_i = x_0 + i\Delta x \quad (12.24)$$

For convenience, we may express Euler's method in compact notation:

$$y(x_{i+1}) = y(x_i) + k_1 \quad (12.25)$$

where the k_1 term is an estimate for the change in y, or Δy_i across the increment in x:

$$\frac{k_1}{\Delta x_i} = f(x_i, y_i) \quad (12.26)$$

The local truncation error of order $O(\Delta x^2)$ in each integration step does not represent the overall global error in the solution for y. Instead, the errors from the n number of steps accumulate to $n \times O(\Delta x^2)$. From Equation (12.21), for n number of steps:

$$x_n = x_0 + n\Delta x \quad \text{or} \quad n = \frac{x_n - x_0}{\Delta x} \quad (12.27)$$

Substitute for n in the accumulated error:

$$nO(\Delta x^2) = \left(\frac{x_n - x_0}{\Delta x}\right) O(\Delta x^2) = (x_n - x_0) O(\Delta x) \quad (12.28)$$

Thus, Euler's method is globally first-order accurate over the domain of integration.

We refer to this type of numerical solution method as "marching" integration because Equations (12.22) "march" or step through the solution. Euler's method is particularly easy to implement in a computer spreadsheet application, such as *Excel*. However, Euler's method is used infrequently because of the relatively large error associated with truncating second order and higher terms from the Taylor series.

Example 12.1 Adiabatic Plug Flow Reactor Simulation

Known/Schematic: We use a tubular, plug flow reactor (PFR), depicted in Figure 1.14, with a second order elementary reaction. The non-isothermal PFR design equation comes from a steady-state mole balance around the reactor:

$$\frac{dX_A}{dV} = k_0 \exp\left(-\frac{E_a}{R_g T}\right) \frac{[C_{A0}(1 - X_A)]^2}{F_{A0}} \quad (12.29)$$

The temperature is a function of conversion calculated from the energy balance around the reactor:

$$T = T_0 - \frac{X_A \Delta H_r}{c_{pA}} \quad (12.30)$$

Find: Solve for the conversion at the end of the reactor, given the tabulated reaction parameters in Figure 12.1.

Analysis: The derivatives of the function in Equation (12.29) become too complicated for us to attempt a Tayler series solution. Instead, we use Euler's method to approximate the solution. Provide the initial conditions and Euler integration steps in an *Excel* worksheet in the top two *rows* as seen in Figure 12.1, respectively. Auto-fill the formulas from the second row down into successive rows until reaching the upper limit of integration, which corresponds to the volume of the PFR. *Excel* automatically updates the relative cell references in Euler's recursive formula to refer to the cells in the previous row.

CHAPTER 12: INITIAL-VALUE PROBLEMS

	A	B	C	D	E	F	G	H	I
1	Molar Feed Rate	F_A0	100	mol/s		V/L	X_A	T/K	-r_A/(mol/L s)
2	Pre-exponential Factor	k_0	55	L/mol s		0	0	=C6	=$C$2*EXP(-$C$3/H2)*($C$5^2)*(1-G2)^2
3	Activation Energy	E_a/R_g	100	K		0.5	=G2+(F3-F2)*I2/$C$1	=$C$6-G3*$C$7/$C$4	=$C$2*EXP(-$C$3/H3)*($C$5^2)*(1-G3)^2
4	Heat Capacity	c_pA	1	J/mol K		1	=G3+(F4-F3)*I3/$C$1	=$C$6-G4*$C$7/$C$4	=$C$2*EXP(-$C$3/H4)*($C$5^2)*(1-G4)^2
5	Initial Concentration	C_A0	1	mol/L		1.5	=G4+(F5-F4)*I4/$C$1	=$C$6-G5*$C$7/$C$4	=$C$2*EXP(-$C$3/H5)*($C$5^2)*(1-G5)^2
6	Initial Temperature	T_0	300	K		2	=G5+(F6-F5)*I5/$C$1	=$C$6-G6*$C$7/$C$4	=$C$2*EXP(-$C$3/H6)*($C$5^2)*(1-G6)^2
7	Heat of Reaction	ΔH_r	-100	J/mol K		2.5	=G6+(F7-F6)*I6/$C$1	=$C$6-G7*$C$7/$C$4	=$C$2*EXP(-$C$3/H7)*($C$5^2)*(1-G7)^2
8	Volume	V	5	L		3	=G7+(F8-F7)*I7/$C$1	=$C$6-G8*$C$7/$C$4	=$C$2*EXP(-$C$3/H8)*($C$5^2)*(1-G8)^2
9	Initial Conversion	X_A0	0	-		3.5	=G8+(F9-F8)*I8/$C$1	=$C$6-G9*$C$7/$C$4	=$C$2*EXP(-$C$3/H9)*($C$5^2)*(1-G9)^2
10						4	=G9+(F10-F9)*I9/$C$1	=$C$6-G10*$C$7/$C$4	=$C$2*EXP(-$C$3/H10)*($C$5^2)*(1-G10)^2
11						4.5	=G10+(F11-F10)*I10/$C$1	=$C$6-G11*$C$7/$C$4	=$C$2*EXP(-$C$3/H11)*($C$5^2)*(1-G11)^2
12						5	=G11+(F12-F11)*I11/$C$1	=$C$6-G12*$C$7/$C$4	=$C$2*EXP(-$C$3/H12)*($C$5^2)*(1-G12)^2

	F	G	H	I
1	V/L	X_A	T/K	-r_A/(mol/L s)
2	0	0	300	39.4092
3	0.5	0.1970	319.7	25.9359
4	1	0.3267	332.7	18.4586
5	1.5	0.4190	341.9	13.8568
6	2	0.4883	348.8	10.8116
7	2.5	0.5424	354.2	8.6858
8	3	0.5858	358.6	7.1399
9	3.5	0.6215	362.1	5.9786
10	4	0.6514	365.1	5.0830
11	4.5	0.6768	367.7	4.3772
12	5	0.6987	369.9	3.8106

Figure 12.1

Euler's solution to an ordinary differential equation in an *Excel* worksheet.

Plot the Euler integration results for reaction conversion and temperature, as shown in Figure 12.2. The effluent solution is 70% converted with a final temperature of 370 K. The solid line represents interpolated values between the integration steps using the cubic spline function **CSPLINE** from *PNMSuite*, with clamped end conditions:

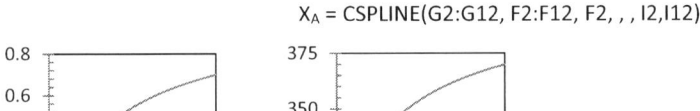

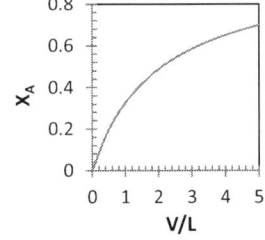

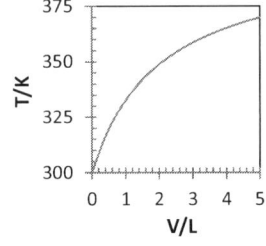

$$X_A = \text{CSPLINE}(G2:G12, F2:F12, F2, , , I2, I12)$$

Figure 12.2

Euler integration results. The solid line represents interpolated values using cubic splines.

Comments: We recommend Euler's method only for obtaining results quickly in a spreadsheet, where solution accuracy is not critical. Euler's method is simple to set up in *Excel*, but requires a relatively large number of integration steps to achieve an accurate solution when compared with higher order methods such as trapezoidal or Runge-Kutta.

We may solve this example by separation of variables and substitution using numerical integration:

$$V = \frac{F_{A0}}{k_0 C_{A0}^2} \int_0^{X_{Af}} \frac{dX_A}{\exp(-E_a/R_g T)(1-X_A)^2} \qquad (12.31)$$

We provide an initial guess for the final conversion and solve the integral using the user-defined function **GKAD**. We then use the **Goal Seek** to find the final conversion to match the volume in Equation (12.31).

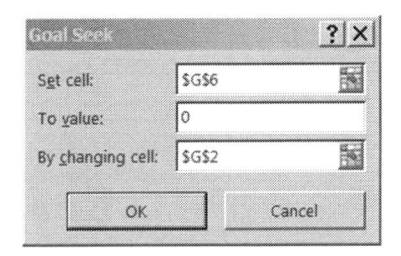

Goal Seek
Set cell: $G$6
To value: 0
By changing cell: $G$2

	A	B	C	D	E	F	G
1	Molar Feed Rate	F_A0	100	mol/s		X_A Lower Limit	0
2	Pre-exponential Factor	k_0	55	L/mol s		X_A Upper Limit	0.672909661
3	Activation Energy	E_a/R_g	100	K		Integrand	1.395612425
4	Heat Capacity	c_pA	1	J/mol K		Integral	2.74999087
5	Initial Concentration	C_A0	1	mol/L		Volume/L	4.999983399
6	Initial Temperature	T_0	300	K		ΔV	1.66006E-05
7	Heat of Reaction	ΔH_r	-100	J/mol K			

□

12.2.1 Stability and Stiffness

"If you're in a bad situation, don't worry, it'll change. If you're in a good situation, don't worry it'll change." – John A. Simone

A stable solution typically exhibits a smooth trend in the relationship between the dependent and independent variables. We recognize unstable solutions by extreme oscillations in the results or sudden divergence from an expected trend. Euler's method is conditionally stable for a local integration step size that satisfies the following stability requirement:

$$\Delta x_i \leq \left| \frac{y_i}{f(x_i, y_i)} \right| \qquad (12.32)$$

Solution stiffness has many definitions. For our purpose, we define "stiff" differential equations as requiring relatively small values for the integration step size Δx for stable solutions.[104] Local regions of the solution may exhibit stiffness while other regions of the solution do not exhibit stiff behavior. We recognize potential for stiffness in regions where small changes in the independent variable, x, result in large changes in the dependent variable, y. Despite the stability criteria, unstable solutions may also be the result of computational round-off errors accumulating with each integration step.

Do not confuse stability with accuracy. A stable Euler solution that meets the criterion in Equation (12.32) has accuracy proportional to the magnitude of the integration step size Δx. We may improve the accuracy as well as the stability by selecting step sizes well below the stability criterion. Beware that computer round-off error also creeps in when using extremely small Euler step sizes, which may also affect Euler integration stability.

To promote stability for larger integration steps, we reluctantly recommend implicit, backward derivative methods that require iterative, trial and error solutions to the difference equations at each integration step. Local iterations at each integration step add significant computational costs to finding an approximate solution to a differential equation. We derive the simplest form of an implicit method by expanding the first-order Taylor series approximation about the (yet to be determined) solution at the far end of the integration step:

$$y(x_i) \cong y(x_{i+1}) - \Delta x_i f(x_{i+1}, y_{i+1}) \qquad (12.33)$$

Upon rearrangement, we obtain an implicit or backward derivative form of Euler's recursive relationship that requires the solution for y_{i+1} at the end of the step on both sides of the equality:

$$y(x_{i+1}) \cong y(x_i) + \Delta x_i f(x_{i+1}, y_{i+1}) \qquad (12.34)$$

Nonlinear forms of Equation (12.34) in terms of y_{i+1} require an iterative solution for y_{i+1} at each integration step. We can use ordinary iteration in *Excel* by enabling iterative calculation in the settings for **File> Options> Formulas**. Do not forget what we learned about the importance of initial guesses and function arrangement in Section 6.6 for achieving a converged solution.

An important technique that controls error while combating moderate stiffness uses variable step sizes through the range of integration. We take smaller steps in stiff regions and larger steps outside of stiff regions of the problem domain. Section 12.5.2 introduces methods that adapt the size of the integration step to control the error and reduce the computational burden.

---

[104] The term "stiff" comes from the characteristics of the solution for the second-order differential equation describing the harmonic oscillation of a stiff spring.

Example 12.2 Backwards (Implicit) Euler's Method for Stiff Equations

Known: Consider the following stiff, first-order, ordinary differential equation integrated over the range $0 \leq x \leq 1$:

$$\frac{dy}{dx} = -5y \qquad (12.35)$$

subject to the initial condition $y = 3$ at $x = 0$. The exact solution of Equation (12.35) is $y = 3e^{-5x}$.

Find: Solve Equation (12.35) using implicit backward and explicit forward Euler's methods with step sizes of 0.25 and 0.1. Compare the results with the exact solution.

Analysis: Implement Euler's methods in an *Excel* worksheet, as seen in Figure 12.3 for $\Delta x = 0.25$. Enter the formulas for explicit and implicit Euler's methods in columns **B** and **C**, respectively. The relatively large integration step size produces an unstable solution with the explicit method:

$$y_{i+1} = y_i + \Delta x(-5y_i) \qquad (12.36)$$

The backward Euler solution for y_{i+1} uses the following recursive formula:

$$y_{i+1} = y_i + \Delta x(-5y_{i+1}) \qquad (12.37)$$

rearranged for

$$y_{i+1} = \frac{y_i}{1 + 5\Delta x}$$

	A	B	C	D
1	x	y Explicit	y Implicit	y Exact
2	0	3	3	3
3	0.25	=B2+0.25*(-5*B2)	=C2/(1+5*0.25)	=3*EXP(-5*A3)
4	0.5	=B3+0.25*(-5*B3)	=C3/(1+5*0.25)	=3*EXP(-5*A4)
5	0.75	=B4+0.25*(-5*B4)	=C4/(1+5*0.25)	=3*EXP(-5*A5)
6	1	=B5+0.25*(-5*B5)	=C5/(1+5*0.25)	=3*EXP(-5*A6)

	A	B	C	D
1	x	y Explicit	y Implicit	y Exact
2	0	3.00000	3.00000	3.00000
3	0.25	-0.75000	1.33333	0.85951
4	0.5	0.18750	0.59259	0.24625
5	0.75	-0.04688	0.26337	0.07055
6	1	0.01172	0.11706	0.02021

Figure 12.3 Forward and backwards Euler's solutions compared with the true (exact) solution.

Plot the integration results, as presented in Figure 12.4. A smaller step size of $\Delta x = 0.1$ solves the stability problem of the forward difference method, but still gives a poor solution. The explicit method with a small step size under predicts the solution, whereas the implicit method, though stable, over predicts the solution.

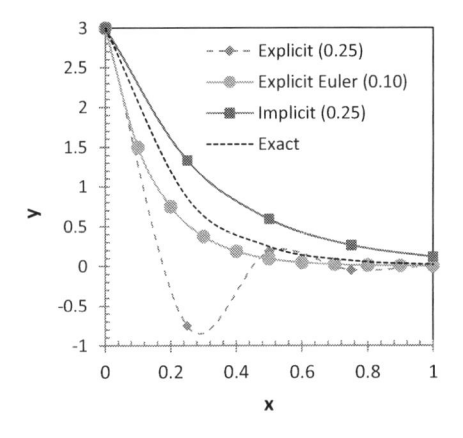

Figure 12.4

Euler's integration result for (12.35) using a step size of 0.25.

Comments: Backward difference integration schemes have a wider range of stability. Smaller step sizes generally improve the accuracy of the solution approximation.

12.2.2 Simultaneous First-order Differential Equations

We may readily adapt marching integration methods to coupled systems of first-order differential equations. For a system of n coupled-equations, Equation (12.23) for Euler's method becomes

$$y_j(x_{i+1}) = y_{j,i} + \Delta x_i f_j(x_i, y_{j,i}) \quad \text{for all } j = 1, 2 \ldots n \quad (12.38)$$

where the subscript j is the index distinguishing the different dependent variables, y_j, while index i represents the step in the independent variable, x_i. To illustrate, consider the smallest system consisting of two-coupled first-order ODEs:

$$\frac{dy_1}{dx} = f_1(x, y_1, y_2) \quad \text{and} \quad \frac{dy_2}{dx} = f_2(x, y_1, y_2) \quad (12.39)$$

subject to initial conditions for y_1 and y_2. The recursive Euler relationships are:

$$y_1(x_{i+1}) = y_{1,i} + \Delta x_i f_1(x_i, y_{1,i}, y_{2,i}) \quad \text{and} \quad y_2(x_{i+1}) = y_{2,i} + \Delta x_i f_2(x_i, y_{1,i}, y_{2,i}) \quad (12.40)$$

The solution procedure follows the method outlined for a single differential equation. We evaluate the derivative functions simultaneously at each integration step in terms of the value for x, and the solutions for the dependent variables from the previous step, beginning at the initial conditions. Implicit methods use similar equations with the functions evaluated at the end of the integration step.

Example 12.3 Model of a Complex Reaction in a Batch Reactor

Known: The following first-order reactions occur in a well-mixed batch reactor.

$$A \xrightarrow{1} 2B \quad (12.41)$$

$$A \underset{3}{\overset{2}{\rightleftharpoons}} C \quad (12.42)$$

$$B \xrightarrow{4} C + D \quad (12.43)$$

The first-order rate constants are $k_1 = 1$ min^{-1}, $k_2 = 0.2$ min^{-1}, $k_3 = 0.05$ min^{-1}, $k_4 = 0.1$ min^{-1}. Initially, only species A is present in the reactor with a concentration of 1 mol/L.

Find: Plot the concentration profiles for species A, B, and C over a 20-minute period.

Analysis: Use Euler's method in an *Excel* worksheet to approximate the solution to the following species mole balances:

$$\frac{dC_A}{dt} = -k_1 C_A - k_2 C_A + k_3 C_C \quad (12.44)$$

$$\frac{dC_B}{dt} = 2k_1 C_A - k_4 C_B \quad (12.45)$$

$$\frac{dC_C}{dt} = k_2 C_A - k_3 C_C + k_4 C_B \quad (12.46)$$

Note that species D does not affect the reaction rate expressions and is left out of the system of mole balances for integration. Add the time steps in column **A**. The worksheet in Figure 12.5 shows the first 0.2 minutes. The named cells have references: **k_1 = B1, k_2 = B2, k_3 = B3, k_4 = B4, dt = D1**. Place the values for the initial conditions in row seven. Use the following formulas with relative cell referencing in columns **B** through **D** for the implementation of Euler's method:

B8 = B7 + dt*(-k_1*B7 - k_2*B7 + k_3*D7)   C8 = C7 + dt*(2*k_1*B7 - k_4*C7)   D8 = D7 + dt*(k_2*B7 - k_3*D7 + k_4*C7)

Auto fill the formulas for the remaining time steps into subsequent cells until $t = 20$ minutes in column **A**.

CHAPTER 12: INITIAL-VALUE PROBLEMS

	A	B	C	D
1	k_1/(min^{-1}) =	1	Δt/min =	0.1
2	k_2/(min^{-1}) =	0.2		
3	k_3/(min^{-1}) =	0.05		
4	k_4/(min^{-1}) =	0.1		

	A	B	C	D
6	t/(min)	C_A/M	C_B/M	C_C/M
7	0	1.000	0.000	0.000
8	0.1	0.880	0.200	0.020
9	0.2	0.775	0.374	0.040

Figure 12.5

Euler's method applied to a system of ODEs in an *Excel* worksheet.

Figure 12.6 has a plot of the concentration profiles in a batch reactor for three interacting species, A, B, and C, generated from Euler's integration method for 20 minutes with 0.1 minute time steps.

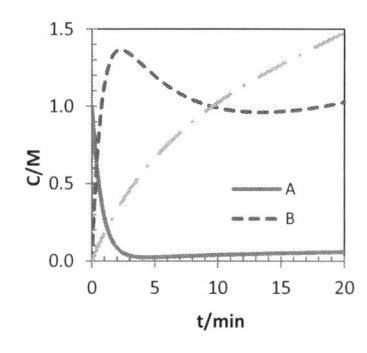

Figure 12.6

Concentration profiles from mass balances for three interacting species, A, B, and C, in a batch reactor.

Comments: The concentration profiles reveal strong interactions among the three chemical species within the first 10 minutes that may affect procedures for starting the reactor or maximizing the concentration of species B.

□

12.3 Trapezoidal Rules for IVPs

"Trapezoid: A device for catching zoids."

Although the implicit backwards Euler's method does not necessarily improve the accuracy compared to the explicit Euler's method, the stability of the solution is unconditional (ignoring round-off errors). As seen in Example 12.2, we can enhance the accuracy of the backward Euler solution by averaging the results from the explicit (forward) and implicit (backward) Euler's methods.

- *Trapezoidal Method* $\quad y_{i+1} = y_i + \dfrac{\Delta x}{2}\left[f(x_i, y_i) + f(x_{i+1}, y_{i+1})\right]$ (12.47)

where $\Delta x = x_{i+1} - x_i$.

Just like Equation (11.9) for integrals, the trapezoidal rule in Equation (12.47) gets its name from the formula for the area of a trapezoid in terms of the derivative functions evaluated at each end of the integration step:

For nonlinear derivative functions involving the dependent variable, the trapezoidal rule in Equation (12.47) is implicit in y_{i+1}, requiring an iterative solution at each integration step. The trapezoidal rule also requires two derivative function evaluations at the first integration step. However, we reuse the derivative evaluated at the end of the previous step for the beginning of the next step – ultimately requiring just one derivative evaluation for each following step. Thus, we gain an improvement in accuracy at practically no additional cost in computation required by the implicit backward Euler method.

The trapezoidal rule increases the accuracy of the solution by one order of magnitude over Euler's method, allowing for larger integration steps. The penalty for the increased accuracy is the added computational burden required to solve the simultaneous implicit equations at each integration step. We may choose from any of the iterative solution methods described in Chapter 6 to solve the implicit equations, but recommend ordinary iteration with Steffensen's method for convergence acceleration. We note two important features of Equation (12.47); it is already in the form required for successive substitution, and Steffensen's method does not require derivative evaluations of the simultaneous nonlinear equations, as required by Newton's method. In summary, the trapezoidal method requires:

1. Evaluate the derivative function at the initial conditions:

$$f_0 = f(x_0, y_0) \tag{12.48}$$

2. Calculate the value of x at the end of the first step. Then, solve the implicit trapezoidal equation and derivative function for f_1 and y_1:

$$x_1 = x_0 + \Delta x \quad \rightarrow \quad f_1 = f(x_1, y_1) \quad \rightarrow \quad y_1 = y_0 + \frac{\Delta x}{2}[f_0 + f_1] \tag{12.49}$$

3. Calculate x at the end of the next step. Solve the implicit trapezoidal equation and derivative function for f_2 and y_2. Note that the derivative function evaluated at x_1 gets recycled from the previous step:

$$x_2 = x_1 + \Delta x \quad \rightarrow \quad f_2 = f(x_2, y_2) \quad \rightarrow \quad y_2 = y_1 + \frac{\Delta x}{2}[f_1 + f_2] \tag{12.50}$$

4. Continue the pattern through the rest of the integration steps to the upper limit of the integration domain:

$$x_n = x_{n-1} + \Delta x \quad \rightarrow \quad f_n = f(x_n, y_n) \quad \rightarrow \quad y_n = y_{n-1} + \frac{\Delta x}{2}[f_{n-1} + f_n] \tag{12.51}$$

We use the result for y from a previous step for the guess to initiate the iterative solution for y at the current step. Enable iteration on circular references in *Excel* **File>Options>Formulas** to solve the implicit trapezoidal rule equation by ordinary iteration in an *Excel* worksheet.

Example 12.4   Comparison of the Trapezoidal Rule with Euler's Methods

Known/Find: Solve the following ordinary, first-order, differential equation in the domain $1 \leq x \leq 2$, subject to the initial condition:

$$\frac{dy}{dx} = \frac{1}{x^2} - \frac{y}{x} - y^2 \qquad y(1) = -1 \tag{12.52}$$

Compare the solutions using three methods including forward and backward Euler's methods and the trapezoidal rule, with the exact solution, $y = -1/x$.

Analysis: First, create a simple *VBA* user-defined function of the right-hand side of Equation (12.52):

```
Public Function f(x As Double, y As Double): f = 1 / x ^ 2 - y / x - y ^ 2: End Function
```

Apply the user-defined function in an *Excel* worksheet to implement Euler's and trapezoidal solutions using 10 integration steps. The formulas in the range **B3:E3** are:

A3=A2+0.1   B3=-1/A3   C3 = C2+0.1*f(A2,C2)

D3 = D2+0.1*f(A2,D3)   E3 = E2+(0.1/2)*(f(A2,C2)+ f(A2,D3))

When we plot the forward and backward Euler results along with the exact solution, as shown in Figure 12.7, we notice that the forward method *over* predicts the solution by approximately the same magnitude that the backward Euler solution *under* predicts the solution. This observation suggests that an average of the Euler methods (represented

CHAPTER 12: INITIAL-VALUE PROBLEMS

by the dashed lines) may produce a better result. The trapezoidal rule, as the average of the forward and backward Euler methods at each integration step (represented by the solid line), does a better job matching the exact solution.

	A	B	C	D	E
1	x	y Exact	y Euler F	y Euler B	y Trap
2	1	-1.000	-1.000	-1.000	-1.000
3	1.1	-0.909	-0.900	-0.918	-0.909
4	1.2	-0.833	-0.817	-0.850	-0.833
5	1.3	-0.769	-0.746	-0.793	-0.768
6	1.4	-0.714	-0.685	-0.744	-0.713
7	1.5	-0.667	-0.632	-0.702	-0.665
8	1.6	-0.625	-0.585	-0.666	-0.623
9	1.7	-0.588	-0.544	-0.634	-0.586
10	1.8	-0.556	-0.507	-0.606	-0.554
11	1.9	-0.526	-0.473	-0.582	-0.524
12	2	-0.500	-0.443	-0.560	-0.498

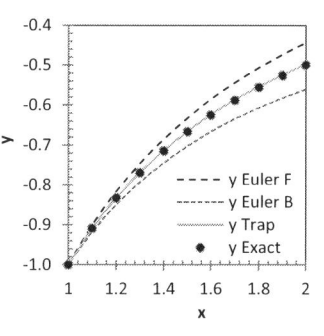

Figure 12.7

Comparison of Euler's method with the trapezoidal rule for Equation (12.52).

Comments: The trapezoidal rule improves the solution with larger integrations steps. Both Euler and trapezoidal methods are easy to remember and implement in an *Excel* worksheet.

□

Like Euler's method, we may extend the trapezoidal method to systems of differential equations.

Example 12.5   Plug Flow Reactor Design

Known/Find: Repeat the problem in Example 12.1. Use the implicit form of the trapezoidal rule with larger step sizes and compare the results with Euler's method.

Analysis: Employ the iteration feature of *Excel* to solve for the semi implicit solution at each integration step in the trapezoidal rule. In Figure 12.8, we obtain essentially the Euler result, but with the larger step size of 2.5 L.

	A	B	C	D	E	F
1	V/L	X_A	T/K	k/(L/mol·s)	C_A/(mol/L)	$(-r_A)$/(mol/L·s)
2	0	0	300	=55*EXP(-100/C2)	=1-B2	=D2*E2^2
3	2.5	=B2+2.5*(F2+F3)/(200)	=300+B3*100	=55*EXP(-100/C3)	=1-B3	=D3*E3^2
4	5	=B3+2.5*(F3+F4)/(200)	=300+B4*100	=55*EXP(-100/C4)	=1-B4	=D4*E4^2

	A	B	C	D	E	F
1	V/L	X_A	T/K	k/(L/mol·s)	C_A/(mol/L)	$(-r_A)$/(mol/L·s)
2	0.00	0.000	300.000	39.409	1	39.409
3	2.50	0.583	358.304	41.606	0.417	7.234
4	5.00	0.716	371.586	42.023	0.284	3.393

Figure 12.8

Excel worksheet for solving the PFR design equations by the implicit trapezoidal method showing the formulas and results.

Comments: Euler's method gives $X_A = 0.98$ (~36% error) at $V = 5$ L using the same large step size of 2.5 L. Although the implicit method is more stable, smaller integration steps are still required for higher accuracy. We can convert the problem into an *explicit* formula in *V* by inverting the differential equation:

$$\frac{dV}{dX_A} = \frac{F_{A0}}{-r_A} \qquad (12.53)$$

then integrate from $X_A = 0$. The trapezoid recursive formula for Equation (12.53) is *explicit* in *V*:

$$V_{i+1} = V_i + \frac{\Delta X_A F_{A0}}{2}\left[\frac{1}{-r_{A,i}} + \frac{1}{-r_{A,i+1}}\right] \qquad (12.54)$$

Now the integration method no longer requires the solution to an implicit equation at each integration step. Figure 12.9 shows the trapezoidal solution to the inverted equation using smaller step sizes.

	A	B	C	D	E	F
1	X_A	V	T	k	C_A	$-r_A$
2	0	0	300	=55*EXP(-100/C2)	=1-A2	=D2*E2^2
3	0.2	=B2+(A3-A2)*(1/F2+1/F3)/2	=300+A3*100	=55*EXP(-100/C3)	=1-A3	=D3*E3^2
4	0.4	=B3+(A4-A3)*(1/F3+1/F4)/2	=300+A4*100	=55*EXP(-100/C4)	=1-A4	=D4*E4^2
5	0.6	=B4+(A5-A4)*(1/F4+1/F5)/2	=300+A5*100	=55*EXP(-100/C5)	=1-A5	=D5*E5^2
6	0.8	=B5+(A6-A5)*(1/F5+1/F6)/2	=300+A6*100	=55*EXP(-100/C6)	=1-A6	=D6*E6^2

	A	B	C	D	E	F
1	X_A	V	T	k	C_A	$-r_A$
2	0	0	300	39.40922208	1	39.40922208
3	0.2	0.006420539	320	40.23885959	0.8	25.75287014
4	0.4	0.017081085	340	40.98538494	0.6	14.75473858
5	0.6	0.03886076	360	41.66058206	0.4	6.66569313
6	0.8	0.113000775	380	42.27412896	0.2	1.690965159

Figure 12.9

Trapezoidal rule solution of Equation (12.54)

□

A *VBA* macro **STIFFODE**, listed in Figure 12.10, employs the implicit trapezoidal or optional backward (implicit) Euler method for solving a stiff system of equations. The user-form **ODEFIXSTEP** also has options for the stiff methods. The macros use Steffensen's method of ordinary iteration (see Section 6.3.2) with convergence acceleration and relaxation to solve the system of implicit expressions in Equation (12.47) for each variable. The algorithm uses the explicit modified Euler method from Section 12.4 to generate initial guesses to initiate ordinary iteration on the nonlinear equations. Input boxes prompt the user for an optional relaxation factor to promote convergence. Relaxation may become important for some problems using larger integration steps. To use the macro:

- Specify in cells on an *Excel* worksheet the lower limit of integration, initial conditions, and derivative functions for the first-order differential equations in terms of the cells containing the values for initial conditions.
- Specify the upper integration limit, number of integration steps, the number of print steps, relative convergence tolerance for Steffensen's method, and a relaxation factor.
- Specify the top-left location of the columns for integration output.

There is an option for displaying the intermediate calculations, which is helpful when getting a feel for how the macro works, but slows down the calculations.

Illustration: Setup an *Excel* worksheet for the macro **STIFFODE** for approximating the solution to the following system of differential equations and corresponding initial conditions in the domain $0 \leq x \leq 1$:

$$\frac{dy_1}{dx} = 3y_1 + \frac{2y_2}{x+1} \qquad y_1(0) = 0 \qquad \text{and} \qquad \frac{dy_2}{dx} = 2xy_1 + y_2 \qquad y_2(0) = 1$$

1. Specify the lower and upper integration limits in the range **B1:B2**.
2. Specify the initial conditions in the range **E1:E2**.
3. Specify the derivative functions in the range **H1:H2**. Note that the formulas for the differential equations reference the initial conditions for the independent and dependent variable.

	A	B	C	D	E	F	G	H
1	x init	0		Init y1	0		dy1/dx	=3*E1+2*E2/(B1+1)
2	x final	1		Init y2	1		dy2/dx	=2*B1*E1+E2

4. Run the macro **STIFFODE** and select the ranges for the initial and final values for *x*, initial values for *y*, and the functions for *dy/dx* in the input boxes. For the following results, use the default settings for the number of integration (1000) and print steps (10), and relaxation factor (0.1). Select cell **A4** for the macro output.
5. Use the macro **GRAPHYXGUIDE** to plot the numerical results in a *y* vs *x* plot for visual inspection.

CHAPTER 12: INITIAL-VALUE PROBLEMS

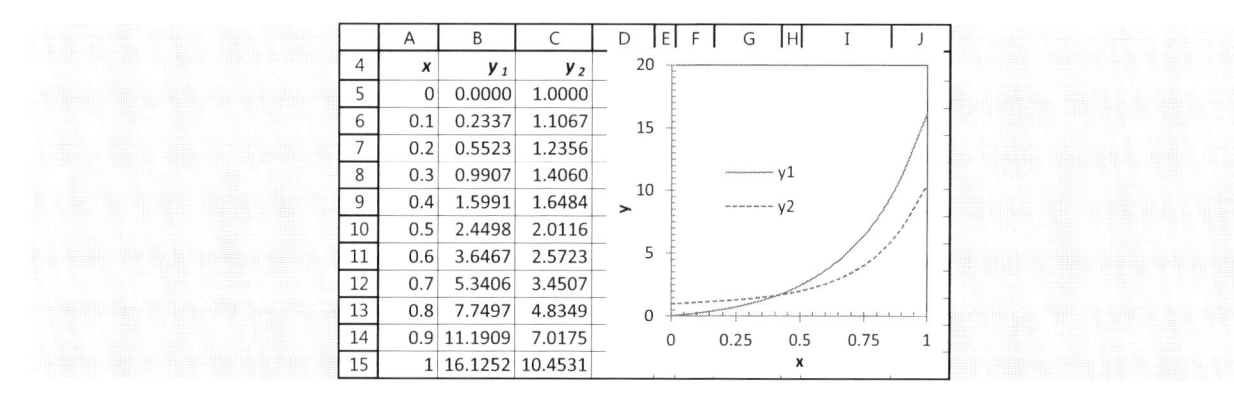

	A	B	C
4	x	y₁	y₂
5	0	0.0000	1.0000
6	0.1	0.2337	1.1067
7	0.2	0.5523	1.2356
8	0.3	0.9907	1.4060
9	0.4	1.5991	1.6484
10	0.5	2.4498	2.0116
11	0.6	3.6467	2.5723
12	0.7	5.3406	3.4507
13	0.8	7.7497	4.8349
14	0.9	11.1909	7.0175
15	1	16.1252	10.4531

Because of the large computational requirements and relatively slow speed of computation, an alternative version named **STIFFODE_sub** is included in the *PNMSuite* that uses the sub procedures in *Example 12.9*, sub **INITC** to code the initial conditions and control parameters, and sub **DYDX** to code the derivative functions for the ordinary differential equations. The much faster macro **STIFFODE_sub** does not require repeated references to worksheet ranges for calculating the derivative functions.

```
Public Sub STIFFODE()
' Implicit (backward) Euler or trapezoidal 2nd order fixed-step numerical integration
' method for solving stiff systems of 1st-order ordinary differential equations. User
' defines the derivatives in terms of the cells containing the initial conditions in
' an Excel worksheet. Steffensen's solves the implicit functions iteratively
' method of convergence acceleration.  The user may use a relaxation factor to promote
' convergence for some problems using larger step sizes.
' *** Required User-defined Functions and Sub Procedures from the PNM Suite ***
' JUSTABC_123 = renames worksheet with allowed object naming characters
' ROWCOL = transpose arrays if needed for user layout on worksheet
' *** Nomenclature ***
Dim boxtitle As String ' = string with title for message and input boxes
Dim d As Double ' = denominator in Aitkin acceleration
Dim dx As Double, dx1 As Double, dx2 As Double, dxf As Double, _
    dxf1 As Double ' = step size in x
Dim dxp As Double ' = print step size
Dim dy() As Double ' = derivative function
Dim fr As Range ' = range of derivative functions
Dim g() As Double ' = guesses for implicit solution
Dim i As Long, j As Long ' = loop index for integration and variables
Dim im As Integer, im1 As Integer ' = backward Euler/implicit trapezoidal (im1 = im-1)
Dim intmethod As String ' integration method (case 2 = Trapezoid, 1 = Backward Euler)
Dim n As Long ' = number of equations
Dim ni As Long ' = number of integration steps
Dim np As Long ' = number of print steps
Dim pc As Double ' = percent complete
Dim pcol As Long, prow As Long ' = print column, prow = print row
Dim ps As Double ' = print step
Dim res As Double ' = residual in y for implicit solution iterations
Dim sc As Integer ' = show calculations
Dim small As Double ' = small number to avoid division by zero
Dim tol As Double ' = convergence tolerance for backward Euler iterations
Dim w As Double, w_1 As Double ' = relaxation parameter, w_1 = 1-w
Dim x As Double ' = independent variable
Dim xfinal As Double, xinit As Double ' = final x/save the initial x value
Dim xp As Double ' = print variable in x
Dim xr As Range ' = range of lower integration limit
Dim y() As Double ' = dependent variables
Dim yinit() As Double ' = save the initial y conditions
Dim yr As Range ' = range of dependent variables
Dim yrows As Long ' = number of rows in yr
Dim z() As Double ' = Euler step for improved Euler method
'*************************************************************************************
small = 0.000000000001: boxtitle = "STIFFODE" ' Set small value & title
im = MsgBox(prompt:="Default method is Implicit Trapezoidal." & vbNewLine & _
```

```
                            "Use Backward Euler?", Buttons:=vbYesNo, Title:=boxtitle)
im = im - 5 ' change from 6 = Yes, 7 = No to 1 for Euler, 2 for Trapezoid
Select Case im
    Case 1
        intmethod = "Backward Euler"
    Case 2
        intmethod = "Trapezoidal"
End Select
With Application
IntLim: ' Label for inputs
Set xr = .InputBox(Title:=boxtitle, Type:=8, prompt:="Cell with INITIAL x:" & _
                   vbNewLine & "(used in derivative functions)")
xinit = xr ' save initial x
xfinal = .InputBox(Title:=boxtitle, Type:=1, prompt:="Final x:")
If xfinal < xinit Then
    MsgBox prompt:="Set x final > x init.", Buttons:=vbExclamation, Title:=boxtitle
    GoTo IntLim
End If
y_input: 'Get ranges for initial conditions and ODEs
Set yr = .InputBox(prompt:="Range of INITIAL conditions, y:" & vbNewLine & _
                   "(must be in contiguous cells)", Title:=boxtitle, Type:=8)
n = yr.Count ' Number of equations/redimension arrays
Set fr = .InputBox(prompt:="Range of DERIVATIVE function(s), dy/dx." & vbNewLine & _
                   "(use initial conditions for x and y," & vbNewLine & _
                   "must be in contiguous cells):", Title:=boxtitle, Type:=8)
If n <> fr.Count Then
    MsgBox Buttons:=vbExclamation, Title:=boxtitle, prompt:="Match IC's & ODE's."
    GoTo y_input
End If
ni = .InputBox(Title:=boxtitle, Default:=1000, prompt:="INTEGRATION steps:", Type:=1)
np = .InputBox(Title:=boxtitle, Default:=10, prompt:="PRINT steps:", Type:=1)
ni = Int(Abs(ni)): np = Int(Abs(np)) ' Number of steps must be positive integers
np = .Min(ni, np): dxf1 = xfinal - xinit: dx = dxf1 / ni: ps = dxf1 / np ' step size
tol = 0.00001 ' convergence tolerance
w = .InputBox(Title:=boxtitle, Type:=1, Default:=0.1, _
              prompt:="Iteration RELAXATION factor, 0<w<1:")
w = .Max(.Min(w, 1), 0): w_1 = 1 - w ' control the value of w to be 0<w<1
sc = MsgBox(prompt:="SHOW calculations?", Title:=boxtitle, _
            Buttons:=vbYesNo + vbQuestion + vbDefaultButton2)
If sc = 7 Then .ScreenUpdating = False
prow = output.row: pcol = output.Column: yrows = yr.Rows.Count ' row/column for output
ReDim y(1 To n) As Double, yinit(1 To n) As Double, dy(1 To n) As Double, _
      z(1 To n) As Double, g(1 To n, 0 To 2) As Double
Worksheets(ws).Cells(prow, pcol) = "x" ' Headings for output columns x & y
For i = 1 To n
    With Worksheets(ws).Cells(prow, pcol + i)
        .Value = "y" & i
        .Characters(Start:=2, Length:=Len(CStr(i))).Font.Subscript = True
    End With
Next i
With Range(Worksheets(ws).Cells(prow, pcol), Worksheets(ws).Cells(prow, pcol + n))
    .HorizontalAlignment = xlRight: .Font.Bold = True: .Font.Italic = True
End With
prow = prow + 1
For i = 1 To n: yinit(i) = yr(i): Next i ' initial conditions
dx1 = dx: dx2 = dx / im: im1 = im - 1 ' initial step sizes and index for method
With Worksheets(ws)
    .Cells(prow, pcol) = xinit ' write initial conditions to spreadsheet
    Range(.Cells(prow, pcol + 1), .Cells(prow, pcol + n)) = yinit
End With
x = xinit: y = yinit: xp = xinit + ps: dxf = xfinal - x ' initialize variables
Do While dxf / ps > small ' Integration steps. Check for integration end
    For i = 1 To n: dy(i) = fr(i): z(i) = y(i) + dx * dy(i): Next i ' Euler Step
    xr = x + dx: Call ROWCOL(n, z, yr, yrows)
    For i = 1 To n ' Modified Euler step for initial guesses for yr
        z(i) = y(i) + dx * (dy(i) + fr(i)) / 2: g(i, 0) = z(i) ' Save initial guesses
    Next i
    Call ROWCOL(n, z, yr, yrows) ' Put results on worksheet to update derivatives
    Do ' Successive substitution Iterations, set residual sum to 0 for each iteration
```

```
            res = 0
            For j = 1 To 2
                For i = 1 To n ' Successive substitution & save iterations
                    yr(i) = w_1 * yr(i) + w * (y(i) + dx2 * (im1 * dy(i) + fr(i)))
                    g(i, j) = yr(i)
                Next i
            Next j
            For i = 1 To n  ' Steffensen's Delta^2 upgrade of last three iterations
                d = g(i, 2) - 2 * g(i, 1) + g(i, 0)
                yr(i) = g(i, 0) - ((g(i, 1) - g(i, 0)) ^ 2) / (d + small): g(i, 0) = yr(i)
                ' Get the maximum relative residual for current iteration
                res = .Max(Abs(yr(i) - g(i, 2)) / (Abs(yr(i)) + small), res)
            Next i
        Loop While res > tol ' Check for convergence
        For i = 1 To n: y(i) = yr(i): Next i ' save y for next step
        x = xr: dxp = xp - x ' Upgrade x and dxp for the next integration step
        If Abs(dxp) < Abs(dx) Then ' Output intermediate results at print step
            dx = dxp: dx2 = dx / im
        End If
        If dxp / ps <= small Then ' print to worksheet when reaching print step
            prow = prow + 1
            With Worksheets(ws)
                .Cells(prow, pcol) = x
                Range(.Cells(prow, pcol + 1), .Cells(prow, pcol + n)) = y
            End With
            xp = xp + ps: dx = dx1: dx2 = dx / im ' reset integration and print step sizes
        End If
        dxf = xfinal - x
        If Abs(dxf) < Abs(dx) Then ' Check for end
            dx = dxf: dx2 = dx / im
        End If
        pc = (x - xinit) / dxf1 ' Display progress in status bar at bottom of worksheet
        .StatusBar = boxtitle & " " & VBA.Format(pc, "0%")
    Loop
    xr = xinit: dx = dx1: Call ROWCOL(n, yinit, yr, yrows) ' Reset initial conditions
End With
End Sub ' STIFFODE
```

Figure 12.10 *VBA* macro STIFFODE for backward Euler or second order implicit trapezoidal fixed step integration of first-order, ordinary differential equations.

12.4   Improved Euler's Method with Extrapolation*

An explicit form of the trapezoidal rule uses a forward Euler step to predict the solution, represented by z at $x + \Delta x$. The derivative function is then evaluated at $x + \Delta x$ from the Euler prediction for $y(x + \Delta x)$. We use the average of the derivatives to improve the solution for $y(x + \Delta x)$. This modified Euler method requires alternating derivative function evaluations between Euler and trapezoidal steps:

1. Euler Prediction Step: $z_{i+1} = y_i + \Delta x f(x_i, y_i)$ (12.55)

2. Trapezoidal Correction Step: $y_{i+1} = y_i + \dfrac{\Delta x}{2}\left[f(x_i, y_i) + f(x_{i+1}, z_{i+1})\right]$ (12.56)

The explicit trapezoidal rule requires two derivative function evaluations for each integration step. The solution to systems of differential equations follows the same approach in Section 12.1 for Euler's method. An *explicit* trapezoidal step provides roughly the same accuracy an *implicit* trapezoidal step, $O(\Delta x^2)$, without the trial-and-error solutions at each integration step. However, we lose some range of stability in the choice of step size.

Example 12.6 Predator-Prey Model

Known: The simple predator-prey model[105] has application in a variety of disciplines, including process control, biology, chemical reaction kinetics, and population dynamics:

$$\text{Predator:} \quad \frac{dx}{dt} = -x(1-y) \qquad \text{Prey:} \quad \frac{dy}{dt} = 2y(1-x) \qquad (12.57)$$

Find: Integrate the coupled system of first-order, ordinary differential equations in the domain $0 < t < 15$ for the initial conditions: $y = 0.5$, $x = 3$.

Analysis: Use the second-order, explicit trapezoidal rule in an *Excel* worksheet, shown in Figure 12.11, with the help of *VBA* user-defined functions for each of the differential equations:

```
Public Function dxdt(x As Double, y As Double): dxdt = -x * (1 - y): End Function ' Predator
Public Function dydt(x As Double, y As Double): dydt =  2 * y * (1 - x): End Function ' Prey
```

The worksheets show the first two integration steps of the explicit trapezoidal rule for predator-prey dynamics, including formulas with a named cell **dt = $B$1**.

	A	B	C	D	E	F	G
1	Δt =	0.1			Euler Step		Trapezoid Step	
2	t	x	y	dx/dt	dy/dt	dx/dt	dy/dt
3	0	0.5	3	=dxdt(B3,C3)	=dydt(B3,C3)	=dxdt(B3+dt*D3,C3+dt*E3)	=dydt(B3+dt*D3,C3+dt*E3)
4	=A3+dt	=B3+0.5*dt*(D3+F3)	=C3+0.5*dt*(E3+G3)	=dxdt(B4,C4)	=dydt(B4,C4)	=dxdt(B4+dt*D4,C4+dt*E4)	=dydt(B4+dt*D4,C4+dt*E4)
5	=A4+dt	=B4+0.5*dt*(D4+F4)	=C4+0.5*dt*(E4+G4)	=dxdt(B5,C5)	=dydt(B5,C5)	=dxdt(B5+dt*D5,C5+dt*E5)	=dydt(B5+dt*D5,C5+dt*E5)

	A	B	C	D	E	F	G
1	Δt =	0.1		Euler Step		Trapezoid Step	
2	t	x	y	dx/dt	dy/dt	dx/dt	dy/dt
3	0	0.500	3.000	1.000	3.000	1.38	2.64
4	0.1	0.619	3.282	1.413	2.501	1.925034892	1.693595416
5	0.2	0.786	3.492	1.958	1.495	2.592916642	0.13327601

Figure 12.11

Solution to the predator prey model by the explicit trapezoidal rule in an *Excel* worksheet.

Experimentation with the step size in the named cell **$B$1** indicates that the solution becomes stable for $\Delta t \leq 0.1$. The complete integration results plotted in Figure 12.12 show the dynamic relationship between the predator and prey populations. A phase plane illustrates the cyclic nature of the predator-prey dynamics.

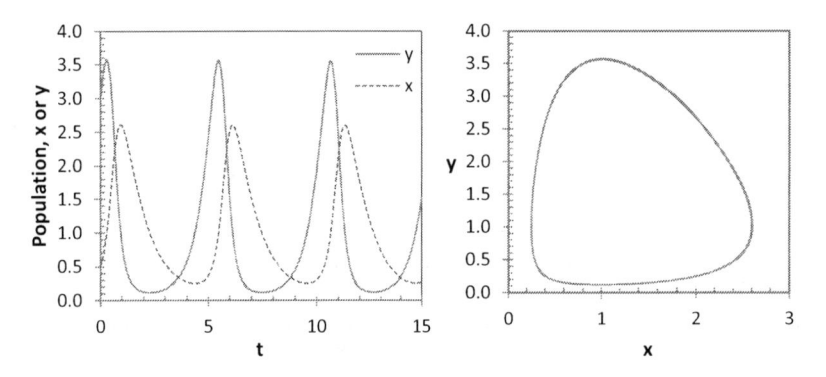

Figure 12.12

Population and phase-plane of the periodic predator-prey dynamics.

We may also integrate the system of differential equations by the implicit trapezoidal method using the macro **STIFFODE**. Set up an *Excel* worksheet as shown in Figure 12.13. The lower and upper integration limits are in cells

---

[105] Historically, the wolf-moose population dynamics on Isle Royale in Lake Superior is a wonderful example, see: http://www.isleroyalewolf.org

CHAPTER 12: INITIAL-VALUE PROBLEMS

B1:B2. The range **B3:B4** contains the initial conditions. We reuse the VBA user-defined functions for the derivatives in the range **B5:B6**.

$$B5 = dydt(B4, B3) \qquad B6 = dxdt(B4, B3)$$

STIFFODE
Click on cell with LOWER integration limit (Initial x):
$B$1

STIFFODE
UPPER integration limit (Final x):
=$B$2

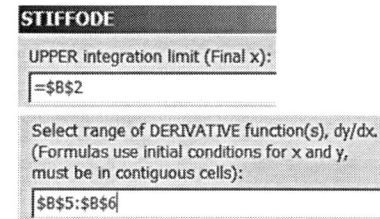

Select range of INITIAL conditions, y:
(must be in contiguous cells)
$B$3:$B$4

Select range of DERIVATIVE function(s), dy/dx.
(Formulas use initial conditions for x and y, must be in contiguous cells):
$B$5:$B$6

Because the macro is doing all of the computational work, we use the default value of 1000 integration steps with 100 print steps to capture the trends in the results (Try plotting the solution with just 10 print steps. What is missing?). Finally, we direct the output to cell **D1**. The worksheet shows the results from the first four integration steps.

	A	B	C	D	E	F
1	t low	0		t	y	x
2	t up	15		0	3	0.5
3	y init	3		0.15	3.392053	0.696448
4	x init	0.5		0.3	3.554303	1.013449
5	dy/dt	3		0.45	3.31624	1.468373
6	dx/dt	1		0.6	2.665602	1.986822

Figure 12.13

Excel worksheet setup for the macro STIFFODE.

Comments: We observe in Figure 12.12 that the predator population is out of phase with the prey population. For the conditions of this problem, the model predicts an initial increase in both prey and predator, following by a sharp decline in prey population as the hungry predator population increases, and vice versa.

□

A common mistake made when setting up the explicit form of the trapezoidal method is to solve the problem first by Euler's method over the range for the independent variable, followed by application of the trapezoidal rule. It is important to recognize that the Euler step starts from the trapezoidal result from the previous iteration, not from the previous Euler result. The improved Euler method of Hanna (1988) retains the second-order global solution accuracy with only a *single* derivative function evaluation per step. To implement the improved method, first evaluate the function for the derivative using the initial conditions:

$$f_0 = f(x_0, y_0) \qquad (12.58)$$

Then, integrate with the following recursive formula for $i = 0, 1, 2 \ldots$ until reaching the upper limit of integration:

$$z_{i+1} = y_i + \Delta x f_i \quad \rightarrow \quad x_{i+1} = x_i + \Delta x \quad \rightarrow \quad f_{i+1} = f(x_{i+1}, z_{i+1}) \quad \rightarrow \quad y_{i+1} = y_i + \frac{\Delta x}{2}[f_i + f_{i+1}] \qquad (12.59)$$

As done previously, we may extend the *improved* Euler method to systems of differential equations.

We used Romberg's implementation of Richardson's extrapolation to upgrade integral solutions in Section 11.2.4. We apply Richardson's extrapolation to the global solutions from multiple trajectories of improved Euler or trapezoidal methods for higher-order accuracy. Hanna and Sandall (1995) recommend three improved Euler trajectories with two extrapolations as a compromise between accuracy and computational efficiency. If we use integration steps that differ by a factor of two, we may apply Equation (11.42) to extrapolate the results from two trajectories, and then extrapolate the results from the first two extrapolations:

$$y_1 = \frac{4y(\Delta x/2) - y(\Delta x)}{3} \pm O(\Delta x^4) \quad y_2 = \frac{4y(\Delta x/4) - y(\Delta x/2)}{3} \pm O(\Delta x^4) \quad \rightarrow \quad y_4 = \frac{16y_2 - y_1}{15} \pm O(\Delta x^6) \qquad (12.60)$$

We illustrate the improved Euler method with extrapolation by example.

Example 12.7 Improved Euler's Method with Richardson's Extrapolation

Known/Find: Repeat *Example 12.4* using improved Euler's method with extrapolation, as shown below.

Analysis: Column **L** contains the exact solution for comparison with the numerical approximations using the improved Euler method. Columns **B:E** display the formulas for the solution using the original integration step size of $\Delta x = 0.1$ with a maximum global relative error of 1.3%. The solution displayed in columns **F:H** cuts the integration step size in half, $\Delta x = 0.05$ with a maximum global relative error of 0.34%. We used Richardson's extrapolation of the two improved Euler's method results in column **K** to produce a solution with a maximum global relative error of 0.02%, which corresponds to an error approximation in $y \pm O(\Delta x^4) = \pm 0.0001$.

	A	B	C	D	E
1	x	z	f	y(Δx)	%Error
2	1	-1	=f(A2,B2)	-1	=ABS(D2-L2)/ABS(L2)
3	1.05				
4	1.1	=D2+(A4-A2)*C2	=f(A4,B4)	=D2+(A4-A2)*(C4+C2)/2	=ABS(D4-L4)/ABS(L4)
5	1.15				
6	1.2	=D4+(A6-A4)*C4	=f(A6,B6)	=D4+(A6-A4)*(C6+C4)/2	=ABS(D6-L6)/ABS(L6)

	F	G	H	I	J	K	L
	z	f	y(Δx/2)	% Error	y Extrap	% Error	y Exact
	-1	=f(A2,F2)	-1	=ABS(H2-L2)/ABS(L2)	-1	=ABS(L2-J2)/ABS(L2)	=-1/A2
	=H2+(A3-A2)*G2	=f(A3,F3)	=H2+(A3-A2)*(G3+G2)/2				
	=H3+(A4-A3)*G3	=f(A4,F4)	=H3+(A4-A3)*(G4+G3)/2	=ABS(H4-L4)/ABS(L4)	=(4*H4-D4)/3	=ABS(L4-J4)/ABS(L4)	=-1/A4
	=H4+(A5-A4)*G4	=f(A5,F5)	=H4+(A5-A4)*(G5+G4)/2				
	=H5+(A6-A5)*G5	=f(A6,F6)	=H5+(A6-A5)*(G6+G5)/2	=ABS(H6-L6)/ABS(L6)	=(4*H6-D6)/3	=ABS(L6-J6)/ABS(L6)	=-1/A6

	A	B	C	D	E	F	G	H	I	J	K	L
1	x	z	f	y(Δx)	%Error	z	f	y(Δx/2)	% Error	y Extrap	% Error	y Exact
2	1.000	-1.00000	1.00000	-1.00000	0.00%	-1.00000	1.00000	-1.00000	0.00%	-1.00000	0.0000%	-1.00000
3	1.050					-0.95000	0.90929	-0.95227				
4	1.100	-0.90000	0.83463	-0.90827	0.09%	-0.90680	0.82852	-0.90882	0.03%	-0.90901	0.0092%	-0.90909
5	1.150					-0.86740	0.75802	-0.86916				
6	1.200	-0.82481	0.70148	-0.83146	0.22%	-0.83126	0.69617	-0.83280	0.06%	-0.83325	0.0099%	-0.83333
7	1.250					-0.79800	0.64160	-0.79936				
8	1.300	-0.76132	0.59774	-0.76650	0.35%	-0.76728	0.59321	-0.76849	0.10%	-0.76915	0.0103%	-0.76923
9	1.350					-0.73883	0.55011	-0.73991				
10	1.400	-0.70673	0.51555	-0.71084	0.48%	-0.71240	0.51155	-0.71336	0.13%	-0.71421	0.0110%	-0.71429
11	1.450					-0.68779	0.47691	-0.68865				
12	1.500	-0.65928	0.44931	-0.66260	0.61%	-0.66481	0.44568	-0.66559	0.16%	-0.66659	0.0120%	-0.66667
13	1.550					-0.64330	0.41743	-0.64401				
14	1.600	-0.61766	0.39516	-0.62037	0.74%	-0.62314	0.39178	-0.62378	0.20%	-0.62492	0.0133%	-0.62500
15	1.650					-0.60419	0.36844	-0.60478				
16	1.700	-0.58086	0.35031	-0.58310	0.87%	-0.58635	0.34712	-0.58689	0.23%	-0.58815	0.0147%	-0.58824
17	1.750					-0.56953	0.32761	-0.57002				
18	1.800	-0.54807	0.31275	-0.54995	1.01%	-0.55364	0.30970	-0.55408	0.26%	-0.55546	0.0164%	-0.55556
19	1.850					-0.53860	0.29323	-0.53901				
20	1.900	-0.51867	0.28097	-0.52026	1.15%	-0.52435	0.27804	-0.52473	0.30%	-0.52622	0.0183%	-0.52632
21	1.950					-0.51083	0.26400	-0.51118				
22	2.000	-0.49216	0.25386	-0.49352	1.30%	-0.49798	0.25101	-0.49830	0.34%	-0.49990	0.0203%	-0.50000

Comments: The improved Euler's method gives results with accuracy on par with the implicit trapezoidal method. Each extrapolation increases the accuracy by two orders of magnitude. We have learned to check the accuracy of numerical integrals by repeating the solution using half interval sizes. In this example, by halving the step size, we may trust the solution in the second decimal place. With two solutions and relatively little additional effort, we should always apply one Richardson's extrapolation to get free precision improvement, down from the second to the fourth decimal place in this example. We can always reduce the error by taking smaller integration step sizes.

□

Hanna and Sandall (1995) developed sophisticated programs that implements explicit and implicit improved Euler methods for one, two, or three solution trajectories with extrapolation using the Richardson's error estimate to adjust in integration step size and control the solution error. In the next section, we introduce alternative Runge-Kutta methods that give fourth-order accuracy at the same level as extrapolation methods with similar computational costs, but less bookkeeping.

12.5 Runge-Kutta Higher Order Methods

We may reduce the error associated with Euler's method by decreasing the step size in the independent variable. Unfortunately, as in the case of Euler's method, reducing the error by decreasing the step size often requires extremely large numbers of integration steps, which translates into many derivative function evaluations. Computer round-off errors creep into the solution for the large number of iterative calculations required for an accurate solution. The explicit trapezoidal rule reduces the error, permitting larger integration steps, by averaging the derivative function evaluations at both ends of the integration step. Following this logic, if two derivative function evaluations are better than one, then three, four, or more derivative function evaluations distributed across an integration step will reduce the error even further, permitting larger step sizes with improved accuracy. This is the basis for the variety of nth-order Runge-Kutta methods (Butcher 1996). The Runge-Kutta methods have two primary objectives:

1. Use explicit integration steps to avoid iterative solutions, which are computationally costly in terms of derivative function evaluations.
2. Reduce the local integration error while using larger integration steps to decrease the overall number of derivative evaluations, lower the computational burden at each integration step, and mitigate the effects of round-off error.

An nth order Runge-Kutta formula for a single step has the recursive formula:

$$y_{i+1} = y_i + a_1 k_1 + a_2 k_2 + \ldots a_n k_n \qquad (12.61)$$

where the k terms are the step integrations for y evaluated at various points between the start and end of an integration step x_i to $x_i + \Delta x_i$:

$$k_1 = \Delta x f(x_i, y_i) \qquad k_2 = \Delta x f(x_i + p_2 \Delta x, y_i + q_{21} k_1) \qquad k_3 = \Delta x f(x_i + p_3 \Delta x, y_i + q_{31} k_1 + q_{32} k_2) \quad \ldots \quad (12.62)$$

The Runge-Kutta method finds the corresponding coefficients $a_1, a_2, \ldots, p_2, p_3, \ldots, q_{21}, q_{31}, q_{32}\ldots$ such that Equation (12.61) is equivalent to an nth-order Taylor series expansion about point i. In the spirit of pragmatism, we skip over the considerable algebra needed to determine values for $a, p, q \ldots$ and derive a version of Runge's fourth-order method from the pattern set by first-order Euler's and second-order explicit trapezoidal rules. Higher order Runge-Kutta methods have complicated derivations that we replace with an analogy to low order methods (Kunz 1957).

Recall that for each step in Euler's method, we approximate the average value of the derivative across an integration step by the value of the derivative function evaluated with the conditions at the beginning of the step, $f(x_i, y_i)$. We may find a better approximation for the average value of the derivative (slope) over the interval by evaluating the derivative at the midpoint conditions of the step, $f(x_{i+1/2}, y_{i+1/2})$, as illustrated in Figure 12.14. Usually, the slope of the curve at the midpoint of the step in Figure 12.14 gives a better approximation for the average slope of the interval compared with the slopes at either end of the interval. To evaluate the derivative function at the midpoint we may need some approximation for the dependent variable at the midpoint. Our experience with the explicit trapezoid method gives us some direction.

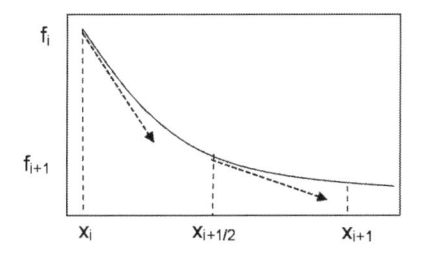

Figure 12.14

Derivative approximations at the start and midpoint of an integration step.

The explicit trapezoidal rule is an example of a second-order Runge-Kutta method. We approximate the derivative function at the midpoint of a whole integration step by taking a half Euler step, $y_{i+1/2} = y_i + k_1/2$:

$$\frac{k_1}{\Delta x} = f(x_i, y_i) \quad \text{and} \quad \frac{k_2}{\Delta x} = f\left(x_i + \frac{\Delta x}{2}, y_i + \frac{k_1}{2}\right) \tag{12.63}$$

This gives an improved approximation for y_{i+1} over the interval $x_i + \Delta x_i$:

$$y_{i+1} = y_i + k_2 \tag{12.64}$$

Equation (12.64) is an alternative *second-order* accurate Runge-Kutta marching integration method (*RK2*). The magnitude of the global error in *RK2* is $O(\Delta x^2)$ as with the explicit trapezoidal rule. As expected, we improve the accuracy of our solution at each integration step by increasing the number of derivative function evaluations appropriately averaged over the interval.

In the classical fourth-order accurate Runge-Kutta method (*RK4*), we use the second-order Runge-Kutta step to approximate the value of the derivative function, at the interval midpoint again:

$$\frac{k_3}{\Delta x} = f\left(x_i + \frac{\Delta x}{2}, y_i + \frac{k_2}{2}\right) \tag{12.65}$$

A fourth derivative function evaluation at the end of the integration step uses the third result to estimate the value of the dependent variable at $x + \Delta x$:

$$\frac{k_4}{\Delta x} = f(x_i + \Delta x, y_i + k_3) \tag{12.66}$$

We now have four approximations for the change in the dependent variable Δy_i, over the integration step Δx: k_1, k_2, k_3, and k_4. We obtained each of the four approximations from an explicit difference equation. The superiority of the approximations evaluated at the midpoint is evident in Figure 12.14. We take two half steps across the interval for a proper weighted average of the four integral k terms. Note that both k_2 and k_3 apply equally to each of the half steps, as shown in Figure 12.15. Each half step uses an average of all k terms that apply to the step:

$$y_{i+1/2} = y_i + \frac{1}{2}\left[\frac{k_1 + k_2 + k_3}{3}\right] \quad \text{and} \quad y_{i+1} = y_{i+1/2} + \frac{1}{2}\left[\frac{k_2 + k_3 + k_4}{3}\right] \tag{12.67}$$

Combining the two half steps for a full integration step gives the following fourth-order accurate classical Runge-Kutta difference equation:

$$y_{i+1} = y_i + \frac{1}{6}\left[k_1 + 2(k_2 + k_3) + k_4\right] \tag{12.68}$$

where the k terms at the midpoint of the integration step are weighted double those at the beginning and ends of the interval because they approximate the slope in both of the half-step intervals.

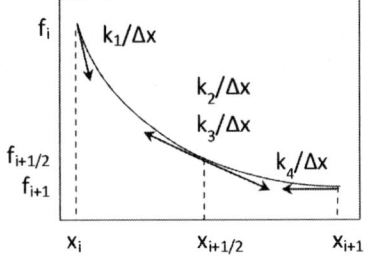

Figure 12.15

Derivative weighting in Runge-Kutta fourth order Equation (12.68).

Note that Runge's fourth-order Equation (12.68) collapses to Simpson's 1/3 rule in Equation (11.49) when the derivative function only involves the independent variable, x. The fourth-order Runge-Kutta methods have the following stability criterion for the integration step size:

CHAPTER 12: INITIAL-VALUE PROBLEMS

$$\Delta x_i < \left| \frac{1.4 y_i}{f(x_i, y_i)} \right| \qquad (12.69)$$

Illustration: Implementation of the fourth-order Runge-Kutta method in an *Excel* worksheet is tedious and error prone. Consider the solution to the initial value problem in Equation (9.1) with $C_0 = 2$ and $k = 0.5$ using an integration step of $\Delta t = 1$. Compare the exact solution with the results using each of the first-order Euler's, implicit second-order trapezoidal and fourth-order Runge-Kutta methods in an *Excel* worksheet. To simplify the implementation in the worksheet, we use a simple *VBA* user-defined function to evaluate the differential equation:

```
Public Function f(C): k = 0.5: f = -k * C: End Function
```

Formulas are shown for the first few integration steps. The implicit trapezoidal rule required iterative calculations enabled. The integration step size is in a named cell dt in cell **F1**

	A	B	C	D	E	F
1	k 0.5		C₀ 2			Δt 1
2	t	C_Exact	C_Euler	C_Trap	C_RK4	k₁
3	0	=Co	=Co	=Co	=Co	=dt*f(E3)
4	=A3+dt	=Co*EXP(-k*A4)	=C3+dt*f(C3)	=D3+0.5*dt*(f(D3)+f(D4))	=E3+(F3+2*(G3+H3)+I3)/6	=dt*f(E4)
5	=A4+dt	=Co*EXP(-k*A5)	=C4+dt*f(C4)	=D4+0.5*dt*(f(D4)+f(D5))	=E4+(F4+2*(G4+H4)+I4)/6	=dt*f(E5)

	G	H	I	J	K	L
1				Euler	Trap	RK4
2	k₂	k₃	k₄	Error	Error	Error
3	=dt*f(E3+F3/2)	=dt*f(E3+G3/2)	=dt*f(E3+H3)	=(C3-B3)/B3	=(D3-B3)/B3	=(E3-B3)/B3
4	=dt*f(E4+F4/2)	=dt*f(E4+G4/2)	=dt*f(E4+H4)	=(C4-B4)/B4	=(D4-B4)/B4	=(E4-B4)/B4
5	=dt*f(E5+F5/2)	=dt*f(E5+G5/2)	=dt*f(E5+H5)	=(C5-B5)/B5	=(D5-B5)/B5	=(E5-B5)/B5

	A	B	C	D	E	F	G	H	I	J	K	L
1	k 0.5		C₀ 2		Δt 1					Euler	Trap	RK4
2	t	C_Exact	C_Euler	C_Trap	C_RK4	k₁	k₂	k₃	k₄	Error	Error	Error
3	0	2.0000	2.0000	2.0000	2.0000	-1.0000	-0.7500	-0.8125	-0.5938	0.00%	0.00%	0.00%
4	1	1.2131	1.0000	1.2000	1.2135	-0.6068	-0.4551	-0.4930	-0.3603	-17.56%	-1.08%	0.04%
5	2	0.7358	0.5000	0.7200	0.7363	-0.3682	-0.2761	-0.2991	-0.2186	-32.04%	-2.14%	0.08%
6	3	0.4463	0.2500	0.4320	0.4468	-0.2234	-0.1675	-0.1815	-0.1326	-43.98%	-3.20%	0.12%
7	4	0.2707	0.1250	0.2592	0.2711	-0.1355	-0.1017	-0.1101	-0.0805	-53.82%	-4.24%	0.16%
8	5	0.1642	0.0625	0.1555	0.1645	-0.0822	-0.0617	-0.0668	-0.0488	-61.93%	-5.27%	0.20%
9	6	0.0996	0.0313	0.0933	0.0998	-0.0499	-0.0374	-0.0405	-0.0296	-68.62%	-6.29%	0.24%
10	7	0.0604	0.0156	0.0560	0.0606	-0.0303	-0.0227	-0.0246	-0.0180	-74.13%	-7.30%	0.28%
11	8	0.0366	0.0078	0.0336	0.0367	-0.0184	-0.0138	-0.0149	-0.0109	-78.67%	-8.30%	0.32%
12	9	0.0222	0.0039	0.0202	0.0223	-0.0111	-0.0084	-0.0091	-0.0066	-82.42%	-9.28%	0.36%
13	10	0.0135	0.0020	0.0121	0.0135	-0.0068	-0.0051	-0.0055	-0.0040	-85.51%	-10.26%	0.40%

The Runge-Kutta method improves the approximate solution over the trapezoidal and Euler's methods by at least two orders of magnitude, but requires complicated formulation in an *Excel* worksheet. Extending the Runge-Kutta fourth-order method to systems of ordinary differential equations in an *Excel* worksheet quickly becomes onerous. Fortunately, we have *VBA* macros in the *PNMSuite* to simplify the application of Runge-Kutta methods in *Excel*.

In practical application, we recommend the following Runge-Kutta-Gill (1951) version that gives numerical results with slightly higher accuracy by reducing the effects of round-off errors:

$$y_{i+1} = y_i + \frac{1}{6}\left\{ k_1 + 2\left[\left(1 - \frac{1}{\sqrt{2}}\right)k_2 + \left(1 + \frac{1}{\sqrt{2}}\right)k_3\right] + k_4 \right\} \qquad (12.70)$$

$$k_1 = \Delta x f(x_i, y_i) \qquad (12.71)$$

$$k_2 = \Delta x f\left(x_i + \frac{\Delta x}{2}, y_i + \frac{k_1}{2}\right) \qquad (12.72)$$

$$k_3 = \Delta x f\left[x_i + \frac{\Delta x}{2}, y_i + \left(\frac{1}{\sqrt{2}} - \frac{1}{2}\right)k_1 + \left(1 - \frac{1}{\sqrt{2}}\right)k_2\right] \quad (12.73)$$

$$k_4 = \Delta x f\left[x_i + \Delta x, y_i - \frac{k_2}{\sqrt{2}} + \left(1 + \frac{1}{\sqrt{2}}\right)k_3\right] \quad (12.74)$$

The VBA macro **RK4G**, with input boxes, and the user-form **ODEFIXSTEP** solve a system of first-order, ordinary differential equations by the fourth-order Runge-Kutta-Gill method. The worksheet setup, user input requirements, and output code are essentially identical to those required by **STIFFODE** in Figure 12.10. To use the macro, the user must populate worksheet cells and ranges for the control parameters, integration limits, initial conditions, and derivative functions in terms of the cells containing the initial conditions for the dependent variables and lower integration limit of the independent variable.

Example 12.8 Batch Enzyme Catalyzed Reactions

Known: Three first-order differential equations describe the enzyme-catalyzed reactions of a substrate in a batch reactor depicted in Figure 1.24, according to the reaction mechanism in Equations (1.29) and (1.30):

$$\frac{dS}{dt} = -k_1 S[E_0 - (ES)] + k_2(ES) \quad (12.75)$$

$$\frac{d(ES)}{dt} = k_1 S[E_0 - (ES)] - (k_2 + k_3)(ES) \quad (12.76)$$

$$\frac{dP}{dt} = k_3(ES) \quad (12.77)$$

where E_0, S, (ES), and P are the concentrations of initial enzyme, substrate, enzyme-substrate complex, and product species, respectively. The system of first-order differential equations is subject to the following initial conditions and reaction rate constants:

$S_0 = 0.1\,M$ $\quad\quad E_0 = 0.01\,M$ $\quad\quad (ES)_0 = P_0 = 0$

$k_1 = 40\dfrac{\ell}{mol \cdot min}$ $\quad\quad k_2 = 5\dfrac{1}{min}$ $\quad\quad k_3 = 0.5\dfrac{1}{min}$

Find: Plot the changes in concentrations with time for the substrate, intermediate complex, and product.

Assume: Constant temperature, well mixed tank

Analysis: To use the VBA macro **RK4G**, place the initial conditions (y) in contiguous cells in a row or column on an Excel worksheet. Formulate the derivative functions in a contiguous row or column of cells. Define the formulas for the derivative functions in terms of the values for the dependent variables, located in the worksheet range for y, as shown in Figure 12.5. According to the stability criterion in Equation (12.69), the step size must be smaller than $\Delta t < 3.5$ min. Use a conservative integration step size of 0.1 to avoid instability in the numerical solution (compared with 0.0001 required for a stable Euler's method).

	A	B	C	D
1	Number of integration steps	1000		
2	Number of print steps	10		
3	Initial t (x)	0		
4	Final t (xf)	100		
5	Initial Conditions S, ES, P (y)	0.1	0	0
6	Derivative Functions (dydx)	=-40*B5*(0.01-C5)+5*C5	=40*B5*(0.01-C5)-5.5*C5	=0.5*C5

Figure 12.16

Excel worksheet formulas for solving a system of three ordinary differential equations using the VBA macro RK4G.

Use the default values of 1000 integration and 10 print steps. Direct the output of the results to the worksheet, beginning in cell **A8**, as shown in Figure 12.17. Plot the solution to the system of first-order differential equations for the three dependent variables with the macro **GRAPHYXGUIDE**.

	A	B	C	D
8	x	y₁	y₂	y₃
9	0	0.1	0	0
10	10	0.077293352	0.003604563	0.019102086
11	20	0.061096979	0.003082963	0.035820058
12	30	0.047475163	0.002573287	0.049951549
13	40	0.036299222	0.002095137	0.06160564
14	50	0.027351109	0.00166523	0.070983662
15	60	0.020348552	0.001294492	0.078356956
16	70	0.014978943	0.000986892	0.084034165
17	80	0.010932207	0.000740187	0.088327607
18	90	0.007925354	0.000547864	0.091526782
19	100	0.005716084	0.000401334	0.093882582

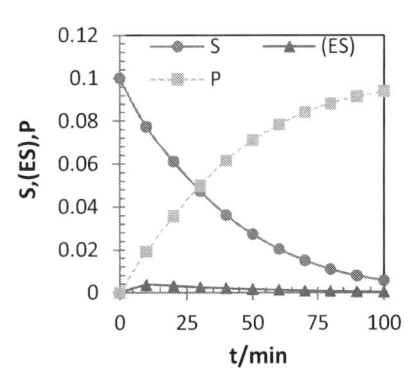

Figure 12.17

Integration results for using the *Excel VBA* macro RK4G. Markers are used to distinguish between the lines for the dependent variable trajectories.

Comments: The enzyme-substrate complex concentration (y_2) never reaches appreciable levels. The model of this reacting system is a candidate for a simplifying assumption, such as pseudo-steady-state concentration for (*ES*) (Menten and Michaelis 1913) and (Briggs and Haldane 1925).

□

Runge-Kutta methods come in a variety of flavors for approximating solutions to first-order differential equations. The fourth-order accurate Runge-Kutta methods are popular based on their balance between improved accuracy and computational cost, when compared with the lower-order accurate Euler and trapezoid methods that require significantly smaller integration steps. Third-order Runge-Kutta methods have issues with numerical stability, whereas higher order accurate methods tend towards an unfavorable tradeoff with computational efficiency (W. Press, S. Teukolsky and S. Vetterling, et al. 1992).

12.5.1 Integration Accuracy versus Precision

"[Some facts] can be seen more clearly by an example than by a proof." – Leonhard Euler

We need to check the accuracy of fixed-step solutions by experimenting with the number of integration steps. Check the accuracy of numerical integration by comparing the precision of the results using successively smaller integration step sizes, in a way similar to the bisection test for the accuracy of integral solutions, as described in Section 11.2.2. If the solution at decreasing step sizes does not change within the degree of precision required by the problem, we may be inclined to trust the result. Variable step-size integration methods are available to control the local integration error, which takes the guesswork out of picking an appropriate integration step size.

Slow computation time is the primary disadvantage when accessing ranges on the worksheet for evaluating the variables and functions of integration. To speed up the computation, we may modify the macros with a sub procedure for evaluating the differential equations in the code, instead of using the worksheet to calculate the functions.

Example 12.9 Speedup Numerical IVP Solutions with User-defined Functions

Known/Find: The *VBA* macro **RK4G** uses an *Excel* worksheet to evaluate variables and functions of integration for systems of ordinary differential equations. To speed up the calculations, modify the macro with a sub procedure for evaluating the differential equations. Test the modified macro using the problem from *Example 12.8*.

Analysis: Create three new macros, **RK4G_sub**, **INITC**, and **DYDX**, for implementing Runge-Kutta-Gill and specifying initial conditions and functions of the differential equations. Start from the code for the macro **RK4G**, available from the *PNMSuite*. Code the initial conditions and functions for the differential equations from *Example 12.8*:

```
Private Sub INITC(tol, im, n, ni, np, w, output, xi, xf, yi)
' tol = integration relative error control tolerance
' im = 1 = implicit Euler, 2 = implicit trapezoidal methods (used for STIFFODEs only
' n = number of equations              ni = initial number of integration steps
' np = number of print steps           w = relaxation factor used by STIFFODEs
' output = range with cell location for top left corner of output
' xi,xf = initial/final values for independent variable,  yi() = initial y conditions
'****************************************************************************
' Used with subs STIFFODE_s, RK45ODE_s, or RK45DP_s.
' Initialize the parameters for the system of ODEs.  The user must assign the range
' for the worksheet output, the number of equations, the number of print steps, and
' the error control tolerance.  Specify the integration limits for the independent
' variable and the initial conditions for the dependent variables of integration.
'****************************************************************************
Set output = Range("A17") ' Set output location on worksheet
n = 3: ni = 1000 ' Number of differential equations & Initial number of integration steps
np = 10: tol = 0.00001 ' number of print steps, relative integration error control tolerance
w = 0.1 ' relaxation factor for STIFFODE only
im = 2 ' 1 for Euler, 2 for Trapezoidal method (STIFFODEs only)
xi = 0: xf = 100 ' Assign initial and final values for independent variable
ReDim yi(1 To n) As Double ' Redimension size of yi array to match problem size
yi(1) = 0.1: yi(2) = 0: yi(3) = 0' Assign the initial conditions for dependent variables
End Sub

Private Sub DYDX(n, x, y, f)
' f() = array of derivative functions
' n = number of equations    x = independent variable y() = array of dependent variables
'****************************************************************************
' Used with subs STIFFODE_s, RK45ODE_s, or RK45DP_s.
' Evaluate the derivative functions dy/dx = f(x,y) in terms of x and y.
' The user must declare any new variables in a DIM statement.
' *** Nomenclature ***
Dim e0 As Double ' = initial enzyme concentration
Dim es As Double ' = concentration of enzyme-substrate complex
Dim k1 As Double, k2 As Double, k3 As Double ' = rate constants
Dim p As Double, s as Double ' = product concentration, substrate concentration
'****************************************************************************
k1 = 40: k2 = 5: k3 = 0.5: e0 = 0.01 ' define model parameters
s = y(1): es = y(2): p = y(3) ' provide variable names for the dependent variables
f(1) = -k1 * s * (e0 - es) + k2 * es ' set the expressions for derivative function array
f(2) = k1 * s * (e0 - es) - (k2 + k3) * es: f(3) = k3 * es
End Sub
```

In the sub procedure **RK4G_sub** we delete the input boxes, call sub **INITC** to assign the initial conditions, and replace references to xr, yr, and fr with x, y, and f from sub **INITC**. We then call the sub **DYDX** in place of referencing cells on the worksheet to calculate the derivative functions at each integration step:

```
Do While dxf / ps > small ' RK integration step. Check for the end of integration
    Call dydx(x, y, f)
    For i = 1 To n: k1(i) = dx * f(i): y1(i) = y(i) + 0.5 * k1(i): Next i
    Call dydx(x + 0.5 * dx, y1, f)
    For i = 1 To n: k2(i) = dx * f(i): y1(i) = y(i) + c * k1(i) + a * k2(i): Next i
    Call dydx(x + 0.5 * dx, y1, f)
    For i = 1 To n: k3(i) = dx * f(i): y1(i) = y(i) - d * k2(i) + b * k3(i): Next i
    Call dydx(x + dx, y1, f)
```

CHAPTER 12: INITIAL-VALUE PROBLEMS

```
        For i = 1 To n: k4(i) = dx * f(i): Next i
        For i = 1 To n ' Put it all together for the integration step
            y1(i) = y(i) + (k1(i) + 2 * (a * k2(i) + b * k3(i)) + k4(i)) / 6
        Next i
        y = y1: x = x + dx: dxp = xp - x ' Keep solution for next integration step
        If Abs(dxp) < Abs(dx) Then dx = dxp ' Output at print step
        If dxp / ps <= small Then
            prow = prow + 1: Cells(prow, pcol) = x
            Range(Cells(prow, pcol + 1), Cells(prow, pcol + n)) = y: xp = xp + ps: dx = dx1
        End If
        dxf = xf - x ' Check for end of integration
        If Abs(dxf) < Abs(dx) Then dx = dxf
    Loop
```

Set the number of print steps to 50 and get the number of integration steps from **Cell B1** on the worksheet. We arrive at the solution in significantly faster time when compared with **RK4G**. We can now explore the effect of the number of integration steps. For illustration, modify the worksheet to run the macro automatically whenever the user changes the number of integration steps defined in **Cell B1**. We use the following code in the Microsoft *Excel* Objects/Sheet1.

```
Private Sub worksheet_Change(ByVal target As Range)
    Select Case target.Address
        Case "$B$1": Call RK4ODEs: End Select
End Sub
```

We then plotted the results using 250, 300, 350, and 400 integration steps in Figure 12.18 to show how the numerical solution to the differential equations behaves as the integration step size decreases. With relatively larger steps, the solution for the reactive intermediate enzyme-substrate complex *es* is unstable, which is an indication of a stiff set of equations – a good candidate for the macro **STIFFODE_sub**. At larger steps, with ni > 400, the solution becomes stable, and stops changing in the fifth decimal place.

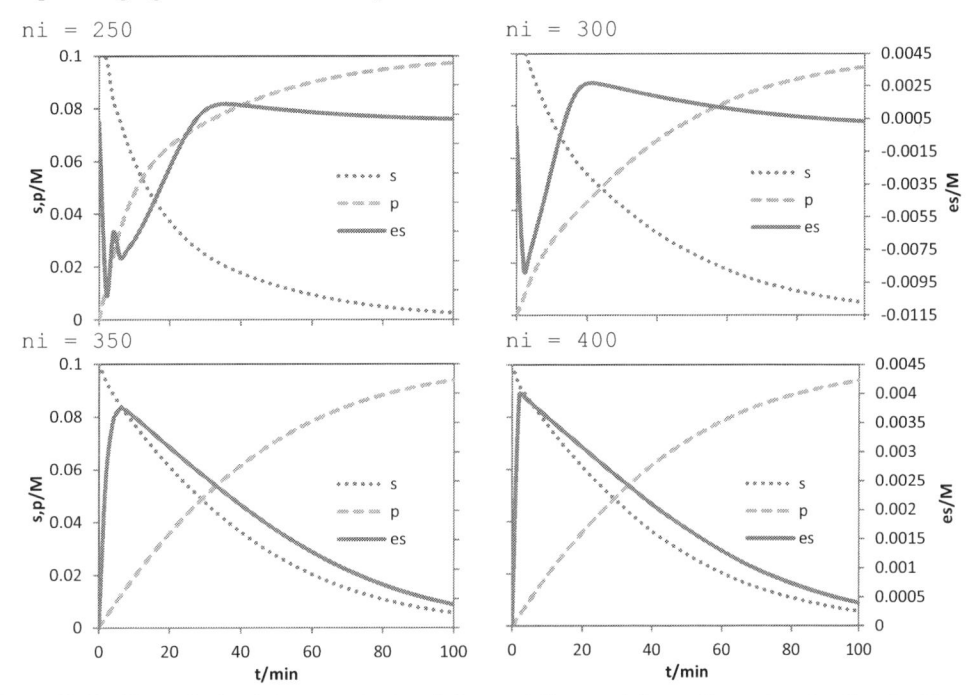

Figure 12.18 Numerical solutions to the batch fermentation model equations showing the significance of the number of steps (or step size) in achieving a stable solution.

Comments: The solution method using *VBA* procedures instead of worksheet evaluations is significantly faster, but requires the user to access the *VBE* and modify the code for the initial conditions and derivative functions.

12.5.2 Variable Step Runge-Kutta Methods

> *"Enjoying success requires the ability to adapt. Only by being open to change will you have a true opportunity to get the most from your talent."* – Baseball Hall of Famer, Nolan Ryan

The classical Runge-Kutta methods employ a uniform step size. While particular regions of the solution may require small integration steps to control error and stability, other regions may not require such a small step. If we allow the integration step size to vary, we can judiciously take small steps when necessary to control stability and error in the solution, and larger steps in stable regions. A variable step-size integration method may improve the computational efficiency of the solution by further reducing the overall number of derivative function evaluations.

The stability criterion in Equation (12.69) provides some guidance on the choice of step size, but does not consider the error in each integration step. We may estimate local integration errors from the difference between fourth-order and fifth-order accurate Runge-Kutta steps. You yell, "Foul! This will now require four plus five, or a total of nine derivative function evaluations for one step!" The trick is to find pairs of 4th and 5th order difference equations that share the derivative function evaluations, similar to the Gauss-Kronrod method of adaptive quadrature. Fehlberg was the first to find a scheme that only required six derivative evaluations. The *PNMSuite* workbook has four such schemes, including Fehlberg (1970), Merson (1957), Cash-Karp (1990), and Dormand-Prince (1980).

Variable step versions of the fourth-order Runge-Kutta routine have a common formula for weight averaging the derivative function evaluations in a step:

$$k_{i1} = \Delta x_i f(x_i, y_i) \tag{12.78}$$

$$k_{ij+1} = \Delta x_i f\left(x_i + a_j \Delta x_i, y_i + \sum_{l=1}^{j} b_{jl} k_l\right) \quad j = 1, 2, \ldots 6 \tag{12.79}$$

where *i* and *j* are indexes for the variables and Runge-Kutta approximations, respectively. We use the results from the Runge-Kutta steps to calculate fourth and fifth-order accurate approximation for the solution at the end of the step:

$$y_{i+1} = y_i + \sum_{j}^{4,5,\text{ or } 6} c_j k_{i,j} \tag{12.80}$$

where c_j are weight factors. We then estimate the error in the step from the difference between the fourth and fifth order accurate solutions:

$$\Delta y_{i+1} = \sum_{j=1}^{4,5,\text{ or } 6} \Delta c_j k_{ij} \tag{12.81}$$

We select the initial step size to give smaller integration steps than print steps, starting with a value smaller than 0.1. The absolute relative error in the solution is:

$$\varepsilon_{calc} = \left|\frac{\Delta y_{i+1}}{y_{i+1} + \delta}\right| \tag{12.82}$$

where we use the small number δ to avoid division by zero. We adjust the new step size from the old value to reflect the fourth-order accuracy of the solution:

$$\left(\frac{\Delta x_{new}}{\Delta x_{old}}\right)^4 = \left(\frac{tol}{\varepsilon_{calc}}\right) \quad \text{or} \quad \Delta x_{new} = \Delta x_{old} \left(\frac{tol}{\varepsilon_{calc}}\right)^{1/4} \tag{12.83}$$

To control the integration step size for systems of differential equations, use the maximum estimated error for any of the dependent variables. We note that our variable step algorithm requires five derivative function evaluations for each integration step – more if the step size requires downward adjustment to satisfy the error criterion before taking the next step. This may not feel like a gain in computational efficiency compared with the classical fourth-order fixed-step method. However, if we compare a case where the total number of integration steps using a variable

CHAPTER 12: INITIAL-VALUE PROBLEMS

step method is 50% fewer than the fixed-step method, we will find that we use ~60% fewer derivative-function evaluations, *and* control the error – which represents a significant gain in efficiency!

If the implementation of the fixed-step Runge-Kutta method in an *Excel* worksheet is impractical, then we certainly will not try to set up the variable-step method directly in an *Excel* worksheet. Instead, we provide general-purpose *VBA* macros **RK45ODE**, **RK45DP**, and user-form **RK45VS** for solving systems of first-order, ordinary differential equations using a variable-step (adaptive) method. **RK45ODE** has three options: Cash-Karp (default), Fehlberg, and Merson. **RK45DP** uses the parameters required in Equation (12.80) discovered by Dormand and Prince.

The Dormand-Prince adaptation of the Runge-Kutta scheme uses seven derivative approximations, but only requires six function evaluations by reusing the last function evaluations at the start of the next step.[106] The *VBA* code for the macro **RK45DP** listed in Figure 12.19 contains the unique Runge-Kutta constants for Dormand-Prince. An alternative, faster version **RK45DP_sub** uses the sub procedures **INITC** and **DYDX** listed in *Example 12.9* to provide the initial conditions and user-coded derivative functions to speed up the computations, in place of referencing ranges on the worksheet.

The worksheet setup is similar to the requirements for the user-form **ODEFIXSTEP**. One difference among the variable step macros and the fixed-step methods is the absence of the requirement to specify the number of integration steps. Instead, we use an input box for specifying the relative integration error tolerance. The number of integration steps is automatically determined to control the error. The variable step integration macros use a default maximum number of integration steps of 10^5 to terminate the integration if unreasonably small steps are required. In this case, we have the option of increasing the maximum number of integration steps. However, we recommend using a stiff integrator, like the implicit trapezoidal method when the variable step method requires unusually small steps.

```
Public Sub RK45DP()
' Runge-Kutta 4-5th order variable step numerical integration method using Dormand-
' Prince parameters. This program solves a set of first order ode's subject to the
' initial conditions, y(i) over the range of independent variable from x0 to xfinal.
' The user specifies the initial conditions and the derivative functions in contiguous
' ranges on the Excel worksheet.  Derivative functions reference initial conditions.
' *** Nomenclature ***
Dim a1 As Double, a2 As Double, a3 As Double, a4 As Double, _
    a5 As Double, a6 As Double, b11 As Double, b21 As Double, b31 As Double, _
    b41 As Double, b51 As Double, b22 As Double, b32 As Double, b42 As Double, _
    b52 As Double, b33 As Double, b43 As Double, b53 As Double, b44 As Double, _
    b54 As Double, b55 As Double, b61 As Double, b62 As Double, b63 As Double, _
    b64 As Double, b65 As Double, b66 As Double, c11 As Double, c12 As Double, _
    c13 As Double, c14 As Double, c15 As Double, c16 As Double, c17 As Double, _
    c21 As Double, c22 As Double, c23 As Double, c24 As Double, c25 As Double, _
    c26 As Double, c27 As Double, dc1 As Double, dc2 As Double, dc3 As Double, _
    dc4 As Double, dc5 As Double, dc6 As Double, _
    dc7 As Double ' = Dormand-Prince coefficients (ii represents integers)
Dim boxtitle As String ' = string title for message and input boxes
Dim del As Double ' = normalized scaled error in integration step
Dim dely As Double ' = relative error in y
Dim dx As Double ' = distance in x to end of integration
Dim dxf As Double ' = final step size
Dim dxfl As Double ' = range of x integration
Dim dxp As Double ' = distance in x to end of print step
Dim dy() As Double ' = derivatives at beginning of integration step
Dim fr As Range ' = range of derivative functions
Dim h As Double, hmin As Double ' = step size in x, hmin = minimum step in x
Dim i As Long ' index for variables
Dim k1() As Double, k2() As Double, k3() As Double, k4() As Double, _
    k5() As Double, k6() As Double, k7() As Double ' = Runge-Kutta steps
Dim n As Long ' = number of equations
Dim np As Long ' = number of equally-spaced print steps
Dim pc As Double ' = percent complete
Dim pcol As Long, prow As Long ' = column/row number for printout
Dim ps As Double ' = print parameter
Dim sc As Integer ' = show calculations
Dim small As Double ' = small number avoids division by zero
```

---

[106] Variable step Dormand-Prince is the Runge-Kutta variant of ODE solver used by Matlab, Octave, and SciLab.

```vb
Dim tol As Double ' = step size error control tolerance
Dim x As Double ' = initial value for x
Dim xfinal As Double, xinit As Double ' = final value for x, xinit = save initial x
Dim xp As Double ' = next print x
Dim xr As Range ' = range of lower integration limit
Dim y() As Double, y1() As Double, y2() As Double, y3() As Double, y4() As Double, _
    y5() As Double, y6() As Double ' = dependent variables 1-6 order
Dim yinit() As Double ' = save the initial y conditions
Dim yr As Range ' = range of initial conditions used for dependent variables y
Dim yrows As Long ' = number of rows in range yr
'****************************************************************************
small = 0.000000000001: boxtitle = "RK45DP"
a1 = 0.2: a2 = 0.3: a3 = 0.8: a4 = 8 / 9: a5 = 1: a6 = 1 ' Dormand-Prince constants
b11 = 0.2: b21 = 3 / 40: b22 = 9 / 40
b31 = 44 / 45: b32 = -56 / 15: b33 = 32 / 9
b41 = 19372 / 6561: b42 = -25360 / 2187: b43 = 64448 / 6561: b44 = -212 / 729
b51 = 9017 / 3168: b52 = -355 / 33: b53 = 46732 / 5247
b54 = 49 / 176: b55 = -5103 / 18656
b61 = 35 / 384: b62 = 0: b63 = 500 / 1113: b64 = 125 / 192: b65 = -2187 / 6784
b66 = 11 / 84
c11 = 5179 / 57600: c12 = 0: c13 = 7571 / 16695: c14 = 393 / 640
c15 = -92097 / 339200: c16 = 187 / 2100: c17 = 1 / 40
c21 = 35 / 384: c22 = 0: c23 = 500 / 1113: c24 = 125 / 192
c25 = -2187 / 6784: c26 = 11 / 84: c27 = 0
dc1 = c11 - c21: dc2 = c12 - c22: dc3 = c13 - c23: dc4 = c14 - c24
dc5 = c15 - c25: dc6 = c16 - c26: dc7 = c17 - c27
With Application
IntLim: ' Integration limits & Initialize the control variables
Set xr = Application.InputBox(Title:=boxtitle, Type:=8, prompt:="Cell with x INITIAL:")
x = xr: xinit = x ' initial x
xfinal = .InputBox(Title:=boxtitle, Type:=1, prompt:="x FINAL:")
If xfinal < x Then
    MsgBox prompt:="x Final must be > x Initial.", _
                   Buttons:=vbExclamation + vbOKOnly, Title:=boxtitle
    GoTo IntLim ' Try again
End If
dxf1 = xfinal - xinit ' range of x integration
h = (xfinal - x) * 0.001: hmin = Abs(xfinal - x) * small ' initial/minimum step size
np = .InputBox(Title:=boxtitle, prompt:="Print steps:", Default:=10, Type:=1)
np = Abs(Int(np)) ' number of print steps
If np = 0 Then
    MsgBox prompt:="Print steps must be > 0.", Buttons:=vbExclamation, Title:=boxtitle
    GoTo IntLim
End If
ps = (xfinal - x) / np ' print step size
If ps < h Then
    MsgBox prompt:="Too many print steps.  Must be < 1E8.", _
           Buttons:=vbExclamation, Title:=boxtitle
    GoTo IntLim
End If
yinput: ' Label, Get ranges for ICs and ODE's
Set yr = .InputBox(prompt:="Range of INITIAL CONDITIONS (y):" & vbNewLine & _
                   "(must be in contiguous cells)", Title:=boxtitle, Type:=8)
n = yr.Count ' Number of equations
Set fr = .InputBox(prompt:="Range of DERIVATIVE functions dy/dx:" & vbNewLine & _
                   "(formulas use cells for x,y initial conditions, " & vbNewLine & _
                   "must be in contiguous cells)", Title:=boxtitle, Type:=8)
If n <> fr.Count Then
    MsgBox Buttons:=vbExclamation, Title:=boxtitle, _
           prompt:="Number of initial conditions and ODE's must match."
    GoTo yinput
End If
tol = .InputBox(prompt:="ERROR Tolerance", Title:=boxtitle, Default:=0.00001, Type:=1)
tol = .Max(tol, small * 10)
sc = MsgBox(prompt:="SHOW calculations?", Title:=boxtitle, _
            Buttons:=vbYesNo + vbQuestion + vbDefaultButton2)
If sc = 7 Then .ScreenUpdating = False
ReDim k1(1 To n) As Double, k2(1 To n) As Double, k3(1 To n) As Double, _
      k4(1 To n) As Double, k5(1 To n) As Double, k6(1 To n) As Double, _
```

CHAPTER 12: INITIAL-VALUE PROBLEMS

```
        k7(1 To n) As Double, dy(1 To n), y(1 To n) As Double, y1(1 To n) As Double, _
        y2(1 To n) As Double, y3(1 To n) As Double, y4(1 To n) As Double, _
        y5(1 To n) As Double, y6(1 To n) As Double, yinit(1 To n) As Double
' Get the row and column for the output
prow = output.row: pcol = output.Column: yrows = yr.Rows.Count
Worksheets(ws).Cells(prow, pcol) = "x" ' Format header for table of output
For i = 1 To n
    With Worksheets(ws).Cells(prow, pcol + i)
        .Value = "y" & i
        .Characters(Start:=2, Length:=Len(CStr(i))).Font.Subscript = True
    End With
Next i
With Range(Worksheets(ws).Cells(prow, pcol), Worksheets(ws).Cells(prow, pcol + n))
    .HorizontalAlignment = xlRight: .Font.Bold = True: .Font.Italic = True
End With
prow = prow + 1
Worksheets(ws).Cells(prow, pcol) = xinit ' write initial conditions to spreadsheet
For i = 1 To n: y(i) = yr(i): yinit(i) = y(i): Next i
Range(Worksheets(ws).Cells(prow, pcol + 1), Worksheets(ws).Cells(prow, pcol + n)) = y
xp = ps + x ' Next x value for printing the output.
Do While (xfinal - x) / ps >= small ' Check for end & replace initial conditions
    For i = 1 To n: dy(i) = fr(i): Next i ' save derivative evaluations at step start
    Do ' Loop for variable step size adjustment. Runge-Kutta integration steps
        For i = 1 To n: k1(i) = h * dy(i): y1(i) = y(i) + b11 * k1(i): Next i
        xr = x + a1 * h: Call ROWCOL(n, y1, yr, yrows)
        For i = 1 To n
            k2(i) = h * fr(i): y2(i) = y(i) + b21 * k1(i) + b22 * k2(i)
        Next i
        xr = x + a2 * h: Call ROWCOL(n, y2, yr, yrows)
        For i = 1 To n
            k3(i) = h * fr(i)
            y3(i) = y(i) + b31 * k1(i) + b32 * k2(i) + b33 * k3(i)
        Next i
        xr = x + a3 * h: Call ROWCOL(n, y3, yr, yrows)
        For i = 1 To n
            k4(i) = h * fr(i)
            y4(i) = y(i) + b41 * k1(i) + b42 * k2(i) + b43 * k3(i) + b44 * k4(i)
        Next i
        xr = x + a4 * h: Call ROWCOL(n, y4, yr, yrows)
        For i = 1 To n
            k5(i) = h * fr(i)
            y5(i) = y(i) + b51 * k1(i) + b52 * k2(i) _
                + b53 * k3(i) + b54 * k4(i) + b55 * k5(i)
        Next i
        xr = x + h: Call ROWCOL(n, y5, yr, yrows)
        For i = 1 To n
            k6(i) = h * fr(i)
            y6(i) = y(i) + b61 * k1(i) + b63 * k3(i) _
                + b64 * k4(i) + b65 * k5(i) + b66 * k6(i)
        Next i
        Call ROWCOL(n, y6, yr, yrows)
        del = 0 ' Error control with scaled dependent variables
        For i = 1 To n
            k7(i) = h * fr(i)
            y5(i) = y(i) + c21 * k1(i) + c22 * k2(i) + c23 * k3(i) _
                + c24 * k4(i) + c25 * k5(i) + c26 * k6(i) + c27 * k7(i)
            dely = Abs(dc1 * k1(i) + dc2 * k2(i) + dc3 * k3(i) _
                + dc4 * k4(i) + dc5 * k5(i) + dc6 * k6(i) + dc7 * k7(i))
            del = .Max(del, dely / (Abs(y5(i)) + small))
        Next i
        ' Check for error in integration step (w/ safety factor)
        If del < tol Or Abs(h) < hmin Then Exit Do
        h = 0.84 * h * (tol / del) ^ 0.25
    Loop
    y = y5: x = x + h: xr = x ' Keep best solution for next step
    If (xp - x) / ps <= small Then ' Control step size for print control
        prow = prow + 1
        With Worksheets(ws)
            .Cells(prow, pcol) = x
```

```
                Range(.Cells(prow, pcol + 1), .Cells(prow, pcol + n)) = y5
            End With
        xp = xp + ps
    End If

    h = 0.87 * h * (tol / (del + small)) ^ 0.2 ' Adjust step size (w/ safety factor)
    h = .Max(Abs(h), hmin) * Sgn(h) ' Check minimum step size, keep sign of step
    dxp = xp - x ' Control step size to avoid integrating beyond upper limit
    If Abs(dxp) < Abs(h) Then h = dxp
    pc = (x - xinit) / dxf1: .StatusBar = boxtitle & ": " & VBA.Format(pc, "0%")
Loop
xr = xinit: Call ROWCOL(n, yinit, yr, yrows) ' Replace initial conditions
End With
End Sub ' RK45DP
```

Figure 12.19 *VBA* macro RK45DP variable step Runge-Kutta-Dormand-Prince integration with error control for a system of 1st-order ordinary differential equations.

Example 12.10 Model PI Control of a Well-mixed Heated Tank

Known: A flow-through tank uses an electric heater to control the temperature. The feed stream to the tank consists of water at a rate of 100 kg/s and temperature of 55°C. A thermocouple with a response time constant $\tau_{tc} = 5$ seconds measures the tank effluent temperature, T_{tc}, as shown in the schematic. The heater uses proportional integral (*PI*) control to maintain the temperature at the set point value of 75°C:

$$q = q_{ss} + K\left(T_{sp} - T_{tc} + \frac{\Delta}{\tau}\right) \tag{12.84}$$

where the parameters and variables are defined in an *Excel* worksheet shown in Figure 12.20.

	A	B	C	D
1	Parameters	Cell Name	Value	Units
2	Density	rho	1000	kg/m3
3	Specific Heat	c_p	4.2	kJ/kg °C
4	Mass Flow Rate	m	100	kg/s
5	Tank Volume	V	1	m3
6	Set Point Temperature	T_sp	75	°C
7	Initial Feed Temperature	T_0	55	°C
8	Feed Step Change	T_step	30	°C
9	Steady State Heat Rate	q_ss	8400	kW
10	Controller Gain	K	4	kW/°C
11	Integral Time Constant	tau	2	sec
12	Thermal Couple Time Constant	tau_tc	5	sec
13	Initial Time	t_o	0	sec
14	Final Time	t_f	200	sec
15	Inlet Temperature	T_in	55	°C

Figure 12.20

Excel worksheet with control parameters.

Schematic: A well-mixed, heated, flow through tank with a *PI* temperature controller uses an immersed electric heating coil. The mass flow rates of the influent and effluent, $\dot{m}$, are considered constant.

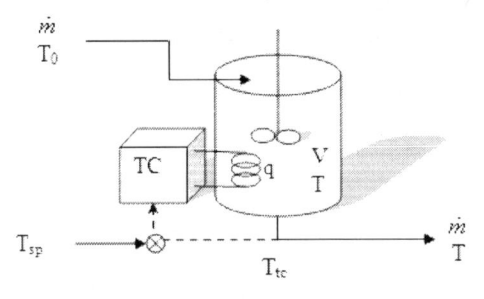

CHAPTER 12: INITIAL-VALUE PROBLEMS

Find: Plot the tank and thermocouple temperature responses to a step change of –30°C in the feed temperature for controller gains of 4, 30, 200, and 1000 kW/°C. Integrate over a period of 200 seconds with the step change occurring after 10 seconds.

Assumptions: Steady-state mass flow rate in and out of the tank, well-mixed, constant properties

Analysis: Calculate the steady-state heating rate from an energy balance:

$$q_{ss} = \dot{m}c_p \left(T_{sp} - T_0\right) \tag{12.85}$$

A dynamic energy balance around the system of the water in the tank (*rate of accumulation = rate in - rate out*) includes the advection terms in and out of the tank and the heat transfer rate (q) from the immersed heater:

$$\rho V c_p \frac{dT}{dt} = q + \dot{m}c_p \left(T_0 - T\right) \tag{12.86}$$

where T = effluent temperature, °C. Assume a first-order model for the thermocouple temperature:

$$\frac{dT_{tc}}{dt} = \frac{T - T_{tc}}{\tau_{tc}} \tag{12.87}$$

The differential equation for the temperature deviation Δ used for integral control includes the difference between the set point and thermocouple values:

$$\frac{d\Delta}{dt} = T_{sp} - T_{tc} \tag{12.88}$$

Initialize all temperatures at the desired set point, then use *Excel* and the VBA macro **RK45ODE** to solve the system of first-order, ordinary differential Equations (12.86), (12.87), and (12.88). Use named cells for the model parameters in the *Excel* worksheet shown in Figure 12.20. The cell names for column **C** are listed in column **B** on the *Excel* worksheet. Specify the initial conditions and heat rate with Equation (12.84) in row **18**. The formula for the heat rate in worksheet cell **E18** uses values in the named cells:

E18 = qss + K*(Tsp – Ttc + Δ/tau)

The formulas for the derivative functions in row **19** of the worksheet use the parameter cell names, including the cell names for the initial conditions when referencing the model variables:

B19= (q + m*cp*(Tin - T))/ (rho*V*cp)   C19 = (T - Ttc)/tautc   D19 = Tsp – Ttc

	A	B	C	D	E
17		T	T_{tc}	Δ	q
18	Initial Conditions & Heat Rate	75	75	0	8400
19	Derivative Functions	0	0	0	

To force a step change in the inlet temperature at 10 seconds, use the *Excel* worksheet function **IF()** in cell **C15**:

C15 = IF(to<10, T0, T0 - Tstep)

The VBA macro **RK45ODE** prompts the user for the lower and upper integration limits. Change the number of print steps from the default value of 10 to a value of 100 and use the default value of 10^{-3} for error control. Select the range of cells **B18:D18** for the initial conditions and **B19:D19** for the derivative functions. Finally, select the location for the output of the integration results being careful not to write over content on the worksheet.

Figure 12.21 displays the first 20 seconds of the integration results for a gain of $K = 200$ and thermal couple time constant of 50 seconds using 100 print steps. Notice the step change at 10 seconds.

	A	B	C	D
22	x	y₁	y₂	y₃
23	0	75	75	0
24	2	75	75	0
25	4	75	75	0
26	6	75	75	0
27	8	75	75	0
28	10	74.818929	75	0
29	12	69.414533	73.96	0.7499
30	14	64.99486	71.66	4.9885
31	16	61.386797	68.81	14.47
32	18	58.448638	65.83	29.836
33	20	56.064046	62.96	51.077

Figure 12.21

Solution to a system of ordinary differential equations from the *VBA* macro RK45ODE.

Generate time-temperature profiles to compare the temperature responses with the step change for the four cases of controller gain. To simplify the calculations, hard code the references to the required input information by replacing the input boxes with range references or numerical values for each of the following parameters:

```
Set xr = Range("C13"): xfinal = Range("C14"): np = 100
Set yr = Range("B18:D18"): Set fr = Range("B19:D19"): tol = 0.001
```

Add a button to the worksheet to call the modified *VBA* macro **RK45ODE** after making changes to the process gain on the worksheet:

Run

```
Private Sub CommandButton1_Click() : Call RK45ODE : End Sub
```

Figure 12.22 has plots of the temperature responses to a -30°C step change in the feed temperature using controller gains of 4, 30, 200, and 1000 in Equation(12.84). Use the *VBA* user-defined function **CSPLINE** with clamped end conditions to functionalize the integration results by interpolation.

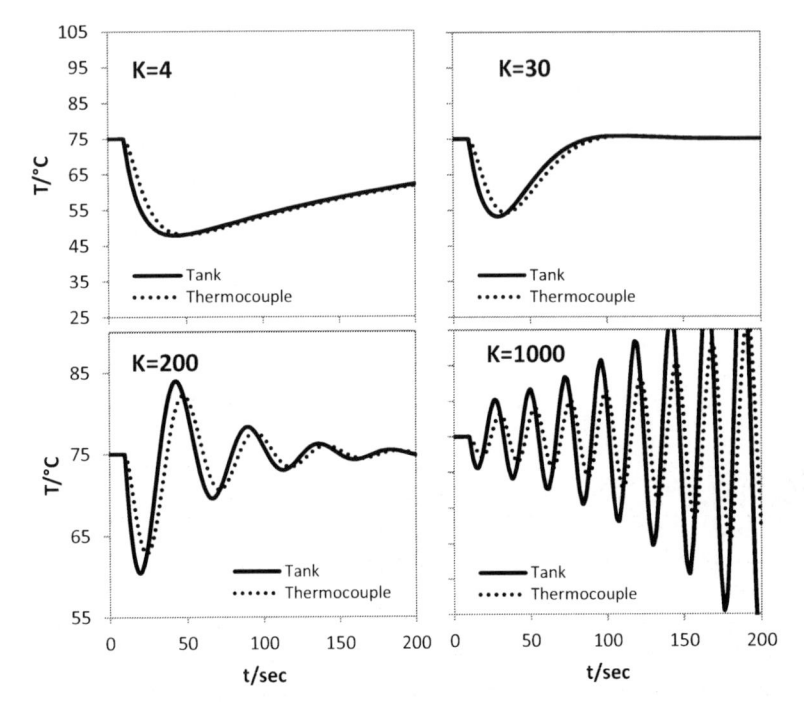

Figure 12.22

Temperature responses to a negative 30°C step change in the feed temperature at 10 seconds using controller gains of 4, 30, 200, and 1000.

Smaller gains $K < 1000$ move the temperature towards the desired steady-state set-point temperature. A gain of $K = 30$ shows the most stable response (critically damped). A gain of $K = 1000$ shows an unstable response as the

CHAPTER 12: INITIAL-VALUE PROBLEMS

temperatures diverge out of control. We observe for $K = 1000$ that the thermocouple temperature does not coincide well with the high and low temperatures, which contributes to the loss of control.

Comments: The sensitivity of the model parameters yields interesting results. Figure 12.23 plots of the temperature responses to a $-30°C$ step change in the feed temperature using a controller gain of 200 with a larger thermocouple time constant of 50 seconds. A formerly stable response now becomes unstable and out of control, as evidenced by the increasing magnitude of the oscillations and offset temperature readings for both tank and thermocouple.

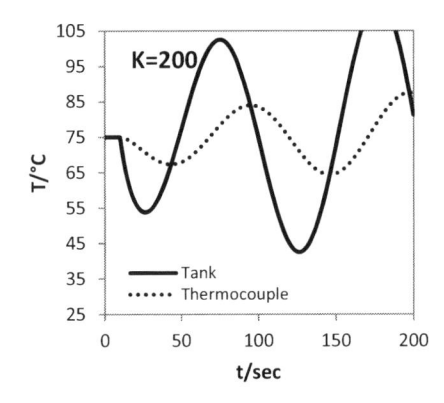

Figure 12.23

Temperature responses to a $-30°C$ step change in the feed temperature using a controller gain of 200 with a larger thermocouple time constant of 50 seconds.

The oscillating numerical solutions plotted in Figure 12.22 provide an opportunity for a reminder to use caution when selecting larger print steps. On the one hand, we want to reduce the size of the table of numerical integration results to minimize clutter on our *Excel* worksheet. On the other hand, we must not choose a print step size that hides the true nature of the solution. To illustrate, we selected a larger print step size of 20 seconds for the case of $K = 1000$ and graphed the solution as shown in Figure 12.23 to show how a step size that is too larger can mislead us in our interpretation of the results.

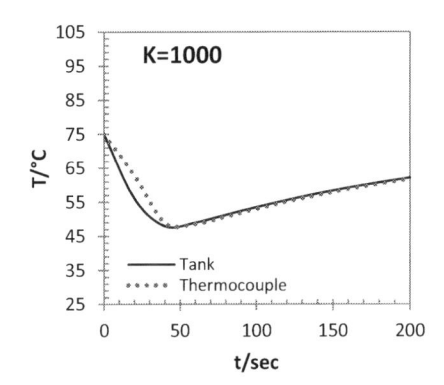

Figure 12.24

Illustration of the potential pitfall when selecting large print step sizes that miss the trend in the results.

The graph uses 20-second print steps for displaying the numerical solution with a controller gain of 1000. Compare the graph of the same solution using print steps of 2 seconds in Figure 12.22. The incorrect display hides the oscillations and indicates a misleading conclusion that the tank and thermocouple temperatures converge after 50 seconds.

□

12.6 *Adams-Bashforth-Moulton Multi-Step Method**

Euler, trapezoid, and Runge-Kutta techniques are single step integration methods that start from the end of the previous step. Multistep methods are generally able to take larger integration steps by using interpolation formulas to extrapolate to the end of the next step from the solutions at several previous steps. We illustrate the method with the Adams-Bashforth-Moulton predictor-corrector method that has favorable stability properties (Moulton 1926). The derivation uses Lagrange interpolating polynomials to integrate the derivative function over several integration steps:

$$y_{i+1} = y_i + \int_{x_i}^{x_{i+1}} f(x, y) dx \tag{12.89}$$

where $f(x, y)$ is the function of the derivative in the differential Equation (12.2). We integrate Lagrange's interpolating polynomial for an explicit fourth-order prediction formula named for Adams and Bashforth. In a fashion similar to the improved Euler's method in Section 12.4, we then correct the predicted solution using the value at the end of the current integration step.

We derive Adams' *predictor* formula with the following third-order Lagrange polynomial using four equally spaced points at x_{i-3}, x_{i-2}, x_{i-1}, and x_i, illustrated in Figure 12.25:

$$f(x) \cong \frac{(x-x_{i-3})(x-x_{i-2})(x-x_{i-1})}{6\Delta x^3} f_i - \frac{(x-x_{i-3})(x-x_{i-2})(x-x_i)}{2\Delta x^3} f_{i-1}$$
$$+ \frac{(x-x_{i-3})(x-x_i)(x-x_{i-1})}{2\Delta x^3} f_{i-2} - \frac{(x-x_i)(x-x_{i-2})(x-x_{i-1})}{6\Delta x^3} f_{i-3} \quad (12.90)$$

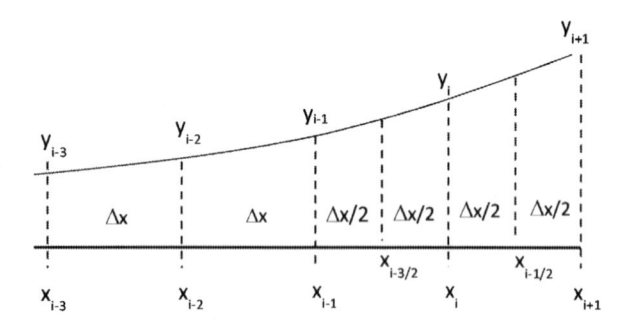

Figure 12.25

Points of interpolation and extrapolation for the Adams-Bashforth-Moulton multipoint method.

Let $U = x - x_i$ for

$$x - x_{i-3} = U + x_i - x_{i-3} = U + 3\Delta x$$
$$x - x_{i-2} = U + x_i - x_{x-2} = U + 2\Delta x \quad (12.91)$$
$$x - x_{i-1} = U + x_i - x_{i-1} = U + \Delta x$$

Substitute from Equation (12.91) into the Lagrange polynomial of Equation (12.90):

$$f(x) \cong \frac{(U+3\Delta x)(U+2\Delta x)(U+\Delta x)}{6\Delta x^3} f_i - \frac{(U+3\Delta x)(U+2\Delta x)U}{2\Delta x^3} f_{i-1}$$
$$+ \frac{(U+3\Delta x)U(U+\Delta x)}{2\Delta x^3} f_{i-2} - \frac{U(U+2\Delta x)(U+\Delta x)}{6\Delta x^3} f_{i-3} \quad (12.92)$$

The Adams-Moulton *corrector* formula comes from the following third-order Lagrange polynomial using the four points at x_{i-2}, x_{i-1}, x_i, x_{i+1}.

$$f(x) \cong -\frac{(x-x_{i+1})(x-x_{i-2})(x-x_{i-1})}{2\Delta x^3} f_i + \frac{(x-x_{i+1})(x-x_{i-2})(x-x_i)}{2\Delta x^3} f_{i-1}$$
$$+ \frac{(x-x_{i+1})(x-x_i)(x-x_{i-1})}{6\Delta x^3} f_{i-2} - \frac{(x-x_i)(x-x_{i-2})(x-x_{i-1})}{6\Delta x^3} f_{i+1} \quad (12.93)$$

Let $x - x_{i+1} = U + x_i - x_{i+1} = U - \Delta x$ for

$$f(x) \cong -\frac{(U-\Delta x)(U+2\Delta x)(U+\Delta x)}{2\Delta x^3} f_i + \frac{(U-\Delta x)(U+2\Delta x)U}{2\Delta x^3} f_{i-1}$$
$$- \frac{(U-\Delta x)U(U+\Delta x)}{6\Delta x^3} f_{i-2} + \frac{U(U+2\Delta x)(U+\Delta x)}{6\Delta x^3} f_{i+1} \quad (12.94)$$

We now approximate the solution to Equation (12.89) by integrating the Lagrange polynomials:

CHAPTER 12: INITIAL-VALUE PROBLEMS

Predictor
$$f_i = f(x_i, y_i) \tag{12.95}$$

$$z_{i+1} = y_i + \frac{\Delta x}{24}(55f_i - 59f_{i-1} + 37f_{i-2} - 9f_{i-3}) \tag{12.96}$$

Corrector
$$f_{i+1} = f(x_{i+1}, z_{i+1}) \tag{12.97}$$

$$y_{i+1} = y_i + \frac{\Delta x}{24}(9f_{i+1} + 19f_i - 5f_{i-1} + f_{i-2}) \tag{12.98}$$

An alternative corrector formula may use an implicit form in y_{i+1} through the derivative function $f_{i+1} = f(x_{i+1}, y_{i+1})$. The predictor formula provides a good approximation for y_{i+1} to initiate the trial-and-error solution method. We may improve the solution for stability by iterating on y_{i+1} starting from the predictor result. However, a better approach may be to reduce the integration step size to improve the accuracy because the implicit form of Adams-Bashforth-Moulton may have a smaller range of stability than the implicit Euler or trapezoidal methods.

From Taylor series analysis, we estimate the error in the Adams-Bashforth-Moulton approximations:

$$E_{i+1} \cong -\frac{19}{720}\Delta x^5 \frac{d^5 y}{dx^5} \tag{12.99}$$

For sufficiently small steps, the fifth derivative is relatively constant, such that:

$$E_{i+1} \cong -\frac{19}{720}(y_{i+1} - z_{i+1}) \tag{12.100}$$

Unfortunately, multistep methods are not self-starting. We resort to Runge-Kutta fourth-order solutions for the first three steps along with the initial conditions to provide the four points needed by the extrapolation formulas in Equations (12.96) and (12.98). Programming multistep methods becomes complicated when we need to control the step size and select print steps.

A *VBA* sub procedure **ABMODE** is available in the *PNMSuite*. The macro requires the same worksheet setup as **RK4G**. A user-form named **ODEFIXSTEP** is available to select from among the implicit Euler, trapezoidal, Runge-Kutta-Gill, or Adams-Bashforth-Moulton methods. The next example demonstrates the *implicit* Adams-Bashforth-Mouton method in an *Excel* worksheet.

Example 12.11 Adams-Bashforth-Moulton Multistep Method of Integration

Known/Find: Reconsider the problem in Example 12.2 using a multistep method of solution. Approximate the solution to Equation (12.35) with the implicit Moulton method using Equation (12.98), with a step size of $\Delta x = 0.1$. Compare the results with the exact solution.

Analysis: We use *Excel* with iteration enabled for our computational platform. We obtain the first three integration steps using the macro **RK4G**. The Runge-Kutta solution is in columns **A** through **F** in Figure 12.26. Column **B** contains the integration limits, initial value for *y* and the function of the differential equation, as required by the macro **RK4G**:

$$B5=-5*B4$$

The values in the range **H3:H6** for implementing the Adams-Moulton method were copied from the Runge-Kutta solution in the range **D3:D6**. We then apply the Adams-Moulton implicit formula beginning in row seven using iteration to complete the integration, as shown in Figure 12.26. The worksheet formulas in cells **H7** and **I7** use the implicit formula in Equation (12.98):

H7 =H6+0.1*(9*I7+19*I6-5*I5+I4)/24   I7 =-5*H7   J7 =Abs(E7-H7)/E7

Finally, the last columns house the solution from the Adams-Bashforth-Moulton method using the macro **ABMODE**, which significantly underperforms because of the relatively large size of the fixed integration step.

	A	B	C	D	E	F	G	H	I	J	K	L	M	N
1	RK4ODE	- Runge-Kutta-Gill			Exact	RK4		Adams-Moulton				Adams-Bashforth-Moulton		
2	x_0	0	x	y_1	y	% Error		y	$f(y)$	% Error		x	y_1	%Error
3	x_1	1	0	3	3	0.00%		3	-15	0.00%		0	3	0.00%
4	y_0	3	0.1	1.820313	1.819592	0.04%		1.820313	-9.10156	0.04%		0.1	1.82031	0.04%
5	dy/dx	-15	0.2	1.104513	1.103638	0.08%		1.104513	-5.52256	0.08%		0.2	1.10451	0.08%
6			0.3	0.670186	0.66939	0.12%		0.670186	-3.35093	0.12%		0.3	0.67019	0.12%
7			0.4	0.406649	0.406006	0.16%		0.405924	-2.02962	0.02%		0.4	0.40343	0.64%
8			0.5	0.246743	0.246255	0.20%		0.245933	-1.22967	0.13%		0.5	0.24275	1.42%
9			0.6	0.149716	0.149361	0.24%		0.148974	-0.74487	0.26%		0.6	0.14642	1.97%
10			0.7	0.090844	0.090592	0.28%		0.090245	-0.45123	0.38%		0.7	0.08815	2.70%
11			0.8	0.055121	0.054947	0.32%		0.054667	-0.27334	0.51%		0.8	0.05302	3.51%
12			0.9	0.033446	0.033327	0.36%		0.033116	-0.16558	0.63%		0.9	0.03193	4.21%
13			1	0.020294	0.020214	0.40%		0.020061	-0.1003	0.76%		1	0.01922	4.91%

Figure 12.26 Comparison of the results using RK4G with Adams-Moulton and Adams-Bashforth-Moulton integration of Equation (12.35).

Comments: For this example, the fourth-order Runge-Kutta-Gill method gives higher accuracy when compared with each of the multistep methods of Adams-Bashforth-Moulton. However, Adams' implicit method is much simpler to implement directly in an *Excel* worksheet to produce results that are also fourth-order accurate. Of course, we need three steps to get Adams' method started, so we may ask, "Why not just complete the integration using Runge-Kutta?" The answer has to do with stability. In some problems, the implicit Adams' methods allow us to take larger integration steps where Runge-Kutta needs much smaller steps to control stability.

□

12.7 Differential-Algebraic Systems*

Engineering models often involve coupled systems of differential and non-linear equations.[107] Consider the following simple system involving a first-order ordinary differential equation and a nonlinear equation:

$$\frac{dy_1}{dx} = f(x, y_1, y_2) \tag{12.101}$$

$$g(x, y_1, y_2) = 0 \tag{12.102}$$

where x is the independent variable, and y_1 and y_2 are two dependent variables. One approach is to solve equation (12.102) for y_2 by a trial and error technique, such as Newton's method, at each integration step. A more efficient approach converts the equation into a first-order differential equation by applying the chain rule[108] to Equation (12.102):

$$\frac{\partial g}{\partial y_1}\frac{dy_1}{dx} + \frac{\partial g}{\partial y_2}\frac{dy_2}{dx} = 0 \tag{12.103}$$

Rearrange Equation (12.103) for dy_2/dx with substitution from Equation (12.101) for dy_1/dx:

$$\frac{dy_2}{dx} = -\frac{\left(\frac{dy_1}{dx}\right)\left(\frac{\partial g}{\partial y_1}\right)}{\left(\frac{\partial g}{\partial y_2}\right)} = -f\frac{\left(\frac{\partial g}{\partial y_1}\right)}{\left(\frac{\partial g}{\partial y_2}\right)} \tag{12.104}$$

---

[107] Linear equations may be simply rearranged explicitly for the unknown variable and substituted into the differential equation.

[108] For $g(x,y)$ where $x(t)$ and $y(t)$ are continuous, the chain rule gives: $\frac{dg}{dt} = \frac{\partial g}{\partial x}\frac{dx}{dt} + \frac{\partial g}{\partial y}\frac{dy}{dt}$

CHAPTER 12: INITIAL-VALUE PROBLEMS

This gives a coupled system of first-order differential Equations (12.101) and (12.103), subject to initial conditions for y_1 and y_2. Determine the initial condition for Equation (12.103) by solving the algebraic Equation (12.102) once for y_2 at the initial condition for y_1.

Example 12.12 Differential-Algebraic Equations for a Plug Flow Reactor

Known: Consider an adiabatic plug flow reactor, described in Example 12.1 in terms of the mass and energy balances with a first-order reaction:

$$\frac{dX_A}{dV} = \frac{k_0 \exp\left(-\frac{E_a}{R_g T}\right)\left[C_{A0}(1-X_A)\right]^2}{F_{A0}} \tag{12.105}$$

$$g(X_A, T) = 0 = X_A \Delta H_r + \int_{T_0}^{T} c_p dT \tag{12.106}$$

where X_A is the reactant conversion and T is the temperature in the reactor. These balances assume constant heat of reaction, ΔH_r. C_{A0} and T_0 are the feed concentration and temperature, respectively. For this example, use a polynomial function of temperature for the heat capacity:

$$c_p = a + bT + cT^2 + dT^3 \tag{12.107}$$

Substitute for c_p from Equation (12.107) into Equation (12.106), followed by integration, for the following algebraic form of the energy balance:

$$g = X_A \Delta H_r + a(T-T_0) + \frac{b}{2}(T^2 - T_0^2) + \frac{c}{3}(T^3 - T_0^3) + \frac{d}{4}(T^4 - T_0^4) = 0 \tag{12.108}$$

Find: Given the conditions displayed in the *Excel* worksheet in Figure 12.27, calculate the final conversion and temperature in the reactor. Solve the material and energy balances using the Runge-Kutta-Gill method of integration.

Assumptions: Steady state, plug flow.

Analysis: Transform Equation (12.106) into a first-order differential equation:

$$\frac{dT}{dV} = -\frac{\left(\frac{\partial g}{\partial X_A}\right)\left(\frac{dX_A}{dV}\right)}{\left(\frac{\partial g}{\partial T}\right)} = -\frac{\Delta H_r}{c_p}\left(\frac{dX_A}{dV}\right) = -\left(\frac{\Delta H_r}{a+bT+cT^2+dT^3}\right)\left(\frac{-r_A}{F_{A0}}\right) \tag{12.109}$$

subject to the initial condition: $T = T_0$ at $V = 0$. Set up an *Excel* worksheet for the VBA macro **RK4G**:

	A	B	C	D	E
1	Parameters			Initial Conditions	
2	V/L	5		V/L	0
3	F$_{A0}$/(mol/s)	100		X$_A$	0
4	C$_{A0}$/(mol/L)	1		T/K	300
5	ΔH$_r$/(J/mol)	-10000			
6	k$_0$/(1/s)	55		Derivative Functions	
7	(E/R)/K	100		dX$_A$/dV	0.394092221
8	k/(1/s)	39.40922		dT/dV	52.61718882
9	c$_{pA}$/(J/mol K)	74.898			

Figure 12.27

Excel worksheet for solving a system of ordinary differential equations using the VBA macro RK4G.

Define the *Excel* formulas for the reaction rate constant and heat capacity in terms of the temperature located in the cell for the initial condition:

	A	B
8	k/(1/s)	=B6*EXP(-B7/E4)
9	c_pA/(J/mol K)	=18.3+0.4721*E4-0.001339*E4^2+0.000001314*E4^3

Next, define the functions for the differential equations in terms of the cells containing the initial conditions, as well as the function for the derivative of conversion in the case of the energy balance:

	D	E
7	dX_A/dV	=B8*(B4*(1-E3))^2/B3
8	dT/dV	=-B5*E7/B9

The inputs to the *VBA* macro **RK4G** are from the worksheet:

Input Box Prompt:	Worksheet Reference	Input Box Prompt:	Worksheet Reference
Initial x:	E2	Number of Print Steps:	10
Final x:	B2	Initial Conditions, y:	E3:E4
Number of Integration Steps:	1000	Derivative Functions, dy/dx = f(x, y):	E7:E8

Map the solution output in Figure 12.28 to the system of differential equations: $V = x$, $X_A = y_1$, $T = y_2$.

	A	B	C
11	x	y_1	y_2
12	0	0	300
13	0.5	0.166324	322.1364
14	1	0.287091	338.1324
15	1.5	0.378264	350.1695
⋮			
19	3.5	0.591062	378.1273
20	4	0.623604	382.3837
21	4.5	0.651417	386.017
22	5	0.675454	389.1534

Figure 12.28

Solution to the system of differential equations using the macro RK4G. $V = x$, $X_A = y_1$, $T = y_2$.

Compare the results graphically using a labeled vertical axis for *x* and a secondary axis for *T*. To create a second *y*-axis, select the *T* data series for the second axis on the chart. Click in the ribbon on the **Format>Format Selection** in the **Current Selection** group, as seen in Figure 12.29 to display the **Format Data Series** dialog box. On the **Series Options** > **Plot Series On**, click **Secondary Axis**:

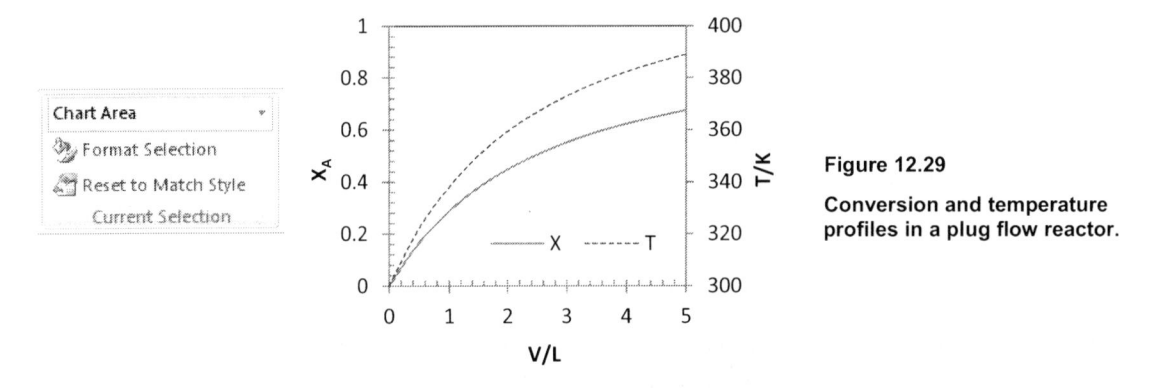

Figure 12.29

Conversion and temperature profiles in a plug flow reactor.

Comments: The temperature increase follows the conversion for the exothermic reactor. By converting the energy balance to a differential equation, we avoid the inconvenience of iteratively solving the energy balance for temperature at each integration step.

12.8 Newton's Method with Continuation for Root-finding (Homotopy)*

We combine our tools for solving differential equations with those for solving systems of linear equations into a robust homotopic method for finding roots to nonlinear equations called Newton's method with continuation, or Homotopy (Hanna and Sandall 1995). Numerical continuation may require a bit more effort to set up, but tends to find roots when other methods fail to converge from poor initial guesses for the roots. We derive the method for a single equation in one variable with an example before demonstrating the method for multivariable problems.

We begin our search method by introducing a new variable, λ, into our nonlinear equation to create a new function in two unknowns:

$$H(x,\lambda) = 0 = f(x) - (1-\lambda)f(g) \tag{12.110}$$

where g is an initial guess for the root to the function. With $\lambda = 0$, the new function in Equation (12.110) is zero for any initial guess $x = g$: $H(g,0) = f(g) - (1-0)f(g) = 0$. As the continuation variable λ increases from zero to one, we get back our original equation, $f(x) = 0$.

We next differentiate the right side of Equation (12.110) with respect to λ:

$$0 = \frac{\partial f(x)}{\partial x}\frac{dx}{d\lambda} + f(g) \tag{12.111}$$

Then rearrange Equation (12.111) for a first-order, ordinary differential equation, using a guess for the root as the initial condition:

$$\frac{dx}{d\lambda} = \frac{-f(g)}{\left[\partial f(x)/\partial x\right]} \quad \text{and} \quad x = g \quad at \quad \lambda = 0 \tag{12.112}$$

The root to the equation $f(x) = 0$ is the solution to the differential equation for x at $\lambda = 1$. For convenience, we may use a numerical approximation for the derivative in the denominator of the right-hand-side of Equation (12.112).

Example 12.13   Probability of Bioreactor Sterilization by Heat Treatment

Known: Bioreactors require sterile conditions to avoid contamination with undesirable microbes (Shuler and Kargi 2002). The probability of culture contamination is:

$$P_c(t) = 1 - P_e(t) = 1 - \left[1 - \exp(-k_d t)\right]^{N_0} \tag{12.113}$$

where t is the time of sterilization, P_e is the probability of microbe extinction, k_d is the first order death rate constant, and N_o is the initial number of contaminating microbes at the beginning of sterilization, $N_o = 1.7 \times 10^{11}$.

Find: Given the death constant of $k_d = 1$ min^{-1} at 121°C and $k_d = 61$ min^{-1} at 140°C, calculate the holding time at each temperature to maintain the probability of extinction greater than 99%.

$$P_e = \left[1 - \exp(-k_d t_h)\right]^N \tag{12.114}$$

Analysis: Try the macro **REGULAFALSI** to get the root by the modified regula falsi method. Cell **B1** contains the initial value for the lower bracket. The function in cell **D1** references cell **B1** for the independent variable $k_d t_h$. Set the upper bracket to 31 in the input box. After just seven iterations, the regula falsi method found the root in cell **B1** within a relative convergence tolerance of 10^{-5}.

	A	B	C	D
1	kdtlow =	30	f =	=(1-EXP(-B1))^(170000000000)-0.99

	A	B	C	D
1	kdtlow =	30.4592572	f =	9.38E-06

Alternatively, try Newton's method. Define the function for Newton's method as follows:

$$f(t_h) = 0 = \left[1 - \exp(-k_d t_h)\right]^N - 0.99 \tag{12.115}$$

The derivative of the extinction function required by Newton's method is

$$\frac{df}{dt_h} = Nk_d \exp(-k_d t_h)\left[1 - \exp(-k_d t_h)\right]^{N-1} \tag{12.116}$$

Newton's method fails to converge for the sterilization time, even when starting from the solution!
Compare the solution using Newton's method with the method of bisection in an *Excel* worksheet.

	A	B	C	D	E	F	G	H	I	J	K
1	N_0	1.70E+11				Sterilization time					
2	k_d/min^{-1}	1				t/min =	30.4592				
3	P_e	0.99		1							
4											
5		Newton's Method				Bisection					
6	Iteration	$k_d t$	f	f'		$k_d t_{low}$	f_{low}	$k_d t_{up}$	f_{up}	$k_d t_{bisect}$	f_{bisect}
7	0	30.46	9.4E-06	0.00994		30	-0.00578	31	0.00417	30.5	0.0004
8	1	30.4591	-9.3E-06	0.00995		30	-0.00578	30.5	0.0004	30.25	-0.0023
9	2	30.46	9.4E-06	0.00994		30.25	-0.0023	30.5	0.0004	30.375	-0.00087
10	3	30.459	-9.3E-06	0.00995		30.375	-0.00087	30.5	0.0004	30.4375	-0.00021
11	4	30.46	9.4E-06	0.00994		30.4375	-0.00021	30.5	0.0004	30.4688	0.0001
12	5	30.459	-9.3E-06	0.00995		30.4375	-0.00021	30.4688	0.0001	30.4531	-6.5E-05
13	6	30.46	9.4E-06	0.00994		30.4531	-6.5E-05	30.4688	0.0001	30.4609	9.4E-06
14	7	30.459	-9.3E-06	0.00995		30.4531	-6.5E-05	30.4609	9.4E-06	30.457	-2.8E-05
15	8	30.46	9.4E-06	0.00994		30.457	-2.8E-05	30.4609	9.4E-06	30.459	-9.3E-06
16	9	30.459	-9.3E-06	0.00995		30.459	-9.3E-06	30.4609	9.4E-06	30.46	9.4E-06
17	10	30.46	9.4E-06	0.00994		30.459	-9.3E-06	30.46	9.4E-06	30.4595	9.4E-06
18	11	30.459	-9.3E-06	0.00995		30.459	-9.3E-06	30.4595	9.4E-06	30.4592	9.4E-06
19	12	30.4599	9.4E-06	0.00994		30.459	-9.3E-06	30.4592	9.4E-06	30.4591	-9.3E-06
20	13	30.459	-9.3E-06	0.00995		30.4591	-9.3E-06	30.4592	9.4E-06	30.4592	-9.3E-06
21	14	30.4599	9.4E-06	0.00994		30.4592	-9.3E-06	30.4592	9.4E-06	30.4592	-9.3E-06
22	15	30.459	-9.3E-06	0.00995		30.4592	-9.3E-06	30.4592	9.4E-06	30.4592	-9.3E-06

Even with a good initial guess for the root, Newton's method falls into an oscillating pattern. The bisection method converges after 15 iterations to the root, as illustrated in Figure 12.30.

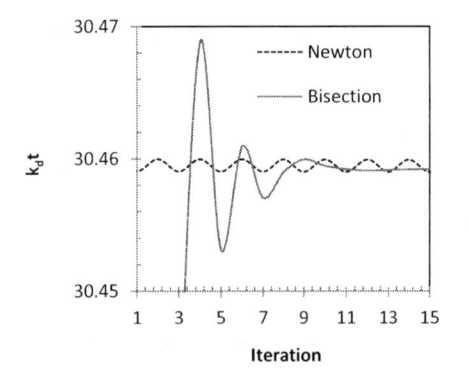

Figure 12.30

Convergence of bisection method compared with oscillations of Newton's method around the root.

Equation (12.113) is difficult to solve for k_d or t_h by indirect methods using function derivatives because of the wide differences in orders of magnitude of the relative terms.

Next, try the method of continuation, which does not require brackets on the root. Substitute the function and derivatives from Equations (12.115) and (12.116) into Equation (12.112):

CHAPTER 12: INITIAL-VALUE PROBLEMS

$$\frac{d(k_d t_h)}{d\lambda} = \frac{-\left[\left[1-\exp(-g)\right]^N - 0.99\right]}{N\exp(-k_d t_h)\left[1-\exp(-k_d t_h)\right]^{N-1}} \tag{12.117}$$

Set up an *Excel* worksheet for solution of the differential equation using the *PNMSuite* macro **RK45DP** from Chapter 12, with an initial guess of 25 for $k_d t_h$. We use just three print steps because we are only interested in the result at $\lambda = 1$. A small integration error tolerance of 10^{-12} is needed to get the same result found using bisection. Be careful to refer to the *Excel* worksheet cell **B4** (not **B3**) in response to the input box for the initial condition.

	A	B	C	D	E
1	λ lower limit	0		x	y₁
2	λ upper limit	1		0	25
3	Initial guess fo kt	25		0.5	26.34977
4	Initial kt	25		1	30.45836
5	d(kt)/dλ	4.021651			

Cell **B5** on the *Excel* worksheet contains the formula for the derivative of the function:

B5 = (0.99 - (1-EXP(-B3)) ^1.7E11)/ (1.7E11*EXP(-B4)*(1-EXP(-B4)) ^ (1.7E11-1))

Comments: Newton's method with continuation produces the same result found with the brute force bisection and regula falsi methods, even when starting from a poor initial guess for $k_d t_h$.

□

We may extend the continuation approach to *n* simultaneous equations, with one equation similar to Equation (12.118) for each variable:

$$\sum_{j=1}^{n} \frac{\partial f_i(x)}{\partial x_j} \frac{dx_j}{d\lambda} = -f_i(g) \qquad i = 1, 2 \ldots n \tag{12.118}$$

We derive the method of multivariable continuation with a pair of coupled nonlinear equations:

$$f_1(x_1, x_2) = 0 \qquad \text{and} \qquad f_2(x_1, x_2) = 0 \tag{12.119}$$

Add the continuation variable using initial guesses for the roots:

$$f_1(x_1, x_2) = (1-\lambda) f_1(g_1, g_2) \qquad \text{and} \qquad f_2(x_1, x_2) = (1-\lambda) f_2(g_1, g_2) \tag{12.120}$$

Differentiate each equation with respect to λ:

$$\frac{\partial f_1(x_1, x_2)}{\partial x_1} \frac{dx_1}{d\lambda} + \frac{\partial f_1(x_1, x_2)}{\partial x_2} \frac{dx_2}{d\lambda} = -f_1(g_1, g_2) \tag{12.121}$$

$$\frac{\partial f_2(x_1, x_2)}{\partial x_1} \frac{dx_1}{d\lambda} + \frac{\partial f_2(x_1, x_2)}{\partial x_2} \frac{dx_2}{d\lambda} = -f_2(g_1, g_2) \tag{12.122}$$

Note that the coefficients of the λ derivatives come from the Jacobian matrix (similar to Equation (6.103) for Newton's method). Use matrix algebra to solve for the system of differential equations, $dx_1/d\lambda$ and $dx_2/d\lambda$ at each integration step:

$$\begin{vmatrix} \frac{dx_1}{d\lambda} \\ \frac{dx_2}{d\lambda} \end{vmatrix} = \begin{vmatrix} \frac{\partial f_1(x_1, x_2)}{\partial x_1} & \frac{\partial f_1(x_1, x_2)}{\partial x_2} \\ \frac{\partial f_2(x_1, x_2)}{\partial x_1} & \frac{\partial f_2(x_1, x_2)}{\partial x_2} \end{vmatrix}^{-1} \cdot \begin{vmatrix} -f_1(g_1, g_2) \\ -f_2(g_1, g_2) \end{vmatrix} \tag{12.123}$$

Larger systems follow the same pattern. Although the method of continuation is computationally expensive when compared with Newton's method for finding roots to nonlinear equations, it is our most powerful tool for difficult problems with sensitivity to initial guesses, or otherwise have problems converging (Hanna and Sandall 1995). Besides, we do not have a "bisection" type method for coupled equations.

To solve the differential equations, we use numerical approximation methods from Chapter 12. The macro **NEWTCON** employs the Cash-Karp method of integration and Gaussian elimination with maximum column pivoting for finding roots to nonlinear equations by Newton's method with continuation. The user-form **ROOTS** includes an option for Newton's method of continuation. The worksheet setup has the same requirements as the macro **QUASINEWTON**. To get higher precision in the solution for the roots, restart from the previous solution as many times as necessary, or use a smaller tolerance for the integration error.

Example 12.14 Steady-state Reactor Analysis with the Method of Continuation

Known/Find/Analysis: Solve for the steady-state concentrations in Example 6.14 using Newton's method with continuation, starting from poor initial guesses for the concentrations at values of 10 mol/L. Use the same *Excel* worksheet used for Newton's method, but replace the initial guesses for the values in the range **B9:B12** with 10. Add the formulas for the system of nonlinear equations in the range **C9:C12**. First try the quasi-Newton method in the macro **ROOTS** to find that it quickly diverges, as shown in Figure 12.31.

	A	B	C	B	C
8		Concentrations	Mole Balances	Concentrations	Mole Balances
9	A	10	-1582.455532	-2.995054833	#NUM!
10	B	10	-2500	4.001163569	-1439.441541
11	C	10	3632.455532	-5.996218402	#NUM!
12	D	10	3500	-15.9813829	1439.441541
13		SSR =	34198898.7	SSR =	#NUM!

Figure 12.31

Divergent solution using Newton's method starting from initial guesses of 10 for each variable.

Next, reset the initial guesses to 10 and use the user-form **ROOTS** with the option for the method of continuation to obtain the correct roots shown in Figure 12.32.

	A	B	C
8		Concentrations	Mole Balances
9	A	0.318865812	0
10	B	0.783883977	0
11	C	0.534981835	0
12	D	0.491579272	0
13		SSR =	0

Figure 12.32

Converged solution using continuation starting from initial guesses of 10.

Comments: We recommend the robust Newton's method with continuation for finding roots only when we run into difficulty achieving convergence with other, generally more efficient methods. The penalty for using continuation is slow convergence based on the high computation requirements of solving differential equations. When using the method of continuation for large systems, use larger integration error control with multiple restarts, or exercise patience to allow the slow *VBA* macro to finish the integration with higher precision settings for error control.

□

12.9 Variable Step Quadrature*

With the exception of the versatile cubic spline and lowly trapezoidal rule[109], most integration methods such as Simpson's, Romberg, and Gauss require function evaluations at specific points between the integration limits. The accuracy of these methods is proportional to the length of the interval between points. We do not need small interval sizes for relatively flat integrand functions. To avoid wasting computational effort, we use integration techniques that employ variable interval sizes. Adaptive Gaussian quadrature is the most popular and efficient method when it works. The variable step Runge-Kutta fourth order marching integration methods are also relatively efficient for approximating solutions to integrals.

Consider a single-variable integral set equal to a dependent variable y:

$$y = \int_a^b f(x)\,dx \tag{12.124}$$

Differentiate both sides of Equation (12.124) to give a first-order, ordinary differential equation with the initial condition set to zero at the lower integration limit:

$$\frac{dy}{dx} = f(x) \quad \text{and} \quad y(a) = 0 \quad \text{for} \quad a \leq x \leq b \tag{12.125}$$

The solution of the integral in Equation (12.124) is just the final numerical solution for $y(b)$ at the upper limit of integration. For instance, a trapezoidal integration step gives:

$$y_{i+1} - y_i = \int_{x_i}^{x_{i+1}} f(x)\,dx = \frac{1}{2}(x_{i+1} - x_i)[f(x_i) + f(x_{i+1})] \tag{12.126}$$

The sum of the integration steps over the range of $a < x < b$ yields an approximate solution to the integral:

$$y_{bi} - 0 = \int_a^b f(x)\,dx = \frac{1}{2}\sum_{i=1}^{n-1}(x_{i+1} - x_i)[f(x_i) + f(x_{i+1})] \tag{12.127}$$

We recognize the result in Equation (12.127) as the composite trapezoidal rule derived in Equation (11.16). We may extend this approach to higher order Runge-Kutta integration methods. The Runge-Kutta-Merson method with variable integration steps is similar to Simpson's method on a local level and particularly efficient on a computational level for solving integrals with regions of flat and steep integrand functions. This method of quadrature also becomes important when we need to integrate the results from the solution to a differential equation.

Illustration: The following integral is a good candidate for demonstrating the potential advantage of a variable step integration technique.

$$I = \int_0^{0.99} \frac{dx}{1-x}$$

Solve the integral using the variable step Runge-Kutta-Merson method and compare the results with those derived from fixed and adaptive quadrature schemes. A plot of the integrand function shows a long, flat section followed by an exponentially increasing steep section. Solve this integral by the Trapezoidal rule, Simpson's 1/3 composite rule, Romberg, adaptive Simpson, adaptive Gauss-Kronrod, and user-form **RK45VS** (Merson).

---

[109] Without end corrections

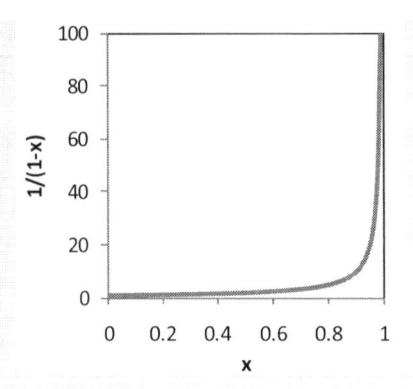

Specify the problem for the *VBA* user-form **RK45VS** in an *Excel* worksheet using Merson's coefficients and provide the required user input.

	A	B
1	x init	0
2	x final	0.99
3	y init	0
4	dy/dx	=1/(1-B1)

B1 = Lower integration limit
B2 = Upper integration limit
B3 = Range of initial conditions
B4 = Range of ODE functions
% Error control = 1E-8

Compare the results for the number of integrand evaluations using relative error tolerances of 10^{-7}.

Integration Method	Integration Result	Number of Integrand Function Evaluations
Exact Analytical Solution	4.6051702	-
Composite Trapezoidal Rule without End Corrections	4.6051710	32600
Simpson's 1/3 Composite Rule	4.6051703	2300
Romberg Integration	4.6051702	2049
RK45VS (Merson)	4.6051703	655
Adaptive Simpson Quadrature	4.6051702	225
Adaptive Gaussian Quadrature	4.6051702	195

Simpson and Romberg methods both require uniform sized intervals. For this example, these methods waste computational effort by putting an inordinate number of intervals where the function is relatively flat. The user-form **RK45VS** approach with Merson's variable integration step method uses large integration steps in the flat region and reserves the small integration steps for the region where the function changes rapidly with small changes in *x*. The superiority of the adaptive Simpson and Gauss-Kronrod methods of quadrature for this problem is evident from the substantially smaller number of integrand function evaluations required to achieve a result with the same degree of precision.

12.10 Nonlinear Parameter Estimation in IVPs*

Parameter estimation from experimental data representing a time-series is a challenging problem for models involving nonlinear differential equations that require a numerical solution, such as Runge-Kutta methods. We learned techniques based on Newton's method of least-squares regression in Chapter 9 for adjusting model parameters to minimize the sum of squared residuals between the model predictions and experiments. Newton's methods require the Jacobian of partial derivatives of the minimization objective function with respect to the model parameters. Approximating the Jacobian of a numerical solution from marching integration becomes computationally intensive, particularly for the *Excel* computational environment. To reduce the computational complexity of nonlinear parameter estimation involving initial-value problems, we recommend the following steps:

1. Interpolate the experimental data series with the **BSPLINE** user-defined function to smooth noise in the dependent variable data. Alternatively, try a least-squares fitted rational function to interpolate the data that includes uncertainty estimates for the interpolating function values.

2. Provide initial guesses for the differential equation model parameters based on good engineering judgement for the physics of the problem.

3. Calculate the minimization objective by integrating the squared residuals from the difference between the numerical solution for the model and the values from the B-spline interpolation of the time series data:

$$Min\,OF = \int_{t_0}^{t_{final}} \left[y_{Bspline}(t) - y_{model}(t) \right]^2 dt \qquad (12.128)$$

We may use high precision methods, such as Romberg integration from Chapter 11, for solving the integral in Equation (12.128). Alternatively, we may use a variable-step quadrature method to solve the integral as described in Section 0 by converting the integral in Equation (12.128) into a first-order, ordinary differential equation:

$$\frac{d(OF)}{dt} = \left[y_{Bspline}(t) - y_{model}(t) \right]^2 \qquad (12.129)$$

subject to the initial condition $OF = 0$ at $t_{initial}$. We integrate Equation (12.129) simultaneously with the first-order differential model equations for y to obtain a value for the objective function in Equation (12.128), which corresponds to the numerical solution for OF from Equation (12.129) at t_{final}.

4. Apply a direct method of minimization, such as Powell's or the Nelder and Mead's Simplex algorithm from Chapter 7, to find the model parameters in the differential equation that give the best fit of the experimental data series defined by Equation (12.129). Nonlinear parameter estimation may require a global minimization method such as the **Solver's** *Evolutionary* genetic algorithm, Simulated Annealing, Luus-Jaakola, PSO, or Firefly, to escape a local optimum and find the true minimum.

5. The last step is not really a step – just a warning that nonlinear parameter estimation requires significant computational resources to arrive at the solution. We suggest using a dedicated computer for this type of analysis for extremely large problems, or scheduling the application for a lengthy period when you will be away from your computer!

Nonlinear parameter estimation also involves calculating the expanded uncertainty in each of the ODE model parameters. The jackknife method of regression uncertainty analysis, presented in Section 0, overcomes the problems of derivative approximations with the Gauss-Newton or Levenberg-Marquardt regression analysis and the requirement for a large (>10,000) number of regressions with Monte Carlo uncertainty analysis. Nevertheless, for a large data series, the jackknife method will require a large number of regressions, which can take a long time in the *Excel* VBA environment. We demonstrate nonlinear parameter estimation in an ODE model with an example.

Example 12.15 Nonlinear Kinetic Parameter Estimation from Fermentation Data

Known: A continuous-flow fermentation reactor was used to determine the kinetic parameters for the specific growth rate according to the following Monod growth model:

$$\mu = \frac{\mu_o S}{K + S} \qquad (12.130)$$

where μ_o is the maximum specific growth rate, K is the equilibrium constant, and S is the concentration of the growth substrate. The experimental data for the substrate-concentration versus time series are provided in columns **I** and **J** of the *Excel* worksheet, as shown in Figure 12.33. The cell mass yield is $Y = 0.1$. The reactor volume, feed flow rate, and concentrations of substrate and cell mass are $V = 1000$, $Q = 250$, $S_0 = 1400$ and $X_0 = 800$, respectively.

Find: Use a method of nonlinear parameter estimation to determine the Monod reaction kinetic parameters for a batch fermentation reactor.

Schematic: The steady volumetric feed rate Q to the reactor with volume V contains initial concentrations of substrate and cell mass, S_0 and X_0.

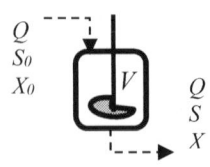

Assumptions: well-mixed reactor

Analysis: Derive the first-order differential equations for the unsteady-state material balances for the substrate and cell mass concentrations and the objective function (Elnashaie and Uhlig 2007):

$$V\frac{dS}{dt} = Q(S_o - S) - \frac{V}{Y}\left(\frac{\mu_o SX}{K+S}\right) \tag{12.131}$$

$$V\frac{dX}{dt} = Q(X_o - X) + V\left(\frac{\mu_o SX}{K+S}\right) \tag{12.132}$$

$$\frac{d(OF)}{dt} = \left[BSPLINE(S_{data}, t_{data}, t) - S\right]^2 \tag{12.133}$$

subject to the initial conditions at $t = 0$, $S = 1000$, $X = 400$, and $OF = 0$. Put the initial conditions in the range **B11:B12** and the formulas for the differential equations in the range **B13:B15**:

    B14=(B2/B1)*(B3-B11)-B6*B11*B12/(B5*(B7+B11))
    B15=(B2/B1)*(B4-B12)+B6*B11*B12/(B7+B11)
    B16=(B11-BSPLINE(J2:J26,I2:I26,B9))^2

Provide initial guesses for the kinetic parameters: $\mu = 0.01$ and $K = 100$ in cells **B6** and **B7**, respectively. Solve the system of first-order ordinary differential equations using the Runge-Kutta-Cash-Karp 4th-order variable step integration method for $0 \le t \le 24$ in columns **E**, **F**, and **G**. The final value for the objective function in cell **G26** is used by the simplex minimization routine to adjust the Monod kinetic parameters to achieve the best fit of the model equations to the substrate data, as shown in Figure 12.33. To implement the solution, we modified copies of the macros **RK45ODE**, **SIMPLEXNLP**, and **JACKKNIFE** to solve the specific system of ODEs in Equations (12.131) to (12.133), minimize the objective function calculated from the numerical solution to Equation (12.133) at the final value for t, and estimate the uncertainty. Follow these steps to implement our method of nonlinear parameter estimation:

1. Copy the macros and rename the copies to **EX12_16OF** for **RK45ODE**, **EX12_16MIN** for **SIMPLEXNLP**, and **EX12_16U** for **JACKKNIFE**.

2. Replace all of the input boxes with direct references to the worksheet ranges for the initial conditions, differential equations, and output in **EX12_16OF**. For example, replace the input box for the initial conditions:

    ```
    Set xr = Range("B9") ' Cell with INITIAL x
    ```

3. Add a **Call EX12_16OF** statement before each reference to the range of the objective function in **EX12_16MIN**. For example, in the first loop to randomly initialize the simplex vertices, we add the call statement before the assignment of the value in the range variable **fr** to the vector element **f**:

    ```
    Call EX12_16OF
    f(i) = fr: nfe = nfe + 1 ' set the initial f vector
    ```

4. Adjust the error control and convergence tolerances in the macros as needed for your own purposes.

5. Run the macro **EX12_16MIN** and select the objective function in cell **G26**. Select the minimization variables in the range **B6:B7**.

CHAPTER 12: INITIAL-VALUE PROBLEMS

6. Modify **EX12_16U** with a prompt for the range of independent *t* data and calls for **EX12_16MIN** in the jackknife loop used to generate the regressed values for the model parameters for each subset of data that removes one data point at a time. The objective function is the same as step 5. We modified the formula in the jackknife loop for the **BSPLINE** function that interpolates the subset of experimental data for the *S* vs *t* series.

```
For i = 1 To n ' Get regression parameters for the jackknifed data regression
Select Case i ' remove a data pair from the n data points at i index from the series
Case 1
   Range("B17").Formula = "=BSPLINE(" & ydatar(2).Address & ":" & ydatar(n).Address & "," & _
               xdatar(2).Address & ":" & xdatar(n).Address & "," & Range("B9").Address & ")"
Case n
   Range("B17").Formula = "=BSPLINE(" & ydatar(1).Address & ":" & ydatar(n - 1).Address & "," & _
               xdatar(1).Address & ":" & xdatar(n - 1).Address & "," & Range("B9").Address & ")"
Case Else
   Range("B17").Formula = "=BSPLINE((" & ydatar(1).Address & ":" & ydatar(i - 1).Address & "," & _
               ydatar(i + 1).Address & ":" & ydatar(n).Address & "),(" & xdatar(1).Address & _
               ":" & xdatar(i - 1).Address & "," & xdatar(i + 1).Address & ":" & _
               xdatar(n).Address & ")," & Range("B9").Address & ")"
   End Select
   Call EX12_16MIN: Next i
```

The workbook with the *VBA* programs for this example is available for download from the book's companion website. First, run the macro **EX12_16MIN** to get the best-fit parameters, and then run the macro **EX12_16U** to obtain uncertainty estimates for the model prediction and parameters.

	A	B	C	D	E	F	G	H	I	J
1	V	1000		Model t	Model S	Model X	Model OF		t_{data}	S_{data}
2	Q	250		0	1000	400	0		0	1000
3	S_o	1400		1	1022.355421	495.0921134	17.53387916		1	1019
4	X_o	800		2	1027.348573	570.3916524	34.97026236		2	1027
5	Y	0.1		3	1021.445943	630.0141225	43.05193662		3	1019
6	μ	0.01753		4	1009.142399	677.218806	47.49605165		4	1009
7	K	59.4132		5	993.5037791	714.5875115	49.06850381		5	992
8				6	976.5705775	744.1656718	50.14327583		6	976
9	t_o	0		7	959.6557586	767.573889	50.35675406		7	961
10	t_f	24		8	943.5628259	786.0961928	50.64938208		8	942
11	Init S	1000		9	928.7442732	800.7499139	53.46393203		9	926
12	Init X	400		10	915.4157438	812.3410262	54.49939769		10	916
13	Init OF	0		11	903.6374687	821.5079942	65.03017278		11	909
14	dS/dt	33.8237		12	893.3716828	828.7565216	87.64227742		12	898
15	dX/dt	106.6176		13	884.5225224	834.4870963	91.555595		13	882
16	dOF/dt	0		14	876.9632501	839.0168263	92.04154388		14	879
17	BSPLINE	1000		15	870.5543938	842.5967524	96.59147194		15	873
18	OF	117.4636		16	865.1554461	845.4255743	98.54145288		16	865
19				17	860.6320554	847.6605315	98.61180198		17	861
20				18	856.8601151	849.42603	99.9083519		18	854
21				19	853.7277598	850.8204781	101.4741018		19	854
22				20	851.135988	851.9217045	103.7861561		20	848
23				21	848.9984182	852.7912501	109.3924404		21	847
24				22	847.2405284	853.4777677	113.2797077		22	845
25				23	845.7986189	854.0197146	114.0673295		23	847
26				24	844.6186559	854.4474834	117.4636331		24	847

$u_Y = \pm$	2.3E+0	$U_{Y,95\%} = \pm$	4.7E+0
$u_1 = \pm$	5.0E-5	$U_{1,95\%} = \pm$	1.1E-4
$u_2 = \pm$	2.7E+0	$U_{2,95\%} = \pm$	5.6E+0
$R^2 =$	0.9989	Adj $R^2 =$	0.9988
SSR =	1.17E+2	DoF =	23
AAD =	1.85E+0	$t_{95\%} =$	2.07

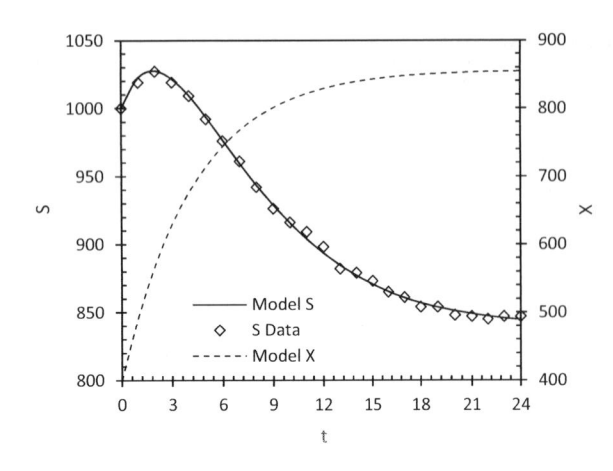

Figure 12.33 Experimental data and model-parameter estimation results with uncertainty for unsteady composition of a continuous flow fermentation reactor.

Comments: The uncertainty analysis by the jackknife method gives a model prediction for the substrate concentration of ± 4.7 (95% confidence). The results from the nonlinear kinetic parameter estimation are shown on the worksheet for modeling the specific growth rate, with corresponding uncertainty:

$$\mu_o = 0.01753 \pm 0.00011 \, (0.63\% \text{ at } 95\% \text{ confidence}) \qquad K = 59.4 \pm 5.6 \, (9.4\% \text{ at } 95\% \text{ confidence})$$

The specific growth rate has a relatively large sensitivity to the equilibrium constant. We were able to estimate the kinetic parameters with a local search method. Some parameter estimation problems may require a stochastic global search method, such as Luus-Jaakola, to find the best-fit model parameters.

☐

12.11 Epilogue on Initial-value Problems

> "Baseball players do not need to be able to solve the non-linear differential equations which govern the flight of the ball. They just catch it." – Paul Ormerod

First-order differential equations are a staple of engineering modeling, problem solving, and analysis. We rely on ordinary differential equations to help us understand and predict behaviors and performance of our engineered systems. Because most real-world problems are nonlinear, we resort to highly accurate numerical methods to find solutions to our models. Fortunately, we have several options for reaching solutions in *Excel* worksheets directly or indirectly using sophisticated *VBA* macros.

Table 12.2 summarizes the numerical methods for solving initial-value problems. We control the accuracy of the solution with judicious selection of the integration step size. Low-order methods, such as Euler and trapezoidal solutions, require relatively small integration steps to control the error. We recommend Euler's method only for quick calculations directly in an *Excel* spreadsheet. Euler's method works well for linear, or nearly linear, solutions, but has serious stability issues for stiff, nonlinear problems using reasonable step sizes. We should use a variable-step method, such as Runge-Kutta-Cash-Karp, to obtain higher precision results while controlling the error in the solution. When encountering a stiff problem, try the implicit trapezoidal rule directly in *Excel*, or use one of the stiff ODE macros. Recall that the trapezoidal rule requires more integration steps for accuracy, when compared with higher order methods, such as Runge-Kutta for non-stiff problems. Several of the *VBA* sub procedures allow the user to select the print step size different from the integration step size. We must be careful to select a print step size that captures the behavior of the solution. If we select a large print step to reduce the size of the output, we risk hiding the true nature of the solution. Consider the numerical results displayed in Figure 12.22 that show periodic, oscillatory behavior. A large print step size may conceal the nature of the solution between the print points.

Table 12.2 Comparison of IVP numerical marching integration methods.

Method	Strengths	Weaknesses
Euler	• Easy to remember and simple implementation in an *Excel* worksheet. • Backward (implicit) Euler method may be more stable than implicit trapezoidal method for some problems.	• Low accuracy • Requires small integration steps • Unstable for large steps • No estimate of error - must compare solutions with different step sizes to estimate accuracy
Trapezoid	• Easy to remember and simple to implement in an *Excel* worksheet using iteration on circular references. • Implicit method is more stable.	• Lower integration accuracy relative to Runge-Kutta • No estimate of error - must compare solutions with different step sizes to estimate accuracy
Improved Euler	• One derivative evaluation per step (per ordinary differential equation) • Implicit (backward) method is more stable than the trapezoidal method.	• Requires small integration steps for higher accuracy • No variable integration step method
Fourth-order Runge-Kutta-Gill fixed step	• Higher, fourth-order accuracy for fewer integration steps.	• No estimate of error - must compare solutions with different step sizes to estimate accuracy • Difficult to implement in an *Excel* worksheet without VBA
Variable Step Runge-Kutta	• Fourth-order accuracy • Error control for high precision solutions of moderately stiff problems	• May require a large number of steps for stiff problems
Multistep Adams-Bashforth-Moulton	• Implicit form has relatively improved stability for stiff problems when compared with explicit fixed step Runge-Kutta methods • Accurate with larger step sizes	• Non-self-starting - requires Runge-Kutta method for the first three integration steps. • Complicated algorithms for adaptive integration step sizes • Lower region of stability than backward Euler or implicit Trapezoidal methods for stiff problems

Table 12.3 summarizes the VBA macros for solving IVPs in *Excel* worksheets. We may use interpolation methods from Chapter 10 to functionalize then interpolate the tabulated results from our methods of integrating differential equations. Where necessary, we recommend rational least squares functions of the results for greater portability of the solution among analysis tools, such as *Excel*.

Table 12.3 Summary of VBA macros, user-forms, and the methods of input for differential equations.

Macro	Inputs	Integration Method
ABMODE	Macro with input boxes	Adams-Bashforth-Moulton method
NEWTCON	Macro with input boxes	Newton's method of continuation (homotopy) for finding roots
ODEFIXSTEP_UsrFrm	Show user-form macro	Select from several fixed-step methods
RK45DP	Macro with input boxes	Variable step fourth/fifth order Dormand Prince method
RK4G	Macro with input boxes	Runge-Kutta-Gill fixed-step method
RK45ODE	Macro with input boxes	Variable step integration methods (Merson, Fehlberg, Cash-Karp)
RK45VS_UsrFrm	Show user-form macro	Select from variable-step, four/fifth order methods of Fehlberg, Merson, Cash Karp, and Dormand-Prince
STIFFODE	Macro with input boxes	Select from backward Euler and implicit trapezoidal methods

Chapter 13 Boundary-value Problems

"Nature is an infinite sphere of which the center is everywhere and the circumference is nowhere." – Blaise Pascal

A few common differential equations describe a wide variety of natural phenomena involving transport phenomena with chemical reaction (Bird, Stewart and Lightfoot 2007). Despite the same form of differential equations, the conditions at the boundaries makes the problem unique. For example, the same energy balance described in Equation (1.68) describes both temperature profiles for adiabatic and heated surface boundary conditions depicted in Figure 1.7 and Figure 1.8.

> **SQ3R Focused Reading Questions**
> 1. How do boundary conditions make a problem unique?
> 2. What are the strengths and weaknesses of shooting versus finite difference methods of solution?
> 3. How can I check the accuracy of a finite difference solution?
> 4. How does a symmetric boundary condition simplify a problem?
> 5. How can I convert a second-order problem into a system of first-order differential equations?

For our purposes, boundary-value problems (*BVP*) consist of second order, ordinary differential equations (*ODE*) with known conditions prescribed at the boundaries of the domain. The general form of an ordinary, two-point, and second-order differential equation involves a combination of terms that include first- and second-order derivatives and the dependent and independent variables:

$$\frac{d^2 y}{dx^2} = R\left(x, y, \frac{dy}{dx}\right) \qquad (13.1)$$

where *R* represents a general function of some combination of the independent and dependent variables. The solution to the second-order differential equation is subject to a pair of split boundary conditions.

Boundary conditions specify values for the derivative of the dependent variable:

$$b_0 = \alpha_0 \left.\frac{dy}{dx}\right|_0 - \beta_0 (y - \gamma_0) = 0 \quad at \ x = x_0 \quad and \quad b_n = \alpha_n \left.\frac{dy}{dx}\right|_n - \beta_n (y - \gamma_n) = 0 \quad at \ x = x_n \qquad (13.2)$$

The unique boundary conditions define the parameters α, β, and functions γ. The γ term may involve a more complicated function of dy/dx, y, and x, as required by a particular problem. Equations (13.1) and (13.2) are general expressions that allow for a variety of combinations of coefficients and functions to match most problems that arise in modeling momentum, heat, and mass transfer with chemical reaction. The split boundary conditions in Equations (13.2) become the Dirichlet type for $\alpha = 0$ or $dy/dx = 0$. In the case of Neumann boundaries, set $\beta = 0$.

One example of a boundary-value problem is the model of steady-state diffusion with second-order reaction in a spherical catalyst pellet:

$$\frac{d^2 C}{dr^2} = -\frac{2}{r}\frac{dC}{dr} + \frac{k}{D}C^2 \qquad (13.3)$$

with split boundary conditions at the pellet center and surface:

- *Neumann* $\qquad \dfrac{dC}{dr} = 0 \ \ at \ r = 0$ $\qquad (13.4)$

- *Dirichlet* $\qquad C = 1 \ at \ r = R$ $\qquad (13.5)$

where C is the concentration of the diffusing molecule, r is the dimension of radius, D is the coefficient of diffusion, and k is the reaction rate constant. Comparing the general form in Equations (13.1) and (13.2) to the example in Equations (13.3), (13.4), and (13.5) gives $R = -(2/r)dC/dr + (k/D)C^2$; $\alpha_0 = 1, \beta_0 = 0$ at $r = 0$ and, $\alpha_n = 0, \beta_n = 1, \gamma_n = 1$ at $r = 1$.

Numerical methods for obtaining good approximate solutions to boundary-value problems consist of hybrids of numerical techniques presented in previous chapters:

- The shooting method uses marching integration techniques, like Runge-Kutta, for solving systems of first-order differential equations combined with a technique for finding the root of a nonlinear function (the unknown initial condition in this case) to eliminate the residual at a boundary.

- The finite-difference method approximates the solution at discrete points between the boundaries by replacing the derivatives with finite-difference approximations. The result is a system of algebraic equations. We then use matrix algebra or fixed-point iterative methods for finding roots to the large systems of linear or nonlinear equations that are the result of discretization.

- The method of orthogonal collocation uses a polynomial trial function for the solution at optimal collocation points generated from the zeroes of orthogonal Legendre polynomials (similar to Gauss quadrature). We get the polynomial coefficients from the solution to the system of equations derived by substituting the trial function into the boundary-value problem and boundary conditions.

13.1 Shooting Method

"Arriving at one goal is the starting point to another." – John Dewey

The shooting method involves transforming a second-order, ordinary-differential equation into a system of two, first-order (initial-value) ordinary-differential equations. To make the transformation, we introduce a new variable for the first derivative:

$$z = \frac{dy}{dx} \tag{13.6}$$

We then substitute the new variable z for dy/dx in Equation (13.1) to transform the second order ordinary differential equation into a first-order initial value problem:

$$\frac{dz}{dx} = R(x, y, z) \tag{13.7}$$

Now the transformed problem seeks the solution to the coupled pair of first-order ordinary differential equations (13.6) and (13.7) with an *undetermined* initial condition y_0 or z_0 at $x = x_0$. The lower boundary condition at x_0 defines one of the initial conditions for either y or z, depending on the problem:

$$b_0 = \alpha_0 z_0 - \beta_0 (y_0 - \gamma_0) \quad at \quad x = x_0 \tag{13.8}$$

The *known* boundary condition at the opposite end serves as our target:

$$b_n = \alpha_n z_n - \beta_n (y_n - \gamma_n) \quad at \quad x = x_n \tag{13.9}$$

The shooting method derives its name from an analogy with target practice. A shooter, whether hunting, playing darts, or basketball, aims at a target by selecting an initial angle of trajectory towards the target. Consider the illustration of a basketball player "shooting hoops" in Figure 13.1. The initial conditions of direction and force applied by the shooter shape the trajectory of the basketball towards the goal. If the ball misses the target, the shooter adjusts the initial conditions for angle and velocity for the succeeding trajectories until reaching the target. In Figure 13.1, the first trajectory is too low. The shooter compensates in the next throw by aiming slightly higher. However, the second trajectory is too high. An interpolation between the first and second shots, not too low or high, hits the target.[110] In a fashion similar to the basketball analogy, with the numerical shooting method we "aim" at the target boundary condition by guessing the unknown initial condition for either y or z at the lower limit of integration, and then calculate a trajectory toward the target by numerical solution of the pair of differential equations.

The shooting method is an example of creative problem solving – the joining together of two or more ideas.[111] The method consists of an initial-value integration technique, such as Runge-Kutta from Section 12.5, to generate the trajectory, and a nonlinear root-finding technique, such as the secant method from Section 6.2.7, to automate the adjustments for the "aim" or unknown initial condition.

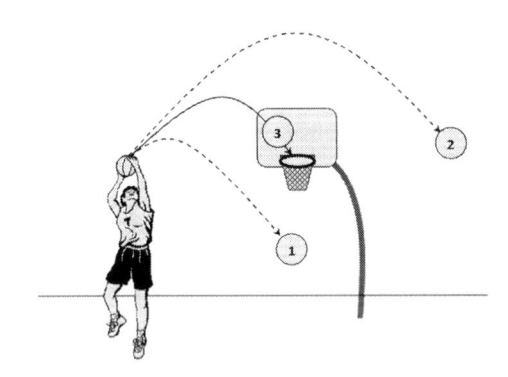

Figure 13.1

A basketball player makes two shooting adjustments to get the ball through the hoop on the third shot.

In summary, the shooting method cycles through the following steps:

1. Guess the *unknown* initial condition for y_0 or z_0 at the lower boundary x_0.
2. Use a marching integration method, such as Euler's, trapezoidal, or Runge-Kutta, to calculate a numerical solution to the set of first-order differential equations at x_n.
3. Compare the numerical *solution* at the upper boundary location x_n with the *known* value for the boundary condition for either y_n or z_n. The residual difference between the known value and the calculated value at the second boundary becomes the nonlinear "function" for finding the root of the unknown boundary condition.
4. If we started with the correct guess for the undetermined initial condition y_0 or z_0 at the lower boundary, the numerical solution at x_n will match the known condition for the dependent variables at the upper boundary.
5. Otherwise, upgrade the guess for the unknown initial condition until the solution converges on the target upper boundary condition within an acceptable tolerance.

We may use any root-finding method to determine the values of the boundary conditions, including **Goal Seek** and **Solver** in an *Excel* worksheet. When the solution becomes unstable, try switching directions and integrating from the upper boundary condition back towards the lower boundary condition.

Example 13.1 Diffusion and Reaction in a Spherical Pellet

Known: The dimensionless concentration profile for a reactant in a spherical catalyst pellet is modeled as a second-order, ordinary differential equation:

$$\frac{d^2c}{dr^2} = \phi c^2 \qquad (13.10)$$

with boundary conditions

$$\frac{dc}{dr} = 0 \quad at \quad r = 0 \qquad and \qquad c = 1 \quad at \quad r = 1 \qquad (13.11)$$

---

[110] Just like the children's story, *Goldilocks and the Three Bears* ... the porridge was too cold, too hot, and then just right!
[111] Green Engineering: Reduce, Reuse, Recycle your numerical methods.

CHAPTER 13: BOUNDARY-VALUE PROBLEMS

Schematic: A spherical, porous catalyst pellet has a radial dimension: $0 \leq r \leq 1$.

Find: Plot the concentration profile in the spherical pellet with Thiele modulus $\phi = 10$.

Assumptions: Homogeneous properties, symmetric concentration profile about the pellet center

Analysis: First, transform the problem into a pair of first-order differential equations:

$$\frac{dc}{dr} = z \quad \text{and} \quad \frac{dz}{dr} = \phi c^2 \tag{13.12}$$

The transformed boundary conditions become

$$z = 0 \quad at \quad r = 0 \quad \text{and} \quad c = 1 \quad at \quad r = 1 \tag{13.13}$$

At this stage of the solution process, the initial condition for c at $r = 0$ is unknown. Use good engineering judgement to guess the value for the undetermined condition at the lower boundary as $c \cong 0.5$ at $r = 0$ to set-up the worksheet for the solution to the system of equations in *Excel*. Integrate the differential equations by the implicit trapezoidal rule with a uniform integration step size of $\Delta r = 0.1$. Our guess for the initial concentration at the pellet center seems reasonable because by definition, the dimensionless concentration ranges between values of zero and one. Knowing that the implicit trapezoidal rule requires iterative solutions, we enable iteration in *Excel* before formulating the implicit equations on the worksheet in Figure 13.2 to avoid a circular reference warning.

	A	B	C
1	r	c	z
2	0	0.27771881725543	0
3	0.1	=B2+0.1*(C2+C3)/2	=C2+0.1*10*(B2^2+B3^2)/2
4	0.2	=B3+0.1*(C3+C4)/2	=C3+0.1*10*(B3^2+B4^2)/2
5	0.3	=B4+0.1*(C4+C5)/2	=C4+0.1*10*(B4^2+B5^2)/2
6	0.4	=B5+0.1*(C5+C6)/2	=C5+0.1*10*(B5^2+B6^2)/2
7	0.5	=B6+0.1*(C6+C7)/2	=C6+0.1*10*(B6^2+B7^2)/2
8	0.6	=B7+0.1*(C7+C8)/2	=C7+0.1*10*(B7^2+B8^2)/2
9	0.7	=B8+0.1*(C8+C9)/2	=C8+0.1*10*(B8^2+B9^2)/2
10	0.8	=B9+0.1*(C9+C10)/2	=C9+0.1*10*(B9^2+B10^2)/2
11	0.9	=B10+0.1*(C10+C11)/2	=C10+0.1*10*(B10^2+B11^2)/2
12	1	=B11+0.1*(C11+C12)/2	=C11+0.1*10*(B11^2+B12^2)/2

Figure 13.2

Implicit trapezoid formulas for integrating differential equations for the shooting method.

Starting the integration with $c = 0.5$ at $r = 0$, the calculated value for the concentration at the pellet surface ($r = 1$) is $c = 6.9$. Adjust the initial guess and shoot again for the known boundary condition $c = 1$ at $r = 1$. Use the **Solver** as shown in Figure 13.3 to find the initial condition in cell **B2** required to match the solution to the known boundary condition at $r = 1$.

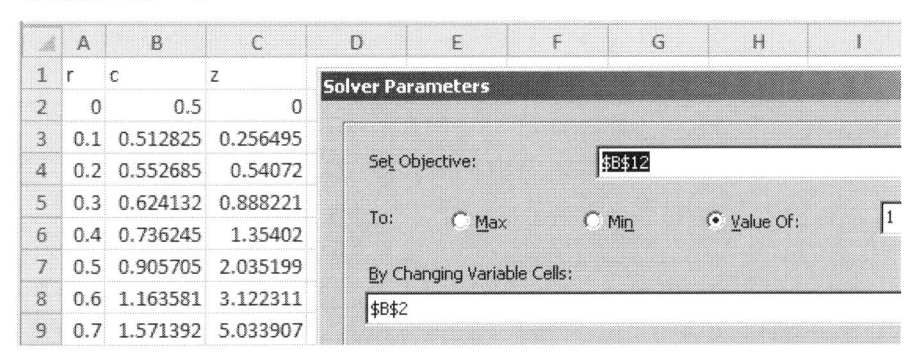

	A	B	C
1	r	c	z
2	0	0.5	0
3	0.1	0.512825	0.256495
4	0.2	0.552685	0.54072
5	0.3	0.624132	0.888221
6	0.4	0.736245	1.35402
7	0.5	0.905705	2.035199
8	0.6	1.163581	3.122311
9	0.7	1.571392	5.033907

Figure 13.3

Solution to Equations (13.12) with $c_o = 0.5$, and Solver parameters for finding the unknown boundary condition.

The converged solution for the initial condition of concentration at the pellet center is $c = 0.28$. Figure 13.4 includes a plot of the concentration and derivative profiles.

	A	B	C
1	r	c	z
2	0	0.2777188	0
3	0.1	0.2816299	0.07822157
4	0.2	0.2935898	0.160976758
5	0.3	0.3143122	0.253470315
6	0.4	0.3451065	0.362415619
7	0.5	0.3880908	0.497272108
8	0.6	0.446569	0.672291306
9	0.7	0.5256926	0.910179587
10	0.8	0.6336574	1.249116762
11	0.9	0.7839724	1.757183978
12	1	1.0000591	2.564549446

Figure 13.4

Solution for the concentration profile in a catalyst pellet using the Solver to find the unknown initial condition in the shooting method.

Comments: Larger values for ϕ require smaller integration steps for a converged solution of the differential equations by the trapezoidal rule. We may use the results for dc/dr at the pellet surface to calculate the diffusive flux of the reacting molecule into the pellet for a nominal rate of reaction.

□

The secant method, described in Section 6.2.7, is convenient for automatically upgrading the guesses for the unknown initial condition at x_0 because it does not require derivatives of the root function. The "function" in the secant formula used by the shooting method is simply the residual between the numerical solution and the upper boundary condition at x_n:

$$f = b_n \quad at \quad x = x_n \tag{13.14}$$

To begin the search, the secant method requires two guesses, g_1 and g_2, for the unknown initial condition. Depending on the nature of the problem, the guess may be for the unknown y_0 or z_0. Use the residuals from these trajectories to upgrade the guess for the unknown boundary condition, as illustrated in Figure 13.5. With the recursive secant formula in Equation (6.57), calculate an upgraded value for the unknown initial condition from the results of the first two trajectories:

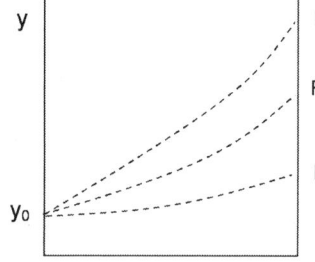

$$g_3 = g_2 - \frac{b_2[g_1 - g_2]}{b_1 - b_2} \tag{13.15}$$

Figure 13.5

Trajectories for secant iterations in the shooting method. Define the residuals by the boundary condition at the upper limit of integration.

A *VBA* macro and user-form **SHOOTING** are available in the *PNMSuite* for solving second-order boundary-value problems by the shooting method. To set up an *Excel* worksheet for either macro:
1. Specify the initial conditions for x, y and z in worksheet cells.
2. Add formulas in a range of cells for dy/dx and dz/dx using references to the cells for the initial conditions for instances of x, y, and z in the differential equations.

The program prompts for the cell locations of the boundary conditions. The **SHOOTING** macro employs the fixed-step Runge-Kutta-Gill fourth-order or implicit trapezoidal integration (stiff) methods for a system of two ordinary differential equations, and the secant method for finding the unknown initial condition.

Example 13.2 Concentration Profile in an Immobilized Enzyme Pellet

Known: A second-order boundary-value equation describes diffusion with reaction in a sphere of immobilized enzyme assuming Michaelis-Menten kinetics:

$$\frac{d^2s}{dr^2} + \frac{2}{r}\frac{ds}{dr} - \frac{9\theta^2 s}{1+Ks} = 0 \tag{13.16}$$

subject to the boundary conditions

$$\frac{ds}{dr} = 0 \quad \text{at } r = 0 \quad \text{and} \quad s = 1 \text{ at } r = 1 \tag{13.17}$$

where $K = 5$ and $\theta = 2$, and s and r are the dimensionless substrate concentration and radial coordinate, respectively.

Find: Plot the concentration profile in the pellet.

Assumptions: Steady state and constant properties, Michaelis-Menten kinetics.

Analysis: Transform the second order differential equation and boundary conditions into a set of first-order differential equations:

$$\frac{ds}{dr} = z \quad \text{and} \quad \frac{dz}{dr} = -\frac{2z}{r} + 9\theta^2 \frac{s}{1+Ks} \tag{13.18}$$

$$z = 0 \quad \text{at} \quad r = 0 \quad \text{and} \quad s = 1 \text{ at } r = 1 \tag{13.19}$$

Add formulas for the differential equations in an *Excel* worksheet, as shown in Figure 13.6, and then define the independent variable in terms of the cell with the lower limit of integration. Guess a value of $s = 0.5$ for the missing dimensionless substrate concentration at the pellet center, $r = 0$.

	A	B
1	Pellet center, r	0
2	Pellet surface, r	1
3	Initial s	0.211502132633863
4	Initial z	0
5	ds/dr	=B4
6	dz/dr	=IF(B1=0,-2+36*B3/(1+5*B3),-2*B4/B1+36*B3/(1+5*B3))

Figure 13.6

Set up differential equations in *Excel* for the shooting method. The user-form SHOOTING prompts the user for information about the setup and location of the output.

Note how the second term in Equation (13.18) is indeterminate at $r = 0$. In the limit that $r \to 0$ the term becomes -2. Use the logical **IF( )** worksheet function in *Excel* to automate the selection of the appropriate form of the equation at $r = 0$ or $r > 0$. Add the initial conditions to cells **B3:B4** for the dependent variables.

The *VBA* user-form **SHOOTING** prompts for information about the integration and location of the output. In this example, equation one refers to the variable and derivative function of s. The second equation refers to the variable and derivative function of z. We know the boundary condition at $r = 1$ for s, but we do not know the value for s at the pellet center, $r = 0$. The shooting method searches for the unknown initial condition for s at $r = 0$ until the solution for s matches the known condition at $r = 1$. Use the default 1000 integration and 10 print steps, respectively, with a convergence tolerance of 10^{-6}. Locate the output on the worksheet beginning at cell **D1** where it does not overwrite cells with setup information.

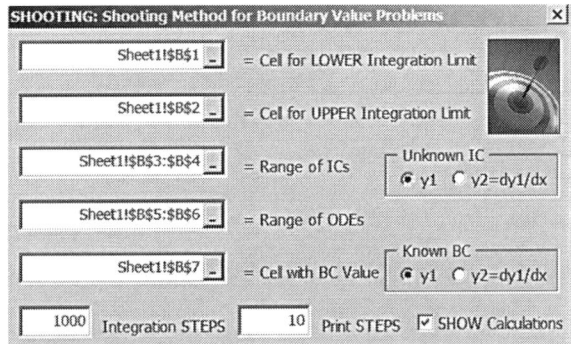

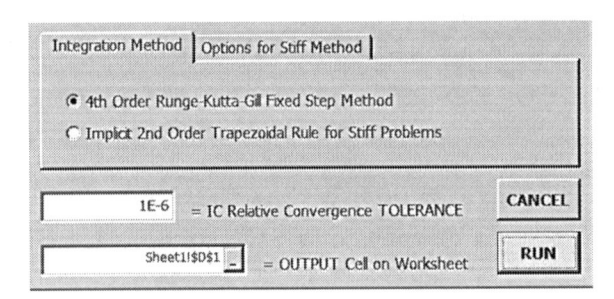

The numerical solution gives $s = 0.21$ at $r = 0$ after five secant iterations. The flux of substrate to the surface of the immobilized enzyme particle is proportional to the gradient in the substrate concentration at $r = 1$, which is the value of the new variable, $z(1) = ds/dr|_{r=1} = 1.8$. Figure 13.7 includes a plot of the substrate concentration with the derivative profile on the secondary axis.

	D	E	F
1	x	y_1	y_2
2	0	0.211502	0
3	0.1	0.217696	0.124395
4	0.2	0.236584	0.254848
5	0.3	0.269042	0.396526
6	0.4	0.316385	0.553019
7	0.5	0.380198	0.726039
8	0.6	0.462144	0.91553
9	0.7	0.563808	1.120108
10	0.8	0.686597	1.33762
11	0.9	0.831682	1.565653
12	1	1	1.801884

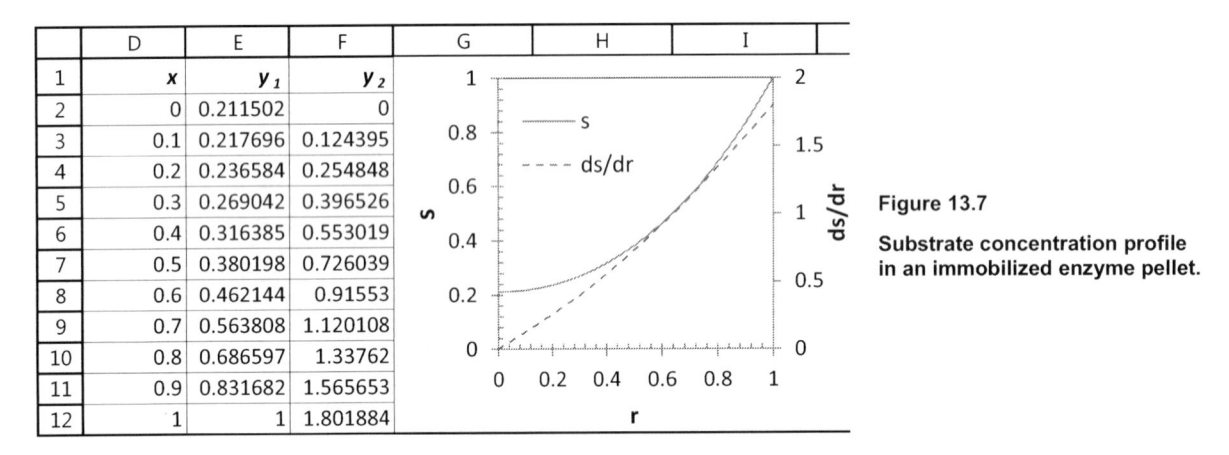

Figure 13.7
Substrate concentration profile in an immobilized enzyme pellet.

Comments: The rate of reaction is proportional to the slope of the concentration profile at the surface. How does the diameter of the sphere affect the concentration profile? Because we know the derivatives at the endpoints, we may use clamped cubic spline to interpolate the concentration profile and integrate an average concentration in the pellet:

$$\bar{s} = \frac{\int_0^1 sr^2 dr}{\int_0^1 r^2 dr} = 3\int_0^1 sr^2 dr \qquad (13.20)$$

The derivative of the integrand in the numerator is

$$\frac{d(sr^2)}{dr} = 2rs + r^2 \frac{ds}{dr} \qquad (13.21)$$

Create additional columns for the integrand and derivative, then apply the user-defined function **CSPLINE** to evaluate Equation (13.20) in the worksheet for the average dimensionless concentration in cell **J1 = 0.66**. The arguments include the lower and upper integration limits, separated by -1 for the anti-derivative:

	G	H	I	J
1	sr^2	$d(sr^2)/dr$	Average s =	=CSPLINE(D2:D12,G2:G12,D2,-1,D12,"C",H2,H12)*3
2	=E2*D2^2	=E2*2*D2+F2*D2^2		
3	=E3*D3^2	=E3*2*D3+F3*D3^2		

CHAPTER 13: BOUNDARY-VALUE PROBLEMS

The shooting method also applies to *systems* of *first*-order equations with split boundary conditions. In the next example, we consider the energy balance around a counter-current heat exchanger.

Example 13.3 Countercurrent Double Pipe Heat Exchanger Analysis

Known: A counter flow double-pipe heat exchanger uses water to cool an oil stream. Oil flows through the center pipe and water flows in the annulus between the pipes. An energy balance around the water includes heat transfer with the surrounding air. The oil energy balance only involves heat transfer with the water (Himmelblau 1996):

$$\dot{m}_w c_{pw} \frac{dT_w}{dx} = U_i \pi D_i (T_{oil} - T_w) + U_o \pi D_o (T_{air} - T_w) \tag{13.22}$$

$$\dot{m}_{oil} c_{poil} \frac{dT_{oil}}{dx} = U_i \pi D_i (T_{oil} - T_w) \tag{13.23}$$

The heat exchanger parameters are tabulated with values and definitions in an *Excel* worksheet shown in Figure 13.8. The heat capacities are provided as polynomial functions of temperature:

$$c_{pw}/(J/mol \cdot K) = 18.30 + 0.4721T - 1.339 \times 10^{-3} T^2 + 1.314 \times 10^{-6} T^3 \tag{13.24}$$

$$c_{poil}/(J/mol \cdot K) = -31.42 - 0.009761T + 2.354 \times 10^{-3} T^2 - 3.093 \times 10^{-6} T^3 \tag{13.25}$$

Schematic: Oil enters the heat exchanger at position 1 on the left side and exits at position 2 on the right side. Water flows in the opposite direction from position 2 to 1.

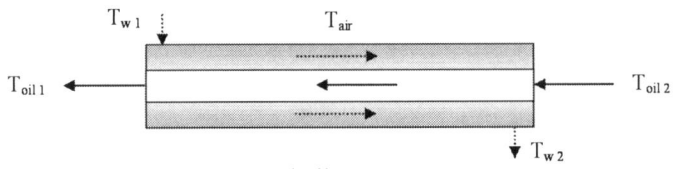

Find: Calculate the exit temperatures of the water and oil streams.

Assumptions: Steady-state operation, plug flow, constant heat transfer coefficients

Analysis: Use the shooting method to find the outlet temperatures. Start by guessing the oil outlet temperature at position one. Integrate for the fluid temperatures at position two and compare the calculated result with the oil inlet temperature. Adjust the oil outlet temperature using the secant method to match the oil inlet temperatures.

To simplify the set-up in an *Excel* worksheet, create VBA user-defined functions for the heat capacity of each fluid. The functions divide the heat capacity by the molecular weight to convert to specific heat:

```
Public Function cpw(T As Double) As Double    ' Specific heat of water, J/kg K
    cpw = (18300 + 472.1 * T - 1.339 * (T ^ 2) + 0.001314 * (T ^ 3))/ 18.016
End Function

Public Function cpoil(T As Double) As Double   ' Specific heat of oil, J/kg K
    cpoil = (-31421 - 9.761 * T + 2.354 * (T ^ 2) - 0.003093 * (T ^ 3)) / 86.17
End Function
```

Use named cells in an *Excel* worksheet as shown in Figure 13.8 to specify the initial and boundary conditions, differential equations, and integration limits. The cell formulas for the differential equations are:

C11 = (Ui*PI()*Di*(C11 - C10) + Uo*PI()*Do*(Tair - C10)) / (mw*cpw(D19))

C12 = (Ui*PI()*Di*(C11 - C10)) / (moil*cpoil(C11))

Find the oil outlet temperature using the VBA user-form **SHOOTING**, with the following inputs where the energy balances for water and oil are shooting equations one and two, respectively. Use the default numbers of integration and print steps, and convergence tolerance. Direct the output to cell **F1** and plot the temperature profiles:

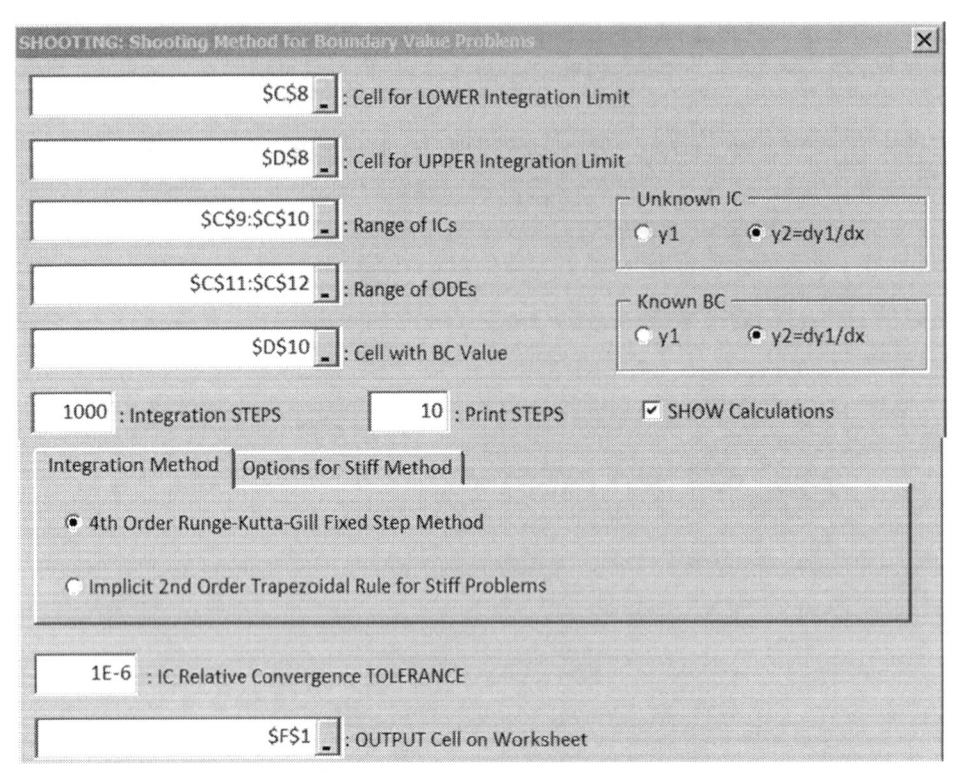

	A	B	C	D	E	F	G	H
1	Water Flow Rate	m_w	2	kg/s		x	y_1	y_2
2	Oil Flow Rate	m_{oil}	1	kg/s		0	290	295.0800047
3	Pipe Inner Diameter	D_i	0.15	m		1	292.107267	298.8162793
4	Pipe Outer Diameter	D_o	0.2	m		2	294.7659411	303.6278829
5	Internal Heat Transfer Coefficient	U_i	1450	W/m K		3	298.144301	309.7624046
6	External Heat Transfer Coefficient	U_o	112	W/m K		4	302.4258942	317.447729
7	Air Temperature	T_{air}	295	K		5	307.7940903	326.8573071
8	Length	x	0	10	m	6	314.4122662	338.081447
9	Water Temperature	T_w	290		K	7	322.4044729	351.1162355
10	Oil Temperature	T_{oil}	295.0800047	400	K	8	331.8423831	365.8753219
11	Derivative of Water Temperature	dT_w/dx	1.881838782		k/m	9	342.7425741	382.2203633
12	Derivative of Oil Temperature	dT_{oil}/dx	3.279795413		k/m	10	355.0747372	400

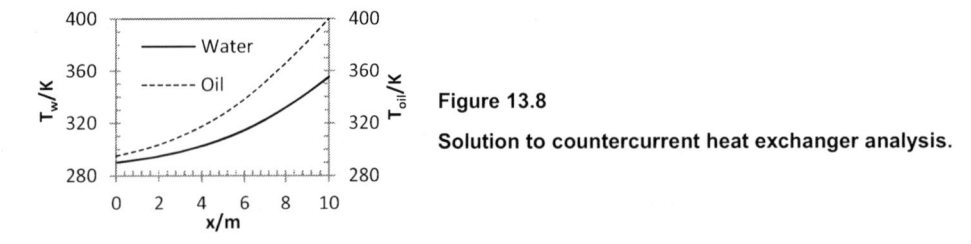

Figure 13.8

Solution to countercurrent heat exchanger analysis.

The outlet temperature for oil is the solution for the initial condition in cell **C10**, also found in cell **H2**. The water outlet temperature is the solution to the differential equation at $x = L$ in cell **G12**.

Comments: Traditional methods for finding the outlet temperature use either the effectiveness/number of transfer units or log-mean temperature difference. However, these require average values for the specific heat and do not properly account for variation of properties with temperature, or heat loss to the surroundings.

□

13.2 Finite-difference Method

The finite-difference method uses the derivative approximations summarized in Table 5.3 to convert an ordinary differential equation into a system of simultaneous algebraic equations. We then solve the systems of algebraic finite-difference equations by matrix algebra for linear systems or iterative methods for nonlinear systems, introduced in Chapter 4 and Chapter 6, respectively.

To implement the finite-difference method, divide the range for the independent variable x into n intervals of equal size, Δx, similar to the example shown in Figure 13.9.

$$\Delta x = x_{i+1} - x_i = x_i - x_{i-1} \tag{13.26}$$

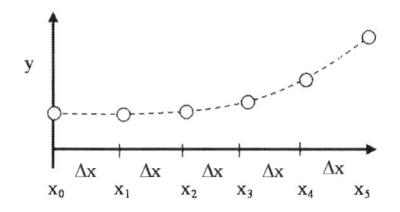

Figure 13.9

Discretize the range of the independent variable into uniform intervals for finite-difference approximations at each node.

Refer to the points between intervals as nodes.[112] Replace the derivatives in the differential equation with corresponding finite-difference approximations in terms of the variables at neighboring nodes. *Central difference* approximations of the first and second derivatives at node i are:

$$\left.\frac{dy}{dx}\right|_{x_i} \cong \frac{y_{i+1} - y_{i-1}}{2\Delta x} \quad \text{and} \quad \left.\frac{d^2y}{dx^2}\right|_{x_i} \cong \frac{y_{i+1} - 2y_i + y_{i-1}}{\Delta x^2} \tag{13.27}$$

Substitute the finite-difference results for the first and second-order derivatives into Equation (13.1):

$$\left[\frac{y_{i+1} - 2y_i + y_{i-1}}{\Delta x^2}\right] + R\left(x_i, y_i, \frac{y_{i+1} - y_{i-1}}{2\Delta x}\right) = 0 \tag{13.28}$$

Construct similar finite-difference equations for each of the interior nodes at x_i, $1 \leq i \leq n-1$. Determine the values of y_0 and y_n from the finite-difference forms of the boundary conditions, using forward and backward difference approximations at the left and right boundaries, respectively:

$$b_0 \cong \left(\frac{-3y_0 + 4y_1 - y_2}{2\Delta x}\right) - \beta_0(y_0 - \gamma_0) = 0 \quad \text{and} \quad b_n \cong \left(\frac{y_{n-2} - 4y_{n-1} + 3y_n}{2\Delta x}\right) - \beta_n(y_n - \gamma_n) = 0 \tag{13.29}$$

The result is a system of $n + 1$ simultaneous equations in $n + 1$ unknowns for y at each node.

In the case of linear systems, we may solve Equation (13.28) for the value of the dependent variable explicitly at node i in terms of the values for y at the neighboring nodes and the size of the interval in x between nodes:

$$y_i = f(\Delta x, x_i, y_{i-1}, y_{i+1}) \tag{13.30}$$

The explicit form of the boundary conditions in Equation (13.29) become:

$$y_0 = \frac{4y_1 - y_2 + 2\Delta x \beta_0 \gamma_0}{3 + 2\Delta x \beta_0} \quad \text{and} \quad y_n = \frac{y_{n-2} - 4y_{n-1} + 2\Delta x \beta_n \gamma_n}{2\Delta x \beta_n - 3} \tag{13.31}$$

As illustrated below, the solution for y_i represents an average value for y across the interval half way from the neighboring values for y_{i-1} and y_{i+1}.

---

[112] We recommend equally spaced nodes for convenience. However, we may strategically use smaller node spacing to improve the accuracy in regions where the solution is changing rapidly. See Section 5.3.3 for non-uniform node spacing in finite-difference derivative approximations.

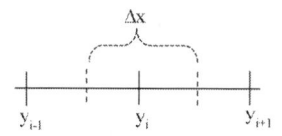

From Taylor series analysis in Chapter 5, we learn that the accuracy of the solution improves as we increase the number of nodes n (reduce the interval size Δx).

The finite-difference method spreads the errors of the derivative approximations over the range of integration unlike the shooting method that compounds the error with each integration step. However, second-order derivative approximations are less accurate than fourth-order derivative approximations employed by the higher order marching integration techniques, such as Runge-Kutta schemes. Use more nodes or five-point derivative approximations from Table 5.4 in finite-difference solutions to achieve a level of accuracy similar to fourth-order shooting methods.

Example 13.4   Catalyst Pellet Reactant Concentration Profile

Known/Find: Solve the problem in Example 13.2 by the finite-difference method.

Analysis: The procedure consists of discretizing the independent variable into a large number of equally spaced intervals. Calculate the length of a finite interval by dividing the range $0 \leq r \leq 1$ between the limits of integration by the number of intervals:

$$\Delta r = \frac{1-0}{n} \tag{13.32}$$

Use a forward finite-difference approximation of the derivative at the left boundary, node $i = 0$:

$$\frac{-3s_0 + 4s_1 - s_2}{2\Delta r} = 0 \tag{13.33}$$

Set the value of s at the right boundary, node $i = n$, according to the boundary condition:

$$s_n = 1 \quad \text{at} \quad r = 1 \tag{13.34}$$

Use central difference derivative approximations in the differential equation for the interior nodes, $i = 1, 2 \ldots n - 1$:

$$\frac{s_{i-1} - 2s_i + s_{i+1}}{\Delta r^2} + \frac{2}{i\Delta r}\left(\frac{s_{i+1} - s_{i-1}}{2\Delta r}\right) - \frac{9\theta^2 s_i}{1 + Ks_i} = 0 \tag{13.35}$$

Rearrange the finite-difference equations for the variable at each node for solution by fixed-point iteration:

$$s_0 = \frac{4s_1 - s_2}{3} \tag{13.36}$$

$$s_i = \frac{1}{2}\left[s_{i-1} + s_{i+1} + \frac{(s_{i+1} - s_{i-1})}{i} - \frac{9\Delta r^2 \theta^2 s_i}{1 + Ks_i}\right] \quad \text{for } i = 1, 3, \ldots n-1 \tag{13.37}$$

$$s_n = 1 \tag{13.38}$$

Calculate the independent variable at each node from the node index and interval size:

$$r_i = i\Delta r \tag{13.39}$$

This gives $n + 1$ simultaneous equations in $n + 1$ unknowns s at each node. To set up an *Excel* worksheet, add the formulas for the implicit finite-difference equations at each node in an *Excel* worksheet as shown in Figure 13.10 for solution using the macro **STEFFENSEN** from the *PNMSuite*.

1. Use named cells for Δr, θ, and K in the range **B1:B3**.
2. Place the values for r in column **B** according to Equation (13.39) beginning with cell **B6**.

CHAPTER 13: BOUNDARY-VALUE PROBLEMS 595

3. Provide initial guesses for each unknown value of $s = 0$ in column **C**.
4. Construct formulas for the implicit functions in column **D** using named cells for Δr (delr), θ (theta), and K. Use **AutoFill** to add the formula in cell **D7** into the remaining cells with relative cell references:

   B6 = A6*delr   D6 = (4*C7 - C8)/3   D7 = 0.5*(C6 + C8 + (C8 - C6)/ (A7) - 9*(delr*theta) ^ 2*C7/ (1 + K*C7))

5. Fix the boundary condition at node 10 to $s = 1$.

	A	B	C	D
1	Δr =	0.1		
2	θ =	2		
3	K =	5		
4				
5	Node	r	s	F(s)
6	0	0	0.212643533	0.2126455
7	1	0.1	0.218917021	0.21891877
8	2	0.2	0.237731597	0.23773304
9	3	0.3	0.270073796	0.27007505
10	4	0.4	0.317271666	0.31727275
11	5	0.5	0.380921425	0.38092232
12	6	0.6	0.462699255	0.46269995
13	7	0.7	0.564202346	0.56420284
14	8	0.8	0.686842742	0.68684303
15	9	0.9	0.831797007	0.83179711
16	10	1	1	

Figure 13.10

Excel **worksheet set up for solving finite-difference equations by the macro STEFFENSEN.**

The solution using the macro **STEFFENSEN** converges after 77 iterations with a relative tolerance of 10^{-5}.

Alternatively, we may set up the worksheet by employing the user-defined function **FDERIV** in place of the first and second derivatives in Equation (13.16), as seen in Figure 13.11. We use a three-point forward difference approximation for the first derivative at the boundary condition $r = 0$. The rest of the formulas use three-point central difference approximations for the first and second derivatives. Use the **Solver** to find the values of s in column **C** that minimize the sum of squares of the formulas in column **D**. The sum of squares is located in cell **D33**.

	A	B	C	D
22	Node	r	s	Diffusion with Reaction Equation
23	0	0	0.212664	=FDERIV(B23:B25,C23:C25,1)
24	1	=B23+delr	0.218935	=FDERIV(B23:B25,C23:C25,2)+(2/B24)*FDERIV(B23:B25,C23:C25,0)-9*(theta^2)*C24/(1+K*C24)
25	2	=B24+delr	0.237749	=FDERIV(B24:B26,C24:C26,2)+(2/B25)*FDERIV(B24:B26,C24:C26,0)-9*(theta^2)*C25/(1+K*C25)
26	3	=B25+delr	0.270090	=FDERIV(B25:B27,C25:C27,2)+(2/B26)*FDERIV(B25:B27,C25:C27,0)-9*(theta^2)*C26/(1+K*C26)
27	4	=B26+delr	0.317286	=FDERIV(B26:B28,C26:C28,2)+(2/B27)*FDERIV(B26:B28,C26:C28,0)-9*(theta^2)*C27/(1+K*C27)
28	5	=B27+delr	0.380934	=FDERIV(B27:B29,C27:C29,2)+(2/B28)*FDERIV(B27:B29,C27:C29,0)-9*(theta^2)*C28/(1+K*C28)
29	6	=B28+delr	0.462709	=FDERIV(B28:B30,C28:C30,2)+(2/B29)*FDERIV(B28:B30,C28:C30,0)-9*(theta^2)*C29/(1+K*C29)
30	7	=B29+delr	0.564209	=FDERIV(B29:B31,C29:C31,2)+(2/B30)*FDERIV(B29:B31,C29:C31,0)-9*(theta^2)*C30/(1+K*C30)
31	8	=B30+delr	0.686847	=FDERIV(B30:B32,C30:C32,2)+(2/B31)*FDERIV(B30:B32,C30:C32,0)-9*(theta^2)*C31/(1+K*C31)
32	9	=B31+delr	0.831799	=FDERIV(B31:B33,C31:C33,2)+(2/B32)*FDERIV(B31:B33,C31:C33,0)-9*(theta^2)*C32/(1+K*C32)
33	10	=B32+delr	1	=SUMSQ(D23:D32)

Figure 13.11 Finite difference solution to the boundary-value problem with the user-defined function FDERIV for the first and second derivatives.

Comments: Finite-difference solutions do not directly produce approximate values for the derivatives at the boundaries like the shooting method. Instead, we calculate the derivatives from the solution using the formulas for finite-difference derivative approximations programmed in the user-defined functions **FINDIF** or **FDERIV**.

□

"Life never presents us with anything which may not be looked upon as a fresh starting point, no less than as a termination" – Andre Gide

Symmetric Boundary Conditions: Models with symmetry in the dependent variable at a boundary use a Neumann boundary condition with the first derivative set to zero. Instead of a forward or backward finite-difference approximation for the derivative employed in Example 13.4, we may take advantage of any symmetry and use a central difference approximation by adding a false node that mirrors the image on the opposite side of the boundary node, as illustrated in Figure 13.12.

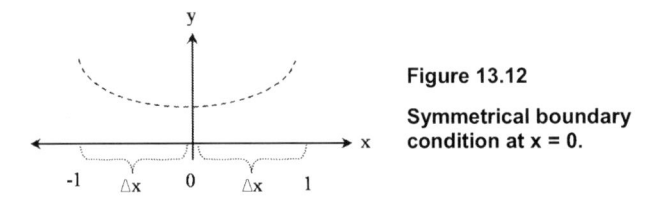

Figure 13.12

Symmetrical boundary condition at x = 0.

We use the false node in the central finite-difference formulas for the differential equation at the symmetric node. For example, at node 0:

$$\left.\frac{dy}{dx}\right|_{x_0} \cong \frac{y_1 - y_{-1}}{2\Delta x} = 0 \quad or \quad y_1 = y_{-1} \tag{13.40}$$

Substitute for y_{-1} into the second-order differential Equation (13.28) at node 0, which reduces to the following:

$$2\left[\frac{y_1 - y_0}{\Delta x^2}\right] + R(x_0, y_0) = 0 \tag{13.41}$$

We demonstrate the application of false nodes in the following example.

Example 13.5 Finite-difference Solution for Diffusion in a Catalyst Pellet

Known/Find: Repeat Example 13.1 using the finite-difference method with a false node at the left boundary.

Analysis: Divide the range into 10 equally spaced intervals. For comparison, we first use the forward-difference approximation at the boundary condition $r = 0$:

$$\frac{-3c_0 + 4c_1 - c_2}{2\Delta r} = 0 \tag{13.42}$$

At $r = 1$, the boundary condition is simply $c_{10} = 1$. Rearrange each equation implicitly in terms of c at the node:

For $i = 0$
$$c_0 = \frac{4c_1 - c_2}{3} \tag{13.43}$$

The finite-difference form of Equation (13.10) for the nodes between boundary conditions is

$$\frac{c_{i+1} - 2c_i + c_{i-1}}{\Delta r^2} = \phi c_i^2 \tag{13.44}$$

Rearrange Equation (13.44) into an explicit form at the interior nodes $i=1\ldots9$:

$$c_i = \frac{c_{i+1} + c_{i-1}}{\Delta r^2 \phi c_i + 2} \tag{13.45}$$

The *Excel* worksheet in Figure 13.13 shows the finite-difference implicit formulas using iteration with circular references enabled. From the plot we see how the slope goes to zero at the center of the pellet where $r = 0$.

CHAPTER 13: BOUNDARY-VALUE PROBLEMS

	A	B
1	r	c
2	0	=(4*B3-B4)/3
3	0.1	=(B2+B4)/(10*B3*(A3-A2)^2+2)
4	0.2	=(B3+B5)/(10*B4*(A4-A3)^2+2)
5	0.3	=(B4+B6)/(10*B5*(A5-A4)^2+2)
6	0.4	=(B5+B7)/(10*B6*(A6-A5)^2+2)
7	0.5	=(B6+B8)/(10*B7*(A7-A6)^2+2)
8	0.6	=(B7+B9)/(10*B8*(A8-A7)^2+2)
9	0.7	=(B8+B10)/(10*B9*(A9-A8)^2+2)
10	0.8	=(B9+B11)/(10*B10*(A10-A9)^2+2)
11	0.9	=(B10+B12)/(10*B11*(A11-A10)^2+2)
12	1	1

	A	B
1	r	c
2	0	0.281484
3	0.1	0.285561
4	0.2	0.297793
5	0.3	0.318893
6	0.4	0.350162
7	0.5	0.393693
8	0.6	0.452723
9	0.7	0.532249
10	0.8	0.640103
11	0.9	0.788931
12	1	1

Figure 13.13 *Excel* worksheet finite-difference solution to a catalyst pellet problem.

An alternative form of the boundary condition at $r = 0$ uses the centered difference approximation for the first derivative with a false node. In this case, we assume a false node to the left of the first node (or the cell in the row above the boundary at the first node in Figure 13.14):

$$\frac{c_1 - c_{-1}}{2\Delta r} = 0 \tag{13.46}$$

Rearrange for the concentration at the false node:

$$c_{-1} = c_1 \tag{13.47}$$

Now, the finite-difference approximation for the differential equation at the first node becomes

$$\frac{2(c_1 - c_0)}{\Delta r^2} = \phi c_0^2 \tag{13.48}$$

We can solve for the concentration at the center of the pellet to give an implicit expression at node 0:

$$c_0 = c_1 - \frac{\phi c_0^2 \Delta r^2}{2} \tag{13.49}$$

We simply include the false node in our system of equations and use a central-difference approximation for the second derivative at node 0. The *Excel* worksheet in Figure 13.14 shows the first five formulas and solution using a false node for the symmetric condition at the left boundary, $r = 0$. The rest of the formulas are the same as seen in Figure 13.13.

	A	B
1	r	c
2	-0.1	=B4
3	0	=(B2+B4)/(10*B3*(A3-A2)^2+2)
4	0.1	=(B3+B5)/(10*B4*(A4-A3)^2+2)
5	0.2	=(B4+B6)/(10*B5*(A5-A4)^2+2)
6	0.3	=(B5+B7)/(10*B6*(A6-A5)^2+2)

	A	B
1	r	c
2	-0.1	0.285916764
3	0	0.281942194
4	0.1	0.285916764
5	0.2	0.298066174
6	0.3	0.319099928

Figure 13.14

Finite-difference formulas and converged cell values using a false (mirror) node for the boundary condition at the pellet center, $r = 0$. (only first 5 nodes displayed)

Comments: We observe slight differences between the solutions from the two different finite-difference representations of the left boundary condition due to slightly more accurate central difference derivative approximations. Thus, we prefer using a false (mirror) node for symmetric boundary conditions (*e.g.*, $dy/dx = 0$).

□

13.3 Orthogonal Collocation on Finite Elements*

"It's always good to take an orthogonal view of something. It develops ideas." – Ken Thompson

Orthogonal collocation is an important method of weighted residuals for solving "stiff" boundary value problems with steep gradients, such as those encountered in chemical reaction engineering (Finlayson 1980). The applications of collocation methods reach far into many branches of engineering, including solid mechanics and dynamics, as well as fluid flow and heat transfer. With the method of weighted residuals, we substitute a polynomial trial function for y into the differential Equation (13.1) and search for the values for the coefficients of the polynomial that minimize the integral of the residual over the domain of integration, similar to the method of continuous least-squares described in Section 11.8:

$$\text{Min}\left\{\text{Residual}=\int_{x_0}^{x_n}\left[\frac{d^2y}{dx^2}-R\left(x,y,\frac{dy}{dx}\right)\right]^2 dx\right\} \quad (13.50)$$

where the dependent variable y and its derivatives are polynomial functions of the independent variable, x:

$$y = a_0 + a_1 x + a_2 x^2 + a_3 x^3 \ldots \qquad \frac{dy}{dx} = a_1 + 2a_2 x + 3a_3 x^2 \ldots \qquad \frac{d^2 y}{dx^2} = 2a_2 + 6a_3 x \ldots \quad (13.51)$$

Determining the polynomial coefficients that minimize the integral in Equation (13.50) subject to boundary conditions like Equation (13.2) is a challenging numerical problem. To avoid the complex integration imbedded in a minimization routine, we use an orthogonal Legendre polynomial for the trial function to replace the integral in Equation (13.50) with the algebraic residuals at the polynomial collocation points. Similar to Gauss quadrature, the collocation points are the zeroes of shifted orthogonal Legendre polynomials to avoid Runge's (1901) phenomena. Collocation points for shifted Lagrange polynomials up to order seven are available in the *VBA* user-defined array functions **A_OC**, **X_OC**, and **Y_OC** listed in Figure 13.15 from the *PNMSuite*. In place of minimizing the integral in Equation (13.50), we determine the trial polynomial coefficients that minimize just the algebraic residuals at the collocation points, including the residual boundary conditions at the integration limits.

```
Public Function A_OC(y As Variant)
' y = range of guesses for dependent variable at each collocation point
'******************************************************************************
' User-defined array function that requires CNTL-SHIFT-ENTER. Returns the n number of
' coefficients for the (n-1)th-order (2<=n-1<=7)Legendre orthogonal polynomial trial
' function corresponding the values for y at the n collocation points xj:
' y(xj) = A1 + A2*xj... An*xj^(n-1).
' To use the user function, select a column range of cells containing y-points
' corresponding to the collocation points with size equal to the order (n+1) of the
' polynomial trial function
' *** Nomenclature ***
Dim j As Integer, k As Integer ' = index variables for row and column loops
Dim n As Integer ' = order + 1 of trial polynomial (number of collocation points)
Dim q() As Variant ' = matrix of x terms at each collocation point for trial function
Dim x() As Variant ' = vector of collocation points from trial polynomial of order n-1
'******************************************************************************
n = Application.Caller.Rows.Count ' get the size of the trial function
If n < 3 Or n > 8 Then ' error message if out of range
    MsgBox "Number of points in COLUMN is limited to 3 <= n <= 8.": Exit Function
End If
ReDim a(1 To n) As Variant, q(1 To n, 1 To n) As Variant, x(1 To n) As Variant
x(1) = 0: x(n) = 1 ' get the collocation points in vector x
Select Case n
Case 3
    x(2) = 0.5
Case 4
    x(2) = 0.2113248654: x(3) = 0.7886751346
Case 5
    x(2) = 0.1127016654: x(3) = 0.5: x(4) = 0.8872983346
Case 6
```

CHAPTER 13: BOUNDARY-VALUE PROBLEMS

```vb
            x(2) = 0.0694318442: x(3) = 0.3300094783: x(4) = 0.6699905218: x(5) = 0.9305681558
    Case 7
            x(2) = 0.0469100771: x(3) = 0.230765345: x(4) = 0.5: x(5) = 0.7692346551
            x(6) = 0.953089923
    Case Else
            x(2) = 0.0337652429: x(3) = 0.1693953068: x(4) = 0.380690407: x(5) = 0.6193095931
            x(6) = 0.8306046933: x(7) = 0.9662347571
    End Select
    For j = 1 To n ' rows for collocation points -> get the matrix of x terms at points
        For k = 1 To n: q(j, k) = x(j) ^ (k - 1): Next k ' columns for polynomial terms
    Next j
    With Application: A_OC = .MMult(.MInverse(q), y): End With ' return coefficients A
End Function ' A_OC

Public Function X_OC(Optional xlo As Double = 0, Optional xup As Double = 1)
' xlo = (optional) lower limit of integration (default = 0)
' xup = (optional) upper limit of integration (default = 1)
'*******************************************************************************
' Array function that requires CNTL-SHIFT-ENTER. Provides n collocation points for a
' polynomial of order 2<=n-1<=7. To use the function, select a column range of cells
' with size equal to the order of the polynomial trial function.
' *** Nomenclature ***
Dim n As Integer ' = number of terms in trial function polynomial (includes zeroes)
Dim s As Double ' = scale factor for integration variable
Dim x() As Variant ' = vector of collocation points
'*******************************************************************************
n = Application.Caller.Rows.Count ' Get the size of the selected range of cells.
If n < 3 Or n > 8 Then ' error message if out of range
    MsgBox "Number of points in COLUMN is limited to 3 <= n <= 8.": Exit Function
End If
s = xup - xlo ' Set the scale factor for collocation points
ReDim x(1 To n): x(1) = xlo: x(n) = xup ' Set the first and last collocation points
Select Case n ' Set the interior collocation points between limits of integration.
Case 3
        x(2) = xlo + 0.5 * s
Case 4
        x(2) = xlo + 0.2113248654 * s: x(3) = xlo + 0.7886751346 * s
Case 5
        x(2) = xlo + 0.1127016654 * s: x(3) = xlo + 0.5 * s: x(4) = xlo + 0.8872983346 * s
Case 6
        x(2) = xlo + 0.0694318442 * s: x(3) = xlo + 0.3300094783 * s
        x(4) = xlo + 0.6699905218 * s: x(5) = xlo + 0.9305681558 * s
Case 7
        x(2) = xlo + 0.0469100771 * s: x(3) = xlo + 0.230765345 * s: x(4) = xlo + 0.5 * s
        x(5) = xlo + 0.7692346551 * s: x(6) = xlo + 0.953089923 * s
Case Else
        x(2) = xlo + 0.0337652429 * s: x(3) = xlo + 0.1693953068 * s
        x(4) = xlo + 0.380690407 * s: x(5) = xlo + 0.6193095931 * s
        x(6) = xlo + 0.8306046933 * s: x(7) = xlo + 0.9662347571 * s
End Select
X_OC = Application.Transpose(x) ' Use CTRL-SHIFT-ENTER to add the values to column
End Function ' X_OC

Public Function Y_OC(x As Double, a As Variant, Optional dy As Integer = 0) As Double
' a = range of cells containing trial polynomial coefficients
' dy = (optional) argument for returning first or second derivative of y;
'         1 = first derivative, 2 = second derivative
'*******************************************************************************
' Calculate y, y' = dy/dx, or y" = dy2/dx2 at x from a trial polynomial
' y = a(1) + a(2)*x + a(3)*x^2 + ... a(n-2)*x^(n-2) + a(n)*x^(n-1)
' y' = a(2) + 2*a(3)*x + ... (n-2)*a(n-2)*x^(n-3) + (n-1)*a(n)*x^(n-2)
' y" = 2*a(3) + ... (n-3)*(n-2)*a(n-2)*x^(n-4) + (n-2)*(n-1)*a(n)*x^(n-3)
' *** Nomenclature ***
Dim k As Integer ' = index of coefficients in polynomial; k-1=order of polynomial term
Dim n As Integer ' = number of coefficients in polynomial (includes zeroes)
'*******************************************************************************
n = a.Count
Select Case dy
Case 1 ' calculate first derivative
```

```
        Y_OC = a(2)
        For k = 3 To n: Y_OC = Y_OC + (k - 1) * a(k) * x ^ (k - 2): Next k
Case 2 ' calculate second derivative
        Y_OC = 2 * a(3)
        For k = 4 To n: Y_OC = Y_OC + (k - 1) * (k - 2) * a(k) * x ^ (k - 3): Next k
Case Else ' otherwise, calculate y
        Y_OC = a(1)
        For k = 2 To n: Y_OC = Y_OC + a(k) * x ^ (k - 1): Next k
End Select
End Function ' Y_OC
```

Figure 13.15 User-defined array functions for implementing the method of orthogonal collocation in *Excel*. A_OC generates the trial function coefficients for specific values of the dependent variable at the collocation points. Requires CTRL SHIFT ENTER.

We outline how to apply the method of orthogonal collocation in *Excel* for a one-dimensional boundary-value problems then follow up with examples. The interested reader should consult advanced treatments of this topic for application to more complicated geometries:

1. Scale (shift) the independent variable x in the differential equation for a domain of integration, $0 \leq x^* \leq 1$:

$$x^* = \frac{x - x_0}{x_n - x_0} = \frac{x - x_0}{\Delta x} \quad (13.52)$$

where x_0, and x_n are the values for the independent variable at the boundaries (integration limits), and Δx is the length of the domain of the independent variable. Scaling the independent variable has the added benefit of reducing round-off errors and promoting convergence.

The scaled boundary-value problem in Equation (13.1) and boundary conditions in Equations (13.2) become:

$$\frac{d^2 y}{dx^{*2}} - \Delta x^2 R\left(x^* \Delta x + x_0, y, \frac{1}{\Delta x}\frac{dy}{dx^*}\right) = 0 \quad (13.53)$$

$$\alpha_0 \left.\frac{dy}{dx^*}\right|_0 - \Delta x \beta_0 (y - \gamma_0) = 0 \text{ at } x^* = 0 \quad \text{and} \quad \alpha_n \left.\frac{dy}{dx^*}\right|_n - \Delta x \beta_n (y - \gamma_n) = 0 \text{ at } x^* = 1 \quad (13.54)$$

where $R(\ )$ represents a general function of x^*, y, and dy/dx^*.

2. Decide on the order N of the Legendre polynomial used for the trial function and generate the collocation points in an *Excel* worksheet with the user-defined array function **X_OC**. The function determines the order of the polynomial trial function from the length of the range of cells selected for the collocation points:

$$y(x) = \sum_{k=0}^{N} a_k \left(x^*\right)^k \quad (13.55)$$

The results are the first and last points $x^*=0$ and $x^*=1$, and up to six interior x collocation points. Thus, the maximum number of trial polynomial values and coefficients is limited to eight (for an order seven polynomial). To use the array function **X_OC**:

   a) Select a range of contiguous cells in a column with a total number of cells equal to the order of the polynomial trial function plus one (between 3 and 8).
   b) With the range selected, type the name of the user-defined array function: = **X_OC()**. The user-defined array function **X_OC(Optional x initial, Optional x final)** has optional arguments for the lower and upper limits for x. The default values are 0 and 1, respectively.
   c) Type the keyboard combination *CNTL SHIFT ENTER* to replace the values in the selected range with the collocation points.

3. We may elect to solve for the polynomial coefficients directly, or solve for the solution to the differential equation at each collocation point and calculate the coefficients. The choice of solution method is a matter of convenience. We show both ways in the following examples. In our experience, the numerical solution is more stable when solving for the values of the dependent variable at the collocation points. In this case, we provide initial estimates on the worksheet for the solution of the dependent variable y_j at each collocation

CHAPTER 13: BOUNDARY-VALUE PROBLEMS

point x_j^*, then calculate the values for the trial polynomial coefficients, a, at each point with the user-defined array function **A_OC**. The trial polynomial coefficients are calculated from the $j = 0...N$ collocation points in the vector x^* and y:

$$a = q^{-1} y \qquad (13.56)$$

where

$$q = \begin{vmatrix} 1 & x_0 & x_0^2 & \cdots & x_0^N \\ 1 & x_1 & x_1^2 & \cdots & x_0^N \\ 1 & x_2 & x_2^2 & \cdots & x_0^N \\ \vdots & \vdots & \vdots & \ddots & \vdots \\ 1 & x_N & x_N^2 & \cdots & x_N^N \end{vmatrix}$$

The user-defined function **A_OC** uses the *Excel* worksheet functions for matrix inversion and multiplication introduced in Section 4.4. To generate the coefficients, select a range of cells in a column for the results, then type = **A_OC(y)** where the argument **y** is the range of values in a column for y. Type CNTL SHIFT ENTER for the result.

4. Generate values for the first and second-derivatives of the dependent variable y with the user-defined function **Y_OC**. Note that **Y_OC** is *NOT* an array function. The arguments for **Y_OC** include the value for the scaled dependent variable x^* and the range of values for polynomial coefficients **a** from Step 3. **Y_OC** uses an optional argument **dy** to return either the first (**1**) or the second (**2**) derivative of the trial polynomial function.

5. Substitute the trial polynomial function from Equation (13.55) into the differential Equation (13.53) and evaluate the residual at each of the $(1 < j < N - 1)$ interior collocation points, x_j^*. For example:

$$\text{Residual}_j = \left.\frac{d^2 y}{dx^{*2}}\right|_j - \Delta x^2 R\left(x_j^*, y_j, \frac{1}{\Delta x}\left.\frac{dy}{dx^*}\right|_j\right) \qquad j = 1, 2 ... N - 1 \qquad (13.57)$$

where

$$y_j = a_0 + a_1 x_j^* + a_2 x_j^{*2} + \sum_{k=3}^{N} a_k x_j^{*k} \qquad (13.58)$$

$$\left.\frac{dy}{dx^*}\right|_j = a_1 + 2a_2 x_j^* + \sum_{k=3}^{N} k a_k x_j^{*(k-1)} \qquad (13.59)$$

$$\left.\frac{d^2 y}{dx^{*2}}\right|_j = 2a_2 + \sum_{k=3}^{N} k(k-1) a_k x_j^{*(k-2)} \qquad (13.60)$$

6. At the lower and upper integration limits, use the residuals at the boundary conditions. For example:

$$\text{Residual}_0 = \alpha_0 a_1 - \Delta x \beta_0 (a_0 - \gamma_0) = 0 \qquad (13.61)$$

$$\text{Residual}_N = \alpha_n \left(a_1 + 2a_2 + \sum_{k=3}^{N} k a_k\right) - \Delta x \beta_1 (y_1 - \gamma_n) = 0 \qquad (13.62)$$

7. Use any of the root finding or minimization methods from Chapter 7 or Chapter 9 to solve for the values y_j at each collocation point, or directly for the coefficients of the polynomial trial function that minimize the sum of squared residuals for the system of $N + 1$ Equations (13.57), (13.61), and (13.62).

8. Interpolate the solution for y between collocation points using the user-defined function **Y_OC** with the final solution for the trial polynomial coefficients.

When compared with the finite-difference method, the method of orthogonal collocation uses the values at all of the points of discretization to approximate each of the function and derivatives, not just the two or three points at the left and right intervals of a collocation point.

Example 13.6 Concentration Profile in an Immobilized Pellet (Revisited)

Known/Find: Repeat Example 13.2 using the method of orthogonal collocation in an *Excel* worksheet.

Analysis: Follow these steps to implement the method of orthogonal collocation in an *Excel* worksheet to determine the trial polynomial coefficients for this example:

1. Provide guesses for the coefficients, *a*, in a seventh order trial polynomial in column **A**.
2. Use array function **X_OC** to get the scaled collocation points for the independent variable *r*. Select the range **B2:B9**, type **X_OC()**, then *CNTL SHIFT ENTER* for the scaled values x^* in column **B**.
3. Use the *VBA* user-defined function **Y_OC** to calculate values for *s*, *ds/dr*, and d^2s/dr^2 in columns **C**, **D**, and **E** from the trial function and guessed coefficients from step one.
4. Use the polynomial trial-function values for *s* and its derivatives to evaluate the residuals of the boundary conditions and differential equation from Equations (13.16) and (13.17) in column **F**, as shown by the formulas in Figure 13.16.
5. Calculate the sum of the squared residuals in cell **F8** and use the **Solver**, or alternative regression method such as Levenberg-Marquardt, to find the values for the trial function parameters, *a*, that minimize the sum or squared residuals (*SSR*). When using the **Solver** for this type of problem, be careful to change the convergence tolerance to 10^{-8} and uncheck the default option **Make Unconstrained Variables Non-Negative**.
6. Finally, plot the solution at the collocation points, located in columns **B** and **C**, for *s* versus *r*.

	A	B	C	D	E	F	G	H
1	a	r	s	ds/dr	d^2s/dr^2	Residual	K =	5
2	0.2115	0	=y_OC(B2,$A$2:$A$9)	=y_OC(B2,$A$2:$A$9,1)	=y_OC(B2,$A$2:$A$9,2)	=D2	θ =	2
3	0.0007	0.0337	=y_OC(B3,$A$2:$A$9)	=y_OC(B3,$A$2:$A$9,1)	=y_OC(B3,$A$2:$A$9,2)	=E3+(2/B3)*D3-(9*C3*$H$2^2)/(1+$H$1*C3)		
4	0.6070	0.1693	=y_OC(B4,$A$2:$A$9)	=y_OC(B4,$A$2:$A$9,1)	=y_OC(B4,$A$2:$A$9,2)	=E4+(2/B4)*D4-(9*C4*$H$2^2)/(1+$H$1*C4)		
5	0.0422	0.3806	=y_OC(B5,$A$2:$A$9)	=y_OC(B5,$A$2:$A$9,1)	=y_OC(B5,$A$2:$A$9,2)	=E5+(2/B5)*D5-(9*C5*$H$2^2)/(1+$H$1*C5)		
6	0.1862	0.6193	=y_OC(B6,$A$2:$A$9)	=y_OC(B6,$A$2:$A$9,1)	=y_OC(B6,$A$2:$A$9,2)	=E6+(2/B6)*D6-(9*C6*$H$2^2)/(1+$H$1*C6)		
7	0.0533	0.8306	=y_OC(B7,$A$2:$A$9)	=y_OC(B7,$A$2:$A$9,1)	=y_OC(B7,$A$2:$A$9,2)	=E7+(2/B7)*D7-(9*C7*$H$2^2)/(1+$H$1*C7)		
8	-0.1567	0.9662	=y_OC(B8,$A$2:$A$9)	=y_OC(B8,$A$2:$A$9,1)	=y_OC(B8,$A$2:$A$9,2)	=E8+(2/B8)*D8-(9*C8*$H$2^2)/(1+$H$1*C8)		
9	0.0555	1	=y_OC(B9,$A$2:$A$9)	=y_OC(B9,$A$2:$A$9,1)	=y_OC(B9,$A$2:$A$9,2)	=C9-1		
10					SSR =	=SUMSQ(F2:F9)		

	A	B	C	D	E	F	G	H
1	a	r	s	ds/dr	d^2s/dr^2	Residual	K =	5
2	0.2115	0.0000	0.2115	0.0007	1.2141	7.24E-04	θ =	2
3	0.0007	0.0338	0.2122	0.0419	1.2253	-1.83E-05		
4	0.6071	0.1694	0.2294	0.2138	1.3229	4.61E-05		
5	0.0423	0.3807	0.3060	0.5217	1.6134	-5.54E-05		
6	0.1863	0.6193	0.4802	0.9538	2.0028	4.57E-05		
7	0.0534	0.8306	0.7286	1.4065	2.2625	-2.64E-05		
8	-0.1567	0.9662	0.9405	1.7213	2.3746	7.88E-06		
9	0.0555	1.0000	1.0000	1.8019	2.4005	-3.34E-07		
10					SSR =	5.33E-07		

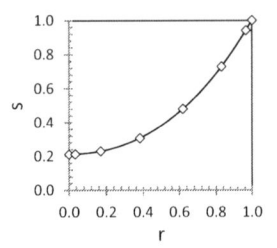

Figure 13.16

Excel worksheet setup for the solution to a boundary value problem by orthogonal collocation for diffusion and reaction in a catalyst pellet.

Comments: The orthogonal collocation results agree with those from the shooting method. Although we can use an interpolation scheme, such as cubic spline to interpolate the results from the shooting method, the method of orthogonal collocation produces the parameters *a* for the polynomial trial function directly for interpolating the solution to the differential equation:

$$s = 0.2115 + 7.240 \times 10^{-4} r + 0.6071 r^2 + 0.04229 r^3 + 0.1863 r^4 + 0.05336 r^5 - 0.1567 r^6 + 0.05552 r^7$$

The user-defined function **Y_OC** is also available for calculating *s*, d*s*/d*r*, or d^2s/dr^2 at any position *r* from the solution for the coefficients of the trial function.

□

CHAPTER 13: BOUNDARY-VALUE PROBLEMS

For stiff problems with steep gradients, the solution may require application of orthogonal collocation to piecewise smaller finite elements within the range of integration (Finlayson 1980). We must shift the independent variable j for each element i and incorporate the scaled variable into the differential equation applied to each element:

$$x_{i,j}^* = \frac{x_j^* - x_{i,j}^*}{x_{i+1,j}^* - x_{i,j}^*} = \frac{x^* - x_{i,j}^*}{\Delta x_i^*} \quad (13.63)$$

where x_i^* and x_{i+1}^* are the starting and ending positions for element i and Δx_i^* is the width of the integration domain in each scaled element. The rescaled residuals of the differential equation takes the following form at the collocation points within each element:

$$\text{Residual}_{i,j} = \frac{d^2 y}{dx_{i,j}^{*2}} - \Delta x_i^{*2} R\left(x_{i,j}^* \Delta x_i^*, y, \frac{1}{\Delta x_i^*} \frac{dy}{dx_{i,j}^*}\right) \quad (13.64)$$

For example, if we divide the range of integration into two finite elements, each having width $\Delta x_i^* = 0.5$, using five collocation points in each element, the scaled values for the independent variable become:

Element i	1	1	1	1	1	2	2	2	2	2
Collocation Point j	0	1	2	3	4	0	1	2	3	4
x_i^*	0	0.112701665	0.5	0.887298335	1	0	0.112701665	0.5	0.887298335	1
x^*	0	0.056350833	0.25	0.443649167	0.5	0.5	0.556350833	0.75	0.943649167	1

For the case of two finite elements, we split the original boundary conditions between the solutions for two trial functions. The trial function uses boundary condition at the lower limit for the first finite element, whereas the boundary condition at the upper limit is used in the trial function for the second finite element. We need two more boundary conditions, one for each element's trial function at the shared point in the middle. Similar to interpolating splines, we find the boundary conditions by requiring continuity in the trial functions and their derivatives at the interior boundary (knot) between the two elements:

$$y_{i=1, j=4} = y_{i=2, j=0} \quad \text{and} \quad \left.\frac{dy}{dx^*}\right|_{i=1, j=4} = \left.\frac{dy}{dx^*}\right|_{i=2, j=0} \quad (13.65)$$

This gives a system of simultaneous equations for the two sets of trial polynomial coefficients. Following this pattern, we may use any number of finite elements in our problem domain.

Excel's **Solver** is convenient for solving a system of equations in an *Excel* worksheet, although we may apply any method, such as Newton's method and optimization techniques, like Levenberg-Marquardt. Try the VBA macro **SCALIT** with the **Solver** if the values of the trial function coefficients differ by several orders of magnitude. Once we have the unique polynomials for each element, we must be careful to match our position in x in an element with the corresponding polynomial. Use the worksheet function **IF()** to select the polynomial from the appropriate element.

We may also apply the methods of orthogonal collocation to finding approximate solutions to elliptical and parabolic partial differential equations (Rice and Do 1995).

Example 13.7 Concentration Profile in a Pellet with a Large Thiele Modulus

Known/Find: Repeat *Example 13.1* using the method of orthogonal collocation with $\phi = 300$. Compare the solution with one and two finite elements in an *Excel* worksheet.

Analysis: With a large Thiele modulus, the concentration decreases towards zero near the center of the pellet. This also produces a steeper gradient in the concentration profile near the pellet surface. For comparison, solve the problem using a single element with a 7th-order polynomial trial-function, then again using two finite elements, each with 7th order polynomial trial-functions.

Apply the user-defined array function **X_OC** to add the values for the radial positions r in column **A**. Start with initial guesses for the concentrations at each collocation point in column **B**. Calculate the polynomial coefficients in column **C** with the user-defined array function **A_OC**. Employ the user-defined function **Y_OC** in the worksheet to create ranges for $c' = dc/dr$, and $c'' = d^2c/dr^2$ in columns **D**, and **E**. Column **F** contains the residuals from the boundary conditions (rows **2** and **9**) and differential equation (rows **3** to **8**).

	A	B	C	D	E	F
1	r	c	a	c'	c''	Residual
2	=x_oc()	0.5	=a_oc(B2:B9)	=y_oc(A2,$C$2:$C$9,1)	=y_oc(A2,$C$2:$C$9,2)	=D2
3	=x_oc()	0.5	=a_oc(B2:B9)	=y_oc(A3,$C$2:$C$9,1)	=y_oc(A3,$C$2:$C$9,2)	=E3-300*B3^2
4	=x_oc()	0.5	=a_oc(B2:B9)	=y_oc(A4,$C$2:$C$9,1)	=y_oc(A4,$C$2:$C$9,2)	=E4-300*B4^2
5	=x_oc()	0.5	=a_oc(B2:B9)	=y_oc(A5,$C$2:$C$9,1)	=y_oc(A5,$C$2:$C$9,2)	=E5-300*B5^2
6	=x_oc()	0.5	=a_oc(B2:B9)	=y_oc(A6,$C$2:$C$9,1)	=y_oc(A6,$C$2:$C$9,2)	=E6-300*B6^2
7	=x_oc()	0.5	=a_oc(B2:B9)	=y_oc(A7,$C$2:$C$9,1)	=y_oc(A7,$C$2:$C$9,2)	=E7-300*B7^2
8	=x_oc()	0.5	=a_oc(B2:B9)	=y_oc(A8,$C$2:$C$9,1)	=y_oc(A8,$C$2:$C$9,2)	=E8-300*B8^2
9	=x_oc()	0.5	=a_oc(B2:B9)	=y_oc(A9,$C$2:$C$9,1)	=y_oc(A9,$C$2:$C$9,2)	=B9-1

	A	B	C	D	E	F
1	r	c	a	c'	c''	Residual
2	0	0.5	0.5	3.41061E-13	-1.98952E-12	3.41061E-13
3	0.033765243	0.5	3.41061E-13	2.78337E-13	-1.72638E-12	-75
4	0.169395307	0.5	-9.9476E-13	1.13445E-13	-7.21967E-13	-75
5	0.380690407	0.5	1.3074E-12	1.03127E-13	5.4816E-13	-75
6	0.619309593	0.5	-1.13687E-13	3.3801E-13	1.26517E-12	-75
7	0.830604693	0.5	-2.27374E-13	5.92795E-13	9.78999E-13	-75
8	0.966234757	0.5	0	6.79371E-13	2.14122E-13	-75
9	1	0.5	0	6.82121E-13	-5.68434E-14	-0.5

The quasi-Newton option in the user-form **ROOTS** from the *PNMSuite* workbook was able to solve for the concentrations in column **B** that make the residuals zero, or $f(x) = 0$, in column **F**.

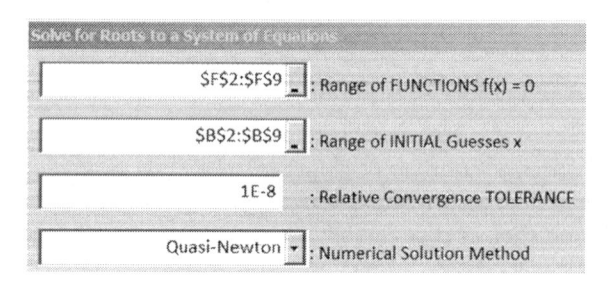

Solve for Roots to a System of Equations

$F$2:$F$9 : Range of FUNCTIONS f(x) = 0

$B$2:$B$9 : Range of INITIAL Guesses x

1E-8 : Relative Convergence TOLERANCE

Quasi-Newton : Numerical Solution Method

	A	B	C	D	E	F
1	r	c	a	c'	c''	Residual
2	0	0.022462977	0.022462977	-2.22045E-16	-4.987577969	-2.22045E-16
3	0.033765243	0.020768632	-2.22045E-16	-0.071693495	0.129400827	2.35922E-14
4	0.169395307	0.027489679	-2.493788984	0.154374317	0.226704743	1.74444E-13
5	0.380690407	0.031812515	34.89227863	-0.107913633	0.30361084	1.04822E-12
6	0.619309593	0.07818499	-160.3091648	0.576811895	1.833867791	3.97371E-12
7	0.830604693	0.200193907	328.8800298	0.803275821	12.02328015	9.25837E-12
8	0.966234757	0.642460272	-310.4180616	8.186042034	123.8265605	1.54898E-11
9	1	1	110.426244	13.328086	183.6171116	0

The solution converged on the roots after eight iterations:

$$c = 0.02246 - 2.494r^2 + 34.89r^3 - 160.3r^4 + 328.9r^5 - 310.4r^6 + 110.4r^7 \tag{13.66}$$

To assess the quality of the solution, we generated a plot of the trial polynomial function in Figure 13.17 with the user-form **QYXPLOT**. The trial function over a single element has suspicious behavior, particularly in the region for $0.2 < r < 0.4$ that exhibits local maxima and minima. This boundary value problem is a good candidate for two or more finite elements.

CHAPTER 13: BOUNDARY-VALUE PROBLEMS

To add a second finite element, scale the independent variable following Equation (13.63) and substitute into the differential equation and boundary conditions:

$$z_1 = 2r \quad \text{and} \quad z_2 = 2r - 1 \tag{13.67}$$

$$\frac{d^2 c_1}{dz_1^2} = 300 (\Delta r)^2 c_1^2 \quad \begin{cases} \dfrac{dc_1}{dz_1} = 0 & z_1 = 0 \; (r = 0) \\ c_1 = c_2 & z_1 = 1 \; (r = 0.5) \end{cases}$$

$$\frac{d^2 c_2}{dz_2^2} = 300 (\Delta r)^2 c_2^2 \quad \begin{cases} \dfrac{dc_1}{dz_1} = \dfrac{dc_2}{dz_2} & z_2 = 0 \; (r = 0.5) \\ c_2 = 1 & z_2 = 1 \; (r = 1) \end{cases} \tag{13.68}$$

where $\Delta r = 0.5$ in the case of two finite elements. The spreadsheet employs the user-defined function **X_OC**, **A_OC**, and **Y_OC** to generate the collocation points for z_1 and z_2, and the corresponding values for a, c' and c''. Note how the user-defined array function **X_OC** uses the optional arguments for the integration limits in x to break up the domain into two elements $0 \leq x \leq 0.5$ and $0.5 \leq x \leq 1$. Implement the boundary conditions in rows **2**, **9**, **10**, and **17**. Find the values for the coefficients of the trial polynomial function with the quasi-Newton option in the user-form **ROOTS** and plot the results in Figure 13.17 with intermediate values using the user-form **QYXPLOT**. The solution converges after nine iterations.

	A	B	C	D	E	F	G
1	r	z	c	a	c'	c''	Residual
2	=x_OC(0,0.5)	=x_OC()	0.0226316734441503	=a_oc(C2:C9)	=y_OC(B2,$D$2:$D$9,1)	=y_OC(B2,$D$2:$D$9,2)	=E2
3	=x_OC(0,0.5)	=x_OC()	0.0226535366754759	=a_oc(C2:C9)	=y_OC(B3,$D$2:$D$9,1)	=y_OC(B3,$D$2:$D$9,2)	=F3-(0.5^2)*300*C3^2
4	=x_OC(0,0.5)	=x_OC()	0.0231873877179419	=a_oc(C2:C9)	=y_OC(B4,$D$2:$D$9,1)	=y_OC(B4,$D$2:$D$9,2)	=F4-(0.5^2)*300*C4^2
5	=x_OC(0,0.5)	=x_OC()	0.0255342102372887	=a_oc(C2:C9)	=y_OC(B5,$D$2:$D$9,1)	=y_OC(B5,$D$2:$D$9,2)	=F5-(0.5^2)*300*C5^2
6	=x_OC(0,0.5)	=x_OC()	0.0308934401178881	=a_oc(C2:C9)	=y_OC(B6,$D$2:$D$9,1)	=y_OC(B6,$D$2:$D$9,2)	=F6-(0.5^2)*300*C6^2
7	=x_OC(0,0.5)	=x_OC()	0.0390745681035148	=a_oc(C2:C9)	=y_OC(B7,$D$2:$D$9,1)	=y_OC(B7,$D$2:$D$9,2)	=F7-(0.5^2)*300*C7^2
8	=x_OC(0,0.5)	=x_OC()	0.0469127216577762	=a_oc(C2:C9)	=y_OC(B8,$D$2:$D$9,1)	=y_OC(B8,$D$2:$D$9,2)	=F8-(0.5^2)*300*C8^2
9	=x_OC(0,0.5)	=x_OC()	0.049295832336265	=a_oc(C2:C9)	=y_OC(B9,$D$2:$D$9,1)	=y_OC(B9,$D$2:$D$9,2)	=C9-C10
10	=x_OC(0.5,1)	=x_OC()	0.0492958305810629	=a_oc(C10:C17)	=y_OC(B10,$D$10:$D$17,1)	=y_OC(B10,$D$10:$D$17,2)	=E9-E10
11	=x_OC(0.5,1)	=x_OC()	0.0515103768445781	=a_oc(C10:C17)	=y_OC(B11,$D$10:$D$17,1)	=y_OC(B11,$D$10:$D$17,2)	=F11-(0.5^2)*300*C11^2
12	=x_OC(0.5,1)	=x_OC()	0.0654810036270019	=a_oc(C10:C17)	=y_OC(B12,$D$10:$D$17,1)	=y_OC(B12,$D$10:$D$17,2)	=F12-(0.5^2)*300*C12^2
13	=x_OC(0.5,1)	=x_OC()	0.097666891984372	=a_oc(C10:C17)	=y_OC(B13,$D$10:$D$17,1)	=y_OC(B13,$D$10:$D$17,2)	=F13-(0.5^2)*300*C13^2
14	=x_OC(0.5,1)	=x_OC()	0.182461399233597	=a_oc(C10:C17)	=y_OC(B14,$D$10:$D$17,1)	=y_OC(B14,$D$10:$D$17,2)	=F14-(0.5^2)*300*C14^2
15	=x_OC(0.5,1)	=x_OC()	0.390229401264757	=a_oc(C10:C17)	=y_OC(B15,$D$10:$D$17,1)	=y_OC(B15,$D$10:$D$17,2)	=F15-(0.5^2)*300*C15^2
16	=x_OC(0.5,1)	=x_OC()	0.796084708452755	=a_oc(C10:C17)	=y_OC(B16,$D$10:$D$17,1)	=y_OC(B16,$D$10:$D$17,2)	=F16-(0.5^2)*300*C16^2
17	=x_OC(0.5,1)	=x_OC()	0.999999964276601	=a_oc(C10:C17)	=y_OC(B17,$D$10:$D$17,1)	=y_OC(B17,$D$10:$D$17,2)	=C17-1

	A	B	C	D	E	F	G
1	r	z	c	a	c'	c''	Residual
2	0	0	0.022631673	0.022631673	7.1063E-09	0.038292276	7.1063E-09
3	0.016882621	0.033765243	0.022653537	7.1063E-09	0.001296109	0.038488709	4.8803E-09
4	0.084697653	0.169395307	0.023187388	0.019146138	0.006616977	0.040324125	3.85119E-09
5	0.190345204	0.380690407	0.02553421	0.000842717	0.015896858	0.048899692	5.29398E-10
6	0.309654797	0.619309593	0.03089344	0.001400281	0.029915111	0.071580349	7.25706E-10
7	0.415302347	0.830604693	0.039074568	0.008901027	0.049019682	0.11451164	-4.7408E-11
8	0.483117379	0.966234757	0.046912722	-0.008008397	0.067699809	0.165060259	-3.17193E-10
9	0.5	1	0.049295832	0.004382385	0.073553011	0.181980777	1.7552E-09
10	0.5	0	0.049295831	0.049295831	0.073553012	-0.906442457	-9.43446E-10
11	0.516882621	0.033765243	0.051510377	0.073553012	0.063812973	0.198998918	-1.10897E-09
12	0.584697653	0.169395307	0.065481004	-0.453221228	0.140966858	0.321582138	2.66307E-10
13	0.690345204	0.380690407	0.097666892	7.514814902	0.178101264	0.715411633	-7.67267E-10
14	0.809654797	0.619309593	0.182461399	-34.31379586	0.604098644	2.496912166	5.49217E-10
15	0.915302347	0.830604693	0.390229401	71.55349018	1.641531924	11.42092392	-9.6575E-10
16	0.983117379	0.966234757	0.796084708	-68.77704228	5.144718243	47.53131473	-1.39404E-09
17	1	1	0.999999964	25.35290541	7.031906901	64.997459	-3.57234E-08

Combine the results for the two trial polynomial functions in the two elements using **IF(r < 0.5, c_1, c_2)**:

$$c_1 = 0.02263 + 0.01915(2r)^2 + 0.0008427(2r)^3 + 0.0014(2r)^4$$
$$+ 0.008901(2r)^5 - 0.008008(2r)^6 + 0.004382(2r)^7 \quad \text{for} \quad 0 \le r < 0.5$$

and

$$c_2 = 0.0493 + 0.07355(2r-1) - 0.4532(2r-1)^2 + 7.515(2r-1)^3$$
$$- 34.31(2r-1)^4 + 71.55(2r-1)^5 - 68.78(2r-1)^6 + 25.35(2r-1)^7 \quad \text{for} \quad 0.5 \le r \le 1$$

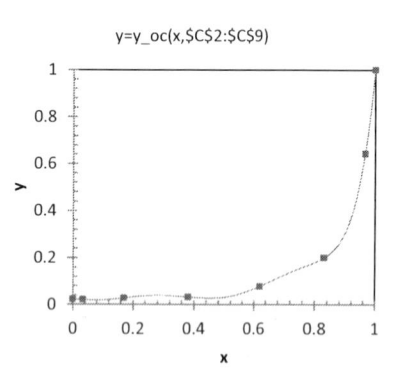

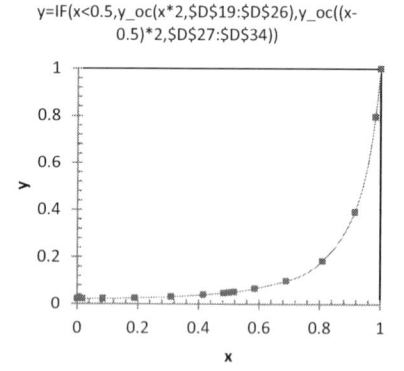

Figure 13.17 QYXPLOT of polynomial trial functions for the orthogonal collocation solution of Equation (13.10) with $\phi = 300$ and one or two finite elements. Markers show collocation points.

Comments: The solution with two elements appears to behave reasonably and matches the trend in the solution obtained from the shooting method in Figure 13.4.

□

13.4 Nonlinear Parameter Estimation in BVPs*

In Section 12.8 we developed the method of nonlinear parameter estimation using marching integration techniques to integrate the objective function for models involving initial-value problems. For second-order boundary-value problems, we turn to the method of orthogonal collocation to interpolate the value for y_{model} in the objective function and integrate Equation (12.128) using the Gauss-Kronrod user-defined function. By employing orthogonal collocation to approximate the solution to the second-order differential equation, we transform the original problem to a problem involving the solution to constrained minimization. We may select from several numerical minimization methods and constraining techniques from Chapter 7 to find the parameters in both the trial function and model. The **Solver** is particularly useful for searching for the model parameters and values of the dependent variable at each of the collocation points in the second-order differential equation:

$$\text{Min}(OF) = \int_0^1 \left[y_{Bspline}(x) - y_{OC}(x) \right]^2 dx^* \tag{13.69}$$

where

$$x = x^* \Delta x + x_0$$

subject to the following $N+1$ equality constraints at the collocation points:

- *Interior Points*: $\left.\dfrac{d^2 y}{dx^{*2}}\right|_j - \Delta x^2 R\left(x_j^*, y_j, \dfrac{1}{\Delta x}\left.\dfrac{dy}{dx^*}\right|_j\right) = 0 \qquad j = 1, 2 \ldots N-1$ (13.70)

- *Left Boundary Condition*: $\alpha_0 a_1 - \Delta x \beta_0 (a_0 - \gamma_0) = 0$ (13.71)

- **Right Boundary Condition:** $\alpha_n \left(a_1 + 2a_2 + \sum_{k=3}^{N} k a_k \right) - \Delta x \beta_1 (y_1 - \gamma_n) = 0$ (13.72)

In summary, apply the following steps for nonlinear parameter estimation in boundary-value problems:
1. Use your best engineering judgement to guess values for the model parameters.
2. Select the order of the polynomial trial function and generate the dimensionless collocation points.
3. Provide initial guesses for the solution of the boundary-value problem y at the collocation points and calculate the corresponding trial function coefficients, a.
4. Interpolate the experimental data for y versus x using the user-defined function **BSPLINE**.
5. Set up the solution to the objective function integral in Equation (13.69) using a numerical method from Chapter 11, such as **GKRAD**.
6. Formulate the constraints in the worksheet as defined by Equations (13.70) through (13.72).
7. Use the **Solver** or another optimization method to find the model parameters and y's at the collocation points that minimize the objective function while satisfying the constraints.

13.5 Epilogue on Boundary-value Problems

The shooting, finite difference, and collocation methods of solution, compared in Table 13.1, do not introduce completely new techniques, rather they combine methods from previous chapters applied to solving second-order, ordinary differential equations with split boundary conditions. The choice of solution method is usually a matter of convenience, as dictated by the nature of the problem at hand.

Table 13.2 contains a summary of the inputs required for the user-defined functions developed for this chapter.

Table 13.1 Comparison of BVP integration methods

Method	Strengths	Weaknesses
Shooting Method	• Higher accuracy when using high order variable step integration schemes. • Simple implementation. • Gives derivative approximations at the boundaries.	• Must convert the second order ordinary differential equation into a system of first order differential equations.
Finite Difference	• Spreads the error across the range of integration. • Simple implementation. • Useful for interpolating data that fits boundary value models.	• Requires a large number of nodes for good accuracy when using second order accurate finite difference derivative approximations. • Does not directly provide values for the derivatives at the boundaries.
Orthogonal Collocation (on Finite Elements)	• Portable polynomial expression for the solution. • Higher accuracy when applied across finite elements. • Uses all collocation points for interpolation, which spreads the error across the domain of the independent variable.	• May have lower accuracy using one element for highly nonlinear functions. • Requires extraordinary effort to implement in a spreadsheet.

Table 13.2 Summary of VBA macros, user-defined functions (UDF) and user-forms for BVPs and the methods of input.

Macro	Inputs	Integration Method
A_OC	UDF array arguments	Returns the orthogonal trial polynomial coefficients.
SHOOTING SHOOTING_UsrFrm	Macro input boxes Show user-form macro	Fixed-step fourth-order Runge-Kutta-Gill with Secant method.
X_OC	UDF array arguments	Orthogonal collocation points.
Y_OC	UDF arguments	Evaluate trial function and derivatives at collocation points.

Chapter 14 Partial Differential Equations

"We have the only cookbook in the world that has partial differential equations in it."
– Nathan Myhrvold, Microsoft Chief Technology Officer and Chef

We extend the numerical methods for boundary-value problems to find numerical solutions of partial differential equations (PDEs) that involve multiple dimensions in space or time. Common examples of partial differential equations come from modeling transient fluid, heat, and mass transport in multidimensional geometries. Following the pattern established in Equation (13.1), we may expand these forms of boundary-value problems to three dimensions, or higher order differential equations as required by our system. In this chapter, we present four numerical methods for approximating solutions to partial differential equations in *Excel*:

- The finite-difference method for solving partial differential equations approximates the solution at discrete points between the boundaries by replacing the derivatives with finite-difference approximations for elliptic equations. The result is a system of algebraic equations. We then use fixed-point iterative methods for finding roots to the large systems of equations that are the result of discretization.

- The method of lines and Crank-Nicolson method for initial-value-boundary-value type partial differential equations use finite-difference methods for derivative approximations in one dimension to convert the problem into a system of coupled first-order, ordinary differential equations. We solve the system of first-order, ordinary differential equations using marching integration techniques, such as the trapezoidal or Runge-Kutta methods, presented in Chapter 12.

- The method of orthogonal collocation uses a polynomial trial function for the solution in the space dimensions at optimal collocation points generated from the zeroes of orthogonal Legendre polynomials. We get the polynomial coefficients from the solution to the system of equations derived by substituting the trial function into the boundary-value problem and boundary conditions. Like the method of lines, this approach converts the higher-order partial differential equation into a system of ordinary first-order differential equations, one for each collocation point, which require the numerical methods from Chapter 12 for solution.

> **SQ3R Focused Reading Questions**
> 1. How are partial differential equations different from ordinary differential equations?
> 2. What numerical methods are recycled into new methods for approximating solutions to PDEs?
> 3. How are the method-of-lines and Crank-Nicolson similar or different?
> 4. When should the numerical method of orthogonal collocation be used?
> 5. How can I check the accuracy of a numerical solution to partial differential equations?

Because of limitations in the *Excel* and *VBA* computing platform, we limit our coverage to simple systems with symmetric geometries that fit nicely within Cartesian, cylindrical, or spherical coordinate systems. Other, more advanced techniques for solving partial differential equations in complex geometries include finite-volume and finite-element methods. Many partial differential equation solvers are available from commercial providers to develop and solve model equations with complicated geometries and boundary conditions. Sophisticated applications have tools for preprocessing design, equation formulation, and post-processing the numerical results using graphics and analysis.

14.1 Finite Difference Solutions of Elliptic Equations

"Think left and think right and think low and think high. Oh, the thinks you can think up if only you try!" – Dr. Seuss

Elliptic second-order partial differential equations (*PDE*) have the general form:

$$\frac{\partial^2 y}{\partial x^2} + \frac{\partial^2 y}{\partial w^2} = P\left(w, x, y, \frac{\partial y}{\partial x}, \frac{\partial y}{\partial w}\right) \qquad (14.1)$$

subject to two general forms of mixed boundary conditions for each dimension x and w:

$$b_{x0} = \alpha_{x0} \left.\frac{\partial y}{\partial x}\right|_{x0} - \beta_{x0}(y - \gamma_{x0}) = 0 \qquad \text{at} \quad x = x_0 \qquad (14.2)$$

$$b_{xn} = \alpha_{xn} \left.\frac{\partial y}{\partial x}\right|_{xn} - \beta_{xn}(y - \gamma_{xn}) = 0 \qquad \text{at} \quad x = x_n \qquad (14.3)$$

$$b_{w0} = \alpha_{w0} \left.\frac{\partial y}{\partial w}\right|_{w0} - \beta_{w0}(y - \gamma_{w0}) = 0 \qquad \text{at} \quad w = w_0 \qquad (14.4)$$

$$b_{wn} = \alpha_{wn} \left.\frac{\partial y}{\partial w}\right|_{wn} - \beta_{wn}(y - \gamma)_{wn} = 0 \qquad \text{at} \quad w = w_n \qquad (14.5)$$

We first encountered a simple example of a finite-difference solution to an elliptical equation in Example 4.5. In two-dimensional problems, we require two indices to locate the nodes on a grid. Use the indices i and j for the x and z dimensions, respectively. Figure 14.1 shows the relative node placements for Cartesian coordinates. Use the finite-difference form of Equation (14.1) at node i, j:

$$\left(\frac{y_{i-1,j} - 2y_{i,j} + y_{i+1,j}}{\Delta x^2}\right) + \left(\frac{y_{i,j-1} - 2y_{i,j} + y_{i,j+1}}{\Delta z^2}\right) + P\left(x_i, z_j, y_{i,j}, \frac{y_{i+1,j} - y_{i-1,j}}{2\Delta x}, \frac{y_{i,j+1} - y_{i,j-1}}{2\Delta z}\right) = 0 \qquad (14.6)$$

where the interval lengths in the x and z dimensions are not necessarily equal. The finite-difference form of the boundary conditions is the same as Equation (13.29).

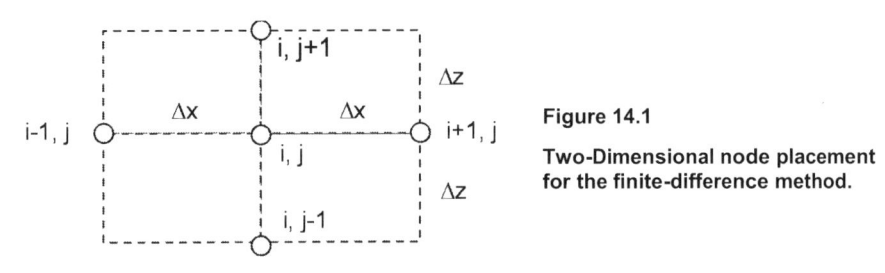

Figure 14.1
Two-Dimensional node placement for the finite-difference method.

Example 14.1 Steady-state Heat Equation in Two Dimensions

Known: Consider the following elliptical equation for steady-state temperature in a two-dimensional square:

$$\frac{\partial^2 T}{\partial x^2} + \frac{\partial^2 T}{\partial y^2} = 0 \qquad (14.7)$$

subject to the boundary conditions shown in the schematic.

Schematic: The x and y-axis run along the bottom and left side of the square, respectively.

```
                    T = 125°C  at y = 1
    ∂T
    ── = 0  at x = 0              T = 25°C  at x = 1
    ∂x

                    T = 5°C  at y = 0
```

Find: Use *Excel* to solve for the temperatures in the *yx* space by the finite-difference method with uniform spacing $\Delta x = \Delta y = 0.1$. Plot the temperature profile in a surface plot.

Assumptions: Constant properties, steady-state conduction

Analysis: The finite-difference method replaces the derivatives with finite-difference approximations, where *i* is the *x*-dimension index, and *j* is the *y*-dimension index:

$$\left(\frac{T_{i-1,j} - 2T_{i,j} + T_{i+1,j}}{\Delta x^2}\right) + \left(\frac{T_{i,j-1} - 2T_{i,j} + T_{i,j+1}}{\Delta y^2}\right) = 0 \tag{14.8}$$

For the case where $\Delta x = \Delta y$, Equation (14.8) reduces to an explicit expression for $T_{i,j}$ in terms of the temperatures of the surrounding nodes:

$$T_{ij} = \frac{1}{4}\left(T_{i+1,j} + T_{i-1,j} + T_{i,j+1} + T_{i,j-1}\right) \tag{14.9}$$

Apply the boundary condition at $x = 0$ using a false node $T_{-1,j} = T_{1,j}$ to simplify Equation (14.9):

$$T_{1j} = \frac{1}{4}\left(2T_{1,j} + T_{1,j+1} + T_{1,j-1}\right) \tag{14.10}$$

Set up the finite-difference solution in an *Excel* worksheet:

1. Enable iteration in **File> Options>Formulas>Enable Iterative Calculations**. Change the number of iterations and convergence settings as needed.

2. Define the *y* versus *x* coordinates of the nodes by setting the *x* and *y* positions in column **A** and row **13**, as shown in Figure 14.2.

3. Specify the boundary conditions at $x = 0$, $x = 1$, and $y = 1$. The boundary condition at $y = 0$ uses a formula like Equation (14.10) for column **B** between the two boundaries:

    B3= (2*C3 + B2 + B4)/4

4. The interior nodes use a formula like Equation (14.9) for the cells in rows **3** to **11** and columns **C** to **K**:

    C3 = (D3 + C4 + B3 + C2)/4

5. Select the cell with the formula and use the fill handle to drag the formula with relative referencing into the rest of the cells in the *y* versus *x* range. *Excel* iterates towards convergence of the temperature at each node.

CHAPTER 14: PARTIAL DIFFERENTIAL EQUATIONS

	A	B	C	D	E	F	G	H	I	J	K	L	M
1	y												
2	1	125.00	125.00	125.00	125.00	125.00	125.00	125.00	125.00	125.00	125.00	75.00	
3	0.9	111.52	111.43	111.15	110.62	109.74	108.30	105.88	101.55	93.06	74.14	25.00	
4	0.8	98.21	98.05	97.54	96.59	95.05	92.59	88.67	82.27	71.56	53.49	25.00	
5	0.7	85.22	85.02	84.36	83.17	81.26	78.34	73.93	67.29	57.42	43.27	25.00	
6	0.6	72.65	72.43	71.73	70.46	68.49	65.59	61.41	55.55	47.55	37.19	25.00	
7	0.5	60.54	60.32	59.65	58.46	56.66	54.10	50.59	45.95	40.04	32.92	25.00	
8	0.4	48.85	48.66	48.08	47.08	45.59	43.55	40.90	37.61	33.74	29.46	25.00	
9	0.3	37.54	37.39	36.95	36.18	35.08	33.62	31.85	29.85	27.86	26.17	25.00	
10	0.2	26.53	26.43	26.13	25.62	24.91	24.01	23.02	22.10	21.66	22.38	25.00	
11	0.1	15.71	15.66	15.51	15.26	14.92	14.51	14.10	13.88	14.30	16.67	25.00	
12	0	5.00	5.00	5.00	5.00	5.00	5.00	5.00	5.00	5.00	5.00	15.00	
13		0	0.1	0.2	0.3	0.4	0.5	0.6	0.7	0.8	0.9	1	x

Figure 14.2

Solution to the finite-difference approximation of two-dimensional heat equation described by Equation (14.7). Select the range of cells with temperature values for conditional formatting.

6. Color-code the cells according to the temperature values to visualize the relative hot and cold regions of the square. To do this, select the range of cells containing temperature values. On the **Home** tab, in the **Styles** group, select **Conditional Formatting>Color Scales>Red-Yellow-Green Color Scale** as seen in Figure 14.3. There are several color scale options, as well as custom features. Explore the other options under conditional formatting to learn about the rich features available for accenting worksheets. Particularly, note the **Conditional Formatting>Clear Rules** option to reset the appearance of the formatted cells if necessary.

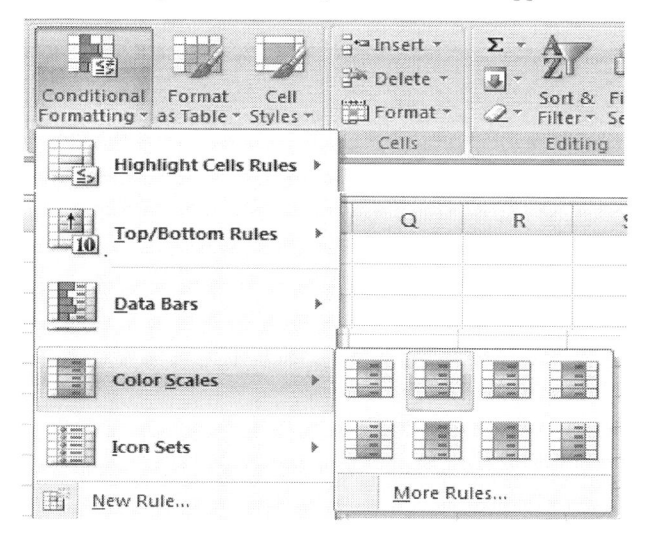

Figure 14.3

Specify conditional formatting for selected cells.

Notice that *Excel* provides a range of shades among the three colors that gives an appearance of interpolation, as seen in Figure 14.4.

1	125.00	125.00	125.00	125.00	125.00	125.00	125.00	125.00	125.00	125.00	75.00
0.9	111.52	111.43	111.15	110.62	109.74	108.30	105.88	101.55	93.06	74.14	25.00
0.8	98.21	98.05	97.54	96.59	95.05	92.59	88.67	82.27	71.56	53.49	25.00
0.7	85.22	85.02	84.36	83.17	81.26	78.34	73.93	67.29	57.42	43.27	25.00
0.6	72.65	72.43	71.73	70.46	68.49	65.59	61.41	55.55	47.55	37.19	25.00
0.5	60.54	60.32	59.65	58.46	56.66	54.10	50.59	45.95	40.04	32.92	25.00
0.4	48.85	48.66	48.08	47.08	45.59	43.55	40.90	37.61	33.74	29.46	25.00
0.3	37.54	37.39	36.95	36.18	35.08	33.62	31.85	29.85	27.86	26.17	25.00
0.2	26.53	26.43	26.13	25.62	24.91	24.01	23.02	22.10	21.66	22.38	25.00
0.1	15.71	15.66	15.51	15.26	14.92	14.51	14.10	13.88	14.30	16.67	25.00
0	5.00	5.00	5.00	5.00	5.00	5.00	5.00	5.00	5.00	5.00	15.00
	0	0.1	0.2	0.3	0.4	0.5	0.6	0.7	0.8	0.9	1

Figure 14.4

Conditional formatting shows the temperature profile by color shading in the cells. See the *Excel* example file from the book's web site for a color view.

Comments: The color formatting allows the viewer to visualize the temperature distribution. A surface plot shows the gradients in the temperature profile, or the direction of heat flux. To create a surface plot:

1. Select only the cells containing the temperature data in the *Excel* worksheet, as seen in Figure 14.2.
2. From the **Insert** tab, select **Other Charts>Surface>Wire Frame Surface** to create a surface chart of the temperature profile, as seen in Figure 14.5.
3. Right-click on the chart and select **3-D Rotation** from the popup menu to display user controlled options for rotating the chart for viewing from different angles. Regions of the surface plot with steeper temperature gradients indicate higher heat transfer rates.

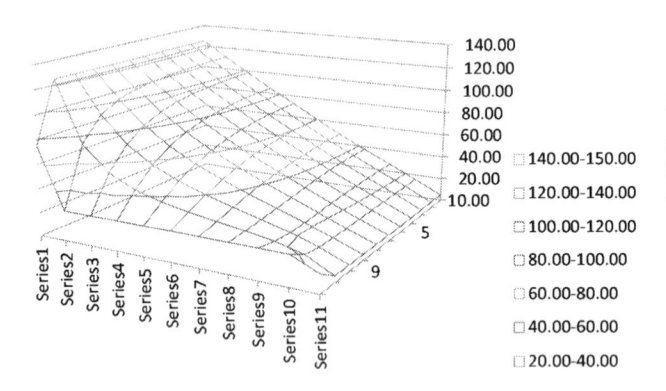

Figure 14.5

Surface plot of the temperature profile in a rectangular solid. See the *Excel* example file for a color view.

□

14.2 Parabolic Equations

Parabolic partial differential equations of the form:

$$\frac{\partial y}{\partial t} = P\left(x, y, t, \frac{\partial y}{\partial x}, \frac{\partial^2 y}{\partial x^2}\right) \tag{14.11}$$

are subject to the general boundary conditions in Equations (13.2) as well as the following initial conditions:

$$y = y(x) \qquad \text{for all } x \text{ at } t = 0 \tag{14.12}$$

In this case, the boundary conditions may also be functions of the variable for the time dimension.

We present three numerical techniques for approximating solutions to parabolic partial differential equations:

1. Method of lines
2. Crank-Nicolson method
3. Orthogonal collocation

14.2.1 Method of Lines

> *"The real voyage of discovery consists not in seeking new landscapes, but in having new eyes."* — Marcel Proust

The method of lines is a hybrid technique involving both marching integration and the finite-difference method for solving parabolic problems, such as Equation (14.11). To implement the method of lines, discretize the x-dimension by the finite-difference method into n intervals. This gives a system of $n-1$ first-order ordinary differential equations in the independent variable t, and two algebraic difference equations for the boundary conditions:

$$\alpha_0 \left(\frac{-3y_0 + 4y_1 - y_2}{2\Delta x}\right) - \beta_0 \Delta y_0 = 0 \tag{14.13}$$

CHAPTER 14: PARTIAL DIFFERENTIAL EQUATIONS

$$\frac{dy_i}{dt} = P\left(x_i, y_i, t, \frac{y_{i+1} - y_{i-1}}{2\Delta x}, \frac{y_{i-1} - 2y_i + y_{i+1}}{\Delta x^2}\right) \quad i = 1, 2 \ldots n-1 \tag{14.14}$$

$$\alpha_n\left(\frac{y_{n-2} - 4y_{n-1} + 3y_n}{2\Delta x}\right) - \beta_n \Delta y_n = 0 \tag{14.15}$$

$$x_i = i\Delta x \quad \text{for } i = 0, 1, 2 \ldots n \tag{14.16}$$

where Δy_0 and Δy_n are the residuals between the calculated and known y values at the boundaries. When the β terms in the boundary conditions are nonlinear functions of y_0 or y_n, use the method described in Section 12.7 to convert the expressions into first-order differential equations.

Solve the set of first-order, ordinary differential equations by a marching integration technique, such as a variable step Runge-Kutta method. The method of lines also works for multi-dimensional equations by tracking the second dimension with an additional index. One unfortunate feature of the method of lines is that the integration method becomes less stable as the number of nodes in the spatial coordinate increases, or as the size of the interval in the finite-difference method decreases, $\Delta x \to 0$ (Hanna and Sandall 1995). Therefore, care is required to balance the solution accuracy with stability. One method of improving stability shrinks the marching integration step size. The method of lines becomes stable only when the step size in the t-dimension is sufficiently smaller than the square of the size of each interval in the x-dimension:

$$2\Delta t \leq \Delta x^2 \tag{14.17}$$

We may elect to enforce stability in the variable step Runge-Kutta 4-5th method by fixing the upper limit on the step size accordingly. For example, add the following line to the *VBA* macro **RK45DP**:

```
h = 0.87 * h * (ec / (del + small)) ^ 0.2 : h = .Min(h, 0.5*delx^2)
```

Example 14.2   One-dimensional, Unsteady-state Heat Equation

Known: Consider the following transient form of the heat equation for thermal conduction through a solid wall of thickness $L = 0.1$ m:

$$\frac{\partial T}{\partial t} = \alpha \frac{\partial^2 T}{\partial x^2} \tag{14.18}$$

where the thermal diffusivity $\alpha = 2 \times 10^{-6}$ m²/s. The wall has an initial temperature of $T_{init} = 100°$C. The left side of the wall is maintained at the initial temperature while the right side of the wall is exposed to 25°C air, with a heat transfer coefficient of $h_t = 100$ W/m²·°C. The thermal conductivity of the wall material is $\kappa = 3$ W/m·°C. Under these conditions, the boundary conditions for this problem take the following form:

$$T = T_{init} \quad \text{at } x = 0 \quad \text{and} \quad -\kappa \frac{dT}{dx} = h_t(T - T_{air}) \quad \text{at } x = L \tag{14.19}$$

Schematic: Refer to Figure 1.7.

Find: Calculate the temperature of the wall exposed to air after ten minutes using the method of lines.

Assumptions: Constant properties

Analysis: Define the parameters for the method of lines.

- Finite step size in the x dimension:
$$\Delta x = \frac{L}{n} \tag{14.20}$$

- T at node 0:
$$T_0 = T_{init} \tag{14.21}$$

- Energy balance for nodes 1 to $n-1$ using central difference formulas:

$$\frac{dT_i}{dt} = \alpha \left(\frac{T_{i-1} - 2T_i + T_{i+1}}{\Delta x^2} \right) \quad \text{for } i = 1, 2 \ldots n-1 \tag{14.22}$$

- Surface energy balance at node n using backward finite-difference for the derivative in Equation (14.19):

$$-\kappa \left(\frac{T_{n-2} - 4T_{n-1} + 3T_n}{2\Delta x} \right) = h_t (T_n - T_{air}) \tag{14.23}$$

The boundary condition at node n is defined explicitly in terms of T_n by rearranging Equation (14.23):

$$T_n = \frac{\kappa(4T_{n-1} - T_{n-2}) + 2\Delta x h_t T_{air}}{3\kappa + 2\Delta x h_t} \tag{14.24}$$

- Initial conditions for each interior node in the x-dimension:

$$T_i = T_{init} \quad \text{for } i = 1, 2, 3 \ldots n-1 \tag{14.25}$$

Arrange an *Excel* worksheet for the solution using 10 intervals in the x-dimension, as shown in Figure 14.6. Specify the initial conditions for each differential equation in row seven. Formulas for the derivative functions in row eight use the corresponding variables in row seven.

	A	B	C	D	E	F	G	H	I	J	K	L
1	α	2.0E-06		Δx	0.01							
2	k	3		t init	0							
3	h	100		t final	600							
4	Tair	25										
5	n	0	1	2	3	4	5	6	7	8	9	10
6	x	0	0.01	0.02	0.03	0.04	0.05	0.06	0.07	0.08	0.09	0.1
7	T	100	100	100	100	100	100	100	100	100	100	86.4
8	dT/dt		0	0	0	0	0	0	0	0	-0.27	

Figure 14.6

Excel worksheet for the method of lines solution to the transient heat equation.

Formulate the boundary condition at $x = 0.1$ in terms of the dependent variable at the nodes to the left, *e.g.*:

L7 = ($B$2*(4*K7 - J7) + 2*$E$1*$B$3*$B$4)/ (3*$B$2 + 2*$E$1*$B$3)

Try the user-form **RK45VS** with Fehlberg's coefficients to solve the system of first-order, ordinary differential equations with output directed to cell **A10**. Provide the required information to the form and use the default values for tolerance and print steps:

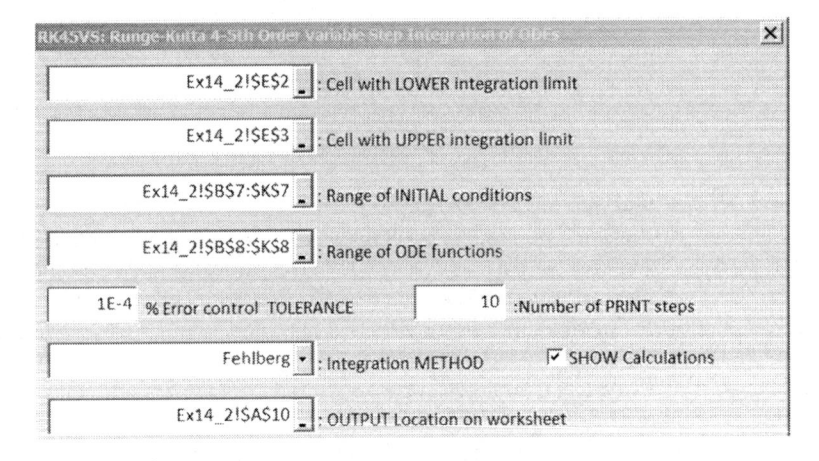

```
RK45VS: Runge-Kutta 4-5th Order Variable Step Integration of ODEs

  Ex14_2!$E$2   : Cell with LOWER integration limit
  Ex14_2!$E$3   : Cell with UPPER integration limit
  Ex14_2!$B$7:$K$7  : Range of INITIAL conditions
  Ex14_2!$B$8:$K$8  : Range of ODE functions
  1E-4   % Error control TOLERANCE    10  :Number of PRINT steps
  Fehlberg  : Integration METHOD   ☑ SHOW Calculations
  Ex14_2!$A$10  : OUTPUT Location on worksheet
```

Complete the temperature profile by adding the constant values for the temperatures in columns **B** and **L** for the left wall and right fluid boundary conditions, respectively. To add the temperatures, select the range of cells containing the numerical results. Click and hold on the border and move the cells to the right one column to make room for the

CHAPTER 14: PARTIAL DIFFERENTIAL EQUATIONS

left boundary. Fill the range **B10:B21** with 100. Use the formula in Equation (14.24) to complete the temperature in the range **L10:L21** at the right boundary using **AutoFill**:

L12 = ($B$2*(4*K12 - J12) + 2*$E$1*$B$3*$B$4)/ (3*$B$2 + 2*$E$1*$B$3)

Add color scales, as shown in Figure 14.7, to the range of solution cells using **Red-Yellow-Green** conditional formatting, as described in Example 14.1. The solution for the outside wall surface temperature decreases to 54.2°C after 10 minutes.

	A	B	C	D	E	F	G	H	I	J	K	L
10	x		y_1	y_2	y_3	y_4	y_5	y_6	y_7	y_8	y_9	
11	0	100	100.0	100.0	100.0	100.0	100.0	100.0	100.0	100.0	100.0	86.4
12	60	100	100.0	100.0	100.0	100.0	100.0	99.9	99.7	98.8	96.1	89.0	75.4
13	120	100	100.0	100.0	100.0	100.0	99.8	99.5	98.5	96.1	91.2	82.7	69.9
14	180	100	100.0	99.9	99.8	99.4	98.5	96.6	93.1	87.2	78.3	66.1
15	240	100	99.9	99.7	99.4	98.6	97.2	94.6	90.4	83.9	74.9	63.4
16	300	100	99.8	99.5	98.8	97.7	95.8	92.6	87.8	81.1	72.1	61.1
17	360	100	99.6	99.1	98.2	96.7	94.3	90.7	85.6	78.6	69.9	59.3
18	420	100	99.4	98.6	97.4	95.6	92.8	88.9	83.5	76.6	67.9	57.7
19	480	100	99.2	98.1	96.7	94.5	91.5	87.2	81.7	74.7	66.2	56.4
20	540	100	98.9	97.6	95.8	93.4	90.1	85.7	80.0	73.0	64.7	55.2
21	600	100	98.6	97.1	95.1	92.4	88.8	84.2	78.5	71.5	63.4	54.2

Figure 14.7

Method of lines solution to the unsteady-state, one-dimensional heat equation. The x-column is time. The y-columns are temperature (°C) at positions x. See the *Excel* example file from the book's website for a color view.

Comments: The method of lines solution in this example tends to become unstable for more than 20 intervals in the *x*-direction, which limits the precision of the solution.

□

14.2.2 Crank-Nicolson and Dufort-Frankel Implicit Methods

The Crank Nicolson method is a cross between the trapezoidal rule applied to the time derivative in Equations (14.11) and the finite difference method applied to the space derivative. A weighted average of forward and backward differences gives the following iterative equation for one spatial dimension:

$$\frac{y_i^k - y_i^{k-1}}{\Delta t} = \omega P\left(x_i^k, y_i^k, t_i^k, \frac{y_{i+1}^k - y_{i-1}^k}{2\Delta x}, \frac{y_{i-1}^k - 2y_i^k + y_{i+1}^k}{\Delta x^2}\right)$$

$$+ (1-\omega) P\left(x_i^{k-1}, y_i^{k-1}, t_i^{k-1}, \frac{y_{i+1}^{k-1} - y_{i-1}^{k-1}}{2\Delta x}, \frac{y_{i-1}^{k-1} - 2y_i^{k-1} + y_{i+1}^{k-1}}{\Delta x^2}\right) \quad (14.26)$$

where the weight factor varies between $0 \leq \omega \leq 1$. For $\omega = 0$, Equation (14.26) is an explicit form of Euler's method, similar to the method of lines. Otherwise, the equation becomes implicit in the solution at each step k in the independent variable t. When $\omega = 0.5$, the method becomes unconditionally stable in t as with the implicit trapezoidal rule (ignoring round-off issues). Notice that we must not confuse the stability of the integration technique with the stability of the solution method. The advantages of the weighted average method include larger step sizes in t, unconditional integration stability for $0.5 \leq \omega \leq 1$, and faster convergence when iterative solution methods, such as Steffensen's, are employed. The computational penalty in this scheme is nearly double the number of calculations required for each iteration. We may also solve multidimensional problems using Crank-Nicolson by adding weight-averaged expressions for the derivatives in each dimension.

An alternative approach of Dufort and Frankel substitutes a central approximation for the time derivative (Woolfson and Pert 1999):

$$\frac{y_i^k - y_i^{k-2}}{2\Delta t} = P\left(x_i^{k-1}, y_i^{k-1}, t_i^{k-1}, \frac{y_{i+1}^{k-1} - y_{i-1}^{k-1}}{2\Delta x}, \frac{y_{i-1}^{k-1} - 2y_i^{k-1} + y_{i+1}^{k-1}}{\Delta x^2}\right) \quad (14.27)$$

Equation (14.27) is not self-starting and requires an initial time step for y_i^{k-1}. To improve the stability of the solution for larger time steps, we replace y_i^{k-1} with $\left(y_i^k + y_i^{k-2}\right)/2$, which changes the method from an explicit solution to an implicit problem requiring an iterative solution for y_i^k at each time step.

If needed for solution stability, try the following upwind form of the finite difference derivative approximation (such as the backward difference approximation in Table 5.3) for partial differential equations that do not involve second order derivatives in the spatial dimension, e.g.:

$$\frac{\partial T}{\partial t} = a\frac{\partial T}{\partial x} \quad \text{becomes} \quad \frac{dT_i}{dt} = \alpha\left(\frac{3T_i - 4T_{i-1} + T_{i-2}}{2\Delta t}\right) \tag{14.28}$$

Example 14.3 Solution of the Transient Heat Equation by Crank-Nicholson

Known: Consider a rectangular wall initially at temperature $T_{init} = 50\ °C$ with thermal diffusivity $\alpha = 0.001\ m^2/s$. The adiabatic left side at $x = 0$ has no heat flux through the wall. The right surface temperature at $x = L$ instantly increases to $T_s = 100\ °C$.

Find: Calculate the time versus temperature profile at $x = 0$.

Schematic: An infinite vertical wall has thickness $L = 1$ m, with T_0 and T_s initial and outside surface temperatures, respectively.

```
|////|                                              |
|////|         T_init = 50 °C                       |
|////|  Adiabatic                                   | T_s = 100 °C
|////|  Zero Heat Flux                              |
|////|                                              |
     x = 0                                          x = L
```

Assumptions: Constant thermal diffusivity

Analysis: The Heat Equation governs the energy balance with the following initial and boundary conditions:

Heat Equation:
$$\frac{\partial T}{\partial t} = \alpha \frac{\partial^2 T}{\partial x^2} \tag{14.29}$$

Initial Conditions:
$$T = T_{init} \quad \begin{cases} t = 0 \\ 0 \le x \le L \end{cases} \tag{14.30}$$

Boundary Conditions:
$$\frac{\partial T}{\partial x} = 0 \quad \begin{cases} t > 0 \\ x = 0 \end{cases} \quad \text{and} \quad T = T_s \quad \begin{cases} t > 0 \\ x = L \end{cases} \tag{14.31}$$

The Crank-Nicolson finite-difference form of Equation (14.29) is

$$\frac{T_i^k - T_i^{k-1}}{\Delta t} = \frac{\alpha}{2\Delta x^2}\left[\left(T_{i-1}^k - 2T_i^k + T_{i+1}^k\right) + \left(T_{i-1}^{k-1} - 2T_i^{k-1} + T_{i+1}^{k-1}\right)\right] \tag{14.32}$$

or

$$T_i^k = T_i^{k-1} + \frac{1}{2}Fo\left[\left(T_{i-1}^k - 2T_i^k + T_{i+1}^k\right) + \left(T_{i-1}^{k-1} - 2T_i^{k-1} + T_{i+1}^{k-1}\right)\right]$$

where we define the dimensionless Fourier number as $Fo = \Delta t \alpha / \Delta x^2$. The Crank-Nicolson method remains numerically stable for $Fo \le 0.5$. However, computer round-off error may cause the method to become unstable, even at smaller values of the Fourier number.

CHAPTER 14: PARTIAL DIFFERENTIAL EQUATIONS

Use a central-difference form of the first-order derivative at the left boundary with a false temperature node to the left of the node at $x = 0$:

$$0 = \left.\frac{\partial T}{\partial x}\right|_{x=0} \cong \frac{T_{-1} - T_{+1}}{2\Delta x} \qquad (14.33)$$

The node with subscript -1 refers to a fictitious mirror image node outside the region of integration, as illustrated in Figure 14.8.

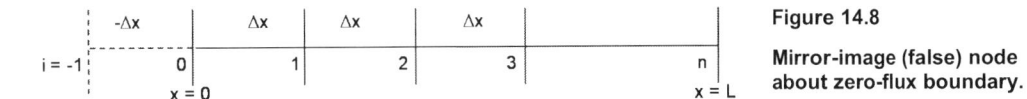

Figure 14.8

Mirror-image (false) node about zero-flux boundary.

Incorporation of a false node into the solution allows us to use the finite-difference form of the second order differential equation at the zero-flux boundary, $x = 0$. Remember that the false node technique does not apply to boundary conditions with non-zero Neumann boundary conditions. For the boundary condition in Equation (14.31), the false node produces a simple relationship between the nodes just before and after the boundary node:

$$T_{-1} = T_1 \qquad (14.34)$$

Solve the Crank-Nicolson equations in an *Excel* spreadsheet, shown in Figure 14.9 using named cells for the parameters for α, Δx, Δt, and *Fo*. Place the false nodes in column **B**. Column **C** contains the solution for T at the adiabatic surface. Solve the system of equations using successive substitution by enabling *Excel*'s **Iteration** with a maximum of 1000 iterations.

	A	B	C	D	E
1	α =	0.001		Fo =	=alpha*delt/(delx^2)
2	Δx =	0.1			
3	Δt =	5			
4	Node i	-1	0	1	2
5	x =	-0.1	0	=C5+$B$2	=D5+$B$2
6	0	=D6	50	50	50
7	=A6+$B$3	=D7	=C6+Fo*(B6-2*C6+D6+B7-2*C7+D7)/2	=D6+Fo*(C6-2*D6+E6+C7-2*D7+E7)/2	=E6+Fo*(D6-2*E6+F6+D7-2*E7+F7)/2
8	=A7+$B$3	=D8	=C7+Fo*(B7-2*C7+D7+B8-2*C8+D8)/2	=D7+Fo*(C7-2*D7+E7+C8-2*D8+E8)/2	=E7+Fo*(D7-2*E7+F7+D8-2*E8+F8)/2
9	=A8+$B$3	=D9	=C8+Fo*(B8-2*C8+D8+B9-2*C9+D9)/2	=D8+Fo*(C8-2*D8+E8+C9-2*D9+E9)/2	=E8+Fo*(D8-2*E8+F8+D9-2*E9+F9)/2

	A	B	C	D	E	F	G	H	I	J	K	L	M
1	α =	0.001		Fo =	0.5								
2	Δx =	0.1											
3	Δt =	5											
4	Node i	-1	0	1	2	3	4	5	6	7	8	9	10
5	x =	-0.1	0	0.1	0.2	0.3	0.4	0.5	0.6	0.7	0.8	0.9	1
6	0	50	50	50	50	50	50	50	50	50	50	50	100
7	5	50	50	50	50	50	50	50.01	50.09	50.51	52.94	67.16	100
8	10	50	50	50	50	50	50.02	50.11	50.49	52.14	58.33	74.26	100
9	15	50	50	50	50	50.02	50.09	50.36	51.35	54.51	62.64	78.25	100
10	20	50	50	50	50.02	50.07	50.24	50.83	52.55	56.92	65.97	80.85	100
11	25	50.01	50.01	50.01	50.05	50.16	50.5	51.48	53.91	59.15	68.59	82.71	100
12	30	50.03	50.02	50.03	50.1	50.3	50.87	52.26	55.31	61.14	70.72	84.12	100
13	35	50.07	50.04	50.07	50.19	50.51	51.32	53.11	56.67	62.91	72.47	85.24	100
14	40	50.12	50.08	50.12	50.31	50.78	51.84	54.01	57.98	64.48	73.95	86.15	100
15	45	50.21	50.14	50.21	50.47	51.1	52.41	54.91	59.21	65.89	75.22	86.91	100
16	50	50.32	50.22	50.32	50.67	51.46	53.02	55.8	60.36	67.15	76.33	87.56	100

Figure 14.9

Excel worksheet with Crank-Nicolson formulas and results. The solution uses conditional formatting. See the *Excel* example file for a color view.

Figure 14.10 shows a surface plot of the temperature profile. **Series1** corresponds to the adiabatic boundary temperature. **Series9** gives the constant temperature at the opposite boundary. As time increases, the temperature in the wall becomes uniform as indicated by the decreasing number of isotherms. Observe how the finite-difference results also maintain the prescribed boundary conditions.

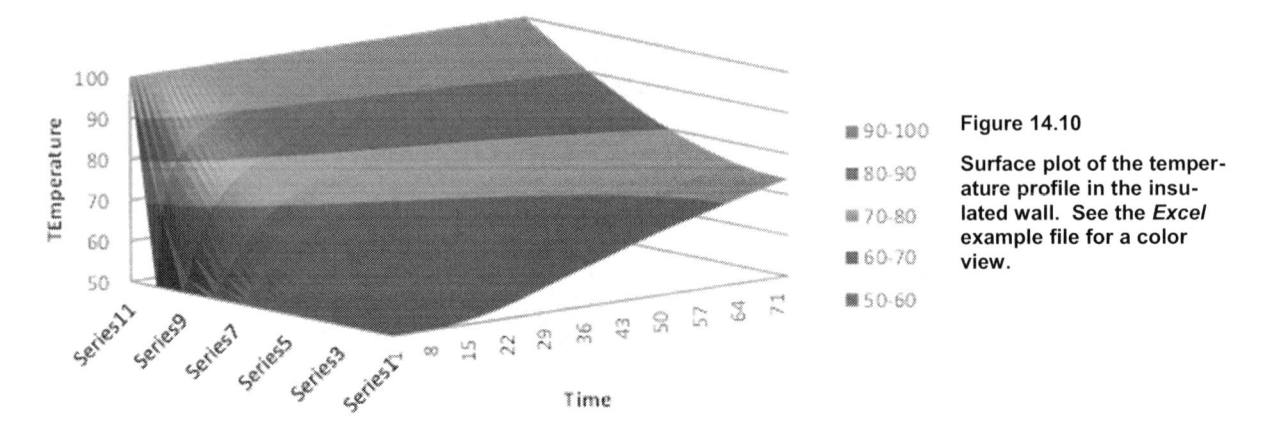

Figure 14.10

Surface plot of the temperature profile in the insulated wall. See the *Excel* example file for a color view.

Figure 14.11 is a plot of the temperature-versus-time profile at the adiabatic surface. Although the adiabatic surface temperature increases with time, the temperature gradient across the adiabatic surface remains zero for no heat transfer.

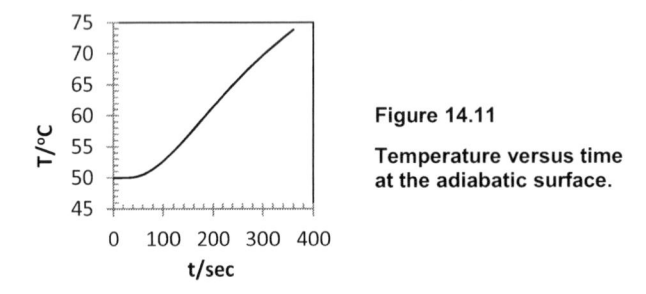

Figure 14.11

Temperature versus time at the adiabatic surface.

Comments: *Excel* provides a convenient platform for solving partial differential equations with simple rectangular, cylindrical, or spherical geometries. The solution tabulated in the formula cells are available immediately for graphing and further analysis. We recommend specialized software for solving partial differential equations in more complex geometries.

□

"There is always an easy solution to every human problem: neat, plausible, and wrong." – H.L. Mencken

14.2.3 Orthogonal Collocation for Parabolic PDEs*

We may extend the method of orthogonal collocation described in Section 13.3 to the solution of stiff parabolic partial differential equations (Finlayson 1980). Similar to the method of lines, the orthogonal collocation method transforms a partial differential equation like Equation (14.11) into a system of simultaneous, first-order differential equations in the dependent variable, one at each of the collocation points. The result is an interpolating polynomial at each integration step for the dependent variable that is exact at the collocation points. Although conceptually simple, the implementation of the method of orthogonal collocation applied to partial differential equations becomes complicated for boundary conditions involving derivatives. We solve for the dependent variable at each collocation point after each integration step. We make extensive use of *Excel's* built-in worksheet matrix functions for the linear algebra required by this technique.

We set the vector of values for the dependent variable at each collocation point from Equation (13.58) equal to the product of a matrix of the power series in x^* with a vector of the trial function parameters:

$$\begin{vmatrix} y_0 \\ y_1 \\ \vdots \\ y_N \end{vmatrix} = \begin{vmatrix} 1 & x_0 & \cdots & x_0^N \\ 1 & x_1 & \cdots & x_1^N \\ \vdots & \vdots & \ddots & \vdots \\ 1 & x_N & \cdots & x_N^N \end{vmatrix} \cdot \begin{vmatrix} a_0 \\ a_1 \\ \vdots \\ a_N \end{vmatrix} \qquad (14.35)$$

For convenience, we dropped the * superscript notation for the scaled value of x in Equation (13.52). In compact matrix notation, Equation (14.35) may be written:

$$Y = ZA \qquad (14.36)$$

where Y is the vector of values for the dependent variable y and Z is the matrix of terms in the power series of the trial function for the independent variable at each collocation point, x. The vector A represents the trial function coefficients:

$$A = Z^{-1}Y \qquad (14.37)$$

To accommodate the general form of a boundary-value problem, we may need the first and second derivatives at each collocation point. We express the first and second derivatives of the power series for the trial function at each of the $N+1$ collocation points in matrix form:

$$\begin{vmatrix} \frac{dy}{dx}\bigg|_0 \\ \vdots \\ \frac{dy}{dx}\bigg|_N \end{vmatrix} = \begin{vmatrix} 0 & 1 & 2x_0 & \cdots & Nx_0^{N-1} \\ 0 & 1 & 2x_1 & \cdots & Nx_1^{N-1} \\ 0 & 1 & 2x_2 & \cdots & Nx_2^{N-1} \\ \vdots & \vdots & \vdots & \ddots & \vdots \\ 0 & 1 & 2x_N & \cdots & Nx_N^{N-1} \end{vmatrix} \cdot \begin{vmatrix} a_0 \\ a_1 \\ a_2 \\ \vdots \\ a_N \end{vmatrix} \qquad \begin{vmatrix} \frac{d^2y}{dx^2}\bigg|_0 \\ \vdots \\ \frac{d^2y}{dx^2}\bigg|_N \end{vmatrix} = \begin{vmatrix} 0 & 0 & 2 & \cdots & (N-1)Nx_0^{N-2} \\ 0 & 0 & 2 & \cdots & (N-1)Nx_1^{N-2} \\ 0 & 0 & 2 & \cdots & (N-1)Nx_2^{N-2} \\ \vdots & \vdots & \vdots & \ddots & \vdots \\ 0 & 0 & 2 & \cdots & (N-1)Nx_N^{N-2} \end{vmatrix} \cdot \begin{vmatrix} a_0 \\ a_1 \\ a_2 \\ \vdots \\ a_N \end{vmatrix} \qquad (14.38)$$

We use compact matrix notation for the first and second derivatives in Equations (14.38) with substitution from Equation (14.37) for A:

$$D = UA \qquad S = WA \qquad (14.39)$$

where D and S represent the vectors of the first and second derivatives at each collocation point, and U and W represent the matrices of the corresponding first and second derivatives of the x-terms in the power series, respectively. In this way, we have transformed the boundary-value part of the problem into a system of simultaneous algebraic equations in matrix form. The system of first-order, ordinary differential equations at the interior collocation points becomes:

$$\frac{dy_i}{dt} = P\left(t, X, Y, D_i^T, S_i^T\right)_i \qquad i = 1, 2 \ldots (N-1) \qquad (14.40)$$

where the D_i^T and S_i^T vectors are formed from the transpose of each row i in Equations (14.39) of the matrices D and S, respectively. The function P of the parabolic partial differential Equation (14.11) uses the row i elements from the vectors X, Y, D_i^T and S_i^T as defined by the specific problem. The simultaneous first-order differential equations in Equation (14.40) are subject to the initial conditions for Y and the boundary conditions located at collocation points x_0 and x_N.

For linear Dirichlet type boundary conditions, we may eliminate y_0 and y_N from the equations for the interior points. For nonlinear Dirichlet boundary conditions in y, Equation (14.40), coupled with the algebraic equations for the boundary conditions, represents a system of differential-algebraic equations solved by the method in Section 12.7. Neumann type boundary conditions require special treatment of the trial function at the collocation point on the boundary. Solve for the value of the dependent variable y at a boundary node from Equation (14.37). Table 14.1 lists the formulas for possible combinations of boundary conditions from Equation (13.2) at the collocation ends in terms of Z^1 and Y. Use the worksheet array functions **SUM** and **MMULT** to evaluate the products of the sub matrices and vectors, Z^1 and Y.

Table 14.1 Boundary conditions in terms of the vectors of Z^{-1} and Y for an orthogonal collocation solution to parabolic PDEs. Compare Equation (13.2).

α	β	y_0 at $x_0 = 0$	α	β	y_N at $x_N = 1$
α_0	0	$\dfrac{1}{-Z_{1,0}^{-1}}\sum_{j=1}^{N} Z_{1,j}^{-1} \cdot y_j$	α_N	0	$\left(\sum_{i=1}^{N}\sum_{j=1}^{N-1} Z_{i,j}^{-1} \cdot y_j\right)\left(-\sum_{i=1}^{N} Z_{i,N}^{-1}\right)^{-1}$
0	β_0	γ_0	0	β_N	γ_N
α_0	β_0	$\left(\dfrac{\beta_0 \gamma_0}{\alpha_0} + \sum_{j=1}^{N} Z_{1,j}^{-1} \cdot y_j\right)\left(\dfrac{\beta_0}{\alpha_0} - Z_{1,0}^{-1}\right)^{-1}$	α_N	β_N	$\left(\dfrac{\beta_N \gamma_N}{\alpha_N} + \sum_{i=1}^{N}\sum_{j=1}^{N-1} Z_{i,j}^{-1} \cdot y_j\right)\left(\dfrac{\beta_N}{\alpha_N} - \sum_{i=1}^{N} Z_{i,N}^{-1}\right)^{-1}$

We recommend a Runge-Kutta variable step method from Chapter 12 for solving the orthogonal collocation system of first-order, ordinary differential equations. We may also apply orthogonal collocation on finite elements to the solution of parabolic partial differential equations being careful to impose the boundary conditions of continuity across a common collocation point between two adjoining elements. We illustrate the orthogonal collocation method applied to parabolic partial differential equations with an example. Although the method of orthogonal collocation is relatively simple, its implementation in *Excel* requires careful bookkeeping and is not for the faint-hearted. You may need to dust off your vector and matrix skills from Chapter 4 before proceeding to the following example.

Example 14.4 Solve the Transient Heat Equation by Orthogonal Collocation

Known/Find: Repeat *Example 14.3* using the method of orthogonal collocation.

Analysis: Select a sixth-order polynomial trial-function. With $N = 6$ we need seven collocation points. For convenience, we programmed a simple *VBA* macro **XiZADS**, listed in Figure 14.12, to create the matrices X, Z^{-1}, A, U, W, D, and S in an *Excel* worksheet, according to Equations (14.37) and (14.38).

```
Public Sub XiZADS()
' Generate the vector for the collocation points X and the matrices of the power
' series terms at the collocation points, Z inverse, D = dY/dX, and S = d2Y/dX2.
' *** Nomenclature ***
Dim i As Integer, j As Integer ' = index variables
Dim n As Integer, n1 As Integer ' = order of trial function
Dim output As Range, outrange As String ' = range for output on worksheet
Dim ow As Integer ' = check for overwriting cells on the worksheet
Dim pcol As Integer, prow As Integer ' = column and row numbers for output to sheet
Dim u() As Double ' = first derivative of trial functions at collocation points
Dim w() As Double ' = second derivative of trial functions at collocation points
Dim ws As String ' = name of worksheet
Dim x() As Double ' = collocation points
Dim z() As Variant, zi() As Variant ' = array of trial functions at collocation points
Dim zr As Range ' = top left cell for output location
'*****************************************************************************
With Application
n = .InputBox(prompt:="ORDER of Trial Polynomial, 2 <= N <= 7:", Default:=2, Type:=1)
n = CInt(.Min(.Max(2, n), 7)) ' limit n to possible integers between 2 and 7
Set zr = .InputBox(prompt:="Location for OUTPUT on Worksheet:", Type:=8)
n1 = n + 1: prow = zr.Row: pcol = zr.Column ' set the number of rows and output
ReDim u(1 To n1, 1 To n1) As Double, w(1 To n1, 1 To n1) As Double, _
      x(1 To n1) As Double, z(1 To n1, 1 To n1) As Variant, zi(1 To n1, 1 To n1)
x(1) = 0#: x(n1) = 1# ' Set the first and last collocation points
Select Case n ' Set the interior collocation points between limits of integration.
Case 2: x(2) = 0.5
Case 3: x(2) = 0.2113248654: x(3) = 0.7886751346
Case 4: x(2) = 0.1127016654: x(3) = 0.5: x(4) = 0.8872983346
Case 5: x(2) = 0.0694318442: x(3) = 0.3300094783: x(4) = 0.6699905218
        x(5) = 0.9305681558
```

CHAPTER 14: PARTIAL DIFFERENTIAL EQUATIONS

```
Case 6: x(2) = 0.0469100771: x(3) = 0.230765345: x(4) = 0.5: x(5) = 0.7692346551
        x(6) = 0.953089923
Case 7: x(2) = 0.0337652429: x(3) = 0.1693953068: x(4) = 0.380690407
        x(5) = 0.6193095931: x(6) = 0.8306046933: x(7) = 0.9662347571
End Select
For i = 1 To n1 ' loop through the matrices for X, Z, U, and W
    For j = 1 To n1: z(i, j) = x(i) ^ (j - 1): Next j
    For j = 2 To n1: u(i, j) = (j - 1) * x(i) ^ (j - 2): Next j
    For j = 3 To n1: w(i, j) = (j - 1) * (j - 2) * x(i) ^ (j - 3): Next j
Next i
zi = .MInverse(z) ' invert the z array
End With ' Application
Cells(prow, pcol) = "X" ' Headings
Cells(prow, pcol + 1) = "Y": Cells(prow, pcol + 2) = "dY/dt"
Cells(prow, pcol + 3) = "t initial": Cells(prow + 1, pcol + 3) = "t final"
With Cells(prow + 1 + n1, pcol)
    .Value = "Z-1" ' Output the results
    .Characters(Start:=2, LENGTH:=2).Font.Superscript = True
End With
Cells(prow + 2 + 2 * n1, pcol) = "U": Cells(prow + 3 + 3 * n1, pcol) = "W"
For i = 0 To n ' put arrays on the worksheet
    Cells(prow + i + 1, pcol) = x(i + 1)
    For j = 0 To n
        Cells(prow + i + 2 + n1, pcol + j) = zi(i + 1, j + 1)
        Cells(prow + i + 3 + 2 * n1, pcol + j) = u(i + 1, j + 1)
        Cells(prow + i + 4 + 3 * n1, pcol + j) = w(i + 1, j + 1)
    Next j
Next i
Cells(prow + 1 + n1, pcol + n1) = "A"
Range(Cells(prow + 2 + n1, pcol + n1), Cells(prow + 2 + n1 + n, pcol + n1)) _
    .FormulaArray = "= MMult(" _
    & Range(Cells(prow + 2 + n1, pcol), Cells(prow + n + 2 + n1, pcol + n)) _
    .Address & "," & Range(Cells(prow + 1, pcol + 1), Cells(prow + n + 1, pcol + 1)) _
    .Address & ")"
Cells(prow + 2 + 2 * n1, pcol + n1) = "D"
Range(Cells(prow + 3 + 2 * n1, pcol + n1), Cells(prow + 3 + 2 * n1 + n, pcol + n1)) _
    .FormulaArray = "= Mmult(" & Range(Cells(prow + 3 + 2 * n1, pcol), _
    Cells(prow + n + 3 + 2 * n1, pcol + n)).Address & "," & _
    Range(Cells(prow + 2 + n1, pcol + n1), Cells(prow + 2 + n1 + n, pcol + n1)) _
    .Address & ")"
Cells(prow + 3 + 3 * n1, pcol + n1) = "S"
Range(Cells(prow + 4 + 3 * n1, pcol + n1), Cells(prow + 4 + 3 * n1 + n, pcol + n1)) _
    .FormulaArray = "= Mmult(" & Range(Cells(prow + 4 + 3 * n1, pcol), _
    Cells(prow + n + 4 + 3 * n1, pcol + n)).Address & "," & _
    Range(Cells(prow + 2 + n1, pcol + n1), Cells(prow + 2 + n1 + n, pcol + n1)) _
    .Address & ")"
zr.Select ' add comment to first output cell
With ActiveCell
    .ClearComments: .addcomment
    With .Comment
        .Visible = False
        .Text Text:="PNMSuite Macro: " & "XiZADS" & vbNewLine _
            & "ORDER of Trial Polynomial Function, N = " & n & vbNewLine _
            & VBA.Date & " at " & VBA.Time
        .Shape.TextFrame.AutoSize = True
    End With
End With
End Sub ' XiZADS
```

Figure 14.12 *VBA* sub procedure **XiZADS** for generating vectors of collocation points and matrices of trial function power series and their derivatives for setting up a solution to a parabolic partial differential equation by the method of orthogonal collocation.

The results from the macro **XiZADS** include the formulas from Equations (14.37) and (14.38) for the vectors of A, D and S in the ranges **H10:H16**, **H18:H24**, and **H26:H32** on the worksheet for live calculations in terms of the values for Y at each integration step in t. The macro also puts labels for the integration limits of t, and initial values for Y and the functions of the derivatives dY/dt at each collocation point. Add values for the range of initial conditions for temperature, T, at each collocation point in the range **B2:B7**. The temperatures corresponds to the vector Y in Equations (14.36) and (14.37). The temperature at the last collocation point is the boundary condition $T = 100°C$ in cell **B8**.

	A	B	C	D	E	F	G	H
1	X	Y	dY/dt	t initial		0		
2	0	50	0	t final		1000		
3	0.04691	50	0.720848					
4	0.230765	50	-0.31536					
5	0.5	50	0.375					
6	0.769235	50	-1.05124					
7	0.95309	50	14.64575					
8	1	100	0					
9	Z^{-1}							A
10	1	0	0	0	0	0	0	50
11	-31	34.69972	-5.03152	2.133333	-1.50942	1.707884	-1	-50
12	240	-335.984	134.1735	-61.8667	44.82986	-51.1524	30	1500
13	-770	1165.631	-626.136	388.2667	-303.983	356.222	-210	-10500
14	1190	-1870.6	1160.966	-866.133	767.0795	-941.316	560	28000
15	-882	1416.448	-956.569	806.4	-799.014	1044.735	-630	-31500
16	252	-410.197	292.5972	-268.8	292.5972	-410.197	252	12600
17	U							D
18	0	1	0	0	0	0	0	-50
19	0	1	0.09382	0.006602	0.000413	2.42E-05	1.36E-06	32.22879
20	0	1	0.461531	0.159758	0.049155	0.014179	0.003926	-55.9811
21	0	1	1	0.75	0.5	0.3125	0.1875	93.75
22	0	1	1.538469	1.775166	1.820692	1.750674	1.616015	-186.608
23	0	1	1.90618	2.725141	3.463073	4.125775	4.718681	654.8046
24	0	1	2	3	4	5	6	1550
25	W							S
26	0	0	2	0	0	0	0	3000
27	0	0	2	0.28146	0.026407	0.002065	0.000145	720.8484
28	0	0	2	1.384592	0.639032	0.245777	0.085075	-315.364
29	0	0	2	3	3	2.5	1.875	375
30	0	0	2	4.615408	7.100663	9.103461	10.50405	-1051.24
31	0	0	2	5.71854	10.90056	17.31536	24.75465	14645.75
32	0	0	2	6	12	20	30	24000

For convenience, name the ranges as follows:

$$X = A2:A8 \qquad T = B2:B8 \qquad A = H10:H16 \qquad Zinv = A10:G16$$

We do not need the values for the first derivatives in the vector D for this problem. The trial function parameters in vector A come from the product of the matrix Z^{-1} with the temperature matrix T according to Equation (14.37).

Use the boundary condition at the first collocation point to calculate the temperature at the left boundary. The boundary condition requires the first derivative of the trial function evaluated at collocation point x_0:

$$\left.\frac{dT}{dx}\right|_{x=0} = \frac{d}{dx}\left(a_0 + a_1 x_0 + a_2 x_0^2 + a_3 x_0^3 + a_4 x_0^4 + a_5 x_0^5 + a_6 x_0^6\right)\bigg|_{x_0=0} = a_1 = 0 \qquad (14.41)$$

To get a single parameter of the trial polynomial, evaluate the vector product of the corresponding row from the matrix Z^{-1} with the T vector, according to Equation (14.37). In this case, for a_1, we evaluate the vector product of the second row of the Z^{-1} matrix with the T vector:

$$a_1 = 0 = Z_{1,0}^{-1} \cdot T_0 + Z_{1,1}^{-1} \cdot T_1 + Z_{1,2}^{-1} \cdot T_2 + Z_{1,3}^{-1} \cdot T_3 + Z_{1,4}^{-1} \cdot T_4 + Z_{1,5}^{-1} \cdot T_5 + + Z_{1,6}^{-1} \cdot T_6 = \sum_{j=0}^{N} Z_{1,j}^{-1} \cdot T_j \qquad (14.42)$$

where the subscripts in the Z^{-1} matrix refer to the row and column (starting with zero for the first row or column). Solve Equation (14.42) for the temperature at the left node T_0 at $x = 0$:

$$T_0 = \frac{Z^{-1}_{1,1} \cdot T_1 + Z^{-1}_{1,2} \cdot T_2 + Z^{-1}_{1,3} \cdot T_3 + Z^{-1}_{1,4} \cdot T_4 + Z^{-1}_{1,5} \cdot T_5 + + Z^{-1}_{1,6} \cdot T_6}{-Z^{-1}_{1,0}} = \frac{1}{-Z^{-1}_{1,0}} \sum_{j=1}^{N} Z^{-1}_{1,j} \cdot T_j \quad (14.43)$$

Using this technique, we may calculate T_0 from the product of the sub vector from the second row of the Z^{-1} matrix with the sub vector of corresponding temperatures, divided by the negative first term in the second row in the *Excel* worksheet. Use a conditional **IF** statement to force the initial value for the temperature to 50°C at the initial time zero:

B2 = IF(E1=0, 50, MMULT(B11:G11, B3:B8) / (-A11))

Note that the vector product does not include the first term in either of the Z^{-1} row or T column. Remember to use the key combination *CNTL SHIFT ENTER* to evaluate the worksheet and user-defined array functions.

The P functions for the first-order differential equations dT/dt in the range **C3:C7** at the interior collocation points come from the values for the second derivatives in the range **H27:H31** calculated from the product of W matrix with the A vector, each multiplied by $\alpha = 0.001$. At the right most boundary, the temperature is constant at 100°C so the derivative of T with respect to t is zero at x_N. With the initial conditions in the range **B3:B8** and the functions of the differential equations in the range **C3:C8**, use the macro **RK45VS** to solve for the temperatures at each collocation point over a period of 1000 seconds. Use the following cell references for the requisite inputs with default values for tolerance and number of print steps:

Lower Integration Limit = J1   Upper Integration Limit = J2
Range of Initial Conditions = E3:E8   Range of ODEs = G3:G8

For the last step of the solution, insert the column temperatures T_0 at the first collocation point for each time using a calculation similar to Equation (14.43). It is necessary to transpose the range of temperatures from the numerical solution arranged in rows at each time step:

B37 =MMULT($B$28:$G$28, TRANSPOSE(C37:H37))/ (-$A$28)

	A	B	C	D	E	F	G	H
35	t	T_0	T_1	T_2	T_3	T_4	T_5	T_6
36	0	50	50	50	50	50	50	100
37	100	52.5373	52.61991	54.56658	63.21917	80.29905	95.82123	100
38	200	61.3858	61.48545	63.7774	72.34212	85.998	97.08349	100
42	600	85.5151	85.55438	86.45635	89.75756	94.86368	98.93327	100
43	700	88.6822	88.71298	89.41775	91.99714	95.98683	99.16633	100
44	800	91.1569	91.18095	91.73163	93.74701	96.86434	99.3486	100
45	900	93.0905	93.1093	93.53955	95.11428	97.54993	99.49125	100
46	1000	94.6013	94.61599	94.95217	96.18257	98.08567	99.60238	100

A plot of the adiabatic surface temperature shows a steady increase that tapers off as the surface approaches the fixed temperature of 100°C at the opposite boundary.

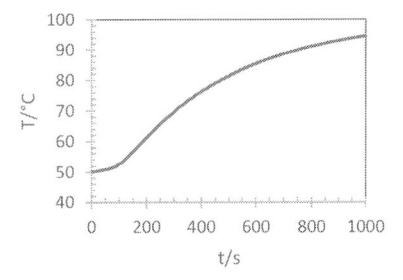

Comments: The solution for the first 350 seconds matches the finite difference solution found in *Example 14.3*. We may use bivariate cubic spline from Section 10.5 to interpolate the two-dimensional results as needed. Once set up for this purpose, we may apply an *Excel* workbook for general orthogonal collocation to any *PDE* problem by changing the formulas for P at the interior points and boundary conditions.

□

14.3 Epilogue on Partial Differential Equations

The field of numerical solutions to partial differential equations is vast, with on-going applications and research. Entire books cover this topic alone. We have only scratched the surface by presenting a few simple techniques, compared in Table 13.1, for implementation in *Excel*. Interested students should consider methods of finite volume for better treatment of boundary conditions and finite elements for handling unusual complex geometries.
Table 13.2 contains a summary of the inputs required for the user-defined function developed for this chapter.

If our problems involve complex systems of partial differential equations, we recommend investing in dedicated software for these types of problems. There are several sophisticated software applications for fluid dynamics, heat and mass transfer, structures, solids, *etc.*, that include advanced algorithms to promote convergence, speed, and flexibility in geometries beyond the simple Cartesian systems used for examples in this chapter.

Table 14.2 Comparison of PDE integration methods.

Method	Strengths	Weaknesses
Crank-Nicolson	• Higher stability for stiff problems. • Simple implementation in a spreadsheet.	• Lower accuracy method uses second order finite difference derivative approximations.
Finite Difference	• Spreads the error across the range of integration. • Simple implementation. • Useful for interpolating data that fits boundary value models.	• Requires a large number of nodes for good accuracy when using second order accurate finite difference derivative approximations. • Does not directly provide values for the derivatives at the boundaries.
Method of Lines	• Possesses all of the advantages of higher order integration of first order differential equations for high accuracy.	• Numerical stability requires a few nodes in the second order variable, which reduces accuracy.
Orthogonal Collocation (on Finite Elements)	• Portable expression for the approximate solution. • Higher accuracy when applied across several finite elements. • Uses all the collocation points for interpolation.	• May have lower accuracy using one element for highly nonlinear functions. • Requires extraordinary effort to implement in a spreadsheet.

Table 14.3 Summary of VBA macro for PDEs and the method of input.

Macro	Inputs	Integration Method
XiZADS	Macro input boxes	Generates a vector of collocation points and matrices of power series for a trial polynomial and its first and second derivatives at the collocation points.

Chapter 15 Review

"Twice and thrice over, as they say, good is it to repeat and review what is good." – Plato

We have encountered a variety of numerical techniques and examples using *Excel* and *VBA*. Some of the topics may be a review of material from our foundational math and engineering courses; others may be completely new to us. Several methods may be immediately applicable in our engineering courses or on the job, whereas others may seem irrelevant to our current needs, but may become useful down the road. Methods that we use frequently will become second nature, while those used infrequently may require review for proper implementation. As we implement numerical methods, pay attention to the following common features:

- Iterative, or "trial and error," methods require good starting points to initiate the solution procedure. Use good engineering judgment to select appropriate initial values.
- Taylor series approximations work better near the points of expansion. We must use good judgment when selecting initial guesses for roots or step sizes for intervals of integration.
- Promote convergence by careful arrangement of equations in terms of variables.
- Not one size fits all – some methods work better, depending on the nature of the problem. We rely on our experience with the methods to find a suitable solution technique.
- Estimate errors in numerical solutions from Taylor series analysis.
- Report the level of reliability in results from an analysis of uncertainty.
- *Excel*'s **Solver** is a versatile tool for tackling a variety of problems – but has limitations. Do not forget alternative methods available using *VBA* functions and macros available in the *PNMSuite*.
- Exploit *VBA* to enhance *Excel*'s capabilities for engineering problem solving and documentation of complex mathematical formulas.
- *Excel* and *VBA* have excellent help files, with examples – we reference them when stuck, or just to expand our expertise. The internet is replete with examples of *VBA* programming.

We recommend the following practices to promote good habits of problem solving:

- Consistently follow the expert problem-solving steps outlined in Table 1.2 of Section 1.3.
- Use graphical analysis to assist with problem solving, understanding, and validation.
- Carefully document solutions for understanding, archiving, trouble-shooting, and collaboration.
- Analyze the sensitivity of solutions to variability in the problem parameters.
- Verify results with experiments or process data when, or wherever, possible.

The appendices to this chapter contain tables of worksheet functions, *VBA* keywords, user-defined functions, sub procedures, and user-forms introduced for implementing numerical methods in *Excel*.

We have treated the topics in this book for the most part as independent units to build up our skill level and diversify our set of tools in our problem solving toolbox. This may be a good time to go back and review some of the topics on mathematical modeling in Chapter 1. In practice, we will find that many problems require some combination of numerical methods to arrive at a complete solution. In the following example, we illustrate how to combine numerical methods through an analysis of the performance of a multicomponent distillation column. To conduct the analysis, we employ methods of least-squares regression, interpolation, numerical quadrature, and finding roots to nonlinear equations. With our set of tools now readily available, we are able to implement the solution in an *Excel* worksheet by selecting functions and macros modified from *PNMSuite*, as well as new custom *VBA* programming.

Example 15.1 Fenske-Underwood-Gilliland Shortcut Distillation Method

Known: We use an eight-tray distillation column for separating a binary mixture of methanol (m) and water (w). Cooling water is available to the condenser at 72°C. The sub cooled feed enters the column at 25°C with the following component flow rates:

Component	Feed Flow Rate/(kmol/hr)	Distillate Mole Fraction, y
Methanol	50	0.95
Water	200	0.05

Schematic: The distillation column is equipped with a total condenser and partial reboiler. The total and component molar flow rates in the feed, distillate, and bottoms are F, D, B and f, d, and b, respectively. The mole fractions in the feed, distillate, and bottoms are z, y, and x, respectively.

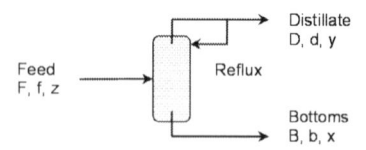

Find: Use the Fenske-Underwood-Gilliland (*FUG*) short cut method to calculate (a) the recovery of methanol in the distillate with 95% methanol, and (b) the reboiler duty.

Assumptions: Fenske-Underwood-Gilliland assumptions (Seader, Henley and Roper 2011), no pressure drop across the column

Analysis: The Fenske and Underwood equations require the cube-root geometric average of the methanol to water relative volatility in the top tray at the conditions of the distillate, the feed tray, and the reboiler:

$$\bar{\alpha} = \sqrt[3]{\alpha_D \alpha_F \alpha_B} \qquad (15.1)$$

where the relative volatility is proportional to the ratio of vapor pressures:

$$\alpha = \frac{(\gamma P^v)_m}{(\gamma P^v)_w} \qquad (15.2)$$

and where γ and P^v are the component activity coefficients and pure species vapor pressures, respectively. To estimate the column pressure, calculate the dew point pressure in the condenser at the temperature of the cooling water:

$$P = \left(\frac{y_m}{\gamma_m P_m^v} + \frac{y_w}{\gamma_w P_w^v}\right)^{-1} \qquad (15.3)$$

To implement the Fenske-Underwood-Gilliland (*FUG*) short cut method, start by guessing the molar flow rate of methanol in the distillate, d_m. Then, iterate on the following steps until reaching a converged solution:

1. Calculate the mole fractions and rates of methanol and water in the distillate and bottom product streams from mass balances:

$$d_w = \frac{y_w}{y_m} d_m \qquad b_m = f_m - d_m \qquad b_w = f_w - d_w \qquad (15.4)$$

$$B = b_M + b_W \qquad x_M = b_M / B \qquad x_W = 1 - x_M$$

2. Calculate the bubble point temperatures in the feed and bottom stages from the modified Raoult's law vapor-liquid equilibrium relationships (Note that the vapor pressure P^v is a function of temperature):

CHAPTER 15: REVIEW

$$P - \gamma_m z_m P_m^v - \gamma_w z_w P_w^v = 0 \quad \text{and} \quad P - \gamma_m x_m P_m^v - \gamma_w x_w P_w^v = 0 \quad (15.5)$$

3. Calculate the thermal condition of the sub cooled feed stream from an energy balance around the feed tray:

$$q = 1 + \frac{1}{\Delta H^{vap}} \int_{T_F}^{T_b} c_p dT \quad (15.6)$$

where T_b and T_F are the bubble point and feed temperatures, ΔH^{vap} is the latent heat of vaporization at the bubble point temperature, and c_p is the heat capacity of the feed.

4. Get the minimum reflux ratio from Underwood's equations. Solve the left equation for θ by an iterative, trial-and-error technique, such as Newton's method:

$$q + \frac{\bar{\alpha} z_m}{\bar{\alpha} - \theta} + \frac{\alpha z_w}{1 - \theta} - 1 = 0 \quad \text{then} \quad R_{min} = \frac{\bar{\alpha} y_m}{\bar{\alpha} - \theta} + \frac{\alpha y_w}{1 - \theta} - 1 \quad (15.7)$$

5. Get the minimum number of stages from Gilliland's data correlated in Equation (9.151):

$$X = \frac{R - R_{min}}{R + 1} \qquad N_{min} = N(1 - Y) - Y \quad (15.8)$$

A common heuristic assumes an economic reflux ratio of $R = 1.2 R_{min}$.

6. Use Fenske's equation to calculate an upgraded value for the methanol molar flow rate in the bottom product:

$$b_m = f_m \left[1 + \left(\frac{d_w}{b_w} \right) \bar{\alpha}^{N_{min}} \right] \quad (15.9)$$

7. Calculate the Fenske molar flow rate of methanol in the distillate from a mass balance:

$$d_m = f_m - b_m \quad (15.10)$$

Implementation of the *FUG* solution method in an *Excel* workbook requires correlations for the vapor pressures, latent heats of vaporization, and heat capacities for methanol and water. We obtained functions to correlate experimental data from least-squares regression (AIChE 2012). We used van Laar models of activity coefficients (Perry 2003) and programmed them as user-defined functions in *VBA*:

```
Public Function H_vM(t As Double) As Double ' Methanol latent heat vaporization, J/kmol
Dim Tr As Double: Tr = t / 512.5
H_vM = 32615000# * (1 - Tr) ^ (-1.0407 + 1.8695 * Tr - 0.60801 * Tr ^ 2)
End Function

Public Function c_pM(t As Double) As Double ' Methanol liquid heat capacity, J/kmol K
c_pM = 256040# - 2741.4 * t + 14.777 * t ^ 2 - 0.035078 * t ^ 3 + 0.000032719 * t ^ 4
End Function

Public Function P_vM(t As Double) As Double ' Methanol vapor pressure, Pa
P_vM = Exp(82.718 - 6904.5 / t - 8.8622 * Log(t) + 0.0000074664 * t ^ 2)
End Function

Public Function H_vW(t As Double) As Double ' Methanol latent heat vaporization, J/kmol
Dim Tr As Double: Tr = t / 647.096
H_vW = 56600000# * (1 - Tr) ^ (0.61204 - 0.6257 * Tr + 0.3988 * Tr ^ 2)
End Function

Public Function c_pW(t As Double) As Double ' Methanol liquid heat capacity, J/kmol K
c_pW = 276370# - 2090.1 * t + 8.125 * t ^ 2 - 0.014116 * t ^ 3 + 0.0000093701 * t ^ 4
End Function

Public Function P_vW(t As Double) As Double ' Water vapor pressure, Pa
P_vW = Exp(73.649 - 7258.2 / t - 7.3037 * Log(t) + 0.0000041653 * t ^ 2)
End Function

Public Function gamma_M(x As Double) As Double
```

```
' Methanol-water activity coefficient, x = mole fraction of methanol
Dim A12 As Double, A21 As Double: A12 = 0.8041: A21 = 0.5619
gamma_M = Exp(A12 * (A21 * (1 - x) / (A12 * x + A21 * (1 - x))) ^ 2)
End Function

Public Function gamma_W(x As Double) As Double
' Methanol-water activity coefficient, x = mole fraction of water
Dim A12 As Double, A21 As Double: A12 = 0.8041: A21 = 0.5619
gamma_W = Exp(A21 * (A12 * (1 - x) / (A12 * (1 - x) + A21 * x)) ^ 2)
End Function
```

Set up an *Excel* worksheet for analysis as follows. Theta (θ) in Underwood's equation was calculated using the user-defined worksheet function **ROOT** from the *PNMSuite*. The initial guess for θ was set as the mean value between one and the average relative volatility:

C6=ROOT(D6, E6)   D6=G20*D12/ (G20-E6) +D13/ (1-E6)-(1-I13)   E6 = Average(1, G20)

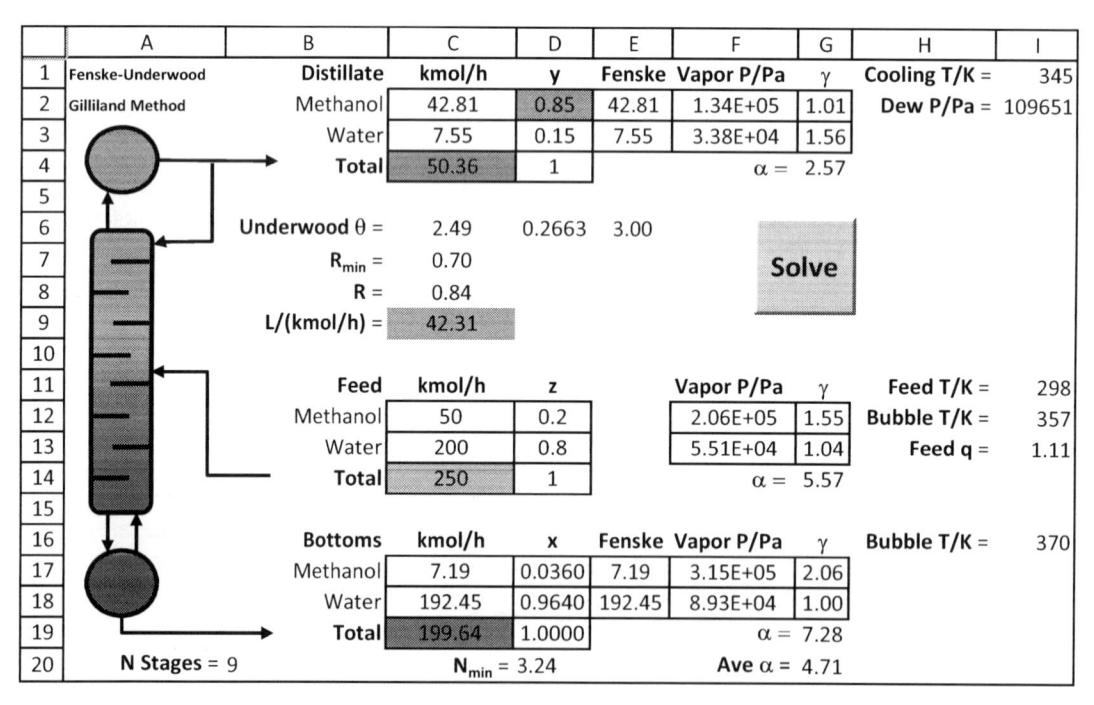

	A	B	C	D	E	F	G	H	I
1	Fenske-Underwood	Distillate	kmol/h	y	Fenske	Vapor P/Pa	γ	Cooling T/K =	345
2	Gilliland Method	Methanol	42.81	0.85	42.81	1.34E+05	1.01	Dew P/Pa =	109651
3		Water	7.55	0.15	7.55	3.38E+04	1.56		
4		Total	50.36	1		α =	2.57		
5									
6		Underwood θ =	2.49	0.2663	3.00				
7		R$_{min}$ =	0.70				Solve		
8		R =	0.84						
9		L/(kmol/h) =	42.31						
10									
11		Feed	kmol/h	z		Vapor P/Pa	γ	Feed T/K =	298
12		Methanol	50	0.2		2.06E+05	1.55	Bubble T/K =	357
13		Water	200	0.8		5.51E+04	1.04	Feed q =	1.11
14		Total	250	1		α =	5.57		
15									
16		Bottoms	kmol/h	x	Fenske	Vapor P/Pa	γ	Bubble T/K =	370
17		Methanol	7.19	0.0360	7.19	3.15E+05	2.06		
18		Water	192.45	0.9640	192.45	8.93E+04	1.00		
19		Total	199.64	1.0000		α =	7.28		
20	N Stages = 9		N$_{min}$ = 3.24			Ave α =	4.71		

We calculated the feed and bottoms bubble-point temperatures using modified versions of the *VBA* macro **REGULAFALSI** from the *PNMSuite*. The cell addresses for the temperatures and functions were hard-coded directly into the macros. For example, in the case of the bubble-point temperature of the bottoms product:

```
Set x_lowr = Range("i16") ' LOWER bracket value (used for x in function)
x_lowr = Range("I1").Value + 1: x_up = x_lowr + 100 ' UPPER bracket value
Set fr = Range("j16") ' Function uses x in cell for low bracket
```

Use the cooling water temperature as the basis for the initial guesses to bracket the root. The formula for the bubble point calculation is

$$J16=I2-D17*F17*G17-D18*F18*G18$$

Calculate the *q*-value by integrating the heat capacity functions using the user-defined *VBA* function **TREX** from the *PNMSuite*, with the integration limits in the range **N12:O12**:

M12 = c_pM(N12)   L12 = TREX(M12, N12, O12)

Program a sub procedure to automate the iterative solution method using Wegstein's approach in Equation (6.13) for accelerating convergence. To speed up the calculations further, we turned off screen updating. The color of the cell containing the distillate mole fraction of methanol changes from red to green when converged:

```
Public Sub FUG()
Dim bm As Range, dm As Range, dm1 As Double, dmf As Range, dmf1 As Double, w As Double
Range("D2").Interior.Color = vbRed: Application.ScreenUpdating = False
Set bm = Range("c17"): Set dm = Range("c2"): Set dmf = Range("e2")
w = 0  ' iterate on the bottoms light key flow rate
Do Until Abs((dmf - dm) / (dm + 0.00000001)) < 0.000001
    dmf1 = dmf: dm1 = dm: dm = (dmf - w * dm) / (1 - w)  ' Wegstein acceleration
    Call BPREGULAFALSI: Call FREGULAFALSI  ' solve for bubble pt T in reboiler and feed
    w = (dmf - dmf1) / (dm - dm1)  ' Wegstein acceleration parameter
Loop
Application.ScreenUpdating = True: Range("D2").Interior.Color = vbGreen
End Sub
```

We added a control button to the worksheet to launch the macro for recalculating the *FUG* method after making changes to any of the parameters or corresponding settings. We put a simple macro in the worksheet object to automate the calculations when changing the value of the cell containing the mole fraction of methanol in the distillate:

| ```
Private Sub CommandButton1_Click()
    Call FUG
End Sub
``` | ```
Private Sub Worksheet_Change(ByVal Target As Range)
Select Case Target.Address
Case "$D$2"
    Call FUG
End Select
End Sub
``` |
|---|---|

We also investigated the sensitivity of methanol recovery in the distillate and reboiler duty over a range of values for the mole fraction of methanol in the distillate. To generate the values for the plots on the worksheet, we used a simple *VBA* macro that looped through a range of mole fractions.

```
Public Sub RecoveryAndDuty(): Dim i As Integer, y As Double: y = 0.84
For i = 1 To 11
    y = y + 0.01: Range("D2") = y
    Cells(i + 24, 1) = y: Cells(i + 24, 2) = Range("C2") / Range("C12")
    Cells(i + 24, 3) = (Range("c14") + Range("c9") - Range("c19")) _
        * (Range("d17") * H_vM(Range("i16")) + Range("d18") * H_vW(Range("i16")))
Next i
End Sub
```

As the purity of methanol in the distillate increases, the distillate recovery and reboiler duty decrease exponentially by removing most of the feed in the bottoms product stream. Although the reflux rate increases, the total distillate decreases, such that the liquid rate of reflux back to the top tray is lower, and the boil up rate is lower as well.

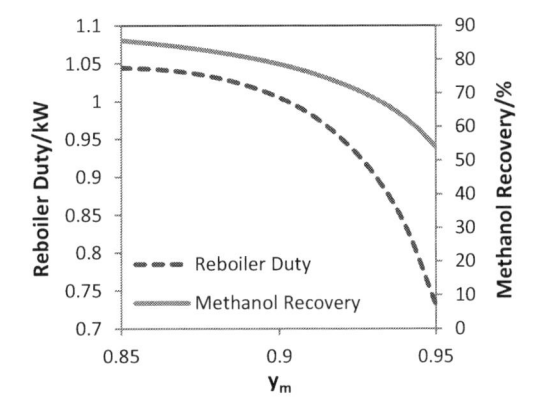

Some final remarks are in order. The cartoon of the distillation column on the worksheet was created using drawing tools from the ribbon tab **Insert>Shapes**. According to the model, we are unable to get higher purity of methanol in the distillate with any quantifiable recovery. Attempts to increase y_m cause the model calculations to fail. To prevent a user from typing a value above 95%, the allowable values in cell **D2** were restricted to $0.7 \leq y_m \leq 0.95$ using **Data Validation** in the **Data Tools** group on the **Data** tab of the *Excel* ribbon.

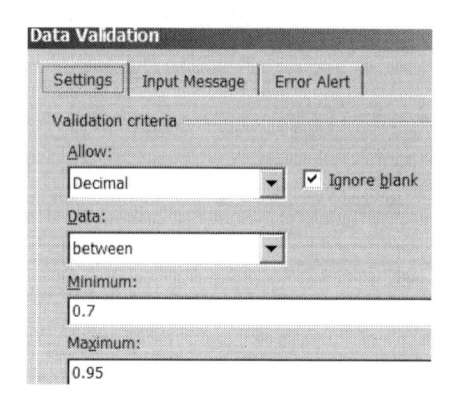

Comments: Short-cut methods get us in the ballpark for design *and* operation. If the results appear promising, we must follow up with a more rigorous analysis of the distillation column using a simulator, such as *ChemSep* (2015).

This example stretched us a little further. We needed to customize our existing methods and macros to arrive at a solution. To implement the solution, we used iteration with Wegstein's method of acceleration, the secant method, regula falsi, Romberg integration, and both linear and nonlinear least-squares regression with interpolation.

□

We covered a lot of ground over the course of this book on practical numerical methods. Perhaps the examples have ignited a spark within you to continue down the path towards mastering numerical methods, as well as *Excel* and *VBA*. The appendices to this chapter summarize several useful *Excel* and *VBA* functions and macros for numerical methods. For new coders, *Excel* with *VBA* can serve as a springboard to other popular computational tools. Multiple resources are available in technical libraries, and on the internet. There we will discover a community of numerical technicians eager to help. Do not be shy about asking for help. Experts know best that they gain as much, if not more from their interaction with us, as we get from them.[113] Onward!

"I went to a bookstore and asked the sales clerks, 'Where's the self-help section?'
They said if they told me, it would defeat the purpose." – George Carlin

---

[113] Personal conversation with 3M's corporate scientist Art Fry, co-inventor of the Post-it Note®.

15.1 Appendix A: Excel Worksheet Functions

The following *Excel* worksheet functions are used to implement numerical methods described in this book.

Worksheet Function	Description
ABS(x)	Absolute value of x
AND(Logical1, Logical2,...)	Returns true when the complete list of logical arguments are all true
AVERAGE(range)	Average (mean) value of range
CEILING(x,m)	Rounds a number x up to the nearest multiple of m
COMBIN(m,k)	Binomial coefficients m = the number of items k = number of selections from the m items
CONFIDENCE(a, s, n)	Confidence interval for a normal-distribution (t = 1.96) a = significance level, *e.g.*, for 95% confidence, a = 1-0.95 = 0.05 s = standard deviation n = degrees of freedom
CONFIDENCE.T(a, s, n)	Confidence interval for a t-distribution a = significance level, *e.g.*, for 95% confidence, a = 1-0.95 = 0.05 s = standard deviation n = degrees of freedom
CONVERT(x, from_unit, to_unit)	Limited unit conversions x = value of parameter from_unit = current unit of parameter x to_unit = new units
CORREL(x, y)	Correlation coefficient for two ranges x and y (must be the same size)
COS(x)	Cosine of x (where x is in radians)
COSH(x)	Hyperbolic cosine of x
COUNT(x)	Number of cells in the range x
COUNTIF(x, s)	Number of cells that contain s in the range x. The argument s can be a number or string in quotation marks. Use * as a wild card in the string for s to indicate anything.
DEGREES(r)	Converts r from radians to degrees
DEVSQ(x)	Sum of squares of deviations of data points from their mean in range x
EVEN(n)	Returns an integer for n rounded up to the nearest even integer
EXP(x)	Exponential of x
FLOOR(x,m)	Rounds a number x down to the nearest multiple of m
FORMULATEXT(c)	Displays the text of the formula in the cell address, c.
GEOMEAN(x1, x2, x3...)	Calculate the geometric mean of a range of values.
HLOOKUP(v, A, row)	Horizontal lookup in a worksheet table. Looks for a value in the top row of a table or array (equal to or smaller than the lookup value) and returns the value in the corresponding column of the row specified. v = lookup value A = range of cells containing the lookup data in the first row (finds the value) row = row number in A from which the matching value is returned
IF(x, y, z)	Logical if statement for controlling a calculation. If x is true then y, otherwise z.
IFS(w, x, y, z)	For IFS (Office 2016 and later), if w is true then x, else if y is true then z, and so on.
INDEX(ref, row, col)	Get the address of a cell at the relative row or column in a range of cells. ref = reference to one or more cell ranges row = number of row in ref from which to return a reference col = (optional) number in the column in reference from which to return a reference
INDIRECT(s)	Returns the worksheet address in the string s. Useful for changing range references on the fly.
INT(x)	Returns the integer portion of x.
INTERCEPT(y, x)	Returns the intercept of linear least-square regression of data in ranges y = dependent data range x = independent variable
LINEST(y, x, c, s)	Multiple linear least-squares regression of data y = range of dependent data x = range of independent data (multivariable) c = optional constant (FALSE = zero intercept)
LN(x)	Natural logarithm of x
LOG(x, [b])	Log base b of x. The parameter b is optional. The default base is 10.

Worksheet Function	Description
LOG10(x)	Log base 10 of x
LOGINV(p, x, s)	Cumulative log normal distribution p = probability x = mean s = standard deviation
MATCH(v, A, type)	Find the relative row or column that contains a value just below or above the match value. Range must be in ascending order. v = lookup value A = array of cells being searched type = (optional) 1 (largest value less than or equal to v (default), 0 (first value equal to v), -1 (default smallest value greater than or equal to v)
MAX(x, y)	Maximum value between x and y
MDETERM(A)	Determinant of a square matrix in range A
MIN(x, y)	Minimum value between x and y, or in a range
MINVERSE(range)	Array function for the inverse of range
MMULT(x, y)	Matrix multiplication of x with y
MROUND(x,m)	Rounds x to the nearest multiple of m.
NORMINV(p, x, s)	Cumulative normal distribution for probability p, with mean x and standard deviation s, mean zero, standard deviation of one.
NORMSINV(p)	Cumulative unit normal distribution for probability p
ODD(n)	Returns an integer for n rounded up to the nearest odd number
OR(Logical1, Logical2, …)	Returns True when at least one of the logical arguments is true
PI()	Constant π = 3.14159 …
QUARTILE(r,q)	Returns the q-level quartile of a data range r
RADIANS(d)	Converts d from degrees to radians
RAND(seed)	Pseudo random number with uniform distribution between 0 and 1. Use seed to control the starting value, otherwise, leave the argument blank.
REPLACE(a, b, c)	Replace occurrences of b in a with c.
ROUND(x, n) ROUNDDOWN(x, n) ROUNDUP(x, n)	Round x to the nth decimal place. The function rounds to the integer value for x when the optional argument is missing. Use negative integer values for n when rounding to the digit left of the decimal place. Round x down to nearest n decimal value. Round x up to nearest n decimal value.
SIGN(x)	+1 for x > 0, 0 for x = 0, -1 for x < 0.
SIN(x)	Sin of x (where x is in radians)
SINH(x)	Hyperbolic sin of x
SLOPE(y, x)	Returns the slope of a linear least-squares regression. y = range of cells containing the dependent data x = range of cells containing the independent data
SQRT(x)	Square root of x
STDEV(range) STDEV.S(range)	Sample standard deviation of range
STEYX(y, x)	Standard error of linear fit for y dependent and x independent data
SUMPRODUCT(x, y)	Vector dot product of corresponding values from two or more ranges, x and y (must be the same length)
SUMSQ(x)	Sum of the squares of the elements of the range x
SUMXMY2(x, y)	Sum of the squares of the difference between x and y elements. x and y are ranges with the same number of terms.
TAN(x)	Tangent of x (where x is in radians)
TANH(x)	Hyperbolic tangent of x
TEXT(Value)	Format the value to text in a specific number format
T.INV(α, v)	Student t-statistic, single, left tailed inverse of a t-distribution (*Excel* 2013 or higher). α = fraction not covered, *e.g.* for 95% confidence, α = 0.05 v = degrees of freedom
TINV(α, v) T.INV.2T(α, v)	Student t-statistic, two-tailed inverse of a t-distribution. T.INV.2T is new for *Excel* 2013 or higher. α = fraction not covered, *e.g.* for 95% confidence, α = 0.05 v = degrees of freedom
TRANSPOSE(x)	Array function. Transpose a contiguous range of cells on an *Excel* worksheet

CHAPTER 15: REVIEW

Worksheet Function	Description
TREND(y, x, x$_i$, c)	Interpolate y vs x data at point x in line y = mx + b y = known y values (dependent variable) x = known x values (independent variable) x$_i$ = point of interpolation c = False for zero intercept
UCASE(s)	Change s from lower to upper case characters.
VLOOKUP(v, A, col)	Vertical lookup in a worksheet table. Looks for a value in the first column of a table A and returns the value in the same row. v = lookup value A = range of cells containing the look uo data in the first column col = column number in A from which the matching value is returned

15.2 Appendix B: VBA Functions and Keywords

The following *VBA* functions and keywords are used to implement numerical methods created for this book.

VBA Keyword(s)	Description of *VBA* keywords used in *PNMSuite*
`Abs(x)`	Absolute value of x
`range.Address`	Object property to get the address of a range variable
`Areas`	Active cell ranges in a SELCTION object of an active worksheet
`Array(i,j)`	Two dimensional array
`Atn(x)`	Arctangent of x, radians
`Beep`	Audible beep sound generated by the computer
`Boolean`	Boolean variable type declaration
`Buttons:=`	Display a button or icon in a message or input box
`Call`	Call a sub procedure (optional)
`Application.Caller`	Refers to the program selection (*e.g.*, range of cells for counting).
`CDbl(x)`	Convert x to double precision type
`Cells(row, col)`	Reference Cell at row number and column number

`.Character.Font`
`.Subscript(Start:=,`
`Length:=)= True`
	Format a part of a string as subscript (or superscript)

`CInt(x)`	Converts argument x to integer type
`CLng(x)`	Converts argument x to long integer type
`Columns("A:Z").EntireCol-` `umn.AutoFit`	Fit the width of columns A through Z to the contents
`.Columns`	Object property that refers to columns in an array or range of cells
`Const x as Type`	Define x as a universal constant as Type (Single, Double, String, etc.) at the top of a module. Place in a procedure to limit the constant to a single procedure.
`.ConvertFor-` `mula(f,xlA1,xlA1,Absolute)`	Application object property to convert references in a string formula f to absolute references
`Cos(x)`	Cosine of radians x (same for *Excel* worksheet function)
`range.Count`	Object property that counts the elements of a range
`CStr(x)`	Convert x to a string
`VBA.Date`	Date from the computer clock
`Default:=`	Default value displayed in an InputBox dialog box
`Dim`	Declare a variable type or dimension an array

`Do While x ... Loop`
`Do Until x ... Loop`
`Do ... Loop While x`
`Do ... Loop Until x`
`While x ...`
	Loop while x is true. x may be a Boolean variable, or a logical expression that returns True or False.

`Double`	Double precision variable type declaration
`End`	Stop all *VBA* running in the workbook
`Err.Description`	*VBA* error message
`Evaluate(f)`	Evaluates a string formula f
`Exit`	Exit a loop, function, or sub procedure

VBA Keyword(s)	Description of VBA keywords used in PNMSuite
Exp(x)	Exponential function of x
False	Boolean value
.Font.Colorindex = 1	Object property that colors the font in a cell according to the color index
For ... Next	Loop control
Format()	Format the output (In Apple Mac version of *Excel*, use *VBA*.Format)
.Formula	Range object property that refers to the cell formula
GoSub label label:... : Return	Sub procedure between label and Return statements inside a sub procedure. (MS Office 2013 or higher)
GoTo Label	Go to the line that starts with the Label:
If Then Else Endif	Logic Program flow control
Input	Open property for importing data from a file
Application.InputBox()	Dialog box to collect information from the user
InStr(f, s)	Returns the integer location of string s within string x, counting from 1 starting at left.
Int(x)	Returns integer part of x
Integer	Integer variable type
.Interior.Color = vbGreen	Cell object property that changes the background color of a cell (e.g., green)
IsError(x)	Boolean operation returns TRUE if x is an error, False, otherwise.
IsMissing(x)	Checks for an optional argument x in a procedure, x must be Variant type.
IsObject(x)	Checks if the argument x is an object, returns Boolean True or False
Label:	Line label (replace Label with distinct name)
Len(a)	Number of characters in variable a
Like(a, b)	Used to compare two strings, returns Boolean True or False
Log(x)	Natural logarithm of x (*Excel* worksheets use LN). To calculate a logarithm in base 10 use log(x)/log(10#). Recall that the default value of the *Excel* worksheet function log() is base 10.
Long	Long integer variable type declaration
Mid(a, m, n)	Returns a **Variant** (**String**) containing a specified number of characters from a string from position m to n.
x Mod n	Returns remainder (decimal) value from a real number, x-n*Int(x/n)
MsgBox ""	Display text information in a popup box
.Name	Object name
Not	Not true
.NumberFormat()	Format the output of a number to a cell or MsgBox
.Offset(row, col)	Apply to the cell located relative to the current active cell by row and columns
On Error statement	Program flow executes *statement* after error, e.g. statement may be a GoTo label.
Open	Open a data file for input or output
Option	Set options for a module. Use Base 1 to set the starting element of arrays to 1, Explicit to require variable type declaration
Optional	Optional arguments in user-defined functions. Limited to Variant data type when using IsMissing to check for the argument.
Output	Property for Open statement for output to a file
Print	Write to a file
Private	Limits constants, or procedures for use within a module or procedure.
Prompt:=	String " " for InputBox dialog
Public	Declaration of variables, constants or procedures for use anywhere in a project
Randomize	Statement to initialize the pseudo random number generator
Range	Range variable type
Range("A1:B2")	Reference cells addresses on worksheet
ReDim ReDim Preserve	Redimension arrays. Use the optional PRESERVE key word to save the array contents when resizing an array.
Replace(a, b, c)	Replace every occurrence of b in a with c.
Resume Next	Resume next calculation
Rnd(s)	Uniform random number between 0 and 1. s is an optional seed. Use Randomize before using Rnd to reseed the pseudo-random number generator. See the *VBA* help for additional information on the argument.
.Rows	Object property that refers to rows in an array or range
Application.Run(s,a)	Evaluate a *VBA* user-defined function string name s given the argument(s) a (separated by commas)

CHAPTER 15: REVIEW

VBA Keyword(s)	Description of VBA keywords used in PNMSuite
`Application.ScreenUpdating`	Set to True or False to turn on or off worksheet updates. VBA runs faster when set to False.
`Select Case x…` `Case x…` `End Select`	Case program flow control, where x is replaced by the value we specify for the case. We may also use comparisons for case, e.g. Case Is < or Case Is >, etc.
`SELECTION`	Object for selected cell ranges on a worksheet
`Set`	Statement used before a Range or InputBox to assign the range address to a range variable
`Sgn(x)`	Returns 1 for x > 0, 0 for x = 0, -1 for x < 0.
`Sin(x)`	Sin of radians x (same for Excel worksheet function)
`Single`	Single precision variable type declaration
`SolverOptions MaxTime:=100, Iterations:=100, Estimates:=1, derivatives:=1, Scaling:=True, SearchOption:=1, Convergence:=0.001`	VBA **Solver** function options. Refer to Table 6.3 for definitions of **Solver** options. Use Record Macro to create simple VBA code to experiment with **Solver** options.
`SolverOk SetCell:=Range(Y), MaxMinVal:=2, ByChange:=Range(X)`	**Solver** set up. Y is the cell reference for the objective function. X is the range of cells for the optimization variables.
`SolverSolve UserFinish:=True`	False causes the function to display the summary message box. True ends the **Solver** function.
`Application.Speech.Speak "text"`	Read aloud string text (in quotes). MS Windows, only.
`Sqr(x)`	Square root of x (Excel worksheets use SQRT)
`Static`	Retains variable values
`Application.StatusBar`	Show information in the worksheet **Status Bar** (bottom)
`Step`	Set the step size in a FOR…NEXT loop
`String`	String variable type
`.Substitute(y, x, j)`	Modifies a string object by replacing the jth occurrence of string x in string y.
`Tab(n)`	Tab n spaces between elements of a print statement
`Tan(x)`	Tangent of radians x
`ThisWorkbook.Path`	Returns a string of the path containing the open workbook.
`Thisworkbook.Worksheets .Add .Delete (Before:=Worksheets(1)).Name = "name"`	Add/delete a new worksheet to the workbook before or after a specified worksheet (numbered starting at 1).
`Title:=`	String in quotation marks " " for InputBox title
`Timer`	Returns the time elapsed in seconds since 12:00 AM
`VBA.Time`	Time from the computer clock
`.Transpose`	Object property to transpose an array
`True`	Boolean value
`Type:=`	Data type for InputBox. Use 1 for Number, 2 for String, 8 for Range
`TypeOf x Is Range`	Checks the object type of x
`Ubound(A, optional rc)`	Get the number of rows (default rc = 1) or columns (rc = 2) in array A
`Ucase(f)`	Make the string f all upper case
`.Value`	Object property to refer to the value of an object, such as a cell or range
`Variant`	Variant variable type
`VbNewLine`	Visual Basic carriage return to put text output on next line
`WorksheetFunction.x`	Use Excel built-in worksheet function x in VBA
`Application.Calculation = xlCalculationAutomatic (= xlCalculationManual)`	Turn auto calculation on (Automatic) or off (Manual)
`&`	Concatenate, or piece together strings and numbers
`;`	Separate elements of a print statement
`:`	Separate VBA statements in the same line, or at the end of a line label
`_`	Underscore to continue on the next line
`" "`	Use quotes around string values
`>`	Greater than
`<`	Less than
`<>`	Not equal to
`=`	Equals

VBA Keyword(s)	Description of VBA keywords used in PNMSuite
`<= or >=`	Less than or equal to, greater than or equal to
`+`	Addition operation (same for Excel worksheet operation)
`-`	Subtraction operation (same for Excel worksheet operation)
`*`	Multiplication operation (same for Excel worksheet operation)
`/`	Division operation (same for Excel worksheet operation)
`\`	Integer divide, rounds the result to nearest integer, e.g. 7\3=2
`^`	Exponent (we must add a space before and after in the code)
`E`	Short hand for power of 10. (e.g. 1.2E-3 = 0.0012)

15.3 Appendix C: VBA User-defined Functions

The following user-defined functions and user-defined array functions are available in the *PNMSuite* workbook for application of numerical methods in *Excel* worksheets. User-defined functions require specific values for the arguments. We must provide the arguments in the same order and type listed:

- To employ a function from the suite, first open the workbook **PNMSuite.xlsm** and enable the macros.
- To use a function in another workbook, prepend path and the workbook name **PNMSuite.xlsm!** before the name of the user-defined function. For example, when the two workbooks are in the same folder

PNMSuite.xlsm!FDERIV(A1:A3, B1:B3, 1)

- To list the required arguments, type the name of the function in an *Excel* worksheet cell and the opening parenthesis after an equal sign, e.g., "**= Function_Name(...**", then type the keyboard combination *CTRL SHIFT A* to fill in the list or required arguments for the function.

Function	Description and arguments of User-defined functions in the *PNMSuite*
`AAD(Y, Y_model)`	Average absolute deviation for two matching ranges Y and Y_model
`A_OC(Y)`	Array function that returns the coefficients of a polynomial trial function for orthogonal collocation. Y = range of values for the dependent variable in the Nth order polynomial at the collocation points.
`AKIMA(Y, X, X`$_i$` [, q, X`$_u$`])` `ASPLINE(Y, X, X`$_i$` [, q, X`$_u$`])`	Returns an interpolated approximation of smooth data using Akima's Hermite cubic spline interpolation method. Default is interpolation. ASPLINE uses Bica's optimized endpoints. Y = range of dependent data X = range of independent data X_i = interpolation value of X or lower integration limit q = (optional) -1 = quadrature, 1 = first derivative, 2 = second derivative X_{up} = (optional) upper integration limit.
`ASPLINE2D(F, X, Y, X`$_i$`, Y`$_i$`)`	Returns bivariate interpolation of two-dimensional data in a grid with Akima's method. F = range of dependent data X = range of x independent data Y = range of y independent data X_i = interpolation value in x Y_i = interpolation value in y
`BSPLINE(Y, X, Xi)`	Returns an interpolated value from Cubic B-spline with fixed ends. Y = range of dependent data X = range of independent data X_i = interpolation value of X or lower integration limit
`CHAUVENET(Y)`	Chauvenet's statistical test for outliers in a normally distributed data set. Returns the value of an outlier, or the string "No Outlier" if none is found. Y = range of values in the data set
`CSPLINE(Y, X, X`$_i$` [, q, X`$_u$`,` `"CS", Alpha, Beta])`	Returns an interpolated approximation of smooth data using cubic spline interpolation with a choice of end conditions. Default is "not-a-knot" end conditions. Y = range of dependent data X = range of independent data X_i = interpolation value of X or lower integration limit q = (optional) -1 = quadrature, 1 = first derivative, 2 = second derivative X_{up} = (optional) upper integration limit. "CS" = (optional) end conditions, "N" = natural, "P" = parabolic, "K" = not-a-knot. Alpha = (optional) slope, (1st derivative) at first data point for clamped end conditions Beta = (optional) slope or 1st derivative at last data point for clamped end conditions

CHAPTER 15: REVIEW

Function	Description and arguments of User-defined functions in the *PNMSuite*
CSPLINE2D(F, X, Y, X$_i$, Y$_i$)	Returns a bivariate interpolation of two-dimensional data in a grid. Ends use not-a-knot conditions. F = range of dependent data X = range of x independent data Y = range of y independent data X$_i$ = interpolation value in x Y$_i$ = interpolation value in y
DENSITY_H2O(T)	Returns the density of water in the temperature range $0 < T < 100°C$ T = Temperature in degrees C.
DREX(F, X [, DV])	Returns a derivative approximation of a formula in a cell using the method of Richardson's extrapolation on finite difference derivative approximations. F = cell containing the formula for the dependent variable (function) X = cell containing the independent variable DV = (optional) integer value for derivative type: 1 = first, 2 = second
FDERIV(F, X [, D])	Returns a derivative approximation of smooth data using a three-point, even or uneven spacing, 2nd-order finite-difference method. F = range of dependent data (3 points) on the worksheet X = range of independent data (3 points) on the worksheet D = (optional) first order forward(1), central (0), or backward (-1), second order (2) difference derivative approximation
FINDIF(F, X, X$_i$ [, D])	Returns a derivative approximation of smooth data using a three-point, even or uneven spacing, 2nd-order finite-difference method. F = range of dependent data (all points) on the worksheet X = range of independent data (all points) on the worksheet X$_i$ = point for derivative approximation must match a value in X D = (optional) first order forward(1), central (0), or backward (-1), second order (2) difference derivative approximation
GKAD(F, XLOW, XUP [, TOL])	Gauss-Kronrod quadrature of a single integral F = Cell reference for formula of Integrand function, integral variable in Worksheet Cell XLOW XLOW = Cell reference for lower limit of integration and independent variable XUP = Upper limit of Integration TOL = (optional) relative convergence tolerance
GRUBBS(Y [, A])	Grubbs' statistical test for outliers in a normally distributed data set. Returns the value of an outlier, or string "No Outlier" if none is determined. Y = range of data A = (optional) significance level (Default = 0.05 for 95% confidence)
JUSTABC_123(S)	Removes non-alphanumeric characters from an *Excel* worksheet name (except underscore). S = string for name of worksheet
LSOLVE(A, B)	Array function (requires CTRL SHIFT ENTER). Solves a linear system of equations arranged in matrix order: Ax=B by matrix inversion and multiplication. The function checks for consistent sizes of arrays and the determinant. A = coefficient matrix B = constant vector
KOWABUNGA(S)	Reads string text to the speaker of the computer. MS Windows only. S = (optional) string text, in quotes.
KSPLINE(Y, X, X$_i$ [, q, Xup])	Returns constrained natural cubic spline interpolation value for smooth data. Y = range of dependent data X = range of independent data X$_i$ = interpolation value of X or lower integration limit q = (optional) -1 (quadrature), 1 = first derivative X$_{up}$ = (optional) upper integration limit.
KSPLINE2D(F, X, Y, X$_i$, Y$_i$)	Bivariate interpolation of dependent data according to independent data in a xy two-dimensional grid using constrained cubic spline with natural end conditions. F = range of dependent data corresponding to X and Y X = range of x values Y = range of y values X$_i$ = value of X point of interpolation Y$_i$ = value of Y point of interpolation
LAGRANGE(Y, X, X$_i$ [, P])	Returns an interpolated value from a polynomial fit of the data derived from Lagrange's method. Y = range of dependent data X = range of independent data X$_i$ = interpolation value of X P = (optional) order of polynomial for interpolation from a subset of the X vs Y data set

Function	Description and arguments of User-defined functions in the *PNMSuite*
LINTERP(Y, X, Xi)	Linear interpolation of data Y = range of dependent data X = range of independent data X_i = interpolation value of X
LINTERP2D(F, X, Y, X_i, Y_i)	Bilinear interpolation of dependent data according to independent data in a xy two-dimensional grid. F = range of dependent data corresponding to X and Y X = range of x values Y = range of y values X_i = value of X point of interpolation Y_i = value of Y point of interpolation
MPEX(F, XLOW, XUP [, TOL])	Integrates an *Excel* worksheet formula between the limits of integration using Romberg's method of Richardson's extrapolation on the composite midpoint rule. F = Cell reference for integrand formula defined in terms of the cell reference for XLOW XLOW = Cell reference for independent variable of integration. The initial value is the lower integration limit. XUP = value of the upper limit of integration. TOL = (optional) relative convergence tolerance
NEWTON(F, X [, TOL])	Returns the root of a single equation defined in an *Excel* worksheet cell formula by the quasi-Newton method. F = cell with formula in terms of cell with x X = cell with independent variable TOL = (optional) relative convergence tolerance, default = 10^{-8}.
NEWTONPOLY(Y, X, X_i)	Returns an interpolated value from a polynomial fit of the data derived from Newton Divided Difference method. Y = range of dependent data X = range of independent data X_i = interpolation value of X
NRDWH()	Returns a normally (Gaussian) distributed random deviate centered at zero.
PADEROOT(F, X [, D, TOL])	Root finding by rational Padé interpolation. F = address for formula of the nonlinear function X = address of the cell with the independent variable, set to the initial guess for the root. TOL = (optional) relative convergence tolerance
POLYROOTS(A)	Returns an array of real and imaginary roots of a polynomial using Bairstow's method. A = (n-1) long range of polynomial coefficients on the worksheet (row or column) starting with the zero-order term increasing to the highest order term: : $a_0 + a_1x + a_2x^2 + ... a_nx^n$.
RATLIN(Y, X, X_i [, P]) RATLINBS(Y, X, X_i)	Returns an interpolated value from smooth data using constrained natural cubic spline interpolation. The alternative user-defined function RATLINBS employs the Bulirsch-Stoer algorithm for the case where the numerator and denominator have the same order polynomial. Y = range of dependent data X = range of independent data X_i = interpolation value of X P = (optional) argument for the order of the polynomial in the numerator (RATLIN only)
RATLSF(X, P, Q, [, I])	Returns the value of a least-squares rational function evaluated at X. Use Ps and Qs from interpolation. X = value of independent variable P = range of values for numerator polynomial coefficients ($P_0 + P_1x + P_2x^2 + ...$) Q = range of values for denominator polynomial coefficients ($Q_1x + Q_2x^2 + ...$) I = (optional) Boolean argument for intercept (False = zero intercept). Default is True
REGFAL(F, X, XUP [,TOL])	Returns a root to a nonlinear function by the method of regula falsi. F = worksheet cell with formula for function, in the form f(x) = 0 X = worksheet cell with independent variable in the formula F and initial lower bracket. XUP = initial upper bracket of the root TOL = (optional) relative convergence tolerance, default = 10^{-8}.
RGAS(P, V, N, T)	Returns values for the ideal gas constant with various units P = string for pressure units. Select from: atm, psi, Pa, mmHg V = string for volume units. Select from: ft3, liter N = string for mole units. Select from: mol, lbmol T = string for temperature units. Select from: K, R
RNDWH()	Pseudo random number generators. Returns a uniformly distributed random value between 0 and 1. No arguments are required. RNDWH has a cycle of 10^{36}.

CHAPTER 15: REVIEW

Function	Description and arguments of User-defined functions in the *PNMSuite*
ROOT(F, X1 [,X2, X3, TOL])	Returns a root to a nonlinear function using Newton, secant, or Muller's methods. F = worksheet cell with formula for function, in the form f(x) = 0 X1 = worksheet cell with independent variable in the formula F. X2 = (optional) additional guess for the root required by Secant or Muller's method X3 = (optional) additional guess for the root required by Muller's method Tol = (optional) relative convergence tolerance, default = 10^{-8}.
ROULETTE(Z, [X])	Returns the index from array Z of probability scores by the roulette algorithm that randomly selects elements proportional to probability. An optional argument X to return the indexed value for the X array
SAMPLE(P)	Simple random sample from a population. Returns an array of values representing a random subset from the population range P.
SHANKS(B)	Accelerate convergence of a series by Shanks transformation provided in the argument as the range B.
SIGFIGS(X, SF)	Returns the value of a number X rounded off to SF significant figures. The default value of the optional argument SF is two.
SIMP(F, XLOW, XUP [, TOL])	Integrates an *Excel* worksheet formula between the limits of integration using an adaptive Simpson 1/3 rule. F = Cell reference for integrand formula defined in terms of the cell reference for XLOW XLOW = Cell reference for independent variable of integration. Value is initially set to the lower integration limit. XUP = value of the upper limit of integration. TOL = (optional) relative convergence tolerance. The default is 10^{-10}.
SIMSPSONDATA(Y, X [, ERROR])	Simpson's rules for integrating equally spaced data. Automatically selects between 1/3 and 3/8 rules depending on the number of data pairs. Applies one Richardson's extrapolation to improve that accuracy if possible (requires data factorable by 4). Y = range of dependent data X = range of independent data ERROR = (optional) Boolean for reporting error estimates in a message box. Default = False
SPOOLED(S,N)	Pooled sample standard deviations S = range of standard deviations N = range of sample sizes
STDEW(x)	Standard deviation of the range x using Welford's formula
STINEMAN(Y, X, X_i)	Returns an interpolated value from smooth data using constrained Stineman interpolation. Y = range of dependent data X = range of independent data X_i = interpolation value of X or lower integration limit
STINEMAN2D(F, X, Y, X_i, Y_i)	Returns a bivariate interpolated value from smooth data using Stineman interpolation. F = Range of dependent Data X = range of independent data Y = range of independent data X_i = interpolation value of X Y_i = interpolation value of Y
TRAP(F, XLOW, XUP [,N])	Integrates an *Excel* worksheet formula between the limits of integration using the composite trapezoidal method. F = Cell reference for integrand formula defined in terms of the cell reference for XLOW XLOW = Cell reference for independent variable of integration. The initial value is the lower integration limit. XUP = value of the upper limit of integration. N = (optional) number of intervals, default = 100
TRAPDATA(YDATA, XDATA [, ERROR])	Integrates a data set by the composite trapezoidal rule. For an even number of equally spaced intervals factorable by two, the function performs one Richardson's extrapolation by integrating every other data pair. YDATA = range of dependent data XDATA = range of independent data ERROR = (optional) Boolean argument to display the error estimate (if possible), default is True.
TREX(F, XLOW, XUP [, TOL])	Integrates an *Excel* worksheet formula between the limits of integration using Romberg's method of Richardson's extrapolation on the composite trapezoidal rules. F = Cell reference for integrand formula defined in terms of the cell reference for XLOW XLOW = Cell reference for independent variable of integration. The initial value is the lower integration limit. XUP = value of the upper limit of integration. TOL = (optional) relative convergence tolerance
TRINV(P, A, B, C)	Inverse of a triangular distribution P = probability, 0 < P < 1 A = (optional) lower limit, default = 0 B = (optional) upper limit, default = 1 C = (optional) mode, default = 0.5

Function	Description and arguments of User-defined functions in the *PNMSuite*
UDF(F, X, X_new)	Template to evaluate a worksheet formula in *VBA*. F = cell with worksheet formula X = cell with independent variable used in the worksheet formula X_new = value of x for evaluation of the worksheet formula in *VBA*
UNITS(V, From_unit, To_unit)	Convert a numerical value between compatible unit systems. V = numerical value with dimensions set by From_unit From_unit = string (in quotes) of the units for V To_units = string (in quotes) of the numerical value returned by UNITS
USIGFIGS(U, SF)	Returns the value of a number U rounded UP to SF significant figures. The default value of the optional argument SF is one.
VNORMRR(x_1, x_2)	Returns the sum of squared relative residuals between two ranges x_1 and x_2.
WAVERAGE(X, S)	Weighted average X = range of averages S = range of standard deviations
X_OC([XLOW, XUP])	Array function that creates a column of collocation points on an *Excel* worksheet. Select the range in a column and type CNTL SHIFT ENTER. This function has no argument, but requires the empty parentheses. XLOW = (optional) lower integration limit (default = 0) XUP = (optional) upper integration limit (default = 1)
Y_OC(X, D [,DY])	Returns the value, first derivative, or second derivative of the orthogonal polynomial with coefficients D at the collocation point X. X = collocation point D = range of polynomial coefficients DY = (optional) argument to return the first (1) or second (2) derivative of the polynomial.

15.4 Appendix D: VBA Sub Procedures (Macros)

The following *VBA* macros are available in the *PNMSuite* workbook for application of numerical methods in *Excel* worksheets. *PNMSuite* macros (sub procedures) interact with values and worksheet formulas in an *Excel* worksheet.

- To use a macro, open the workbook *PNMSuite.xlsm* and enable the macros.
- Run a macro from the ribbon tab **View>Macros>View Macro>Run**.

The following table of macros gives a brief description of the macro and required inputs from the worksheet or user. Before running a macro, be sure to have the worksheet setup with the requisite information. The following table lists the required worksheet setup and additional information for input boxes:

Cell = any unique cell on an *Excel* worksheet.
Range = any unique, contiguous range of cells on an *Excel* worksheet (may be a row or column of cells)

Macro	Description of *VBA* Macros in the *PNMSuite*
ABMODE	Adams-Bashforth-Moulton single step predictor-corrector solution to a system of ordinary differential equations. See RK4G for the required setup and inputs.
BAIRSTOW	Finds the roots to a polynomial of order n in the form: $a_0 + a_1x + a_2x^2 + \ldots a_nx^n$. **Range 1** = range of coefficients starting with constant, ordered from a_0 through a_n. **Range 2** = location of output
BISECTION	Root of a single function by the bisection method. **Cell 1** = lower bracket **Cell 2** = cell containing function in terms of variable in Cell1. Upper bracket, Convergence tolerance
BOOTSTRAP	Apply the bootstrap method for calculating the expanded uncertainty of a statistic of a data set. **Cell 1** = address of the function **Range 1** = range of the data set on the worksheet **Range 2** = address for the output of results P = level of uncertainty in the confidence interval M = number of bootstrap simulations ($\geq$ 1000)
CLRNDX	Create a range of cells on an *Excel* worksheet named Color_Index containing the index and corresponding color.

CHAPTER 15: REVIEW

Macro	Description of VBA Macros in the *PNMSuite*
CROUT	Solution to a linear system of equations by Crout reduction. CROUT uses input boxes. **Range 1** = range of square array of coefficients **Range 2** = range vector of constants **Range 3** = vector of roots
EIGENVI	Solution of Eigen problem for eigenvalues and eigenvectors of a real, square matrix by the method of interpolation. **Range 1** = range of square matrix constants **Range 2** = cell with top left location of results on the worksheet.
FIREFLY FIREFLYDM	Firefly algorithm and Davis modified algorithm for multivariable, multimodal global minimization by random direct search. **Cell 1** = number of fireflies **Cell 2** = initial brightness level **Cell 3** = absorption factor **Cell 4** = randomization factor **Cell 5** = best randomization factor **Cell 6** = objective function **Range 1** = variables of optimization **Range 2** = Initial $\pm$ search region
GAUSSELIM	Solution to a linear system of equations by Gaussian elimination with maximum column pivoting. **Range 1** = matrix of coefficients **Range 2** = vector or constants **Range 3** = location for solution vector on the worksheet.
GAUSSKRONROD	Gauss-Kronrod quadrature of a single integral. **Cell 1** = lower limit of integration **Cell 2** = Integrand function, integral variable in Worksheet Cell 1 **Cell 3** = Integration result Upper integration limit
GAUSSKRONROD2D	Double integration by Gauss-Krondrod adaptive integration for error control. **Cell 1** = integrand function in terms of cells with x and y variables. Cells 2 and 5. **Cell 2** = x variable **Cell 3** = lower x limit of integration (may be a function of cell with y variable) **Cell 4** = upper x limit of integration (may be a function of cell with y variable) **Cell 5** = y variable **Cell 6** = output for result Lower and upper y limits of integration, numbers of x and y intervals
GAUSSLEGENDRE	Gauss-Legendre 10-point (5 node) quadrature of a single integral **Cell 1** = lower limit of integration **Cell 2** = Integrand function, integral variable in Worksheet Cell 1 **Cell 3** = Integration result Upper integration limit
GAUSSNEWTON	Least-squares regression by Gauss-Newton method with uncertainty analysis **Range 1** = range of model parameters **Range 2** = range of dependent data **Range 3** = range of model calculations **Cell 1** = cell with sum of squared residuals **Cell 2** = location of top left corner of output range Convergence Tolerance
GENETIC	Minimization of an objective function by the evolutionary genetic algorithm subject to constraints on the variables of optimization. **Range 1** = Objective function defined in terms of the range of variables **Range 2** = range of variables of optimization (genes) **Range 3** = range of minimum constraints on the variables of optimization **Range 4** = range of maximum constraints on the variables of optimization Mutation rate, population size, number of generations
GRAPHYXGUIDE	Graph of a range of xy data following the recommended graphing guidelines from Section 2.3.4. **Range 1** = range of data with x data in first column and y data in columns 2, 3 … The macro uses the first row headings for the axis labels.

Macro	Description of VBA Macros in the PNMSuite
JACKKNIFE	(Windows only) Estimates model parameter uncertainty in least square regression. Requires the **Solver**. **Range 1** = range of cells containing regression parameters **Range 2** = range of cells containing dependent data **Range 3** = range of cells containing model data **Cell 1** = sum of squared residuals **Cell 2** = worksheet location for upper left corner of output Convergence tolerance and show calculations
JACOBIAN	Approximation of the Jacobian for a system of nonlinear equations. **Range 1** = independent variables in contiguous row or column of cells **Range 2** = functions in contiguous row or column of cells. Formulas use Range 1 cells for independent variables.
JITTER	Uncertainty propagation in mathematical equations. **Cell 1** = function. Formula uses cells in Worksheet Range 1 for independent variables. **Range 1** = independent variables in contiguous row or column of cells. **Range 2** = standard uncertainties of variables in contiguous row or column of cells **Range 3** = degrees of freedom for standard uncertainties corresponding to variables in contiguous row or column of cells **Range 4** = systematic uncertainties corresponding to variables in contiguous row or column of cells. **Cell 2** = location of upper left corner of results. Correlation coefficients (optional)
KRONROD	Single integration by Gauss (10 pt)-Kronrod (15 pt) method. **Cell 1** = lower limit of integration **Cell 2** = integrand function. Formula uses Worksheet Cell 1 for integration variable. **Cell 3** = integration result Upper limit, Number of intervals (even) only required by SIMPSON
KRONROD2D	Double integration by Gauss-Kronrod method. **Cell 1** = cell with x variable **Cell 2** = cell with lower x(y) integration limit **Cell 3** = cell with y variable **Cell 4** = cell with upper x(y) integration limit **Cell 5** = cell with integral function in terms of Cell 1 and Cell 3 **Cell 6** = location of double integral result Lower and upper y integration limits
LEVENBERGM	Nonlinear least-squares regression by the method of Levenberg-Marquardt with uncertainty analysis. **Range 1** = range of model parameters **Range 2** = range of dependent data **Range 3** = range of model calculations **Cell 1** = cell with sum of squared residuals **Cell 2** = location of top left corner of output range Convergence Tolerance
LUUSJAAKOLA	Luus-Jaakola minimization method for multimodal global optimization. **Cell 1** = formula for objective function in terms of Range 1 **Range 1** = range of cells containing initial values for variables **Range 2** = range of search regions for variables in Range 1 Number of passes, number of iterations, number of random variable values, contraction factor, convergence tolerance, option for using best value, show calculations
MC2INT	Double integration (quadrature) by Monte Carlo method. **Cell 1** = formula for integrand in terms of Cell 1 and Cell 2 **Cell 2** = variable x **Cell 3** = lower x integration limit, may be function of y **Cell 4** = upper x integration limit, may be function of y **Cell 5** = variable y **Cell 6** = cell for output of integration result Lower and upper y integration limits, show calculations
MIDPOINT	Integration of a function by the composite midpoint rule **Cell 1** = lower limit of integration **Cell 2** = integrand function in terms of Cell 1 for the variable of integration. **Cell 3** = output cell upper limit of integration, number of intervals

Chapter 15: Review

Macro	Description of VBA Macros in the *PNMSuite*
NEWTCON	Solution to a system of nonlinear equations by Newton's method of continuation. Partial derivatives in the Jacobian are approximated numerically by finite-difference. Uses RK45CK method of integration. **Range 1** = initial guesses for roots **Range 2** = system of equations arranged in the form f(x) = 0 Integration error tolerance
OUTLIER	Tukey test for outliers in a range of data. Colors cells red that contain outliers. **Range 1** = range of data for outlier test in contiguous cells in a row or column on a worksheet
PADE	Accelerate convergence of a power series with a Padé rational function. Similar to RATLINPQ **Range 1** = range of power series (polynomial) coefficients, starting with the lowest order term. **Cell 1** = beginning of column containing the P's and Q's for the rational function coefficients
PDERIV	Calculates the first or second partial derivatives of a formula in a cell on a worksheet by finite-difference with Richardson's extrapolation. **Cell 1** = value of independent variable at derivative **Cell 2** = dependent variable function using Worksheet Cell 1 for value of independent variable.
POWELL	Unconstrained minimization of a multivariable objective function by Powell's method (also used for root-finding of nonlinear equations). **Range 1** = range of cells containing initial values for variables **Cell 1** = cell with formula for objective function in terms of Range 1 Select from Golden Section and Quadratic Interpolation, Convergence tolerance and show calculations
PSO	Particle swarm optimization (PSO) with Latin hypercube initialization and adaptive intertia for global minimization. **Cell 1** = cell with formula for objective function in terms of Range 1 **Range 1** = range of cells containing the variables of optimization **Range 2** = range of cells containing the corresponding lower limits on the variables **Range 3** = range of cells containing the corresponding upper limits on the variables Number of cycles, population size, and maximum cycles with no change in the objective function for convergence.
QYXPLOT	Quick y vs x scatter plot of a formula in one dimension in an *Excel* worksheet **Cell 1** = formula y= f(x) in terms of x in cell 2 **Cell 2** = value of independent variable x Minimum and maximum scale values for x-axis.
QUASINEWTON	Solution to a system of nonlinear functions by quasi Newton's reduced step method. Partial derivatives are approximated numerically by finite-difference. **Range 1** = initial guesses for roots **Range 2** = system of equations arranged in the form f(x) = 0 Relative convergence tolerance
RATLS RATLSGN	Rational Least-squares regression. The macro has options for weighting factors and fixed uncertainty in the dependent variable. RATLS calls the sub procedure RATLSGN for the nonlinear regression analysis by the Gauss-Newton method. **Range 1** = range of independent data **Range 2** = range of dependent data **Range 3** = range for rational function evaluated for each data point. **Cell 1** = location of top-left cell for results output on the worksheet. Weight factors and fixed uncertainty in y (dependent data)
RATLINPQ RATLS	RATLINPQ returns the coefficients in a rational polynomial fit of smooth data for interpolation. RATLS returns the coefficients in a low-order rational polynomial least squares regression fit of data to a linearized rational function. RATLS is may only be used for modeling data located in an *Excel* worksheet in the *PNMSuite*. **Range 1** = range of independent data, x **Range 2** = range of dependent data, y **Range 3** = range of rational function in terms of x **Cell 1** = worksheet location for upper left corner of output Order of the numerator, GaussNewton input requirements
REGULAFALSI	Root of a single function by the method of regula falsi. *Required Worksheet Setup* **Cell 1** = lower bracket **Cell 2** = cell containing function in terms of variable in Cell1. Upper bracket and Convergence tolerance
RELAXIT	Gauss-Seidel iteration with relaxation for the solution to a system of equations. The equations must be defined implicitly, one for each variable in terms of the initial conditions. **Range 1** = range of cells containing initial guesses for the roots **Range 2** = range of cells containing the implicit functions with formulas using the cells for initial guesses **Range 3** = range of relaxation parameters Convergence tolerance and Show Calculations

Macro	Description of VBA Macros in the *PNMSuite*
RIDDERS	Ridders' algorithm of Richardson's extrapolation applied to finite difference derivative approximations of a function. **Cell 1** = range of function for dependent variable **Cell 2** = range of value for independent variable **Cell 3** = range for calculated result for the derivative Order of the derivative, Initial estimate for the finite difference interval Δx
RK45DP RK45DP_sub	4-5th order variable step integration of a system of first-order, ordinary differential equations by the Runge-Kutta-Dormand Prince method. Formulas for the derivatives are defined in terms of the cells containing the initial conditions. **Cell 1** = lower integration limit **Range 1** = range of cells containing the initial conditions for the variables of integration **Range 2** = range of cells containing formulas for the derivative functions in terms of Range 1 **Cell 2** = worksheet location from upper left corner of output Upper integration limit, Number of print steps and Show Calculations, Integration error control tolerance
RK4G RK4G_sub	Classical 4th order constant step integration of a system of first-order, ordinary differential equations by Runge-Kutta-Gill method. Formulas for the derivatives are defined in terms of the cells containing the initial conditions. **Cell 1** = lower limit of integration for independent variable **Range 1** = initial conditions in contiguous row or column of cells **Range 2** = derivative functions using cells from Worksheet Range 1 for dependent variables and Worksheet Cell 1 for independent variable. **Cell 2** = top left corner of results output. Integration error tolerance, Upper limit of integration
RK45ODE	Runge-Kutta 4-5th order variable step integration of a system of first-order, ordinary differential equations by Cash Karp, Fehlberg, Merson methods. Formulas for derivatives are defined in terms of cells containing initial conditions. **Cell 1** = lower integration limit **Range 1** = range of cells containing the initial conditions for the variables of integration **Range 2** = range of cells containing formulas for the derivative functions in terms of Range 1 **Cell 2** = worksheet location from upper left corner of output Upper integration limit, Number of print steps and Show Calculations, Integration error control tolerance
ROMBERGM ROMBERGT ROMBERGI	Romberg's method using the composite midpoint or trapezoidal rule numerical quadrature of a single integral with Richardson's extrapolation. Use ROMBERGM or ROMBERGI for improper integrals. **Cell 1** = integrand function. Formula uses Worksheet Cell 2 for integration variable. **Cell 2** = lower limit of integration **Cell 3** = integration result
ROOTSGN ROOTSLM	Solve for roots to a system of nonlinear equations by the methods of Gauss-Newton or Levenberg-Marquardt. **Range 1** = initial guesses for roots **Range 2** = system of equations arranged in the form f(x) = 0 Relative convergence tolerance
ROWSHUFFLE	Randomly shuffle the order of rows in a range on an Excel worksheet. Range 1 = range of cells to randomize (shuffle) the order.
SCALIT	(Windows only) Optimization or least-squares regression with automatic parameter scaling using the **Solver** to minimize the sum of squared residuals. **Cell 1** = cell containing formula for minimization function in terms of Range 3 **Range 1** = range of cells for variables of optimization **Range 2** = range of cells containing scale factors **Range 3** = range of model parameters = product of corresponding cells in Range 1 and Range 2 Convergence tolerance
SECANT	Finds a root to a function, f(x) = 0 by the secant method. **Cell 1** = the formula for the function in terms of a cell with the variable **Cell 2** = variable x at the initial guess for the root Second guess for the root Maximum iterations, Relative convergence tolerance
SENSITIVITY	Calculates the partial derivative values for sensitivity coefficients. **Cell 1** = cell with formula for the function in terms of Range 1 **Range 1** = range of cells containing the variables **Cell 2** = Location of upper left corner of output to the worksheet.

Chapter 15: Review

Macro	Description of VBA Macros in the *PNMSuite*
SHOOTING	Solve a boundary value problem by the shooting method. Integration uses RK4 method or implicit trapezoidal rule. **Cell 1** = contains the lower limit of integration **Range 1** = range of cells containing the initial conditions **Range 2** = range of cells containing the formulas for the differential equations in terms of the cells with the initial conditions for the variables **Cell 2** = upper left corner of outputs on the worksheet Equation number with unknown initial condition, Equation number with known boundary condition (target), Value of known boundary condition, Upper integration limit, Number of integration and print steps, Convergence tolerance and Show calculations options.
SIMANN	Global minimization by simulated annealing. **Cell 1** = formula for minimization objective function **Range 1** = range of cells with minimization variables **Range 2** = range of cells with standard deviations of minimization variables Number of temperature reduction steps, Number of equilibrium cycles, upper and lower probabilities
SIMPLEXNLP a.k.a., BLOB	Downhill simplex method of minimization for multivariable objective functions. The convergence tolerance are simplex rule parameters are set in the code if modification is required. **Cell 1** = formula for objective function **Range 1** = independent variables in formula for objective function.
SIMPSON	Simpson's composite 1/3 rule for quadrature of a single integral function. **Cell 1** = lower limit of integration **Cell 2** = integrand function. Formula uses Worksheet Cell 1 for integration variable. **Cell 3** = integration result Upper limit, Number of intervals (even) only required by SIMPSON
SIMPSONADAPT	Adaptive quadrature with Simpson's 1/3 rule. User specifies the error tolerance directly in the code. **Cell 1** = integrand formula in terms of Cell 2 **Cell 2** = integration variable and lower limit of integration Upper limit of integration and relative error tolerance.
SIMPSON2D	Double integration by Simpson's composite 1/3 rule. **Cell 1** = integrand function in terms of cells with x and y variables. Cells 2 and 5. **Cell 2** = x variable **Cell 3** = lower x limit of integration (may be a function of cell with y variable) **Cell 4** = upper x limit of integration (may be a function of cell with y variable) **Cell 5** = y variable **Cell 6** = output for result Lower and upper y limits of integration, numbers of x and y intervals
SPIDER	Generates a sensitivity spider plot of an optimized objective function versus plus or minus percent changes to the variables of optimization. **Cell 1** = objective function **Cell 2** = Heading or label for objective function **Range 1** = variables of optimization **Range 2** = range of headings for the variables Percent change in the variables
STEFFENSEN	Aitken's Δ^2 method of acceleration of ordinary iteration for the solution to a system of equations. The equations must be defined implicitly, one for each variable in terms of the initial conditions. **Range 1** = range of cells containing initial guesses for the roots **Range 2** = range of cells containing the implicit functions with formulas using the cells for initial guesses Convergence tolerance and Show Calculations
STIFFODE STIFFODE_sub	Solves a stiff set of first-order ordinary differential equations by either the implicit Euler or trapezoidal methods with fixed integration steps. Uses Steffensen's method to solve the system of implicit equations. **Cell 1** = lower limit of integration for independent variable **Range 1** = initial conditions in contiguous row or column of cells **Range 2** = derivative functions using cells from Worksheet Range 1 for dependent variables and Worksheet Cell 1 for independent variable. **Cell 2** = top left corner of results output. Ordinary iteration tolerance, (optional relaxation factor), Upper limit of integration, number of print/integration steps.
STUDENTIZER	Studentizes a range of regression residuals for identifying potential outliers. **Range 1** = independent variable data **Range 2** = regression residuals Number of model parameters
SWAP	Swaps the contents of two active cell ranges on the active worksheet. The cell ranges must be the same length.

Macro	Description of VBA Macros in the *PNMSuite*
TEARING	Solve a system of equations by the tearing method. The worksheet setup is the same as for QUASINEWTON. **Cell 1** = torn equation **Cell 2** = torn variable, lower bracket **Range 1** = n-1 non torn functions **Range 2** = n-1 non torn variables Upper bracket on the torn variable
TORNADO	Generates a sensitivity tornado chart of an optimized objective function versus plus or minus percent changes to the variables of optimization. **Cell 1** = objective function **Cell 2** = Heading or label for objective function **Range 1** = variables of optimization **Range 2** = range of headings for the variables Percent change in the variables
TRAPEZOID	Composite trapezoidal integration without end correction. **Cell 1** = lower limit of integration **Cell 2** = integrand function in terms of Cell 1 for the variable of integration. **Cell 3** = output cell upper limit of integration, number of intervals
UNITS_TABLES	Creates an *Excel* worksheet with unit strings permitted in the user-defined function UNITS for unit conversions.
UNMCLHS	UNMLHS applies Uncertainty propagation by Monte Carlo simulations with Latin Hypercube sampling. Both methods allow for variable correlation. **Cell 1** = function. Formula uses cells in Worksheet Range 1 for independent variables. **Range 1** = independent variables in contiguous row or column of cells. **Range 2** = standard uncertainties of variables in contiguous row or column of cells **Range 3** = systematic uncertainties corresponding to variables in contiguous row or column of cells. **Cell 2** = location of upper left corner of results. Correlation coefficients (optional)
UNSOLVER	(Windows only) Estimates model parameter uncertainty in least squares regression. Requires the **Solver**. **Range 1** = range of cells containing regression parameters **Range 2** = range of cells containing dependent data **Range 3** = range of cells containing model data **Cell 1** = sum of squared residuals **Cell 2** = worksheet location for upper left corner of output Convergence tolerance and show calculations
XIZADS	Generate vector of collocation points and matrices of power series at each collocation point for the points, first derivatives, and second derivatives.
ZPLOT	Creates a normalized sorted residual plot for least-squares model assessment and calculates Filliben's critical R^2. **Range 1** = range of residuals

Utility sub procedures for building macros for numerical methods. See the VBA code in *PNMSuite* for the required arguments and details of application.

Macro	Description
ASPLINEA	Akima spline interpolation in one direction
AX	Product of square coefficient matrix with the vector x.
AXISCALE	Calculates graphing scale axis parameters
BUBBLESORT	Sort a column of an array in ascending order by the bubble sorting method
BUBBLESORTXY	Sort a pair of arrays in ascending order by the bubble sort method
CHOLESKY	Returns the lower diagonal matrix of a positive-definite symmetric matrix.
CSPLINEC	Cubic spline interpolation in one dimension
DYDX	Sub procedure to evaluate derivative functions for methods of marching integration
GAUSSE	Solve a simultaneous system of linear equations by Gaussian elimination
GAUSSPIVOT	Solves a system of linear equations by Gaussian elimination with maximum column pivoting
HEAPSORT	Sorts an array in ascending order by the method of heap sorting (regional sort push down).
HEAPSORTXY	Sorts a pair of arrays (x & y of the same length) according to the elements of x.
INITC	Initial conditions for marching integration procedures
ITRAP	Implicit trapezoidal rule for marching integration
JACOBI	Returns the Jacobian matrix
KSPLINEK	Constrained cubic spline interpolation in one dimension
MATINV	Invert a matrix

CHAPTER 15: REVIEW

Macro	Description
MATMLT	Matrix product
MATTRN	Transpose a matrix
MIDPT	Composite midpoint rule for numerical quadrature
Q_13	Single integral solution by Simpson's 1/3 rule
QUADRADF	Solves for the real and imaginary roots of a quadratic function
RICHARDSON	Richardson's extrapolation for derivative evaluations
RKG_4	Solve a system of first-order differential equations by the fixed –step Runge-Kutta-Gill method
ROWCOL	Checks for the orientation of an array with one element in rows or columns
STINEMANS	Stineman interpolation in one dimension
THOMAS	Solve a system of linear simultaneous equations represented by a tridiagonal matrix.

15.5   Appendix E: VBA User-forms

The user-forms described in the following table are available in the *PNMSuite* workbook for application of numerical methods in *Excel* worksheets. *PNMSuite* user-forms interact with values and worksheet formulas in an *Excel* worksheet. Show user-form macro names end with _UsrFrm to distinguish them from sub procedures that apply similar numerical methods. As a matter of convenience, we frequently refer to the user-forms by their show user-form macro names throughout the text. To use a user-form:

- Open the workbook *PNMSuite.xlsm* and enable the macros.
- Run a show user-form macro with the suffix _UsrFrm from the ribbon tab **View>Macros>View Macro>Run**.

The following table of show user-form sub procedures gives a brief description of the form and required inputs from the worksheet or user. Before running a show user-form macro, be sure to have the worksheet setup with the requisite information.

Show Form Macro	Description of *VBA* Show user-form macros in the *PNMSuite*
BOOTSTRAP_UsrFrm	Bootstrap resampling method with replacement for calculating the expanded uncertainty of a statistic of a data set. See the macro BOOTSTRAP for the require user input.
INTEGRAL_UsrFrm	Adaptive integration, including Gauss-Kronrod, Simpson, Romberg methods using the composite midpoint or trapezoidal rule numerical quadrature of a single integral with Richardson's extrapolation, or limit transformation for improper integrals. The status bar displays the convergence error of integration. See the macros ROMBERG, GAUSSKRONROD, etc., for required worksheet setup.
LINSYS_UsrFrm	Solves for the roots to simultaneous linear equations by Crout reduction with scaled row pivoting (default) or Gaussian elimination with maximum column pivoting. Both methods employ one iteration of error correction. See the macros CROUT or GAUSSELIM for worksheet set up.
LSREGRESS_UsrFrm	Nonlinear least-squares regression by Gauss-Newton or Levenberg-Marquardt with uncertainty analysis. See the macros GAUSSNEWTON and LEVENBERGM for required worksheet set up.
ODEFIXSTEP_UsrFrm	Solves a system of first-order, ordinary differential equations by the fixed-step methods of backward Euler, trapezoid, Runge-Kutta-Gill, or Adams-Bashforth-Moulton. See the macros RK4G and ABMODE for required worksheet setup and inputs.
OPTIMIN_UsrFrm	Optimization for the minimum (minimization), including Powell, downhill Simplex, Firefly, and Luus-Jaakola. See the macros POWELL, SIMPLEXNLP, FIREFLY, and LUUSJAAKOLA for required set up.
QYXPLOT_UsrFrm	Quick y vs x scatter plot of a formula in one dimension in an *Excel* worksheet. See the macro QYXPLOT for required worksheet setup and inputs.
REGRESSUN_UsrFrm	Stochastic least-squares regression uncertainty by either the jackknife or the Monte Carlo methods using the **Solver** to minimize the sum of squared residuals. See the macros JACKKNIFE and UNSOLVER for the required worksheet setup and inputs.
RK45VS_UsrFrm	Variable step integration methods for initial value problems, including Merson, Fehlberg, Cash-Karp, Dormand-Prince. See the macros RK45ODE and RK45DP for required worksheet setup and inputs.
ROOTS_UsrFrm	Solution to a system of nonlinear equations by Newton's reduced step method of continuation. Partial derivatives in the Jacobian are approximated numerically by finite-difference. Uses RK45CK method of integration. See the macro QUASINEWTON for required worksheet setup and inputs.
SHOOTING_UsrFrm	Solution to a 2nd order boundary value problem by the shooting method. Integration uses RK4 method or implicit trapezoidal rule. See the macro SHOOT for required worksheet setup and inputs.
UNCERTAINTY_UsrFrm	Uncertainty analysis by Jitter or Monte Carlo methods of uncertainty propagation. See the macros JITTER and UNMCLHS for required worksheet setup and inputs.

References

Abramowitz, M., and I.A. Stegun. 1970. *Handbook of Mathematical Functions*. New York: Dover.
AIChE. 2012. "Design Institute for Physical Properties." *DIPPR Project 801*. http://www.knovel.com/web/portal/browse/display?_EXT_KNOVEL_DISPLAY_bookid=1187&VerticalID=0.
Aitken, Alexander. 1926. "On Bernoulli's numerical solution of algebraic equations." *Proceedings of the Royal Society of Edinburgh* (46): 289-305.
Akima, H. 1970. "A New Method of Interpolation and Smooth Curve Fitting Based on Local Procedures." *Journal of the Association for Computing Machinery* 17 (1): 590-602.
Al-ameeri, R.S., and A.S. Said. 1985. "A Simple Formula for the Gilliland Correlation in Multicomponent Distillation." *Separation Science and Technology* 20 (7&8): 565.
Alper, J.S., and R.I. Gelb. 1990. "Standard Errors and Confidence Intervals in Nonlinear Regression: Comparison of Monte Carlo and Parameteric Statistics." *Journal of Physical Chemistry* (94): 4747-4751.
Aris, R. 1993. "Ends and Beginnings in Mathematical Modeling of Chemical Engineering Systems." *Chemical Engineering Science* 2507-2517.
Baker, G.A., and P. Graves-Morris. 1996. *Pade Approximants*. Cambridge: Cambridge University Press.
Beers, K.J. 2007. *Numerical Methods for Chemical Engineering: Applications in Matlab*. Cambridge: Cambridge University Press.
Bell, S. 1999. "A Beginner's Guide to Uncertainty Measurement." *Measurement Good Practice Guide No. 11 (Issue 2)*. Vol. 11. no. 2. Teddington, Middlesex: National Physical Laboratory, August. 2.
Bellman, R.E., and R.E. Kalaba. 1965. *Quasilinearization and Nonlinear Boundary-value Problems*. New York: Elsevier.
Bender, C.M., and S.A. Orszag. 1999. *Advanced Mathematical Methods for Scientists and Engineers*. New York: Springer.
Berendsen, H.J. 2011. *A Student's Guide to Data and Error Analysis*. Cambridge: Cambridge University Press.
Bica, A.M. 2014. "Optimizing at the end-points the Akima's interpolation method of smooth curve fitting." *Computer Aided Geoometric Design* 31: 245-257.
Bieler, P.S., U. Fisher, and K. Hungerbuhler. 2003. "Modeling the Energy Consumption of Chemical Batch Plants - Top-down Approach." *Ind. Eng. Chem. Res.* 42: 6135-6144.
Billo, J. 2006. *Excel for Scientists and Engineers*. New York: Wiley.
Bird, R.B., W.E. Stewart, and E.N. Lightfoot. 2007. *Transport Phenomena*. 2nd. New York: John Wiley & Sons, Inc.
Blom, G. 1958. *Statistical Estimates and Transformed Beta-Variables*. New York: Wiley.
Boehm, B. 1981. *Software Engineering Economics*.
Bourg, D.M. 2006. *Excel Scientific and Engineering Cookbook*. Sebastopol, California: O'Reilly Media, Inc.
Box, G.E.P., and M.E. Muller. 1958. "A note on the generation of normal derivatives." *Annals of Mathematical Statistics* 28: 610-611.
Bradley, E., and G. Gong. 1983. *Statistician* 37: 36.
Briggs, G.E., and J.B.S. Haldane. 1925. "A note on the kinematics of enzyme action." *Biochem J* 19 (2): 338-339.
Bunday, B.D. 1984. *Basic Linear Programming*. London: Edward Arnold.
Burden, R.L., and J.D. Faires. 1985. *Numerical Analysis*. Boston: Prindle, Weber & Schmidt.
Burton. 1998. "US engineering trends." *ASEE Prism* 18-21.
Butcher, J.C. 1996. "A history of Runge-Kutta methods." *Applied Numerical Mathematics* 20: 247-260.
Carnahan, B., H.A. Luther, and J.O. Wilkes. 1969. *Applied Numerical Methods*. New York: Wiley.
Carroll, Lewis. 1939. *The Complete Works of Lewis Carroll*. London: Nonesuch Press.
Cash, J.R., and A.H. Karp. 1990. "A variable order Runge-Kutta method for initial value problems with rapidly varying right-hand sides." *ACM Transactions on Mathematical Software* 16: 201-222.
CCPS. 2009. *Guidelines for Developing Quantitative Safety Risk Criteria*. New York: Wiley.
Chapra, S.C. 2003. *Power Programming with VBA/Excel*. Upper Saddle River, New Jersey: Pearson Education, Inc.
Chapra, S.C., and R.P. Canale. 2002. *Numerical Methods for Engineers, 4th ed*. New York: McGraw-Hill.
2015. *ChemSep Modeling Separation Processes*. Accessed August 11, 2017. http://www.chemsep.com/.
Chen, J.J.J. 1987. "Comments on improvements on a replacement for the logarithmic mean." *Chemical Engineering Science* 42 (10): 2488-2489.
Chong, E.K.P., and S.L. Zak. 1996. *An Introduction to Optimization*. New York: Wiley & Sons.
Civan, F. 2011. "Correlate Data Effectively." *Chemial Engineering Progress* 107 (2): 35-44.
Co, T.B. 2008. "Use Spreadsheets to Estimate Modeling Parameters." *Chemical Engineering Progress* 45-50.
Coleman, H.W., and W.G. Steele. 1989. *Experimentation and Uncertainty Analysis for Engineers*. New York: Wiley.
Constantinides, A., and N. Mostoufi. 1999. *Numerical Methods for Chemical Engineers with MATLAB Applications*. Upper Saddle River: Prentice Hall.
Cornell, L.W., and R.E. Montonna. 1933. *Industrial Engineering and Chemistry* 25: 1131-1335.
Coronell, D.G., and M.H. Hariri. 2008. "The Chemical Engineer's Toolbox: A Glass Box Approach to Numerical Problem Solving." *Chemical Engineering Education* 43 (2): 1-7.
Davis, R.A. 2018. "New and Improved Gilliland Correlation Equations for Shortcut Distillation Column Design." *Separation Science and Technology*.
—. 1992. "The Separation of Carbon Dioxide from Methane by Facilitated Transport in Liquid Membranes." *PhD Dissertation*. Santa Barbara, California: University of California, June. 90.
de Levie, R. 2004. *Advanced Excel for Scientific Data Analysis*. New York: Oxford University Press.

REFERENCES

de Levie, R. 1999. "Estimating Parameter Precision in Nonlinear Least Squares with Excel's Solver." *Jounral of Chemical Education* 76 (11): 1594-1598.
de Levie, R. 2000. "Spreadsheet Calculation of the Propagation of Experimental Imprecision." *Journal of Chemical Education* 77 (4): 534-535.
Deb, K. 2001. *Multi-objective Optimization Using Evolutionary Algorithms*. Chichester, UK: Wiley.
Deb, K., and A. Saha. 2010. "Finding Multiple SOlutions for Multimodal Optimization Problems Using Multi-Objective Evolutionary Approach." *Proceedings of the 12th annual conference on Genetic and evolutionary computation*. New York: GECCO. 447-454.
Delahunty, M.D., and J.P.G. Mack. 1993. "A general method of curve fitting and error analysis using a spreadsheet: determination of the binding constants of tight binging ligands in variable volume assays." *Computer Applications in the Biosciences* 9 (2): 127-131.
Derringer, G., and R. Suich. 1980. "Simultaneous Optimziation of Several Response Variables." *Journal of Quality Technology* 12 (4): 214-219.
Donnelly, Jr., R.A. 2004. *The Complete Idiots Guide to Statistics*. Indianapolis: Alpha Books.
Dormand, J.R., and P.J. Prince. 1980. "A family of embedded Runge-Kuttta formulae." *Journal of Computational Appied Mathematics*, 19-26.
du Plessis, B.J. 2007. "Using Spreadsheets as Curve Fitting Tools." *Chemical Engineering* (5): 66-69.
Duncan, T.M., and J.A. Reimer. 1998. *Chemical Engineering Design and Analysis*. Cambridge: Cambridge.
Dunn, P.F. 2005. *Measurement and Data Analysis for Engineering and Science*. New York: McGraw-Hill.
Edgar, T. 2003. "Computing Through the Curriculum." *CACHE Newsletter*, Fall 2003 ed. http://www.che.utexas.edu/cache/newsletters/fall2003_cover.html.
Eduljee, H.E. 1975. *Hydrocarbon Processing* 120.
Efron, B. 1979. "Bootstrap methods: Another look at the jackknife." *The Annals of Statistics* 7 (1): 1-26.
Ellison, S.L.R., and A. Williams, . 2012. *EURACHEM/CITAC Guide Quantifying Uncertainty in Analytical Measurement*. 3rd. Eurachem/CITAC.
Elnashaie, S., and F. Uhlig. 2007. *Numerical Techniques for Chemical and Biological Engineering Using MATLAB*. Springer.
Englezos, P., and N. Kalogeraki. 2001. *Applied Parameter Estimation for Chemical Engineers*. New York: Marcel-Dekker.
Fadeev, D.K., and V.N. Fadeeva. 1963. *Computational Methods of Linear Algebra*. San Fancisco: W.H. Freeman.
Farhadi, M., P. Azadi, and N. Zarinpanjeh. 2009. *Chemical Enginering Education* 43 (1): 65-69.
Fausett, L. 2002. *Numerical Methods using Mathcad*. Upper Saddle River: Prentice Hall.
Fehlberg, E. 1970. "Klassische Runge-Kutta-Formeln vierter und niedrigerer Ordnung mit Schrittweiten-Kontrolle und ihre Anwendung auf Wärmeleitungsprobleme." *Computing* 6: 61-71.
Felder, R.M., and R.W. Rousseau. 1986. *Elementary Principlesof Chemical Processes*. 2nd. New York: Wiley.
Feller, S.E., and C.F. Blaich. 2001. "Error Estimates for Fitted Parameters: Applicaiton to HCl/DCl Vibrational-Rotational Spectroscopy." *Journal of Chemical Education* 78 (3): 409-412.
Fenton, J.D. 2010. *Numerical methods*. Notes, Institute of Hydraulic and Water Resources Engineering, Vienna University of Technology, Vienna: Fenton.
Filliben, J.J. 1975. "The Probability Plot Correlation Coefficient Test for Normality." *Technometrics* 17 (1): 111-117.
Finlayson, B. 1980. *Nonlinear Analysis in Chemical Engineering*. New York: McGraw-Hill.
Fister, I., I. Fister, X. Yang, and J. Brest. 2013. "A comprehensive review of firefly algorithms." *Swarm and Evolutionaly Computation* 13: 34-46.
Floyd, R.W. 1964. "Algorithm 245-Treesort 3." *Communications of the ACM* 7 (12): 701.
Galilei, G. 1914. *The Two New Sciences*. Translated by H. Crew and A. di Silvio. Macmillan.
Gear, C.W. 1971. *Numerical Initial Value Problems in Ordinary Differential Equations*. Englewood Cliffs: Prentice-Hall.
Gehman, H.W. 2003. "Columbia Accident Investigation Board Report, vol 1." Accessed October 25, 2013. http://www.nasa.gov/columbia/home/CAIB_Vol1.html.
Gerald, C.F., and P.O. Wheatley. 2004. *Applied Numerical Analysis*. 7. New York: Pearson Education.
Gill, S. 1951. "A process for the step-by-step integration of differential equations in an automatic digital computing machine." *Proceedings of the Cambridge Philosophy Society*, 96-108.
Gilliland, E.R. 1940. *Inustrial Engineering Chemistry* 32: 1220.
Goddard, D.L. 2001. *Journal of Engineering Education* 90 (1): 119.
Goldberg, D.E. 2006. *The Entrepreneurial Engineer*. New York: Wiley.
Good, P.I., and J.W. Hardin. 2006. *Common Errors in Statistics*. New York: Wiley.
Goodstein, D.L. 1985. *States of Matter*. New York: Dover.
Gossage, J. 2016. "Plotting McCabe-Thiele Diagrams in Microsoft Excel for Non-ideal Systems." *ASEE's 123rd Annual Conference & Exposition*. New Orleans: ASEE. 15092.
Grcar, J.F. 2011. "Mathematicians of Gaussian Elimination." *Notices of the AMS* 58 (6): 782-792.
Haaland, S.E. 1983. *Transactions ASME JFE* 105: 89.
Hall, B.D., and R. Willink. 2001. "Does Wlech-Satterthwaite make a good uncertainty esitmate?" *Metrologia* 38: 9-15.
Hall, K.R., D.J. Kirwin, and O.L. Updike. 1975. "Reporting Precision of Experimental Data." *Chemical Engineering Education* 24-30.
Hall, S. 2012. *Rules of Thumb for Chemical Engineers*. 5th. Amsterdam: Elsevier.
Hanna, O.T. 1988. *Computers and Chemical Engineering* 12: 1083.
Hanna, O.T., and O.C. Sandall. 1995. *Computational Methods in Chemical Engineering*. Upper Saddle River: Prentice-Hall.
Harrington, C., and T. Zakrajsek. 2017. *Dynamic Lecturing - Research-based Strategies to Enhance Lecture Effectiveness*. Sterling: Stylus Publishing.
Harris, D.C. 1998. "Nonlinear Least-squares Curve Fitting with Microsoft Excel Solver." *Jounral of Chemical Education* 75 (1): 119-121.
Hassan, R., B. Cohanim, and O. de Weck. 2005. "A Comparison of Particle Swarm Optimization and the Genetic Algorithm." *46th AIAA/ASME/ASCE/AHS/ASC Structures, Structural Dynamcics & Materials Conference*. Austin, Texas: AIAA. 1-13.
Hesterberg, T. 2014. "What Teacheres Should Know about the Bootstrap: Resampling in trhe Undergraduate Statistics Curriculum." November 19. Accessed June 2, 2016. https://www.amstat.org/education/pdfs/ResamplingUndergradCurriculum.pdf.
Himmelblau, D. M. 1996. *Basic Principles and Calculations in Chemical Engineering*. 6th Edition. Upper Saddle River: Prentice Hall.
Holman, J.P. 2001. *Experimental Methods for Engineers*. 7. New York: McGraw-Hill.
Incropera, F.P., and D.P. DeWitt. 2002. *Fundamentals of Heat and Mass Transfer*. 4. New York: Wiley.

Jacobi, C.G.J. 1846. "Über ein leichtes Verfahren, die in der Theorie der Säkularstörungen vorkommenden Gleichungen numerisch aufzulösen." *Crelle's Journal* 30: 51-94.
Kahan, W.M. 1980. "Handheld calculator evaluates integrals." *Hewlett-Packard Journal* 23-32.
Kirkup, L. 2002. *Data Analysis with Excel: An Introduction for Physical Scientists*. Cambridge: Cambridge University Press.
Kirkup, L., and B. Frenkel. 2006. *An Introduction to Uncertainty in Measurement*. Cambridge: Cambridge University Press.
Klinkenberg, A. 1948. *Inustrial Engineering and Chemistry Research* 40: 1992-1994.
Koomey, J. 2008. *Turning Numbers into Knowledge*. Oakland: Analytics Press.
Kowalczyk, L.S. 1963. "There are no "Small" Mathematical Errors in Engineering Work." *Chemical Engineering Education* 2 (2): 28-32.
Kronrod, A.S. 1965. *Nodes and weights of quadrature formulas. Sixteen-place tables*. New York: Consultants Bureau.
Kruger, C.J.C. 2011. *Constrained Cubic Spline Interpolation for Chemical Engineering Applications*. Accessed October 28, 2011. http://www.korf.co.uk/spline.pdf.
Kuku, S., and B. Karamani. 2011. "Using Excel and VBA for Excel to Learn Numerical Methods." *1st International Symposium on Computing in Informatics and Mathematics (ISCIM 2011)*. Tirana-Durres. 365-376.
Kumar, K.V., and S. Sivanesan. 2006. "Equilibrium Data, Isotherm Parameters and Process Design for Partial and Complete Isotherm of Methylene Blue onto Activated Carbon." *Journal of Hazardous Materials* B134: 237-244.
Kunz, K.S. 1957. *Numerical Analysis*. New York: McGraw-Hill.
Levenberg, K. 1944. "A Method for the Solution of Certain Non-linear Problems in Least Squares." *The Quarterly of Applied Mathematics* 2: 164-168.
Levenspiel, O. 1984. *Engineering Flow and Heat Exchange*. New York: Plenum Press.
Levenspiel, O. 2002. "Modeling in Chemical Engineering." *Chemical Engineering Science* 4691-4696.
—. 2007. *Rambling Through Science and Technology, 2nd ed.* Lulu.com.
Lewis, R., A. Moshfeghian, and S. Madihally. 2006. "Engineering Analysis in the Chem-E-Car Competition." *Chemical Engineering Education* 66-72.
Liengme, B.V. 2009. *A Guide to Microsoft Excel 2007 for Scientists and Engineers*. San Diego: Academic Press.
Lilley, David G. 2010. "NME: Some Useful Excel/VBA Codes for Numerical Methods in Engineering." *48th AIAA Aerospace Sciences Meeting Including the New Horizons Forum and Aerospace Exposition*. Orlando: American Institute of Aeronautics and Astronautics. 15956-15974.
Linga, P., N. Al-Saifi, and P. Englezos. 2006. "Comparison of the Luus-Jaakola Optimization and Gauss-Newton Methods for Parameter Estimation in Ordinary Differential Equation Models." *Industrial Engineering and Chemistry Research* (45): 4716-4725.
Luus, R., and T. Jaakola. 1973. "Optimization by Direct Search and Systematic Reduction of the Size of Search Region." *AIChE Journal* 19: 760-766.
Lwin, Y. 2000. "Chemical Equilibrium by Gibbs Energy Minimization on Spreadhseets." *International Journal of Engineering Edcuation* 16: 335.
Lynch, D.T., and S.E. Wanke. 1991. "Reactor Design and Operation for Gas-phase Ethylene Polymerization using Ziegler-Natta Catalysts." *Canadian Journal of Chemical Engineering* 69 (1): 332-339.
Lyon. 1998. *All You Wanted to Know About Mathematics But Were Afraid to Ask: Mathematics for Science Students*. Cambridge: Cambridge University Press.
Marini, F., and B. Walczak. 2015. "Particle Swarm Optimization (PSO). A Tutorial." *Chemometrics and Intelligent Laboratory Systems* 149: 153-165.
Marquardt, D. 1963. "An Alorithm for Least-Squares Estimation of Nonlinear Parameters." *SIAM Journal of Applied Mathematics* 11 (2): 431-441.
McCullough, B.D. 2008. "Microsoft Excel's 'Not the Witchmann-Hill' random number generators." *Computational Statistics and Data Analysis* 52: 4587-4593.
Menten, L, and M I Michaelis. 1913. "Die Kinetik der Invertinwirkung." *Biochem Z* 333-369.
Merson, R.H. 1957. "An operational method for the study of integration processes." *Proceedings Symposium on Data Processing*. Salisbury.
Meyer, E.F. 1997. "A Note on Covariance in Propagation of Uncertainty." *Journal of Chemical Education* 74 (11): 1339-1340.
Michalewicz, Z., and D.B. Fogel. 2000. *How to Solve It: Modern Heuristics*. Berlin: Springer.
Moin, P. 2001. *Fundamentals of Engineering Numerical Analysis*. Cambridge: Cambridge University Press.
Molokanov, Y.K., T.P. Korabline, N.I. Mazuriana, and G.A. Nikiforov. 1972. "An Approximation Method for Calculating the Basic Parameters of Multicomponent Fractionation." *International Chemical Engineering* 78 (21): 209.
Montgomery, D.C. 2009. *Design and Analysis of Experiments*. New York: Wiley.
Montgomery, D.C., G.C. Runger, and N.F. Hubele. 1998. *Engineering Statistics*. New York: Wiley.
Morgan, M.G., and M. Henrion. 1990. *Uncertainty: A Guide to Dealing with Uncertainty in Quantitiatic Risk and Policy Analysis*. Cambridge: Cambridge University Press.
Moulton, F.R. 1926. *New methods in exterior ballistics*. Chicago: University of Chigaco Press.
Nelder, J.A., and R. Mead. 1965. "A Simplex Method for Function Minimization." *Computer Journal* 7 (4): 308-313.
Nickabadi, A., M.M. Ebadzadeh, and R. Safabakhsh. 2011. "A novel particle swarm optimization algorithm with adaptive inertia weight." *Applied Soft Computing Journal* 11 (4): 3658-3670.
NIST. 2012. *Engineering Statistics Handbook*. April 1. http://www.itl.nist.gov/div898/handbook/.
Novak, I. 2011. *ScienceL A Many Splendored Thing*. Singapore: World Scientific.
Nyasulu, F.W., and R. Barlag. 2008. "Gas Pressure Sensor Monitored Iodide Catalyzed Decomposition Kinetics of Hydrogen Peroxide: An Initial Rate Approach." *Chemistry Education* 13 (4): 227-230.
Ogren, P., B. Davis, and N. Guy. 2001. "Curve Fitting, Confidence Intervals and Envelopes, Corrleations, and Monte Carlo Visualizations for Multilinear Problems in Chemistry: A General Spreadsheet Approach." *Journal of Chemical Education* 78 (6): 827-836.
Park, S.W., B.S. Choi, and J.W. Lee. 2007. "Breakthrough Data Analysis of Adsorption of Toluene Vapor in a Fixed Bed of Granular Activated Carbon." *Separation Science and Technology* 2221-2233.
Parulekar, S.J. 2006. "Numerical Problem Solving Using Mathcad in Undergraduate Reaction Engineering." *Chemical Engineering Education* 40 (1): 14.

REFERENCES

Perry. 2003. *Chemical Engineer's Handbook.* 6. Edited by Perry, Robert H Perry and Don Green. New York: McGraw-Hill.
Polya, G. 1985. *How to Solve It: A New Aspect of Mathematical Method.* 2nd. Princeton, New Jersey: Princeton University Press.
Press, W.H., S.A. Teukolsky, and W.T., Flannery, B.P. Vetterling. 2007. *Numerical Recipes, 3rd ed.* Cambridge: Cambridge University Press.
Press, W.H., S.A. Teukolsky, S.A. Vetterling, and B.P. Flannery. 1992. *Numerical Recipes in Fortran 77, 2nd ed.* Cambridge: Cambridge University Press.
Ramirez, W.F. 1997. *Computational Methods for Process Simulation.* Oxford: Butterworth-Heinmann.
Rao, S. 2002. *Applied Numerical Methods for Engineers and Scientists.* Upper Saddle River, New Jersey: Prentice Hall.
Rasmuson, A., B. Adersson, L. Olsson, and R. Andersson. 2014. *Mathematical Modeling in Chemical Engineering.* Cambridge: Cambridge University Press.
Reklaitis, G.V. 1983. *Introduction to Material and Energy Balances.* New York: John Wiley & Sons.
Rice, R. G., and Duong D. Do. 1995. *Applied Mathematics and Modeling for Chemical Engineers.* new York: John Wiley & Sons.
Richardson, L.F. 1910. "The approximate arithmetical solution by finite differences of physical problems involving differential equations, with an application to the stress in a masonry dam." *Philos. Trans. Roy. Soc. Ser. A* 210: 307-357.
Ridders, C.J.F. 1982. *Advances in Engineering Software* 4 (2): 75-76.
Riggs, J.A. 2001. *Chemical Process Control.* Lubbock, Texas: Ferret Publishing.
Riggs, J.B. 1994. *An Introduction to Numerical Methods for Chemical Engineers.* 2nd. Lubbock: Texas Tech University Press.
Rosenbrock, H.H. 1960. "An automatic method for finding the greatest of least value of a function." *Computational Journal* 3 (3): 175-184.
Runge, C. 1901. "Über empirische Funktionen und die Interpolation zwischen äquidistanten Ordinaten." *Zeitschrift fur Mathematik und Physik* 46: 224-243.
Schmidt, L. 1998. *The Engineering of Chemical Reactions.* New York: Oxford University Press.
Seader, J.D., E.J. Henley, and D.K. Roper. 2011. *Separation Process Principles: Chemical and Biochemical Operations.* 3rd. New York: John Wiley & Sons Inc.
Seife, C. 2010. *Proffiness.* New York: Viking.
Shacham, M., and N. Brauner. 2002. "Numerical solution of non-linear algebraic equations with discontinuities." *Computers and Chemical Engineering* 29: 1449-1457.
Shuler, M, and F Kargi. 2002. *Bioprocess Engineering.* 2nd Edition. Upper Saddle River, New Jersey: Prentice Hall.
Stephen, E.A., D.R. Bowman, W.J. Park, B.J. Sill, and W.W. Ohland. 2011. *Thinking Like an Engineer.* Upper Saddle River: Pearson.
Stineman, R.W. 1980. "A Consistently Well-Behaved Method of Interpolation." *Creative Computing* 6 (7): 54-57.
Stoer, J., and R. Bulirsch. 1980. *Introuduction to Numerical Analysis.* New York: Springer Verlag.
Su, Y.S. 2008. "It's easy to produce chartjunk using Microsoft Excel 2007 but hard to make good graphs." *Computational Statistics and Data Analysis* 52: 4594-4601.
Taylor, J.R. 1982. *An Introduction to Error Analysis.* Sausalito: University Science Books.
Thanoon, F.H. 2015. "Robust Regression by Least Absolute Deviations Method." *International Journal of Statistics and Applications* 5 (3): 109-112.
Thompson, H.B. 1987. "Good Numerical Techniques in Chemistry: The Quadratic Equation." *Journal of Chemical Education* 1009-1010.
Tufte, E.R. 1983. *The Visual Display of Quantitative Information.* Chesshire: Graphics Press.
Tukey, J.W. 1977. *Exploratory Data Analysis.* Boston: Addison-Wesley.
Tyagi, A., and C.T. Haan. 2001. "Reliability, Risk, and Uncertainty Analysis Using Generic Expectation Functions." *Journal of Environmental Engineering* 127 (10): 938-945.
1997. *U.S. Guide to the Expression of Uncertainty in Measurements.* ANSI, Boulder, Colorado: NCSL International.
Urbaniec, K. 1986. *Optimal Design of Process Equipment.* Chichester: Ellis Horwood, Ltd.
Van Wylen, G.J., and R.E. Sonntag. 1965. *Fundamentals of Classical Thermodynamics.* New York: John Wiley & Sons.
Vasquez, V.R., and W.B. Whiting. 2006. "Accounting for Both Random Errors and Systematic Errors in Uncertainty Propagation Analysis of Computer Models Involving Experimental Measurements with Monte Carlo Methods." *Risk Analysis* 25 (6): 1669-1681.
Vesiland, P.A. 1999. "How to Lie with Engineering Graphics." *Chemical Engineering Education* 304-309.
Walas, S.M. 1990. *Chemical Process Equipment: Selection and Design.* Elsevier.
Wankat, P.C. 2017. *Separation Process Engineering.* Boston: Prentice Hall.
Wegstein, J.H. 1958. "Accelerating convergence of iterative processes." *Comm. ACM* 1 (6): 9-13.
Welford, B.P. 1962. "Note on a Method for Calculating Corrected Sums of Squares and Products." *Technometrics* 4 (3): 419-420.
Wichmann, B.A., and I.D. Hill. 2005. "Generating good pseudo-random numbers." *Computational Statistics and Data Analysis* 51 (3): 1614-1622.
Wisniak, J., and A. Polishuk. 1999. "Analysis of residuals - a useful took for phase equilibrium data analysis." *Fluid Phase Equilibria* 164: 61-82.
Woolfson, M.M., and G.J. Pert. 1999. *An Introduction to Computer Simulation.* Oxford: Oxford University Press.
Yang, X.-S. 2009. *Firefly Algorithms for Multimodal Optimization.* Edited by O. Watanabe and T. Zeugmann. Vol. LNCS 5792. Berlin, Heidelberg: Springer-Verlag.
Zill, D.G. 1982. *A First Course in Differential Equations with Applications.* Boston: Prindle, Weber & Schmidt.

"Most people are other people. Their thoughts are someone else's opinions, their life a mimicry, their passions quotations." – Oscar Wilde

Index

A

A_OC, 598, 607, 636
AAD, 437, 440, 636
ABMODE, 569, 583, 640
`Abs`, 633
ABS, 631
absolute cell referencing, 40
accelerated bracketing, 267
accumulation, 534
accuracy, 30, 315, 318, 496, 540
accuracy, integration, 505, 540, 544, 557
activated carbon, 411
ActiveX Controls, 111
Adams-Bashforth-Moulton, 567
adaptive Gaussian quadrature, 577
adaptive Romberg, 532
`add`, 635
`Address`, 633
adiabatic, 616
adjusted R², 388, 389
adjusting the worksheet display, 38
Aitken's Δ², 231
AKIMA, 464, 484
AKIMAB, 636
Alignment, 64
analog, 326
Analysis ToolPak, 70
Analysis Tools, 71
analytical solution, 3, 4, 6, 167, 254, 496, 534
AND, 631
animate, 114
Antoine, 374
Application, 97
ARCSIN, 100
area, integration, 485, 488
`Areas`, 633
Aris' rules, 22
Aristotle, 317
ARITHMETIC, 107
arithmetic operations, 97
array, 106
`Array`, 633
arrays, 461
Arrhenius, 4, 32, 307, 381, 382, 517, 534
ASPLINE, 465, 484, 512, 636
ASPLINE2D, 478, 484

`ASPLINEA`, 646
assumptions, 15
asymmetric, 350, 421
asymmetric confidence interval, 358
asymmetry, 359
asymptotic, 9, 429, 436
asymptotic analysis, 373
asymptotic limit, 195, 320
`Atn`, 633
Auto Fill, 40, 41, 46, 85
`AutoFit`, 633
Automatic Calculation, 71, 141
automatic recalculation, 37
automatic scaling, 378
auto-scaling, 421
average, 316, 318
AVERAGE, 72, 103, 187, 318, 479, 631
average absolute deviation, 436
average of absolute residuals, 437
AX, 646
Axes, 381
Axis Titles, 57
`AXISCALE`, 125, 646

B

backward derivative, 540
backward difference, 179, 593, 596, 615
Bairstow, 240
BAIRSTOW, 241, 243, 250, 327, 430, 434
basketball, 586
batch distillation, 238, 472
batch reactor, 31, 361, 492, 542, 556
`Beep`, 633
Bernoulli, 17
bias, 270, 336, 356, 479
bias error, 317, 318, 325
bilinear interpolation, 476
bimodal, 253
bins, 350
bisection, 205
BISECTION, 206, 250
bisection rule, 496
blob, 268
boiling point, 415
Boltzmann probability function, 271
`Boolean`, 633
BOOTSTRAP, 320, 323, 640

boundary, 534, 584
boundary condition, 17, 19, 20, 451, 466, 584, 585, 586, 593, 596, 612
boundary-value problems, 584
boxes, 55
Box-Muller transformation, 102
bracket, 208, 263
Bracketing, 260
break text, 64
Briggs-Haldane, 10, 14
B-spline, 481
BSPLINE, 482, 579, 636
bubble sort, 109
BUBBLESORT, 646
BUBBLESORTXY, 646
Bulirsch-Stoer, 473
Buttons, 633
by reference, 81
by value, 81
ByChange, 635

C

Calculation, 635
calculator, 37, 364
calculus, 534
calibration, 325, 426, 498
Call, 633
Caller, 110, 633
cancel, 330
Case, 90, 635
catalyst, 584, 586
CDbl, 633
CEILING, 631
cell comment, 134
cell naming, 40
Cells, 108, 633
centering, 389
central difference, 179, 593
chain rule, 535
Character, 633
Chart Tools, 62, 364, 381, 427
charting, 55
CHAUVENET, 320, 394, 440, 636
ChemE Car, 26
chemical process safety, 314
chemical reaction energy, 15
CHOLESKY, 166, 646
Cholesky decomposition, 151, 356
chromosome, 286
CInt, 633
circular reference, 159, 162, 219, 544, 587, 596
clamped cubic spline, 537
Clear Rules, 611
CLng, 633
CLRNDX, 135, 142, 640
coefficient matrix, 152
coefficient of determination, 387
Colebrook, 213

Color, 634
Colorindex, 135, 634
Columns, 633
COMBIN, 631
combine, 39
combined uncertainty, 316, 318, 328, 330, 344
comment, 25, 65, 82, 87
Comment, 65
comparison operators, 88
concatenate, 108
concentration polarization, 348
conditional formatting, 611, 615
Conditional Formatting, 66
conditionally stable, 540
conduction, 13, 19, 159
confidence, 315
CONFIDENCE, 631
confidence interval, 24, 320, 328, 350, 356, 389, 421, 428, 528
confidence level, 390
CONFIDENCE.T, 631
conjugate directions, 266
conservation, 4, 11, 534
conservation of energy, 18
conservation of mass, 362
Const, 633
constant VBA, 84
constant vector, 152
constitutive equations, 11, 19
constrained cubic spline interpolation, 466
constraint, 291
constraints, 253, 294
continuous function, 270
continuous least-squares regression, 530
contour plot, 194, 259, 266, 300, 306
convergence, 157, 158, 164, 214, 226, 236, 247, 405, 615
Convergence, 635
convergence tolerance, 157
conversion, 299, 510, 539
CONVERT, 50, 631
ConvertFormula, 633
CORREL, 336, 631
correlated variables, 356
correlation, 318, 341, 350, 356, 388, 391, 405, 415, 421, 425, 426
correlation coefficient, 336, 356, 389, 436
Cos, 633
COS, 631
COS(a), 45
COSH, 631
Count, 110, 633
COUNT, 631
COUNTIF, 631
covariance, 336
covariance matrix, 389
coverage factor, 320, 338
Crank Nicolson, 615
Create from Selection, 40
Create names from values, 162

creative problem solving, 586
critical correlation coefficient, 392
critical value, 390
crossover, 286, 287
CROUT, 148, 166, 641
Crout reduction, 148
CROUTLU, 148
CROUTLUX, 148
CROUTRED, 153, 641
CSPLINE, 453, 456, 461, 484, 512, 533, 539, 566, 636
`CSPLINE, integration`, 513
CSPLINE2D, 478, 484, 533, 636, 637
`CSPLINEC`, 646
`CStr`, 633
CSTR, 329
CTRL SHIFT ENTER, 42, 152
CTRL+BREAK, 86
cubic B-spline, 481
cubic spline, 151, 429, 450, 512
cubic spline, 452
cubic spline clamped end conditions, 453
cubic spline parabolic end conditions, 452
cumulative normal distribution, 391
cumulative probability, 355

D

Darcy friction factor, 90
Darcy's law, 384
dart board, 527
data acquisition, 497
Data Analysis, 71
Data Analysis add-in, 323
Data Bars, 103
data transformation, 429
Data Validation, 65
`Date`, 633
Debug.Print, 100
debugging, 84, 135
`Default`, 633
Define Name, 108
Defined Names, 40
degrees of freedom, 21, 143, 316, 319, 320, 326, 338, 358, 390, 441
DEGREES(r), 45, 631
`delete`, 635
DENSITY_H2O, 461, 637
dependent variable, 316, 470, 534, 543, 584
derivative, 34, 178, 254, 461
derivative approximation, 461, 466
derivative of an integral, 535
`derivatives`, 635
`Description`, 633
Descriptive Statistics, 71, 319, 323
Design Mode, 111, 113
desirability functions, 304
determinant, 153
deterministic, 24, 527

Developer, 76, 114
DEVSQ, 631
dew point, 221
diamond, 402
differential algebraic equations, 570
differential element, 19
differential equation, 534, 537
differential method, 34
differentiation, 429, 436, 450
diffusion, 585
digital, 326
digital filter, 479
`Dim`, 84, 633
dimensionless, 22
direct search, 251, 258, 266, 270
Dirichlet, 619
Dirichlet boundary condition, 584
discontinuity, 527
discontinuous, 276
discontinuous function, 350
discrete, 503, 538
discretize, 258, 612
dispersion, 17
distillation, 315
distributed, 19
Dittus-Boelter, 333
divergence, 405, 540
division by zero, 43, 157, 158, 204, 214, 239, 276, 429, 430, 560
`Do`, 633
`Do While`, 633
document, 25, 53
`Double`, 633
DOUBLE, 83
double integral, 519, 522, 525, 528
double pipe heat exchanger, 591
double precision, 7, 8, 83, 91, 99
downhill simplex, 268
drawing tools, 55
drawings, 534
DREX, 191, 193, 198, 637
DYDX, 547, 558, 561, 646

E

`E`, 636
economic pipe diameter, 310
economics, 315
eigenproblems, 154
eigenvalues, 154
eigenvectors, 154
EIGENVI, 154, 155, 166, 641
Einstein, 2
electrical energy, 15
electrical heating, 15
elementary reactions, 4, 31
elliptical, 609
`Else`, 634
empirical, 10

empirical data modeling, 361
Enable Iterative Calculations, 163, 218, 610
End, 633
End Select, 635
Endif, 634
endothermic, 15
energy balance, 18, 32, 584
energy equation, 46
engineering judgment, 31
EntireColumn, 633
enzyme, 382, 556, 589
equally-spaced, 497, 504
equally-spaced data, 532
equation arrangement, 226
equation sequence, 226
equilibrium, 12, 13, 21, 295, 345
Err, 633
error, 65
Error, 634
error bars, 427
Error Bars, 428
error, integration, 505, 537, 540, 553, 560
Esc, 86
escape, 86
Estimates, 635
ethics, 30, 314
Euler's method, 537, 555, 615
Euler's method for systems, 542
Evaluate, 633
EVEN, 631
events, 132
Evolutionary solver, 290
Exit, 633
Exit For, 86
Exit Sub, 137
exothermic, 15, 315
Exp, 634
EXP, 631
expanded uncertainty, 338, 357, 428
expected value, 318, 328, 330, 350
experiment, 23, 336
experimental errors, 317
expert problem solvers, 25
explicit, 5
exponential function, 304, 364, 380, 429
exponential growth phase, 386
exponential model, 362
extrapolation, 34, 429, 456

F

F9 key, 37
False, 634
false node, 596, 597, 617
false position, 208
Fanning friction factor, 18
FDERIV, 185, 186, 198, 449, 595, 637
Fick's law, 19
Fick's second law, 20

filter, 479
Filter, 398
filtration, 384
FINDIF, 186, 449, 637
finite elements, 603
finite volume, 19
finite-difference, 151, 182, 183, 194, 257, 340, 341, 362, 406, 466, 503, 585, 593, 594, 608, 609, 612, 613, 617
FIREFLY, 284, 313, 641
Firefly Algorithm, 283
FIREFLY_ANIMATION, 285, 641
FIREFLYDM, 285, 313, 641
first derivative, 467
first-order derivative, 12, 617
first-order function, 144
first-order ordinary differential equation, 534, 561, 570, 585, 608, 612
first-order polynomial, 22
first-order reaction, 4, 5, 32, 315, 361
first-order Taylor series, 179
First-Order-Plus-Dead-Time (FOPDT) model, 423
fitness, 287
fitness score, 287
fixed error, 318
fixed uncertainty, 316, 324, 354, 356
flash, 170
floating-point, 157
FLOOR, 631
flux, 20
Font, 633, 634
FOPDT, 423
For, 634
forecasting, 24
Form controls, 111
Format, 135, 634
FORMAT, 134
Format Cells, 49
Formula, 634
Formula Auditing, 44, 46
Formulas, 45, 140, 540
formulas, *Excel*, 37
FORMULATEXT, 44, 631
forward difference, 179, 593, 596, 615
Fourier analysis, 70
Fourier's law, 19
Freeze Panes, 39
FREQUENCY, 103
Freundlich isotherm, 469
friction factor, 213
function library, *Excel*, 45
function procedure, VBA, 79
fundamental theorem of calculus, 486, 488
fuzz, 314

G

gain, 423, 566
Galileo, 317
gas absorption, 487

Gauss quadrature, 598
GAUSSE, 646
GAUSSELIM, 146, 147, 166
Gaussian elimination, 145, 151, 405
Gaussian quadrature, 513
GAUSSKRONROD, 515, 533, 641
GAUSSKRONROD2D, 641
GAUSSKRONRON2D, 525
GAUSSLEGENDRE, 514, 533, 641
Gauss-Newton, 372, 405, 411
GAUSSNEWTON, 407, 440, 641
GAUSSPIVOT, 147, 148, 641, 646
Gauss-Seidel, 158, 163, 615
gene, 286
generalized reduced gradient, 290
Generalized Reduced Gradient, 290
generation, 14, 15
GENETIC, 288, 289, 313, 641
genetic algorithm, 290
genetic evolutionary algorithm, 291
GEOMEAN, 631
geometric mean, 382
Gibbs free energy, 295
GIGO, 24
Gilliland correlation, 437
GKAD, 516, 519, 533, 637
global error, 538, 551, 554
global optimum, 253, 259, 270, 275, 291
global variable, 83
GN, 407, 410
Goal Seek, 67, 91, 211, 235, 348
Goal Seek, Shooting Method, 586
Golden Section, 263, 267
goodness of fit, 336, 379, 436
GoSub, 634
GoSub `label`...Return, 96
GoSub... Return, 288
GoTo, 85, 634
gradient, 618
graphical analysis, 258
graphical integration, 486, 519
graphical model, 9
graphical solution, 200
graphing, 247
GRAPHYXGUIDE, 60, 124, 142, 379, 543, 641
GRG, 290
grid, 258
growth phases, 386
GRUBBS, 320, 394, 637
guess, 7, 257, 305, 588
Guide to the Expression of Uncertainty in Measurements, 317
GUM, 317

H

Haaland, 90
HEAPIFY, 109
HEAPSORT, 109, 357, 421, 646

HEAPSORTXY, 109, 448, 453, 456, 467, 646
heat capacity, 397
Heat Equation, 613, 616, 73
heat transfer coefficient, 333
Henri, 4, 9, 11, 15
Henry's constant, 487, 509
Hermite interpolation, 463
Hermite polynomial interpolation, 463
heteroscedasticity, 402
heuristic, 3, 333
Hide, 67
histogram, 24, 357
HISTOGRAM, 102, 350
HLOOKUP, 631
homogeneous, 12
homoscedasticity, 402
hourglass, 402
hypothesis test, 389

I

ideal gas, 3, 28, 61, 198
identity matrix, 152
`If`, 85, 634
IF, 51, 85, 631
`If-Then`, 88
ignition temperature, 202
ill-conditioned, 21
Illustrations, 55
Immediate window, 100, 138
implicit, 5, 204
implicit equation, 5, 157, 165, 225, 587
implicit integration, 540
implicit, trapezoidal rule, 543
improper integrals, 494, 495
improved Euler, 551
independent variable, 316, 436, 470, 485, 528, 534, 584
indeterminate, 429
INDEX, 51, 66, 445, 631
INDIRECT, 52, 53, 631
indirect optimization, 366
indirect search, 251, 253, 254, 256, 257
INITC, 547, 558, 561, 646
initial condition, 534, 542, 544, 551, 571, 585, 586, 588
initial guess, 157, 158, 164, 200, 205, 212, 225, 231, 247, 257, 372, 373, 405, 540
initial-value problem, 534, 585
input, 316
Input, 131, 634
InputBox, 118, 634
Insert Function, 81
insight, 3
InStr, 634
Int, 634
INT, 52, 631
Integer, 634
INTEGRAL, 641
integral method, 34
INTEGRAL_UsrFrm, 500, 510, 515, 518, 533, 647

integrand, 486, 513
integration, 429, 436, 450, 467, 485, 557, 586; spline, 512
integration accuracy, 551
intercept, 366, 382, 426, 429
INTERCEPT, 631
INTERCEPT, *Excel*, 367
interest, 201
`Interior`, 634
interior logarithm, 301
interpolation, 429, 441, 485
Interpolation Method for Eigenproblems, 154
intersection, 39
interval, 488, 503, 537, 593, 613
interval size, 496, 497
interval, integration, 514
inverse, 381
inverse barrier, 301
inverse interpolation, 457
inverse matrix, 152
invert, 470
Invert the derivative, 535
irreversible, 4
`IsError`, 634
`IsMissing`, 634
`IsObject`, 634
isotherm, 411
iteration, 7, 406, 540, 544
Iteration, 617
iteration, *Excel*, 540
`Iterations`, 635
iterative, 315, 615
`ITRAP`, 646

J

JACKKNIFE, 405, 440, 642
jackknife resampling method, 404, 414
JACOBI, 227, 232, 646
Jacobi's method, 157
Jacobian, 226, 231, 418, 575
JACOBIAN, 227, 250, 642
JITTER, 337, 341, 344, 348, 358, 360, 414, 416, 422, 642
jitter method, 341
`JUSTABC_123`, 89, 138, 142, 637

K

kinetics, 361
knot, 450, 451, 466
KOWABUNGA, 111, 142, 637
Kronrod, 515
KRONROD, 515, 533, 642
KRONROD2D, 525, 526, 642
KSPLINE, 467, 468, 484, 512, 637
KSPLINE2D, 478, 484
`KSPLINEK`, 646

L

L2INTERP, 637, 638
`Label`, 634
lag phase, 386
Lagrange, 476, 503, 567
LAGRANGE, 448, 484
Lagrange polynomial, 184, 448, 504
Lagrangian multipliers, 294
Langmuir isotherm, 469
language of engineering, 534
Latin Hypercube Sampling, 354
law of conservation, 11
Law of Propagation of Uncertainty, 336, 350, 414
Layout, 381
least absolute deviation, 437
least-squares, 363, 418
least-squares regression, 364, 405, 421, 441, 479, 515
Legendre polynomial, 513, 598
`Len`, 634
level of significance, 389
LEVENBERGM, 419, 440, 642
Levenberg-Marquardt, 372
Levenspiel, 2
`Like`, 634
limits, integration, 488, 494, 496, 503, 513, 515, 528, 586
line break, 82
`Line Input`, 131
linear, 380
linear algebra, 145
Linear approximation, 6
linear equations, 143
linear function, 143, 350
linear interpolation, 443, 488
linear least-squares regression, 70, 366, 367, 382
linear transforms, 383
linearity, 199, 380
linearization, 6, 144, 167, 197, 362, 373, 382, 383, 386, 405
lines, 55
LINEST, 371, 389, 401, 430, 433, 443, 631
Lineweaver-Burk, 383
LINSYS_UsrFrm, 147, 148, 166, 647
LINTERP, 443, 444, 484, 638
LINTERP2D, 477, 478, 484
LN, 631
local optimum, 253, 257, 258, 259, 266, 270
`Log`, 634
LOG, 97, 98, 631
log normal distribution, 359
log normal probability density function, 359
log plot, 380
Log10, 97, 98
LOG10, 631
logarithmic, 381
logarithmic functions, 429
logarithmic linearization, 365
logical functions, 51
LogInv, 358, 359
LOGINV, 632

logistic, 402, 428
Long, 634
lookup functions, 51
loop, 86
Loop Until, 633
Loop While, 633
LSOLVE, 153, 154, 166, 637
LSREGRESS, 410, 411
LSREGRESS_UN, 414
LSREGRESS_UsrFrm, 407, 419, 440, 647
Luus-Jaakola, 270, 421
LUUSJAAKOLA, 276, 278, 313, 642

M

macro-enabled, 76
Macros, 79
magnitude, 248
Make Unconstrained Variables Nonnegative, 292
marching integration, 538, 542, 554, 577, 585, 586, 594, 612
mass transfer, 20
MATCH, 51, 66, 444, 453, 477, 632
mathematical models, 2, 3, 11, 15, 17, 23, 294, 361, 441
mathematical operations, 38
MATINV, 646
MATMLT, 647
matrix solution improvement, 146
MATTRN, 647
MAX, 632
maximum, 254
Maximum Change, 68, 163, 218
Maximum iterations, 218
Maximum Iterations, 163
maximum uncertainty, 197, 338
MaxMinVal, 635
MAXRES, 164
MaxTime, 635
MC2INT, 529, 533, 642
McLaurin series, 169
MDETERM, 152, 153, 632
mean, 316, 318, 319, 350, 359
measurand, 316
measurement model, 318
measurements, 50, 314, 315, 316, 317, 318, 319, 325, 335, 340, 359, 441, 497
mechanistic, 361
megaphone, 402
membrane, 13
method of continuation, 573
method of lines, 608, 612, 613
Metropolis algorithm, 271
Michaelis-Menten, 10, 14, 382, 589
Michalewic, 259
Mid, 634
midpoint, 553
MIDPOINT, 494, 533, 642
midpoint rule, 494, 514

MIDPT, 647
MIN, 632
minimize, 363, 366, 372
minimum, 254
MINV, 110
MINVERSE, 110, 152, 632
mirror image, 596, 617
mistakes, 317
Mistakes, 317
mixing effects, 17
mixing zones, 16
MMULT, 152, 632
Mod, 634
MODE, 72
model of uncertainty, 318
model parameters, 363
model selection, 429
modeling steps, 23
modified Euler method, 549
module, 78
Module, 77
mole balance, 362
Monod, 32
Monte Carlo, 321
Monte Carlo integration, 527
Monte Carlo methods, 25
Monte Carlo regression uncertainty, 421
Monte Carlo simulations, 331, 351, 354
Monte Carlo uncertainty propagation, 350
moving around in an *Excel* worksheet, 38
moving average, 479
MPEX, 501, 533
MREX, 638
MROUND, 632
MsgBox, 81, 116, 634
Muller, 219
multidimensional optimization, 257
multimodal, 253
multiple integrals, 525, 528
multiple linear least-squares regression, 367, 389
multiple roots, 200
multiple steady-states, 327
multistep method, 567
multivariate interpolation, 476
mutation, 287, 288

N

Name, 634
Name a Range, 40
Name Box, 61
name cells, 40
Name Manager, 40, 61
natural selection, 286
natural spline, 456
Nelder and Mead, 268
Neumann, 619
Neumann boundary condition, 596
Neville, 473

NEWTCON, 250, 576, 643
NEWTON, 213, 250, 638
Newton's law of viscosity, 19
Newton's method, 212, 257, 405
Newton-Cotes, 485
NEWTONPOLY, 443, 446, 448, 484, 637, 638
Next, 634
node, 258, 596, 613
noise, 363, 387, 436, 479, 497
nondimensionalization, 22
non-elementary reactions, 5, 32
nonlinear equations, 5, 6, 16, 167, 199, 201, 203, 205, 217, 225, 226, 235, 252, 257, 405, 585, 608
nonlinear functions, 4, 200, 350, 383, 534
nonlinear least-squares regression, 236, 306, 364, 372, 373, 386
nonlinear parameter estimation, 606
Nonlinear parameter estimation, 579
nonlinearity, 6
normal distribution, 319, 330, 336, 354, 359, 368, 391, 392, 421
normal random deviates, 356
normal random numbers, 356
normalize, 306, 513
normalized sorted residual plot, 392
NORMDIST, 355
NORMINV, 71, 355, 632
NORMSINV, 102, 350, 391, 392, 632
Not, 634
NRDWH, 102, 142, 638
null hypothesis, 390
NumberFormat, 634
numerical methods, 3, 6, 167, 200, 535
Nusselt number, 333

O

object, 78
objective function, 251, 253, 257, 306
observation, 319
Occam's razor, 2, 436
ODD, 632
ODEFIXSTEP_UsrFrm, 546, 556, 569, 583, 647
Offset, 634
On, 634
On Error GoTo, 137
On Error Resume Next, 137
Open, 634
OPTIMIN_UserFrm, 313
OPTIMIN_usrfrm, 267, 277
OPTIMIN_UsrFrm, 647
optimization, 21, 257, 363
Option, 634
Option Base, 106
Option Explicit, 84
Optional, 634
OR, 632
order, 22

order of magnitude, 22, 284, 354, 380, 436
order of operations, 97
ordinary differential equation, 6
ordinary iteration, 204, 225, 544
origin, 429
orthogonal collocation, 585, 608, 618
oscillation, 157, 470, 540
OUTLIER, 394, 643
Output, 634
over specified, 21, 441
overshoot, 470

P

packed, 114, 509
PADE, 174, 638, 643
Padé, 173, 223
Padé approximation, 173, 536
PADEROOT, 223, 250
parabolic partial differential equations, 21, 612, 618
parallel, 380
Parameter estimation, 578
parameter scaling, 305
parameter space, 24
parameters, 316, 366
parentheses, 38, 81, 97, 110
partial derivative, 194, 366
Particle swarm optimization, 280, 643
password, 78, 139
Path, 635
pathname, 131
patterns, 391
PDERIV, 191, 198, 643
penalty function, 301
percent uncertainty, 334, 357
permeate, 14
Personal.xlsb Macro Workbook, 76
perturbation, 194, 306
phase plane, 550
physical reality, 21
PI, 632
PI(), 45
Picard's method, 225
PICNIC, 135
piecewise, 466
pipe flow, 213
plot, 56
plug flow, 16, 538, 571
plug-flow reactor, 510
PNMSuite, 78, 139
point-slope, 204, 208, 443
poles, 429, 471
polynomial, 6, 11, 199, 364, 366, 387, 399, 429, 433, 436, 450, 466, 470, 503, 513
polynomial, roots, 240
POLYROOTS, 638, 640
population, 286, 319
positive-definite correlation coefficient, 356
positive-definite square matrix, 151

Powell, 237
POWELL, 267, 313, 643
POWELL_ANIMATION, 643
Powell's method, 264, 266
Prandtl number, 333
precision, 3, 315, 318, 325, 328, 334, 421, 498
precision, integration, 557
predator-prey, 550
predictor-corrector, 567
present value, 201
Preserve, 634
PRESERVE, 107
Print, 131, 634
Private, 80, 634
probability, 518
probability density function, 319, 325, 350, 528
probability distribution, 314
problem solving steps, 25
procedural errors, 317
procedures, VBA, 79
profile, 618
programming, 534
Project window, 78
Prompt, 634
propagation of uncertainty, 329
propagation of uncertainty, integration, 498
Properties, 113
Propeties window, 78
Protection, 78
pseudo random number generator, 379
pseudo-first-order, 32
pseudo-steady-state, 15
PSO, 280, 282, 312, 313, 643
Public, 80, 634
PUBLIC, 83
PV, 201
p-values, 389

Q

Q_13, 647
QUADRADF, 243, 647
quadratic equation, 6, 40, 258, 302
quadratic exterior, 301
quadratic formula, 6
quadratic interpolation, 267, 443
QUARTILE, 632
QUASINEWTON, 232, 576, 643
quasi-Newton's method, 217
quotation marks, 82
quoted value, 326
QYXPLOT, 127, 142, 327, 379, 456, 643
QYXPLOT_frm, 125
QYXPLOT_UsrFrm, 124, 647

R

R^2, 387, 389
Rachford Rice, 216

RADIANS(d), 45, 632
RAND, 45, 101, 379, 632
random, 421, 528
Random, 379
random deviates, 350
random errors, 317, 319, 340, 391
random normal deviate, 350
random number, 350
Random Number Generation, 71
random numbers, 379, 399
random sample, 354
random scatter, 390
random uncertainty, 316, 318, 354, 356
randomize, 336
Randomize, 634
Range, 634
RANGE, 83, 107
range variables, Excel, 107
ranking, 148
rate constants, 12
rate law, 5, 14, 16, 31, 369, 370, 372, 506
rate of convergence, 158
rate of return, 201
rational function, 429
rational functions, 364, 373, 429, 434, 537
rational interpolation, 473
rational Padé functions, 470
RATLF, 638
RATLIN, 471, 484, 638
RATLINBS, 473, 484, 638
RATLINPQ, 472, 484, 643
RATLS, 430, 433, 440, 643
RATLSF, 430, 433, 440, 472
RATLSGN, 643
Rayleigh equation, 238
reaction kinetics, 361
reaction mechanism, 556
reactor design, 33
reactors-in-series, 16
readability, 325
recalculate, 37, 79
recipe for uncertainty analysis, 340
Record Macro, 79, 93
recursive, 212, 257, 542
ReDim, 634
Redlich-Kwong, 3, 61, 68
reduced-step Newton's method, 232
Redundancy, 302
References, 141
REGFAL, 210, 250, 638
Regression, 377, 389, 397, 443
Regression Analysis, 371, 401, 433
regression sum of squares, 387
Regression tool, 389, 398
REGRESSION_UsrFrm, 647
REGRESSUN, 405
REGRESSUN_UN, 414, 421
REGRESSUN_UsrFrm, 405, 422, 440
regula falsi, 208

REGULAFALSI, 209, 640, 643
relative cell referencing, 40
relaxation, 164, 406
RELAXIT, 164, 166, 643
reliability, 24, 318, 319
Replace, 634
REPLACE, 632
replicated experiments, 318, 319, 325, 328, 340, 349, 350
residence time, 506
residual plots, 389, 390, 402, 414, 415, 434
residuals, 146, 160, 163, 390, 436, 588
residue, 14
resolution, 325, 326
response surface, 194, 262, 306
Resume, 634
Reynolds number, 333
RGAS, 50, 142, 638, 666
RICHARDSON, 647
Richardson's exptrapolation, 189
Richardson's extrapolation, 183, 189, 191, 195, 496, 498, 499, 505, 507
Ridders, 191
RIDDERS, 191, 644
risk, 314
RK2, 554
RK4, 554
RK45DP, 561, 583, 613, 644
RK45DP_sub, 561, 644
RK45ODE, 561, 565, 583, 644
RK45VS_UsrFrm, 561, 583, 614, 647
RK4G, 556, 583, 644
RK4G_sub, 558
RKG_4, 647
RKVS, 564
Rnd, 634
RND, 101
RND3WH, 142
RND4WH, 101, 142, 277, 638
Robust, 437
Romberg integration, 499
ROMBERGI, 499, 501, 533
ROMBERGM, 500, 501, 533, 644
ROMBERGT, 499, 501, 533, 644
ROOT, 221, 250, 639
root of a nonlinear function, 235
roots, 200, 212, 252, 585, 608
ROOTS, 237
ROOTS_UsrFrm, 232, 234, 250, 267, 576, 647
ROOTSGN, 237, 250, 411, 644
ROOTSLM, 237, 250, 419, 644
roughness, 213
ROULETTE, 96, 639
roulette wheel, 96
roulette wheel algorithm, 287, 288
Round, 98
ROUND, 49, 632
ROUNDDOWN, 52, 339, 632
round-off error, 8, 22, 52, 83, 305, 540, 553, 616

ROUNDUP, 52, 328, 632
ROWCOL, 108, 142, 647
Rows, 110, 634
ROWSHUFFLE, 379, 440, 644
RSS, 387
Run, 94, 634
RUN_QUASINEWTON, 232
Runge Phenomena, 443, 598
Runge-Kutta, 585, 613
Runge-Kutta fourth order, 588
Runge-Kutta, fixed step, 554
Runge-Kutta, variable step, 560
Runge-Kutta-Dormand-Prince, 560
Runge-Kutta-Gill, 555
Runge-Kutta-Merson, 560, 577

S

saddle, 254
safety, 30, 314
safety factors, 3
SAMPLE, 110, 142, 639
scale-up, 315
scaling, 247, 305, 421, 433, 436, 537, 603
Scaling, 635
scaling the independent variable in ODEs, 23
SCALIT, 305, 306, 313, 378, 439, 644
scientific notation, 49
ScreenUpdating, 635
SearchOption, 635
secant, 219
SECANT, 220, 250, 644
secant method, 219, 588
second derivative, 254, 257, 466, 467
second order derivative, 22
second order reaction, 32
second y-axis, 572
Secondary Axis, 57
second-order, 6, 22, 32, 168, 554, 585
second-order differential equation, 584
second-order partial differential equation, 19, 609
second-order reaction, 32
Select, 635
Select Data Source, 57
SELECTION, 635
SELECTPRO, 96
semi-log, 381
sensitivity, 194, 306, 310, 330, 567
Sensitivity, 297
SENSITIVITY, 195, 198, 337, 644
sensitivity analysis, 262, 306, 344
sensitivity coefficient, 194, 197, 316, 333, 336, 340, 344, 405
separability, 302
sequence of calculations, 7, 38, 42, 85, 97, 340
Set, 635
SetCell, 635
Sgn, 635
SHANKS, 173, 198, 639

Shanks transformation, 172, 536
Shapes, 55
SHOOTING, 588, 607, 645
shooting method, 585, 594
`SHOOTING_UsrFrm`, 588, 647
short cuts, 38
Show Formulas, 44, 46
SIGFIG, 99
`SIGFIGS`, 99, 142, 329, 461, 639
sigmoid, 434
SIGN, 632
significance level, 320, 390
significant digits, 7
significant figures, 8, 48, 60, 326, 334
`SIMANN`, 272, 273, 313, 645
`SIMP`, 510, 533, 639
simple, 436
Simplex linear programming, 290
SIMPLEX_ANIMATION, 269
`SIMPLEXNLP`, 268, 269, 645
simplifying assumptions, 25
`SIMPSON`, 507, 514, 515, 533, 645
Simpson's 1/3 rule, 503, 514
Simpson's 3/8 rule, 504
Simpson's composite rule, 504, 514, 522
Simpson's rules, 503
`SIMPSON2D`, 522, 523, 645
`SIMPSONADAPT`, 533, 645
`SIMPSONDATA`, 507, 508, 533
SIMSPSONDATA, 639
simulated annealing, 270
simulation, 354, 400
simultaneous differential equations, 555
simultaneous equations, 544
`Sin`, 635
SIN, 632
SIN(a), 45
`Single`, 635
`SINGLE`, 83
single precision, 7, 8, 99
singular matrix, 153
SINH, 632
slack variable, 298
slope, 366, 382, 426
SLOPE, 632
SLOPE, *Excel*, 367
smooth, 436, 441, 461, 470, 479, 497, 515
smoothing, 429, 443
Soave-Redlich-Kwong, 193
Solver, 70, 73, 141, 235, 246, 290, 297, 375, 378, 457
Solver, least-squares regression, 404
Solver Options, 236
Solver, least-squares regression, 403
Solver, nonlinear least-squares regression, 372
Solver, parameter scaling, 305
Solver, Shooting Method, 586
`SolverOk`, 635
`SolverOptions`, 635
`SolverSolve`, 635
sort, 356, 379, 467
Sort, 398
sorted residual plots, 392, 414, 434
sorting, 357, 421
Sparklines, 59, 390
specific growth rate, 33, 187
`Speech.Speak`, 111, 635
speed up, 99
SPIDER, 309, 311, 313, 645
spider plot, 309
splines, 537
split boundary conditions, 584
SPOOLED, 639
`Sqr`, 635
SQRT, 427, 632
square matrix, 152
SSR, 363, 366, 387, 406
SSX, 426
stability, 540, 543, 613, 615
stability, integration, 560
stable, 540, 560, 613, 615, 616
standard devation, 271
standard deviation, 271, 316, 319, 328, 359
Standard deviation, 639
standard uncertainty, 316, 319, 320, 328, 356, 389
standardized residual, 394
standardized residuals, 394
`Static`, 84, 635
stationary phase, 386
statistical analysis, 47
statistical functions, 47
Status Bar, 132, 209, 232, 267
`StatusBar`, 635
STDEV, 72, 103, 319, 632
STDEW, 271, 313, 639
steady-state, 12, 13, 15, 423, 424
Steffensen, 544
STEFFENSEN, 225, 231, 250, 594, 645
Steffensen's method, 546
step, 537, 538, 540, 543, 553, 557, 560, 561
`Step`, 635
step size, 613, 615
step, print, 560
sterilization, 517, 573
STEYX, 632
stiffness, 540
STIFFODE, 546, 549, 583, 645
STIFFODE_sub, 547, 559
STINEMAN, 472, 475, 484, 639
Stineman rational interpolation, 474
STINEMAN2D, 478, 484
STINEMANS, 647
stochastic, 24, 96, 251, 270, 275, 527
Stochastic, 527, 647
Stop Recording, 80
`String`, 635
STRING, 83
strings, 82

INDEX

Student t-statistic, 320
studentized residual, 395
STUDENTIZER, 395, 440, 645
sub procedure, *VBA*, 79
Subscript, 633
Substitute, 105, 635
successive approximation, 7
successive substitution, 157, 204, 225
sufficiently large, 379
sum of squared residuals, 245, 252, 363, 366, 372, 387, 405, 406
sum of the squared residuals, 404, 421
sum of the squares of the deviations, 426
SUMPRODUCT, 632
SUMSQ, 246, 632
SUMXMY2, 376, 632
surface plot, 259, 260, 279, 300
swap, 117
SWAP, 117, 645
Swiss Army Knife, 450
symbolic, 3, 4, 11, 204
symmetric, 356, 514, 596
Symmetric Boundary Conditions, 596
symmetry, 596
systematic errors, 317, 324
systematic uncertainty, 316, 354, 356
systems of first-order differential equations, 542

T

T.INV, 632
Tab, 635
Tan, 635
TAN, 632
TAN(a), 45
TANH, 632
Target Cell, 236
Taylor series, 167, 197, 212, 220, 231, 257, 350, 405, 489, 503, 536, 537, 540
team, 252
TEARING, 248, 646
tearing method, 248
Template, 60, 106
TEXT, 632
text boxes, 55
Text Wrapping, 64
Then, 634
theory, 436
thermal conductivity, 19, 613
thermal diffusivity, 613, 616
Thiele modulus, 587
ThisWorkbook, 132, 635
THOMAS, 151, 166, 647
Thomas algorithm, 151, 452
Three-D Rotation, 612
Time, 635
Timer, 101, 635
TIMER, 100

TINV, 320, 338, 341, 390, 632
Title, 635
titration, 184, 462
tolerance, 515
tornado, 309
TORNADO, 309, 311, 313, 646
tornado chart, 309
trace cells in formulas, 44
tracer, 506
trajectory, 586
transfer units, 487
transformation, 17, 34, 102, 144, 151, 152, 169, 195, 204, 247, 294, 298, 382, 383, 389, 405, 429, 495, 527, 585
transformed data, 365
transient, 12, 178, 613
transport phenomena, 11
transpose, 108
Transpose, 110, 635
TRANSPOSE, 632
trap, 31
TRAP, 490, 492, 501, 533, 639
TRAPDATA, 498, 639
TRAPEZOID, 490, 533, 646
trapezoidal rule, 488, 496, 514, 543, 555, 615
trapezoidal rule with end-corrections, 490, 505
trapezoidal rule, explicit, 549
TREND, 427, 443, 445, 477, 633
Trend line, 187, 362, 364, 382, 385, 433, 436
TREND, *Excel*, 367
trend, residual plot, 390
TREX, 501, 533, 639
trial and error, 5, 67, 171, 200, 213, 231, 247, 252, 540, 570, 625
trial function, 598
tridiagonal, 151
TRINV, 358, 359, 360, 639
True, 635
truncated Taylor series, 6, 167, 331, 535
t-statistic, 320, 390
t-table, 320
t-Test: Two Sample Assuming Unequal Variances, 72
Tufte, 55
turbulent, 333
Type, 633, 635
Type A uncertainty, 317, 319
Type B uncertainty, 317, 325, 339
TypeOf, 635

U

Ubound, 635
Ucase, 635
UCASE, 633
UDF, 640
ultrafiltration, 348
uncertainties in measurements, 317
uncertainty, 3, 21, 24, 197, 314, 442
uncertainty analysis, 314
uncertainty, integration, 497

uncertainty, least-squares regression, 404, 405
UNCERTAINTY_UsrFrm, 341, 356, 360, 647
unconditional stability, 615
undershoot, 470
underspecified, 21
unequally-spaced, 512
unequally-spaced data, 183, 532
Unhide, 67
uniform distribution, 350, 354
uniform random numbers, 356
unimodal, 253, 263
unique, 21
unit conversions, 50; uncertainty, 334
UNITS, 50, 142, 640, 666
UNITS_TABLES, 95, 142, 646, 666
universe, 319
unknown boundary condition, 588
UNMCLHS, 356, 360, 646
UNSOLVER, 421, 422, 440, 646
unstable, 540, 616
upwind, 616
User-defined, 81, 140
user-defined function, *VBA*, 79, 81, 636
UserFinish, 635
USIGFIGS, 329, 360, 640

V

validate, 24
validation, 24, 65, 379, 383, 390, 442
Value, 107, 635
vapor pressure, 415
vapor-liquid equilibrium, 20, 433
variable step, 540
variable step Runge-Kutta, 577
variable transformation, 299
variable, *VBA*, 82
Variant, 99, 110, 635
VARIANT, 82, 107
VBA, 75, 534
VBE, 76
vbNewLine, 135
VbNewLine, 635
verify, 24
view factor, 524
View Macros, 76
Visual Basic Editor, 76
Visual Basic for Applications, 75

VLE, 433
VLOOKUP, 51, 633
VNORMRR, 158, 166, 640

W

WAVERAGE, 640
Wegstein, 165, 204
weighted average, 164, 189, 554, 560, 615
weighted harmonic mean, 466
weighted least-squares regression, 402
weighted regression, 404, 421
weighted sum, 513
weighted uncertainty, 349
Welch-Satterthwaite, 320, 339, 341, 389
Wend, 633
What if …?, 24
What-if Analysis, 67
While, 633
Wire Frame Surface, 612
Workbook_Open, 132
Worksheet_Change, 132
WorksheetFunction, 97, 635
worksheets, 635
worst case scenario, 330
Write, 131

X

X_OC, 598, 607, 640
XIZADS, 620, 621
xlCalculationAutomatic, 635
xlCalculationManual, 635
XLSTART, 76
XZUW, 624, 646

Y

Y_OC, 607, 640
yx plot, *Excel*, 56

Z

zero intercept, 366
Zoom to Selection, 39
ZPLOT, 392, 393, 394, 440, 646
z-score, 392

Summary of Key Equations for Practical Numerical Methods in Excel Worksheets

Laws of Conservation	Rate In – Rate Out + Rate of Generation = Rate of Accumulation
Linear Algebra: MMULT(MINVERSE(A),B), CTRL SHIFT ENTER	$x = A^{-1}b$
First Derivatives	$y_i' \cong \dfrac{y_{i+1} - y_{i-1}}{2\Delta x}$ $\quad$ $y_0' \cong \dfrac{-3y_0 + 4y_1 - y_2}{2\Delta x}$ $\quad$ $y_n' \cong \dfrac{3y_n - 4y_{n-1} + y_{n-2}}{2\Delta x}$
Second Derivative – Central Finite Difference	$y_i'' \cong \dfrac{y_{i+1} - 2y_i + y_{i-1}}{\Delta x^2}$
Secant Root	$x_{i+1} = x_i - f(x_i)[x_i - x_{i-1}]/[f(x_i) - f(x_{i-1})]$
Newton Root	$x_{i+1} = x_i - f(x_i)/f'(x_i)$
Solver Roots (Minimize the Sum of Squared Functions): SUMSQ	$Objective = Min\left(f_1^2 + f_2^2 + f_3^2 \ldots\right)$
Measurement Uncertainty (Fixed, Random, Combined)	$u_z = \pm \delta/\sqrt{3}$ $\quad$ $u_r = s/\sqrt{n}$ $\quad$ $u_x = \sqrt{u_z^2 + u_r^2}$
Law of Propagation of Uncertainty	$U_y = t_v \sqrt{\left(\dfrac{\partial y}{\partial x_1} u_{x1}\right)^2 + \left(\dfrac{\partial y}{\partial x_2} u_{x2}\right)^2 \ldots}$
Least-squares Regression (Minimize Sum of Squared Residuals): SUMXMY2	$Min\left[\sum (y_{\text{data}} - y_{\text{model}})^2\right]$
Logarithmic Linearization	$y = A\exp(Bx) \rightarrow \ln y = \ln A + Bx$
Linear Lagrange Interpolation	$y(x) \cong \left(\dfrac{x - x_{i+1}}{x_i - x_{i+1}}\right) y_i + \left(\dfrac{x - x_i}{x_{i+1} - x_i}\right) y_{i+1}$
Quadratic Interpolation	$y(x) \cong \dfrac{(x - x_i)(x - x_{i+1}) y_{i-1}}{(x_{i-1} - x_i)(x_{i-1} - x_{i+1})} + \dfrac{(x - x_{i-1})(x - x_{i+1}) y_i}{(x_i - x_{i-1})(x_i - x_{i+1})} + \dfrac{(x - x_{i-1})(x - x_i) y_{i+1}}{(x_{i+1} - x_{i-1})(x_{i+1} - x_i)}$
Trapezoidal Integral	$\int_a^b f(x)dx \cong \dfrac{(b-a)}{2}\left[f(a) + f(b)\right]$
Simpson's 1/3 Integral	$\int_{x_0}^{x_2} f(x)dx \cong \dfrac{\Delta x}{3}\left[f(x_0) + 4f(x_1) + f(x_2)\right]$ $\quad$ $\Delta x = x_2 - x_1 = x_1 - x_0$
Simpson's 3/8 Integral	$\int_{x_0}^{x_3} f(x)dx \cong \dfrac{3\Delta x}{8}\left[f(x_0) + 3f(x_1) + 3f(x_2) + f(x_3)\right]$ $\quad$ $\Delta x = x_3 - x_2 = x_2 - x_1 = x_1 - x_0$
Richardson's Extrapolation	2nd Order $I \cong \dfrac{4I_2 - I_1}{3}$ $\quad$ 4th Order $I \cong \dfrac{16I_2 - I_1}{15}$
Euler's ODE	$y_{i+1} = y_i + \Delta x f(x_i, y_i)$
Trapezoidal ODE: Iteration on Circular References	$y_{i+1} = y_i + 0.5\Delta x\left[f(x_i, y_i) + f(x_{i+1}, y_{i+1})\right]$
Shooting Method	$\dfrac{d^2 y}{dx^2} = f\left(x, y, \dfrac{dy}{dx}\right)$ $\Rightarrow$ $\dfrac{dy}{dx} = z$ $\quad$ and $\quad$ $\dfrac{dz}{dx} = f(x, y, z)$

Conversion Factors Available in the VBA User-defined Functions UNITS and RGAS

VBA user-defined function: **UNITS(nu**mber, "from_unit", "to_unit")
- Use strings for the units (subscipts/superscripts to right of unit), e.g., $lb_m.ft/s^3.°F$ is typed with quotes as: "lbm.ft/s3.F"
  - Example Mass Transfer Coefficient Units Conversion: 2.4578 = UNITS(12,"kmol/m2.s","lbmol/ft2.s")
- Use the macro UNITS_TABLES to generate an *Excel* worksheet with the available unit strings for UNITS.
- For combined units, multiply or divide the numerical value by the user-defined function, with one as the first argument: UNITS(1, "from_unit","to_unit"). Note: use dF, dC, or dK for units of temperature change.

Quantity	Equivalent Values and Names of Units Available in the user-defined function UNITS[114]
Mass	1 kg = 1000 g = 0.001 t (or metric ton) = 2.20462 lb_m 1 lb_m = 5 x 10^{-4} ton = 453.593 g = 0.453593 kg
Length	1 m = 100 cm = 1000 mm = 39.37 in = 3.2808 ft = 1.0936 yd = 0.0006214 mi 1 ft = 0.33333 yd = 0.3048 m = 30.48 cm
Volume	1 m^3 = 1000 L (or liter)= 10^6 cm^3 = 35.3145 ft^3 = 264.17 gal 1 ft^3 = 28317 cm^3 = 1728 in^3 = 7.4805 gal = 0.028317 m^3 = 28.317 L (or liter)
Force	1 N = 1 kg.m/s^2 = 0.22481 lb_f 1 lb_f = 32.174 lbm.ft/s^2 = 4.4482 N
Pressure	1 atm = 1.01325 x 10^5 N/m^2 (or Pa) = 1.01325 bar = 760 mmHg (or torr) @ 0 °C = 14.696 lb_f/in^2 (or psi) = 2116.2 lb_f/ft^2 = 68087 $lb_m/ft.s^2$ = 33.9 ft H_2O @ 4 °C = 406.8 in H_2O = 29.921 in Hg @ 0 °C
Energy	1 J = 1 N.m = 2.7778 x 10^{-7} kW.hr = 0.23901 cal = 0.7376 ft.lb_f = 9.486 x 10^{-4} Btu = 3.7251x10^{-7} hp.hr
Power	1 W = 1 J/s = 0.23901 cal/s = 0.7376 ft.lb_f/s = 9.486 x 10^{-4} Btu/s = 1.341 x 10^{-3} hp
Density	1 kg/m^3 = 1 g/L = 0.001 g/cm^3 = 0.062428 lb_m/ft^3 = 3.61275 x 10^{-5} lb_m/in^3 = 0.00834547 lb_m/gal
Viscosity	1 Pa.s (or kg/m.s) = 10 P (poise or g/cm.s) = 10^3 cP (or centipoise) = 0.67197 lb_m/ft.s = 2419.1 lb_m/ft.hr = 0.020886 $lb_f.s/ft^2$
Thermal Conductivity	1 W/m^2.K (or kg.m/s^3.K) = 4.0183 $lb_m.ft/s^3.°F$ = 0.12489 lb_f/s.°F = 0.0023901 cal/s.cm.K = 0.57780 Btu/hr.ft.°F
Diffusivity	1 m^2/s = 10^4 cm^2/s (St or stoke) = 10^6 cSt (or centistoke) = 38750 ft^2/hr
Heat Transfer Coefficient	1 W/m^2.K (or kg/s^3.K) = 10^{-4} W/cm^2.K = 10^3 g/s^3.K = 2.3901 x 10^{-5} cal/cm^2.s.K = 1.2248 $lb_m/s^3.°F$ = 0.038068 lb_f/ft.s.°F = 0.17611 Btu/ft^2.hr.°F
Mass Transfer Coefficient	1 kg/m^2.s = 0.1 g/cm^2.s = 0.20482 lb_m/ft^2.s = 0.0063659 $lb_f.s/ft^3$ 1 kmol/m^2.s = 0.1 mol/cm^2.s = 0.20482 lbmol/ft^2.s
Temperature	K = °C + 273.15 = (°F + 459.67)/1.8 = R/1.8 °C = (°F - 32)/1.8 °F = 1.8 °C + 32 R = °F + 459.67 For Heat Capacity & Heat Transfer Δ°F = ΔR = 1.8Δ°C = 1.8ΔK

Ideal Gas Constant

$$R_g = PV/nT \qquad R_g = 8.31441 \text{ J/(mol-K)} = 1.9872 \text{ cal/(mol-K)}$$

User-defined function returns gas constant for different unit systems (in quotes): **RGAS("P_unit", "V_unit", "n_unit", "T_unit")**

V	T	n	atm	psia	mmHg	Pa
ft^3	K	mol	0.00290	0.0426	2.20	293.3
		lbmol	1.31	19.31	999	1.332 x 10^5
	R	mol	0.00161	0.02366	1.22	162.65
		lbmol	0.730	10.73	555	7.4 x 10^5
liter (L or l)	K	mol	0.08205	1.206	62.4	8315
		lbmol	37.2	547	28,300	3.773 x 10^6
	R	mol	0.0456	0.670	34.6	4613
		lbmol	20.7	304	15,700	2.093 x 10^6

Standard Temperature and Pressure (STP)	Ideal Gas Molar Volume
0 °C, 1 atm	22.413 L/mol
32 °F, 1 atm	359.017 ft^3/lbmol
60 °F, 1 atm	379.463 ft^3/lbmol

Use *kmol* units with volume units of m^3.

---

[114] Adapted from (Felder and Rousseau 1986), (Bird, Stewart and Lightfoot 2007), CRC *Handbook of Chemistry and Physics*, 64th Ed, F-197

Made in the USA
Lexington, KY
08 January 2019